GAMMA-RAY BURSTS

Related Titles from the AIP Conference Proceedings Subseries on Astronomy and Astrophysics

523   Gravitational Waves: Third Edoardo Amaldi Conference
Edited by Sydney Meshkov, June 2000, 1-56396-944-0

522   Cosmic Explosions: Tenth Astrophysics Conference
Edited by Stephen S. Holt and William W. Zhang, June 2000, 1-56396-943-2

516   26th International Cosmic Ray Conference: ICRC XXVI, Invited, Rapporteur, and Highlight Papers
Edited by Brenda L. Dingus, David B. Kieda, and Michael H. Salamon
May 2000, 1-56396-939-4

515   GeV-TeV Gamma Ray Astrophysics Workshop: Towards a Major Atmospheric Cherenkov Detector VI
Edited by Brenda L. Dingus, Michael H. Salamon, and David B. Kieda, May 2000, 1-56396-938-6

510   The Fifth Compton Symposium
Edited by Mark L. McConnell and James M. Ryan, March 2000, 1-56396-932-7

499   Small Missions for Energetic Astrophysics: Ultraviolet to Gamma-Ray
Edited by Steven P. Brumby, December 1999, 1-56396-912-2

471   Solar Wind Nine: Proceedings of the Ninth International Solar Wind Conference
Edited by Shadia Rifai Habbal, Ruth Esser, Joseph V. Hollweg, Philip A. Isenberg, May 1999, 1-56396-865-7

433   Workshop on Observing Giant Cosmic Ray Air Showers from $>10^{20}$ eV Particles from Space
Edited by John F. Krizmanic, Jonathan F. Ormes, and Robert E. Streitmatter, June 1998, 1-56396-788-X

428   Gamma-Ray Bursts: 4th Huntsville Symposium
Edited by Charles A. Meegan, Robert D. Preece, and Thomas M. Koshut, May 1998, 1-56396-766-9

410   The Fourth Compton Symposium
Edited by Charles D. Dermer, Mark S. Strickman, and James D. Kurfess, December 1997, 2 vol. set, 1-56396-659-X

385   Robotic Exploration Close to the Sun: Scientific Basis
Edited by Shadia Rifai Habbal, February 1997, 1-56396-618-2

To learn more about these titles, or the AIP Conference Proceedings Series, please visit the webpage http://www.aip.org/catalog/aboutconf.html

GAMMA-RAY BURSTS

5th Huntsville Symposium

Huntsville, Alabama   18–22 October 1999

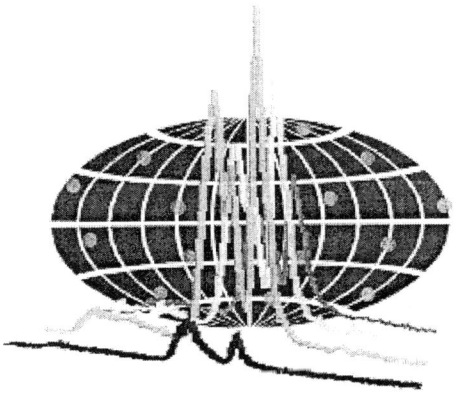

EDITORS
R. Marc Kippen
Robert S. Mallozzi
University of Alabama in Huntsville

Gerald J. Fishman
NASA/Marshall Space Flight Center

Ⓒ **CD-ROM INCLUDED**

Melville, New York, 2000
AIP CONFERENCE PROCEEDINGS ■ VOLUME 526

Editors:

R. Marc Kippen
Robert S. Mallozzi
Gerald J. Fishman

SD50, Space Science Department
NASA/Marshall Space Flight Center
Huntsville, AL 35812
USA

E-mail: marc.kippen@msfc.nasa.gov
robert.mallozzi@msfc.nasa.gov
gerald.j.fishman@msfc.nasa.gov

The articles on pp. 431–440, 570–574, 671–680, and 731–735 were authored by U. S. Government employees and are not covered by the below mentioned copyright.

Authorization to photocopy items for internal or personal use, beyond the free copying permitted under the 1978 U.S. Copyright Law (see statement below), is granted by the American Institute of Physics for users registered with the Copyright Clearance Center (CCC) Transactional Reporting Service, provided that the base fee of $17.00 per copy is paid directly to CCC, 222 Rosewood Drive, Danvers, MA 01923. For those organizations that have been granted a photocopy license by CCC, a separate system of payment has been arranged. The fee code for users of the Transactional Reporting Service is: 1-56396-947-5/00/$17.00.

© 2000 American Institute of Physics

Individual readers of this volume and nonprofit libraries, acting for them, are permitted to make fair use of the material in it, such as copying an article for use in teaching or research. Permission is granted to quote from this volume in scientific work with the customary acknowledgment of the source. To reprint a figure, table, or other excerpt requires the consent of one of the original authors and notification to AIP. Republication or systematic or multiple reproduction of any material in this volume is permitted only under license from AIP. Address inquiries to Office of Rights and Permissions, Suite 1NO1, 2 Huntington Quadrangle, Melville, N.Y. 11747-4502; phone: 516-576-2268; fax: 516-576-2450; e-mail: rights@aip.org.

L.C. Catalog Card No. 00-105301
ISBN 1-56396-947-5 set
ISBN 1-56396-957-2 CD-ROM
ISSN 0094-243X
Printed in the United States of America

Contents

Preface xviii
Dedication: Jan van Paradijs 1946–1999 xxi

I. GENERAL PROPERTIES

A Gamma-Ray Burst Bibliography, 1973–1999 3
    K. Hurley

GRB DETECTION AND CLASSIFICATION

Derivation of a Sample of Gamma-Ray Bursts from BATSE
DISCLA Data 8
    M. Schmidt
A Search for Non-triggered Gamma-Ray Bursts in the BATSE
Continuous Records: The Current Status 13
    B. E. Stern, Y. Tikhomirova, D. Kompaneets, M. Stepanov,
    A. Berezhnoy, and R. Svensson
A Search for Gamma-Ray Bursts in the GRBM/BeppoSAX Database 18
    C. Guidorzi, F. Frontera, L. Amati, E. Costa, and M. Feroci
What Can BeppoSAX Tell Us about Short GRBs: An Update from the
Subsecond GRB Project 23
    G. Gandolfi, M. J. S. Smith, A. Coletta, G. Celidonio, L. Di Ciolo,
    A. Paolino, G. Tarei, G. Tassone, J. M. Muller, E. Costa, M. Feroci,
    F. Frontera, and L. Piro
MeV Measurements of γ-Ray Bursts by CGRO-COMPTEL:
Revised Catalog 28
    A. Connors, C. A. Young, K. Bennett, W. Collmar, W. Hermsen,
    R. M. Kippen, L. Kuiper, M. McConnell, R. Miller, J. M. Ryan,
    V. Schönfelder, O. R. Williams, and C. Winkler
Properties of Gamma-Ray Burst Classes 33
    J. Hakkila, D. J. Haglin, R. J. Roiger, R. S. Mallozzi, G. N. Pendleton,
    and C. A. Meegan
Unsupervised Induction and Gamma-Ray Burst Classification 38
    R. J. Roiger, J. Hakkila, D. J. Haglin, G. N. Pendleton, and R. S. Mallozzi

DETECTION BIASES

A Summary of Biases in the BATSE Burst Trigger 43
    C. Meegan, J. Hakkila, A. Johnson, G. Pendleton, and R. Mallozzi
The Fluence Duration Bias 48
    J. Hakkila, C. A. Meegan, G. N. Pendleton, R. S. Mallozzi, D. J. Haglin,
    and R. J. Roiger

A Monte Carlo Simulation of the Peak Counts Bias and the Slow Riser
Bias in the BATSE Trigger... 53
    A. Johnson, C. Meegan, and J. Hakkila

DISTRIBUTIONS

Luminosities, Space Densities and Redshift Distributions
of Gamma-Ray Bursts.. 58
    M. Schmidt

Constraining the Luminosity Function of GRBs from Time Dilation,
Brightness Distribution and Redshift Data............................. 63
    M. Deng and B. E. Schaefer

Rest Frame Properties of Gamma-Ray Bursts............................. 68
    I. G. Mitrofanov, D. S. Anfimov, M. L. Litvak, A. B. Sanin, M. S. Briggs,
    W. S. Paciesas, G. N. Pendleton, R. D. Preece, and C. A. Meegan

LogN-LogF Distribution of BATSE Bursts Based on the Average Flux
during High Power Emission 73
    D. S. Anfimov, I. G. Mitrofanov, M. L. Litvak, A. L. Sanin, M. S. Briggs,
    W. S. Paciesas, G. N. Pendleton, R. D. Preece, and C. A. Meegan

Connection between Spectral Lags and Peak Luminosity in GRBs 78
    J. P. Norris, G. F. Marani, and J. T. Bonnell

A Simple BATSE Measure of GRB Duty Cycle 83
    J. Hakkila, R. D. Preece, and G. N. Pendleton

The Energy Distribution of GRBs....................................... 87
    T. Piran, R. Jimenez, and D. Band

The Pure Physical Parameters of BATSE Gamma Ray Bursts................ 92
    A. B. Sanin, I. G. Mitrofanov, D. S. Anfimov, M. L. Litvak, M. S. Briggs,
    W. S. Paciesas, G. N. Pendleton, R. D. Preece, and C. A. Meegan

Non-isotropic Angular Distribution for Very Short-Time
Gamma-Ray Bursts?... 97
    D. B. Cline, C. Matthey, and S. Otwinowski

Testing the Intrinsic Randomness in the Angular Distributions
of Gamma-Ray Bursts... 102
    A. Mészáros, Z. Bagoly, I. Horváth, L. G. Balázs, and R. Vavrek

On the Clustering of GRBs on the Sky.................................. 107
    S. K. Sethi, S. G. Bhargavi, and J. Greiner

II. PROMPT GRB EMISSION

SPECTRAL STUDIES

Gamma-Ray Burst Spectroscopy 115
    R. D. Preece

Wide-Band Spectroscopy of GRB 990510.................................. 125
    M. S. Briggs, R. D. Preece, J. van Paradijs, J. in't Zand, J. Heise,
    E. Kuulkers, and C. Kouveliotou

Track Jumps in Hardness-Intensity Correlations of GRB Pulse Decays 130
    L. Borgonovo and F. Ryde
Shape of the Decay Phase of Gamma-Ray Burst Pulses 135
    F. Ryde and R. Svensson
Average Spectral Parameters of High-Power Emission in BATSE
Gamma-Ray Bursts 140
    D. S. Anfimov, I. G. Mitrofanov, M. L. Litvak, A. L. Sanin, M. S. Briggs,
    W. S. Paciesas, G. N. Pendleton, R. D. Preece, and C. A. Meegan
Cosmological Signatures in Intensity and Break Energy Correlations 145
    N. M. Lloyd, V. Petrosian, and R. S. Mallozzi
The Role of the BATSE Instrument Response in Creating the GRB
E-Peak Distribution 150
    J. J. Brainerd, G. N. Pendleton, R. S. Mallozzi, M. S. Briggs,
    and R. D. Preece
Synchrotron Emission as the Source of GRB Spectra, Part II:
Observations 155
    N. M. Lloyd, V. Petrosian, and R. D. Preece
Study of the Characteristics of GRB Energy Spectra 160
    B. M. Belli
GRB 920229: Evidence for a Sharp Spectral Break? 165
    M. S. Briggs and R. D. Preece
Spectra of a Recent Bright Burst Measured by CGRO-COMPTEL:
GRB 990123 170
    C. A. Young, A. Connors, K. Bennett, W. Collmar, W. Hermsen,
    R. M. Kippen, E. D. Kolaczyk, L. Kuiper, M. McConnell, R. Miller,
    J. M. Ryan, V. Schönfelder, O. R. Williams, and C. Winkler
X-Ray Excesses in GRB Spectra 175
    R. D. Preece, J. R. Espley, and M. S. Briggs
A New Method for Studying the Hardness-Intensity Correlation
in Gamma-Ray Bursts 180
    F. Ryde, L. Borgonovo, and R. Svensson
Time Resolved GRB Spectroscopy 185
    M. Tavani, D. Band, and G. Ghirlanda

TEMPORAL STUDIES

Unusual Properties in the Time Profiles of Bright GRBs 190
    F. Quilligan, K. J. Hurley, B. McBreen, L. Hanlon, P. Duggan,
    and D. Watson
Generic Difference between Early and Late Stages of BATSE
Gamma-Ray Bursts 195
    M. L. Litvak, I. G. Mitrofanov, M. S. Briggs, W. S. Paciesas,
    G. N. Pendleton, R. D. Preece, and C. A. Meegan

The Duration Errors of Gamma-Ray Bursts 200
    I. Horváth, E. E. Fenimore, J. P. Norris, P. Mészáros, and Z. Bagoly
Power Density Spectra of Gamma-Ray Bursts 205
    A. M. Beloborodov, B. E. Stern, and R. Svensson
GRB Time-Dilation Measurements Corrected for Trigger Bias 210
    J. T. Bonnell, J. P. Norris, G. F. Marani, and R. J. Nemiroff
Rise and Decay Time of Subpeaks in Short Duration Bursts 215
    V. Gupta, P. Das Gupta, and P. N. Bhat
Aperiodic Properties of Gamma-Ray Bursts 220
    A. S. Pozanenko and V. M. Loznikov
Evidence for and Implications of Turbulence in GRB Time Profiles 225
    Y. Yan
The Technique of Emission Time Estimation for BATSE GRBs 230
    M. L. Litvak, I. G. Mitrofanov, D. S. Anfimov, A. B. Sanin,
    M. S. Briggs, W. S. Paciesas, G. N. Pendleton, R. D. Preece,
    T. M. Koshut, G. J. Fishman, C. A. Meegan, and J. P. Lestrade
The Emission Time Signature of BATSE GRBs 235
    I. G. Mitrofanov, M. L. Litvak, D. S. Anfimov, A. B. Sanin, M. S. Briggs,
    W. S. Paciesas, G. N. Pendleton, R. D. Preece, T. M. Koshut, G. J. Fishman,
    C. A. Meegan, and J. P. Lestrade

OBSERVATIONS AT OTHER ENERGIES

Evidence for TeV Emission From GRB 970417a 240
    J. E. McEnery, R. Atkins, W. Benbow, D. Berley, M. L. Chen, D. G. Coyne,
    B. L. Dingus, D. E. Dorfan, R. W. Ellsworth, D. Evans, A. Falcone, L. Fleysher,
    R. Fleysher, G. Gisler, J. A. Goodman, T. J. Haines, C. M. Hoffman,
    S. Hugenberger, L. A. Kelley, I. Leonor, M. McConnell, J. F. McCullough,
    R. S. Miller, A. I. Mincer, M. F. Morales, P. Nemethy, J. M. Ryan, B. Shen,
    A. Shoup, C. Sinnis, A. J. Smith, G. W. Sullivan, T. Tümer, K. Wang,
    M. O. Wascko, S. Westerhoff, D. A. Williams, T. Yang, and G. B. Yodh
Fast X-Ray Transients and Their Relation to GRBs 245
    V. A. Arefiev, K. N. Borozdin, and W. C. Priedhorsky
LOTIS Upper Limits and the Prompt OT from GRB 990123 250
    G. G. Williams, D. H. Hartmann, H. S. Park, R. A. Porrata, E. Ables,
    R. Bionta, D. L. Band, S. D. Barthelmy, T. Cline, N. Gehrels,
    D. H. Ferguson, G. Fishman, R. M. Kippen, C. Kouveliotou, K. Hurley,
    R. Nemiroff, and T. Sasseen
Preliminary Results from the TAROT Experiment 255
    M. Boër, J.-L. Atteia, M. Bringer, S. Chaty, A. Klotz, F. Morand,
    H. Pedersen, C. Pollas, and D. Toublanc
First Results from the Burst Observer and Optical Transient Exploring
System Station 1 (BOOTES-1) 260
    A. J. Castro-Tirado, J. Soldán, M. Bernas, P. Páta, R. Hudec, T. M. Sanguino,
    B. de la Morena, J. A. Berná, A. de Ugarte, J. Gorosabel, J. M. Más-Hesse,
    and A. Giménez

Optical Observations of GRBs: EN, BART, and OMC 265
    R. Hudec, J. Soldán, V. Hudcová, J. Florian, M. Nekola,
    O. Broz, M. Bernas, P. Páta, F. Hroch, A. J. Castro-Tirado,
    M. Más-Hesse, A. Giménez, E. Palazzi, N. Masetti,
    and G. Pizzichini

An Extended Search for Transient Events in the COBE/DMR Database
Associated with Cosmic Gamma-Ray Bursts 270
    L. J. Beathley, Jr., J. G. Stacy, T. R. Bontekoe, P. D. Jackson,
    and C. Winkler

III. AFTERGLOW, HOSTS, AND SUPERNOVA CONNECTIONS

The Afterglows of Gamma-Ray Bursts 277
    S. R. Kulkarni, E. Berger, J. S. Bloom, F. Chaffee, A. Diercks,
    S. G. Djorgovski, D. A. Frail, T. J. Galama, R. W. Goodrich,
    F. A. Harrison, R. Sari, and S. A. Yost

RADIO AFTERGLOW

A Coordinated Radio Afterglow Program 298
    D. A. Frail, S. R. Kulkarni, M. H. Wieringa, G. B. Taylor,
    G. H. Moriarty-Schieven, D. S. Shepherd, R. M. Wark,
    R. Subrahmanyan, D. McConnell, and S. J. Cunningham

OPTICAL AFTERGLOW

Optical/Multiwavelength Observations of GRB Afterglows 303
    T. J. Galama

Recent Optical/Near-IR Observations of GRBs 313
    A. J. Castro-Tirado, J. Gorosabel, J. Greiner, S. Klose, V. Mohan,
    R. Sagar, I. Bond, N. Rattenbury, P. Yock, F. Vrba, A. Henden,
    C. Luginbuhl, A. Guarnieri, M. R. Zapatero-Osorio, J. Zhu,
    R. Hudec, S. Guziy, A. Shlyapnikov, E. Palazzi, N. Masetti,
    F. Frontera, E. Costa, M. Feroci, and L. Piro

The Detection of Linear Polarization in the Afterglow of GRB 990510
and its Theoretical Implications 318
    D. Lazzati, S. Covino, and G. Ghisellini

Polarimetric Studies of Gamma-Ray Burst Afterglows 323
    S. Klose, B. Stecklum, and O. Fischer

Submillimeter Observations of GRB Counterparts 326
    I. A. Smith, J. van Paradijs, R. P. J. Tilanus, T. J. Galama, P. J. Groot,
    P. Vreeswijk, E. Rol, C. Kouveliotou, R. A. M. J. Wijers, and N. Tanvir

Color Indices of Optical Afterglows of GRBs 329
    V. Šimon, G. Pizzichini, and R. Hudec

Photometric Study of the Improved GRB 970815 Error Box 334
    J. Gorosabel, A. J. Castro-Tirado, N. Benítez, M. R. Zapatero-Osorio,
    A. Campos, J. Trapero, E. Sánchez, and N. Metcalfe

**CCD Observations of the QSO 4C49.29 in the Error Box
of GRB 960720** 339
    S. Guziy, A. Shlyapnikov, and R. Hudec

Optical Study of the Counterpart to GRB 990712 342
    J. Gorosabel, A. J. Castro-Tirado, P. Saizar, N. J. Rattenbury, I. A. Bond,
    P. Yock, J. Hearnshaw, P. M. Kilmartin, Y. Muraki, T. Nakamura,
    K. Ohnishi, M. Reid, T. Saito, and S. Noda

Early Search for the Optical Afterglow from GRB 990806 347
    N. J. Rattenbury, I. A. Bond, A. J. Castro-Tirado, J. Gorosabel,
    J. Hearnshaw, P. M. Kilmartin, Y. Muraki, T. Nakamura, K. Ohnishi,
    M. Reid, T. Saito, M. Feroci, and P. Yock

GRB Optical Searches: Results from UKSTU Plates 352
    R. Hudec

Connections between Parameters of GRB Afterglows 355
    G. M. Beskin, C. Bartolini, G. Cosentino, A. Guarnieri, S. Lodi,
    and A. Piccioni

**What Have We Learned from Optical Detections of GRBs: Feasibility
of Independent Searches?** 360
    R. Hudec

X-RAY/GAMMA-RAY AFTERGLOW

X-Ray Afterglow of Gamma-Ray Bursts 365
    E. Costa

**Constraints to the Nature of the Central GRB Engine from
a Comparative Analysis of X-Ray Properties of Afterglows** 375
    G. Stratta, L. Piro, G. Gandolfi, L. A. Antonelli, E. Costa, M. Feroci,
    P. Soffitta, F. Frontera, J. Heise, and the BeppoSAX GRB Team

Search for X-Ray Afterglows from Gamma-Ray Bursts in the RASS 380
    J. Greiner, D. H. Hartmann, W. Voges, T. Boller, R. Schwarz,
    and S. V. Zharykov

BATSE Observations of Gamma-Ray Burst Tails 385
    V. Connaughton

**Possible Evidence for Soft Glowing Emission from Gamma-Ray Bursts
Detected by the APEX Experiment on Phobos-2** 390
    D. A. Litvine, I. G. Mitrofanov, and A. S. Kosyrev

**Evidence for Early High-Energy Afterglow: BATSE Observations
of GRB980923** 394
    T. W. Giblin, J. van Paradijs, C. Kouveliotou, V. Connaughton,
    R. A. M. Wijers, M. S. Briggs, R. D. Preece, and G. J. Fishman

GRB Spectral Hardness and Afterglow Properties 399
    J. Hakkila, T. Giblin, and R. D. Preece

HOST GALAXIES

Host Galaxies Have 'Normal' Luminosities 404
    B. E. Schaefer
The Host of GRB 970828: Another Subluminous Galaxy? 409
    J. Gorosabel and A. J. Castro-Tirado

SUPERNOVA CONNECTIONS—OBSERVATIONAL

The GRB/SN Connection: An Improved Spectral Flux Distribution
for the SN-Like Component to the Afterglow of GRB 970228,
the Non-detection of a SN-Like Component to the Afterglow of GRB 990510,
and GRBs as Beacons to Locate SNe at Redshifts $z \approx 4-5$ 414
    D. E. Reichart, D. Q. Lamb, and F. J. Castander
The GRB/Supernova Connection 419
    B. E. Schaefer and M. Deng
Are There Unrecognized Optical Afterglows of GRBs among
Observed SNe? 424
    R. Hudec, V. Hudcová, E. Palazzi, N. Masetti, and G. Pizzichini

IV. THEORY

PROMPT EMISSION MECHANISMS

External Shock Model for Gamma-Ray Bursts during
the Prompt Phase 431
    C. D. Dermer
Spectral Modeling of GRB Pulses 441
    H. Papathanassiou
Time Profiles and Spectral Evolution of GRB Pulses 446
    E. Liang, A. Crider, M. Böttcher, and I. Smith
Constraints on the Internal Shock Model from the Temporal Analysis
of GRBs 450
    M. Spada, A. Panaitescu, and P. Mészáros
A Plasma Instability Theory of Prompt Gamma-Ray Burst Radiation 455
    J. J. Brainerd
Emission of Cosmological Gamma-Ray Bursts in the GeV–TeV Energy
Domain 460
    E. V. Derishev, V. V. Kocharovsky, and Vl. V. Kocharovsky
Synchrotron Radiation as the Source of GRB Spectra, Part I: Theory 465
    N. M. Lloyd and V. Petrosian
Comparisons between Analytic and Numerical Calculations
of GRB Spectra 470
    C. D. Dermer, M. Böttcher, and J. Chiang

Evaluating Spectral Functions Used to Test the Synchrotron Shock Model 475
   A. Crider
Redshifts and Compton Attenuation 480
   J. J. Brainerd and R. D. Preece
An Eigenfunction Method for Particle Acceleration at Ultra-relativistic
Shocks 485
   A. W. Guthmann, J. G. Kirk, Y. A. Gallant, and A. Achterberg
The Synchrotron Spectrum of Fast Cooling Electrons Revisited 489
   J. Granot, T. Piran, and R. Sari
Flow-Field Dependent Variation Method for Complex Relativistic Fluids 494
   G. A. Richardson, T. J. Chung, G. R. Karr, and G. N. Pendleton
A Unified Picture for the Various Total Energies of GRBs 499
   T. Totani

AFTERGLOW EMISSION MECHANISMS

Beaming and Jets in Gamma-Ray Bursts 504
   R. Sari
Photospheres, Comptonization and X-Ray Lines in Gamma-Ray Bursts 514
   P. Mészáros
Modeling the Iron Line in GRB Afterglows 519
   M. Böttcher
Particle Acceleration at Ultra-relativistic Shocks and the Spectra
of Relativistic Fireballs 524
   Y. A. Gallant, A. Achterberg, J. G. Kirk, and A. W. Guthmann
Afterglow Lightcurves from Beamed Outflows 530
   A. Majczyna, T. Bulik, R. Moderski, and M. Sikora
The Patchy Shells Model 535
   T. Piran and P. Kumar
Hydrodynamics and Radiation from a Relativistic Expanding Jet
with Applications to GRB Afterglow 540
   J. Granot, M. Miller, T. Piran, and W.-M. Suen
Early GRB Afterglows from Relativistic Blast Waves in General
Radiative Regimes 545
   M. Böttcher and C. D. Dermer
Optical Flashes and Radio Flares in GRB Afterglows: Numerical Study 550
   S. Kobayashi and R. Sari

CENTRAL ENGINE

Gamma-Ray Bursts: The Central Engine 555
   S. E. Woosley
2D Hydrodynamic Simulations of Relativistic Jets from Collapsars 565
   E. Müller, M. A. Angel Aloy, J. Mª Ibáñez, J. Mª Martí,
   and A. MacFadyen

Gamma-Ray Bursts via Pair Plasma Fireballs from Heated
Neutron Stars 570
  J. D. Salmonson, J. R. Wilson, and G. J. Mathews
Asymmetric Neutron Star Coalescences: Implications for GRBs 575
  S. Rosswog, M. B. Davies, F.-K. Thielemann, and T. Piran
Mass-Loss from a Magnetically Driven Wind Emitted by a Disk Orbiting
a Steller-Mass Black Hole 579
  F. Daigne and R. Mochkovitch
Fireballs from Collapse of Neutron Stars Induced by Primordial
Black Holes 584
  E. V. Derishev, V. V. Kocharovsky, and Vl. V. Kocharovsky
Gamma-Ray Bursts from Rapidly Spinning Neutron Stars 589
  H. C. Spruit
Red Hole Gamma-Ray Bursts: A New Gravitational Collapse Paradigm
Explains the Peak Energy Distribution and Solves the GRB Energy Crisis 594
  J. S. Graber
Electric GRBs 599
  L. Körtvélyessy
A Ritzian Interpretation of Variable Stars 603
  R. S. Fritzius

SUPERNOVA CONNECTIONS—THEORETICAL

The Connection between Supernovae and Gamma-Ray Bursts: On the
Distribution of the Circumstellar Matter 608
  R. A. Chevalier
The Type Ib/c Supernova, Gamma-Ray Burst, Soft Gamma-Ray
Repeater, Magnetar Connection 617
  J. C. Wheeler, P. Höflich, L. Wang, and I. Yi
Properties of Hypernovae: SNe 1997ef, 1998bw, and 1997cy 622
  K. Nomoto, K. Maeda, T. Nakamura, K. Iwamoto, T. Suzuki,
  P. A. Mazzali, M. Turatto, I. J. Danziger, and F. Patat
Is Nova Sco 1994 (GRO 1655–40) a Relic of a GRB? 628
  G. E. Brown, C.-H. Lee, H. K. Lee, and H. A. Bethe
Collapsars 633
  A. MacFadyen

PROGENITORS

Looking for GRB Progenitors 638
  K. Belczyński, T. Bulik, and B. Rudak
Making Accretion Disks around Black Holes: GRB Progenitors 643
  C. L. Fryer and W. Zhang
Distribution of Binary Mergers around Galaxies 648
  T. Bulik and K. Belczyński

GRBs AS COSMOLOGICAL PROBES

The Most Distant Gamma-Ray Bursts 653
    D. H. Hartmann, A. I. MacFadyen, and S. E. Woosley
Gamma-Ray Bursts as a Probe of the Very High Redshift Universe 658
    D. Q. Lamb and D. E. Reichart
Gamma-Ray Burst Lensing Limits on Cosmological Parameters 663
    R. J. Nemiroff, G. F. Marani, J. P. Norris, J. T. Bonnell, C. A. Meegan, and K. C. Hurley

V. INSTRUMENTATION

FUTURE FLIGHT EXPERIMENTS

The Swift Gamma-Ray Burst MIDEX 671
    N. Gehrels
Observing Gamma-Ray Bursts with INTEGRAL 681
    C. Winkler
The INTEGRAL Burst Alert System 686
    S. Mereghetti, S. Brandt, D. Jennings, J. Borkowski, and R. Walter
Observations of Gamma-Ray Bursts with MAXI on the International Space Station 691
    N. Kawai, A. Yoshida, T. Mihara, H. Negoro, Y. Shirasaki, I. Sakurai, M. Matsuoka, K. Torii, S. Ueno, M. Sugizaki, H. Tomida, W. Yuan, H. Tsunemi, E. Miyata, and M. Yamauchi
MARGIE: A Gamma-Ray Burst Ultra-long Duration Balloon Mission 696
    D. Band, J. Matteson, M. Cherry, J. G. Stacy, P. Altice, T. G. Guzik, S. C. Kappadath, J. Buckley, P. Hink, J. Macri, M. McConnell, J. Ryan, T. O'Neill, and A. Zych
A New X-Ray Telescope for Monitoring the X-Ray Afterglows of GRBs 701
    R. Hudec, A. Inneman, L. Pina, and P. Gorenstein
Estimation of GRB Detection by FiberGLAST 706
    S. Phengchamnan, K. Aisaka, M. Atac, W. R. Binns, J. H. Buckley, M. L. Cherry, D. Cline, P. Dowkontt, J. W. Epstein, M. H. Finger, G. J. Fishman, T. G. Guzik, P. L. Hink, M. H. Israel, S. C. Kappadath, G. R. Karr, R. M. Kippen, J. Macri, R. S. Mallozzi, M. L. McConnell, Y. Pischalnikov, W. S. Paciesas, T. A. Parnell, G. N. Pendleton, R. D. Preece, G. A. Richardson, K. Rielage, J. M. Ryan, J. G. Stacy, T. O. Tümer, D. B. Wallace, and R. B. Wilson

CURRENT FLIGHT EXPERIMENTS

A Robust Filter for the BeppoSAX Gamma-Ray Burst Monitor Triggers 711
    M. Feroci, C. L. Bianco, F. Lazzarotto, A. Mattei, G. Ventura, E. Costa, and F. Frontera

GRB Localization with the BeppoSAX Gamma-Ray Burst Monitor 716
    B. Preger, E. Costa, M. Feroci, L. Amati, and F. Frontera

Response Function of the Gamma-Ray Burst Monitor (GRBM) Onboard the BeppoSAX Satellite 721
    F. Calura, M. Rapisarda, F. Frontera, E. Montanari, C. Guidorzi,
    L. Amati, M. Feroci, E. Costa, and P. Collina

Progress Incorporating the NEAR Mission into the Interplanetary GRB Network 726
    T. L. Cline, S. Barthelmy, P. Butterworth, T. McClanahan,
    D. Palmer, J. Trombka, K. Hurley, R. Gold, R. M. Kippen,
    C. Kouveliotou, D. Frederiks, S. Golenetskii, and E. Mazets

GRB Coordinates Network (GCN): A Status Report 731
    S. D. Barthelmy, T. L. Cline, P. Butterworth, R. M. Kippen,
    M. S. Briggs, V. Connaughton, and G. N. Pendleton

GROUND-BASED EXPERIMENTS

Super-LOTIS Early Time Optical Counterpart Measurements 736
    H. S. Park, R. A. Porrata, G. G. Williams, E. Ables, D. L. Band,
    S. D. Barthelmy, R. M. Bionta, T. L. Cline, D. H. Ferguson,
    G. J. Fishman, N. Gehrels, D. Hartmann, K. Hurley, C. Kouveliotou,
    C. A. Meegan, R. Nemiroff, and W. Pereira

Rapid, Deep GRB Observations with the U.S. Naval Observatory 1.3-m Wide-Field Telescope 741
    F. J. Vrba, H. C. Harris, B. Canzian, A. A. Henden, S. E. Levine,
    C. B. Luginbuhl, D. H. Hartmann, M. C. Jennings, and R. M. Kippen

TAROT-2: A Versatile Large Observatory for Optical Transients 746
    M. Boër

Milagro: A TeV Observatory for Gamma-Ray Bursts 751
    B. L. Dingus, R. Atkins, W. Benbow, D. Berley, M. L. Chen, D. G. Coyne,
    D. E. Dorfan, R. W. Ellsworth, D. Evans, A. Falcone, L. Fleysher, R. Fleysher,
    G. Gisler, J. A. Goodman, T. J. Haines, C. M. Hoffman, S. Hugenberger,
    L. A. Kelley, I. Leonor, M. McConnell, J. F. McCullough, J. E. McEnery,
    R. S. Miller, A. I. Mincer, M. F. Morales, P. Nemethy, J. M. Ryan, B. Shen,
    A. Shoup, C. Sinnis, A. J. Smith, G. W. Sullivan, T. Tümer, K. Wang,
    M. O. Wascko, S. Westerhoff, D. A. Williams, T. Yang, and B. G. Yodh

Searching for Optical and High-Energy Transients from Gamma-Ray Bursts Simultaneously with Cerenkov Telescopes 756
    A. Piccioni, C. Bartolini, G. Beskin, S. Biryukov, D. Eichler, D. Faiman,
    and A. Guarnieri

VI. SOFT GAMMA REPEATERS

X-RAY/GAMMA-RAY OBSERVATIONS

The 4.5 ± 0.5 Soft Gamma Repeaters in Review 763
    K. Hurley

BeppoSAX and Ulysses Data on the Giant Flare from SGR 1900+14.........771
    M. Feroci, K. Hurley, R. Duncan, C. Thompson, E. Costa,
    and F. Frontera

The Hard Side of SGR 1900+14 776
    P. M. Woods, C. Kouveliotou, J. van Paradijs, M. S. Briggs,
    K. Hurley, E. Göğüş, R. D. Preece, T. W. Giblin, C. Thompson,
    and R. C. Duncan

Spin Period and Burst Rate History of SGR 1900+14......................781
    P. M. Woods, C. Kouveliotou, J. van Paradijs, M. H. Finger,
    C. Thompson, R. C. Duncan, K. Hurley, T. Strohmayer, J. Swank,
    and T. Murakami

Deep Searches for Pulsations in SGR 1627−41 and 0526−66 786
    T. Murakami, M. Ando, K. Hurley, P. Li, C. Kouveliotou,
    P. Woods, D. H. Hartmann, I. Smith, M. Tsujimoto,
    T. Strohmayer, and A. Yoshida

Soft Gamma Repeaters as Relaxation Systems 791
    D. M. Palmer

Statistical Properties of SGR 1900+14 Bursts...........................796
    E. Göğüş, P. M. Woods, C. Kouveliotou, J. van Paradijs,
    M. S. Briggs, R. C. Duncan, and C. Thompson

OPTICAL/INFRARED OBSERVATIONS

Optical Imaging of the SGR 1627−41 Error Box during the SGR Activity in June 1998...801
    A. J. Castro-Tirado, N. Lund, D. Pinfield, and S. Covino

Rapid Optical Follow-up Observations of SGR Events with ROTSE-I 804
    R. Balsano, C. Akerlof, S. Barthelmy, J. Bloch, P. Butterworth,
    D. Casperson, T. Cline, S. Fletcher, G. Gisler, J. Hills, R. Kehoe,
    B. Lee, S. Marshall, T. McKay, A. Pawl, W. Priedhorsky,
    N. Seldomridge, J. Szymanski, and J. Wren

Search for Photometric Variability in the Vicinity of SGR 1900+14 and Discovery of a High-Mass Cluster 809
    F. J. Vrba, C. B. Luginbuhl, A. A. Henden, H. H. Guetter,
    and D. H. Hartmann

Soft Gamma-Ray Repeaters in Clusters of Massive Stars 814
    I. F. Mirabel, Y. Fuchs, and S. Chaty

ISO Observation of a Fraction of the SGR 1801−23 Error Box.............818
    A. J. Castro-Tirado, L. Metcalfe, and R. Laureijs

Optical/Near-IR Observations of SGR 1900+14 during the May−June and Aug−Sep 1998 Active Periods......................................821
    A. J. Castro-Tirado, S. Beckwith, D. Kelson, T. Kerr, C. Lázaro,
    and S. Madruga

NIR Spectroscopic Observations of the SGR 1900+14 M Stars..............825
    E. W. Guenther, S. Klose, and F. J. Vrba

THEORY

Physics in Ultra-strong Magnetic Fields 830
    R. C. Duncan
Magnetic Field Limits on SGRs 842
    R. E. Rothschild, D. Marsden, and R. E. Lingenfelter
Environmental Influences in SGRs and AXPs 847
    D. Marsden, R. Lingenfelter, R. Rothschild, and J. Higdon
Relativistic Compton Scattering in Ultra-strong Magnetic Fields 852
    P. L. Gonthier, R. M. Costello, C. L. Mercer, A. K. Harding, and M. G. Baring
Nuclear Equation of State and Internal Structure of Magnetars 857
    I.-S. Suh and G. J. Mathews
New QED Calculations for Processes in Strong Magnetic Fields 862
    D. Leahy and L. Semionova
Electric Magnetars 867
    L. Körtvélyessy

VII. TECHNIQUES AND MISCELLANY

Attributes of GRB Pulses: An Improved Bayesian Blocks Algorithm for Binned Data 873
    J. D. Scargle, J. Norris, G. Marani, and J. Bonnell
A GRB Tool Shed 877
    D. J. Haglin, R. J. Roiger, J. Hakkila, G. N. Pendleton, and R. Mallozzi
Properties of Karhunen-Loeve Expansion of Astronomical Images in Comparison with Other Integral Transforms 882
    P. Páta, M. Bernas, A. J. Castro-Tirado, and R. Hudec
Fast and Simple Data Compression 887
    M. Nekola and M. Bernas
Accuracy of Press Reports on Gamma-Ray Astronomy 890
    B. E. Schaefer, R. J. Nemiroff, and K. Hurley

Symposium Participants 895
Author Index 901

PREFACE

Since the initial discovery of multiwavelength afterglow emission from the gamma-ray burst (GRB) of 27 February 1997, more than a dozen additional bursts with afterglow have been observed. These hallmark events have led to more progress in our understanding of the GRB phenomenon than the multitude of bursts observed during thirty years of previous research. We now observe that all GRBs with measured redshifts originate from cosmological distances, and most are seen with their host galaxies. These discoveries have enabled a remarkably detailed (but still incomplete) theoretical framework.

The Fifth Huntsville Gamma-Ray Burst Symposium was held during the week of 18–22 October 1999 to discuss recent developments in GRB research. Like previous Huntsville GRB Symposia (held biennially since 1991), much of the discussion was devoted to recent observations and ongoing analyses of archival data. However, there were also renewed, lively discussions related to theoretical developments, reflecting the progress afforded by recent afterglow discoveries.

Even in this new era of GRB research, where much activity is concentrated on afterglow data, there is still a large effort being devoted to prompt GRB emission. In this area, confidence in the cosmological distance scale has led to several attempts to correlate burst properties with luminosity and to search for other signatures that might indicate the intrinsic nature of the burst mechanism. There is continued interest in burst classification, and in the clear distinction between short and long bursts. Afterglow has thus far been detected only from long-duration bursts. In terms of new observations, the discovery of prompt optical emission from GRB 990123 has motivated searches for similar events. There is also tantalizing evidence from the Milagro experiment for prompt TeV emission from one burst.

New afterglow detections continue at a rate of about one burst per month, thanks to coordinated multiwavelength observing programs (now including optical polarimetry). Afterglow emission is generally well behaved (compared to the usually chaotic prompt emission) and described remarkably well by the "standard" relativistic fireball shock model. The most enlightening results are related to departures from the standard behavior, including events with rapid or changing decay rate, which are interpreted as possible evidence for jets, circumstellar winds, or supernova emission. Of further interest is the transition from prompt emission to afterglow, where some progress has been made using gamma-ray and X-ray data. One of the most important uses of afterglow observations has been to identify host galaxies, which show evidence of active star formation and thus hint at massive star progenitors. The most provocative afterglow result—and the source of much speculation—is the possible association of GRB 980425 with the bright, relatively nearby supernova 1998bw.

Because the fireball shock model appears to fit the afterglow data rather well, much theoretical activity is now being devoted to explaining the prompt emission and its association with the central energy "engine." There have been many attempts to describe the rich temporal and spectral variety in the gamma-ray data with various incarnations of shock acceleration, with mixed results and total energy estimates that vary by several orders of magnitude. The prevailing central engine models are compact object mergers (usually evoked to explain short-duration bursts) and supernova-like events known as "collapsars" or "hypernovae." The theoretical similarity between GRBs and supernovae has prompted speculation about a real connection, especially in view of recent observational evidence. If GRBs do have massive star progenitors, they might be a sensitive tracer of star formation in the high-redshift universe.

As usual, the most interesting new results are unfortunately lacking in confidence due to the small number of observations and limited sensitivity. New space missions like HETE-2 (2000) and Swift (2003), and new ground-based instruments, are eagerly anticipated. The Swift mission is particularly promising in that it will attempt to measure both prompt and afterglow properties for hundreds of bursts, and determine redshifts in many cases. In the more distant future, a next generation GRB observatory is envisioned that will detect and measure redshifts for thousands of bursts, and use them to probe the young universe. Early ideas for such a mission were discussed at a special evening session of the Symposium.

Preceding the Gamma-Ray Burst Symposium was a one-day workshop on soft gamma repeater (SGR) observations and theory. There has been remarkable progress in this field due to the recent discovery of quiescent X-ray pulsations from two SGRs, the discovery of a new SGR source, and the observations of a giant flare from SGR 1900+14. Association with supernova remnants and changes in the pulse period indicate a highly magnetized neutron star ("magnetar") origin. However, an alternative interpretation is that SGRs are powered by interactions between the normal magnetic field of a neutron star and strong stellar winds.

In this volume we have assembled the written papers from both oral and poster presentations at the Symposium. The papers have been arranged into topical sections to aid the reader, although many papers could fit into several categories.

This Symposium was a success due to the efforts of many people. In particular, we thank our fellow local organizer V. Connaughton for her invaluable assistance, S. Zarger and other staff members of the Universities Space Research Association for their logistical support, and D. Dooling for his production of web-based stories and organizing press activities. The contributions of the Scientific Organizing Committee—J. Van Paradijs (UoA & UAH, Chair), S. E. Woosley (UCSC), G. J. Fishman (NASA/MSFC), L. Piro (IAS/CNR), N. Gehrels (NASA/GSFC), C. Kouveliotou (USRA), S. R. Kulkarni (CalTech), P. Mészarós (Penn State), R. M. Kippen (UAH), and R. A. M. J. Wijers (SUNY-Stony Brook)—are gratefully acknowledged. We are also grateful to our sponsors, the Compton Observatory Science Support Center, and The Curry Foundation.

Huntsville, Alabama, USA *R. Marc Kippen*
March 2000 *Robert S. Mallozzi*
  Gerald J. Fishman

Jan van Paradijs
1946 – 1999

DEDICATION

These Proceedings are dedicated to the memory of Jan van Paradijs, who passed away less than a month after the Fifth Huntsville Gamma Ray Burst Symposium. Jan was the Chairman of the Scientific Organizing Committee (SOC) for this Symposium. In this position, he selected the other members of the SOC and organized the selection of the entire program. Much of this work was done during the time he was terminally ill and in considerable pain. Jan wanted desperately to attend the symposium, but could not because of his failing health. He sent a tape-recorded greeting that was played for the attendees during the opening session on October 19. Jan died on November 2 in Amsterdam, with his wife, Chryssa Kouveliotou, at his side. Chryssa was also a member of the Scientific Organizing Committee for this Symposium. She was with Jan throughout his illness in Huntsville and in Amsterdam.

We were all saddened by the loss of our colleague and friend. But we are grateful for the work that he did and for the knowledge that he gave to us and to the world. Jan was one of the most productive multiwavelength astronomers in high-energy astrophysics. He had a primary role in the rapid series of discoveries in the past few years that provided the long-sought breakthrough in the study of gamma-ray bursts. For this, Jan will be long remembered.

Those in our field recognized Jan as one of the most knowledgeable, productive, and wide-ranging astrophysicists of our time. He had a unique talent for providing insight concerning many varied objects in high-energy astrophysics. He was a master of both observational and theoretical astrophysics. Until just a few days before his death, he was working with colleagues and students in Amsterdam, Huntsville, and other locations throughout the world. He had just completed a Perspectives article on the link between GRBs and supernovae (*Science*, 22 October 1999) and had done extensive work on a review article on GRB afterglows with Chryssa and his former student, Ralph Wijers (*Ann. Rev.*, in press, 2000).

Soon after Jan married Chryssa, he began an extensive collaboration with the BATSE Team in Huntsville. Early on, he recognized the potential of the Italian-Dutch BeppoSAX mission for GRB investigations. The Dutch-provided WFC on BeppoSAX was capable of rapid, precise GRB locations, and Jan was prepared to take advantage of that capability several years before the launch of BeppoSAX.

The story of the GRB 970227 optical counterpart discovery began with Chryssa receiving a message while she was at a conference in Tokyo that a gamma-ray burst had just been detected and precisely located by BeppoSAX. She called Jan in Huntsville, who immediately began working on an observing program with graduate students Paul Groot and Titus Galama in Amsterdam. The observations were made with the Westerbork Radio Telescope, the William Hershel Telescope, and the Issac Newton Telescope and then promptly analyzed, interpreted, and reported by the team, led by Jan.

Jan usually had four or five papers in progress and several key observations at powerful ground-based or satellite-based observatories about to take place. Simultaneously, he wrote comprehensive review articles, gave invited lectures, edited books, planned symposia, and had major research efforts going on two continents. Many proposals or ideas for proposals were always in the works. Yet, above all else, he considered his teaching and advising role to have the highest priority, and this role was the most satisfying to him. The time that he spent with his students was sacred, and he had their undivided attention.

An interesting short biography (obituary) of this remarkable individual was written by Jan's longtime colleague, collaborator, and friend at the University of Amsterdam, Ed van den Heuvel (*Nature*, 16 December 1999).

Jan will be most remembered for what he would have wanted us to remember him by. His students and their students will continue his legacy into the future: from advisor to student, for many academic generations of astrophysicists to come. For those of us in the GRB field, Jan's memory will live on.

Huntsville, Alabama, USA *Gerald J. (Jerry) Fishman*
March 2000

I. GENERAL PROPERTIES

A Gamma-Ray Burst Bibliography, 1973–1999

Kevin Hurley

UC Berkeley Space Sciences Laboratory
Berkeley, CA 94720-7450

Abstract. On the average, one new publication on cosmic gamma-ray bursts enters the literature every day. The total number now exceeds 4100. I present here a complete bibliography which can be made available electronically to interested parties.

INTRODUCTION

I have been tracking the gamma-ray burst literature for about the past twenty-one years, keeping the authors, titles, references, and key subject words in a machine-readable form. The present version updates previous ones reported in 1994, 1996, and 1998 [1-3]. In its current form, this information is in a Microsoft Word 97 "doc" format. My purpose in doing this was first, to be able to retrieve rapidly any articles on a given topic, and second, to be able to cut and paste references into manuscripts in preparation. The following journals have been scanned on a more or less regular basis starting with the 1973 issues:

Advances in Physics*
Annals of Physics*
Astronomical Journal*
Astronomische Nachrichten*
Astronomy and Astrophysics (letters, main journal, and supplement series)*
Astronomy and Astrophysics Review*
Astronomy Letters* (formerly Soviet Astronomy Letters)
Astronomy Reports*(formerly Soviet Astronomy)
Astrophysical Journal (letters, main journal, and supplements)*
Astrophysical Letters and Communications
Astrophysics and Space Science*
ESA Bulletin*
ESA Journal*
IEEE Transactions on Nuclear Science*

Journal of Astrophysics and Astronomy*
Monthly Notices of the Royal Astronomical Society*
Nature
Nuclear Instruments and Methods in Physics Research Section A*
Observatory*
Physical Review (main journal A and letters)*
Proceedings of the Astronomical Society of Australia*
Publications of the Astronomical Society of Japan*
Publications of the Astronomical Society of the Pacific*
Reports on Progress in Physics*
Science*
Scientific American
Sky & Telescope

The asterisks indicate journals which are scanned using the on-line version of Current Contents. In addition, the following journals have been scanned, but in many cases less regularly, particularly in the past:

Annals of Geophysics
Astrofizika
Bulletin of the American Astronomical Society
Bulletin of the American Physical Society
Chinese Astronomy
Cosmic Research
Journal of Atmospheric and Terrestrial Physics
Journal of the British Interplanetary Society
Journal of the Royal Astronomical Society of Canada
Progress in Theoretical Physics
Solar Physics
Soviet Physics

The above lists are not exhaustive. For example, where theses, newspaper articles, or internal reports have come to my attention, I have included them, too. To be included, an article had to have something to do with gamma-ray burst theory, observation, or instrumentation, or be closely related to one of these topics (e.g., merging neutron stars, AXPs, SGRs, the Bursting Pulsar), and must have been published in some form. With only a few exceptions, preprints which were never published have not been included.

ORGANIZATION OF THE BIBLIOGRAPHY

The overall organization is chronological by year. Within a given year, articles published in journals are listed first, in alphabetical order by first author. Then come theses and conference proceedings articles. The latter are listed in the order in which they appear in the proceedings.

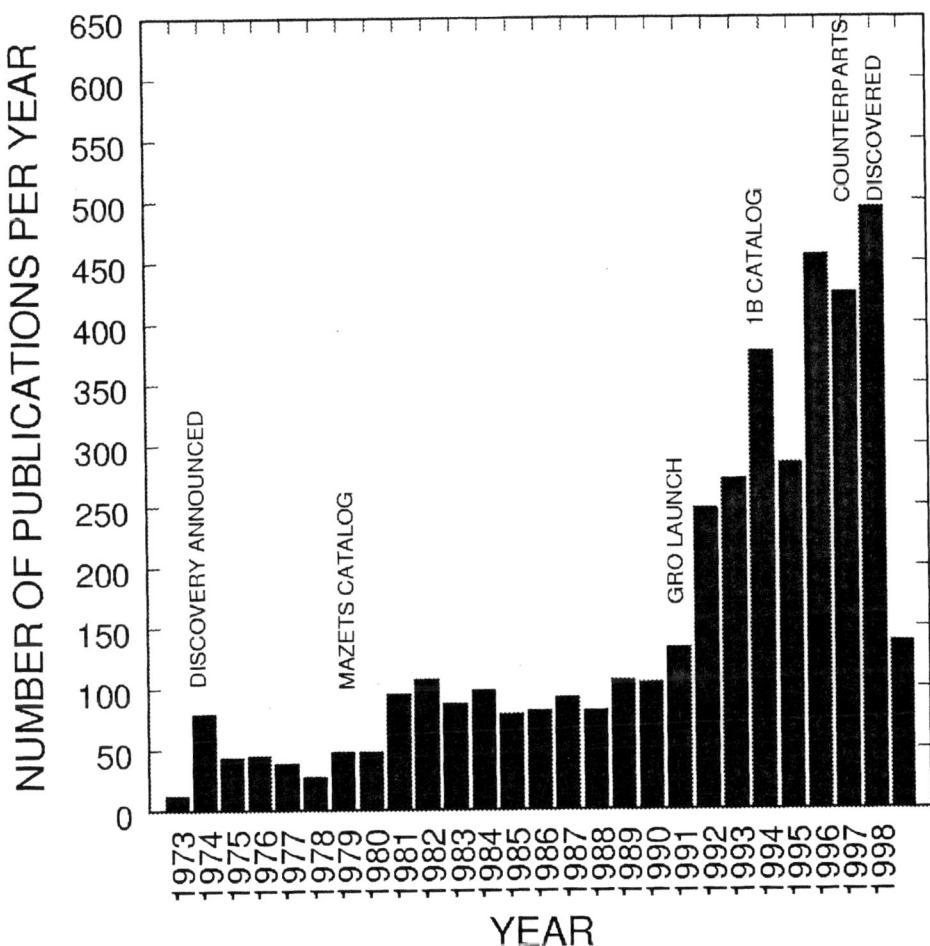

FIGURE 1. Number of gamma-ray burst publications as a function of year. The cutoff date is mid-1999.

The entries are numbered consecutively, so that paper copies which are kept on file can be retrieved quickly. However, to avoid having to renumber this entire file when a new article is added, numbers are skipped at the end of each year and reserved for later inclusion. The complete author list follows, as it appears in the journal, along with the title, journal, volume number, page number, and year. A line containing key words follows this. These are generally not the same key words as the ones listed in the journal, nor are they taken from the title or any particular list. Rather, they are meant to reflect the true content of the article, and provide

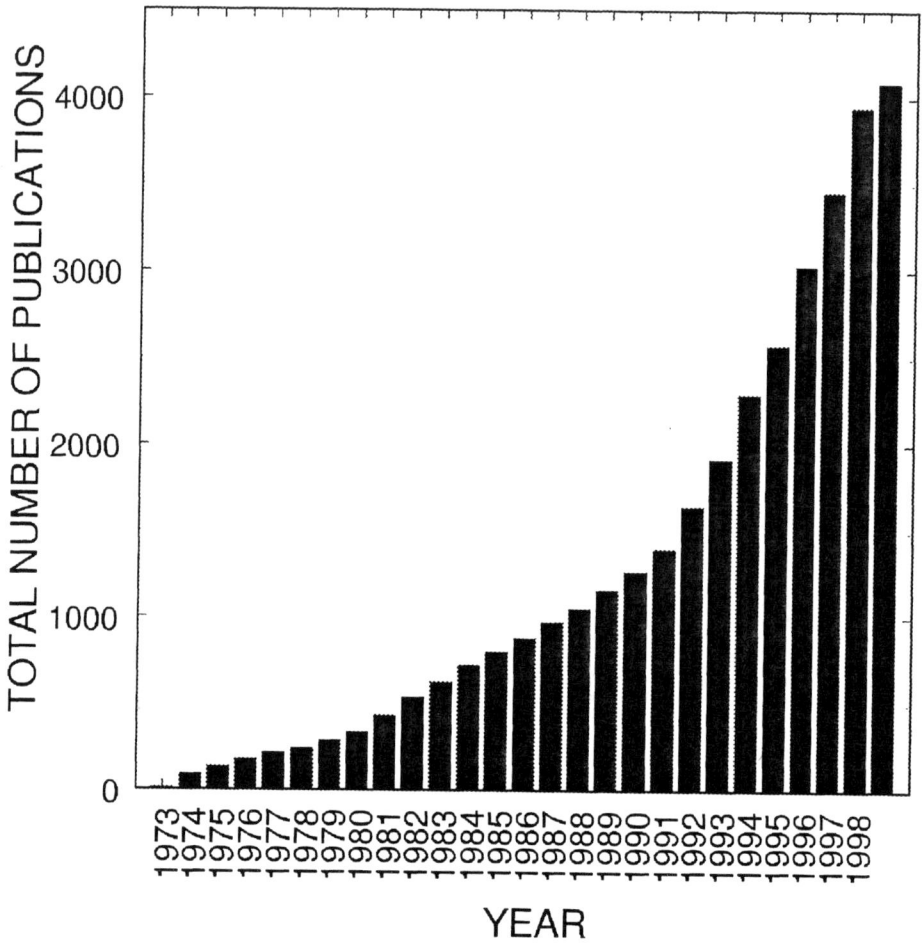

FIGURE 2. The cumulative number of publications by year.

a list of machine-searchable topics. In general, however, key words have not been included for conference proceedings articles.

A FEW INTERESTING STATISTICS

The number of articles published each year since 1973 is shown in Figure 1. Starting with a modest article per month in 1973, it began to exceed one per day in 1994. Several milestones are indicated as the probable causes of sudden increases in the publication rate. The apparent decreases in the rates in 1995 and 1997 are in fact due to a 2 year periodicity in the publications caused by the influx of a large number of articles from the Huntsville Workshop proceedings. In keeping with this publication rate, the bibliography is updated on an approximately daily basis. Note that there are still about as many papers published as there are gamma-ray bursts. The cumulative total is shown in Figure 2.

The sheer volume of the literature has necessitated the development of a program which can search for and extract particular articles. I have written such a program in Microsoft Word Basic (a variant of the BASIC programming language). It allows one to extract all articles between two dates whose entries contain a particular key phrase and write it to a separate file.

AVAILABILITY

The IPN web site http://ssl.berkeley.edu/ipn3/index.html contains a version of this bibliography. More up-to-date versions in plain ASCII, "doc", and "rich text" (rtf) formats can be made available to interested parties as time permits. Please contact me at

khurley@sunspot.ssl.berkeley.edu

to request copies, and indicate your preference for the format. I would appreciate it if users would communicate errors and omissions to me.

ACKNOWLEDGMENTS

This work was carried out under JPL Contract 958056 and CGRO guest investigator grant NAG5-7810.

REFERENCES

1. Hurley, K., in *Gamma Ray Bursts - Second Workshop*, AIP Conference Proceedings 307, eds. G. Fishman, J. Brainerd, & K. Hurley, New York, 1994, p. 726.
2. Hurley, K., in *Gamma Ray Bursts - 3rd Huntsville Symposium*, AIP Conference Proceedings 384, eds. C. Kouveliotou, M. Briggs, & G. Fishman, New York, 1996, p. 985.
3. Hurley, K., in *Gamma Ray Bursts - 4th Huntsville Symposium*, AIP Conference Proceedings 428, eds. C. Meegan, R. Preece, & T. Koshut, New York, 1998, p. 87.

Derivation of a Sample of Gamma-Ray Bursts from BATSE DISCLA Data

Maarten Schmidt

California Institute of Technology, Pasadena, California 91125

Abstract. We have searched for gamma-ray bursts (GRBs) in the BATSE DISCLA data over a time period of 5.9 years. We employ a trigger requiring an excess of at least 5σ over background for at least two modules in the 50–300 keV range. After excluding certain geographic locations of the satellite, we are left with 4485 triggers. Based on sky positions, we exclude triggers close to the sun, to Cyg X-1, to Nova Persei 1992 and the repeater SGR 1806−20, while these sources were active. We accept 1013 triggers that correspond to GRBs in the BATSE catalog, and after visual inspection of the time profiles classify 378 triggers as cosmic GRBs. We denote the 1391 GRBs so selected as the "BD2 sample". The BD2 sample effectively represents 2.003 years of full sky coverage for a rate of 694 GRBs per year. Euclidean V/V_{max} values have been derived through simulations in which each GRB is removed in distance until the detection algorithm does not produce a trigger. The BD2 sample produces a mean value $\langle V/V_{max} \rangle = 0.334 \pm 0.008$.

INTRODUCTION

We have been engaged for several years in an effort to derive a homogeneous sample of gamma-ray bursts (GRBs) from the continuous data stream transmitted by the Burst and Transient Source Experiment (BATSE). In particular, we have used the DISCLA data which provide the counts for each of the eight BATSE detectors in channels 1–4 on a time scale of 1024 ms. The resulting *BD2 sample* of 1391 GRBs covers a time period of 5.9 years [3,4]. It includes 378 GRBs that are not in the BATSE catalog maintained on the BATSE web site. The BD2 sample has been used for a derivation of the characteristic luminosity and local space density of GRBs [4].

We will briefly review the procedures used to construct the BD2 sample, and analyze the differences between the BATSE catalog and the BD2 sample in some detail.

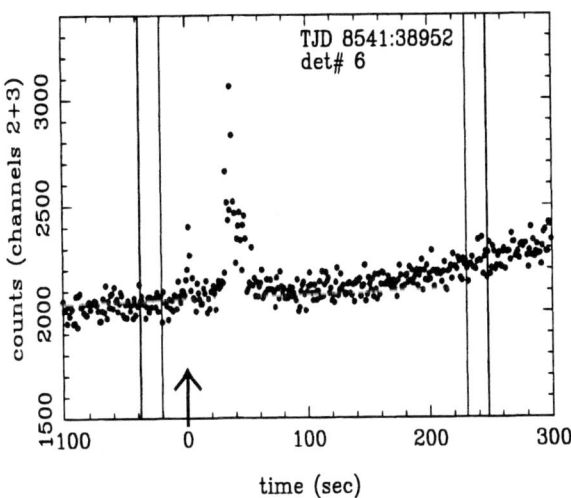

FIGURE 1. The test for the presence of a burst in time interval 0.000 − 1.024 s involves a linear interpolation between two 17.408 s averages of the background, one ending at time −20.48 s, the other starting at time 230.4 s.

DERIVATION OF THE BD2 SAMPLE

For most of the time period covered by the BATSE catalog, the on-board trigger mechanism required that the counts in channels 2 and 3 covering the energy range 50–300 keV exceed the background by 5.5σ on a time scale of 64, 256 or 1024 ms in two of the eight detectors. In our search based on DISCLA data, we used channels 2+3, and required an excess of 5σ above background on a time scale of 1024 ms in two detectors.

The BATSE on-board trigger employs a background averaged over 17.408 s that is updated every 17.408 s. We have taken advantage of the archival nature of the DISCLA data to use a background that is derived by linear interpolation from two averages taken over 17.408 s, as shown in Figure 1. The interval of 20.48 s between the first background average and the test time is intended to alleviate the problem of detecting slowly rising GRBs [1].

We searched for defects in the DISCLA data, since these could lead to false triggers or affect the background. For the time period of TJD 8365–10528, we found around 151,000 defects, ranging from checksum errors that affected only one 1024 ms bin, to gaps caused by transmission problems, passage through the South Atlantic Anomaly, etc. Around each of these defects, we set up an exclusion time window such that the defect does not affect the background estimation.

The initial search yielded 7536 triggers. The geographic coordinates of the satellites at the time of trigger showed strong concentrations over W. Australia and over Mexico and Texas [3]. We then outlined geographic exclusion zones to avoid

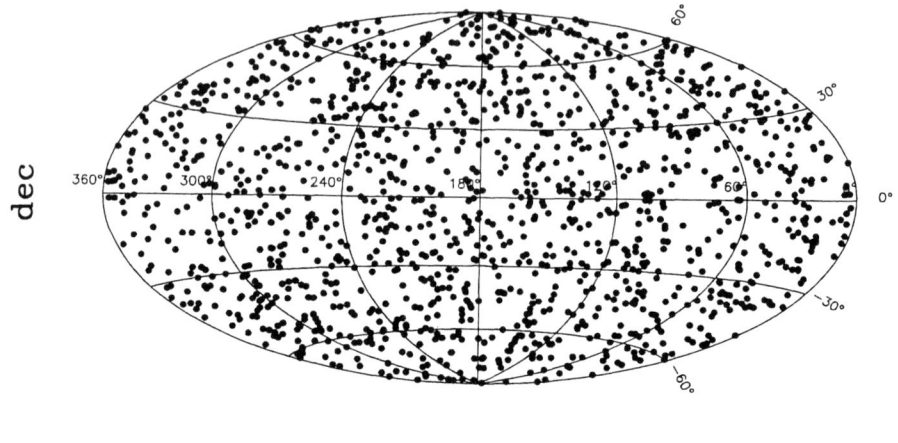

FIGURE 2. Equatorial coordinates of 1391 GRB in the BD2 sample

the trigger concentrations. With these exclusions in place, we were left with 4485 triggers. For each of these we derived celestial coordinates based on the relative response in all eight detectors and the orientation of the satellite, and other properties such as the duration, the hardness ratios, V/V_{max}, etc.

The equatorial coordinates of these triggers showed clear concentrations that were identified as Cyg X-1, Nova Persei 1992, and solar flares along the ecliptic [3]. We excluded from consideration all triggers whose positions were within 23 deg of these sources while they were active. Most of the remaining triggers were either magnetospheric events or cosmic GRBs. For a more detailed description of the selection procedure, the reader is referred to [3]. We ended up with a sample of 1422 GRBs [3] which we call the BD1 sample.

Subsequently, we carefully investigated all our bursts that were either not in the BATSE catalog (for an example, see Fig. 3), or whose positions or times agreed poorly with the catalog data. We eventually rejected 31 of the sources in the BD1 sample (most of which were parts of long bursts or identified as the repeater SGR 1806−20), resulting in the BD2 sample of 1391 GRBs [4]. We show in Table 1 an updated accounting of the classification of all 4485 triggers.

The most important property among those derived for the GRBs in the BD2 sample is V/V_{max}. Since redshifts are not know for most GRBs, V/V_{max} has to be evaluated in Euclidean space. It has usually been assumed that $V/V_{max} = (C_{max}/C_{min})^{-3/2}$, where C_{min} is the minimum detectable burst signal, and C_{max} the maximum amplitude. Instead, we derive V/V_{max} of each GRB by carrying out a simulation in which we move the source out in Euclidean space, and at each step apply the detection algorithm to see whether the reduced burst is still detected [3]. In the process of moving out the source, it may get detected later

TABLE 1. Classification of 4485 Triggers

Description	reject	accept
Within 23 deg of the sun when active	963	
Within 23 deg of CygX-1 when active	827	
Within 23 deg of Nova Persei when active	418	
Within 230.4 s of BATSE-listed burst		1013
Profile inspected: accepted as GRB		378
Profile inspected: rejected	818	
Part of a preceding long burst	18	
Soft repeater 1806−20	6	
Near sun, soft spectrum	44	
Total number of GRBs in the BD2 sample:		1391

and later (depending on its time profile) and some burst signal may be included in the background. In some cases, the C_{max} part of the profile is never detected before the source disappears as it is being removed. Using C_{max} to derive V/V_{max} therefore tends to lead to an underestimate of V/V_{max}.

Using the procedure outlined above, the BD2 sample produces a mean value $\langle V/V_{max}\rangle = 0.334 \pm 0.008$ [4]. The deviation from the value 0.5 expected for a uniform space distribution reflects to first order the effect of using Euclidean space in its derivation rather than a relativistic cosmological model. Hence, the Euclidean $\langle V/V_{max}\rangle$ is effectively a distance indicator, allowing derivation of the characteristic luminosity of GRBs [4].

COMPARISON OF THE BATSE CATALOG AND THE BD2 SAMPLE

Given that the BD2 sample was derived independently of the BATSE catalog, it is of interest to compare the two data sets.

1) The BD2 sample produces an all-sky rate of 694 per year, the BATSE catalog yields ~ 690 per year [2]. One might expect a larger rate in the BD2 sample since its limiting S/N is 5.0, while for the BATSE catalog it is 5.5. However, the BD2 sample is limited to detections at the time scale 1024 ms, while the BATSE catalog also includes time scales of 256 and 64 ms.

2) The BD2 sample has 378 GRB not in the BATSE catalog, and the BATSE catalog has 130 GRB, detected at a time scale of 1024 ms, that are not in the BD2 sample. These differences are related to the different depth of the two data sets (5σ vs. 5.5σ), but they are also influenced by items (3)–(6) below.

3) The $\sim 151{,}000$ time exclusion windows used in the derivation of the BD2 sample are independent from those used in the BATSE search.

4) Following a BATSE trigger, the detection mechanism was disabled until data could be transmitted to the ground. After a BD2 detection, we disabled the software trigger for 230 s.

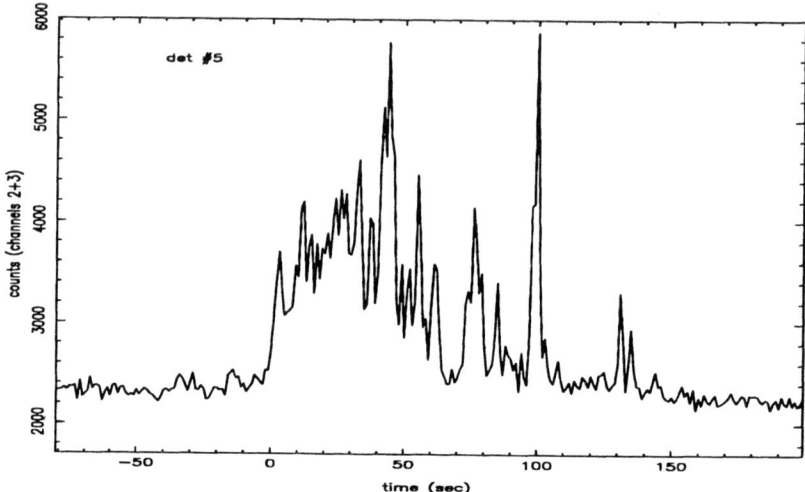

FIGURE 3. Example of a gamma-ray burst in the BD2 sample (TJD 8868:05731) that is not listed in BATSE catalog

5) The trigger criteria for the BATSE database were changed a number of times, see [2].

6) In general, if the background stretches used in two searches are different, then if a source is detected precisely at the limiting S/N in one of the searches, the probability that the other search finds the source is around 50%. The backgrounds used in the BATSE search and the BD2 sample are independent when the BATSE background stretch is separated by less than 3 s from the time of detection and partly dependent when the separation is larger than 3 s. Since we would find a substantial number of GRBs that are not in the BD2 by carrying out a search using different background stretches than those of Figure 1, we prefer to call the resulting collection of sources a sample. It should be stressed that the different samples so found are all statistically equivalent to each other, as long as one background choice is not to be preferred over the other.

REFERENCES

1. Higdon, J. C., and Lingenfelter, R. E., in *Gamma-Ray Bursts*, AIP Conference Proc. 384, New York: AIP, 1996, p. 402.
2. Meegan, C. A., et al. in *Gamma-Ray Bursts*, AIP Conference Proc. 428, New York: AIP, 1998, p. 3.
3. Schmidt, M., *A&A Suppl.* **138**, 409 (1999).
4. Schmidt, M., *ApJ* **523**, L117 (1999).

A Search for Non-Triggered Gamma-Ray Bursts in the BATSE Continuous Records: The Current Status

Boris E. Stern[*†‡], Yana Tikhomirova[†‡], Dmitri Kompaneets[†], Mikhail Stepanov[¶], Alexander Berezhnoy[¶], and Roland Svensson[‡§]

[*] Institute for Nuclear Research, Russian Academy of Sciences, Moscow 117312, Russia
[†] Astro Space Center of Lebedev Physical Institute, Moscow, Profsoyuznaya 84/32, 117810, Russia
[¶] Skobeltsyn Institute of Nuclear Physics, Moscow State University, Moscow 119899, Russia
[‡] Stockholm Observatory, SE-133 36 Saltsjöbaden, Sweden
[§] Institute for Theoretical Physics, University of California, Santa Barbara, CA 93106, USA

Abstract. We present our current results of an off-line search for non-triggered gamma-ray bursts (GRBs) in the BATSE daily records (obtained with 1024 ms time resolution) covering about 8.6 years of observations. We found 1713 non-triggered and 1963 triggered GRBs. The scan was done with artificial test bursts added to the data which allowed us to measure the efficiency of the search. We extended the $\log N - \log P$ distribution down to $P \sim 0.1$ ph/s/cm^2. Previous indications of a turnover of the $\log N - \log P$ distribution at small P are not confirmed.

INTRODUCTION

Many gamma-ray bursts which were too weak to trigger the Burst and Transient Source Experiment (BATSE) on the *Compton Gamma-Ray Observatory* (CGRO) [1] or were missed due to other reasons (data readouts, etc.) can be confidently identified in the BATSE daily records which cover the full period of CGRO operation. The search for non-triggered bursts can substantially extend the GRB sample and, even more importantly, can extend the $\log N - \log P$ distribution of GRBs to smaller peak fluxes, P. The first sample of non-triggered GRBs was found by Rubin et al. [2]. A systematic search for non-triggered GRBs was initiated by Kommers et al. ([3], hereafter K97). Recently, Kommers et al. ([4], [5], hereafter K98 and K99) completed their data scan of 6 years of observations. A similar search, with trigger criteria close to those of the BATSE on-board trigger, was done by Schmidt [6].

Here we report intermediate results of the search for non-triggered bursts in the BATSE daily records started by our group in 1997 [7–9].

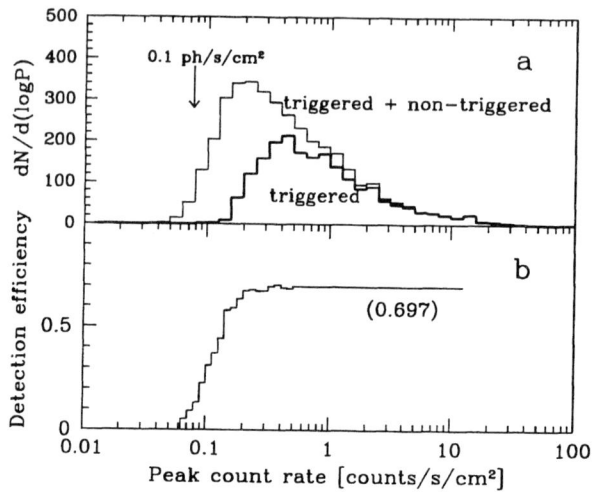

FIGURE 1. a) Peak count rate distribution, $dN/d(\log P)$, of the 1963 BATSE-triggered GRBs (thick histogram) and of all 3676 GRBs detected in our scan (thin histogram). The average ratio between the peak count rate and the photon flux in the 50–300 keV range is 0.75 with a dispersion of 0.05. The arrow shows the count rate corresponding to an average photon flux of 0.1 ph/s/cm². b) The efficiency of the off-line burst detection defined as the fraction of test bursts detected in our scan versus the peak count rate, P. The efficiency is normalized to the total number of events occurring above the Earth's horizon.

THE DATA SCAN

We used the 1024 ms time resolution BATSE data (DISCLA) from the Goddard Space Flight Center archive [1].

The procedure of our scan is described in detail in [8]. The first advance of our scan as compared to K97 and K98 is a more selective off-line trigger, which reduces the number of false triggers to an easily manageable level. The second advance is the test burst technique (i.e., we add artificial bursts to the data and then detect them together with the real bursts), which allows us to control the scan and to measure the detection efficiency function.

We scanned 3130 days (Apr. 23, 1991 to Nov. 17, 1999, i.e., 8.6 years) of the BATSE daily records. A few days were missed because of technical problems (corrupted data). The last 500 days were scanned with an improved procedure (a more convenient visual control) which slightly increased the efficiency of the scan.

We revised our GRB sample stored during the scan using stronger classification criteria and removed ~ 40 events as being of non-GRB origin or because the events had insufficient significance level.

---

[1] ftp://cossc.gsfc.nasa.gov/pub/data/batse/daily/

RESULTS

Our sample contains 1713 non-triggered events that were classified as GRBs, as well as 1963 BATSE-triggered GRBs. K98 and K99 scanned 2200 days (Dec. 9, 1991 to Dec. 16, 1997) and found 873 non-triggered GRBs as well as 1392 BATSE-triggered events.

We also found about 5250 of the 10083 test bursts added to the data. About 30% of the test bursts fell into data gaps, and $\sim 20\%$ were lost being in the peak flux range of low detection efficiency.

The peak count rate distribution of the events found in the scan is presented in Figure 1a. The detection efficiency of the scan is shown in Figure 1b. Note that the BATSE trigger missed some strong events due to readout dead time.

The resulting $\log N - \log P$ distribution in absolute units is presented in Figure 2 in comparison with the results of K98. K98 also detected events below 0.2 ph/s/cm^2. They, however, only presented data having an efficiency larger than 0.8. The efficiency curve used by K98 is a sharper function of the peak flux as compared

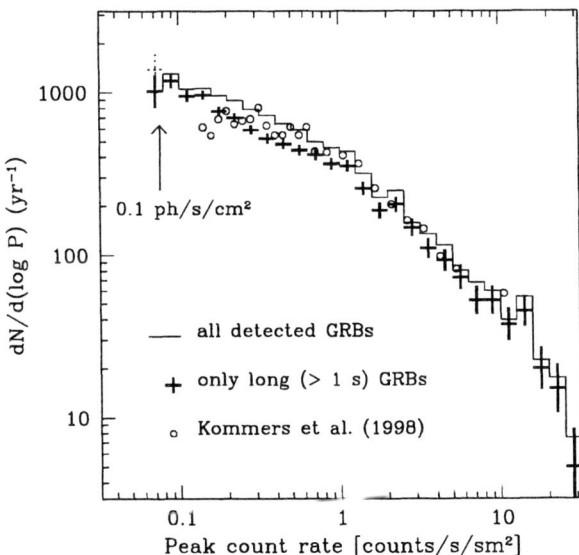

FIGURE 2. The differential $\log N - \log P$ distribution corrected for the efficiency in absolute units for all 3676 GRBs detected in our scan (histogram) and for the 3028 events in our sample with a duration longer than 1 s. The corresponding distribution in absolute units for the 2265 GRBs (of any duration) found by K98 is shown by circles (the K98 data were transformed to the count rate by applying the factor 0.75, see the caption of Fig. 1). The leftmost data point is affected by threshold bias arising from errors in the peak count rates. Its true (preliminary) position estimated by a forward folding fit with a realistic error/efficiency matrix is shown by the dotted cross. The bias for the second data point is less than 10%.

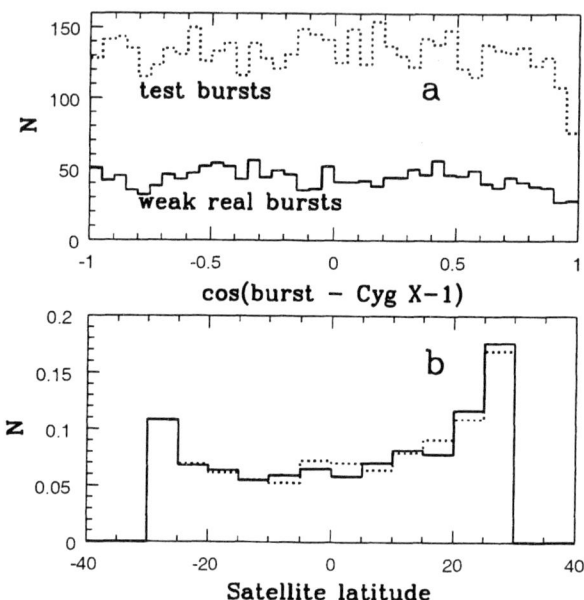

FIGURE 3. a) The angular distribution of detected events with respect to the Cyg X-1 direction. b) The distribution of detected events over CGRO-latitude at the time of detection. The asymmetric shape is caused by the South Atlantic Anomaly. Solid and dotted line histograms represent real bursts and test bursts, respectively.

to our curve.

Our whole sample is not homogeneous because it contains short events where the peak flux estimate in 1 s resolution is wrong. To eliminate the corresponding bias we removed all events consisting of one 1.024 s bin (i.e., the second brightest bin has a peak count rate below 0.5 of that in the brightest bin). The resulting $\log N - \log P$ for long bursts (Fig. 2) cannot be fitted in the simplest cosmological scenario using the standard candle assumption for a non-evolving parent population [8]. The general catalog of all (triggered and non-triggered) GRBs that we found and all numerical data for the distributions presented here are available via WWW or anonymous ftp [2].

Test distributions

Plotting hardness ratios vs. peak fluxes (see [7-8]) shows a brightness - hardness correlation in approximate agreement with Nemiroff et al. [10] and demonstrates that the new weak bursts have a distribution that smoothly extends the distribution

---

[2] http://www.astro.su.se/groups/head/grb_archive.html
ftp://ask.astro.su.se/pub/grb

of stronger GRBs (the non-GRB events are softer on average) and indicates that non-triggered bursts are intrinsically identical to triggered GRBs.

The detected events show no excess towards the Sun, no significant excess in the direction of Cyg X-1 (see Fig. 3a), and have a reasonable distribution in geocentric coordinates (uniform above the Earth's horizon). The polar angular distribution is consistent [9] with the sky exposure function given in [1]. The distribution of detected events over the CGRO-latitude at the time of detection is shown in Figure 3b. It is sensitive to the possible contribution of ionospheric events in our sample as their rate sharply increases with latitude. The distribution over the CGRO-latitude should be compared to that for test bursts as the latter accounts for the dependence of the detection efficiency on latitude. There is no excess of real events compared to test bursts at high latitudes. An approximate 1σ upper limit for the contamination of the sample by events of ionospheric origin at latitudes above $\pm 20°$ is 50 events of $\sim$ 3700 GRBs.

Acknowledgments: This research made use of data obtained through the HEASARC Online Service provided by NASA/GSFC. We are grateful to Robert Preece for valuable discussions and suggestions. This work was supported by the Swedish Natural Science Research Council, the Swedish Royal Academy of Science, the Wennergren Foundation for Scientific Research, a Nordita Nordic Collaboration grant, the Swedish Institute, and NSF grant PHY94-07194.

REFERENCES

1. Meegan, C., et al., in *Gamma Ray Bursts*, eds. C. A. Meegan, R. D. Preece, & T. M. Koshut, AIP Conf. Proc. **428**, New York, 1998, p. 3.
2. Rubin, B. C., Horack, J. M., Brock, M. N., et al., in *Compton Gamma Ray Observatory*, eds. M. Friedlander, N. Gehrels, & D. J. Macomb, AIP Conf. Proc., New York, 1992, p. 719.
3. Kommers, J. M., Lewin, W. H. G., Kouveliotou, C., et al., *Astrophys. J.*, **491**, 704 (1997). (K97)
4. Kommers, J. M., Lewin, W. H. G., Kouveliotou, C., et al., astro-ph/9809300 (1998). (K98)
5. Kommers, J. M., Lewin, W. H. G., Kouveliotou, C., et al., "Current Non-triggered Supplement to the BATSE Gamma-Ray Bursts Catalogs", http://space.mit.edu/BATSE/ (1999). (K99)
6. Schmidt, M., these proceedings.
7. Stern, B. E., Tikhomirova, Y., Stepanov, M., et al., *A&A Suppl. Ser.* **138**, 413 (1999).
8. Stern, B. E., Tikhomirova, Y., Stepanov, M., et al., *Astrophys. J.*, submitted, astro-ph/9903094 (1999).
9. Stern, B. E., Tikhomirova, Y., Stepanov, M., et al., in *Gamma Ray Bursts: The First Three Minutes*, eds. J. Poutanen, & R. Svensson, ASP Conf. Ser. **190**, 1999, pp. 253–261.
10. Nemiroff, R. J., Norris J., Bonnell, J. T., et al., *Astrophys. J. Lett.* **435**, L133 (1994).

A Search for Gamma-Ray Bursts in the GRBM/BeppoSAX Database

C. Guidorzi[1], F. Frontera[1,2], L. Amati[2],
E. Costa[3], and M. Feroci[3]

[1] *Physics Department, University of Ferrara*
Via Paradiso 11, 44100 Ferrara, Italy
[2] *Istituto Tecnologie e Studio Radiazioni Extraterrestri, CNR*
Via Gobetti 101, 40129 Bologna, Italy
[3] *Istituto Astrofisica Spaziale, CNR*
Via Fosso del Cavaliere, 00133 Roma, Italy

Abstract. Preliminary results on a systematic analysis of the data from the Gamma-Ray Burst Monitor on-board BeppoSAX are presented. The purpose of this analysis is the identification of true celestial gamma-ray bursts. The study considers events that are recognized or not by the on-board trigger system logic. As the on-board logic triggers also spurious events which mimic GRBs (e.g., events due to high energy particles, atmospheric events, etc), it becomes necessary to impose additional constraints in order to recognize true GRBs. These further conditions will be the subject of this contribution.

THE GRBM EXPERIMENT ON-BOARD BEPPOSAX

The Gamma-Ray Burst Monitor (GRBM) [1–3], on-board BeppoSAX [4] has already shown good capabilities in the detection of gamma-ray bursts (GRBs). The GRBM consists of four CsI(Na) detector units called "Lateral Shields" (LS), because they surround the PDS experiment [1], forming a square well (each slab has 1135 cm^2 geometrical area and is seen by two PMTs).

The GRBM data

The data continuously stored and transmitted by the GRBM consist of the following classes:

Housekeeping Data (HK) . These include the count rates of each LS in two energy bands: GRBM band and AntiCoincidence (AC) band (integration time = 1 sec).

GRBM band : 40 - 700 keV (nominal values);

AC band : > 100 keV (nominal value).

High Time Resolution Data (HTR). These include the count rates of each slab in the GRBM energy band with a short integration time. These data are available only for those events that have been recognized by the on-board trigger logic. HTR data stored on-board are the following:

1. Ratemeters for 8 s before the trigger time (1024 channels of integrated counts in 7.8125 ms bins, 9 bit counters).

2. Ratemeters for 10 s after the trigger time (20480 channels of integrated counts in 0.48828125 ms bins, 6 bit counters).

3. Ratemeters for 88 s after the first 10 s (11264 channels of integrated counts in 7.8125 ms bins, 9 bit counters).

THE ON-BOARD TRIGGER CONDITIONS

The ratemeters in the 40-700 keV energy band are used to detect the occurrence of a candidate event. All the parameters for determining the on-board trigger conditions can be set from the ground station by command. Whenever the counting rate, accumulated on 7.8125 to 4000 ms, for at least two slabs exceeds the level of $n\sigma$ (n can be set to 4, 8 or 16 by command and σ is the standard deviation of the background statistics), the GRBM logic recognizes that event. The background level is continuously determined from the count rate averaged on a time interval from 8 to 128 s. Then, the temporal profile of the GRB candidate is stored as illustrated above.

SYSTEMATIC ANALYSIS OF THE GRBM DATA

The systematic analysis of the on-board triggered events (some of which are GRB candidates) has shown some relevant features:

1. Presence of many spurious events (crossing of high energy charged particles, atmospheric phenomena, etc.)

2. No identification by the on-board trigger logic of weak events, that are recognized as GRB events by visual inspection.

3. Loss of GRB triggers occurred during the transmission of the HTR data to the on-board tape recorder. Often this transmission is due to false triggers (e.g., particles that crossed two slabs). In these cases, a flag reports on the occurrence of a burst-like event.

Item 1 above means that only a small fraction of the on-board triggered events are GRB candidates. From item 2 we realize that many weak GRBs are not detected by the on-board logic; finally, from item 3 it turns out that we lose HTR data for some strong GRBs that occur during the dead time of the GRBM data transmission. As consequence of these limitations, too restrictive trigger conditions would increase the number of the undetected GRB candidates; on the other side, too weak trigger conditions would increase the false events. We decided to perform a systematic analysis of the ratemeter data of the GRBM in order to catch GRBs present in the data. A similar analysis has already been done for the BATSE/CGRO archive [5]. In the next section we describe the conditions set by us to recognize GRBs. We call these conditions "software trigger conditions" (SWTC), to distinguish them from the on-board trigger conditions (OBTC). GRB candidates that satisfy the SWTC are called "SW triggers", while the on-board trigger events are called "on-board triggers".

SW TRIGGER CONDITIONS

As in the case of the OBTC, for each slab the background level is continuously computed by a moving average of the counting rate on a 100 s time interval. A SW trigger happens when the counting rate exceeds the level of $n\sigma$ with

$$\begin{cases} n = 6 & (40 - 700 \text{ keV}) \\ n = 3 & (> 100 \text{ keV}) \end{cases} \quad (1)$$

in at least one of the following cases:

1. For at least two slabs in both the energy bands.

2. For a single slab in both the energy bands, provided that the slab axis is directed towards sky.

3. For at least three slabs in the 40-700 keV energy band.

The lower number of σ requested for the > 100 keV energy band is due to the fact that in this band there are much less spikes than in the 40-700 keV energy band. The > 100 keV energy band is of key importance for recognizing a true GRB because of the well known GRB spectral hardness. Condition 2 permits us to catch weak GRBs, that could be above our sensitivity limits only in one slab. Finally, condition 3 permits us to reject all the high energy charged particle crossing events.

AVAILABLE DATA FOR EACH SW TRIGGER

For each SW trigger event detected the following data are extracted and/or computed:

1. The SW trigger time.

TABLE 1. Some of these GRBs have been detected also by BATSE/CGRO.

GRB	SW Trigger (UTC)	BATSE # trig	GRBM Peak(σ) (SW thr = 6σ)	AC Peak(σ) (SW thr = 3σ)	On-board thr (σ)	Dur (s)
960802	22:07:01	5559	7 (LS1)	5 (LS1)	8	8
980421	22:44:08	6698	8 (LS1)	4 (LS1)	4	14
960913	23:05:20	5604	9 (LS2)	9 (LS2)	8	18
			12 (LS3)	13 (LS3)	8	
970311	08:43:12	xxxx	10 (LS1)	8 (LS1)	4	0.1
			7 (LS2)	7 (LS2)	4	
980831	16:34:40	xxxx	36 (LS1)	5 (LS1)	4	0.4
990412	01:48:34	xxxx	14 (LS3)	10 (LS3)	4	1
			12 (LS2)	11 (LS2)	4	

2. The max count rate (expressed in σ) in the energy band(s) and slab(s) used for the trigger.

3. The event duration.

4. The pointing direction of the slab(s) used for the trigger.

5. The peak flux (counts/s) in the energy band(s) and slab(s) used for the trigger.

6. The fluence (counts) in the energy band(s) and slab(s) used for the trigger.

7. Detection of the same event by other missions (in particular, BATSE/CGRO, Ulysses).

PRELIMINARY RESULTS

In Fig. 1 we show the temporal profiles of some SW triggered GRBs, some of which have been detected also by BATSE (in this case, the trigger number is reported). The GRBM data obtained are given in TABLE 1. For each GRB the peak fluxes (expressed in σ, integration time = 1 s) in the two energy bands are reported only for those slabs, in which the SW trigger conditions have been satisfied. The GRBM on-board threshold (σ) and the GRB duration are also reported. The first GRB, i.e., GRB960802, triggered only the SWTC in the single slab #1 and the OBTC in no slab at all. The case of GRB980421 is slightly different: it triggered both SWTC and OBTC in the single slab #1, but the on-board logic requires at least two slabs, therefore it has been recognized only by the SW logic. The GRB960913 has been recognized by both SWTC and OBTC, but HTR data for it are not available, because it occurred 3'10" after a spurious event, whose data transmission was taking place at the GRB SW trigger time. Finally, we report the HTR temporal profile of one short GRB, that matched both SWTC and OBTC.

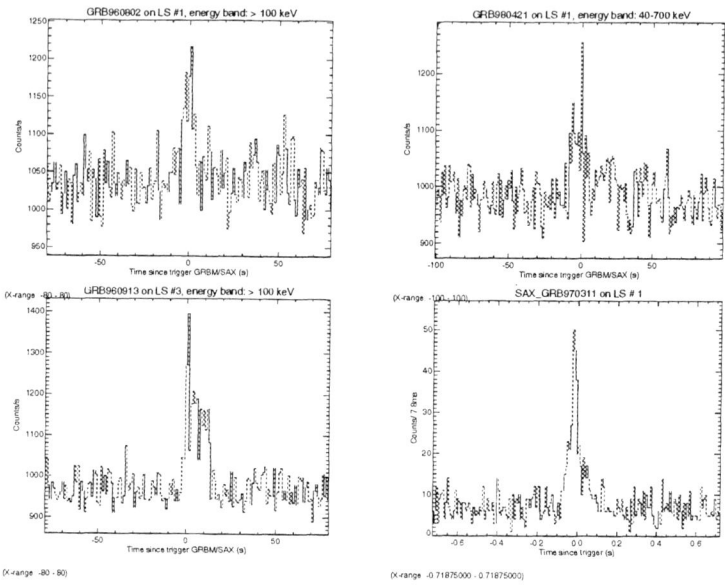

FIGURE 1. Example GRBs identified with our SW trigger logic system.

CONCLUSIONS

The systematic analysis of the GRBM/BeppoSAX database is still in progress. Nevertheless, an appropriate SW trigger logic system has shown good capabilities in detecting GRB candidates otherwise ignored by the on-board logic. We can summarize the main results in the following points:

1. Detection of weak GRBs, that are visible only in one slab.
2. Rejection of the false on-board triggered events.
3. Automatic comparison with other GRB catalogs.

Once the SWTC is refined for the search for GRBs, this SW trigger logic could be implemented for a prompt analysis of the raw data. This should permit us to identify the occurrence of GRBs not detected by the on-board logic soon after the event.

REFERENCES

1. Frontera, F., Costa, E., Dal Fiume, D., et al., *A&AS* **122**, 357 (1997).
2. Feroci, M., Frontera, F., Costa, E. et al., *Proc. SPIE* **3114**, 186 (1997).
3. Pamini, M. et al., *Il Nuovo Cimento* **13C**, 337 (1990).
4. Boella, G., Butler, R.C., Perola, G.C., et al., *A&AS* **122**, 299 (1997).
5. Kommers, J. M., Lewin, W. H., et al., *ApJ* **491**, 704 (1997).

What Can BeppoSAX Tell Us About Short GRBs: An Update from the Subsecond GRB Project

G. Gandolfi[1], M. J. S. Smith[2,4], A. Coletta[2], G. Celidonio[2],
L. Di Ciolo[2], A. Paolino[2], G. Tarei[2], G. Tassone[2], J. M. Muller[2,4],
E. Costa[1], M. Feroci[1], F. Frontera[3,5], and L. Piro[1]

[1] *IAS/CNR, Via del Fosso del Cavaliere, Roma, Italy*
[2] *BeppoSAX-SOC/TELESPAZIO, Via Corcolle 19, Roma, Italy*
[3] *TESRE/CNR, Via Gobetti 101, Bologna, Italy*
[4] *SRON, Sorbonnelaan 2, Utrecht, The Netherlands*
[5] *Dipartimento di Fisica, Universitá di Ferrara, Via del Paradiso 12, Ferrara, Italy*

Abstract. We present some statistical considerations on the BeppoSAX hunt for subsecond GRBs at the Scientific Operation Center. Archival analysis of a BATSE/SAX sub-sample of bursts indicates that the GRB Monitor is sensitive to short (≤ 2 sec) events, that are in fact $\approx 22\%$ of the total. The non-detection of corresponding prompt X-ray counterparts to short bursts in the Wide Field Cameras, in about 3 years of operations, is discussed: with present data no implications on the X-to-γ-ray spectra of short vs. long GRBs may be inferred. Finally, the status of searching procedures at the SOC is reviewed.

INTRODUCTION

The nature and the origin of short bursts, namely the events with a duration ≤ 2 s, that clearly represent a separate class from longer ones (Kouveliotou et al. 1993), is one of the most interesting and puzzling open problems in GRB studies. The celestial distribution of these events is clearly isotropic and their gamma-ray spectra are predominantly harder than those of longer bursts, while the two classes have the same peak flux range. Many questions naturally arise when considering the issue and comparing the phenomenology with the successful paradigm of Fireball models: are short bursts caused by a different emission mechanism? Do they also originate in extragalactic sources? Is their LogN-LogS Euclidean (Tavani 1998)? Are there other duration subclasses (Cline 1999)? Do they possess afterglow signatures like longer bursts? At present, no prompt counterpart detection at any wavelength is available and the recent afterglow discoveries refer only to the longer bursts, leaving a conspicuous gap in our knowledge of the phenomenon. Hence, the identification and localization of at least some event of this kind, that we call "subsecond bursts", remains a key objective of GRB research, an objective that can only be pursued at the moment by means of the BeppoSAX satellite and, in the near future, by means of missions like HETE-II, INTEGRAL and Swift.

We present here some statistical considerations on BeppoSAX triggering sensitivity to short bursts with GRBM. Finally, the non-detection of subsecond events in the sample of WFC-selected GRBs is analyzed and the new real-time procedure for on-ground triggering at the Scientific Operation Centre is briefly summarized.

ESTIMATION OF GRBM EFFICIENCY IN TRIGGERING SHORT BURSTS

The basic procedure for GRB counterpart detection at the BeppoSAX Scientific Operation Centre, performed since the very beginning of the mission, relies on the visual check of PDS Lateral Shield (LS) and WFC rate meters around GRBM trigger times (Coletta 1998). This procedure is recently being improved and extended, discriminating as possible the short, particle induced, false GRBM events from real bursts (Feroci 1999), monitoring the BATSE triggers, and performing an additional on-ground search for excess of counts in Lateral shields and WFCs below the trigger threshold (Smith 1999; Gandolfi 1999). In this work we address the general problem of the BeppoSAX sensitivity to GRBs with duration less than 2 s and we try to test the significance of experimental results in the field, i.e., the non-detection of subsecond bursts at quick look level. The first point to evaluate is the efficiency of GRBM in triggering short events, since the fundamental method used to identify GRBs — the only one systematically applied to the data for a long time — is the check of the counts at trigger times both in WFC and LS lightcurves with 1 s bins. If the sample of BeppoSAX triggers doesn't include the majority of short bursts, we would expect *a priori* a very low probability of finding subsecond X-ray counterparts, because the general SOC monitoring of WFC transients at a quick look level has a poor sensitivity to excess counts that last less than 8 s (the standard bin adopted to scan lightcurves on an orbital basis).

In order to estimate the GRBM sensitivity in the duration range of interest with respect to BATSE, we have selected the subsample of common triggers in the period 10/11/1996 – 15/9/1998 (BeppoSAX/SOC Catalog, 1999; BATSE Current Catalog, 1999). This guarantees the reality of SAX events (and automatically discriminates false triggers), that are also analyzed in high time resolution mode (8 ms bin) in order to obtain the T90 (i.e., the time interval during which they emit the 90% of their fluence). We find 111 common triggers, with 24 events under the 2 s threshold that corresponds to a 22% of the total. Comparing this value with the 26% of the BATSE duration distribution and assuming that the probability is described by a binomial, we conclude that the number of short bursts detected is compatible with completeness with a probability of 15%, but is almost certainly affected by a bias for very brief events. In fact, the bins with GRBs under 0.5 s

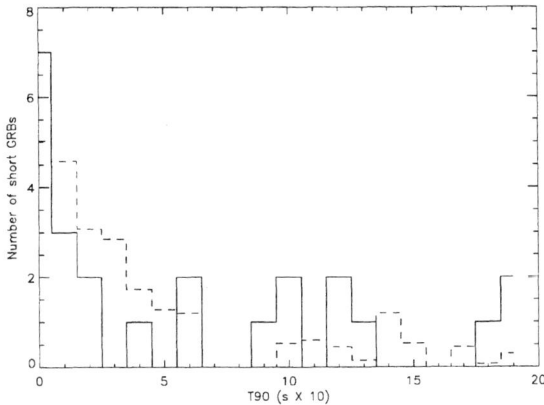

FIGURE 1. Duration distribution of short GRBs. The solid line is the common BeppoSAX/BATSE sample distribution, the dashed one is the normalized 4B BATSE distribution, for comparison.

exhibit a slight deficit of the order of 25% with respect to BATSE 4B duration distribution: coherently with expectations, the triggering efficiency is probably poor in that range due to the small fluence of the shortest events (Fig. 1). An offline analysis of BATSE positions, WFC's field of view and rate meter counts at the relevant trigger times confirms the non-detection of a WFC counterpart in each of the 24 cases.

INCREASING THE SENSITIVITY WITH ON-GROUND TRIGGERING: THE BEPPOSAX SUBSECOND BURSTS PROJECT

The efficiency of the GRBM strongly depends on the spectral properties of the burst and its maximum flux in the 1 s on-board trigger time bin. Other selection effects that could affect GRB detection are mainly geometrical and depend on the effective area illuminated by the burst: in case of an event exactly on the axis of the LS (this is the most favorable condition to catch the counterpart in WFC field of view if the involved lateral shield is unit 1 or unit 3) the area is minimum and the probability of triggering the GRBM is consequently minimized. For this reason, an on-ground triggering procedure (part of the BeppoSAX subsecond GRB Project) was implemented at the SOC last summer (Gandolfi 1999), with the aim of increasing the global detection efficiency of short and long-faint bursts. This semi-automatic procedure relies only on the comparison between lateral shields and WFC rate meters in the whole orbit with a 1 s bin. The new trigger condition, which has been empirically chosen in order to achieve the best sensitivity to short events without overloading the amount of quick look work and is certainly more effective with respect to the on-board one, selects events that have a 3σ excess of counts in the lateral shield and at the same time a 3.5σ excess in the corresponding WFC rate meter bin. The WFC threshold guarantees that all the bursts detectable in the celestial image, that is those with a global one-bin fluence of at least 70 counts in the 2–10 keV range (Smith 1999), are selected. Interesting examples of short events detected by the new semi-automatic routine, are GRB991014 and GRB991106: the first is slightly too long to be considered a subsecond event, the latter is probably a X-ray/GRB without a detected signal in the GRBM. Other candidate short bursts have been found, but the excess of WFC counts never satisfied the above condition and the transients were in fact not detectable in the corresponding WFC celestial image. The detection of a number of such events, even if it doesn't allow localization of the burst, could help to constrain the spectral properties of short GRBs with statistical significance.

EXPERIMENTAL RESULTS AND IMPLICATIONS

The present number of GRBs with prompt X-ray counterparts detected by BeppoSAX is 27 and none are classified as short events. Is this result compatible with the expectations? If we assume completeness in our GRBM detections in the subsecond duration range (and we have seen that we are at least almost complete) we would expect to detect the 26% of short bursts also in the sample of GRBs with revealed counterparts, i.e., ∼7 bursts. This is not the case, but we must remember that just 17 of these events have been triggered by GRBM, the only reliable detection method if the X-ray counterpart has a very low fluence (the GRB quick look procedure guarantees an optimal analysis of WFC, at 1 s or less resolution, only around trigger times). Hence any statistical consideration should be based on the GRBM trigger catalog, which has been — and is being — carefully inspected in real time by Duty Scientists at SOC. Furthermore, the most stringent limit to subsecond counterpart identification is the WFC rate meter sensitivity in the 1 s bin inspected, which decreases with the decreasing duration of the transient. Considering the duration distribution of short GRBs and the X-ray counterpart peak flux distribution for long bursts (Frontera 1999; various IAUCs), we can estimate a rough global detection efficiency: with a minimum 3σ signal in the 1 s bin of WFC rate meter, in the hypothesis of an average offset of 10 degrees from the center of the field of view and of a Crab-like spectrum for the source, we should miss about the 50% of expected events. The conversion from peak flux to fluence that is compared to the threshold in counts for the bin (∼36 in a standard WFC empty field), is done assuming a constant and maximum emission during T90. The expected number of subsecond counterparts in the GRBM-selected sample is therefore $N_{xs} = f_{\gamma s} \times \eta \times N_x$, where $f_{\gamma s}$ is the percentage of short gamma bursts, η the estimated global detection efficiency for short X-ray counterparts and N_x the global number of X-ray counterparts detected by BeppoSAX GRBM. We find with all the above assumptions a value of 2 subsecond events. The non-detection is not statistically significant, because considering again the binomial distribution, the probability $P(x, n, p)$ (with $p = f_{\gamma s} \times \eta$ the probability of detecting the counterpart of a short GRB instead of a long one, $x = 0$ the number of GRBM-selected counterpart detections, and $n = 17$ the dimension of the sample) corresponds to $P(0, 17, 0.13) \sim 0.09$. This result strongly encourages increasing the external trigger capability at quick look level in order to maximize the probability of finding a subsecond counterpart. In fact, extending to candidate events not selected by GRBM the real-time standard GRB procedure, with its robust and tested efficiency, will surely increase the chance of identifying low fluence events.

CONCLUSIONS

At present, no sure prompt X-ray counterparts of subsecond GRBs have been noticed in BeppoSAX WFCs celestial images and we have shown that GRBM triggered detections in this range are compatible with completeness, with a probable slight deficit at shortest durations. This implies, taking into account WFC sensitivity limits and assuming an X to γ peak ratio similar to that of longer GRBs, that 2 events out of a total of 17 triggered with revealed counterpart, should be subsecond bursts. The non detection is clearly not sufficient to test on a statistical basis any spectral hypothesis (i.e., the validity of the assumption on X to γ peak ratio). Furthermore, we can systematically rely on non-triggered detections (that represents the 63% of the whole sample of discovered counterparts) since only about 6 months, thanks to the new on-ground triggering routine implemented at SOC. No inference can be made on the basis of the entire counterpart catalog (i.e., $\sim$ 3-4 events of 27), because untriggered events with a prompt counterpart could have been easily missed at a quick look level with the old standard GRB procedure. On the other side, archival analysis of the GRBM trigger catalog and of the WFC rate meters of the whole mission are in progress, and the optimization of quick look procedures to identify and analyze the high time resolution data has now been obtained within the frame of the BeppoSAX Subsecond Bursts Project.

REFERENCES

1. Coletta, A., et al., in *Proc. of the Active X-Ray Sky Workshop*, Roma, *Nuclear Physics B* **69/1-3**, 712 (1998).
2. Cline, D.B., et al., *astro-ph/9905346* (1999).
3. Feroci, F., et. al., these proceedings.
4. Gandolfi, G., et. al., *http://www.ias.rm.cnr.it/ias-home/sax/subsec.html* (1999).
5. Kouveliotou, C., et al., *ApJ* **413**, L101 (1993).
6. Smith, M.J.S., et al., *A&ASS* **138**, 561 (1999).
7. Tavani, M., *ApJ* **497**, L21 (1998).
8. BeppoSAX/SOC GRBM Trigger Catalog, private communication (1999).
9. Paciesas, W.S., et al., 4B BATSE Catalog, *http://cossc.gsfc.nasa.gov/cossc/batse*

MeV Measurements of γ-Ray Bursts by CGRO-COMPTEL: Revised Catalog

A. Connors[1], C. A. Young[2], K. Bennett[3], W. Collmar[4],
W. Hermsen[5], R. M. Kippen[6], L. Kuiper[5], M. McConnell[2],
R. Miller[2], J. M. Ryan[2], V. Schönfelder[4], O. R. Williams[3],
and C. Winkler[3]

[1] *Eureka Scientific*
[2] *Space Science Center, University of New Hampshire, Durham NH, USA*
[3] *Astrophysics Division, ESTEC, NL-2200 AG Noordwijk, NL*
[4] *Max-Planck-Institut fur Extraterrestrische Physik, D-85740 Garching, FRG*
[5] *SRON-Utrecht, Sorbonnelaan 2, NL-3584 Utrecht, NL*
[6] *CSPAAR, University of Alabama in Huntsville, Huntsville AL, USA*

Abstract. The imaging COMPTEL telescope has accumulated 0.1–30 MeV spectra, time-histories, and positions of more than forty γ-ray bursts within its ∼ 3 sr field of view in the eight years since its launch. CGRO-COMPTEL measures in both imaging "telescope" and single detector "burst spectroscopy" mode. In an ongoing collaboration with BACODINE/GCN, bursts are imaged automatically, with localizations relayed to a global network of multiwavelength observers in near real time (∼ 10 minutes). We have updated our burst search procedure in two ways: 1) using more sensitive search algorithms; and 2) using data from more detectors. The first are double change-point algorithms. With these we can find regions of significant excess flux with no assumptions on the wide range of burst time-scales (e.g., rise-times or decay-times) or intensities, and only one adjustable parameter (the time-averaged count-rate of the detectors). This makes it simpler to combine information on burst time-histories from the larger effective area (but cruder time bins) burst spectroscopy detectors, and hence better pinpoint the best times for imaging each burst. We report the eight bursts detected during 1998–1999.

CGRO-COMPTEL Rapid Burst Response

The imaging Compton telescope COMPTEL, on board the Compton γ-Ray Observatory, has special capabilities for measuring transient events such as γ-ray bursts [1]. Not only are they detected in double-scatter "imaging" or "telescope" mode (0.75–30 MeV, 0.125 μs timing); but two NaI detectors act as independent 0.1–1.6 MeV and 0.6–10 MeV spectrometers (single-scatter "burst-mode"). Upon receiving an on-board burst trigger from BATSE, these burst spectrometers read out spectra at a faster cadence (every ~1 s for 6 s; then every 4 s for several minutes), before reverting to the default "background" integration time of 140 s. COMPTEL's Rapid Burst Response (COMPTEL RBR) [2], an ongoing project with BACODINE [3,4] to broadcast ~ 1° COMPTEL positions of MeV-bright bursts in near-real time ($\geq$ 7 minutes), had so far made use of only the telescope data. Here we report on upgrading the process to incorporate the lower time resolution but higher effective area burst spectroscopy data. These contain effectively no imaging information but do allow us to more accurately constrain both ends of the burst light curve and whether the burst was visible above 0.5 MeV at all. This can improve the signal-to-background ratio for imaging and reduce the false trigger rate. However, properly including these data forced us to rederive methods for finding the start and end time of a burst (or other transient) from first principles.

To Catch a Burst: Change Points + Bayes

Previous methods of determining burst (or flare) start and end times ranged from finding it by eye (BATSE-LOCBURST [5]) to requiring the counts per pre-set time bin to be greater than $n\sqrt{counts}$ above a running average for the background (BACODINE [3]). COMPTEL had used a Negative Double Difference (NDD) algorithm, on the telescope data alone. NDD weights and smooths the data within specified time windows; numerically determines a second derivative; and checks to see if this curvature is beyond threshold. All parameters were empirically determined and carefully crafted to the telescope count rates and likely burst time scales [2]. Unfortunately, upgrading to include burst-mode data was difficult with NDD due to the involved parameter-tuning, which change drastically with countrate and burst duration. By contrast, this was simple with change-point models. Scargle ([6] "Bayesian Blocks") first pointed out this simplicity, using it on GRB light-curves with drastically varying time-structures. Models for these are built up one at a time from piece-wise constant components. The 'change points' are the times at which these components switch, and are estimated by a straightforward Bayesian likelihood calculation. (change points are one of a number of aproaches well-known to statisticians but not to astronomers.) For catching γ-ray bursts, we use double change points: one each for the burst start and end.

Building the Algorithm

In theory, a properly constructed Bayesian likelihood ratio should be the best measure of the "distance" between two hypotheses ([7] and references therein). We built and tested likelihood ratios for several different kinds of change-point algorithms, using the standard Bayesian calculus. These all compared models of three segments delineated by two change points (i.e., two background segments separated by a burst block) with models of only one segment and no change points (i.e., background only). Here we briefly sketch the process (for details see [8]). We used constant or exponential rates μ_i for the segments (see below). For all, the sampling statistic assumed a Poisson process (binned or unbinned), given the model: $\exp(-\mu_i \delta t_i)(\mu_i \delta t_i)^{Y_i}/Y_i!$, with Y_i the counts in the ith time bin with width δt_i. The priors on the average rates were chosen to be exponential, with scale factor β given by the inverse of the long-term average detector count-rate $\langle r \rangle$: $\pi(\mu_i|I) = \exp(-\beta\mu_i)/\beta$, with $\beta = 1/\langle r \rangle$. The prior on the exponential model scale factor was a broad Gaussian centred at zero; while that for the change-point times was uniform over the interval sampled.

The burst models tested included: **1)** Three constant pieces (1st background, burst, 2nd background) versus one (only background); **2)** Three constant pieces with rate before and after the burst constrained to be equal; **3)** Three exponential pieces; **4)** Three exponential pieces with the rate before and after the burst constrained to be the same exponential model; **5)** Two exponential pieces plus one constant "burst block" in the middle; and **6)** Two exponential pieces plus one constant "burst block" in the middle, with the rate before and after the burst constrained to be the same exponential model. The marginalized posterior for these was compared to those for two background models: **1)** One constant piece (for Models 1 and 2); and **2** One exponential piece (for Models 3–6).

Let Y_i represent the total counts and T_i the total livetime in the ith segment, with t_0 and t_1 be the burst start and end times respectively (i.e., the end of the 0th and 1st segments; with segments 0 and 2 being background and 1 the burst). Then, the Bayes likelihood ratio for t_0, t_1 (after marginalizing over the unknown rates μ_0, μ_1, μ_2, and dividing by the similarly marginalized likelihood for no change points) can be written as the ratio of the priors (on the rates and change points) times the ratio of the (marginalized) sampling statistics. For Model 1 (constant background, burst, background) this is:

$$\lambda_{3C}(t_0, t_1) = \frac{1}{\beta} \frac{\beta^3}{(N_T - 3)(N_T - 2)} \times \frac{(\beta + T_T)^{Y_T+1}}{(Y_T!)} \prod_{i=0}^{2} \frac{Y_i!}{(\beta + T_i)^{Y_i+1}}$$

For Model 2 (constant background, burst, same background) it is similar, but the product is taken over only two segments(0+2 and 1). Bayes likelihood ratios for the exponential models were the same but were multiplied by terms for marginalizing over the exponential scale parameter. Finally, to find global (or total) Bayes odds for each model, we marginalized over all change points.

Tests and Results

These were compared on BACODINE-generated COMPTEL data for 44 BATSE burst triggers in COMPTEL's field of view from Jan 1, 1999 though May 31, 1999. For COMPTEL detections, we required that the total Bayes Odds ratio $>10:1$, and that the odds at the maximum likelihood change points exceed $\sim 25\%$ of this.

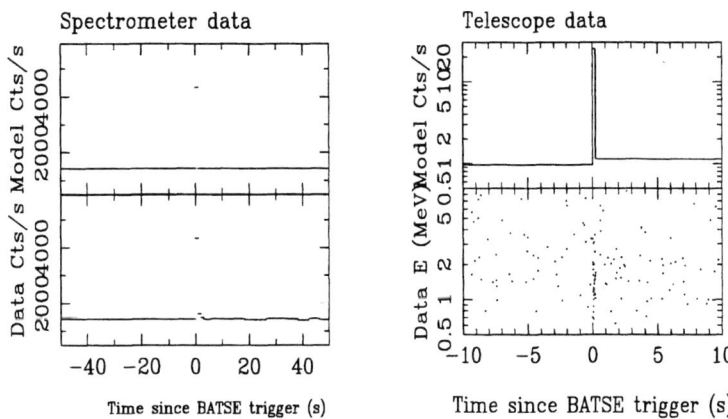

FIGURE 1. GRB980706: Light-curves and Model 1 for burst (*left*) and telescope (*right*) data.

TABLE 1. Results from Automated Timing Analysis

GRB DAY	Burst Spectrometer				Telescope Events			
	Log Bayes Odds		Best times (s)		Log Bayes Odds		Best times (s)	
	Total	Peak	Start[a]	Length	Total	Peak	Start[a]	Length
980124	1092.	1095.	+4.1	6.1	69.2	29.7	+4.2	3.3
980329	462.3	465.6	+3.1	7.1	5.64	10.8	+7.3	0.63
980706	823.1	826.5	+0.0	1.0	69.9	28.1	+0.0	0.23
980828	57.16	60.58	+0.0	2.0	4.49	9.99	+0.8	1.43
990105	167.7	171.1	+0.0	54.4	40.2	1.25	−0.2	0.88
990123	3645.	3649.	+22.2	20.1	64.7	24.8	+20.0	25.39
990728	199.0	202.3	+2.1	40.3	50.5	58.1	−1.9	31.64
990915	754.9	758.3	+0.0	10.1	67.1	26.7	+0.1	1.61

[a] With respect to BATSE trigger time.

We found: 1) Using an exponential to model a burst gave an indeterminant endtime. 2) Background variations gave too many false triggers unless the 3-piecewise-constant or exponential models were used. 3) Models 1 and 6 worked best; the exponential model for the background worked about as well as the 3-constant-components model, but was rather slower. Hence, Model 1 was preferred. 4) Using the change points determined from burst-spectrometer data to set a *window* in which to search for burst start and end times in the telescope data reduced the false trigger rate; increased the signal-to-noise ratio; and increased the speed of the search. 5) We also added a "minimum COMPTEL imaging" criterion: that the burst block must have at least 10 events.

After testing, Model 1 was run on all bursts for which we had BACODINE datasets, from 1996 through the present. For NDD: 8 false triggers, 9 real bursts found, 3 missed. For Model 1: 0 false triggers, 10 real bursts found, 2 missed; plus higher significance detections (and better position contours) for several of the bursts. We illustrate the change-point algorithm with GRB 980706, a very short

TABLE 2. Results from Automated Imaging Analysis

GRB DAY	Peak likelihood ratio	COMPTEL R.A ° 2000	Decl	COMPTEL Azimuth °	Zenith	2σ error °
980124	67.28	285.61	78.69	208.48	21.88	1.57
980329[a]	**	**	**	**	**	**
980706	60.22	161.82	57.53	67.51	46.75	1.84
980828	27.11	141.86	22.44	91.83	21.69	3.00
990105	21.67	307.73	1.28	320.92	22.09	3.12
990123	99.88	225.41	44.96	277.72	53.49	2.00
990728	241.4	211.96	−57.83	27.69	4.09	0.82
990915	61.97	96.89	71.68	280.24	50.69	2.50

[a] Below threshold for the automated algorithm, but was imaged by hand.

burst missed by the NDD algorithm but found by the CP algorithm (Fig. 1).

In sum, through a confluence of Bayesian methods, a classical statistics tool (change points), and knowledge of the COMPTEL instrument, we constructed more robust "burst-catching" algorithms. We have eight new candidates for the COMPTEL burst catalog. We show their timing results and preliminary positions in Tables 1 and 2. These bursts will go through standard COMPTEL processing before final acceptance.

ACKNOWLEDGMENTS

AC acknowledges the hospitality of Wellesley College and UNH; her collaborators in AstroStatistics (esp. J. Scargle, V. Kashyap, T. Loredo, E. Kolaczyk, A. Siemiginowska, and D. van Dyk); and NASA contract NAG5-7984. The COMPTEL project is supported in part through NASA contract NAS 5-26646, DARA grant 50 QV 90968, and the Netherlands Organization for Scientific Research (NWO).

REFERENCES

1. Schönfelder, V. et al., *ApJS* **86**, 629 (1993).
2. Kippen, R.M., et al, *ApJ* **492**, 246, (1998).
3. Barthelmy, S.D., et al, AIP Conf. Proc. **384**, ed C. Kouveliotou, M.S. Briggs & G.J. Fishman, New York, 1996, pp. 580.
4. Barthelmy, S.D., et al, in AIP Conf. Proc. **428**, ed C.A. Meegan, R.D. Preece, & T.M. Koshut, New York, 1998, pp. 99.
5. Kippen, R.M., et al, in AIP Conf. Proc. **428**, ed C.A. Meegan, R.D. Preece, & T.M. Koshut, New York, 1998, pp. 119.
6. Scargle, J., *ApJ* **504**, 405 (1998).
7. Loredo, T.J., in *Statistical Challenges in Modern Astronomy I*, ed G.J. Babu and E.D. Feigelson, Springer Verlag, New York, 1992.
8. Connors, A. et al., in preparation.

Properties of Gamma-Ray Burst Classes

Jon Hakkila*, David J. Haglin*, Richard J. Roiger*,
Robert S. Mallozzi[†], Geoffrey N. Pendleton[†],
and Charles A. Meegan[‡]

*Minnesota State University, Mankato, Minnesota 56001
[†] University of Alabama, Huntsville, Alabama 35812
[‡] NASA/MSFC, Huntsville, Alabama 35899

Abstract. The three gamma-ray burst (GRB) classes identified by statistical clustering analysis [7] are examined using the pattern recognition algorithm C4.5 [8]. Although the statistical existence of Class 3 (intermediate duration, intermediate fluence, soft) is supported, the properties of this class do not need to arise from a distinct source population. Class 3 properties can easily be produced from Class 1 (long, high fluence, intermediate hardness) by a combination of measurement error, hardness/intensity correlation, and a newly-identified BATSE bias (the fluence duration bias). Class 2 (short, low fluence, hard) does not appear to be related to Class 1.

INTRODUCTION

GRB spectral and temporal properties overlap, providing a continuum of burst characteristics. Some of this overlap is intrinsic in nature, while much is due to instrumental and observational biases. In addition to this overlap, there is clustering indicative of classes within the parameter space defined by GRB attributes. In particular, there are two long-recognized GRB classes [2,5] based on duration (divided at roughly 2 seconds) and spectral hardness. A statistically significant third class has been identified using statistical clustering analysis [7].

Can effects attributable to a source population be separated from instrumental effects? To answer this, we have applied computer science pattern recognition algorithms to learn why bursts cluster in some parameter spaces. For this analysis, we have used the supervised decision tree classifier C4.5 [8]. Supervised classifiers establish rules for previously identified patterns, and must be trained by representative class members.

ANALYSIS

The three GRB classes identified by statistical clustering techniques [7] can be found from three significant classification attributes: 50 to 300 keV fluence, T90 duration, and HR321 hardness ratio (the fluence in the 100 to 300 keV band divided by the fluence in the 25 to 100 keV band). The properties of the three classes in terms of these attributes are demonstrated in Table 1.

TABLE 1. Statistical clustering classes, from 3B GRBs.

Attributes	Class 1 (Long)	Class 2 (Short)	Class 3 (Intermediate)
T90	long	short	intermediate
Fluence	large	small	intermediate
Hardness	intermediate	hard	soft

C4.5 was trained on the three GRB classes using five fluences, two durations, three peak fluxes, and three hardness ratios. C4.5 produced a decision tree containing IF THEN ELSE branches for placing each GRB in the appropriate class; these branches were *pruned* to remove branches containing less than four GRBs. Rules were then generated for each class based on the pruned branches. C4.5 identifies outliers with poorly defined rules that often contain few GRBs. Statistical methods find that outliers are not closely bound to the class (cluster) centers. C4.5 rules identified a number of GRBs as having peculiar hardness ratios; these resulted from large individual channel fluence errors. The GRBs with the largest 10% relative errors (error divided by measurement) were subsequently removed from the database. The remaining 3B GRBs were reclassified using C4.5; the resulting rules were used to classify 4B Catalog GRBs and thus increase the database size.

Class 3 Spectral Hardnesses

C4.5 verified that the three GRB classes resulted primarily from the attributes of spectral hardness, duration, and fluence. With the larger classification database, the dependence on spectral hardness could be examined in terms of the spectral fitting parameters α, β, and E_{peak} [1]. Using only these three attributes, C4.5 was able to accurately classify most of the 4B GRBs. The rules generated by C4.5 were able to cleanly separate Class 2 from Class 1, but could not delineate Class 3 from Class 1 (85% of Class 3 bursts were assigned to Class 1).

Upon further examination, Class 3 GRBs were found to have E_{peak} values similar to Class 1 bursts of the same 1024 ms peak flux (Figure 1). The correlation between E_{peak} and peak flux has been interpreted as cosmological redshift [6].

Class 3 Fluences and Durations

Since at least one of the three defining characteristics of Class 3 actually represents a data correlation, we hypothesized that Class 3 GRBs actually belong to Class 1. We decided to see if Class 3 fluences and durations could be explained in terms of Class 1 attributes. This could be the case if some instrumental or sampling bias made Class 1 GRBs appear to be shorter and fainter than they should be.

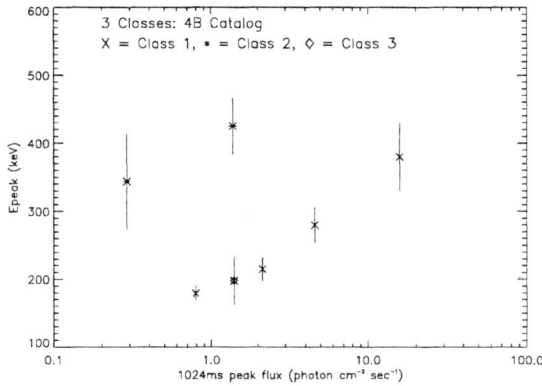

FIGURE 1. E_{peak} vs. p1024 for the Three GRB Classes.

Figure 2 is a plot of fluence vs. 1024 ms peak flux for each of the three GRB classes, and is limited to GRBs detected when BATSE had one homogeneous set of trigger criteria. There are distinct regions outside of which no GRBs are found. GRBs with 1024 ms peak fluxes less than 0.2 photons cm^{-2} second^{-1} are not detected, since this is below BATSE's minimum detection threshold. GRBs do not have fluences less than what would be found in their time-integrated 1024 ms peak fluxes, since this is the shortest timescale on which this peak flux can be measured.

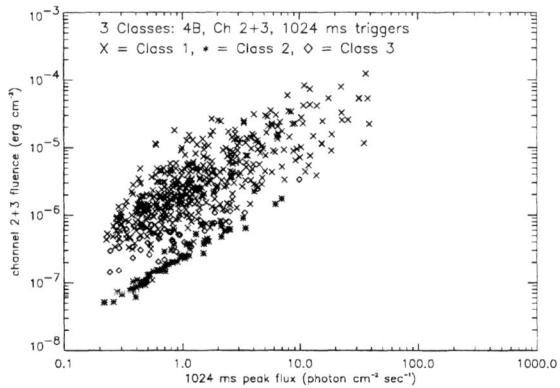

FIGURE 2. Fluence vs. p1024 for the Three GRB Classes.

Figure 3 overlays log(T90) contours for Class 1 GRBs on the fluence vs. 1024 ms peak flux space. The contours demonstrate that GRBs can be modeled as a series of pulses, with pulses containing most of the fluence and interpulse separations primarily defining the duration. Most Class 2 bursts are single-pulsed events as measured on the 1024 ms timescale. This helps define the characteristics of the third distinct region outside of which no GRBs are found: high fluence, faint Class 1 GRBs are missing, whereas low fluence faint, Class 1 GRBs are present. Since a bias favoring detection of GRBs with few photons over those with many photons seems unlikely, we suspect a bias capable of underestimating fluence relative to peak flux.

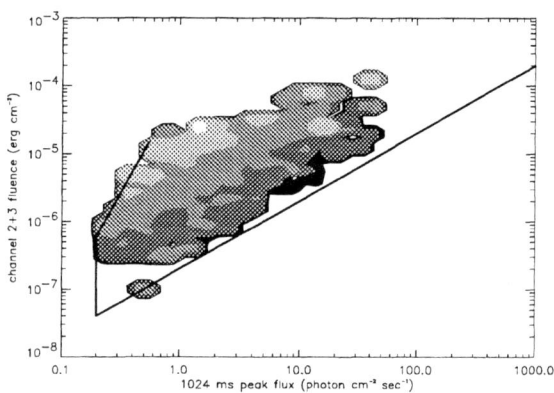

FIGURE 3. Fluence vs. p1024 for Class 1 GRBs; contours indicate regions of constant log(T90).

We have dimmed a number of bright GRBs to where they just trigger in order to study their measured properties as they fade into background. Each burst's peak flux is dimmed, and the time history is "noisified" with a Poisson background. The peak flux and fluence are then re-measured. These actions have been performed ten times on five bright bursts with a range of temporal structures.

One problem quickly became apparent during the analysis: the time interval bounding the fluence measurement (the *fluence duration* [4]) strongly influenced the amount of fluence measured. If the same fluence duration interval was used for undimmed and dimmed measurements, then the fluence-to-peak flux ratio did not change as a GRB was dimmed. If, however, the fluence duration interval shortened to account for faint pulses disappearing into the background and becoming unrecognizable, then the fluence-to-peak flux ratio decreased as the burst dimmed (see Figure 4). This bias becomes stronger near the trigger threshold.

Fluence durations taken from BATSE Catalogs provide supportive evidence for this mechanism. The durations used to calculate fluence of faint Class 1 GRBs are shorter than those of bright Class 1 GRBs [4].

CONCLUSIONS

A mechanism exists whereby some Class 1 (Long) GRBs can develop Class 3 (Intermediate) characteristics via a combination of the hardness intensity relation

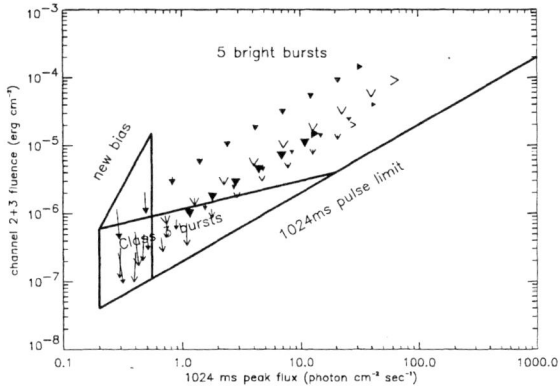

FIGURE 4. Five bright Class 1 GRBs, decremented in peak flux, noisified, with remeasured fluences and peak fluxes. It has been assumed that the GRB duration is measured from identifiable pulses, which become harder to recognize as the peak flux becomes fainter.

and the fluence duration bias. Faint Class 1 GRBs are most likely to develop Class 3 characteristics, but it is possible for even bright GRBs with appropriate time histories and spectral features to develop these characteristics. Class 3 (Intermediate) GRBs do not therefore appear to represent a separate source population, although they cluster in the duration, fluence, hardness, attribute space. Class 2 (Short) GRBs do appear to represent a separate source population. We were unable to find a mechanism by which faint Class 1 GRBs could develop Class 2 characteristics.

GRB population studies can benefit from use of AI classifiers. There are many other attributes developed by the community that could be included for future study. To this end, we are designing a web-based AI tool for GRB classification [3] that includes supervised and unsupervised AI classifiers [9].

REFERENCES

1. Band, D. L., et al., *ApJ* **413**, 281 (1993).
2. Cline, T. L., and Desai, U. D., *Proc. 9th ESLAB Symp.*, pp. 37-45 (1974).
3. Haglın, D. J., et al., this conference.
4. Hakkila, J., et al., this conference.
5. Kouveliotou, C., et al., *ApJ* **413**, L101 (1993).
6. Mallozzi, R. S., et al., *ApJ* **454**, 597 (1995).
7. Mukherjee, S., et al., *ApJ* **508**, 314 (1998).
8. Quinlan J. R., *Machine Learning* **1**, 81 (1986).
9. Roiger, R. J., et al., this conference.

Unsupervised Induction and Gamma-Ray Burst Classification

Richard J. Roiger[†], Jon Hakkila[†], David J. Haglin[†],
Geoffrey N. Pendleton[‡], and Robert S. Mallozzi[‡]

[†]*Minnesota State University, Mankato, MN, 56001*
[‡]*University of Alabama in, Huntsville, AL, 35899*

Abstract. We use ESX, a product of Information Acumen Corporation, to perform unsupervised learning on a data set containing 797 gamma-ray bursts taken from the BATSE 3B catalog [5]. Assuming all attributes to be distributed log-normally, Mukherjee et al. [6] analyzed these same data using a statistical cluster analysis. Utilizing the logarithmic values for T90 duration, total fluence, and hardness ratio HR321 their results showed the instances formed three classes. Class I contained long/bright/intermediate bursts, class II consisted of short/faint/hard bursts and class III was represented by intermediate/intermediate/soft bursts.

When ESX was presented with these data and restricted to forming a small number of classes, the two classes found by previous standard techniques [1] were determined. However, when ESX was allowed to form more than two classes, four classes were created. One of the four classes contained a majority of short bursts, a second class consisted of mostly intermediate bursts, and the final two classes were subsets of the Class I (long) bursts determined by Mukherjee et al. We hypothesize that systematic biases may be responsible for this variation.

INTRODUCTION

Induction-based learning [4] attempts to extract interesting patterns from data. These patterns form concept classes with each class containing data instances. When the induction is unsupervised, the learning model has no *a priori* class knowledge. Rather, the learning algorithm uses one or more statistical or symbolic (machine learning) evaluation functions to cluster instances into concept classes.

Mukherjee et al. [6] performed a statistical cluster analysis on a data set containing 797 gamma-ray bursts taken from the BATSE 3B catalog [5]. Assuming all attributes to be distributed log-normally, and utilizing the logarithmic values for T90 duration, total fluence, and hardness ratio HR321 their results showed the instances formed three classes. Class I contained long/bright/intermediate bursts, class II consisted of short/faint/hard bursts and class III was represented by intermediate/intermediate/soft bursts. Table 1 shows the mean and standard deviation values for the three classes. Table 2 offers a best defining rule for each class, as determined by ESX [7]. The rule for class I bursts indicates that 82.72% of the bursts in this class have a log T90 value between 0.70 and 2.66 and a log Fluence between −5.77 and −3.11. The rule also shows that we can be at least 97% confident that a burst with these characteristics is a class I burst. Table 2 shows that the class III rule does not cover its instances as well as the rules for classes I and II.

TABLE 1. Mean and Standard Deviations for the Classes found by Mukherjee et al. [6]

	Class I Long	Class II Short	Class III Intermediate	Domain
Number of Bursts	486	203	107	796
[a] Log T50 (mean)	1.13	−0.80	0.33	0.53
(sd)	0.45	0.41	0.26	0.93
Log T90 (mean)	1.55	−0.42	0.71	0.93
(sd)	0.40	0.44	0.32	0.94
Log Fluence (mean)	−5.21	−6.37	−6.11	−5.63
(sd)	0.59	0.57	0.37	0.77
[a] Log P256 (mean)	0.21	0.14	−0.08	0.15
(sd)	0.48	0.38	0.33	0.45
[a] Log HR32 (mean)	0.20	0.51	0.09	0.26
(sd)	0.27	0.27	0.40	0.33
Log HR321 (mean)	0.43	0.70	0.35	0.49
(sd)	0.23	0.26	0.39	0.30

[a] Attributes log T50, log P256, and log HR32 were not used in the final analysis since each had a high correlation with its respective counterpart (log T90, log fluence, and log H321).

TABLE 2. Representative ESX Rules for the Three Classes found by Mukherjee et al. [6]

Class I (Long Bursts)	$0.70 \leq \log T90 \leq 2.66$ and $-5.77 \leq \log \text{Fluence} \leq -3.11$:rule accuracy 97.34% :rule coverage 82.72%
Class II (Short Bursts)	$-1.55 \leq \log T90 \leq 0.41$:rule accuracy 90.87% :rule coverage 98.03%
Class III (Intermediate Bursts)	$0.46 \leq \log T90 \leq 0.96$ and $0.17 \leq \log T50 \leq 0.55$:rule accuracy 79.17% :rule coverage 53.27%

In this paper we use ESX [7], a machine learning model and product of Information Acumen Corporation, to perform unsupervised learning on these same data for the purpose of comparative analysis. We chose ESX for this research since ESX explains its behavior has been shown to perform well in several real-world environments [7].

METHOD

The machine learning component of ESX is an induction-based sequential learning model that creates a concept hierarchy [2] from a set of input instances. ESX uses knowledge contained in its concept hierarchy to generate a set of production rules to help define and explain what has been discovered. Supervised as well as unsupervised learning is supported.

ESX accepts data in the form of instances represented in attribute-value format. When learning is unsupervised, ESX takes one of two possible actions for each newly presented instance: (1) place the new instance into an existing cluster, or (2) create a new conceptual cluster containing the instance as its only member.

In addition, ESX allows the user to set a learning parameter so as to encourage or discourage the creation of new clusters. For a given domain, a best value for this parameter can be determined experimentally.

RESULTS

For our first experiment, we set the ESX learning parameter so as to restrict the formation of new classes. As a result, ESX clustered the data into the two classes found by previous standard techniques [1]. Table 3 shows a representative rule for each class. Notice that both clusters are well-defined.

TABLE 3. Representative Rules Taken from the Two Class ESX Clustering

Class I (Long Bursts)	$0.54 \leq \log T90 \leq 2.66$:rule accuracy 98.03% :rule coverage 96.99%
Class II (Short Bursts)	$-1.55 \leq \log T90 \leq 0.38$:rule accuracy 98.14% :rule coverage 90.95%

For our second experiment, we allowed ESX to form a best set of three or more clusters. The results of this experiment showed the formation of four clusters. One of the four clusters contained a majority of intermediate bursts (class 1); a second cluster consisted of mostly short bursts (class 2). The remaining two clusters (classes 3 and 4) were subsets of the Mukherjee class I bursts. The class mean and standard deviation values for each of the six burst attributes are shown in Table 4.

Table 5 offers representative rules for each of the four clusters. Figures 1 and 2 as well as Table 4 indicate that class 3 contains mostly long/soft bursts and class 4 contains long/bright bursts. The following rule represents a covering rule for the cluster formed by combining the class 3 and class 4 bursts.

$$1.19 \leq \log T90 \leq 2.66$$
:rule accuracy 90.26%
:rule coverage 92.68%

TABLE 4. Mean and Standard Deviations for the ESX Four Class Clustering

	Class 1 Intermediate	Class 2 Short	Class 3 Long/Soft	Class 4 Long/Bright	**Domain**
Number of Bursts	182	205	195	215	796
Log T50 (mean)	0.44	−0.78	1.27	1.18	0.53
(sd)	0.44	0.44	0.37	0.44	0.93
Log T90 (mean)	0.85	−0.41	1.67	1.62	0.93
(sd)	0.37	0.46	0.32	0.38	0.94
Log Fluence (mean)	−5.87	−6.36	−5.50	−4.84	−5.63
(sd)	0.45	0.59	0.37	0.61	0.77
Log P256 (mean)	0.04	0.13	−0.07	0.48	0.15
(sd)	0.43	0.38	0.22	0.51	0.45
Log HR32 (mean)	0.11	0.54	−0.03	0.38	0.26
(sd)	0.27	0.30	0.24	0.16	0.33
Log HR321 (mean)	0.36	0.73	0.24	0.59	0.49
(sd)	0.27	0.29	0.21	0.14	0.30

TABLE 5. Representative Rules Taken from the Four Class ESX Clustering

Class 1 (Intermediate)	$0.29 \leq \log T90 \leq 1.20$:rule accuracy 75.00% :rule coverage 74.18%	$0.21 \leq \log T50 \leq 0.63$ and $0.29 \leq \log T90 \leq 1.09$:rule accuracy 89.02% :rule coverage 40.11%
Class 2 (Short)	$-1.55 \leq \log T90 \leq 0.42$ and $-1.92 \leq \log T50 \leq -0.02$:rule accuracy 93.20% :rule coverage 93.66%	$-7.80 \leq \log \text{Fluence} \leq -6.63$ and $-1.92 \leq \log T50 \leq -0.02$:rule accuracy 95.95% :rule coverage 34.63%
Class 3 (Long/Soft)	$1.19 \leq \log T90 \leq 2.66$:rule accuracy 90.26% :rule coverage 92.68%	$0.02 \leq \log \text{HR321} \leq 0.08$:rule accuracy 77.36% :rule coverage 21.03%
Class 4 (Long/Bright)		$-4.85 \leq \log \text{Fluence} \leq -3.11$:rule accuracy 90.91% :rule coverage 51.16%

CONCLUSIONS

We used ESX to cluster data from about 797 gamma-ray bursts. When restricted to forming a small number of classes, two classes were determined. However, when allowed to form more than two classes, four classes were created. Two of the clusters were similar to the class II and class III bursts determined by Mukherjee et al. [6]. Taken together, the two remaining clusters represent the class I Mukherjee et al. bursts. ESX differentiated the class I bursts by brightness and hardness. The separation of long bursts into two classes may be due in part to the fact that ESX makes no *a priori* assumptions about data distribution.

We hypothesize that systematic effects may cause some class I bursts to take on class III characteristics [3]. Systematic biases may explain why class I bursts have been separated into two groups by ESX. Our future work will focus on testing these hypotheses with the help of additional induction-based techniques.

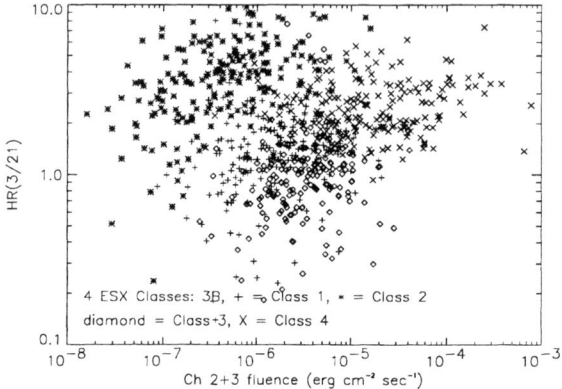

FIGURE 1. 3/21 Hardness Ratio vs. Ch 2 + 3 Fluence

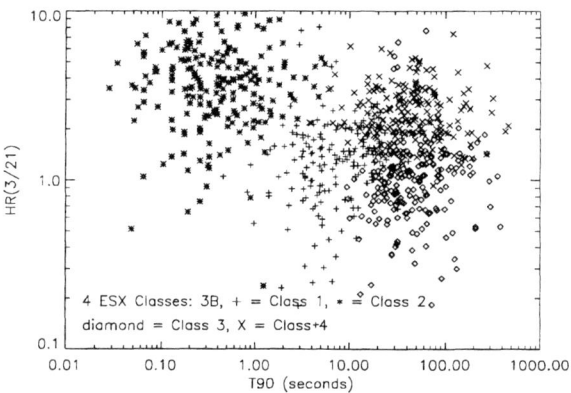

FIGURE 2. 3/21 Hardness Ratio vs. T90 Duration

REFERENCES

1. Cline, T.L., and Desai, U.D., in *Proc. 9th ESLAB Symp.*, 1974, pp. 37–45.
2. Gennari, J.H, et al., *Artificial Intelligence* **40**, 11 (1989).
3. Hakkila, J., et al., these proceedings.
4. Langley, P., and Simon H.A., *Communications of the ACM* **38(11)**, 55 (1995).
5. Meegan, C.A., et al., *ApJS* **106**, 65 (1996).
6. Mukherjee, S., et al., *ApJ* **508**, 314 (1998).
7. Roiger, R.J., et al., in *Proc. Federal Data Mining Symposium and Exposition '99*, 1999, pp. 109–120.

A Summary of Biases in the BATSE Burst Trigger

Charles Meegan*, Jon Hakkila†, Audress Johnson#,
Geoffrey Pendleton‡, and Robert Mallozzi‡

*NASA/MSFC, Huntsville, AL 35812
†Minnesota State University, Mankato, MN 56001
#University of Alabama, Tuscaloosa, AL 35487
‡University of Alabama, Huntsville, AL 35899

Abstract. Interpreting the BATSE burst intensity distribution requires an understanding of the biases introduced by the on-board trigger algorithm. We present a classification and description of these biases.

BATSE TRIGGER ALGORITHM

A burst trigger is generated by BATSE whenever the count rates in a specified energy range, typically 50 to 300 keV, exceed threshold in two or more of the eight detectors. The thresholds are specified in terms of the number of standard deviations, typically 5.5, above the background rate, which is recomputed every 17.408 seconds. Thresholds are determined independently for three integration times: 64 ms, 256 ms, and 1024 ms. Thresholds in effect at the time of each trigger are recorded in the data stream. During the readout of data from a trigger, the thresholds are increased to permit a stronger burst to overwrite the current readout.

PEAK FLUX DISTRIBUTION

The physical parameter of a burst that corresponds most closely to the on-board threshold is peak flux, defined specifically in terms of the energy range and integration time of the trigger algorithm, with units photons cm^{-2} s^{-1}. Peak luminosity is a similarly defined measure of intrinsic brightness, with units photons s^{-1}. By specifying the peak flux in this way, it is possible to determine efficiency corrections based on the conversion between peak flux and peak counts. Since there is no simple relationship between the peak rates on different integration times, the three different trigger timescales must be thought of as separate experiments. Figure 1 shows a plot of the peak flux distribution from the 4Br catalog [1] for bursts that

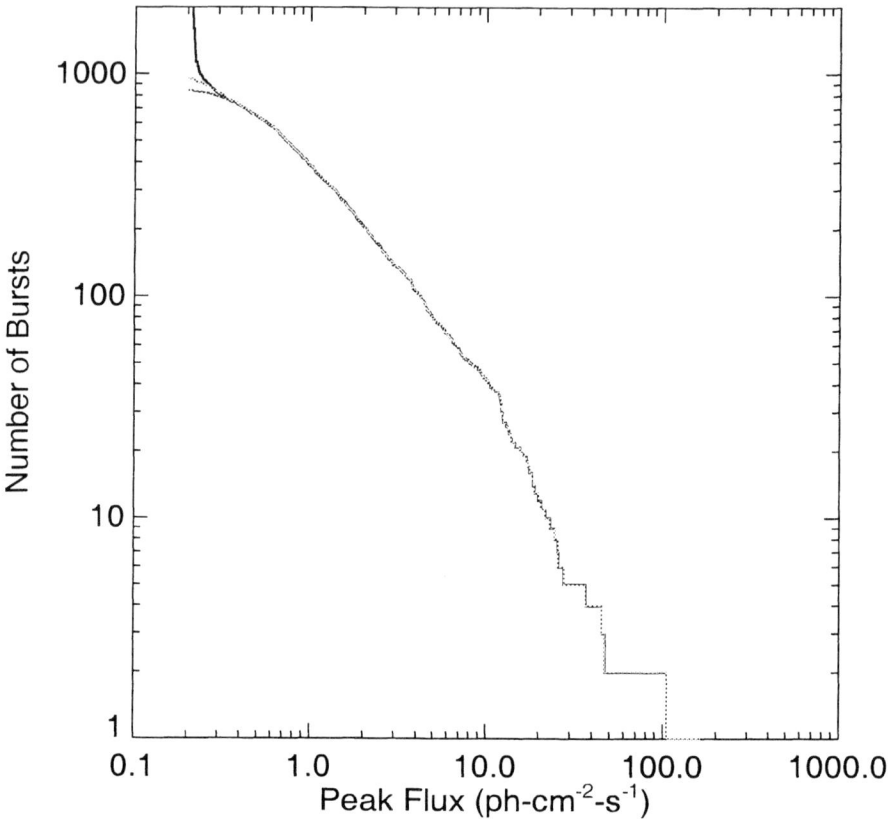

FIGURE 1. Integral brightness distribution for BATSE bursts for the 1024 ms trigger timescale. The upper and lower curves use an efficiency correction assuming all bursts have $E_{peak} = 267$ keV and 1391 keV, respectively. The middle curve represents a realistic range of E_{peak} values.

exceeded the 1024 ms threshold. This curve has efficiency corrections applied as described below. The three curves that diverge near the threshold represent different assumptions about the spectra. Bursts are included only for time intervals during which the thresholds were set at 5.5σ in the 50 to 300 keV energy range. Similar distributions can be generated for the other trigger timescales but the 1024 ms plot shows the greatest deviation from homogeneity, implying that it may be the most sensitive timescale and provide the most stringent test for models. In presenting such a plot, the implied claim is that the BATSE sample is a representative peak-flux-limited sample of gamma-ray bursts. Biases are discussed below in terms of this plot.

TYPES OF BIASES

Trigger biases can be categorized in terms of a series of questions about the peak flux distribution shown in Figure 1. The first question is: how accurately does this distribution reflect the true peak flux distribution of bursts? Errors in this distribution arise from inaccuracies in the efficiency calculation and from errors in the on-board determination of the peak count rate. We refer to these errors as "completeness bias".

The next question is: how representative is the BATSE sample of a true peak-flux-limited sample of bursts? Even if the correction for efficiency is accurate, this correction may be different for bursts with different characteristics. Bursts near threshold would then not be a representative sample of all bursts of that peak flux. We refer to these errors as "sample biases."

The last question is: how representative is the BATSE sample of a volume-limited burst sample? If the peak luminosity of bursts (defined similarly to the peak flux) were a standard candle, then a peak-flux-limited sample would also be a volume-limited sample. We now know that peak flux is not a standard candle, so a volume-limited sample of bursts may have quite different characteristics than the BATSE sample. We refer to these effects as "Malmquist-like biases".

The discussion of each of these biases presupposes the definition of peak flux given above, which necessarily incorporates a specific integration time. Although the 1024 ms timescale may appear inappropriate for bursts whose entire duration is less than 1024 ms, the peak flux and the efficiency correction remain well-defined. This is because the trigger depends on the integrated number of counts in a 1024 ms interval without regard to the distribution of those counts within that interval. If one wishes to determine the burst distribution using a different intensity parameter, e.g. fluence, the efficiency calculation becomes much more difficult, introducing additional biases.

The three types of biases are discussed in separate sections below.

COMPLETENESS BIAS

Trigger Efficiency

Trigger efficiency is the probability that a burst of a specific peak flux will generate a BATSE burst trigger. Errors in the trigger efficiency are one source of completeness bias. The following factors are currently taken into account in the efficiency calculation: detector geometry relative to the burst direction, detector response matrices, which relate photons at true energy E to counts at measured energy E', variations in background rate, variations in burst spectral properties, and increased counts due to scattering by the Earth's atmosphere.

Current peak flux distributions are significantly improved over previous ones as a result of incorporating atmospheric scattering. The largest source of error is uncertainty in the distribution of the spectral properties of faint bursts, and occurs only very close to the BATSE flux threshold, as can be seen in Figure 1.

Peak Counts Measurement

Since BATSE triggers on peak counts, errors in the on-board estimate of a burst's peak count rate will lead to a trigger bias. These can arise from the following effects.

"Slow Risers." As originally noted by Higdon & Lingenfelter [2], a burst with a slow risetime may have some of its flux included in the background calculation. This results in an overestimate of the background rate and hence an underestimate of the peak count rate. If this underestimate falls below the threshold, a burst that should have triggered will not.

Photon Statistics. Statistical fluctuations in the count rates introduce not only a dispersion in the measured rates, but also a systematic bias. The bias results from the fact that the observed peak in a series of count rates is more likely to be a positive than a negative fluctuation. There is then a systematic overestimate of the peak rate. This type of bias has been addressed by in't Zand & Fenimore [3] and Lamb et al. [4].

Sliding vs. Stepped Window. The on-board trigger algorithm uses stepped time windows over which the counts are integrated. The ground measurement of the peak flux steps the window in 64 ms intervals for all three timescales. Thus, the on-board measurement will be an underestimate of the peak on the 256 ms and 1024 ms timescales.

The biases due to the slow riser and photon statistics effects are the subject of an investigation described by Johnson et al. [5] in these proceedings. These biases are important because they even affect the measurement of $V/V_{\max}$.

SAMPLE BIAS

The most important sample bias results from the fact that high energy photons have an appreciable probability of Compton scattering in the detector and generating a count in a lower energy channel. BATSE is therefore more sensitive to triggering on a burst with a hard spectrum than on a burst with a soft spectrum having the same peak flux. Consequently, the sample of bursts near the BATSE threshold will be overpopulated with hard bursts. Note that this effect is opposite to the common belief that BATSE is biased against hard bursts. The hypothesized bias against hard bursts is actually a Malmquist-like bias, discussed in the next section.

Considerations of sample biases are complicated by the fact that burst temporal and spectral properties may be correlated with peak flux, either because of redshift and time dilation, or because of correlations with intrinsic burst luminosity. Therefore, the difference in observed properties between strong and weak bursts is a convolution of trigger biases and real effects.

MALMQUIST-LIKE BIASES

Malmquist-like biases are very different from the previous biases considered here, and probably should not be classified as a trigger bias since the BATSE sample does not purport to be a volume-limited sample of bursts. However, there is interest in determining the characteristics of volume-limited burst samples, which requires accounting for Malmquist-like biases. Such biases can exist whenever the luminosity, defined using the same energy and integration time as the trigger threshold, is not a standard candle.

The most often discussed bias of this type is a bias against bursts with hard spectra. The motivation for this hypothesis is the surprisingly narrow range of values of E_{peak}, a characteristic energy indicating the position of the break in the power law spectrum. BATSE will exhibit a Malmquist-like bias against bursts with high E_{peak} if there is an anticorrelation between E_{peak} and peak luminosity in the 50–300 keV energy range. A model that produces such an anticorrelation is one in which the bolometric luminosity is assumed to be a standard candle. Calculations using such a model permit a significant population of undetected hard bursts [6,7]. It must be emphasized, however, that these conclusions are strongly model dependent. We will not know the extent of Malmquist-like biases in the BATSE sample until we know how burst properties correlate with peak luminosity.

REFERENCES

1. Paciesas, W.S. et al., *ApJS* **122**, 465 (1999).
2. Higdon, J.C. & Lingenfelter, R.E., in *Gamma-Ray Bursts: Third Huntsville Symposium*, AIP Conf. Proc. **384**, 402 (1995).
3. in't Zand, J.J.M. & Fenimore, E.E., in *Gamma-Ray Bursts: Second Workshop*, AIP Conf. Proc. **307**, 692 (1994).
4. Lamb, D.Q., Graziani, C. & Smith, I.A. *ApJ* **413**, L11 (1993).
5. Johnson, A. et al., these proceedings.
6. Cowen, E., Piran, T. & Narayan, R., *ApJ* **500**, 888 (1998).
7. Lloyd, N.M. & Petrosian, V., in *Gamma-Ray Bursts: 3rd Huntsville Symposium*, AIP Conf. Proc. **384**, 233 (1995).

The Fluence Duration Bias

Jon Hakkila*, Charles A. Meegan‡, Geoffrey N. Pendleton†,
Robert S. Mallozzi†, David J. Haglin*, and Richard J. Roiger*

*Minnesota State University, Mankato, Minnesota 56001
‡NASA/MSFC, Huntsville, Alabama 35812
†University of Alabama, Huntsville, Alabama 35899

Abstract. The fluence duration bias causes fluences and durations of faint gamma-ray bursts to be systematically underestimated relative to their peak fluxes. Using Monte Carlo analysis, we demonstrate how this effect explains characteristics of structure of the fluence vs. 1024 ms peak flux diagram. Evidence of this bias exists in the BATSE fluence duration database, and provides a partial explanation for the existence of burst class properties.

INTRODUCTION

The *fluence duration bias* is an instrumental bias causing some gamma-ray burst fluences and durations to be underestimated relative to their peak fluxes. The fluence duration bias does not manifest itself by altering the trigger rate, but instead alters measured burst properties. Elsewhere in this conference [2] we present evidence that the class of Intermediate bursts identified by statistical clustering analysis [3] can be produced from the hardness vs. intensity correlation and the fluence duration bias. We also demonstrate how the bias can be responsible for decreasing fluences and durations of the longest low peak flux Class 1 bursts. In this paper, we describe the fluence duration bias in more detail.

AN EXAMPLE

Figure 1 demonstrates the time history of a bright, Class 1 (Long) BATSE burst (trigger 2831) as measured in the 50 to 300 keV range on the 1024 ms timescale. This burst is complex with an overall duration in excess of 180 seconds.

Figure 2 is a Monte Carlo simulation of what this burst might look like if its 1024 ms peak flux were reduced in intensity to 15% of its measured value (Poisson fluctuations have been added to the reduced signal). If the reduced burst duration is assumed to be identical to that of the unreduced burst, then its measured fluence-to-peak flux ratio is unchanged from the actual value of 19.4 (we measure the result in

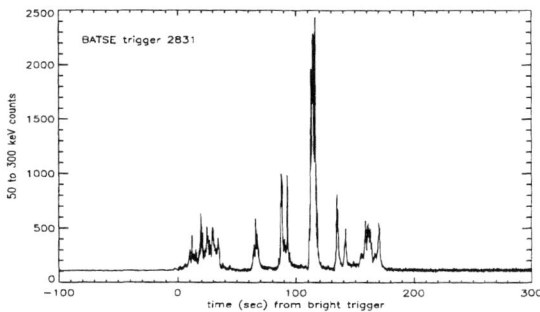

FIGURE 1. BATSE trigger 2831.

terms of the fluence-to-peak flux ratio, because Poisson fluctuations can also cause a burst's peak flux to change). If, however, the reduced burst duration is determined from "recognizable pulses" (pulses that are clearly visible above background; our algorithm assumes that the first and last peaks larger than 4σ above background bound the burst duration because there is no formal algorithm used by a human operator), then the average fluence-to-peak flux ratio drops slightly to 94% of its actual value.

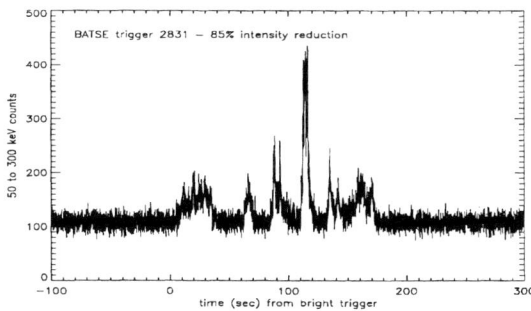

FIGURE 2. BATSE trigger 2831, reduced in intensity by 85%.

Figure 3 shows what the burst might look like if reduced to 2% of its actual value. Most of the burst fluence is confined to a temporal span of roughly 20 seconds. Our "recognizable pulse" algorithm finds that the burst is still considerably longer than this single pulse, but that the total burst duration is still underestimated for the purpose of measuring fluence. The fluence-to-peak flux ratio for the burst in question is only 61% of its actual value. This underestimate is even larger when the burst is reduced to a value closer to the trigger threshold.

It is difficult to accurately model the process by which the fluence duration interval is chosen, since human interaction plays an important role. We suspect

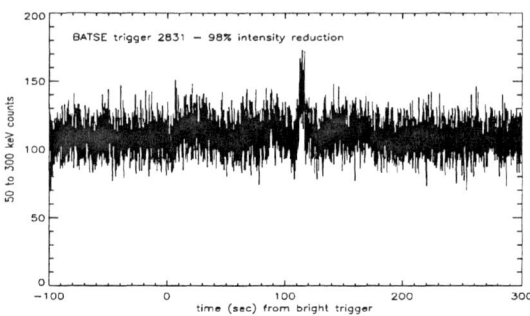

FIGURE 3. BATSE trigger 2831, reduced in intensity by 98%.

that the actual amount of the bias is less than the amount described here, since the human eye and mind are good at removing patterns from noise. Nonetheless, there is evidence that the bias is present, and that it is large enough to cause a depletion in the number of small peak flux, high fluence bursts as well as being responsible for producing some Class 3 burst characteristics from Class 1 bursts.

EVIDENCE FOR THE FLUENCE DURATION BIAS IN THE 4B CATALOG

Fluence appears to be one of BATSE's most accurately measured quantities because its statistical measurement errors are typically only ±5%. However, there is no intensity-dependent component to this measurement error, as might be expected from Figures 1, 2, and 3. It should be mentioned that there are no BATSE bursts with fluences less than zero (as might be expected if background dominated the fluence measurement), and few with fluences less than the fluence found in the 1024 ms peak flux.

The formal fluence error is kept small in part by fitting the background for faint bursts with high-order polynomials. Unfortunately, this process can introduce systematic underestimates of burst fluence by overestimating background [1]. *The fluence error can also be reduced by decreasing the fluence duration.* Figure 4 plots fluence durations for available bursts in the 4B Catalog. The sample has been limited to Class 1 bursts detected using the same trigger criteria (because Class 2 and Class 3 bursts are clearly shorter than the Class 1 bursts, and because different trigger criteria might alter the composition of the sample in a heterogeneous way).

Figure 4 indicates that there are few long Class 1 fluence durations near BATSE's detection threshold (1024 ms peak fluxes slightly greater than BATSE's 0% efficiency of 0.2 photons cm^{-2} second^{-1}). This is strong evidence for the existence of the fluence duration bias, and it indicates that the magnitude of the effect apparently strengthens for fainter bursts.

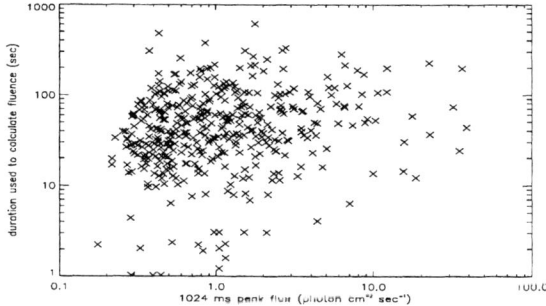

FIGURE 4. Fluence Duration vs. 1024 ms Peak Flux from BATSE 4B Data.

Figures 5, 6 and 7 demonstrate that the fluence duration bias is more difficult to cleanly delineate when peak flux and/or trigger timescales are shorter than 1024 ms (the effect is likewise more pronounced when longer timescales are used). We attribute this to the lower signal-to-noise ratio of shorter timescale measurements, making intensities measured on these timescales less accurate measures with larger intrinsic scatter than 1024 ms. The fluence duration bias is still present on shorter timescales; the scatter of these measures just makes it harder to recognize.

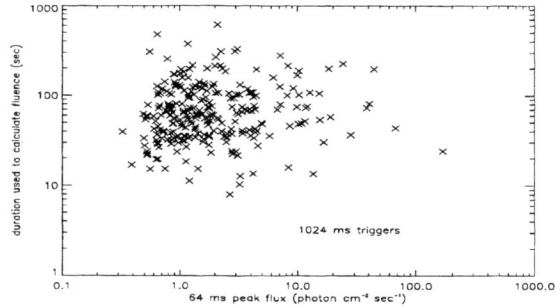

FIGURE 5. Fluence duration vs. 64 ms peak flux for Class 1 bursts triggering on the 1024 ms timescale.

CONCLUSIONS

Monte Carlo modeling of bursts with different temporal structures indicates that fluence duration is easy to underestimate, particularly for faint bursts. This causes some burst fluences and durations to be underestimated. Some bursts, such as trigger 2831, have temporal structures more susceptible to this bias than others. The strength of the bias is hard to judge for an individual burst, as it depends both

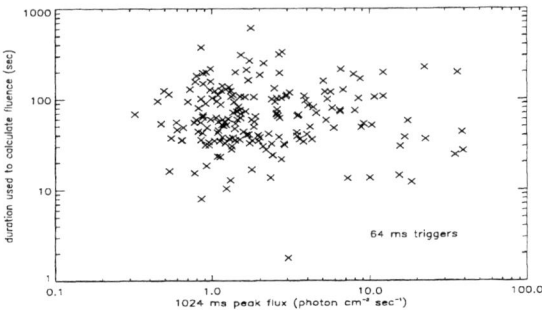

FIGURE 6. Fluence duration vs. 1024 ms peak flux for Class 1 bursts triggering on the 64 ms timescale.

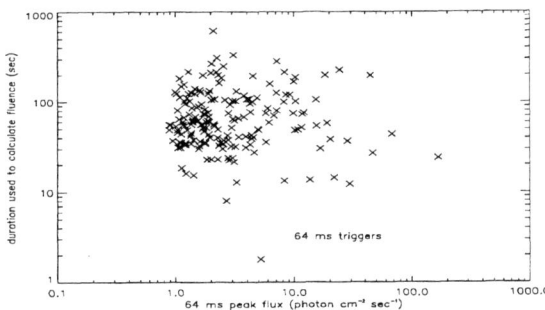

FIGURE 7. Fluence duration vs. 64 ms peak flux for Class 1 bursts triggering on the 64 ms timescale.

on burst temporal morphology and on how the human operator selects a fluence duration interval. The magnitude of the bias depends both on the time intervals chosen for the peak flux and trigger flux, since the fluence underestimate must be made relative to a "fixed" brightness measure. The fluence duration bias appears capable of producing observed characteristics of the fluence vs. 1024 ms peak flux diagram, and of making some Class 1 bursts (primarily faint ones) take on Class 3 characteristics. We currently studying this effect in greater detail.

REFERENCES

1. Bonnell, J. T., et al., *ApJ* **490**, 79 (1997).
2. Hakkila, J., et al., these proceedings.
3. Mukherjee, S., et al., *ApJ* **508**, 314 (1998).

A Monte Carlo Simulation of the Peak Counts Bias and the Slow Riser Bias in the BATSE Trigger

Audress Johnson[1], Charles Meegan[2], and Jon Hakkila[3]

[1] *University of Alabama at Tuscaloosa, Tuscaloosa, AL 35487*
[2] *NASA/MSFC, Huntsville, AL 35812*
[3] *Minnesota State University, Mankato, MN 56001*

Abstract. We have investigated two biases in the BATSE 1024 ms burst trigger using Monte Carlo simulations of burst time histories. The peak counts bias, which arises from statistical fluctuations in the count rates, results in an average overestimate of the on-board peak rate by ~8%. The slow riser bias, which arises from errors in the on-board calculation of background, results in an average underestimate of the on-board peak rate by ~5%.

INTRODUCTION

BATSE has produced the most extensive catalog of gamma-ray bursts. The importance of this catalog for global studies of bursts has motivated efforts to better characterize the biases in the on-board burst trigger algorithm. This algorithm generates a burst trigger when the count rates in two or more of the eight BATSE detectors are above a varying trigger threshold. Rates are tested on three timescales: 64 ms, 256 ms, and 1024 ms. The nominal energy range is 50 keV to 300 keV. The detector thresholds are specified in terms of the number of standard deviations (typically 5.5) above background, and are recomputed every 17.408 seconds. A trigger efficiency is provided by the BATSE team which is used to construct burst intensity distributions. Residual biases in this efficiency correction must be understood to properly interpret BATSE intensity distributions. An overview of these biases is provided elsewhere in these proceedings [1]. Here, we describe a simulation to measure two of the biases: the peak counts bias, and the slow riser bias [2].

The peak counts bias arises from statistical fluctuations in the count rates. The peak rate is more likely to be overestimated because the peak in a series of trials is more likely to be a positive than a negative statistical fluctuation.

The slow riser bias arises from a possible systematic error in the on-board determination of the background rate. If a burst exhibits a sufficiently slow increase in count rate, some burst flux may be included in the calculation of background, artificially increasing the background. The burst peak rate is then underestimated, since it is determined after subtracting the background.

These two biases were investigated by selecting strong bursts, reducing their intensity to near trigger threshold, and measuring the deviations in peak rates as determined by the on-board trigger algorithm. Forty bursts with 1024 ms peak fluxes above 6 photons cm^{-2} s^{-1} were selected for initial analysis. We used count rates for 1024 ms time intervals, 50 to 300 keV, for each of the four detectors that observe each burst. After subtracting background models from the rates, the burst count rates were scaled to produce a burst with a peak flux of 0.3 photons cm^{-2} s^{-1}, based on the cataloged value of the bursts' peak flux. This value of the reduced intensity corresponds to a trigger efficiency of 50% on the 1024 ms timescale. The count rates for the reduced burst were examined to find the peak rate above background reached by the second brightest detector over the time interval of the burst. The effect causing the bias was then applied and the new peak rate found. The ratio of this new peak rate to the previous unbiased peak rate was recorded.

Throughout this paper, the term peak rate refers always to the rate in the second brightest detector, since this is the rate that determines whether a burst triggers, and the biases we wish to measure are due to errors in the on-board measurement of this rate.

Note that it is not necessary to test each trial for burst triggering. Although the trigger probability depends sensitively on the reduced peak flux of the burst, the ratio of the biased to the unbiased peak rate is not so sensitive, so our technique will allow us to find revised trigger probabilities without requiring a fine grid of burst intensities. Our results will be tied to trigger probabilities through the already known trigger efficiency as a function of true peak flux.

PEAK COUNTS BIAS

To determine the peak counts bias, Poisson noise was applied to the burst rates, which were then examined to determine the new peak in the second brightest detector. Note that this peak is not necessarily at the same time or in the same detector as in the original burst. Ten trials with different random noise were performed for each of the forty bursts. For each trial, we found the ratio of the new peak count rate to the peak count rate of the burst before applying noise. Figure 1 shows a histogram of the number of trials as a function of the ratio of the observed to the true peak rate. The distribution has an average of 1.079 and a standard deviation of 0.172. The offset of the distribution from unity is a measure of the peak counts bias for the 1024 ms burst trigger.

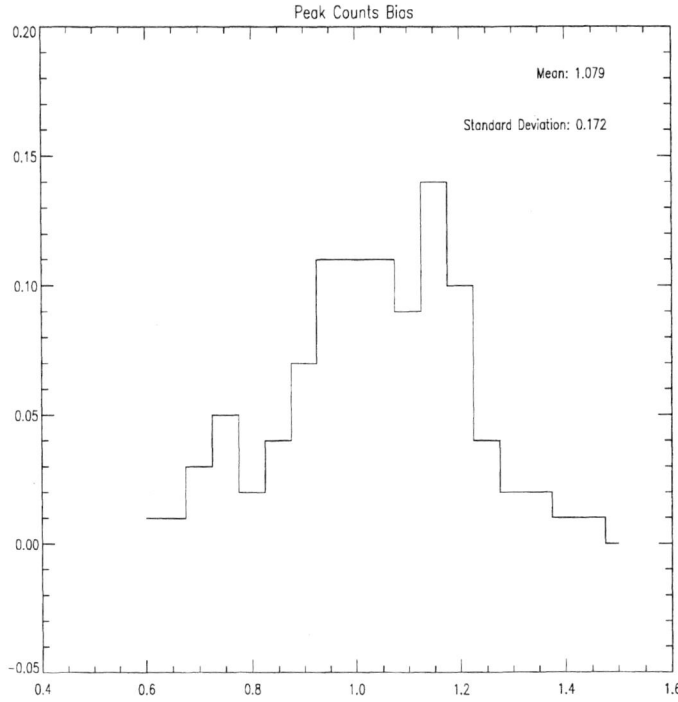

FIGURE 1. The Peak Counts Bias. This shows the distribution of errors in the measured peak rate due to statistical fluctuations in the rates.

SLOW RISER BIAS

To investigate the slow riser bias, we examined variations in the peak rate caused by differences in the subtracted background. Instead of using the background model as in the investigation of the peak counts bias, we computed the background using a sample of the preburst data, as is done on-board to determine burst triggers. For each sample burst we computed 17 backgrounds, each phased differently with respect to the peak. This background was then subtracted from the total count rate to determine the burst peak rate. Thus, the true peak rate was the rate above the best fit background model, while the biased peak rate was the rate above the 17 second preburst background sample as would be computed on-board. The ratio of peak rate to true peak rate was then determined, as before. Figure 2 shows the distribution of this ratio for 17 trials each of the 40 bursts. The distribution has an average of 0.950, and a standard deviation of 0.162. The offset from unity is a measure of the slow riser bias for the 1024 ms burst trigger.

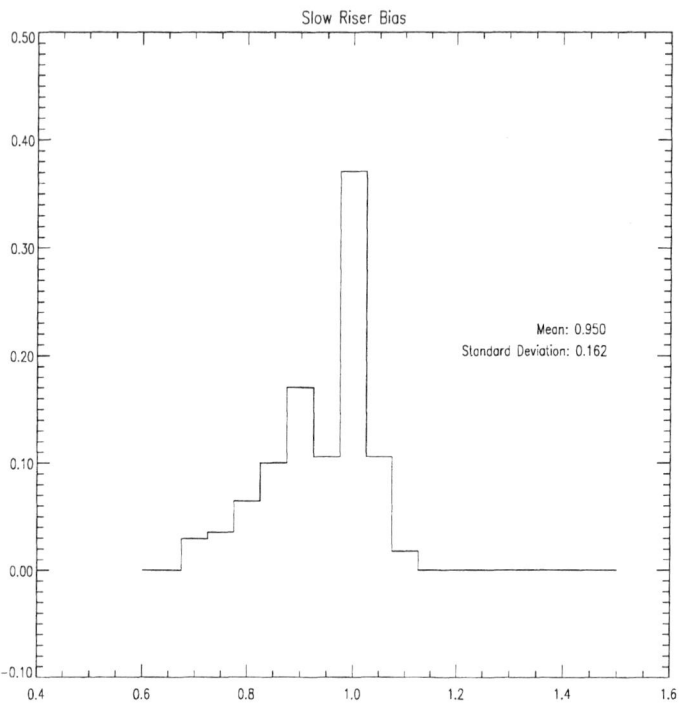

FIGURE 2. The Slow Riser Bias. This shows the distribution of errors in the measured peak rate due to background measurement errors.

CONCLUSION

We have obtained estimates of the peak counts bias and the slow riser bias in the BATSE trigger using Monte Carlo simulations of strong bursts reduced to rates near trigger threshold. Since these biases depend on the temporal structure of bursts, the accuracy of these results is contingent on the temporal similarity of weak and strong bursts. Only the 1024 ms trigger has been considered so far. The effects considered result primarily in a broadening of the trigger efficiency and, since the error distributions are not centered at 1, a bias. Both effects clearly can act in either direction. The peak rate shows fluctuations of about 20%, as would be expected given a 5.5σ threshold, with an average overestimate of the peak rate of $\sim 8\%$. For the slow riser bias, the broadening is due to the varying slopes of the background rate; rising backgrounds lead to underestimates of the peak rate and vice versa. The slow riser bias itself appears as a tail extending to low values of the peak rate ratio, resulting in an average underestimate of the peak rate by $\sim 5\%$.

These biases are not large and will partially cancel since they act in opposite directions. Consequently, we do not expect that they will significantly affect the BATSE burst intensity distribution.

FUTURE WORK

Since they may be correlated, we plan to examine the simultaneous application of the peak counts and the slow riser biases. We will increase the statistical significance of the results by using more bursts. We will perform runs with the bursts reduced to different equivalent peak fluxes. We also expect to perform similar analyses using the 64 ms and the 256 ms trigger timescales. Ultimately, we plan to convolve these results into an improved trigger efficiency correction factor.

REFERENCES

1. Meegan, C., et al., these proceedings.
2. Higdon, J.C. & Lingenfelter, R.E., in *Gamma-Ray Bursts: Third Huntsville Symposium*, AIP Conf. Proc. **384**, 1995, pp. 402.

Luminosities, Space Densities and Redshift Distributions of Gamma-Ray Bursts

Maarten Schmidt

California Institute of Technology, Pasadena, California 91125

Abstract. We use the BD2 sample of gamma-ray bursts (GRBs) based on 5.9 years of BATSE DISCLA data with a variety of models of the luminosity function to derive characteristic GRB luminosities, space densities and redshift distributions. Previously published results for an open universe and modest density evolution of the GRBs showed characteristic peak luminosities around 5×10^{51} ergs s^{-1} in the 50 – 300 keV band if the emission is isotropic, and local space densities around 0.2 Gpc^{-3} y^{-1}. In this paper, we illustrate for several luminosity function models the predicted distributions of peak flux, luminosity and redshifts. We use the luminosity function models also to address the connection between supernovae and GRB. If all supernovae of type Ib/c harbor a GRB, the beaming fraction would have to be in the range $10^{-5} - 10^{-3.5}$. We find that GRB 980425, if correctly identified with SN 1998bw, has to be part of a population different from that of the bulk of GRBs.

INTRODUCTION

The luminosity function and cosmological evolution of extragalactic objects are usually derived from observed samples that are complete above a given flux limit and have measured redshifts. In the case of gamma-ray bursts (GRBs), there is no well defined complete sample with redshifts available at the present time. Under these circumstances, we have found it useful to invert the process, to *assume* a luminosity function, and to derive intrinsic properties of GRBs such as their characteristic peak luminosity L^* and local space density [9]. In this paper, we illustrate the predictions for some of these luminosity function models. We also use these models to address the issue of a connection between supernovae and GRBs.

MODELS OF THE GRB LUMINOSITY FUNCTION

We start with a brief description of the methodology. For a detailed description of the models, the reader is referred to [9]. We used luminosity functions that were broken power laws with characteristic luminosity L^*, with different slopes and

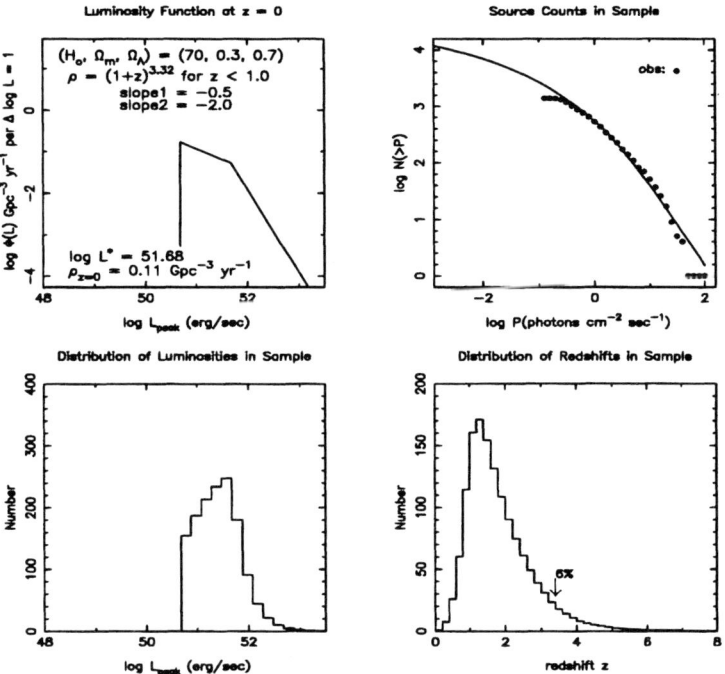

FIGURE 1. Assuming a luminosity function extending from $L^*/10$ to $100L^*$ characterized in the upper left panel and the text, the other panels show the predicted luminosity distribution, source counts and redshift distributions for the BD2 sample of GRBs

different extents above and below L^*. In addition, we assumed density evolution rising to a factor of 10 at $z = 1$ and constant for $z > 1$. We used the BD2 sample of 1391 GRB which was derived from 5.9 years of BATSE DISCLA data on a timescale of 1024 ms [9,10]. For a given luminosity function model, the observed euclidean value of $V/V_{max} = 0.334 \pm 0.008$ allowed a robust determination of L^*. The total number of GRBs in the BD2 sample provided the normalization of the luminosity function. For a given cosmology and evolution, the value of L^* ranged over a factor of 6 and the local space density over a factor of 2. Typical values[1] for an open universe with $q_o = 0.1$ were a characteristic peak luminosity $L^* = 5 \times 10^{51}$ ergs s^{-1} and a local space density around 0.2 Gpc^{-3} yr^{-1}. Beaming reduces the peak luminosity and increases the density by the same factor.

We illustrate in this paper the results based on several different luminosity functions. We are using a cosmological model that is a flat accelerating universe with a matter density $\Omega_m = 0.3$ and cosmological constant $\Omega_\Lambda = 0.7$ [1]. We assume a Band type GRB spectrum [2] with $\alpha = -1$, $\beta = -2$, and break energy $E_o = 150$

---

[1] These values are based on fluxes corrected for a scale error, see footnote in [9]. I thank J. Brainerd for supplying the BATSE detector response and atmospheric scattering matrices.

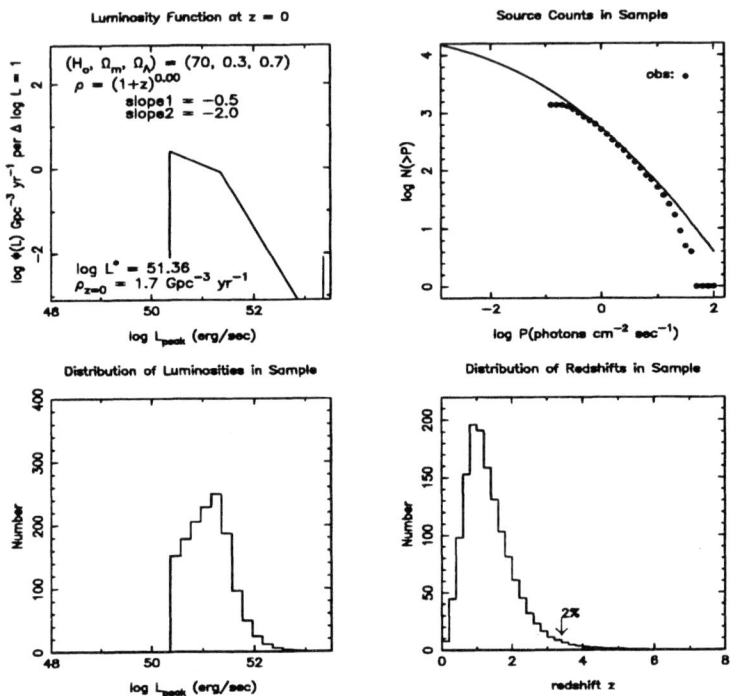

FIGURE 2. Predictions based on a GRB luminosity function with no density evolution

keV. In Figure 1, we show the results for a typical model of the luminosity function, extending from $L^*/10$ to $100L^*$. The density evolution $\rho = (1+z)^{3.32}$ amounts to a factor of 10 at $z = 1$ and is then constant at $z > 1$. The local density is 30% lower than that for an open universe with $q_o = 0.1$ [9]. As shown in Figure 1, the predicted $N(> P)$ distribution agrees well with the observations. The largest GRB redshift so far observed is 3.42 for GRB 971214 [7]. Based on the expected redshift distribution, the probability of finding a redshift of $z = 3.4$ or larger in the BD2 sample is 6%. The median redshift is 1.5 and the largest redshift among the 1391 GRB in the BD2 sample is $z = 9.3$ according to this model.

In Figure 2, we show the results if there is no density evolution. Compared to the previous case, L^* is down by a factor of 2 and the local space density up by a factor of 15. The predicted source counts fall below the observed points. The redshift distribution has only 2% of GRBs at a redshift of 3.4 or above.

THE SUPERNOVA CONNECTION

If the light curves of GRB afterglows show evidence for a supernova of Type Ib/c [4], the question arises whether all such supernovae could be associated with GRBs.

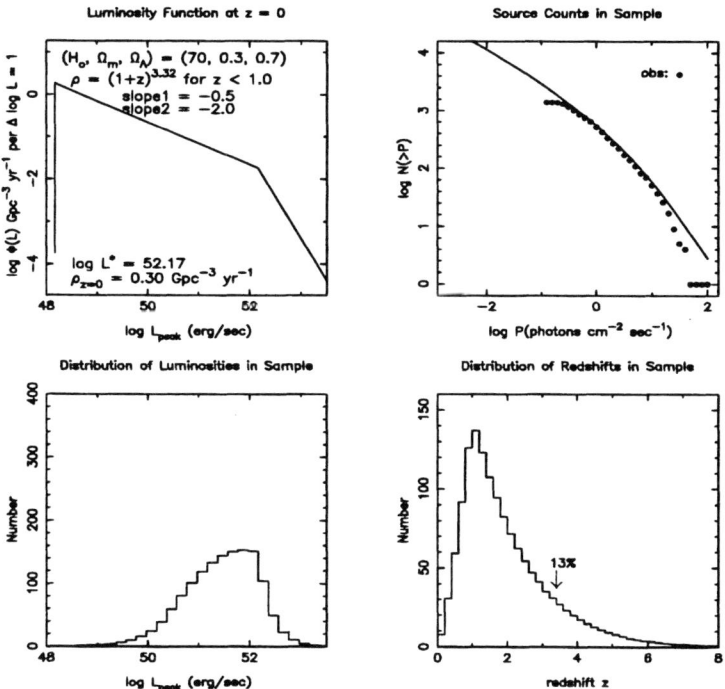

FIGURE 3. Predictions based on a luminosity function extending from $L^*/10000$ to $100L^*$

In the following, we update a discussion of this issue by Lamb [8]. If the gamma-ray emission is beamed over a fraction f_{beam} of the celestial sphere, the local GRB rate is roughly (0.1–3) f_{beam}^{-1} Gpc^{-3} yr^{-1}, where the range mostly reflects the effect of density evolution (up to a factor of 10 at $z = 1$), or no density evolution [9]. The rate of Type Ib/c supernovae [5] in spiral galaxies is $\sim 10^{-13}$ L$_\odot^{-1}$ yr^{-1}. With a luminosity density of $\sim 10^8$ L$_\odot$ Mpc^{-3} [3] for spirals, this corresponds to $\sim 10^4$ SN Gpc^{-3} yr^{-1}. If $f_{beam} = 10^{-2}$, then only 1 in (1000-30) SN Type Ib/c could have an associated GRB. If every Type Ib/c supernova harbors a GRB, f_{beam} would have to be as small as $10^{-5} - 10^{-3.5}$.

If GRB 980425 is associated with the SN 1998bw, its peak luminosity is $\log L = 46.7$ [6]. We explored a luminosity function model that extends a factor of 10^4 below L^* (Fig. 3). Even though the luminosity function reaches down to around $\log L \sim 48$, the distribution of *observed* luminosities predicts only one GRB to have $\log L < 48.5$, or a probability of around 0.1%. Given this low probability, the identification of GRB 980425 with SN 1998bw can only be understood if it represents a separate population of low-luminosity bursts. Assuming that 30% of the observed GRBs are of low luminosity, we show in Figure 4 the expected distributions for the remaining GRBs assuming that they have a luminosity function

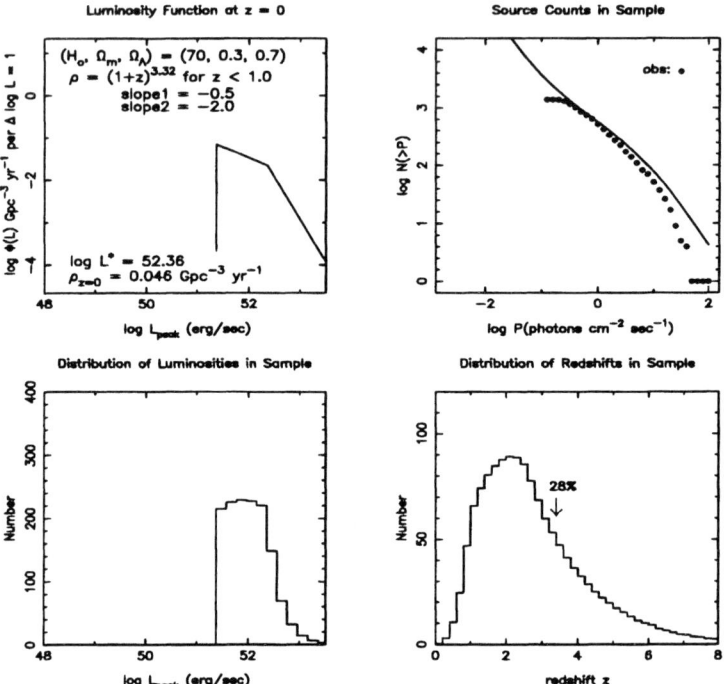

FIGURE 4. Assuming that 30% of observed GRB are of low luminosity, we show predicted distributions for the remaining GRBs for a given model of the luminosity function.

identical in shape to that of Figure 1. The fit of $N(>P)$ to the observed distribution is poor, indicating that at most 30% of observed GRBs can have a low luminosity.

REFERENCES

1. Bahcall, N. A., Ostriker, J. P., Perlmutter, S., and Steinhardt, P. J., *Science* **284**, 1481 (1999).
2. Band, D. L., et al., *ApJ* **413**, 281 (1993).
3. Binggeli, B., Sandage, A. and Tammann, G. A., in *Ann. Rev. Astron. Astroph.* **26**, 509 (1988).
4. Bloom, J. S., et al., *Nature* **401**, 453 (1999).
5. Cappellaro, E., et al., *A&A* **322**, 431 (1997).
6. Galama, T. J., et al., *A&A Suppl.* **138**, 465 (1999).
7. Kulkarni, S. et al., *Nature* **393**, 35 (1998).
8. Lamb, D. Q., *A&A Suppl.* **138**, 607 (1999).
9. Schmidt, M., *ApJ* **523**, L117 (1999).
10. Schmidt, M., *A&A Suppl.* **138**, 409 (1999).

Constraining the Luminosity Function of GRBs from Time Dilation, Brightness Distribution and Redshift Data

Ming Deng and Bradley E. Schaefer

Department of Physics, Yale University, New Haven, CT 06520-8120

Abstract. We constrain the luminosity function of gamma ray bursts (GRBs) by testing several cosmological models versus a combination of (a) observed time dilation in the peak-to-peak time scales of GRB light curves, (b) the $LogN - LogP$ number counts relation, and (c) GRB luminosities directly determined from recent redshift measurements and apparent GRB brightness. A power law luminosity function $\phi(L) = C \cdot L^{-\beta}$ ($L_{min} < L < L_{max}$) is examined within three cosmological models, with cosmological parameters $(\Omega_0, \Omega_\Lambda) = (1.0, 0.0)$, $(0.3, 0.0)$, or $(0.3, 0.7)$. It is found that (a) GRB models with constant comoving density rate do not fit the data while models with GRB density evolution tracing the observed star formation rate can accommodate all the observed data; (b) the width $K = L_{max}/L_{min}$ of the luminosity function is found to be $K > 100$ at the 1 σ level; (c) the *intrinsic* average luminosity $<L>$ is found to be $2 \sim 6 \times 10^{51}$ ergs $\cdot$ s^{-1} at the 3σ level, this implies that bursts at the BATSE 100% efficiency threshold have typical red shifts of $<z_{0.42}> = 2 \sim 3$.

INTRODUCTION

In the past, a lot of research efforts have been undertaken [1–3] to use time dilation or the $LogN - LogP$ relation to study the distance scale to gamma ray bursts (GRBs) assuming a standard candle luminosity of GRBs, or using some of these relations to constrain the GRB luminosity function and density evolution [4–7]. With the recent direct redshift measurements of GRBs, a distribution of the GRB luminosities can be inferred. The diverse luminosities determined from their redshifts and peak flux can now be compared to the constraints of the time dilation and $LogN - LogP$ relation.

In this paper, we test several cosmological models with broad luminosity functions to see if they can simultaneously reproduce (a) the time dilation data, (b) the $LogN - LogP$ curve, and (c) the diverse distribution of redshift measurements vs. brightness. The analysis is performed for models with either constant comoving density (CCD) rate of GRBs, or models with GRB comoving density rate tracing the star formation rate (SFR) [5,6,8].

DATA PREPARATION

To measure the time dilation effect in the cosmological models, we used the peak-to-peak time scales τ_i [9] of BATSE GRBs. The average peak-to-peak time scales are determined for GRBs in 6 different brightness groups, from $P_{dim} = 1.0 - 1.4$ photons $\cdot$ cm$^{-2}\cdot$ s^{-1} to $P_{bright} > 7.7$ photons $\cdot$ cm$^{-2}\cdot$ s^{-1}. The peak-to-peak time scales τ_is are averaged for bursts within each of the 6 brightness groups. The use of peak-to-peak time scales avoids the effect of pulse width/energy correlations which might biases for some other time scales. Nevertheless, the measured time dilation factors in the peak-to-peak time scales agree well with results derived from other time scales [10,11].

For comparison of the number counts vs. brightness relation of the cosmological model and observed $LogN - LogP$ relation, the data from the current (early 1999) BATSE catalog are used. The catalog contains 1838 bursts from trigger 105 to trigger 7361. The peak flux values P_{1024} in the catalog are integrated over 1024 ms time scales, with the BATSE trigger efficiency dropping significantly below 100% when the peak flux P_{1024} is less than a threshold level of 0.42 photons $\cdot$ cm$^{-2}\cdot$ s^{-1}. The GRBs with peak flux values below this threshold level are eliminated, thus the actual number of GRBs used is 1437. These GRBs are divided into 25 bins according to their peak flux P_{256} values, and the number counts is calculated for each of these brightness bins. The resulting number counts vs. peak flux values is used for comparison with cosmological models.

There are now 10 GRB redshifts measured. Together with the peak flux measured by BATSE, the redshifts measured can be used to estimate the luminosity L values of the GRBs assuming isotropic emission. In Table 1, the luminosities of these 10 GRBs are listed along with other properties for a cosmology with $\Omega_0 = 0.3, \Omega_\Lambda = 0.7$ and a Hubble constant of 65 km $\cdot$ s$^{-1}\cdot$ Mpc^{-1}, the estimated luminosity values span over 6 orders of magnitude, from 1.8×10^{53} photons $\cdot$ s^{-1} (GRB 980425) to 3.2×10^{59}

TABLE 1. Properties of the GRBs with redshift measurements.

GRB	P_{1024} photons $\cdot$ cm$^{-2}\cdot$ s^{-1}	Redshift	L photons $\cdot$ cm$^{-2}\cdot$ s^{-1}	Host R
970228	3.3^a	0.695	8.2×10^{57}	24.96
970508	0.969 ± 0.05	0.835	3.8×10^{57}	25.65
970828	4.9	0.958	2.7×10^{58}	24.41
971214	1.95 ± 0.05	3.418	2.4×10^{59}	25.56
980425	0.96 ± 0.05	0.0085	1.8×10^{53}	~ 13
980613	0.63 ± 0.05	1.096	4.8×10^{57}	23.85
980703	2.42 ± 0.06	0.966	1.3×10^{58}	22.43
990123	16.42 ± 0.12	1.600	3.2×10^{59}	23.73
990510	8.16 ± 0.08	1.619	1.6×10^{59}	> 27.0
990712	$\sim 3.0^b$	0.430	2.3×10^{57}	21.71

^{a} From TGRS value.
^{b} Scaled from BeppoSAX peak flux.

photons · s⁻¹ (GRB 990123). Such an extreme diversity is primarily caused by the low luminosity event GRB 980425. However, even without it, the luminosity values still span more than 2 decades.

ANALYSIS

To constrain the luminosity function of GRBs, an analytic form of luminosity function has to be adopted to investigate how the observed effects vary with the parameters of the luminosity function. In this paper, a range limited power luminosity function $\phi(L)$ is adopted:

$$\phi(L) = C \cdot L^{-\beta}, L_{min} < L < L_{max} \qquad (1)$$

There are two parameters of this luminosity function, the average luminosity $<L>$ and width K. These two parameters are defined as

$$<L> = \frac{\beta-1}{\beta-2} \cdot \frac{1-K^{2-\beta}}{1-K^{1-\beta}} \cdot L_{min}, K = L_{max}/L_{min} \qquad (2)$$

To generate simulated data from the cosmological models with the luminosity function, a distribution of redshift values are sampled from the volume vs. redshift in the cosmological model, taken into account of the density evolution of GRBs. The L values are sampled from this luminosity function. For each burst with sampled z and L, the peak flux, P, is then derived. (For purposes of K-corrections, we adopt the reasonable approximation that the typical burst count spectrum is shaped as E^{-2}). Bursts with peak flux values less than the 100% BATSE efficiency threshold $P = 0.42$ photons · s⁻¹· cm⁻² are eliminated as it is done with the real BATSE data set. The remaining peak flux P, luminosity L, and redshift z distributions form the simulated data sample to be compared with the observed distributions. The $LogN - LogP$ relation of the simulated data is calculated by dividing the bursts into 25 brightness bins as is done with the real data.

To simulate the time dilation effect in the cosmological models, a simulated distribution of GRB time scales are created. Since GRBs with the same brightness have widely ranging durations and peak-to-peak time scales, any time dilation effect can only be detected statistically. We simulate the observed τ_i as the product of the time dilation factor and a τ_i randomly sampled from a distribution of 157 bright GRB time scales. The time dilation factor $(1+z)/(1+z_{bright})$ is determined by the redshift of the simulated data and the redshift z_{bright} of the sampled bright GRB. The simulated bursts are then divided into 6 brightness groups using the same brightness ranges as the observed distribution. The average time dilation factor of each brightness group can be calculated relative to the brightest burst group. This will produce 6 average time dilation factors that can be compared to the results of the BATSE data.

The distribution of the simulated luminosity L values are weighted by a smooth efficiency function that crudely accounts for the detection efficiency of the optical

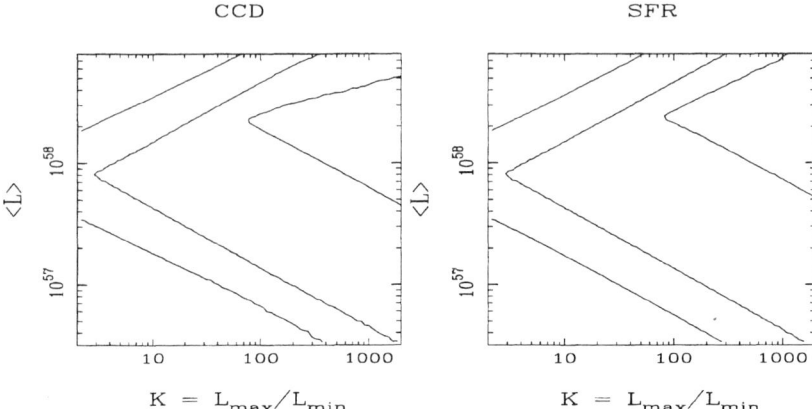

FIGURE 1. The acceptable region of $<L>$ and K from the redshift data alone. The three contour levels have KS probabilities of 99.73%, 95.4%, and 68.3% respectively from left to right. The cosmological parameters are $\Omega_0 = 0.3, \Omega_\Lambda = 0.7$, with the GRB density rate either having no evolution (left) or tracing star formation rate(right).

counterparts. The simulated distribution is compared to the observed distribution of 9 luminosity estimates using a Kolmogorov-Smirnov test. Such comparisons are made for 3 cases of cosmological models: $(\Omega_0, \Omega_\Lambda) = (1.0, 0.0), (0.3, 0.0)$ and $(0.3, 0.7))$ with the GRB density rate either having no evolution (CCD) or tracing the star formation rate (SFR). Several luminosity functions with different power law index β are tested. In Figure 1, the constraints on the two parameters of the luminosity function $<L>$ and K are shown as acceptable region within contour levels for 2 of the 6 cases considered with $\beta = 1.5$. It is clear from the graphs that $K > 100$ at 68.3% (1σ) level. For the other four cases of cosmological models tested, the results are similar.

The simulated time dilation and $LogN - LogP$ relation are compared to the observed relations using χ^2 tests. Again, comparisons are made for the above 6 cases of cosmology and various β. In Figure 2, the combined χ^2 test of time dilation and $LogN - LogP$ are displayed as acceptable regions $<L>$ and K with reasonable χ^2 values for the same 2 cases as in Figure 1. There are two conclusions that can be drawn from these graphs. First, there are no acceptable regions for the no evolution models, the minimum χ^2_{MIN} for the $<L>$ and K searched is 62.7. This is a bad fit χ^2 that is beyond the 99.73% for a 28 degree χ^2 variable. Second, The star formation rate models produces reasonable fits and the acceptable fit $<L> = 1 \sim 2 \times 10^{58}$ photons $\cdot$ s^{-1} or $3 \sim 6 \times 10^{51}$ erg $\cdot$ s^{-1} at 99.73% (3σ) level. For other cosmological models, it is found that CCD models consistently do not fit, while the variation of cosmological parameters slightly affect the acceptable range of $<L>$ in SFR models. In all cases, acceptable fits of SFR models have $<L>$ ranges from $0.7 \sim 2 \times 10^{58}$ photons $\cdot$ s^{-1} or $2 \sim 6 \times 10^{51}$ erg $\cdot$ s^{-1} at 99.73% (3σ) level.

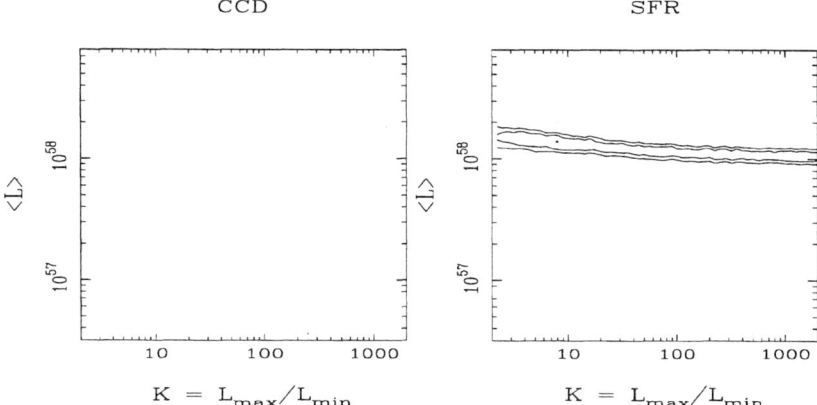

FIGURE 2. Constraint from the time dilation and $LogN - LogP$ relation. The acceptable region of $<L>$ and K is obtained by a χ^2 test of the simulated time dilation and number counts relation vs. the observed relations. The three contour levels corresponds to probabilities of 99.73%, 95.4%, and 68.3% for a χ^2 variable that has 28 degrees of freedom. The cosmological parameters are $\Omega_0 = 0.3, \Omega_\Lambda = 0.7$, with the GRB density rate either having no evolution (left) or tracing star formation rate (right). *There is no acceptable solution within the CCD model.*

CONCLUSION

The results of the simulation tests show that the GRB comoving density rate is not constant at different redshifts. If the GRB comoving density rate follows the star formation rate, then the GRB luminosity function is found to have a width of $K > 100$ at 1σ level, and an average luminosity $<L> = 0.7 \sim 2 \times 10^{58}$ photons $\cdot$ s^{-1} or $<L> = 2 \sim 6 \times 10^{51}$ erg $\cdot$ s^{-1} is required for all 3 cases of cosmological models $(\Omega_0, \Omega_\Lambda) = (1.0, 0.0)$, $(0.3, 0.0)$ and $(0.3, 0.7)$ to be compatible with observed time dilation, number counts and redshift data.

REFERENCES

1. Fenimore, E. E. et al., *Nature* **357**, 140 (1992).
2. Rutledge, R. E., Hui, L., & Lewin, W. H. G., *MNRAS* **276**, 753 (1995).
3. Wijers, A. M. J., et al., *MNRAS* **294**, L17 (1998).
4. Reichart, et al., *ApJ* **483**, 597 (1997).
5. Totani, T., *ApJ* **486**, L71 (1997).
6. Totani, T., *ApJ* **511**, 41 (1999).
7. Krumholz, M. and Thorsett, S. E., *ApJ* **506**, L81 (1998).
8. Madau, P., Pozzetti, L., and Dickinson, M., *ApJ* **498**, 106 (1998).
9. Deng, M. and Schaefer, B. E., *ApJ* **502**, L109 (1998).
10. Norris, J. P., et al., *ApJ* **439**, 542 (1995).
11. Bonnell, J. T., et al., *ApJ* **490**, 79 (1997).

Rest Frame Properties of Gamma-Ray Bursts

Igor G. Mitrofanov[1], Dmitrij S. Anfimov[1], Maxim L. Litvak[1],
Anton B. Sanin[1], Michael S. Briggs[2], William S. Paciesas[2],
Geoffrey N. Pendleton[2], Robert D. Preece[2], and Charles A. Meegan[3]

[1] *Space Research Institute, Profsojuznaya str. 84/32, 117810 Moscow, Russia*
[2] *Department of Physics, University of Alabama in Huntsville, Huntsville, AL 35899*
[3] *NASA/Marshall Space Flight Center, Huntsville, AL 35812*

Abstract. The small *reference* group of BATSE gamma-ray bursts with measured red-factors of optical afterglows is compared with the large *comparison* group of brightest BATSE bursts. These two groups are shown to be similar both in respect with distributions of measured E_p parameters in the Observer frame and also in respect with frame-independent distributions of newly implemented cosmological invariant parameters (CIP). Using the known values of red-factors Z for the *reference* group, the distribution of $E_p^{(RF)}$ is built for the rest frames of their sources. De-redshifted statistics of $E_p^{(RF)}$ are compared with observed distributions of E_p for the *comparison* group of bursts. From this comparison the collective estimation of red-shifts $Z = 2.5 - 3.5$ has been performed for the group of brightest BATSE bursts.

INTRODUCTION

The strongest direct argument in favor of the cosmological paradigm of gamma-ray bursts (GRBs) has recently been presented by the detection of red-shifted spectral lines of afterglowing optical counterparts of gamma-ray bursts. There are 5 BATSE events with measured red-shifts of fading optical counterparts [1]. These events could are considered below the *reference* group of bursts, whose observed physical signatures could be transformed into the rest frames of emitters.

The bursts with found afterglows belong to the group of so-called long bursts with duration time T_{50} more than 2.5 seconds. The current 4B database contains about several hundred of such events, and 248 of them have the peak flux brighter than the peak flux $F_{max}^{(64)} = 2.65$ ph cm^{-2}s^{-1} of the dimmest event GRB 971214 of the *reference* group with known red shifts. We define the sample of 209 BATSE bursts with $T_{50} > 2.5$ sec and peak fluxes greater than $F_{max}^{(64)} = 2.65$ ph cm^{-2}s^{-1}, as the *comparison* group of GRBs.

The subject of this paper is to compare between the *reference* group of bursts with Z-measured afterglows and the *comparison* group of bright bursts which belong to the −3/2 part of counting statistics.

THE CONSISTENCY BETWEEN *REFERENCE* AND *COMPARISON* GROUPS OF GRBS

Gamma-ray bursts are known to be highly variable transients, and nobody actually could measure the instantaneous energy spectra of photons at each preselected moment of a burst. To compare two groups of GRBs we have to implement the definite spectral parameter to each event, and then to compare statistics of these parameters for these two groups.

We suppose that the best spectral representation of a burst is the spectrum integrated over the full set of time bins selected by the emission time criteria (see definition of emission time in [2]). The first, it represents the moments of the most powerful emission, which would be similarly selected either for brighter or for dimmer events. The second, it would be similarly selected either for stretched and for non-stretched time profiles. It is a well determined signature both for complex multi-pulse time histories and for simple single-pulse events. And finally, it represents time intervals of bursts with the highest signal-to-noise ratio.

Accepting this parameterization, we built up the summed spectrum over the time intervals of high power emission for each BATSE burst of the *comparison* and *reference* groups. We used the MER data type with an appropriate calibration matrix for counts to photons deconvolution. We used the Band model parameterization of each spectrum. And, we implemented one spectral parameter for each burst defined as the peak energy E_p of νF_ν spectrum of high power emission.

The distribution of spectral parameters for the *comparison* group is presented in the Figure 1 (left). The observed statistics $[E_p^{(COMP)}]$ agrees with the log-normal distribution, and the log-normal mean value is $\left\langle E_p^{(COMP)} \right\rangle_{LN} = 250 \pm 10$ keV. We tried to find the probability that 5 observed values $E_p^{(REF)}$ of *reference* group and values $E_p^{(COMP)}$ of *comparison* group belongs to the same distribution using the Pearson criterion and the Student criterion. According to Pearson the probability is 0.40 (5 degrees of freedom) and according to the Student criterion the probability is 0.27 (212 degrees of freedom).

Formally speaking, the consistency of spectral parameters of these two groups in the Observer frame does not prove the similarity of out-bursting sources. One may imagine the situation that two different kinds of sources emit softer and harder gamma-rays in their rest frames, but the difference of spectral parameters in k times is occasionally compensated by opposite difference of cosmological red-shifts.

One could prove the internal consistency of out-bursting emitters provided the *reference* and the *comparison* groups of bursts would be compared by the parameter determined in the rest frames of emitters. We have implemented such parameter as the product of emission time τ_{50} and spectral parameter of high power emission E_p:

$$CIP = \tau_{50} \cdot E_p. \quad (1)$$

This parameter does not depend by definition on the *geometrical* cosmological transformation from the Observer frame into the rest frame of the emitter with a red-shift Z; the factor $1/(1+Z)$ of time stretching is compensated by the factor $(1+Z)$ of energy red-shift (see [3]). So we call it the Cosmological Invariant Parameter (CIP). The statistics of CIP values of GRBs characterizes the intrinsic properties of emitters in their co-moving frames.

The *comparison* group has the distribution of CIP values that is well fitted by the log-normal law (Figure 1, right). Again, we tried to find the probability that 5 CIP values of *reference* group and CIP values of *comparison* group belongs to the same distribution using the same criteria. According to Pearson the probability is 0.15 (5 degrees of freedom) and according to the Student criterion the probability is 0.05 (212 degrees of freedom).

COLLECTIVE ESTIMATION OF RED-SHIFT FACTORS OF GAMMA-RAY BURSTS

We may transform the spectral parameters $E_p^{(REF)}$ of gamma-ray bursts with measured red-shifts of afterglows into the rest frames of emitters:

$$E_p^{(rest)} = E_p \cdot (1 + Z). \quad (2)$$

The statistics of $E_p^{(rest)}$ may be used for collective estimation of red-shifts for the comparison group of bursts with measured spectral parameters E_p. If we assume that all sources of bursts belong to the same class of emitters, we could estimate the best fitting red-shift factor $(1+Z)$ of this group from the condition of consistency between the reference red-shifted statistics $[E_p^{(rest)}]$ and the comparison red-shifted statistics $[(1+Z) \cdot E_p^{(em)}]$. We used different statistical criteria to find the limits of Z values in accordance to 1σ, 2σ and 3σ levels of consistency (Figure 2).

DISCUSSIONS

We found that the group of gamma-ray bursts with measured red-shifts of optical afterglows is consistent in two aspects with the main sample of long and bright BATSE gamma-ray bursts. The first, the consistency is proved for statistics of measured spectral parameters of these two groups for summed time intervals of high power emission. It points out that they are the same observational phenomenon. The second, the consistency is also proved for the statistics of cosmological invariant parameters of bursts. It points out that the intrinsic properties of emitters are also non-distinguishable for these two groups. Therefore, we conclude that bursts with detected optical afterglows could be considered as a randomly selected sub-set from the general sample of long bright events.

Following this assumption, we have compared the statistics of red-shifted spectral parameters for the *reference* group with observed statistics for the *comparison* group. We find significant inconsistency between them. We implemented the free transformation factor of spectral parameters of the *comparison* group to achieve the consistency with red-shifted reference values. The best fitting value represent so-called collective estimation of the red-shift factor for the sample of bright long bursts. The best fitting value of $Z^* = 3.0 \pm 0.5$ points out the large cosmological distances for these bursts. The 3σ limits of Z corresponds to 0.9-7.2.

We know that the counting statistics of bright gamma-ray bursts with peak flux $F_{peak} > 10$ ph cm^{-2} s^{-1} corresponds to the slope of $-3/2$ of the $logN/logF$ distribution. We found that bursts with a peak flux above $F_{peak} > 10$ ph cm^{-2} s^{-1} have large average red-shifts of ~ 2.2. This red-shift corresponds to strongly non-flat space, and the slope of $-3/2$ cannot be the result of the basic properties of the space. We need to consider the models with non-standard emitters and possible $Z-$related evolution to explain this slope. We may build the particular model of emitters with some definite values of emitters distribution and evolution, but the slope of $-3/2$ will not be preselected for this model by either physical or geometrical properties of the emitting population. We have to consider the question of why the combination of different independent physical distributions of cosmological emitters has resulted to the signature of $-3/2$ which is generic for standard emitters in the flat space.

If we assume the collective estimation of Z for the *comparison* group of bursts, we may evaluate the most probable values of spectral parameters in their rest frames $\sim \left\langle (1+Z^*) \cdot E_{\mathrm{p}}^{(COMP)} \right\rangle = 992^{+41}_{-39}$ keV. It is found to be very close to the energy of the electron-positron rest mass. It might be a chance coincidence. However, we believe that any future model of gamma-ray burst emission has to deal with the fact that the spectral peak of emission is around the energy of electron-positron annihilation.

ACKNOWLEDGMENTS

The work in USA was supported by NASA project CGRO-98-120 of the Compton Observatory Guest Observations Program. The work in Russia was supported by the RFBR grant 98-02-17380 and INTAS grant 96-0315.

REFERENCES

1. Briggs M.S. et al., *ApJ* **524**, 82, 1999.
2. Mitrofanov I.G. et al., *ApJ* **522**, 1069, 1999.
3. Mitrofanov I.G. et al., *ApJ* **523**, 192, 1998.

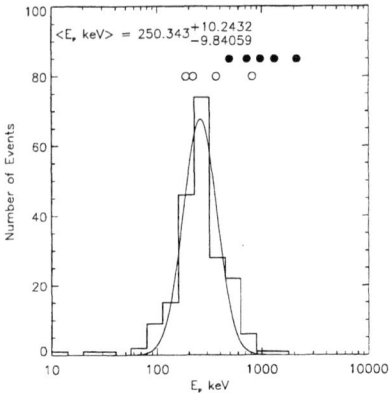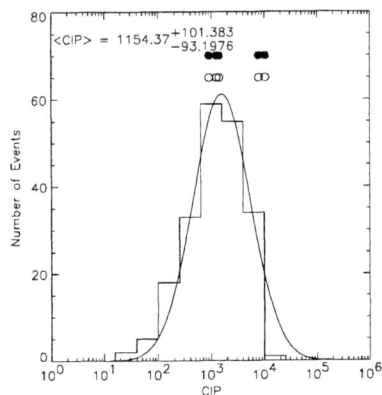

FIGURE 1. The distribution of E_p for all bursts (**left**) and distribution of CIP for all bursts (**right**) are shown. The E_p values of the *reference* group in the Observer frame (○) and the Rest frame (●) are shown.

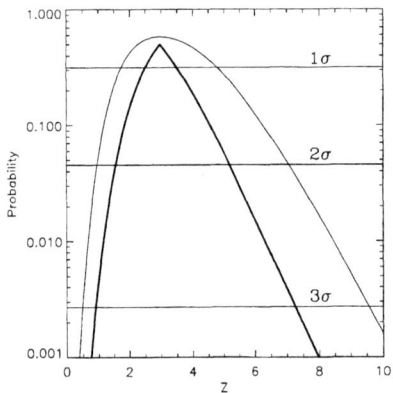

FIGURE 2. The probabilities that 5 $E_p^{(rest)}$ values and $E_p^{(COMP)}$ values of comparison group belongs to the same distribution according to the Pearson (fine line) and Student (thick line) criteria, as functions of red-shift of sources.

LogN-LogF Distribution of BATSE Bursts Based on the Average Flux During High Power Emission

Dmitrij S. Anfimov[1], Igor G. Mitrofanov[1], Maxim L. Litvak[1], Anton L. Sanin[1], Michael S. Briggs[2], William S. Paciesas[2], Geoffrey N. Pendleton[2], Robert D. Preece[2], and Charles A. Meegan[3]

Abstract. A physically-based flux characteristic for gamma-ray bursts is proposed that consists of the photon flux in a particular energy range positioned at the peak of the energy spectrum accumulated during the sum of emission time intervals. LogN-LogF distributions are investigated using these fluxes for bright BATSE bursts with peak fluxes on the 64 ms time scale larger than 2 ph/cm^2·s.

INTRODUCTION

Throughout the first period of gamma-ray burst (GRB) investigations, the most important question concerned the distance scale of GRB sources. Recently, much progress has been achieved by finding the optical afterglows of several bursts. The observed values of red-shifts indicate that burst sources might be placed at large cosmological distances.

However, if we restrict ourselves to the gamma-ray data, the signature of a cosmological origin of GRBs can only be determined from studies of the brightness-frequency distribution *logN − LogF* of GRBs. For a cosmological model with standard candle sources, we expect a −3/2 slope for the brightest events from the flat region of space, and we should see a downward deviation from the −3/2 law for dimmer bursts, as they are thought to come from curved space with large cosmological effects. Indeed, the observed curve for the total sample of BATSE and PVO events has the slope of −3/2 for bright bursts, and declines downward from this law for dimmer events [1].

Gamma-ray bursts with large measured z are bright bursts, and some of them are positioned on the *logN − logF* curve at the Euclidean portion (with a −3/2 slope). Therefore, there is a disagreement between the optical measurements of red-shifts for particular bursts with afterglows and the concept of the flatness of space, where the brightest bursts are positioned according to the −3/2 slope of brightness-frequency statistics. We might consider what are the physical reasons for this disagreement, but we also have to examine the possibility that the peak flux used for the curve is biased by instrumental effects, and that one should implement a pure physical parameter for the flux evaluation in order to resolve the shape of the real $LogN - LogF$ curve with clear physical sense (see [2]).

DEFINITION OF HIGH POWER EMISSION FLUX

For an estimation of the photon flux, we need two values: the fluence at a fixed energy channel and the time interval corresponding with the fluence accumulation. We suggest determining these two values using the technique of high-power emission [3,4]. The emission time τ_{50} of GRBs represents the sum of the intervals of high-power emission that contribute 50% of the total fluence. The spectrum of high-power emission intervals represents the energy distribution of photons accumulated during these intervals. The peak energy E_p of the $\nu F(\nu)$ spectrum represents the region of the highest spectral density emission. We define the average flux of the high-power emission of GRBs to be the photon fluence accumulated during the emission time intervals in the spectral range $(0.5E_p; 2E_p)$ around E_p, divided by the emission time. This flux is a pure physical parameter F_{PPP} of the observed emission; it depends upon neither the instrument spectral range, nor the instrument time resolution.

BRIGHTNESS-FREQUENCY STATISTICS FOR BRIGHT SET OF BATSE GRB

According to the implemented definition, we calculate F_{PPP} for bright BATSE bursts with an estimated peak flux of photons above 2 ph/cm²·s on the 64 ms time scale. There are 308 events in the BATSE 4B catalog [5] with 16-channel MER data available, which we need to calculate the energy spectra of high-power emission. Differential and integral distributions for the high-power emission flux are presented in Figures 1 and 2 for the long and short classes of bursts, respectively. It has been shown previously [6] that these two classes have not only different durations and spectral hardnesses, but also different slopes for their $logN - LogF$ distributions.

We used the differential presentation of brightness-frequency statistics to find the best-fitting power indices. We know that these indices depend on the selected brightness range between the lower-limit flux and the brightest detected flux that is used for fitting procedure. The best-fit power-low indices are presented in Figure 3 for 236 long and 72 short bursts with $\tau_{50} > 0.6$s and $\tau_{50} < 0.6$s, respectively, for different values of the lower limit of flux F_{PPP}.

CONCLUSIONS

The brightness-frequency statistics were generated using the F_{PPP} parameter as a measure of brightness for the two classes of long and short GRBs (Figures 1 and 2). It was shown that power-law indices for these two brightness-frequency curves are different from the value of $-3/2$ that is thought to correspond to a homogeneous distribution of standard emitters in a flat Euclidean space (Figure 3). On the other hand, the brightness-frequency statistics of short and long bursts do not show any significant differences between the best-fitting indices of a power-law approximation. The brightness-frequency statistics represent the luminosity function of emitting objects and their distribution in space. The consistency of these two curves for classes of long and short events points out that we have no evidence to associate them with different outbursting astronomical objects. We may assume that any difference between emission parameters of bursts between short and long classes of events may be associated rather with different types of outbursts than with distinct populations of emitters.

ACKNOWLEDGEMENTS

The work in USA was supported by NASA grant CRO-96-173. The work in Russia was supported by RFBR grant 96-02-1882.

REFERENCES

1. Fenimore E. E and Bloom J. S, *ApJ* **453**, 25 (1995).
2. Sanin A. B. et al., these proceedings.
3. Litvak, M. L. et al., these proceedings.
4. Mitrofanov, I. G. et al., these proceedings.
5. Meegan C. A. et al., 1997, in "Gamma-Ray Bursts, 4th Huntsville Symposium", eds. Meegan C. A., Preece R. D. & Koshut T. M., (AIP: New York), 428, p. 3 (1997).
6. Pizzichini G., in Proc. of the XXIV ICRC Conf., p. 81 (1995).

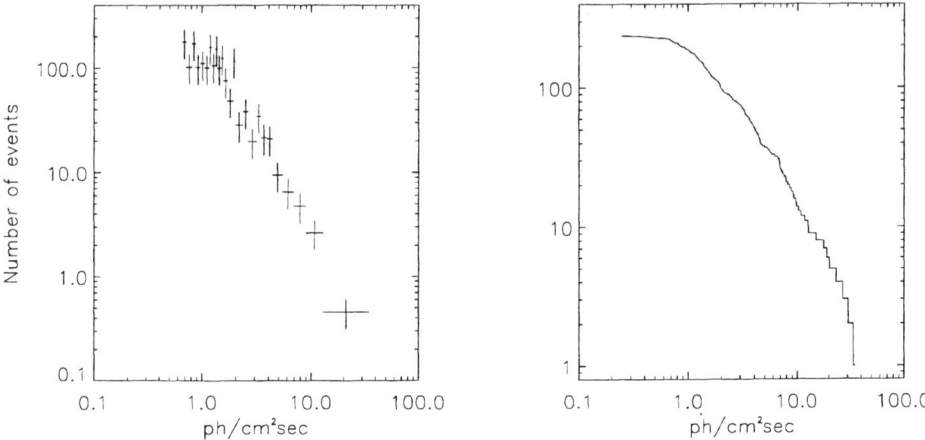

FIGURE 1. *Left panel*: Differential distribution for events with τ_{50} greater than 0.6 s. *Right panel*: Integral distribution for same events.

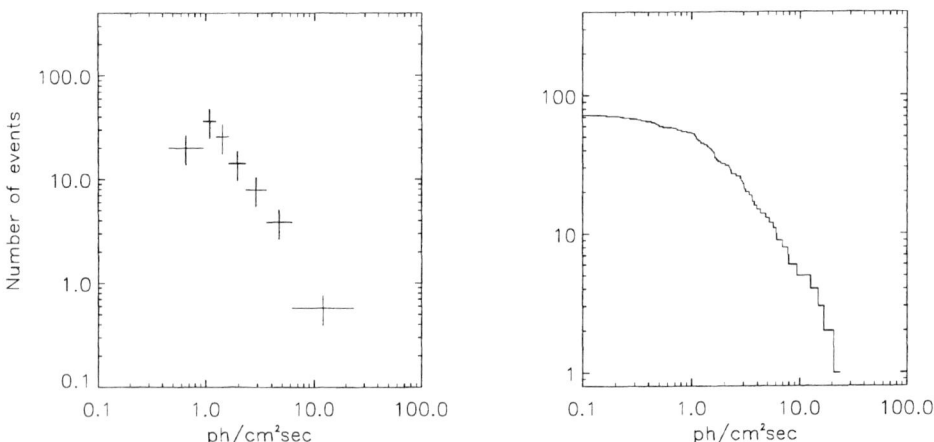

FIGURE 2. *Left panel*: Differential distribution for events with τ_{50} less than 0.6 s. *Right panel*: Integral distribution for same events.

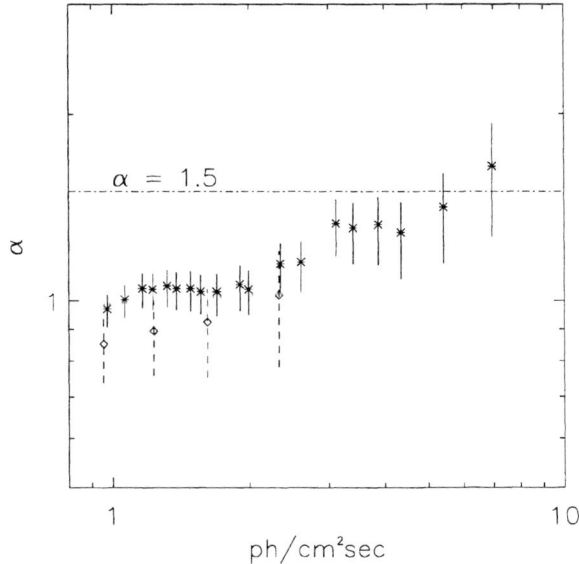

FIGURE 3. *Solid line*: Average power-law index α as function of flux for events greater then 0.6 s. *Dashed line*: the same for events shorter than 0.6 s.

Connection between Spectral Lags and Peak Luminosity in GRBs

J. P. Norris[1], G. F. Marani[1,2], and J. T. Bonnell[1,3]

[1] NASA/Goddard Space Flight Center, Greenbelt, MD 20771
[2] National Research Council
[3] Universities Space Research Association

Abstract. For the set of six gamma-ray bursts with redshifts observed by BeppoSAX and CGRO/BATSE, their isotropic peak luminosities and spectral lags are anti-correlated. In this small set of bursts, the relationship is approximated by a power-law, $L_{53} \approx 1.3 \times (\tau/0.01\ s)^{-1.15}$. While GRB 980425 (if associated with SN 1998bw) would appear to extend this trend qualitatively, it falls below the power-law relationship by a factor of several hundred. The same underlying anti-correlation, but apparently convolved with redshift, appears to be manifest in the 30% brightest BATSE bursts – dimmer bursts tend to have longer spectral lags. These results appear to lend at least empirical meaning to the GRB pulse paradigm: short spectral lag and therefore narrow pulse width are somehow related to high luminosity.

INTRODUCTION

The phenomena of spectral evolution in gamma-ray bursts (GRBs) are two distinct, observed effects: (1) pulse peaks migrate to later times and become wider at lower energies; and (2) burst spectra tend to soften as the event progresses, such that the individual, evolving pulses have impressed upon them an envelope which governs a global spectral decay. This global spectral behavior is arguably related to burst temporal asymmetry on long timescales [1,2]. Progressively longer decays are now detected from hard X-ray energies to wavelengths stretching into the radio bands, and must be manifestations of radiation transfer, and of dissipation of the burst energy preceding and during the afterglow phase. The physical explanation for pulse evolution is probably more closely connected to the primal energy generation, rather than dissipation, mechanism. While understanding pulse spectral behavior is a paramount goal, only empirical relationships have been established so far. From analysis of many individual pulses in bright, long BATSE bursts, Norris and colleagues [3] proposed that a "pulse paradigm" operates in GRBs: among pulses where shape is relatively well determined, the rise-to-decay ratio is unity or less; as this ratio decreases, pulses tend to be wider, the pulse centroid is shifted to later times at lower energies, and pulses tend to be spectrally softer.

Many investigators have observed that pulses within a given burst tend to exhibit a large degree of self-similarity, or little variation in pulse width. In particular, Fenimore, Ramirez-Ruiz, and Wu [4] demonstrated quantitatively via autocorrelation analysis that different intervals within GRB 990123 exhibit comparable pulse widths. Ramirez-Ruiz and Fenimore [5] arrived at the same conclusion for bright bursts in general via two analysis approaches. However, for lack of a Rosetta Stone, the physical import of the pulse paradigm has remained undeciphered. This situation appears changed by the detection of several optical and radio counterparts to GRBs, which have disparate implied isotropic luminosities and total energies. Here we analyze cross correlation lags between energy bands for two important GRB samples: six events with associated redshifts that are detected by BeppoSAX and BATSE; and 372 bright, long BATSE bursts. For the bursts with redshifts, spectral lag is anti-correlated with isotropic peak luminosity, following a power-law relationship. For the large sample, bursts with lower peak flux tend to have longer lags. These results suggest that eventual calibration of GRB luminosities may be possible, allowing GRBs to be used as cosmological tools.

ANALYSIS AND RESULTS

We considered all GRBs (1) detected by BeppoSAX with associated optical or radio afterglows that yielded redshift determinations, and (2) with usable BATSE data through May 1999. This sample includes six bursts: GRBs 970508, 970828, 971214, 980703, 990123, and 990510. The larger sample of 372 bursts is defined by: T_{90} duration > 2 s; BATSE peak flux (50–300 keV) > 2 photons cm^{-2} s^{-1}; and peak count rate (> 25 keV) > 7000 count s^{-1}. The 4-channel time profiles, background fits, and T_{90} duration estimates were constructed following procedures described in Bonnell et al. [6].

For the sample of six bursts with redshifts we computed lags using the cross-correlation function (CCF) between BATSE channels 1 and 3. The analysis was performed for portions of the time profiles with count rates down to 10%, 30%, and 50% of peak intensity (> 25 keV) for the sample with redshift, in order to search for possible dependence of lag on intensity. For the large sample only the portions above 10% of peak intensity were analyzed. The detailed considerations for computation of spectral lag and treatment of exceptions are described at length in a complete analysis [7]. Here, the peak flux threshold has been lowered from 4 to 2 photons cm^{-2} s^{-1}, and bursts retained that have little or no emission above 300 keV, resulting in the addition of ~ 200 bursts. The CCF lag coordinate was determined from the median of 101 realizations, each one obtained by adding Poisson-distributed noise to the original time profiles. The peak in the CCF was determined by over-resolving the time dimension of the model fit by a factor of 32 and fitting the region near the CCF peak with a quadratic or cubic.

Results for the small sample with associated redshifts are illustrated in Figure 1. The dashed line is a power-law fit which yields $L_{53} \approx 1.3 \times (\tau/0.01\ s)^{-1.15}$, where

L_{53} is the luminosity in units of 10^{53} ergs s^{-1} (assuming an $[\Omega_m,\Omega_\Lambda] = [0.3,0.7]$ universe), and τ is the spectral lag. A variation of $\sim$0.10 in the fitted spectral index results from fitting the CCF lags computed for the different signal inclusion levels relative to the peak intensity (in Figure 1: 0.1 – diamonds; 0.3 – triangles; 0.5 – squares). Only the lags for GRB 990123 depend significantly on the portion of the time profile analyzed; including the latter, less intense half of the burst in the CCF results in a longer lag by a factor of $\sim$4 (diamond and triangle symbols). GRB 980425 (not shown in Figure 1, see reference [7]), while exhibiting a qualitatively similar signature – long spectral lag ($\sim$ 4 s), and very low peak luminosity for a GRB ($\sim 10^{47}$ ergs s^{-1}) – falls below the fitted power-law relationship for the six bursts at cosmological distances by a factor of $\sim$ 400–700, depending on the precise power-law index adopted. We conclude that either GRB 980425 is truly a different kind of GRB, "on a different branch of the GRB HR diagram," or else it is not associated with SN 1998bw.

Figure 1 suggests a nascent "luminosity–color" relationship for GRBs, similar

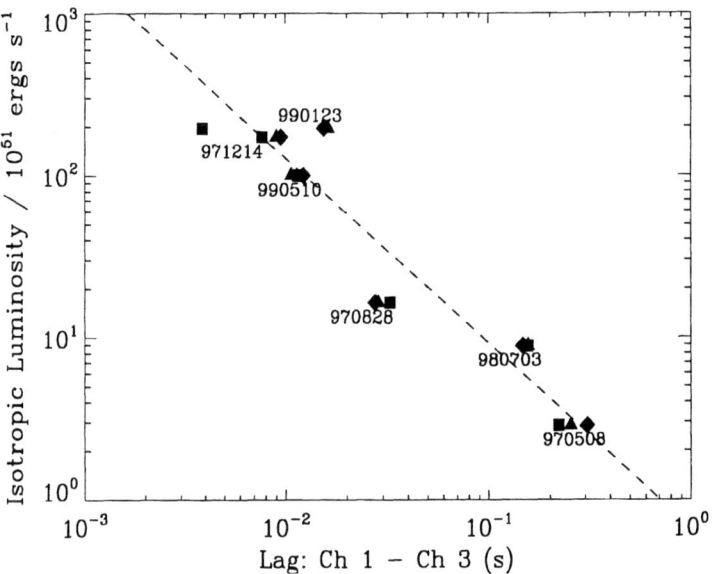

FIGURE 1. CCF lag, corrected for time dilation, versus (would-be) isotropic peak luminosity, for the subset of six bursts with associated redshifts. Symbols represent CCF lags computed for included signal intervals above different thresholds relative to peak intensity: $\times$ 0.1 (diamonds), $\times$ 0.3 (triangles), $\times$ 0.5 (squares). Individual GRBs are labeled in YYMMDD format. Dashed line is a power-law fit to the lags, yielding $L_{53} \approx 1.3 \times (\tau/0.01 \text{ s})^{-1.15}$.

to the HR diagram for main sequence stars, or the period-luminosity relationship for Cepheids. Ramirez-Ruiz and Fenimore [8] obtain a similar relationship between a variability factor and luminosity. If corroborated with subsequent measurements, the lag–luminosity relationship would constitute another rung on the cosmic distance-scale ladder.

Figure 2 illustrates results for the sample of the 30% brightest BATSE bursts. Note the near absence of negative lags for very bright bursts, with peak flux > 10 photons cm^{-2} s^{-1}. A few negative lags begin to appear at lower peak fluxes, but the values are not significantly different from zero or small positive lag, as evidenced by our inspection of bootstrap errors from the 101 realizations per burst. Thus most lags are short, implying that most *detected* GRBs occupy the high end of their luminosity distribution. However, it is therefore likely that a *volume-limited* GRB sample – shorn of a bias for detecting luminous bursts – would manifest a different luminosity distribution, possibly peaked towards lower luminosities [9].

Also, we observed a profound trend while computing the CCF lags: bursts with long (short) lags tend to have one or a few wide (many narrow) pulses. This tendency is more evident for the dimmer bursts, below peak fluxes of ~4 cm^{-2} s^{-1}, where long lags become prevalent. Thus the pulse paradigm previously reported

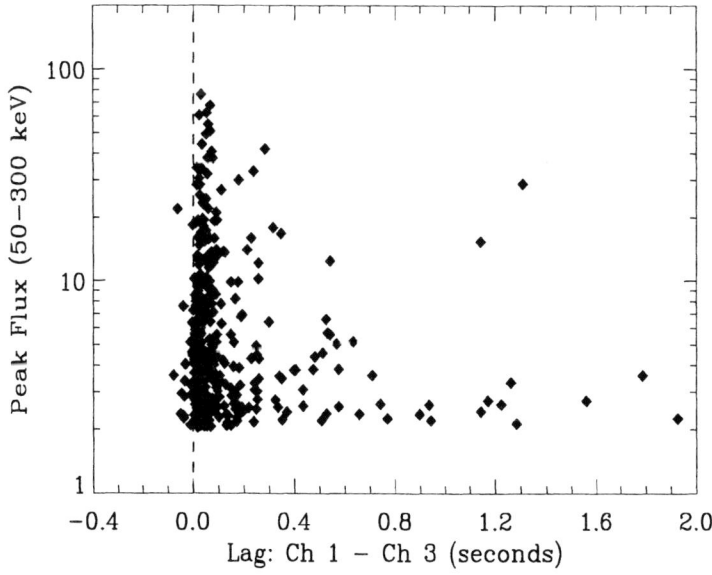

FIGURE 2. CCF lags between BATSE channels 1 and 3 versus peak flux (50–300 keV) for the sample of 372 long ($T_{90} > 2$ s) bright (peak flux > 2 photons cm^{-2} s^{-1}) bursts.

only for relatively bright bursts [3], appears to extend to dimmer bursts – assuming that the integral measure of CCF spectral lag indeed reflects pulse widths. Hence wide pulses appear to be essentially monolithic structures, organized in time and energy, even in dim bursts where sparse counting statistics do not support measurements of the individual pulses. While significant short timescale fluctuations can be superposed on wide pulses, nevertheless the long lags found for wide pulses argue strongly that they are not composites of many narrow, crowded pulses, which would have a resultant short composite lag.

The expectation is that spectral lags for the general population of GRBs would translate into luminosities, once the lags are corrected for the one expected extrinsic effect, redshift. This single parameter enters as four factors, twice on each axis of Figure 2: on the ordinate, once as distance and once due to the finite bandpass in which peak flux is measured; on the abscissa, once as time dilation, and again for the finite bandpasses of the two channels that are lagged. It is conceivable that a straightforward, iterative algorithm could deconvolve this single parameter from the scatter plot of lag versus peak flux, to yield a more "one-to-one" relationship in the large sample, as appears in Figure 1. Almost decidedly, a good physical understanding of the lag-luminosity relationship would still be required to ensure that evolution does not complicate the picture significantly (see Drell, Loredo, and Wasserman [10] for an excellent discussion on possible effects of cosmic evolution in type Ia supernovae). If such a program could be realized for GRBs, they might be considered part of the toolbox for investigating cosmological parameters in the high redshift universe.

REFERENCES

1. Link, B., Epstein, R. I., and Priedhorsky, W. C., *ApJ* **408**, L81 (1993).
2. Nemiroff, R. J., et al., *ApJ* **423**, 432 (1994).
3. Norris, J. P., et al., *ApJ* **459**, 393 (1996).
4. Fenimore, E. E., Ramirez-Ruiz, E., and Wu, B., *ApJ* **518**, L73 (1999).
5. Ramirez-Ruiz, E., and Fenimore, E. E., *A&A Supp.* **138**, 521 (1999).
6. Bonnell, J. T., Norris, J. P., Nemiroff, R. J., and Scargle, J.D., *ApJ* **490**, 79 (1997).
7. Norris, J. P., Marani, G. F., and Bonnell, J. T., *ApJ*, accepted (2000).
8. Ramirez-Ruiz, E., and Fenimore, E. E., these proceedings.
9. Fenimore, E. E., private communication.
10. Drell, P. S., Loredo, T. J., and Wasserman, I., *ApJ*, accepted (1999).

A Simple BATSE Measure of GRB Duty Cycle

Jon Hakkila*, Robert D. Preece[†], and Geoffrey N. Pendleton[†]

*Minnesota State University, Mankato, Minnesota 56001
[†]University of Alabama, Huntsville, Alabama 35899

Abstract. We introduce a definition of gamma-ray burst (GRB) duty cycle that describes the GRB's efficiency as an emitter; it is the GRB's average flux relative to the peak flux. This GRB duty cycle is easily described in terms of measured BATSE parameters; it is essentially fluence divided by the quantity peak flux times duration.

Since fluence and duration are two of the three defining characteristics of the GRB classes identified by statistical clustering techniques (the other is spectral hardness), duty cycle is a potentially valuable probe for studying properties of these classes.

INTRODUCTION

The term *duty cycle* in astrophysics is defined as "the fraction of time a pulsed beam is on" [3]. This term is more appropriate in describing periodic emitters such as pulsars than it is for non-periodic, one-time emitters such as gamma-ray bursts (GRBs). GRB emission consists of pulses varying in intensity over a burst's duration, and is thus more conducive to a definition recognizing the continuous nature of GRB emission than to one limited as either "on" or "off." A more appropriate definition of GRB duty cycle should describe the effectiveness of a GRB as an emitter during the time that it emits. We therefore define the *GRB duty cycle* as the average flux relative to its peak flux. This duty cycle definition can be described in terms of measured BATSE parameters; it is essentially fluence divided by the quantity peak flux times duration.

Fluence and duration are two of the three defining characteristics of the three GRB classes identified by statistical clustering techniques [5]. Spectral hardness is the third. Fluence (time-integrated flux) incorporates information about duration in its definition. Overlapping information appears to be contained in fluence and duration [1]. For this reason, duty cycle (as defined here) is a potentially valuable probe for studying properties of the three GRB classes.

Properties of the three GRB classes as determined from statistical clustering techniques [5] are demonstrated in Table 1.

TABLE 1. Statistical clustering classes, from 3B GRBs.

Attributes	Class 1 (Long)	Class 2 (Short)	Class 3 (Intermediate)
T90	long	short	intermediate
Fluence	large	small	intermediate
Hardness	intermediate	hard	soft

DUTY CYCLE DEFINITION

We define the duty cycle (DC) in terms of BATSE parameters as

$$DC = \frac{S_{23}}{A \cdot P_{64} \cdot T_{90}}. \quad (1)$$

Here, S_{23} is the channel 2+3 fluence (time-integrated flux between 50 and 300 keV), T_{90} is the duration spanning 90% of the GRB emission, and P_{64} is the 64 ms peak flux. A is a constant for converting photon counts to energy (assuming a diagonal detector response matrix as a first-order approximation).

The peak flux used in this calculation must be measured on the shortest available timescale (64 ms) in order to avoid arbitrarily smoothing out the maximum value of the peak flux. A peak flux underestimate produces a corresponding duty cycle overestimate.

GRBs with T_{90} durations less than 64 ms have had their T_{90} values set to 64 ms, so that their durations are not given a different temporal resolution than that of their peak flux measure.

ANALYSIS OF GRB CLASSES

Our database consists of non-overwriting GRBs in the BATSE 4Br Catalog [6] triggering on 1024 ms peak flux in the 50 to 300 keV range with the trigger threshold set 5.5σ above background. These criteria prevent trigger biases [4] from influencing our conclusions. We have also removed GRBs with large relative measurement errors in each of the four-channel fluences, T_{90}, and P_{64} so that measurement error does not bias our conclusions.

The GRBs have been assigned to a class using the supervised decision tree classifier C4.5 [7]. The technique is described in more detail elsewhere [2].

We obtain an average value of $A \approx 2.24 \times 10^{-7}$ ergs photon^{-1} by integrating typical bright GRB spectra (soft power law index $\alpha \approx -1$) over the 50 to 300 keV trigger energy range.

Many Class 2 (Short) GRBs have duty cycles $DC > 1.0$, as their harder spectra lead to an underestimate of A and to a corresponding overestimate of DC. We obtain a separate value of $A \approx 2.80 \times 10^{-7}$ ergs photon^{-1} for these GRBs (using $\alpha \approx 0$), and recalculate their duty cycles. No attempt is made to account for Class

·3 (Intermediate) spectra, which have similar spectral components to those of faint Class 1 (Long) bursts [2].

Figure 1 is a plot of DC vs. hardness ratio $HR321$ (100 to 300 keV fluence divided by 25 to 100 keV fluence). This hardness ratio represents the third delineating attribute of the three GRB classes.

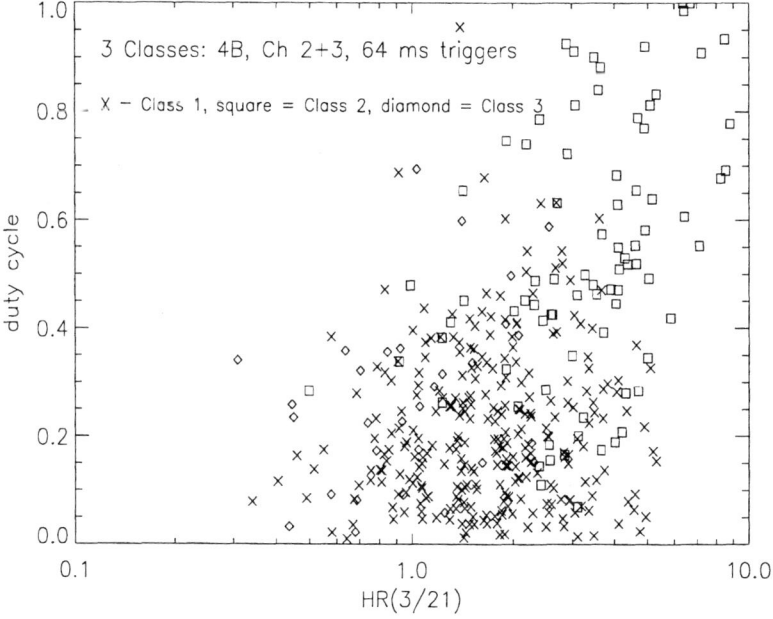

FIGURE 1. GRB Duty Cycle vs. hardness ratio HR321.

We note the following characteristics of GRBs, and of the three GRB classes (based upon Figure 1), as they pertain to the duty cycle:

- There are no efficient, soft GRBs.
- There are no inefficient, hard GRBs.
- Class 2 (Short) GRBs are generally efficient, with duty cycles of $DC \geq 0.1$.
- Class 1 (Long) GRBs are rarely efficient, with duty cycles of $DC \leq 0.7$.
- Class 3 (Intermediate) GRBs blend into the Class 1 (Long) GRBs in this plot.

CONCLUSIONS

The duty cycle measure as defined here is fairly effective and easy to calculate. The simplifying assumption of a diagonalized detector response matrix is not completely valid. A small correction factor is needed to account for excess high-energy photons from hard GRBs (primarily those belonging to Class 2). Nonetheless, this approach allows preprocessed BATSE attributes to be incorporated directly into the duty cycle calculation, without requiring the use of data in a less processed form.

Despite the aforementioned problem, Class 2 (Short) is well delineated from Class 1 (Long) in a plot of duty cycle vs. HR321. Class 2 GRBs are harder, more efficient emitters than Class 1 GRBs.

Class 3 (Intermediate) does not appear to be distinct from Class 1 (Long) on the basis of the duty cycle attribute. This result is in agreement with the findings of our artificial intelligence study [2].

REFERENCES

1. Bagoly, Z., et al., *ApJ* **498**, 342 (1998).
2. Hakkila, J., et al., these proceedings.
3. Hopkins, J. *Glossary of Astronomy and Astrophysics*, Chicago: University of Chicago Press, p. 54 (1980).
4. Meegan, C. A., et al., these proceedings.
5. Mukherjee, S., et al., *ApJ* **508**, 314 (1998).
6. Paciesas, W. S., et al., *ApJS* **122**, 465 (1999).
7. Quinlan, J. R., *Machine Learning* **1**, 81 (1986).

The Energy Distribution of GRBs

Tsvi Piran*, Raul Jimenez[†], and David Band[‡]

*Racah Institute for Physics, The Hebrew University, Jerusalem, Israel
[†] Institute for Astronomy, University of Edinburgh, Blackford Hill, Edinburgh EH9–3HJ, UK
[‡] X-2, Los Alamos National Laboratory, Los Alamos, NM, USA 87545

Abstract. We analyze the distribution of total energy of bursts with optical afterglows. Our sample contains eleven bursts for which there are BATSE data allowing detailed fits to the spectra. Six of these bursts have measured redshifts while five have host galaxies whose redshifts are unknown. Using a new technique based on the distribution of magnitudes of observed high redshift galaxies, we bracket the expected redshifts of these host galaxies. We perform a maximum likelihood fit for the energy distribution of these GRBs. Assuming that the total energy emitted has a log-normal distribution, we find that the average isotropic energy emitted is $E_\gamma = 1.5^{+1.5}_{-0.6} \times 10^{53}$ergs (for $H_0 = 65$ km s^{-1} Mpc^{-1}, $\Omega_m = 0.3$ and $\Omega_\Lambda = 0.7$) with a standard deviation $\sigma_\gamma = 1.2^{+0.6}_{-0.4}$. The corresponding distribution of X-ray afterglow energy (for seven bursts) is significantly narrower with $\sigma_x = 0.5^{+0.4}_{-0.25}$ and $E_x = 3.8^{+2}_{-1.5} \times 10^{51}$ergs, in agreement with the prediction of the patchy shell model of Kumar and Piran [1]. We also give a table with the detailed fits to the spectra of the 11 GRBs with optical afterglow with BATSE data.

The energy and luminosity functions of GRBs are among the most interesting still unknown parameters of these objects. Clearly a good estimate of the total energy emitted in these bursts could shed a lot of light on possible sources. Until recently the absence of a measured burst distance hindered the calculation of the total energy radiated by a burst. However we have now determined the redshifts of some of the host galaxies associated with a number of bursts. We have expanded this sample by estimating the redshifts of host galaxies without measured redshifts based on their apparent magnitudes. With this sample of bursts with estimated energy emission we can calculate the distribution of energy emitted. One has to remember though that this estimate is valid only for the subclass of GRBs with observed optical afterglows.

GRB DATA

We consider the sample of GRBs with an observed optical afterglow and with a detected host galaxy with or without a measured redshift. To obtain a uniform

TABLE 1. Main Properties of the Sample of Gamma Ray Bursts.

Name	N_0 10^{-2}	α	β	E_0 (keV)	F_γ a,b	slope$_x$ b	F_x	t (s)	R_{gal}	z_{obs}	z_{est}
970228	...	...	...	...	...	...	...	...	24.6	0.695	0.8
970508	0.17	−1.19	−1.83	480.84	3.17	−1.2	0.39	23.44	25.8	0.835	1.2
970828	1.15	−0.70	−2.07	229.74	96	−1.0	1.82	146.59	24.5	0.958	0.8
971214	0.72	−0.78	−2.57	155.96	9.44	−1.6	1.31	45.45	25.5	3.412	1.2
971227	1.06	−1.44	−4.20	112.03	1.21	...	...	6.94	25.0	...	1.2
980326	2.11	−1.33	−4.34	77.19	0.92	...	...	4.01	25.3	...	1.2
980329	2.58	−0.96	−2.43	235.65	55.1	−1.34	0.56	50.15	26.3	...	1.2
980425	0.47	−1.27	...	161.20	4.96	...	...	37.41	14.3	0.0085	0.01
980519	0.49	−1.35	...	315.94	10.3	...	...	56.35	24.7	...	0.8
980613	...	...	...	...	...	−1.6	0.55	...	24.4	1.096	0.8
980703	0.44	−1.31	−2.40	370.26	22.6	−1.67	1.27	102.37	23.0	0.967	0.6
990123	2.62	−0.90	−2.48	549.51	268	−1.6	8.98	104.61	24.3	1.600	0.8
990506	1.51	−1.37	−2.15	449.78	194	...	...	220.38	...	1.2	...
950510	0.80	−1.28	−2.67	174.24	22.6	...	...	103.84	...	1.619	...
990712	...	...	...	...	...	...	...	...	22.0	0.43	0.40

a 20–2000 keV
b 10^{-6} erg · cm^{-2}

estimate of the energy fluence we consider only those bursts detected by BATSE. For these bursts we obtain a best fit to the spectrum of the entire burst, which is described by the Band function [2]. The parameters of the bursts, including the parameters of the Band function are given in Table 1.

Only six out of the eleven BATSE bursts with an optical afterglow and a host galaxy have a measured redshift (mostly measured for the associated host). In order to increase the number of relevant bursts and to improve the statistics, we implement here a method to determine the redshift of those host galaxies for which there are currently no spectroscopic redshifts.

In addition to the energy emitted by the burst (mostly in γ-rays), significant emission occurs during the first few hours of the afterglow (mostly in X-rays). We estimate this energy (excluding the prompt X-ray emission) using the observed late (few hours) X-ray flux and assuming this emission decays as a power law with the index "slope$_x$" shown in the above table over the period 100 to 10^5 seconds after the burst.

REDSHIFT DETERMINATION

The use of photometry to derive galaxy redshifts has flourished during the past 5 years (e.g., [3–6]). It is now common to determine redshifts to an accuracy of 10% (e.g., [7]) by using multi–band photometry (usually 4 to 6 bands). It is therefore fair to say that multi-band photometry is an economical way of determining redshifts. Unfortunately, in the case of GRB hosts one is limited, in most cases, to observations in one band (typically R).

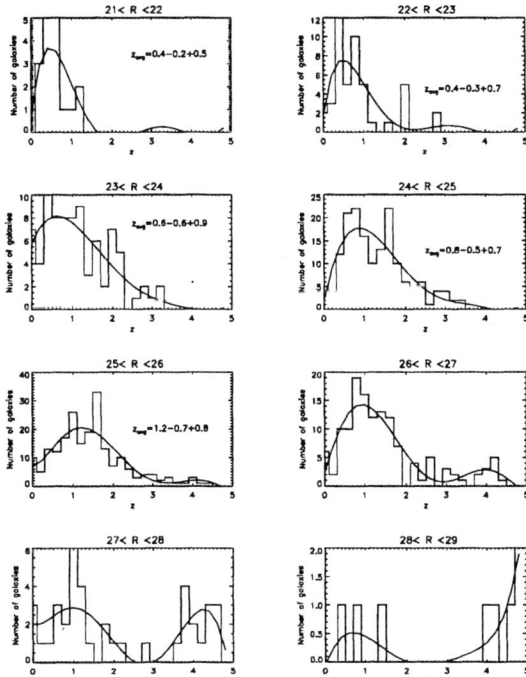

FIGURE 1. The redshift distribution of galaxies in the R band from the HDF (see text). The average redshift (z_{avg}) has been obtained from the maximum of the polynomial fit (solid curves) to the observed data and the errors reflect the redshift for which the solid curve encloses 68% of the area.

In principle it may seem futile to try to determine the redshift of a galaxy from a single band. Galaxies span a large range in stellar masses (i.e., luminosities) and therefore it is possible to find galaxies of a given magnitude at virtually any redshift (c.f., [8]). On the other hand, since the volume element dV/dz peaks at around $z = 1 - 2$ (for reasonable cosmologies) and galaxies formed stars at a higher rate at $z > 1$ (and therefore were brighter), one expects galaxies are more likely to inhabit the $1 < z < 2$ region for not too faint (or too bright) magnitudes. In this paper we use the Hubble Deep Field (HDF) observations to calibrate an approximate redshift-magnitude relationship. In doing so we assume that GRB hosts belong to the general galaxy population.

In Figure 1 we show the redshift distribution of the HDF [8] galaxies as a function of apparent magnitude R. We have transformed the HST F606W magnitudes into cousins R magnitudes. First inspection shows that for $R < 26$ galaxies are systematically more common at $z < 2$ than at higher redshifts, as expected. Note that the average redshift increases with galaxy magnitude.

A concern is that the HDF is *not* a volume limited sample, as illustrated by

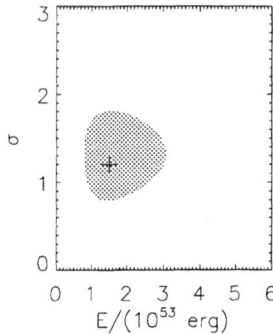
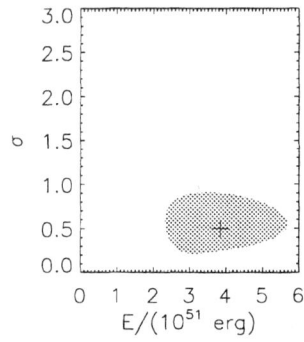

FIGURE 2. The 1σ confidence region in the 20–2000 keV band (left panel). Right panel: same as left panel but for the X-ray energy. In this case we have integrated the X-ray light curve between 100 and 10^5 seconds.

Figure 1 in [8]. Clearly the HDF is missing only very bright galaxies at $z < 0.5$, which will not affect our redshift estimates as GRB hosts are much fainter in general (see Table 1). Note that while Figure 1 in [8] might lead you to suspect that a large number of $z > 1$ galaxies may be missing, luminosity evolution shows this is not the case. At $z > 1$ galaxies were forming many more stars than at $z < 1$ (e.g., [9,10]), and thus they were more luminous (by 2-3 magnitudes [10]) than nearby galaxies. Thus we conclude that for $R < 26 - 27$ the HDF represents a fair sample of the Universe. A statistical redshift-magnitude relationship is justified by the relative paucity of $z < 0.5$ galaxies in the redshift distributions for faint magnitudes $24 < R < 27$.

To evaluate the error of our redshift estimate, we compare the estimated redshifts z_{est} with the spectroscopic redshifts z_{obs} for those bursts where both are available (see Table 1). The differences between these two redshifts are comparable to the widths of the redshift distributions in Figure 1. The major exception is GRB971214, where the estimated redshift is much lower than the spectroscopic one. Therefore we conclude that we can estimate redshifts (*statistically*) from one band within a 40% error.

RESULTS AND CONCLUSIONS

We assume that the probability for a GRB to have an energy E_i in the interval E_i to $E_i + dE_i$ is given by a log–normal distribution with two parameters E_{av} and $\sigma(\log E)$. We use a maximum likelihood method to determine these two parameters. For half of our samples the redshift is only known within 40%. To reflect this uncertainty we modify the log–normal probability distribution to allow for errors in the redshift (energy) determination by simply assuming that these errors are normally distributed. Then the likelihood function that we maximize is

$$ln\mathcal{E}(E_{av},\sigma) = \sum_i ln\left[\frac{1}{2\pi\sigma_i\sigma}\int \frac{\exp\left[-\frac{1}{2}\left[\frac{(ln(E)-ln(E_{av}))^2}{\sigma^2} + \frac{(E_i-E)^2}{\sigma_i^2}\right]\right]}{E}dE\right], \quad (1)$$

where σ_i is the error due to the redshift uncertainty of the burst i; for the bursts with spectroscopic redshifts $\sigma_i=0$ and the integral over E becomes the evaluation of a δ function.

For the gamma-ray band we estimate the characteristic isotropic energy, E_γ, and its spread, σ_γ, in the 20–2000 keV band (Figure 2, left panel). The preferred value for E_γ is 1.5×10^{53} erg with $\sigma=1.2$. Recall that this is the standard deviation of a log-normal distribution, thus this value corresponds to a spread of about two and a half decades in energy. Note that there is no need to K–correct the data in this case because the 20–2000 keV band already contains most of the energy of the GRB.

We also perform a similar analysis for the X–ray energy between 100 and 10^5 seconds after the burst. The right panel of Figure 2 shows the 1σ confidence contour. Note that the *variance ($\sigma \sim 0.5$) is significantly smaller than that for the gamma ray band*. This is in agreement with the theoretical predictions of the patchy shell model [1].

We have determined the luminosity function of GRBs with observed optical afterglow. In agreement with the predictions of the patchy shell model [1] the γ-ray energy emitted during the GRB has a significantly wider distribution than the X-ray energy emitted during the afterglow. It is interesting to compare our results with the luminosity function determined for the whole GRB sample. Schmidt [11] finds for a broken power law distribution that the characteristic energy (the position of the break) is 1.2×10^{53} ergs. This is very close to our average energy. The average energy in Schmidt's distribution is about factor of five smaller than its characteristic energy. These results are compatible since the subsample of bursts with observed afterglow, which we examine, is much brighter than the whole sample.

REFERENCES

1. P. Kumar and T. Piran, *astro-ph* 9909014 (1999), and these proceedings.
2. D. Band et al., *ApJ* **413**, 281–292 (1993).
3. A. J. Connolly et al., *AJ* **110**, 2655 (1995).
4. M. J. Sawicki, H. Lin, and H. K. C. Yee, *AJ* **113**, 1 (1997).
5. S. P. Driver et al., *ApJL* **496**, L93 (1998).
6. Arnouts et al., *astro-ph* 9902290 (1999).
7. I. Csabai, A. Connolly, A.S. Szalay, and T. Budavari, *astro-ph* 9910389 (1999).
8. S. Driver, *astro-ph*, 9909469 (1999).
9. C. C. Steidel et al., *ApJ* **519**, 1 (1999).
10. A. F. Heavens and R. Jimenez, *MNRAS* **305**, 770 (1999).
11. M. Schmidt, *astro-ph* 9908206 (1999).

The Pure Physical Parameters of BATSE Gamma Ray Bursts

Anton B. Sanin[1], Igor G. Mitrofanov[1], Dmitrij S. Anfimov[1], Maxim L. Litvak[1], Michael S. Briggs[2], William S. Paciesas[2], Geoffrey N. Pendleton[2], Robert D. Preece[2], and Charles A. Meegan[3]

[1] *Space Research Institute, Profsojuznaya str. 84/32, 117810 Moscow, Russia*
[2] *Department of Physics, University of Alabama in Huntsville, Huntsville, AL 35899*
[3] *NASA/Marshall Space Flight Center, Huntsville, AL 35812*

Abstract. The concept of Pure Physical Parameters (PPP, or P^3) is presented, as the instrument-independent parameterization for the statistical search for basic physical signatures of gamma-ray bursts. The emission time of photon light curves and the peak of photon spectra of high power emission are proposed for each burst characterization, as P^3 parameters for time variability and for spectral energy of gamma-rays. The P^3 parameters are presented for 6 gamma-ray bursts in their rest frames with known red-shifts of optical afterglows.

INTRODUCTION

The measurements of gamma-rays are known to be affected by the detector properties. The measured energies of counts do not equal to energies of detected photons. Also, a detector has different sensitivity for gamma-rays of different energies. The counts-based time history of a GRB at some energy range is distinguishably different from the actual light curve of incoming photons. To study the physics of gamma-ray bursts, we need to describe the variable flux of GRBs by instrument-free pure physical parameters. We have to deconvolve the measured time history of counts into the original light curves of photons by robust and non-biased procedures.

After the discovery of the red-shift of optical afterglows for several GRBs, we know that bursts have very broad spread of intrinsic luminosity. One can not assume any longer that bright and dim bursts correspond to closer and to more distant sources, respectively. It is quite probable that bright and dim bursts are produced by intrinsically strong and intrinsically weak emitters at large red-shifts. To build a physical model, we need to know the parameters of emission in the rest frames of outbursting sources. However, for de-red-shiftization of gamma-ray bursts with known Z we need a set of pure physical parameters, which could be transformed from the observer frame into the rest frame in accordance with the model of the expanding Universe.

DEFINITION OF PURE PHYSICAL PARAMETERS OF GRBS

We suggest to implement the following P^3 parameters of GRBs:
(1) the emission time $\tau(E_p)$ (see [1–3]) at the spectral range around a peak energy E_p,
(2) the number of separate pulses at time history N (see [4,5]),
(3) the peak energy E_p of νF_ν spectrum of photons measured at moments of emission time $\tau(E_p)$ (see [6]),
(4) the spectral range δE_p around a peak energy E_p,
(5) the flux, as a number of photons at the spectral range δE_p at moments of emission time $\tau(E_p)$ (see [7]),
(6) the cosmological invariant parameter $CIP = E_p \cdot E_p(\tau(E_p))$ (see [8]).

The first parameter (1) of emission time represents the total integrity of emission moments, when a pre-defined fraction of the total fluence of counts has been produced with the highest power [1–3]. Thus, to provide the P^3 measure of emission variability one has to apply the concept of emission time onto deconvolved photons-based time profiles. The group of connected time intervals of high power emission could be attributed as the signature of a separate pulse. The number N of these groups within emission time intervals could be related to the number of separate pulses (see [4-5]). Parameter (2) N is P^3 measure of a burst variability. The ratio $\delta \tau = \tau/N$ is the average duration of a separate peak of emission.

The energy spectra are known to be highly variable, and one has to define time intervals for spectrum integration, which would be determined by the physical but not by the instrumental criteria. To evaluate the P^3 spectral measure of bursts (3), we have to sum up the data for all time intervals contributing into the emission time.

Parameters (1) and (3) provide a self-consistent measure of the time-and-spectral signature of gamma-rays bursts. It has been shown that the couple of these parameters may be calculated by an iterative procedure, which starts from selection of emission time intervals in counts, then building the spectrum of photons for these intervals, then determining the peak energy E_p (3) and the energy range δE_p (4) for this spectrum, then re-selecting time bins for time history of photons in the range δE_p, and then determining the spectrum of photons for new bins, etc. This procedure converges well (Scheme 1).

Parameters of flux (5) for a burst detected by different instruments would corresponded to the same value provided these instruments are well-calibrated. The difference in their sensitivity would only result in different errors.

Finally, the product (6) of time-dimensional parameter (1) and energy-dimensional parameter (2) is not changed by cosmological transformation. The average cosmological invariant parameter has already been implemented as the average signature of intensity groups of bursts (see [6]). The statistics of P^3 invariant parameters for individual bursts allows us to compare them in their rest frames.

We believe that statistics of parameters (1) – (6) will allow us to resolve the physical difference between different samples of bursts selected by different criteria. This comparison could be performed for several particular cases. This analysis is now in progress (see [2,3,6,7,10]).

PURE PHYSICAL PARAMETERS FOR GBRS WITH MEASURED RED-SHIFTS OF OPTICAL AFTERGLOW

There are 6 BATSE gamma-ray bursts with measured red-shifts of optical afterglow [9], which we may use to build the full set of P^3 parameters. For these bursts the P^3 parameters could be transformed from the observer frame into rest frames of emitters (see also [10]).

We do not see any clear correspondence between smaller/larger values of the total luminosity and smaller/larger values of re-redshifted P^3 parameters of total emission time $\tau(E_p)$ (Figure 1, left), of peak-related emission time $\delta\tau = \tau/N$ (Figure 1, right)) and of spectral peak energy E_p (Figure 2, left). Also, we do not see any relation between the total luminosity of these bursts with their cosmological invariant parameters (Figure 2, right).

These results could be compared with the similar comparison of bursts with known red-shift based on so-called "filling factors" of their time history in counts [11]. The strong correlation has been found between the "filling factors" and intrinsic luminosities. On the contrary, we have no evidence that intrinsic properties of GRBs presented by P^3 parameters (1) – (6) manifest a correlation with the intrinsic luminosities of gamma-ray bursts.

CONCLUSIONS

For a spread of intrinsic luminosity about 3 orders of magnitude for 6 bursts with known red-shifts, de-redshifted values of emission times $\tau(E_p)$ changes in about 20 times (Figure 1, left), and, even more, de-redshifted values of spectral peaks E_p changes in about 2 times only (Figure 2, left). We do not know other examples of astronomical transients which vary the total energy at so broad a dynamical range with so little changes of the time of emission and the spectral energy of radiated photons.

We believe that more Z-anchored bursts will be known soon from HETE-2 and existing instruments. We hope that a much larger number of them will produce better knowledge of intrinsic properties of emission, and we suggest that P^3 parameters of bursts will be used in these studies.

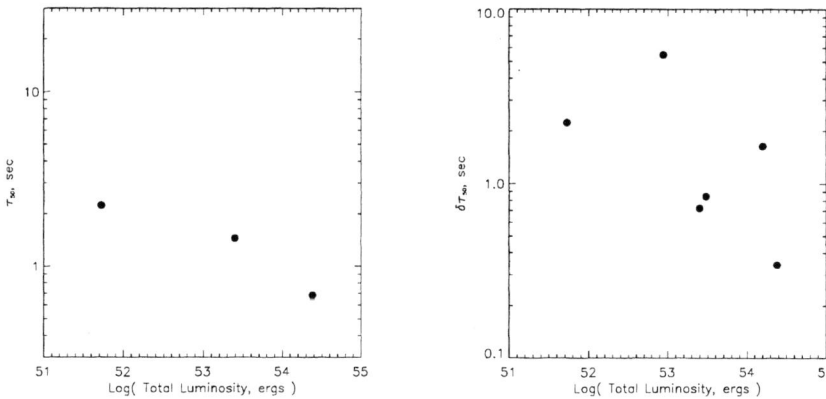

FIGURE 1. Left: The values of de-redshifted emission time versus the total luminosity estimations for 6 BATSE gamma-ray bursts with known red-shifts of their optical afterglows. **Right**: The values of de-redshifted average emission time of peaks versus the total luminosity estimations for 6 BATSE gamma-ray bursts with known red-shifts of their optical afterglows.

τ_{50}^γ, sec   15.616   ↘   12.800   ↘   12.800
                                  ↗   ↗   ↓
E_p, keV   766.2   814.2   814.2

Scheme 1. The results of a converging numerical procedure to calculate emission time τ_{50}^γ for photons-based time profiles at the spectral range $[0.1 \cdot E_p, E_p]$, and the spectral peak energy E_p for the spectrum νF_ν accumulated along τ_{50}^γ.

ACKNOWLEDGMENTS

The work in the USA was supported by NASA project CGRO-98-120 of the Compton Observatory Guest Observations Program. The work in Russia was supported by the RFBR grant 98-02-17380 and INTAS grant 96-0315.

REFERENCES

1. Mitrofanov, I.G. et al., *ApJ* **522**, 1069 (1999).
2. Litvak, M.L. et al., these proceedings.
3. Mitrofanov, I.G. et al., these proceedings.
4. Mitrofanov, I.G. et al., *ApJ* **504** (1998).
5. Pozanenko, A.S. et al., in *Gamma-Ray Bursts: Fourth Huntsville Gamma-Ray Burst*

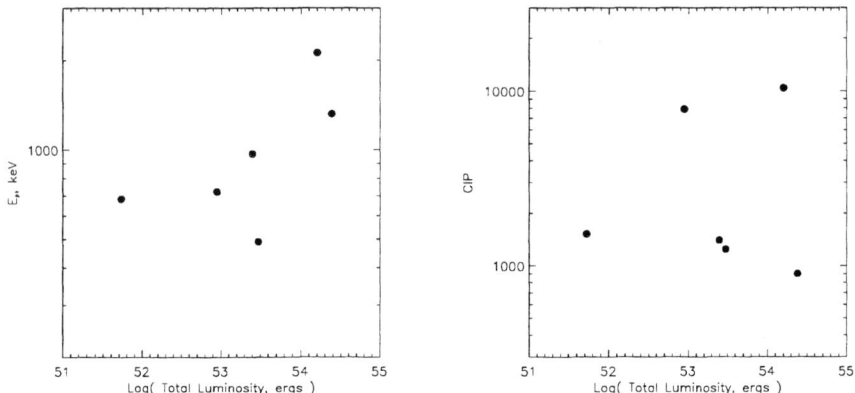

FIGURE 2. **Left**: The values of de-redshifted spectral peak energy versus the total luminosity estimations for 6 BATSE gamma-ray bursts with known red-shifts of their optical afterglows. **Right**: The values of cosmological invariant parameters versus the total luminosity estimations for 6 BATSE gamma-ray bursts with known red-shifts of their optical afterglows.

Symposium eds. Ch. A. Meegan, R. D. Preece & T. M. Koshut, New York: AIP, pp. 59-62 (1997).
6. Anfimov, D.S. et al., these proceedings (1).
7. Anfimov, D.S. et al., these proceedings (2).
8. Mitrofanov, I.G. et al., *ApJ* **523**, 192 (1999).
9. Briggs, M.S. et al., *ApJ* **524**, 82 (1999).
10. Mitrofanov, I.G. et al., these proceedings.
11. Ramirez-Ruiz, E. et al., these proceedings.

Non-Isotropic Angular Distribution for Very Short-Time Gamma-Ray Bursts?

David B. Cline, Christina Matthey, and Stanislaw Otwinowski

University of California, Los Angeles
Department of Physics and Astronomy, Box 951457
Los Angeles, CA 90095-1547 USA

Abstract. While the bulk of the gamma-ray bursts (GRBs) are now believed to be from cosmological distances, the origin of very short-time GRBs is still not known. In the past, we have shown that GRBs with time duration less than 100 ms may form a separate class of GRBs based on the hardness and time distribution of these events. We also showed that the ln N - ln S distribution is consistent with the expectation of a quasi-Euclidean distribution of sources. In this letter, we report the study of the angular location of these GRBs that show a strong deviation from isotropy within the Galactic coordinates of $+180^0 <$ longitude $< +90^0$ and $-30^0 <$ latitude $< 30^0$. We have studied the rest of the GRBs and do not find a similar deviation. This further indicates that the very-short GRBs likely form a separate class of GRBs, most likely from sources of Galactic or near solar origin.

INTRODUCTION

The bulk of the GRBs is now considered to come from cosmological distances; of course, this refers mainly to long-duration bursts. It is well known that there are at least two distinct classes of GRBs: one with time duration in the 10s of seconds and one with time duration of about one second [5]. Both of these classes display a completely isotropic angular distribution and a $\langle V/V_{max} \rangle$ much less than 0.5, which is in concert with cosmological sources [6]. Much less is known about GRBs with a much shorter time duration ($\sim$50 msec on average). We have carried out an extensive study of these GRBs and have concluded that [4,2]:

(A) The hardness distribution is very different from the long-duration GRBs,

(B) The $\langle V/V_{max} \rangle$ is consistent with 0.5 or a quasi-Euclidean distribution of sources.

These results are in press for publication [4]. In this letter, we present the results of our study of the angular distribution of the very short GRBs and find that this distribution deviates sharply from an isotropic distribution. We compare this with that of the short GRBs with a time duration of about 1 second; we then speculate on the possible cause of the asymmetry for the very-short GRBs.

DIFFERENT CLASSES OF GRBS

Figure 1 shows the time distribution T90 for all GRBs from the BATSE detector up to Nov. 1998. Hereafter we will use this data. We divide the GRBs into three classes in time duration: long, L ($\tau >1$ s); short, M (1 s $> \tau > 0.1$ s); and very short, S ($\tau < 100$ ms). We use the duration time of T90 for all of this analysis. Henceforth in this letter, we confine the discussion to the M and S classes of GRBs. Since these events are adjacent in time, it is important to contrast the behavior.

We note that the short bursts are strongly consistent with a $C_p^{-3/2}$ spectrum, indicating a Euclidean source distribution, as was shown previously [2]. In the medium (from 100 ms to 1 s) time duration, the ln N - ln S distribution seems to be non-Euclidean; in the long duration ($\tau > 1$ s) bursts, the situation is more complicated as we have shown recently [3]. The $\langle V/V_{max} \rangle$ for the S, M, and L class of events is, respectively, 0.52 ± 0.06, 0.36 ± 0.02, 0.31 ± 0.01.

One possibility for explaining these effects is that a big part of the short bursts may come from a local Galactic source. This explanation is likely not viable for the medium time bursts, since $\langle V/V_{max} \rangle$ is 0.36 ± 0.02, indicating a likely cosmological source. The longer bursts are clearly mainly from cosmological distance.

We assume that the S GRBs constitute a separate class of GRBs and fit the time distribution in Fig. 1 with a three-population model. The fit is excellent but does not in itself give significant evidence for a three-population model.

We now turn to the angular distributions of the S and M GRBs. In Fig. 2A, we show this distribution for the very short bursts. We can see directly that this is not an isotropic distribution. To ascertain the significance of the anisotropy, we break up the Galactic map into eight equal probability regions. In Fig. 2B, we show the distribution of events in the eight bins; clearly one bin has a large excess. To determine the statistical probability for such a deviation, we calculate the Poisson probability distribution for the eight bins with a total of 42 events. This distribution is also shown in Fig. 2B. The probability of observing 19 events in a single bin is 1.6×10^{-5}; we consider this a very significant deviation from an isotropic distribution. We neglect non-uniformity in sky exposure.

To contrast the distribution of the S GRBs and to test for possible errors in the analysis, we plot the same distributions for the M sample in Figs. 3A and 3B. As can be seen from Fig. 3A, this distribution is consistent with isotropy. Fig. 3B shows the same analysis as Fig. 2B, indicating that there are no bins with a statistically significant deviation from the hypothesis of an isotropic distribution. Of the 42 very short S GRBs, there are 19 in the excess region. We have studied events from S and M samples and can find no real differences between the properties of the excess events and those outside of the excess region.

To further contrast the GRBs in the S and M regions, we have calculated V/V_{max}

for each event using the C_p values from the BATSE data. As we have previously noted [4], this distribution, for the S events is totally consistent with $\langle V/V_{max}\rangle = 1/2$ for a local distribution. In contrast, the same distribution for the M events indicate a $\langle V/V_{max}\rangle$ much less than 1/2 consistent with the same mean values for the L (long) events, which is now widely interpreted as being due to the cosmological sources for those GRBs. It is probable that the M events are also from cosmological sources; however the S events appear to come from local sources.

We can make a preliminary conclusion based on Figs. 2A, 2B, and $\langle V/V_{max}\rangle = 0.52 \pm 0.06$ that the S events are likely from Galactic, or possibly more local (solar neighborhood), sources. This is the first convincing evidence of GRBs that are probably at non-cosmological distances.

SOURCES OF SHORT BURSTS

Independent of a direct association of the GRB events with specific Galactic sources, it would be even harder to explain the distributions in Fig. 2A as being due to extra-galactic or even cosmological sources. The value of the $\langle V/V_{max}\rangle$ and the location asymmetry would seem to strongly support a Galactic origin of the sources for the S GRBs. It is also clear that this distinction would equally argue for a separate class of S GRBs from the M or L classes shown in Fig. 1.

We believe a future study of the S GRB population so a possible shorter time distribution will be fruitful. First there is evidence that we may only have detected a small fraction of the short bursts [8].

The bulk of the very short bursts identified here all have time duration at or below the BATSE 64-ms integration time. We therefore believe that the BATSE trigger is likely an inefficient method of identifying such events; we also believe that many weak bursts may have been missed. Recently the issue of detecting very short GRBs with time duration of $\tau < 64$ ms (the smallest BATSE trigger time scale) has been raised [8]. They show that the detector efficiency will drop sharply for bursts below 64 ms. Since several GRBs have been detected with bursts less than 64 ms, it is likely that there is a significant population of short bursts that have been missed. Here we will not attempt to determine the detector efficiency as a function of but simply refer to the work of [8]. They state that there could be as many missed GRBs of 1-ms duration as the number that have been detected at 10 s.

We now turn to a possible population of sources in the Galaxy or near solar neighborhood that would follow a pattern as shown in Fig. 2A. While the excess is in the vicinity of the Galactic plane, it is offset from the Galactic center and very asymmetric. We have found no specific set of Galactic plane sources with this distribution. We have also studied the nearby star population and found no pattern that fits the distribution found on Fig. 2A. One explanation for the very short bursts that we have offered before primordial black hole (PBH) evaporation [1,2] might not naturally give such a distribution; however, we cannot exclude the possibility that the PBHs are clustered in the Galactic plane in a manner such as is seen in Fig. 2A. We have also looked at a possible Oort cloud explanation (i.e., comets, comets colliding with PBHs, or comets colliding with each other), and there is no obvious association [7].

In conclusion, we have carefully studied a class of very short-duration GRBs (less than 100 ms) and find that they

1) have a harder energy spectrum than the bulk of GRBs,
2) have a $\langle V/V_{max} \rangle$ consistent with 1/2, and
3) display a non-isotropic angular distribution.

We believe that this can indicate a separate class of GRBs that are most likely from Galactic sources.

REFERENCES

1. Cline D. B., and Hong W. P., *ApJ* **401**, L57 (1992).
2. Cline D. B., Sanders D. A., and Hong W. P., *ApJ* **486**, 169 (1997).
3. Cline D. B., Matthey C., and Otwinowski S., Wide Field Surveys in Cosmology (Proc. 14th IAP Astrophysics Colloq., Institut d'Astrophysique de Paris, CNRS) eds. S. Colombi, Y.Mellier, B. Raban (Editions Frontieres, Paris), pp. 374-378 (1998).
4. Cline D. B., Matthey C., and Otwinowski S., *ApJ* in press (1999).
5. Kouveliotou C., et al., *ApJ* **413**, L101 (1993).
6. Meegan, C. A., et al., *ApJS* **106**, 65 (1996).
7. Maoz, E. *ApJ* **414**, 877, (1993).
8. Nemiroff, R. J., et al., *ApJ* **46**, L13 (1998).

BATSE GRB EVENTS (SINCE APRIL 21, 1991 TILL NOVEMBER 18, 1998)

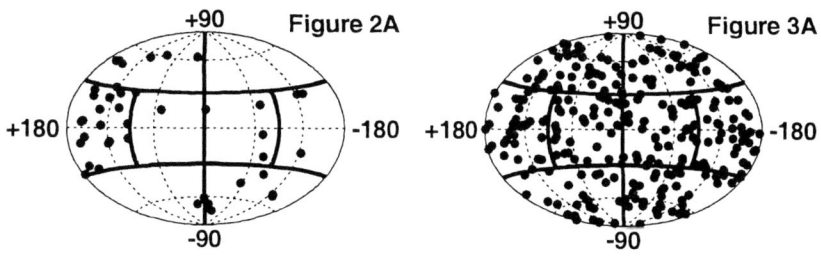

Galactic Coordinates, Aitoff projection

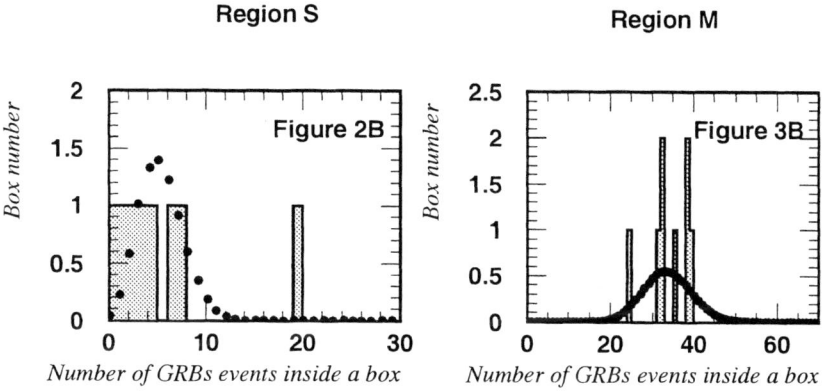

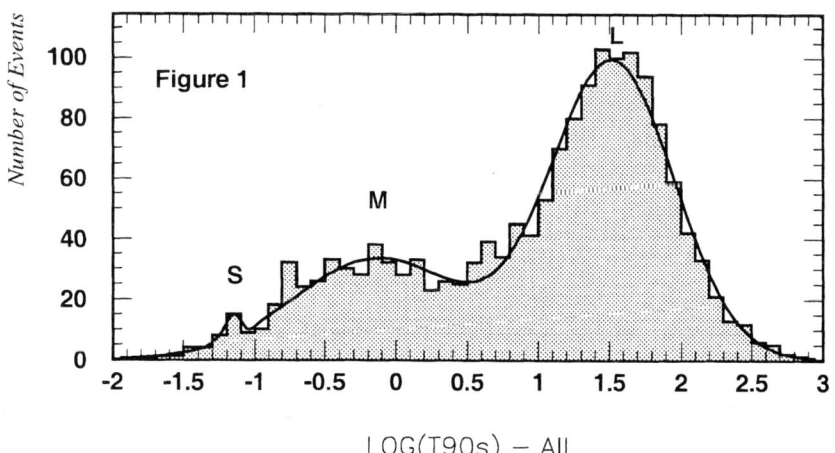

LOG(T90s) — All

Testing the Intrinsic Randomness in the Angular Distributions of Gamma-Ray Bursts

Attila Mészáros[1], Zsolt Bagoly[2], István Horváth[3],
Lajos G. Balázs[4], and Roland Vavrek[4]

[1] *Department of Astronomy, Charles University, V Holešovičkách 2, CZ-180 00 Prague 8, Czech Republic*
[2] *Laboratory for Information Technology, Eötvös University, Pázmány Péter sétány 1/A, H-1518 Budapest, Hungary*
[3] *Department of Physics, BJKMF, Box 12, H-1456 Budapest, Hungary*
[4] *Konkoly Observatory, Box 67, H-1505 Budapest, Hungary*

Abstract. The counts-in-cells and the two-point angular correlation function method are used to test the randomness in the angular distributions of both the all gamma-ray bursts collected in the BATSE Catalog, and also their three subclasses ("short", "intermediate", "long"). The methods *eliminate* the non-zero sky-exposure function of BATSE instrument. Both tests *suggest* intrinsic non-randomness for the intermediate subclass; for the remaining three cases we use only the correlation function method. The confidence levels are between 95% and 99.9%. Separating the GRBs into two parts ("dim half" and "bright half", respectively) we obtain the result that the "dim" half shows a non-randomness at the 99.3% confidence level from the counts-in-cells test.

INTRODUCTION

Recently, two different articles [1,2] simultaneously suggest that the earlier separation of gamma-ray burts (GRBs) - detected by BATSE - into short and long subclasses is incomplete. These articles show that the earlier long subclass alone should be further separated into a new "intermediate" subclass (2 s $< T_{90} <$ 10 s) and into a "truncated long" subclass ($T_{90} >$ 10 s). Therefore, in what follows, the long subclass will contain only the GRBs with $T_{90} >$ 10 s, and the *intermediate subclass* will be considered as a new subclass (2 s $< T_{90} <$ 10 s). The "short" subclass is defined by $T_{90} <$ 2 s (for definition of T_{90} see [3]).

Fully independently, Balázs et al. [4,5] suggest that GRBs are distributed anisotropically on the sky. In addition, they show that the short subclass shows an anisotropy, but the intermediate + long subclasses do not show this. The different behavior is confirmed at the 99.7% confidence level [5]. It is difficult to explain such behavior of subclasses by the instrumental effects alone. Hence, some **intrinsic** anisotropy should exist. A recent study [6], which is based on the spherical harmonic analysis, shows that just the GRBs of the "intermediate subclass" **have an intrinsic anisotropy** at the confidence level of 97%.

Here we shortly describe and summarize the new results of two further tests. The details of these tests will be published elsewhere [7,8].

GRBs will be taken between trigger numbers 0105 and 6963 from the current BATSE Catalog [3] having defined T_{90} (i.e., all GRBs detected up to August 1996 having measured T_{90}). From them we exclude, similarly to [9,4,5], the faintest GRBs having a peak flux (on 256 ms trigger) smaller than 0.65 photon/(cm^2-s). The 1284 GRBs obtained in this way define the "all" class. From them there were 339 GRBs with $T_{90} < 2$ s (the "short" subclass), 181 GRBs with 2 s $< T_{90} < 10$ s (the "intermediate" subclass) and 764 GRBs with $T_{90} > 10$ s (the "long" subclass). We will study the all classes and the three subclasses separately.

COUNTS-IN-CELLS TEST

We separate the sky in declination into $m_{dec} > 1$ stripes having the same "effective" area ($4\pi/m_{dec}$ steradian). This means that the boundaries of stripes are the declinations, which ensure that the probability to have a GRB in a stripe is *for any stripe* the same ($= 1/m_{dec}$). Because the sky-exposure function of BATSE is *not* depending on right ascension, this may easily be done by the convenient choices of declinations. We also separate the sky in right ascension α into $m_{ra} > 1$ stripes. Hence, we separated the sky into $M = m_{dec} \times m_{ra}$ areas ("cells") having the same "effective" size $4\pi/M$ steradian.

If there are N GRBs on the sky, then $n = N/M$ is the mean of GRBs at a cell. Let n_i, $i = 1, 2, ..., M$ be the observed number of GRBs at the ith cell ($\sum_{i=1}^{M} n_i = N$). Then

$$var_M = (M-1)^{-1} \sum_{i=1}^{M} (n_i - n)^2 \qquad (1)$$

defines the observed variance. For the given cell structure with M cells, the measured variance var_M should be identical to the theoretically expected value $n(1 - 1/M)$.

Q cell structures may be probed for the same sample of GRBs. We will choose $m_{dec} = 2, 3, \ldots, 8$ and $m_{ra} = 2, 3, \ldots, 16$, i.e., it will be $Q = 105$. In the coordinate system with axes $x = 1/M$ versus $y = \sqrt{var_M/n} = (var/mean)^{1/2}$ the Q values of $(var/mean)^{1/2}$ define Q points, where $j = 1, 2, ..., Q$). The theoretical expectation is verified by least squares estimation ([10], Chapter 5.3.1.). Our estimator is the dispersion

$$\sigma_Q = \sum_{j=1}^{Q} (y_j - \sqrt{1 - 1/M}\,)^2 \,. \qquad (2)$$

We throw 1000-times randomly N points on the sphere, and repeat the above calculation leading to σ_Q for every simulated sample. Then we compare the size of the σ_Q obtained from this simulation with σ_Q obtained from the actual GRB positions. Let ω be the number of simulations, when the obtained σ_Q is bigger than the actual value of σ_Q. Then $(100 - \omega/10)$ is the confidence level in percentage.

TWO-POINT ANGULAR CORRELATION FUNCTION TEST

Given N_D GRBs on the sky, there are $N_D(N_D - 1)/2$ angular distances among them. These measured distances are binned into bins with binwidth 4 degrees. Then, using a Monte Carlo simulation, there are distributed randomly N_R points on the sky, where $N_R \gg N_D$. (In our case, $N_R = 1000\ N_D$.) Then the $N_R(N_R-1)/2$ angular distances are binned in the same manner. In addition, the $N_R N_D$ random data pairs are also binned in the same manner.

In [11] the following formula is proposed in order to obtain the $w(\theta)$ correlation function:

$$w(\theta) = \frac{<DD>}{<RR>} - 2\frac{<RD>}{<RR>} + 1 , \qquad (3)$$

where $<...>$ means a normalized mean (see [11]).

The 1σ uncertainty in $w(\theta)$ for the given bin is given by $\delta w(\theta) = n_{iDD}^{-1/2}$ [11]. This formula allows to test the zero expectation value of $w(\theta)$. For any bin be calculated the dimensionless value x from the relation $|w(\theta) - x\delta w(\theta)| = 0$. Then $x\sigma$ is the probability that $w = 0$ holds for the given bin. Because $w = 0$ must occur for any bin, simply the biggest value of $x\sigma$ defines the "Poissonian" confidence level [11].

We will verify the confidence levels by Monte Carlo simulations, too. We will take - instead of the the actual BATSE positions - randomly generated N_D positions, and then we will repeat the above procedure. This simulation will be done 500 times. If there are ω such simulations, when the absolute value of w_{sim} is bigger than the absolute value of w_{act}, then $(500 - \omega)/5$ is the confidence level in percentage. As the final confidence level the smaller value will be taken.

We exclude the non-uniform sky exposure function of BATSE instrument as follows. Assume that the Monte Carlo simulation defines a point at a given position. We generate for any such point an additional random number between 0 and 1, too. If this number is bigger than the actual value of sky-exposure function at that point, then this point is omitted from the Monte Carlo sample.

Note that this test is usually more effective than the counts-in-cells one, because usually the "distance-based" tests are more sensitive than "cell-based" tests ([10], Chapter 2.6).

THE RESULTS

The counts-in-cells tests give $\omega = 287$ ($\omega = 80$, $\omega = 36$, $\omega = 440$) for all (short, intermediate, long) GRBs. Hence, *the rejection of null-hypothesis of randomness is confirmed for the intermediate subclass* on the 96.4% confidence level. For the short and long subclasses, respectively, and also for all GRBs the null-hypothesis cannot be rejected at the $> 95\%$ confidence level.

The calculated four correlation functions give the following results:

1. For the short subclass there is an essential departure from zero for $\theta = (14 \pm 2)$ degrees. We *have* a 99.2% confidence level for the non-randomness.

2. In the case of intermediate subclass the "suspicious" angles are the values $\theta = (6 \pm 2)$, $\theta = (50 \pm 2)$, and $\theta = (90 \pm 2)$ degrees. For $\theta = (6 \pm 2)$ degrees we *have* a 99.8% confidence level for non-randomness.

3. In the case of long subclass for the angle $\theta = (94 \pm 2)$ degrees we *have* a 99.0% confidence level for non-randomness.

4. In the case of all GRBs the following angles are "suspicious": $\theta = (50 \pm 2)$, $\theta = (94 \pm 2)$, $\theta = (130 \pm 2)$ and $\theta = (150 \pm 2)$ degrees. For $\theta = (94 \pm 2)$ we *have* a 99.8% confidence level for non-randomness.

FIRST CONCLUSION: The intrinsic non-randomness **is confirmed** on the confidence level $> 95\%$ for **the intermediate subclass** by both methods; the angular correlation function (counts-in-cells) method gives a 99.8% (96.4%) confidence level. This supports the results of [6].

SECOND CONCLUSION: For the remaining two sub-classes and for all GRBs the null-hypotheses of intrinsic randomness are rejected only by the angular correlation function method on the confidence levels between 99% and 99.9%. The counts-in-cells test, similarly to [6], did not reject the null-hypothesis of randomness for these cases.

The peak flux $= 2$ photons/(cm^{2}s) (on 0.256s trigger) is practically identical to the medium. Therefore, we consider the GRBs having smaller (bigger) peak flux than 2 photons/(cm^{2}s) as the "dim" ("bright") subclass ("half") of the intermediate subclass. There are 92 GRBs (89 GRBs) at the "dim half" ("bright half").

The 105 "var/mean" tests for these two parts give the results that *the "dim half" has an intrinsic non-randomness* on the 99.3% confidence level; the "bright half" can well be random. The sky distribution of 92 dim GRBs is shown on Figure 1.

THIRD CONCLUSION: The intrinsic non-randomness of **the "dim half" of the intermediate subclass of GRBs is confirmed** at the 99.3% confidence level by the counts-in-cells method.

ACKNOWLEDGMENTS

The authors acknowledge the supports from the following grants: GAUK 36/97, GAČR 202/98/0522, Domus Hungarica Scientiarium et Artium (A.M.); OTKA T024027 (L.G.B.); F029461 (I.H.).

REFERENCES

1. Horváth, I., *ApJ* **508**, 757-759 (1998).
2. Mukherjee, S., Feigelson, E. D., Babu, G.J., Murtagh, F., Fraley, C., and Raftery, A. *ApJ* **508**, 314-327 (1998).
3. Meegan, C. A., et al., *Current BATSE Gamma-Ray Burst Catalog*, http://gammaray.msfc.nasa.gov/batse, 1998.
4. Balázs, L. G., Mészáros, A., and Horváth, I., *A&A* **339**, 1-6 (1998).
5. Balázs, L. G., Mészáros, A., Horváth, I., and Vavrek, R., *A&A Suppl.* **138**, 417-418 (1999).
6. Mészáros, A., Bagoly, Z., and Vavrek, R. *A&A* in press (1999).
7. Mészáros, A., Bagoly, Z., Horváth, I., Balázs, L.G. & Vavrek, R., *ApJ* submitted (1999).
8. Bagoly, Z., Mészáros, A., Horváth, I., Balázs, L.G. & Vavrek, R., *ApJ* submitted (1999).
9. Pendleton, C. N., et al., *ApJ* **489**, 175-198 (1997).
10. Diggle, P.J., 1983, *Statistical Analysis of Spatial Point Patterns*, Academic Press, London (1983).
11. Landy, S.D., and Szalay, A.S., *ApJ* **412**, 64-71 (1993).

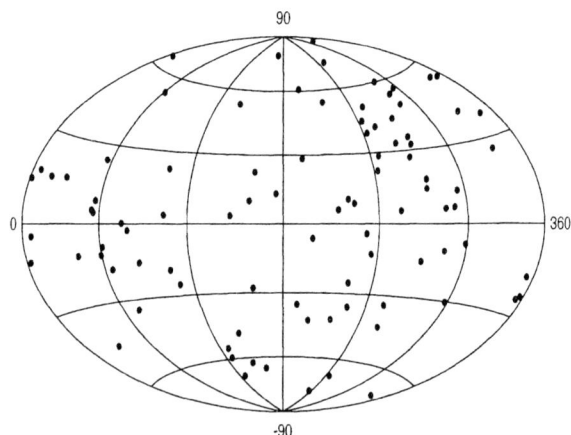

FIGURE 1. Sky distribution of 82 "dim" GRBs of the intermediate subclass in equatorial coordinates.

On the Clustering of GRBs on the Sky

Shiv K. Sethi[1], S. G. Bhargavi[2], and Jochen Greiner[3]

[1] Mehta Research Institute, Chhatnag Road, Jhusi
Allahabad 221 506, India
[2] Indian Institute for Astrophysics, Korumangla
Bangalore 560034, India
[3] Astrophysikalisches Institut Potsdam
D-14482 Potsdam, An der Sternwarte 16, Germany

Abstract. The two-point correlation of the 4th (current) BATSE catalog (2494 objects) is calculated. It is shown to be consistent with zero at nearly all angular scales of interest. Assuming that GRBs trace the large scale structure in the universe we calculate the angular correlation function for the standard CDM (sCDM) model. It is shown to be $\leq 10^{-4}$ at $\theta \simeq 5°$ if the BATSE catalog is assumed to be a volume-limited sample up to $z \simeq 1$. Combined with the error analysis on the BATSE catalog this suggests that nearly 10^5 GRBs will be needed to make a positive detection of the two-point angular correlation function at this angular scale.

INTRODUCTION

Recent optical identification of gamma-ray bursts (GRBs) has established the cosmological origin of GRBs and redshifts have been measured in 9 cases. (For a comprehensive list of references on this subject see [1]). However, the physical origin of these bursts, their environment, and their relationship with other astrophysical objects still remains an unsolved puzzle. If these bursts are associated with the underlying large scale structure in the universe, then they should show clustering in their positions on the sky as expected of cosmological objects.

One way to search for the clustering is to determine the two-point angular auto-correlation function of the burst positions [2–4]. We compute this quantity for the 4th (current) BATSE catalog (2494 objects) in the next section. In §3 we calculate the two-point correlation function from existing, viable, theoretical models of structure formation. Section 4 summarizes the main results.

TWO-POINT CORRELATION FUNCTION

Given a two-dimensional distribution of N point objects in a solid angle Ω, the two-point angular correlation function is defined using the relation [5]

$$n_{\rm DD} = n\, d\Omega\, N(1 + w(\theta)). \tag{1}$$

Here $n_{\rm DD}$ is the total number of pairs between angular separation θ and $\theta + d\theta$; $n = N/\Omega$; and $d\Omega$ is an infinitesimal solid angle centered around θ. $w(\theta)$ is the two-point correlation function. It measures the excess of pairs over a random Poisson distribution at a given separation θ. Eq. (1) is not very convenient for estimating the two-point correlation function and several alternative estimators of the two-point angular correlation function have been suggested. We experimented with several estimators [6–9]. The advantage of using either [8] or [9] is that the error on the two-point function is nearly Poissonian; the leading term in the error for the other two estimators is $\propto 1/N$, which can dominate over the Poisson term for large bin size [9]. In this paper we report results using the estimator given by Landy and Szalay [9]:

$$\tilde{w}(\theta) = \frac{n_{\rm DD} - 2n_{\rm DR} + n_{\rm RR}}{n_{\rm RR}}. \tag{2}$$

Here $n_{\rm DD}$ is the number of pairs (for a given θ) in the GRB catalog, $n_{\rm RR}$ is the number of pairs in a mock, random, isotropic sample, and $n_{\rm DR}$ is the catalog-random pair count. The variance of $\tilde{w}(\theta)$ is given by

$$\delta \tilde{w}(\theta)^2 \simeq \frac{1}{n_{\rm DD}}. \tag{3}$$

In Figure 1 (left panel) we show the angular correlation function with 1σ error bars for the current BATSE (2494 objects) catalog. We also plot the 1σ errors (Eq. (3)). The main conclusions of our analysis are the following:

1. The two-point angular correlation function is consistent with zero on nearly all angular scales of interest.

2. From Figure 1 (left panel) it is seen that at several angular scales a 1σ detection of the correlation function seems to be possible. To make a definitive statement about a detection we need to take into account several uncertainties in our analysis. One of the dominant sources of uncertainty is the heterogeneity of the sample with respect to the error in angular positions of the GRBs (the localization uncertainty varies from $\simeq 1°$–$10°$). This means that errors at $\theta \leq 10°$ are much larger than seen in Figure 1 (left panel). Another major source of uncertainty comes from the anisotropic exposure function of the BATSE instrument, which results in a non-zero correlation function even for a completely isotropic intrinsic distribution[1]. Though it is possible that some of the signal at large angular scales is not an artifact, more careful analysis would be required to confirm it.

---

[1] For more details see http://gammaray.msfc.nasa.gov/batse/grb/catalog/

THEORETICAL PREDICTIONS

The two-point angular correlation function can be related to the two-point three-dimensional correlation function $\xi(r)$ using Limber's equation (for details see [5]). If we assume that the GRBs constitute a volume-limited sample up to a distance $r_{\max}$ and that the comoving number density of objects is constant, the Limber's equation reduces to

$$w(\theta) = \frac{\int_0^{r_{max}} \int_0^{r_{max}} r_1^2 r_2^2 dr_1 dr_2 \xi(r_{12}, z_1, z_2)}{[\int_0^{r_{max}} r^2 dr]^2}. \qquad (4)$$

Here

$$r_{12}^2 = r_1^2 + r_2^2 - 2r_1 r_2 \cos\theta, \qquad (5)$$

and r is the coordinate distance in an isotropic, homogeneous universe. The two-point correlation function is related to the power spectrum $P(k)$ of the density fluctuations as

$$\xi(r,t) = b^2 \frac{1}{2\pi^2} \int_0^\infty k^2 dk P(k,t) \frac{\sin(kr)}{kr}. \qquad (6)$$

The bias factor b denotes the clustering of visible matter relative to the dark matter. While its absolute value is still uncertain, the relative bias between nearby rich clusters of galaxies and optically-identified galaxies is $\simeq 5$. And hence if GRBs originate in clusters rather than ordinary galaxies their correlation can be 25 times larger. In this paper, we use the linear perturbation theory predictions for $P(k,t)$. We have checked that for the angular scales of interest ($\theta \geq 5°$) it is a reasonable assumption. We use the BBKS fit [10] for the linear power spectrum of the standard CDM (sCDM) model and some of its variants. We normalize the power spectrum requiring $\sigma_8 = 0.7$. The time dependence of linear power spectrum is $P(k,t) \propto (1+z)^{-2}$, which is also the time dependence of the two-point correlation function. It should be noted that in general the two-point correlation function depends on both the separation between two points and their redshifts, as indicated in Eq. (4). However, the two-point correlation function is negligible for points separated by a large enough redshift difference. Therefore, for most purposes $\xi(r,t) \propto (1+z)^{-2}$, where z refers the redshift of any of the two points.

In Figure 1 (right panel) we show the theoretically predicted angular two-point correlation function for sCDM model. The bias b is taken to be one. If observed GRBs constitute a complete sample up to $z \simeq 1$ and they are assumed to be associated with highly biased structures like rich clusters, the value of correlation function is $\leq 10^{-4}$ at $5°$. This is the smallest angular scale at which information is possible in the BATSE catalog. At larger angles the correlation function typically scales as θ^{-1}.

CONCLUSIONS AND SUMMARY

The two-point correlation function of the 4th (current) BATSE catalog (2494 objects) is consistent with zero at nearly all angular scales that can be probed in the BATSE catalog. This result is consistent with theory if the GRBs are assumed to trace the dark matter distribution with some bias and are a complete sample up to $z \simeq 1$.

When can a detection of the two-point correlation function become possible? The error in the two-point correlation function scales as $n_{\rm DD}^{-1/2}$ (Eq. 3) and $n_{\rm DD} \propto N^2$, N being the number of objects in the catalog. Therefore the error in estimating the two-point correlation function scales as $1/N$. Theory suggests that the value of correlation function at $\theta = 5°$ is $\leq 10^{-4}$ if the GRB sample is assumed to be complete up to $z \simeq 1$. We check that $n_{\rm DD}$ at $\theta \simeq 5°$ is $\simeq 10^{-2}$ times the total number of pairs ($\simeq N^2/2$) in the GRB sample. This would suggest that a detection might become possible at this angular scale when the number of objects in the sample exceeds 10^5.

Future surveys like HETE-II and SWIFT will localize the GRBs to a few arc-minutes. This means smaller angular scales could be probed. And as the theoretically-predicted two-point correlation function scales as $\sim \theta^{-1}$, the probability of detection will increase. SWIFT will detect nearly 1000 objects over a period of 3 years with an angular resolution $\leq 1''$. However, though the two-point

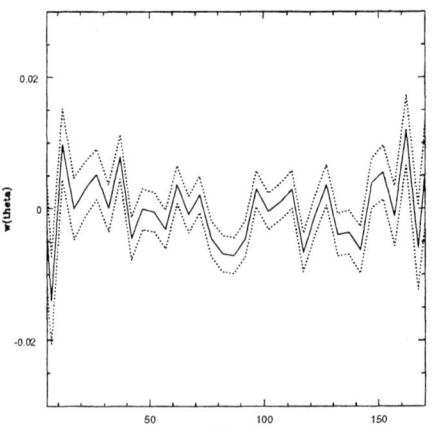 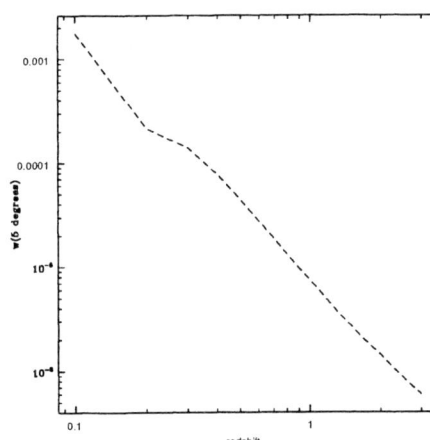

FIGURE 1. *Left Panel*: The two-point angular correlation function for the BATSE catalog (2494 objects) and the 1σ error bars are shown. The *solid* line corresponds to the two-point correlation function. The *dotted* lines show the 1σ errors given by Eq. (3). *Right Panel*: Theoretical prediction for the two-point angular correlation function is shown for the sCDM model as a function of depth (redshift) of the sample. The quantity plotted is the absolute value of the two-point correlation function at $\theta = 5°$.

correlation function is large at these angular scales, the average separation between 1000 objects on the complete sky is $\simeq 6°$. Therefore as long as $w(\theta) \leq 1$, the probability of finding an object within a few arcseconds of the other is negligible. It is possible that $w(\theta) \gg 1$ at sub-arcsecond scales. However, detailed analysis, taking into account the non-linear correction to the power spectrum of density perturbation, is needed to make precise theoretical predications for the future surveys.

ACKNOWLEDGEMENTS

JG is supported by the German Bundesministerium für Bildung, Wissenschaft, Forschung und Technologie (BMBF/DLR) under contract No. 50 QQ 9602 3. RJG acknowledges a travel grant from DFG (KON 1973/1999 and GR 1350/7-1) to attend this conference. SKS thanks S. Bharadwaj for pointing out an error. We thank D. H. Hartman for valuable comments on the anisotropic exposure function of BATSE.

REFERENCES

1. Greiner, J., http://www.aip.de/People/JGreiner/grbgen.html (1999).
2. Blumenthal, G. R., Hartmann, D. H., & Linder, E. V., *AIP* **307**, p.117, New York (1994).
3. Hartmann, D. H. & Blumenthal, G. R., *ApJ* **342**, 521 (1989).
4. Lamb, D. Q. & Quashnock, J. M., *ApJ* **415**, L1 (1993).
5. Peebles, P. J. E., *Large Scale Structure of the Universe*, Princeton University Press, Princeton (1980).
6. Peebles, P. J. E. & Hauser, M. G., *ApJS* **28**, 19 (1974).
7. Davis, M. & Peebles, P. J. E., *ApJ* **267**, 465 (1983).
8. Hamilton, A. J. S., *ApJ* **417**, 19 (1993).
9. Landy, S. D. & Szalay, A. S., *ApJ* **412**, 64 (1993).
10. Bardeen, J. M., Bond, J. R., Kaiser, N., & Szalay, A. S. , *ApJ* **304**, 15 (1986).

II. PROMPT GRB EMISSION

Gamma-Ray Burst Spectroscopy

Robert D. Preece

Dept. of Physics, University of Alabama in Huntsville, Huntsville, AL 35899

Abstract. Recent findings from the continuum spectroscopy analyses of gamma-ray bursts can be used to constrain proposed emission models. All the observed richness of behavior comes despite the simplicity of the canonical four-parameter functional form that is so successful for fitting spectra in the BATSE energy band. The distributions of time-resolved fitted spectral parameters, now available for bright bursts in the first BATSE Spectroscopy Catalog, are quite revealing. I will review the issues pertaining to the current status of GRB spectroscopy, with an emphasis on observations that must be accounted for by theory.

I INTRODUCTION

The history of the field of gamma-ray burst (GRB) continuum spectroscopy was reviewed very effectively by David Band for the Third Huntsville Symposium [2], so I will not cover that topic here. Rather, it is important to examine the questions left open to theory by the more recent, high statistics, studies of burst continuum spectroscopy made possible by the continuing long-duration mission of BATSE [29]. The open questions revolve around the intrinsic distributions of the continuum spectral parameters. The technique behind spectral model fitting has been presented before [14,6]; it should be noted here that the results are fairly robust, despite the 'obliging' nature of the data.

II THE FOUR-PARAMETER CONTINUUM

The first look at the BATSE data revealed an astonishing fact: although each burst's time history is unique, most burst's spectra are quite well characterized by a model that has at most four spectral parameters, including the overall normalization. The most famous of these models is the 'GRB' function [1] that was first used to fit the average spectra of bright BATSE bursts, using Spectroscopy Detector (SD) data. Next, the same spectral form was successfully applied to time-resolved spectroscopy of SD data [15]. The model can be described as two asymptotic power-law components, characterized by their spectral indices α and β, corresponding to low and high energies, respectively. The two power laws are joined continuously and smoothly, at a break energy E_{break}, marking the onset of the high-energy power-law portion. The function characterizing the GRB model can also be re-parameterized so that the fitted value of E_{break} is re-cast into the energy of the peak of the function in $\nu\mathcal{F}_\nu$ (but only if $\beta < -2$), which is called E_{peak}.

Several questions remain, the first of which addresses the empirical nature of the function: What physics might correspond to the observed continuum? Related to this is the question of whether the chosen function is unique, which can be restated: Are there related functions that obtain better fits to the data and are more closely allied to some physical mechanism? The next question considers what shape the continuum takes outside of the 'normal' burst energy range of 25–2000 keV. A good example of the range of validity of the four-parameter continuum can be seen in Figure 1, which has a departure of the data rate from the model rate only in the lowest channel.

Smoothly Broken Power Law. Ryde [31] has discussed another empirical spectral function, the smoothly-broken power law that has the advantage of better numerical stability over the GRB function, which has the disadvantage of a large intrinsic curvature that can mask the true spectral index of the low-energy power law [27]. Another form of this function made an earlier appearance in the BATSE spectral analysis software WINGSPAN, and is used extensively in the BATSE spectroscopy catalog [29] (see Fig. 1).

Alternative Spectral Models. Some effort has been put into finding models that are 'better' than the commonly-used GRB function in some sense; either because they are physically-motivated or because they perform better in goodness of fit tests using available data. A model based on synchrotron shocks was put forward by Tavani [33,34]. In its most basic form the synchrotron shock model (SSM) predicts a low-energy power-law index that is related to the emission physics: a $-2/3$ photon index results from the single-particle emissivity and is appropriate in cases where extensive cooling of the emitting particles can be neglected, while $-3/2$ is more appropriate for bursts, where considerable cooling is likely. Cohen et al. [10] have compared the low-energy predictions of the SSM with time-averaged spectra of 10 bright BATSE bursts, finding rough agreement. Liang and Crider [21,12] have presented a model that involves saturated synchrotron self-Compton and have successfully fit several bursts. Interestingly, this model requires a soft component to be present in GRB spectra, which has actually been observed (as discussed in the next section).

Another physically-motivated spectral model was derived by Brainerd to answer the important question of the narrowness of the observed E_{break} distribution [3]. In the Compton attenuation model (CAM), observed spectra arise from power-law spectra that have had most of their flux Compton-scattered out of the line of sight by an optically-dense medium. It is important to note that while addressing the E_{break} problem, the CAM also predicted the X-ray excess that was later observed in many spectra. In addition, the model predicted the large observed red-shifts that have been observed in GRB host galaxies, since the red-shift distribution is directly related to the E_{break} distribution. The intrinsic (or rest-frame) break energy

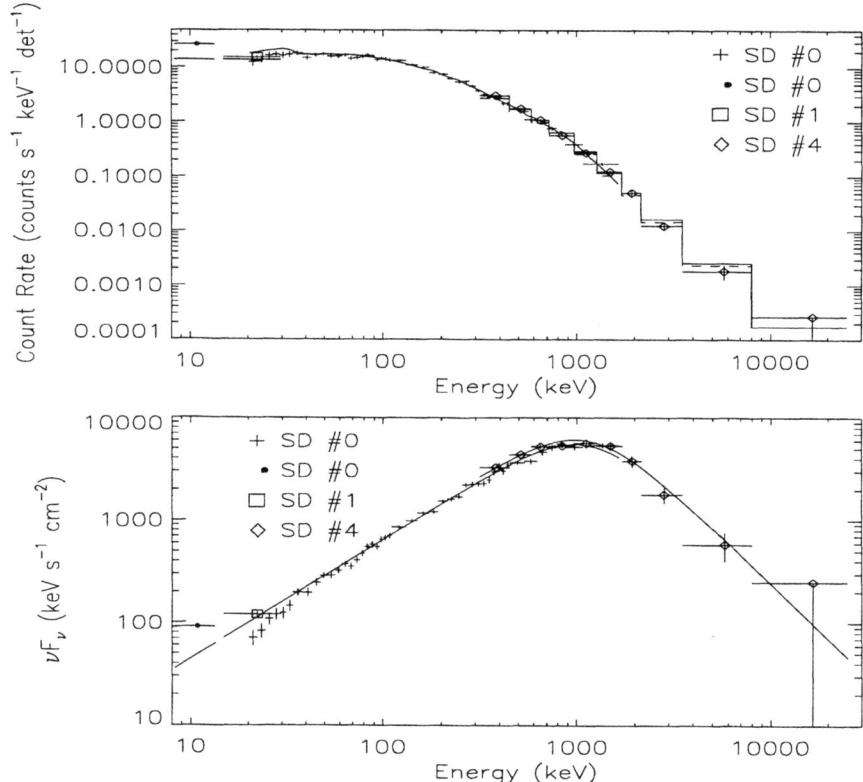

FIGURE 1. Smoothly-broken power-law fit to a 5 s spectral accumulation for GRB990123. The offset where the BATSE SD data overlap is an adjustment for differences between the absolute calibration of the two detectors. At low energies, a single SD discriminator point lies considerably above the fitted model, an example of an X-ray excess (*filled circle*). There is no calibration offset between this discriminator point and the higher-energy data from the same detector (*crosses*).

in burst spectra comes from the process of relativistic Compton scattering, and is thus a constant in the model. The CAM has been fit to a number of burst spectra with good success [4]; although there are other bursts where the fit is poor [8].

X ray Excesses. Although there are reports of X-ray precursors and tails accompanying GRB emission time histories [23], it is quite another thing to investigate the X-ray emission simultaneous with the GRB emission. The reason is simple: to properly account for all the instrumental effects, the relevant data sets must be jointly deconvolved, using the appropriate detector response matrices (DRMs). When this is done, for spectra integrated over the duration of each burst, the lower-energy data (defined as falling between 8 – 20 keV in the BATSE SD discriminators) are consistent with the assumed model to within 5σ for 85% of 86 bursts studied [26]. In the remainder, an excess in the low-energy data was observed over the model prediction. This excess is quite large in some cases, as seen in Figure 1. Similar results with fewer bursts were obtained in the study of *Ginga* data [32], with one burst out of 22 showing a considerable excess in the three lowest channels. In at least one event observed by both *BeppoSAX* and BATSE (GRB970111), an X-ray excess is confirmed in a joint deconvolution of the spectra (although this result is based upon a preliminary calibration of the WFC).

The evolution of the X-ray excess is another interesting question. While general statements about this subject await further study [30], it was found that the X-ray excess observable in the two main pulses of GRB 990123 is not observable in the tail of the emission that starts 44.93 s after the BATSE trigger [7]. No excess was observable in weak emission before the main peak as well, and there is a secondary flare on the falling portion of the first main pulse that also does not show an excess.

Two Spectral Breaks. Comprehensive information on the soft X-ray continuum properties comes from *Ginga* [32], where 22 bursts were fitted with the GRB function. Within small-number statistics, the resulting distribution in E_{break} agrees with the BATSE distribution, except at the lowest end, where *Ginga* observed seven bursts with very low E_{break} values, between 3 and 10 keV (values unobservable by BATSE). In fact, some of these seem to be true E_{peak} values, since they report $\beta < -2$. With small number statistics, one must be cautious when interpreting this result as indicating a hitherto unknown part of the E_{break} distribution. In at least one case, GRB910429, the data for the two instruments can be combined (Fig. 2). In the *Ginga* data, a 4 keV E_{peak} is reported; however, the BATSE data, which are much more sensitive over the energies where the two instruments overlap, indicates that *Ginga* slightly underestimates the fitted value for β and that there is another spectral break at the higher energy of ~ 100 keV. Confirmation of this result awaits an actual joint fit of the data from the two detectors. What is quite likely, for the *Ginga* results, is that there are *two* distributions of break energies, corresponding to two spectral breaks in burst continua, with the higher-energy break identified with E_{peak}. This possibility can be explored with available BATSE spectral data, and indeed, some evidence for two spectral breaks can be found. Indications of two spectral breaks have also been observed in afterglow spectra [17]. Of course, many bursts seem to have only a single spectral break; one such example (GRB990510) is discussed in these proceedings [8].

III SPECTRAL PARAMETER DISTRIBUTIONS

The Low-Energy Power-Law Index. The distribution of observed values of α, corrected for an upward bias that is due to the functional form of the GRB model [28], is presented in Figure 3. Not only does it peak at the value -1, which is not predicted by any simple emission model, but it also has too many large-value points to accommodate the strict upper limit of $-2/3$ imposed on the index by the synchrotron shock model [33]. Other recent studies of the low-energy index were done for bursts observed by *Ginga* [32] and BeppoSAX [16].

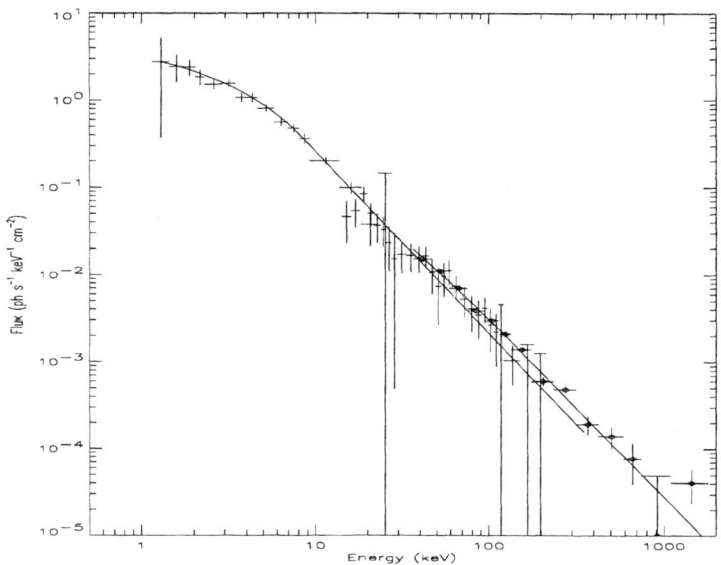

FIGURE 2. Two spectral fits from the same time interval of GRB910429 are overlaid. They consist of *Ginga* data, covering 2–200 keV, and BATSE SD data from 14–1800 keV. The two break energies are ~ 4 and 173 keV.

Optically-thin synchrotron emission can be ruled out completely for $\sim 30\%$ of all spectra, and many bursts are observed to have smooth evolution from $\alpha > -2/3$ (crossing the 'line of death' for the SSM) to $< -2/3$. Synchrotron self-absorption (SSA) allows for spectra with $\alpha = +3/2$. However, there is the problem that such regions could rapidly become optically thick, due to the dependence of the pair-creation rate on photon densities [24]. If the low energy power law is due to SSA, there should be a region of the spectrum that *does* satisfy $\alpha = -2/3$, up to another spectral break at the characteristic synchrotron energy. Although there are spectra with more than one spectral break, the observed intermediate spectral index is not consistent with this scheme. This raises the possibility that $\nu_{SSA} > \nu_{synch}$, which certainly says that the source is quite optically thick, in which case, the emission should be thermal at the high end, which is not observed.

The Compton attenuation model [3] assumes that a high optical depth scattering region lies between us and the emission region. The low-energy index in this model arises from the interplay between the optical depth of the scatterers and the power-law index of the original unscattered spectrum, which can be observed as the high-energy power-law component. Finally, reprocessing can take the form of saturated synchrotron self-Compton [21], which seems to require fairly strong magnetic fields to be effective for GRBs.

A recent paper [9] speculates that the relative motion of different potions of

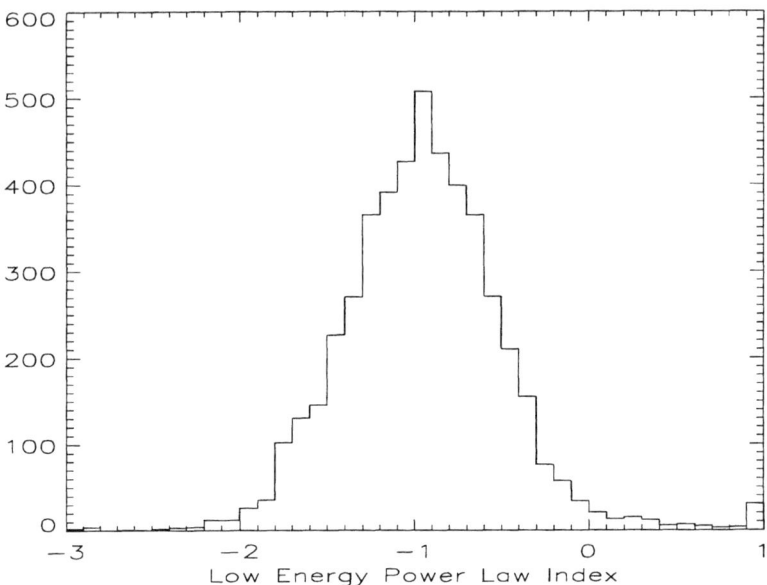

FIGURE 3. Low-energy power-law index distribution for the entire sample of 5500 spectral fits from the BATSE spectroscopy catalog.

a blast wave, as seen from a fixed, external observer, contributes to a Doppler spreading of the fundamental emission such that the low-energy portion of observed spectra ought to be confined to $-5/4 < \alpha < -3/4$, as observed. Clearly, this should be confirmed with actual models for the emission, since only a monochromatic energy source has been fully modeled.

The High-Energy Power-Law Index. As can be seen from the distribution (Fig. 4), the most probable values for the high-energy power-law index β is -2.2, with only a fraction of about 16% greater than -2, where there is no peak in the $\nu \mathcal{F}_\nu$ (power density) spectrum, E_{peak}, but rather only a fitted break energy. In addition, there are some spectra within bursts where only an upper limit on β can be determined, as indicated by the lowest bin of the histogram in Figure 4. For six out of the sample of 156 bursts in the BATSE spectroscopy catalog, the three-parameter variant of the GRB model, which is essentially an exponentially-attenuated power law, was used for all of the spectra in each. The general behavior of β over time in single bursts is from hard to soft [27]. This is in sharp contrast to the behavior of the other spectral parameters, especially E_{peak}, which may have a correlation with the burst intensity in addition to a general hard-to-soft trend [15]. Only 12% of the sample had a significant correlation between β and E_{peak}. Finally, β changes over the entire time history of a burst by an average of -0.4. When the BATSE values for β are compared with results from COMPTEL, covering a higher energy band, it is BATSE that underestimates the power-law index [18]. There

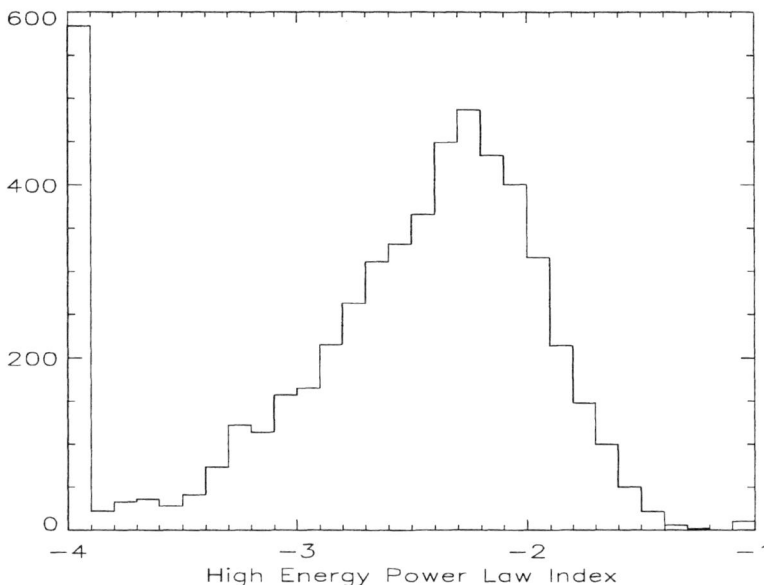

FIGURE 4. High-energy power-law index distribution for the BATSE spectral catalog sample. All spectra that either do not have β values or that have values < -4 (555/5500) have been included in the lowest bin.

may a systematic effect that leads to the underestimation of the power-law index when statistics are poor that should be investigated further. The data available from EGRET observations of bursts indicate the high-energy power law continues to very high energies, with no evidence for a high-energy break [13,20].

The Break Energy Distribution. Within the context of the typical four parameter spectral continuum models investigated for gamma-ray bursts, the break energy is the only part of the spectrum that reflects the relative motion between the source of emission and the observer. This relative motion could be a combination of the relativistic recession velocity of the source due to the expansion of the universe and the relativistic bulk motion of the emission source toward the observer. Here, 'break energy' stands in for many concepts: it is the GRB model's E_{peak} parameter, regardless of whether $\beta < -2$; it is also the break energy for broken and smoothly broken power-law models; finally, it is the characteristic energy from the exponentially attenuated power law form of the GRB model. The observed distribution of break energies (loosely referred throughout as E_{break}) is presented in Figure 5. It has a roughly Gaussian shape in log(Energy) (in other words, it is log-normal) with a peak very close to 250 keV and a full width at half maximum that extends from 130 keV to 440 keV, or roughly a factor of four. A log-normal distribution is not expected a priori, and a narrow distribution runs contrary to common sense, except within the context of certain models, such as the CAM. The considerable

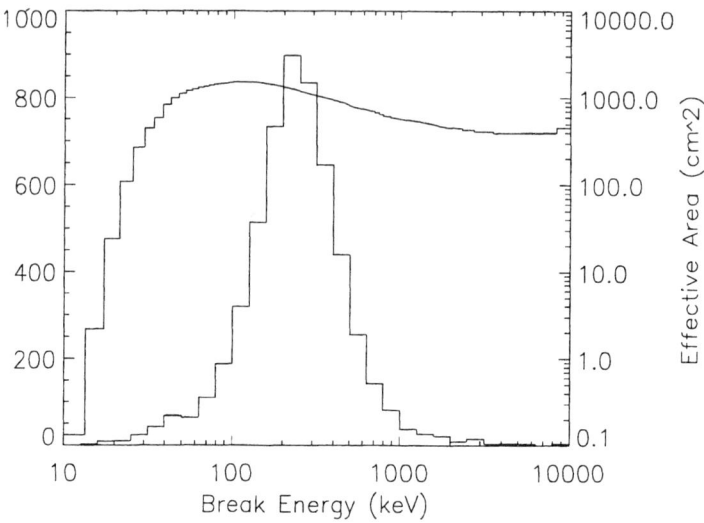

FIGURE 5. Break energy distribution for the entire sample of 5500 spectral fits from the BATSE spectroscopy catalog. Overplotted is the BATSE LAD total effective area (*right-hand axis*).

spectral evolution found within single bursts alone [15] must contribute much of the observed width. Relative distances to the sources within the sample population of bright bursts must also contribute, leaving seemingly no room for the bulk Lorentz motion that must be present in the shock model so that the emission region does not become optically thick to the highest emerging photons from pair production. The restriction of a physical parameter to within a factor of four presents serious problems for the most commonly-accepted models for burst emission (outside the CAM), so the correspondence between the observations and the intrinsic, physical distribution should be made explicit.

The peak of the E_{break} distribution is confined entirely within the BATSE energy pass band, so a question arises concerning the reality of the tails of the distribution. Certainly, corrections must be made for the changes in the instrumental response with energy. Figure 5 indicates this qualitatively, with the low-energy slope of the distribution of E_{break} values covered by the most sensitive region in the response. Above that, the response drops by only a factor of two, while the the distribution tails off to roughly zero. The answer to the difficult question of how to properly correct for instrumental effects depends upon a number of assumption that may be made about correlations between bursts' peak intensities and E_{break} [11,22,5]. The high-energy tail of the E_{break} distribution was investigated using archival data from SMM [19]. The technique used in this analysis consisted of an off-line trigger of the entire archived data, using a pass band covering 800 keV – 10 MeV. Once

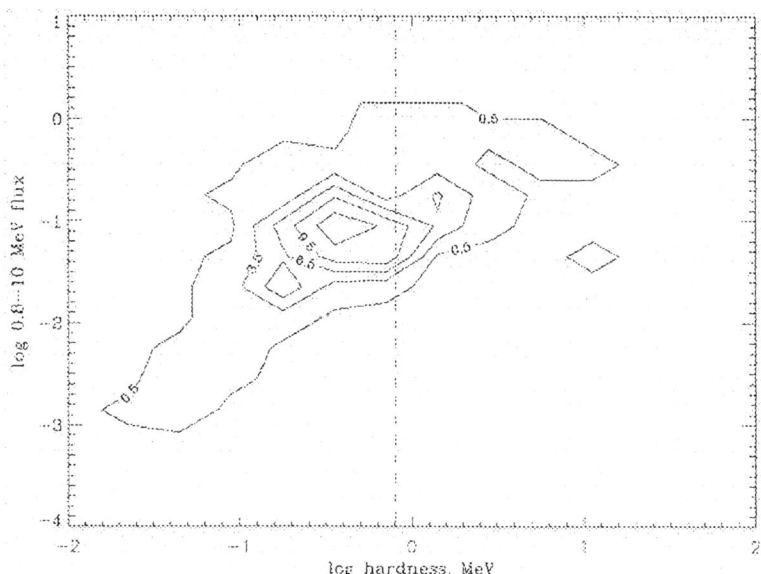

FIGURE 6. Distribution of 138 GRBs detected by SMM during 1980–1989, as a function of spectral hardness and 0.810 MeV flux in the form of a contour plot. *Dashed line*: Hardness value $H_0 = 0.8$ MeV (from [19]).

a triggering event was discovered, the total spectrum was analyzed and E_{break} determined. The resulting distribution does not peak within the energy pass band picked, but instead a rather broad peak centered near 500 keV *below* the chosen energy pass band (see Fig. 6). This peak energy is different than the BATSE result; however, the distribution has not been corrected for the SMM detector response in this energy range. Given that the lowest energy observable in SMM is roughly 300 keV, the agreement with the BATSE distribution is actually quite remarkable. More importantly, no large population of high-E_{break} events was discovered.

IV CONCLUSIONS

What more is needed before some of the outstanding questions can be reliably answered? First of all, there needs to be a comprehensive study of broad-band spectroscopy in a large sample of bursts (including the set of short bursts not covered here). At the low-energy end of the continuum, joint spectral fits between BATSE and instruments with X-ray observing capability, such as will fly on HETE-2 and Swift as well as the WFC data set, can elucidate the properties of the X-ray excess and the second population of spectral breaks. Future missions like AGILE and GLAST will answer the question of the high-energy cut-off in GRB spectra; always assumed to exist but never observed. These will also improve our understanding of the temporal behavior of the high-energy power-law component. Theory must account for the narrow E_{break} distribution, at least as well as does the CAM. Synchrotron shock proponents must explain the peak at $\alpha \approx -1$, as well as answer tough questions about pair production saturation, if they would invoke SSA to resolve the 'line of death' issue.

REFERENCES

1. Band, D. L., et al., *ApJ* **413**, 281 (1993).
2. Band, D. L., in *Gamma-Ray Bursts: 3rd Huntsville Symposium*, edited by Kouveliotou, Briggs & Fishman, AIP, New York, 1996 pp. 123.
3. Brainerd, J. J., *ApJ* **428**, 21 (1994).
4. Brainerd, J. J., et al., *ApJ* **501**, 325 (1999).
5. Brainerd, J. J., et al., *ApJ*, submitted (2000).
6. Briggs, M. S., in *Gamma-Ray Bursts: 3rd Huntsville Symposium*, edited by Kouveliotou, Briggs & Fishman, AIP, New York, 1996 pp. 133.
7. Briggs, M. S., et al., *ApJ* **524**, 82 (1999).
8. Briggs, M. S., et al., these proceedings.
9. Cen, R., *ApJ* **517**, L113 (1999).
10. Cohen, E., et al., *ApJ* **488**, 330 (1997).
11. Cohen, E., Piran, T., & Narayan, R., *ApJ* **500**, 888 (1998).
12. Crider, A., et al., *ApJ* **497**, L39 (1997).
13. Dingus, B. L., *Ap. & S. S.* **231**, 187 (1995).
14. Fenimore, E. E., et al., *ApJ* **335**, L71 (1988).
15. Ford, L. A., et al., *ApJ* **439**, 307 (1995).
16. Frontera, F., et al., *ApJ*, submitted, astro-ph9911228, (2000).
17. Galama, T., et al., *ApJ* **500**, L97 (1998).
18. Hanlon, L. O., et al., *Ap. & S. S.* **231**, 157 (1995).
19. Harris, M. J, & Share, G. H., *ApJ* **494**, 724 (1998).
20. Hurley, K., et al., *Nature* **372**, 652 (1994).
21. Liang, E. P., et al., *ApJ* **497**, L35 (1997).
22. Lloyd, N. M., & Petrosian, V., *ApJ* **511**, 550 (1999).
23. Murakami et al., *Nature* **350**, 592, (1991).
24. Papathanassiou, H., *A&AS* **138**, 525 (1999).
25. Pendleton, G. N., et al., *ApJ* **489**, 175 (1997).
26. Preece, R. D., et al., *ApJ* **473**, 310 (1996).
27. Preece, R. D., et al., *ApJ* **496**, 849 (1998).
28. Preece, R. D., et al., *ApJ* **506**, L23 (1998).
29. Preece, R. D., et al., *ApJ Supp.* **126**, in press (2000).
30. Preece, R. D., Espley, J. R., & Briggs, M. S., these proceedings.
31. Ryde, F., *Astrophys. Lett. and Comm.*, in press, astro-ph/9811462 (1999).
32. Strohmayer, T. E., et al., *ApJ* **500**, 873 (1997).
33. Tavani, M., *Ap. & S. S.* **231**, 181 (1995).
34. Tavani, M., *ApJ* **466**, 768 (1996).

Wide-Band Spectroscopy of GRB 990510

Michael S. Briggs[†], Robert D. Preece[†], Jan van Paradijs[†¶1],
Jean in 't Zand[*], John Heise[*], Erik Kuulkers[*§],
and C. Kouveliotou[‡]

[†]*Department of Physics, University of Alabama in Huntsville*
[¶]*Anton Pannekoek Institute, University of Amsterdam*
[*]*Space Research Organization Netherlands*
[§]*Astronomical Institute, Utrecht University*
[‡]*Universities Space Research Association*

Abstract. Observations made with detectors of BeppoSAX and BATSE provide spectral coverage of the prompt burst emission from 2 keV to 1800 keV. The light curve exhibits very strong spectral evolution, while the integrated spectrum is well-fit by the Band GRB function.

INTRODUCTION

Numerous observations were made of both the prompt and afterglow emission of GRB 990510. The observations of the afterglow provided the best evidence to date that the emission occurs in a jet and the first evidence of optical polarization (Covino et al. 1999, Wijers et al. 1999). The optical afterglow was detected 0.35 days after the GRB (Vreeswijk et al. 1999a); absorption lines place a lower-limit on the redshift of 1.619 (Vreeswijk et al. 1999b). Further optical observations (Beuermann et al. 1999, Harrison et al. 1999, Israel et al. 1999, Stanek et al. 1999) provide a well-sampled multi-color light curve of the afterglow emission. The light curve exhibits an achromatic break at about one day, which is evidence that the outflow is collimated in a jet (Beuermann et al. 1999, Harrison et al. 1999). The decay of the X-ray afterglow, as observed with the BeppoSAX Narrow Field Instruments from 8 to 44 hours after the burst, can be understood if the cooling frequency lies between the optical and X-ray bands (Kuulkers et al. 2000).

The prompt burst emission was observed with at least three spacecraft: BATSE (Kippen 1999), BeppoSAX (Amati et al. 1999) and Ulysses (Hurley & Barthelmy 1999). Herein, we report on the light curve and spectrum using the observations of the BeppoSAX Wide Field Camera (WFC) #2 and of several BATSE Large Area Detectors (LADs) and Spectroscopy Detectors (SDs). Together, these observations

---

[1] deceased

provide high spectral and temporal resolution from 2 to 1800 keV. While gamma-ray bursts have been extensively observed in the gamma-ray band and observed in the X-ray band, simultaneous observations of both bands are rare. A high-energy resolution time-integrated spectrum of GRB 980329 was obtained by in 't Zand et al. (1998) using the BeppoSAX WFC (2 to 20 keV) and the BeppoSAX Gamma-Ray Burst Monitor (GRBM) (70 to 650 keV). For eight GRBs, Frontera et al. (2000) combine high energy resolution WFC data (2 to 26 keV) with two channel data from the GRBM (40 to 700 keV).

The empirical Band 'GRB' function, which blends a low- and high-energy power law with an exponential (Band et al. 1993), has been quite successful at fitting GRB spectra. Of the four spectral parameters of the Band function, E_{peak}, the energy at which the spectrum peaks in νF_ν units, has received the most attention. The BATSE team has reported that the distribution of E_{peak} is remarkably narrow, extending between 80 keV and 800 keV (Mallozzi et al. 1995, Preece et al. 2000); others have questioned this (Piran & Narayan 1996, Lloyd & Petrosian 1999). Possible deviations from the Band function that have been reported include "X-ray excesses" above the Band function fit (Preece et al. 1996), and the possibility of a second break in the X-ray range or perhaps that E_{peak} will be observed in the X-ray range (Strohmayer et al. 1998).

RESULTS

GRB 990510 exhibits spectacular spectral evolution (Figure 1). For example, the brightest peak in the 2 to 4 keV band (at 18 s relative to the BATSE trigger) is weak in the 21 to 62 keV band and is not detected above 330 keV. The burst begins approximately simultaneously in all bands, excepting that there is little emission above 330 keV until 42 s: the first complex of peaks, from 0 to 32 s, consists of No-High Energy emission (NHE) peaks (Pendleton et al. 1997)

The burst lasts longer at lower energies, with significant 2 to 4 keV flux to at least 105 s. In comparison, the 21 to 62 keV emission appears to end at 75 s and the > 300 keV at 62 s. The greater duration of many GRBs in X-rays is a well-known phenomenon (e.g., Wheaton et al. 1973, Laros et al. 1984, Yoshida et al. 1989, Murakami et al. 1992, Frontera et al. 2000). An early description of this phenomena was "X-ray tails", of which one interpretation was black body emission (e.g., Laros et al. 1984, Yoshida et al. 1989). An alternative view is that the X-rays lagging the gamma-rays is an aspect of the overall hard-to-soft evolution almost universally seen in GRBs, albeit modulated by a hardness-intensity correlation when new peaks occur (Ford et al. 1995). Another interpretation is that the X-ray tail is the birth of the afterglow (e.g., Frontera et al. 2000).

The spectral evolution trends of the overall event are mirrored in the trends in pulse complexes and in pulses. For example, the second pulse complex begins approximately simultaneously in all four bands at $\approx$ 40 s, while the duration of the complex ranges from about 65 s in the 2 to 4 keV band down to only about 20 s in

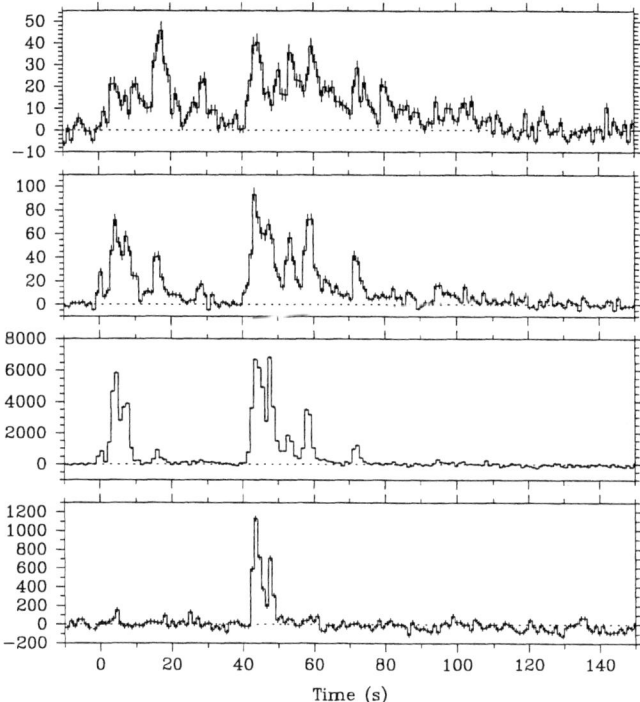

FIGURE 1. Light curves in four energy bands: the units are counts s^{-1} received by the detectors. The times are referenced to the BATSE trigger time of 31746.297 s UT. The count rate data of each detector have been binned to 1 s or 1.024 s resolution and background subtracted based upon a low-order polynomial fit to pre- and post-burst data. The WFC data are direct count rates rather than mask-correlated imaging data. **top**: 2.0 to 4.1 keV as observed with WFC #2, **next to top**: 7.9 to 12.5 keV as observed with WFC #2, **next to bottom**: 21 to 62 keV as observed with three BATSE LADs, **bottom**: > 330 keV as observed with three BATSE LADs.

the >300 keV band. This is an aspect of the well-known hard-to-soft evolution of GRB pulses (Norris et al. 1986, Ford et al. 1995).

We have only begun to explore this spectral evolution with detailed modeling. Herein we report the time-integrated spectrum (Figure 2). The data are well fit by the canonical Band GRB function, with $\chi^2 = 120.4$ for 140 degrees-of-freedom. The parameter values are $E_{\text{peak}} = 143 \pm 13$ keV, low-energy index $\alpha = -1.31 \pm 0.03$ and high-energy index $\beta = -2.5 \pm 0.3$. The fluence above 20 keV is 3.1×10^{-5} erg cm^{-2}; the energy range 2 to 20 keV contributes an additional 4.6×10^{-6} erg cm^{-2}. The isotropic-equivalent energy release above 2 keV is 1.4×10^{53} ergs (for $z = 1.619$, $H_0 = 65$ km s^{-1} Mpc^{-1}, $\Omega_0 = 1.0$ and $\Lambda = 0$). The evidence for beaming indicates that the energy of the burst is lower by ~ 300 (Harrison et al. 1999).

The low-energy break reported at the conference was an artifact of the initial WFC calibration. With an improved calibration, the spectrum is perfectly fit by

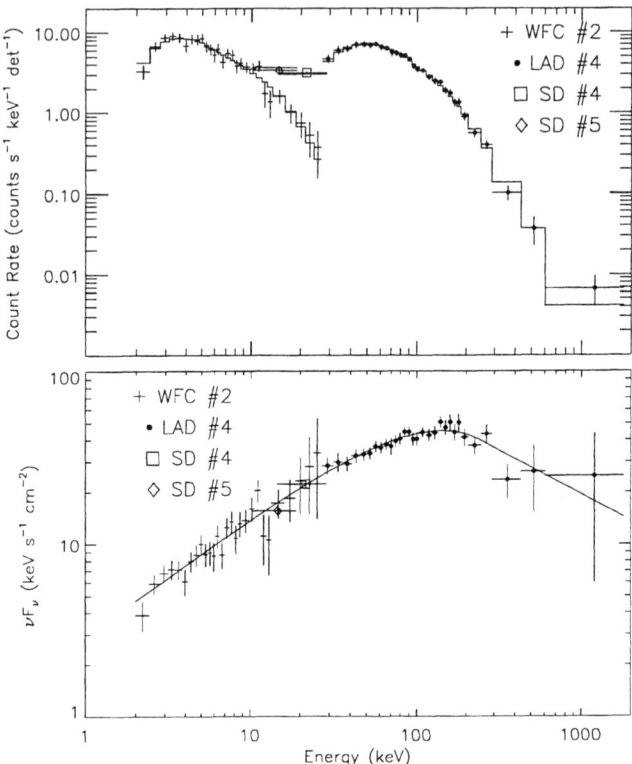

FIGURE 2. Spectral data and fits using the Band GRB function (parameter values given in the text) for the interval from 0 to 112 s relative to the BATSE trigger time. The WFC data are from the source image obtained via correlating with the mask. The BATSE data are background-subtracted counts received by the detectors. The LAD data have been rebinned above 100 keV for clarity. **top:** Count-rate data with ±1σ error bars and the best-fit photon model convolved through the detector response models to produce the best-fit count-rate models (histograms). **bottom:** Best-fit Band GRB model in νF_ν units. The data have been transformed to these units in a model-dependent manner.

the Band GRB function. There are functions that do not fit, e.g., the Compton attenuation model (Brainerd 1994). The revised calibration and detector response function expands the usable energy to 2 to 26 keV, models the time variations of the gain, reduces the systematic errors in the energy boundaries of the channels, and incorporates photon escape effects. The calibration has been tested using ≈50 long observations of the Crab at various off-axis angles.

The data from the two instruments agree perfectly and no relative normalization is necessary. The agreement of the SD low-energy discriminator points with the overlapping WFC data helps validate their calibration (Preece et al. 1996). Examination of the SD low-energy discriminator data showed 15% of the GRBs to have excess low-energy emission above the Band GRB function (Preece et al. 1996). Clearly, GRB 990510 falls into the majority of bursts lacking X-ray excesses.

The Band GRB function is now tested with high-spectral resolution data over almost 3 decades in energy. The time-integrated spectrum of GRB 990510 exhibits no deviations from this empirical function and has a typical E_{peak} value. We look forward to applying the Band GRB function to time-resolved analysis of GRB 990510 and to other GRBs observed with BATSE and the BeppoSAX WFCs.

REFERENCES

1. Amati, L., Frontera, F., Costa, E., et al., GCN #317 (1999).
2. Band, D. L., Matteson, J., Ford, L., et al., *ApJ* **413**, 281 (1993).
3. Beuermann, K., Hessman, F. V., Reinsch, K., et al., *A&A* **352**, L26 (1999).
4. Brainerd, J. J., *ApJ* **428**, 21 (1994).
5. Covino, S., Lazzati, D., Ghisellini, G., et al., *A&A* **348**, L1 (1999).
6. Ford, L. A., Band, D. L., Matteson, J. L., Briggs, M. S., et al., *ApJ* **439**, 307 (1995).
7. Frontera, F., Amati, L., Costa, E., et al., *ApJS*, in press, astro-ph/9911228 (2000).
8. Harrison, F. A., Bloom, J. S., Frail, D. A., et al., *ApJ* **523**, L121 (1999).
9. Hurley, K. & Barthelmy, S., GCN #309 (1999).
10. Israel, G. L., Marconi, G., Covino, S., et al., *A&A* **348** (1999).
11. Kippen, R. M., GCN #322 (1999).
12. Kuulkers, E., Antonelli, L. A., Kuiper, L., et al., *ApJ*, submitted (2000).
13. Laros, J. G., Evans, W. D., Fenimore, E. E., et al., *ApJ* **286**, 681 (1984).
14. Lloyd, N. M. & Petrosian, V., *ApJ* **511**, 550 (1999).
15. Mallozzi, R. S., Paciesas, W. S., Pendleton, G. N., et al., *ApJ* **454**, 59 (1995).
16. Murakami, T., Inoue, H., van Paradijs, J., et al., in *Gamma-Ray Bursts*, ed. C. Ho, R. I. Epstein, & E. E. Fenimore, Cambridge Univ. Press, 1992, pp. 238.
17. Norris, J. P., Share, G. H., Messina, D. C., et al., *ApJ* **301**, 213 (1987).
18. Pendleton, G. N., Paciesas, W. S., Briggs, M. S., et al., *ApJ* **489**, 175 (1997).
19. Piran, T. & Narayan, R., in *Gamma-Ray Bursts: 3rd Huntsville Symposium*, ed. C. Kouveliotou, M. S. Briggs & G. J. Fishman, AIP Conf. Proc. **384**, New York, 1996, pp. 233.
20. Preece, R. D., Briggs, M. S., Pendleton, G. N., et al., *ApJ* **473**, 310 (1996).
21. Preece, R. D., Briggs, M. S., et al., *ApJS* **125**, in press, astro-ph/9908119 (2000)
22. Stanek, K. Z., Garnavich, P. M., Kaluzny, J., et al., *ApJ* **522**, L39 (1999).
23. Strohmayer, T. E., Fenimore, E. E., Murakami, T. & Yoshida, A., *ApJ* **500**, 873 (1998).
24. Vreeswijk, P. M., Galama, T. J., Rol, E., et al., GCN #310 (1999a).
25. Vreeswijk, P. M., Galama, T. J., Rol, E., et al., GCN #324 (1999b).
26. Wheaton, Wm. A., Ulmer, M. P., Baity, W. A., et al., *ApJ* **185**, L57 (1973).
27. Wijers, R. A. M. J., Vreeswijk, P. M., Galama, T. J., et al., *ApJ* **523**, L33 (1999).
28. Yoshida, A., Murakami, T., Itoh, M., et al., *PASJ* **41**, 509 (1989).
29. in 't Zand, J. J. M., Amati, L., Antonelli, L. A., et al., *ApJ* **505**, L119 (1998)

Track Jumps in Hardness-Intensity Correlations of GRB Pulse Decays

Luis Borgonovo and Felix Ryde

Stockholm Observatory, SE-133 36 Saltsjöbaden, Sweden

Abstract. The hardness-intensity correlation (HIC) during the decay phase of gamma-ray burst pulses can often be described by a simple power law. The pulse decays in multi-pulse bursts present power-law indices that are equal to within the errors. We find that in some pulses the HIC changes to a parallel power law, a feature that we graphically call a *track jump*. The use of the index constancy is proposed as an auxiliary tool for pulse identification, and examples of its application are given.

INTRODUCTION

The understanding of the spectral and temporal evolution of gamma-ray bursts (GRBs) renders us important clues to the underlying processes giving rise to the phenomenon. The evolution has been studied both over the entire burst, giving the overall behavior, and over individual pulse structures (see, e.g., [1]). One important relation that characterizes the spectral evolution is that between some measure of the "hardness" of the spectrum and the instantaneous flux (or intensity), usually referred to as the Hardness-Intensity Correlation (HIC). It was first studied in GRBs by Golenetskii et al. [2], and later by others like Kargatis et al. [3,4]. They found that in many cases, during the decay phase of strong individual pulses, the HIC could be described by a simple power law relation $F \propto E_{pk}^{\gamma}$, evolving from hard to soft. F represents the energy flux, and E_{pk} is the energy at which the power is at its maximum, i.e., the peak in the EF_E-spectrum, that is used as a measure of the spectral hardness. In this work we will discuss cases where the HIC can be divided into two power laws of identical indices. These cases show what we will refer to graphically as a *track jump*.

We conducted a study [5] on multi-pulse bursts over a complete sample. We showed that, although the HIC indices present a broad distribution among different bursts, their values are equal to within the errors when comparing pulses within a burst. The probability of a random coincidence was negligible ($p < 10^{-5}$).

The measure of the flux $F \equiv \int F_E dE$ can be affected by the finite energy band of the detector, especially when the spectrum is broad or peaks close to the boundaries of the energy range. The existence of a separate soft component and/or additional weak pulses could also affect the analysis, changing the shape of the spectrum. This motivated us to introduce an alternative representation of the flux, given by the value of EF_E at $E=E_{pk}$, which we will denote by ϕ. A more detailed discussion can be found in these proceedings [6]. Since ϕ is roughly proportional to the total flux, the HIC can again be described by a power law as $\phi \propto E_{pk}^{\eta}$. In general, better correlation coefficients are found with this method. In addition, we observed a better constancy of the η power law indices among pulses of the same burst than when we compared the corresponding γ indices.

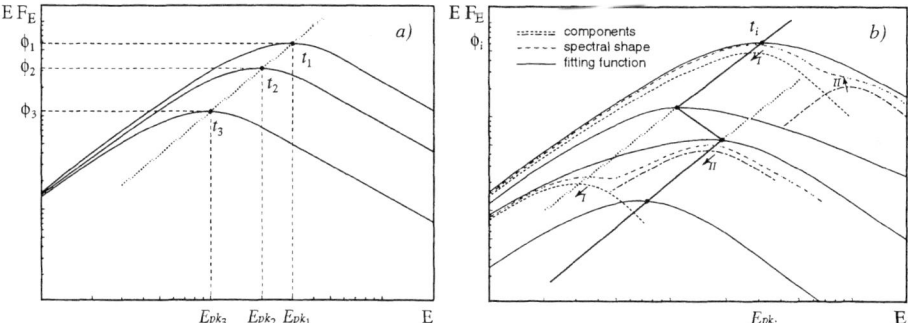

FIGURE 1. a) During pulse decays, in the EF_E representation, the spectral peaks evolve in time following a power law. b) The *track jumps* found in some cases can be explained with a two-component model. The figure shows a qualitative sketch. Each component evolves in a similar way, following approximately parallel tracks when decaying. However, the second is delayed in time and the "jump" occurs when it overcomes the first component. See the text for further details.

Figure 1a illustrates the observed evolution of $\phi(E_{pk})$ along a *track* during the decay phase of a single pulse. When many pulses are present in the light curve of a burst, although there is no apparent general trend during the rising phases, they will follow almost parallel tracks during the decay phases. To actually observe this behavior, it is necessary that the pulses do not overlap significantly. Also, they need to be bright and long enough to establish the direction of the *track*. As an example, the overall evolution of GRB 970420 (trigger #6198) is shown in Figure 2.

TRACK JUMPS

Our work was conducted on data taken by the Burst and Transient Source Experiment (BATSE) on board the *Compton Gamma-Ray Observatory*. We used the high-energy resolution burst data (HERB) which were reduced using standard procedures. The spectra were modeled using the empirical Band et al. [7] function.

Pulses are a common feature in the light curves of GRBs and appear to be their fundamental constituent [8,9]. An approach frequently employed is to identify each pulse as a single physical event. Depending on the model chosen, this event could

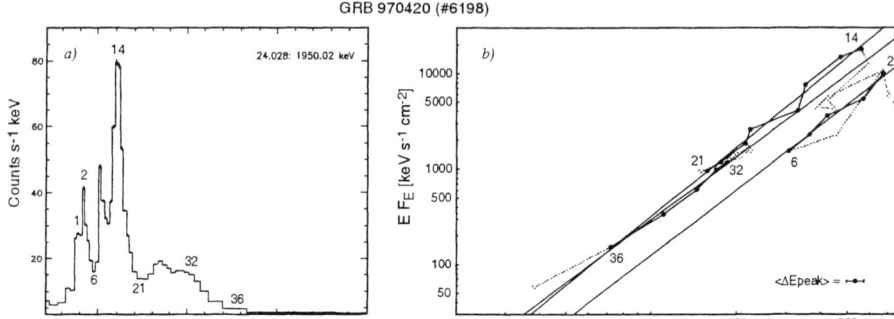

FIGURE 2. *a*) Count time history of GRB 970420 (#6198) showing the binning used to study the spectral evolution. The second pulse (at bin #8) actually consists of two close-lying pulses, which are seen in the 64-ms data. *b*) The peaks of the instantaneous spectra at successive times are represented by dots. Error bars are excluded for better visibility, and only a typical average value is shown separately. The numbers refer to the time-bins in (*a*). The decaying phases (dark dots) follow very good power laws with indices $\eta_1 = 2.25 \pm 0.20$, $\eta_2 = 2.40 \pm 0.15$, and $\eta_3 = 2.20 \pm 0.15$, equal to within the estimated errors.

be the collision between two expanding *internal* shells or the "activation" of a region on a single *external* shell. To validate this reductionistic method, it is essential to find common trends between pulses, such as the HIC that we describe here. But GRB light curves very often show a complex overlap of pulses and to attempt to separate them requires additional assumptions.

We show in Figure 3 the spectral evolution of GRB 950104 (#3345). Initially, the second decay has approximately the same HIC power law index as the first one. It then makes a change into a parallel track. Checking the 64-ms count time history, one can see that this *track jump* coincides with the presence of a small secondary peak which is hidden by the low resolution binning. The significance of this peak is somewhat marginal (3.2σ), but the agreement with the direction of the first pulse track makes it a more compelling case. The power law indices obtained for the tracks are $\eta_1 = 0.77 \pm 0.05$, $\eta_2 = 0.79 \pm 0.03$, and $\eta_3 = 0.76 \pm 0.05$, equal to within the estimated errors. We show in the same way the case of GRB 970925 (#6397). Again, we have modeled the decay evolution with two power laws. There is now no clear indication of a secondary peak, although a small "bump" coincides with the *track jump*. The measured indices, $\eta_1 = 0.77 \pm 0.15$ and $\eta_2 = 0.86 \pm 0.05$, are again equal to within the errors.

As a likely explanation of the *track jumps*, we illustrate a two spectral component model in Figure 1b. The components follow identical directions during their decay phases. At some point the second component overcomes the first, the fitting routine finds a better χ^2 value shifting the Band function parameters to peak at a higher E_{pk}, and the jump occurs. Note that due to the rather broad spectral shape, the change in the values of the asymptotic slopes of the Band function is an artifact of the fitting. The fitting routine finds the peaks by minimizing the χ^2 merit function. The errors are estimated evaluating the confidence region around the χ^2 minima. Since the spectral shape has a local maximum close to the absolute one, this produces an overestimation of $E_{\rm pk}$ errors. The alignment of the data points along the tracks proves that the shape features are real and not a random occurrence (see Figure 3).

Other workers [4,10] have fixed the parameters that control the asymptotic behavior of the Band function, to reduce the estimated errors of $E_{\rm pk}$. Such a solution provides more "stable" $E_{\rm pk}$ values, but a case like GRB 950104 (#3345) would be missed. It smoothes out the *track jump* in such a way that a single, steeper track is obtained for the whole decay.

The simplest interpretation of the *track jump* is to assume that it is produced by an overlapping secondary pulse. From this point of view, this feature is present also in cases like that shown in Figure 2. The only difference is that there we can resolve the individual pulses in the light curve. We therefor propose to use the constancy of the HIC power law index among pulses of the same burst as an auxiliary method for pulse identification in overlapped pulses [5]. To identify *track jumps* with some confidence, the distance between tracks should be larger than the observed dispersion along individual tracks, with a minimum number of points on each track required to produce a reasonable fit. We found few cases in our BATSE sample that meet these requirements. However, future instrumentation will allow higher time-spectral resolution and these criteria will then have a wider application.

ACKNOWLEDGMENTS

This research made use of data obtained through the HEASARC Online Service provided by NASA/GSFC. We are grateful to the GROSSC for support. We are indebted to R. Svensson for useful discussions. We gratefully acknowledge support from the A.-G. and H. Crafoord Fund, the G. and E. Kobbs Stipend Fund, the Swedish Natural Science Research Council, and the Swedish National Space Board.

REFERENCES

1. Ryde F., *ASP Conf. Series* **190**, 103 (1999).
2. Golenetskii, S. V. et al., *Nature* **306**, 451 (1983).
3. Kargatis, V. E. et al., *ApJ* **422**, 260 (1994).
4. Kargatis, V. E. et al., *A&SS* **231**, 177 (1995).
5. Borgonovo, L., and Ryde, F., in preparation.
6. Ryde, F., Borgonovo, L., and Svensson, R., these proceedings.
7. Band, D. et al., *ApJ* **413**, 281 (1993).
8. Norris, J. P. et al., *ApJ* **459**, 393 (1996).
9. Stern, B., and Svensson, R., *ApJ* **469**, L109 (1996).
10. Liang, E. P., and Kargatis, V. E., *Nature* **381**, 495 (1996).

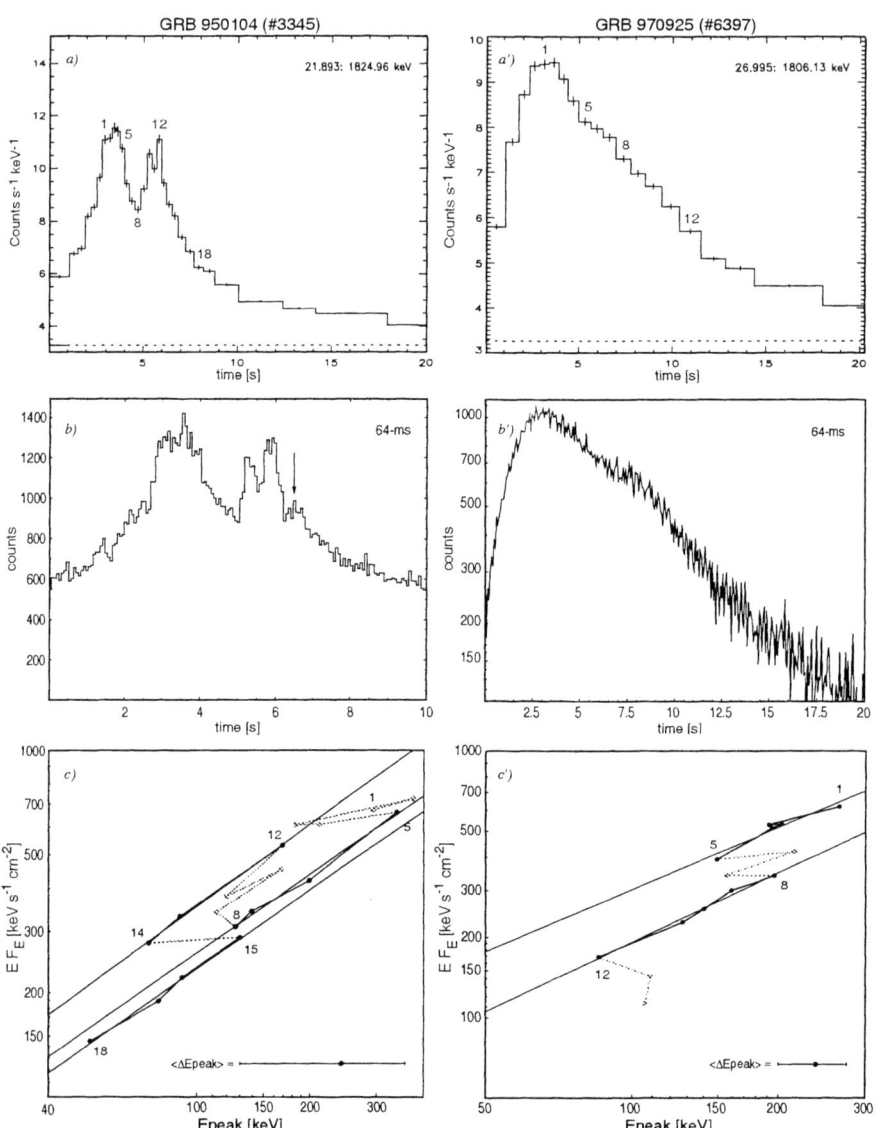

FIGURE 3. Spectral evolution of GRB 950104 (#3345) and GRB 970925 (#6397). *a)* Count time histories showing the binning used to study the spectral evolution ($S/N = 30$) and the background level (dashed line). *b)* The same using 64-ms time resolution data. *c)* The peaks of the instantaneous spectra at successive times are represented by dots. The numbers refer to the time-bins in (*a*). Error bars are excluded for better visibility, and only a typical average value is shown separately. The fitting procedure largely overestimates the errors in these cases. See discussion in the text.

Shape of the Decay Phase of Gamma-Ray Burst Pulses

Felix Ryde[*] and Roland Svensson[*†]

[*]Stockholm Observatory, SE-133 36 Saltsjöbaden, Sweden
[†]Institute for Theoretical Physics, University of California, Santa Barbara, CA 93106, USA

Abstract. We study the decay phase of GRB pulses belonging to the large subsample for which two important correlations between spectral hardness and intensity, and spectral hardness and photon fluence, respectively, are valid. These pulses exhibit a common temporal behavior, i.e., the spectral hardness and intensity of the decay phase of the pulses follow a reciprocal function in time.

INTRODUCTION

The spectral evolution for the decay phase of pulses in the light curve of gamma-ray bursts (GRBs) is described by two well-established empirical relations: i) the hardness-intensity correlation (HIC) [1], and ii) the hardness-photon fluence correlation (HFC) [2]. The HIC relates the instantaneous photon flux, $N(t)$ (in units of photons cm^{-2} s^{-1}), and a measure of the "hardness", e.g., the peak energy $E_{\rm pk}(t)$, of the instantaneous $E^2 N_{\rm E}(E,t)$ spectrum:

$$E_{\rm pk}(t) = E_{\rm pk,0} \left[\frac{N(t)}{N_0}\right]^\delta, \qquad (1)$$

where δ is the correlation index. The HFC relates the spectral hardness and the time-integrated photon flux:

$$E_{\rm pk}(t) = E_{\rm pk,0} e^{-\Phi(t)/\Phi_0}, \qquad (2)$$

where $\Phi(t)$ is the photon fluence $\equiv \int_0^t N(t')\, dt'$ (in units of cm^{-2}) integrated from the time of $E_{\rm pk,0}$, and Φ_0 is the exponential decay constant.

In Ryde & Svensson [3] the combination of these two correlations was studied. Neither the HIC nor the HFC includes any explicit time dependence of the spectral evolution. However, combined they do, as the fluence is the time integral of the flux. This was used to find a compact and quantitative description of the time evolution of the decay phase of a GRB pulse. This gives constraints on the pulse profiles, which should reflect the underlying physical mechanisms.

DESCRIPTION OF THE DECAY PHASE

The combination of the two empirical relations, given by equations (1) and (2), makes the description of the temporal behavior complete. This can be expressed as differential equations either for $E_{\mathrm{pk}}(t)$ as

$$-\frac{dE_{\mathrm{pk}}(t)}{dt} = \frac{N_0}{\Phi_0(E_{\mathrm{pk},0})^{1/\delta}} E_{\mathrm{pk}}^{1+1/\delta}, \qquad (3)$$

or for $\Phi(t)$ as

$$\frac{d\Phi(t)}{dt} = N_0 e^{-\Phi(t)/\delta\Phi_0}, \qquad (4)$$

or for $N(t)$ as

$$-\frac{dN(t)}{dt} = \frac{1}{\delta\Phi_0} N^2(t). \qquad (5)$$

Integrating these results, the time evolution of the peak energy will be given by

$$E_{\mathrm{pk}}(t) = \frac{E_{\mathrm{pk},0}}{\left[1 + \frac{N_0}{\delta\Phi_0}t\right]^\delta}, \qquad (6)$$

the time evolution of the fluence by

$$\Phi(t) = \delta\Phi_0 \ln\left[1 + \frac{N_0}{\delta\Phi_0}t\right], \qquad (7)$$

and finally, the time-evolution of the intensity will be given by

$$N(t) = \frac{N_0}{1 + \frac{N_0}{\delta\Phi_0}t}, \qquad (8)$$

or equivalently by

$$\frac{1}{N(t)} = \frac{1}{N_0} + \frac{t}{\delta\Phi_0}. \qquad (9)$$

Equation (8) is an important result, as it describes how the intensity declines with time in the decay phase of a GRB pulse belonging to the subgroup of pulses for which both the two empirical relations, equations (1) and (2), are valid. The reciprocal, instantaneous intensity is a linear function in time, shown in equation (9). This should be compared to the generally discussed exponential decay, e.g., in the terms of a FRED (fast rise, *exponential* or *stretched exponential* decay) often used to characterize single pulses within GRBs (e.g., Norris et al. [4]). Our result, on the other hand, is an analytical result following directly from the two empirical relations (1) and (2).

As shown by equation (7), the fluence increases logarithmically with time. This divergent behavior must eventually change, when the emission of radiation changes behavior, terminates, or shifts out of the observed spectral range.

The hardness, represented by the peak energy, E_{pk}, also declines reciprocally with time, but it is stretched by the δ-power. A comparison between the intensity decline and the E_{pk} decline is shown, for instance, in Ford et al. [6].

The time evolution is fully defined through four parameters describing the empirical relations, for instance, by N_0, $F_{pk,0}$, Φ_0, and δ.

BATSE SAMPLE

We have studied a large, complete sample of 83 GRB pulses observed by the Burst and Transient Source Experiment (BATSE) on the *Compton Gamma-Ray*

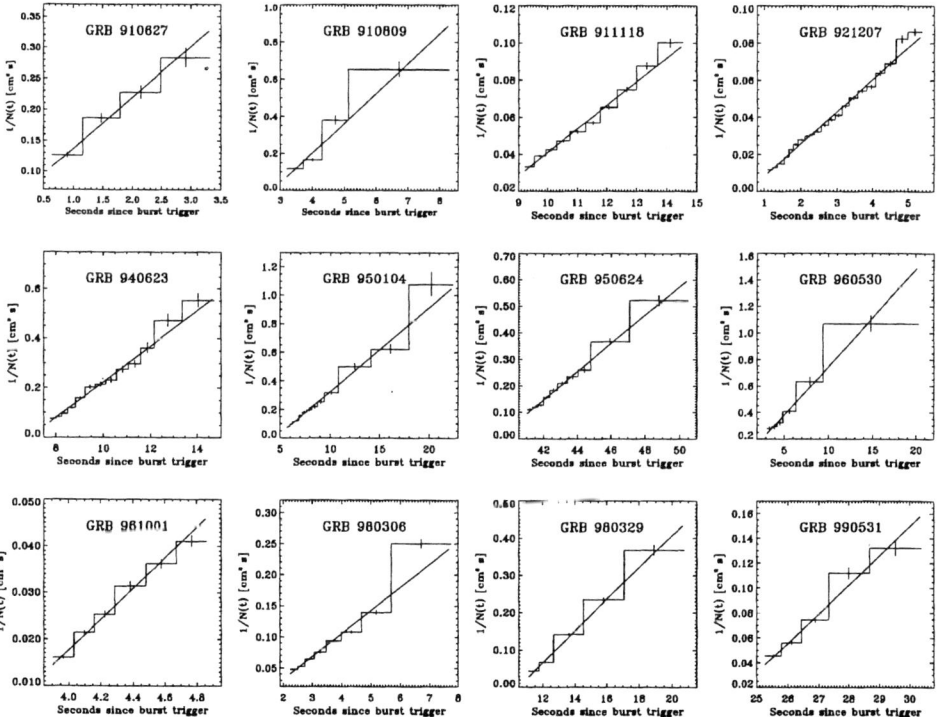

FIGURE 1. Observed background-subtracted photon flux light curves of the decay phases of a few cases in the studied sample. The linear behavior of $1/N(t)$ is shown. See the text for details.

TABLE 1. The decay parameters for a sample of bright BATSE burst pulses. The most prominent pulse in the light curves is studied.

Burst Name	N_0 [cm^{-2} s^{-1}]	$E_{\rm pk,0}$ [keV]	τ [s]	δ	Φ_0[a] [cm^{-2}]	Time[b] [s]
GRB 910627 (#451)	9.3 ± 0.5	440 ± 250	1.3 ± 0.2	1.8 ± 1.2	6.7 ± 3.2	2.69
GRB 910809 (#660)	13.7 ± 0.8	1060 ± 360	0.46 ± 0.05	1.09 ± 0.25	5.6 ± 1.0	5.12
GRB 911118 (#1085)	32.1 ± 0.4	123.2 ± 2.4	2.46 ± 0.07	0.63 ± 0.03	128 ± 7	5.31
GRB 921207 (#2083)	99.5 ± 1.3	474 ± 9	0.58 ± 0.01	0.90 ± 0.02	64.1 ± 1.5	4.29
GRB 940623 (#3042)	16.3 ± 0.4	523 ± 68	0.85 ± 0.04	0.78 ± 0.09	18.4 ± 2.6	6.98
GRB 950104 (#3345)	14.0 ± 0.4	140 ± 18	1.22 ± 0.07	0.41 ± 0.10	36 ± 9	16.77
GRB 950624 (#3648)	10.5 ± 0.2	300 ± 15	1.80 ± 0.08	1.07 ± 0.05	16.9 ± 0.9	9.47
GRB 960530 (#5478)	4.1 ± 0.2	187 ± 14	3.3 ± 0.4	0.86 ± 0.11	15.2 ± 2.3	17.15
GRB 961001 (#5621)	69.1 ± 1.8	440 ± 60	0.44 ± 0.03	1.1 ± 0.2	26 ± 5	0.96
GRB 980306 (#6630)	24.7 ± 0.6	253 ± 8	1.10 ± 0.04	1.01 ± 0.04	27.9 ± 1.2	5.50
GRB 980329 (#6665)	34.3 ± 1.0	240 ± 24	0.68 ± 0.03	0.21 ± 0.09	103 ± 36	9.60
GRB 990531 (#7592)	25.9 ± 0.8	138 ± 7	1.65 ± 0.13	0.69 ± 0.09	62 ± 8	5.12

[a] As fitted from equation (2).
[b] Length of the time interval during which the decay was studied.

Observatory (CGRO), see [3]. We find that the behavior described above is common ($\sim 45\%$). There are also several examples of varying behavior of the HIC within single pulse structures. In Ryde & Svensson [5], other behaviors that do not follow the common description (equations 1, 2) are studied. According to the prescription, a pulse profile that does not show this time behavior will, of necessity, have a different HIC or HFC. The light curve is much better suited for statistical analysis as the noise (counting statistics) is relatively small (compared to the HIC or the HFC).

Figure 1 shows several cases which follow the description. They are presented as $1/N(t)$ to clearly emphasize the reciprocal linear behavior arrived at in equation (9). The data are rebinned to a signal-to-noise-ratio of at least 30. These fits to the data give the parameters N_0 and the time constant $\tau \equiv \delta \Phi_0 / N_0$. Corresponding fits to the time evolution of $E_{\rm pk}(t)$ [equation (6)] give $E_{\rm pk,0}$ and δ. The results of the fits are given in Table 1. Over a time τ, the light curve decreases by a half. For most of the cases, the decay phase persists over more than 5τ with the pulse in GRB 980329 (#6665) covering 14τ. Note that the time variable in Figure 1 is counted from the trigger time of the burst, while the time variable t in the text is counted from the peak of the pulse.

DISCUSSION

We have studied a subgroup of GRB pulses, for which the two empirical relations are valid and show that for these the decay phase of the pulse should be (with the characteristic time scale of the decay, the time constant, $\tau \equiv \delta \Phi_0 / N_0$)

$$N(t) = \frac{N_0}{(1 + t/\tau)}, \tag{10}$$

$$E_{\text{pk}}(t) = \frac{E_{\text{pk},0}}{(1 + t/\tau)^\delta}, \tag{11}$$

where the initial values are $(N_0, E_{\text{pk},0})$ and the number of additional parameters is limited to two: the time constant τ $[N(t = \tau) = N_0/2]$, and the HIC index δ. The peak energy has a similar time dependence as the intensity, differing only by the power law index δ. The exponential decay constant of the HFC is given by $\Phi_0 \equiv N_0\tau/\delta$. The formulation given above is a condensate of the HIC and the HFC, each of which has been proven to be valid in most sufficiently bright GRB-pulses.

ACKNOWLEDGMENTS

This research made use of data obtained through the HEASARC Online Service provided by NASA/GSFC. We are also grateful to the GROSSC for support. We acknowledge support from the Gustaf and Ellen Kobbs Stipend Fund at Stockholm University, the Swedish National Space Board, the Swedish Natural Science Research Council (NFR), and NSF grant PHY94-07194.

REFERENCES

1. Golenetskii, S. V., Mazets, E. P., Aptekar, R.L., and Ilyinskii, V. N., *Nature* **306**, 451 (1983).
2. Liang, E. P., and Kargatis, V. E., *Nature* **381**, 495 (1996).
3. Ryde F., and Svensson R., *ApJ* **529**, L13 (2000).
4. Norris, J. P., Nemiroff, R. J., Bonnell, J. T., Scargle, J. D., Kouveliotou, C., Paciesas, W. S., Meegan, C. A., and Fishman, G. J., *ApJ* **459**, 393 (1996).
5. Ryde, F., and Svensson, R., *ApJ*, in preparation (2000).
6. Ford, L. A. et al., *ApJ* **439**, 307 (1995).

Average Spectral Parameters of High-Power Emission in BATSE Gamma-Ray Bursts

Dmitrij S. Anfimov[1], Igor G. Mitrofanov[1], Maxim L. Litvak[1], Anton L. Sanin[1], Michael S. Briggs[2], William S. Paciesas[2], Geoffrey N. Pendleton[2], Robert D. Preece[2], and Charles A. Meegan[3]

[1] *Space Research Institute, Profsojuznaya str. 84/32, 117810 Moscow, Russia*
[2] *Department of Physics, University of Alabama in Huntsville, Huntsville, AL 35899*
[3] *NASA/Marshall Space Flight Center, Huntsville, AL 35812*

Abstract. Photon spectra are investigated that were accumulated during the times of high emission in gamma-ray bursts. The corresponding spectral parameters are obtained for the sample of bright bursts in the BATSE 4B catalog with peak flux more than 2 phot/(cm^2 s) on the 64 ms time scale. The difference in spectra corresponding to the short and long classes of bursts is investigated. The evolution of spectral parameters is studied along successive time intervals of high-power emission and around the moment of primary peaks.

INTRODUCTION

GRBs are known to have a total duration from milliseconds up to thousands of seconds. We may attribute this time scale to the total duration of the outbursting stage of some source. The well-known time parameters T_{50} and T_{90} represent the measure of this duration. A burst usually has very randomized time profile, and the radiated power is distributed over one or more pulses that are often separated by low-emission interpulses. We have suggested [1-3] defining a complementary temporal parameter, called the emission time τ_{50}, which represents the total duration of high-power emission intervals, during which a pre-defined fraction of the total fluence (e.g., 50%) has been produced.

Selection of time intervals of high-power emission defines the total of those moments when a source has generated gamma rays with the highest power. This approach allows us to implement a number of observational parameters that might work as very generic signatures of burst emission. Indeed, the sum of these intervals creates the emission time parameter [1-3]. The sum of photon energy spectra from the emission time intervals describes the physical gamma-ray generation mechanism. Finally, the total fluence divided by the emission time corresponds to the average power of emission [4]. This article is devoted to describing general properties of spectral parameters during high-power emission of bright BATSE gamma-ray bursts.

ENERGY SPECTRA OF HIGH-POWER EMISSION OF GAMMA-RAY BURSTS

Deconvolution of count spectra into photon energy spectra is based on the 'Band model' photon spectrum [5], which describes the spectrum by a low-energy power index α, a high-energy power index β and the peak energy E_p, which is defined as the peak of the spectrum in the $\nu \cdot F_\nu$ representation. The β parameter is not usually well-determined by the deconvolution procedure, so we will use the α and E_p parameters for describing the spectral signatures of high-power emission of GRBs.

In previous studies, the time-averaged spectra have been integrated over an burst's entire time profile. In this case, a one-pulse burst would be represented by the spectrum of the pulse, while a multi-pulse event would be represented by the mixture of pulse spectra and the broad inter-pulse valleys between them. Therefore, the integral spectra of different bursts could represent different stages of the emission process. Another approach uses the spectrum at the main peak of a burst as the spectral signature of an individual event. However, the duration of a peak spectrum is determined by the fixed time resolution of the instrument, either 64 or 2048 ms, without any adjustment to the actual time history of a burst. In this case, a burst with a broad primary pulse would be represented by a spectrum at the peak, while another burst with a narrow primary pulse would be represented by a spectrum of the entire pulse added with interpulse wings around it.

The spectrum of high-power emission manifests the energy distribution of photons emitted at a high power level, when 50% of the total fluence has been produced. We believe that the high-power emission spectra can be used for studying the generic spectral signature of emitters of the main sample of many gamma-ray bursts at similar physical conditions; i.e., the conditions under which they produce the highest power emission.

We use 16 energy channel BATSE MER and CONT data, accumulated with 16 ms and 2048 ms time resolution, respectively. The total sample of these bursts from the 4B catalog, with complete sets of usable data, contains 312 events. We calculate emission time values for these bursts based upon time profiles of counts in the broad spectral range 50–300 keV.

The scatter-plot distribution for these events of the emission time versus spectral peak energy is presented in Figure 1. A signature of bimodality in the emission time is well seen as a gap between the left-hand cloud of short bursts and right-hand cloud of long events. When presented as a one-parameter distribution, the emission time shows the bi-modality as two peaks [2]. The dashed line indicates a boundary between the two classes of long (LB) and short bursts (SB). There are 240 LB and

72 SB in our data base. Short and long classes of GRBs are different not only in τ_{50}. The spectral signatures of the high-power emission are also different for the short and long classes. The SB class has larger values in the spectral parameter E_p than the LB class. The distributions of E_p values for LB and SB are shown in Figure 2.

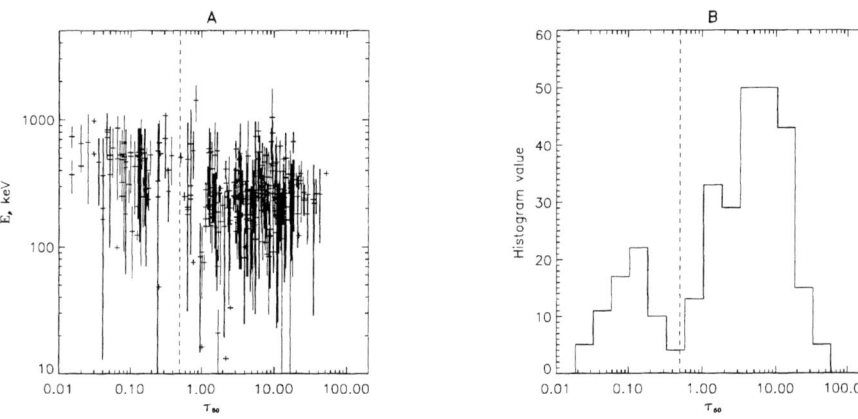

FIGURE 1. A: τ_{50} vs. E_p for all bursts. B: The distribution of τ_{50}.

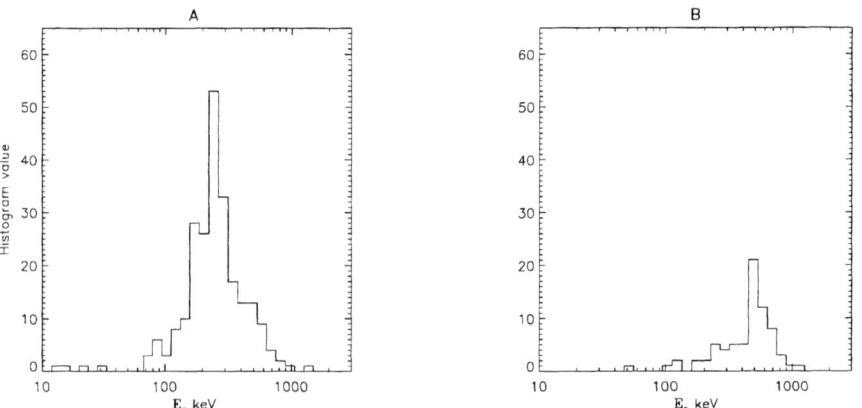

FIGURE 2. A: E_p distribution for events with $\tau_{50} > 0.6$ s. B: E_p distribution for events with $\tau_{50} < 0.6$ s.

SPECTRAL EVOLUTION OF HIGH-POWER EMISSION OF LONG BURSTS AROUND THE PRIMARY PEAKS

One may explain the difference in the emission spectra between long and short bursts as resulting from the difference of their time histories: for long events we integrate many peaks with high-power emission, but for short events we take into account the single peak only. One could suspect that primary peaks of long multi-peak events have the same spectral properties as single peaks of short events, and that the observed softness of E_p for the LB class (Figure 1) results from the dominant contribution of minor peaks of long bursts.

To check this possibility, we can compare the average spectral parameters of the SB class with the average spectral parameters for time intervals of LB events selected by the criteria of high-power emission around the primary peak of each of these events. Among bursts of the LB class, we selected events with a total emission time longer than 6.0 s. We divided the total emission time of these bursts into 8 time intervals around the primary peak, corresponding to the following intervals of high-power emission: $\pm[3.0; 1.4]$ s, $\pm[1.4; 0.6]$ s, $\pm[0.6; 0.2]$ s and $\pm[0.2; 0.0]$ s. The average value of the spectral parameters E_p and α for these intervals are presented in Figure 3. We see peaks in both the α and E_p average spectral parameters at the moment $t = 0.0$ s relative to primary peaks in the time histories. We next compare these average values with similar average spectral parameters for the SB class. There is a significant difference: for the SB class, the average spectral parameters E_p and α are larger. Average values for β are between -2.5 and -2.3, and do not show a difference between LB and SB. We compare the average spectral parameters for bursts with $\tau_{50} > 6.0$ s with similar parameters for bursts with $\tau_{50} > 1.4$ s (see Figure 3). There is no significant difference of high-power emission around the

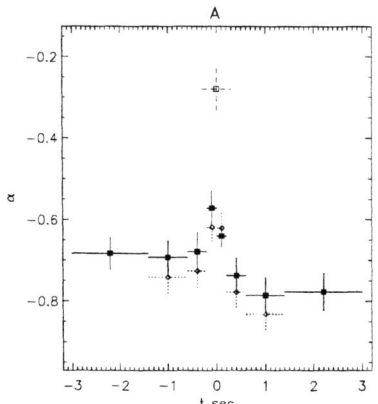
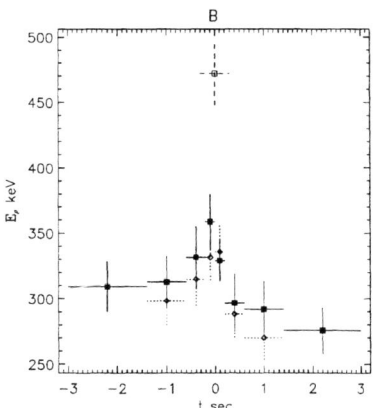

FIGURE 3. A: α vs. t (s) around burst peak emission for events with $\tau_{50} > 3.0$ s (*solid*), $\tau_{50} > 1.4$ s (*dotted*), $\tau_{50} < 0.6$ s (*dashed*). B: The same as in A, for E_p vs. t (s).

primary peaks for bursts of LB class with longer and shorter emission time.

At the moment of highest peak emission, some physical transition from the early stage of burst development into the late stage of relaxation may take place [6]. In this case, we expect large changes of the average spectral parameters around this moment. The time-irreversible evolution of the average spectral parameters is quite observable for emission time intervals on both sides of the moment of primary peaks (Figure 3). We confirm that main peaks are physically-selected points of gamma-ray bursts [6].

CONCLUSIONS

The photon spectra of high-power emission of GRBs has been suggested as an effective tool for GRB investigations. It allows us to build instrument-independent spectral parameters for each burst that describe the moments of high-power emission. Using the emission time as a temporal measure of bursts, and the spectral peak energy of high-power emission as a spectral measure, we have confirmed the previously-known effect that short GRBs are systematically harder than long GRBs. The effect was first seen in the distribution of hardness ratio of GRBs vs. T_{50} by Belli [7].

We found that the average spectral parameters during the highest power emission vary around the moments of primary peaks. When emission goes through the main peak to the later portion of the emission, it becomes softer. The high-power emission of GRBs is thus a time-irreversible process.

ACKNOWLEDGEMENTS

The work in USA was supported by NASA grant CRO-96-173. The work in Russia was supported by RFBR grant 98-02-17380 and by INTAS project 96-0315.

REFERENCES

1. Litvak, M. L., et al., these proceedings.
2. Mitrofanov, I. G., et al., these proceedings.
3. Mitrofanov, I. G., *ApJ* **522**, 1069 (1999).
4. Anfimov, D. S., these proceedings.
5. Band, D. L., et al., *ApJ* **413**, 281 (1993).
6. Litvak, M. L., et al., these proceedings.
7. Belli, B. M., in *Gamma-Ray Bursts, 4th Huntsville Symposium*, eds. Meegan, Preece & Koshut, AIP 428, New York, 1997, pp. 82.

Cosmological Signatures in Intensity and Break Energy Correlations

Nicole M. Lloyd[1], Vahé Petrosian[1], and Robert S. Mallozzi[2]

(1) Stanford University, Stanford, California 94305
(2) University of Alabama in Huntsville, Huntsville, AL 35899

Abstract. We examine the association between the peak of the νF_ν spectrum of gamma ray bursts, E_p, and the burst's energy fluence. Because bursts near the detector threshold are not usually able to provide reliable spectral parameters, we analyze the results for the brightest bursts in which we can better understand the selection effects relevant to E_p and burst strength. We find that there is a strong correlation between total fluence and E_p. We discuss these results in terms of both cosmological and intrinsic effects. In particular, we show that for realistic distributions of the burst parameters, cosmological expansion alone cannot account for the correlation between E_p and total fluence; the observed correlation is likely a result of an intrinsic relation between the burst rest-frame peak energy and the total radiated energy, as expected in an internal shock model.

INTRODUCTION

The measured redshifts to 8 GRBs have helped provide insight into the energetics of bursts, but they have not helped us learn much of the cosmological distribution of GRBs as a whole; the "luminosity functions" of these bursts are broad, and mask any cosmological signature. Another useful way to gain insight into the distribution and energetics of GRBs is to study correlations between various burst spectral properties. For example, a burst that is further away will be weaker (assuming little luminosity evolution) and its spectrum redshifted; thus one would expect a positive correlation between, say, the peak of the νF_ν spectrum, E_p, and fluence if cosmological effects were dominant. Intrinsic effects could either wash out or intensify such a correlation.

However, the degree of the correlation can be affected by various detector selection biases. Many low intensity GRBs near the detection threshold will not provide adequate spectral fits. This then will introduce a bias against bursts with peak counts near the threshold. Similarly, there may exist biases against high and low values of E_p. In this paper, we examine the correlation between GRB total fluence, F_{tot}, and E_p, accounting for such biases. Our main goal is to determine to what extent the correlation is due to cosmological redshift and to what extent it is a signature of an intrinsic relation between radiated power and spectral features.

DATA

Our sample consisted of a set of bursts from the 4B catalog [5] which have 16 channel CONT (continuous) data. Using these data, the bursts are fit to a Band Spectrum [1]. The spectrum is parameterized by a low energy photon index, α, a high energy photon index, β, the peak of the νF_ν spectrum, E_p, and a normalization, A (ph/cm^2/s/keV). Because the brightest bursts give the best fits, most bursts which have spectral fits have $C_{max}/C_{min} \gg 1$ (where C_{max} is the maximum counts in the detector trigger energy range and C_{min} is the threshold). Getting a spectral sample complete in C_{max}/C_{min} is essential in understanding the selection effects against E_p and burst strength (which can significantly alter the correlation results). To circumvent this situation, we truncate the available data at 5×10^{-5} erg/cm^2 (that is, we only analyze bursts above this threshold). This allows us to get a complete sample in terms of C_{max}/C_{min}, without reducing our spectral sample to an unreasonably small size. For further discussion of the truncation process and selection effects against E_p and burst strength, see Lloyd and Petrosian [3].

RESULTS

Figure 1 shows the average E_p versus binned values of F_{tot}, for the whole sample without any consideration of completeness or detector bias (dotted histogram), as well as our more limited but complete sub-sample (solid histogram) which we analyze in this paper. There is a significant correlation between total fluence and E_p for the whole sample and sub-sample. However, the functional dependence the correlation is much stronger for high (solid histogram) rather than low values of the fluence. Again, in what follows we shall use the truncated sub-sample (bursts above 5×10^{-5} erg/cm^2) to search for cosmological signatures; the flatter functional form at low F_{tot} may be due to selection effects.

COSMOLOGICAL?

A positive correlation between an observed burst strength and E_p follows the trend expected from cosmological effects; bursts which are further away have lower intensity as well as a lower (redshifted) value of E_p. We now test to see if the observed correlations between F_{tot} and E_p can be attributed *fully* to these effects. Total fluence can be related to the total radiated energy and the redshift of the burst, without any need for the so-called K-correction or correction for duration bias: $F_{tot} = \mathcal{E}_{\rm rad}/(\Omega_b[d_\mathcal{E}(\Omega_i, z)]^2)$. $\mathcal{E}_{\rm rad}$ is the total radiated energy (in the gamma-ray range), Ω_b is the beaming solid angle, and $d_\mathcal{E}(\Omega_i, z)$ is a modified luminosity

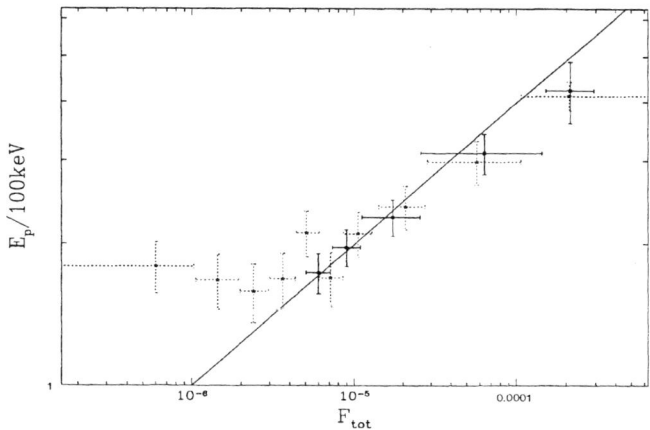

FIGURE 1. Average peak energy, $\overline{E_p}$ vs. total fluence, F_{tot} for the whole (dashed histogram) and more complete sub (solid histogram) spectral sample. The solid line is a least squares fit to the data, $\log(E_p) = 0.28 \pm 0.04 \, \log(F_{tot})$, which is in agreement with the functional form of the correlation we obtain from our statistical methods.

distance dependent on redshift and cosmological parameters, Ω_i. We also define E_{po} as the value of the peak of the νF_ν spectrum in the rest frame of the burst, so that the observed $E_p = E_{po}/(1+z)$.

Assuming isotropic emission and that the intrinsic parameters $\mathcal{E}_{\rm rad}$ and E_{po} are uncorrelated and do not evolve with cosmic time, the distribution function Ψ of redshift and intrinsic burst parameters becomes separable: $\Psi(E_{po}, \Omega_b, \mathcal{E}_{\rm rad}, z) = \phi(\mathcal{E}_{\rm rad})\zeta(E_{po})\rho(z)$. Changing variables from $\mathcal{E}_{\rm rad}$ to F_{tot} and E_{po} to E_p, and given a cosmological model and functional forms for the distributions of intrinsic burst parameters, we can then compute average value of E_p as a function of F_{tot} (e.g., see [2]).

To see if the observed correlation is purely cosmological in origin, we carry out the following tests. We assume various plausible models for the functions ζ, ϕ, and ρ, and compute the expected cosmological relation $\overline{E_p(F_{tot})}$. We then remove this correlation from the data by the transformation $F'_p = E_p/\overline{E_p(F_{tot})}$, and see if we are left with any correlation between the observed F_{tot} and E'_p distributions. Only if none remains (i.e., $|\tau| < 1$, where τ indicates the significance of the correlation between F_{tot} and E_p), can we attribute the correlation between F_{tot} and F_p to cosmological effects alone.

1) For a *standard candle radiated energy*, $\mathcal{E}_*$, i.e. $\phi = \delta(\mathcal{E}_{\rm rad} - \mathcal{E}_*)$, and a Gaussian in E_{po} with mean Q and dispersion σ_Q, we find that the correlation *can* be removed ($|\tau| < 1$) for $\mathcal{E}_* = (5 \pm 2) \times 10^{52}$ ergs. This is evident in Figure 2a, where we plot τ as a function of $\mathcal{E}_*$. We note that this result is independent of the rate evolution ρ, and is fairly insensitive to the value of Q and σ_Q of the E_{po} distribution. However, we know from the bursts with measured redshifts that a delta function in burst

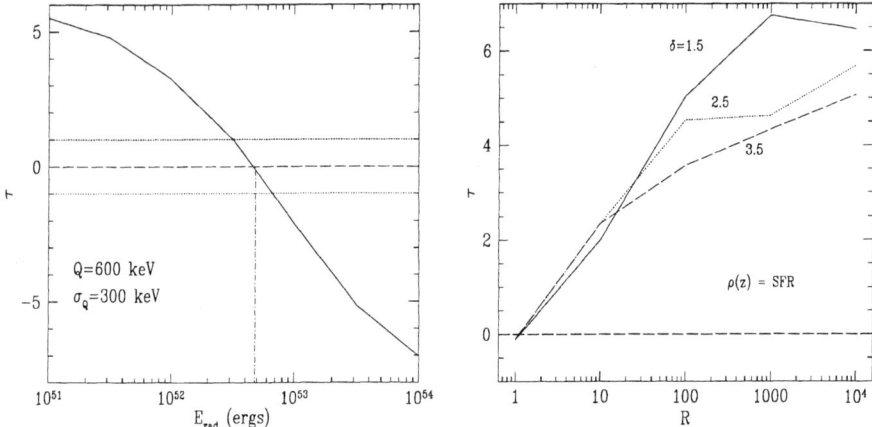

FIGURE 2. (a) **Left Panel:** Significance of the correlation, τ, vs. radiated energy, $\mathcal{E}_{\rm rad}$, for the model given in case 1 (standard candle energy). Here, we find that the correlation is removed for a standard candle energy of 5×10^{52} erg. (b) **Right Panel:** Significance of the correlation, τ, vs. $R \equiv \mathcal{E}_{\rm max}/\mathcal{E}_{\rm min}$, for the model in case 2 (power law luminosity function) in the text. The curves are shown for three different values for the luminosity function index β, and for a density following the star formation rate.

radiated energy is not a good approximation. Hence we try a more realistic model for the distribution of burst parameters.

2) We can choose a *power law distribution in the radiated energy*, $\phi \propto \mathcal{E}_{\rm rad}^{-\delta}$, with upper and lower cutoff $\mathcal{E}_{\rm max}$ and $\mathcal{E}_{\rm min}$. Figure 2b plots τ vs. $R \equiv \mathcal{E}_{\rm max}/\mathcal{E}_{\rm min}$ for a density following the star formation rate (our results are qualitatively similar for a constant density). Here we have chosen $\sqrt{\mathcal{E}_{\rm min}\mathcal{E}_{\rm max}} = 5 \times 10^{52}$ erg. In general, for a broader, more realistic "luminosity function", the correlation due cosmological expansion is smeared out and becomes too weak to explain the strong observed correlation. The natural interpretation in this case is that the observed correlation between F_{tot} and E_p is either caused by the cosmological evolution of one or more of the parameters or is due to an intrinsic correlation between E_{po} and $\mathcal{E}_{\rm rad}$.

INTRINSIC?

One possible explanation is that the observed correlation is explained by evolutionary effects. A discussion of this is found in [2]. We believe that a more likely possibility is an intrinsic correlation between E_{po} and $\mathcal{E}_{rad}$. For a model as in case 2 where $\phi \propto \mathcal{E}_{\rm rad}^{-\delta}$, $\mathcal{E}_{\rm min}/\mathcal{E}_{\rm max} = 10^{2.5}$ (a somewhat broad luminosity function, within the range of the dispersion evident in bursts with measured redshifts), $\delta = 2.5$, and ρ = SFR, we find that the observed correlation can be explained if the mean of the E_{po} distribution, Q, is related to the radiated energy as $Q \propto \mathcal{E}_{\rm rad}^{0.47 \pm 0.08}$ (similar results are found for constant rate density). These results can be used to test models of GRB emission, most of which predict some sort of relation between these quantities. A discussion of this is given in [2]. We find that the internal shock model can explain an intrinsic correlation more naturally.

CONCLUSIONS

We find significant correlations between the burst fluence and E_p. For brighter bursts, the data are well understood and the correlation can be tested against models of its origin. For a plausible, broad GRB luminosity function (e.g., a power law distribution $\phi(\mathcal{E}_{\rm rad}) \propto \mathcal{E}_{\rm rad}^{-\delta}$), we find that neither for a constant co-moving rate density nor for rates proportional to the star formation rate, can the correlation be solely attributed to cosmological expansion. The expected correlation is essentially "washed out" by the broad luminosity function. We conclude that the observed correlation between F_{tot} and E_p must be primarily due to an intrinsic correlation between the total radiated energy, $\mathcal{E}_{\rm rad}$, and the rest frame peak energy, E_{po}. This kind of correlation seems to be a natural consequence of optically thin synchrotron emission by a power law distribution of electrons with Lorentz factors greater than some minimum cutoff, γ_m. Using the results from a detailed modeling of the emission in an external or internal shock model, we find that the internal shock model can explain the strong intrinsic correlation more simply.

REFERENCES

1. Band, D., et al., *ApJ* **413**, 281 (1993).
2. Lloyd, N.M. et al., to appear in *ApJ* (2000); astro-ph 9908191.
3. Lloyd, N.M. & Petrosian, V., *ApJ* **511**, 550 (1999).
4. Mallozzi, R.S., et al., *ApJ* **454**, 597 (1995).
5. Paciesas, W.S. et al., *ApJS* **122**, 465 (1999).

The Role of the BATSE Instrument Response in Creating the GRB E-Peak Distribution

J. J. Brainerd, G. N. Pendleton, R. S. Mallozzi,
M. S. Briggs, and R. D. Preece

University of Alabama in Huntsville, Huntsville, AL 35899

Abstract. Of the properties possessed by gamma-ray bursts, the most constraining on the physics of the emission region is the narrow distribution of E-peak (E_p), the photon energy at which the νF_ν curve is a maximum. The importance of this observation for gamma-ray burst theory has motivated us to study the influence of BATSE instrumental effects on the observed distribution. Two aspects of measuring the E-peak distribution were investigated: the derivation of the value of E-peak for a given burst, and the triggering of the BATSE instrument on bursts of different E-peak. We find that instrumental effects cannot account for the observed distribution. We use a log-normal E-peak distribution to model the observations, finding the number of bursts per unit decade must decrease rapidly as one moves away from the observed peak of the E-peak distribution for burst spectra that are normalized to a constant total flux.

INTRODUCTION

The distribution of E-peak values for bursts observed with the BATSE instrument is narrowly distributed. The question is whether the distribution is intrinsic to the gamma-ray bursts, or whether it is produced by instrumental effects. We examine the second hypothesis in this article. We find that the observed narrowness is intrinsic to the gamma-ray bursts. A more detailed discussion of our work has been submitted to the Astrophysical Journal [2].

ANALYSIS

Instrumental effects come into play in two ways. First, the character of the detector response and the finite number of counts in a spectrum produce an error in the derived value of E_p. Second, the energy range over which the instrument can observe a burst introduces a dependence on E_p into the efficiency of burst detection. We show below that these effects do not produce the observed distribution.

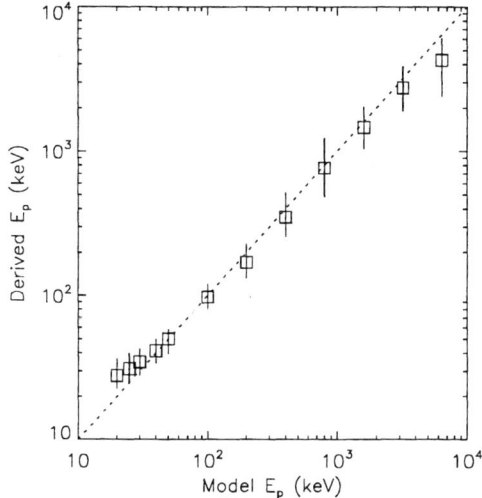

FIGURE 1. Derived E_p versus true E_p for $\alpha = -0.5$ and $\beta = -2.5$. The average E_p of fits to model spectra are given as squares. The solid lines give the positive and negative standard deviations away from E_p.

We first examine the accuracy with which E_p is determined with BATSE. It has already been shown that the detector response matrices are accurate [3]. The goal of this analysis is to determine whether the deconvolution technique of forward fitting introduces a systematic effect.

We first generate a model count spectrum with statistical fluctuations by folding a model photon spectrum through the detector response matrix, adding a background model based upon the observed background, and randomly choosing counts from the resulting count distribution function. We use the Gamma-Ray Burst Spectral Form [1] for the model photon spectrum in this study. It is given by

$$N_E = \begin{cases} N_p \left(\frac{E}{E_p}\right)^\alpha e^{-(2+\alpha)\left(\frac{E}{E_p}-1\right)}, & \text{for } E < \frac{\alpha-\beta}{2+\alpha} E_p; \\ N_p \left(\frac{\alpha-\beta}{2+\alpha}\right)^{\alpha-\beta} e^{2+\beta} \left(\frac{E}{E_p}\right)^\beta, & \text{otherwise.} \end{cases} \quad (1)$$

One hundred model count spectra are produced for each value of E_p in which we are interested. A value of E_p is then calculated for each model spectrum by forward fitting equation (1) to each, employing χ^2 minimization, and allowing the four free-parameters to minimize to their best values. For some spectra, no model can be fit. These bursts are dropped from the ensemble.

Results of this analysis are shown in Figures 1 and 2 for two different spectral models. The value of N_p is adjusted with E_p to make the photon flux at 100 keV independent of E_p. The value of N_p is chosen so that the count rate is twice the trigger threshold. From these figures, one sees that the value of E_p is correctly derived between 50 keV and 2 MeV. For bursts with a large value of α, so that

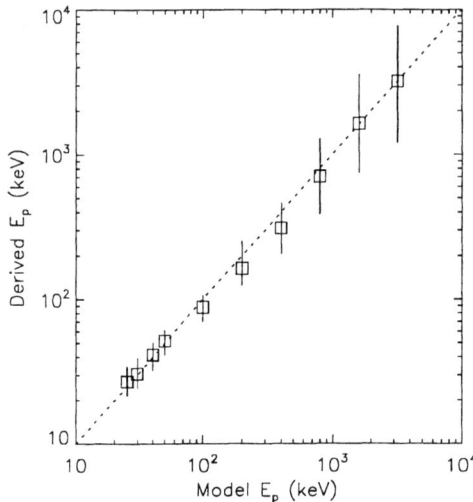

FIGURE 2. Derived E_p versus true E_p. For $\alpha = -1.0$ and $\beta = -2.5$. Same as Fig. 1.

the break in the spectrum is strong, the variation in E_p is relatively narrow. For a smaller value of α, the errors become larger, but remain centered about the true value. We therefore conclude that the value of E_p is not systematically skewed to a characteristic value.

To examine the dependence of the burst trigger on E_p, we simulate the triggering of the BATSE instrument on bursts with a distribution of E_p values. In the simulation, model burst spectra are passed through all eight BATSE detectors for randomly chosen burst positions and CGRO orientations relative to the earth. The number of bursts observed above a given flux limit in the trigger energy range is then determined as a function of E_p.

To derive an expected E_p distribution, both the intrinsic E_p distribution and the peak-flux distribution must be modeled. Two intrinsic E_p distributions are examined. These are a power law, given by

$$K(E_p) = K_c \left(\frac{E_p}{250\,\text{keV}}\right)^k, \qquad (2)$$

and a log-normal with a high-end power law, given by

$$K(E_p) = \begin{cases} K_0 \, e^{-\frac{1}{2}w \ln^2\left(\frac{E_p}{E_k}\right)}, & \text{for } E_p < E_k \, e^{\frac{k_h}{w}}; \\ K_0 \, e^{\frac{k_h^2}{2w}} \left(\frac{E_p}{E_k}\right)^{-k_h}, & \text{otherwise.} \end{cases} \qquad (3)$$

These distributions are assumed to be independent of the burst luminosity. The peak-flux distribution is assumed to be a broken power-law with an index of -1.8

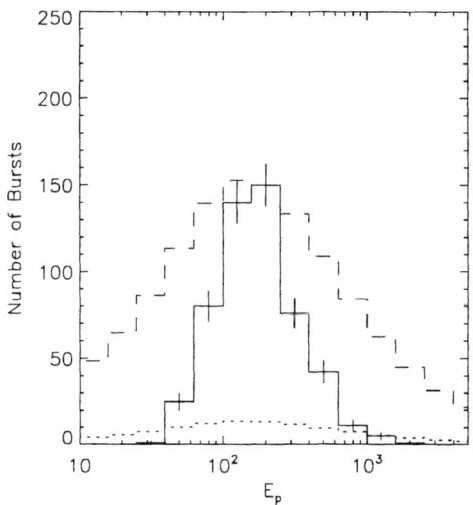

FIGURE 3. Model fit for a power law distribution of E_p (eq. 2) to the observed E_p distribution. The spectral model has $\alpha = -0.5$ and $\delta = -2.5$. The solid line gives the data, the dotted line the model derived through chi-square minimization, and the dashed line gives the best fit model renormalized to the peak of the observed distribution.

for low peak-flux and an index of -2.5 at high peak-flux. The burst spectra are normalized to give a total power that is independent of E_p for bursts at the break in the peak-flux distribution.

Figure 3 shows the best fit to the BATSE data of the simulated triggering of bursts with the E_p distribution given by equation (2). The threshold is set so that most bursts are on the -1.8 portion of the peak-flux distribution. The data are the subset of bursts with E_p determined to better than 25%. The best fit gives a power law index of $k = -0.77 \pm 0.12$. The fit has $\chi^2 = 438.4$ for 8 degrees of freedom. This model is clearly a poor fit to the data.

Figure 4 shows the best fit produced using equation (3) as the E_p models distribution. The model is fit to the data used in Figure 3. The fit has $\chi^2 = 7.8$ for 6 degrees of freedom; the probability of χ^2 being larger than this is $\approx 25\%$, so the model provides a good fit to the data.

The derived value w is $\approx 2.77 \pm 0.19$, which implies the e^{-1} points occur at a factor of 2.5 above and below the energy at which the peak occurs. The value of k_h is 2.28 ± 0.21. The peak of the distribution in E_p is at 124 ± 5 keV, which places the maximum in $\log E_p$ at 200 keV. The best fit intrinsic model is therefore a narrow distribution centered about 200 keV in $\log E_p$. For these values, the number of burst per unit decade falls rapidly as one moves away from 200 keV.

CONCLUSIONS

We find that accurate values of E-peak are derived when applying forward-fitting to the burst spectra of BATSE for values between 50 keV and 2 MeV. We also find

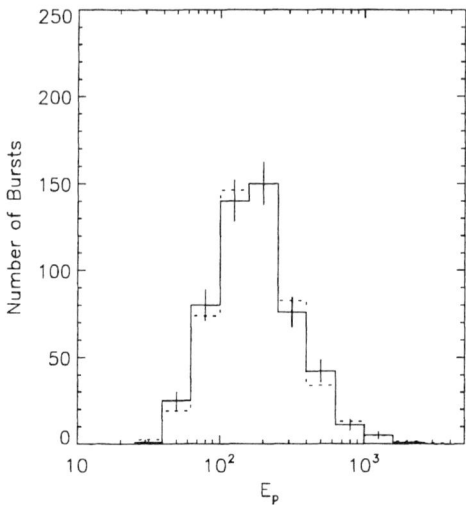

FIGURE 4. Model fit for a log-normal power-law-tail distribution of E_p (eq. 3) to the observed E_p distribution. Similar to Fig. 3.

that the dependence of the detector efficiency on E_p cannot produce the observed E_p distribution. The narrowness of the observed E_p distribution is therefore a consequence of the physics within the gamma-ray burst.

A log-normal distribution for E_p with a high-end power law provides a good fit to the observed distribution. A fit of this model to the observed distribution gives a distribution with a maximum in $\ln E_p$ at 200 keV. The half-maxima of this distribution are at approximately 80 keV and 500 keV. The number of bursts per unit decade falls on either side of the distribution maximum, so most gamma-ray bursts have E_p values that are within the BATSE energy range.

REFERENCES

1. Band, D. et al., *ApJ* **413**, 281–292 (1993).
2. Brainerd, J. J. et al., *ApJ*, submitted (1999).
3. Pendleton, G. N. et al., *Nuclear Instrum. & Methods in Phys. Res.* **364**, 567 (1995).

Synchrotron Emission as the Source of GRB Spectra, Part II: Observations

Nicole M. Lloyd[1], Vahé Petrosian[1], and Robert D. Preece[2]

[1] *Stanford University, Stanford, California 94305*
[2] *University of Alabama in Huntsville, Huntsville, AL 35899*

Abstract. We test the models of synchrotron emission presented in Part I of this series (Lloyd & Petrosian, these proceedings [8]) against the distributions and evolution of GRB spectral parameters (particularly the low energy index, α). With knowledge of the E_p distribution and the correlation between α and E_p presented in [8], we show how to derive the expected distribution of α from fits to optically thin synchrotron spectra, and compare this with the observed distribution. We show that there is no difficulty explaining bursts below the "line of death", $\alpha < -2/3$, and that these bursts indicate that the spectrum of accelerated electrons must flatten or decline at low energies. Bursts with low energy spectral indices that fall above this limit are explained by the synchrotron self-absorption frequency entering the lower end of the BATSE window. Finally, we discuss a variety of spectral evolution behavior seen in GRBs and explain this behavior in the context of synchrotron emission from internal shocks.

INTRODUCTION

The high and low energy spectral indices of GRB spectra contain important information about GRB physics. The high energy photon index, β, usually reflects the steepness of the underlying particle energy distribution, in any non-thermal emission model. The asymptotic value of the low energy photon index, α, however, varies from model to model and can distinguish between different scenarios for GRB emission. The peak and dispersion in the distribution of the index α are difficult to explain in the usual simple synchrotron model (SSM - optically thin synchrotron emission from a power law distribution of electrons with some minimum cutoff). Preece et al. [13] point out flaws in the SSM primarily based on a significant fraction of bursts above the "line of death" value of $-2/3$. Others [7,3] have pointed out that the evolution of α throughout the time history of the GRB is difficult to explain by synchrotron emission in simple GRB emission models (i.e. external shock models). Indeed all of these phenomena must be explained by any GRB emission mechanism.

In this paper, we show that synchrotron emission can accommodate both the distribution and temporal evolution of GRB spectral parameters. We focus particularly on the low energy spectral index, α, because - again - synchrotron models make definite predictions about the value of α. As shown in [8], there is a strong correlation between the value of E_p and the value of the "asymptote", α, as determined by a Band function [1] fit to the data from BATSE, limited to 25 keV to about 1.5 MeV. We can use this relationship and knowledge of the E_p distribution to determine the resultant α distribution. Finally, we give examples of spectral evolution in GRBs and show how this is consistent with synchrotron emission from internal shocks.

THE α DISTRIBUTION

As shown in [8], there exists a relationship between the values of α and E_p obtained from spectral fits - the lower the value of E_p (i.e. as at moves closer toward the low energy edge of the BATSE window), the lower (softer) the value of α. Given the mean and dispersion of the observed E_p distribution [14], we can test if the peak and dispersion in the observed α distribution can be attributed to this correlation. We approximate the correlation between α and E_p by a simple analytical function: $\log(E_p) = h(\alpha)$ (the function $h(\alpha)$ depends on the specifics of the synchrotron model; see Figure 2 in [8]). We then approximate the E_p distribution, $f(\log(E_p))$ by a Gaussian in $\log(E_p)$, with a mean and dispersion representative of the observed distribution. [It is important to point out that there has been considerable controversy over whether the observed distribution of E_p is real or suffers from selection bias [9,2]. In the past, we have estimated the selection bias in E_p without accounting for the non-diagonality of the detector response matrices (DRMs), which allow for photons from higher energies (outside the BATSE band) to scatter to low energies (into the BATSE band). The DRMs reduce the selection bias, but it is still not clear to what degree since there does not exist a complete sample (in terms of brightness) of bursts with spectral fits (see, e.g., [10]). As a result, we use the most conservative form of the E_p distribution in our analysis - the raw observed BATSE distribution.] The distribution of α is then obtained from the relation

$$g(\alpha) = f(\log(E_p)) \frac{dh(\alpha)}{d\alpha}. \qquad (1)$$

Figure 1 compares the resultant α distributions for a sharp ($q = \infty$, right solid curve), intermediate ($q = 2$, middle short-dashed curve), and flat ($q = 0$, left long-dashed curve) cutoff to the electron distribution with the observed distribution obtained from fits to the BATSE data using the Band spectral form. Several conclusions can be reached from this comparison. First, given a distribution in q, an instantaneous optically thin spectrum can easily accommodate bursts with $\alpha < -2/3$ or below the line of death, where most of the bursts are located. The second conclusion is that the electron energy distribution below the turnover energy E_* must be falling off, or at least flat ($q \leq 0$). Otherwise, the SSM would predict too many bursts with α less than about -1.5. This restriction will become stronger for a more realistic and broader distribution of E_p.

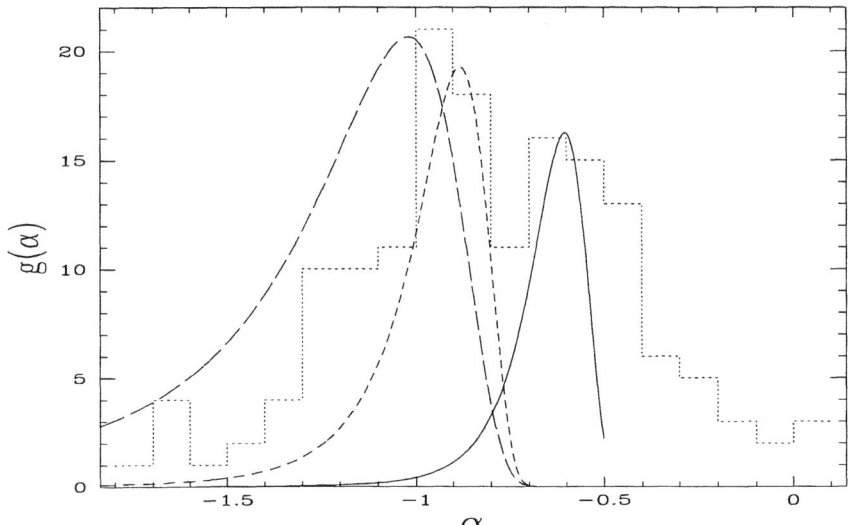

FIGURE 1. Simulated distributions of α for a sharp ($q = \infty$, right solid curve), intermediate ($q = 2$, middle short-dashed curve), and flat ($q = 0$, left long-dashed curve) cutoff to the electron distribution, and the observed distribution (dotted histogram).

As, pointed out by others [12], the SSM fails to explain the bursts with $\alpha > -2/3$, above the "line of death". We believe these bursts that lie beyond our simulated distributions may be physically explained by a self-absorption cutoff entering the BATSE window (however, see also [5]). As shown in [8], for some bursts synchrotron self-absorption is necessary to provide a good spectral fit to the data. However, we usually do not see the sharp $\alpha = 3/2$ ($\nu_{min} < \nu \ll \nu_{abs}$) or $\alpha = 1.0$ ($\nu \ll \min[\nu_{min}, \nu_{abs}]$) cutoff; the average value of α above the line of death is about 0. There are two things that could give a low α when the absorption cutoff is present in the BATSE window. One is due to the correlation discussed above (as E_p moves closer to the lower edge of the BATSE window, a lower (softer) value of α is measured). But there is an additional factor when two breaks are present. If $\nu_{min} > \nu_{abs}$ and our fits place $E_p \propto \nu_{min}$, then when the self-absorption frequency enters into the BATSE window, the Band spectrum cannot accommodate this additional break. As a result, the low energy index ends up being a weighted average (depending on the relative values of ν_{min} and ν_{abs} of the optically thin ($-2/3$) and optically thick (1) asymptotes. Support for this idea comes from the GINGA data [15]; we believe their low E_p values not measured by BATSE are due to this absorption break ν_{abs} and not ν_{min}.

SPECTRAL EVOLUTION

GRB spectra are known to vary throughout the duration of the burst, and can even vary during individual pulses. If the above explanation for the α distribution is correct, we would then expect a strong positive correlation in the time histories of α and E_p for most bursts. This relation should reflect the $h(\alpha)$ curves shown in Figure 2 of [8] unless the parameters describing the distribution of the accelerated particles (q, E_*, p, in [8]) or the magnetic field vary during a pulse or from pulse to pulse during a burst. Depending on the nature of these variations, the expected correlation could be strengthened or weakened.

There have been many studies of the time evolution of spectral parameters (e.g. [11,4,3,12]). Crider et al. [3] investigate the behavior of the low energy spectral index α for a sample of 30 BATSE GRBs. They find that 18 of these bursts show hard-to-soft evolution of α, while 12 exhibit "tracking" of the burst time profile, $\alpha(t) \propto A(t)$, where $A(t)$ is the amplitude of the photon spectrum. All of these bursts show a strong correlation between α and the peak energy, E_p, as a function of time. Recently, Preece et al. [14] published a catalog of spectral data with high time resolution. We have examined a sample of 46 bursts from this data set, and find a variety of behaviors for $\alpha(t)$ and $E_p(t)$: (a) both α and E_p "track" the flux in time. This can be explained by the correlation between α and E_p discussed above (this, then, implies an intrinsic correlation between E_p and the flux throughout this burst); (b) the parameter α tracks the flux; E_p varies on the same timescale as the flux, but is in an envelope of hard-to-soft evolution. This can be explained by a sharpening of the cutoff of the electron distribution from pulse to pulse. Note the correlation between α and E_p is seen within each individual pulse; (c) the parameter α appears to evolve from hard (above the "line of death") to soft, while E_p fluctuates around fairly high (~ 500 keV) values. This can be explained by a transition from an optically thick (self-absorption) to optically thin regime throughout the duration of the burst.

All of the observed spectral behaviors we have encountered be explained by i) the correlation between α and E_p, and ii) regarding each pulse as an independent emission episode (as one would expect for internal shocks, for example), which allows for evolution of the smoothness of the particle energy distribution cutoff as well as the optical depth in the shock. We will discuss this in more detail in a future publication.

CONCLUSIONS

We conclude that synchrotron emission from internal shocks can reproduce both the distribution and temporal evolution of GRB spectral parameters. Depending on the conditions at the GRB, synchrotron spectra can have different values for the low energy asymptote of its spectrum, and the apparent correlation between E_p and α (which results from fitting over a finite bandpass) can explain the peak and

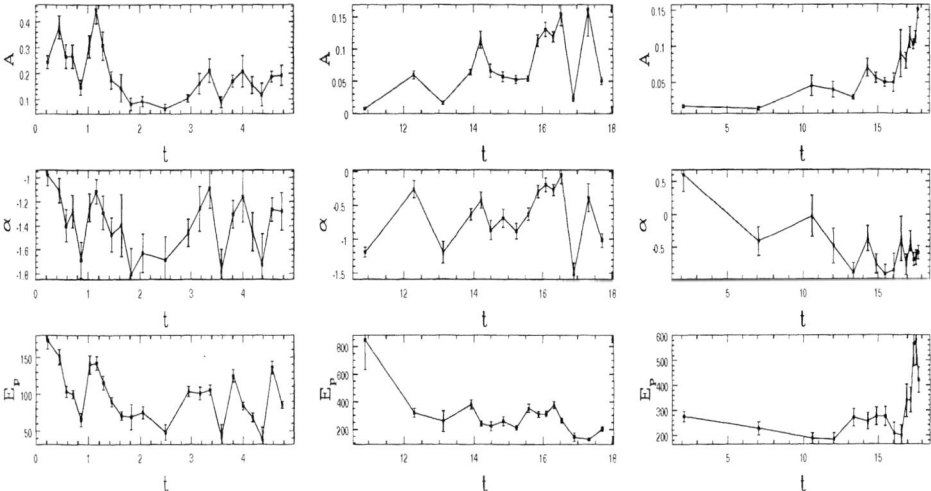

FIGURE 2. Three examples of spectral evolution of α, E_p, and the normalization A discussed in the text [(a), left; (b), middle; (c), right]

dispersion of the α distribution. In addition, we conclude that the electron energy distribution must flatten or decline at low energies; otherwise, we would see many more bursts with $\alpha < 1.5$. Finally, allowing for variation of internal parameters from pulse to pulse as in an internal shock model, synchrotron emission is consistent with the variety of spectral evolution see in GRBs.

REFERENCES

1. Band, D., et al., *ApJ* **413**, 281 (1993).
2. Cohen, E., Piran, T., & Narayan, R., *ApJ* **500**, 888 (1998).
3. Crider, A., et al., *ApJ* **479**, L39 (1997).
4. Ford, L.A., et al., *ApJ* **439**, 307 (1995).
5. Granot, J. et al., these proceedings.
6. Harris, M.J. & Share, G. H, *ApJ* **494**, 724 (1998).
7. Liang, E.P. & Kargatis, V.E., *Nature* **381**, 49 (1996).
8. Lloyd, N.M. & Petrosian, V., these proceedings.
9. Lloyd, N.M. & Petrosian, V., *ApJ* **511**, 550 (1999).
10. Lloyd, N.M., et al., to appear in *ApJ* (1999); astro-ph 9908191.
11. Norris, J.P., et al., *ApJ* **301**, 213 (1986).
12. Preece, R.D., et al., *ApJ* **496**, 849 (1998).
13. Preece, R.D., et al., *ApJ* **506**, L23 (1998).
14. Preece, R.D., et al., to appear in *ApJS* (1999); astro-ph 9908119.
15. Strohmeyer, T.E., et al., *ApJ* **500**, 873 (1998).

Study of the Characteristics of GRB Energy Spectra

B. M. Belli

Istituto di Astrofisica Spaziale CNR, 00133 Roma, Italy

Abstract. We analyzed the GRB energy spectra obtained from the fluence values reported in the BATSE catalog, averaging the data over contiguous HR ranges. When submitted to this type of analysis, these GRB spectra show a continuous evolution with the spectral hardness, that can be represented as a shift of the E^2 (power emitted per keV, that can be fitted with a synchrotron law) versus higher values of the maximum emission energy. For the highest values of the spectral hardness considered range we found an indication of power excess at the lowest energies (25-50 keV) suggesting the presence of a secondary maximum of emission.

INTRODUCTION

The energy spectrum study gives one of the most direct approaches to investigate the physical process at the origin of gamma-ray bursts (GRBs). Many authors have studied and are studying the spectra generally of the most intense events that provide the possibility of a very detailed analysis of the spectral characteristics [1,8,5]. Here we study the GRB energy spectra in their totality, looking for their general characteristics and trying to give a scenario as clear as possible. The data used for the analysis are the fluences reported in the BATSE catalog: four for each event relative to the four intervals of energy in which the total energy range observed by the experiment has been divided. These very synthetic spectra are available for all the recorded events including for the weakest ones. In this way all the observed events have been taken into account. In a previous paper [2] GRBs have been divided in two classes, I and II, in the duration hardness-ratio plane. The two classes are at the right and at the left of the straight line $HR = 2\ D^{0.5}$ respectively, in logarithmic scale. Class I is principally composed of the longest events and Class II of the shortest ones. Pendleton et al. [6] have divided GRBs by two spectral types: high energy events that have photons of energy over 300 keV (HE) and the no-high energy events without photons of energy over 300 keV (NHE). We studied separately both the high energy events and the no-high energy ones. The NHE events have been further divided in Class I and Class II, the HE have been divided into a large number of groups, to have as much as possible homogeneous classes of events.

SPECTRAL SHAPES

In this section we study how the GRB averaged energy spectra change varying the hardness ratio. As previously stated, the energy spectra which we use here are the very concise ones provided by the BATSE catalog. We analyze how the fluence values vary in the first and the fourth interval, for a given value of the ratio of the fluences relative to the third and the second interval (HR). To reduce the error bars, sometimes very large, we divided the GRBs into subsets relative to contiguous hardness-ratio ranges and calculated the average of the fluences for each range. The fluence mean performed is a mean weighted using as weights the inverse of the square of the variances given by the catalog. Besides, the partial averaged fluences have been divided by the energy range width to which refer, to obtain fluences per keV, and then multiplied by the energy of the center of the interval to obtain the emitted energy power per keV-E^2 (erg cm^{-2}). Operating in this manner, i.e. taking the half point energy to represent the averaged photon energy, we certainly introduce systematic errors, but we do not *a priori* know the exact spectral shape, for a more precise calculation. However all these energy spectra have been obtained with the same procedure and for this can be useful to put in evidence the shape evolution with HR.

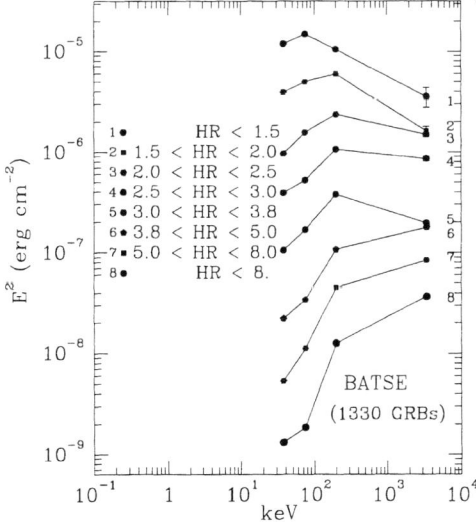

FIGURE 1. Averaged energy square density E^2 (erg cm^{-2}), for the 1330 current BATSE catalog HE events, for the different HR ranges reported in the figure. The number of the events for each ranges is about 200. The spectra have been shifted between them for clarity.

Fig. 1 shows for 1330 HE events of the BATSE current catalog the averaged values of E^2 obtained by $fl1, fl2, fl3, fl4$ (erg cm^{-2}) of 8 hardness-ratio intervals, consisting of about two hundred events for each one, from the softer to the harder ones. The statistical error bars are all smaller than the dimensions of the marks but for the highest energy interval. For this reason the obtained errors are sufficiently small to allow to observe variability in the ordinate values. Fig. 1 clearly indicates that the GRB spectra vary with continuity, going from the softer to the harder ranges of HR.

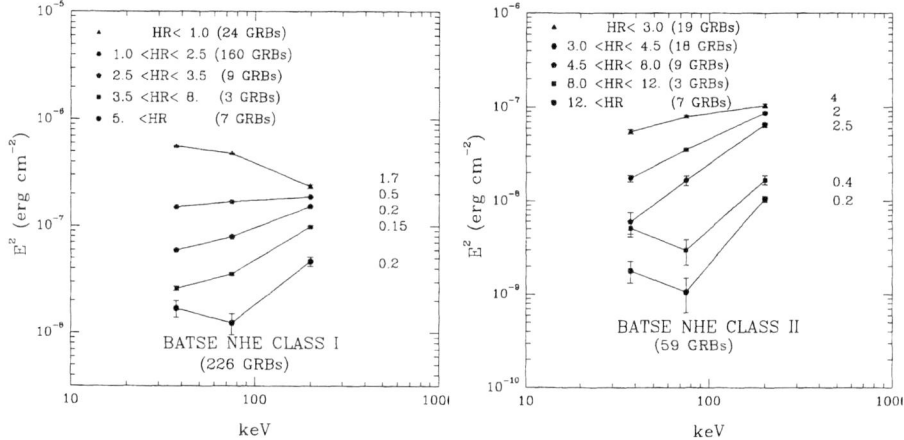

FIGURE 2. Averaged energy square density E^2 (erg cm^{-2}), on the left for Class I and on the right for Class II NHE events, for the different HR indicated ranges. The number of the events and the shift factor are also reported for each HR range.

SPECTRAL SHAPES FOR SEPARATED SUBSETS

The analysis of the subsets of Class I and Class II of NHE events, reported in Fig. 2 shows a possible excess of energy power in the 25-50 keV range for the hardest groups. The significance is low, but this behavior is present in the two classes. Analogously for HE events, if we further divide the first and last set of events in Fig. 1 in four subsets, we find an excess of power in the last subset of the hardest group, with the 98% of confidence, (Fig. 3 and Table 1).

DISCUSSIONS

It is very interesting to have a complete picture of the energy spectra variability for GRBs, even if we can only give a rough outline of it. The HR value determines

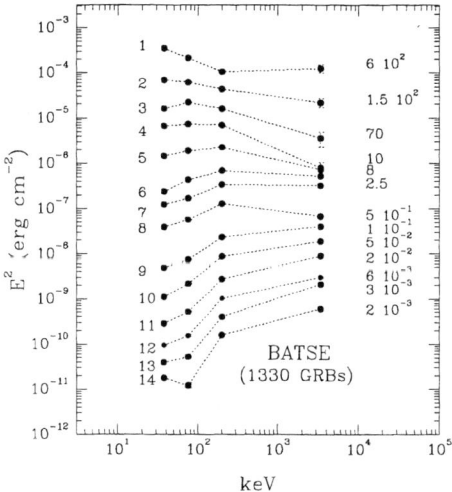

FIGURE 3. The first and the last four spectra of this graph have been obtained dividing the first and last spectra of Fig. 1 in four subsets of about 40 events for each one. Also here the data points are shifted by a factor indicated at the right of the figure. Table 1 reports the HR ranges and the photon number for each spectrum, starting from the top.

TABLE 1. BATSE 1330 HE GRBs in 14 HR Ranges and Relative Numbers of Events.

HR range	GRB number	HR range	GRB number
0.00 - 0.85	30	3.00- 3.80	215
0.85 - 1.15	30	3.80- 5.00	193
1.15 - 1.35	26	5.00- 8.00	225
1.35 - 1.50	22	8.00- 9.20	37
1.50 - 2.00	128	9.20-11.80	37
2.00 - 2.50	178	11.80-14.80	18
2.50 - 3.00	158	14.80 -∞	26

the averaged spectrum shape, that is to say that the spectrum shape changes as a function of HR. For the emitted power spectra E^2 we find an evolution of the spectral shape with HR that can be described as a shift of the power function (interpreted as a synchrotron emission) along the energy range observable by the BATSE experiment as HR increases, from a decreasing curve to a gradually increasing curve with a value of the maximum emission energy E_p higher. This behavior continues to be present even if we divide the HR range into much more than eight intervals.

The correlation we found between the HR and the E_p has been already observed by other researchers and this gives some confirmation of the validity of the present results. Briggs et al. in a recent paper [3] found in the very strong event of 1999 January 23th an increase of photons at the lowest energies of the spectrum which should go in the same direction of our results. Also other events present a similar behavior [4]. However, it is necessary to remark that the results of the present analysis could contain artifacts caused by the procedure used to obtain the fluence values from the photons detected with the Large Area Detectors. In fact, the fluence values have been obtained fitting, for necessity, the data with a simple power law (4B Revised BATSE Catalog), whereas best fits are generally obtained with two power laws. We think that this circumstance could some how alter the spectral shapes but not generate their evolution. Instead artifacts could play a role in the presence of the excess at the lower energies found in some events [5]. The present analysis, if performed on more suitable data, could bring results posing important constraints on the models of the origin of GRBs [7].

REFERENCES

1. Band, D. et al., *ApJ* **413**, 281 (1993).
2. Belli, B. M., *Ap&SS* **231**, 43 (1995).
3. Briggs, M. S. et al., *ApJ*, submitted (1999).
4. Briggs, M. S. et al., these proceedings.
5. Preece, R. D. et al., *ApJS*, in press (1999).
6. Pendleton, G. N. et al., *ApJ* **489**, 175 (1997).
7. Rees, M. J. and Meszaros, P., *MNRAS* **258**, 41 (1992).
8. Tavani, M., *ApJ* **466**, 768 (1996).

GRB 920229:
Evidence for a Sharp Spectral Break?

Michael S. Briggs and Robert D. Preece

Department of Physics
The University of Alabama in Huntsville, Huntsville, AL 35899

Abstract. It has been reported that observations made with BATSE of GRB 920229 show the continuum spectrum to be best-fit with a broken power law and that this fit demonstrates a sharp cutoff above 200 keV. The sharpness of this reported cutoff, $\Delta E/E = 18\%$, is difficult to explain within the current paradigm for GRBs.

We report our analysis of the continuum spectrum of GRB 920229. While a broken power law provides an excellent fit to the data, we find that the now standard Band GRB function provides an equally good fit. Furthermore, a broken power-law fit gives a moderate change in slope at the break rather than the reported dramatic change. Our analysis shows that the spectrum has a gradual cutoff typical of No-High Energy (NHE) events.

INTRODUCTION

Using BATSE data, both Scargle, Norris & Bonnell (1998) and Walker & Schaefer (1998) reported that GRB 920229 has the sharpest known temporal feature in a GRB. In a recent paper, Schaefer & Walker (1999) further report that GRB 920229 has "by far the sharpest known spectral break or continuum feature". Intrigued by this unusual spectrum, we examined the data ourselves.

The BATSE burst datatype most suitable for this spectral analysis is the 128 energy channel High-Energy Resolution (HERB) datatype of the Large Area Detectors (LADs). The temporal resolution of the HERB data is insufficient to resolve the short spike; following Schaefer & Walker, we analyze a 1.920 s interval which includes most of the flux of the event. We use the data of LADs 1 & 5 because they are the only two LADs with good viewing angles, 23° and 51°, respectively. Our background spectrum is created by averaging the remaining ≈ 80 s of HERB data.

We use the standard forward-folding technique: a parameterized photon model is assumed, the model is multiplied the the Detector Response Matrix to produce a count rate model, the count rate model is compared to the count rate data using χ^2, and the photon model parameters are optimized to minimize χ^2. While for clarity the graphs show the data rebinned into broad channels, the actual fits have been done using the full channel resolution of the data. We simultaneously fit the data of the two LADs with a common model—the data of the two detectors are in excellent agreement and no normalization difference is necessary.

SPECTRAL MODELS

We use three models to fit the HERB data: a broken power law, the Band 'GRB' function, and a power law times an exponential, $dN/dE = AE^{-\lambda}\exp(-E/E_0)$, which is referred to as the 'Comptonized' model. Because the Comptonized model is a special case of the Band GRB function, and because the two models give virtually identical results, for brevity we show only the results obtained with the broken power-law model and the Comptonized model.

Figure 1 shows the data and the model fits in the count rate units of the data. The broken power-law model has one more parameter than the Comptonized model, but its χ^2 is smaller by only 1.9. The residual plots (Figure 2) are virtually indistinguishable. On these grounds, we find that both models provide equally acceptable fits to the data. Figure 3 shows the deconvolved spectra in νF_ν units.

Schaefer & Walker concluded that the data significantly favored the broken power model based upon broad energy bands in which they found significant differences between the observations and the predictions of the Comptonized model. We do not reproduce these findings (Figure 2): the residuals for the Comptonized model show no wide-band trends nor do they differ from the residuals of the broken power law.

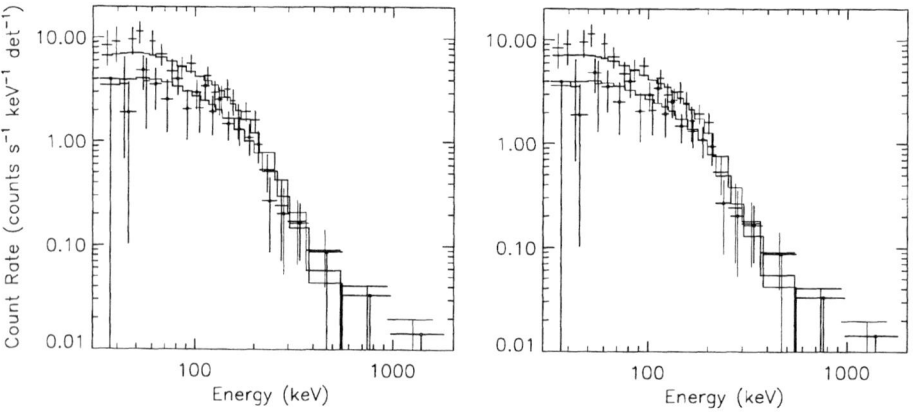

FIGURE 1. Model fits to the count-rate data of LADs 1 & 5. The count-rate data points are shown with crosses ($\pm 1\sigma$) or as T-symbols (2σ upper-limits), while the models are depicted with histograms. The data points of LAD 5 are designated with dots. **Left**: Comptonized model (power law with exponential cutoff), $\chi^2 = 173.9$ with 216 degrees-of-freedom, **Right**: broken power law, $\chi^2 = 172.0$ with 215 degrees-of-freedom.

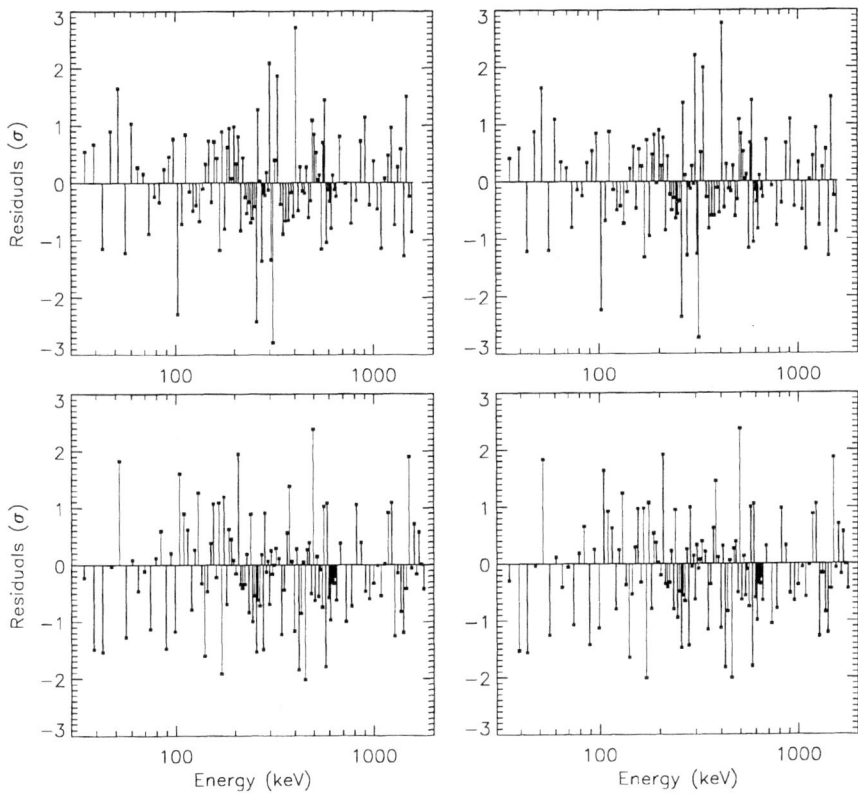

FIGURE 2. Count-rate residuals in units of σ for each channel. **Left**: Comptonized model, **Right**: broken power law. **Top**: LAD 1, **Bottom**: LAD 5.

As an example of a wide-energy band residual, Schaefer & Walker state that the GRB function (which they agree reduces for this spectrum to a power law times an exponential) "predicts that the burst photons [counts] in channel 4 [i.e., above 300 keV] should be ~1500, whereas only 120 are observed." An earlier sentence identifies 120 counts as the background-subtracted counts in channel 4 for the *entire* burst. Testing of the spectral fit should be based on the same data as the fit, i.e., the 1.92 s interval of the high-energy resolution HERB data, however, the results for the two intervals should be very similar because only a small fraction of the burst flux occurs outside the 1.92 s interval. Above 300 keV, we find 1436 observed counts in LADs 1 and 5, of which 1378.7 are predicted by the background model, leaving 57.3 ± 37.9 counts to be attributed to the source. Propagating the Comptonized photon model through the detector response model, our best fit predicts 88.2 source counts. Our observed (57.3 ± 37.9) and model source counts (88.2) agree well. Considering the differing time intervals, there is acceptable agreement between

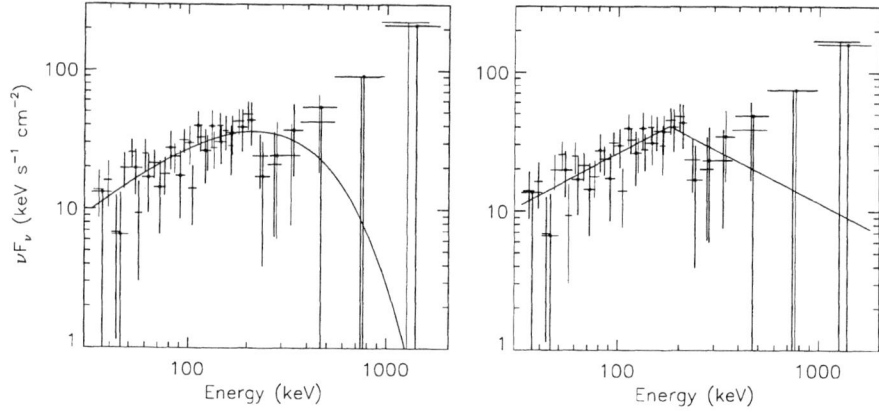

FIGURE 3. Deconvolved spectra in νF_ν units. **Left**: the Comptonized model (power law with exponential cutoff): $\lambda = -0.78 \pm 0.34$, $E_0 = 180 \pm 75$ keV, **Right**: Broken power law: $\lambda_L = -1.24 \pm 0.16$, $\lambda_H = -2.74 \pm 0.64$ and $E_{\text{break}} = 180 \pm 40$ keV.

our background-subtracted observed counts (57.3) and those of Schaefer & Walker (120). There is a significant discrepancy between the model source counts from the best-fit Comptonized models: 88.2 compared to ~1500 (Schaefer & Walker).

We find that the broken power law and the Comptonized model (along with the Band GRB function) explain the data equally well. This short burst has a very low fluence (4.7×10^{-7} erg cm^{-2}) compared to those we typically analyze spectroscopically, so there are few counts and it is not surprising that the data cannot distinguish among these three models.

EVIDENCE FOR A SHARP BREAK?

From their broken power-law fit, Schaefer and Walker conclude that the burst has a remarkably sharp high-energy cutoff. We disagree with this interpretation for two reasons, either of which suffices:

- Because several models are equally consistent with the data, conclusions should not be reached that are derived only from the broken power-law fit,
- Even if the broken power-law fit is used, the values we obtain for the spectral indices show no evidence for an unusually sharp break: Schafer & Walker obtained power-law indices of -1.3 ± 0.1 and -5.5 ± 1.0 with a break at 200 keV, while we find values of -1.24 ± 0.16 and -2.74 ± 0.64 with a break of 180 ± 40 keV.

Schaefer & Walker report a break scale of $\Delta E = 39$ keV ($\Delta E/E = 18\%$), where ΔE is a Half-Width. We find a Full-Width at Half-Maximum of 390 keV (0.8 decades) for the broken power-law model and 480 keV (0.9 decades) for the Comptonized model.

The most discrepant spectral parameter is the high-energy index λ_H of the broken power-law model (-2.74 ± 0.64 versus -5.5 ± 1.0). Our analysis shows that while the error region of λ_H is asymmetrical, the value of -5.5 is excluded at the 2σ level. Because essentially the same data were used for both analyses, this difference cannot be understood as a 2σ statistical fluctuation and must be due to a difference in the analysis procedures. Our analysis has been made using WINGSPAN and variances for χ^2 calculated from the model. We obtain similar results using likelihood but obtain results similar to Schaefer & Walker by using χ^2 with data variances. Schaefer (private communication) reports that χ^2 with model variances was used within BSAS, so the cause of the discrepancies remains unknown.

CONCLUSIONS

On two crucial points, our findings differ from those of Schaefer & Walker:
- contrary to their conclusion that the data favor a broken power law, we find that a broken power law, the 'Comptonized' function, and the Band GRB function are equally consistent with the data,
- we find a gradual cutoff rather than a sharp spectral break, even if the burst is modeled with a broken power law. The data are inconsistent with the reported spectral scale $\Delta E/E = 18\%$ and indicate a FWHM of almost one decade.

GRBs with little flux above 300 keV are well known and have been named No-High Energy (NHE) bursts (Pendleton et al. 1997). GRB 920229 is formally an HE event because it has a few too many counts above 300 keV (Pendleton, private communication), but the paucity of high-energy counts and the evidence for a gradual cutoff beginning at a few hundred keV show it to be similar to NHE GRBs. The 'Comptonized' model is a special case of the Band function which commonly fits No-High Energy (NHE) spectra. Of the 2028 GRBs with BATSE catalog fluence values by 1999 November 29, 1275 have larger channel 3 (100–300 keV) fluence values than GRB 920229 and therefore have comparable or better statistics. Of these, fully 823 are softer in the sense of channel 4 (> 300 keV) fluence divided by channel 3 fluence. The implications are that the spectrum of this burst has a fairly common cutoff that is typical of NHE-like events and that this spectrum imposes no unique requirements on GRB theory.

REFERENCES

1. Pendleton, G. N., Paciesas, W. S., Briggs, M. S., et al., *ApJ* **489**, 175–198, (1997).
2. Scargle, J. D., Norris, J., & Bonnell, J., in AIP Conf. Proc. **428**, ed. C. A. Meegan, R. D. Preece & T. M. Koshut, 1998, pp. 181–185.
3. Schaefer, B. E. & Walker, K. C., *ApJ* **511**, L89–L92 (1999).
4. Walker, K. C. & Schaefer, B. E., in AIP Conf. Proc. **428**, ed. C. A. Meegan, R. D. Preece & T. M. Koshut, 1998, pp. 266–270.

Spectra of a Recent Bright Burst Measured by CGRO-COMPTEL:GRB 990123

C. A. Young[1], A. Connors[2], K. Bennett[3], W. Collmar[4],
W. Hermsen[5], R. M. Kippen[6], E. D. Kolaczyk[7], L. Kuiper[5],
M. McConnell[1], R. Miller[1], J. M. Ryan[1], V. Schönfelder[4],
O. R. Williams[3], and C. Winkler[3]

[1] *Space Science Center, University of New Hampshire, Durham NH, USA*
[2] *Eureka Scientific*
[3] *Astrophysics Division, ESTEC, NL-2200 AG Noordwijk, NL*
[4] *Max-Planck-Institut fur Extraterrestrische Physik, D-85740 Garching, FRG*
[5] *SRON-Utrecht, Sorbonnelaan 2, NL-3584 Utrecht, NL*
[6] *CSPAAR, University of Alabama in Huntsville, Huntsville AL, USA*
[7] *Department of Mathematics and Statistics, Boston University, Boston MA, USA*

Abstract. CGRO-COMPTEL measures gamma-ray burst positions, time-histories and spectra in the 0.1–30 MeV energy range, in both imaging "telescope" and single detector "burst spectroscopy" mode. GRB 990123, one of the most recent bright bursts seen by COMPTEL, was caught in the optical while the gamma-ray emission was ongoing. The burst spectral shape can be characterized by a peak in $\nu - F_\nu$ just below 1 MeV and a power-law tail above (photon index ~ -2.4), and flattening below. There is also spectral evolution by downward movement of the peak and/or softening of the power laws. We present light-curves, time resolved spectra and an image map for this burst.

INTRODUCTION

GRB 990123 is the only gamma-ray burst to be simultaneously observed in optical wavelengths. The Burst and Transient Source Experiment (BATSE) on CGRO triggered the Compton Telescope (COMPTEL) on CGRO at 35216.121 s UT on 23 January 1999 (BATSE trigger #7343). The burst was located at a zenith angle (w.r.t. CGRO and COMPTEL) pointing of 58.4°. The gamma-ray emission lasted for about 100 s with the > 1 MeV emission seen by COMPTEL most significant between 18 s and 46 s after the BATSE trigger. The Robotic Optical Transient Search Experiment (ROTSE) detected optical emission during and after the gamma-ray emission. ROTSE made 6 measurements starting 22.2 s after the BATSE trigger.

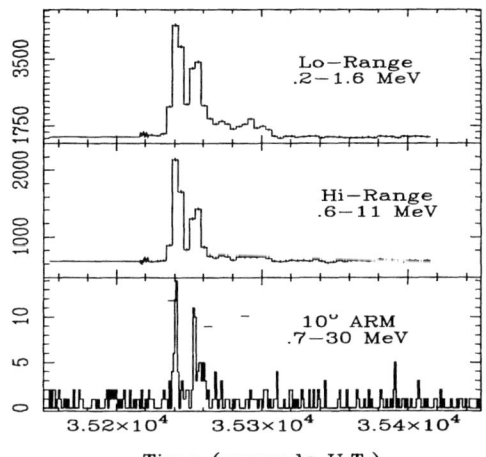

FIGURE 1. GRB 990123 (BATSE #7343) in the COMPTEL low (0.2–1.6 MeV) and high (0.6–11 MeV) range burst mode, plus the 0.7–30 MeV telescope mode light-curves. The ROTSE m_v are indicated with the telescope mode.

COMPTEL detected the second and third peaks of GRB 990123 in both its imaging telescope ("double scatter"; 0.75–30 MeV) and non-imaging burst-spectroscopy ("burst mode"; 0.3–1.5 MeV and 0.6–10 MeV; see Schönfelder et al. 1993) modes. Telescope position constraints were broadcast (via GCN/BACODINE) about 10 minutes after burst onset and preliminary light-curves were posted soon thereafter (Young et al. 1999; Connors et al. 1999).

In the imaging telescope mode, COMPTEL provides detailed information on individual time-tagged photons with $\frac{1}{8}$ ms time resolution. Because of the small effective area and limited telemetry of the telescope mode, however, only ~200 burst events were recorded. In spectroscopy mode the effective area was roughly two orders of magnitude greater, and the dead time was negligible. The COMPTEL light-curves are displayed in Fig. 1.

TELESCOPE MODE

The spectra from the two modes were handled differently. For the 0.75–30 MeV telescope data, one selects only events whose event circle falls within a certain angle of the source position ("angular resolution measure", or ARM; Schönfelder et al. 1993), both reducing background and providing a nearly diagonal response (e.g., Kippen et al. 1998). For the telescope spectrum displayed in Fig. 2, a 10° ARM limit was used. The 32.768 s integration interval of (35229.452 s, 35262.22 s) was chosen both to cover all the significant gamma-ray emission and to allow an accurate live time calculation. The background data were taken 15 orbits prior to the burst (see Kippen et al. 1998). The data were fit to a simple power-law via a forward folding technique. The best-fit is $2.0 \pm 0.4 (E/1\,\text{MeV})^{-3.33 \pm 0.4}$ photons/cm^2-s-MeV, giving a total fluence of $(0.98 \pm 0.5) \times 10^{-4}$ ergs/cm^2 (0.75–30 MeV). However, a turnover at or below 1 MeV is also consistent with the telescope data.

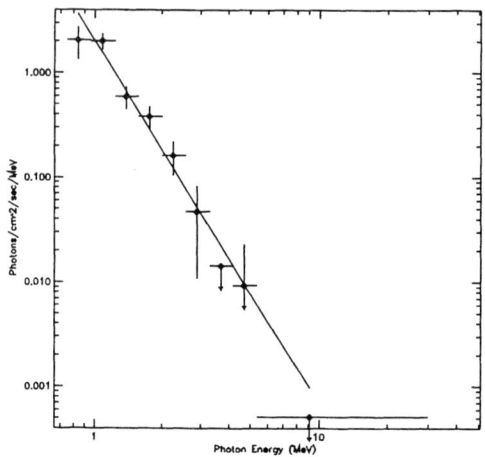

FIGURE 2. The best-fit power-law spectrum is $2.0 \pm 0.4 (E/1\,\text{MeV})^{-3.33 \pm 0.4}$ photons/cm^2-s-MeV, giving a total fluence of $(0.98 \pm 0.5) \times 10^{-4}$ ergs/cm^2 (0.75–30 MeV). However, a turnover at or below 1 MeV is also consistent with these telescope data.

BURST MODE

The single detector count spectra obtained in "spectroscopy mode" were processed as follows: the background was estimated from a spectrum of 140 s duration starting 202 s prior to the BATSE trigger (at 35216 s). Eight high range detector (0.6–10.0 MeV) spectra (4 s integration time each) covering a 32 s time interval (35230.2 s, 35262.3 s), were background subtracted and summed. The spectral fitting of these time-averaged data was performed by convolving a trial photon spectrum with the detector response matrix to produce a count spectrum. We used a χ^2 statistic to compare the model spectrum with the data. We first assumed a single power law for the spectrum between 0.9 and 5.0 MeV. As shown in Fig. 1 there is no significant signal above 4 MeV. The best fit parameters for the single power law in this energy range are: normalization = (1.37 ± 0.10) photons/[cm^2-sec-MeV] at 1 MeV; and index = (-2.63 ± 0.16). The fluence (0.9–5 MeV, 32 sec) is 7.86×10^{-5} erg/cm^2. We note a clear break (Briggs et al. 1999) in the spectrum below 0.9 MeV, where the spectrum becomes flatter. Also, preliminary analysis of the low range (0.3–1.5 MeV) spectroscopy data covering the same 32 s time interval indicates a single power law with index -2.0.

Due to evidence for spectral evolution (Briggs et al. 1999) with a break in COMPTEL's energy range, one goal of this analysis was to produce time resolved spectra of this burst. Traditional spectral analysis techniques (like those applied above) use a parametric approach assuming a particular model and approximately account for the Poisson nature of the data using χ^2 type approximations or data transformations. These techniques thus require high statistics and so don't allow for fine time binning of our data. We have begun the implementation of a newly developed technique that addresses some of these shortcomings enabling the use of finer time resolution. Nowak and Kolaczyk (1999) have presented a method for deconvolving spectra that is both non-parametric and explicitly handles Poisson data. We

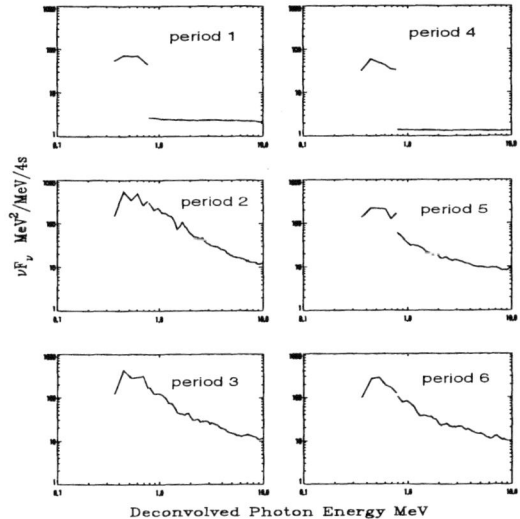

FIGURE 3. Time resolved $\nu - F_\nu$ spectra of 6 consecutive 4 s intervals of GRB 990123, showing qualitative evidence for spectral evolution. The deconvolution was performed using a non-parametric approach.

divided the low and high range burst data into seven 4 s time intervals. Using this non-parametric technique we deconvolved these spectra with the instrument response. Though only preliminary, Fig. 3 shows qualitative spectral structure consistent with the previous analysis with a spectral break below 1 MeV. The time periods displayed don't show any spectral evolution but they are consistent with the comparable times shown in Briggs et al. (1999).

IMAGE MAP

GRB 990123 was so bright in MeV gamma-rays that COMPTEL was able to image it in a 10 sigma detection. (The preliminary COMPTEL detection, broadcast about 11 minutes after the BATSE trigger, was 8.2 sigma.) Fig. 4 shows the combined COMPTEL and BATSE 1, 2, and 3 sigma location contours. The IPN, BeppoSAX X-ray, prompt Optical (ROTSE), fading optical, and radio counterpart locations are included on the map.

CONCLUSIONS

For GRB 990123, observed in the optical during the gamma-ray emission, one notes that the optical flux peaks after the brightest gamma-ray portion and is not a simple extrapolation of the MeV flux. Some have noted this may be a signature of self-absorption; while others suggest this shows the optical to have come from a separate component (Briggs et al. 1999). Traditional analysis techniques show consistent results for the burst and telescope data from COMPTEL. A preliminary non-parametric analysis is consistent with other methods and shows qualitative

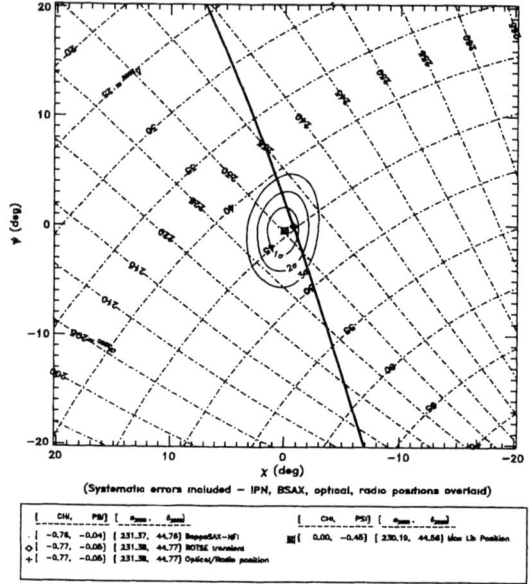

FIGURE 4. The combined COMPTEL and BATSE Huntsville 1, 2 and 3 sigma location contours (systematic uncertainties included), along with the IPN timing arc (Hurley, Feroci et al GCN 222), and indications of the BeppoSAX X-ray, prompt optical (ROTSE), fading optical, and radio counterparts.

evidence for a spectral break just under 1 MeV. Further investigation of this new technique will allow for quantitative results.

ACKNOWLEDGMENTS

The COMPTEL project is supported in part through NASA grant NAS 5-26646, DARA grant 50 QV 90968, and the Netherlands Organization for Scientific Research (NWO). AC is supported in part through the hospitality of Wellesley College and NASA grant NAG5-7984. MM is supported in part through NASA grant NAG5-7829.

REFERENCES

1. Akerlof, C. W., et al., GCN 205 (1999).
2. Briggs, M. S., et al., *ApJ* **524**, 82 (1999).
3. Connors, A., et al., GCN 230, 237 (1999).
4. Hurley, K., et al., GCN 222 (1999).
5. Kippen, R. M., et al., *Adv. Space Res.* **22** (7), 1097-1100 (1998).
6. Loredo, T. J., in *Statistical Challenges in Modern Astronomy*, eds. G. J. Babu & E. D. Fiegelson, Springer-Verlag, New York, 1999.
7. Nowak, R. D., Kolaczyk, E. D., *IEEE Trans. Inform. Theory*, Sept. (1999).
8. Schönfelder, V., et al., *ApJS* **86**, 629 (1993).

X-ray Excesses in GRB Spectra

R. D. Preece*, J. R. Espley[†], and M. S. Briggs*

*University of Alabama in Huntsville, Dept. of Physics, Huntsville, AL 35899
[†]University of Virginia, Dept. of Astronomy, Charlottesville, VA, 22903

Abstract. The observation of X-ray excesses greater than 5 sigma above an assumed model spectrum in approximately 15% of 86 bright bursts observed by BATSE has been verified in recent work. We have looked at a further 64 events, and observe the same behavior of a 15% excess rate in the statistically-independent sample. We have examined the time-history of excesses for 10 bursts that had an overall greater than 8 sigma excess, in order to compare the behavior with other spectral parameters. In addition, we have looked at 17 bursts that exhibit no global excess, to determine the expected temporal variations in the BATSE low-energy data.

INTRODUCTION

Previous work [2] has shown that while standard spectral models have been largely successful at energies greater than 20 keV, there is evidence that these models are insufficient to correctly model spectral behavior of some bursts in the 5–20 keV regime. The present work strongly confirms this previous work using the four years worth of data that have accumulated since the earlier work. More importantly, we also performed an analysis of the spectral evolution of the ten bursts with the highest overall low-energy excess and compared these excesses with various model parameters.

X-RAY EXCESSES IN BURSTS AFTER GRB951203

In order to examine the low-energy behavior of gamma-ray bursts in the most recent four years of BATSE observations, we combine several data sets from the BATSE Large Area and Spectroscopy Detectors (LADs and SDs). First, we determine the brightest bursts from this time period in order to have a reasonable signal to noise ratio. We require that the peak flux be greater than 10 photons cm^{-2}s^{-1} in the 50–300 keV range and that the total fluence be at least 4×10^{-5} erg cm^{-2}; we find that 65 bursts satisfy these criteria. For these 65 bursts, we choose data sets that give the greatest range of total energy coverage. Typically, we choose one LAD data set, providing high-statistics data with energy coverage from 30–1800 keV, and discriminator data from one of the SDs, giving us coverage in the energy range from 5–20 keV. Occasionally, we use high energy resolution data from the SDs (SHERB) that covers the same energy range as the LAD data, when those data is not available (typically because of inadequate time coverage). The exact energy range covered depends on the gain setting of the particular SD used. Our choice of SD is dictated by the lowest available energy coverage and also by the constraint that the detector zenith angle be less than 60° in order to avoid an excessive detector response dependence on the azimuthal angle [3].

Having chosen our data sets, we perform a time-averaged joint fit of our combined data to one of several standard spectral models: the "GRB" spectral model [1], a smoothly broken power law, and the Comptonized model [4]. For each burst, we find that a smoothly broken power law consistently yields the best fit. In order to quantify the difference between the observed and predicted count rate for the SD data (the low-energy point), we take the difference between the two and divide by the standard error, derived from the variance of the model. We call this quantity the sigma residual.

In 10 out of 65 bursts (15%), the time-averaged sigma residual in the low-energy SD discriminator data was greater than 5σ, despite a good fit to the higher-energy LAD data. This percentage is exactly in keeping with the previous work [2] that found 12 out 86 bursts had similar excesses, giving a total of 22 out of 151 bursts with significant excesses. We also find that 2 bursts (3%) have an X-ray deficit greater than 5σ.

TEMPORAL EVOLUTION OF THE EXCESSES

From the data set of 22 bursts mentioned above, we chose the 10 bursts with the highest sigma residuals (8σ or more) and did an analysis of how the excesses evolved with time. Additionally, we did a similar analysis on the two bursts in the set with a deficit of 5σ or more, and as a control set, on 17 bursts with residuals close to zero.

Our procedure follows closely the one outlined above for the general analysis but differs in two significant ways. First, to include the widest available energy range, we fit as many data sets as possible that meet our energy coverage and angle requirements from additional BATSE detectors, even if they overlap with the data used for the time-averaged analysis. We often end up with as many as four data sets to be included in the joint fit. The second significant difference was that we divide each time history into segments of roughly equivalent signal-to-noise ratios, as defined by the LAD data energy interval. In order to accomplish this, we must determine time intervals common to every data set. When a satisfactory set of time intervals has been found, we perform joint spectral fits to all the data in each time interval. This enables us to track the sigma residual as a function of the spectral and temporal parameters throughout each burst. In most cases, the smoothly broken power law model yields the best fit (as in the general analysis)

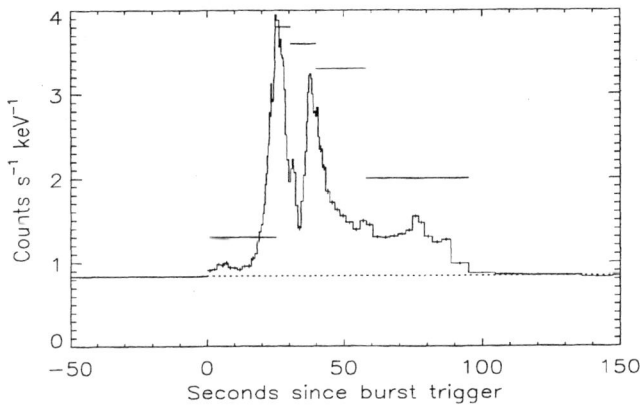

FIGURE 1. GRB990123 time history in the SHERB data. The 5 time bins chosen for the time-resolved analysis with identical signal to noise ratios are indicated by horizontal lines.

but occasionally the GRB model produces the closest fit, especially when β, the high-energy power law index, is poorly constrained.

Figure 1 shows a typical result. BATSE trigger #7343 (GRB990123) had a sigma residual of 14.9 in our analysis of its time-averaged spectra. We divided its time history into 5 separate time bins, based upon equal signal-to-noise ratios in the SHERB data that we were using for our principal data set. An additional SHERB data set extends the energy coverage to higher energies and two SD discriminator data points cover the low-energy regime without a break below the lowest SHERB data point. The resulting energy coverage is as follows: 8–13, 14–29, 30–1650, and 220–5000 keV. Figure 2 shows a typical joint fit to these 4 data sets for the first of the five intervals. The model used is the smoothly broken power law. The joint fits show a clear departure in the amount of X-ray flux in the SD#0 and SD#4 data points from the fitted model, as determined by the best joint fit to all the data. Typically, we see that the excess dies away with time (see Fig. 3).

PRELIMINARY RESULTS

Further analysis is underway, but a few preliminary results have already been seen. No obvious correlations exist between the sigma residual and other spectral parameters, such as the low-energy power-law index α, the high-energy power-law index β, flux or time-integrated fluence. This holds for both bursts with excess as well as those with a deficit.

However, if we select all spectra from bursts with large (> 3) low-energy excesses and plot their fitted E_{peak} values (Fig. 4), the distribution is clearly harder than the scaled E_{peak} distribution from the BATSE Spectroscopy Catalog [4]. Additionally, there is also an indication that the shorter intervals for the spectral accumulation

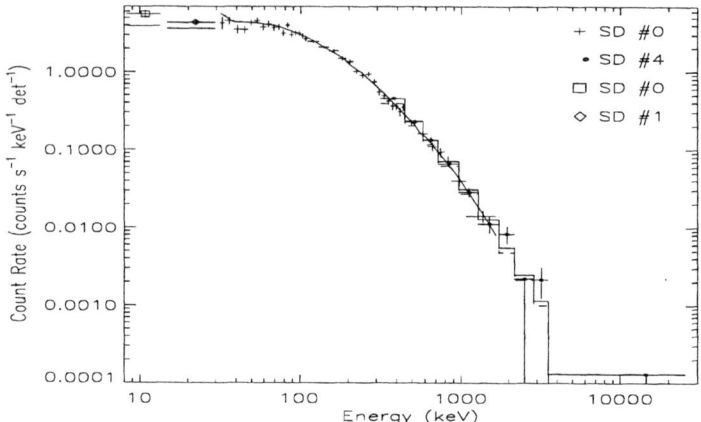

FIGURE 2. Spectral data for GRB990123 in one time-averaged spectrum corresponding to the first time bin shown in Figure 1. The four included data sets each have their own symbol including error bars and the solid line represents the model fit to the data.

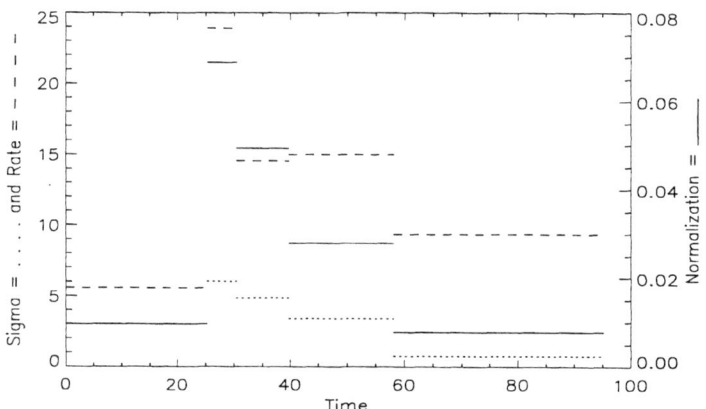

FIGURE 3. Sigma residuals and model count rates for GRB990123 binned into 5 time bins. The right-hand axis displays the normalization of the fitted model (i.e., the model rate at 100 keV), as an indication of the high-energy flux history.

in the sample have the greatest excesses. In Figure 5, the spectra with $> 5\sigma$ excess are found in spectral accumulations of 1–10 s. It is clear that some additional X-ray spectral component is needed to explain this excess in 15% of the bursts and it is hoped that additional analysis will shed some light as to what form this component should take.

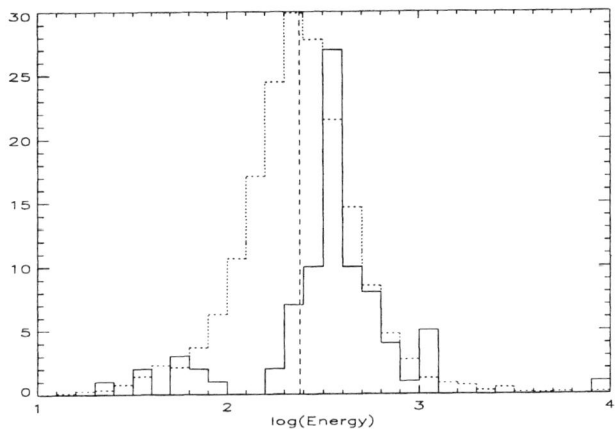

FIGURE 4. Histogram of fitted E_{peak} values for the general BATSE Spectroscopy Catalog (*dotted line*) with bursts having a large ($> 3\sigma$) low-energy excess (*solid line*).

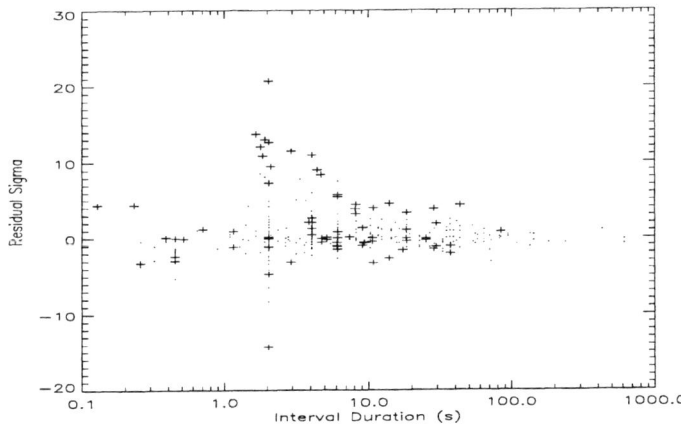

FIGURE 5. Plot of sigma residual verses duration of the spectral accumulation interval in seconds for all spectra from the 29 bursts studied in detail. Spectra from those 10 bursts with overall $\sigma > 5$ are marked with crosses.

REFERENCES

1. Band, D. L., et al., *ApJ* **413**, 281 (1993).
2. Preece, R. D., et al., *ApJ* **473**, 310 (1996).
3. Pendleton, G. N., et al., *Nucl. Inst. Meth.* **364**, 567 (1995).
4. Preece, R. D., et al., *ApJS* **126**, in press (1999).

A New Method for Studying the Hardness-Intensity Correlation in Gamma-Ray Bursts

Felix Ryde*, Luis Borgonovo*, and Roland Svensson*[†]

*Stockholm Observatory, SE-133 36 Saltsjöbaden, Sweden
[†]Institute for Theoretical Physics, University of California, Santa Barbara, CA 93106, USA

Abstract. We introduce a new method to study the hardness-intensity correlation (HIC) within gamma-ray bursts, in which the intensity is represented by the peak value of the EF_E spectrum. We compare it to the traditional method where the intensity over a finite energy range is used. This new method gives stronger correlations and is useful in studying several aspects of the HIC.

INTRODUCTION

Gamma-ray bursts (GRBs) often exhibit strong spectral evolution (e.g., [1] and references therein). This can be quantified by studying the evolution of the hardness of the spectrum versus its intensity. Such hardness-intensity correlations (HICs) can often be described with a power law, especially over the decay phase of individual pulse structures in the GRB light curve [2–7]:

$$F \propto E_{\text{pk}}^{\gamma}, \qquad (1)$$

where F is the energy flux and E_{pk} is the peak energy of the EF_E spectrum. A general problem, encountered in HIC studies, arises from the limited spectral coverage of the used detector which might affect the assigned measure of the bolometric flux, i.e., the energy-integrated flux. A second problem, when one aims at studying the correlation over single pulses in the light curve, arises from the fact that the observed spectra may contain contributions from other pulses which even could be unresolved. Furthermore, additional soft components [8] could also affect the measured flux value and thus weakening the correlations.

We therefore introduce a new representation of the HIC which might resolve some of these complications. The GRB spectrum is often peaked in an EF_E representation [9] at some energy E_{pk} in the γ-ray band. The value of EF_E at E_{pk} is a representation of the energy-integrated flux as it is proportional to the total flux. We denote this quantity by ϕ and study the HIC correlation as

$$\phi \propto E_{\text{pk}}^{\eta}, \tag{2}$$

where η is the new HIC index. Our discussion is limited to the cases where the peak actually exists within the studied band $[E_{\min}, E_{\max}]$. Representing the spectrum by the Band et al. model [9], this means that the low- and high-energy power-law indices must have $\alpha > -2$ and $\beta < -2$. In the most common case, where $E_{\max} > (\alpha - \beta) E_{\text{pk}}/(\alpha+2)$, the proportionality between ϕ and F=$\int F_E dE$ becomes

$$k(\alpha, \beta, y_{\min}, y_{\max}) = \frac{F}{\phi} = \frac{e^{(\alpha+2)}}{(\alpha+2)^{\alpha+2}} \times \tag{3}$$
$$\left[\Gamma(\alpha+2)\{P(\alpha+2, \alpha-\beta) - P(\alpha+2, y_{\min})\} + \frac{(\alpha-\beta)^{\alpha-\beta} y_{\max}^{\beta+2} - (\alpha-\beta)^{\alpha+2}}{(\beta+2) e^{\alpha-\beta}} \right],$$

where $y_{\min} = (\alpha+2) E_{\min}/E_{\text{pk}}$ and $y_{\max} = (\alpha+2) E_{\max}/E_{\text{pk}}$. $\Gamma(\alpha+2)$ and $P(\alpha+2, y)$ are the gamma function and the incomplete gamma function, respectively (e.g., Press et al. [10]). In the case that α and β do not have a strong dependence on E_{pk}, the only E_{pk} dependence in $k(\alpha, \beta, y_{\min}, y_{\max})$ is in $y_{\min}$ and $y_{\max}$. In particular, when the integration is chosen over the whole energy range from 0 to ∞, there will be no dependence at all. The ϕ-value can be interpreted as integrating over the whole energy range, or at least over a range for which $y_{\min} \ll 1$ and $y_{\max} \gg 1$, and should, under some circumstances, be a better representation of the bolometric flux for the study of the HIC.

COMPARING THE TWO METHODS

To compare the methods, represented by equations (1) and (2), we study a sample of 45 GRB pulse decays, introduced by [6], observed by the Burst and Transient Source Experiment (BATSE) on the *Compton Gamma-Ray Observatory*. These pulses have good power-law HICs, as measured by the square of the correlation coefficients, R^2. The light curve for each burst in the sample was rebinned to achieve a signal-to-noise ratio of at least 30. The spectrum for each time bin was fitted with the Band et al. [9] model, allowing a deconvolution to find the energy spectrum and the peak energy. The energy spectrum was integrated over the energy band of the BATSE detector ($\sim 25-1900$ keV), for the strongest illuminated large area detector (LAD), to find the instantaneous energy flux.

Then, for each pulse decay we analyzed the spectral evolution and determined both the γ values, according to the F-method, Equation (1), and the η values, according to the ϕ-method, Equation (2). In Figure 1, a typical example of a HIC in a strong GRB pulse is shown. Graphically, in a plot of $\log EF_E$ versus $\log E$, the correlation describes how the peak of the spectrum moves following a straight track during the pulse decay from hard to soft. In general, the HICs follow a power law better when the ϕ-method is used. When comparing the coefficients given by the two different methods, R_γ^2 and R_η^2, we found that in 85% of the cases R_η^2 is greater than R_γ^2, implying that the new method, in general, gives better correlations.

The relation between the two HIC indices γ and η for all pulse decays in the sample is shown in Figure 2. A linear fit to these data gives the proportionality constant 0.95 ± 0.07. This shows that there is a good average correspondence between the two methods. A scatter around an exact correspondence between the two methods is to be expected. As mentioned above, additional flux components can be contributing to the estimated flux, affecting the resulting HIC index. The ϕ-method is less dependent on this. Furthermore, the exact correspondence between

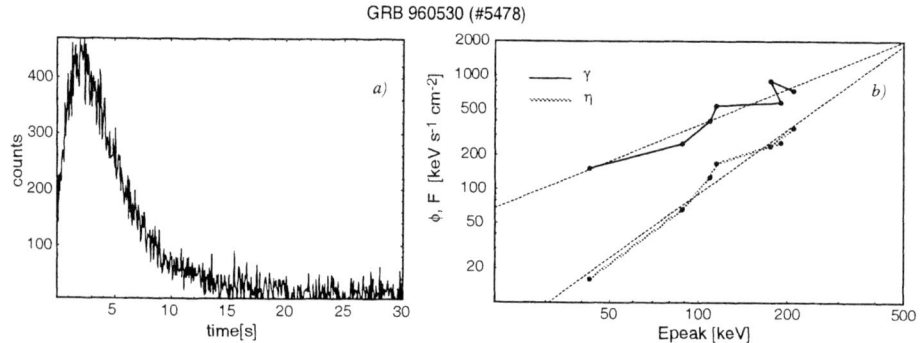

FIGURE 1. The HIC shown as measured by the two different methods, taking as example GRB 960530 (#5478). a) The count time history shows a single, long pulse. b) The HIC, during the decay phase of the pulse. In this case, we found $R^2_\gamma = 0.884$ and $R^2_\eta = 0.967$. In general, the ϕ-method results in better power-law correlations.

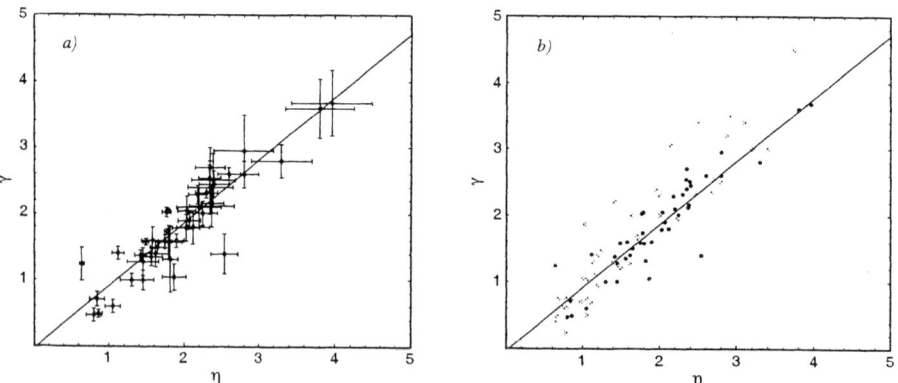

FIGURE 2. The relation between the two HIC indices γ and η. a) The correlation for the sample of 45 pulse decays. The straight line shows a linear fit to the data which gives a proportionality constant of 0.95 ± 0.07. The error bars given by the HIC fits are displayed. b) The dark dots represent the same pulses as shown in (a), but without error bars. Additional shorter pulses and weaker correlated cases are also included (grey dots). Note that the scatter across the line is larger for the pulse decays with weaker HICs.

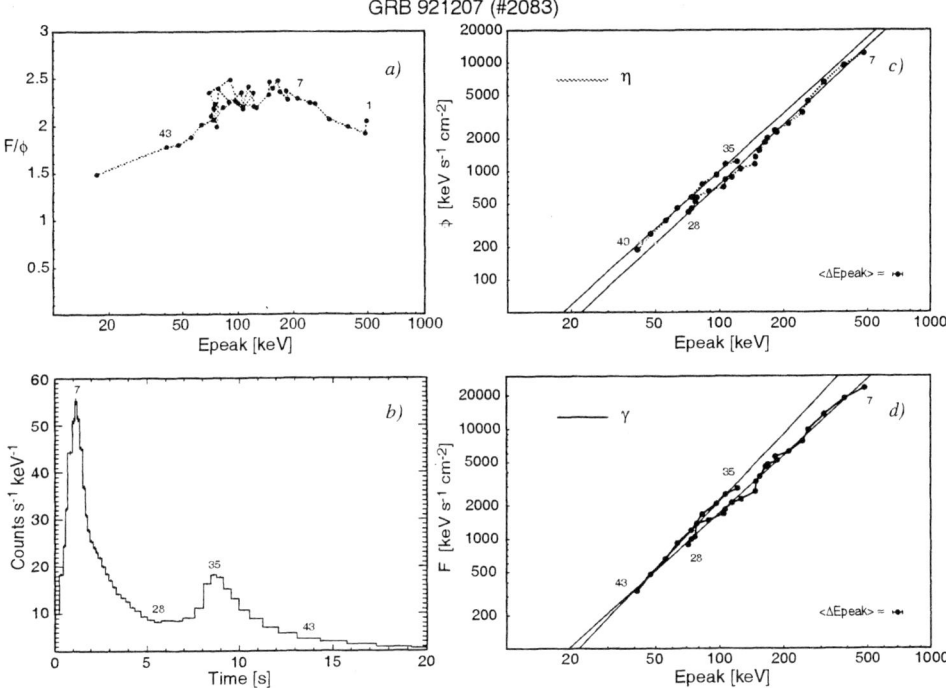

FIGURE 3. a) The time evolution of the ratio $k = F/\phi$, in equation (3), for the case GRB 921207 (#2083). The integration of the flux was made over the energy range 24–2000 keV. The F-measure will suffer from window problems at high and low energies. b) The count time history showing the time binning used to study the spectral evolution. A few bin numbers are shown. c) The HIC measured with the ϕ-method. Error bars are excluded for clarity, and only a typical average value is shown separately. The numbers refer to the time-bins in (b). We found, for each pulse decay, the power-law indices $\eta_1 = 1.78 \pm 0.04$ and $\eta_2 = 1.77 \pm 0.06$. Only the decay phases are shown here. d) Same as (c), now using the F-method. The corresponding indices are $\gamma_1 = 1.74 \pm 0.04$ and $\gamma_2 = 2.03 \pm 0.07$. Note that, as in general, the similarity between the HIC indices of pulses within a burst is more evident using the ϕ-method.

the power-law indices measured by the two methods is only valid in the case that α and β do not vary with E_{pk}, which they do in many bursts. The F-method also suffers from the finite detector range and might not be measuring the actual bolometric γ-ray flux. In fact, the behavior in Figure 3a, with a decline in k at high and low peak energies, is expected due to this [6]. Furthermore, the flux F is found from integrating the deconvolved spectrum. This spectrum is model-dependent since the deconvolution is based on the model spectrum. This could introduce additional scatter into the HIC relation.

Note that in studying the HICs, the measured errors of E_{pk} and the energy flux measure, as found in the fitting procedure, were not used. The data points were simply compared to the power-law behavior. One reason for doing this is that these errors are often overestimated, see discussion in [6].

DISCUSSION

The fact that the power-law correlations that we find using the ϕ-method are better than the ones given by the F-method, could indicate that the F-method suffers from effects which are not part of the correlation. The ϕ-method thus provides an efficient way of studying detailed features in the observed HICs. The two methods are compared, in detail, in Borgonovo & Ryde [6], where, for instance, the distribution of the indices γ and η are given both between different bursts and within bursts. When comparing pulses within a burst, HIC indices are similar, and in particular, η indices are equal to within the estimated errors (see Figure 3). Furthermore, Borgonovo & Ryde [7] uses this method to analyze a feature, that we call *track jumps*, found in some HICs.

ACKNOWLEDGMENTS

This research made use of data obtained through the HEASARC Online Service provided by NASA/GSFC. We are also grateful to the GROSSC for support. We acknowledge support from the Anna-Greta and Holger Crafoord Fund, the Gustaf and Ellen Kobbs Stipend Fund at Stockholm University, the Swedish Natural Science Research Council, the Swedish National Space Board, and NSF grant PHY94-07194. FR wishes to thank the BATSE team for their hospitality during his visit to NASA's Marshall Space Flight Center.

REFERENCES

1. Ryde, F., *ASP Conf. Series* **190**, 103 (1999).
2. Golenetskii, S. V., Mazets, E. P., Aptekar, R. L. and Ilyinskii, V. N., *Nature* **306**, 451 (1983).
3. Kargatis, V. E., Liang, E. P., Hurley, K. C., et al., *ApJ* **422**, 260 (1994).
4. Kargatis, V. E., et al., *A&SS* **231**, 177 (1995).
5. Ryde, F., and Svensson, R., *ApJ* **529**, L13 (2000).
6. Borgonovo, L., and Ryde, F., to be submitted.
7. Borgonovo, L., and Ryde, F., these proceedings.
8. Preece, R. D., Briggs, M. S., Pendleton, G. N., et al., *ApJ* **473**, 310 (1996).
9. Band, D., et al., *ApJ* **413**, 281 (1993).
10. Press, W. H., Teukolsky, S. A., Vetterling, W. T., and Flannery, B. P., *Numerical Recipes in Fortran* 2nd Ed., Cambridge Univ. Press, Cambridge, 1992.

Time Resolved GRB Spectroscopy

Marco Tavani*, David Band†, and Giancarlo Ghirlanda*

*Istituto Fisica Cosmica CNR, Milan (Italy)
†X-2, Los Alamos National Laboratory, Los Alamos, NM 87545

Abstract. We present the main results of a study of time-resolved spectra of 43 intense GRBs detected by BATSE. We considered the 4-parameter Band model and the Optically Thin Synchrotron Shock model (OTSSM). We find that the large majority of time-resolved spectra of GRBs are in remarkable agreement with the OTSSM. However, about 15% of *initial GRB pulses* show an apparent low-energy photon suppression. This phenomenon indicates that complex radiative conditions modifying optically thin emission may occur during the initial phases of some GRBs.

INTRODUCTION

We study a sample of 43 GRBs selected for the high quality of their time-resolved spectra obtained with the BATSE Spectroscopy Detectors (sensitive in the energy range 25–1800 keV). The time over which each spectrum was accumulated was varied so that the signal-to-noise ratio was greater than 15 (in the hard X-ray energy band). These data provide excellent temporal resolution: in many cases we obtain more than 10 spectra per burst with accumulation times as short as 256 ms.

SPECTRAL MODELS

We fitted each time-resolved GRB spectrum with two models: (1) the Band model [1], and (2) the Optically Thin Synchrotron Shock Model (OTSSM) [3,4]. The (purely phenomenological) four-parameter Band model [1] consists of two power-law components (of spectral indexes α and β) joined smoothly by an exponential roll-over near a break energy E_0.

$$N(E) = A \left(\frac{E}{100\,\text{keV}} \right)^{\alpha} \exp\left(-\frac{E}{E_0}\right) \qquad \text{for } E \leq (\alpha - \beta) E_0 \quad (1)$$

$$N(E) = \left[A \left(\frac{(\alpha - \beta) E_0}{100\,\text{keV}} \right)^{\alpha-\beta} \exp(\beta - \alpha) \right] \left(\frac{E}{100\,\text{keV}} \right)^{\beta} \qquad \text{for } E \geq (\alpha - \beta) E_0 \quad (2)$$

We used the (three-parameter) OTSSM of Refs. [3,4]. We performed an independent spectral fitting for the Band and OTSSM models for each of the time-resolved spectra of all GRBs of our sample. For each GRB we obtain 4 (3) best-fit parameters as a function of time representing the complete spectral evolution.

RESULTS

We find GRB spectral evolutions of two types: (1) a "tracking behavior", with spectral parameters in approximate one-to-one correspondence with the changing energy flux, and (2) a "hard-to-soft evolution", with spectral parameters evolving independently of the energy flux (see, e.g., ref. [2]).

Fig. 1 shows the distribution for *all* collected time-resolved spectra of the low-energy spectral index α. A few bursts show values $\alpha \geq -2/3$, typically during the initial-rising part of their most intense pulses. The high-energy spectral index β is less constrained, and in some cases varies substantially over consecutive spectra within the same burst. The β distribution (Fig. 2, left panel) is peaked near -2 for the Band model representation, and is broader for the OTSSM fits. Break energies E_0 are typically well below 500 keV. Interestingly, we find that the OTSSM provides a very good representation of time-resolved spectral data. Fig. 2 (right panel) show the cumulative distribution of the reduced χ^2 for the Band and OTSSM models.

DISCUSSION

We studied 43 GRBs from the BATSE spectral archive selected by their large signal-to-noise ratios. We collected information for a total of 1046 spectra.

Our results indicate that the OTSSM is quite successful in describing the majority of GRB spectra. Fig. 3 shows the spectral evolution of the remarkable GRB 990123, demonstrating the validity of the OTSSM for very intense bursts. However, violations of the simple OTSSM are apparent in about 15% (at 3σ level) of our time resolved spectra. These violations (typically with a low-energy index $\alpha > -2/3$) always occur at the beginning of major GRB pulses (as in Fig. 4).

The OTSSM was derived [3] for idealized plasma and hydrodynamic conditions that are most likely valid far from the central source. Several plasma and dynamic conditions (probably involving emission sites close to a central object) may produce the apparent suppression of soft photons at the beginning of some GRB pulses.

REFERENCES

1. Band, D., Matteson, J., Ford, L., et al., *ApJ* **413**, 281 (1993).
2. Ford, L.A., et al., *ApJ* **439**, 307 (1995).
3. Tavani, M.; *ApJ* **466**, 768 (1996).
4. Tavani, M., Ghirlanda, G., & Band, D., in preparation (2000).

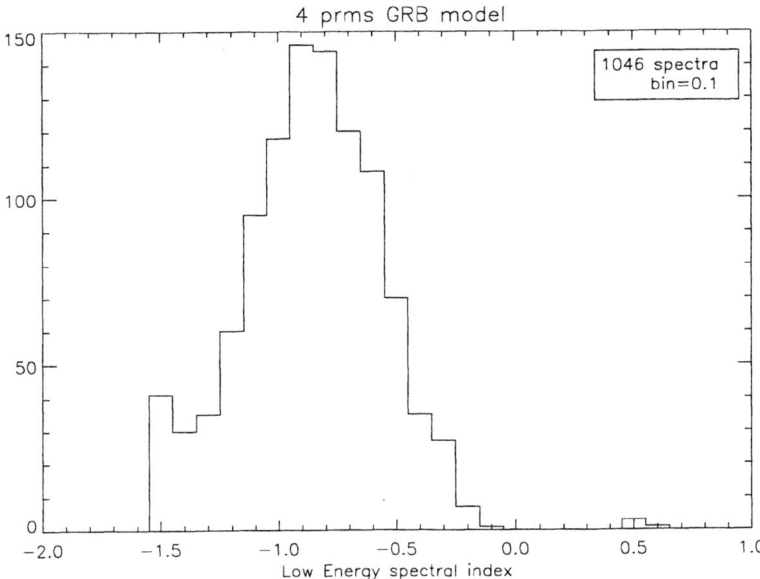

FIGURE 1. Low-energy (α) spectral index distribution from all the time-resolved spectra

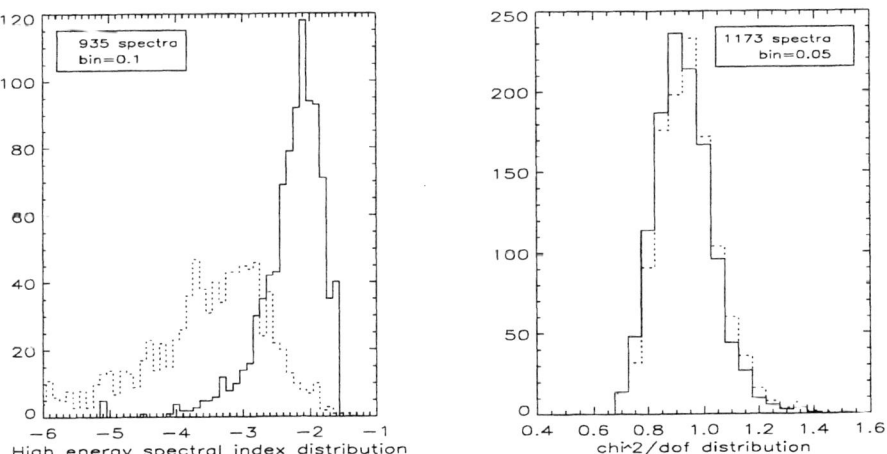

FIGURE 2. *Left panel:* High energy spectral index β distributions for the Band model (*solid line*) and the OTSSM (*dotted line*). *Right panel:* Reduced χ^2 distributions.

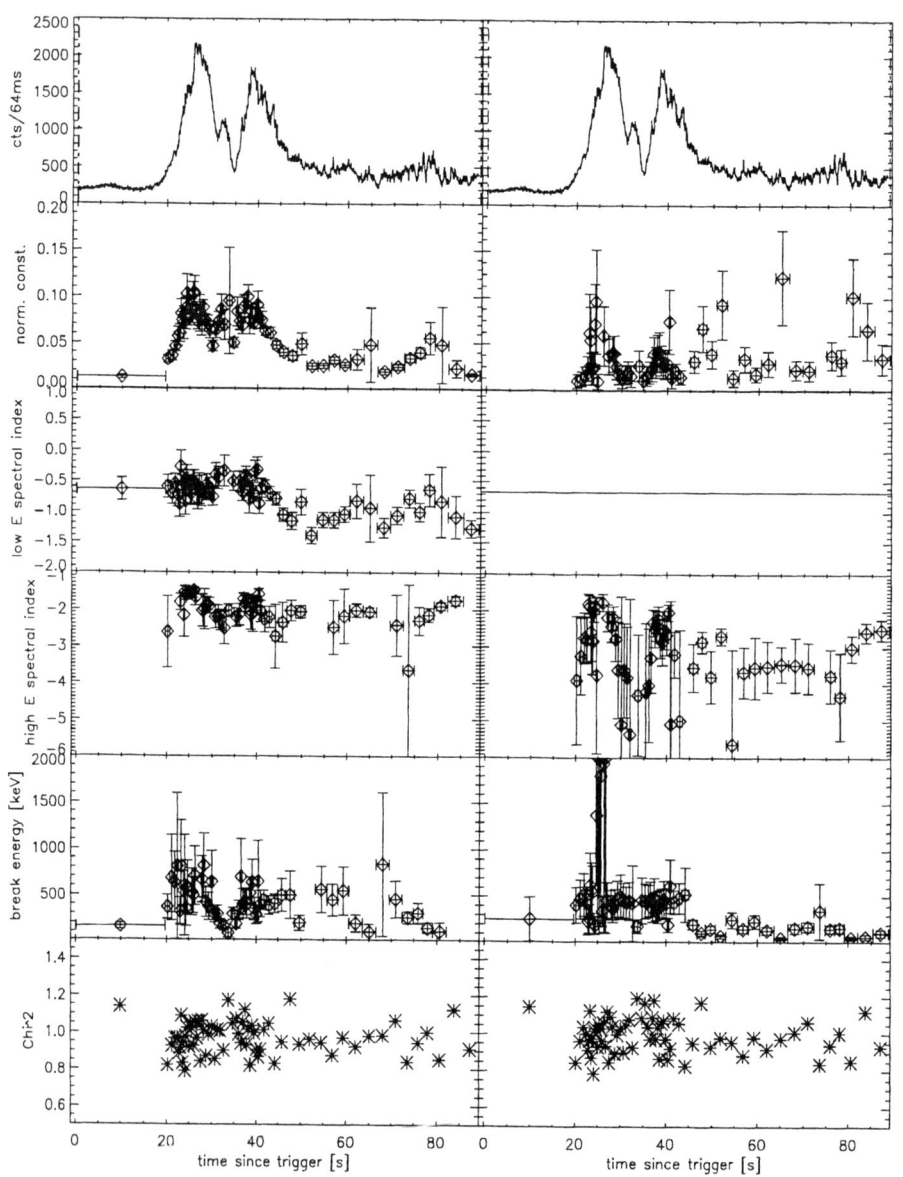

FIGURE 3. GRB 990123 spectral evolution of the 4-parameter Band model (*left column*) and the 3-parameter OTSSM (*right column*). The α parameter is fixed in the OTSSM.

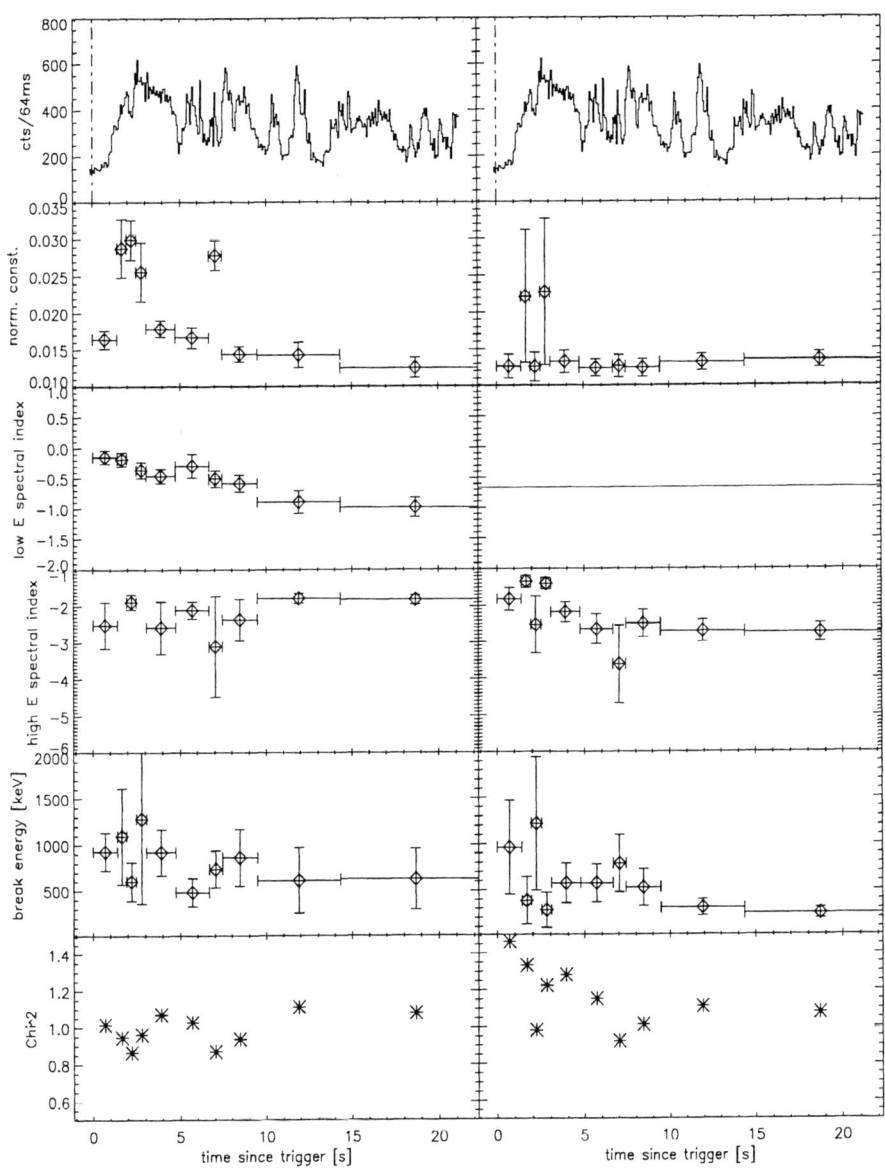

FIGURE 4. GRB 910814 spectral evolution of the 4-parameter Band model (*left column*) and the 3-parameter OTSSM (*right column*). The α parameter is fixed in the OTSSM.

Unusual Properties in the Time Profiles of Bright GRBs

Fergus Quilligan, Kevin J. Hurley, Brian McBreen,
Lorraine Hanlon, Paul Duggan, and Darach Watson

Physics Department, University College, Dublin, Ireland.

Abstract. The 200 brightest GRBs with T90 > 2 sec were selected from the BATSE 4B catalog and analyzed using wavelets. The rise times, fall times, FWHM, peak heights and peak areas were measured for statistically significant peaks in the sample of 200 GRBs and are consistent with lognormal distributions.

The properties of peaks before and after the strongest peak in a GRB were compared for three different categories of bursts, based on the number of peaks in the burst. There is no statistically significant change between the distributions of the rise times, fall times and FWHM of the peaks before and after the strongest peak. Given the spectral evolution observed in GRBs, the lack of temporal evolution is surprising, and should be addressed in any shock scenario within the fireball model.

Furthermore, we found that the GRBs with large numbers of peaks to have narrower and faster peaks and larger fluences than the simpler GRBs. This result can be explained by GRBs with more peaks having shocks with a higher average value of Γ.

INTRODUCTION

The light curves of gamma ray bursts (GRBs) are irregular and complex and statistical studies are necessary to identify the physical properties of the emission mechanism. One of the first studies [1] revealed that lognormal distributions can adequately describe some of the properties of GRBs. Subsequent studies [2–4] have confirmed the applicability of lognormal distributions in accounting for the wide range in the observed properties of GRBs. A variety of techniques has been used to quantify the temporal properties of GRBs and the results have significantly constrained models of this phenomenon [5–7].

In the framework of the internal shock model, the rapid variability in the GRB time profile is due to emission from multiple shocks in a relativistic wind [8–10]. The instabilities in the wind leads to shocks which convert a fraction of the bulk kinetic energy to internal energy remote from the central engine. A turbulent magnetic field then accelerates electrons which radiate by synchrotron emission and inverse Compton scattering, generating the GRB. The work presented here expands on the earlier analysis and provides new insight into the mechanism which generates GRBs.

DATA PREPARATION

The data were obtained from the BATSE 4B catalog [11]. A subset of this catalog was selected ($T_{90} > 2$ s and $P_{256ms} > 3.28$ ph cm^{-2} s^{-1}) yielding a sample of 200 long bright bursts. These bursts were processed with a multiscale edge detection wavelet with a denoising routine, and peaks were then selected using a simple algorithm [3]. The analysis included a method of measuring the degree of separation of the peaks in the GRB. The technique used determines the percentage of the total height of the peak which is above the higher of the two minima on either side of the maximum, and the peaks are considered isolated at that percentage level.

Each GRB was divided into two sections to include the peaks that occurred before and after the strongest peak in the burst. For this part of the analysis only GRBs with more than 2 peaks were included, resulting in a reduced sample of 161 GRBs. The first half (pre-main peak) of the GRB was then compared with the second half (post-main peak). The bursts were also split by the number of peaks in the burst into three categories which we label the 'simple' bursts (3-15 peaks), the 'average' bursts (15-35 peaks), and the 'complex' bursts (35-85 peaks) [12]. Table 1 contains a summary of the GRB properties used in this analysis.

RESULTS AND DISCUSSION

A detailed comparison has been made between the distributions of the properties of the peaks in the first half and second half of the bursts for the three categories of GRBs. The distributions of fall times, rise times and full widths at half maximum (FWHM) are presented for these categories in Figures 1, 2 and 3 respectively. The

TABLE 1. Summary of the properties of the bursts used in the first half and second half analysis.

GRB Category	Simple	Average	Complex
No. of Peaks per GRB	3-15	15-35	35-85
No. of GRBs	112	36	13
Total No. of Peaks	775	773	641
Number in 1st half/2nd half	310/353	353/384	303/325
No. of Isolated Peaks at 50% level 1st half / 2nd half	144/163	142/209	68/136
No. of Isolated Peaks at 75% level 1st half / 2nd half	60/84	54/69	26/36
Mean T90 (sec)	45	69	75
Mean Total Fluence (ergs/cm^2)	0.3×10^{-4}	0.7×10^{-4}	2.3×10^{-4}

results of Kolmogorov-Smirnov (KS) tests are included in the figures. The KS probability is a measure of whether the two distributions (first half/second half) are compatible with being drawn from the same parent distribution. There is no statistically significant difference between the first half and second half of the GRB in any of the three categories. The mechanism that generates a GRB is remarkably constant throughout the burst. This result is compatible with the constancy of the pulse widths observed by Ramirez-Ruiz & Fenimore [13]. Given the spectral evolution observed in GRBs, the lack of temporal evolution is surprising, and should be addressed in any shock scenario within the fireball model.

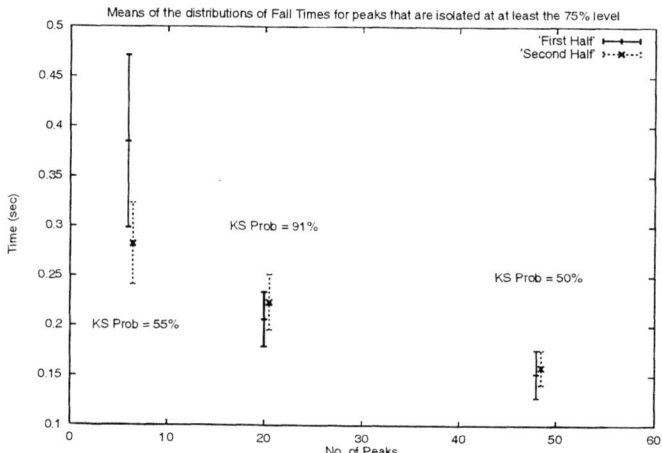

FIGURE 1. The means of the distributions of fall times for peaks in the first half and second half for each GRB category, for peaks that are isolated at least at the 75% level. Each data point represents a distribution of values and the bars shown represent the error on the mean value and should only be taken as a qualitative guide. More quantitative information is obtained from the Kolmogorov-Smirnov (KS) test. The KS probability is given in the figure and is a measure of whether the two distributions (first half/second half) are compatible with being drawn from the same parent distribution.

However, it is also evident from Figures 1-3 that as the number of peaks in a burst increases, the rise times, fall times and the FWHM all decrease by a factor of about 3. This behavior could result from collisions between internal shocks, if we assume the bulk Lorentz factor, Γ, increases with the number of peaks in the GRB. The average increase in Γ is $\sqrt{3}$ because the peak width varies as Γ^{-2} [13]. The power output is proportional to Γ^3 [9] for an isotropically emitting blob. The mean fluence in the three categories of GRBs reveals an increase of a factor of 8 (Table 1) between simple and complex bursts. This result is also consistent with complex bursts having a higher value of Γ by a factor of $\sqrt[3]{8} = 2$

There is an indication (3% probability) that the rise times of peaks may be faster

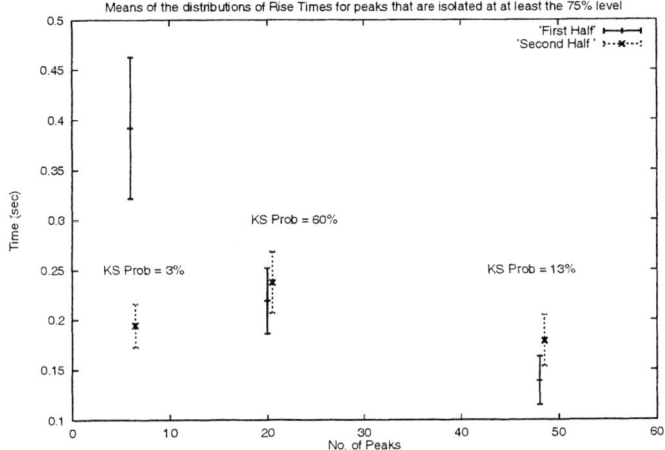

FIGURE 2. The means of the distributions of rise times for peaks in the first half and second half for each GRB category for peaks that are isolated at least at the 75% level. The notation is the same as that used in Fig. 1.

in the second half of GRBs with 3-15 peaks (Fig. 2). If confirmed, a variation of this type can be explained as a clearing out effect that slows the rise times during the initial part of a GRB [14]. This effect should be more apparent for bursts that are emitted at a lower value of Γ, and is compatible with the sample of GRBs having the smallest number of peaks.

CONCLUSIONS

The peaks in a large sample of GRBs have been selected using wavelets and the properties of the peaks have been compared for GRBs in three categories. There is no statistically significant difference in the rise times, fall times and full widths at half maximum of the peaks in the first half and second half of GRBs. Given the spectral evolution observed in GRBs, the lack of temporal evolution is surprising, and should be addressed in any shock scenario within the fireball model.

Furthermore, the rise times, fall times and FWHM of peaks in GRBs decrease as the number of peaks increases. GRBs with more peaks have faster and narrower peaks, and larger fluences. This result can be explained by GRBs with more peaks having shocks with a higher average value of Γ.

REFERENCES

1. McBreen, B., Hurley, K.J., Long, R. and Metcalfe, L., *MNRAS* **271**, 662 (1994).

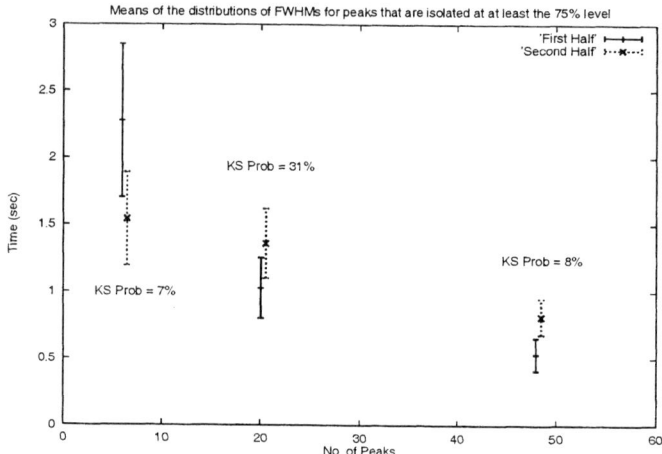

FIGURE 3. The means of the distributions of FWHM for peaks in the first half and second half for each GRB category, for peaks that are isolated at least at the 75% level. The notation is the same as that used in Fig. 1.

2. Li, H. and Fenimore, E.E., *ApJ* **469**, L115 (1996).
3. Hurley, K.J., McBreen, B., Quilligan, F., Delaney M. and Hanlon, L., 1998, "Wavelet Analysis and Lognormal Distributions in GRBs," in *Gamma Ray Bursts*, eds. C.A. Meegan et al., AIP Conference Proceedings 428, New York, pp 191-195 (1998).
4. Quilligan, F., Hurley, K.J., McBreen, B., Hanlon, L. and Duggan, P., *A&AS* **138**, 419-420 (1999).
5. Norris, J.P. et al, *ApJ* **459**, 393 (1996).
6. Mitrofanov, I., *Ap & Space Sci.* **231**, 103 (1995).
7. Beloborodov, A.M., Stern, B. E. & Svensson, R., *ApJ* **508**, L25 (1998).
8. Rees, M.J., Meszaros, P., *ApJ* **430**, L99 (1994).
9. Piran, T., *Phys. Rep.* **314**, 575 (1999).
10. Spada M., Panaitescu, A., Meszaros, P., astro-ph/9908097 (1999).
11. Fishman, G.J. and Meegan, C.A., *ARAA* **33**, 415 (1995).
12. Quilligan, F, Hurley, K.J., McBreen, B., Hanlon, L., in preparation (2000).
13. Ramirez-Ruiz, E. and Fenimore, E., *A&AS* **138**, 521-522 (1999).
14. Dermer, C.D. and Mitman, K., *ApJ* **513**, L5 (1997).

Generic Difference between Early and Late Stages of BATSE Gamma-Ray Bursts

Maxim L. Litvak[1], Igor G. Mitrofanov[1], Michael S. Briggs[2],
William S. Paciesas[2], Geoffrey N. Pendleton[2], Robert D. Preece[2],
and Charles A. Meegan[3]

[1] *Space Research Institute, Profsojuznaya str. 84/32, 117810 Moscow, Russia*
[2] *Department of Physics, University of Alabama in Huntsville, Huntsville, AL 35899*
[3] *NASA/Marshall Space Flight Center, Huntsville, AL 35812*

Abstract. The concept of early and late stages (ES and LS) of gamma-ray bursts is implemented for the statistical analysis of the large sample of BATSE events. The primary peak is proposed to be the boundary between the early and late stages of emission. Significant differences are found between ES and LS: the early stage is shorter than the late one, it has the harder emission, and it becomes shorter in respect to all burst time histories for groups with decreasing intensity.

INTRODUCTION

Gamma-ray bursts are known to have very different time profiles. Some of them have complex multi-pulse time histories with broad interpulse valleys between pulses, and others are one-pulse events with well-seen rise and decay. On the other hand, each burst, as a transient phenomenon, has a beginning and an end. There is a key question in respect to the mechanism of gamma-ray burst emission: is there an envelope of all-burst temporal evolution from the beginning to the end which determines the generic properties of emission along a light curve? Does a separate pulse possess an individual behavior independently whether it is at an early or at a late stage, or there is some generic law which predetermines the individual properties of pulses depending on the moment in the light curve?

Therefore, we check in this paper for the existence of a signature of all-burst evolution. We shall test the difference between the early and late stages of bursts, as the evidence for all-burst evolution.

EMISSION OF BURSTS BEFORE AND AFTER THE PRIMARY PULSE

For each burst we may define the totality of time intervals (so-called emission time), which contribute the pre-determined fraction of the total fluence with the highest counts rate [1,8]. We suggest to use the fraction of 50% of fluence for the emission time determination. The all-burst emission time τ_{50} is shorter than (or equal to) the classical duration parameter T_{50}, because for τ_{50} the fluence is influenced by time intervals with the highest count rate. The parameter of emission time describes how long it takes for the source to generate 50% of the total fluence with the highest flux. This parameter does not depend on interpulse intervals with lower count rate.

We define two fractions of the total emission time, as it is divided by the primary peak. The corresponding time intervals of high power emission before and after the primary peak could be defined as emission time of early stage $\tau_{50}^{(ES)}$ and late stage $\tau_{50}^{(LS)}$, respectively. The sum of these two values $\tau_{50}^{(ES)} + \tau_{50}^{(LS)}$ equals the total burst emission time by definition.

Taking into account that short and long bursts might be associated with different physical populations, it is necessary to study these sets separately [8]. In this section we compare early and late stages of the total sample and for short and long bursts separately in terms of temporal and spectral properties. In the next section we compare early and late stages in terms of intensity properties only for long bursts. Short bursts will be considered in a future paper because finite time resolution and considerable problems from noise produced effects and triggering effects require special investigation.

Temporal properties: We may characterize each burst by the early-to-late ratio $\Theta = \tau_{50}^{(ES)}/\tau_{50}^{(tot)}$, which is the fraction of early stage in respect to the total burst emission. The distribution of Θ values is presented in Figure 1 for 1029 BATSE bursts with $F_{max}^{(64)} > 1$ ph cm^{-2} s^{-1} taken from the current BATSE catalog [5]. It has a non-symmetric shape with the maximum at ~ 0.3 and with the mean value $\langle\Theta\rangle = 0.429 \pm 0.007$. If the position of primary peaks is randomly distributed along time histories of long bursts, one should suppose a flat distribution of fraction ratios with the most probable value at 0.5. The observed distribution shows that primary peaks have a tendency to be at the beginning of bursts. This means that the moment of the primary peak could be pre-determined by the process of emission. This result confirms the effect found by the peak alignment method that early stages of long bursts are predominately shorter than the late stages after them [2].

The distribution of Θ values for 745 long bursts with $T_{90} > 2$ s demonstrates that $\langle\Theta\rangle = 0.434 \pm 0.009$. For 284 short bursts with $T_{90} < 2$ s, the value is equal 0.421 ± 0.013. Comparing these results, we conclude that in both cases there is about the same ratio between duration of the early stage and that of the late stage.

Spectral properties: The average hardness ratio for long bursts at the early stage is $\langle HR_{ES}^{long}\rangle = 1.19 \pm 0.02$. This is significantly harder than the average hardness ratio $\langle HR_{LS}^{long} \rangle = 1.01 \pm 0.02$ at the late stage. It is in good agreement with the effect of the difference of average HR at rise front and back slope of ACE [2]. The values of HR for short bursts during the early and late stages are practically the same. The average HR value at the early stage of $\langle HR_{ES}^{short} \rangle = 1.29 \pm 0.02$ is larger than the average HR value at the late stage of $\langle HR_{LS}^{short} \rangle = 1.24 \pm 0.02$, but this difference is not significant.

One may conclude that the classes of short and long bursts demonstrate the difference between ES and LS. But for long bursts this difference is seen both in terms of emission time durations and in terms of the HR. In the case of short events, this difference is based only on the different durations of the early and late stages.

INTENSITY-DEPENDENT EFFECTS OF THE PRIMARY PEAK POSITION OF LONG BURSTS

One may suspect that separation into early and late stages of emission is different for bursts with different intensities [3,4]. We study below this effect for the class of long bursts.

For this purpose we divided the sample of 745 long BATSE events into 4 groups of intensity with decreasing peak fluxes (Table 1). Each of these groups includes a large number of events (~ 185), which could be enough for conclusive results.

We build the distributions of the Θ parameter and estimate the mean values $\langle \Theta_i \rangle$, $i = 1 - 4$ for each intensity group. We applied the most robust method for measuring these values, which takes into account the known noise-produced bias and the effect of non-unique selection of primary peaks.

The different statistical noise of bright and dim gamma-ray bursts is known to lead to two biases of the emission time distributions:

(a) with decreasing S/N level the all-burst emission time goes down [1].

(b) with decreasing S/N level the separation between early and late stages of emission is determined not only by the actual moment of the primary peak in the light curve, but by a random moment of the largest positive fluctuation also.

To take into account effect (a), we produce artificially dimmed reference groups $i' = 1' - 3'$ from the real groups 1,2,3 which have the same set of events, but with the same S/N ratio as events in the dimmest group 4 (the details of this procedure can be found in [3]). In this case we receive four groups $1', 2', 3', 4$ with equal S/N ratio which allow us to make a robust comparison between $\langle \Theta_i \rangle$ for different intensity groups and to avoid possible systematic errors caused by effect (a).

To take into account effect (b), we evaluate for each pulse the probability that this pulse corresponds to the true primary peak in the emission light curve. The sum of these probabilities equals 1 by definition. For a given burst we calculate the values of $\tau_{50}^{(ES)}$ and $\tau_{50}^{(LS)}$ several times in respect to each of these pulses, and then average them with weighting by their probabilities.

BATSE is a triggering instrument, and a possible instrumental effect is the increasing fraction of missed slow-risers for groups with decreasing intensity (for example, see [7]). However, in the present study we avoid this problem using relatively bright bursts with peak flux $F_{max}^{(64)}$ above the level of 1 ph cm^{-2} s^{-1}.

The $\langle \Theta_i \rangle$ values are presented in Table 1 and in Figure 1. It is seen that $\langle \Theta_i \rangle$ is decreasing with decreasing intensity which points out that the fraction of early stages is decreasing with decreasing intensity.

To check the significance of this effect, we tried to fit $\langle \Theta_i \rangle$ vs. intensity with a constant function. The best fit is shown by the dashed line at the Figure 1. The probability of good agreement between the data and the best fit is less than 1%.

CONCLUSIONS

We may conclude that the primary peaks of long gamma-ray bursts with $T_{90} > 2$ s separate two physically distinguishable stages of emission: an early stage and a late stage. Around the primary peak, the shorter early stage of harder gamma-rays (ES) is transforming into the longer late stage of softer emission (LS).

This suggests that there is some generic change of emission process around the primary peak moment. The time history of long gamma-ray bursts is time irreversible. The average early-to-late ratio is 0.434±0.009 and the difference in the HR between ES and LS is 0.18±0.03.

Moreover, the difference between early and late stages of long bursts is found to be intensity-dependent. We found that the early-late ratio decreases with decreasing intensity of bursts. We have shown that this effect is not a result of missing of slow-risers and we believe that it is associated with a generic property of gamma-ray bursts.

If brighter and dimmer bursts would be produced by intrinsically stronger and weaker sources, the difference between early and late stages of bursts should be attributed to intrinsic properties of emitters. Stronger sources, on average, should rise up faster and decay more quickly than weaker sources. It does not look similar to other well-known cases of astronomical explosions, where stronger cataclysms are longer than weaker ones, but gamma-ray bursts have never been found to be a usual phenomenon.

ACKNOWLEDGEMENTS

The work in USA was supported by NASA project CGRO-98-120 of the CGRO Guest Observations Program. The work in Russia was supported by RFBR grant 98-02-17380 and by INTAS project No. 96-0315.

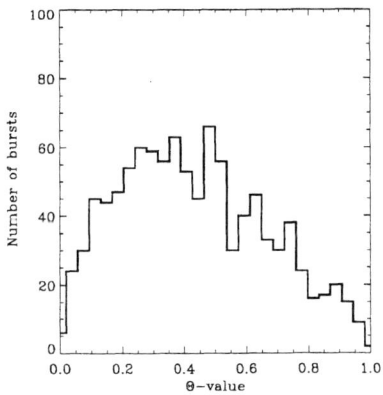 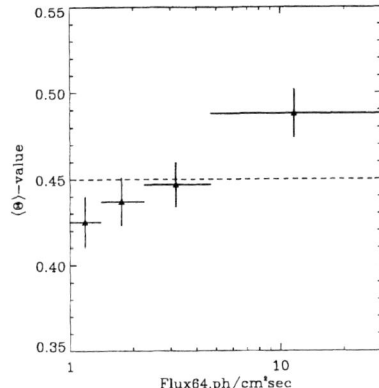

FIGURE 1. The distribution of Θ for all bursts (left) and average values of Θ for different intensity groups of long bursts (right).

TABLE 1.

Intensity group	Peak Flux (ph/cm^2−s)	Number of events	$\langle \Theta_i \rangle$
1	> 4.70	185	0.488±0.014
2	2.26-4.70	186	0.447±0.013
3	1.41-2.26	185	0.437±0.014
4	1.00-1.41	189	0.425±0.015

REFERENCES

1. Mitrofanov I. G., et al., *ApJ* **522**, 1069 (1999).
2. Mitrofanov, I. G., et al., *ApJ* **459**, 570 (1996).
3. Mitrofanov, I. G., et al., *ApJ* **523**, 610 (1999).
4. Stern B. E., et al., *ApJ* **464**, L111 (1996).
5. Current BATSE Catalog, electronically available
   http://gammaray.msfc.nasa.gov/batse/grb/catalog/
6. Norris, J. P., in "Gamma-Ray Bursts, 4th Huntsville Symposium", eds. Meegan, Preece & Koshut, (AIP: New York), **428**, 176 (1997a).
7. Higdon, J. C., & Lingenfelter, R. E., in "Gamma-Ray Bursts, 3rd Huntsville Symposium", eds. Kouveliotou, Briggs & Fishman, (AIP: New York), **384**, 402 (1996).
8. Mitrofanov et al., these proceedings.

The Duration Errors of Gamma-Ray Bursts

Istvan Horváth[1], Edward E. Fenimore[2], Jay P. Norris[3], Peter Mészáros[4], and Zsolt Bagoly[5]

[1] *Department of Physics, ZMN University, BJKMF, POB 12, H-1456 Budapest, Hungary*
[2] *MS D436, Los Alamos National Lab., Los Alamos, NM 87545*
[3] *NASA/Goddard Space Flight Center, Greenbelt, MD 20771*
[4] *Dept. of Astronomy & Astrophysics, Pennsylvania State University, University Park, PA 16802*
[5] *Laboratory for Information Technology, Eötvös University, Pázmány Péter sétány 1/A, H-1518 Budapest, Hungary*

Abstract. We have recalculated the durations of GRBs using a method which calculates the parabolical background fit using the standard background and burst intervals. This analysis indicates that the errors associated with the durations of the short bursts are larger than previously published in gamma-ray burst catalogs. Also the error bars are asymmetric, i.e. non-gaussian.

INTRODUCTION

Many burst models make prognostications that should be observable in a GRB time history [1–5]. Therefore time history and the length of the bursts are very important quantities, e.g. in the classification of the bursts based on the duration, either one believes two [6] or three [7,8] classes. In the 4B catalog [9] there are 1637 GRBs of which 1234 have a listed duration. Here we discuss a more detailed definition of GRB durations, going beyond the T_{50} and T_{90} definitions. Using a χ^2 method the background subtraction is detailed in Section 2. In Section 3 we discuss the errors derived for the durations, which are larger than previously published values [9], and are asymmetric (non-gaussian).

THE BACKGROUND FIT

The measurement of GRB durations is complicated for various reasons. Its usual definition [10] is through T_{90}, which is the time it takes to observe 90% of the total background-subtracted S_{tot} counts in a burst, starting and ending when 5% and 95% of S_{tot} have been observed. To calculate T_{90} we need a background fit.

From [11] we have background and burst interval definitions. This makes our task simpler, because it is very difficult to define a burst interval and the background intervals. It is additionally useful because we use the same interval definitions as used in the other papers [12,13]. Although [11] also contains background fits, we have also made our own independent background fits. There are no significant difference between the two fits.

For more than half of the bursts the linear background fit is not good. Therefore we have made parabolic background fits. Using the 1024 ms scale data we made a parabolic fit using the χ^2 method. For the best fit we have the smallest χ^2 (we call it χ^2_{best}). The best parabolic background fit for the added counts of the four Channels has three numbers a, b, c, because the fit is $ax^2 + bx + c$ (on the 1024 ms scale). For a, b and c we are looking for the intervals where χ^2 is less than $\chi_{lim} = \chi^2_{best} + 3.5$, which is within the 1σ (68%) level [14,15]. In this volume (which is rectangular in the a, b, c 3D space) we calculate χ^2 $100 \times 100 \times 100$ times (10^6 different points). Out of these million numbers, typically 10-20 thousand have χ^2 less than χ_{lim}. Using these 10-20 thousand points means some 10 thousand different fits where $\chi^2 < \chi_{lim}$. There is no statistical difference between them, according to χ^2 method. After this we can calculate N1, N2, N3, T_{90} and T_{50}. Here N1 is photon count in the first background interval; N3 is photon count in the burst interval; N2 is photon count in the second background interval. In all the three cases the backgrounds were subtracted. For example in BATSE trigger number 0543 we have the numbers:

543	11217				
-1618.00	-169.51	1278.00			
-2122.00	-188.97	1744.00			
81049.00	81854.48	82660.00			
1.9840	1.9923	2.0480	4.4160	4.5104	4.6080

The first number is the trigger number, the second is how many good ($\chi^2 < \chi_{lim}$) parabolic fits were made. Therefore we have 11217 potentially different N1, which is the difference between the background fit and an observed datum on the first background interval. The mean of these numbers is -169.5140. The biggest of these N1 is 1278, the smallest is -1618. Remember that in this interval the fitted background has already been subtracted. The second line is the same for N2. The third line contains the numbers of the burst's interval. At that point we have 11217 numbers which are photon counts above the background fit (total observed burst's photon counts). We have 11217 numbers between 81049 and 82660. The mean of them is 81854.48. The next line contains the calculated duration. The 11217 good background fits imply a burst's duration. The numbers in this line are the following: sT_{50}, T_{50}, bT_{50}, sT_{90}, T_{90} and bT_{90}. The T_{50} (whose definition is similar to T_{90} in the first paragraph in this section) is a mean of the 11217 different T_{50}. The sT_{50} and bT_{50} mean the 1σ limits. For T_{90} the numbers have the same meaning. Therefore our T_{90} is 4.51 sec and it is a 68% probability that T_{90} is between 4.416s and 4.608s. In the 4B catalog these numbers are 4.662s and 5.066s. Therefore our results look the same as what is published in the 4B catalog, except that our error is twice smaller.

THE RESULTS

For various reasons, the overlap between the bursts with both a 4B duration and a duration calculated by us with the above method consists of 930 bursts. There is a big difference between the two calculations. At the 3σ level only 21 (40%) of the population have the same duration (T_{90}). 22 (17%) have less and 57 (43%) have larger duration than the 4B catalog describes. Therefore, although our calculation has a general agreement with the 4B durations, our calculated T_{90}s mostly are larger than the 4B durations. Our calculated duration errors are also different than the 4B catalog.

For long bursts (using the definition in [6], long means $T_{90} > 2s$) 58% of our errors are larger than the 4B errors. 30% of our errors are twice as large as the

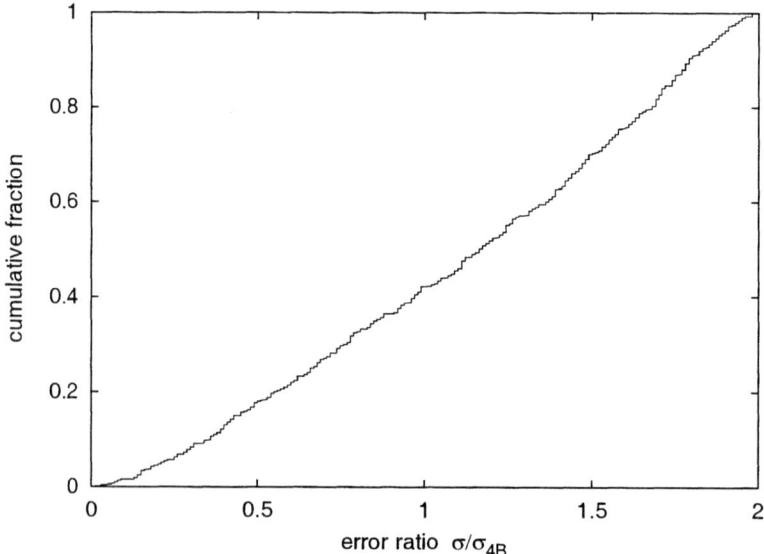

FIGURE 1. The σ/σ_{4B} duration error ratio distribution. For error ratios > 1 the $2 - \sigma_{4B}/\sigma$ transformation is used to transform the ratio from the $[1, \infty]$ interval to $[1, 2]$.

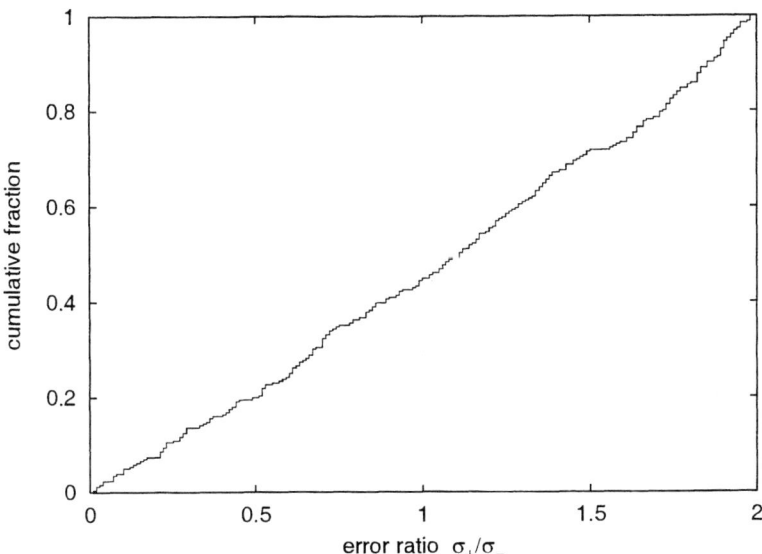

FIGURE 2. The σ_+/σ_- error ratio shows the duration error asymmetricity. Here the $2-\sigma_-/\sigma_+$ transformation is also applied.

4B errors and only 17% are less than half of the 4B errors. Figure 1 shows the error ratio (σ/σ_{4B}) distribution. If $\sigma_{4B} > \sigma$ the ratio is between zero and one. If $\sigma > \sigma_{4B}$ the ratio is bigger than one (between one and infinity). To make the viewpoint symmetric, one should use the ratio between 0 and 1. If it is larger than one let us have the opposite ratio (σ_{4B}/σ). We have chosen a simple way to show this. On the horizontal axis there is a ratio. On the left side (between 0 and 1) there is a ratio if σ/σ_{4B} less than a unity. On the right side (between 1 and 2) we have drawn the $2 - \sigma_{4B}/\sigma$ function if σ/σ_{4B} bigger than a unity. On the vertical axis there is the cumulative fraction. One means all the 930 bursts, while 0.5 means half of them (465 GRBs). One can find the numbers mentioned above by checking the distribution function values at 0.5 and 1 and 1.5 (0.17; 0.42; 0.70).

From our calculations, for each duration we obtain two different error bars, since we calculate both the plus and minus one σ levels. Subtracting them from the mean of the durations, we can obtain the errors of the duration. Let us call them plus and minus errors. Of course in this calculation the two numbers are different. The question is whether they differ just accidentally or not. The answer is that the difference is real. One of them is typically larger then the other.

Figure 2 shows a function similar to that in Figure 1. Here one can find the cumulative distribution of the σ_+/σ_- (σ_+ is the plus error bar, σ_- is the minus error bar). Above one we use again the $2-\sigma_-/\sigma_+$ transformation on the horizontal axis. 55% of the plus errors are larger than the minus ones. 29% of them are larger than twice the minus ones, and 19% are smaller than half of the minus ones. Therefore nearly half of the bursts have a factor of two difference between the two (plus (σ_+) and minus (σ_-)) duration errors.

ACKNOWLEDGEMENTS

This research was supported in part through NASA NAG5-2857, OTKA 29461. Useful discussion with L. G. Balazs, E. Feigelson, T. M. Koshut, J. Nousek, J. D. Scargle are appreciated.

REFERENCES

1. Usov, V. V., *Nature* **357**, 472 (1992).
2. Begelman, M. C., et al. *MNRAS* **265**, L13 (1993).
3. Mészáros, P., and Rees, M. J., *ApJ* **405**, 278 (1993).
4. Katz, J. I., *ApJ* **422**, 248 (1994).
5. Panaitescu, A., and Mészáros, P., *ApJ* **492**, 683 (1998).
6. Kouveliotou, C., et al. *ApJ* **413**, L101 (1993).
7. Mukherjee, S., Feigelson, E. D., Babu, G.J., Murtagh, F., Fraley, C., and Raftery, A., *ApJ* **508**, 314 (1998).
8. Horváth, I., *ApJ* **508**, 757 (1998).
9. Paciesas, W. S., et al. *ApJ Suppl.* **122**, 465 (1999).
10. Koshut, T. M., et al. *ApJ* **463**, 570 (1996).
11. CGRO Space Science Center, BATSE database, *cossc.gsfc.nasa.gov*.
12. Bonnell, J. T., Norris, J. P., Nemiroff, R. J., and Scargle, J. D., *ApJ* **490**, 79 (1997).
13. Lee, T., and Petrosian, V., *ApJ* **470**, 479 (1996).
14. Press, W. H., Flannery, B. P., Teukolsky, S. A., and Vetterling, W. T., *Numerical Recipes*, Cambridge University Press, Cambridge (1992).
15. Lampton, M., Margon, B., and Bowyer, S., *ApJ* **208**, 177 (1976).

Power Density Spectra of Gamma-Ray Bursts

Andrei M. Beloborodov, Boris E. Stern, and Roland Svensson

Stockholm Observatory, Saltsjöbaden SE-133 36, Sweden

Abstract. The longest gamma-ray bursts (GRBs) display power-law power density spectra (PDSs). Shorter bursts have PDSs more strongly affected by statistical fluctuations. The underlying power law can then be reproduced with high accuracy by averaging the PDSs for a large sample of bursts. It indicates that different GRBs are random realizations of the same stochastic process. The power-law PDS provides a new sensitive tool for studies of gamma-ray bursts. In particular, we calculate the PDSs of bright bursts in separate LAD energy channels. The average PDS flattens in the hardest channel ($h\nu > 300$ keV) and steepens in the softest channel ($h\nu < 50$ keV). The average PDS of bolometric light curves has a slope $\alpha \approx -5/3$ and a sharp high frequency break at ~ 1 Hz.

INTRODUCTION

Contrary to the complicated diverse behavior in the time domain, long GRBs show a simple behavior in the Fourier domain (Beloborodov, Stern, & Svensson 1998). Their PDS is a power law of index $\alpha \approx -5/3$ (with a high frequency break at ~ 1 Hz) plus standard (exponentially distributed) statistical fluctuations superimposed onto the power law. The $-5/3$ slope and the 1 Hz break characterize the process randomly generating the diverse light curves of GRBs. Intriguingly, the PDS slope coincides with the Kolmogorov law

Here, we illustrate the power law seen in the longest bursts, present the average PDS for an extended sample of GRBs, and compare the PDSs in separate energy channels. We quantify the difference of the temporal structure between the channels in terms of the PDS slope and compare the results with previous studies of the autocorrelation function (ACF). An extended description of this work is given in Beloborodov (1999) and Beloborodov, Stern, & Svensson (2000).

In our analysis, we use GRB light curves with 64 ms resolution obtained by BATSE in the four LAD energy channels, I–IV: (I) $20-50$ keV, (II) $50-100$ keV, (III) $100-300$ keV, and (IV) $h\nu > 300$ keV. The background is subtracted using linear fits to the 1024 ms data. Our sample consists of 527 GRBs with $T_{90} > 20$ s and the peak flux in channels II+III, $C_{\text{peak}} > 100$ counts per 64 ms time bin.

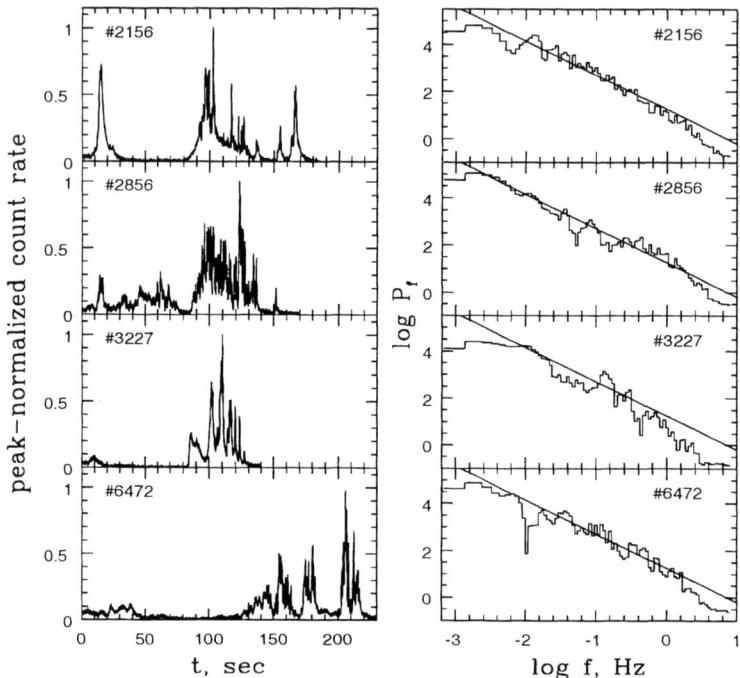

FIGURE 1. The light curves (in channels II+III) and their PDSs for the four brightest bursts in the sample with $T_{90} > 100$ s. The light curves are peak-normalized and, correspondingly, the PDSs are normalized by C_{peak}^2. We smoothed the PDSs on the scale $\Delta \log f = 0.04$ before the plotting. The straight lines show the fit to the average of the 4 PDSs, $\log P_f = 1.25 - 1.45 \log f$.

INDIVIDUAL POWER DENSITY SPECTRA

The four brightest bursts with $T_{90} > 100$ s in our sample have trigger numbers 2156, 2856, 3227, and 6472. To simplify the comparison of different bursts, we use peak-normalized light curves, $C(t)$. Their Fourier transform, C_f, is therefore normalized by C_{peak}, and the PDS, $P_f = C_f C_f^*$, is normalized by C_{peak}^2. The light curves and their PDSs are shown in Figure 1. The light curves are very different while the PDSs are similar. They can be described as a single power law, $\log P_f = A + \alpha \log f$ ($A \approx 1$ and $\alpha \approx -1.5$) with super-imposed fluctuations $\Delta P_f / P_f \sim 1$. In spite of the large ΔP_f, the power-law behavior can be seen in each burst due to the power-law extending over more than two decades in frequency.

The PDS analysis was previously performed for a number of individual bursts. Belli (1992) analyzed GRBs detected by the Konus experiment and Giblin, Kouveliotou, & van Paradijs (1998) studied individual BATSE bursts. The power-law PDS can be observed in their longest complex bursts as well.

THE AVERAGE PDS OF GRBS

In most of the GRB models (e.g., in the internal shock model), different bursts are produced by one physical mechanism of a stochastic nature, i.e., an individual burst is a random realization of the same standard process. With such an approach, individual GRBs are like pieces of a "puzzle", and the features of the standard engine can be probed with statistical methods applied to a large ensemble of GRBs.

The simplest statistical characteristic of the PDSs is the average PDS. The average PDS, $\bar{P}_f$, for the 527 peak-normalized light curves in LAD channels (II+III) is shown in Figure 2. For comparison, we also plot the average PDS for light curves normalized by $\sqrt{\Phi}$, where Φ is the burst fluence. The average PDS then has a different amplitude. Nevertheless, $\bar{P}_f$ again follows a power-law with approximately the same slope. This provides evidence that the self-similar behavior with $\alpha \approx -5/3$ is an intrinsic property of GRBs, rather than an artifact of the averaging procedure. This interpretation is also supported by the fact that we observe the same power-law behavior for the longest individual bursts (see Fig. 1). Note also that the distribution of P_f around $\bar{P}_f$ is self-similar with respect to shifts in f (see Beloborodov, Stern, & Svensson 1998). The self-similar behavior breaks at ~ 1 Hz.

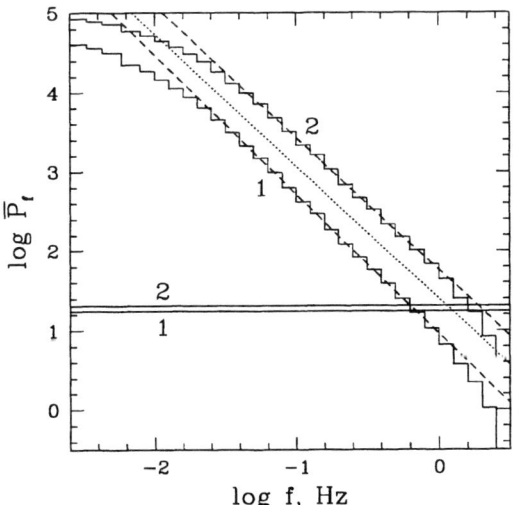

FIGURE 2. The average PDS (in LAD channels II+III) for the full sample of 527 GRBs. The histograms marked as 1 and 2 correspond to the peak-normalization and the $\sqrt{\Phi}$-normalization, respectively. The horizontal lines show the corresponding normalized Poisson levels averaged over the sample. *Dashed lines* are the power-law fits in the range $-1.4 < \log f < -0.1$: $\log \bar{P}_f = 0.96 - 1.75 \log f$ in the peak-normalized case and $\log \bar{P}_f = 1.76 - 1.67 \log f$ in the $\sqrt{\Phi}$-normalized case. The *dotted line* shows the $-5/3$ slope.

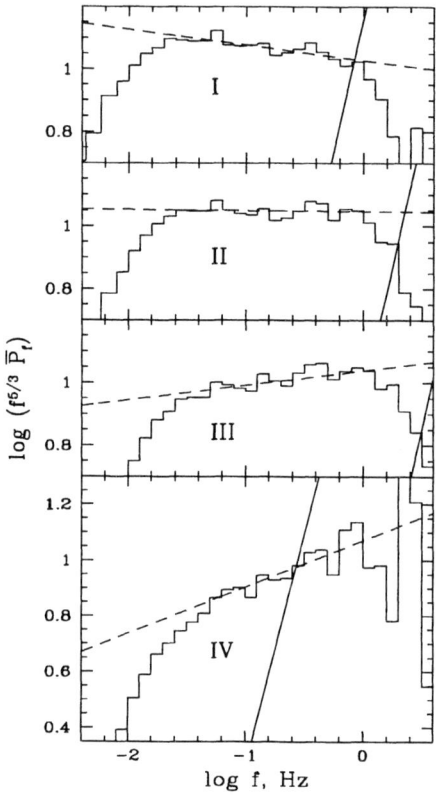

FIGURE 3. The average PDS for the sample of 152 GRBs in separate energy channels, I–IV. *Dashed lines* show the power-law fits (see Table 1). *Solid lines* show the Poisson level.

PDS AND ACF IN SEPARATE LAD CHANNELS

The average PDSs for the brightest 152 bursts in LAD channels I–IV are shown in Figure 3. The differences in the slopes are clearly seen. We fitted the PDSs by power laws, $\log \bar{P}_f = A + \alpha \log f$, in the range $-1.6 < \log f < 0$. Channel IV is fitted in the range $-1.3 < \log f < -0.1$. A and α of the fits are listed in Table 1.

TABLE 1. Fitting parameters for the average PDS and ACF

Channel	PDS slope, α	PDS amplitude, A	ACF index, β	ACF width, τ_0
I	−1.72	1.03	0.73	14.0
II	−1.67	1.05	0.67	10.7
III	−1.60	1.06	0.63	7.3
IV	−1.50	1.07	0.60	5.1

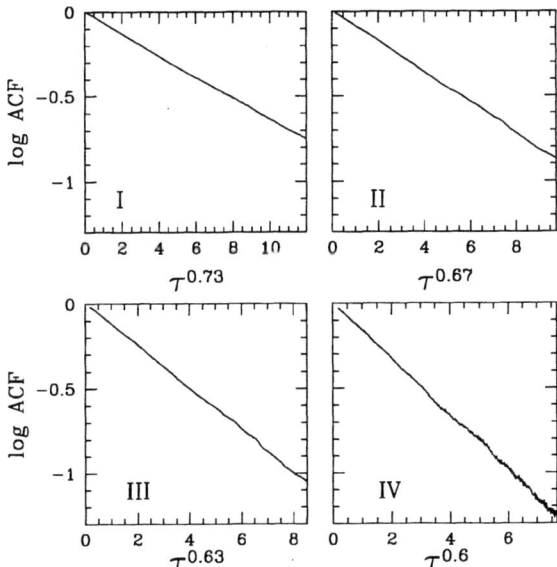

FIGURE 4. The average ACFs in channels I–IV are plotted against τ^β which should give a straight line for a stretched exponential.

The autocorrelation function (ACF) contains the same information as the PDS since one is the Fourier transform of the other (the Wiener-Khinchin theorem). The average ACF, $\bar{A}(\tau)$, for our sample of 152 bright GRBs is shown in Figure 4. In each channel, the ACF is perfectly fitted by a stretched exponential: $\bar{A}(\tau) = \exp(-[\tau/\tau_0]^\beta)$. The parameters τ_0 and β are listed in Figure 2. Note the approximate relation between the ACF index, β, and the PDS slope, α, $\beta \approx -(1 + \alpha)$. τ_0 measures the ACF width. The ACF gets narrow at high energies, in agreement with previous studies (Fenimore et al. 1995).

REFERENCES

1. Belli, B. M., *ApJ* **393**, 266 (1992).
2. Beloborodov, A. M., in *Gamma-Ray Bursts: The First Three Minutes*, eds. J. Poutanen & R. Svensson, ASP Conf. Ser. 190, San Francisco, 1999, pp. 47.
3. Beloborodov, A. M., Stern, B. E., and Svensson, R., *ApJ* **508**, L25 (1998).
4. Beloborodov, A. M., Stern, B. E., and Svensson, R., *ApJ*, in press, astro-ph/0001401 (2000).
5. Fenimore, E. E., et al., *ApJ* **448**, L101 (1995).
6. Giblin, T. W., Kouveliotou, C., and van Paradijs, J., in *Gamma-Ray Bursts: 4th Huntsville Symposium*, eds. C. A. Meegan, R. D. Preece, & T. M. Koshut, AIP Conf. Proc. 428, New York, 1998, pp. 241.

GRB Time-Dilation Measurements Corrected for Trigger Bias

Jerry T. Bonnell[1], Jay P. Norris[1], Gabriela F. Marani[1], and Robert J. Nemiroff[2]

[1] *NASA Goddard Space Flight Center, Greenbelt, MD 20771, USA*
[2] *Department of Physics, Michigan Technological University, MI 49931, USA*

Abstract. We measure time-dilation factors (TDFs) for 1158 long gamma-ray bursts (GRBs) using a peak-alignment method. Our standard corrections for noise bias and spectral redshift are implemented. Also for the first time we account for trigger bias against detecting relatively short, low signal-to-noise GRBs using a retriggering procedure which approximates the on-board BATSE algorithm on the 1024-ms timescale. This procedure implicitly addresses any bias that would be introduced when comparing bright bursts with dim bursts. The retriggering step significantly reduces the relative TDFs between the 10% brightest bursts and the 30% dimmest bursts. These corrected TDFs are compared with Monte Carlo simulations of a distribution of GRBs adopting standard cosmologies, GRB luminosity functions, and GRB rate-density evolution with cosmic time. Statistically, the corrected GRB TDFs are consistent with the expected extrinsic cosmological time dilation.

INTRODUCTION

When corrected for noise bias and systematic effects, BATSE gamma-ray burst (GRB) durations and time profiles have been demonstrated (e.g. [1–3]) to show an anticorrelation with their 50-300 keV peak fluxes – dimmer burst time profiles are longer or stretched when compared to brighter bursts. Norris et al. [4] initially suggested that this effect was consistent with the signature of cosmological time dilation. The sources of long GRBs are now known to arise at cosmological distances, but it has been argued (e.g. [5–7]) that this apparent time dilation could well be an intrinsic property of the GRBs themselves and not the naturally expected extrinsic effect of general relativity.

Is the observed GRB time-dilation consistent with cosmology? We have examined the question by performing detailed peak-aligned time-dilation measurements for 1158 long GRBs, determining the relative stretching or time-dilation factors (TDFs) for a range of GRB brightness groups. Our standard corrections for noise bias and spectral redshift are implemented. In addition, for the first time we have applied a correction for trigger bias against relatively short, low signal-to-noise ratio (S/N),

TABLE 1. GRB Peak Fluxes (50-300 keV) and Peak-Aligned TDFs.

Peak Flux (ph cm^{-2} s^{-1})	TDFs ($\pm 1\sigma$) errors
> 7.77	1.00
3.81 – 7.77	$1.26^{+0.12}_{-0.12}$
2.36 – 3.81	$1.28^{+0.15}_{-0.12}$
1.71 – 2.36	$1.64^{+0.17}_{-0.16}$
1.30 – 1.71	$1.66^{+0.21}_{-0.16}$
1.06 – 1.30	$1.90^{+0.18}_{-0.10}$
0.85 – 1.06	$1.41^{+0.16}_{-0.17}$
0.69 – 0.85	$2.00^{+0.33}_{-0.25}$
0.55 – 0.69	$1.57^{+0.21}_{-0.17}$
0.30 – 0.55	$2.04^{+0.33}_{-0.30}$

bursts in measuring time-dilation factors. This significant correction reduces TDFs by factors of $\sim 0.6 - 0.9$ for the dimmest 3 classes. As a result we find that

$$TDF_{corr} = (1 + z_{\text{dimmest}})/(1 + z_{\text{brightest}}) \approx 2 \qquad (1)$$

when comparing the dimmest 10% and brightest 10% of GRBs. We also find that the measured TDF vs. peak flux relation is consistent with extrinsic cosmological time dilation when standard cosmologies, a GRB luminosity distribution, and number density evolution are considered in Monte Carlo simulations of GRBs distributed to large redshifts.

ANALYSIS

We divide the 1158 long BATSE GRBs ($T_{90} > 2$ s), through January 23, 1999 into ten peak-flux groups with ≈ 116 GRBs per group. The 50-300 keV peak-flux divisions are given in Table 1. The brightest group, spanning from 7.77 to 183 ph cm^{-2} s^{-1}, is a factor of 25-100 brighter than the dimmest group.

In this work we require that all GRBs trigger on a 1024-ms timescale when time profiles are compared after S/N equalization. Most of the dimmest BATSE GRBs trigger on this timescale, however, some of the brightest bursts (group 1), with which the nine dimmer groups' time-dilation measurements are to be compared, would not have triggered BATSE if they had been detected at a S/N typical of the dimmest bursts profiles. Therefore, we S/N equalize the eight-detector time profiles of group 1 GRBs successively to the S/N appropriate to the peak-intensity levels of the nine dimmer groups.

For a bright burst to survive at each dimmer level, we require that at least two of the eight-detector time profiles exceed a 5.5σ level above a preceding background (17 1.024-s accumulations), similar to the on-board BATSE requirements. The

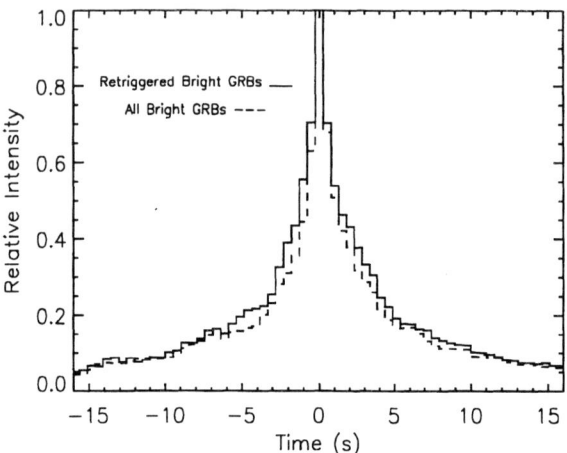

FIGURE 1. Average peak-aligned GRB time-profiles for bright GRBs before (dashed) and after (solid histogram) retriggering on a 1024-ms timescale. The retriggered average time profile is systematically broader.

number of detectors actually examined is drawn from the distribution of the number of triggered detectors per peak-flux group.

For example, Figure 1 illustrates the net effect of this requirement on the average peak-aligned profiles for group 1 (brightest) bursts when retriggered at the S/N appropriate for group 10 (dimmest) GRBs. The average profile for the 112 bright GRBs of group 1 is shown as the dashed histogram. The solid histogram is the average profile for the 61 GRBs from this group which pass the retriggering requirement. The average retriggered bright GRB profile is systematically broader (stretched) when compared the total group 1 profile (all bright GRBs). Previous comparisons of bright and dim GRBs which did not account for this systematic effect would therefore over-estimate the relative peak-aligned TDFs.

For each peak-flux group, the GRB time profiles (0.5-s resolution) are co-added in registration at the time of maximum intensity. The widths of the averaged peak-aligned profiles for the nine dimmer groups, at several intensity points within a $(-12, +24)$-bin window of the average peak are ratioed to the same widths for the average peak-aligned profile of group 1 – dimmed and S/N equalized to each peak-flux groups' levels. Only those bright bursts (group 1) which retriggered the simulated BATSE trigger at each dimmer group's peak intensity level are included in the width measurement. For each time-dilation measurement, dim group to group 1, the average of the ratios from the several power points yields the measured (relative) time-dilation factor. Bootstrap errors for the observed time-dilation factors (TDFs) are generated by producing a parent population of dimmed and S/N equalized profiles for the bright bursts, and drawing from this population 500 times, thence estimate the 1σ errors.

GRB time profiles are energy-dependent: individual pulses and durations are shorter at higher energy. Therefore in our fixed spectral band of observation, we record time profiles of bursts from higher redshifts, downshifted into our spectral band (25 keV to ≈ 1 MeV). We estimate the correction for this redshift narrowing of the average profiles by comparing the original bright bursts' time profiles with the same redshifted and time dilated. Trial TDFs, which reproduce the observed TDFs, are then estimates of the actual $TDF = (1 + z_{\text{dim}})/(1 + z_{\text{brightest}})$. These redshift-corrected TDFs versus peak fluxes are shown in Table 1 and illustrated in Figure 2 where they are compared directly with Monte Carlo expectations from cosmological modeling. The modeling considers standard cosmological parameters and includes estimates of the GRB luminosity distribution and rate-density evolution with cosmic time.

MONTE CARLO UNIVERSE

The Monte Carlo universe of GRBs considered here accounts for GRB spectral diversity, rate-density evolution, and luminosity distributions in a universe where $\Omega_M = 0.3$, $\Omega_\Lambda = 0.7$, $H_o = 65$ km s^{-1} Mpc^{-1}.

The GRB spectral diversity was based on the spectral fits of Band et al. [8]. Since these spectral shapes correspond to bursts at z > 0, a K-correction is applied following Fenimore and Bloom [2]. The GRB rate density distribution was alternately based on three models: constant comoving density (CCD) with $\rho(z) = \rho_o$; the star formation rate deduced from the rest-frame UV luminosity density (SFR-UV) from Madau et al. [9]; and star formation rate tracking the total output of radio-loud AGN (SFR-AGN) from Hughes et al. [10]. The GRB luminosity L was also drawn from three different distributions: standard candles; a negative power-law luminosity function (e.g. [11]); and a positive power-law luminosity function where $\Phi(L) = L^{-\beta}$, if $L_{\min} < L < L_{\max}$.

Briefly, the Monte Carlo code works as follows: given ρ_o and L, we create a Monte Carlo universe for long bursts, randomly picking $0 < z < z_{\max}$, and one K-corrected spectrum. We then compute the apparent duration of the burst and the 50-300 keV peak flux. This peak flux indicates to which brightness group the simulated burst belongs. We repeat these steps thousands of times. For each group, the time-dilation measurement is given by the ratio of averages of the logarithms of the bursts' durations. We accept the values of ρ_o and L if the fits can reproduce the observed $\langle V/V_{max}\rangle$ test and total number of events per year, for long bursts with peak fluxes P = 1 ph cm^{-2} s^{-1} (i.e., $\langle V/V_{\max}\rangle = 0.37 \pm 0.02$, ≈ 100 GRBs/yr).

CONCLUSIONS

For the first time, we have applied a correction for trigger bias against relatively short, low S/N bursts in computing time-dilation measurements. This important correction makes TDFs a factor of ≈ 0.6-0.9 smaller for the dimmest 3 classes.

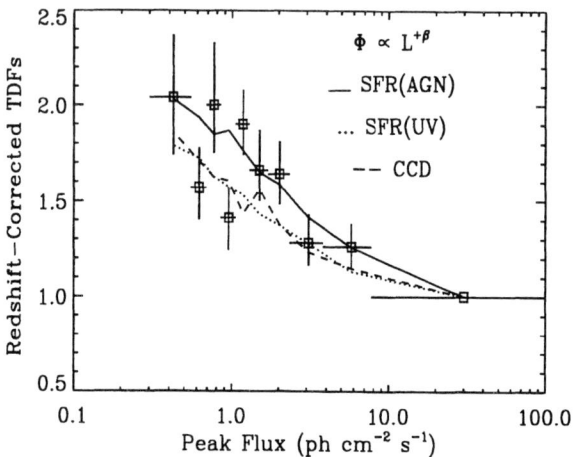

FIGURE 2. Corrected GRB time-dilation measurements are compared with a Monte Carlo simulation in a universe where where $\Omega_M = 0.3$, $\Omega_\Lambda = 0.7$, $H_o = 65$ km s^{-1} Mpc^{-1}. A power-law GRB luminosity function and three different GRB rate-density distributions were adopted.

The resulting general trend of corrected TDFs with peak fluxes are well fitted by cosmological models. In general we find that all the observed time-dilation measurements are within 2σ confidence level, for combinations of a standard candle or broad functions for the GRB luminosity distribution, and constant comoving or star formation models for the GRB rate-density distribution. This would imply that the detected time dilation in GRBs is consistent with being attributable to cosmological effects. However, we note that using a negative power-law luminosity function, and adopting the SFR-UV model for the GRB rate-density distribution yields the worst fit, in some cases, $\approx 3\sigma$ away from the observed values.

REFERENCES

1. Bonnell, J. T., Norris, J. P., Nemiroff, R. J., and Scargle, J.D, *ApJ* **490**, 79 (1997).
2. Fenimore, E. E. and Bloom, J., *ApJ* **453**, 25 (1995).
3. Deng, M. and Schaefer, B. in AIP Conf. Proc. **428**, Gamma-Ray Bursts, eds. C. A. Meegan, R. D. Preece and T. M. Koshut (New York: AIP) 251, (1998).
4. Norris, J. P. et al., *ApJ* **424**, 540 (1994).
5. Mitrafanov, I. G. et al., *ApJ* **459**, 570 (1996).
6. Stern, B. E., Poutanen, J. and Svensson, R., *ApJ* **489**, L41 (1997).
7. Brainerd, J. J., *ApJ* **487**, 21 (1996).
8. Band, D. et al, *ApJ* **413**, 281 (1993).
9. Madau, P. et al., *ApJ* **297**, L17 (1998).
10. Hughes et al., *Nature* **394**, 241 (1998).
11. Kommers, J. M. et al., *ApJ* **491**, 704 (1997).

Rise and Decay Time of Subpeaks in Short Duration Bursts

Varsha Gupta[1], Patrick Das Gupta[2], and P. N. Bhat[3]

[1] *Department of Physics & Astrophysics, University of Delhi, Delhi 110 007, India*
[2] *Tata Institute of Fundamental Research, Homi Bhabha Road, Colaba, Mumbai 400 005, India*

Abstract. Temporal profiles of subpeaks detected in GRBs, belonging to a sample of 65 short duration bursts selected from the 3B catalog, have been studied by fitting the former with lognormal functions since most subpeaks exhibit fast rise and slow decay. We present statistical evidence for a systematic increase in the ratio of rise to decay time with peak position. Using weighted correlation analysis we find that decay time displays a strong anti-correlation with peak position. On the other hand, no significant correlation between rise time and instant of peak maximum has been observed. If subpeaks in a burst arise due to emission from distinct shells then those scenarios in which later shells either suffer less deceleration or have larger speeds so that they catch up with slower matter earlier, are favored as they provide the required explanation for the observed decay time versus peak position anti-correlation.

TEMPORAL CHARACTERISTICS OF SUBPEAKS

We identify the subpeaks present in each burst belonging to a sample of 65 short duration GRBs taken from the 3B catalog, according to a systematic procedure [1-3]. Each subpeak is fitted with a lognormal function of the following form,

$$C_j = \begin{cases} \frac{N}{(j-k)\sqrt{2\pi}\sigma} \exp\left[-(\log(j-k)-\mu)^2/2\sigma^2\right] & j > k, \\ 0 & j \leq k. \end{cases} \quad (1)$$

C_j being the expected photon count in the jth time bin and k is the time bin at which the subpeak starts. The best fit in case of each burst correspond to low χ^2 (except in the case of the burst with BATSE trigger no. 2614 which, therefore, is excluded from the analysis that follows) tracing out the observed time profile rather closely.

Rise time Tr (decay time Td) of a peak in a GRB is taken to be the time interval during which the value of the fitted lognormal function increases (decreases) from 5% (95%) to 95% (5%) of its maximum value. The symmetry of a peak is characterized by a parameter r_{rd} defined as,

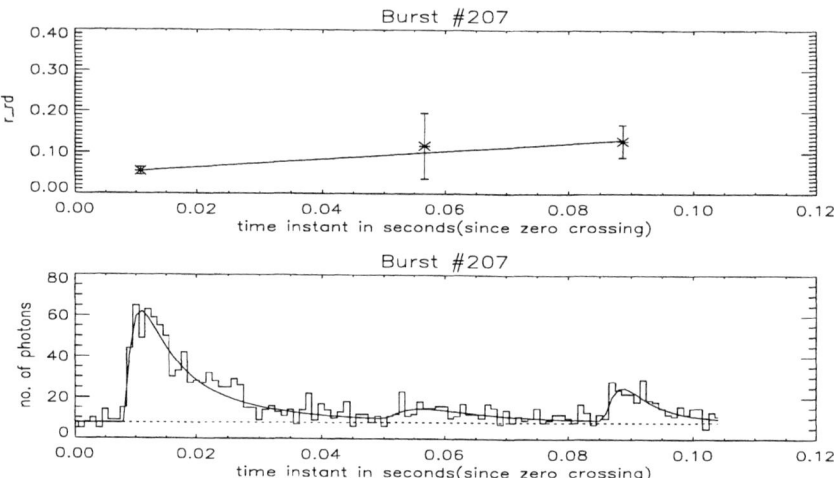

FIGURE 1. Lower part of the figure shows the time history of the GRB with BATSE trigger no. 207 with fitted lognormal functions. Top portion provides a plot of r_{rd} along with 1σ error against peak position - the straight line fit is obtained from χ^2 minimization.

$$r_{rd} \equiv \frac{Tr}{Td} \quad (2)$$

that is manifestly redshift independent, and is found to lie in the range ~ 0.02 to ~ 0.8. In many instances, it is observed that the first peak tends to be more asymmetric than the subsequent ones, leading one to fit the r_{rd} versus peak position plots with straight lines obtained by minimizing the corresponding χ^2. A few of these plots are displayed in Figures 1 and 2. Except in 13 cases, the slopes of all straight line fits are positive suggesting a tendency of peaks evolving towards increasing values of r_{rd}. The distribution of slopes is shown in left box of Figure 3. The cumulative behavior of r_{rd} corresponding to all 161 peaks detected in 64 short duration bursts has been studied using a weighted correlation analysis of r_{rd} versus peak position. The observed degree of correlation is ~ 0.17 corresponding to a probability of 0.03 of its being due to pure chance. Right box of Figure 3 shows a scatter diagram of r_{rd} against peak position.

To understand the feature described above, we subject both Tr as well as Td to weighted correlations against peak position. It is found that while Tr does not show a significant correlation, the decay time is strongly anti-correlated with the peak position, the degree of correlation being ~ -0.43, corresponding to a chance probability of 6×10^{-8}. It is evident then that the propensity of r_{rd} to increase with peak position is essentially due to an associated decrease in the value of decay time, in a statistical sense.

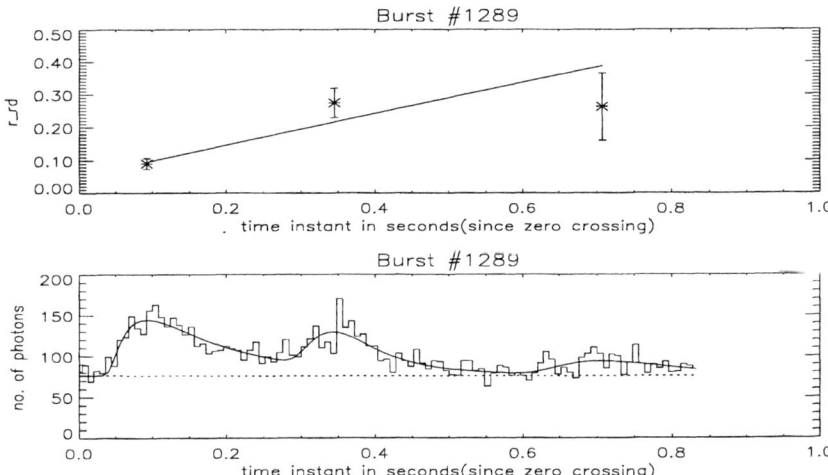

FIGURE 2. Bottom portion provides the time history of the burst with BATSE trigger no. 1289 superposed with the fitted function. The peak symmetry parameter with 1σ error against peak position has been plotted in the upper part of the figure.

THEORETICAL IMPLICATIONS

Assuming that a subpeak in a burst is due to emission of gamma photons from a distinct shocked shell, we consider a shell created at the instant t_n^c, and thereafter billowing through the ISM with highly relativistic speed. For $\Gamma > 100$, even when the shell deceleration is sufficiently small, the associated Lorentz factor can change by a large factor. Therefore, in the present study we include effects of time varying Lorentz factor, otherwise our analysis closely follows previous work [4–7].

Suppose the nth subpeak in a burst is due to emission from such a shell between the instants t_n and $t_n + \Delta t_n$, the rapid rise in the observed count rate may be attributed to those photons emitted directly along the line of sight so that the rise time is

$$Tr_n \approx \frac{\Delta t_n}{2\Gamma^2(\bar{v}_n)} \approx \frac{\Delta \tau_n \bar{\Gamma}_n}{2\Gamma^2(\bar{v}_n)} \qquad (3)$$

where $\Gamma(\bar{v}_n)$ is the Lorentz factor corresponding to the time averaged speed,

$$\bar{v}_n \equiv \frac{\int_{t_n}^{t_n+\Delta t_n} v_n(t)dt}{\Delta t_n} \qquad (4)$$

while $\Delta \tau_n$ is the time interval of emission evaluated in the comoving frame corresponding to the shell and $\bar{\Gamma}_n$ is given by

$$\bar{\Gamma}_n \equiv \frac{\int_{\tau_n}^{\tau_n+\Delta \tau_n} \Gamma(\tau)d\tau}{\Delta \tau_n} \qquad (5)$$

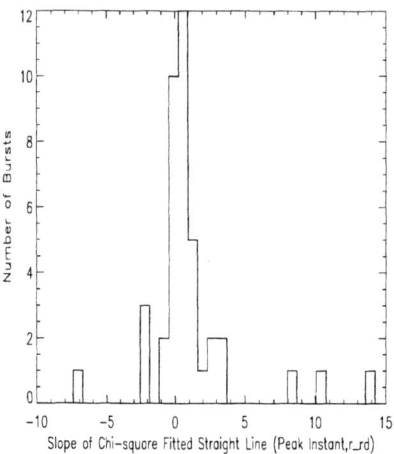 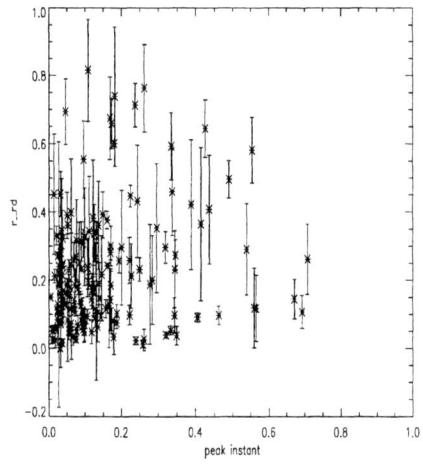

FIGURE 3. Left box shows the distribution of slopes of linear fits corresponding to r_{rd} versus peak position for 41 GRBs and the right box shows the scatter diagram of r_{rd} against peak position for 64 bursts.

The decaying tail of the peak is largely constituted of photons emitted at time $t_n + \Delta t_n$ from those regions of the shell within an angle $\sim 1/\Gamma'_n$, where Γ'_n is the Lorentz factor of the shell at that instant. Hence, the decay time is given by

$$Td_n \approx \frac{t_n + \Delta t_n - t_n^c}{2\Gamma'^2_n} \approx Tr_n \left(\frac{\Gamma(\bar{v}_n)}{\Gamma'_n}\right)^2 + (\tau_n - \tau_n^c)\frac{\bar{\Gamma}_n^-}{2\Gamma'^2_n} \quad (6)$$

where $\tau_n - \tau_n^c$ is the time interval as measured in the comoving frame of the shell during which latter remains 'inactive', while $\bar{\Gamma}_n^-$ is the time averaged Lorentz factor of the shell prior to its 'gamma active' phase, evaluated in its rest frame.

If we assume that $\Delta \tau_n$ does not vary systematically from peak to peak [7] then it is expected that rise time does not show any significant correlation with peak position since the ratio of $\bar{\Gamma}_n$ to $\Gamma^2(\bar{v}_n)$ is least likely to change with n even when v_n differs. The significant anti-correlation observed between decay time and peak position could be due to a systematic decrease with n either in the parameter κ defined as

$$\kappa \equiv \frac{\bar{\Gamma}_n^-}{\Gamma'^2_n} \quad (7)$$

or in the magnitude of $\tau_n - \tau_n^c$.

In the first alternative, where $\tau_n - \tau_n^c$ does not have a definite variation, the physical cause of a steady decrease in κ could be due to the tendency of later shells suffering less deceleration as they plow through a rarer ambient medium because of sweeping away of matter by previous shells. The other alternative, in which $\tau_n - \tau_n^c$

tends to decrease with peak position can naturally arise in internal shock models where later peaks are produced due to interaction between faster wind catching up with the slower one.

DISCUSSIONS

The parameter r_{rd} is observed to increase with peak position in most cases implying that subsequent peaks in a burst tend to become more symmetric. The origin of this behavior can be attributed to the observation that decay time exhibits a strong anti-correlation with peak position while rise time does not show any such trend. The latter naturally follows if the proper time interval of emission that results in a subpeak does not systematically change from peak to peak.

If each subpeak is due to emission from a distinct shell in a multiple wind scenario then the systematic decrease in decay time with peak position could possibly be due to a combination of two distinct physical mechanisms: (1) less deceleration experienced by subsequent shells as the latter speed through a wind-swept, rarefied ambient medium, and (2) a systematic decrease in the proper time interval between generation of the wind and subsequent formation of a radiative shocked shell upon former's catching up with slowly moving plasma.

REFERENCES

1. Bhat, P. N. et al. AIP Conference Proceedings 307, Gamma Ray Bursts, Second Workshop, Huntsville, AL, pp. 953-957 (1993).
2. Bhat, P. N. et al. AIP Conference Proccedings 384, Gamma Ray Bursts, Third Huntsville Symposium, Huntsville, AL , pp. 197-201 (1995).
3. P.N.Bhat, Varsha Gupta and Patrick Das Gupta, Fifth Compton Symposium, in preparation (1999).
4. E.E.Fenimore, C.D.Madras and S.Nayakshin, *ApJ* **473**, 998-1012 (1996).
5. Shiho Kobayashi, Tsvi Piran and Re'em Sari, *ApJ* **490**, 92-98 (1997).
6. Tsvi Piran, *Phys. Rep.* **314**, 575-667 (1999).
7. E.F.Fenimore, E. Ramirez-Ruiz and Bobing Wu, *ApJ* **518**, L73-L76 (1999).

Aperiodic Properties of Gamma-Ray Bursts

Alexei S. Pozanenko and Vladimir M. Loznikov

Institute for Space Research, Moscow 117810, Russia

Abstract. Aperiodic properties of Gamma-Ray Burst (GRB) time profiles were investigated on the basis of BATSE catalog. Power density spectra (PDS) obtained with the FFT procedure for each GRB were averaged for all GRBs. We analyze the continuum and high-frequency part of the spectrum. The spectra averaged for different brightness groups of GRBs are found to be different. The hardness of PDS monotonously changes with brightness of bursts, which may point to the absence of "standard candle" model of Gamma-Ray Bursts.

INTRODUCTION

One of the methods used for GRB time profile investigation is a study of power density spectra of GRB time profiles. However, the PDS of individual bursts are very different [1,2]. Let us assume that each burst is a random realization of the same standard burst engine. Then one can study the whole sample of bursts by summing the PDS of each burst in the same manner as used for investigation of X-ray variables, summing different moments of observation of the same source to improve statistical significance of PDS. The difference is that GRBs are transient events and the duration of the events ranges from milliseconds up to hundreds of seconds. We have postulated that the burst engine is the same for all bursts. Hence, the duration is an inherent parameter of burst and we include all bursts in the present consideration. On the other hand, short bursts ($T_{90} < 2$ s) may represent a different population of GRBs (because of the bimodal distribution of T_{90}) and those burst should be considered separately.

DATA ANALYSIS AND RESULTS

For the present analysis, we use events from the BATSE catalog [3], up to burst number 5624, which have 64 ms data from the BATSE Large Area Detectors (DISC-SC+PREB). There are 815 bursts for which T_{90} values exist and $T_{90} > 2$ s. Time profiles are constructed from the sum of discriminator channels 2 and 3, which corresponds to the 50–300 keV energy range. Background subtraction is performed in each discriminator channel separately, using the interpolation of a fitted polynomial model.

Before averaging individual PDS, we should decide about normalization of PDS and subtraction of Poisson noise from the PDS. Each PDS has a different amplitude and shape, which are due to the intrinsic variability and intensity of the time profile of a particular burst. To provide equal weight of each PDS, we use normalization of each PDS for the same value. In other words, we use the normalization when the total power of every burst is equal to 2. This choice of normalization is appropriate for shape investigation of averaged PDS and is discussed elsewhere [5].

The averaging procedure is the following. For each time profile we remove the "trend" associated with background. Then PDS is calculated on the same time interval $T = 64$ (ms) $\times$ 8192 (bins) with the standard FFT procedure and Poisson noise level is determined (e.g., [4]). The total power of PDS is normalized by 2. The accordingly normalized Poisson noise level is subtracted from the normalized PDS and the remainder is used for averaging.

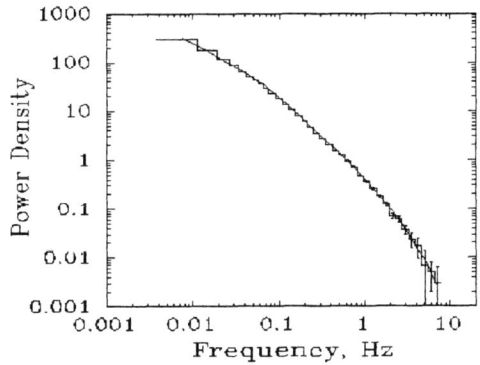

FIGURE 1. The averaged PDS of 815 GRBs ($T_{90} > 2$ s) is shown in logarithmically spaced bins.

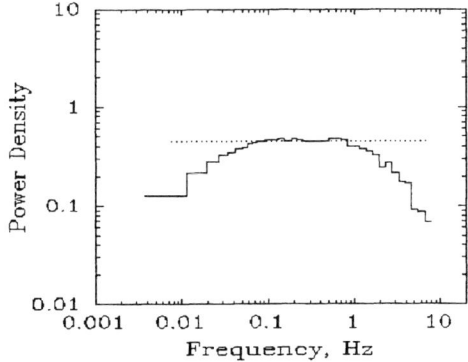

FIGURE 2. The same as in Figure 1 but PDS is multiplied by $\nu^{1.6}$ (dotted line). Statistical errors are not shown.

The averaged PDS of 815 bursts is presented in Figures 1 and 2. One can observe three features of the spectrum. The left break is associated with the finite duration of bursts, or more precisely, with the absence of long-duration bursts in the sample. Indeed, the logarithmic mean of the long mode of the T_{90} distribution is around 10 seconds. The right part of the spectrum (high frequency tail) represents intrinsic variability of GRB sources, while the central part of the spectrum represents both variability of the source and properties of the duration distribution. One can also observe that the central part of the spectrum can be approximated by a power law. Qualitatively, the spectrum corresponds to the spectrum obtained for averaged PDS for only long bursts ($T_{90} > 20$ s) [6].

For quantitative analysis of the PDS we use two analytical laws (1) an exponential power law $P_f \sim \nu^{-a} \exp(\nu/\nu_1)$ and (2) Band's law [7] successfully used for GRB energy spectra, approximation $P_f \sim \nu^{-b} \exp(\nu/\nu_0)$, if $(a - b)\nu_0 \geq \nu$ and $P_f \sim \nu^{-a} \exp(\nu/\nu_1)[(a - b)\nu_0]^{(a-b)} \exp(b - a)$, if $(a - b)\nu_0 < \nu$. Both analytical approximations describe the observed spectra fairly well.

TABLE 1. Definition of brightness groups

Brightness group	Peak Flux F_{64} (ph/cm^2-s)	Number of GRBs
1	> 4.0	171
2	1.7–4.0	195
3	1.0–1.7	199
4	0.5–1.0	250

TABLE 2. Approximation by exponential power law averaged PDS of simulated time profiles

Brightness group	a	ν_1 Hz
1	1.43 ± 0.02	2.74 ± 0.14
2	1.36 ± 0.07	2.10 ± 0.10
3	1.54 ± 0.05	3.53 ± 0.19
4	1.51 ± 0.04	2.63 ± 0.11

To compare averaged PDS of different burst groups, we subdivide all bursts into four different brightness groups (Table 1). Before comparing results of actual brightness groups, we investigate the possible bias of averaged PDS parameters against signal-to-noise reduction with Monte-Carlo simulation. Time profiles of the brightest group (1) are artificially noisified into dimmer groups 2, 3 and 4. The artificially noisified time profiles are then used as samples for the calculation of averaged PDS for each new group 2, 3 and 4. One should stress that for Monte-Carlo simulation we use only time profiles of the brightest group (1). The artificial noisification is a rather standard method and described in detail in [8]. Results of the approximation of averaged PDS by the exponential power law are presented in Table 2. There is no significant difference in parameters against the brightness group. Therefore, if a significant difference will be found in PDS parameters of actual brightness groups, then it should be ascribed to GRB properties of these groups.

Results obtained for the actual brightness groups are presented in Tables 3 and 4. The power law approximates PDS of any brightness group in more than 1.5 decades of frequency. Moreover, the power-law index a monotonically steepens while the burst brightness decreases. The change of power-law index a is significant and cannot be explained by bias or systematic errors of the applied procedure. On the other hand, parameters ν_1 and ν_0 do not depend on burst brightness.

TABLE 3. Power-law parameters for averaged PDS

Brightness group	a	ν_1 Hz
1	1.43 ± 0.02	2.74 ± 0.14
2	1.51 ± 0.03	3.26 ± 0.27
3	1.64 ± 0.05	2.99 ± 0.48
4	1.70 ± 0.04	3.29 ± 0.73

TABLE 4. Band function parameters for averaged PDS

Brightness group	b	a	ν_0 Hz	ν_1 Hz
1	0.71 ± 0.04	1.43 ± 0.02	0.14 ± 0.02	2.88 ± 0.19
2	0.82 ± 0.05	1.53 ± 0.03	0.15 ± 0.03	3.50 ± 0.31
3	0.91 ± 0.16	1.61 ± 0.05	0.12 ± 0.06	2.94 ± 0.54
4	0.96 ± 0.14	1.68 ± 0.06	0.11 ± 0.05	3.36 ± 0.85

DISCUSSION

The power-law PDS found in a wide frequency range means that there is no specific time scale in this range. Hence, not only do observed time profiles consist of random realizations of emission episodes, but the burst itself (and its temporal parameters, e.g., duration) is a random realization of the same process. Thus, from the point of view of PDS, one can consider every part of time profile (e.g., single pulse) as a separate burst. This is the way to compare intrinsic variability of bursts of different modes of the bi-modal T_{90} distribution.

If GRBs are cosmological, then the power-law index of PDS should not be changed if properties of the burst engine remain the same because cosmological transformation does not change the PDS slope. Monotonic steepening the power-law index a with decreasing burst brightness indicates a model of the GRB source with evolution.

The averaged PDS has a high-frequency break described in approximation laws by the parameter ν_1. There is a specific time scale ~ 0.4 s corresponding to the frequency ν_1. If the specific time scale is an intrinsic property of GRBs then it should evolve with redshift z as $(1+z)$ in case of cosmological GRB origin. The observed parameter ν_1 does not change with brightness. The absence of that change is rather strange and might be explained by evolution of the specific time scale with cosmological redshift $\sim (1+z)^{-1}$ exactly compensated by cosmological time dilation of $(1+z)$. If this is the case, it would strongly restrict models of GRB emission. Otherwise, the brightness of a burst does not represent the distance of GRB sources and might be strong evidence for a broad luminosity function. However, more careful investigation of ν_1 is necessary.

Different models of time profiles may provide the power law of PDS. Models either involve pulses of the same form or any shape arriving at random times. In latter case, the duration of pulses must follow a power law (e.g., [9]). Using the specific pulse and time structure parameters [8,10] one can find the model of GRB time profile which will satisfy the observed PDS [5].

REFERENCES

1. Belli, B.M., *ApJ* **393**, 266 (1992).
2. Giblin, W.T., et al., in AIP Conf. Proc. 428, 241 (1998).
3. Meegan, C.A., et al., Catalog of BATSE gamma-ray bursts (1998).
4. Leahy, D.A., *ApJ* **266**, 160 (1983).
5. Pozanenko, A.S., et al., in preparation (1999).
6. Beloborodov, A.M., et al, astro-ph/9807139 (1998).
7. Band, D., *ApJ* **413**, 281 (1993).
8. Mitrofanov, I.G., et al., *ApJ* **504**, 925 (1998).
9. Press, W.H., *Comments Astrophys.* **7**, 103 (1978).
10. Pozanenko, A.S., et al., in AIP Conf. Proc. 428, 59 (1998).

Evidence for and Implications of Turbulence in GRB Time Profiles

Yuan Yan

Department of Physics and Astronomy
Mississippi State University, MS 39762

Abstract. Previous work has shown that turbulent processes are present at GRB sites [15,16]. Here we present a cycle analysis for GRB time profiles. Our analyses of time profiles show intermittency, a characteristic of turbulence. We present arguments that these characteristics of turbulence imply that the relaxation phase of the GRB phenomenon occurs under adiabatic conditions.

PREVIEW

Our previous study [16] found indications of a time scaling property in GRB time profiles. We also examined the Fourier spectra from 125 long duration GRB time profiles [17] ($T_{90} > 65$ s), though earlier studies on the spectra reached no significant results [9]. The examined decay exponent of the low frequency part of the spectra shows that the Fourier spectra $F(f) \propto 1/f^s$ decay, with an average exponent $s = 0.63 \pm 0.13$. The value $s = 0.63$ is consistent with the time domain result of Stern et al. [15] The link that is responsible for this consistency between the results from the time and frequency domains is the Kolmogorov turbulence theory of 1941 [8].

CYCLE STATISTIC ANALYSIS

The cycle statistic algorithm used here to test the intermittency of GRB time profiles is based on Hurst's R/S scaling analysis as in reference [7]. The calculation begins with a partition process which takes 1024 data points from each long duration GRB time profile (64-ms bins). The partition process is a binary cascade. That is, on a cascade stage b, the GRB time profile is broken into 2^b pieces. Through calculation of accumulated departure and rescaled range R/S [7] for each piece, the $\overline{(R/S)_n}$ value is obtained on scale n for the whole data set. After the binary cascade process a special "V-Statistic" is composed to expose the intermittency:

$$V_n = \frac{\overline{(R/S)_n}}{\sqrt{n}} \tag{1}$$

A plot of V_n vs. n from a long duration GRB time profile is shown in Figure 1a. The plateaux are where $dV_n/dn \equiv 0$. In this methodology, the plateaux are easily found along the dV_n/dn vs. n plot (Figure 1b). Some of the long-duration GRB time profiles from the 3B catalog have been tested with the V_n vs. n plot. The dV_n/dn zeros show that there are cyclic "statistical" properties in the GRB time profiles.

This appears to be a paradox: on the one hand, there are no significant spectral peaks in the Fourier spectra of GRB time profiles, i.e., with the fewest of exceptions, periodicities are not found in GRB time profiles [9]. On the other hand, according to this V_n cycle analysis, there are cyclic phenomena embedded in those profiles. The answer to this paradox is that the signature of turbulence in these profiles is not overwhelming. This leads to statistical cycles that do not show up in a standard Fourier Spectrum. The V_n analysis is more sensitive to these than the Fourier analysis. As characteristics similar to fully developed turbulence, the Fourier spectrum of a chaotic dynamic system is flat. The noisy waveforms of fully developed turbulence in time domain shows breaks or gaps, an intrinsic property of fully developed turbulence. That is intermittency. Between the mean duration and the difference between the associated parameter and its critical value which causes the intermittent transition, there are scaling relationships [12].

A TURBULENCE APPROACH

Ruelle and Takens [13], and Eckmann [6] show that the dynamics of turbulence is composed of an infinite number of Fourier components. However, the quasi-periodic underlying dynamics can be exposed through a study of the deterministic modes. Each mode is composed of a sum of the "minimum number of rationally independent frequencies" [14]. The corresponding time domain "modes" are expected to be the cycles observed from the V_n analysis. This cycle picture provides us with a clue to the quasi-periodic dynamical processes that give rise to gamma ray bursts.

Some of the cycles shown on the V_n plot for GRB time profiles (Figures 1a, 1b) are possibly explained as follows. First of all, the longest cycle of about 10^3 data points is due to the data being truncated to 1024 data points (≈ 65 s). The second and third longest cycles are both due to 'intermittency' in the (presumed) turbulence model for gamma ray bursts. The intermittent property comes from the sparse filling of the eddies in fluid turbulence. That is, the daughter eddies usually only fill a fraction of the space which the parent eddies occupied. In GRB processes, the breakdown reflects the fact that the particle production-annihilation chain reaches a dead end in the reactions. These sudden breaks give the turbulent signal its burst like appearance.

There are several other plateaux in the V_n curve. The shortest-duration cycles correspond to the highest and second highest dynamo modes in the turbulence

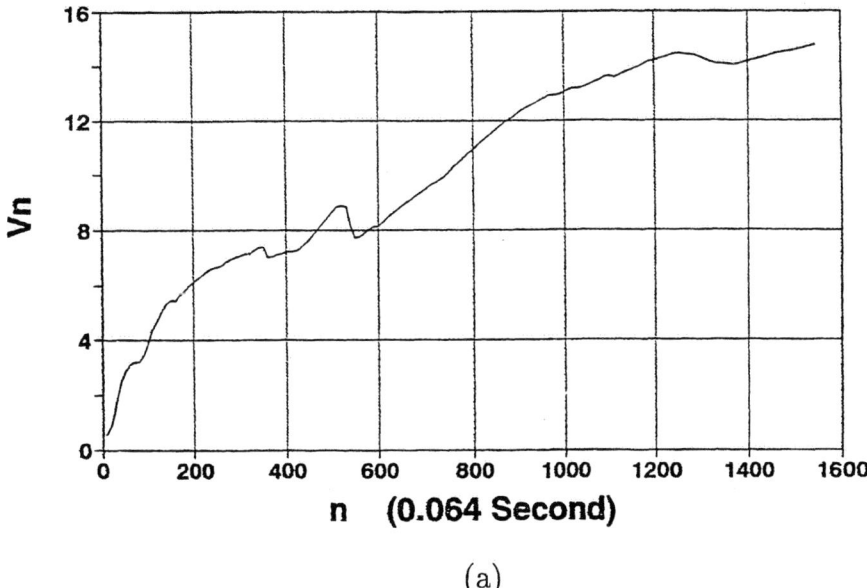

(a)

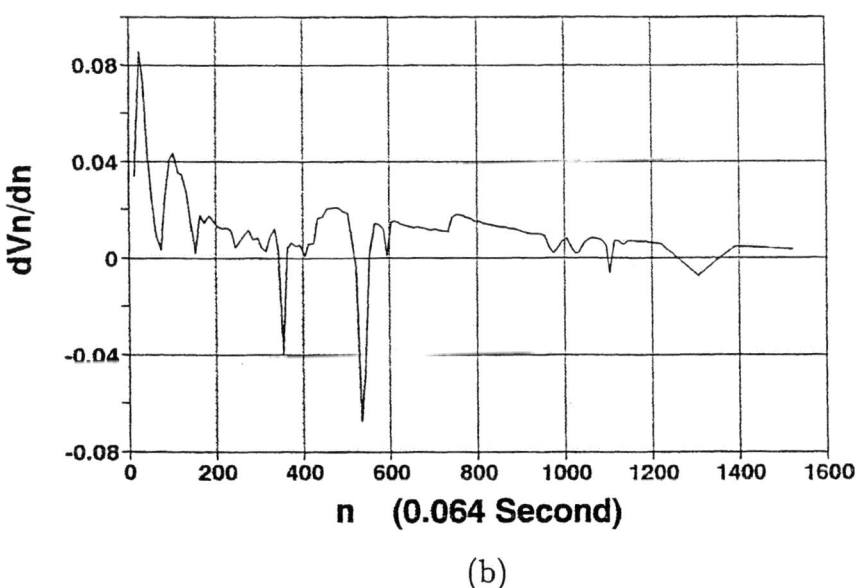

(b)

FIGURE 1. V-Statistics for a GRB time profile, (a) V_n vs. n; (b) dV_n/dn vs. n.

[13]. This is caused by vortices in turbulent flow. The curl motion is composed of sub-curl motions which have different velocities (different directions and angular speeds). These observations are quickly pushing us into the regime of nonlinear dynamics. Specifically, if the total number of modes or cycles is an integer power of 2, then this shows that a bifurcation process is present in the turbulent process. It often happens that there are a non-2^n or odd number of plateaux, in which case we are seeing a transition to chaos in GRB dynamo. If there are 3 non-factorable (non-reducible) cycles, then the GRB process is chaotic [10]. In this special case the attractor in the phase diagram is strange.

Since the vortex motion is composed of many modes of curl velocity, the circular frequency ω of a vortex can be written as the superposition from each mode or vector base ω_n:

$$\omega = a_1\omega_1 + a_2\omega_2 \cdots + a_n\omega_n \quad , \tag{2}$$

where $a_1, \cdots a_n$ are constants, read as eigenvalues. At the GRB site, there are only a few modes of ω_n which need to be included in the analysis due to the less-known number of degrees of freedom of the system. The number of V_n plateaux provides evidence that there are several modes in the turbulent GRB process. It is known that the number of degrees of freedom for a flow motion increase from a laminar flow to a turbulent flow according to the transition from translational motion to additional curl motion represented as ω_n [6]. In phase space, this forms an n-manifold torus and is referred to as the classic Ruelle-Takens-Newhouse scenario [6,13]. If $n \geq 3$, a strange attractor can be formed; if $n \geq 4$, then one can take the $n \to \infty$ approximation. Then an infinite number of eigenvalues become unstable simultaneously [4]. The strange attractor has fractal dimension and is itself a signature of chaos. The infinite number of spectral peaks means that broad-band, noise-like Fourier spectrum, which turbulence usually exhibits, will appear [6].

An interesting aspect of the V_n plot of GRB time profiles is that there are usually three or more plateaux on it. This means that the number of modes in the motion of a GRB process shows *in situ* turbulent processes. Furthermore, fundamental to the GRB turbulent scenario is a transition to chaos.

A PERSPECTIVE

Previous work by the author shows that the Fourier spectrum from GRB time profiles has a characteristic $1/f^s$ with $s = 0.63 \pm 0.13$. This scaling exponent s remains invariant only on the time scale 0.66 s to 16.4 s. Beloborodov et al. [3] show that the Kolmogorov 5/3 law is valid for the PDS from GRB time profiles on time scales from 1.0 s to 50.0 s. These results present a clear way to study GRB physical processes. That is, on the one hand, the inertial scale characteristic shows that there is 'developed turbulence' on the GRB sites. It can be referred to as the e^+e^- pair annihilation-photon production process in relativistic outflow. On the other hand, on the limited inertial scale, the fact that the scaling law is not adapted to all time scales shows that the decaying tail phase for gamma ray bursts is just like turbulence, which is usually considered to be locally isolated [8]. When considering a self organized criticality state in a dynamic system [1,2], the state must have self-similar fractal structure, and the frequency spectrum from the system must have the characteristic of $1/f$ noise. Furthermore, the scaling law for the $1/f$ spectrum must propagate over all temporal/spatial scales etc.; there is no "inertial scale" being reserved - scale free - for a self organized criticality state [1,2]. Since the self organized criticality must be established under a 'transportation' process, the system with self organized criticality is an open one. However, a turbulent system with the characteristic of 'inertial scale' is locally isolated, so the 'eddies' in the outflow of GRB must be locally isolated as well. In other words, on the temporal or spatial 'inertial scale,' there is no massive energy or mass transportation at the 'eddies' in the relativistic outflow. The e^+e^- pair production-annihilation and the multi-photon upscattering procedures [5] should occur in an environment opaque to γ-ray photons. The decaying tail phase can be considered an adiabatic process. The resulting observed gamma ray bursts are a result of the photons 'catching up' with each other from the relativistic outflow [11], so the temporal (frequency) scaling of the counts for the γ-ray photons is the spatial scaling of the processes at the GRB outflow. Further study could follow a calculation on intermittency scaling from GRB time profiles due to reference [12].

REFERENCES

1. Bak, P. et al., *PRL* **59**, 381, Cambridge Univ. Press, N.Y. (1987).
2. Bak, P., *How Nature Works*, Springer-Verlag, N.Y. (1996).
3. Beloborodov, A. et al., *ApJL* **508**, L25 (1998).
4. Bohr T. et al., *Dynamical Systems Approach to Turbulence*, Cambridge Univ. Press, N.Y.
5. Burke, D. et al., *PRL* **79**, 1626 (1997).
6. Eckmann, J-P., *RMP* **53**, 643 (1981).
7. Feder, J., *Fractals*, Plenum Press, N.Y. (1988).
8. Frisch, U., *Turbulence*, Cambridge Univ. Press, N.Y. (1995).
9. Kouveliotou, C., et al., AIP Conf. Proc. **265**, 299 (1992).
10. Li, T. et al., *Am. Math. Month.* **82**, 985 (1975).
11. Meszaros, P., astro-ph/9904038 (1999).
12. Ott, E., *Chaos in Dynamical System*, Cambridge Univ. Press, N.Y. (1993).
13. Ruelle, D. et al., *Comm. Math. Phy.* **20**, 167 (1971).
14. Ruelle, D., *Chaotic evolution and strange attractors*, Cambridge Univ. Press, N.Y. (1989).
15. Stern, B. et al., *ApJL* **469**, L109 (1996).
16. Yan, Y. et al., AIP Conf. Proc. **384**, 111 (1995).
17. Yan, Y., *JMAS* **43** (1998).

The Technique of Emission Time Estimation for BATSE GRBs

Maxim L. Litvak[1], Igor G. Mitrofanov[1], Dmitrij S. Anfimov[1],
Anton B. Sanin[1], Michael S. Briggs[2], William S. Paciesas[2],
Geoffrey N. Pendleton[2], Robert D. Preece[2], Thomas M. Koshut[3],
Gerald J. Fishman[3], Charles A. Meegan[3],
and John Patrick Lestrade[4]

[1] *Space Research Institute, Profsojuznaya str. 84/32, 117810, Moscow, Russia*
[2] *Department of Physics, University of Alabama in Huntsville, Huntsville, AL 35899*
[3] *NASA/Marshall Space Flight Center, Huntsville, AL 35812*
[4] *Department of Physics, Mississippi State University, MS 93762*

Abstract. The technique of estimation of the *emission time* τ_{50}, introduced as a new temporal parameter [1,2], is described.

INTRODUCTION

We continue discussion devoted to the concept of a new burst temporal parameter, the *emission time*. The emission time is defined as sum of most intensive time bins that contribute 50 percent of total fluence for a given burst [1,2]. In this paper we consider possible effects which may influence the estimation of the emission time parameter.

NOISE-PRODUCED EFFECTS ON THE EMISSION TIME FOR INDIVIDUAL EVENTS

The decreased signal-to-noise ratio of dimmer bursts may bias temporal parameters because it leads to a larger number of erroneously selected pulses and valleys between them.

The bursts GRB 940217 and GRB 940206 are two rather good examples of the variety of burst time profiles. The burst GRB 940217 has a complex time profile with 11 or more separate pulses. The fluence of this burst is due to several separate pulses, and the interpulse intervals between them contribute a large part to its total duration. The other example, GRB 940206, has a time profile whose duration is

TABLE 1. GRB940206 and GRB940217 at Reduced S/N Ratio.

GRBs:	940206			940217		
S/N:	100σ	50σ	15σ	200σ	50σ	15σ
$\langle T'_{50}\rangle/T_{50}$	0.96 ± 0.02	0.88 ± 0.02	0.78 ± 0.04	1.00 ± 0.003	0.99 ± 0.03	0.90 ± 0.14
$\langle \tau'_{50}\rangle/\tau_{50}$	0.99 ± 0.01	0.95 ± 0.01	0.85 ± 0.03	0.99 ± 0.01	0.97 ± 0.03	0.82 ± 0.07

associated with the pulse width. Koshut et al. [4] have used them to illustrate the noise-produced effects on the parameters T_{50} and T_{90}. We will also use these two events to estimate the noise-produced effects for the emission time τ_{50} in comparison with similar effects for the duration time T_{50}. The PREB and DISCSC data, from the BATSE 4B Catalog [3], were used for measurements of gamma ray emission times. The procedure for calculating the emission time τ_{50} is described in [1,2].

Originally, GRB 940217 and GRB 940206 have signal-to-noise ratios S/N=350 and 206, respectively. We reduce the time histories of these bursts down to smaller signal-to-noise levels: S/N=200, 50 and 15 for GRB 940217 and S/N=100, 50 and 15 for GRB 940206. We use the same levels of S/N as were used by Koshut et al. [4]. We create 500 noisified realizations for each burst at each level of S/N, and for each realization we calculate the parameters T_{50} and τ_{50}.

As the signal-to-noise ratio decreases, the average values of T_{50} are become smaller for noisified realizations of bursts. Table 1 presents the ratio of the average values of the temporal parameters for noisified realizations to their original values. The errors of these values correspond to the standard deviation of the noisified parameters divided by their original value. At the level of S/N=15, the downward shift of the average T_{50} is 10% for GRB 940217 and 22% for GRB 940206. Moreover, the distributions of the measured values of T_{50} become broader for dimmer realizations (Table 1). In the S/N=200 case of GRB 940217 (100 for GRB 940206) the relative standard deviation of T_{50} is equal to 0.3% (1.8% for GRB940206), but for S/N=15 the relative standard deviation is equal to 14% (4% for GRB940206).

The emission time values of τ_{50} also decrease with decreasing S/N levels (Table 1) but it is more stable to noise biases because only most intensive time bins are selected for estimation of emission time.

NOISE-PRODUCED EFFECTS ON THE EMISSION TIME FOR DIFFERENT INTENSITY GROUPS OF GRBS

The possible anti-correlation between emission time and intensity may result from time-stretching of dimmer bursts with respect to the brighter bursts. That is why it is necessary to estimate distortions of shifting factors between bright and dim groups caused by noise-produced effects. To do this we have taken 745 GRBs with $T_{90} > 2$ s, and $F_{64}^{(max)} > 1.0$ phot cm^{-2} s^{-1} from the current BATSE catalog.

TABLE 2.

Intensity group	Peak Flux (ph cm^{-2} s^{-1})	Number of events	Y_{noise}
1	> 7.6	121	1.00
2	3.6-7.6	121	1.04
3	2.3-3.6	119	1.07
4	1.7-2.3	127	1.08
5	1.3-1.7	125	1.12
6	1.0-1.3	132	1.16

It is rather bright bursts which belong to the "long" class of GRBs. From one side they are sufficiently bright to exclude non-triggering effects, from other side they are weak enough to study noise-produced biases [2]. All these bursts were divided into six intensity groups (see Table 2). Using the brightest group (number 1) as a prototype, we created 5 artificial groups with higher noise in accordance with smaller fluxes of groups 2-6. Each realization for group 1 is transformed from the original bright bursts of group 1 by simulating the noise which corresponds to the peak flux level of the comparison group 2-6. The Monte Carlo transformation procedure includes the following steps:

a) For each original burst of reference group 1, some counterpart event is randomly selected inside the testing group i.

b) The ratio of peak fluxes $\lambda = F_{\max}^{(1)}/F_{\max}^{(i)}$ is estimated between the peak fluxes of the original burst and its weaker counterpart.

c) The time profile of the original bright burst is divided by the factor λ, and represents an artificially-dimmed version of the original burst we are simulated using Poisson statistics:

$$D_j = \frac{C_{\text{S},j}}{\lambda} + C_{\text{B},j}, \qquad (1)$$

where $C_{\text{S},j}$, $C_{\text{B},j}$ are the signal and background counts accumulated during the jth time bin for a burst from the reference group.

The best stretching coefficients Y_{noise} have been estimated between distributions for the artificial reference groups ($i = 2$–6) and the emission time distribution for the original reference group.

One can see the largest effect occurs in case of the 6th artificial group ($i = 6$). The distribution for this group must be shifted by a factor of 1.16 to fit the distribution of the reference group ($i = 1$). The noise-produced factors of ~ 1.2 (Table 2) contribute into the dispersion factors σ of the emission time distribution [2]. This contribution is smaller than σ_l, which shows that the factor of dispersion σ_l is also a result of the intrinsic difference of time profiles (see next section). The estimated noise-produced stretching should be taken into account also for comparison of different intensity groups (see [2]). Therefore, if the observed stretching factor between the distribution of emission time for the actual observed dim group and the reference group is equal to 1, it should be interpreted as evidence for real stretching on the order of Y_{noise} between the dim and reference groups.

CHOICE STATISTIC

We know that gamma-ray bursts have very different time profiles, and we have to use a large sample to build a histogram that accurately represents the emission time distribution. Indeed, the direct comparison of average emissivity curves for different samples with ~ 150 events has shown [1] that they are much different than would be expected from the sample variance within each sample only. This suggests that these samples are not large enough to be the *representative samples*, and we have to estimate the errors of stretching factors using *random choice statistics* to make any conclusive estimations of stretching factors between emission time distributions for distinct samples of ~ 100 bursts (see details in [1]).

In order to estimate the errors in our procedure, we compare subpopulations of the total sample of 745 bursts described in the previous paragraph. We select 100 bursts randomly from the total sample and then another random group of 100 bursts from the remaining 645 bursts. We determine the best stretching factor Y between these two groups of 100 bursts. We repeat this process to generate 10^4 pairs of samples of 100 bursts and 10^4 simulated values of Y. For 68% of the realizations, the stretching factors are below 1.15. Therefore, the stretching factor between emission time distributions has the error $\delta Y/Y = 0.15$ according to random choice statistics. This value should be used to estimate the errors of stretching factors between the groups under comparison.

CONCLUSIONS

The technique of the estimation of the emission time is discussed. The comparison between standard T_{50} and T_{90} temporal parameters and emission time parameters is done. The noise produced effects are considered as well for individual bursts and for large groups of GRBs. The choice statistic technique is applied to distributions of emission time parameters to take into account possible errors due to non-presentativity of samples of GRBs.

ACKNOWLEDGMENTS

The work in USA was supported by NASA project CGRO-98-120 of CGRO Guest Observations Program. The work in Russia was supported by RFBR grant 98-02-17380 and by INTAS project No.96-0315.

REFERENCES

1. Mitrofanov I.G. et al., *ApJ* **522**, 1069 (1999).
2. Mitrofanov I.G. et al., these proceedings.
3. Paciesas W.S., et al., BATSE 4B Catalog (1999).
4. Koshut T.M., et al., *ApJ* **463**, 570 (1996).

The Emission Time Signature of BATSE GRBs

Igor G. Mitrofanov[1], M.L. Litvak[1], Dmitrij S. Anfimov[1],
Anton B. Sanin[1], Michael S. Briggs[2], William S. Paciesas[2],
Geoffrey N. Pendleton[2], Robert D. Preece[2], Thomas M. Koshut[3],
Gerald J. Fishman[3], Charles A. Meegan[3],
and John Patrick Lestrade[4]

[1] *Space Research Institute, Profsojuznaya str. 84/32,117810,Moscow, Russia*
[2] *Department of Physics, University of Alabama in Huntsville, Huntsville, AL 35899*
[3] *NASA/Marshall Space Flight Center, Huntsville, AL 35812*
[4] *Department of Physics, Mississippi State University, MS 93762*

Abstract. The *emission time parameter* τ_{50} is introduced as a new temporal parameter of gamma-ray bursts.

INTRODUCTION

We implement the concept of a new burst temporal parameter, *emission time*, complementary to the classical parameters of T_{50} and T_{90}. It is defined as the time of emission τ_ξ of a fixed fraction of fluence, $\xi\%$, detected from the highest flux level. It excludes all low-emission intervals of a burst and characterizes the phase of peak power emission.

PARAMETER OF BURST EMISSION TIME

The PREB and DISCSC data, obtained with BATSE, were used for measurements of gamma-ray burst emission times [1]. The procedure for calculating the emission time τ_ξ is as follows [2]. Each time bin in the burst profile is ranked by its background-subtracted intensity (measured in counts s^{-1}). Beginning with the time bin associated with the largest intensity, the flux of each successively weaker time bin is summed until $\xi\%$ of the fluence has been observed (we denote as f_ξ the flux of the time bin which last contributed to this fluence). The value τ_ξ is then calculated by summing the durations of each time interval with flux above f_ξ which contributed to $\xi\%$ of the fluence. The selection of time intervals that contribute to

τ_{50} ($\xi = 50$) is shown for BATSE events GRB94206 and GRB 940217 in Figures 1 and 2.

The optimal choice of a fraction ξ for evaluating the emission time τ_ξ is a compromise between the demand to have smaller statistical errors of counts in selected time bins, which tends towards choosing a smaller ξ and a correspondingly higher threshold f_ξ, and the opposite demand to a have a larger number of time bins for better measurement of the emission time, which tends towards choosing a larger ξ and a correspondingly lower threshold f_ξ. Selection of fractions of 30-50% of total fluence looks to be the most feasible for the emission time definition.

It is convenient to use the criteria $\xi = 50\%$, giving an emission time τ_{50}, because the same fraction of fluence is used for the estimation of the duration time T_{50}.

For a burst with a complex multi-pulse time history, the two parameters τ_{50} and T_{50} could be quite different. For the burst GRB 940206 with a single-pulse time profile, the values $T_{50} = 9.02$ s and $\tau_{50} = 8.13$ s are quite close (Figure 1). On the other hand, GRB 940217 has $T_{50} = 66.05$ s and $\tau_{50} = 18.5$ s (Figure 2).

DISTRIBUTION OF BURSTS OVER THE EMISSION TIME

There are 1029 bright bursts with $F_{64}^{(\max)} > 1.0$ phot cm^{-2} s^{-1} which belong to the current BATSE Catalog [1] and for which PREB and DISCSC data are available. The distribution of τ_{50} for sample of 1029 BATSE bright bursts is presented in Figure 3. It consists of a bimodal double-peaked curve, with an evident gap at $\tau_{50} \sim 0.4$ s. The fraction of 284 short burst was processed using BATSE time-tagged events (TTE) data with high temporal resolution. We have performed a fit to the distribution using a function that represents the sum of two log-normal Gaussians (Figure 3):

$$F(\tau_{50}) = \frac{A_s}{\sigma_s} \exp\left[-\left(\frac{\log \tau_{50} - \log \mu_s}{\sigma_s}\right)^2\right] + \frac{A_l}{\sigma_l} \exp\left[-\left(\frac{\log \tau_{50} - \log \mu_l}{\sigma_l}\right)^2\right]. \quad (1)$$

A similar formula was used by Koshut et al. [4] to perform a non-linear fit of the $T_{50/90}$ distributions. The best fit parameters $\mu_s, \sigma_s, \mu_l, \sigma_l$ for the τ_{50} distribution are as follows:

$\mu_s = 0.09 \pm 0.05$ s, $\sigma_s = 1.7 \pm 0.6$, $\mu_l = 4.6 \pm 0.9$ s, $\sigma_l = 1.4 \pm 0.2$.

The parameters μ_s and μ_l represent the log-normal mean values for two peaks of τ_{50}. Factors σ_s and σ_l represent the dispersion of these peaks which could result from contributions from both the intrinsic difference of events and from noise (see [3]).

There are two peaks in the T_{90} curve: one below and one above $T_{90} \sim 2$ s (see [4]). Furthermore, there are two samples of bursts that contribute to the short-τ and long-τ peaks of the τ_{50} distribution. The following question arises: is there a direct correspondence between the samples of bursts making up the peaks in the

T_{50} and emission time distributions? To respond to this question we calculated the distribution of emission times separately for the sample of 745 bright bursts with $T_{90} > 2.0$ s. It is found that this sample perfectly corresponds to the long-τ peak of the emission time distribution (Figure 3).

Therefore, we may conclude that the bimodality of the T_{50} and T_{90} distributions and the τ_{50} distribution represents the same two classes of gamma-ray bursts. There are two distinct classes of bursts that have different duration time $T_{50,90}$, and emission time τ_{50} parameters. We call them long bursts (LB) and short bursts (SB).

INTENSITY-DEPENDENT COMPARISON OF EMISSION TIME DISTRIBUTIONS

The sample of bright bursts with $F_{64}^{(\mathrm{max})} > 1.0$ phot cm^{-2} s^{-1} with long duration time $T_{90} > 2.0$ s contains 745 events. They correspond to the right peak of the curve of emission time distribution (Figure 3), and they are thought to represent the LB class of bursts, as defined above. Below we consider the question of whether there is a difference between the distributions of emission time for two well-separated intensity groups within the LB sample of 745 bright bursts with long emission times. The possible anti-correlation between emission time and intensity might be interpreted, if it were observed, as the time-stretching of dimmer bursts with respect to the brighter events.

We have divided all these bursts into six equal intensity groups 1-6 (see Table 1). The brightest group (1) will be called the reference group and the other groups will be mentioned as comparison groups (2-6).

For each comparison group i=2-6, the stretching factors $Y^{(i)}$ were estimated with respect to reference group 1 as the best shift factor between emission time distributions for group i and group 1.

We know that such comparison should be done between given comparison groups i=2-6 and an artificially noisified reference group [3]. In this situation, the artificially noisified reference group has the same signal-to-noise ratio as the comparison group. The best fitting factors taking into account this issue are presented in Table 1. In order to estimate the errors in our procedure, we use a choice statistic technique described in [3]. According to this method, the stretching factor between emission time distributions has the error $\delta Y/Y = 0.15$.

The dependence of Y versus peak flux is presented in Figure 4. One can see the monotonic increasing of stretching factors with decreasing average peak flux. The single exception concerns group 3, where the factor Y_3 falls below the Y_2 value. The largest value is achieved for group 6 and is equal to 1.86±0.28. It is concluded that we see significant stretching of the average emission time duration with decreasing burst intensity.

TABLE 1.

Intensity group	Peak Flux $(ph/cm^2 sec)$	Number of events	Stretching factors
1	> 7.6	121	1.00
2	3.6-7.6	121	1.58±0.23
3	2.3-3.6	119	1.36±0.20
4	1.7-2.3	127	1.63±0.24
5	1.3-1.7	125	1.85±0.28
6	1.0-1.3	132	1.86±0.28

CONCLUSIONS

The *emission time* parameter τ_{50} is implemented as complementary to the well known parameters of *duration time* T_{50}, and T_{90}. The clocks of emission time τ_{50} and duration time T_{50} correspond to two different properties of outbursting sources. The measured intervals for these parameters add different sets of time bins.

The physical properties of GRBs were investigated based on an estimation of emission time. This study show two principal results:

1. The distribution of emission time values for short and long bursts demonstrate the same bimodality property as in the case of the standard T_{50}, T_{90} distributions. This result confirms the fact all GRBs may be divided into two different classes: short burst and long bursts.

2. Significant stretching is found between emission time distributions created for different intensity groups, in agreement with similar results obtained by different methods (see [5-7]).

ACKNOWLEDGMENTS

The work in USA was supported by NASA project CGRO-98-120 of CGRO Guest Observations Program. The work in Russia was supported by RFBR grant 98-02-17380 and by INTAS project No.96-0315.

REFERENCES

1. Paciesas W.S., et al., BATSE 4B catalog (1999).
2. Mitrofanov I.G. et al., *ApJ* **522**, 1069 (1999).
3. Litvak M.L. et al.,these proceedings.
4. Koshut T.M. et al., *ApJ* **463**, 570 (1996).
5. Mitrofanov I.G. *ApJ* **523**, 610, (1999).
6. Norris, J. P., 4th Huntsville Symposium, (AIP: New York), **428**, 176 (1998).
7. Stern, B. E., *ApJL* **464**, L111 (1996).

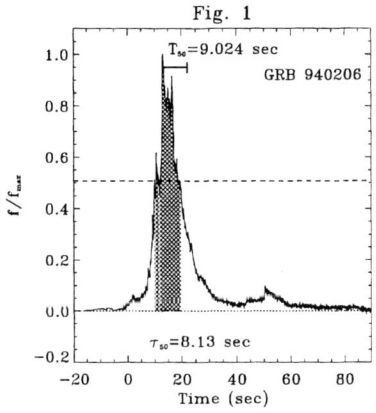
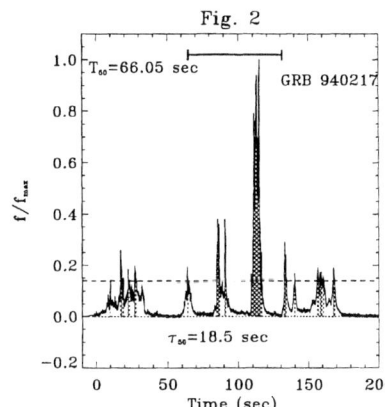

FIGURE 1. The selection of emission time intervals is shown for GRBs 940206. The dashed line corresponds to the level of 50 % of fluence.

FIGURE 2. The selection of emission time intervals is shown for GRBs 940217. The dashed line corresponds to the level of 50 % of fluence.

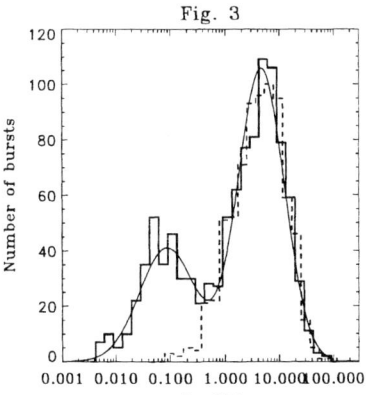
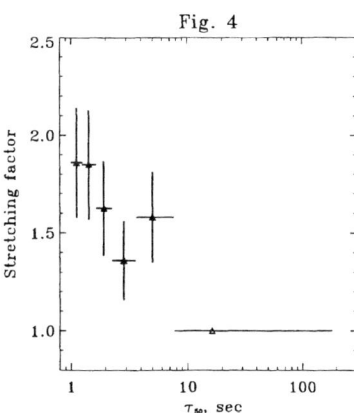

FIGURE 3. The distribution of of τ_{50} for 1029 BATSE bursts with $F_{max}^{(64)} > 1$ ph cm^{-2}s^{-1} and its Gaussian fit are shown. The dashed line corresponds to long BATSE bursts with $T_{90} > 2$ s.

FIGURE 4. The stretching factors are shown for six intensity groups selected from 745 long BATSE bursts with $T_{90} > 2$ s.

Evidence for TeV Emission From GRB 970417a

J.E. McEnery[1], R. Atkins[1], W. Benbow[2], D. Berley[3,10],
M.L. Chen[3,11], D.G. Coyne[2], B.L. Dingus[1], D.E. Dorfan[2],
R.W. Ellsworth[5], D. Evans[3], A. Falcone[6], L. Fleysher[7], R. Fleysher[7],
G. Gisler[8], J.A. Goodman[3], T.J. Haines[8], C.M. Hoffman[8],
S. Hugenberger[4], L.A. Kelley[2], I. Leonor[4], M. McConnell[6],
J.F. McCullough[2], R.S. Miller[8,6], A.I. Mincer[7], M.F. Morales[2],
P. Nemethy[7], J.M. Ryan[6], B. Shen[9], A. Shoup[4], C. Sinnis[8],
A.J. Smith[9], G.W. Sullivan[3], T. Tumer[9], K. Wang[9], M.O. Wascko[9],
S. Westerhoff[2], D.A. Williams[2], T. Yang[2], and G.B. Yodh[4]

(1) University of Utah, Salt Lake City, UT 84112, USA
(2) University of California, Santa Cruz, CA 95064, USA
(3) University of Maryland, College Park, MD 20742, USA
(4) University of California, Irvine, CA 92697, USA
(5) George Mason University, Fairfax, VA 22030, USA
(6) University of New Hampshire, Durham, NH 03824, USA
(7) New York University, New York, NY 10003, USA
(8) Los Alamos National Laboratory, Los Alamos, NM 87545, USA
(9) University of California, Riverside, CA 92521, USA
(10) Permanent Address: National Science Foundation, Arlington, VA 22230, USA
(11) Now at Brookhaven National Laboratory, Upton, NY 11973, USA

Abstract. Milagrito, a detector sensitive to γ-rays at TeV energies, monitored the northern sky during the period February 1997 through May 1998. With a large field of view and high duty cycle, this instrument was used to perform a search for TeV counterparts to γ-ray bursts. BATSE detected 54 GRBs within the field of view of Milagrito during this period. This paper describes the results of an analysis to search for TeV emission correlated with BATSE detected bursts. Milagrito detected an excess of events coincident both spatially and temporally with GRB 970417a, with chance probability 2.8×10^{-5} within the BATSE error radius. No other significant correlations were detected. Since 54 bursts were examined, the chance probability of observing an excess with at least this significance in any of these bursts is 1.5×10^{-3}.

OBSERVATIONS AND ANALYSIS

Milagro, a new type of TeV γ-ray observatory sensitive at energies above 100 GeV, with a field of view of over one steradian and a high duty cycle, began operation in February 1999, near Los Alamos, NM. A predecessor of Milagro, Milagrito [7], operated from February 1997 to May 1998. During this time interval, 54 γ-ray bursts (GRBs) detected by BATSE [1] were within Milagrito's field of view (less than 45° zenith angle).

A search was conducted in the Milagrito data for an excess of events above the background coincident with each of these γ-ray bursts. For each burst, a circular search region was defined by the BATSE 90% confidence interval, which incorporates both the statistical and systematic position errors [2]. The size of this 90% confidence interval ranged from 4° to 26° for the 54 GRBs in the sample. The search region was tiled with an array of overlapping 1.6° radius bins spaced 0.2° apart. The number of events falling within each of the 1.6° bins was summed for the duration of the burst defined by the T90 interval reported by BATSE. T90 ranged from 0.1 seconds to 195 seconds for the 54 bursts examined.

For each GRB, the angular distribution of background events on the sky was characterized using two hours of data surrounding each burst [3]. This distribution was normalized to the number of events (N_{T90}) detected by Milagrito over the entire sky during T90. The resulting background data were also binned in the same 1.6° bins as the initial data. Each bin in the actual data was compared to the corresponding bin in the background map. The Poisson probability of a background fluctuation giving rise to an excess at least as large as that observed was calculated and the bin with lowest probability was taken as the candidate position of a TeV γ-ray counterpart to the BATSE burst. The background and signal counts in this bin were used to calculate a fluence or fluence upper limit for each burst.

RESULTS

The flux sensitivity of Milagrito to γ-ray bursts depends on the zenith angle and duration of the burst, and on the instrument conditions at the time. During the lifetime of the Milagrito detector, data were taken with three different water depths (0.9 m, 1.5 m and 2.0 m). In addition, for the period February 1997 through the end of March 1997 a considerable amount of snow collected on the cover of the pond. Detector simulations were used to obtain effective area as a function of zenith angle for an assumed $E^{-2.0}$ spectrum for each of these configurations. These were then used to calculate flux upper limits for each burst. Flux upper limits in the range $10^{-6} - 10^{-8}$ $\gamma/\text{cm}^2/\text{s}$ were obtained for 53 of the 54 bursts in the sample.

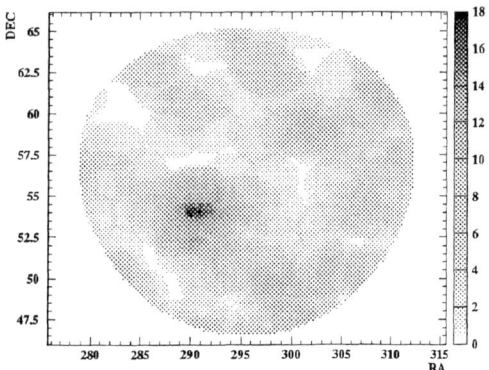

FIGURE 1. Number of events recorded by Milagrito during T90 in the BATSE 90% error radius for GRB 970417, each bin contains the number of events detected by Milagrito within a 1.6 degree radius.

One of the 54 bursts, GRB 970417a, shows a substantial excess above background in the Milagrito data [6]. The BATSE measurements indicate this burst is weak, with a fluence in the 50–300 keV energy range of 1.5×10^{-7} ergs/cm^2 and has a T90 of 7.9 seconds. BATSE determined the burst position to be RA = 295.66°, DEC = 55.77°. The 90% positional uncertainty was 9.4°. The 1.6° radius bin with the largest excess in the Milagrito data is centered at RA = 289.89° and DEC = 54.0°, corresponding to a zenith angle of 21°. This position is 3.8° away from the position reported by BATSE; well within the BATSE 1-σ position error of 6.2°. The uncertainty in the position of the TeV candidate is approximately 0.5°. Figure 1 shows the number of counts in this search region for the array of 1.6° bins. The bin with the largest excess has 18 events with an expected background of 3.46 ± 0.11. The Poisson probability for observing a signal at least this large due to a background fluctuation is 2.89×10^{-8}.

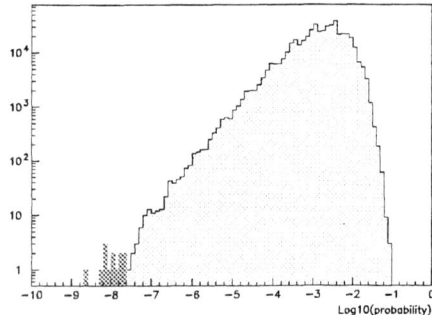

FIGURE 2. The distribution of minimum probabilities for the ensemble of simulated data-sets for GRB 970417a.

To obtain the significance of this result one must account for the size of the search region. The probability of obtaining the observed significance anywhere within the entire search region was determined by Monte Carlo simulations. A set of simulated signal maps was made by randomly drawing N_{T90} events from the background distribution. Each map was searched, as before, for a significant excess within the search region defined by BATSE. The probability of the observation in the actual data being due to a fluctuation in the background, after accounting for the size of the search region, is given by the ratio of the number of simulated data sets with probability less than that observed for the actual data to the total number of simulated data sets. The distribution of the probabilities for 4.65×10^6 simulated data sets is shown in Figure 2; thirteen of which had Poisson probability less than 2.89×10^{-8}. We therefore find that the chance probability of such a detection within the entire 9.4° search region for GRB 970417a to be 2.8×10^{-5}. The probabilities for each of the other 53 bursts in the sample were obtained using the same method, the distribution of these probabilities, after correcting for the size of the search region, is shown in Figure 3. 54 bursts were examined. Therefore the chance probability of observing such a significant excess due to fluctuations in the background for any of these bursts is 1.5×10^{-3}.

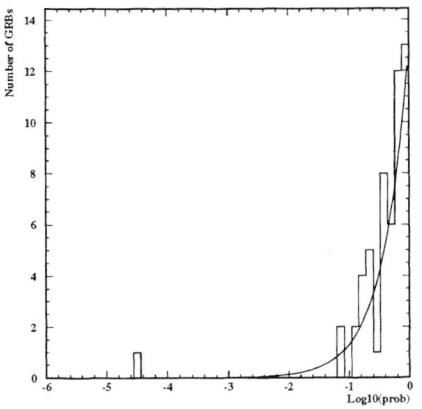

FIGURE 3. The distribution of probabilities, corrected for the size of the search region for the 54 GRBs in the sample. The solid curve shows the probability distribution expected from background data.

Although the initial search was limited to T90, for GRB 970417a longer time intervals were also examined. To allow for the positional uncertainty of the excess observed by Milagrito, the radius of the search bin was increased to 2.2° for this search. A search for TeV γ–rays integrated over long time intervals of one hour, two hours and a day after the GRB start time did not show any significant excess.

DISCUSSION

If the observed excess of events in Milagrito is indeed associated with GRB 970417a then it represents the highest energy photons yet detected from a GRB in coincidence with the sub-MeV emission. The following discussion assumes that the excess observed by Milagrito was due to TeV γ-rays from GRB 970417a. The TeV spectrum and maximum energy of emission are difficult to determine from Milagrito data [7]. Monte Carlo simulations of γ-ray initiated air showers show that the effective area increases slowly with energy, so that the energy threshold is undefined. Figure 4 shows the implied fluence of this observation above 50 GeV as a function of upper cutoff energy for a range of power-law input spectra.

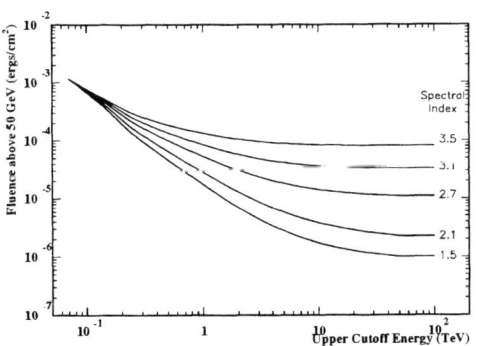

FIGURE 4. Implied fluence (> 50 GeV) as a function of high-energy cutoff for five assumed differential spectral indices

Some information about the energies of the events observed for GRB 970417a can be obtained by considering the response of the summed untriggered counting rate of the individual PMTs in Milagrito. Detector simulations of the effect on PMT counting rates of γ-ray induced air-showers indicate that these rates are more sensitive than the standard shower data at energies below a few hundred GeV, but are only sensitive to very large fluxes [7]. No excess was observed in these rates, which implies that the air-showers detected by Milagrito were probably due to γ-rays at energies above 500 GeV.

High-energy γ-rays from sources at cosmological distances will be absorbed via electron-positron pair production with infrared photons in the intergalactic medium. Several studies find that the opacity due to pair production for above 500 GeV γ-rays exceeds one for redshifts larger than 0.3 [4,5]. Thus, if Milagrito has indeed detected high-energy photons from GRB 990417a, it must be from a relatively nearby object.

To summarize, an excess of events with chance probability 2.8×10^{-5} coincident both spatially and temporally with the BATSE emission for GRB 970417a was observed by Milagrito. The chance probability that an excess of this significance would be observed from the entire sample of 54 bursts is 1.5×10^{-3}. The spectrum must extend with no cutoff to at least a few hundred GeV. The inferred TeV fluence (> 50 GeV) from this result at least an order of magnitude greater than the sub-MeV fluence and the emission extends to at least several hundred GeV. If the observed excess from GRB 970417a is not an unlikely fluctuation of the background, then a GRB bright at TeV energies has been identified. A search for other coincidences with BATSE, to verify this result, will be continued with the current instrument, Milagro, which has increased sensitivity to TeV γ-ray bursts between 0.1 and 10 TeV.

ACKNOWLEDGMENTS

This research was supported in part by the National Science Foundation, the U.S. Department of Energy (Office of High Energy Physics and Office of Nuclear Physics), Los Alamos National Laboratory, the University of California, the Institute of Geophysics and Planetary Physics.

REFERENCES

1. Paciesas, W. S., et al., astro-ph-9903205 (1999).
2. Briggs, M. S., et al., *Astrophys. J. Supp.* **122(2)**, 503 (1999).
3. Smith, A. J., et al., to appear in *GeV-TeV Astrophysics: Toward a Major Atmospheric Cherenkov Telescope VI*, 1999.
4. Salamon, M. H., and Stecker, F. W., *Astrophys. J.* **493**, 547 (1998).
5. Primack, J. R., et al, *Astroparticle Physics* **11**, 93 (1999).
6. Atkins, R., et al., *Astrophys. J.*, submitted (2000).
7. Atkins, R., et al., *Nucl. Inst. and Methods*, accepted (1999).

Fast X-ray Transients and their Relation to GRBs

Vadim A. Arefiev*, Konstantin N. Borozdin[†], and William C. Priedhorsky[†]

*Space Research Institute, 117810 Moscow, Russia
[†]NIS-2, Los Alamos National Laboratory, Los Alamos, NM 87545

Abstract. Fast X-ray Transients (FXTs) are short duration X-ray sources that have timescales of minutes to hours. They have been observed by many instruments, from HEAO-1 to BeppoSAX, however, they are still unexplained. Due to their wide range of observational characteristics, it is suspected that FXTs are a heterogeneous collection of objects involving different emission mechanisms. The relationship between FXTs and gamma-ray bursts (GRBs) is particularly intriguing. The results of BeppoSAX observations confirm that some but not all fast X-ray transients are counterparts of gamma-ray bursts. A fraction of "classical" FXT are believed to be counterparts of GRBs, based largely on time coincidences with GRB events. Other FXTs might be "gamma-ray silent" GRBs. We discuss the statistics and distribution of FXTs, and compare their characteristics with GRB prompt counterparts and afterglows.

INTRODUCTION

The detection of X-ray afterglows with BeppoSAX was a very important step in understanding of the gamma-ray burst phenomenon. These observations brought to a life a new wave of interest to the observation of GRBs in X-rays. It is evident now, that if not all, then at least most GRB sources, whatever they are, emit X-rays as well as gamma-rays, and in copious amounts [1,2]. Studies of this X-ray emission can help us to answer whether GRB emission is beamed or isotropic, and whether the prompt X-ray emission and afterglow are generated by the same mechanism or by different ones. Even today, we know surprisingly little about X-ray emission from GRBs to answer these questions. The X-ray counterparts and X-ray afterglows of GRB show temporal properties that put them in the class of of Fast X-ray Transients (FXT). FXT events were detected by many X-ray surveys and all-sky monitors, from Vela to BeppoSAX and ASM/RXTE. The results of historical experiments have been recently reanalyzed [3,4] to better understand differences in beaming between the two bands. The object of this work is to better understand event statistics and progenitor populations.

FXT STATISTICS, SKY MAP AND PROGENITORS

The Sky Survey Instrument on Ariel-V detected 27 FXTs during its 5 years in orbit [5]. 10 and 8 FXTs, respectively, were reported by the HEAO A-1 and A-2 surveys [6,7]. 7 FXT were discovered by Watch/Granat [8]. 22 X-ray counterparts of GRB were detected with Ginga [9], X-ray emission from 23 GRBs has been studied by BeppoSAX [2]. In addition, the BeppoSAX team reported 9 X-ray flashes, or GRB-like events without corresponding GRBs [10]. So most FXTs seen by BeppoSAX were associated with GRBs. It is quite probable that some of the FXTs seen by other instruments are GRB counterparts or afterglows also, but the majority of GRB-associated events in the whole FXT sample is worth further inquiry, because the FXTs seen by various instruments have been selected with a range of criteria.

To date, FXTs have shown no significant deviation from isotropy on the sky. This conclusion is limited by small number statistics. The dipole moment of the distribution towards Galactic Center is 0.02, and the quadrupole moment is 0.31, compared to 0.00 and 0.33 for an isotropic distribution. We note, however, that the majority of the surveys were biased against FXT from the Galactic plane or Galactic bulge, due to source confusion. The localization accuracy of the archival instruments was poor, so that any X-ray burst from a crowded region would be tend to be attributed to to a known X-ray source. Hence, even if there were a population of FXT with galactic distribution, it would be difficult to distinguish.

Some FXTs have been identified with various types of nearby flaring stars. For example, a large fraction (6 out of 27) of Ariel's FXTs was identified as RS CVn binary systems [5]. The localization error boxes from Ariel-V differ by two orders of magnitude from source to source. For some sources the identification with RS CVn systems looks reliable, while in other cases statistical coincidence cannot be excluded. For GRBs, identifications can draw on temporal as well as spatial coincidence. The WATCH/Granat all-sky monitor observed 95 events which, due to the 8 to 100 keV response band of the experiment [11,8], can be classified as either FXT or GRB. None of 47 localized events was, however, identified with a RS Cvn system, so it is evident that this sample of FXTs is quite different from one observed with Ariel-V. The share of GRB-associated events in the integrated FXT database remains unknown.

LOG N–LOG S

Based on the events seen by several experiments (Ariel-V, HEAO-1 A-1 and A-2 surveys, 7 FXT observed by WATCH/Granat) [5–8] we have tried to reconstruct the Log N–Log S curve for FXT events. The result is presented in Figure 1. However, this result depends significantly on several assumptions. For the HEAO-1 surveys, which are the basis for our log N–log S relationship at lowest fluences (time integrated flux in units of erg cm^{-2}), the real fluences were not measured, but

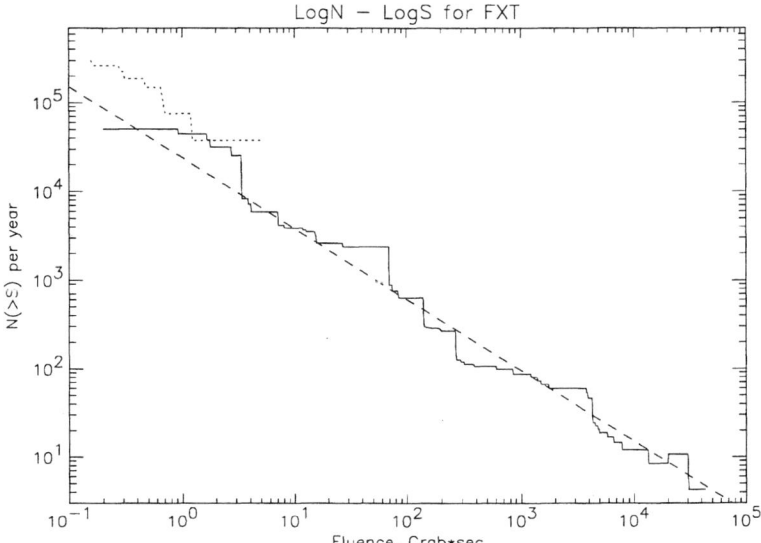

FIGURE 1. LogN–LogS distribution of FXTs observed by HEAO A1,A2; Ariel-V; WATCH (converted to a number of yearly events in whole sky). Solid line - average FXT distribution, dotted line - FXT distribution under assumption of minimum event durations for HEAO-1 A-2 survey, dashed line - a power law approximation with slope −0.8

can be estimated only by assuming an event duration which is poorly constrained. For example, for the HEAO-1 A-2 survey the duration of detected events is within the range from 60 to 2000 s. We assumed a geometric mean of 346.4 s in this case. To demonstrate the impact of this assumption, we also show in Figure 1 (dotted histogram) an alternative Log N–Log S for HEAO-1 A-2, calculated under the assumption that each event lasted the minimum possible time (60 sec - see [7]). We accepted the correction factors that were used by the original authors to calculate total number of the events per year from their observed event rates. These factors are also duration-dependent, and we calculated them for the assumed duration (which was about 55 thousands of 346s-events per year for HEAO-1 A-2). Estimations given by different surveys (and different authors) differ significantly for the same range of fluences, and we applied the geometric mean again to obtain an average histogram. Hence for our average curve (solid line) we tried to avoid extremes and follow the middle path as described above, but it doesn't guarantee the precise results given an accuracy of the measurements. Still, we note that our average curve has a power law slope close to −0.8, which was measured by [5] from Ariel-V alone and confirmed also by [3].

We estimated the fraction of FXTs in the Ariel-V survey that could correspond to GRBs. For this analysis we took the log N–log S distribution of long GRBs (with T_{90} longer than 2 s) reported in the BATSE 4B Catalog, and simulated their observation with Ariel-V. Critical to this analysis are two parameters: the fraction of GRBs that emit in the X-ray band, and the ratio between X-ray and gamma-ray fluxes Based on results from BeppoSAX and WATCH [2,11], we assumed that all GRBs with duration longer than 1-2 sec emit X-rays. We left the fluence ratio Rx/Rg, as a free parameter. Strohmayer et al. [9] obtained an average value of 0.24 for 22 bright Ginga bursts. However, for most bursts this value was close to 0.03, and the average value was much higher due to few events with Rx/Rg close to unity or even greater. Figure 2 demonstrates that this parameter is crucial to the fraction of GRBs in this FXT database. A better estimate of the ratio requires its measurement for a large samples of GRBs. Nonetheless, we cannot rule out the possibility that most FXTs are related to gamma-ray bursts. If so, FXTs would be a powerful new observational tool to understand the gamma-ray burst phenomenon.

WHAT IS NEEDED TO BETTER UNDERSTAND FXTS

The uncertainty of our analysis reflects our ignorance about the FXT phenomenon and its relation with GRBs. A large, uniform sample of FXTs would define their distribution on the sky, and highlight any Galactic components. For-

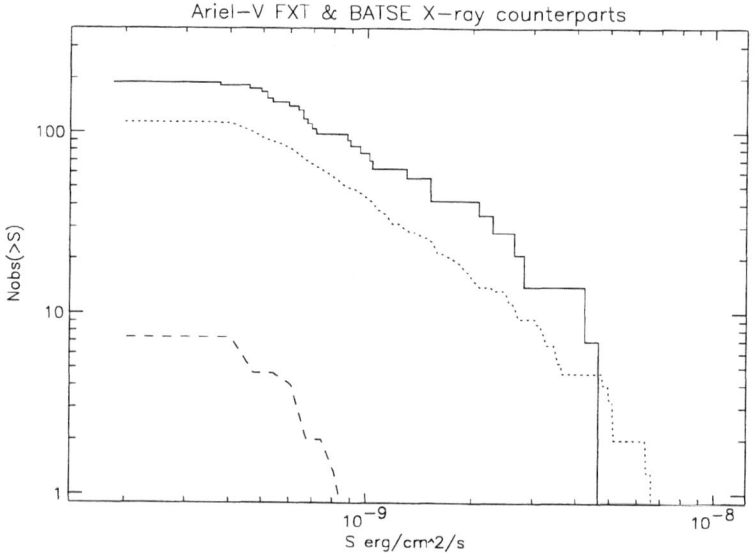

FIGURE 2. Ariel-V FXT and GRB X-ray counterparts. Solid line - Ariel-V FXTs (converted to number of events in whole sky per year). Dotted line - simulated observations of BATSE GRBs by Ariel-V; prompt X-ray ratio Rx/Rg=0.24 (Ginga average) and afterglow time decay slope −1.3 (BeppoSAX average). Dashed line - the same as dotted line, but for Rx/Rg=0.03.

tunately, accurate position measurements with simple instruments are easier in the X-rays than in the gamma-rays. Precise locations and real-time event announcements would allow the identification of counterparts in other wavebands. In short, we need to do for FXTs the job done by BATSE and BeppoSAX for GRBs. Both these goals can be achieved by placing a sensitive wide field-of-view X-ray all-sky monitor on a platform with a fast data link to the ground. The design for this experiment already exists [12,13], and flight-qualified hardware is on the shelf [14]. By adding a gamma-ray detector on the same platform, we can obtain simultaneous observations of FXTs in both bands. With these data, we can go far beyond the limits of archival surveys, and hope to finally understand the connection between FXTs and GRBs.

REFERENCES

1. Piro, L., Amati, L., Antonelli, L. et al., *A&A* **331**, L41 (1998).
2. Frontera, F., Amati, L., Costa, E. et al., preprint **astro-ph/9911228**, (1999).
3. Grindlay, J. E., *ApJ* **510**, 710 (1999).
4. Woods, E. & Loeb, A., *ApJ* **523**, 187 (1999).
5. Pye, J. & McHardy, I., *MNRAS* **205**, 875 (1983).
6. Ambruster, C. & Wood, K., *ApJ* **311**, 258 (1986).
7. Connors, A., Serlemitsos, P., Swank, J., *ApJ* **303**, 769 (1986).
8. Castro-Tirado, A., Brandt, S., Lund, N. et al., *A&A* **347**, 927 (1999).
9. Strohmayer, T., Fenimore, E., Murakami, T. et al., *ApJ* **500**, 873 (1998).
10. in't Zand, J. et al., these proceedings.
11. Sazonov, S., Sunyaev, R., Terekhov, O. et al., *A&A SS* **129**, 1 (1998).
12. Holt, S.S. & Priedhorsky, W.C., *Space Sci. Rev.* **45**, 269 (1987).
13. in't Zand, J., Priedhorsky, W. C., Moss C. E., Fenimore, E. E., Black, J. K., Kelley, R. L., Stilwell, D. E., Birsa, F. B., Borozdin, K. N., and Arefiev, V. A., *Proc. SPIE* **2279**, 458 (1994).
14. Borozdin, K. N., Priedhorsky, W. C., Arefiev, V. A. et al., *Proc. Small Mission Workshop, Los Alamos* (1999).

LOTIS Upper Limits and the Prompt OT from GRB 990123

G. G. Williams*, D. H. Hartmann*, H. S. Park[†], R. A. Porrata[†],
E. Ables[†], R. Bionta[†], D. L. Band[¶], S. D. Barthelmy[§], T. Cline[§],
N. Gehrels[§], D. H. Ferguson[‡], G. Fishman[⋆], R. M. Kippen[⋆],
C. Kouveliotou[⋆], K. Hurley[♯], R. Nemiroff[b], and T. Sasseen[||]

*Dept. of Physics and Astronomy, Clemson University, Clemson, SC 29634
[†]Lawrence Livermore National Laboratory, Livermore, CA 94550
[¶]Los Alamos National Laboratory, Los Alamos, NM 87545
[§]NASA Goddard Space Flight Center, Greenbelt, MD 20771
[‡]Dept. of Physics, California State University, Hayward, CA 94542
[⋆]NASA Marshall Space Flight Center, Huntsville, AL 35812
[♯]Space Sciences Laboratory, University of California, Berkeley, CA 94720
[b]Dept. of Physics, Michigan Technological University, Houghton, MI 49931
[||]Dept. of Physics, University of California, Santa Barbara, CA 93106

Abstract. GRB 990123 established the existence of prompt optical emission from gamma-ray bursts (GRBs). The Livermore Optical Transient Imaging System (LOTIS) has been conducting a fully automated search for this kind of simultaneous low energy emission from GRBs since October 1996. Although LOTIS has obtained simultaneous, or near simultaneous, coverage of the error boxes obtained with BATSE, IPN, XTE, and BeppoSAX for several GRBs, image analysis resulted in only upper limits. The unique gamma-ray properties of GRB 990123, such as very large fluence (top 0.4%) and hard spectrum, complicate comparisons with more typical bursts. We scale and compare gamma-ray properties, and in some cases afterglow properties, from the best LOTIS events to those of GRB 990123 in an attempt to determine whether the prompt optical emission of this event is representative of all GRBs. Furthermore, using LOTIS upper limits in conjunction with the relativistic blast wave model, we weakly constrain the GRB and afterglow parameters such as density of the circumburster medium and bulk Lorentz factor of the ejecta.

INTRODUCTION

The ultimate reward for the Gamma-Ray Burst Coordinates Network (GCN) [1] came when the Robotic Optical Transient Search Experiment (ROTSE) detected prompt optical emission from GRB 990123 [2]. Although this discovery marks another milestone in comprehending the physics of GRBs, bright optical transients

(OTs) may be the exception rather than the rule. Both LOTIS and ROTSE have unsuccessfully attempted to detect these predicted flashes on many occasions [3–8]. Although some of the non-detections may be attributed to large extinction, GRB 990123 demonstrated that the progenitor is not always obscured.

OBSERVATIONS & ANALYSIS

During more than 1100 nights of possible observations (since October 1996), LOTIS has responded to 127 GCN triggers. Of these, 68 triggers were unique GRB events; a rate of approximately one unique GRB event every 16.5 days. The quality of the LOTIS "coverage" for a given event depends on five factors: observing conditions, LOTIS response time, difference between the initial and final coordinates, size of the final error box, and the duration of the GRB. Table 1 lists 13 events for which LOTIS achieved good coverage.

First we compare GRB 990123 with the LOTIS upper limits to test whether the flux of the prompt optical emission scales with some gamma-ray property. Here and throughout the analysis we neglect extinction effects. The first row in Table 1 lists the properties of GRB 990123 [9,2]. The columns display the UTC date of the burst, the BATSE trigger number, the 64 ms and 1024 ms peak fluxes (50 - 300 keV), and the gamma-ray fluence (>20 kev) of each event. The last three columns are the scaled magnitudes,

$$m_{\rm GRB} = m_{\rm GRB990123} - 2.5 \log \left(\frac{X_{\rm GRB}}{X_{\rm GRB990123}} \right), \quad (1)$$

TABLE 1. LOTIS GRB events with good coverage and predictions for the scaled magnitudes of the prompt optical emission.

Date	BATSE Trig.	F_p (64 ms) (γ cm^{-2} s^{-1})	F_p (1024 ms)	$S/10^{-7}$ (erg cm^{-2})	$m_{F;64}$	$m_{F;1024}$	m_S
990123	7343	16.96	16.41	3000	9.0	9.0	9.0
961017	5634	4.22	1.98	5.07	10.5	11.3	15.9
961220	5719	1.93	1.60	18.11	11.4	11.5	14.5
970223	6100	19.41	16.84	968	8.9	9.0	10.2
970714	6307	1.89	1.32	17.09	11.4	11.7	14.6
970919	6388	1.10	0.77	22.49	12.0	12.3	14.3
971006	6414	2.08	1.79	258	11.3	11.4	11.7
971227	6546	3.32	2.11	9.25	10.8	11.2	15.3
990129	7360	5.88	4.99	585	10.2	10.3	10.8
990308	7457	2.02	1.26	164	11.3	11.8	12.2
990316	7475	3.87	3.67	529	10.6	10.6	10.9
990413	7518	3.78	2.57	68.13	10.6	11.0	13.1
990803	7695	16.99	12.19	1230	9.0	9.3	10.0
990918	7770	5.69	3.17	25.21	10.2	10.8	14.2

where $m_{\text{GRB990123}} = 9.0$, the peak magnitude of GRB 990123, and X_{GRB} and $X_{\text{GRB990123}}$ are the gamma-ray peak flux and fluence values for those events.

The LOTIS sensitivity varies depending on observing conditions but in general a conservative limiting magnitude is $m \approx 11.5$ prior to March 1998 (upgrade to cooled CCD) and $m \approx 14.0$ following that date. Table 1 shows that the scaled prompt optical emission for both peak flux and fluence is often brighter than the LOTIS upper limits which suggests that these simple relationships are not valid.

Briggs et al. [9] show that the optical flux measured during GRB 990123 is not consistent with an extrapolation of the burst spectrum to low energies. However Liang et al. [10] point out that the extrapolated tails rise and fall with the optical flux. A low-energy enhancement would produce an upward break which might account for the measured optical flux during GRB 990123. It is important to determine if there is a low-energy upturn in the spectrum since it would establish whether or not the optical and gamma-ray photons are produced by the same electron distribution. The LOTIS upper limits can be used to constrain a low-energy enhancement assuming it is common to all GRBs.

For the events listed in Table 1 we fit the gamma-ray spectra during the LOTIS observations to the Band functional form [11]. In a few cases the low-energy extrapolation is near the LOTIS upper limit. The solid line in Figure 1 shows the Band fit to GRB 971006 and its extrapolation to low energies. Fits to the spectra of GRB 990123 during the first (short dash), second (dash-dot), and third

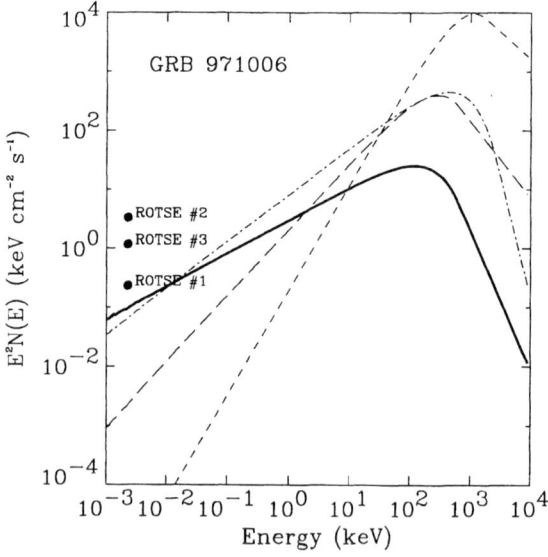

FIGURE 1. Extrapolated spectrum of GRB 971006 during the LOTIS observation and GRB 990123 during ROTSE detections.

(long dash) ROTSE observations and the corresponding ROTSE detections (filled circles) are also shown. The extrapolation of GRB 971006, predicts an $m \approx 12.4$ optical flash. Even a slight upward break in the spectrum would have produced a detectable OT. We conclude that the LOTIS upper limits support the hypothesis that the low-energy emission is produced by a different electron distribution than the high energy-emission.

Finally, we attempt to use the LOTIS upper limits and the external reverse shock model to constrain the physical properties of the GRB blast wave. Sari and Piran [12] show that the fraction of the energy that is emitted in the optical band depends on the values of the cooling frequency and the characteristic synchrotron frequency. For the external reverse shock these frequencies are given by

$$\nu_c = 8.8 \times 10^{15} \text{Hz} \left(\frac{\epsilon_B}{0.1}\right)^{-3/2} E_{52}^{-1/2} n_1^{-1} t_A^{-1/2}, \tag{2}$$

$$\nu_m = 1.2 \times 10^{14} \text{Hz} \left(\frac{\epsilon_e}{0.1}\right)^2 \left(\frac{\epsilon_B}{0.1}\right)^{1/2} \left(\frac{\gamma_0}{300}\right)^2 n_1^{1/2}, \tag{3}$$

where ϵ_e and ϵ_B are the fraction of equipartition energy in the electrons and magnetic field, E_{52} is the total energy in units of 10^{52} erg, n_1 is the density of circumburster medium in cm^{-3}, γ_0 is the initial Lorentz factor, and t_A is the duration of the emission in seconds.

Sari and Piran assume the frequency dependencies modify the fluence of a moderately strong GRB, i.e. 10^{-5} erg cm^{-2}. In this analysis we compare the afterglow

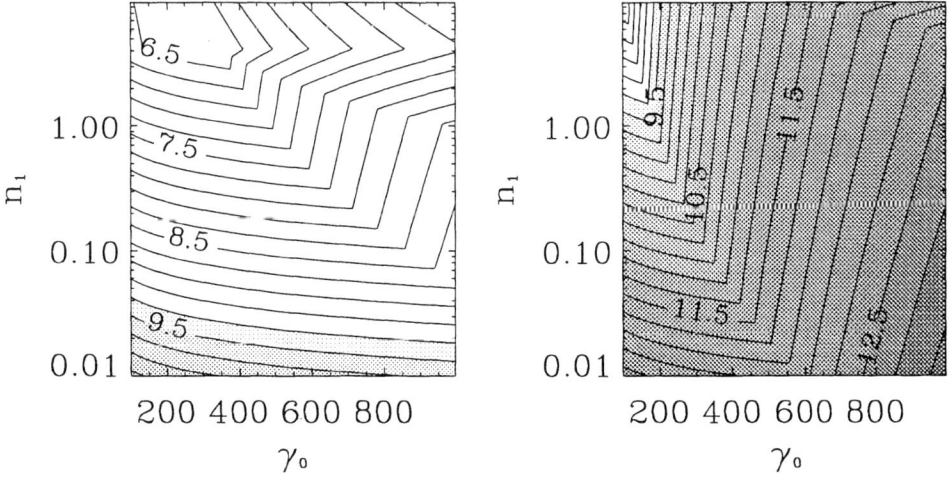

FIGURE 2. Predicted magnitude of the prompt optical flash for $E_{52} = 3.5$, $\epsilon_e = 0.12$, and $\epsilon_B = 0.089$ (left panel) and $E_{52} = 0.53$, $\epsilon_e = 0.57$, and $\epsilon_B = 0.0082$ (right panel).

properties of GRB 970508 found by Wijers and Galama [13] to those found by Granot et al. [14]. Therefore we use a fluence of 3.1×10^{-6} erg cm^{-2} emitted over the entire LOTIS integration time of $t_A = 10.0$ s. The index of the electron power-law distribution is set to p = 2.2.

Figure 2 shows contour plots of the predicted magnitude of the prompt OT for GRB 970508 as a function of n_1 and γ_0. GRB 970508 could not be observed by LOTIS or ROTSE since it occurred during the day. Values of $E_{52} = 3.5$, $\epsilon_e = 0.12$, and $\epsilon_B = 0.089$ from Wijers and Galama are used in the left panel and values of $E_{52} = 0.53$, $\epsilon_e = 0.57$, and $\epsilon_B = 0.0082$ from Granot et al. are used in the right panel. The right panel demonstrates the effect of altering the total energy and the distribution of energy to the electrons and the magnetic field. The smaller values of E_{52} and ϵ_B shift the contours to the upper left while the larger ϵ_e steepens the breaks in the contours. The increased shading corresponds to a decreasing detection probability. However for nearly all values of n_1 and γ_0 shown the predicted OT could have been detected by the upgraded LOTIS system.

Wijers and Galama find a circumburster medium density of $n_1 = 0.030$ which predicts an $m = 9.0 - 9.5$ optical flash nearly independent of the initial Lorentz factor. Granot et al. find a considerably higher vlaue of $n_1 = 5.3$, which predicts an $m = 8.7 - 12.4$ OT which is very dependent on the initial Lorentz factor. The LOTIS upper limits mildly favor the GRB blast wave values determined by Granot et al. since dim OTs are predicted over a larger range of initial Lorentz factors.

REFERENCES

1. Barthelmy, S. D., et al., "The GRB Coordinates Network (GCN): A Status Report", in *Gamma-Ray Bursts 4th Huntsville Symposium*, edited by C. A. Meegan, R. D. Preece, & T. M. Koshut, AIP Conf. Proc. 428, Woodbury, 1998, pp. 99-103.
2. Akerlof, C. W., et al., *Nature* **398**, 400-402, (1999).
3. Park, H. S., et al., *ApJ* **490**, 99-108 (1997).
4. Park, H. S., et al., *ApJ* **490**, L21-L24 (1997).
5. Williams, G. G., et al., "First Year Results from LOTIS", in *Gamma-Ray Bursts 4th Huntsville Symposium*, edited by C. A. Meegan, R. D. Preece, & T. M. Koshut, AIP Conf. Proc. 428, Woodbury, 1998, pp. 837-841.
6. Williams, G. G., et al., *ApJ* **519**, L25-L29 (1999).
7. Schaefer, B. E., et al., *ApJ* **524**, L103-L106 (1999).
8. Kehoe, R., et al., submitted to *Proc/ of the 1999 May Symposium of the Space Telescope Institute*, (1999); (astro-ph/9909219).
9. Briggs, M. S., et al., *ApJ* **524**, 82-91 (1999).
10. Liang, E. P., et al., *ApJ* **519**, L21-L24 (1999).
11. Band, D. L., et al., *ApJ* **413**, 281-292 (1993).
12. Sari, R., and Piran, T., *ApJ* **520**, 641-649 (1999).
13. Wijers, R. A. M. J., and Galama, T. J., *ApJ* **523**, 177-186 (1999).
14. Granot, J., et al., *ApJ*, submitted (1999); (astro-ph/9808007).

Preliminary Results from the TAROT Experiment

M. Boër[1], J.-L. Atteia[1], M. Bringer[1], S. Chaty[1], A. Klotz[1],
F. Morand[2], H. Pedersen[3], C. Pollas[3], and D. Toublanc[1]

[1] *Centre d'Etude Spatiale des Rayonnements (CESR/CNRS), BP 4346,
31028 Toulouse Cedex 4, France*
[2] *Observatoire de la Côte d'Azur, 2130 route de l'Observatoire, Caussols,
06460 St Vallier de Thiay, France*
[3] *NBIFaG, Copenhagen University Observatory*

Abstract. TAROT-1 has been operating routinely since July 1999. During that time 6 alerts have been observed in good weather conditions, leading to limits in magnitude of about R = 14 to 17 for the early afterglow of the gamma-ray burst sources. Future prospects include the replacement of the present camera by a more sensitive one, a larger field of view, and an improved readout time.

INTRODUCTION

Since the GCN [1] sends timely positions of GRB sources, various attempts have been made to catch optical flashes from GRBs. The discovery of the optical transient (hereafter OT) associated with GRB 990123 [2] demonstrated that OTs are a tool to understand the physics of GRBs, and they are potentially a means to provide an early accurate position of the source, and to make subsequent spectroscopy and photometry of the afterglow on the short and long term.

The Télescope à Action Rapide pour les Objets Transitoires [3] (Rapid Action Telescope for Transient Objects, TAROT-1) has been in operation for about one year now. The major technical parameters of TAROT-1 are summarized in Table 1. After a test phase, the reliability of the system was greatly enhanced, and no major software and/or hardware problems have been noticed since June 1999. In short, TAROT is operational every clear night. The connection with the GCN is quite reliable and fast, since the travel time of a GCN packet is typically 250 ms, with little loss. We get an observable alert every 2 weeks on average, though the weather during the month of November has been rather bad, and no alert could be observed. Between two alerts TAROT makes various routine observations, which may help in understanding the background of variable or transient events, and ensure the availability of the experiment.

TABLE 1. TAROT-1 main technical characteristics

Optics	
Aperture	25 cm, f/3.2
Field of view	2° × 2°
Optical Resolution	20 μm
Mount	
Mount type	Equatorial
Slew speed	Max. 80°/s
Traking speed	Adjustable α, δ 0°/s – 40°/s
Maximum acceptable wind speed	80 km/h
Detector (present)	
Size	1024 × 1270 pixels
Pixel size	15 μm, 3.2″
Readout speed	20 s
Readout noise	≤ 30 e$^-$
Limiting magnitude	R = 17
Detector (planned 1st Quarter 2000)	
Size	2048 × 2048 pixels
Pixel size	14μm, 3.1″
Readout speed	1 s
Readout noise	10 e$^-$
Exposure characteristic times	
Typical integration time (alert)	10 – 20 s
Typical integration time (routine)	300 s
Minimum integration time	0.1 s
Maximum integration time	300 s

OBSERVATIONS

In case of an alert from BATSE, TAROT starts a mosaic observation, as shown in Figure 1. With the present camera, the full process takes about 25 min, for a covered area of 5° × 5°, while for the new camera an area of 10° × 10° will be covered in less than 5 min. If the origin of the alert allows a reduced area to be covered, then we adapt our mosaic, or we make direct observations. The data are then processed and a list of sources is produced, matched with the USNO A2.0 catalog. The analysis of a full image takes about 1 min, introducing little delay. As soon as a complete mosaic has been processed, an e-mail is sent to a pager to awake the duty scientist, whose role is to check the results, and to have a close look at the unmatched objects in order to see whether one of them can be associated to the GRB source.

We observed seven GRB sources as of December 7th, 1999. While in one case (the first one) we had FITS compatibility problems (which were fixed subsequently), in the others we were unable to find any source which could be associated with the GRB. All the "transient" sources are either asteroids or variable stars, or appear only on one frame. Table 2 summarizes the limits we have.

Figure 1 illustrates the process of taking mosaics with BATSE locations. The

TABLE 2. TAROT-1 GRB observations, as of December 7th, 1999

GRB date	BATSE trigger #	Limit	Remarks
990208	7688	N/A	FITS problem
990807	7703	R > 14	
990903	7750	R > 14	IRAS 23024+6729 identified as Mira variable
990915	7766	R > 14	Clouds on 3rd mosaic
990917	7769	R > 15	
991002	7784	R > 16	SN 99eb not detected
991115	7858	N/A	Cloudy, not enough data

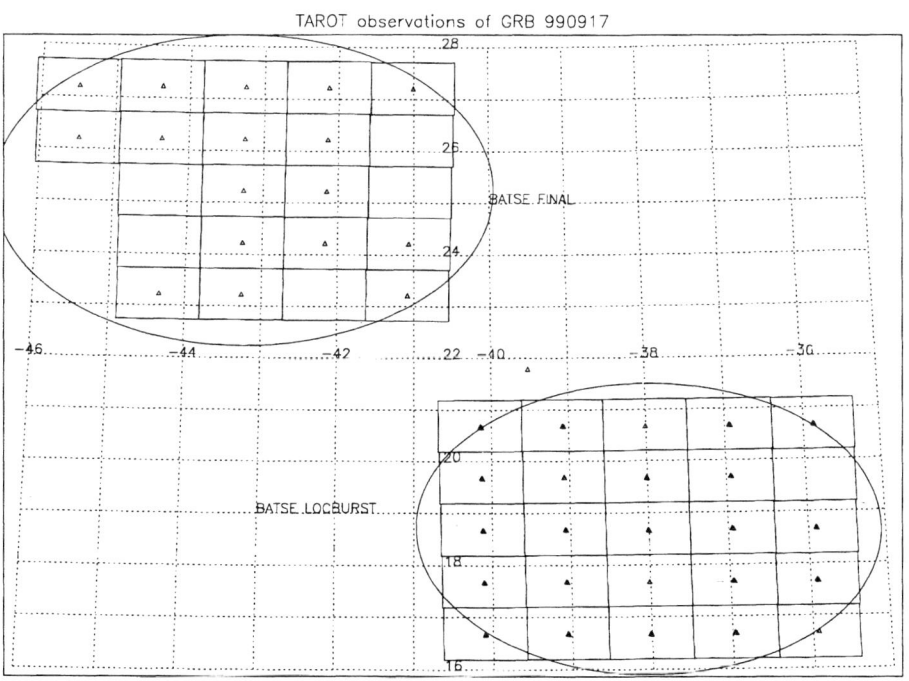

FIGURE 1. TAROT observations of the GRB 990917 source

limits of the various frames are shown, as well as the BATSE error box, including the systematic error. TAROT uses first the BATSE ORIGINAL and FINAL notices, and starts to acquire images. As soon as the LOCBURST position is available, the system starts a new mosaic centered on the new position, interrupting the last exposure of the previous mosaic. We note that this position may be offset from the FINAL position by an angle larger than the quoted error. Several comments should be made from the data we acquired already:

- The initial, fast locations given by the GCN for BATSE may have somewhat large errors. This means that we have to span a larger error area over the sky (e.g., $15° \times 15°$), in order to be sure to catch the right location. This will be easier with the new, $2° \times 2°$ camera, because of its quick readout and larger area covered.

- As soon as a better position arrives, e.g., BATSE LOCBURST, the width of the mosaic is reduced.

- We react also to alerts given, through the GCN socket connection, by the BeppoSAX, ALEXIS, and RXTE satellites. In these cases, as obvious, only frames centered on the burst source position are taken.

- TAROT reaction times, from the coordinate reception to the beginning of the exposure, are between 1 and 2 seconds, depending weakly on the angular amplitude of the telescope move.

CONCLUSION AND PERSPECTIVES

TAROT-1 is already one of the most sensitive and autonomous instruments devoted to the search of gamma-ray burst sources at optical wavelengths. The limits we can reach (R = 17 [10σ] in 10 s, and 19 for a 1 min exposure) will probably be enhanced with the new camera, but are already among the most stringent limits on the prompt counterpart and early afterglow. They probably mean that the optical luminosity of GRB 990123 was quite exceptional.

The perspectives for TAROT-1 are both in space and on the ground. The HETE-II satellite, due to launch in January 2000, will give timely positions, with an accuracy that matches well the TAROT field of view. On ground, the camera we are assembling and testing at present will enter into operations by the beginning of the next year, giving a faster response, a larger field of view, reduced deadtimes, etc. As an example, with this camera it will be possible to make a 1 s sampling of the light curve, provided the GRB prompt counterparts are bright enough. On the mid-term, TAROT-1 will be able to work with various space missions, from SWIFT to BALLERINA, including GLAST and INTEGRAL, because of its large field of view.

We also plan to enhance the reliability of our data reduction system and to implement a fast data management system that will have to be able to process the data, to recognize quickly if there is a GRB source optical candidate in our data, and to send the position to other instruments. Already, TAROT alone has the good field/sensitivity compromise to be able to detect the optical flash from a GRB source and the early afterglow for several hours, providing a link with larger facilities.

ACKNOWLEDGMENTS

The Télescope à Action Rapide pour les Objets Transitoires (TAROT-1) has been funded by the Centre National de la Recherche Scientifique, Institut National des Sciences de l'Univers (CNRS/INSU)

REFERENCES

1. Barthelmy, S., et al., these proceedings.
2. Akerlof, C., et al., *Nature* **398**, 400 (1999).
3. Boër, M., et al., *A&ASS* **138**, 579 (1999).

First Results from the Burst Observer and Optical Transient Exploring System Station 1 (BOOTES-1)

A. J. Castro-Tirado[1,2], J. Soldán[3], M. Bernas[4], P. Páta[4], R. Hudec[3], T. M. Sanguino[5], B. de la Morena[5], J. A. Berná [6], A. de Ugarte[7], J. Gorosabel[1], J. M. Más-Hesse[1], and A. Giménez[1,2]

[1] *Laboratorio de Astrofísica Espacial y Física Fundamental (LAEFF-INTA), P.O. Box 50727, E-28080, Madrid, Spain*
[2] *Instituto de Astrofísica de Andalucía (IAA-CSIC), P.O. Box 03004, E-18080 Granada, Spain*
[3] *Astronomical Institute, Academy of Sciences of Czech Republic, 251 65 Ondrejov, Czech Republic*
[4] *Czech Technical University, Faculty of Electronic Engineering, Department of Radioelectronics, 16627 Prague, Czech Republic*
[5] *Centro de Experimentación de El Arenosillo (CEDEA-INTA), Mazagón, Huelva, Spain*
[6] *Departamento de Física, Ingeniería de Sistemas y Teoría de la Señal, Universidad de Alicante, Alicante, Spain*
[7] *Facultad de Ciencias Físicas, Universidad Complutense, Madrid, Spain*

Abstract. The Burst Observer and Optical Transient Exploring System (BOOTES) is considered as a part of the preparations for ESA's INTEGRAL satellite, and is currently being developed in Spain, in collaboration with two Czech institutions. It makes use of two sets of wide-field cameras, 240 km apart, and two robotic 0.3-m telescopes. The first observing station (BOOTES-1) is located in Huelva (Spain) and the first light was obtained in July 1998. During the test phase, it has provided rapid follow-up observations with the wide-field cameras for 19 GRBs detected by BATSE aboard *CGRO*, and narrow-field imaging for 6 bursts. Limiting magnitudes for any GRB optical afterglow are I $\sim$ 13 and R $\sim$ 16.5, a few minutes after the events.

INTRODUCTION

After years of search, the first optical counterparts to gamma-ray bursts (GRBs) have been found in 1997, beginning 3–20 hr after the onset of the high-energy events. A power-law decline with the flux $F \propto t^\alpha$ is usually observed, where α is in the range -1.1 to -2.0 (for a review, see [1] and references therein). Under the assumption of the bursts being a repeating phenomenon, archival plates have been used in order to look for optical transient emission in the smallest GRB error

TABLE 1. Features of the BOOTES Wide Field Cameras

Lens	50 mm @ f/2.0	18 mm @ f/2.8
CCD	1534 × 1020 pixels	1534 × 1020 pixels
field of view	16° × 11°	40° × 28°
angular resolution	0.63'	1.58'
limiting magnitude (300-s)	I ~ 13	R ~ 11

TABLE 2. Features of the BOOTES Robotic Telescope

Telescope	D = 0.30 m @ f/3.3
CCD	1530 × 1020 pixels
field of view	49' × 33'
angular resolution	1.9"
limiting magnitude (60-s)	R ~ 16.5

boxes, and about 50 candidates have been identified so far, but most of them were rejected as they turned out to be plate defects [2].

INSTRUMENTATION

The first observing station of the *Burst Observer and Optical Transient Exploring System* (named BOOTES-1) is located at El Arenosillo (Huelva), a dark-sky site in Spain owned by the Instituto Nacional de Técnica Aeroespacial (INTA). Wide-field and narrow-field cameras are part of the BOOTES instrumentation.

The Wide Field Cameras

The wide field cameras (WFC) consist of commercial 50-mm wide-field lenses attached to two ST8 CCD cameras. Each pair of the BOOTES wide-field cameras is mounted atop a 0.3-m Schmidt-Cassegrain telescope, allowing long integrations of a previously selected region. The four cameras monitor the same region of the sky, both in the I- and V-bands. The typical limiting magnitude is I ~ 11 for an integration time of 30 s, and I ~ 13 for 300 s (see Table 1). A third ultra wide-angle 18-mm lens is mounted atop the telescope, providing a 40° × 28° field of view. The system is operated by commercial PCs. It is planned to work most of the time as discussed in [3,4], allowing discrimination against flashing objects closer than one million km, thus ruling out satellite glints and other atmospheric and near-Earth events. When information on a GRB position is obtained from the GCN (the GRB Coordinates Network [5]), the corresponding GRB error box is imaged by the cameras.

TABLE 3. Follow-up observations by the BOOTES WFC in 1998-99.

GRB	Data taken after	Error box coverage	Lim. mag. (I)
980808	66-min	95%	13
980810	125-min	100%	13
981009	255-min	100%	13
981203	35-min	100%	13
981205	30-min	100%	13
990321	14-min	75%	13
990713	5-min	50%	13
990720	302-min	100%	13
990724	3-min	70%	13
990725	4-min	35%	13
990802	6-min	100%	13
990822	59-min	70%	13
990903	317-min	100%	13
990904	235-min	100%	13
990905	34-min	100%	13
991002	55-min	100%	13
991014	420-min	100%	13
991115	37-min	100%	10
991219	4-min	100%	10

The Robotic Telescope

The Robotic Telescope is based on a previous design [6]. The CCD camera (Narrow-Field Camera, NFC) at the Cassegrain focus performs rapid follow-up observations of events detected by *BATSE*, *BeppoSAX*, *Rossi-XTE* and future GRB detectors like *HETE-II*, *SWIFT* and *BALLERINA*. Thus, the BOOTES telescope is able to slew immediately and take deep frames at the GRB positions. Selected objects (variable stars, nearby galaxies, bright QSOs, etc.) will be regularly monitored, searching for flaring behavior. The first telescope unit is already placed at the BOOTES-1 station, under a special enclosure which is opened automatically, according to weather conditions. The main features are shown in Table 2.

SCIENTIFIC OBJECTIVES AND FIRST RESULTS

The main goal is the observation of GRB error boxes simultaneously to the GRB occurrence ($\sim$ 6 GRBs per year). The faint transient emission that has been detected for hours after the event seems to be a consequence of the expanding remnant produced by the GRB. This provides information about the surrounding medium, but not about the burster itself. Although these optical counterparts are not brighter than 19 mag a few hours after the burst, transient emission *simultaneous* to the event is expected to be more intense. In fact, an extrapolation of the γ-ray power-law spectrum indicates that the simultaneous optical flash should lie

TABLE 4. Follow-up observations by the BOOTES NFC in 1999.

GRB	Data taken after	Imaged region (deg)	Lim. mag. (R)
990802	27-min	3.5 × 3.3	17.0
990903	371-min	4.5 × 4.7	16.5
990904	235-min	3.5 × 3.3	16.5
990905	28-min	4.5 × 4.7	16.0
991115	71-min	5.5 × 5.3	16.0
991219	25-min	3.5 × 3.3	16.0

in the range 10–15 mag, depending on the burst intensity. Current theories predict that there should be an optical flash reaching a red broad-band magnitude R ∼ 10, or brighter, as has been seen for the first time in GRB 990123 [7]. Follow-up observations by the BOOTES-WFC have been performed for 19 bursts so far, with delays ranging from 3 min to 7 hr. In all cases, the limiting magnitude for any GRB optical afterglow emission is I = 13 (See Table 3). The BOOTES-NFC has also allowed imaging of 5 bursts as soon as the Huntsville/Bacodine/Locburst positions became available, starting ∼ 25 min after some events. By means of a spiral-shape mosaic of 35 to 99 frames, the imaged region varies from 3.5° × 3.3° to 5.5° × 5.3° down to a limiting magnitude of R ∼ 16.5 (Table 4).

Another objective is the observation of the sky in the I and V filters, as a part of the preparations for the ESA's satellite project INTEGRAL (the International Gamma-Ray Laboratory), in which Spain and the Czech Republic are involved with the Optical Monitoring Camera (OMC [8]). The preparation includes the test of technologies, data processing, ground based observational network, etc. For instance, weekly monitoring of the Galactic plane (Fig. 1) has led to a pre-discovery image of Nova Aquilae 1999 No. 1, and images obtained during the Leonid Storm on 17 Nov 1998 allowed a precise determination of the radiant [9].

An additional goal is the monitoring of several objects (bright AGNs/QSOs, old GRB positions, etc.) looking for *recurrent* transient optical emission arising from these sources [10]. There are hints that sudden and rapid flares occur. This will be achieved by means of the BOOTES-NFC.

More information can be found at http://www.laeff.esa.es/∼ajct/BOOTES.

Acknowledgements. We are very grateful for the support given by the Space Sciences Division (DCE) at INTA, through the project IGE 4900506. We thank J. Torres for his encouragement, M. Iríbar, J. Martín-Francía, F. Souvrier and J. M. Vilaplana for their hospitality and help at El Arenosillo. We are also indebted to E. López and V. Gallardo (SMA) and to the GCN team (in particular to S. Barthelmy) for rapidly distributing the GRB positions obtained by BATSE. This work is partially supported by a Spanish CICYT grant ESP95-0389-C02-02. The Czech contribution is supported by the Ministry of Education and Youth of the Czech Republic, projects ES02 and ES36.

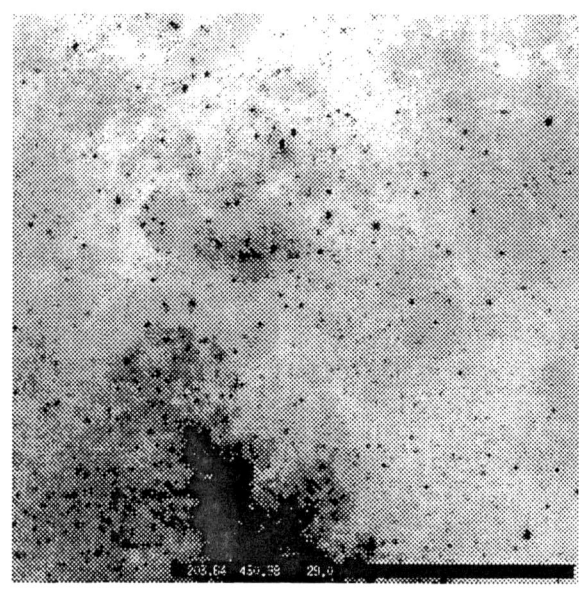

FIGURE 1. A 8° × 8° field close to the galactic center, as imaged by the BOOTES WFC in July 1999.

REFERENCES

1. Castro-Tirado, A. J., *A&SS* **263**, 15, astro-ph/9903187, (1999).
2. Hudec, R., *Astrophys. Lett. Comm.* **28**, 359 (1993).
3. Bernas, M., Pata, P., Hudec, R., Soldán, J., Rezek, T., & Castro-Tirado, A. J., in *AIP Conf. Proc. 428*, eds. Meegan, C. A., Preece, R. D. and Koshut, T. M., 1998, pp. 864.
4. Castro-Tirado, A. J., Hudec, R. & Soldan, J., in *AIP Conf. Proc. 384*, eds. Kouveliotou, C., Briggs, M. F. and Fishman, G. J., 1996, pp. 814.
5. Barthelmy, S., et al., in *AIP Conf. Proc. 428*, eds. Meegan, C. A., Preece, R. D. and Koshut, T. M., 1998, pp. 99.
6. Soldán, J., Hudec, R., Nemcek, M. & Rezek, T., in *AIP Conf. Proc. 428*, eds. Meegan, C., Preece, R. D. and Koshut, T. M., 1998, pp. 855.
7. Akerlof, C., et al., *Nature* **398**, 400 (1999).
8. Más-Hesse, J.M., et al., *Physica Scripta* **T77**, 44 (1998).
9. Trigo, J. M., Castellano-Roig, J. & Castro-Tirado, A. J., WGN 27:5, issue of IMO, 1999, pp. 258.
10. Hudec, R., et al., in *ASP Conf. Ser. 110*, eds. Miller H. R., Webb, and Noble, J. C., 1996, pp. 129.

Optical Observations of GRBs: EN, BART, and OMC

Rene Hudec[1], Jan Soldan[1], Vera Hudcova[1], Jan Florian[1], Martin Nekola[1], Ondrej Broz[1], Martin Bernas[2], Petr Pata[2], Filip Hroch[3], Alberto J. Castro-Tirado[4], Miguel Mas-Hesse[4], Alvaro Gimenez[4], Eliana Palazzi[5], Nicola Masetti[5], and Graziella Pizzichini[5]

[1] *Astronomical Institute, CZ-251 65 Ondrejov, Czech Republic, e-mail: rhudec@asu.cas.cz*
[2] *Faculty of Electronic Engineering, Czech Technical University Prague, Czech Republic*
[3] *Masaryk University Brno, Czech Republic*
[4] *LAEFF-INTA, Madrid, Spain*
[5] *TESRE, CNR, Bologna, Italy*

Abstract. We report on the ongoing projects at the Astronomical Institute Ondrejov. The EN (European Fireball Network) is based on 11 photographic stations and has already provided simultaneous optical data for 100 GRB triggers. A summary and discussion of the obtained results are given as well as discussion of detected candidates. The BART (Burst Alert Robotic Telescope) is a 25 cm aperture remotely controlled system in test operation. The OMC (Optical Monitoring Camera), prepared by a wide international collaboration for the INTEGRAL satellite (launch 2001), will also have capability for the detection of optical transients and optical afterglows of GRBs, assuming that their rate is higher than the GRB rate (caused by different beaming).

EN: IMPROVED ALL-SKY MONITORING

The European Fireball Network, EN, is an all-sky photographic patrol with 11 stations in the Czech Republic. The stations are equipped with high-quality Fish-Eye lenses with photographic records (sky diameter 8 cm in the focal plane). Typical exposure times amount to 3 hr. The network provides simultaneous optical data for various projects, and represents complete sky monitoring (180° diameter field of view) with limiting magnitude up to 11 for stars.

The main goal of the network is to provide sky monitoring to detect and study meteors. The use of the sky archive is, however, much more general. For example, providing real-time and pre-burst data (sky monitoring) in visible light is extremely important for investigations of GRBs. This kind of information may be provided only by patrol programs. No follow-up experiment will be able to provide data for times before GRBs as well as for the first few seconds during GRB onsets. This is very crucial since some theories predict optical/UV emissions preceding GRBs.

The stations are equipped with fish-eye lenses imaging the whole visible sky hemisphere on photographic emulsion. This allows a large fraction of GRBs to be observed (so far nearly 100) with limiting magnitudes between 6 and 11 at times before, during and after the GRB triggers. This still represents one of the faintest real-time and pre-burst magnitudes of GRBs.

Parameters of the recent system: (1) Optics: Fish-Eye Objective F-Distagon 3.5/30; (2) Detector: Planfilm FOMAPAN 400 ASA or 100 ASA (panchromatic emulsion) 90 mm × 120 mm, sky diameter 80 mm; (3) Typical exposure time: 3 hr for guided cameras, whole night for fixed cameras; (4) 2 stations equipped with guided and fixed cameras; (5) 9 stations equipped with fixed cameras; (6) Sensitivity for a brief 1 s flash 2–3 mag, for stars up to mag 11; (7) Response limited to the red light above 400 nm.

Preferences: (1) Large sky coverage (full visible hemisphere); (2) Large fraction of observation time: 2,400 to 6,000 sr hr for one station/year; (3) Multiplicity of data to eliminate background triggers easily; (4) Classification of detected triggers by paralax; (5) Simultaneous and pre-burst optical data (limits) for GRBs.

The recent mechanical and emulsion improvements will result in improved limiting magnitudes (up to mag 6 for stars on trailed plates and mag 12 on pointed plates). The network will be operated as a fully remote controlled network — without any human assistance — two years from now. The total number of plates in the Ondrejov plate archive is around 110,000. Only two such patrols are operated in the world, namely the variable star patrol at the Sonneberg Observatory and the meteor patrol at the Ondrejov Observatory.

Summary of GRB related results

(1) No optical emission above mag 5 (1 s duration assumed) or mag 13 (full exposure time) or $L\gamma/Lo > 100...300$ has been detected for a few GRBs. The faintest limit (320) exists for GRB 830313 [1]. (2) No optical emission above magnitudes 0...3 (1 s duration assumed) or 4–11 (full exposure time) or $L\gamma/Lo > 0.1...10$ has been detected for many (100) GRBs. (3) Optical Transient (OT) was detected on the plate taken ~7 hr after GRB790929 inside its error box [2]. (4) The preliminary BeppoSAX-related results are follows:

- GRB960720 11:37 UT

GRB at daytime, plates available for night before and after:
960719/20 7 plates 20:40–01:27 UT lim mag 4–11 for stars
960720/21 8 plates 20:45–01:30 UT lim mag 4–11 for stars
Nothing 10 hr before GRB limit 11, nothing 9 hr after GRB limit 11

- GRB970111 09:44 UT

GRB at daytime, plates available for the night after
970111/12 1 plate 16:45–20:15 UT lim mag 4
Nothing 7 hr after GRB limit 4

- GRB970228 02:28 UT

GRB at nighttime but below the local horizon at the time of event, plates available for the night of GRB
970227/28 2 plates 17:56–21:58 UT lim mag 4
970227/28 9 plates 00:50–04:32 UT lim mag 4–10 but GRB below horizon
970228/0301 23 plates 17:56–01:46 lim mag 4–11
Nothing 4.5 hr before GRB lim mag 4, nothing 13.5 hr after GRB lim mag 10

- GRB980326 12:20 UT

980324/25 5 plates 18:37–03:41 UT lim mag 9
980325/26 6 plates 18:34–03:40 UT lim mag 10 OT?
980326/27 5 plates 18:35–03:35 UT lim mag 10
Suspicious object at the GRB position on the plate A 4515 (18:34-21:40 UT), i.e., 24 hr before GRB (in further study)

- GRB980329 03:44 UT

980329/30 5 plates 18:42–3:20 UT lim mag 9
980330/31 3 plates 01:47–03:29 UT lim mag 9
980331/0401 5 plates 22:49–03:24 UT lim mag 9
Nothing down mag 9 15 hr after the GRB

- GRB980519 12:20 UT

980517/18 1 plate 22:31–01:34 UT lim mag 9
980518/19 1 plate 22:02–01:34 UT lim mag 9
980519/20 3 plates 21:16–01:27 UT lim mag 9 OT?
980520/21 2 plates 20:25–01:30 UT lim mag 10
980521/22 1 plate 21:14–00:00 UT lim mag 10
Suspicious object at the GRB position on the plate 4A 2331 (21:16–01:27 UT), i.e., 9 hr after GRB (in further study)

- GRB980613 04:551 UT

980614/15 2 plates 21:08–01:05 UT lim mag 10
980615/16 1 plate 21:00–23:10 UT lim mag 8

- GRB 980703 04:24 UT

980630/0701 1 plate 00:03–01:07 lim mag 6
980702/03 2 plates 21:07–01:09 GRB outside
980703/04 1 plate 21:39–01:06 GRB outside

- GRB981220 03:42 UT

GBR at night and visible position, but cloudy at the GRB time
981216/17 3 plates 17:55–04:40 UT lim mag 6-9
981217/18 6 plates 16:20–05:57 UT lim mag 8–10
981218/19 11 plates 16:20–05:11 UT lim mag 6–10
981219/20 5 plates 21:27–05:32 UT lim mag 5–6
981220/21 1 plate 01:11–05:33 UT lim mag 2?
981221/22 1 plate 16:31–20:10 UT lim mag 6
Nothing 14 min before GRB lim mag 5.5, nothing 31 hr before lim mag 10, nothing 26 hr before lim mag 9

- GRB981226 23:29 UT

981223/24 3 plates 16:31–05:31 UT lim mag 8
981225/26 6 plates 16:25–04:02 UT lim mag 3?
981228/29 3 plates 16:30–05:35 UT lim mag 6.5

- GRB990123 15:25 UT

99011/22 3 plates 16:53–05:30 lim mag 9.5
990122/23 7 plates 18:50–05:30 lim mag 9.5
990123/24 4 plates 17:20–05:24 lim mag 3?
Nothing 10 hr before lim mag 9.5

- GRB990506 11:54 UT

990503/04 4 plates 19:49–02:05 UT lim mag 6
990505/06 10 plates 19:52–02:07 UT lim mag 5.5
990506/07 5 plates 20:29–02:00 UT lim mag 4
990507/08 7 plates 19:58–01:56 UT lim mag 5.5
Nothing 10 hr before lim mag 5, nothing 12 hr before lim mag 5.5

- GRB990520 08:35 UT

990517/18 3 plates 20:18–01:37 UT lim mag 9
990518/19 3 plates 20:25–01:36 UT lim mag 9.5
990519/20 11 plates 20:19–01:34 UT lim mag 8
990521/22 2 plates 20:22–01:26 UT lim mag 7
Nothing 9 hr before lim mag 9

The recently detected optical transients (OTs) related to GRBs are generally considered as optical afterglows (OAs), while the direct optical emission of bursts has been detected in only one case. The data presented here hence represent valuable limits for the direct optical luminosity of GRBs in question: the direct optical emission seems not to exceed V magnitude 10.

ROBOTIC SYSTEMS BART AND BOOTES

BART – Burst Alert Robotic Telescope: This system is in test operation in Ondrejov starting Sept. 1999. It consists of a remotely controlled, small-aperture CCD telescope (25 cm aperture) with a wide-field CCD camera, limiting magnitudes 18 over a narrow field of view and 13 over a wide field of view, follow-up observations of GRBs with delays of 1 min and more, monitoring of selected targets.

BOOTES – Burst Observer and Optical Transient System: This device is in test operation in El Arenosilo, Spain, since Aug. 1998. A second station 200 km away is in preparation. The system is based on modified BART: 30 cm remotely controlled telescope with CCD camera 1530×1020 pixels, FOV $49' \times 33'$, with angular resolution $1.9''$, and limiting magnitude $V \sim 19$ (60 s exposure).

The device is equipped with two WF cameras (50-mm f/1.2 lenses, CCD 1534×1020 pixels, FOV $16° \times 11°$, angular resolution $0.63'$, limiting magnitude $V \sim 14$ (in 300 s). The scientific goals include rapid response to GRB alerts, monitoring of selected triggers such as variable blazars, QSOs, etc.

OMC INTEGRAL

The OMC (Optical Monitoring Camera) represents one of four experiments onboard the ESA's INTEGRAL (International Gamma-Ray Astrophysics Observatory) satellite. The device is a result of collaboration of scientific institutes in Spain, UK, Belgium, Ireland and the Czech Republic. The OMC is based on a high quality 50 mm aperture objective with a 1024×2048 pixels CCD camera, FOV $5° \times 5°$, with limiting magnitude of $V \sim 19$.

Scientific objectives of the OMC: (1) To monitor the optical emission of high-energy targets within the FOV simultaneously with the gamma- and X-ray instruments; (2) Especially important for blazars and quasars: 30 EGRET sources, violent variability at all frequencies; (3) To provide simultaneous and calibrated standard V filter photometry of the high-energy sources; (4) To provide optical photometry and astrometry of optical counterparts of HE sources detected by other instruments; (5) To monitor selected optical variable sources inside the FOV (up to 100 targets per pointing); (6) To search for flaring objects; (7) To provide long-time optical light curves of selected triggers. The possibility to detect an OA of GRB by chance is, due to the small FOV, negligible. However, the OMC can detect OAs if their rate is much larger than the GRB rate (due to different beaming).

OMC source catalogue: The OMC catalogue includes all objects of interest (variable optical objects such as optical counterparts of HE sources, variable stars, AGNs, QSOs, blazars, etc.). The catalogue will be regularly updated. The searching software (developed by Astronomical Institute Ondrejov) will extract objects of interest inside the FOV for each observing run. Up to 100 triggers can be evaluated per observation.

CONCLUSIONS

There are 3 new small aperture (25 and 30 cm) robotic telescopes (BART and 2 × BOOTES) available for optical GRB alert with moderate limiting magnitudes (15–18). There is all-sky patrol, operating down to mag 12, providing optical real-time and pre-burst data for numerous GRBs. After launch of INTEGRAL in 2001, the OMC will be able to monitor selected objects down to lim mag 17–19.

Despite recent instrumental developments, the photographic patrol data still represent unique databases for real-time (simultaneous and pre-bursts) optical observations of GRBs. The recently operational alert systems can provide rapid follow up optical data (with time delays of order of tens of seconds) for less precisely localized CGRO-BATSE triggers but their magnitude limits in such cases are generally not substantially better (since they have to cover large FOVs) than those of the photographic sky patrols and further, they will never be able to provide data for times during and before bursts. The improved sky patrols can yield valuable real-time and pre-burst optical data in the magnitude range 12–15.

ACKNOWLEDGEMENTS

This work has been supported by the Project of The Czech Ministry of Education and Youth No. ES02, and by the Grant Agency of the Czech Republic, grant 205/99/0145.

REFERENCES

1. Hudec, R., *Astroph. Letters and Communications* **28**, 359 (1993).
2. Borovi_ka, J., Hudec, R. and Dedoch, A., *A&A* **258**, 379 (1992).

An Extended Search for Transient Events in the COBE/DMR Database Associated with Cosmic Gamma-Ray Bursts

Louis J. Beathley, Jr.[1], J. Gregory Stacy[1,2],
Tj. Romke Bontekoe[3], Peter D. Jackson[4],
and Christoph Winkler[5]

[1] *Southern University, Baton Rouge, LA*
[2] *Louisiana State University, Baton Rouge, LA*
[3] *Bontekoe Data Consultancy, The Netherlands*
[4] *Raytheon ITSS Corporation, Greenbelt, MD*
[5] *Astrophysics Division, ESA/ESTEC, The Netherlands*

Abstract. We report on an extension of earlier work to search the archival database of the Differential Microwave Radiometers (DMR) aboard the COBE satellite for evidence of transient events at microwave wavelengths associated with cosmic gamma-ray bursts (GRBs). Over its four-year lifetime the DMR experiment repeatedly surveyed the sky to measure fluctuations in the cosmic microwave background. On a number of occasions at least one of the horns of the DMR was also serendipitously pointing in the direction of a cosmic gamma-ray burst at the moment of burst occurrence. Our original investigation covered the eight-month period April–December 1991 when a number of GRBs were observed by the COBE/DMR at or within a few seconds of the CGRO/BATSE trigger time. Upper limits only were obtained for any simultaneous microwave flux from these events, in the range 7–42 kJy. Most recently, we have extended our search using the additional 24 months of COBE DMR data now publicly available, covering the time period January 1992 through December 1993, and the latest BATSE 4B Catalog of gamma-ray bursts.

SCIENTIFIC MOTIVATION

The first phase of the project described here was initiated prior to the launch of the BeppoSAX satellite in 1997, and before the definitive discovery of afterglow counterpart emission from GRBs originating at cosmological distances [1]. Our original analysis demonstrated that the COBE/DMR instruments have observed GRBs on a number of occasions over the course of the COBE mission. Due to the low point-source sensitivity of the DMR, however, only upper limits to prompt microwave emission from GRBs have been obtained to date [2].

We have been re-motivated to extend our earlier search of the COBE/DMR database following the detection since 1997 of GRB counterparts at x-ray, optical, and radio wavelengths with BeppoSAX and other telescopes. Recent developments include 1) the detection of radio afterglows in approximately nine GRBs [3]; 2) the detection of bright "prompt" optical emission from GRB 990123 [4] during the active gamma-ray phase of the burst; 3) the detection of a radio flare associated with GRB 990123 (and possibly in GRB 970828 [3]); 4) increasing reports that a small fraction of single-pulse GRBs may be associated with unusual, radio-bright supernova events (termed S-GRBS) in relatively *nearby* galaxies (for example, GRB 980425 with SN 1998bw [5,6]); 5) evidence that at least one nearby galaxy (M101, [7]) contains large, bright supernova remnants suggestive of high-energy "hypernova" events that have been proposed as a possible origin of GRBs [8]; 6) the parallel development of detailed theoretical models that can account in a self-consistent manner for (i) the gamma-ray emission of the burst itself, (ii) the prompt, flaring emission observed in the optical and radio bands, and (iii) the more long-term multiwavelength afterglow emission seen in GRBs from days to weeks following the burst.

All of these phenomena have been modeled in terms of internal shocks in relativistic flows (for the burst itself), external shocks interacting with the interstellar medium (for the long-term afterglow emission), and with reverse shocks leading to the prompt low-frequency emission observed in the optical and radio bands [9–11].

CHARACTERISTICS OF THE COBE/DMR

The COBE Differential Microwave Radiometers (DMR) operated at three frequencies: 31.5, 53, and 90 GHZ. For each frequency there were two independent radiometers, or channels. Each channel received the signals from two horns (7° HPBW), with their pointing directions separated by 60°. A differential signal was obtained by switching at a rate of 100 Hz between the two horns of each channel. At any given instant, the DMR observed almost 300 deg^2 of the sky, so that there was a $\sim 0.7\%$ chance that a particular GRB would be observed by the DMR at the time of burst occurrence. With a burst rate of approximately one per day one expects a few GRBs per year to have occurred within the field of view of the DMR at one of its three frequencies. The COBE spacecraft had a spin period of 75 s, implying that a point on the sky took approximately 1.5 s to pass through half of the beam of a DMR horn. As the COBE spin-axis rotated due to orbital motion around the Earth, the six DMR horns would pass successively over a given position on the sky for about one minute, yielding approximately twelve separate chances for a source detection from a particular direction [12–15].

An unfortunate consequence of the large sky coverage of the DMR was a relatively low sensitivity of its instruments to point sources of microwave emission. Further, since the instrument was only passively cooled (to about 140 K), the radiometer noise temperatures were comparatively high. The DMR time-ordered data (TOD) used in the present study were accumulated over 0.5-s integration intervals. A point source passed through half of the DMR beam (7° HPBW) in 1.5 s, resulting in the rms uncertainties given in **Table 1**. The radio afterglows observed in GRBs *to date*, in contrast, are several orders of magnitude lower in intensity (in the μJy to mJy range [3]) compared to the flux limits presently achievable using the DMR TOD datasets.

TABLE 1. RMS Uncertainty in the Measurement of Microwave Emission from a Point Source with the COBE-DMR.

DMR Channel	T_a (mK)	S (kJy) (0.5-s integration)	S (kJy) [a] (1.5-s integration)
31A	56.8	36.4	15.1 (21.0)[b]
31B	58.5	37.5	...
53A	22.5	40.8	17.9
53B	26.3	47.7	...
90A	38.5	213	71.8
90B	29.3	153	...

[a] Averaged over both frequency channels.
[b] After October 4, 1991, the 31B radiometer became very noisy for almost a year, so only channel 31A could be used with an rms uncertainty of 21.0 kJy.

ANALYSIS METHOD

The search for microwave emission from GRBs proceeds as follows. First, a list of BATSE 4B bursts [16] viewable by the COBE DMR over the period 1991–1993 is determined, based on the viewing constraints (due to solar-angle restrictions) of the COBE spacecraft. For each of these individual burst candidates the relevant DMR time-ordered dataset (TOD) is read and searched over an initial temporal window of 20 minutes before and after the burst trigger time. Within this time window, for every 0.5-second minor frame of DMR data, the angle of separation between the burst location and the pointing direction of each DMR horn on the sky is calculated. If the separation distance from a DMR horn is found to lie within an angular tolerance set by the uncertainty of the burst location, the DMR data are selected for further processing. Corrections are applied to the DMR signal for the dipole component of the cosmic microwave background and the microwave emission of the Galaxy, and the DMR response is converted from milliKelvins of antenna temperature to Janskys of detected flux. DMR data associated with selected events are then written to output files containing event times, positions, fluxes, and processing flags, along with computed errors. Finally, the output DMR signal files are searched for exact coincidences in time and DMR pointing direction for individual bursts, and for any associated microwave signal above background levels.

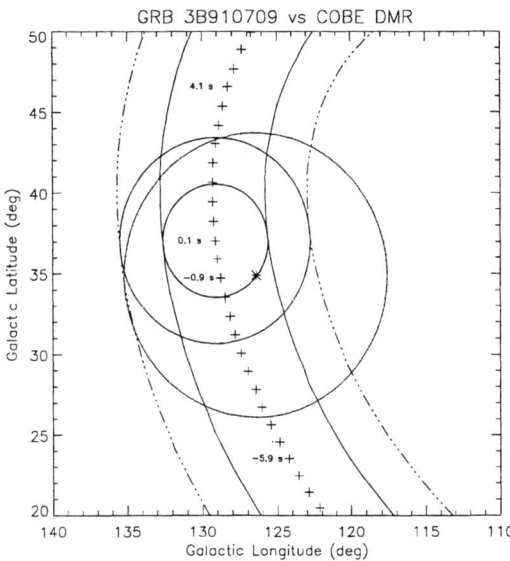

FIGURE 1. A skymap showing the field containing the burst GRB 3B910709, observed by one of the DMR 90 GHz horns for several seconds before, during, and after burst trigger. An asterisk marks the position of the GRB, centered within the BATSE 99%-confidence error circle. Plus symbols mark successive positions of the DMR horn, bounded by the FWHM and 10%-sensitivity levels of the DMR beam.

RESULTS TO DATE AND ONGOING WORK

Over the first phase of this project [2], covering the eight-month interval from 1991 April-December, 210 bursts from the Third (3B) Catalog were viewable by the COBE DMR. For five of these events the DMR was pointing within 7° of the burst position at the exact moment of the burst occurrence (see, for example, **Figure 1**). For another four events the DMR was pointed within 2° of the BATSE positions within 10 seconds of the burst trigger time, while a total of 12 GRBs were observed within 30 seconds of burst triggering. No obvious microwave emission (at 31.5, 53, 90 GHz), with upper limits in the 10–100 kJy range, can be associated with any of these events (see **Figure 2**). By extending an acceptance window in time of up to 20 minutes before and after a burst another 60 GRBs are sampled by the DMR, whose signals have been analyzed to deduce that the "average" GRB produces less than about 7–42 kJy in simultaneous microwave radiation.

We are presently extending our search to the full 32 months of DMR TOD data now publicly available, and using the latest burst information published in the Fourth BATSE (4B) Catalog [16]. Given the four-fold increase in DMR data and time coverage, a corresponding increase in the number of bursts observed by the DMR is expected, along with at least a factor of two improvement in sensitivity limits for microwave emission from an average burst. The results of this ongoing investigation will be the subject of a future report.

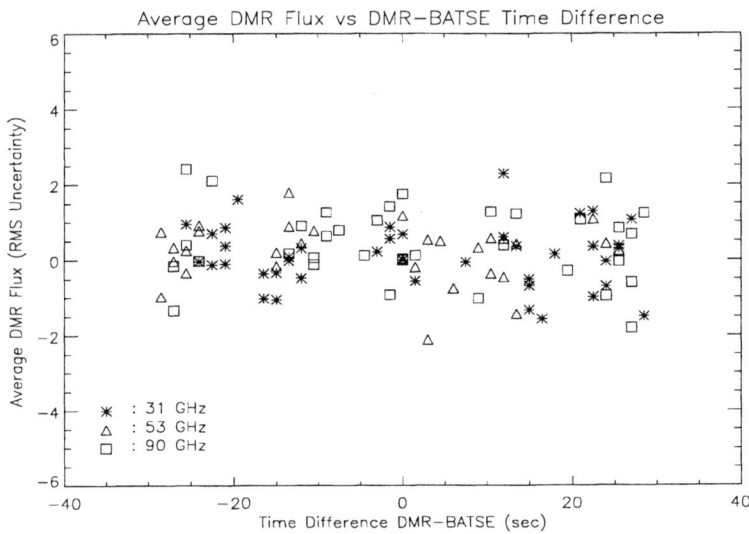

FIGURE 2. A plot of the average DMR flux as a function of time for twelve BATSE GRBs observed within 30 seconds of the burst trigger time.

REFERENCES

1. Costa, E., these proceedings.
2. Jackson, P. D., et al., *Ap. J.* **505**, 1 (1998).
3. Kulkarni, S. R., et al., *Ap. J.* **522**, L97 (1999).
4. Akerlof, C. W., et al., *Nature* **398**, 400 (1999).
5. Galama, T. J., et al., *Nature* **395**, 670 (1998).
6. Kulkarni, S. R., et al., *Nature* **395**, 663 (1998).
7. Wang, Q. D., *BAAS* **31**, 695 (1999).
8. Paczynski, B., *Ap. J.* **494**, L45 (1998).
9. Sari, R., Piran, T., and Narayan, R., *Ap. J.* **497**, L17 (1998).
10. Meszaros, P., and Rees, M. J., *Ap. J.* **476**, 232 (1997).
11. Paczynski, B., and Rhoads, J. E., *Ap. J.* **418**, L5 (1993).
12. Bennett, C. L., et al., *Ap. J.* **391**, 466 (1992).
13. Bennett, C. L., et al., *Ap. J.* **436**, 423 (1994).
14. Bennett, C. L., et al., *Ap. J.* **464**, L1 (1996).
15. Smoot, G., et al., *Ap. J.* **360**, 685 (1990).
16. Paciesas, W., et al., *Ap. J. Suppl.*, **122**, 465 (1999).

III. AFTERGLOW, HOSTS, AND SUPERNOVA CONNECTIONS

The Afterglows of Gamma-Ray Bursts

S. R. Kulkarni*, E. Berger*, J. S. Bloom*, F. Chaffee¶,
A. Diercks*, S. G. Djorgovski*, D. A. Frail†, T. J. Galama*,
R. W. Goodrich¶, F. A. Harrison*, R. Sari*, and S. A. Yost*

*California Institute of Technology, Pasadena, CA 91125, USA
†National Radio Astronomy Observatory, Socorro, NM 87801, USA
¶W. M. Keck Observatory, Kamuela, HI 96743, USA

Abstract. Gamma-ray burst astronomy has undergone a revolution in the last three years, spurred by the discovery of fading long-wavelength counterparts. We now know that at least the long-duration GRBs lie at cosmological distances with estimated electromagnetic energy release of 10^{51}–10^{53} erg, making these the brightest explosions in the Universe. In this article we review the current observational state, beginning with the statistics of X-ray, optical, and radio afterglow detections. We then discuss the insights these observations have given to the progenitor population, the energetics of the GRB events, and the physics of the afterglow emission. We focus particular attention on the evidence linking GRBs to the explosion of massive stars. Throughout, we identify remaining puzzles and uncertainties, and emphasize promising observational tools for addressing them. The imminent launch of *HETE-2* and the increasingly sophisticated and coordinated ground-based and space-based observations have primed this field for fantastic growth.

I INTRODUCTION

GRBs have mystified and fascinated astronomers since their discovery. Their brilliance and their short time variability clearly suggest a compact object (black hole or neutron star) origin. Three decades of high-energy observations, culminating in the definitive measurements of CGRO/BATSE, determined the spatial distribution to be isotropic yet inhomogeneous, suggestive of an extragalactic population (see [14] for a review of the situation prior to the launch of the BeppoSAX mission). Further progress had to await the availability of GRB positions adequate for identification of counterparts at other wavelengths.

In the cosmological scenario, GRBs would have energy releases comparable to those of supernovae (SNe). Based on this analogy, Paczyński & Rhoads [65] and Katz [44] predicted that the gamma-ray burst would be followed by long-lived but fading emission. These papers motivated systematic searches for radio afterglow, including our effort at the VLA [15]. The broad-band nature of this "afterglow" and its detectability was underscored in later work [59,78].

Ultimately, the detection of the predicted afterglow had to await localizations provided by the Italian-Dutch satellite, BeppoSAX [6]. The BeppoSAX Wide Field Camera (WFC) observes about 3% of the sky, triggering on the low-energy (2–30 keV) portion of the GRB spectrum, localizing events to $\sim$ 5–10 arcminutes. X-ray afterglow was first discovered by BeppoSAX in GRB 970228, after the satellite was re-oriented (within about 8 hours) to study the error circle of a WFC detection with the 2–10 keV X-ray concentrators. The detection of fading X-ray emission, combined with the high sensitivity and the ability of the concentrators to refine the position to the arcminute level, led to the subsequent discovery of long-lived emission at lower frequencies [10,77,16].

Optical spectroscopy of the afterglow of GRB 970508 led to the definitive demonstration of the extragalactic nature of this GRB [60]. The precise positions provided by radio and/or optical afterglow observations have allowed for the identification of host galaxies, found in almost every case. Not only has this provided further redshift determinations, but it has been useful in tying GRBs to star formation through measurements of the host star formation rate (e.g., [46,11]). HST, with its exquisite resolution, has been critical in localizing GRBs within their host galaxies and thereby shed light on their progenitors (e.g., [29,41,4]). Observations of the radio afterglow have directly established the relativistic nature of the GRB explosions [16] and provided evidence linking GRBs to dusty star-forming regions. Radio observations are excellent probes of the circumburst medium and the current evidence suggests that the progenitors are massive stars with copious stellar winds. The latest twist is an apparent connection of GRBs with SNe [5]. Separately, an important development is the possible association of a GRB with a nearby (40 Mpc) peculiar SN [30,47].

In this paper we review the primary advances resulting from afterglow studies. §II discusses the statistics of detections to-date, including possible causes for the lack of radio and optical afterglows from some GRBs. In §III we review constraints on the nature of the progenitor population(s), in particular evidence linking some classes of GRBs to SNe. §IV describes the status of current understanding of the physics of the afterglow emission. Here we compare observations to predictions of the basic spherically-symmetric model, and describe complications arising from deviations from spherical symmetry and non-uniform distribution of the circumburst medium. We conclude with speculations of the near- and long-term advances in this field (§V).

We point out that this review has two biases. First, given the concentration of previous review articles on optical and X-ray observations, we emphasize the unique contributions of radio afterglow measurements. Second, this article is intended to also provide a summary of the efforts of the Caltech-NRAO-CARA GRB collaboration, and therefore details our work in particular. This review is in response to review talks given at the 1999 Maryland October meeting (SRK) and the 5th Huntsville GRB meeting (DAF and SRK).

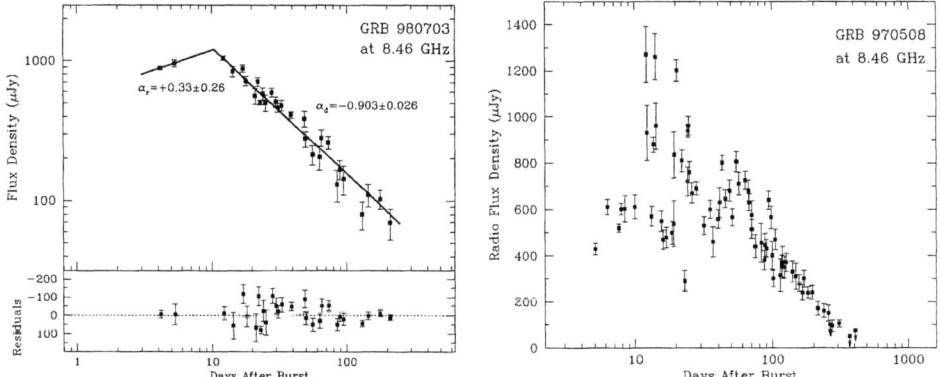

FIGURE 1. *Left:* The radio light curve of GRB 980703. This is a typical afterglow, a rise to a peak followed by a power-law decay. The longer lifetime of the radio afterglow allows us to see both the rise and the fall of the afterglow emission. In contrast, at optical and X-ray emission, most of the times we see only the decaying portion of the light curve. *Right:* The radio light curve of GRB 970508 [21]. The wild fluctuations of the light curve in the first three weeks are chromatic. At later times, the fluctuations become broad-band and subdued. These fluctuations are a result of multi-path propagation of the radio waves in the Galactic interstellar medium. As the source expands (at superluminal speeds) the scintillation changes from diffractive to refractive scintillation. This is analogous to why stars twinkle but planets do not.

II STATISTICS OF AFTERGLOW DETECTIONS

Afterglow emission was first detected from GRB 970228, both at X-ray [10] and optical frequencies [77], but not at radio wavelengths [17]. The first radio afterglow detection came following the localization of GRB 970508 [16]. Figure 1 shows two examples of radio lightcurves. The radio afterglow of GRB 970508 is famous for several reasons: it was the first radio detection, it gave the first direct demonstration of relativistic expansion, and it remains the longest-lived afterglow [21].

Afterglow emission is now routinely detected across the electromagnetic spectrum. BeppoSAX has been joined in studying the X-ray afterglows by the All Sky Monitor (ASM) aboard the X-ray Timing Explorer (XTE), the Japanese ASCA mission, and recently the Chandra X-ray observatory (CXO). A veritable armada of optical facilities (ranging from 1-m class telescopes to the 10-m Keck telescopes) routinely discover and study optical afterglows. HST has been primarily used to make exquisite images of the host galaxies (see above) but in the near future we expect other uses such as UV spectroscopy and identification of underlying SNe. The VLA has led the detection in radio. However, other centimeter-wavelength facilities (the Australia Telescope National Facility, Westerbork Synthesis Radio Telescope, the Ryle Telescope) and millimeter wavelengths (James Clerk Maxwell Telescope, the Owens Valley Millimeter Array, IRAM and the Plateau de Bure

Interferometer) are now regularly contributing to afterglow studies.

Figure 2 summarizes the statistics of afterglow detections. In almost all cases, X-ray emission has been detected, establishing the critical importance of prompt X-ray observations. Optical afterglow appears to be detected in about 2/3 of all well-localized events if sufficiently deep optical images are taken rapidly (i.e., within a day or so of the burst). Radio afterglows are detected in 40% of the cases — far more often than usually assumed. We refer the reader to the Frail et al. [22] for a comprehensive summary of the X-ray/optical/radio afterglow detection statistics. The non-detections are, as discussed below, as interesting as the detections.

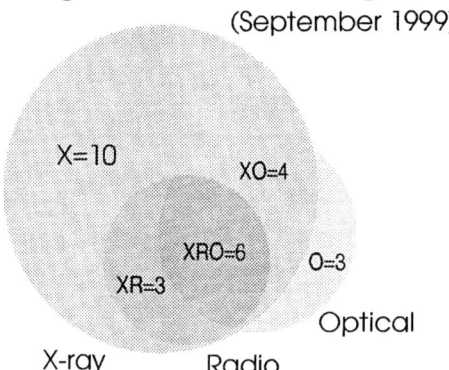

FIGURE 2. *A Venn diagram showing the detection statistics for 26 well-localized GRBs in the Northern and Southern hemispheres. Of the 23 GRBs for which X-ray afterglows have been detected to date, 10 have optical afterglows (XO + XOR) and 9 have radio afterglows (XR + XOR). In total there are 13 optical and/or radio afterglows with corresponding X-ray afterglows.*

Radio Non-detection. The failure to find radio afterglow is most likely due to lack of sensitivity. The brightest radio afterglow to date is that from GRB 991208 (Frail GCN [1] 451) with a peak flux of 2 mJy, a 60-σ detection (at centimeter wavelengths), whereas the weakest afterglow is typically around 5σ. In contrast, at optical and X-ray wavelengths, afterglow emission is routinely detected at hundreds of sigma. If the VLA were to be upgraded by a factor of 10 in sensitivity, then we predict that radio afterglow emission would, like X-ray emission, be detected from most GRBs.

Optical Non-detection. Non-detection at optical wavelengths is more interesting, as it may result in some cases from extinction along the line of sight or within the source. Bad weather as well as rapid fading of the afterglow has certainly hindered some optical searches, which, due to notification delays, typically begin some

---

[1] GCN refers to the GRB Coordinates Network Circular Services. This network is maintained by S. Barthelmy at the Goddard Space Flight Center; see http://gcn.gsfc.nasa.gov/gcn/

hours after the event. Furthermore, low Galactic latitude events may be obscured, or hidden in crowded foregrounds. However, in some cases deep searches have been performed with no success. Here, non-detection likely results from extinction by dust in the burst host galaxy and/or absorption by the intergalactic medium. GRB 970828 [38] is one example, as is the more dramatic case of GRB 980329. This burst was one of the brightest events in the WFC [42]. Searches for optical afterglow emission failed to identify any counterpart. VLA observations identified an unusual radio variable in the field [76]. Soon thereafter, a red afterglow and a bright IR afterglow were identified (Klose GCN 43, Larkin et al. GCN 44). Taylor et al. [76] suggest that the GRB arose in a region with high extinction. Further optical and IR work on this interesting afterglow can be found in [34,64,69].

Optically dim "red" but bright IR afterglows can also result from the GRB being located at high redshift. Intergalactic HI absorption will result in a wavelength cutoff below the Lyman limit, $< 912(1+z)$ Å, where z is the redshift of the source. This effect was originally invoked to explain the faint R-band but bright IR emission from GRB 980329 [27]. We now know, based on recent Keck observations, that the GRB host is blue, incompatible with a high-z origin. Rather, it is more tenable that the host is a typical star-forming galaxy with dusty star-forming regions, and that the GRB occurred in one such region [76]. We are presently carrying out IR spectroscopy of this host to determine the redshift and the star formation rate (SFR). While searching for "R dropouts" may in the future provide an effective method for finding high-redshift events, it is clear that cross-calibrated multi-band photometry of higher quality than currently exists will be required to make this useful.

X-ray Non-detection. The spectra of most GRB events clearly extend into the X-ray band, as established by *GINGA* observations [75]. How the X-ray emission observed during the burst connects to the X-ray afterglow is uncertain, due to sensitivity limitations of wide-field monitors. X-ray afterglow emission appears to be ubiquitous. Observations of the X-ray afterglow are important for two reasons: (i) the observations of the X-ray afterglow by sensitive imaging instruments (e.g., the concentrators aboard BeppoSAX) result in sufficiently precise (arcminute) localization and (ii) a significant (perhaps even a dominant) fraction of the explosion energy appears to be radiated in this band. Of all the SAX bursts, GRB 970111 is peculiar for the absence of X-ray afterglow (admittedly the data were obtained about 17 hours after the burst) [25]. In view of the critical role played by X-ray afterglow in localization of GRBs we regard this non-detection to be worthy of further investigation.

III THE NATURE OF THE PROGENITORS

In almost all cases, a host galaxy has been identified at the location of the fading afterglow. GRB redshifts can be obtained either via absorption spectroscopy (when the transient is bright) or by emission spectroscopy of the host galaxy. In Figure 3

TABLE 1. Basic Properties of Selected GRBs

GRB	α(J2000) (h m s)	δ(J2000) (° ′ ″)	R_{host} (mag)	S ×10^{-6} (erg cm^{-2})	z	References[a]
970228	05 01 47	+11 46.9	25.2[b]	1.7	0.695	Djorgovski et al. GCN 289
970508	06 53 49	+79 16.3	25.7	3.1	0.835	[60,2]
970828	18 08 32	+59 18 52	TBD	74	0.957	[12]
971214	11 56 26	+65 12.0	25.6	11	3.418	[46]
980326	08 36 34	−18 51.4	≳ 27.3	1	...	
980329	07 02 38	+38 50.7	25.4	50	...	
980519	21 22 21	+77 15.7	26.2	25	...	
980613	10 17 58	+71 27.4	24.5	1.7	1.096	Djorgovski et al. GCN 189
980703	23 59 07	+08 35.1	22.6	37	0.966	[11]
981226	23 29 37	−23 55 54	≳ 22	N.A.	...	
990123	15 25 31	+44 46 00	24.4	265	1.600	[48]
990510	13 38 07	−80 29 49	≳ 28	23	1.619	Vreeswijk et al. GCN 324
990712	22 31 53	−73 24 29	21.78	N.A.	0.430	Galama et al. GCN 388
991208	16 33 54	+46 27 21	≳ 25	100	0.706	Dodonov et al. GCN 475
991216	05 09 31	+11 17 07	24.5	256	1.020	Vreeswijk et al. GCN 496

[a] References to redshift determination.
[b] V-band magnitude from HST. All others are R magnitude in the Johnson system.

and Table 1 we summarize the measured redshifts and host galaxy magnitudes. While the distance scale debate is settled (at least for the class of long-duration GRBs, see below) we remain relatively ignorant of the nature of the central engine. Currently popular GRB models fall into two categories: (i) the coalescence of compact objects (neutron stars, black holes and white dwarfs [13,54,61,63]) and (ii) the collapse of the central iron core of a massive star to a spinning black hole, a "collapsar" [85,57]. We now summarize the light shed on the progenitor problem by afterglow studies.

The Location of GRBs Within Hosts. A fundamental insight into the nature of SNe came from their location with respect to other objects within the host galaxy (specifically HII regions and spiral arms) and the morphology of the host galaxy itself (elliptical versus spiral). In a similar manner, we are now making progress in understanding GRB progenitors by measuring offsets with respect to other objects in the host galaxies. The rather good coincidence of GRBs with host galaxies already suggests that they are unlikely to be a halo population (as would be expected in the coalescence scenario [3]). On the other hand, with the possible exception of GRB 970508 [66], they are clearly not associated with galactic nuclei (i.e., massive central black holes). Typical offsets of GRBs from the centroid of their host galaxies are comparable to the half-light radii of field galaxies at comparable magnitudes, suggesting that GRBs originate from stellar populations.

Host Galaxies. Demonstrating a direct link between GRBs and (massive) star formation is more difficult. On the whole, the population of identified hosts seems typical in comparison to field galaxies in the same redshift and magnitude range. The hosts have average luminosities for field galaxies, modulo corrections due to

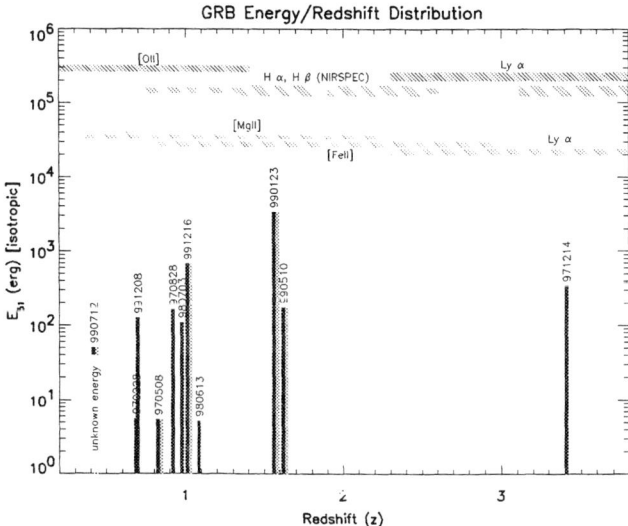

FIGURE 3. *The isotropic gamma-ray energy distribution of GRBs with confirmed redshifts. Bursts indicated in black are those with spectroscopically confirmed emission lines from the host galaxies; bursts indicated by a shaded column (e.g., GRB 990123) are those with absorption line redshifts. The relevant key absorption or emission features are noted at the top of the figure.*

evolution. Their emission line fluxes and equivalent widths are also statistically indistinguishable from the normal field galaxy population. The observed star formation rates, derived from recombination line fluxes (mostly the [O II] 3727 Å line) and from the UV continuum flux range from less than 1 $M_\odot$ yr^{-1} to several tens of $M_\odot$ yr^{-1} — typical of normal galaxies at comparable redshifts (extinction corrections can increase these numbers by a factor of a few, but similar corrections apply to the comparison field galaxy population as well). It will probably be necessary to have a sample of several tens of GRB hosts before a correlation of GRBs with the (massive) star formation rate can be tested statistically. However, below we point to several specific examples that are suggestive of a link between GRBs and star-forming regions.

Association with Starforming Regions. There is evidence showing that GRBs arise from dusty regions within their host galaxies. In this respect, radio observations provide a unique tool for detecting events in regions of high ambient density (as was the case for GRB 980329). An even more extreme example is GRB 970828, where the host was identified based *solely* on the VLA discovery of a radio flare [12]. Interestingly enough, this is the dustiest galaxy in the sample of GRB hosts to-date.

Second, some GRBs appear to be located within identifiable star-forming regions. An example is GRB 990123 [4,28,41]. VLA observations of GRB 980703 [19] are perhaps more convincing. The radio observations can be sensibly interpreted by appealing to free-free absorption from a foreground HII region (which would dwarf the Orion complex). If this interpretation is correct then this would be strong evidence for a GRB being located within a starburst region.

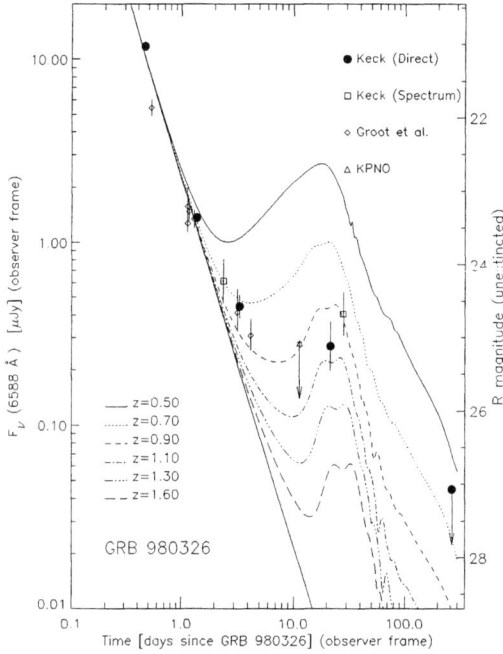

FIGURE 4. *R-band light curve of GRB 980326 and the sum of an initial power-law decay plus Ic supernova light curve for redshifts ranging from $z = 0.50$ to $z = 1.60$ (reprinted by permission from Nature [5] copyright (1999) Macmillan Magazines Ltd.).*

The GRB–SN link. If GRBs arise from the collapse of a massive star, it is an unavoidable consequence that emission from the underlying supernova should be superimposed on the afterglow. Bloom et al. [5] may have made the first detection of a possible SN component in the GRB 980326 lightcurve (Fig. 4). These authors noted that SNe, in contrast to afterglows, have distinctive temporal and spectral signatures: rising to a maximum at $\sim 20(1+z)$ days, with little emission blueward of about 4000 Å in the restframe (and certainly blueward of 3000 Å) owing to a multitude of resonance absorption lines. This discovery has led to other possible SN detections, most notably GRB 970228 [31,68].

The suggestion of a GRB–SN connection is an intriguing one but it has yet to be placed on a firm footing. Important questions are: (i) are all long-duration GRBs accompanied by SNe? (ii) if so, are these SNe of type Ib/c? Ground-based observations are possible in those cases where the afterglow decays rapidly (e.g., GRB 980326) or if high quality optical and IR observations exist (e.g., GRB 970228).

We need more examples to test the GRB–SN link. Future progress will depend on a combination of ground and HST observations. For relatively nearby GRBs, especially those with a rapidly decaying optical afterglow, it would be attractive and feasible to obtain the spectrum of the SN around the time when the flux from the SN peaks. A moderate quality spectrum with SN-like features would have the singular advantage of definitively confirming the SN interpretation (as opposed to alternatives involving re-radiation by dust [80]). However, for most GRBs, we expect HST observations to play a critical role. HST's widely recognized strengths in accurate photometry of sources embedded in galaxies [32] and photometric stability make the detection of a faint SN against the optical afterglow and the host galaxy possible.

Diversity of the Progenitor Population. As was the case with SNe, it is likely naive to think of a single progenitor population. Below, we discuss the two additional classes which show some promise: the mysterious short duration GRBs and a possible class of low luminosity GRBs associated with SNe.

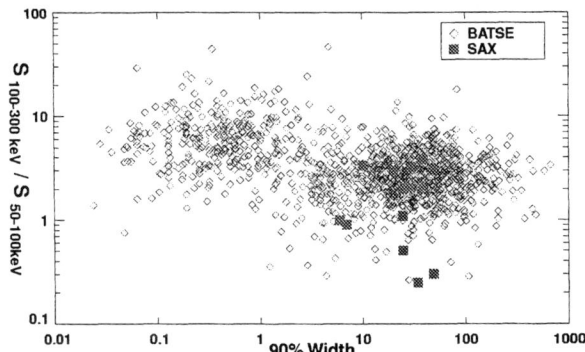

FIGURE 5. *Distribution of duration (T_{90}) vs. spectral hardness for BATSE bursts (diamonds) from the 4B catalog. There is a clear suggestion of two groups of GRBs: short/hard and long/soft events. Events localized by BeppoSAX (solid squares) appear to belong to the long-duration class.*

Short Events. It has been known for some time that the distribution of the duration of GRBs appears to be bimodal [14]; see Figure 5. Furthermore, these two groups may have different spatial distributions [45], with the short bursts being detected out to smaller limiting redshifts. However, we know very little about this class of GRBs since, as noted earlier, all bursts localized by BeppoSAX and RXTE thus far are of long duration (Figure 5). Fortunately, improvements in BeppoSAX and the imminent launch of HETE-2 provide for the first time the opportunity to follow-up short GRBs.

The short duration bursts are difficult to accommodate in the collapsar model, given the long collapse time of the core. However, they find a natural explanation in the coalescence models. How would these bursts manifest themselves? Li & Paczyński [56] speculate that if the short-duration bursts result from NS–NS mergers then they may leave a bright, but short-lived ($\lesssim 1$ day) optical transient. Radio observations provide a complementary tool for determining the nature of the short duration bursts. The low ambient density would result in weak afterglows (since

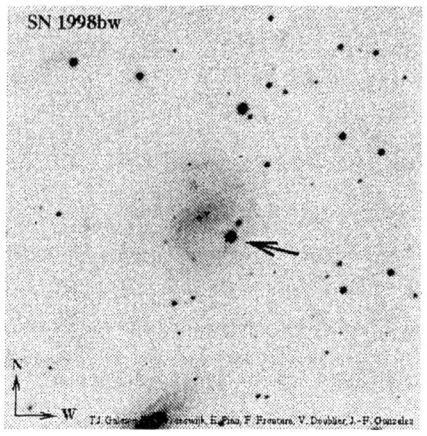

FIGURE 6. *Discovery image of SN 1998bw (reprinted by permission from Nature [30] copyright (1998) Macmillan Magazines Ltd.). The SN is the bright object (marked with an arrow) SW of the nucleus. Relative to typical SNe, this SN is more energetic and appears to have synthesized ten times more Nickel.*

flux $\propto \rho^{1/2}$) that are potentially detectable. Radio observations have additional advantages of a longer lived afterglow, immunity from weather and freedom from the diurnal cycle.

Gamma-Ray Bursts Associated with Supernovae. Observers and theorists alike have been intrigued by the possibility that the bright supernova, SN 1998bw, discovered by Galama et al. [30] in the error circle of GRB 980425 [67], is associated with the gamma-ray event (Figure 6). Kulkarni et al. [47] discovered that the SN had an extremely bright radio counterpart; see Figure 7. We noted that the inferred brightness temperature exceeded the inverse Compton catastrophe limit of 5×10^{11} K and to avoid rapid cooling we postulated the existence of a relativistically expanding blastwave ($\Gamma \gtrsim 2$). This relativistic shock is, of course, in addition to the usual sub-relativistic SN shock. This relativistic shock may have produced the GRB at early times. (We note here that we disagree with the much lower energy estimates of [81]; our recent calculations using the same assumptions as those made in [81] result in an energy estimate similar to that obtained earlier [47] from minimum-energy formulation). The optical modeling of the lightcurve and the spectra show that GRB 980425 was especially energetic [43,86] with an energy release of 3×10^{52} erg and Nickel production of nearly nearly a solar mass.

If GRB 980425 is associated with SN 1998bw, then this type of event is rare among the SAX localizations. GRB 980425 is most certainly not a typical GRB: the redshift of SN 1998bw is 0.0085 and the γ-ray energy release in GRB 980425 is at least four orders of magnitude less than in other cosmologically located GRBs. For this reason, most astronomers (especially those in the GRB field; see Wheeler's foray in experimental sociology [82]) do not believe the association between GRB 980425 and SN 1998bw. On the other hand, as evidenced by the intense interest in and modeling of the radio and optical data of SN 1998bw, this object is of considerable interest to the SN community. Indeed, we believe that the proposed GRB–SN association controversy has muddied the main issue: SN 1998bw is an interesting

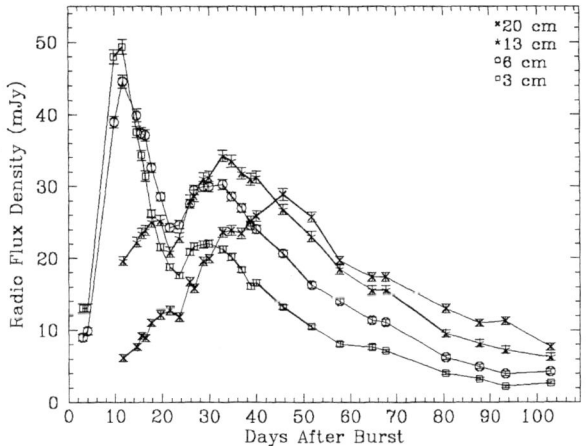

FIGURE 7. *The radio light curve of SN 1998bw at four wavelengths (reprinted by permission from Nature [47] copyright (1998) Macmillan Magazines Ltd.). The peak brightness temperature of SN 1998bw at early times is 10^{13} K, well in excess of the inverse Compton limit of 5×10^{11} K, and can be best understood if the radio emission originates from a relativistic shock ($\Gamma \gtrsim 2$).*

SN in its own right.

What is the true distinguishing feature of SN 1998bw that may connect it to a GRB event? Is it the large energy release, as suggested by several authors [43,33]? We argue that in fact it is the energy *coupled into relativistic ejecta* that most closely connects SN 1998bw to a GRB. In a typical SN, about 10^{51} erg is coupled to the envelope of the star (a small fraction of the total SN energy release of 10^{53} erg). In a GRB, a similar amount of energy (10^{51}–10^{52} erg depending on the event) is coupled to a much smaller ejecta mass, resulting in relativistic outflow. For SN 1998bw, applying the minimum energy formulation to the radio observations we infer the relativistic shell to contain $\sim 10^{50}$ erg. Not only is this uncharacteristic of a typical SN (there exists no evidence for relativistic ejecta in ordinary SN), but it is not dissimilar from the energy implied for GRB outflows. One could therefore envisage a continuum of physical phenomenon between SN 1998bw and cosmological GRBs provided we use the energy in the relativistic ejecta as the basic underlying parameter and not the isotropic gamma-ray release.

IV AFTERGLOW: THE PHYSICS AND ENERGETICS OF THE FIREBALL

One can consider a GRB to be like a SN explosion with a central source releasing energy E_0 (comparable to the mechanical release of energy in an SN). This is the so-called fireball model. The difference between an SN and a GRB is primarily in ejecta mass: 1–10 $M_\odot$ for SNe whereas only 10^{-5} $M_\odot$ for GRBs. The evolution of a GRB is much faster than that of a SN due to two factors: the ejecta expand relativistically and, thanks to the smaller ejecta mass, the optical depth is considerably smaller.

As the ejecta encounter ambient gas, two shocks are produced: a short-lived reverse shock (traveling through the ejecta) and a long-lived forward shock (propagating into the swept-up ambient gas). Afterglow emission is identified with emission from the forward shock. In order to obtain significant afterglow emission, several conditions are necessary. (1) Rapid equipartition of electrons with the shocked protons (which hold most of the energy). (2) Acceleration of electrons to a power-law spectrum (particle Lorentz factor distribution, $dN/d\gamma \propto \gamma^{-p}$). (3) Rapid growth of the magnetic field with energy density in the range of 10^{-2} of that of the protons. Under these circumstances, afterglow emission is dominated by synchrotron emission of the accelerated particles; see [71,79]. The weakness of this model is the assumption of growth in the magnetic field strength to the high values noted above (R. Blandford, pers. comm.).

The theoretically expected afterglow spectrum is shown in Figure 8. Three key frequencies can be identified: ν_a, the synchrotron self-absorption frequency; ν_m, the frequency of the electron with a minimum Lorentz factor (corresponding to the thermal energy behind the shock) and ν_c, the cooling frequency. Electrons which radiate above ν_c cool on timescales equal to the age of the shock. The evolution of these three frequencies is determined by the hydrodynamical evolution of the shock, which in turn is affected by two principal factors: the environment of the GRB and the geometry of the explosion.

The GRB environment. The earliest afterglow models made the simplifying assumption of expansion into a constant density medium. This is an appropriate assumption should the GRB progenitor explode into a typical location of the host galaxy. However, there is increasing evidence tying GRBs to massive stars (see §III). It is well known that massive stars lose matter throughout their lifetime and thus one expects the circumburst medium to exhibit a density profile, $\rho \propto r^{-2}$ where r is the distance from the progenitor. Chevalier & Li [8] refer to these two models as the ISM (interstellar medium) and the wind model respectively. As can be seen from Figure 8 these two models give rise to rather different evolution of the three critical frequencies.

Geometry: Jets versus Spheres. The hydrodynamics is also affected by the geometry of the explosion. Many powerful astrophysical sources have jet-like structure. There is evidence (from polarization observations) indicating asymmetric expansion in SNe [82], so it is only reasonable to assume that GRB afterglows also have jet-like geometry as well. A clear determination of the geometry is essential in order to infer the true energy of the explosion. This is especially important for energetic bursts such as GRB 990123 whose isotropic energy release approaches $M_\odot c^2$.

Let the opening angle of the jet be θ_0. As long as the bulk Lorentz factor, Γ, is larger than θ_0^{-1}, the evolution of the jet is exactly the same as that of a sphere (for an observer situated on the jet axis). However, once Γ falls below θ_0^{-1} then two effects become important. First, for a well defined jet, the on-axis observer sees an edge and thus one expects to see a break in the afterglow emission. Second, the lateral expansion of the jet (due to heated and shocked particles) will start affecting the hydrodynamical explosion.

Wind or ISM? The two key diagnostics to distinguish these two models are the evolution of the cooling frequency (see Figure 8) and the early behavior of the radio emission. In the wind model, the radio emission rises rapidly (relative to the ISM model) and the synchrotron self-absorption frequency falls rapidly with time. Both these result from the fact that the ambient density decreases with radius (and hence in time) in the wind model.

Unfortunately, in general, the current data are not of sufficient quality to firmly distinguish the two models. For example in GRB 980519, the same optical and X-ray data appear to be adequately explained by the jet+ISM model [73] and the sphere+wind model [8]. Including the radio data tips the balance, but only slightly, in favor of the wind model [23]. In our opinion, the best example for the wind model is that of GRB 980329 [20]; see Figure 9. This afterglow exhibits the two unique signatures of the wind model: high ν_a and a rapid rise. Given the importance of making the distinction between the wind and the ISM model, we urge early wide band radio observations (especially at high frequencies).

Energetics. Of all the physical parameters of the fireball, the most eagerly sought parameter is the total energy E_0. By analogy with supernovae, it is E_0 which sets the GRB phenomenon apart from other astrophysical phenomena. Classes of GRBs

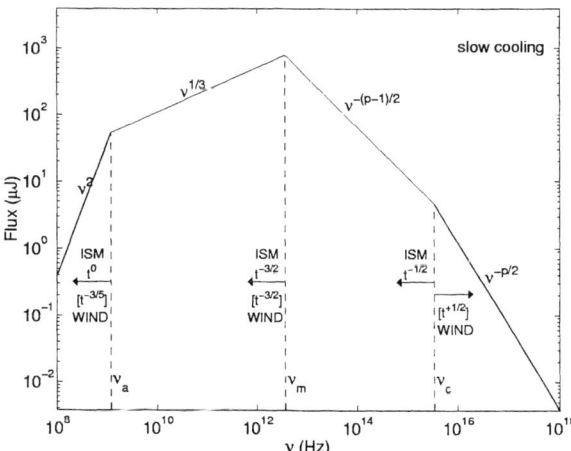

FIGURE 8. *Broad-band spectrum (f_ν) of the afterglow from a spherical fireball with constant density ("ISM" model; see text) and $\rho \propto r^{-2}$ medium ("wind" model; see text). This is representative of the observed spectrum a few days after the burst. Note the distinct evolution of ν_a and ν_c in the two models.*

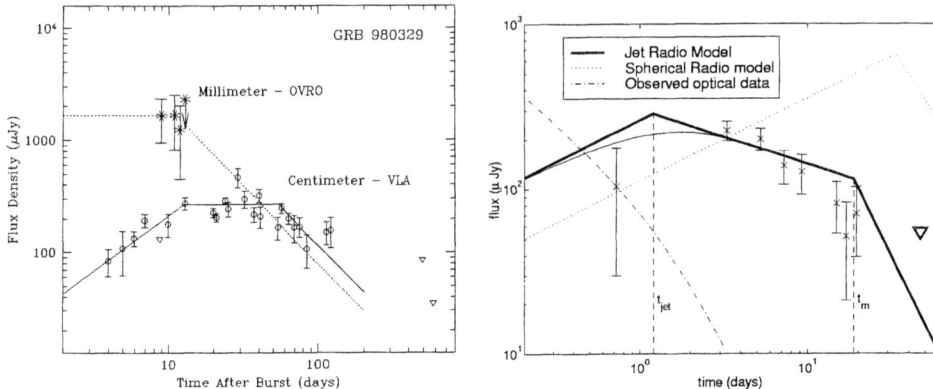

FIGURE 9. *Left:* Radio afterglow of GRB 980329 [20]. The rapid rise of the centimeter flux and the high absorption frequency (signified by the considerable strength of the millimeter emission) offer good support for GRB 980329 expanding into a circumburst medium with density falling as inverse square distance. The lines represents a wind model based on X-ray, optical, IR, mm and cm data. *Right:* Observed and model radio light curves of GRB 990510 [39]. The model predictions for the radio afterglow emission are displayed by the solid line (jet fireball model) and dotted line (spherical fireball model). The observed optical afterglow emission is displayed by the dotted-dashed line; see text for more details.

may eventually be distinguished and ranked by their energy budget; for example, long-duration events, short duration events and supernova-GRBs (see §III).

One approach has been to use the isotropic γ-ray energy as a measure of E_0; see Figure 3. There are three well known problems with such estimates. First, collimation of the ejecta (jets) will result in overestimation of the total energy release. For GRB 990510, where a good case for a jet has been established (Figure 9), the standard isotropic energy estimate is probably a factor of 300 more than the true energy [39]. Second, even after accounting for a possible jet geometry, the efficiency of converting the shock energy into gamma-ray emission is very uncertain. For example, some authors [51] advocate low efficiency ($\sim 1\%$), which would result in an enormous upward correction to the usual isotropic estimates. Third, the bulk Lorentz factor is extremely high during the emission of γ rays and thus the estimates critically depend on assumption of the geometry and granularity [52] of the emitting region. In particular, if the emission is from small blobs [52] then the inferred estimates are grossly in error.

In contrast to this highly uncertain situation, afterglows offer (in principle) more robust methods to evaluate E_0. In view of the importance of determining E_0 we summarize the different methods of determining E_0 from afterglow observations. One approach is to fit a "snapshot" broad-band afterglow spectrum (from radio to X rays) to an afterglow model; this approach was pioneered by Wijers & Galama [83]. The strength of this method is that the estimated E_0 is, in principle, robust.

Specifically, the estimate does not depend on the usually unknown environmental factors (run of density). However, in practice, this method is very sensitive to the values of the critical frequencies (Figure 8) which are usually not well determined. This difficulty explains the wildly differing estimates of E_0 for GRB 970508 [83,35]. Furthermore, this method uses measurements obtained at early times (when the afterglow at high frequencies is bright) with the result that the true source geometry is hidden by relativistic beaming.

A second approach is to model the light curves of the afterglow in a given band, specifically a radio band. The advantages of this method are the photometric stability of radio interferometers and the low Lorentz factor at the epoch of the peak of the radio emission. The disadvantages are two-fold: the sensitivity to the environmental parameters (density) and the assumption of the constancy of the microphysics parameters (electron and magnetic field equipartition factors). Application of this approach to GRB 980703 has resulted in seemingly accurate measures of the fireball parameters [19].

Freedman & Waxman [24] take yet another approach, and estimate the energy release from late time X-ray observations. They show that the X-ray flux is insensitive to the GRB environment, and obtain robust estimates of the fireball energy per unit solid angle: from 3×10^{51} erg to 3×10^{53} erg.

With all the above approaches, however, the possible collimation of the ejecta in jets is still a major uncertainty. This can be addressed by observing the evolution of the afterglow as the "edge" of the jet becomes visible. In most cases no evidence for jets has been seen, with the notable exceptions of GRB 990510 and possibly GRB 990123. In addition, a variety of statistical arguments (the absence of copious numbers of "orphan afterglows") [37,36,70] suggests that, on average, the collimation cannot be extreme, and that for most bursts the opening angle is not less than 0.1 radian. Thus, the total energy for most bursts may be reduced to the range of 10^{50} erg to 3×10^{51} erg, but could easily be much higher in at least some cases.

Possibly the best approach to determining the energetics, which minimizes uncertainties due both to collimation (jets) and to the environment, is to model the afterglow after it becomes non-relativistic. This method builds on the well-established minimum energy formulation and the self-similarity of the Sedov solution. Not only are the ejecta truly non-relativistic, but they are also essentially spherical, as by this time jets will have had sufficient time to have undergone significant lateral expansion. Indeed, we can justifiably call this "fireball calorimetry" [21]. Applying this technique to the long-lived afterglow of GRB 970508 (Figure 1) led to the surprising result that $E_0 \sim 5 \times 10^{50}$ erg — weaker than a standard SN! This is an astonishing result. If true, this result would suggest that it is not E_0 which is the prime distinction between GRBs and SNe but the ejecta mass. However, Chevalier & Li [9] interpret the same data in the wind framework and derive much larger E_0. Clearly, we need more well-studied afterglows with sufficient observations to first distinguish the circumburst environment (wind versus ISM) and then radio observations over a sufficiently long baseline to undertake calorimetry. Nonetheless, one should bear in mind that the current evidence for large energy release in GRBs is not as strong as is usually assumed.

V EPILOGUE AND FUTURE

Clearly, the GRB field is evolving rapidly. Along what direction[s] will this field proceed in the coming years? One way to anticipate the future is by considering analogies from the past.

In §III we already discussed the parallels between the SN field and the GRB field. Here we discuss the numerous parallels with quasar astronomy. First discovered at radio wavelengths, we now study quasars across the electromagnetic spectrum. Although still identified by their gamma-ray properties, we now recognize the tremendous value of pan-chromatic GRB and afterglow studies. In both cases, there was considerable controversy about the distance scale. However, once this issue was settled, it became clear that quasars are the most energetic objects (sustained power) whereas GRBs are the most brilliant. For both, the ultimate energy appears to be related to black holes (albeit of different masses).

The raging issues in GRB astronomy today are the same that fueled quasar studies in the 60's: the spatial distribution, the extraction of energy from the central engine, the transfer of energy from stellar scales to parsec scales, and the geometry of the relativistic outflow (sphere or jet). Astronomers took decades to unify the seemingly diverse types of quasars, and to conclude that there are two types of central engines: radio loud and radio quiet. Likewise, there may well be two types of GRB engines: rapidly and slowly spinning black holes emerging respectively from collapse of a rotating core of a massive star or coalescence of compact objects and the collapse of a massive star. This picture could potentially explain both the cosmologically located GRBs and SN 1998bw. Finally, we can project that in the future, GRBs may be used to probe distant galaxies, just as quasars are used today to study the IGM.

There is a feeling in the astronomical community (outside the GRB community) that the GRB problem is "solved". The truth is that the GRB problem is now getting defined! We now summarize our view of the major issues and anticipated near-term advances. In our opinion the major issues are Diversity, Progenitors and Energy Generation.

As discussed earlier, high energy observations suggest the existence of two classes: short and long-duration bursts. It is possible that afterglow observations may demarcate additional classes. If so, one can contemplate that within a year (assuming abundant localizations by HETE-2) that we will have new GRB designations such as sGRBs (GRBs with late time bump indicative of an underlying SN), wGRBs (GRBs whose afterglow clearly indicates a wind circumburst medium shaped by stellar winds), iGRBs (GRBs which explode in the interstellar medium) and so on.

The broad indications are that GRBs are associated with stars and most likely massive stars. However, we know little beyond this. Comparing the unbeamed GRB event rate of 1.8×10^{-10} yr^{-1} Mpc^{-3} [74] with 3×10^{-5} Type Ibc SN yr^{-1} Mpc^{-3} and 10^{-6} yr^{-1} NS–NS merger Mpc^{-3} [53] shows that GRBs events are extremely rare; here we note that the present data do not support a collimation correction in excess of 100. It will be quite some time before we will be in a position to identify the conditions necessary for a star to die as a GRB.

It is our opinion that SN 1998bw is a major development in the field of stellar collapse. The association (or lack) with GRB 980425 unfortunately has distracted our attention of this important development. The existence of a significant amount of mildly relativistic material, $\sim 10^{50}$ erg [47], is fascinating and it is ironic that none of the models can account for this inferred value whereas most of the theoretical effort has gone into explaining the gamma-ray burst itself (especially considering the uncertain association of GRB 980425 with SN 1998bw). Clearly, SN 1998bw is a rare event, but we are convinced that more such events will be found and accordingly have mounted a major campaign to identify these SNe. The robust signatures of this class are high T_B and prompt X-ray emission since these are necessary consequences of a relativistic ejecta. We note that if these future events are as bright as SN 1998bw then the energy in the relativistic ejecta can be directly measured by VLBI observations of the expanding radio shell.

It is vitally important to make quantitative progress in determining the energy release in GRBs. As discussed in §IV, firm estimates of the energy release require well-sampled broad-band data at early times and densely sampled radio light curves out to late times. This will require a *coordinated* approach and necessarily involve many observatories around the world and in space. The same datasets will also help us understand a profound puzzle: if GRBs indeed arise from the death of massive stars, then why do we not see signatures for a circumburst medium shaped by stellar winds in *all* long-duration GRBs? Even ardent supporters of the wind model [8,9] concede that some GRBs (e.g., GRB 990123, GRB 990510) are due to a jet expanding into a constant density medium.

We now discuss the anticipated returns. True to our tradition as observers, we order the discussion by wavelength regimes!

Radio Observations: Dusty Galaxies, Circumstellar Edges and Reverse Shocks. Perhaps the most exciting use of radio afterglow is in identifying dusty star-forming host galaxies. Such host galaxies are not readily seen at optical wavelengths. Currently, such galaxies are eagerly sought and studied at sub-millimeter wavelengths. However, the sensitivity and localization of such galaxies by sub-millimeter telescopes is poor. In contrast, GRB host galaxies are identified at the sub-arcsecond level. The present radio afterglow detection rate of 40% already places an upper limit on the amount of star-formation in dusty regions, viz. this rate is not larger than that measured from optical observations. This result is entirely independent of the conclusion based on studies in the sub-millimeter regime, or the diffuse cosmic FIR background found in the COBE data. However, the result does rely on two assumptions: (i) GRBs trace star formation and (ii) the GRB explosion and its aftermath does not radically alter the ambient medium (i.e., with a prompt and complete destruction of dust grains along the line of sight).

Radio observations of SNe offer a probe of the distribution of the circumstellar matter. A spectacular example is SN 1980K whose radio flux dropped 14 yr after the explosion [58]. A progenitor star which suffered mass loss with variation in the wind speed could explain the observations. Indeed, one *expects* significant radial structure in the circumburst medium as the progenitor evolves from a blue star to a red supergiant and thence to possibly a blue supergiant, etc. If GRBs come from binary stars which undergo a phase of common envelope evolution [7] then the structure would be even more complicated. Thus, radio observations have the potential (in fortunate circumstances) to give us insight into the mass-loss history of the progenitor star(s).

The prompt optical emission from GRB 990123 [1] has been interpreted to arise from the reverse shock [72]. Far less discussed is the prompt radio emission — a radio flare — also seen from this burst [49]. Sari & Piran [72] suggest that the radio emission also originates from the reverse shock as the electrons cool. Observations related to the reverse shocks are important since it is only through these observations that we have a chance of studying the elusive ejecta. We now have four such examples of radio flares [50] and this represents an order of magnitude better success rate than ROTSE+LOTIS. We urge theorists to pay attention to these new findings. More to the point, radio observations appear to be fruitful for the study of reverse shocks, especially when combined with observations of the prompt optical emission. This bodes well for the coming years given the efforts underway to increase the sensitivity of ROTSE [1].

X-ray Observations: Diversity & Progenitors. *GINGA* identified a number of X-ray rich GRBs. BeppoSAX has found several such examples with some bursts lacking significant gamma-ray emission — the so-called X-ray flashes [40]. We know very little about these X-ray transients. Could they be GRBs in a very dense environment (with red giant progenitors)? We need to take such transients more seriously and intensively followup on such bursts.

Another interesting finding from *GINGA* was the discovery of precursor soft X-ray emission [62]. There is no simple explanation for this phenomenon in the current internal-external shock model. We suggest that the soft X-ray emission precursor is similar to the UV breakout of ordinary SNe. This hypothesis can be confirmed or rejected by obtaining the redshift to such bursts.

The X-ray rich GRB 981226 [26,18] was marked with two additional peculiarities: a precursor emission and afterglow emission which is seemingly undetectable after about 12 hours but then rises rapidly before commencing decay. Above we alluded to the fact that massive stars do not have a single phase of mass loss but instead have a veritable history of mass loss (from birth to death). The X-ray observations of GRB 981226 could be accounted for in a model in which the progenitor has first a red supergiant wind followed by a blue supergiant wind.

Optical Observations: SN Link, Short Bursts & Geometry. The GRB–SN connection is best probed by optical observations. The value of optical observations has already been demonstrated by the current observations of GRB 980326 and GRB 970228. Clearly, more observations are needed to establish this link. Once this link is established then one can undertake detailed spectroscopic studies of the SN with large ground-based telescopes and photometric studies with HST.

Offsets of GRBs and the morphology of the host galaxies will continue to be of great interest. Such observations will help us differentiate whether some GRBs come from nuclear regions or always from star-forming regions. Under the current paradigm, the discovery of GRBs coincident with elliptical galaxies would be a major surprise. On the other hand, one expects short bursts to arise in the halo of their galaxies and thus, in this case, no coincidence is expected. We expect HETE-2 to contribute significantly to these issues. Finally, polarization measurements offer a very convenient way to probe the geometry of the emitting region as has already been demonstrated from the discovery of polarization in GRB 990510 (e.g., [55,84]).

ACKNOWLEDGMENTS

Our research is supported by NASA and NSF. JSB holds a Fannie & John Hertz Foundation Fellowship, AD holds a Millikan Postdoctoral Fellowship in Experimental Physics, TJG holds a Fairchild Foundation Postdoctoral Fellowship in Observational Astronomy and RS holds Fairchild Foundation Senior Fellowship in Theoretical Astrophysics. The VLA is a facility of the National Science Foundation operated under cooperative agreement by Associated Universities, Inc. The W. M. Keck Observatory is operated by the California Association for Research in Astronomy, a scientific partnership among California Institute of Technology, the University of California and the National Aeronautics and Space Administration. It was made possible by the generous financial support of the W. M. Keck Foundation.

REFERENCES

1. Akerlof, C., Balsano, R., Barthelmy, S., et al., *Nature* **398**, 400 (1999).
2. Bloom, J. S., Djorgovski, S. G., Kulkarni, S. R. & Frail, D. A., *ApJ* **508**, L17 (1998).
3. Bloom, J. S., Sigurdsson, S. & Pols, O. R., *MNRAS* **305**, 763 (1999).
4. Bloom, J. S., et al., *ApJ* **518**, L1, (1999).
5. Bloom, J. S., et al., *Nature* **401**, 453 (1999).
6. Boella, G., et al., *A.&A. Suppl. Ser.* **122**, 298 (1997).
7. Brown, G. E., Lee, C. -H., Lee, H. K. & Bethe, H. A., astro-ph/9911458 (1999).
8. Chevalier, R. A. & Li, Z.-Y., *ApJ* **520**, L29, (1999).
9. Chevalier, R. A. & Li, Z. -Y., astro-ph/9908272 (1999).
10. Costa, E., et al., *Nature* **387**, 783 (1997).
11. Djorgovski, S. G., Kulkarni, S. R., Bloom, J. S., Goodrich, R., Frail, D. A., Piro, L & Palazzi, E., *ApJ* **508**, L17 (1998).
12. Djorgovski, S. G. et. al., in preparation, (2000).

13. Eichler, D., Livio, M., Piran, T., & Schramm, D. N., *Nature* **340**, 126 (1989).
14. Fishman, G. J. & Meegan, C. A., *Annu. Rev. Astron. Astrophys.* **33**, 415 (1995).
15. Frail, D. A. & Kulkarni, S. R., *Astrophys. Space Sci.* **231**, 277 (1995).
16. Frail, D. A., Kulkarni, S. R., Nicastro, S. R., Feroci, M., & Taylor, G. B., *Nature* **389**, 261 (1997).
17. Frail, D. A., Kulkarni, S. R., Shepherd, D. S., & Waxman, E., *ApJ* **502**, L119, (1998).
18. Frail, D. A., et al., *ApJ* **525**, L81 (1999).
19. Frail, D. A., Bloom, J. S., Kulkarni, S. R., Sari, R., & Taylor, G. B., in preparation, (2000).
20. Frail, D., Kulkarni, S., Sari, R., Taylor, G., Shepherd, D., Bloom, J., Young, C., Nicastro, L., & Masetti, N., *ApJ* in press, (1999).
21. Frail, D., Waxman, E. & Kulkarni, S. R., *ApJ* in press, (2000).
22. Frail, D. A., Kulkarni, S. R., Wieringa, M. H., et al., astro-ph/9912171, (1999).
23. Frail, D. A., Kulkarni, S. R., Sari, R., et al., astro-ph/9910060, (2000).
24. Freedman, D. L. & Waxman, E., astro-ph/9912214 (1999).
25. Frontera, F., Amati, L., Costa, E., et al., astro-ph/9911228, (1999).
26. Frontera, F., Antonelli, L. A., Amati, L., et al., *ApJ*, submitted (2000).
27. Fruchter, A. S., *ApJ* **512**, L1 (1999).
28. Fruchter, A. S., Thorsett, S., Metzger, M. R., *ApJ* **519**, L13 (1999).
29. Fruchter, A. S., Pian, E., Thorsett, S. E., et al., *ApJ* **516**, 683 (1999).
30. Galama, T. J., et al., *Nature* **395**, 670, (1998).
31. Galama, T. J. et al., *ApJ* in press, (1999).
32. Garnavich, P. M. et al., *ApJ* **493**, L53 (1998).
33. Germany, L., Reiss, D. J., Sadler, E. M., Schmidt, B. P. & Stubbs, C. W., astro-ph/9906096 (1999).
34. Gorosabel, J., Castro-Tirado, A. J., Pedrosa, A., et al., *A.&A.* **347**, L31 (1999).
35. Granot, J., Piran, T. & Sari, R., *ApJ* **527**, 236 (1999).
36. Greiner, J., et al., *A& AS* **138**, 441 (1999)
37. Grindlay, J. E., *ApJ* **510**, 710 (1999)
38. Groot, P. J., Galama, T. J., van Paradijs, J., et al., *ApJ* **493**, L27 (1998).
39. Harrison, F. A., et al., *ApJ* **523**, L121 (1999).
40. Heise, J., talk at the 5th Hunstville GRB conference, (1999).
41. Holland, S. & Hjorth, J., *A&A* **344**, L67 (1999).
42. In't Zand, J. J. M., Amati, L., Antonelli, L. A., et al., *ApJ* **505**, 1191 (1998).
43. Iwamoto, K., et al., *Nature* **395**, 672, (1998).
44. Katz, J. I., *ApJ* **422**, 248 (1993).
45. Katz, J. I. & Canel, L. M., *ApJ* **471**, 915 (1996).
46. Kulkarni, S. R., Djorgovski, S. G., Ramaprakash, A. N., et al., *Nature* **393**, 35 (1998).
47. Kulkarni, S. R., Frail, D. A., Wieringa, M. H., et al., *Nature* **395**, 663 (1998).
48. Kulkarni, S. R., Djorgovski, S. G., Odewahn, S. C., et al., *Nature* **398**, 389 (1999).
49. Kulkarni, S. R., et al., *ApJ* **522**, L97 (1999).
50. Kulkarni, S. R. & Frail, D. A., in preparation, (2000).
51. Kumar, P., astro-ph/9907096 (1999).

52. Kumar, P. & Piran, T., astro-ph/9909014 (1999).
53. Lamb, D. Q., astro-ph/9909026 (1999).
54. Lattimer, J. M. & Schramm, D. N., *ApJ* **192**, L145 (1974).
55. Lazzati, D., Covino, S. & Ghisellini, G., astro-ph/9912247 (1999).
56. Li, L.-X. & Paczynski, B., *ApJ* **507**, L59, (1998).
57. Macfadyen, A. I. & Woosley, S. E., *ApJ* **524**, 262, (1999).
58. Montes, M. J., van Dyk, S. D., Weiler, K. W., Sramek, R. A. & Panagia, N., *ApJ* **506**, 874, (1998).
59. Mészáros, P. & Rees, M. J., *ApJ* **476**, 232 (1997).
60. Metzger, M. R., Djorgovski, S. G., Kulkarni, S. R., Steidel, C. C., Adelberger, K. L., Frail, D. A., Costa, E., & Frontera, F., *Nature* **387**, 879 (1997).
61. Mochkovitch, R., Hernanz, M., Isern, J., & Martin, X., *Nature* **361**, 236–238, (1993).
62. Murakami, T., et al., *Nature* **350**, 592 (1991).
63. Narayan, R., Paczyński, B., & Piran, T., *ApJ* **395**, L83 (1992).
64. Palazzi, E., et al., *A.&A.* **336**, L95 (1998).
65. Paczyński, B. & Rhoads, J., *ApJ* **418**, L5 (1993).
66. Pian, E., Fruchter, A. S., Bergeron, L. E., et al., *ApJ* **492**, L103 (1999).
67. Pian, E. et al., *A&AS* **138**, 463 (1999).
68. Reichart, D. E., *ApJ* **521**, L111 (1999).
69. Reichart, D. E., Lamb, D. Q., Metzger, M. R., et al., *ApJ* **517**, 692 (1999).
70. Rhoads, J. E., *ApJ* **487**, L1 (1997)
71. Sari, R., Piran, T. & Narayan, R., *ApJ* **497**, L17 (1998).
72. Sari, R. & Piran, T., *ApJ* **517**, L109 (1999).
73. Sari, R., Piran, T., & Halpern, J. P., *ApJ* **519**, L17 (1999).
74. Schmidt, M., *ApJ* **523**, L117 (1999).
75. Strohmeyer, T. E., Fenimore, E.E., Murakami, T. & Yoshida, A., *ApJ* **500**, 873 (1998).
76. Taylor, G. B., Frail, D. A., Kulkarni, S. R., et al., *ApJ* **502**, L115 (1998).
77. van Paradijs, J., Groot, P. J., Galama, T., et al., *Nature* **368**, 686 (1997).
78. Vietri, M., *ApJ* **488**, L105 (1997).
79. Waxman, E., *ApJ* **489**, L33 (1997).
80. Waxman, E., & Draine, B. T., astro-ph/9909020 (1999).
81. Waxman, E. & Loeb, A., *ApJ* **515**, 721 (1999).
82. Wheeler, J. C., astro-ph/9912403, (1999).
83. Wijers, R. A. M. J. & Galama, T. J., *ApJ* **523**, 177, (1999)
84. Wijers, R. A. M. J., Vreeswijk, P. M., Galama, T. J., et al., astro-ph/9906346 (1999).
85. Woosley, S. E., *ApJ* **405**, 273, (1993).
86. Woosley, S. E., Eastman, R. G. & Schmidt, B. P., *ApJ* **516**, 788 (1999).

A Coordinated Radio Afterglow Program

D. A. Frail[1], S. R. Kulkarni[2], M. H. Wieringa[3], G. B. Taylor[1],
G. H. Moriarty-Schieven[4], D. S. Shepherd[1], R. M. Wark[3],
R. Subrahmanyan[3], D. McConnell[3], and S. J. Cunningham[3]

[1] *National Radio Astronomy Observatory, Socorro, NM 87801*[1]
[2] *Division of Physics, Mathematics, and Astronomy 105-24, Caltech, Pasadena, CA 91125*
[3] *Australia Telescope National Facility, CSIRO, Epping 2121, Australia*
[4] *Joint Astronomy Centre, 600 A'Ohoku Place, Hilo, HI 96720*

Abstract. We describe a ground-based effort to find and study afterglows at centimeter and millimeter wavelengths. We have observed all well-localized gamma-ray bursts in the Northern and Southern sky since BeppoSAX first started providing rapid positions in early 1997. Of the 23 GRBs for which X-ray afterglows have been detected, 10 have optical afterglows and 9 have radio afterglows. A growing number of GRBs have both X-ray and radio afterglows but lack a corresponding optical afterglow.

INTRODUCTION

BeppoSAX revolutionized gamma-ray burst (GRB) astronomy not only through its discovery of X-ray afterglows but also through the dissemination of accurate and timely GRB positions to ground-based observers, who then conduct searches for afterglows at optical and radio wavelengths. Our collaboration uses the interferometer facilities of the Very Large Array (VLA), the Australia Telescope Compact Array (ATCA), the Very Long Baseline Array (VLBA) and the Owens Valley Radio Observatory (OVRO) Interferometer. At high frequencies, we use single dish telescopes which include the James Clerk Maxwell Telescope (JCMT) and the OVRO 40-m Telescope. All afterglow searches begin with the VLA in the northern hemisphere ($\delta > -45°$, $\sigma_{rms} = 45$ μJy in 10 min, FOV $\simeq 5'$) and the ATCA in the southern hemisphere ($\delta < -45°$, $\sigma_{rms} = 45$ μJy in 240 min, FOV $\simeq 5'$), typically at a frequency of 8.5 GHz, which provides a balance between sensitivity and field-of-view. Follow-up programs at the other radio facilities are begun after a VLA or ATCA transient is discovered.

As with quasars, radio observations provide unique diagnostics complementary to those obtained at X-ray and optical wavelengths. Our collaboration has discovered

---

[1]) The NRAO is a facility of the National Science Foundation, operated under cooperative agreement by Associated Universities, Inc.

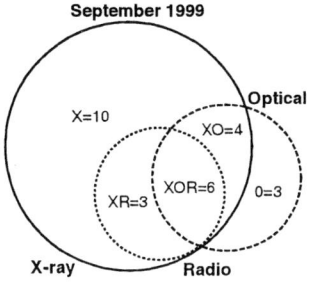

FIGURE 1. A Venn diagram showing the detection statistics for 26 well-localized GRBs. Of the 23 GRBs for which X-ray afterglows have been detected to date, 10 have optical afterglows (XO + XOR) and 9 have radio afterglows (XR + XOR). In total there are 13 optical and/or radio afterglows with corresponding X-ray afterglows (XO + XR + XOR). Only 6 GRBs have afterglows detected in all three bands (XOR).

all known radio afterglows to date, leading to a number of important results: the direct demonstration of relativistic expansion of the ejecta (Frail et al. 1997a), evidence for a reverse shock (Kulkarni et al. 1999), the first true calorimetry of a GRB explosion (Frail, Waxman & Kulkarni 1999), the discovery of optically obscured events (Taylor et al. 1998), the first unambiguous evidence that the ejecta are collimated in jets (Harrison et al. 1999), and the discovery of a possible link between supernovae and GRBs (Kulkarni et al. 1998).

RADIO AFTERGLOW STATISTICS

Since 1997 we have observed 19 GRBs with the VLA and detected a total of eight radio afterglows (see Figure 1, Tables 1 and 2). The peak fluxes (F_{peak}) of the detections range from 1200 μJy to 150 μJy. This small range of F_{peak} values suggests that our ability to detect radio afterglows is severely limited by the sensitivity of the telescope. The "lifetime" (i.e., t_{max}) of the radio afterglows is signal-to-noise limited but it is clear, at least among bursts of comparable brightness, that t_{max} varies substantially. Of special note are the three GRBs (970828, 981226, and 990506) which have no optical counterparts (i.e., XR class). These may represent an important group of GRBs whose optical emission is extincted by dust.

There are 11 GRBs for which a VLA search of the error box failed to detect a radio afterglow (see Table 2). The peak fluxes given in the table are conservative upper limits for a radio afterglow on a time-scale of 1 to 30 days and at frequencies between 1.4 and 8.5 GHz. These non-detections vary in quality depending on the size of the error circle but most observations had sufficient sensitivity to detect radio afterglows with fluxes comparable to those listed in Table 1.

TABLE 1. Radio Afterglow Detections.

GRB	F_{peak} (μJy)	t_{max} (days)	AG Class	Instruments	References
970508	1200	450	XOR	VLA, VLBA, OVRO, JCMT	Frail et al. (1997a)
970828	150	3.5	XR	VLA	Djorgovski et al. (1999)
980329	300	135	XOR	VLA, OVRO, JCMT	Taylor et al. (1998)
980425[a]	50,000	>300	XOR	ATCA, JCMT	Kulkarni et al. (1998)
980519	300	65	XOR	VLA, OVRO	Frail et al. (1999a)
980703	1200	210	XOR	VLA, VLBA, JCMT	Bloom et al. (1998)
981226	170	20	XR	VLA	Frail et al. (1999b)
990123	260	1.2	XOR	VLA, OVRO, OVRO 40-m, JCMT	Kulkarni et al. (1999)
990506	550	<16	XR	VLA	Taylor et al. (1999)
990510	225	20	XOR	ATCA	Harrison et al. (1999)

[a] Related to SN1998bw. We do not include this GRB in the detection statistics.

TABLE 2. VLA Afterglow Non-Detections.

GRB	F_{peak} (μJy)	AG Class	References
970111	<300	X	Frail et al. (1997b)
970228	<50	XO	Frail et al. (1998)
970616	<150	X	IAUC #6691
970815	<50	X	IAUC #6723
971214	<25	XO	Ramaprakash et al. (1998)
971227	<50	X	
980326	<150	O	
980613	<50	XO	
981220	<125	X	GCN #269
990520	<125	X	
990704	<125	X	

TABLE 3. ATCA Afterglow Non-Detections.

GRB	F_{peak} (μJy)	AG Class	References
970402	<300	X	
980109	<550	a	
990217	<175	a	GCN #266
990627	<125	X	GCN #357
990705	<100	XO	GCN #376
990712	<100	O	

[a] GRB 990109 and 990217 were seen only in gamma-rays. No afterglows were detected at any wavelength.

There have been two radio afterglow detections made at the ATCA (see Table 1). The possible relation of GRB 980425 to SN1998bw makes it a rather unusual event, so we do not include it in the detection statistics. The upper limits of the six ATCA non-detections in Table 3 were not sufficient to have detected the weaker radio afterglows in Table 1.

SUMMARY

In summary, our coordinated program has been very successful in detecting radio afterglows from GRBs. In particular:

- Six gamma-ray bursts are seen at X-ray, optical and radio wavelengths (GRB 970508, GRB 980329, GRB 980519, GRB 980703, GRB 990123, GRB 990510) (see Figure 1).

- Of the 23 X-ray afterglows, nine have been detected at radio wavelengths (XOR + XR) for a rate of 39%. At the VLA the detection rate is 8/19 or 42%. The small range in the observed peak flux densities suggests that our ability to detect radio afterglows is mainly limited by the sensitivity of the telescopes (VLA and ATCA).

- Of the 23 X-ray afterglows, ten have been detected at optical wavelengths (XOR + XO) or 43%. The detection rate of well-localized GRBs is comparable at optical and radio wavelengths.

- There exists a growing class (XR) of "dark" GRBs which have X-ray and radio afterglows but no known optical afterglow. These may represent an important group of GRBs whose optical emission is extincted by dust.

REFERENCES

1. Bloom, J. S., Frail, D. A., Kulkarni, S. R., Djorgovski, S. G., Halpern, J. P., Marzke, R. O., Patton, D. R., Oke, J. B., Horne, K. D., Gomer, R., Goodrich, R., Campbell, R., Moriarty-Schieven, G. H., Redman, R. O., Feldman, P. A., Costa, E., and Masetti, N. The Discovery and Broadband Follow-up of the Transient Afterglow of GRB 980703. *ApJ* **508**, L21–L24, (1998).
2. Djorgovski, S. G., et. al. The radio afterglow from GRB 970828 and optical spectroscopy of its host galaxy. ApJ (Letters) submitted, (1999).
3. Frail, D. A., Kulkarni, S. R., Nicastro, S. R., Feroci, M., and Taylor, G. B. The radio afterglow from the gamma-ray burst of 8 May 1997. *Nature* **389**, 261–263, (1997a).
4. Frail, D. A., Kulkarni, S. R., Costa, E., Frontera, F., Heise, J., Feroci, M., Piro, L., Dal Fiume, D., Nicastro, L., Palazzi, E., and Jager, R. Radio Monitoring of the 1997 January 11 Gamma-Ray Burst. *ApJ* **483**, L91–L94, (1997b).
5. Frail, D. A., Kulkarni, S. R., Shepherd, D. S., and Waxman, E. No Radio Afterglow from the Gamma-Ray Burst of 1997 February 28. *ApJ* **502**, L119–L122, (1998).
6. Frail, D. A., Kulkarni, S. R., Sari, R., Taylor, G. B., Shepherd, D. S., Bloom, J. S., Young, C. H., Nicastro, L., and Masetti, N. The radio afterglow from GRB 980519: A test of the jet and circumstellar models. ApJ submitted; astro-ph/9910060, (1999a).

7. Frail, D. A., Kulkarni, S. R., Bloom, J. S., Djorgovski, S. G., Gorjian, V., Gal, R. R., Meltzer, J., Sari, R., Chaffee, F. H., Goodrich, R. W., Frontera, F., and Costa, E. The radio afterglow and the host galaxy of the x-ray rich GRB 981226. ApJ (Letters) in press; astro-ph/9909407, (1999b).
8. Frail, D. A., Waxman, E., and Kulkarni, S. R. A 450-day light curve of the radio afterglow of GRB 970508: Fireball calorimetry. ApJ (Letters) submitted; astro-ph/9910319, (1999).
9. Harrison, F. A., Bloom, J. S., Frail, D. A., Sari, R., Kulkarni, S. R., Djorgovski, S. G., Axelrod, T., Mould, J., Schmidt, B. P., Wieringa, M. H., Wark, R. M., Subrahmanyan, R., McConnell, D., McCarthy, P. J., Schaefer, B. E., McMahon, R. G., Markze, R. O., Firth, E., Soffitta, P., and Amati, L. Optical and Radio Observations of the Afterglow from GRB 990510: Evidence for a Jet. *ApJ* **523**, L121–L124, (1999).
10. Kulkarni, S. R., Frail, D. A., Wieringa, M. H., Ekers, R. D., Sadler, E. M., Wark, R. M., Higdon, J. L., Phinney, E. S., and Bloom, J. S. Radio emission from the unusual supernova 1998bw and its association with the gamma-ray burst of 25 April 1998. *Nature* **395**, 663–669, (1998).
11. Kulkarni, S. R., Frail, D. A., Sari, R., Moriarty-Schieven, G. H., Shepherd, D. S., Udomprasert, P., Readhead, A. C. S., Bloom, J. S., Feroci, M., and Costa, E. Discovery of a Radio Flare from GRB 990123. *ApJ* **522**, L97–L100, (1999).
12. Ramaprakash, A. N., Kulkarni, S. R., Frail, D. A., Koresko, C., Kuchner, M., Goodrich, R., Neugebauer, G., Murphy, R., Eikenberry, S., Bloom, J. S., Djorgovski, S. G., Waxman, E., Frontera, F., Feroci, M., and Nicastro, L. The energetic afterglow of the gamma-ray burst of 14 December 1997. *Nature* **393**, 43–46, (1998).
13. Taylor, G. B. *et al.* The Discovery of the Radio Afterglow from the Optically Dim Gamma-Ray Burst of 1998 March 29. *ApJ* **502**, L115, (1998).
14. Taylor, G. B. *et al.* The gamma-ray burst of 1999 May 6 and the population of optically dim afterglows. In preparation, (1999).

Optical/Multiwavelength Observations of GRB Afterglows

Titus J. Galama[1]

Astronomy, MS 105-24, California Institute of Technology, Pasadena, CA 91125

Abstract. I review γ-ray burst optical/multiwavelength afterglow observations since 1997, when the first counterparts to GRBs were discovered. I discuss what we have learned from multiwavelength observations of GRB afterglows in relation to the 'standard' fireball plus relativistic blast-wave models. To first order, the 'standard' model describes the afterglow observations well, but a wealth of information can be gathered from the deviations of GRB afterglow observations from this 'standard' model. These deviations provide information on the nature of the progenitor and on the physics of GRB production. In particular, I focus on the possible connection of GRBs to supernovae, on jet and circumstellar wind models, on the early-time afterglow, and on the emission from the reverse shock.

I INTRODUCTION

Fireball plus relativistic blast-wave models predict low-energy radiation following GRBs (see e.g., [1]). This radiation has been dubbed the 'afterglow'. The basic model is that of a point explosion: a large amount of energy, $\sim 10^{52-53}$ ergs, is released in a compact region (less than a light millisecond across), which leads to a 'fireball', an optically thick radiation-electron-positron plasma with initial energy much larger than its rest mass that expands ultra-relativistically (see e.g., [2] for an extensive review). The GRB may be due to a series of 'internal shocks' that develop in the relativistic ejecta before they collide with the ambient medium. When the fireball runs into the surrounding medium a 'forward shock' ploughs into the medium and heats it, and a 'reverse shock' does the same to the ejecta. As the forward shock is decelerated by increasing amounts of swept-up material it produces a slowly fading 'afterglow' of X rays, followed by ultraviolet, optical, infrared, millimeter, and radio radiation. As the reverse shock travels through the ejecta it may give rise to a bright optical flash.

Models for the origin of GRBs that (in principle) can provide the required energies, are the neutron star-neutron star (e.g., [3]) and neutron star-black hole

---

[1] The author is supported by the Sherman Fairchild Foundation.

mergers [4–6], white dwarf collapse [7], and core collapses of very massive stars ('failed' supernovae or hypernovae [8,9]).

This review consists of two parts. In the first part, I discuss several confirmations of the relativistic nature of GRB events and discuss the generally good agreement between the 'standard' fireball plus relativistic blast wave model and the observations of GRB afterglows. In the second part, I then proceed to discuss the 'devious' deviations of some GRB afterglows from this standard model, and discuss the wealth of information that we can gather from them. In particular, I discuss what such deviations may tell us about the nature of the progenitor and about the physics of GRB production.

II CONFIRMATION OF THE RELATIVISTIC BLAST-WAVE MODEL

A The Forward Shock

Let us first concentrate on the forward shock and assume slow cooling (the bulk of the electrons do not radiate a significant fraction of their own energy and the evolution is adiabatic); this appears applicable to some observed GRB afterglows at late times ($t > 1$ hr).

The electrons are assumed to be accelerated, in the forward shock, to a power-law distribution of electron Lorentz factors, $N(\gamma_e) \propto \gamma_e^{-p}$, with some minimum Lorentz factor γ_m. Then, the synchrotron spectrum of such a distribution of electrons is a power law with $F_\nu \propto \nu^{1/3}$ up to a maximum, F_m, at the peak frequency ν_m (corresponding to the minimum Lorentz factor γ_m). Above ν_m it is a power law, $F_\nu \propto \nu^{-(p-1)/2}$, up to the cooling frequency, ν_c. Electrons with energies $\gamma_e m_e c^2 > \gamma_c m_e c^2$, where γ_c is the electron Lorentz factor associated with the cooling frequency ν_c, radiate a significant fraction of their energy and thereby cause a spectral transition; above ν_c we have $F_\nu \propto \nu^{-p/2}$. Synchrotron self absorption causes a steep cutoff of the spectrum at low frequencies, $\nu < \nu_a$ ($F_\nu \propto \nu^2$ if $\nu_a < \nu_m$), where ν_a is the synchrotron self absorption frequency. Thus, the spectrum consists of four distinct power-law regimes, separated by three break frequencies: (i) the self absorption frequency, ν_a, (ii) the peak frequency, ν_m, and (iii) the cooling frequency, ν_c (see Fig. 2).

The simplest assumption is that of spherical symmetry and a constant ambient density. For example, if GRBs are the result of the merger of a compact binary system (such as a double neutron star or a neutron star-black hole binary system), then we would expect the fireball to encounter a homogeneous ambient medium. In that case the afterglow can be described by the spectral shape described above combined with the following scalings: $\nu_m \propto t_{obs}^{-3/2}$, $\nu_c \propto t_{obs}^{-1/2}$, $\nu_a \propto t_{obs}^0 = $ constant, and $F_m \propto t_{obs}^0 = $ constant (see [10] and [11] for details).

The First X-Ray and Optical Counterparts

Since both the afterglow spectrum and the temporal evolution of the break frequencies (ν_a, ν_m, and ν_c) are power laws in this model, the evolution of the flux is also a power law in time. For example, for $\nu_m \leq \nu \leq \nu_c$, the decay of the flux is $F_\nu \propto t_{\rm obs}^{-3(p-1)/4}$, and the power-law spectral slope α relates to the spectral slope β as $\alpha = -3/2\beta$. A stringent test of the relativistic blast-wave model came with the discovery of the first X-ray [12] and optical [13] counterparts to GRB 970228. Several authors [14–16] showed that to first order the model describes the X ray and optical afterglow very well (see Fig. 1).

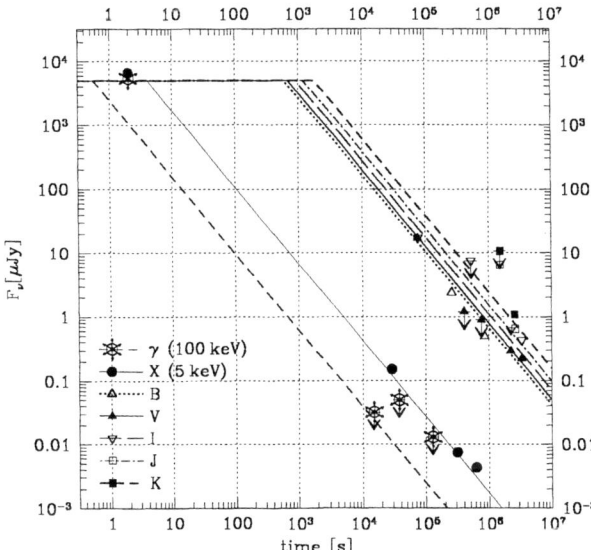

FIGURE 1. The light curves of GRB 970228 from gamma rays to near-infrared (from [14]; reprinted with permission from Blackwell Science Ltd.). To first order, the light curves are power laws and the offsets between them satisfy the expectations from the model.

Detection of absorption features in the optical transient (OT) spectrum of GRB 970508 [17] established that this event was at a redshift $z > 0.835$, showing that GRBs are located at cosmological distances and are thus extremely powerful events. This was also the first GRB with a radio counterpart [18]. The radio light curves (8.5 and 4.9 GHz) show large variations on time scales of less than a day, but these damp out after one month. This finds a viable explanation in interstellar scintillation (irregular plasma refraction by the interstellar medium between the source and the observer). The damping of the fluctuations can then be understood as the effect of source expansion on the diffractive interstellar scintillation. Thus, a source size of roughly 10^{17} cm was derived (at 3 weeks), corresponding to a mildly relativistic expansion of the shell [18].

GRB 970508 remains one of the best observed afterglows: the radio afterglow was visible for at least 368 days (and at 2.5 sigma on day 408.6 [19]), and the optical afterglow up to ~ 450 days (e.g., [20–23]). In addition, millimeter [24], infrared and X-ray [25] counterparts were detected, and it is the first GRB for which a spectral transition in the optical/near-IR range was found [22,26]; this transition is interpreted as the effect of the passage of the cooling frequency through the optical/near-IR passbands. These multiwavelength observations allowed the reconstruction of the broad radio to X-ray spectrum for this GRB [26] (see Fig. 2). It is found that the 'standard' model provides a successful and consistent description of the afterglow observations over nine decades in frequency, ranging in time from the event until several months later [26]. The synchrotron afterglow spectrum of this GRB allows measurement of the electron energy spectrum p, the three break frequencies (ν_a, ν_m and ν_c), and the flux at the peak, F_m. For GRB 970508 the redshift, z, is also known, and all blast wave parameters could be deduced: the total energy (per unit solid angle) $E = 3.5 \times 10^{52}$ erg, the ambient (nucleon) density $n = 0.030$, the fraction of the energy in electrons $\epsilon_e = 0.12$ and that of the magnetic field $\epsilon_B = 0.089$ [11]. The numbers themselves are uncertain by an order of magnitude (see e.g., [27]), but the result shows that the 'standard' model fits the expectations very well.

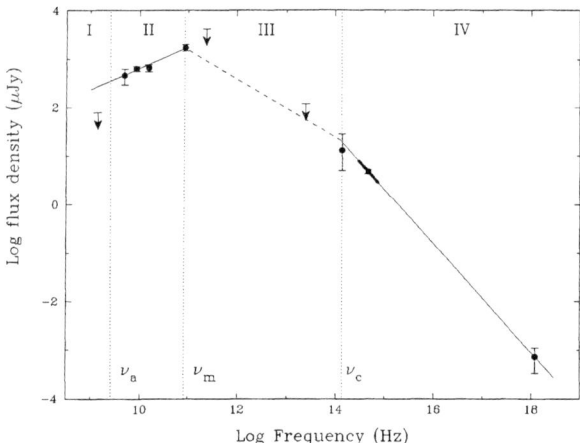

FIGURE 2. The X-ray to radio spectrum of GRB 970508 on May 21.0 UT (12.1 days after the event). The location of the break frequencies ν_a, ν_m and ν_c, inferred from transitions in the light curves and from spectra of the afterglow, are indicated (from [26]).

B The Reverse Shock

The ROTSE telescope obtained its first images only 22 seconds after the start of GRB 990123 (i.e., during the GRB), following a notification received from BATSE aboard the Compton Observatory satellite. The ROTSE observations show that the optical light curve peaked at $m_V \sim 9$ magnitudes some 60 seconds after the event began [28]. After maximum, a fast decay followed for at least 15 minutes. The late-time afterglow observations show a more gradual decline [29-33] (see Fig. 3).

The redshift $z = 1.6$, inferred from absorption features in the OT spectrum, implies that the optical flash would have been as bright as the full moon had the GRB occurred in the nearby galaxy M31 (Andromeda). If one assumes that the emission detected by ROTSE comes from a non-relativistic source of size ct,

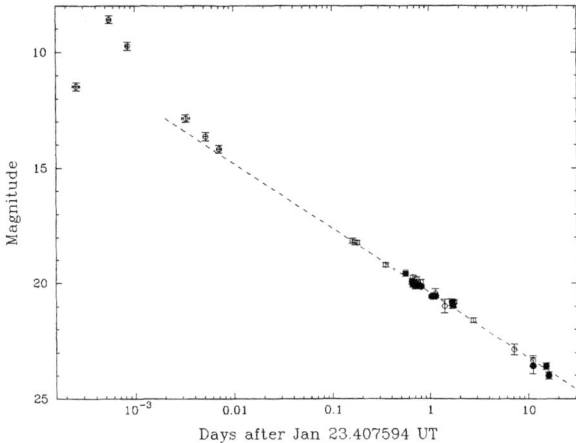

FIGURE 3. R-band light curve of the afterglow of GRB 990123. The ROTSE data show that the optical light curve peaked at $m_V \sim 9$ [28]. The dashed line indicates a power-law fit to the light curve (for $t > 0.1$ days), which has exponent -1.12 ± 0.03 (Reprinted by permission from Nature [29] copyright (1999) Macmillan Magazines Ltd.).

then the observed brightness temperature $T_b \gtrsim 10^{17}$ K of the optical flash exceeds the Compton limit of 10^{12} K, confirming the highly relativistic nature of the GRB source [29].

The ROTSE observations show that the prompt optical and γ-ray light curves do not track each other [28]. In addition, detailed comparison of the prompt optical emission with the BATSE spectra of GRB 990123 (at three epochs for which both optical and gamma-ray information is available) shows that the ROTSE emission is not a simple extrapolation of the GRB spectrum to much lower energies [29,34].

Emission from the reverse shock is predicted to peak near the optical waveband during or just after the GRB [1,35]. The observed properties of GRB 990123 appear to fit this model quite well [29,33,36]. If this interpretation is correct, GRB 990123 would be the first burst in which all three emitting regions have been seen: internal shocks causing the GRB, the reverse shock causing the prompt optical flash, and the forward shock causing the afterglow. The emissions thus arise from three different emitting regions, explaining the lack of correlation between the GRB, the prompt optical and the late-time optical emission [29] (but see [37]).

III DEVIATIONS

As discussed above, the 'standard' model explains the multiwavelength observations of GRB afterglows very well. Now that we have a basic understanding of GRB afterglows it is interesting to consider what we can learn (and what we have learned in the past year) from the observational departures from the 'standard' model.

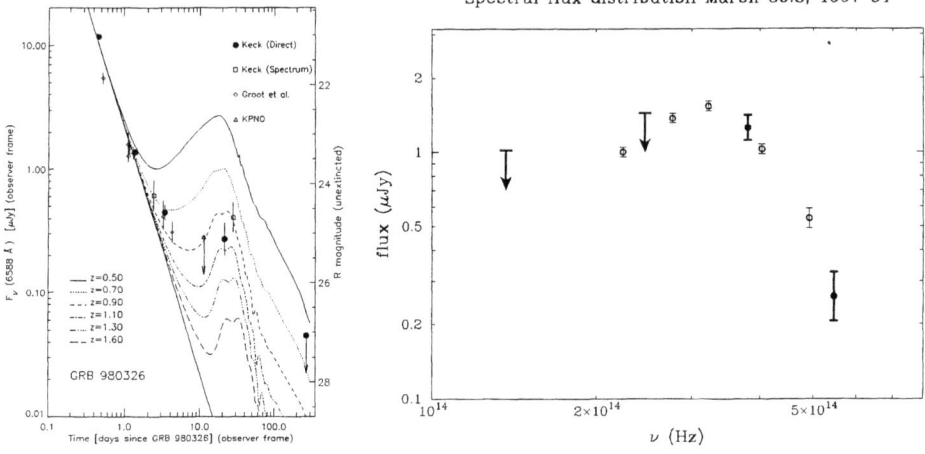

FIGURE 4. Left: R-band light curve of GRB 980326 and the sum of an initial power-law decay plus Ic supernova light curve for redshifts ranging from $z = 0.50$ to $z = 1.60$ (Reprinted by permission from Nature [40] copyright (1999) Macmillan Magazines Ltd.). Right: The broad-band spectrum of the OT of GRB 970228 at March 30.8, 1997 UT (• and upper-limit arrow). Also shown is the spectral flux distribution of SN 1998bw (○) redshifted to the redshift of GRB 970228 ($z = 0.695$). The similarity of the spectral flux distributions is remarkable (from [43]).

A The GRB/Supernova Connection

A direct consequence of the collapsar model is that GRBs are expected to be accompanied by supernovae (SNe).

The first evidence for a possible GRB/SN connection was provided by the discovery by Galama et al. [38] of SN 1998bw in the error box of GRB 980425. The temporal and spatial coincidence of SN 1998bw with GRB 980425 suggest that the two phenomena are related [38,39]. GRB 980425 is most certainly not a typical GRB: the redshift of SN 1998bw is 0.0085 and the corresponding γ-ray peak luminosity of GRB 980425 and its total γ-ray energy budget are about a factor of $\sim 10^5$ smaller than those of 'normal' GRBs. Such SN-GRBs may well be the most frequently occurring GRBs in the Universe.

Bloom et al. [40] realized that the late-time red spectrum and the late-time rebrightnening of the light curve of GRB 980326 are possible evidence that at late times the emission is dominated by an underlying supernova. The authors find that a template supernova light curve, provided by the well-studied type $I_{b/c}$ SN 1998bw provides an adequate description of the observations (see Fig. 4).

In fact, the behavior of GRB 970228 already showed first indications that the standard model was not sufficient to describe the observations in detail [41]. The early-time decay of the optical emission is faster than that at later times and, as the source faded, it showed an unexpected reddening [41]. Indeed, Galama et al.

[41] conclude that although the initial behavior is in agreement with the 'standard' model, the subsequent behavior is harder to explain. It was not until Bloom et al. [40] discussed evidence for a supernova-like emission accompanying GRB 980326 that the behavior of GRB 970228 was better understood. Also for GRB 970228 the late-time light curve and reddening of the transient can be well explained by an initial power-law decay modified at late times by SN 1998bw-like emission [42,43] (see Fig. 4).

The relation between distant GRBs like GRB 980326 and GRB 980425/SN 1998bw is unclear. Is SN 1998bw a different phenomenon or a more local and lower energy equivalent? Are all afterglows consistent with such a phenomenon? The answer to the latter question requires detailed analysis of existing data on GRB afterglows, but more convincing evidence may be provided by future observations of GRB afterglows around the time of the SN emission maximum.

B Collimated Outflow (Jet) and/or Circumstellar Wind Model

If, as suggested by the evidence for a GRB/SN connection (see §III A), at least some GRBs are produced by the core collapse of massive stars to black holes, then the circumburst environment will have been influenced by the strong wind of the massive progenitor star. For a constant wind speed, the circumstellar density falls as $n \propto r^{-2}$, where r is the radial distance. In this so called circumstellar wind model, the afterglow can be described by the same synchrotron spectral shape (see §II A), but with different scalings for the break frequencies and the peak flux: $\nu_m \propto t_{obs}^{-3/2}$, $\nu_c \propto t_{obs}^{+1/2}$, $\nu_a \propto t_{obs}^{-3/5}$, and $F_m \propto t_{obs}^{-1/2}$ (see [45,46] for details).

Due to relativistic beaming, only a small portion of the emitting surface with opening angle $1/\gamma$ is visible. As the fireball evolves γ decreases and the beaming angle will eventually exceed the angular size of the collimated outflow (the size of the jet). In this jet model, we then expect to see an increase in the decay rate. Slightly later, the jet begins a lateral expansion, which causes a further steepening of the light curve. In this case the scalings for the break frequencies are: $\nu_m \propto t_{obs}^{-2}$, $\nu_c \propto t_{obs}^{0} =$ constant, $\nu_a \propto t_{obs}^{-1/5}$, and $F_m \propto t_{obs}^{-1}$ (see for details [47–49]). At late times, when the evolution is dominated by the spreading of the jet, the decay is as fast as $F_\nu(t) \propto t^{-p} \sim t^{-2.2}$, where p is the power-law index of the electron energy spectrum.

Non 'standard' behavior: The optical and X-ray light curves of GRB 970508 show a maximum that is reached around 1 day and is followed by characteristic power-law decaying light curves. The onset of the X-ray flare roughly coincides with that of the optical bump [25]. This behavior is not yet well understood. Panaitescu et al. [44] have tested several possible models to explain the flare: (i) by continued energy injection from the central source, (ii) by ejecta with a range of Lorentz factors, (iii) as the effect of a jet that is observed slightly off-center, and (iv) by the encounter of a shell of dense ambient material.

The afterglow peak flux F_m of GRB 970508 decays with time; it is ~ 1700 μJy at 86 GHz at ~ 12 days, while only ~ 700 μJy at 8.5 GHz at ~ 60 days [18,24,26]. Also, the self-absorption frequency ν_a evolves to lower frequencies. However, in the 'standard' model the peak flux and the self-absorption frequency would remain constant in time. Again, these features have several possible explanations: (i) the effect of collimated outflow, (ii) the effect of a circumstellar wind, or (iii) the transition from an ultra-relativistic to a non-relativistic evolution [19]. Note however, that the 'standard' model and the circumstellar wind model predict a distinctively different evolution of the cooling break ν_c; the observed evolution for GRB 970508 fits the 'standard' model well and is hard to reconcile with the wind model.

Fast decaying afterglows: GRB 980326 was the first example of a rapidly decaying afterglow [50]. Unfortunately, no attempt was made to observe the X-ray afterglow, and the optical spectral information is only sparse. It was not until GRB 980519 that it was decisively found that the rapidly decaying afterglow could not be understood in the terms of the 'standard' model; the relation between the spectral slope and the temporal decay is not as expected from the 'standard' model. The observations can either be explained by a jet [48,51] or by a circumstellar wind model [45]. Radio observations of GRB 980519 are well described by a wind model, but cannot decisively reject the jet model [52]. The reason that it is hard to distinguish the different models is because of the absence of high quality data; afterglows are faint. Future radio observations at early and late times may allow to decisively distinguish the models.

Perhaps the actual light-curve transition (from a regular to a fast decay caused by 'seeing' the edge of the jet) has been observed in the optical afterglow of GRB 990123 [30–32]. However, no evidence for such an increase of the decay rate was found in near-infrared K-band observations [30]. A similar transition was better sampled in afterglow data of GRB 990510; optical observations of GRB 990510, show a clear steepening of the rate of decay of the light between ~ 3 hours and several days [53,54] to roughly $F_\nu(t) \sim t^{-2.2}$. Together with radio observations, which also reveal a transition, it is found that the transition is very much frequency-independent; this virtually excludes explanations in terms of the passage of the cooling frequency, but is what is expected in case of beaming [53]. Harrison et al. [53] derive a jet opening angle of $\theta = 0.08$, which for this burst would reduce the total energy in γ rays to $\sim 10^{51}$ erg.

C The Early Afterglow and the Reverse Shock

The radio observations of GRB 990123 show a brief flare at one day after the event [29,56]. Such radio behavior is unique, both for its early appearance as well as its rapid decline. The flare has been suggested to be due to the reverse shock [33,52]. However, understanding the full evolution still requires interpretation in terms of the forward shock and a jet in addition to the reverse shock. An alternative interpretation in terms of emission by the forward shock only is also consistent with the observations [29]. This interpretation is also not without problems; the spectrum is required to be relatively flat around the maximum. In this interpretation the energy density of the magnetic field is very low $\epsilon_B < 10^{-6}$, similar to what is derived for GRB 980703 [55]. The differences in afterglow behavior may thus reflect variations in the magnetic-field strength in the forward shock [29]. Other possibilities have been put forward: an explanation in terms of the forward shock and a jet [57] and an explanation in terms of the forward shock and a dense ambient medium [58]. Interestingly, observations of the light curve at times between 15 min and several hours could distinguish between some of the models; this is the region of transition from early times, where the emission is believed to be due to the reverse shock, to late times where the emission of the forward shock is dominant. The imminent launch of HETE-2 will provide the unique possibility to study this time window, by providing accurate localizations to the community within minutes after the events.

IV CONCLUSIONS

Although the 'standard' model describes the afterglow observations well, a wealth of information is provided by the deviations of GRB afterglows from the 'standard' model; in particular, by the possible connection of GRBs to supernovae, by possible evidence for collimated outflow and circumstellar winds, by the early-time afterglow and by the emission from the reverse shock.

REFERENCES

1. Mészáros, P. and Rees, M.J., *ApJ* **476**, 232 (1997).
2. Piran, T., *Physics Report*, in press, astro-ph/9810256 (1999).
3. Eichler, D., et al., *Nature* **340**, 126 (1989).
4. Mochkovitch, R., et al., *Nature* **361**, 236 (1993).
5. Lattimer, J.M. and Schramm, D.N., *ApJL* **192**, L145 (1974).
6. Narayan, R., Paczyński, B. and Piran, T., *ApJL* 395, L83 (1992).
7. Usov, V.V., *Nature* **357**, 472 (1992).
8. Woosley, S.E., *ApJ* **405**, 273 (1993).
9. Paczyński, B., *ApJL* **494**, L45 (1998).
10. Sari, R., Piran, T. and Narayan, R., *ApJL* **497**, L17 (1998).
11. Wijers, R.A.M.J. and Galama, T.J., *ApJ* **523**, 177 (1999).
12. Costa, E., et al., *Nature* **387**, 783 (1997).
13. Van Paradijs, J., et al., *Nature* **386**, 686 (1997).
14. Wijers, R.A.M.J., Rees, M.J. and Mészáros, P., *MNRAS* **288**, L51 (1997).
15. Reichart, D.E., *ApJL* **485**, L57 (1997).
16. Waxman, E., *ApJL* **485**, L5 (1997).
17. Metzger, M.R., et al., *Nature* **387**, 879 (1997).
18. Frail, D.A., et al., *Nature* **389**, 261 (1997).

19. Frail, D.A., Waxman, E. and Kulkarni, S.R., *ApJ*, in press, astro-ph/9910319 (2000).
20. Fruchter, A., et al., *ApJL*, submitted, astro-ph/9902236 (1999).
21. Bloom, J.S., et al., *ApJL* **507**, L25 (1998).
22. Galama, T.J., et al., *ApJL* **497**, L13 (1998).
23. Castro-Tirado, A., et al., *Science* **279**, 1011 (1998).
24. Bremer, M., et al., *A&A* **332**, L13 (1998).
25. Piro, L., et al., *A&A* **331**, L41 (1998).
26. Galama, T.J., et al., *ApJL* **501**, L97 (1998).
27. Granot, J., Piran, T. and Sari, R., *ApJL* **527**, 236 (1999).
28. Akerlof, C., et al., *Nature* **398**, 400 (1999).
29. Galama, T.J., et al., *Nature* **398**, 394 (1999).
30. Kulkarni, S.R., et al., *Nature* **398**, 389 (1999).
31. Castro-Tirado, A., et al., *Science* **283**, 2069 (1999).
32. Fruchter, A.S., et al., *ApJL* **519**, L13 (1999).
33. Sari, R. and Piran, T., *ApJL* **517**, L109 (1999).
34. Briggs, M.S., et al., *ApJ* **524**, 82 (1999).
35. Sari, R. and Piran, T., *ApJL* **520**, 641 (1999).
36. Mészáros, P. and Rees, M. J., *MNRAS* **306**, L39 (1999).
37. Liang, E.P., et al., *ApJL* **519**, L21 (1999).
38. Galama, T.J., et al., *Nature* **395**, 670 (1998).
39. Kulkarni, S.R., et al., *Nature* **395**, 663 (1998).
40. Bloom, J.S., et al., *Nature* 401, 453 (1999).
41. Galama, T.J., et al., *Nature* **387**, 479 (1999).
42. Reichart, D.E., *ApJL* **521**, L111 (1999).
43. Galama, T.J., et al., *ApJ*, submitted, astro-ph/9907264 (1999).
44. Panaitescu, A., Mészáros, P., and M. J. Rees, M.J., *ApJ* **503**, 314 (1998).
45. Chevalier, R.A. and Li, Z., *ApJL* **520**, L29 (1999).
46. Chevalier, R.A. and Li, Z., *ApJ*, submitted, astro-ph/9908272 (1999).
47. Rhoads, J.E., *ApJ* **525**, 737 (1999).
48. Sari, R., Piran, T. and Halpern, J.P., *ApJL* **519**, L17 (1999).
49. Panaitescu, A. and Mészáros, P., *ApJ* **526**, 707 (1999).
50. Groot, P.J., et al., *ApJL* **502**, L123 (1998).
51. Halpern, J.P., *ApJL* **517**, L105 (1999).
52. Frail, D.A., et al. *ApJ*, submitted, astro-ph/9910060 (1999).
53. Harrison, F.A., et al., *ApJL* **523**, L121 (1999).
54. Stanek, K.Z., et al., *ApJL* **522**, L39 (1999).
55. Vreeswijk, P.M., et al., *ApJL* **523**, 171 (1999).
56. Vreeswijk, P.M., et al., *ApJL* **522**, L97 (1999).
57. Wang, X.Y., Dai, Z.G. and Lu, T., *MNRAS*, submitted, astro-ph/9906062 (2000).
58. Dai, Z.G. and Lu, T., *ApJL* **519**, L155 (1999).

Recent Optical/Near-IR Observations of GRBs

A. J. Castro-Tirado[1,2], J. Gorosabel[1], J. Greiner[3], S. Klose[4],
V. Mohan[5], R. Sagar[5], I. Bond[6], N. Rattenbury[6], Ph. Yock[6],
F. Vrba[7], A. Henden[7], C. Luginbuhl[7], A. Guarnicri[8],
M. R. Zapatero-Osorio[9], J. Zhu[10], R. Hudec[11], S. Guziy[12],
A. Shlyapnikov[12], E. Palazzi[13], N. Masetti[13], F. Frontera[14]
E. Costa[14], M. Feroci[14], and L. Piro[14]

[1] *Laboratorio de Astrofísica Espacial y Física Fundamental (LAEFF-INTA), P.O. Box 50727, E-28080, Madrid, Spain*
[2] *Instituto de Astrofísica de Andalucía (IAA-CSIC), P.O. Box 03004, E-18080 Granada, Spain*
[3] *Astrophysikalisches Institut, An der Sternwarte 16, D-14482, Potsdam, Germany*
[4] *Tautenburg Landessternwarte, Germany*
[5] *Uttar Pradesh Observatory, Nainital, India*
[6] *University of Auckland, New Zealand*
[7] *U.S. Naval Observatory, Flagstaff, AZ, USA*
[8] *Universita di Bolonia, Italia*
[9] *Instituto de Astrofísica de Canarias (IAC), La Laguna, Spain*
[10] *Beijing Astronomical Observatory, Beiging, China*
[11] *Astrophysics Space Group, Nikolaev Astronomical Observatory, Ukraine*
[12] *Astronomical Observatory, Ondrejov, Czech Republic*
[13] *ITESRE, CNR, Bologna, Italy*
[14] *Istituto di Astrofisica Spaziale (IAS), CNR, Italy*

Abstract. We present the results of optical/near-IR observations carried out in the period Dec 1998-Dec 1999 at Spanish, Indian and New Zealand observatories. Most of the searches proved to be unfruitful, in spite of the rapid follow-up observations, starting 3-24 hr after the events. This indicates that a significant fraction of the bursts occur in high density environments preventing optical and near-IR photons to be detected. In a couple of events, GRB 990123 and GRB 991208 the optical/near-IR afterglows were detected.

GRB 981226

Infrared images (J- and K'-band) were obtained starting 8.4 hr after the GRB and continuing until 3 Jan 1999 with the 3.5-m at the German- Spanish Calar Alto Observatory (CAHA). Optical (VRI-band) images were also obtained at CAHA and Tenerife on Dec 26-27. None of the objects initially reported as possible counterparts [1-3], seem to have significantly changed in brightness over a 9 month time interval. No IR/optical transient shows up at the position of the variable radio source detected at the VLA [4].

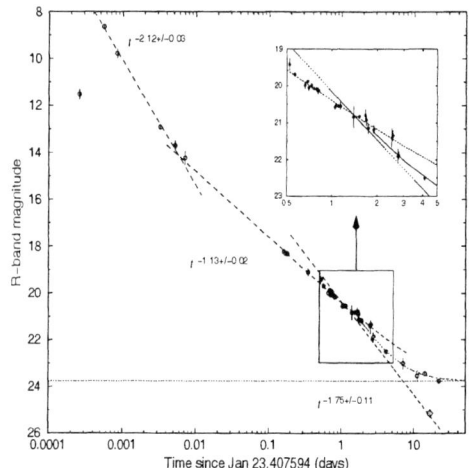

FIGURE 1. The R-band light-curve of the GRB 990123 optical transient. Based on our observations (filled circles) and other data reported elsewhere (empty circles). The doted line is the contribution of the underlying galaxy, with R ~ 23.77 ± 0.10. The three dashed lines are the contribution of the OA, following $F \propto t^\delta$ with $\delta = -2.12$ up to ~ 10 min, $\delta = -1.13$ up to ~ 1.5 d, and $\delta = -1.75$ after that time. The solid line, only drawn after 1.5 d for clarity is the total observed flux (OA + galaxy).

GRB 990123

Broad-band (ultraviolet to near-infrared) observations of the intense gamma-ray burst, GRB 990123, started ~8.5 hours after the event and continued until 18 Feb 1999. When combined with other data, in particular from the Robotic Optical Transient Search Experiment (ROTSE) and the Hubble Space Telescope (HST), evidence emerges for a smoothly declining light curve, suggesting some color dependence, that could be related to a cooling break passing the UV-optical band at about 1 day after the high-energy event. The steeper decline rate seen after 1.5-2 days (Figure 1) may be evidence for a collimated jet pointing towards the observer [5]. UBVRI magnitudes for the host galaxy imply a flat spectral distribution, i.e., a star-forming galaxy. Optical spectroscopic observations revealed that the redshift of the burst source is $1.6 \leq z < 2.05$. Assuming isotropic emission, a Hubble constant of $H_0 = 70$ km s^{-1} Mpc^{-1}, and a cosmological density parameter $\Omega_0 = 0.3$, implies an energy release in γ-rays of $\sim 2.5 \times 10^{54}$ erg, using the BeppoSAX GBM data [6]. This is larger by a factor of 10–100 than measured for any of the four other GRBs for which a redshift is known. Polarimetric observations only implied an upper limit of 1.8% [7].

GRB 990506

B and R-band images obtained at the 2.2-m CAHA on 6 May (starting 15 hr after the event) covering the entire IPN/RXTE error box. I-band images taken at the 0.8-m IAC telescope in Tenerife covering a $\sim$75% fraction of the IPN/RXTE error box. No clearly variable object is detected within the IPN annulus down to $R_c = 23$, based on comparison amongst USNO frames [8]. No optical transient is detected at the position of the variable radio source detected at the VLA [9], coincident with a faint R = 24.8 galaxy [10].

GRB 990520

Optical images (B- and R-band) obtained at the 1.23-m and 2.2-m CAHA on May 20 (starting 21 hr after the burst) and May 21. Infrared (H-band) images were obtained at the 3.5-m CAHA on May 20 and 21. No new sources seen down to the second epoch Digitized Sky Survey (DSS-2) limit (R $\sim$ 21), neither in the optical [11] nor in the near-IR and a similar result when comparing with USNO images [12].

GRB 990627

Images obtained through a wide B and R-band filters at the 0.6-m Mount John University Observatory Telescope (MJUO) on June 28, starting 26 hr after the trigger. No new sources were seen down to the DSS-2 limit.

GRB 990704

R- and B-band images obtained at the 2.2-m CAHA on July 4, starting 3.3 hr after the event and I-band images at the 0.8-m IAC telescope on the same night. No new sources were detected to a limiting magnitude of R = 18 within the BSAX-WFC error box [13].

GRB 990907

Images obtained through a wide R+I-band filter at the 0.6-m MJUO on Sep 9.40 (42 hr after the trigger). No new sources were seen down to the DSS-2 limit [14].

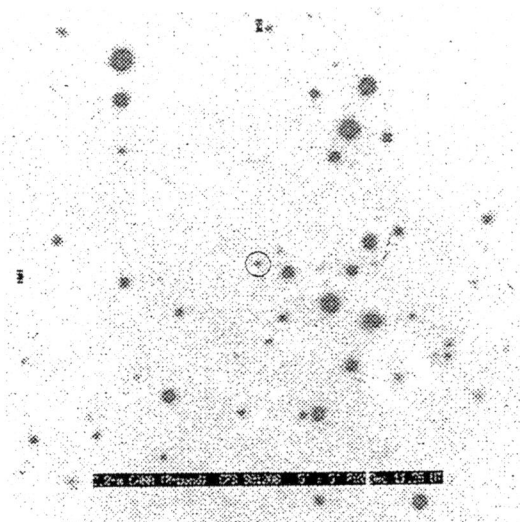

FIGURE 2. The optical counterpart (circle) of GRB 991208 on an R-band image taken at CAHA on Dec 11.27, 1999.

GRB 990908

Images obtained through a wide R+I-band filter at the 0.6-m MJUO on Sep 9.50 (34 hr after the trigger). No new sources were seen within the BSAX-WFC error box down to the DSS-2 limit.

GRB 990915

A significant amount of observing time was devoted to this GRB. 150 frames (60 s Exp. time, no filter) were taken at the 0.8-m Schmidt CAHA on Sep 15 (starting at 0.5 hr after the burst), covering $\sim$15% of the NEAR error box [15]. 25 additional frames (60 s Exp. time, R-band) were obtained at the 1.23-m CAHA (starting 30 hr after the burst) covering an additional $\sim$65% of the NEAR region, and another 10 frames (150 s, R-band) at the 1.23-m CAHA (starting 50 hr after the burst) covered the rest ($\sim$20%) of the NEAR error box. No sources varying more than 0.3 mag were seen to a limiting magnitude of R = 20.5 when comparing with the DSS-1 [16].

GRB 991014

R-band images obtained at the 1.04-m telescope at Uttar Pradesh Observatory (UPSO) on Oct 16.9 (starting 48 hr after the burst). No new sources were seen within the BSAX-NFC error box down to the DSS-2 limit.

GRB 991208

Imagery obtained with the 2.2-m CAHA were taken on Dec 11.25, 1999 (R-band filter). After a visual comparison with the Digital Sky Survey, a new source was clearly detected with R = 19.5 ± 0.1 (Dec 11.27; see Figure 2) [17]. This object coincides with the radio source reported in [18].

A BTA/MPSF 4500 s spectrum of the optical transient was obtained on Dec 14.14, 1999 with the Special Astrophysical Observatory (SAO-RAS) 6-m telescope. The data were obtained with the integral field spectrograph (MPFS). Emission lines at $\lambda = 6350$ Å, 8550 Å and 8470 Å are present, with the most likely identifications of these emission lines being: [OII] 3727 Å, [OIII] 4959,5007 Å at a redshift of $z = 0.707 \pm 0.002$, presumably that of the GRB host galaxy [19].

ACKNOWLEDGMENTS

We are grateful to many colleagues that we cannot include as co-authors in this manuscript because of the limited space, who helped us getting these observations. This work has been partially supported by a Spanish CICYT grant ESP95-0389-C02-02.

REFERENCES

1. Galama, T., et al., GCN Circ. 172 (1998).
2. Castro-Tirado, A. J., et al., GCN Circ. 173 (1998).
3. Wozniack, et al., GCN Circ. 177 (1998).
4. Frail, D., et al., GCN Circ. 195 (1999).
5. Castro-Tirado, A. J., et al., *Sci* **283**, 2079 (1999).
6. Andersen, M., et al., *Sci* **283**, 2075 (1999).
7. Hjorth, J., et al., *Sci* **283**, 2073 (1999).
8. Henden, A., et al., GCN Circ. 300 (1999).
9. Taylor, et al., GCN Circ. 350 (1999).
10. Bloom, J., et al., GCN Circ. 351 (1999).
11. Castro-Tirado, A. J., et al., GCN Circ. 336 (1999).
12. Luginbuhl, C., et al., GCN Circ. 341 (1999).
13. Castro-Tirado, A. J., et al., GCN Circ. 362 (1999).
14. Gorosabel, J., et al., GCN Circ. 415 (1999).
15. Hurley, K., et al., GCN Circ. 412 (1999).
16. Castro-Tirado, A. J., et al., GCN Circ. 416 (1999).
17. Castro-Tirado, A. J., et al., IAU Circ. 7332 (1999).
18. Frail, D., et al., GCN Circ. 451 (1999).
19. Dodonov, N., et al., GCN Circ. 475 (1999).

The Detection of Linear Polarization in the Afterglow of GRB 990510 and its Theoretical Implications

Davide Lazzati[*†], Stefano Covino[*], and Gabriele Ghisellini[*]

[*] *Osservatorio Astronomico di Brera, via E. Bianchi 46, I-23807 Merate (LC) Italy*
[†] *Dipartimento di Fisica, Università degli Studi di Mailano, via Celoria 16, I-20133 Milano, Italy*

Abstract. We present the recent discovery of linear polarization of the optical afterglow of GRB 990510. Effects that could introduce spurious polarization are discussed, showing that they do not apply to the case of GRB 990510, which is then intrinsically polarized. It will be shown that this observation constrains the emission mechanism of the afterglow radiation, the geometry of the fireball and degree of order of the magnetic field. We then present the theoretical interpretations of this observation with particular emphasis on the possibility of observing polarization in beamed fireballs.

INTRODUCTION

Polarization is one of the clearest signatures of synchrotron radiation, if this is produced by electrons gyrating in a magnetic field that is at least in part ordered. For this reason, polarization measurements can provide a crucial test of the synchrotron shock model [11], the leading scenario for the production of the burst and, in particular, the afterglow photons.

Attempts to measure the degree of linear polarization yelded only an upper limit ($\sim 2.3\%$ for GRB 990123 [6]), until the observations of the afterglow of the burst of May 10, 1999. A small but significant amount of polarization was detected ($1.7 \pm 0.2\%$ [2]) ~ 18 hours after the BATSE trigger and confirmed in a subsequent observation two hours later [13].

Even if synchrotron radiation can naturally account for the presence of linearly polarized light in a GRB afterglow, a significant degree of anisotropy in the magnetic field configuration or in the fireball geometry is required. If, in fact, the synchrotron emission is produced in a fully symmetrical set-up, all the polarization components average out, giving a net unpolarized flux. The presence of a partially ordered magnetic field (in causally disconnected domains) has been discussed by Gruzinov & Waxman [5], however their model over-predicts, in its simplest formulation, the observed amount of polarization. Here we discuss a different possibility, in which the asymmetry is provided by a collimated fireball observed off-axis, while the magnetic field is tangled in the plane perpendicular to the velocity vector of the fireball expansion. Indeed, the smooth break in the lightcurve of GRB 990510 [7] has been interpreted as due to a collimated fireball observed slightly off-axis.

GRB 990510 MEASUREMENTS

GRB 990510 was detected by BATSE on-board the Compton Gamma Ray Observatory and by the *Beppo*SAX Gamma Ray Burst Monitor and Wide Field Camera on 1999 May 10.36743 UT [8,3]. Its fluence (2.5×10^{-5} erg cm^{-2} above 20 keV) was relatively high [8]. Follow up optical observations started ~ 3.5 hr later and revealed an $R \simeq 17.5$ [1] optical transient (OT). The OT showed initially a fairly slow flux decay $F_\nu \propto t^{-0.85}$ [7], which gradually steepened; Vreeswijk et al. [12] detected Fe II and Mg II absorption lines in the optical spectrum of the afterglow. This provides a lower limit of $z = 1.619 \pm 0.002$ to the redshift, and a γ-ray energy of $> 10^{53}$ erg, in the case of isotropic emission.

We observed the OT associated with GRB 990510 ~ 18 hours after the gamma-ray trigger at the ESO VLT-Antu (UT1) in polarimetric mode, performing four 10 minute exposures in the R band at four angles (0°, 22.5°, 45° and 67.5°) of the retarder plate [2]. The average magnitude of the OT in the four exposures was $R \sim 19.1$. Relative photometry with respect to all the stars in the field was performed and each couple of simultaneous measurements at orthogonal angles was used to compute the points in Fig. 1 (left panel) (see [2] for details). The parameter $S(\phi)$ is related to the degree of linear polarization P and to the position angle of the electric field vector ϑ by:

$$S(\phi) = P \cos 2(\vartheta - \phi). \tag{1}$$

P and ϑ are evaluated by fitting a cosine curve to the observed values of $S(\phi)$. The derived linear polarization of the OT of GRB 990510 is $P = (1.7 \pm 0.2)\%$ (1σ error), at a position angle of $\vartheta = 101° \pm 3°$. Fig. 1 (left panel) shows the data points and the best fit $\cos \phi$ curve. The statistical significance of this measurement is very high. A potential problem is represented by a "spurious" polarization introduced by dust grains interposed along the line of sight, which may be preferentially aligned in one direction. The normalization of the OT measurements to the stars in the field already corrects for the average interstellar polarization of these stars, even if this does not necessarily account for all the effects of the galactic ISM along the line of sight to the OT (e.g., the ISM could be more distant than the stars, not inducing any polarization of their light). To check this possibility, we plot in Fig. 1 (right panel) the degree of polarization vs. the instrumental position angle for each star and for the OT. It is apparent that, while the position angle of all stars are consistent with being the same (within 10 degrees), the OT clearly stands out. The polarization position angle of stars close to the OT differs by $\sim 45°$ from the position angle of the OT. This is contrary to what one would expect if

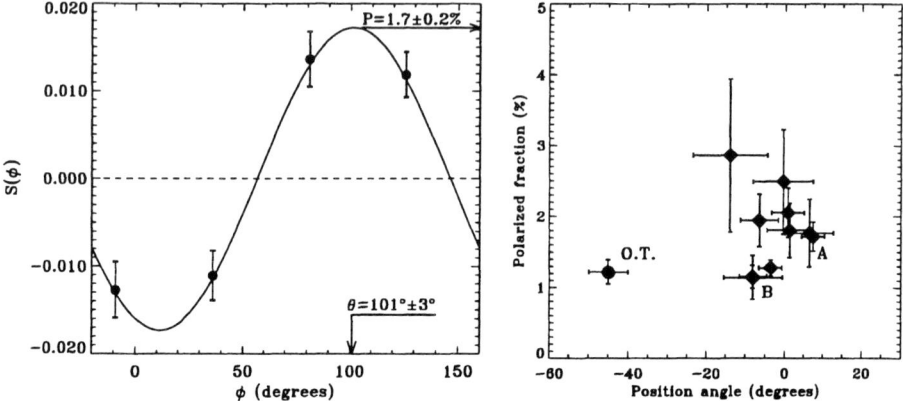

FIGURE 1. Left Panel: our polarization data taken at four different position angles ϕ are fitted with a cosine curve. The amplitude of this curve corresponds to the degree of linear polarization, and its maximum to the polarization position angle. Data are normalized to the average of the stars in the same field. **Right Panel**: The unnormalized degree of polarization vs. the instrumental polarization position angle of the stars in the field and the optical transient. The optical transient clearly stands out (P = $1.2 \pm 0.2\%$). The two stars closest to the OT are labelled A and B.

the polarization of the OT were due to the galactic ISM. Polarization induced by absorption in the host galaxy can be constrained to be $P_{host} < 0.2\%$, due to the lack of any absorption in the optical filters in addition to the local value (see [2] for more details). We therefore conclude that the OT, even if contaminated by interstellar polarization, must be intrinsically polarized to give the observed orientation.

POLARIZATION FROM BEAMED FIREBALLS

We consider a slab of magnetized plasma, in which the configuration of the magnetic field is completely tangled if the slab is observed face on, while it has some some degree of alignment if the slab is observed edge on. Such a field can be produced by compression in one direction of a volume of 3D tangled magnetic field [9] or by Weibel instability [10]. If the slab is observed edge-on, the radiation is therefore polarized at a level, P_0, which depends on the degree of order of the field in the plane. If the emitting slab moves in the direction normal to its plane with a bulk Lorentz factor Γ, we have to take into account the relativistic aberration of photons. This effect causes photons emitted at $\theta' = \pi/2$ in the (primed) comoving frame K' to be observed at $\theta \sim 1/\Gamma$ (see also [10]).

We assume that the fireball is collimated into a cone of semi-aperture angle θ_c, and that the line of sight makes an angle θ_o with the jet axis. As long as $\Gamma > 1/(\theta_c - \theta_o)$, the observer receives photons from a circle of semi-aperture angle

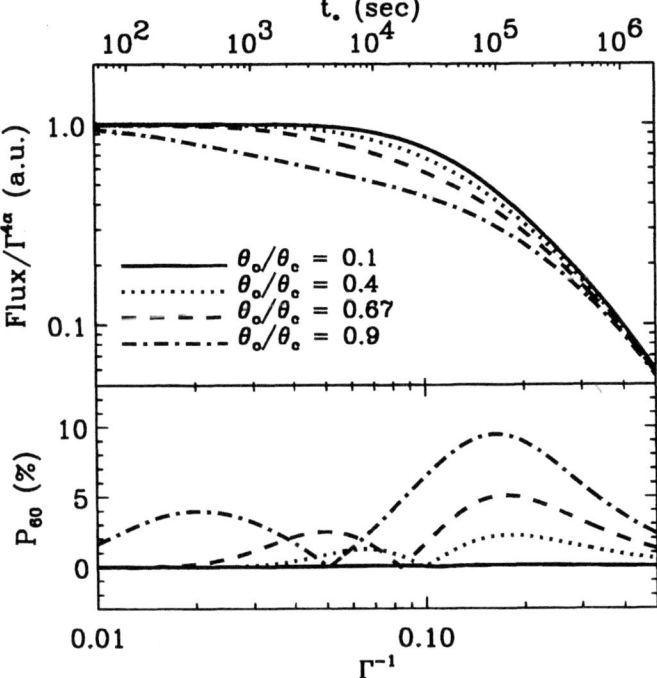

FIGURE 2. Lightcurves of the total flux (upper panel) and of the polarized fraction (bottom panel) for four different choices of the ratio θ_o/θ_c. The cone aperture angle $\theta_c = 5°$, while θ_o is the viewing angle. The higher the ratio θ_o/θ_c, the higher the polarized fraction due to the increase of the asymmetry of the geometrical setup. The actual value of the observed polarization depends linearly upon P_0. In this figure we assumed $P_0 = 60\%$. The lightcurve of the total flux assumes a constant spectral index $\alpha = 0.6$ for the emitted radiation. Note that the the highest polarization values are associated with total flux lightcurves steepening more gently. To calculate the values of the upper x-axis (t_*), we assumed $(t_*/t_0) = (\Gamma/\Gamma_0)^{-8/3}$ with $t_0 = 50$ s and $\Gamma_0 = 100$.

$1/\Gamma$ around θ_o. Consider the edge of this circle: radiation coming from each sector is highly polarized, with the electric field oscillating in radial direction (see [4] for more details). As long as we observe the entire circle, the configuration is symmetrical, making the total polarization vanish. However, if the observer does not see part of the circle, some net polarization survives in the observed radiation. This happens if a beamed fireball is observed off-axis when $1/(\theta_c + \theta_o) < \Gamma < 1/(\theta_c - \theta_o)$.

At the beginning of the afterglow, when Γ is large, the observer sees only a small fraction of the fireball and no polarization is observed. At later times, when Γ becomes smaller than $1/(\theta_c - \theta_o)$, the observer will see only part of the circle centered in θ_o: there is then an asymmetry, and a corresponding net polarized flux. To understand why the polarization angle in this configuration is horizontal, consider that the part of the circle which is not observed would have contributed to the polarization in the vertical direction. At later times, as the fireball slows down even more, a larger area becomes visible. When $\Gamma \sim 1/(\theta_c + \theta_o)$, the dominant contribution to the flux comes from the upper regions of the fireball which are vertically polarized. The change of the position angle happens when the contributions from horizontal and vertical polarization are equal, resulting in a vanishing net polarization. At still later times, when $\Gamma \to 1$, light aberration vanishes, the observed magnetic field is completely tangled and the polarization disappears.

Figure 2 shows the result of the numerical integration of the appropriate equations (see [4] for the detailed discussion). As derived in the above qualitative discussion, the lightcurve of the polarized fraction shows two maxima, with the position angle rotated by 90° between them. It is interesting to note the link with the lightcurve. The upper panel of Fig. 2 shows the lightcurve of the total flux divided by the same lightcurve in the assumption of spherical geometry. As expected, the lightcurve of the beamed fireball shows a break with respect to the spherical one. A larger off-axis ratio produce a more gentle break in the lightcurve, and is associated with a larger value of the polarized fraction. The behavior of the total flux and of the polarization lightcurves allow us to constrain the off-axis ratio ϑ_o/ϑ_c, but is insensitive to the absolute value of the beaming angle ϑ_c. Therefore, even if we could densely sample the polarization lightcurve, the beaming angle could be derived only assuming a density for the interstellar medium, i.e., a relation between the observed time and the braking law of the fireball. On the other hand, the detection of a 90° rotation of the polarization angle of the afterglow would be the clearest sign of beaming of the fireball, especially if associated with a smooth break in the lightcurve. Polarimetric follow-up of afterglows is hence a powerful tool to investigate the geometry of fireballs.

REFERENCES

1. Axelrod T., Mould J., and Schmidt B., GCN 315 (1999).
2. Covino S. et al., *A&A* **348**, L1 (1999).
3. Dadina M. et al., IAUC 7160 (1999).
4. Ghisellini G., and Lazzati D., *MNRAS* **309**, L7 (1999).
5. Gruzinov A., and Waxman E., *ApJ* **511**, 852 (1999).
6. Hjorth J., et al., *Science* **283**, 2037 (1999).
7. Israel G. L. et al., *A&A* **348**, L5 (1999).
8. Kippen R. M., GCN 322 (1999).
9. Laing R. A., *MNRAS* **193**, 439 (1980).
10. Medvedev M. V, and Loeb A., *ApJ* submitted (astro-ph/9904363).
11. Meszàros P., and Rees M. J., *ApJ* **476**, 232 (1997).
12. Vreeswijk P. M., et al., GCN 324 (1999).
13. Wijers R. A. M. J. et al., *ApJ* **523**, L33 (1999).

Polarimetric Studies of Gamma-Ray Burst Afterglows

S. Klose[1], B. Stecklum[1], and O. Fischer[2]

[1] *Thüringer Landessternwarte Tautenburg, D-07778 Tautenburg, Germany*
[2] *Universitäts-Sternwarte Jena, D-07745 Jena, Germany*

Abstract. We report on a target of opportunity program to study GRB afterglows based on observations with the Calar Alto 3.5-m telescope equipped with the near-infrared camera Omega Cass.

INTRODUCTION

Early in 1999 we proposed to perform polarimetric, rapid follow-up observations of GRB afterglows using the Calar Alto 3.5-m telescope equipped with the near-infrared camera Omega Cass. The goal of the project is to measure the degree of linear polarization of afterglows in the K' band. The importance of polarimetric observations has been discussed by various authors (Ghisellini & Lazzati 1999; Gruzinov 1999; Gruzinov & Waxman 1999; Loeb & Perna 1998; Medvedev & Loeb 1999; Sari 1999) and first observations have already been reported (Covino et al. 1999; Hjorth et al. 1999; Rol et al. 2000; Wijers et al. 1999).

Our project was approved as a target of opportunity (ToO) program and the schedule started in May 1999. In its first phase it will last for one year.

TECHNICAL DATA

The near-infrared camera Omega Cass is equipped with a 1024×1024 HAWAII array (see http://www.caha.es/CAHA/Instruments/IRCAM/OCASS/index.html). It can make use of an adaptive optics system which provides diffraction limited resolution in combination with a laser guide star. Given the accuracy of GRB error boxes, we decided to start the polarimetric observations in the wide-field mode (0.3″/pixel), resulting in a field of view of about $5' \times 5'$. This mode can be used to image *BeppoSAX* error boxes when an optical counterpart of a GRB afterglow is not yet known at the time of the near-infrared observations. For imaging small *HETE-2* error boxes, as well as in cases where an optical afterglow is already detected, the pixel scale we will use is 0.12″/pixel and the field of view will be about $2' \times 2'$.

Polarimetric observations are performed using wire-grid analyzers which provide individual images of light, linearly polarized corresponding to the wire-grid orientations. Four images have to be taken through the four offered analyzers which are mounted at position angles of 0, 45, 90, and 135 deg. Data reduction will be performed in a standard manner (cf., Fischer, Stecklum, & Leinert 1998).

OBSERVING CONSTRAINTS

During the summer semester (May - November 1999) Omega Cass was mounted at the 3.5-m telescope for about 50 nights, during the winter semester (November 1999 - May 2000) it will be mounted for about 40 nights. The number of requests for ToO observations is restricted to three bursts per semester. Based on 1 hour integration time we can perform polarimetric measurements of a GRB afterglow, if it is not fainter than about $K' = 18$.

FIRST OBSERVATIONS

GRB follow-up observations have been performed so far from the bursts 990704, 991014, and 991106. During the first observing run (990704) a problem with the instrument occurred and the frames obtained did not have sufficient signal-to-noise ratio for a further data analysis. The other two runs were successful, but the bursts occurred at low Galactic latitudes and no optical afterglow was found in spite of world-wide observations. At the time of this writing, our K'-band frames of the GRB 991014 follow-up observations are not yet reduced. In the case of the burst 991106, a comparison of our reduced K'-band data with deep optical images of the GRB error box did not reveal the GRB afterglow (Stecklum et al. 1999; see also Gorosabel et al. 1999). Unfortunately, the nature of GRB 991106 itself is under question (Gandolfi et al. 1999).

ACKNOWLEDGMENTS

We thank Ulrich Thiele and Markus Feldt, Max-Planck-Institut für Astronomie, Heidelberg, for their help during the early phase of the project and for performing first observations.

REFERENCES

1. Covino, S. et al., *A&A* **348**, L1 (1999).
2. Fischer, O., Stecklum, B., and Leinert, Ch., *A&A* **334**, 969 (1998).
3. Gandolfi, G., et al., *GCN* 448 (1999).

4. Ghisellini, G., and Lazzati, D., *MNRAS* **309,** L7 (1999).
5. Gorosabel, J., et al., *GCN* 447 (1999).
6. Gruzinov, A., *ApJ* **525,** L29 (1999).
7. Gruzinov, A., and Waxman, E., *ApJ* **511,** 852 (1999).
8. Hjorth, J., et al., *Science* **283,** 2073 (1999).
9. Loeb, A., and Perna, R., *ApJ* **495,** 597 (1998).
10. Medvedev, M. V., and Loeb, A., preprint astro-ph/9904363 (1999).
11. Rol, E., these proceedings (2000).
12. Sari, R., *ApJ* **524,** L43 (1999).
13. Stecklum, B., et al., *GCN* 446 (1999).
14. Wijers, R. A. M. J., et al., *ApJ* **523,** L33 (1999).

Submillimeter Observations of GRB Counterparts[1]

I. A. Smith[*], J. van Paradijs[†,‡], R. P. J. Tilanus[||], T. J. Galama[†],
P. J. Groot[†], P. Vreeswijk[†], E. Rol[†], C. Kouveliotou[¶],
R. A. M. J. Wijers[§], and N. Tanvir[**]

[*]*Department of Space Physics and Astronomy, Rice University, MS-108, 6100 South Main, Houston, TX 77005-1892 USA*
[†]*Astronomical Institute 'Anton Pannekoek', University of Amsterdam and Center for High-Energy Astrophysics, Kruislaan 403, 1098 SJ Amsterdam, The Netherlands*
[‡]*Department of Physics, University of Alabama in Huntsville, Huntsville, AL 35899 USA*
[||]*Joint Astronomy Centre, 660 N. Aohoku Place, Hilo, HI 96720 USA*
[¶]*Universities Space Research Association, NASA Marshall Space Flight Center, ES-62, Huntsville, AL 35812 USA*
[§]*Department of Physics and Astronomy, SUNY, Stony Brook, NY 11794-3800 USA*
[**]*Department of Physical Sciences, University of Hertfordshire, College Lane, Hatfield, Herts AL10 9AB, UK*

Abstract. We briefly summarize the results from our ongoing program of submillimeter observations of GRB counterparts using the SCUBA instrument on the James Clerk Maxwell Telescope (JCMT). Full details have already been published elsewhere [1,2].

INTRODUCTION

The sub-millimeter is an important band for counterpart observations because it is where the emission peaks in some bursts in the days to weeks following the burst. Our previous SCUBA observations have shown that they are essential to: 1) determine the breaks in the radio to sub-millimeter to optical spectrum to compare with the shock models, 2) determine the evolution of the sub-millimeter flux, and 3) look for underlying quiescent sources that may be dusty star-forming galaxies at high redshifts. Even when we do not detect the counterpart with SCUBA, our early observations can be combined with the optical and radio fluxes to determine the location of the breaks in the spectrum.

---

[1] This work is dedicated to the memory of Jan van Paradijs.

1997 – 1999 RESULTS

Over the past five semesters we have been performing Target of Opportunity observations of GRB counterparts using SCUBA. So far, we have observed ten bursts. The detailed SCUBA results for the first eight (GRB 970508, 971214, 980326, 980329, 980519, 980703, 981220, and 981226) are described in Smith et al. [1], while GRB 990123 is discussed in Galama et al. [2]. Observations of GRB 990520 were also made in mediocre weather.

Our most interesting result to date is the detection of a fading counterpart to GRB 980329 at 850 μm. While a fading X-ray counterpart [3] and a variable radio source VLA J070238.0+385044 [4] were soon found for this burst, it proved to be difficult to find the infrared counterpart [5]: this indicated that the optical extinction was significant for this source [6,4], and/or the redshift was large. Our sub-millimeter detection shows that this counterpart was similar to the well-studied GRB 970508.

Figure 2 of Smith et al. [1] plots the evolution of the 850 μm SCUBA flux. For a power law decay with the flux density $\propto t^{-m}$ where t is the time since the burst, the best fit power law index is $m = 3.0$. However, m is not tightly constrained: the 90% confidence interval is $m = 1.2$ to $m = 5.3$.

Figure 3 of Smith et al. [1] adds the SCUBA results to the VLA-OVRO results presented in Figure 2 of Taylor et al. [4]. Taylor et al. found that a power law $S_\nu \propto \nu^\alpha$ with $\alpha = +0.9$ gave the best fit to the VLA-OVRO data alone. We found that this extends very well to the SCUBA results. The popular $\nu^{1/3}$ power law [7,8] attenuated by a synchrotron self-absorption component does not seem to give a good description of the shorter wavelength emission for this counterpart. However, the reduced $\chi^2_\nu = 2.6$ for this fit, and the probability that a random set of data points would give a value of χ^2_ν as large or larger than this is $Q = 0.034$. It is therefore not possible to exclude this model, and it will be important to study more bursts to determine the range and evolution of the longer wavelength spectral indices. For example, an $\alpha \sim +1$ VLA-SCUBA power law spectrum is definitely ruled out for GRB 980703, and possibly also for GRB 980519 [1].

We remark that one interesting possibility is that part of the SCUBA flux for GRB 980329 comes from an underlying quiescent sub-millimeter source, such as the dusty star-forming galaxies at high redshifts seen by SCUBA. More sensitive sub-millimeter instruments will be able to determine whether such a quiescent source is present.

ONGOING PROGRAM

Despite a lack of appropriately located bursts in 1999, our program of Target of Opportunity observations using SCUBA is ongoing.

The current efforts to upgrade the sensitivity of SCUBA and the telescope will hopefully allow us to detect more afterglows and monitor their evolution in better detail.

The upcoming launch of *HETE-II* promises to give us more well-localized bursts to study, as well as permitting us to search for the sub-millimeter afterglow emission from bursts whose duration is less than one second.

ACKNOWLEDGMENTS

The James Clerk Maxwell Telescope is operated by The Joint Astronomy Centre on behalf of the Particle Physics and Astronomy Research Council of the United Kingdom, the Netherlands Organization for Scientific Research, and the National Research Council of Canada.

REFERENCES

1. Smith, I. A., et al., *A&A* **347**, 92 (1999).
2. Galama, T. J., et al., *Nature* **398**, 394 (1999).
3. in't Zand, J., et al., *ApJ* **505**, L119 (1998).
4. Taylor, G. B., Frail, D. A., Kulkarni, S. R., Shepherd, D. S., Feroci, M., & Frontera, F., *ApJ* **502**, L115 (1998).
5. Klose, S., Meusinger, H., & Lehmann, H., IAU Circ. 6864 (1998).
6. Larkin, J., et al., GCN 44 (1998).
7. Katz, J. I., *ApJ* **432**, L107 (1994).
8. Waxman, E., *ApJ* **489**, L33 (1997).

Color Indices of Optical Afterglows of GRBs

V. Šimon*, G. Pizzichini[†], and R. Hudec*

*Astronomical Institute, 251 65 Ondřejov, Czech Republic
[†]Istituto di Tecnologie e Studio delle Radiazioni Extraterrestri (CNR), via Gobetti 101, 40129 Bologna, Italy

Abstract. We present results of the study of the comprehensive properties of optical afterglows of gamma-ray bursts (GRBs) using the color indices determined from published multicolor photometry. Suitable data for nine afterglows are available. We found that the color variations, detectable for the well-covered events (usually reddening on late decline), are generally small enough to allow for a comparison of the respective afterglows (especially for $t - T_0 < 10$ days). Most afterglows concentrate at $V - R = 0.4$, $R - I = 0.6$, $B - V = 0.5$.

INTRODUCTION

The color indices of the optical afterglows (OAs) of gamma-ray bursts (GRB) as well as their time evolution can serve as an important parameter reflecting the related physical processes and perhaps even to resolve between the respective models. A comparative study of the colors of the afterglows determines the comprehensive properties of these events and also offers a clue to resolve whether the optical event is related to afterglow, even in those cases when only the optical observations are available without gamma-ray detection (yet hypothetical). This is an innovative approach with quite new ideas and results.

COLLECTION AND ANALYSIS OF THE DATA

The search in the literature and on the Internet archives enabled to collect suitable optical data for nine afterglows. Unfortunately, the published photometry usually comprises unorganized observations in various passbands, in many cases the data in the respective filters come from different nights. Only in a few cases were simultaneous data over UV, visible and IR region obtained. This fact seriously complicates reconstruction of the light curves in the respective passbands and the analysis of the color variations over the whole event.

TABLE 1. Summary of the literature used for the respective afterglows.

GRB970228	Guarnieri et al. (1997), IAU Circ.6618, van Paradijs et al. (1997), Castander and Lamb (1998), Pedichini et al. (1997), IAU Circ.6588, IAU Circ.6631, IAU Circ.6619, Fruchter et al. (1999), IAU Circ.6619
GRB970508	Galama et al. (1998), Zhakirov et al. (1998), Sokolov et al. (1998)
GRB971214	Halpern et al. (1998), Diercks et al. (1998), Kulkarni et al. (1998), GCN 61, IAU Circ. 6793, Gorosabel et al. (1998), Ramaprakash et al. (1998)
GRB980519	Halpern et al. (1999)
GRB980613	GCN 109, 117, 118, 134
GRB980703	Bloom et al. (1998)
GRB990123	Galama et al. (1999)
GRB990510	GCN 316, 318, 319, 321, 323, 325, 328, 329, 330, 331, 332
GRB990712	GCN 389, 391, 402, 403, IAU Circ. 7225

The color indices were determined from plots of the light curves in the respective passbands. It emerged that in most cases the light curves are free of features on the time scale of hours to a few days. Interpolation between the neighboring measurements was therefore possible in many cases — at least one color index was therefore obtained for 9 afterglows. The light curves of two best-covered events (GRB970508 and GRB990510) are displayed in Figure 1a and 1c. The error bars, quoted from the original sources, show that the uncertainty of the respective measurements is rather small (a few times 0.01 mag) on early decline and of the order of 0.1–0.2 mag on the final decline. Figure 2abc shows the color indices $R - I$, $V - R$, $B - V$ of the respective afterglows, plotted as a function of the time interval $t - T_0$ elapsed from the GRB. The scale of the axes is identical for all three plots. Observations of most events lie within 10 days from the corresponding GRB. It can be seen that the color indices of the respective events are scattered even for a very similar $t - T_0$. Evolution of the color indices over the afterglow decline could be followed for GRB970508 (Figure 1b) and GRB990510 (Figure 1d).

GRB970508 (Figure 1ab): *a)* rise to Max. brightness — reddening in $V - R$, $R - I$, $V - I$ (no data for $B - V$); *b)* early decline to 0.8 mag(V) below max. brightness – slight reddening continues in $B - V$, $V - R$, $R - I$, $V - I$; *c)* further decline to 1.5 mag(V) below Max. brightness — tendency for a slight decrease of the color indices; *d)* decline to 2.2 mag(V) below max. brightness — strong reddening in $B - V$, strong decrease of $V - R$, $V - I$, $R - I$; *e)* final decline to 5 mag(V) below Max. brightness — strong reddening in $V - I$, $R - I$; no reddening in $B - V$, $V - R$.

GRB990510 (Figure 1cd): strong reddening of $V - I$ through the decline, defined by OGLE data.

Color-Color Diagrams

Most afterglows concentrate at $V - R = 0.4$, $R - I = 0.6$, $B - V = 0.5$ (Figure 2abc). The color variations are detectable for the well-covered events (usually

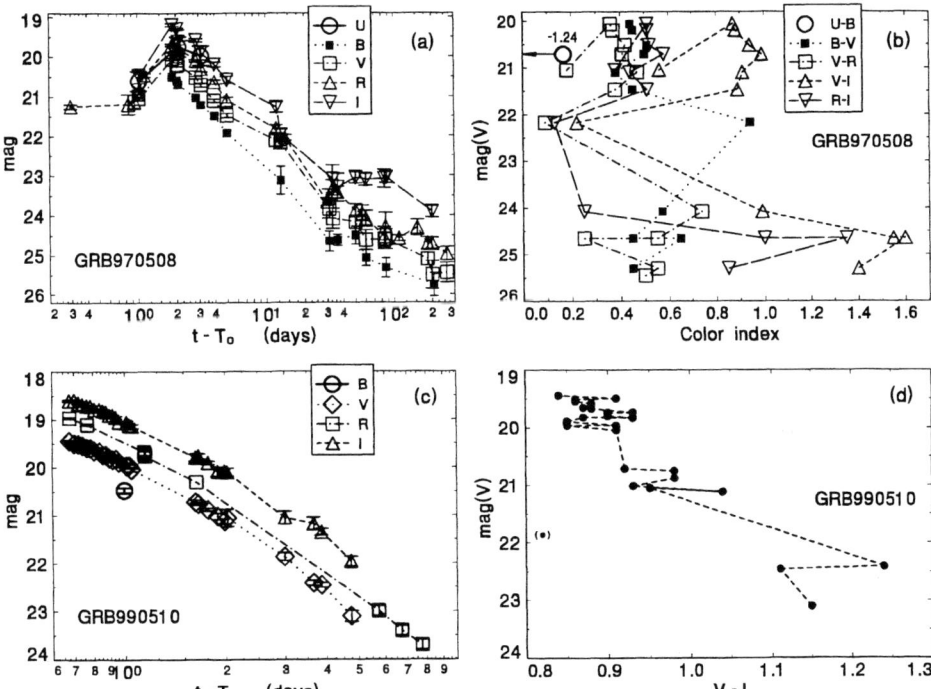

FIGURE 1. Light curves and color variations of the two best-covered afterglows (GRB970508 and GRB990510). The respective points are connected by lines to guide the eye. See text for details.

reddening on late decline). Fortunately, these changes are generally sufficiently small to allow for a comparison of the respective afterglows even when the indices represent different values of $t - T_0$, provided that $t - T_0$ stays within about 10–30 days. The respective color-color diagrams (Figure 2def) have identical scales of the axes to allow a direct comparison of the scatter in the respective plots.

$V - R, R - I$ diagram (Figure 2d): The afterglow of GRB970508 displays a prominent track over the late rise, the maximum and the decline. Notice that most other afterglows lie within this track. However, the position of GRB970228 is strikingly different from all remaining ones. Positions of some main-sequence stars are included — it can be seen that the colors of most afterglows accumulate near mid-G or early K types.

$B - V, V - R$ diagram (Figure 2e): The track of the afterglow of GRB970508 is clearly visible again. All afterglows occupy just a small region of the diagram. We note, however, that there are no data for GRB970228.

$B - V, V - I$ diagram (Figure 2f): The position of GRB970228 is more consistent

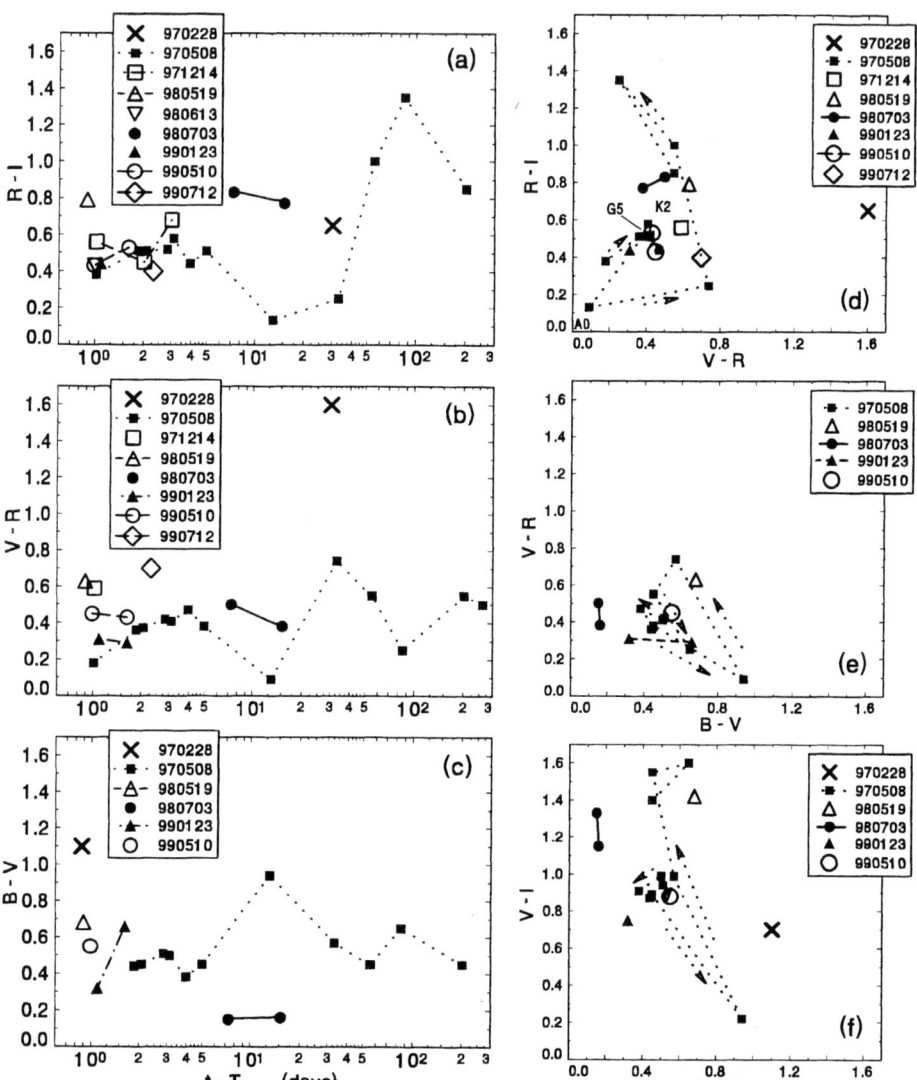

FIGURE 2. (abc) Color indices of the respective afterglows as a function of the time interval $t - T_0$ elapsed from the GRB. (def) Color-color diagrams for the afterglows. The scales of the axes of all three plots are identical to allow for a direct comparison of the scatter in the respective diagrams. The arrows denote the course of the track of the afterglow of GRB970508. See text.

with the remaining afterglows now.

$U - B$ index: Only two afterglows (GRB970508 and GRB990123) allow us to determine the $U - B$ index. This color diagram therefore is not meaningful at present. Nevertheless, the available data suggest that $U - B$ in both events is very negative (-1.2 and -0.9, respectively).

CONCLUSIONS

It is obvious that the color evolution of OAs exists and that even complicated changes are possible during their declines. When available, the afterglows appear quite red in the spectral region of wavelengths longer than corresponds to B-filter and, at the same time, very blue in $U - B$. Most afterglows concentrate in a relatively well localized region of the color diagrams. The color indices $V - R = 0.4$, $R - I = 0.6$, $B - V = 0.5$ appear to be typical. This information can serve to distinguish genuine OAs related to GRBs from another types of objects (background) in optical GRB searches.

The evaluation of the observational data available so far for OAs of GRBs has indicated that there is a need for a suitable strategy. In the optical observations of future OAs of GRBs, we propose to focus on simultaneous multicolor photometry allowing not only the light profiles in particular colors, but also the color indices and their time evolution to be studied.

ACKNOWLEDGMENTS

We acknowledge the support provided by the Grant Agency of the Czech Republic, Grant 205/99/0145 as well as by the CNR/AS CR project.

REFERENCES

1. Bloom J.S., et al., *ApJ*, submitted, astro-ph/9808319 (1998).
2. Castander F.J., and Lamb D.Q., *ApJ*, submitted, astro-ph/9807195 (1998).
3. Diercks A.H., et al., *ApJ* **503**, L105 (1998).
4. Fruchter A.S., et al., *ApJ* **516**, 683 (1999).
5. Galama T.J., et al., *ApJ* **497**, L13 (1998).
6. Galama T.J., et al., astro-ph/9903021 (1999).
7. Guarnieri A., et al., *A&A* **328**, L13 (1997).
8. Halpern J.P., et al., *Nature* **393**, 41 (1998).
9. Halpern J.P., et al., *ApJ*, submitted, astro-ph/9903418 (1999).
10. Kulkarni S. R., et al., *Nature* **393**, 35 (1998).
11. Pedichini F., et al., *A&A* **327**, L36 (1997).
12. Sokolov V.V., et al., *A&A* **334**, 117 (1998).
13. van Paradijs J., et al., *Nature* **386**, 686 (1997).
14. Zhakirov S.V., et al., *A&A* **337**, 356 (1998).

Photometric Study of the Improved GRB 970815 Error Box

J. Gorosabel[1], A. J. Castro-Tirado[1,2], N. Benítez[3],
M. R. Zapatero-Osorio[4], A. Campos[5], J. Trapero[1,6],
E. Sánchez[2], and N. Metcalfe[7]

[1] *LAEFF-INTA, P.O. Box 50727, 28080 Madrid, Spain.*
[2] *Instituto de Astrofísica de Andalucía (IAA-CSIC),
P.O. Box 03004, E-18080 Granada, Spain.*
[3] *Department of Astronomy, University of California,
Berkeley, 601 Campbell Hall, Berkeley, CA 94720, USA.*
[4] *Instituto de Astrofísica de Canarias, Vía Láctea s/n,
38200 La Laguna, Tenerife, Spain.*
[5] *Instituto de Matemáticas y Física Fundamental, Consejo Superior
de Investigaciones Científicas, Serrano 113 bis, E-28006 Madrid, Spain.*
[6] *Universidad SEK, Campus de Santa Cruz La Real,
c/ Cardenal Zúñiga 12, 4003, Segovia, Spain.*
[7] *Department of Physics, University of Durham,
South Road, Durham DH1 LE, UK.*

Abstract. An improved error box of GRB 970815 has been reported very recently (Smith et al. 1999). This error box is consistent with the decaying X-ray source detected by *ROSAT*, which strongly supports the source as the GRB X-ray afterglow. We report a re-analysis of the follow up observations performed in 1997 containing the improved *RXTE* error box. No variable optical/IR source has been found neither within the entire *RXTE* error box nor within the *ROSAT* and *ASCA* X-ray circles. Upper limits for any transient object are $K' \geq 18$ (Aug 17 1997), B $\geq$ 21.5, R $\geq$ 22 (20 Aug 1997), B $\geq$ 22.5 and R $\geq$ 23 (23 Aug 1997). Also we report additional BVRI optical observations performed in 1999 of the *ROSAT* X-ray source aimed to detect the host galaxy of GRB 970815.

INTRODUCTION

GRB 970815 was detected on 15 Aug 1997, UT 12 h 07 min 4 s by BATSE on *CGRO* and ASM on *RXTE* [1]. The burst lasted about 130 s and showed a double-peaked structure, reaching a maximum intensity of almost 2 Crab (2-12 keV) in the second peak. Its position was refined to a 14.6′ × 2.9′ error box (see Fig. 1, the solid line diamond-shape error box). Following the ASM/*RXTE* detection, the *ASCA* satellite performed an observation on 18.68–18.9 Aug, revealing no sources above

10^{-13} erg cm^{-2} s^{-1} (2-10 keV) within the ASM/$RXTE$ error box. However, a nearby X-ray source was seen slightly outside the ASM/$RXTE$ error box [2]. Although some candidates were proposed, optical/IR and radio follow up attempts in order to detect a counterpart were unfruitful [3-7]. A deep HRI/$ROSAT$ observation performed on 20-22 Aug revealed ten X-ray sources outside the ASM/$RXTE$ error box. One of them fell just on the border of the $RXTE$ error box. The flux level of this source was consistent with decay, suggesting that it may be the afterglow of GRB 970815 [8].

Very recently the $RXTE$ team reported a revised error box for GRB 970815 [9], which includes the decaying $ROSAT$ X-ray Source. Once we knew this position, new observations of the content of the error box were planned. Such observations were performed in 1999 with the 2.2-m telescope of Calar-Alto (CAHA). Thus, in this paper we show the results of the observations performed from 1997 to 1999 of the improved $RXTE$ error box and the X-ray sources detected by $ASCA$ and $ROSAT$ satellites.

OBSERVATIONS

The observations are divided in two runs, the first one took place in 1997 and the second one in 1999. Concerning our 1997 observations, optical and IR images of the $RXTE$ box [1] were obtained beginning 56 hr after the burst. The K'-band frames were obtained at the 3.5-m CAHA telescope equipped with OMEGA on 17 Aug. B & R-band frames were taken at the (OSN[1] 1.5-m telescope), CAHA (2.2-m + CAFOS) on 20-21 Aug and La Palma (4.2-m WHT) on 22 Aug. The large field of view of CAFOS allowed us to cover the overall GRB 970815 error box. In 1999, once a new error box was reported, new observations were carried out. This second run is based on BVRI photometry performed with the 2.2-m CAHA telescope on 2 and 3 March 1999.

RESULTS AND CONCLUSION

The main result is that no variable IR/optical source was found neither within the entire $RXTE$ error box nor within the $ROSAT$ and $ASCA$ X-ray error circles [8]. Upper limits for any transient object are K' $\geq$ 18 (17 Aug 1997), B $\geq$ 21.5, R $\geq$ 22 (20 Aug 1997), B $\geq$ 22.5, R $\geq$ 23 (23 Aug 1997) [7]. The optical observations performed in 1999 did not reveal any object inside the $ROSAT$ X-ray circle, giving the following conservative upper limits for the host galaxy: B $\geq$ 23.0, V $\geq$ 23.3, R $\geq$ 23.9 and I $\geq$ 22.8.

---

[1] The telescope of the Observatorio de Sierra Nevada, owned by the Spanish Instituto de Astrofísica de Andalucía in Granada.

TABLE 1. Journal of the GRB 970815 optical/infrared observations.

Date	Time (UT)	Telescope	Field	Filter	Integration time (ks)
20 Aug 1997	00:41	1.5 OSN	4	B	1.2
20 Aug 1997	01:00	1.5 OSN	4	R	3.6
20 Aug 1997	01:12	2.2 CAHA	16	R	1.5
20 Aug 1997	02:24	2.2 CAHA	16	B	0.9
20 Aug 1997	22:08	2.2 CAHA	16	B	1.2
21 Aug 1997	04:05	2.2 CAHA	16	R	0.6
22 Aug 1997	21:00	4.2 WHT	4	B	0.6
22 Aug 1997	21:14	4.2 WHT	4	B	0.3
23 Aug 1997	04:00	4.2 WHT	4	B	0.6
23 Aug 1997	04:14	4.2 WHT	4	R	0.3
2 Mar 1999	19:51	2.2 CAHA	16	B	2.4
2 Mar 1999	20:53	2.2 CAHA	16	R	4.8
2 Mar 1999	23:00	2.2 CAHA	16	V	3.0
3 Mar 1999	00:11	2.2 CAHA	16	I	3.0
17 Aug 1997	20:38	3.5 CAHA	6	K'	0.5
17 Aug 1997	21:49	3.5 CAHA	6	K'	0.5

ACKNOWLEDGMENTS

This work has been partially supported by Spanish CICYT grant ESP95-0389-C02-02.

REFERENCES

1. Smith, D. A., Levine, A. M., Morgan, E. H., and Wood, A., *IAU Circ.* 6718 (1997).
2. Murakami, T., Ueda, Y., Ishida, M., Fujimoto, R., Yoshida, A., and Kawai, N., *IAU Circ.* 6722 (1997).
3. Harrison, T. E., Mcnamara, B. J., Johnson, J. J., Mason, P. A., Balam, D. D., Klemola, A. R., Stanek, K. Z., Sasselov, D. D., Garcia, M. R., and Robinson, C. R., *IAU Circ.* 6721 (1997).
4. Groot, P. J., Galama, T. J., Hurley, K., and Kouveliotou, C., *IAU Circ.* 6723 (1997).
5. Kulkarni, S. R., Odewahn, S. C., Djorgovski, S. G., Frail, D. A., Goodrich, R. F., Maxfield, L. M., and Groot, P. J., *IAU Circ.* 6723 (1997).
6. Adams, M. T., Ward, M., Ma, F., Howell, A., Wang, L., and Wheeler, J. C., *IAU Circ.* 6725 (1997).
7. Castro-Tirado, A. J., Gorosabel, J., Greiner, J., Campos, A., Metcalfe, N., Sanchez, E., Hagen, H. J., and Trapero, J., *IAU Circ.* 6744 (1997).
8. Greiner, J. *IAU Circ.* 6742 (1997).
9. Smith, D. A., Levine, A. M., Bradt, H. V., Remillard, R., Jernigan, J., Hurley, K., Wen, L., Briggs, M., Cline, T., Mazets, E., Golenetskii, S., and Frederics, D., *ApJ in press, (astro-ph/9907332)* (1999).

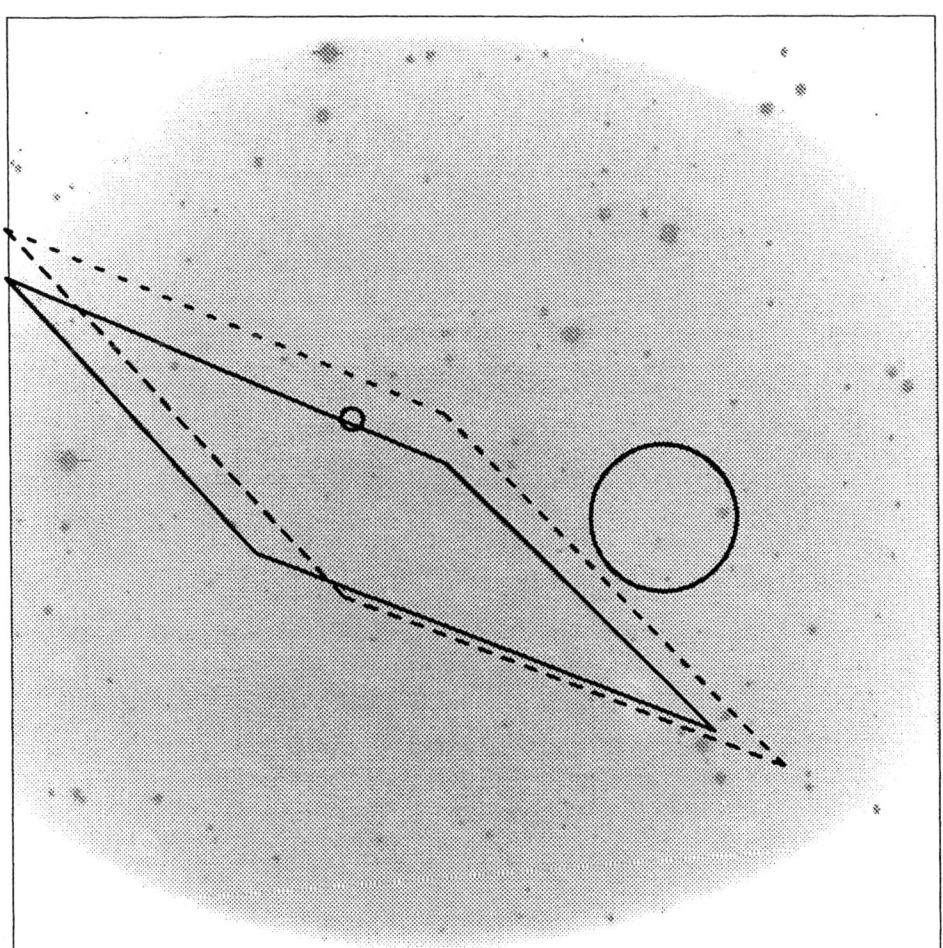

FIGURE 1. An R-band image of the GRB 970815 error box taken by the 2.2-m telescope (+CAFOS) of CAHA on 21.17 Aug 1997. The large field of view provided by CAFOS (diameter of 16′) allowed us to image the previous (solid line, [1]) and the improved (dashed-line, [9]) diamond-shape *RXTE* error boxes. The large circle represents the *ASCA* X-ray source error box (radius=1′) and the small one is the *ROSAT* X-ray circle (radius=10″). As can be seen the *ROSAT* X-ray source is inside the improved *RXTE* error box. A blow-up of the *ROSAT* X-ray error circle can be seen in Fig. 2. North is up and east to the left.

FIGURE 2. A 1500 s R-band image of the *ROSAT* X-ray error box taken by the 2.2-m telescope (+CAFOS) of CAHA on 20–21 Aug 1997. The limiting magnitude is R ∼ 23. The circle represents the HRI/*ROSAT* 10″ radius X-ray error box. No variable IR/optical source was found within the X-ray error box over R ≥ 23. North is up and east to the left.

CCD Observations of the QSO 4C49.29 in the Error Box of GRB960720

Sergei Guziy[1], Aleksey Shlyapnikov[1], and Rene Hudec[2]

[1] Kalinenkov Astronomical Observatory, Nikolaev State University, Nikolskaya 24, Nikolaev, 327030 Ukraine, e-mail: root@aok.mk.ua
[2] Astronomical Institute, CZ-251 65 Ondrejov, Czech Republic, e-mail: rhudec@asu.cas.cz

Abstract. We present preliminary results of optical investigations of the radio-loud QSO 4C49.29 in R band, performed during 1999 year. The data were taken with the 0.7-m telescope at the Kalinenkov Astronomical Observatory of the Nikolaev State University with CCD SBIG ST-7. More details regarding the optical observations and light curves are presented.

INTRODUCTION

Despite recently available identifications of GRBs with objects at another wavelengths, the problem of their physical nature remains urgent. More observational data are needed for better understanding of the GRB phenomenon.

The GRB's hosts seem to be at high redshift. The association with QSOs has been discussed and is not fully excluded at the recent stage [1].

We report on observations of QSO 4C49.29 (J173044.5 + 490626, V = 18.8m), located inside the GRB960720 [2] error box area. The precise localization has been provided by the *Beppo*SAX satellite upon observing the X-ray afterglow, which to within 50" coincides with the position of the QSO 4C49.29. The random probability for such association is only of order of 10^{-4} [3]. This QSO is a rather bright source of radio radiation [4]. The goal of our observations was to study the possible optical activity of the object. Earlier observations [5] have shown that the QSO does not exhibit any extraordinary activity, except for the 2 magnitude amplitude flare found on an archival plate [6].

OBSERVATIONS

CCD photometry of QSO 4C49.29 was obtained on 13 nights between 1999 March and 1999 October. The 0.7-m telescope of the Kalinenkov Astronomical Observatory of the Nikolaev State University was used with CCD SBIG ST-7. The observations were obtained in the R-band, scale of the image 0.66"/pixel. The time of the exposure was from 600–900 s for one frame (three frames were usually taken for one observation).

After the standard reduction (correction for flat field, hot pixels and background) all images were summarized and measured with the help of the program CCDOPS.

The photometric binding was used as well as the photometric standard SA 107 [6]. For the control of accuracy of observations and the changes of magnitudes of the target, five control stars of various brightness were chosen (Fig. 1).

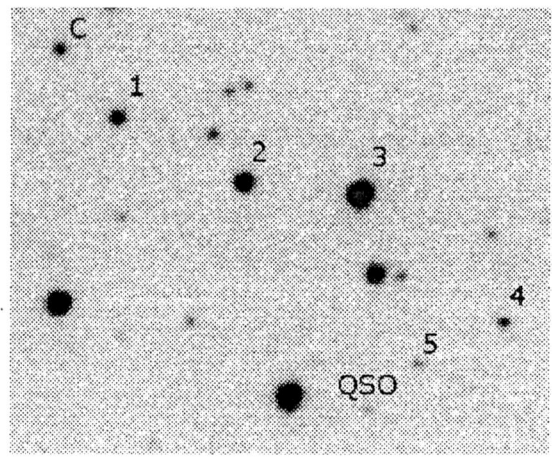

FIGURE 1. The image area with QSO 4C49.29. The figure shows control stars, with the letter C designating the main star of comparison. The size of the image is $2' \times 2'$.

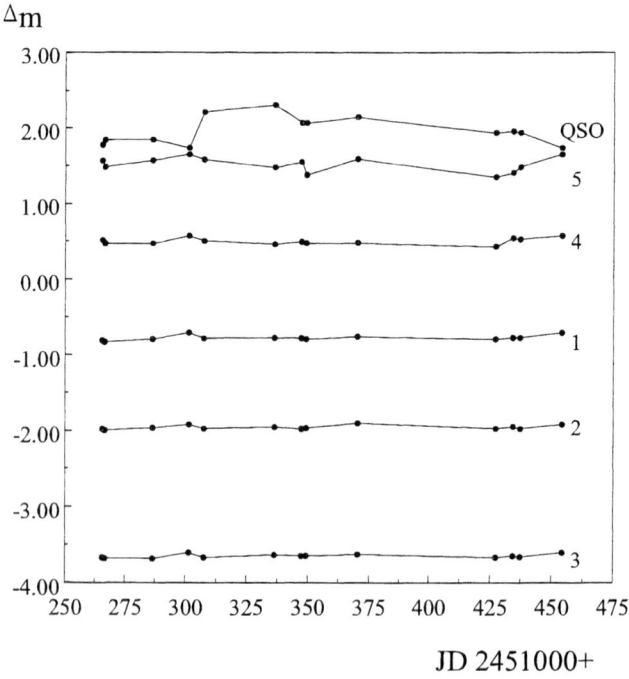

FIGURE 2. The light curves of the QSO and control stars as indicated in Fig. 1.

RESULTS

During the reduction of the observations, one star was accepted as a main star-comparison. Other stars have served as control stars. In Fig. 2 the light curves of the QSO and of control stars for the period of observations are given.

CONCLUSIONS

The QSO 4C49.29 did not show any flare or any other remarkable activity during the time of our observations. The insignificant fluctuations of magnitude are almost consistent with observational errors.

ACKNOWLEDGEMENTS

The study has been supported by the Project KONTAKT ES002 provided by the Ministry of Education of the Czech Republic.

REFERENCES

1. Burenin, V.V., et al., Proc. 3rd Integral Workshop, 1999, in press.
2. In't Zand , J., Heise, J., Hoyng, P. et al., *IAUC* 6569 (1997).
3. Greiner, J., Heise, J., Piro, L. et al., *IAUC* 6570 (1997).
4. Walsh, D., Beckers, J.M., Carswell, R.F., Weymann, R.J., *MNRAS* **211**, 105 (1984).
5. Beskin, G.M., Lyutyj, V.M., Neyzvestnyj, S.I., Pustilnik, S.A., Sahvartsman, V.F., *Astron. Zhurn.* **62**, 432 (1985).
6. Hudec, R. et al., Proc. 3rd Integral Workshop, 1999, in press.

Optical Study of the Counterpart to GRB 990712

J. Gorosabel[1], A. J. Castro-Tirado[1,2], P. Saizar[3], N. J. Rattenbury[4], I. A. Bond[4], P. Yock[4], J. Hearnshaw[5], P. M. Kilmartin[5], Y. Muraki[6], T. Nakamura[7], K. Ohnishi[8], M. Reid[9], To Saito[10], and S. Noda[11].

[1] *LAEFF-INTA, P.O. Box 50727, 28080 Madrid, Spain.*
[2] *Instituto de Astrofísica de Andalucía (IAA-CSIC), P.O. Box 03004, E-18080 Granada, Spain.*
[3] *Facultad de Ingeniería, Universidad de Morón, Argentina.*
[4] *Faculty of Science, University of Auckland, New Zealand.*
[5] *Department of Physics and Astronomy, University of Canterbury, Ernest Rutherford Building, Christchurch, New Zealand.*
[6] *Solar-Terrestrial Environment Laboratory, Nagoya University, Japan.*
[7] *Yukawa Institute, Kyoto University, Japan.*
[8] *Nagano National College of Technology, Japan.*
[9] *Department of Physics, Victoria University of Wellington, New Zealand.*
[10] *Tokyo Metropolitan College of Aeronautics, Japan.*
[11] *Solar-Terrestrial Environment Laboratory, Nagoya University, Furocho,Chikusa-ku, Nagoya 464-8601, Japan.*

Abstract. Quasi-simultaneous BVR-band observations performed from New Zealand and Argentina ∼16 hr after the burst clearly detected the optical counterpart to GRB 990712. Based on these measurements we construct the optical multi-band spectrum. We report that the spectrum between the R and B bands follows a power law $F_\nu \propto \nu^\beta$ with index $\beta = -0.50 \pm 0.16$. The spectrum is consistent with a stretch of an afterglow spectrum between the peak frequency, ν_m, and the cooling break, ν_c. The photon index derived following the model of Sari et al. (1998), $p = 2.36 \pm 0.08$ is compatible with β and the power law decay, α, only if no absorption is introduced. Thus, our results support that GRB 990712 occurred in a low density region, resembling GRB 970508.

INTRODUCTION

GRB 990712 was detected by the Wide Field Cameras of *BSAX* on July 12.6955, showing the strongest X-ray counterpart observed to date [1].
Optical observations performed by the South African Astronomical Observatory revealed a bright object that was pointed out as the optical counterpart to the GRB [2,3]. Later observations revealed the fading optical behavior of the candidate [4,5].

Several weeks after the gamma event, optical observations revealed the host galaxy with a magnitude R $\sim$ 21.7 [6,7]. A spectrum taken with the VLT determined the redshift of the galaxy at $z = 0.430 \pm 0.005$ [8]. In this paper we show the detection of the optical counterpart by means of optical images taken from the Mount John University Observatory (MJUO), New Zealand and the Complejo AStronómico el LEOncito (CASLEO), Argentina.

OBSERVATIONS

The observations at MJUO were performed with the 61-cm Ritchey-Chretien Cassegrain telescope devoted to the Microlesing Observations in Astrophysics project (MOA). It provides a field of view of 0.92° × 1.39° with a pixel scale of 0.81"/pixel (for further details see [9]). On the other hand, the observations at CASLEO are based on the 2.15-m telescope equipped with the focal reductor, which provides a circular field of view with a diameter of 9'.6 and a pixel scale of 0.813"/pixel.

The observations at MJUO were used to construct the R-band light curve, and the results are published elsewhere [10]. Instead, The CASLEO observations were aimed at constructing the multiwavelength spectrum of the afterglow. The multi-wavelength spectrum is based on BVR images, which allow us to derive the spectral index β_{obs}. The images were calibrated observing the standard star LTT 1020 at similar airmass to the GRB field. The extinction coefficients for CASLEO were taken from [11]. The BVR magnitudes derived for the secondary standards used by [3] are consistent with our independent calibration, with a maximum discrepancy in the V filter of 0.04 mag. Fig. 1 shows the counterpart of the GRB in the V-band.

First, we corrected the magnitudes for the Galactic extinction. We made use of the dust extinction maps of [12]. We derived a color excess of $E(B-V) = 0.03$ towards the GRB direction. Once the magnitudes were corrected by Galactic extinction, we extrapolated all the magnitudes to the same observing time. The shifts are based on a fit to the R-band light curve following $F_\nu = F_{host} + C(t-t_o)^\alpha$; obtaining $\alpha = -1.022 \pm 0.06$ and $m_R(\text{host}) = 21.76$ ($\chi^2/dof = 0.45$). We assumed that B, V and R-band fluxes decay following the same α.

Then, the rest frame spectrum of the afterglow between the R and B bands was obtained. Thus, we propose that the spectral flux density is well described by a power law $F_\nu \sim \nu^\beta$ ($\chi^2/dof = 2.16$), as predicted by the afterglow model (see Fig. 2). The derived spectral index is $\beta_{obs} = -0.50 \pm 0.16$. To overcome the lack of published multi-band observations for the host galaxy, many undesirable assumptions have to be made about the host galaxy spectrum. Thus, the host galaxy has not been subtracted in these preliminary results.

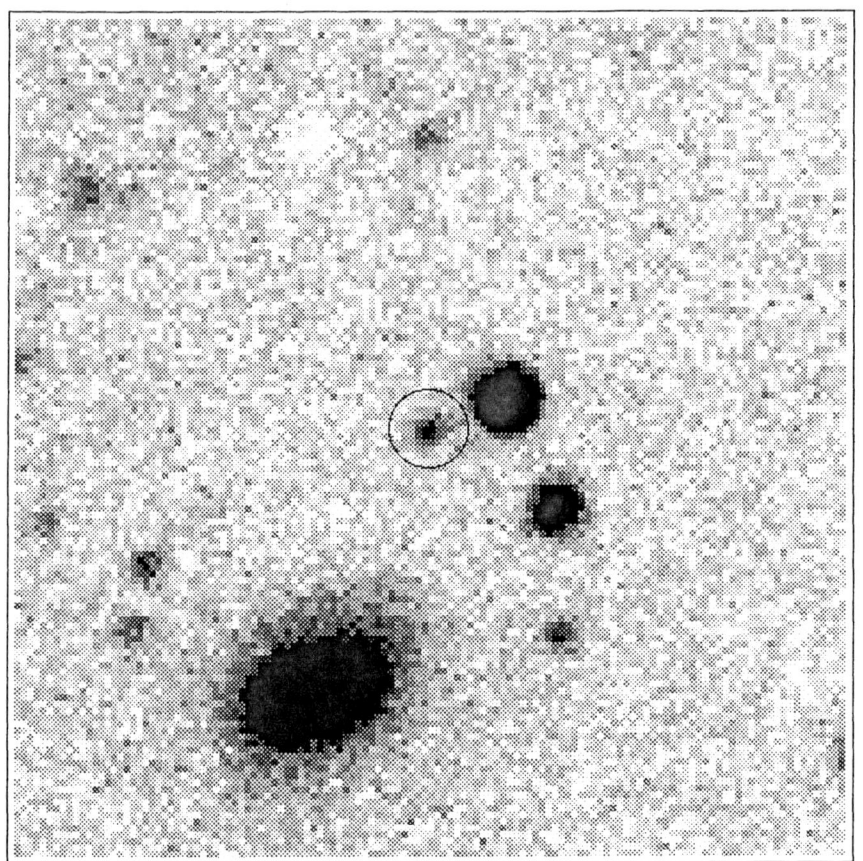

FIGURE 1. The circle shows counterpart of the GRB in this 600 s V-band image. The field of view is $1.7' \times 1.7'$. North is up and East is to the left.

RESULTS AND CONCLUSION

We assume that our data follow the slow-cooling regime of the model described by [13]. This was the case of GRB 970508, for which the cooling time was $t_o = 500$ s [14], and therefore much earlier than the date of our measurements (~ 16 hr after the GRB). In the slow-cooling regime two cases can be distinguished:

- If the peak frequency, ν_m, has passed the BVR bands, but the cooling frequency, ν_c, has not yet; $\nu_m < \nu_{BVR} < \nu_c$:

$$p = -(4\alpha - 3)/3 = 2.36 \pm 0.08$$
$$\beta = -(p-1)/2 = -0.68 \pm 0.04$$

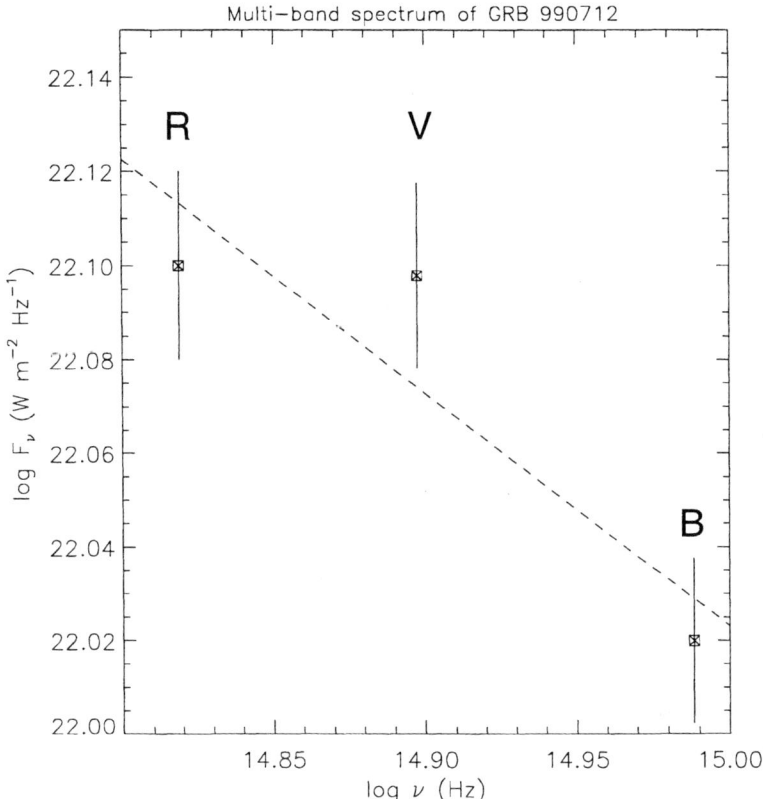

FIGURE 2. The figure shows the rest frame multiwavelength spectrum once the Galactic absorption has been corrected. The measured spectral index $\beta_{obs} = -0.50 \pm 0.16$.

The value predicted for β is marginally consistent with the one measured in our spectrum $\beta_{obs} = -0.50 \pm 0.16$. The inclusion of intrinsic absorption would increase the value β_{obs} and hence the agreement with the theoretical prediction would get worse.

- If both the break frequency ν_m and ν_c are below the BVR bands, $\nu_m < \nu_c < \nu_{BVR}$, the value predicted for β is not consistent with the measured one.

$$p = -(4\alpha - 2)/3 = 2.02 \pm 0.08$$
$$\beta = -p/2 = -1.01 \pm 0.04$$

Thus, our spectrum supports that $\nu_m < \nu_{BVR} < \nu_c$. In this case, no absorption would be required to match our measurements with the prediction of the model.

ACKNOWLEDGMENTS

This work has been partially supported by Spanish CICYT grant ESP95-0389-C02-02.

REFERENCES

1. Heise, J., in't Zand, J., Tarei, G., Torroni, V., Feroci, M., Gandolfi, G., and Palazzi, E., *IAU Circ.* 7221 (1999).
2. Bakos, G., Sahu, K., Menzies, J., Vreeswijk, P., and Frontera, F., *IAU Circ.* 7225 (1999b).
3. Bakos, G., Sahu, S., Menzies, J., Vreeswijk, P., and Frontera, F., *GCN Circ.* 387 (1999a).
4. Hjorth, J., Courbin, F., Cuadra, J., and Minniti, D., *GCN Circ.* 389 (1999a).
5. Thompson, I., Preston, G.W., Bloom, J.S., Harrison, F.A., Kulkarni, S.R., Djorgovski, S.G., and Frail, D.A., *GCN Circ.* 391 (1999).
6. Hjorth, J., Fynbo, J., Dar, A., Courbin, F., and Moller, P.: *GCN Circ.* 403 (1999b).
7. Kemp, J. and Halpern, J., *GCN Circ.* 402 (1999).
8. Galama, T. J., Vreeswijk, P., Rol, E., Kaper, L., Masetti, N., Pian, E., Palazzi, E., and Frontera, F. *GCN Circ.* 388 (1999).
9. Rattenbury, N. J., Castro-Tirado, A.J., Gorosabel, J., Feroci, M., Bond, I.A., Hearnshaw, J., Kilmartin, P.M., Muraki, Y., Nakamura, T., Ohnishi, K., Reid, M., Saito, To, Yock, P., *These Proceedings*.
10. Sahu, K. et al. *ApJ* in press, (2000).
11. Minniti, D., Claria, J. J., and Gomez, M. N., *Ap&SS* **158**, 9 (1989).
12. Schlegel, D. J., Finkbeiner, D. P., and Davis, M., *ApJ* **500**, 525 (1998).
13. Sari, R., Piran, T., and Narayan, R., *ApJL* **497**, L17 (1998).
14. Galama, T. J., Wijers, R. A. M. J., Bremer, M., Groot, P. J., Strom, R. G., Kouveliotou, C., and Van Paradijs, J., *ApJL* **500**, L97 (1998).

Early Search for the Optical Afterglow from GRB 990806

Nicholas J. Rattenbury*, Ian A. Bond*, Alberto J. Castro-Tirado[†],
Javier Gorosabel[‡], John Hearnshaw[||], Pam M. Kilmartin[||],
Yasushi Muraki[¶], Takashi Nakamura[★], Kouji Ohnishi[◊],
Michael Reid[♣], Toshiharu Saito[♡], Marco Feroci[♠], and Philip Yock*

*Faculty of Science, University of Auckland, New Zealand.
[†]IAA-CSIC (Granada) and LAEFF-INTA (Madrid), Spain.
[‡]LAEFF-INTA (Madrid), Spain.
[||]Department of Physics and Astronomy, University of Canterbury, New Zealand.
[¶]Solar-Terrestrial Environment Laboratory, Nagoya University, Japan.
[★]Yukawa Institute, Kyoto University, Japan.
[◊]Nagano National College of Technology, Japan.
[♣]Department of Physics, Victoria University of Wellington, New Zealand.
[♡]Tokyo Metropolitan College of Aeronautics, Japan.
[♠]IAS-CNR (Frascati), Spain.

Abstract. We describe the early optical observations of the GRB 990806 error box by the Japan/New Zealand Microlensing Observations in Astrophysics (MOA) collaboration. We discuss the telescope system, its operation and suitability for rapid observations for gamma-ray burst optical afterglows.

THE MOA PROJECT

The Microlensing Observations in Astrophysics (MOA) project is a joint Japan/New Zealand collaboration formed in 1995. There are around 30 members from over 10 institutions. The project goals are: [1]

- The search for dark matter in the Galactic halo via microlensing observations towards the Magellanic Clouds and the Galactic bulge.

- The search for planetary systems (especially those with sub-Jovian mass planets) via microlensing.

- The study of variable stars in the Magellanic Clouds and Galactic bulge.

- Observations of time critical phenomena including optical counterparts of gamma-ray bursts and other targets of opportunity.

Observatory

The MOA observations are made at the Mount John University Observatory (MJUO), owned and operated by the University of Canterbury. The MJUO is located at Lake Tekapo, in the center of the South Island of New Zealand at 170°28′E, −43°59′, 1030 m above sea level.

Telescope

The telescope used by the MOA collaboration is a modified Boller & Chivens 61-cm Ritchey-Chretien Cassegrain telescope with a focal ratio of $f/6.25$. The field of view afforded by the telescope and camera system is $0.92° \times 1.39°$. This allows prompt coverage of a typical BATSE error box by taking several images in order to perform a mosaic.

Camera

Three 2048×4096 pixel SITe CCDs comprise the MOA-cam2 camera. The pixel size is $15\mu m$, each covering $0.81''$ on the sky at $f/6.25$. The CCDs are thinned, back-illuminated and have a peak QE of 85%. To allow for the modest aperture of the telescope, two very broad band filters (designated MOA-red and MOA-blue) are used with step function transmittances longer and shorter than 630 nm respectively [2].

Viewing Schedule

The MOA observations are generally divided into one of two modes: a patrol mode where survey images of the LMC, SMC and bulge fields are made, and an alert mode where intensive, repeated observations are made of a microlensing event in progress. This alert mode is used also for the investigation of other transient phenomena, such as gamma-ray bursts (GRBs), and recently the preliminary astrometry of Nova Velorum 1999 [3].

Analysis

The microlensing image data are processed and analyzed using an UltraSparc workstation with four 250 MHz CPUs, 384 Mb physical memory and 90 Gb storage space. The workstation is located on site at the B&C dome. The software used to perform photometry on the images obtained with the B&C telescope was written by I. Bond and is based on the image subtraction and difference image photometry procedures developed by Alard et al. [4].

GAMMA RAY BURST OPTICAL AFTERGLOWS

The MOA project offers some advantages in the search for GRB Optical Afterglows (OAs).

- The system has a large field of view enabling a large error box to be scanned with minimum delay.

- The MOA observation schedule already expects and can accommodate regular time critical target-of-opportunity (TOO) observations.

- The data are processed and analyzed on site at the observatory and the software used in the microlensing research can be adapted for the analysis of GRB OAs.

- The isolated location of MJUO in the South Pacific is useful for obtaining continuous light curves of OAs in combination with ESO's La Silla Observatory and other southern observatories.

RECENT RESULTS

Since the addition of the search for GRB OAs as a project goal in mid 1999, MOA has been involved in the search for several GRB OAs. We discuss the results from several recent events here.

GRB 990806

The GRBM and WFC instruments on the BeppoSAX satellite simultaneously detected GRB 990806 on August 6, 14:28:07 UT [5]. The refined co-ordinates from the WFC were $\alpha = 03^h10^m35.5^s, \delta = -68°06'51.4''$. GRB 990806 was detected by the BATSE instrument on the CGRO as trigger 7701 on August 6, 14:28:06 UT. The burst profile as recorded by BATSE showed a FRED structure with $T_{90} = 17.41 \pm 3.84$ sec [6].

The first MOA observations of the GRB 990806 error box began at 1732 UT (0532 Aug 7 NZST). For the following two nights, intensive observations were made of the event location and co-added images were compared to Digital Sky Survey (2nd generation) images. A short log of the MOA observations of the GRB 990806 error box is given in Table 1.

It became clear that there were no obvious objects present in the stacked images that could be the OA of GRB 990806. The $S/N = 3$ magnitude limit of the co-added images is $R = 23$. This early null result is important considering how soon after the GRB event the first images were obtained. A BeppoSAX followup observation of the GRB 990806 WFC error box detected a previously unknown X-ray source [8], and observations using the NFI on BeppoSAX about 8 hours after the burst confirmed the source, 1SAX J0310.6-6806 at $\alpha = 03^h10^m35^s, \delta = -68°06'35''$

TABLE 1. MOA observation log for GRB 990806

Date (UT)	Time	Event
6 Aug	1428	BeppoSAX detects GRB 990806
	1732	First observations at MJUO
	1832	Position refined $\alpha = 3^h10^m35.5^s$ $\delta = -68°06'51.4''$
		3' error circle
7 Aug	1028	Second observation
	1337	Third observation
8 Aug	1519	Forth observation

(error radius 1') [7]. We conclude that significant intrinsic absorption had to be present in order to have not detected the OA of this burst. A speculation at this stage is that the GRB event occurred deep within a host galaxy which obscured any optical signal. The non-detection of an OA for GRB 990806 is consistent within the limits presented by other authors [9–13]. Additional observations of GRB event locations with similar response times could yield important position information, which could then be used to direct follow-up spectroscopic observations.

Other Bursts

GRB 990728

The MOA observations of this event are included to illustrate the advantage of the large $0.92° \times 1.32°$ MOA field of view. GRB 990728 was detected by the BATSE instruments aboard CGRO at 990728, 1103:50 UT. The original event co-ordinates were reported as $\alpha = 14^h09^m09^s, \delta = -56°12'06''$ with a 2° radius error circle. In order to cover such a large section of the sky, a mosaic of 20 fields covering the error circle was imaged. A further 18 fields were added to the mosaic. An improved event location was received on 30 July and further observations were made of the mosaic fields immediately covering this new position. A further improved location was available at 990731, 1430 UT. Searching such a large section of the sky for an optical afterglow presents some challenges, but by utilizing the B&C telescope's wide field of view we can maximize the chances of obtaining an early optical afterglow image, if one exists, identifying it on early images once more accurate position data become available.

GRB 990712

MOA observed the OA for GRB 990712 over three consecutive nights. Image reduction and photometry were performed on site. The MOA results for this event are given elsewhere in these proceedings (Gorosabel et al.).

GRB 990907 and GRB 990908

The MOA telescope has also been used for follow up observations of GRB 990907 and GRB 990908. For GRB 990907, Gorosabel et al. report that no OA was found to the DSS-2 limit within the BSAX NFI error box [14].

CONCLUSION

The wide field of view and target-of-opportunity scheduling of the 61-cm telescope used by the MOA collaboration are well suited for the rapid identification of gamma-ray burst optical afterglows. Several GRB error boxes have been imaged by MOA, most notably GRB 990712 and most recently GRB 990908. A search of the MOA images of the $4'$ error circle associated with GRB 990806 for an optical transient has yielded a null result to $R = 23$. This early limit magnitude for GRB 990806, obtained three hours after the burst event, is a useful constraint. Further early observations of GRB error boxes will likely result in positive OA identifications.

REFERENCES

1. Yock, P.C.M. MOA - The Japan/NZ Gravitational Microlensing Project. In Sato, H., Sugiyama, H. editors, *Black Holes and High Energy Astrophysics*, number 23 in Frontiers Science Series, pages 375-386. Universal Academy Press, 1998.
2. Yanagisawa, T. et al., *Experimental Astronomy*, to be submitted.
3. Green, D. W. E., May 1999. IAU Circular 7176.
4. Alard, C., Lupton, R.H. A Method for Optimal Image Subtraction. *Astrophys. J*, 503, 1998.
5. Piro, L. GCN Circular 392, 1999.
6. Giblin, T. et al., GCN Circular 400, 1999.
7. Frontera, F. et al., GCN Circular 401, 1999.
8. Gandalfi, G. et al., GCN Circular 393, 1999.
9. Schaefer, B. GCN Circular 394, 1999.
10. Greiner, J. et al., GCN Circular 396, 1999.
11. Kemp, J. et al., GCN Circular 397, 1999.
12. Vreeswijk, P. et al., GCN Circular 398, 1999.
13. Prochaska, J. et al., GCN Circular 399, 1999.
14. Gorosabel, J. et al., GCN Circular 415, 1999.

GRB Optical Searches: Results from UKSTU Plates

René Hudec

Astronomical Institute, CZ-251 65 Ondrejov, Czech Republic, e-mail: rhudec@asu.cas.cz

Abstract. We report on the results of investigations of precise GRB localizations provided by the BeppoSAX and RXTE satellites. These positions have been investigated on deep sky plates taken by the UKSTU in Siding Springs, Australia. From the list of 20 triggers known in November 1998, 7 have been covered by these plates. Altogether 107 plates with limiting magnitudes between 19 and 23 have been investigated for optical activity related to these positions. No activity has been found. We will also show that the recently found relation of GRBs and star forming regions in distant galaxies could explain possible repetitions (from the same host galaxy, not from the same GRB) on the time scale of the order of tens of years (or more). In this scenario, optical transients found on archival sky plates close to some GRBs positions could be explained as recurrent events in the particular star forming region.

INTRODUCTION

The operation of BeppoSAX and RXTE satellites has provided a list of 36 reliably and precisely positioned gamma-ray bursts (GRB). The typical localization accuracy amounts to roughly 1...5 arc-min, and is even essentially better if there have been X-ray, optical and/or radio afterglows found.

For a large fraction of these GRBs, low-energy counterpart afterglows have been found, including optical [1]. The optical afterglows (OAs) of GRBs have been found to peak at roughly 18–23 mag and then decrease slowly according to a power law over a time period of weeks to months. In one case, the direct optical emission peaking at about mag 9 and lasting for about 10 min has been found [2].

The archival plates taken by the UKSTU in Siding Springs, Australia, have typical limiting magnitudes of 19...23 and are hence well suited to search for optical emissions related to GRBs, including faint OAs.

The archival search conducted has two major prospects: (i) to search for underlying GRB hosts and/or peculiar objects, and (ii) to search for possible activity (including recurrence).

The recently suggested association of GRBs and star forming regions can account for activity recurrence due not by the recurrent physical model of the central engine, but by the apparently recurrent (however not identical) events in the same galaxy.

THE DATA

We have found altogether 107 high-quality plates covering the positions of 7 BeppoSAX/RXTE precisely positioned GRBs in the UKSTU (United Kingdom Schmidt Telescope Unit) plate archive located at the Royal Observatory Edinburgh, UK. Their limiting magnitudes were in the range 19...23, depending both on the observing conditions, exposure times as well as filter used. The Table 1 gives more details.

For GRB980425, the error boxes of both X-ray sources as well as the SB1998bw position have been examined. The deepest plates show an extended structure in the spiral arm in the host galaxy of the SN1998bw and consistent with its position, perhaps related to the star-forming region responsible for both events.

TABLE 1. GRBs investigated on UKSTU plates.

GRB	OA yes/no	Number of plates analyzed	Optical activity found
970228	y	2	n
970402	n	13	y[1]
970616	n	9	n
980109	n	8	n
980326	y	14	n
980425	SN?	25	n
980515	?	27	n

GRBs INVESTIGATED

Since the positional errors of the triggers studied were usually small, of order of 1 arc-min and/or below, the GRB localizations were investigated both by the Zeiss blink microscope as well as by the plate microscope. This proved to be more effective than the time-consuming scanning of the whole plates. Moreover, the false objects may be effectively eliminated on original plates if compared with digitized files where the 3rd dimension (along the line of sight) is lost. The goal for the study was to search for any kind of optical activity (including possible burst recurrence and light variability of underlying objects/hosts) at the positions of GRB triggers at times before and after GRBs.

Although the recent results on GRBs do not seem to provide any kind of support for recurrence of the triggers, they still cannot rule this out completely. The BATSE GRB catalog (4B) shows trigger overlaps consistent with statistical expectations for non-recurrent triggers but these studies are affected by large errors of events detected by BATSE and they are also unable to detect any kind of repetitions with time scales equal or exceeding the operational time of CGRO, which was about 8 years at the time of the study.

The recent GRB results relate GRBs to faint and distant host galaxies and there are even indications that GRBs are related to star-forming regions [3,4]. This scenario,

---

[1] Variable star BL Circini

however, does not fully exclude the possibility of recurrent events on a long time basis. The recurrent trigger would then simply represent another (unrelated) event in the same galaxy.

It should also be noted that the recurrence and/or non-recurrence of the triggers could be crucial in understanding their physical nature which still — despite recent progress in GRB investigations — remains hidden. This strengthens the value of searches for event repetitions since confirmation of recurrence could strictly support one group of theoretical models explaining GRBs and definitely eliminate others.

In the star-forming scenario, the estimated apparent recurrence would provide an unprecedented possibility to estimate the star forming rate.

CONCLUSIONS

We have investigated 107 high-quality deep UKSTU Schmidt plates covering the positions of 7 precisely positioned GRBs provided by BeppoSAX and RXTE satellites until November 1998. No optical activity exceeding limiting magnitudes of the corresponding plates which amounts to mag 19...23 has been found except the peculiar variable star BL Cir marginally consistent with the error box of GRB970402 [5]. Despite the negative results found, this study provides valuable experience for using high-quality deep optical archival data to search for OAs and OTs of GRBs in general. i.e., independently from satellite experiments.

We also note that the UKSTU plates cover only southern positions. Hence, for the recent 15 GRBs detected by BeppoSAX and RXTE between November 1998 and September 1999, 12 are covered by these plates, i.e., a significant fraction.

ACKNOWLEDGEMENTS

This study has been supported by a grant provided by the Grant Agency of the Czech Republic No. 205-99-0145, by the project KONTAKT ES002 provided by the Ministry of Education and Youth of the Czech Republic and also by the Academic Link between the Astronomical Institute Ondrejov and the University of Westminster, Harrow, UK provided by the British Council in Prague. I also acknowledge the support provided by the UKSTU staff at the ROE.

REFERENCES

1. McNamara, B. J. and Harrison, T. E., *Nature* **396**, 233-236 (1999).
2. Akerlof, C. and McKay, T. A., *IAUC* 7100 (1999).
3. Djorgovski, S. G., Kulkarni, S. R., Bloom, J. S., Goodrich, R., Frail, D. A., Piro, L. and Palazzi, E., *ApJ* **508**, L17-L20 (1998).
4. Bloom, J. S., Djorgovski, S. G., Kulkarni, S. R. and Frail, D. A., *ApJ* **507**, L25-L28 (1998).
5. Hudec, R. et al., Proc. 3rd INTEGRAL workshop, Taormina, Italy, 1999, in press.

Connections between Parameters of GRB Afterglows

G. M. Beskin[1], C. Bartolini[2], G. Cosentino[2],
A. Guarnieri[2], S. Lodi[2], and A. Piccioni[2]

[1] *Special Astronomical Observatory, Nizniji Arhkiz, Russia*
[2] *Dipartimento di Astronomia, Università di Bologna, Italy*

Abstract. The optical light curves of GRBs can be grouped in two families according to their α values. Significant clues to investigate the physics of GRBs could come from the detection of correlations between the parameters of optically identified GRB afterglows.

INTRODUCTION

Since February 28, 1997, thirteen optical afterglows connected with GRBs have been observed. Our understanding of the nature of the GRBs can receive a remarkable support from the search for correlations among the observed properties of the sources in the different spectral bands. For this reason we have collected from the literature observational data on 11 GRBs with optical afterglows. We excluded the latest two cases (GRB 990705 and 990712) because of the scarcity of the data so far published.

THE COLLECTED DATA

The astrophysical quantities we have considered are the following:
i) γ-ray fluxes at the peak in the 40–700 keV and 50–300 keV ranges, durations T, T_{50} and T_{90} in the γ-ray range, number of peaks and fluence in γ range and the hardness ratios in the bands 100–300 keV and 50–100 keV.
ii) X-ray fluxes in the ranges 2–25 keV and 2–10 keV.
iii) Radio fluxes at the maximum, delays between γ-ray and radio detections.
iv) R magnitudes at the observed maximum and at the first observation, delays between the times of maximum in γ-ray and optical ranges, R magnitudes of the host galaxy, optical fluences, time intervals ΔT for which optical observations are available, mean exponents α of the power law for the fading of the optical flux.

TABLE 1. GRB parameters.

GRB	Fluence$_\gamma$ erg cm^{-2}	Fluence$_{opt}$ erg cm^{-2}	R at max. mag	$\Delta T_{\gamma-opt}$ days	α
970228	6.1x10^{-6}	1.58x10^{-9}	20.9	0.87	1.112
970508	3.1x10^6	3.0 x10^{-8}	19.78	1.95	1.126
971214	1.09x10^{-5}	2.22x10^{-10}	21.67	0.48	1.65
971227	9.3x10^{-7}	–	20	0.56	>2
980326	1x10^{-6}	2.72x10^{-10}	21.19	0.42	1.508
980329	5x10^{-5}	5.76x10^{-11}	23.5	0.83	1.813
980519	2.54x10^{-5}	6.73x10^{-10}	20.39	0.65	2.334
980613	1.71x10^{-6}	1.11x10^{-9}	22.9	0.68	0.849
980703	4.59x10^{-5}	7.97x10^{-10}	20.6	0.94	0.995
990123	5.09x10^{-4}	8.38x10^{-9}	18.2	0.16	1.166
990510	2.56x10^{-5}	1.11x10^{-8}	18.24	0.15	1.196

For the optical transients (OTs) that show a variation of slope, the initial and final values α_1 and α_2 were considered as well.

Table 1 reports fluence in the γ–ray and the optical ranges, R magnitude at the maximum, the delay between the times of maximum in the γ–ray and the optical ranges (in days), and the index α.

The fluxes from the original references were converted in erg cm^{-2} sec^{-1} by the following conversion factors: 1 count cm^{-2} sec^{-1} = 6.15 x 10^{-7} erg cm^{-2} sec^{-1}, F_R = 3.371 × 10^{-6} × 10$^{-0.4 m_R}$ [1].

For the 7 GRBs for which the redshifts have been measured, the durations T_γ, T_{opt}, and the delays $\Delta T_{\gamma-opt}$ were transformed to the rest frame of the host galaxy through the well known formula $t_{RF} = t_{obs}/1+z$. Luminosity distances were computed by the following formula:

$$D_L = \frac{c}{H_0 q_0^2}\{q_o z + (q_0 - 1)[(1 + 2q_0 z)^{1/2} - 1]\} \quad (1)$$

assuming H_0 = 65 km/s-Mpc and q_0 = 1/2. These data, along with luminosities and energies in γ–ray and optical ranges and their ratios are reported in Table 2.

SLOW AND FAST AFTERGLOWS

The values of α index reported in Table 1 can be grouped in two sets: those smaller than 1.2 (slow afterglows) and those greater than 1.5 (fast afterglows).

The first group has a mean value of 1.07, the second of 1.86. The standard deviations around these mean values (0.05 for the slow ones and 0.14 for the fast ones) are at least 5 times as small as the difference of the two mean values. This

TABLE 2. Luminosities and energies in optical and γ ranges.

GRB	D_L (Gpc)	L_γ (erg/s)	L_{opt} (erg/s)	E_γ erg	E_{opt} erg	E_γ/E_{opt}
970228	3.62	5.85×10^{51}	2.32×10^{43}	9.56×10^{51}	2.49×10^{48}	3800
970508	4.43	3.92×10^{50}	9.72×10^{43}	7.28×10^{51}	7.08×10^{49}	100
971214	21.4	6.56×10^{52}	3.96×10^{44}	5.97×10^{53}	1.22×10^{49}	48900
980613	5.98	1.66×10^{51}	1.00×10^{43}	7.32×10^{51}	4.77×10^{48}	1500
980703	5.17	4.80×10^{51}	6.25×10^{43}	1.47×10^{53}	2.56×10^{48}	57400
990123	9.11	1.00×10^{53}	1.76×10^{45}	5.05×10^{54}	8.32×10^{49}	60200
990510	9.24	5.12×10^{52}	1.74×10^{45}	2.61×10^{53}	1.13×10^{50}	2300

means that the confidence probability of the hypothesis of a real difference between the two groups of GRBs is higher than 99%.

The distribution of the α coefficients, shown in Figure 1, appears to be bimodal. No α values were found in the interval 1.2–1.5. It has still to be ascertained by better statistics if this is due to the scarcity of the sample or if it is indicative of the existence of two families of GRBs.

TABLE 3. Connections having correlation coefficients greater 0.5.

X axis	Y axis	Corr. Coeff.	Slope	Stand. Dev.
$\text{Log } \Delta T_{\gamma-opt}$	$\text{Log } E_{opt}$	0.96	-1.65	0.21
$\text{Log } L_\gamma$	$Log E_\gamma$	0.88	1.08	0.26
z	$\text{Log } E_\gamma$	0.80	1.85	0.69
$\text{Log } \Delta T_{\gamma-opt}$	$\text{Log } L_{opt}$	0.79	-1.42	0.49
$\text{Log } E_\gamma$	$\text{Log } E_\gamma/E_{opt}$	0.77	0.72	0.27
$\text{Log } L_{opt}$	$Log E_{opt}$	0.77	0.66	0.28
$\text{Log } L_\gamma$	$\text{Log } L_{opt}$	0.75	0.75	0.28
$\text{Log } L_\gamma$	$\text{Log } E_\gamma/E_{opt}$	0.73	0.84	0.35
$\text{Log } E_\gamma/E_{opt}$	$\text{Log } T_{opt}$	0.65	-0.77	0.40
α	$\text{Log } L_\gamma$	0.58	0.16	0.10
α	$\text{Log } T_\gamma$	0.58	-0.35	0.22
$\text{Log } \Delta T_{\gamma-opt}$	$\text{Log } E_\gamma/E_{opt}$	0.53	-1.12	0.79
$\text{Log } L_\gamma$	T_{opt}	0.51	-0.49	0.36
z	$\text{Log } E_{opt}$	0.51	0.54	0.41

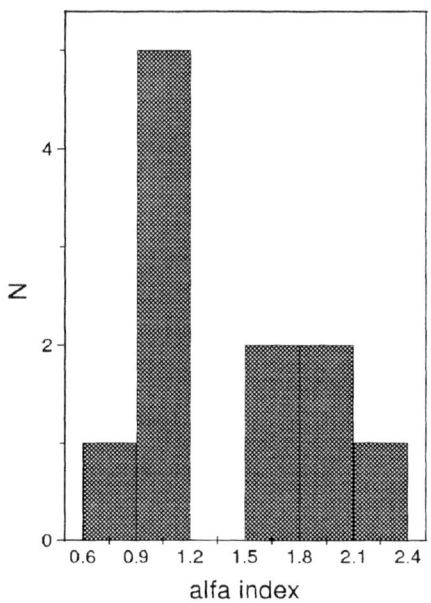

FIGURE 1. Distribution of α coefficients.

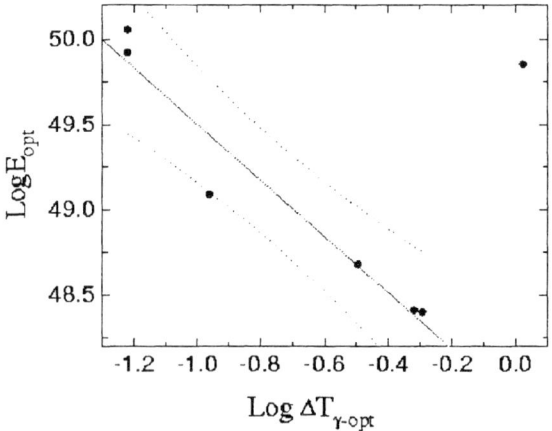

FIGURE 2. Correlation energy vs. opt-gamma delay.

CORRELATIONS BETWEEN PAIRS OF PARAMETERS

Significant clues for investigating the physics of the GRBs could come from the detection of correlations between the quantities presented in Tables 1 and 2.

In 14 cases we found that observed quantities are linearly correlated with coefficients greater than 0.5. They are listed in Table 3.

The correlation coefficient between $\Delta T_{\gamma-opt}$ and E_{opt} was computed without taking into account GRB 970508 (the point on top at right in the Figure 2).

CONCLUSIONS

The optical light curves of GRBs appear to be grouped in two families (slow and fast) according to their α value. High values of α could be explained by a beamed stream of particles crossing thicker interstellar matter. In this case the beaming angle would increase, the spread of the radiation would be higher and the detected flux would drop faster.

Positive correlations between parameters in γ-ray and in optical ranges indicate that the energy release in the optical band is a direct consequence of processes at high energies.

Negative correlations between optical energy or luminosity and time delay between γ-ray and optical flares show that energy of the first stream was dissipated in the environment before enough mass was accumulated in order to start the optical emission. There are at least two reasons for the energy losses: particles interaction with magnetic fields and plasma and spread of the beam. In either case, a long time delay would correspond to fainter optical transients.

Correlations between optical or γ-ray energy of GRBs and z seem to indicate evolution of GRB sample properties with the age of the universe.

These results are only preliminary speculations and require a detailed statistical analysis and theoretical interpretation.

ACKNOWLEDGEMENTS

We thank Prof. G. Setti and Dr. E. Waxman for useful discussions, DrÅ. Chianura and Dr. N. Ialeggio for their help in collecting the observational data. This investigation was supported by the University of Bologna (Funds for selected Research Topics), by the Italian Ministry of Foreign Affairs, by the Russian Fund of Fundamental Research and by Educational-Scientific Center "Cosmion".

REFERENCES

1. Fukugita, M., Shimasaku, K., Ichikawa, T., *PASP* **107**, 945 (1995).

What Have We Learned from Optical Detections of GRBs: Feasibility of Independent Searches?

René Hudec

Astronomical Institute, CZ-251 65 Ondrejov, Czech Republic

Abstract. Recent results indicate that nearly half of GRBs are accompanied by optical emissions. I focus on the feasibility of independent optical searches for GRBs based on sky surveys. I estimate the optical transient and optical afterglow rates based on recent observational results and theoretical predictions (beaming), and compare them with background rates of similar transient phenomena (such as flare stars, flaring AGNs, supernovae, etc.). I show that independent optical searches for GRBs using both ground-based and space experiments are feasible but require a very high level of background elimination. I also propose a strategy for these searches and will show that deep CCD-based wide field surveys and deep high-quality sky patrol plates may be involved in such studies.

INTRODUCTION

The recent detection of optical afterglows (OAs) and optical transients (OTs) of gamma ray bursts allow us to consider optical ground-based independent monitoring of these phenomena. Optical surveys achieving lim mag better than 19–23 for stars and/or ~10 for 1 min exposures may detect OAs and OTs of GRBs. This opens the possibility of independent optical searches. These searches must be of large field of view, i.e., CCD surveys and/or deep patrol plates are suitable. The prospects of such searches are as follows:

(1) The optical surveys may provide a larger sample (due to different beaming) and better localization accuracy (~1 arcmin or better) than are provided by gamma ray satellite detectors (2) The larger sample of OAs and their host galaxies may be crucial for understanding the nature of GRBs (3) The actual rate of OAs can place strong constraints on the afterglow appearance fraction and the initial beaming angle of GRB sources (4) The UV flashes predicted by some theories [1] could be detected and studied. The corresponding delays regarding GRBs could serve to study the nature of the sources. This can be addressed only by surveys, not by follow-up devices since the flashes may preceed the GRBs [1] (5) The optical surveys are cost-effective.

It is expected that sources emit jets from which the gamma-ray emission is more beamed than the subsequent optical afterglow radiation due to the deceleration of the jet by the ambient gas and the corresponding decline in its relativistic beaming with time [2]. Because the shift to lower frequencies accompanies the shift to lower bulk Lorentz factor, the minimum solid angle into which the transient can radiate increases with time. A jet geometry hence implies a higher rate of OA detections: if bursts are highly collimated, the gamma rays will radiate into a small solid angle, the optical light into a larger one, and radio into a still larger one. On the other hand, if bursts emit isotropically, we do not expect OAs unaccompanied by GRBs. The ratio of transients detected hence allows the ratio of the mean solid angle into which transients radiate to be estimated. The GRB rate is known already and the OA rate with characteristics typical for the observed OAs can be estimated by present and future sky survey programs.

THE ESTIMATED BEAMING OF GRBs

Recently, the evidence for collimation in GRBs has grown: (1) $\Delta\Omega/4\pi > 0.11$ for 970508 [3], < 0.003 for 980519 [4], < 0.01 for 990123 [5,6] and < 0.003 for 990510 [7]. The strongest evidence for collimation is in the brightest sources. This evidence is based on the observed profiles of the light curves — steeper slopes and even breaks in the slope as a collimated flow slows and spreads laterally. The number of optically selected OAs could be greater than the number of gamma-ray selected GRBs by a factor of $(\gamma_o/\gamma_a)^2$ where the γ_o is the initial Lorentz factor (the initial gamma-ray emission is beamed to an angle of about $1/\gamma_o$) and the afterglow emission is produced when the fireball has been decelerated to a modest Lorentz factor γ_a. Since typical bursts have $\gamma_o < 10^{2-3}$ [8,9] while the optical afterglow emission occurs at γ_a nearly 10^{1-2}, this could boost the expected OA rate by up to four orders of magnitude relative to the GRB rate (we will call this boosting factor).

THE ESTIMATED RATES OF OAs AND OTs IN REAL EXPERIMENTS

The recent digitization of deep sky patrol plates as well as CCD surveys provide valuable input data for these searches.

Optical Transients (prompt emission): observed rate of GRBs (by BATSE) is about 1 GRB/day, beaming factor 1–10 (probably no more since the time after GRB is small), boosting factor of 1–100, actual GRB fraction with OTs 0.01–1 (brighter than mag 12), estimated OT rate 0.01–100 OTs/day for the whole sky sphere. This means $2.5 \times 10^{-7}...2.5 \times 10^{-3}$ deg^{-2}. Examples: astrograph plate (100 deg^2) $2.5 \times 10^{-5}...2.5 \times 10^{-1}$ OTs per plate (brighter than mag 12).

Optical Afterglows (delayed emission): observed rate of GRBs (by BATSE) is about 1 GRB/day, beaming factor 1–100, boosting factor of 1–10000, actual GRB fraction with OAs 0.5 (brighter than mag 23) and/or 0.05 (brighter than mag 18), estimated OA rate 0.5–5,000 OAs/day for the whole sky sphere lim mag 23 or 0.05–500 OAs/day for the whole sphere lim mag 18. This means $1 \times 10^{-5}...1 \times 10^{-1}$ deg^{-2} lim mag 23 and $1 \times 10^{-6}...1 \times 10^{-2}$ lim mag 18 (not in contrast with results obtained by plate searches). Example: UKSTU plate area of 41 deg^2, 23 mag limit: $4 \times 10^{-4}...4$ objects per plate, 18 mag limit: $4 \times 10^{-5}...4 \times 10^{-1}$ objects per plate.

Constraints from observations [10,11]: < 0.15 events · deg^{-2} lim mag 23, i.e., consistent with above estimations (boosting factor < 10000). Note, however, that the background of SNe, variable AGNs, variable stars, flare stars, etc., is higher. Depending on galactic latitude, their integrated rate may achieve ~1000 to 5000 variable objects per UKSTU plate, lim mag 23 [11].

It seems to be feasible that both flaring (OTs) as well as fading (OAs) optical emission related to GRBs may be detected by optical sky patrols. Although the true rate of these triggers remains unknown, in is very probable that their rate is substantially below the background rate, hence good knowledge of all background triggers must be available as well as a reliable technique for their classification and elimination.

BACKGROUND

Background is represented by false events not related to GRBs but with similar transient behavior. Background sources are represented by SN, AGN flares/brightenings, stellar flares, variable stars, OTs of unknown nature and origin, and by non-astrophysical triggers.

Supernovae: Supernovae, especially those of Type Ia, may represent an important source of confusion due to their occurence rates, rise and decay timescales, and magnitudes. The expected rate of occurrence for faint events (down to R 23 mag) is roughly 2 deg^{-2} or 0.0015 arcmin^{-2} [12,13,14]. But, at least some SNe may be related to GRBs (e.g., SN 1998bw and GRB980425).

AGN flares: The typical flare amplitudes are between 0.5 and 1.3 mag, depending on the AGN type [15]. These are mean values. The spread of particular flare amplitudes is large (0.1–6.7 mag). Recently, there is growing evidence for large amplitude (more than 10 mag) AGN flares. The AGN/QSO surface densities are ~ 20 deg^{-2}, lim mag B/V 20 [16,17,18], ~ 100 deg^{-2}, lim mag B 22.5 [18], most of them variable [19].

Stellar flares: There is growing evidence for large amplitude (5 mag and more) stellar flares. The archival searches have revealed large-amplitude flares (5–9 mag) from otherwise typical dMe flare stars. The rate of analogous stellar flares, however, remains unknown [11].

Variable stars: Variable stars may exhibit light behavior similar to OAs and OTs. Variable stars are observed more commonly in decline than in increase since the declines are typically more slowly than rises, such as delta Cep stars, U Gem stars, flare stars, novae, etc. Example: Y Dra Mira type variable inside GRB910709 detected and positioned by COMPTEL: light variations between 6 and 15 mag, gradual light decrease after maximum (period 322 d). The surface density: estimated rates for VS brighter than 20 mag: 80 deg^{-2} for /bII/ < 20 deg, and 4 deg^{-2} for /bII/ > 40 deg [20], but the discovery probability is ~ 0.1 (using a blink microscope). No statistics for variable stars below 20 mag is available (no systematic surveys).

OTs of unknown nature and origin: There are real OTs of unknown nature but of astrophysical origin detected both on emulsions and CCDs. Examples (real CCD detections): (1) OT 970215: real CCD detection, V 13 mag, nothing down 20 mag on the position, amplitude more than 7 mag [21], (2) OT 950806: real object, detected on 20 CCD frames, peak magnitude I 7.5, amplitude more than 10 mag, nothing down to mag 21, 48 hrs after [22], (3) OT triggers found by SNe searches [10]: mystery events found at a rate of about 0.15 deg^{-2}/per time scale (between 10 min and 3 days) lim mag R 23.5. In three cases, no host galaxy is seen down to mag R 24, in one case, host galaxy is clearly visible. Two events at a low galactic latitude and hence could be flare stars. It should be noted that SN searches reject events with timescales less than 3 min (as cosmic rays) hence can detect OAs but not all OTs. Further, these searches are very limited so far (6 SN runs done, 2–6 deg^2 per run [10]).

Non-astrophysical triggers: There are also OT triggers of non-astrophysical origin especially on very short timescales. They include head-on meteors, satellite glints, aircraft flashes, CCD and electronics defects, photographic faults, etc.

SN versus OA rates: Cosmological GRBs are much rarer than SNe (10^4 to 10^6 times [24]), the GRB afterglows are, however, at peak luminosity 10^3 to 10^4 times brighter than Type II SNe [24]. In Euclidean space, this implies that GRBs are detected from a volume bigger by a factor of about $(10^4)^{3/2} = 10^6$, roughly canceling the factor by which they are rarer than SNe. The average detection rate of SNe and OAs may be comparable at lim mag brighter than mag 18–20 in all colors (U, B, R, K) [24], assuming the OA frequency is the same as that of gamma-ray selected GRBs. If true, this has a significant consequence: in optical searches down to lim mag 18–20, roughly identical numbers of SNe and of OAs are expected. This means that OAs may be among the detected SNe [24].

How to distinguish OAs and SNe (and other background events): (1) Light curve, (2) Peak luminosity (only for objects with known redshift) and (3) Color information: most OAs have $R - I = 0.6 \pm 0.4$, $V - R = 0.4 \pm 0.3$, $B - V = 0.5 \pm 0.3$ [25].

CONCLUSIONS

The recent results of OA and OT searches indicate that GRBs may be monitored and studied by observing their optical emissions, i.e., independent of satellite projects. The rate of OAs may be (significantly) higher than the rate of GRBs due to different beaming. It is feasible to use ground based optical devices to monitor OAs and OTs of GRBs. This opens a new observing window for GRBs. These surveys must be of wide field and high sensitivity. There is, however, a background of false triggers (not related to GRBs, but with similar transient behavior) with poorly known statistics (for faint magnitudes). This background is due to supernovae, AGN/QSOs, stellar flares, variable stars, optical transients of unknown origin and non-astrophysical triggers. The expected total number of optically variable sources of astrophysical origin is large: ~ (0.5–2) in a 5' × 5' error box, depending on galactic latitude, lim mag 23 [11]. The recent searches (e.g., SN searches) and databases (e.g., UKSTU plate collection) may be used to detect OAs — there even may be unrecognized OAs in detected SNe catalogs [24]. The detected OAs and especially OTs (since they will be recorded only once due to their short duration in most cases) must be further studied in detail to eliminate them from background triggers.

Suitable devices/methods include the OMC (Optical Monitoring Camera) on INTEGRAL (5° × 5°, lim mag 19), CCD-based devices and telescopes (ASPA, USNO, ROTSE, OTM, BOOTES,...), digitized plate surveys, as well as digitized deep archival plates (e.g., UKSTU plate collection, Siding Springs Schmidt, 17 000 plates with lim mag 20–23, different filters/colors).

ACKNOWLEDGEMENTS

The investigations of gamma-ray bursts and optical transients are supported by the project KONTAKT ES002 provided by the Ministry of Education and Youth of the Czech Republic and by the grant 205/99/0145 provided by the Grant Agency of the Czech Republic. The investigation of plate defects has been supported by the Academic Link between the University of Westminster and Astronomical Institute Ondrejov provided by the British Council in Prague.

REFERENCES

1. Protheroe, R. J. and Bednarek, W., astro-ph/9904279, *Astropart. Phys.*, accepted (1999).
2. Rhoads, J. E., *ApJ* **487**, L1 (1997).
3. Sari, R., Piran, T. and Halpern, J. P., *ApJ* **519**, L17 (1999).
4. Halpern, J. P., Kemp, J., Piran, T. and Bershady, M. A., *ApJ* **517**, L105 (1999).
5. Anderson, M. I., *Nature* **398**, 400 (1999).
6. Kulkarni, S., et al., *Nature* **398**, 389 (1999).
7. Harrison, F. A., et al., *ApJ* in press, astro-ph/9905306 (1999).
8. Fenimore, E. F., Epstein, R. I. and Ho C., *A&ASS* **97**, 59 (1993).
9. Woods, E. and Loeb, A., *ApJ* **453**, 583 (1995).
10. Schmidt, B., et al., 1999, priv. comm.
11. Hudec, R., Proc. 4th Huntsville GRB conference, AIP Conf. Proc., 1999.
12. Evans, R., et al., *ApJ* **345**, 752 (1989).
13. Pain, R., et al., *ApJ* **473**, 356 (1996).
14. Brainerd, J. J., AIP Conf. Proc., 1998, 428, 545.
15. Smith, A. G., Blazar Continuum Variability ASP Conf Ser 110, 1996, 3.
16. Iovino, A., et al. *A&ASS* **119**, 265-269 (1996).
17. Hartwick, F. D. A. and Scade, D. *ARA&A* **28**, 437 (1990).
18. Trevese, D., et al., *AJ* **98**, 1 (1989).
19. Trevese, D. and Kron, R.G., in *Multi-Wavelength Continuum Emission of AGNs*, eds. Courvoisier, T. and Blecha, A., 1994.
20. Hudec, R. and Wenzel, W., *A&ASS* **120**, C707 (1996).
21. Vidal-Saiz, J., et al., *IBVS* Budapest 4324 (1996).
22. Toth, I., et al., *A&A* **315**, 153 (1996).
23. Woods, E. and Loeb, A., *ApJ* **508**, 760 (1998).
24. Hudec, R., Hudcova, V., et al., 2000, this volume.
25. Simon, V., Pizzichini, G. and Hudec, R., 2000, this volume.

X-ray Afterglow of Gamma-Ray Bursts

E. Costa

Istituto di Astrofisica Spaziale - CNR
Via del Fosso del Cavaliere, 00133 Roma, Italy

Abstract. In almost three years BeppoSAX has detected 27 gamma-ray bursts with the Wide Field Cameras. The majority have been localized in a short time and observed with the Narrow Field Instruments. More bursts detected by various satellite experiments or by the Interplanetary Network have been followed up by X-ray Telescopes, including SAX itself. Around 15 afterglow sources have been unambiguously detected and studied.

Several features have been interpreted in the framework of synchrotron shocks in fireballs, but not all the data may be interpreted in terms of this simplified approach. I discuss the relevance of X-ray afterglow data to answer some open questions.

I INTRODUCTION

A physical explosive phenomenon still to be understood, but likely associated with gravitational collapse, gives rise to two distinct phenomena. The *Main Event* (or gamma-ray burst itself) is peaked energetically in the γ-ray band (where the phenomenon has been historically detected) but is visible as well in the X-ray and (at least in one case) in the optical. The *Afterglow* is mainly peaked in the X-ray band (where the phenomenon has been detected for the first time [6]). In many cases it is also present in the optical, where it was discovered soon after [38] and in a subset of these in the radio band [11].

Some phenomena, already detected before the existence of afterglow sources was clearly established in 1997, like many of those above 50 MeV, are surely related to the Afterglow. We believe that the two phenomena originate from the same event, but are produced in different physical mechanisms and sequenced in time, although the transition from the first to the second is not clear and some overlap may be present.

Besides the very early stages (when it is partially confused with the main event) the afterglow is not as outstanding as the main event. The discovery and study of the GRB Afterglow sources requires a high sensitivity (10^{-13} erg cm^{-2} s^{-1}) and a strategy of quick positioning and pointing, achieved for the first time with BeppoSAX [7] and can be also obtained by combining the capabilities of other missions: An important addition of information derives from the serendipitous detection of GRBs and/or their afterglow from past X/γ-ray missions.

II THE BEPPOSAX SAMPLE

The BeppoSAX sample was built by a baseline procedure for the detection of the afterglow. It foresees: 1) a trigger from the Gamma Ray Burst Monitor or BATSE, 2) an excess in the time series of a Wide Field Camera, and 3) a point-like source in the WFC image of the photons corresponding to this excess not present in images before.

An alternative to step 1 can be: i) a direct detection of the X-ray excess in time series produced on ground at QLA level, confirmed by an excess in the GRBM (even without a trigger); ii) a detection and positioning from XTE/ASM [34]; iii) a detection and positioning from IPN. Actually, IPN has so far only cooperated to improve positioning from other satellites and we still lack a TOO performed on an IPN position only. A trigger from BATSE before BeppoSAX data are received may result in a faster follow-up pointing, by shortening the decision times.

When these conditions occur, a follow-up pointing of the Narrow Field Instruments is usually performed, with some exceptions. Since the crisis of the gyroscopes in May 1997, BeppoSAX is pointed with a widely new pointing procedure, and during an orbit in different phases, the pointing is locked with different combinations of sensors. In some of these combinations the attitude reconstruction of WFCs may be poor. If the error in the quick-look attitude reconstruction is such that the the absence of a detected source could be an ambiguous result, no follow up is activated. NFIs and WFCs are mutually oriented in such a way that a source detected in the field of a WFC can be pointed at with the telescopes. Nonetheless, for particular conditions the pointing was not possible because of the solar angle. Of the 27 GRBs detected so far (October 1999) by BeppoSAX:

- SVP (July 1996 to December 1996): 1 GRB detected off-line, 1 follow up

- AO1 (Jan 1987 to February 1998): 7 GRBs detected, 6 follow ups

- AO2 (March 1988 to February 1999): 9 GRBs detected, 8 follow ups

- AO3 (March 1999 to February 2000): 10 GRBs detected, 7 follow ups

We assume that a source in NFI is a GRB afterglow source when: 1) the detection is significant, 2) the position is consistent with WFC error box, 3) the source was not previously detected by other missions, and 4) The source shows an unambiguous fading. According to these criteria we have from 21 follow ups: 12 unambiguous detections, 1 unambiguous non-detection (GRB990217), and 8 ambiguous cases.

Two more afterglows have been detected by ASCA and by SAX using ASM burst positions.

III SYNCHROTRON RADIATION FROM SHOCKS IN A FIREBALL: THE NEW ORTHODOXY

We can state that the afterglow is nowadays better understood than the main burst itself. Reviewing the theories of GRBs is not the purpose of this paper, but I want to summarize the more widely accepted points of view and the most often debated questions, with the aim of discussing which data on afterglows intrigue, and may challenge in the next few years, the prevailing point of view.

In the years preceding the discovery of the afterglow, it was clear that synchrotron radiation in collisionless shocks was playing a major role in the emission from GRBs. The spectra and their evolution [1], in particular in the wide energy range studied with GRO [36], provided strong supporting evidence, although some other features could not be explained in this model only. The attempt to build scenarios where this emission could be generated, combined with the complexities discovered by BATSE, lead various authors [23,25,20,39] to suggest that two kinds of shocks should be foreseen. External shocks, more straightforward and easy to figure, are produced by the impact of the blast wave on the interstellar medium. Internal shocks, a little bit more elusive, are due to the mutual interaction of different parts of the ejecta themselves (such as two shells expelled with different velocities). This framework turned out to be very suitable to interpret the X-ray afterglow. Actually, a few days after the detection of the afterglow of GRB970228, all the major features were explained in terms of external shocks [41,40]. The theory developed by many contributors arrived to give a good description of the afterglow over a broad band and on several timescales [31]. The application of this model to observational data [16], spanning from long radio wavelengths to X rays, was very successful. The overall spectrum is represented by broken power laws and the major parameters of the electron distributions are expressed via the three cut-off frequencies: ν_a is the self absorption cut-off and is operating only in the first days at very low frequencies; ν_m is the lower edge of the electron energy distribution; ν_c is the frequency corresponding to the energy above which synchrotron radiation is the prevailing cooling mechanism. The evolution of these three frequencies allows for a complete description of the afterglow phenomenology. The detection of linear polarization in the optical afterglow of GRB990510, and of the persistence of the polarization angle [8,42], is the conclusive evidence of the prominent role of synchrotron radiation in the afterglow. This good description of the afterglow is only based on the hypothesis that a huge amount of energy is injected in the ISM in the form of a few baryons with a very high kinetic energy. *Any* central machine that can produce the blast will give the same phenomenology, when impinging on the same medium. Incidentally, the magnetic field is also likely produced in the explosion and not carried from the original object [19]. The major mystery of the nature of the central engine cannot be solved on the basis of the afterglow data only. The afterglow is distinct from the main event, but some of the parameters derived for it (such as the source distance or the Lorentz factor) may be of enormous value to interpret the main event in addition to the direct measurements [13].

I try now to discuss some of the open points on the afterglow physics and the possible contribution of existing or arriving X-ray data.

A The Transition From the Main Event to the Afterglow

One of the most important parameters, and one of the least completely measured, is the burst energy. The decay is well described by a power law and most of the energy is released in the first part. SAX data have a wide gap between the last significant points of the WFC and the first NFI measurements (6–10 hours after).

By analyzing the data (Figure 1) of the *prototypical* GRB970228 [6,12], the SAX GRB group arrived at the conclusion (supported by spectral analysis) that: 1) the afterglow starts soon after the main event (actually 35 s after the trigger) and lasts for at least 10 days; 2) there is a gap with no detectable emission between the two.

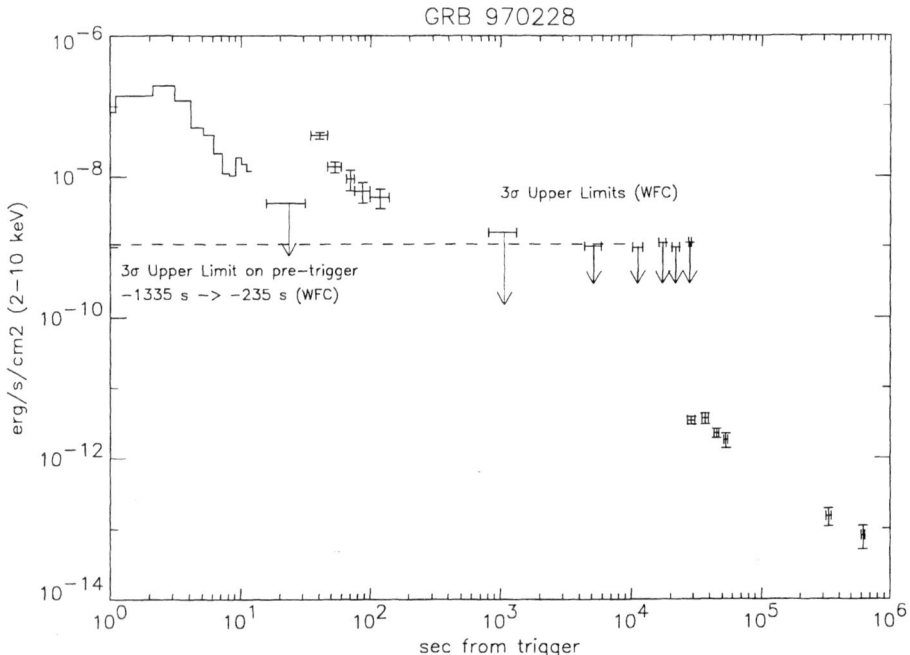

FIGURE 1. Light curve of the GRB970228 including both the main event and the afterglow. The figure combines WFC and NFI data from BeppoSAX and from ASCA in the 2–10 keV band. All data can be fit with a single power law and are consistent with a single energy spectrum from 35 s to 5 days after the burst. If we include ROSAT data (assuming a constant spectrum) this extends to two weeks. In this case the transition from the main event to the afterglow is very clear.

This is a relatively *easy* case: we know exactly when the afterglow starts and we know the spectrum and how it evolves in the WFC data. Hence, we can identify the main event (the first pulse) and have a good hypothesis to interpolate the gap between WFC and NFI. Therefore, we can compute the energy content of the afterglow, compare it with the GRB energy, convert it into absolute units using the red-shift [9], measure the position of the burst within the host, and so on.

For other (less cooperative) bursts it is much more difficult to fix the border line between the main event and the afterglow. In some bursts there is a continuous evolution [27,44,4] and a turning point where the rapidly decaying main event (more or less exponential) is overcome by a slow, power-law decaying afterglow.

The study of this transition is not only relevant for the content of energy but is likely related to the density distribution of the ISM [32].

In a first systematic analysis of 8 GRB/afterglow, by combining WFC, GRBM and NFI BeppoSAX data, Frontera et al. [13] tried to fix with criteria as objective as possible the start of the afterglow interpolated with the NFI data, and computed the energy content in the afterglow that came out to be from 1 to 20% of the total energy content in the Main Event. An even larger energy content (close to the 100%) is present in GRB980613.

These evaluations of the fraction of the energy in the afterglow are still underestimated because while in the Main Event we have a *bolometric* evaluation of the energy (from the spectral model), in the afterglow we only know the flux from 2–10 keV, cut-off by the X-ray optics band pass. More energy may be above this limit and actually in the bright GRB990123, photons were detected up to more than 60 keV, while upper limits to the flux above 10 keV from other bursts are consistent with the extrapolation of the low-energy spectrum, and therefore bring no evidence of any cut-off. Our ignorance of the upper boundary of the flat afterglow spectrum is one of the most severe, indeed.

B Are GRBs or their Afterglows Beamed?

If the GRB emission is concentrated in a beam of opening angle θ_b, the real energy will be much less than that computed with the isotropic assumption: $E = F_{iso} \times \theta_b^2$. This degree of freedom (spanning orders of magnitude) will add to the models. Just as an example, NS-NS merging is viable (with beaming) from the energy point of view and would become very unlikely from the point of view of matching the frequency with the formation rate.

From the condition of compactness we know that the emitting region is in relativistic expansion and, hence, beamed, from our observer frame, into an angle of $\theta_{exp} = 1/\gamma$. When the shell decelerates and $\gamma < 1/\theta_b$ the slope of the decay is expected to change from slow to fast, which is not easy to explain for a fireball with spherical symmetry. More generally, the correlation of spectral slope and decay slope is different in case of narrow beams or jets [33]. The change in slope of the optical decay of GRB990510 can be interpreted with such a beaming effect [15], but other interpretations could be invoked (e.g., the encounter of the blast wave with a changed environment). Since the basic mechanism is kinematic it should be achromatic as well. The presence of the same break in the X-ray decay curve would be an excellent test of this interpretation. Unfortunately, the X-ray afterglow [22] does not show such a break, nor can one be ruled out with high confidence. A longer pointing could give a better response (on intense afterglow only). This might be a good target for future BeppoSAX observations.

In these proceedings L. Piro et al. show the results of the search for the effect of beaming based on the systematic comparison of spectral indexes and temporal decay indexes of the brightest afterglows [29]. The statistics of the BeppoSAX sample do not yet provide conclusive evidence in favour of beaming.

C Is the Detection of Fe in GRBs Real?

One of the few points well established from the theoretical analysis of GRB data is that the source of the X/γ-ray photons is expanding relativistically with a Lorentz factor of 10^2–10^3. One straightforward consequence is that, independent of any particular picture of the emission mechanism, we should not expect any line emitted from the main source of these photons, since it would be shifted and broadened until becoming undistinguishable from the continuum. However, this radiation will interact with the local environment and this interaction will introduce features detectable in our observer frame. Such features are absorption edges due to photons removed from the direct beam and fluorescence or recombination lines following the photoionization of the medium around the source. The only realistic candidate for emission features is Fe (if we except the possibility of Co or Ni of very recent synthesis). Two detections of Fe lines have been reported [28,43]. In both cases (GRB970508 and GRB970828) the evidence is good, but not compelling. The strongest point is that both are associated with a deviation from the power-law decay with a continuum spectral change. The association of an Fe reprocessing feature with an encounter with a dense region is very attractive. The ASCA spectral performance is better, but the case is challenged by a candidate counterpart radio source within a host at $z = 1$, inconsistent with the measured Fe line energy. A confirmation may come from more detections of *rebursting* afterglow, that are for sure a minority.

IV IS THERE A SINGLE CLASS OF GRBS?

A Do GRBs and X-ray Flashes Belong to the Same Class?

J. Heise has communicated at this conference that X-ray flashes (or γ-less gamma-ray bursts) have been detected by the SAX WFC in 3 years, at a rate of around 1 for every 3 GRBs. It is possible that many X-ray transients detected at high galactic latitude by previous satellites could belong to this class. They could be: 1) unusual flare stars; 2) high z GRBs; 3) soft GRBs without any implication on red-shift; 4) supernovae.

The GRB connection is very likely for some of them. In fact, more soft GRBs (such as GRB990704 [10] or GRB991226 [14]) have similar X/γ ratios and, if less luminous, would be probably classified as "X-ray flashes". More detections are needed for a successful follow-up or a more compelling upper limit. The follow-up of these flashes in now included in the BeppoSAX observing program.

B The Quest for the Afterglow of Short Bursts

Several attempts have been performed to classify or group GRBs. In 1972 Kouveliotou et al. [21] found that gamma-ray bursts have a bimodal distribution in duration with a saddle around 2 s. More clustering is evident if the GRBs are scattered in a plot that, besides duration, uses also the hardness ratio. One of the most interesting aspects of these analyses is that the various classes show different log N - log P distributions [2,37]. The short bursts, and possibly the long-soft bursts, have a luminosity distribution consistent with a homogeneous distribution in space. This would leave room for completely different classes and is one of the big open questions in GRB science. Unfortunately, we still do not know whether short bursts have afterglow emission and we do not have a single good position supporting an association with other astronomical objects. For short bursts we are in the same ignorance we had for all the bursts before 1997. The solution can only arrive from observations. The short bursts are a minority, but if no selection effect is present, 2–4 of them should have been detected by the BeppoSAX WFCs. The BeppoSAX [35,18] team is putting a big effort into preparing a follow up of such a GRB, if detected in a WFC. So far none has been detected. They are likely selected negatively from the BeppoSAX WFC analysis based on the 1-s rate meters, but we must admit that something is not completely understood.

C Is a part of GRBs Associated with Supernovae?

The gravitational collapse of a massive star provides many of the ingredients that are found in a GRB scenario and the potential association of GRBs with SN is extremely suggestive. Of course the time scales are completely different and according to the current SN picture a large amount of energy is lost in the form of neutrinos. The association of SN and GRB was suggested by the discovery in the error box of GRB980425 the SN1998bw [17] with unusually bright radio emission. No unambiguous fading X-ray source was found in the error box [26] of the WFC. In absence of a good alternative X-ray/optical or radio transient, the association with this very peculiar SN becomes attractive [26].

A further step forward was provided by the behavior of the afterglow of GRB980326 [3]. The optical decay curve after one month brightens again and eventually disappears below the visibility threshold, without any detected host. This behavior can be interpreted as due to the light curve of a SN similar to SN1998bw superimposed on a regular decaying afterglow. The SN contribution becomes apparent in time and eventually prevails.

Further evidence [30] is provided from the GRB970228. The minor, but significant deviations from the straight power-law decay, combined with a strong reddening, are not easy to explain with a simple fireball, and may be interpreted as the superposition of a SN and the *regular* afterglow. We notice that GRB980425 and GRB980326 are those with the highest X/γ ratio of the sample of the first 16 BeppoSAX GRBs [7]. GRB970228 has a very strong afterglow while in GRB980425 it is possibly missing. If the three associations are correct, we must assume strong beaming, which could also explain the large difference in the intrinsic luminosity.

V ANY COSMOLOGY WITH GRBS?

Since we know that GRBs (at least the long ones) are at cosmological distances ($0.43 < z < 3.4$ detected so far) it is attractive to think that they may become a tool for cosmological observations. GRB990123 at a red-shift of 1.6 has a fluence of 3×10^{-4} erg cm^{-2}. Thus, some of the bursts with a fluence of 10^{-6} to 10^{-7} erg cm^{-2} (and possibly the X-ray flashes) could be GRBs at $z > 10$. With respect to QSOs and galaxies, they could be more easily singled out because of the peculiar time behavior. Moreover, gamma radiation is more penetrating than the optical. It is possible that we miss optical counterparts of some GRBs because of the extreme extinction, while, so far, we could never really detect a significant (not local) absorption column from an afterglow [24].

May GRBs become a primary subject of cosmological observations?

1. GRBs belong to a relatively early Universe. Any picture of this Universe must include them. This is a somewhat trivial statement.

2. Can we use GRBs as a candle? Not as they are. Apparently the intrinsic luminosity's (X, γ, and afterglow) are quite different.

3. Can we identify a subset of GRBs as standard candles (e.g., excluding SN and beamed bursts), or can we find a tight correlation of intrinsic luminosity with some other observable? Some attempts are in progress.

4. Can we measure red-shift from the γ/X rays only, and detect sources not measurable in the Optical/IR band? This cannot be excluded but potentially Fe is the only candidate line and how much Fe is there at $z = 10$?

More conservatively, GRBs are already providing some important information on their environment from spectroscopy of reprocessed radiation, from the study of the transition from the GRB to the afterglow, from the comparison of optical extinction with X-ray absorption, and from the position within the host.

VI THE FUTURE OF AFTERGLOW INVESTIGATION

The BeppoSAX mission is extended to April 2001. The number of GRBs detected in the WFCs should reach 40, with around one half of afterglow sources and a few more X-ray flashes.

A new capability is now developing from the extension of the Interplanetary Network to include the NEAR satellite [5] and making the procedures faster.

The expectations for the near future are focussed on HETE-2, which will be a source of burst positions more frequent than SAX. The γ triggers only should be 2–3 times more frequent due to the larger field of view, but with a selection in luminosity. A further improvement will come from the X-ray cameras, intrinsically not more sensitive than the BeppoSAX WFCs, but equipped with a more complex trigger capability, involving the simultaneous use of data in different wavelengths, and also thanks to an observation program that avoids the crowded and rapidly variable galactic fields. The major impact of HETE-2 will come from the quick dissemination of the GRB positions to X-ray, optical and radio telescopes.

Although a significant improvement in the number of rapid GRB positions is expected, the number and rapidity of X-ray TOO pointings is not very clear, since large observatories will not invest too much time to such a program.

A HETE-2/BepppoSAX combination is likely the fastest (and actually such observations are already approved), but the different constraints on solar angles will limit the frequency of such observations.

A clue to the near future of afterglow studies will be the pointing of Chandra and XMM following BeppoSAX, HETE-2 and IPN positions. The expectations on the detection of low fluxes in the late afterglow, spectral features in absorption and in emission, and, hopefully, the first red-shift detection in X rays only, are important goals, providing an essay of the potential of the recently approved Swift mission. This mission will produce the first large scale survey of afterglow sources.

REFERENCES

1. Band, D., et al., *ApJ* **413**, 281 (1993).
2. Belli, B.M., *ApJ* **479**, L31 (1997).
3. Bloom, J.S., et al., *Nature* **401**, 453 (1999).
4. Burenin, R.A., et al., *A&A* **344**, L53 (1999).
5. Cline, T.L., et al., *A&A* **138**, 557 (1999).
6. Costa, E., et al., *Nature* **387**, 783 (1997).
7. Costa, E., *A&A* **138**, 425 (1999).
8. Covino, S., et al., *A&A* **348**, L1 (1999).
9. Djorgovski, S.G., et al., *GCN* 289 (1999).
10. Feroci, M., in preparation (2000).
11. Frail, D.A., et al., *Nature* **389**, 261 (1997).
12. Frontera, F., et al., *ApJ* **493**, L67 (1998).
13. Frontera, F., et al., *ApJ* in press (2000).

14. Frontera, F., et al., *ApJ*, submitted (2000).
15. Harrison, F., et al., *ApJ* **523**, L121 (1999).
16. Galama, T., et al., *ApJ* **500**, L97 (1998).
17. Galama, T., et al., *Nature* **395**, 670 (1998).
18. Gandolfi, G.G., these proceedngs.
19. Gruzinov, A., & Waxman, E., *ApJ* **511**, 852 (1999).
20. Katz, J.I., *ApJ* **432**, L107 (1994).
21. Kouveliotou, C., et al., *ApJ* **413**, L101 (1993).
22. Kuulkers, E., et al., *ApJ*, submitted (2000).
23. Meszaros, P. & Rees, M.J., *ApJ* **476**, 232 (1997).
24. Owens, A., et al., *A&A* **339**, L37 (1998).
25. Paczynsky, B. & Xu, G., *ApJ* **427**, 708 (1994).
26. Pian, E., et al., *ApJ*, submitted (1999).
27. Piro, L., et al., *A&A* **331**, L41 (1998).
28. Piro, L., et al., *ApJ* **514**, L73 (1999).
29. Piro, L. & Stratta G., these proceedings.
30. Reichart, D.E., *ApJ* **521**, L111 (1999).
31. Sari, R., Piran, T., & Narayan, R., *ApJ* **497**, L17 (1998).
32. Sari, R., 1997, *ApJ* **489**, L37 (1998).
33. Sari, R., Piran, T., & Halpern, J.P., *ApJ* **519**, L17 (1999).
34. Smith, D.A., et al., *ApJ*, in press (2000).
35. Smith, M., et al., *A&A* **138**, 479 (1999).
36. Tavani, M., *ApJ* **466**, 768 (1996).
37. Tavani, M., *ApJ*, **497**, L21 (1998).
38. van Paradijs, J., et al., *Nature* **386**, 686 (1997).
39. Vietri, M., *ApJ* **478**, L9 (1997).
40. Vietri, M., *ApJ* **488**, L105 (1997).
41. Wijers, R.A.M.J., Rees, M.J., & Meszaros, P., *MNRAS* **288**, L51 (1997).
42. Wijers, R.A.M.J., et al., *ApJ*, in press (2000).
43. Yoshida, A., et al., *A&A* **138**, 433 (1999).
44. in't Zand, J., et al., *ApJ* **505**, L119 (1998).

Constraints to the Nature of the Central GRB Engine from a Comparative Analysis of X-ray Properties of Afterglows

G. Stratta[a], L. Piro[a], G. Gandolfi[a], L. A. Antonelli[d], E. Costa[a], M. Feroci[a], P. Soffitta[a], F. Frontera[b,c], J. Heise[e], and the BeppoSAX GRB team

[a] Istituto di Astrofisica Spaziale - CNR - Frascati, Italy
[b] Istituto Tecnologie e Studio Radiazioni Extraterrestre - CNR - Bologna, Italy
[c] Dipartimento di Fisica - Universita' di Ferrara - Ferrara, Italy
[d] BeppoSAX-SDC - Osservatorio Astronomico di Roma
[e] Space Research Organizations in the Netherlands, Utrecht, The Netherlands

Abstract. Afterglows are a powerful source of information on GRBs. In this paper we present a systematic study of 11 afterglows observed with the Narrow Field Instruments on board on BeppoSAX. The intent of this analysis is to: i) disclose the nature of GRB progenitors by the temporal and spectral features imprinted by the nearby environment; ii) disentangle the jet versus spherical expansion of the fireball; iii) test and constrain the fireball model relations. We find no strong evidence of iron lines, except for the cases of GRB970508 (BeppoSAX) and GRB970828 (ASCA). From the average spectral and temporal properties we estimate a lower limit on the opening angle of the jet: $\theta > 12°$.

INTRODUCTION

Two of the most important issues regarding the GRB phenomenon are the nature of GRB progenitor and the energetics involved in this phenomenon. The former could be disclosed by the analysis of the nearby environment, whereas the latter may be estimated by disentangling jet vs. spherical emission. In this paper we discuss these arguments and present a systematic study of X-ray afterglows, in order to search for spectral and temporal features related to the GRB nearby environment and to investigate the compatibly of BeppoSAX data with fireball model predictions for spherical or collimated expansions.

GOALS

Disclose the Nature of GRB Progenitor by Temporal and Spectral Features

The nature of the GRB progenitor and central engine could be disclosed by the analysis of temporal and spectral features imprinted by the nearby environment. In fact, massive and short lived progenitors, such as hypernova, are to be found in star forming regions where the environment is particularly dense and enriched with high metallicity elements. On the contrary, the merging of two neutron stars, or a BH plus NS, could happen far away from the star forming region, in an environment more rarefied and less dense, due to the kick velocity acquired during the SN explosion phase. A possible diagnostic for metallicity and density of the environment is the presence of Fe K-α and Fe K-edge emission lines in the spectra. These are produced by fluorescence from the surrounding matter illuminated by the X-ray photons originating in the afterglow of the GRB [2,7,11]. The formation of a thick shell of material at ~ 1 light-day from the central engine is more easily explained in the case of massive progenitors than in low mass binary mergers. This thick shell could be seen either in the spectra as an iron emission line, or in the X-ray afterglow light curves as an enhancement of the luminosity after about 1 day. Two events, by far, show evidence of these spectral and temporal features: GRB970508 [1] and GRB970828(ASCA) [10].

Test and Constrain the Fireball Model Predictions

In the fireball model, writing afterglow fluxes as $F_\nu(\nu,t) \sim t^{-\alpha}\nu^{-\beta}$, spectral and temporal slopes are linked together by precise relations that depend on the geometry of the fireball expansion [4,5]. Under the hypothesis of an adiabatic, spherical expansion in a constant density medium, there are two possible slopes for the decaying part of the light curve, where the generic synchrotron frequency ν is respectively higher and lower than the cooling frequency ν_c. The temporal index is $\alpha = 3p/4 - 3/4$ for $\nu < \nu_c$ and $\alpha = 3p/4 - 1/2$ for $\nu > \nu_c$, where p is the power law electron density distribution index. The spectral index corresponds to $\beta = (p-1)/2$ for $\nu < \nu_c$ and to $\beta = p/2$ for $\nu > \nu_c$. Thus, the relation between α and β in an adiabatic fireball is either $\alpha = 3\beta/2$ or $\alpha = 3\beta/2 - 1/2$. We have neglected at this stage the presence of stellar winds, a common phenomenon in massive stellar progenitor environments, a scenario still under study. Optical temporal indices observed at the same time as the X-ray afterglows were collected from the literature and compared with the light curves predicted by the fireball model in the adiabatic and in the spherical expansion cases. According to the model, if the critical frequency ν_c is below or above both the X-ray band and the optical band (that is we have a unique regime in both spectral bands) the optical temporal slope must be equal to the X-ray slope. If ν_c lies between the two bands, the X-ray slope follows the decay law expected for $\nu > \nu_c$ and the optical slope that for $\nu < \nu_c$.

Disentangle Jet vs. Spherical Expansion

One of the key problems in GRB physics concerns the amount of energy released in the explosion. The hypothesis that such explosions might be collimated in two jets is very reasonable and afterglow observations still do not discriminate this scenario from the spherical expansion case. However, different relations hold when beaming is assumed. These new relations are $\alpha = 2\beta + 1$ for $\nu < \nu_c$ and $\alpha = 2\beta$ for $\nu > \nu_c$ [5]. Evaluating which one of the expected fireball model relations better describes spectral and temporal data is a valid test of spherical vs. jet expansion. With the same assumptions described in the previous section, we can now relate the slopes in the energy domain to those in the temporal domain for a number of BeppoSAX afterglows.

METHODS

As far as the 7th of October, 1999, the Gamma Ray Burst Monitor and the two Wide Field Cameras on board BeppoSAX have seen 24 events; 19 of them triggered a fast follow up with the Narrow Field Instruments (NFI) and are afterglow candidates. By now, a systematic analysis of spectra and light curves of 11 X-ray afterglows has been carried out (GRB970228, GRB971214, GRB971227, GRB980329, GRB980515, GRB980519, GRB980613, GRB980703, GRB981226, GRB990123, GRB990510). A fit of LECS and MECS spectral data (0.1-10.0 keV), in each first Target Of Opportunity (TOO) observation, has been performed using a power law function, with energy index β, multiplied by an exponential photoelectric absorption factor at low energy. Power law function fits were performed on MECS count rates (1.6-10.0 keV) to determine the decay law for each TOO. A χ^2 test was applied to evaluate the goodness of the fit and to estimate the best fit temporal decay and spectral index parameters. On the basis of the spectral results obtained for GRB970508 [1], first $20ks$ exposure spectral data for afterglows with known redshift (GRB970228, GRB971214, GRB980703, GRB990123) were fitted with the previous spectral model plus a Gaussian function with fixed energy and width. The energy has been fixed at the redshifted value expected for the Fe K-α fluorescent line and Fe K-edge recombination, at 6.4 keV and at 9.28 keV and a width of 0.1 keV (that is, the width of a line originating from material with bulk velocity $0.01(z+1)c$.

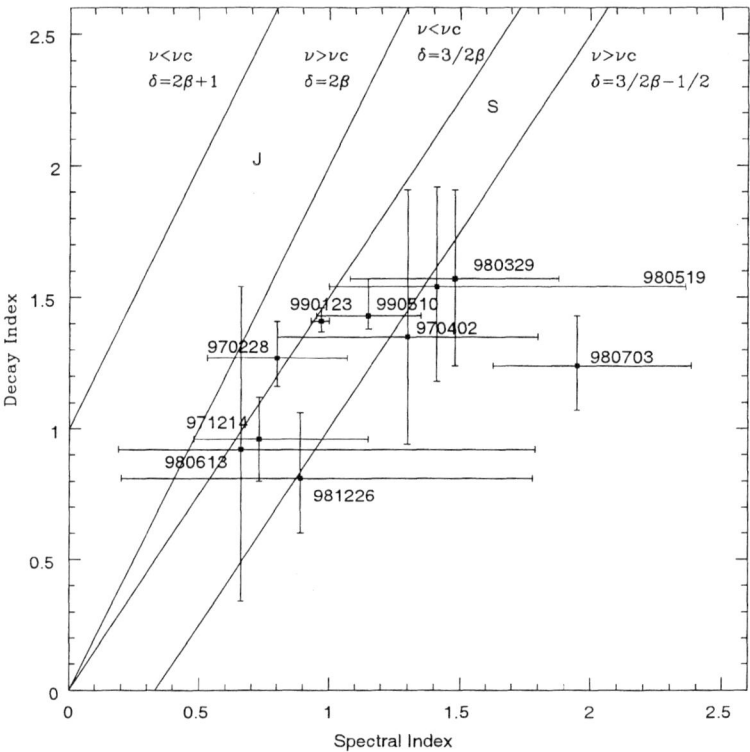

FIGURE 1. Temporal vs. spectral X-ray afterglow index.

RESULTS

Features

No other afterglows in the sample show evidence either of re-bursting or of line presence. Emission line intensity upper limits from Fe K-α fluorescence at 6.4 keV and Fe K-edge recombination at 9.28 keV have been estimated. All are consistent with $\sim$ a few 10^{-5} ph cm^{-2} s^{-1}.

Fireball Model and Jet vs. Spherical Expansion

In Figure 1, decay index vs spectral index (1σ errors) of the analyzed afterglows are compared within the frame of spherical and jet expansion fireball models. The jet expansion case (J) predicts two relations between the energy and the temporal domain, in case of fast or slow cooling, that are different from the spherical expansion ones (S). The sample is not biased in the afterglow decay index range >1.5 as ~95% of all the events detected in WFC were detected by NFI. Afterglows show an average trend, compatible with the spherical expansion fireball model relations. We stress that this does not imply a spherical expansion but does put a lower limit on the opening angle of the jet. It can be no less than the relativistic angle at the moment of the observed X-ray afterglow, that is about 2 days from the gamma event, i.e. when our afterglow observations end. Using the expression that relates the time t when the two angles are equal to the beaming angle, $\theta_{jet} \sim 0.1[\frac{t}{6.2}(\frac{n}{E_{52}})^{1/3}]^{3/8}$ [5] and assuming a constant density $n = 1$ cm^{-3}, a released total energy $E = 10^{-52}$ ergs and a time of ~48 hr from the burst, the collimation angle must be >12°. Hence, if the beaming is present, the collimation can not be very high.

REFERENCES

1. Piro, L., et al., *A&A* **331**, L41 (1998).
2. Meszaros, P. & Rees, M.J., *MNRAS* **299**, L10 (1998).
3. Piro, L., et al., *ApJ* **514**, L73 (1999).
4. Sari, R., Piran, T., & Narayan, R., *ApJ* **497**, L17 (1997).
5. Sari, R., Piran, T., & Halpern, J.P., *ApJ* **519**, L17 (1999).
6. Vietri, M., Perola, C., Piro, L. & Stella, L., astro-ph/9906288.
7. Boettcher, M., astro-ph/9912030, Apj Submitted.
8. Harrison, F.A., et al., astro-ph/9905306.
9. Weth, C., Meszaros, P., Kallman, T., & Rees, M.J., astro-ph/9908243, Apj Submitted.
10. Yoshida, A., et al., 1999, A&ASS, 138, 433, Proc. of the Gamma Ray Burst in the Afterglow Era, F. Frontera & L. Piro eds.
11. Lazzati, D., et al., *MNRAS* **304**, L31 (1999).

Search for X-ray Afterglows from Gamma-Ray Bursts in the RASS

J. Greiner[1], D. H. Hartmann[2], W. Voges[3], T. Boller[3], R. Schwarz[1], and S. V. Zharykov[4]

[1] Astrophysical Institute Potsdam, An der Sternwarte 16, 14482 Potsdam, Germany
[2] Clemson Univ., Dept. of Physics and Astronomy, Clemson, SC 29634, USA
[3] MPI for Extraterrestrial Physics, 85740 Garching, Germany
[4] Special Astrophysical Observatory, 357147 Nizhnij Arkhyz, Russia

Abstract. We report on a search for X-ray afterglows from gamma-ray bursts using the ROSAT all-sky survey (RASS) data. If the emission in the soft X-ray band is significantly less beamed than in the gamma-ray band, we expect to detect many afterglows in the RASS. Our search procedure generated 23 afterglow candidates, where about 4 detections are predicted. Follow-up spectroscopy of several counterpart candidates strongly suggests a flare star origin of the RASS events in many, if not all, cases. Given the small number of events we conclude that the data are consistent with comparable beaming angles in the X-ray and gamma-ray bands. Models predicting a large amount of energy emerging as a nearly isotropic X-ray component, and a so far undetected class of "dirty fireballs" and re-bursts are constrained.

SURVEY DATA AND EXPECTED AFTERGLOW RATE

If afterglow and burst emission are from separate regions one must seriously consider the possibility that prompt γ-ray and delayed X-ray emission are beamed (if at all) differently. If so, one expects X-ray afterglows to be less beamed than GRBs. We describe here our results to test this possibility with a search for X-ray afterglows that were fortuitously detected during the RASS. All technical details and a more thorough discussion are reported in Greiner et al. (1999).

During the RASS, the ROSAT field of view scans a full 360° circle on the sky, covering a source located inside the scan circle for typically 10–30 sec. A source is covered by consecutive telescope scans between two days (near the ecliptic equator) up to 180 days (at the ecliptic poles). Our study relies on the product of exposure in time and coverage in area so that the large exposure at the poles and low equatorial exposure is compensated by the correspondingly small/large solid angles (according to cos[ecliptic latitude]), thus yielding a rather uniform search pattern. Even with a single exposure of 10–30 s duration the sensitivity of ROSAT is sufficient to detect GRB X-ray afterglows for several hours after the burst (Fig. 1)

The fraction, f, of afterglows detectable during the RASS depends critically on three parameters: (1) the fraction of GRBs that have detectable X-ray afterglows, (2) the possible correlation of X-ray flux to γ-ray peak flux (or fluence, or some other characteristic aspect of the GRB itself), (3) the X-ray intensity decay law. It is currently not clear how one should combine all these factors into a proper statistical distribution from which to derive the overall sampling fraction f. We thus simply use the existing database as a representative set of templates and compare this set to the ROSAT PSPC sensitivity. This implies that the RASS would in fact be sensitive enough to detect all GRB afterglows in 3 subsequent scans, and $\sim$80% in 5 scans (see Fig. 1). We adopt a conservative fraction of $f = 0.8$.

The number of detectable X-ray afterglows from GRBs beamed towards us (based on the BATSE detection rate) during the RASS is $N^{agl} = f \times S_R^{agl} \times R_{GRB}$, where $R_{GRB} = 900$ GRBs/sky/yr $\equiv 1$ GRB/(16628$\square° \times$ days) is the rate density of GRBs and S_R^{agl} is the RASS afterglow coverage function. The temporal completeness of the RASS was 62.5% (Voges et al. 1999), so that $S_R^{agl} = 76435$ $\square° \times$ days. Thus, we expect $N^{agl} = 4.6 \times f \sim 3.7$ GRB afterglows to be detected during the RASS.

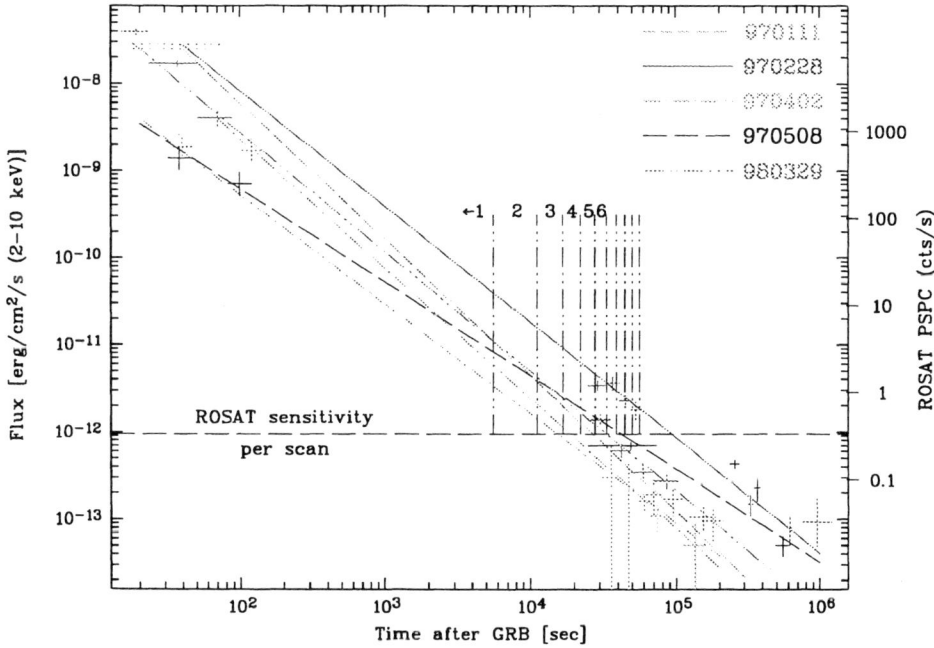

FIGURE 1. Afterglow light curves of some observed GRB X-ray afterglows in the 2–10 keV range (GRB 970111: Feroci et al. 1998; GRB 970228: Costa et al. 1997; GRB 970402: Nicastro et al. 1998; GRB 970508: Piro et al. 1998; GRB 980329: in 't Zand et al. 1998) extrapolated into the ROSAT band (scale on the right). The vertical lines mark the time windows for the possible coverage of a GRB location by ROSAT during its scanning mode.

THE SEARCH FOR AFTERGLOW CANDIDATES

We produced scan-to-scan light curves for all RASS sources with either a count rate larger than 0.05 cts/s or a detection likelihood exceeding 10, resulting in a total of 25,176 light curves. Each of these light curves consists of about 20 to 450 bins spaced at 96 min., with each bin corresponding to 10-30 sec. exposure time. We apply three selection criteria to these light curves: (1) The maximum bin should have a signal-to-noise ratio of S/N>3 above the mean count rate around the maximum. (2) The mean count rate derived from observations obtained until one bin prior to the maximum count rate should be consistent with zero. (3) The mean count rate at times later than those covered by 5 bins past maximum should also be consistent with zero. This suppresses transient sources that have quiescent emission at detectable levels, such as flare stars.

Application of the above listed criteria yields a total of 32 GRB afterglow candidates. We then proceed with additional conditions that proper afterglows should display: (i) Sources with double and multi-peak structures are excluded, because this pattern does not fit "standard" X-ray afterglow behavior (4 sources). (ii) Sources with a rise extending over several bins and zero flux immediately after the peak are removed (2 sources). (iii) Sources with low-level (below the RASS threshold) persistent X-ray emission during serendipituous pointed ROSAT observations were excluded (3 sources). (iv) We correlate the candidate list with optical, infrared, and radio catalogs, and exclude sources with known counterparts (1 source).

The application of these selection steps yields a total of 23 transients as viable X-ray afterglow candidates. About 50% of the light curves display single peaks, i.e., outbursts with just one bin satisfying S/N>3 and otherwise zero count rate. The remainder show decays that more closely resemble GRB afterglow behavior.

To estimate the flare star fraction of the events we obtained optical spectra for six randomly selected sources. All 6 objects are Me flare stars. Three further objects of our sample were optically identified by other groups, and also are flare stars. Based on the optical brightness of these flare stars and the well-known L_X/L_{opt} ratio of 1/50...1/100 the expected X-ray intensity during quiescence is $1\times10^{-14}...2\times10^{-13}$ erg/cm^2/s. This corresponds to ROSAT PSPC count rates of 0.0015...0.03 cts/s and is below the RASS sensitivity, thus consistent with the non-detection outside the X-ray flare (which caused detection during the RASS).

We thus argue that the bulk of the "afterglows" are probably due to X-ray flares from nearby late-type stars, and that the existing data support the notion that the RASS contains at most a few X-ray afterglows from GRBs. This interpretation is consistent with the expected number of afterglows ($N^{agl} = 3.7$). 1RXS J120328.8+024912 is the best candidate for a GRB X-ray afterglow simply due to the fact that the ROSAT error box does not contain a bright ($m < 22$ mag) stellar object though the light curve is single-peaked. While it is difficult to determine the likelihood that a flare of this large amplitude from a position with no optical counterpart could be due to a statistical fluctuation, we note that this event is among the largest amplitude events of our whole sample.

If we argue that the RASS data contain a few afterglows, then the data are obviously consistent with the expected theoretical rate (especially considering the significant uncertainties affecting our estimate of the afterglow expectation value). This implies that GRB afterglows do not have a significantly wider beaming angle in the X-ray band relative to the gamma-ray band. This is to some extent in agreement with predictions of the "standard" fireball model (Meszaros & Rees 1997; Piran 1999), given the fact that we are only sampling a few hours of emission following the GRB. As the fireball slows due to interaction with a surrounding medium the bulk Lorentz factors of the flow decrease and the beaming angle increases. However, the RASS data cover a time interval of $\sim$1–8 hrs after the GRB event. During this time the fireball is expected to decelerate from $\Gamma \gtrsim 100$ to $\Gamma \sim 10$. Thus, the flow is still highly relativistic and the afterglow emission is still far from isotropic.

On the other hand, if we argue that those of the events which are not optically identified are in fact GRB afterglows, then the rate apparently exceeds expectations. However, the enhancement factor is less than a few. Furthermore, the uncertainties are large and the sample is still small. Again we would conclude that the RASS results support consistency between observations and theoretical expectations, with only marginal evidence for less beaming in the X-ray band.

Both points of view basically conclude the same; beaming of GRBs and of their afterglows is, if it exists, comparable. This conclusion supports a similar result (Grindlay 1999) obtained from an analysis of fast X-ray transients observed with *Ariel V* (Pye & McHardy 1983) and earlier instruments. We also emphasize that our results and those discussed by Grindlay (1999) can be used to place constraints on presently undetected GRB populations that preferentially emit in the X-ray band. Dermer & Mitman (1999) pointed out that the initial fireball Lorentz factor, Γ_0, is crucial for determining the appearance of the GRB. Since Γ_0 is related to the ratio of total burst energy to rest mass energy of the baryon load a "clean" (low baryon load and/or large energy) fireball is characterized by Γ_0 in excess of 300 (according to Dermer's definition), while a "dirty" fireball (heavy load) is characterized by a very small Lorentz factor. Dermer & Mitman argue that clean fireballs produce GRBs of very short duration with emission predominantly in the high-energy regime, while dirty fireballs produce GRBs of long duration that preferentially radiate in the X-ray band. These bursts are in fact predicted to be X-ray bright, but have probably not yet been detected by BATSE and similar instruments, because these detectors are "tuned" to events for which Γ_0 falls in the range 200–400 (Dermer & Mitman 1999). The absence of a significant number of X-ray transients in the RASS and the *Ariel* survey thus suggests that the frequencies of "dirty" GRBs relative to bursts with a "normal" baryon load is comparable.

Vietri et al. (1999) drew attention to the "anomalous" X-ray afterglows from GRB 970508 and GRB 970828, which exhibit a resurgence of soft X-ray emission and evidence for Fe-line emission. These authors interpret the delayed "rebursts" in the framework of the SupraNova model (Vietri & Stella 1998) in which the GRB progenitor system creates a torus of iron-rich material. The GRB fireball heats the torus, which cools via Bremsstrahlung, leading to a "reburst" in the X-ray band. The emission pattern of this heated torus should be nearly isotropic, so that one expects many X-ray afterglows that are not accompanied by GRBs. The RASS data place severe constraints on this type of reburst scenario, because these delayed components are predicted (Vietri et al. 1999) to be bright (10^{-4} erg cm^{-2}) and of long duration ($\sim 10^3$ s). The rarity of afterglows in the RASS data suggests that GRBs from "SupraNovae" do not constitute the bulk of the observed GRB population, unless the GRBs are also roughly isotropic emitters (which is in conflict with the correspondingly large energy requirements).

Another constraint can be placed on GRBs related to supernovae (SN). If the association of GRB 980425 with SN1998bw is real (e.g., Galama et al. 1998, Woosley et al. 1999) then such SN-related GRBs would dominate the total GRB rate by a factor of $\sim$1000 due to their low luminosities implied by the small redshift ($z = 0.0085$) of the host galaxy. It can be argued that GRB 980425 was beamed away from us, and we merely saw the less beamed afterglow emission. If this is true, we expect many X-ray afterglows in the RASS data. Again, our results constrain these possibilities, but more quantitative results require detailed simulations.

Acknowledgments: JG and RS are supported by the German Bundesministerium für Bildung, Wissenschaft, Forschung und Technologie (BMBF/DLR) under contract Nos. 50 QQ 9602 3 and 50 OR 9708 6, respectively, and SZh by INTAS N 96-0315. JG acknowledges a travel grant from DFG (KON 1973/1999 and GR 1350/7-1) to attend this conference. DHH expresses gratitude for support and hospitality during visits to the AIP in Potsdam and the MPE in Garching. The *ROSAT* project is supported by BMBF/DLR and the Max-Planck-Society.

REFERENCES

1. Costa E., Frontera F., Heise J., et al., 1997, Nat. **387**, 783
2. Dermer C.D., Mitman K.E., 1999, ApJ **513**, L5
3. Feroci M., Antonelli L.A., Guainazzi M., et al., 1998, A&A **332**, L29
4. Galama T., et al., 1998, Nat. **395**, 670
5. Greiner J., Hartmann D.H., Voges W., Boller T., Schwarz R., Zharykov S.V., 1999, A&A (in press; astro-ph/9910300)
6. Grindlay J.E., 1999, ApJ **510**, 710
7. in 't Zand J.J.M., Amati L., Antonelli L.A., et al., 1998, ApJ **505**, L119
8. Meszaros P., Rees M.J., 1997, ApJ **476**, 232
9. Nicastro L., Amati L., Antonelli L.A., et al., 1998, A&A **338**, L17
10. Piran T., 1999, Phys. Rep. **314**, 575
11. Piro L., Amati L., Antonelli L.A., et al., 1998, A&A **331**, L41
12. Pye J.P., & McHardy I.M., 1983, MNRAS **205**, 875
13. Vietri M., Stella L., 1998, ApJ **507**, L45
14. Vietri M., Perola C., Piro L., Stella L., 1999, MNRAS (subm.; astro-ph/9906288)
15. Voges W., Aschenbach B., Boller Th., et al., 1999, A&A **349**, 389
16. Woosley S.E., Eastman R.G., Schmidt B.P., 1999, ApJ **516**, 788

BATSE Observations of Gamma-Ray Burst Tails

Valerie Connaughton

NASA Marshall Space Flight Center[1]
Huntsville, AL 35812

Abstract. With the discovery of low-energy radiation appearing to come from the site of gamma-ray bursts in the hours to weeks after the initial burst of gamma rays, it would appear that astronomers have finally seen a cosmological imprint made by the burster on its surroundings. I discuss in this paper the phenomenon of post-burst emission in BATSE gamma-ray bursts at energies traditionally associated with prompt emission. By summing the background-subtracted signals from hundreds of BATSE bursts, I find that tails out to 2000 seconds after the trigger may be a common feature of Class I events, but not of the shorter Class II bursts. The tail component appears independent of both the duration and brightness of the burst, and may be softer.

INTRODUCTION

Afterglow radiation at wavelengths from radio to X-rays has now been seen in several gamma-ray bursts (GRB) detected by Beppo-SAX and BATSE (e.g., [4,8,9]). It is clear that there is a frequency-dependent temporal decay to this emission for a particular event, but the afterglow signatures of different bursts exhibit a wide variety of temporal behaviors. Indeed, there does not even appear to be a relationship between the brightness of the initial GRB and the brightness or duration of the afterglow emission.

There are gaps in the temporal coverage of the emission by the various detectors involved in burst observations. The prompt emission at soft gamma-ray energies is strong enough to trigger BATSE and other dedicated GRB detectors. Prompt soft X-ray emission can be seen by the SAX Wide-Field Camera, delayed soft X-rays are seen hours later with the more sensitive Narrow Field Instruments. The relationship between these components is unclear. Shock models for GRBs suggest that the initial photons (prompt emission) are a result of collisions of shells of different Lorentz factor initiated by some kind of central engine, while the afterglow is a result of the external shock formed by these shells colliding with an external medium. It has been suggested in the case of GRB 970228 that if one extends

---

[1] National Research Council Research Associate

the frequency-dependent temporal decay back in time, some of the soft gamma-ray emission which appears to be part of the prompt component is consistent with being afterglow radiation [3]. It is difficult to make this argument for bursts in general: the later peaks in bursts do not seem to be significantly different from the initial ones, and not all bursts are multi-episodic. The argument does raise, however, the question of the nature of any hard component to the afterglow emission. What does it look like, when does it start and how long does it last? Recently, Fenimore et al. [5] have proposed that the prompt and afterglow emission may overlap if a shell of prompt photons is decelerated by the external medium prior to being caught up by other shells produced by the central engine.

In the work presented here, the period of time from the GRB trigger to 2000 seconds later is explored using BATSE continuous data above 20 keV.

BATSE MEASUREMENTS OF GRB TAILS

BATSE receives telemetry with 16 energy channels and 2.048 second time resolution for the $\sim 80\%$ of the time during which the detector voltages are activated and the spacecraft is in contact with a TDRSS satellite. Over the the ~ 5600 second spacecraft period, large variations in background count rates are registered by the BATSE Large Area Detectors (LADs) owing to passage through regions of differing magnetospheric activity. Careful background subtraction is required to extract any subtle signal over these time-scales. A technique using the average of the count rates 15 orbits before and after the orbit of interest has been developed for the purpose of this analysis [2].

For an individual orbit, the sensitivity of this technique is estimated to be of the order of 10^{-9} erg cm^{-2} s^{-1}, or an order of magnitude lower than the BATSE trigger threshold. The technique has been applied in examining the possible presence of a weak tail in the bright gamma-ray burst GRB 980923 [6], where the traditional method of background subtraction by interpolation of intervals before and after the GRB failed owing to the duration of the tail, the variations in background levels, and the faintness of the signal in the tail.

To explore GRB tail emission in general, rather than individual (possibly unique) events, the background-subtracted signals from many bursts were summed, aligning the lightcurves at the peaks of the GRBs as done by Mitrofanov et al. [7] in their analysis of the asymmetry of burst emission. The robustness of the background subtraction to such signal stacking was assessed using test orbits [2] and the resulting test lightcurve was flat to within the BATSE count rate statistics.

Data from 6 CGRO orbits on 3 days without spacecraft reorientation (to maintain a constant detector geometry relative to the burst position) or GRB triggers (apart from the burst of interest) are needed to analyze a burst, leading to a high attrition rate in data selection. The count rates in each 2.048 second bin for which data exist for all three days and during which the burst location is not Earth-occulted are included in the light curve.

RESULTS

Figure 1 shows the combined lightcurve for 400 Class I GRBs. These bursts are all the BATSE GRBs in the first 7500 triggers which (i) satisfy the criteria for background subtraction, (ii) have satisfactory background fits and (iii) have durations greater than 2 seconds. The time axis is logarithmic and in seconds relative to the time of the aligned peaks, t_0, and the rates are summed above 20 keV and given per burst by dividing the total rates in a time bin by the number of bursts contributing to that bin. The time axis starts at $t_0 + 40$ seconds.

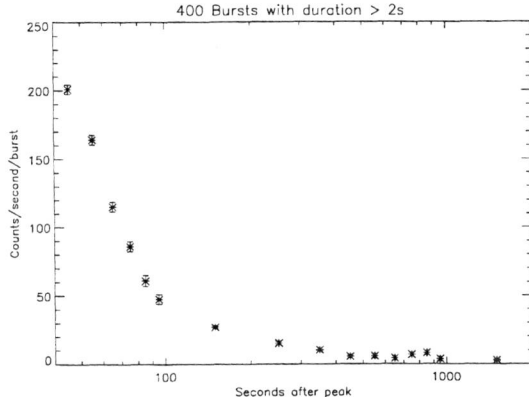

FIGURE 1. Lightcurve for summed background subtracted Class I BATSE bursts after peak alignment, with peak time suppressed.

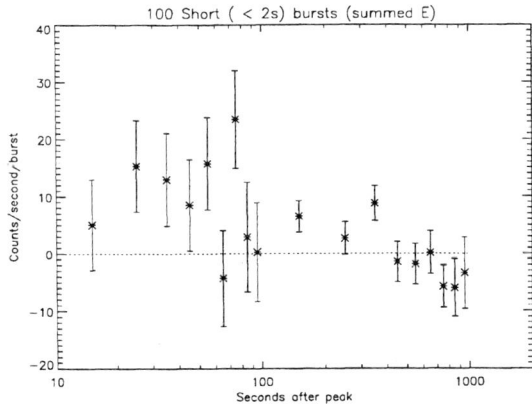

FIGURE 2. Lightcurve for summed background subtracted Class II BATSE bursts after peak alignment, with peak time suppressed.

Figure 2 shows the Class II bursts — those with durations less than 2 seconds and starts at $t_0 + 10$ seconds. A time bin must have data for at least 40 bursts to merit inclusion in either figure. There are regions of the orbit following a trigger which are not well sampled. It is very likely that a burst position will be occulted 2000 seconds after its detection. It can be seen that a tail is seen in the Class I bursts which is totally absent in the Class II bursts. The tail extends to at least the period between 1000 and 2000 seconds after t_0, the final bin for which data are available. It is reasonable to worry about the tail being due to the longest bursts, so the Class I bursts were sorted by duration, and a medium-duration subsample defined as the shortest 50% of Class I events.

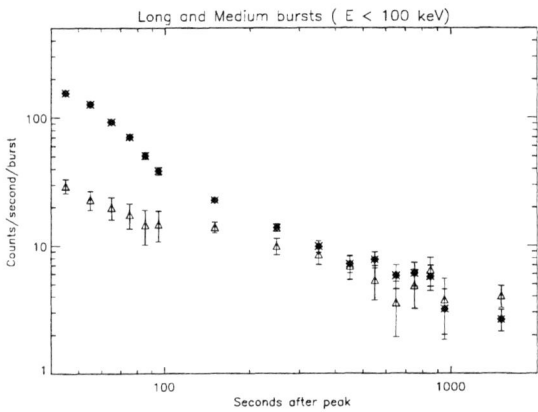

FIGURE 3. Class I BATSE bursts summed and peak-aligned with peak time suppressed. 400 bursts lasting longer than 2 s (stars) and 200 medium bursts (triangles) with 30 s as the dividing duration.

Figure 3 starts at $t_0 + 40$ seconds and shows on a logarithmic scale the count rates per burst in the long and medium Class I bursts between 20 and 100 keV. The 200 medium-duration bursts last between 2 and 30 seconds so that the main emission is deemed to be over by the time of the start of the plot. In Figure 4 the count rates are given as a fraction of the count rates at t_0. At 1000 seconds after the peak of the hypothetical average GRB, for example, the tail is at 1/1000 the flux level at the peak.

The tail is visible in both duration groups, and although the count rates are higher for the longer bursts until $t_0 + 200$ seconds, the contribution to the tail beyond this time is independent of burst duration. This suggests a component distinct from the prompt emission. If the emission seen from the 200 medium-duration bursts is mostly tail emission from $t_0 + 40$s onward, then the average appearance of the tail can be fit by a power law of index 1.6. It is probable, however, that like most burst and afterglow properties, this tail varies considerably in individual bursts so that the actual average power law index may not be a physically meaningful quantity.

Division of the 400 Class I bursts into three brightness-selected groups indicates that the tail is measured in the summed light-curves of even the dimmest group, and that although the magnitude of the tail appears brightest in the bright group, the fraction of the peak count rate in the tail is smaller than in the dimmer groups.

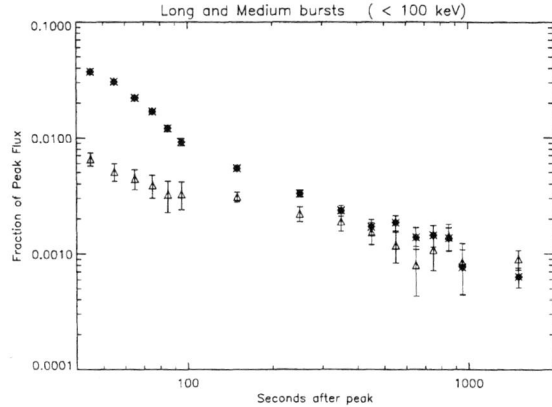

FIGURE 4. Signal as a fraction of peak count rates for summed, peak-aligned Class I BATSE bursts with peak time suppressed. 400 bursts longer than 2 s (stars) and 200 medium bursts (triangles) between 2 s and 30 s.

This again suggests a component independent of the main emission mechanism.

As a measure of the spectrum of the tail relative to the main emission, the channel 2:1 hardness ratio shows that in the complete Class I sample, the type of softening with time seen in the spectra of individual bursts can be seen until about $t_0 + 200$ seconds. After this a discontinuous softening is seen, with the remainder of the lightcurve appearing flat, but statistics are poor. Clearly, the issue of spectral differences between prompt and tail emission is important in establishing distinctions between the two components. This study is probably best performed on individual bursts, where the transition between prompt and tail emission is clearly seen. Giblin et al. [6] and Burenin et al. [1] have found such spectral transitions in the tails of BATSE GRB 980923 and GRANAT GRB 920723, respectively.

It is not clear whether tail components are part of prompt emission or afterglow, but they do seem to be a feature of the long BATSE gamma-ray bursts. No tails are seen in the combined lightcurve from short GRBs; neither have afterglows been measured thus far for short bursts by BeppoSAX.

REFERENCES

1. Burenin, R. A., et al., *Astron. Astrophys. Suppl.* **138**, 443 (1999).
2. Connaughton, V., in preparation.
3. Costa, E., et al., in *Gamma-Ray Bursts: 4th Huntsville Symp.*, eds. C.A. Meegan, R.D. Preece, T.M.Koshut, AIP Conf. Proc. 428, New York, 1998, pp. 409.
4. Costa, E., et al., *Nature* **387**, 783 (1997).
5. Fenimore, E. E., et al., these proceedings.
6. Giblin, T., et al., *Astrophys. J.* **524**, L47 (1999).
7. Mitrofanov, I. G., et al., *Astrophys. J.* **459**, 570 (1996).
8. Piro, L., et al., *Astron. Astrophys.* **331**, L41 (1998).
9. van Paradijs, J., et al., *Nature* **386**, 686 (1997).

Possible Evidence for Soft Glowing Emission from Gamma-Ray Bursts Detected by the APEX Experiment on Phobos-2

Dmitri A. Litvine, Igor G. Mitrofanov, and Alexandr S. Kosyrev

Space Research Institute, Profsojuznaya str. 84/32, 117810, Moscow, Russia

Abstract. Data are presented on gamma-ray bursts detected by the Russian-French APEX experiment on the Phobos-2 interplanetary spacecraft. The stable background conditions of this experiment are shown to be favorable for the search for a long duration component of gamma-ray bursts. Several events are presented with possible evidence for this component.

INTRODUCTION

Recent discovery of X-ray afterglow from gamma-ray bursts (e.g., see [1]) has resulted in one very principal question on the bursts' physics: does the afterglow begin during the burst emission, or does it follow it? This question is rather important for comprehensive understanding of the gamma-ray burst phenomenon. We believe that the final response will be presented by future observations (e.g., by the HETE-2 mission [2]). However, we wish to present below our studies of this question based on data from the APEX experiment on the Russian Phobos-2 interplanetary mission [3].

DESCRIPTION OF THE APEX EXPERIMENT ON THE RUSSIAN PHOBOS-2 INTERPLANETARY MISSION

The Soviet-French APEX experiment on the Russian Phobos-2 spacecraft operated for 8 months from July 1998 until March 1989 [3]. It used large 10 cm × 10 cm cylindrical CsI(Tl) gamma-ray detector with a spectral resolution of 11% at 662 keV. As far as we know, APEX was one of the most sensitive gamma-ray detectors ever flown in interplanetary space.

APEX was developed for measurements of Martian gamma rays [4] and for the detection of gamma-ray bursts and solar flares [3]. During the Phobos-2 mission,

62 triggered events were identified as gamma-ray bursts [3]. The stable background was a very particular property of APEX operation. Except during several strong solar flares, the APEX background was very stable or varied quite predictably. Moreover, the Phobos-2 spacecraft carried several another instruments for monitoring electrons, protons are heavy ions at different energies [5]. The data from these instruments are available to know the local background of charge particles at any required moment of the mission. Therefore, data from the Phobos-2 mission provide the interesting possibility to look for long-duration glowing emission from gamma-ray bursts. We explore this below.

POSSIBLE EVIDENCE OF A GLOWING COMPONENT FOR BRIGHT APEX GAMMA-RAY BURSTS

To search for glowing emission, we use the 17 brightest APEX gamma-ray bursts, with $V/V_{max} < 0.25$, because brighter events could correspond to brighter glowing.

The main type of APEX burst spectral data is M4. For each trigger, the format of M4 data includes 116 energy spectra, with 8 from memory recorded just before a trigger. Each M4 spectrum contains 108 energy channels from 64 keV up to 9.2 MeV. The duration of each M4 spectrum was determined by a time-to-spill algorithm, which corresponds to the accumulation of a pre-determined number of counts in the range 122–1420 keV.

APEX measured the same number of M4 spectra for each burst, and for many of them, several first spectra and several last spectra represent the local background around a burst. These measurements were used to determine the local background model (LBM) (Figure 1).

APEX also continuously measured the spectrum of background at the beginning of each 4-hour interval (M7 data). These spectra are similar to M4, but corresponded to a much larger accumulation number ~70,000. The M7 data allow calculation of ~ 5-min averaged background count rates in the energy range 122–1420 keV. Moreover, the M7 spectra were measured continuously without telemetry until a trigger, when the last spectrum was recorded as M1 data. M1 represents the background count rate 5–10 minutes before a trigger. The M7 and M1 data provide the global background model (GBM) at 122–1420 keV (Figure 2).

To check the existence of a glowing component in the course of APEX gamma-ray bursts, we compare the GBM level (based on M7 and M1) with the LMB background produced by M4 data around a burst. For the sample of 17 bright bursts, we found practically in all cases that levels of LBM are well above the levels of GBM. We illustrate this effect by GRB880925b and GRB881225 (Figure 1).

For better illustration we use a special time scale along the time axis, which has zero moment at the primary peak of a burst and goes to the left and right sides with logarithmic scale. Large squares are M7 and M1 measurements, and dotted lines are GBM models for these bursts (Figure 2). The errors of M7 and M1 measurements are smaller than size of squares. We selected zero-flux time

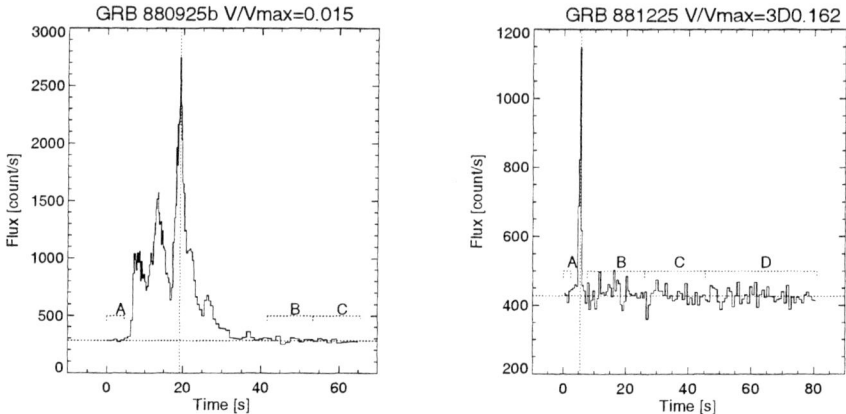

FIGURE 1. *Left*: GRB 880925b time profile according to M4 data at 122–1420 keV. A dotted line represents the LBM level. Intervals A, B, and C have zero flux according to the LBM. *Right*: GRB 881225b time profile according to the same data. Intervals A, B, C and D have zero flux according to the LBM.

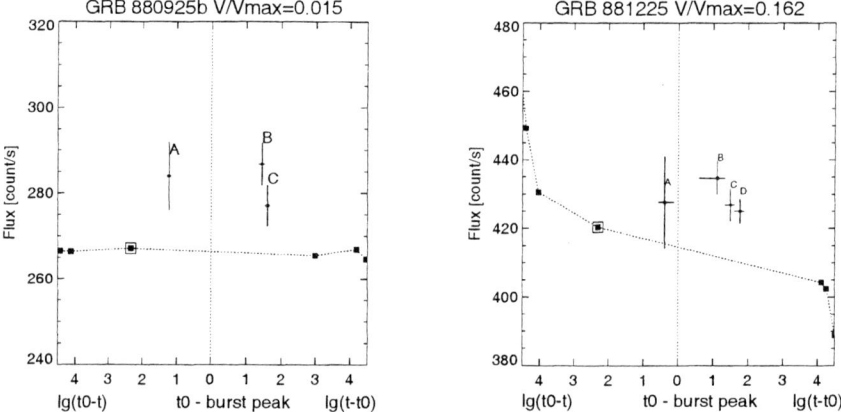

FIGURE 2. *Left*: Long background history for GRB 880925b according to M7 data (solid squares) and M1 data (framed squares) at 122–1420 keV. A dotted line represents the GBM level. Points A, B and C correspond to the intervals shown in Figure 1. *Right*: Long background history for GRB 881225 according to the same data. Points A, B, C and D correspond to the intervals shown in Figure 1.

intervals at standard time profiles of these bursts, which correspond to the level of LBM (Figure 1). For both bursts we find significant excess of LBM intervals in comparison with interpolations of GBM (Figure 2). This excess corresponds to 3.8σ for GRB 880925b and at least 3.1σ for GRB 881225. The energy spectrum of this excess is rather soft, its energy range is 64 keV – 250 keV.

The difference between the LBM and GBM levels could not result from different decoding of the different data types. First, these data were provided by the same Pulse Height Analyzer and were processed by the same digital electronics board. Second, we compared these levels for non-burst triggers, and we found very good agreement between them.

CONCLUSIONS

We suppose that a difference between the local background model (LBM) and the large-scale background interpolation (GBM) for two bright APEX bursts indicates a soft glowing emission component of these gamma-ray bursts.

The measurements of M1 data (Figure 2) at ~ 100 s before a burst do not show the existence of pre-burst glowing emission. The nearest post-burst point of M7 data several hours after a burst also does not show any detectable glowing. Thus, the glowing emission may start within a burst time history and grows up to the level of ~ 0.1–0.3 photons cm^{-2} s^{-1}, then it decreases down to the level of detectability after $\sim 10^3 - 10^4$ s.

We confirm the reported finding of an emission component along a tail of several BATSE gamma-ray bursts [6,7]. However, the BATSE experiment is known to have quite variable background, and the findings of a glowing emission component for APEX bursts is rather important because of the very stable background.

ACKNOWLEDGMENTS

The work in Russia was supported by RFBR grant 98-02-17380. It was supported also by INTAS project No.96-0315.

REFERENCES

1. Costa, E., these proceedings.
2. Ricker, G., et al., these proceedings.
3. Mitrofanov, I., et al., *Planet. Space Sci.* **39(1)**, 23 (1991).
4. d'Uston, C., et al., *Nature* **341**, 598 (1989).
5. Afonin, V., et al., *Planet Space Sci.* **39(1)**, 153 (1991).
6. Connaughton, V., these proceedings.
7. Giblin, T., et al., these proceedings.

Evidence for Early High-Energy Afterglow: BATSE Observations of GRB980923

Timothy W. Giblin[1,2], Jan van Paradijs[1,4,7], Chryssa Kouveliotou[2,3], Valerie Connaughton[5], Ralph A. M. J. Wijers[6], Michael S. Briggs[1,2], Robert D. Preece[1,2], and Gerald J. Fishman[2]

[1] *Department of Physics, University of Alabama in Huntsville, Huntsville, AL 35899*
[2] *NASA Marshall Space Flight Center, SD-50, Huntsville, AL 35812*
[3] *Universities Space Research Association, Huntsville, AL 35806*
[4] *University of Amsterdam, Kruislaan 403, 1098 SJ Amsterdam, The Netherlands*
[5] *NRC, NASA Marshall Space Flight Center, SD-50, Huntsville, AL 35812*
[6] *Department of Physics & Astronomy, SUNY, Stony Brook, NY 11794-3800*
[7] *Deceased*

Abstract. The relation between the prompt γ-ray emission of a Gamma-Ray Burst (GRB) and its afterglow emission is unclear. We present evidence in the BATSE data for a prompt high-energy (25-300 keV) afterglow component associated with GRB980923 that is initiated during the burst. The event exhibits rapid variability lasting ~ 40 s, followed by a smooth power-law emission tail lasting ~ 400 s. The energy spectrum changes abruptly to a cooling synchrotron spectrum when the tail becomes noticeable, consistent with the internal-external shock scenario in the fireball model. This overlap of prompt and afterglow emission illustrates that the external shocks can be generated during the γ-ray emission phase, as also suggested by the simultaneous optical emission of GRB991023.

INTRODUCTION

GRB afterglow measurements at X-ray, optical, and radio wavelengths at early and late times after the prompt γ-ray emission are in good agreement with the relativistic fireball model [1,2]. From an observational viewpoint, one crucial question remains open: what is the relation of the afterglow with the prompt emission? The afterglow is attributed to synchrotron emission from electrons in the forward shock that is produced when a relativistic blastwave encounters the ambient medium. The afterglow peak frequency depends strongly on the electron bulk Lorentz factor (Γ), whereas the start time of this emission relative to the γ-ray emission is uncertain and may highly depend on the ambient medium. It has been suggested

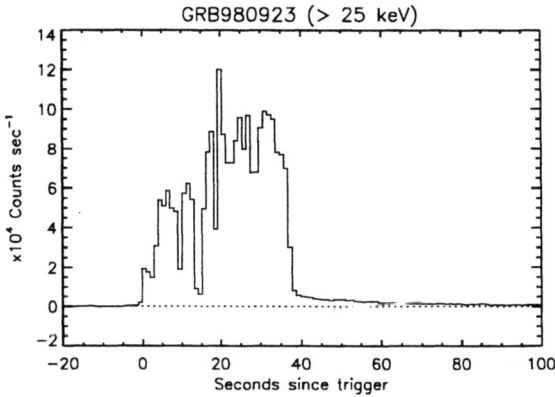

FIGURE 1. Time history of GRB98023 with 1.024 s accumulation intervals.

that the synchrotron frequency can initially peak near $\nu = 10^{19}$ Hz [3], which is within the BATSE energy range. If the external shock occurs at early times, then we might expect to see a smooth power-law decay in soft γ-rays during the burst and/or near the end of the burst. Here, we present the temporal and spectral properties of GRB980923, a burst that shows a very long and smooth γ-ray emission tail after a period of strong variability lasting ~ 40 s. Note that emission tails are a somewhat common burst property and have been observed by several instruments: GINGA [4], PHEBUS [5], OSSE [6], SIGMA [7], APEX [8].

BATSE OBSERVATIONS OF GRB980923

On 1998 September 23 at 20:10:47.5 UT BATSE was triggered on the 64 ms time scale by a long and intense GRB (trigger 7113), shown in Figure 1. The burst was not in the field of view of the BeppoSAX WFC. RXTE/PCA made a single observation in stare mode in response to the BATSE Rapid Burst Response team, but observed no change in count rate. A full scan of the GRB error circle by RXTE/PCA was not possible at the time. Consequently, no follow-up counterpart observations were made at longer wavelengths.

The 50–300 keV peak flux of the burst on the 64 ms time scale is $(1.16 \pm 0.02) \times 10^{-5}$ ergs s^{-1} cm^{-2}, ranking twelfth in brightness among all BATSE GRBs. The fluence of the burst (> 25 keV) is $(4.84 \pm 0.02) \times 10^{-4}$ ergs cm^{-2}, ranking third among all BATSE GRBs. The fluence in the tail (> 40 s) is $\sim 7\%$ of the total burst fluence.

To accurately model the temporal and spectral properties of the tail we used an orbital background subtraction algorithm developed at Marshall Space Flight Center [9].

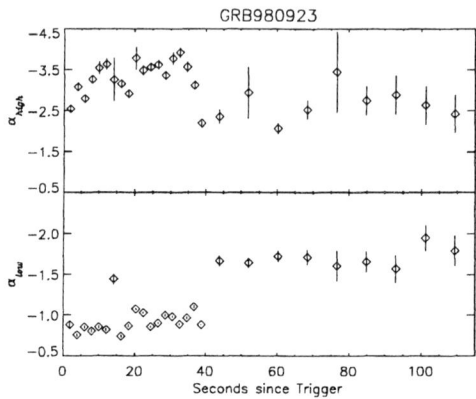

FIGURE 2. Time evolution of the low- and high-energy power-law indices.

BURST EMISSION VS. AFTERGLOW EMISSION

The smooth tail emission is well-described by a power-law function of the form $A(t-t_0)^\beta$. We generated a series of two parameter (A and β) fits in the 25–300 keV range for a range of t_0 values. We find that χ^2 has a minimum at $t_0 = 9.6$ s for $\beta = -1.8 \pm 0.02$. Individual fits to the three lowest discriminator energy channels (25–50, 50–100, 100–300 keV) show that β remains constant [10].

The key to deciphering the burst emission from the afterglow emission lies in spectral evolution. Spectra were generated in 2.048 s intervals using 16-channel CONT data covering the 22–1880 keV energy range. Two time-integrated spectra were also constructed: one for the burst emission (1–40 s) and one for the tail (40–400 s). All spectra were modeled with the GRB function (α, β, E_p) for comparison with the known distribution of GRB spectral parameters [11], and a smoothly broken power-law (SBPL) function ($\alpha_{\text{low}}, \alpha_{\text{high}}, E_b$) for comparison with the relativistic synchrotron spectrum expected from external shocks.

The time-integrated spectrum of the burst proper gives GRB function parameters $\alpha_{\text{GRB}} = -0.61 \pm 0.01$, $\beta_{\text{GRB}} = -2.95 \pm 0.08$, $E_p = 364 \pm 3$ keV, typical for GRBs [11]. From the time-integrated fit of the tail, we obtain $\alpha_{\text{low}} = -1.66 \pm 0.04$, $\alpha_{\text{high}} = -2.20 \pm 0.11$, and $E_b = 203 \pm 59$ keV ($\chi^2/\text{dof} = 1.06$). The GRB function gives nearly identical parameter values ($\alpha_{\text{GRB}} = -1.59 \pm 0.05$, $\beta_{\text{GRB}} = -2.19 \pm 0.14$, $E_p = 271 \pm 52$), also similar to values found in typical GRB spectra [11]. However, the GRB function proved to be less robust than the SBPL in the time-resolved fits. From the time-resolved fits, we find that α_{low} and α_{high} do not vary significantly in the tail (see Figure 2), with weighted average values -1.68 ± 0.03 and -2.09 ± 0.07, respectively. An abrupt change in spectral form is evident where the burst emission drops sharply and the tail of the burst becomes detectable at ~ 40 s. We further find that the break energy decays according to $t^{-0.52}$, indicating that the spectrum retains a constant shape as it shifts in time to lower energies.

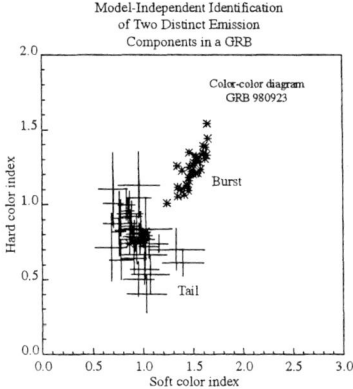

FIGURE 3. Color-color diagram for GRB980923. Plotted are "hard" (100-300 keV)/(50-100 keV) vs. "soft" (50-100 keV)/(25-50 keV) hardness ratios.

Clearly, if we examine only the tail, we are led to believe that it is very "GRB-like". However, when compared with the burst, the spectral parameters indicate two separate types of emission present in GRB980923. The two-emission component hypothesis is further corroborated by a model-independent color-color diagram (CCD) shown in Figure 3. The burst produces a common crescent CCD pattern [12], while the CCD pattern of the tail is clearly decoupled from that of the burst.

DISCUSSION

Is the tail of GRB980923 consistent with the synchrotron spectrum expected for afterglow emission? Initially after the shock, the electrons are in the fast-cooling mode ($\nu_m > \nu_c$) and cool on a dynamical time-scale [13]. When the synchrotron frequency passes below the cooling frequency (i.e., when $\nu_m = \nu_c$) the electrons enter the slow-cooling regime where the expected change in spectral slope across ν_c is $p/2 - (p-1)/2 = 0.5$. From the time-integrated spectral fit, we find a slope change of 0.54 ± 0.12 (0.42 ± 0.09 from the time-resolved fits), in agreement with the slow-cooling synchrotron spectrum. The electron distribution index, p, can be obtained from $p/2 = -(\alpha_{high} + 1)$, leading to $p = 2.4 \pm 0.11$, typical for afterglows. The fast-cooling adiabatic and radiative cases are confidently excluded and the slow-cooling spectrum is described in detail in Giblin et al. [10].

GRB980923 represents a case in which the external shock emission overlaps the GRB emission, similar in principle to the simultaneous optical emission observed in GRB991023. In the case of GRB980923, a large bulk Lorentz factor ($\Gamma \sim 1000$) is expected due to the high observed peak frequency of the prompt afterglow emission.

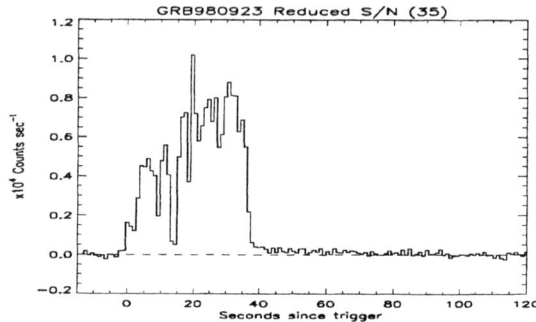

FIGURE 4. Time history of GRB980923 with reduced signal-to-noise.

EARLY HIGH-ENERGY AFTERGLOWS IN MORE BATSE GRBS?

We are currently searching for early high-energy afterglows in the BATSE catalog. Recent studies of the average properties of the late-time behavior of GRBs suggests that high-energy tails may be an intrinsic property of GRBs [14]. If we hypothesize that external shock emissions peak in soft γ-rays in all GRBs, then we are subject to an observational sensitivity bias because we do not see tails in *all* GRBs. The detection in the case of GRB980923 may hinge strongly on the fact that this burst was extremely bright. After reducing the signal-to-noise of the peak to 35 (Figure 4), the burst remains bright but the tail is marginally visible. Results from our extended study in the BATSE data will be presented in a forthcoming paper [15].

REFERENCES

1. Wijers, R. A. M. J., Rees, M. J., and Meszaros, P., *MNRAS* **288**, L51 (1997).
2. Waxman, E., *ApJ* **485**, L5 (1997).
3. Sari, R., and Piran, T., *ApJ* **520**, 641 (1999).
4. Yoshida, *Adv. Space Res.* **15**, 5(89) (1991).
5. Tkachenko, et al., *Nature* **359**, 217 (1995).
6. Matz, S., et al. *MNRAS* **277**, 287 (1995).
7. Burenin, R. A., et al. *A & A* **344**, L53 (1999).
8. Litvine, D., et al., these proceedings (1999).
9. Connaughton, V., et al., in preparation (1999).
10. Giblin, T. W., et al. *ApJ* , 266 (1999).
11. Preece, R. D., *ApJS* **126**, in press (2000).
12. Giblin, T. W., BAAS **30**, 4, 79.06 (1999).
13. Sari, R., Narayan, R., and Piran, T., *ApJ* **497**, L17 (1999).
14. Connaughton, V., et al., these proceedings (1999).
15. Giblin, T. W., et al., in preparation (2000).

GRB Spectral Hardness and Afterglow Properties

Jon Hakkila*, Timothy Giblin†, and Robert D. Preece†

*Minnesota State University, Mankato, Minnesota 56001
†University of Alabama, Huntsville, Alabama 35812

Abstract. A possible relationship between the presence of a radio afterglow and gamma-ray burst spectral hardness is discussed. The correlation is marginally significant; the spectral hardness of the bursts with radio afterglows apparently results from a combination of the break energy E_{break} and the high-energy spectral index β. If valid, this relationship would indicate that the afterglow does carry information pertaining to the GRB central engine.

INTRODUCTION

Early observations of gamma-ray bursts (GRBs) with afterglows led us to hypothesize that GRBs with afterglows are spectrally harder than those without. However, the heterogeneous nature of GRB observations, coupled with the multi-wavelength nature of afterglow observations, has led us to a number of *concerns*:

1. Afterglow observations are heterogeneous: observational biases for detecting radio afterglows are different than those for detecting optical and/or x-ray afterglows.

2. A GRB might have no intrinsic afterglow, or inadequate search conditions might result in no afterglow being detected.

3. Spectral hardness measures are biased by instrument performance: One instrumental dataset should be used.

Our *solutions* to these concerns are as follows:

1. We require GRBs to either have detected radio afterglows or moderately-complete radio afterglow searches. This condition satisfies the first and second concerns.

2. We require that GRBs be observed by BATSE. This satisfies the third concern.

TABLE 1. GRBs with and without radio afterglows.

GRB	Radio afterglow	HR(43/21)	HR21	E_{break} (keV)	α	β
970111	No	1.93 ± 0.03	1.38 ± 0.01	194.2 ± 4.0	-1.14 ± 0.01	-3.65 ± 0.07
970815	No	1.83 ± 0.32	0.93 ± 0.02	100.8 ± 57.9	-0.82 ± 0.40	-3.09 ± 0.58
971214	No	4.31 ± 0.34	1.67 ± 0.05	183.3 ± 36.7	-1.14 ± 0.08	-2.85 ± 0.35
980425	?	1.42 ± 0.38	0.96 ± 0.04	450.6 ± 766	-1.67 ± 0.08	-11.0 ± 39.7
970508	Yes	8.27 ± 1.32	1.09 ± 0.10	137.2 ± 96.9	-1.18 ± 0.24	-1.88 ± 0.25
970828	Yes	5.4 ± 0.2	1.81 ± 0.05	261.0 ± 44.2	-0.09 ± 0.18	-3.72 ± 0.49
980329	Yes	6.05 ± 0.09	1.31 ± 0.01	171.2 ± 4.78	-1.28 ± 0.01	-2.22 ± 0.03
980519	Yes	2.65 ± 0.33	1.05 ± 0.02	812 ± 1308	-1.60 ± 0.03	-10.7 ± 37.3
980703	Yes	6.36 ± 0.50	0.98 ± 0.03	181.8 ± 46.2	-1.32 ± 0.06	-2.04 ± 0.13
990123	Yes	18.0 ± 0.14	1.36 ± 0.01	618.6 ± 13.4	-0.92 ± 0.01	-3.51 ± 0.28
990506	Yes	5.15 ± 0.06	1.25 ± 0.01	258.7 ± 12.1	-1.09 ± 0.03	-2.06 ± 0.05
990510	Yes	1.87 ± 0.09	1.21 ± 0.02	120.8 ± 11.5	-1.43 ± 0.04	-2.49 ± 0.07

Our resulting database is shown in Table 1.

GRB 980425 may or may not have a radio afterglow, depending on its association with supernova SN 1998bw. If there is an association with SN 1998bw, then GRB 980425 has x-ray, optical, and radio afterglows. If there is no association with SN 1998bw, then this GRB has only an x-ray afterglow, with no optical or radio afterglow. The SAX team [1] lists two possible x-ray afterglow sources for GRB 980425; the first one is consistent with SN 1998bw, the second one is not.

The SAX Team indicates that the second afterglow source might have "rebursted", making its classification as a GRB counterpart questionable (this is not standard GRB behavior). However, it should be noted that the flux measurement of the second observation of this source represents less than a 3σ detection, placing the rebursting claim in doubt.

If the first afterglow source is associated with SN 1998bw, then GRB 980425 is significantly less luminous than typical GRBs. As discussed elsewhere in the literature [2], very few BATSE GRBs can have luminosities this small in typical spatial distribution models.

It appears to us that the afterglow source of GRB 980425 is still in question. Because of these doubts, we consider independently the cases where GRB 980425 has and does not have a radio afterglow.

ARE GRBS WITH RADIO AFTERGLOWS DIFFERENT FROM THOSE WITHOUT?

Figure 1 demonstrates that GRBs with radio afterglows appear to be harder than those without. The hardness ratio HR(43/21) (100-1000 keV energy fluence divided by 25-100 keV energy fluence) has been used in this analysis, because it has the largest signal-to-noise ratio available in BATSE 4-channel data and spans the largest spectral range. However, similar results can be obtained from other hardness ratios.

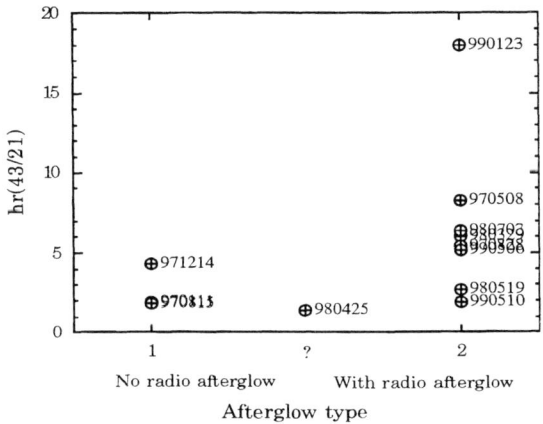

FIGURE 1. HR(43/21) vs. Radio Afterglow Type.

In Table 2 we summarize Student's t-test probabilities that the log[HR(43/21)] distributions of bursts without radio afterglows and bursts with radio afterglows have different means. The significance of a correlation depends strongly on the status of GRB 980425 due to small number statistics. If GRB 980425 is not associated with SN 1998bw, then the correlation between spectral hardness and radio afterglow is more likely.

TABLE 2. Probability that GRBs With and Without Radio Afterglows Have Different Distributional Means.

Status of SN 1998bw	t-Test Probability
GRB980425 NOT associated with SN 1998bw	0.963
GRB980425 associated with SN 1998bw	0.781

DISCUSSION

To determine why this difference in hardness might exist, we checked other hardness ratios in the four-channel data. Hardness ratios involving channels 3 and 4 indicate similar results as obtained in Figure 1. Thus, any spectral differences dependent on radio afterglow type appear to result from the distribution of high energy photons. This is supported by Figure 2, which indicates no correlation of radio afterglow type with hardness ratio HR21.

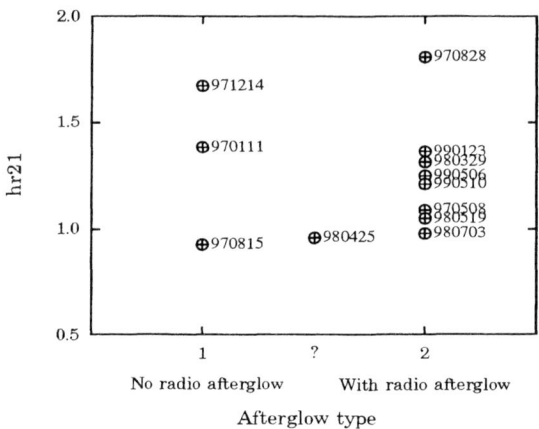

FIGURE 2. HR21 vs. Radio Afterglow Type.

This is also supported by Figure 3, which compares the GRB function spectral parameters E_{break} and β for the bursts in question (GRBs 980425 and 980519 are not plotted due to large β errors). Large values of E_{break}, large values of β, or both produce conditions indicating many high-energy photons. GRBs with radio afterglows tend to occupy a different diagram region than GRBs without radio afterglows. Since E_{break} and β are obtained from time-averaged spectra, we suspect that the diagram regions might be even more distinct if signal-to-noise were better for faint BATSE GRBs.

The possible correlation between GRB spectral hardness and afterglow type can be clarified with additional observations in the future. Resolution of the status of GRB 980425 would also help clarify this issue.

If GRB spectral hardness is an indicator of radio afterglow type, then a direct link between the central engine and the delayed emission is established. Such a link could be very important to the understanding of GRB physics. The Lorentz factor of the expanding external shock could be constrained by this correlation. For this reason, it is as important to determine upper flux limits on afterglow non-detections as it is to provide information on detections.

CONCLUSIONS

There is evidence that GRBs with radio afterglows have harder gamma-ray burst emission than those without. Due to small number statistics, the significance of this correlation depends at present time on whether or not GRB 980425 is associated with supernova SN 1998bw. There is evidence of a similar correlation between

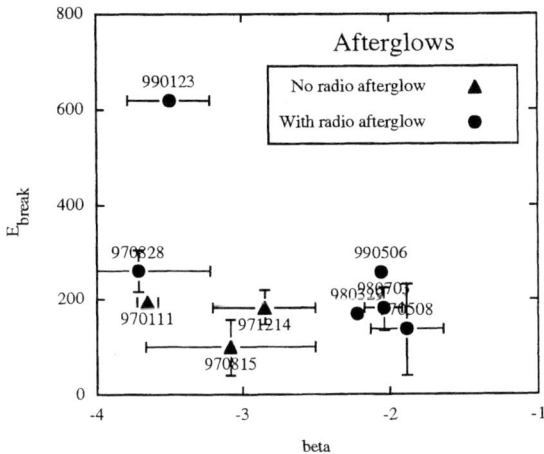

FIGURE 3. E_{break} vs. β for different radio afterglow types.

bursts with optical afterglows, but this is more difficult to document because the literature is less clear on conditions under which an optical search failed to yield an afterglow.

If we assume that a relationship exists between spectral hardness and radio afterglow type, then GRBs with radio afterglows appear to have more high-energy photons ($E \geq 100$ keV) than those without radio afterglows, as determined from the spectral parameters E_{break} and β. Also, roughly 2/3 of BATSE-detected GRBs should produce radio afterglows based on the overall distributions of E_{break} and β. It should be noted that all GRBs producing afterglows of any type belong to the long, bright, soft GRB class.

ACKNOWLEDGEMENTS

We thank Dale Frail, Chip Meegan, Geoff Pendleton, Ralph Wijers, David Haglin, and Chryssa Kouveliotou for valuable discussions. Jon Hakkila acknowledges 1999 NASA/ASEE Summer Faculty Fellowship Program support.

REFERENCES

1. Pian, E., et al., *A & A* submitted.
2. Schmidt, M., this conference.

Host Galaxies Have 'Normal' Luminosities

Bradley E. Schaefer

Yale University, Physics Department, 260 Whitney, New Haven CT 06511

Abstract. The galactic environment of Gamma Ray Bursts can provide good evidence about the nature of the progenitor system, with two old arguments implying that the burst host galaxies are significantly subluminous. New data and new analysis has now reversed this picture: (A) Even though the first two known host galaxies are indeed greatly subluminous, the next eight hosts have absolute magnitudes typical for a population of field galaxies. A detailed analysis of the 16 known hosts (ten with red shifts) shows them to be consistent with a Schechter luminosity function with $R^* = -21.8 \pm 1.0$, as expected for normal galaxies. (B) Bright bursts from the Interplanetary Network are typically 18 times brighter than the faint bursts with red shifts, however the bright bursts do not have galaxies inside their error boxes to limits deeper than expected based on the luminosities for the two samples being identical. A new solution to this dilemma is that a broad burst luminosity function along with a burst number density varying as the star formation rate will require the average luminosity of the bright sample ($>6 \times 10^{58}$ ph $\cdot$ s^{-1} or $>10^{52}$ erg $\cdot$ s^{-1}) to be much greater than the average luminosity of the faint sample ($\sim 10^{58}$ ph $\cdot$ s^{-1} or $\sim 6 \times 10^{51}$ erg $\cdot$ s^{-1}). This places the bright bursts at distances for which normal host galaxies will not violate the observed limits. In conclusion, all current evidence points to GRB host galaxies being normal in luminosity.

INTRODUCTION

In the past, two arguments have been presented that made strong cases that most GRBs appeared either outside normal galaxies or in systematically subluminous hosts [1,2]. The first argument was that GRB970228 and GRB970508 (the first two identified GRB hosts) are in the bottom $\sim 1\%$ of the luminosity weighted Schechter luminosity function, with this result being unlikely unless the GRB hosts are systematically subluminous. The second argument was that a dozen very bright bursts seen with the Interplanetary Network (IPN) have no galaxies in their small error boxes to B magnitudes from 20 to 24, whereas the hosts should have been easily visible if the bursters reside in normal galaxies for the luminosities allowed by LogN-LogP studies.

In the past year, new burst red shifts have greatly changed the situation from that presented in the previous paragraph. Also, I here propose an alternative solution for the lack of sufficiently-bright hosts for the bright bursts. This paper presents these two new analyses, with the conclusion that GRBs reside in galaxies whose luminosities are distributed as a normal Schechter luminosity function.

HOSTS OF FAINT BURSTS

The first two discovered GRB hosts are galaxies at the bottom of the luminosity function. But now we have data for hosts on sixteen GRBs with optical transients or radio transients (OT/RT) to provide arc-second localizations and ten of these have measured red shifts. This much larger sample can answer the question of "What is the luminosity function for the host galaxies of faint bursts?"

An approximate answer to this question can be obtained by merely examining the derived absolute magnitudes of the hosts. The first two GRB hosts are fortuitously the least luminous hosts by about one magnitude. This means that the early argument for subluminous hosts based on GRB970228 and GRB970508 is wrong due to a rather unlikely coincidence. Further, the typical R-band absolute magnitude is around -21, a value which is comparable to the R^* value characteristic of the R-band Schechter luminosity function. R^* is approximately -21.2 mag in the local vicinity for a Hubble constant of 65 km $\cdot$ s^{-1} $\cdot$ Mpc^{-1}. So to first order, the host galaxies of faint GRBs have a normal luminosity.

However, a variety of effects and biases can affect this conclusion: The probability of a detected burst yielding a red shift and an apparent magnitude for the host depends on the burst distance, the burst luminosity, and the host's absolute magnitude. So, for example, the OT/RT sample will be biased towards luminous hosts for which a red shift is more likely to be measured. Also, the OT/RT bursts have a typical red shift of ~ 1, so that effects due to the values of cosmological parameters, the K-corrections for both bursts and hosts, and changes in the galaxy luminosity function will affect the conclusion.

An improved analysis is to model all these effects and biases with a Monte Carlo calculation to produce a simulated catalog of bursts containing subsets with red shifts and host apparent magnitudes. I have adopted a Hubble constant of 65 km $\cdot$ s^{-1} $\cdot$ Mpc^{-1} in a flat Universe with $\Omega_m = 0.3$. I take the burst number density to follow the star formation rate as given by [3]. The burst luminosity function is taken as the usual truncated power law with slope -2, dynamic range of 1000, and a minimum luminosity of 10^{57} ph $\cdot$ s^{-1}, to be consistent with the observed time dilation, red shifts, and LogN-LogP curve [4] and the light curve variability [5]. The host luminosity function was taken to have the shape of the Schechter luminosity function with slope $\alpha = -1$ and a characteristic R-band absolute magnitude, R^*, which is a free parameter. The probability of getting an arc-second position for an observed burst is approximated as rising linearly from zero for P_{256} values ranging between 0.5 to 5.5 ph $\cdot$ s^{-1} $\cdot$ cm^{-2}. The probability of measuring an apparent magnitude for a host galaxy is taken to be 0.7 if the burst has an arc-second position and a host brighter than $R = 25.7$ mag. The K-corrections for the host

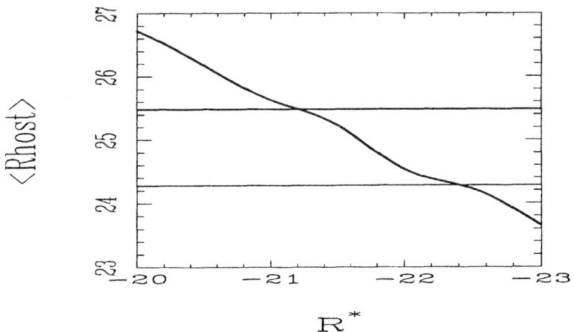

FIGURE 1. The graph compares the results from realistic Monte Carlo simulated burst catalogs (the sloped curve) with the range of uncertainty for the observed values (between the two horizontal lines; 24.88 ± 0.60 mag). The one free parameter in the model is the R^* value which characterizes the host galaxy luminosity. The acceptable range is then $-21.2 > R^* > -22.4$ for the adopted model parameters, although this range is increased to $-20.8 > R^* > -22.8$ when allowance is made for plausible uncertainties in the adopted model parameters. From this, we see that GRB hosts are apparently of normal luminosity and certainly not greatly subluminous on average (despite the first two known hosts being greatly subluminous).

are taken for those of an Sb galaxy as given by [6]. The K-corrections for the burst are taken for a count spectrum varying as E^{-2} [7,8].

A reasonable aggregate parameter for model comparisons is the median for the apparent magnitudes of detected bursts which equals 24.88 mag. (The standard deviation for the distribution in R_{host} of observed bursts is consistent with the model values.) The $<R_{host}>$ value for 16 bursts varies with a standard deviation of 0.60 mag. So the target for the model is 24.88 ± 0.60 mag.

For what values of R^* does the model reproduce the observed distribution of host apparent magnitudes? Figure 1 displays the model predictions as a function of the adopted R^*. An acceptable range of R^* is then from -21.2 to -22.4 mag, with the best value being around -21.8 mag.

However, uncertainties in the model input parameters will enlarge the acceptable range of R^*. This can be quantified by calculating the change in the model $<R_{host}>$ when one input parameter is changed over some plausible range (with the luminosity function shifted such that the observed $<LogL>$ is held constant). Plausible changes in Ω_m, the Hubble constant, the average slope of the GRB count spectrum, the mean intrinsic burst luminosity, and the shape of the burst luminosity function move $<R_{host}>$ by ~0.4 mag. Then, the range of acceptable R^* values increases to from -20.8 to -22.8, so the final model estimate of R^* is -21.8 ± 1.0 mag. This derived R^* value is easily consistent with normal galaxies, yet is inconsistent with greatly subluminous hosts.

HOSTS OF BRIGHT BURSTS

The bursts with optical or radio transients are typically rather faint, with the median P_{256} being only a factor of 3 above the BATSE completeness threshold. These GRBs are greatly fainter than the bursts positioned with the IPN. For a fair comparison, the median of sixteen OT/RT bursts is 18 times fainter than the median of the sixteen IPN bursts with the smallest error boxes [8].

For many reasonable models, the IPN bursts should thus be ~4 times closer than the OT/RT events and then will be substantially immune to many problems that plague the interpretation of the high red shift OT/RT events (uncertainties in the K-corrections, the cosmological parameters, and the galaxy luminosity function). For some purposes, the IPN burst sample might then be more important than the OT/RT sample because the low red shift Universe can be readily interpreted.

Examinations [1,2] of the limits on R_{host} for the IPN GRBs conclude that the hosts can have normal luminosities (i.e., can be drawn from the usual Schechter luminosity function) only if the average burst luminosity is greater than 6×10^{58} ph·s^{-1} ($Log L = 58.8$). This directly contradicts fits to the LogN-LogP curve, the time dilation of burst light curves, as well as the observed luminosities for the OT/RT bursts [4]. Possible solutions to this dilemma were that the GRBs were ejected from their birth galaxy or that the host galaxies are systematically subluminous for some reason. Neither solution now seems plausible.

I would like to point out another solution which fits well with currently popular ideas. The dilemma arises because the bright IPN bursts were plausibly assumed to have the same mean luminosity as the faint OT/RT bursts. However, if GRBs simultaneously have a broad luminosity function and their number density increases greatly with red shift, then the bright bursts will have a much greater average luminosity than will faint bursts. That is, if the OT/RT events have $Log L \approx 58.0$ while the IPN events have $Log L > 58.8$, then the host galaxies of the IPN bursts will have $R_{host} \approx 24$ and be fully consistent with the limits in [8].

To provide a quantitative evaluation of this idea, I have calculated the average luminosity and red shift for bursts with peak fluxes brighter than some threshold for a variety of burst luminosity functions. The required integrals were performed numerically for red shifts over the range 0–6 with bins of 0.01. I have used the same model assumptions that were used in the previous section. The burst luminosity function was taken either as a log-normal distribution or as a truncated power law. The average luminosity was set such that $<Log L>$ for a population observed with $P_{256} > 1$ ph·s^{-1} was 58.34. Figure 2 displays the results for the truncated power law as well as for two log-normal functions of different widths.

Both power law and log-normal distributions give similar results, in that samples of bright bursts will be much more luminous than samples of dim bursts. For broad luminosity functions, the $<Log L>$ for observed bursts is determined by the overall slope of the intrinsic luminosity function; thus for $Log L \sim 58$, the GRB luminosity function must scale close to L^{-2} regardless of the behavior at high and low luminosity. The mean red shift of bright burst samples is much higher than

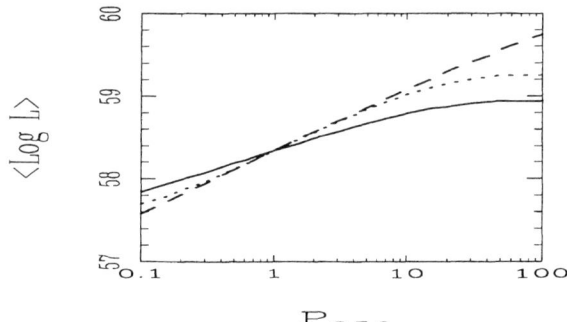

FIGURE 2. Bright bursts are much more luminous than faint bursts. The three curves are for broad logarithmic and log-normal luminosity functions. A comparison between BASTE or OT/RT bursts (threshold $P_{256} = 0.85$ ph·s^{-1}·cm^{-2}) and IPN bursts (threshold $P_{256} = 30$ ph·s^{-1}·cm^{-2}) reveals a factor of $\sim$10 difference in average luminosity. This realization resolves the discrepancy that faint bursts have $<LogL>$ around 58.0 (based on time dilation, the LogN-LogP curve, and the few known red shifts) while bright bursts have $<LogL>$ of >58.8 (based on the lack of host galaxies to deep limits).

would be expected from simple scaling by $P^{-0.5}$ from the red shift of faint bursts.

The primary point of Figure 2 is that the average luminosity of the bright bursts is greatly larger than for the faint bursts. The most important comparison is for bursts with $P_{256} > 0.85$ ph·s^{-1}·cm^{-2} (the BATSE completeness threshold which is relevant for LogN-LogP studies and for the OT/RT bursts) versus bursts with $P_{256} > 30$ ph·s^{-1}·cm^{-2} (for the IPN bursts). For the three broad luminosity functions in Figure 2, the ratio of luminosities for these two thresholds is 8.3, 4.0, and 13.5. That is, the average luminosity of the IPN bursts is roughly an order of magnitude brighter than for the OT/RT and BATSE bursts. This means that for OT/RT and BATSE bursts with $<LogL>\sim 58.0$ [4,5,9,1], the IPN bursts likely have $<LogL>\sim 59.0$. This is completely consistent with the lack of hosts in IPN boxes to deep limits [1].

REFERENCES

1. Schaefer, B. E., *ApJ* **511**, L79 (1999).
2. Band, D. L., Hartmann, D., and Schaefer, B. E., *ApJ* **514**, 862 (1999).
3. Madau, P., Pozzetti, L., and Dickinson, M., *ApJ* **498**, 106 (1998).
4. Deng, M. and Schaefer, B. E., *ApJ* submitted (1999).
5. Ramirez-Ruiz, E. and Fenimore, E. E., *ApJ* submitted (1999).
6. Rocca-Volmerange, B., and Guideroni, B., *A & AS* **75**, 93 (1988).
7. Schaefer, B. E., et al., *ApJS* **92**, 285 (1994).
8. Schaefer, B. E., et al., *ApJ* **492**, 696 (1998).
9. Horváth, I., Mészáros, P., & Mészáros, A., *ApJ* **470**, 56 (1996).

The Host of GRB 970828: Another Subluminous Galaxy?

Javier Gorosabel* and Alberto J. Castro-Tirado*,†

*LAEFF-INTA, P.O. Box 50727, 28080 Madrid, Spain
† Instituto de Astrofísica de Andalucía (IAA-CSIC),
P.O. Box 03004, E-18080 Granada, Spain

Abstract. Optical observations performed with the William Herschel Telescope did not reveal any object with R > 24.5 inside the small X-ray error box determined by ROSAT (Greiner et al. 1997). On the other hand, an X-ray spectrum taken with ASCA revealed an emission line which could be attributed to the redshifted Fe I line at 6.4 keV (Yoshida 1999). Both facts impose severe limits to the luminosity of the host galaxy. Thus, we conclude that the host galaxy of GRB 970828 must be fainter than $M_R = -20.28$, resembling the one related to GRB 970508.

INTRODUCTION

GRB 970828 was discovered with the All-Sky Monitor on the Rossi X-Ray Timing Explorer (ASM/$RXTE$) on 1997 August 28 17h44m36s UT. It provided an elliptical region centered at $\alpha_{2000} = 18^h08^m29^s$, $\delta_{2000} = +59°\ 18'$, with a major axis of 50' and a minor axis of 20' [1,2]. The mean flux measured by the camera #1 of ASM/$RXTE$ experiment was 756 mCrab (2–12 keV). Ninety seconds later when the source was scanned by ASM/$RXTE$ camera #2, the flux was only 238 mCrab. Both measurements were consistent with an exponential law decay with a 30 s time constant.

On the other hand, the Proportional Counting Array on the Rossi X-Ray Timing Explorer (PCA/$RXTE$) started a follow-up observation only 3.6 hr after the gamma-ray event and revealed a faint X-ray source (0.5 mCrab) consistent with the elliptic uncertainty region provided by ASM/$RXTE$ [3]. The source was also consistent with the position provided by BATSE and the Interplanetary Network (IPN) [4]. A further X-ray observation by the $ASCA$ satellite starting 28 hr after the GRB [5], led to a refined position, providing finally an error box of 0'.5 error radius. The flux was $F_{(2-10\ keV)} = 4 \times 10^{-13}$ erg s^{-1} cm^{-2}.

The fact that the source was never seen before, together with its swift decline, supports its identification as the GRB fading X-ray counterpart with a time power-law index $\alpha = -1.4$, being $F \propto t^\alpha$ [6]. This was confirmed by the 61 ksec

ROSAT observation on 3–5 Sep, which detected the source with an unabsorbed flux, $F_{(0.1-2.4\ keV)} = 2.5 \times 10^{-14}$ erg s^{-1} cm^{-2}, consistent with a -1.4 power-law decay [7].

An X-ray spectrum taken with ASCA of the afterglow revealed an emission line which could be attributed to the redshifted Fe I line at 6.4 keV. It would imply a redshift of $z = 0.33$ for the host galaxy [8]. The X-ray spectrum also shows a typical cut-off at low X-ray energies, typical of a dense environment like the star forming regions. Although many optical/near-IR follow up observations were performed, no reliable counterpart candidates were found [9–17].

OBSERVATIONS

Optical and near-IR observations of the RXTE error box [1] were started only 6.5 hr after the event [11]. Table 1 and Table 2 display the log of the optical and near-IR observations, respectively.

TABLE 1. Summary of the optical observations performed for the GRB 970828 error box. Dashes in column 2 denote unfiltered images.

Date	Filter	Exposure time (s)	Telescope
29.03 Aug	R	3600	0.82-m IAC80
29.91 Aug	-	1500	2.2-m CAHA
29.95 Aug	R	3600	0.82-m IAC80
30.09 Aug	-	1500	2.2-m CAHA
30.92 Aug	B	1500	2.2-m CAHA
31.10 Aug	-	600	2.2-m CAHA
3.85 Sep	R	600	4.2-m WHT
3.93 Sep	B	1000	4.2-m WHT
4.91 Sep	R	600	4.2-m WHT
5.92 Sep	R	600	4.2-m WHT
10.89 Sep	R	2400	4.2-m WHT
24.0 Oct	R	4800	2.5-m NOT

TABLE 2. Summary of the near-IR observations performed for the GRB 970828 error box.

Date	Filter	Exposure time (s)	Telescope
13.30 Sep	K	2160	3.8-m UKIRT
13.25 Oct	K	2880	3.8-m UKIRT
16.25 Oct	K	1800	3.8-m UKIRT

The R-band frames were acquired at the 0.82-m IAC80 telescope. Unfiltered, B and R-band frames were taken at 2.2-m Calar Alto (CAHA) telescope. Additional B and R-band images were obtained at La Palma with the WHT and NOT telescopes. The near-IR observations are based on K-band frames taken at the 3.8-m UKIRT telescope.

RESULTS AND CONCLUSION

Our result is that no variable optical/near-IR counterpart was found within the preliminary $1'.5$ radius *ASCA* X-ray source error box [5], and further reduced to a $0'.5$ radius region [6]. At the position of the X-ray source found by *ROSAT* (coincident with the *ASCA* source) we found no variable optical/near-IR objects down to R = 24.5, K = 20. The object found by [17] has not changed significantly in our images.

If the host galaxy is at $z = 0.33$ (as suggested by [8]) and is fainter than R = 24.5 we can impose an upper limit to the luminosity of the host galaxy. As we know from the *ASCA* spectrum the gamma event occurred in a dense medium, i.e., the absorption effects must be taken into account for the luminosity estimate. We assume a spectral index $\beta = -1.56$, similar to GRB 970508 [18], in order to perform the K-correction. The cosmological parameters considered were $q_o = 0.5$ and $H_o = 75$ km s^{-1} Mpc^{-1}. The value of the Galactic hydrogen column density towards the GRB direction is $N_H(\text{Galactic}) = 3.6 \times 10^{20}$ cm^{-2} or $A_v = 0.2$ by the relation $A_v = 5.6 \pm 0.1 \times 10^{-22} N_H(\text{Galactic})$ derived from [19].

Preliminary results of the inferred intrinsic hydrogen column density yields $N_H(\text{host}) = 1.31^{+2.14}_{-1.31} \times 10^{21}$ cm^{-2} [8]. This value has to be increased by a factor of $(1 + z)^{2.6}$ in order to calculate the effective $N_H(\text{host})$ at $z = 0$, since the spectral turnover is reduced by a factor of $(1 + z)$ [20]. Thus, the rest frame value of the intrinsic hydrogen column density is $N_H(\text{host}) = 6.04^{+9.86}_{-6.04} \times 10^{21}$ cm^{-2}.

As shown in Fig. 1, this value of $N_H(\text{host})$ makes the host galaxy to be fainter than $M_R = -20.28$, in order to be consistent with our limit imposed over the R-band magnitude of the host galaxy (R > 24.5). Thus $M_R > M_R^* = -22.2$ and the galaxy would be below the knee of the luminosity function. However, the large error in the determination of $N_H(\text{host})$ makes that the value imposed over M_R ranges from -16 to -26. As it is also clearly shown, if we assume that GRB 970828 had the same intrinsic hydrogen column density and absolute R-band magnitude as GRB 970508 — $N_H(\text{host}) = 1.8 \times 10^{21}$ atom cm^{-2} [21] and $M_R = -19.9$ — the host galaxy of GRB 970828 would have been clearly detected (see square of Fig. 1).

The absence of optical/near-IR afterglow for GRB 970828 is probably due to extinction by dust, as derived from the hydrogen column density of 1.31×10^{21} cm^{-2} measured by *ASCA* [22]. This would support models in which the events occur in high, densely populated regions, perhaps related to star forming regions. These results could be strengthened in the future, when a final value of $N_H(\text{host})$ is published.

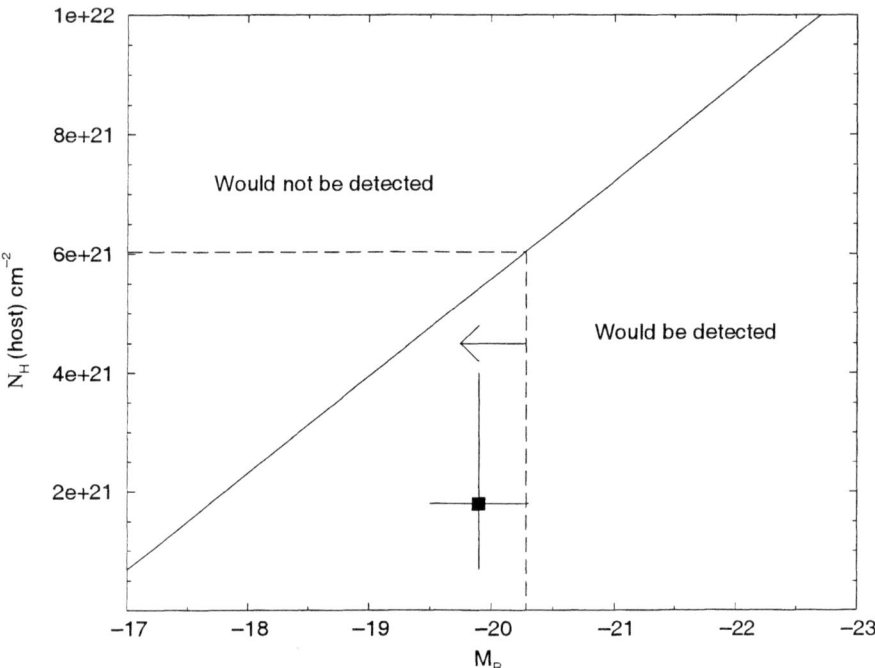

FIGURE 1. The straight line corresponds to values of N_H(host) and M_R that make the host galaxy have an apparent magnitude R = 24.5. The region above the diagonal corresponds to values of N_H(host) and M_R for which R > 24.5 and therefore undetectable in our images. The area under the line shows the values of N_H(host) and M_R that imply R < 24.5. The square shows the place where GRB 970508 host galaxy would be located if its redshift were $z = 0.33$. As it is shown in this case, the GRB 970508 host galaxy would be detected in our images. The intrinsic hydrogen column density of the GRB 970828 host galaxy, indicated by the long-dashed line, implies that the host galaxy must be fainter than $M_R = -20.28$.

ACKNOWLEDGMENTS

This work has been partially supported by Spanish CICYT grant ESP95-0389-C02-02. The authors wish to thank A. Yoshida for valuable information.

REFERENCES

1. Remillard, R., Wood, A., Smith, D., and Levine, A., *IAU Circ.* 6726 (1997).
2. Smith, D., Levine, A., Remillard, R., and Wood, A., *IAU Circ.* 6728 (1997).
3. Marshall, F. E., Cannizzo, J. K., and Corbet, R. H. D., *IAU Circ.* 6727 (1997).
4. Hurley, K., Kouveliotou, C., Fishman, G., Meegan, C., Connaughton, V., van Paradijs, J., *IAU Circ.* 6728 (1997).
5. Murakami, T., Ueda, Y., Ozaki, M., Yoshida, A., Kawai, N., Yamauchi, M., Marshall, F. E., Corbet, R. H. D., and Takeshima, T., *IAU Circ.* 6729 (1997a).
6. Murakami, T., Ueda, Y., Yoshida, A., Kawai, N., Marshall, F. E., Corbet, R. H. D., and Takeshima, T., *IAU Circ.* 6732 (1997b).
7. Greiner, J., Schwarz, R., Englhauser, J., Groot, P.J., Galama, T.J., *IAU Circ.* 6757 (1997).
8. Yoshida, A., Priv. Comm. (1999).
9. Filippenko, A. V., Stern, D., Treffers, R.R., Peng, C. Y., Bond, H.E., and Richmond, M.W., *IAU Circ.* 6729 (1997).
10. Brocato, E., Stiavelli, M., Bond, H.E., Sahu, K., *IAU Circ.* 6729 (1997).
11. Castro-Tirado, A. J., Gorosabel, J., Iglesias, J., Cairos, L. M., Vilchez, J., Mora, A., Gutierrez, C., Licandro, J., Bejar, V., and Greiner, J., *IAU Circ.* 6730 (1997).
12. Djorgovski, S.G., Odewahn, S.C., Steidel, C.C., Adelberger, K.L., Kellog, M., *IAU Circ.* 6730 (1997).
13. Yanagisawa, K., Nakada, Y., Abe, M., Ishibashi, Y., *IAU Circ.* 6731 (1997).
14. Guarnieri, A., Bartolini, C., Masetti, N., Piccioni, A., *IAU Circ.* 6733 (1997).
15. Odewahn, S.C., Djorgovski, S.G., Kulkarni, S.R., *IAU Circ.* 6735 (1997).
16. Stanek, K.Z., Garcia, M.R., Krockenberger, M., *IAU Circ.* 6735 (1997).
17. Klose, S., Eisloeffel, J., and Stecklum, B., *IAU Circ.* 6756 (1997).
18. Bloom, J. S., Djorgovski, S. G., Kulkarni, S. R., and Frail, D. A., *ApJL* **507**, L25 (1998).
19. Predehl, P. and Schmitt, J., *A&AS* **293**, 889 (1995).
20. Morrison, R. and McCammon, D., *ApJ* **270**, 119 (1983).
21. Owens, A., Guainazzi, M., Oosterbroek, T., Orr, A., Parmar, A. N., Costa, E., Feroci, M., Piro, L., Soffitta, P., Dal Fiume, D., Frontera, F., Palazzi, E., Pian, E., Heise, J., int Zand, J. J. M. I. T., Maccarone, M. C., and Nicastro, L. *A&AS* **339**, L37 (1998).
22. Yoshida, A., Namiki, M., Otani, C., Kawai, N., Murakami, T., Ueda, Y., Shibata, R., and Uno, S., in Huntsville Gamma-Ray Burst Fourth Symposium (ed.), Meegan, C.A., Preece, R. D. and Koshut, T. M., AIP 428, p. 441 (1998).

The GRB/SN Connection: An Improved Spectral Flux Distribution for the SN-Like Component to the Afterglow of GRB 970228, the Non-Detection of a SN-Like Component to the Afterglow of GRB 990510, and GRBs as Beacons to Locate SNe at Redshifts z ≈ 4 − 5

Daniel E. Reichart[*], Donald Q. Lamb[*], and Francisco J. Castander[†]

[*]*Department of Astronomy & Astrophysics, University of Chicago, 5640 South Ellis Avenue, Chicago IL, 60637*
[†]*Observatoire Midi-Pyrénées, 14 Av. Edouard Belin, 31400 Tolouse, France*

Abstract. We better determine the spectral flux distribution of the supernova candidate associated with GRB 970228 by modeling the spectral flux distribution of the host galaxy of this burst, fitting this model to measurements of the host galaxy, and using the fitted model to better subtract out the contribution of the host galaxy to measurements of the afterglow of this burst. Furthermore, we discuss why the non-detection of a SN1998bw-like component to the afterglow of GRB 990510 does not necessarily imply that a SN is not associated with this burst. Finally, we discuss how bursts can be used as beacons to locate SNe out to redshifts of $z \approx 4 - 5$.

AN IMPROVED SPECTRAL FLUX DISTRIBUTION FOR THE SN-LIKE COMPONENT TO THE AFTERGLOW OF GRB 970228

The discovery of what appear to be SNe dominating the light curves and spectral flux distributions (SFDs) of the afterglows of GRB 980326 (Bloom et al. 1999) and GRB 970228 (Reichart 1999; Galama et al. 1999) at late times after these bursts strongly suggests that at least some, and perhaps all, of the long bursts are related to the deaths of massive stars. Here, we build upon the results of Reichart (1999) by modeling the SFD of the host galaxy of GRB 970228, fitting this model to measurements of the host galaxy, and using the fitted model to better subtract out the contribution of the host galaxy to measurements of the afterglow of this burst.

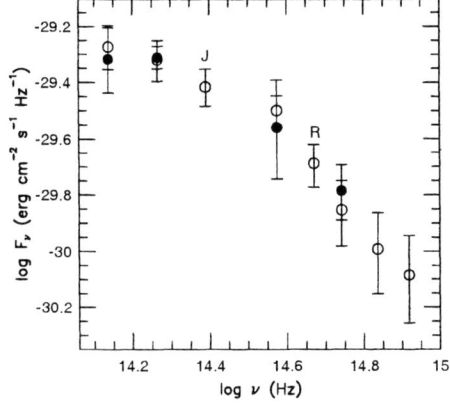

FIGURE 1. The observed (filled circles) and modeled (unfilled circles) K- through U-band SFDs of the host galaxy of GRB 970228.

In Figure 1, we plot the observed SFD of the host galaxy of GRB 970228, as measured with HST/WFPC2, HST/NICMOS2, and Keck I, and converted to the standard bands (Castander & Lamb 1999a; Fruchter et al. 1999). To these measurements and a broadband measurement made with HST/STIS (Castander & Lamb 1999a; Fruchter et al. 1999), which is not plotted, we fit a two-parameter, spectral synthesis model (see Castander & Lamb 1999a for details). The two parameters are the normalization of the SFD, and the age of the galaxy, defined to be the length of time that star formation has been occurring at a constant rate. Taking $A_V = 1.09$ mag for the Galactic extinction along the line of sight (Castander & Lamb 1999b), we find a fitted age of 270^{+460}_{-180} Myr; different values of A_V affect primarily the fitted age, and not the fitted SFD. Furthermore, models in which star formation slows considerably, or ceases, are generally too red to account for the measurements. Finally, we note that the fitted J- and R-band spectral fluxes are perfectly consistent with what one finds simply from linear interpolation between adjacent photometric bands.

In Figure 2 (left panel), we plot the SFD of the afterglow minus the *observed* SFD of the host galaxy from Figure 1. For the SFD of the afterglow, we use the revised K-, J-, and R-band measurements of Galama et al. (1999) and the I- and V-band measurements of Castander & Lamb (1999a; see also Fruchter et al. 1999); all of these measurements were taken between 30 and 38 days after the burst. We have scaled these measurements to a common time of 35 days after the burst, and have corrected these measurements for Galactic extinction along the line of sight (see Reichart 1999 for details). The K-band measurement of the afterglow is consistent with that of the host galaxy (Galama et al. 1999), resulting in an upper limit in Figure 2; J- and R-band measurements of the host galaxy are not available, again resulting in upper limits in Figure 2. As originally concluded by Reichart (1999), this SFD is consistent with that of SN 1998bw, after transforming it to the redshift of the burst, $z = 0.695$ (Djorgovski et al. 1999), and correcting it for Galactic extinction along its line of sight (see Reichart 1999 for details).

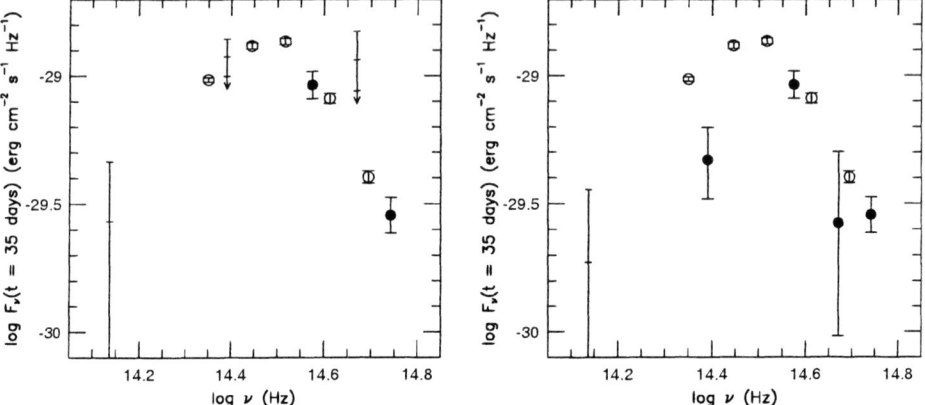

FIGURE 2. The K- through V-band SFDs of the late afterglow of GRB 970228 after subtracting out the observed (left panel) and modeled (right panel) SFDs of the host galaxy from Figure 1 and correcting for Galactic extinction (filled circles and upper limits), and the I- through U-band SFD of SN 1998bw after transforming to the redshift of GRB 970228, $z = 0.695$, and correcting for Galactic extinction (unfilled circles). The K-, J-, and R-band upper limits are 1, 2, and 3 σ.

In Figure 2 (right panel), we plot the same distribution, but minus the *modeled* SFD of the host galaxy from Figure 1. The SN-like component to the afterglow is detected in the J band, and possibly in the R band. The J-band measurement suggests that the SN-like component is $\approx 1/2$ mag fainter, and $\approx 1/2$ of a photometric band bluer, than SN 1998bw; however, this difference in J-band spectral fluxes is significant only at the $\approx 2.5\ \sigma$ level. When possible photometric zero point errors and uncertainties in our spectral synthesis model of the SFD of the host galaxy are included, this difference is significant only at the $\approx 2\ \sigma$ level. However, it is suggestive of what is generally expected: the Type Ic SNe that are theorized to be associated with bursts (e.g., Woosley 1993) are not expected to be standard candles.

THE NON-DETECTION OF A SN-LIKE COMPONENT TO THE AFTERGLOW OF GRB 990510

Given the rapid rate at which the afterglow of GRB 990510 faded, $t^{-2.4}$ (Stanek et al. 1999) or $t^{-2.2}$ (Harrison et al. 1999), at late times after the burst, a SN1998bw-like component to the afterglow, if present, could have dominated the light curve after about a month at red and NIR wavelengths (Figure 3; Lamb & Reichart 1999). However, Fruchter et al. (1999b) find no evidence of such a component, to within a factor of 3 − 7 in brightness, from two broadband *HST*/STIS observations.

Caution is in order before one concludes that a SN was not associated with GRB 990510, or one uses the non-detection of SN1998bw-like components to afterglows

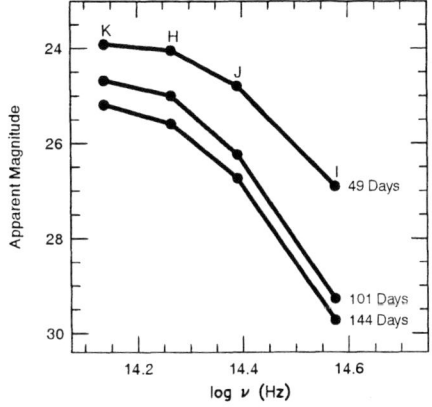

FIGURE 3. The SFD of SN 1998bw, transformed to the redshift of GRB 990510, $z = 1.619$, and corrected for Galactic extinction along the line of site (see Lamb & Reichart 1999 for details). Had NICMOS not run out of cryogen half a year earlier, the presence or absence of a SN component to the afterglow of this burst could have been firmly established.

to place lower limits on the redshifts of bursts. In the case of GRB 990510, the above calculation requires that assumptions be made about (1) the form of the SFD of the afterglow at the times of the observations, since the observations spanned a wavelength range of 300 – 900 nm; (2) how to extrapolate the non-power-law light curve of the early afterglow to the times of the observations; e.g., Stanek et al. (1999) and Harrison et al. (1999) do this differently; (3) the range of the luminosity distribution of SNe associated with bursts, relative to the luminosity of SN 1998bw, since these SNe are not expected to be standard candles; (4) the SFD of SN 1998bw at ultraviolet wavelengths, since the redshift of this burst is $z = 1.619$ (Vreeswijk et al. 1999); (5) whether differences in host galaxy extinction along the SN 1998bw and GRB 990510 lines of sight can be ignored; and (6) the underlying cosmological model.

GRBS AS BEACONS TO LOCATE SUPERNOVAE AT VERY HIGH REDSHIFTS

If bursts are indeed associated with SNe, then the first bursts should have occurred shortly after the first stars formed, at redshifts of $z \approx 15 - 20$ (Ostriker & Gnedin 1996; Gnedin & Ostriker 1997; Valageas & Silk 1999). Lamb & Reichart (1999) show that bursts and their afterglows should be detectable out to these very high redshifts. One way, of many (see Lamb & Reichart 1999 for details), in which bursts can be used to probe the early universe is to use them as beacons to locate SNe at very high redshifts.

In Figure 4, we plot the V-band light curve of SN 1998bw added to the best-fit source-frame V-band light curve of the early afterglow of GRB 970228 from Reichart (1999), transformed to various redshifts and corrected for Galactic extinction (see Lamb & Reichart 1999 for details). We use the V band because SN 1998bw peaked in this band. Clearly, the peak of the SN component of the light curve can be detected, at least from space, out to the K band, which corresponds to a redshift

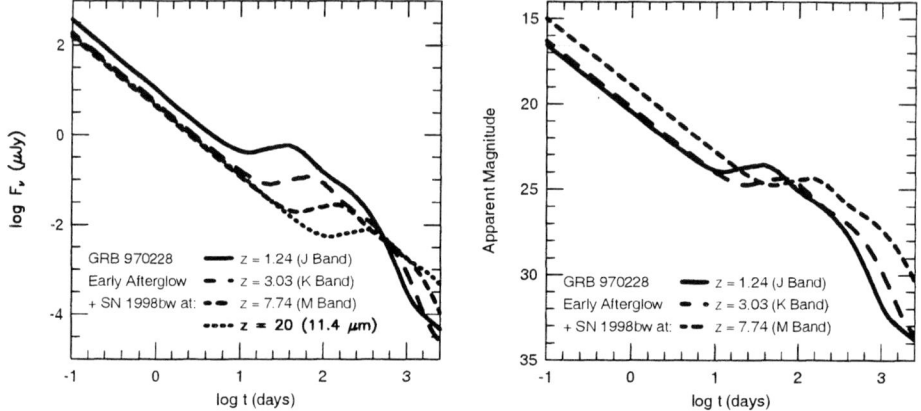

FIGURE 4. V-band light curve of SN 1998bw added to the best-fit source-frame V-band light curve of the early afterglow of GRB 970228 from Reichart (1999), transformed to various redshifts and corrected for Galactic extinction

of $z \approx 3$. The peak cannot be detected at longer wavelengths since such faint magnitudes cannot be reached in the L and M bands. However, from the K band, one should be able to detect SNe like SN 1998bw blueward of the peak out to redshifts of $z \approx 4 - 5$.

REFERENCES

1. Bloom, J. S., et al. 1999, Nature **401**, 453.
2. Castander, F. J., & Lamb, D. Q. 1999a, ApJ **523**, 593.
3. Castander, F. J., & Lamb, D. Q. 1999b, ApJ **523**, 602.
4. Djorgovski, S. G., et al. 1999, GCN Report 289.
5. Fruchter, A. S., et al. 1999a, ApJ **516**, 683.
6. Fruchter, A. S., et al. 1999b, GCN Report 386.
7. Galama, T. J., et al. 1999, ApJ, submitted.
8. Gnedin, N. Y., & Ostriker, J. P. 1997, ApJ **486**, 581.
9. Harrison, F. A., et al. 1999, ApJ **523**, L121.
10. Lamb, D. Q., & Reichart, D. E., ApJ, submitted.
11. Ostriker, J. P., & Gnedin, N. Y. 1996, ApJ **472**, L63.
12. Reichart, D. E., 1999, ApJ **521**, L111.
13. Stanek, K. Z., et al. 1999, ApJ **522**, L39.
14. Valageas, P., & Silk, J. 1999, A&A **347**, 1.
15. Vreeswijk, P. M., et al. 1999, GCN Report 324.
16. Woosley, S. E. 1993, ApJ **405**, 273.

The GRB/Supernova Connection

Bradley E. Schaefer and Ming Deng

Yale University, Physics Department, 260 Whitney, New Haven CT 06511

Abstract. We report here three unsuccessful tries to find a supernova event associated with a Gamma Ray Burst. (A) We examined published and unpublished records for six pre-1997 bursts for optical transients inside the error boxes from 4–30 days after the burst, with observed limits on any supernova of typically $V \sim 19$. For SN1998bw-like SNe, the bursters need only be at red shifts of greater than 0.07 for the transient to be undetected. Thus, the pre-1997 burst data does not usefully constrain model predictions which state that all long GRBs have underlying SNe events. (B) GRB781104 has a bright galaxy at $z = 0.0024$ in the middle of its small error box, so it is possible that this burst is a very low luminosity event with a SN1998bw-like transient visible for more than a year afterwards. We have examined nine deep plates from six-to-ten months after the burst, and these rule out the possibility of a supernova. For example, plate J5215 has no point source transients to a limit of 20.5 mag on 24 July 1979, when a SN1998bw-like event would have appeared at 17.2 mag. (C) We have examined whether observed Type IIn supernovae have positional/temporal coincidences with BATSE bursts. We find no significant correlation.

INTRODUCTION

Supernovae (SNe) have been associated with Gamma Ray Bursts (GRBs) since Colgate's shock breakout model before the discovery of bursts. In recent years, models have proposed that GRBs are the collapse of a massive star which should also produce a supernova [1]. GRB980425 is associated with a bright and unusual Type Ic SN appearing in a nearby bright galaxy [2]. And several other faint bursts with arc-second positions have had inflections in the fading afterglow which might well be from an underlying supernova [3,4]. With the scant available information, it is worthwhile trying to look for more GRB/SN connections.

DO ALL LONG GRBs HAVE SUPERNOVAE?

At the Rome GRB meeting, S. Woosley advanced a model in which the long duration GRBs are caused by the core collapse of a massive fast-rotating star with a massive accretion disk around the forming black hole and a jet which burrows out through the infalling material to appear as a GRB to observers within the beam.

In an excellent display, Woosley then presented over a dozen specific predictions on which his model could be tested. One of his specific predictions was that *all* long duration GRBs should have a supernova (probably like SN1998bw) appear in the following month or so.

This is difficult to test with current events. For the usual distance to SAX GRBs of $z = 1$, the underlying SN would peak at $R = 25.1$ roughly twenty days after the burst. This faint light can easily get swamped by the light from the afterglow. For example, an afterglow at $R = 21$ one day after the burst fading as $T^{-1.1}$ will appear at $R = 24.6$ at the time of the peak SN light. In such cases, the SN would only appear as an inflection in the decline of the afterglow and can never be convincingly identified.

Our immediate reaction to Woosley's prediction was that the pre-1997 events could well place a stricter limit than the post-1997 events. The reason is that the old events were greatly brighter than the SAX events with optical transients. Specifically, the bright IPN GRBs are 18 times brighter on average than the recent OT/GRBs. So with the old events being much brighter and presumably closer, there was a good chance that significant limits could be placed to test Woosley's prediction.

From a wide variety of sources, published and unpublished, we have collected observations taken 4–30 days after a GRB which could show a point source that would have been recognized as a transient like SN1998bw. This means that no new source was visible to the stated limiting magnitude. We were somewhat surprised to find no relevant observations from the 1978–1982 IPN positions, but this can be understood by the delay in producing those error boxes being typically months. The result was only nine observations from six pre-1997 bursts which could plausibly have detected an underlying SN. These bursts have a median P_{256} of 36 ph · s^{-1} · cm^{-2}, which is 14 times brighter than the post-1997 bursts with arcsecond positions. The observations are for GRB920311, GRB920406, GRB930131, GRB930614, GRB930706, and GRB940301 [5–7] all of which are long duration bursts. The limits on any supernova are typically $V \sim 19$ from 4–26 days after the burst, although the deepest is a $B = 22.0$ limit 5.0 days after GRB940301.

The basic analysis was in asking the questions "How far away must the burst be such that a SN1998bw-like event would have been fainter than the observed limit?" and "At this distance, what would be the lower limit on the burst peak luminosity?". This lower limit can then be compared to the burst peak luminosities derived from LogN-LogP or time dilation or observed red shift measures. A variety of corrections were made for extinction in our Milky Way galaxy, K-corrections for the flux from the supernova, the time dilation applied to the SN1998bw light curve, and the K-corrections for the GRB peak luminosity.

For all nine observations of the six pre-1997 bursts, our limits on the red shift are poorer than $z = 0.07$ and our limits on the burst peak luminosity are poorer than 10^{57} ph · s^{-1}. That is, if the six bursters are more distant than $z = 0.07$ (or more luminous than 10^{57} ph · s^{-1}), then any underlying SN would be fainter than the observed limits. This is to be compared to expected red shifts of $z > 0.5$ and luminosity for bright bursts of $> 10^{58}$ ph·s^{-1}. Our conclusion is that the pre-1997 observations can not realistically address Woosley's prediction that all long GRBs should have underlying SNe.

DID GRB781104 HAVE AN EVENT LIKE SN1998bw?

Another way of stating the objective of this paper is to say "Can we find any more SN1998bw-like events?" SN1998bw/GRB980425 is unique in having a bright nearby galaxy in the middle of the error box. Perhaps we can use this trait to find SN/GRB events? That is, can we find a GRB error box with a bright galaxy for which we can seek to find evidence of a supernova within one-or-two years after the event?

A compilation of the bright IPN events [8] might be the best sample to find bright galaxies and perhaps supernovae because the IPN bursts are bright and hence presumably close. Indeed, one event, GRB781104, stands out as having a bright nearby galaxy in the center of the error box [8]. This event has a T_{90} of around 20s and is thus one of the long GRBs. The V = 15.26 galaxy is MCG-04-47-011 at a red shift of $z = 0.0024$. Could GRB781104 have spawned a supernova like SN1998bw? If so, then the supernova would have peaked at around twelfth magnitude in mid-November 1978 and should still be visible on deep plates showing the galaxy for over a year later.

We found plates from three sources (the UK Schmidt, E. Helin's asteroid search on the 18" Palomar Schmidt, and the Harvard Damons). These images have all been examined and there is no evidence for any transient source (see Table 1). The most restrictive limit is for plate J5215 which shows no transient to 3.3 mag deeper than a SN1998bw-like event should have appeared. The lack of any transient to this depth proves that GRB781104 did not have a SN1998bw-like event.

Another try is to realize that SN1998bw was very bright in the radio and that the GRB781104 galaxy is even closer. However, the first post-burst radio map is

TABLE 1. Observed Magnitude Limits for a SN1998bw-like Event for GRB781104.

Plate	Date	Observed Magnitude	Expected Magnitude
DSB585	25 May 1979	> 15.5	16.2
DSB615	29 May 1979	> 16.9	16.2
J5215	24 Jul 1979	> 20.5	17.2
...	25 Jul 1979	> 16.5	17.2
R5244	2 Aug 1979	> 19.5	17.3
J5250	3 Aug 1979	> 18.5	17.3
...	21 Aug 1979	> 16.5	17.6
J5354	20 Sep 1979	> 21.0	18.1
UJ5360	21 Sep 1979	> 20.5	18.1

with a deep VLA look in 1985 [9]. Indeed, a radio source was found at the position of the galaxy, but it was extended and likely just the galaxy itself. And at this late time, even SN1998bw would have been below the threshold of the VLA data.

Nevertheless, we are left with the conclusion that GRB781104 did not have a SN1998bw-like event. Either GRB781104 was an extremely low-luminosity burst with no following supernova or a galaxy as bright as MCG-04-47-011 appears in an IPN error box by an unlikely chance.

This idea can be turned around. That is, the *lack* of a bright galaxy in error boxes can indicate that the GRB was *not* like SN1998bw. This can be used to place an upper limit on the fraction of GRBs that are like SN1998bw. For this test, we can only use small GRB boxes for which the probability of chance coincidence with a bright galaxy is much smaller than the final fraction of events. The only two samples for this test are the 23 bright IPN bursts and the 17 SAX bursts. Out of these 40 events, 2 of them (GRB781104 and GRB980425) have bright nearby galaxies in the middle of the box. So I conclude that the fraction of all GRBs with SN1998bw-like events is $\sim 5\%$ or perhaps smaller.

ARE TYPE IIn SUPERNOVAE ASSOCIATED WITH GRBs?

Type Ic (and possibly Type Ib) SNe are associated with GRBs both due to GRB980425/SN1998bw and due to theoretical models. Perhaps a necessary condition for this class of explosions is a very massive progenitor star with its outer envelope stripped. Another class of supernova that might then be of interest for burst studies is the Type IIn events. These are thought to be produced by massive star in the act of ejecting a very thick wind. Terlevich and Fabian [10] has proposed that the GRB991002 produced the Type IIn SN1999eb (with the first detection on 29 September) based solely on positional and temporal coincidence (with no explanation for why the burst would be seen long after the core collapse).

The problem with such identifications is that there are many GRBs (typically with $\sim 5°$ positional accuracy) and there are many Type IIn SNe (typically with uncertainties in the date of core collapse of a month), so that positional/temporal coincidences are expected. (GRB980425/SN1998bw had a coincidence good to an arc-minute and better than one day, so the *a posteriori* connection is significant.) So merely identifying one coincidence in the eighth year of BATSE is not enough to establish a connection.

We have addressed the problem by broadly seeking coincidences between all BATSE GRBs and all identified Type IIn events between April 21, 1991 and 9 September, 1999. For the data of time and positions, we have taken the BATSE list of GRBs and the CFA catalog of supernovae (http://cfa-www.harvard.edu/cfa/ps/lists/RecentSupernovae.html). There are 2515 GRBs on the BATSE list and 27 TypeIIn discovered during the period (http://merlino.pd.astro.it/ supern/snean.txt). We have been generous in seeking supernovae whose discovery dates coincide within 30 days after the GRB and

TABLE 2. Correlations Between GRBs and Type IIn Supernovae.

BATSE Trigger	Supernova	Angular Separation	GRB $1-\sigma$ Error Box	Time Separation
1675	1992an	8.4°	5.19°	29
3403	1995G	9.2°	3.94°	17
6166	1997bs	6.0°	8.95°	7
7207	1998et	8.3°	1.49°	6

whose positions agree to within 10°. This resulted in 4 pairs of possible matches (see Table 2). None of these bursts has additional positional information, for example from the IPN.

Of these four coincidences, two SNe are within the 2-sigma GRB error boxes and one SN is within the 1-sigma GRB error boxes. The expected number of matches between the 27 identified Type IIn SNe and BATSE bursts detected at a rate of roughly one per day with our 10° and 30 day window is 5.4. However, with both the supernovae discoveries and BATSE bursts being clumped in time and position, any exact calculation of the expected number of coincidences will be problematic. Instead, we have made an identical search except that we require the SN to be before the GRB, and we find two coincidences and take this to be the random coincidence rate. It is clear that the number of GRB/IIn coincidences is not significant.

ACKNOWLEDGMENTS

Ken Lawrence scanned the Palomar 18" plates taken by Eleanor Helin. Dorrit Hoffleit served as courier for the Damon plates from Harvard with the help of Alison Doane. Sue Tritton prepared the UK Schmidt plates.

REFERENCES

1. MacFadyen, A. and Woosley, S. E., *ApJ* **524**, 262 (1999).
2. Galama, T. et al., *Nature* **395**, 670 (1998).
3. Bloom, J. S. et al., *Nature* **401**, 453 (1999).
4. Reichart, D. et al., *ApJ* **521**, L111 (1999).
5. Schaefer, B. E. et al., *ApJ* **422**, L71 (1994).
6. McNamara, B. J. et al., *ApJS* **103**, 173 (1996).
7. Barthelmy, S. D., Palmer, D. M., and Schaefer, B. E., in *Gamma Ray Bursts, edited by G. J. Fishman, J. J. Brainerd, and K. Hurley, AIP Conf. Proc.* **307**, New York, 392 (1994).
8. Schaefer, B. E. et al., *ApJS* **118**, 353 (1998).
9. Schaefer, B. E. et al., *ApJ* **340**, 455 (1989).
10. Terlevich, R. and Fabian, A. IAU Circ, 7269 (1999).

Are There Unrecognized Optical Afterglows of GRBs Among Observed SNe?

René Hudec[1], V. Hudcová[1], Eliana Palazzi[2], Nicola Masetti[2], and Graziella Pizzichini[2]

[1]*Astronomical Institute, CZ-251 65 Ondrejov, CZ-251 65 Ondrejov, Czech Republic, e-mail: rhudec@asu.cas.cz*

[2]*TESRE, CNR, Bologna, Italy*

Abstract: The observational properties of optical afterglows of GRBs resemble to some extent those of supernovae. This means that there may be unrecognized afterglows especially among poorly investigated SNe. We present a study based on cross-correlations of GRB 4B and SNe catalogues and focused on search for such unrecognized afterglows. We present a list of possible candidates and show that the main limiting factors so far in this type of study are the poorly known parameters (light curves etc.) of particular SNe.

INTRODUCTION

The recent detections of optical afterglows of Gamma Ray Bursts (GRBs) confirm the low energy emission of these events. There have been optical afterglows observed for 12 GRBs so far (from subset of 24 with good optical data), hence limited information is available now for their light curves, magnitudes, as well as other parameters. The common properties of optical afterglows (henceforth OA) can be summarized as follows: (1) The OA peaks at roughly mag 18–22 (2) They exhibit power law declines mostly with mean index of −1.2 (3) They seem to be related to faint host galaxies (but not in all cases) (4) Not all of the GRBs have an associated OA. These general properties and behavior of OA resembles to some extent those of supernovae, especially by peak magnitudes, transient behavior, and also by mean decline rates. Furthermore, the rate of OA may be much larger than the GRB rate due to different beaming. This factor is unknown but may reach up to four orders of magnitude [1]. It is hence obvious that some of the detected and poorly in detail investigated supernovae may, in fact, represent unrevealed OA. The NS mergers are also expected to produce OT events that could be, in principle, found among detected SNe [2]. There are, however, also indications that there may be even direct physical relations between (some) GRBs and (some) SNe, as illustrated by the positional and temporal coincidence of SN1998bw with GRB980425, and redshifted SNe probably detected in some OAs.

This is why we have decided to study the possible correlations of GRBs and SNe in more detail. In this search, we have taken all detected SNe into account, including faint and poorly investigated events, since such may represent unrevealed OAs. Our approach is to study in detail the coincidences found and SN in question rather than make pure statistical conclusions (which are difficult due to large positional inaccuracies of GRBs and another influences such as the incompleteness of SN data bases). We note that the incompleteness of the SNe catalogues is large and that only $10^{-3}...10^{-4}$ of all SN down to mag 23 are included.

DATA

The 4B GRB catalog available on the www has been used as the source of GRBs parameters and positions. This database has been cross-correlated with the list of supernovae compiled by us over the whole time interval of the BATSE 4B catalog from reports published in the IAUC as well as with the IAU list of detected SNe. All events occurring before Sept 27, 1999 have been taken into account. Since the OAs are known to peak no more than one day (with only one exception for GRB980508) after the corresponding GRB and then declining gradually, and since the first SN observations are usually delayed after their explosions, we have searched for correlations between GRBs and supernovae detected within the particular GRB error boxes up to 60 days after the gamma ray events.

PRELIMINARY RESULTS

The subset (SNe detected up to 30 days after GRB) of correlations found is listed in the Table 1. The given errors represent both statistical as well as systematical errors. Altogether, 2521 GRBs and 804 SN detected between April 1991 and September 1999 have been analyzed. The number of correlations found is consistent with the assumption of unrelated samples, as was confirmed by two independent methods.

SN1990aj and SN1993R

Moreover, we have considered the possible correlations with GRBs of SN1990aj and SN1993R [3], based on their spectra similar to SN1998bw. The SN1993R is included in the above mentioned Table and may be related to GRB 930524, although this correlation is rather weak due to the large error box of this GRB. For the SN1990aj, no reliable GRB catalogs are available. Our search based on the archival gamma-ray observations of the HEASARC Master Catalogs (http:// heasarc.gsfc.nasa.gov/cgi-bin/W3Browse/w3browse.pl) has revealed only one catalogue of GRBs detected in 1990 with data from the GRANAT satellite. None of the 13 GRBs with known locations occurring between Dec. 1989 and Dec. 1990 and detected by GRANAT is spatially coincident with SN1990aj [4]. On the other hand, the GRANAT catalogue does not represent a complete data set of GRBs detected at times close to the SN1990aj and also the sensitivity was much worse than that of BATSE. Other GRBs detected by GRANAT/WATCH in the given period as well as those detected by GRANAT/Phebus have no coordinates available [5]. We hence cannot either confirm or exclude the connection of SN1990aj and GRBs.

Comparison with previous searches

Our search is more general than those of Kippen et al. (1998) [6] and Wang and Wheeler (1998) [7]: we consider all known SNe without any restriction. The study of Kippen et al. [6] is restricted to brighter events only (brighter than mag 17) but this is in contrast with BeppoSAX OA statistics (only the SN1998bw was brighter than 17 mag, all other 12 OAs were fainter) while Wang and Wheeler restrict to SNe of known Ia, Ib types (but the OA may be also among weak and unclassified SN).

The comparison with results of Wang and Wheler [7] and Kippen et al. [6] shows that they have found two objects which have seemingly escaped our cross-correlation. While SN1992at was optically detected more than one month after the corresponding BATSE trigger (GRB920628) and is slightly (0.3 deg) outside the BATSE error box, SN1995ac was excluded by our cross-correlation because is about 1 degree outside the BATSE error box (the distance between the SN1995ac and GRB950917 reported by Wang and Wheeler is wrong).

SN EVENTS BEFORE GRBs?

For GRBs with beams completely off the line of sight, theoretical predictions have been made that the GRB events may be delayed by weeks compared to the SN events, and the GRB should be spatially displaced in the sky by light-weeks [8]. The subsequent afterglows will be further delayed and spatially displaced from the SN. The vast majority of GRB/SN events should belong to this class. Combining the time delay and the angular separation between the SN and GRB (probably only for nearby ones in practice) should allow a determination of the angle of the jet relative to the line of sight. This is why we have cross-correlated the catalogues also for GRBs following SNe with time windows of up to 60 days. The number of correlations is again consistent with unrelated samples.

DISCUSSION

The statistical significance of results of cross-correlations between the GRBs and SN is affected and limited by the following: (1) the positional uncertainty of a quite large fraction of the GRBs is rather high (2) the dates of SN explosions and/or peaks are unknown in most cases (3) the results of optical SN searches do not represent a full and homogeneous sample so that many (and even a large majority) of the SN may be missed especially at faint magnitudes (4) there is no systematic sky patrol survey at low magnitudes below 15.

For the correlations found, many of the related SNe represent poorly investigated events with poorly known light curves, decline rates, color indices and no or limited spectral information hence the decision whether they could be related to the GRBs in question is difficult. The recent detections of faint OA of GRBs strongly support and justify further, more extended and more complete searches for faint new and variable optical objects, especially at low magnitudes (18 and less). Some of such information

TABLE 1. Preliminary results of SN – BATSE GRB correlations for SNe detected within 30 days after GRBs and within their error boxes.

SN	Supernovae α_{2000}	δ_{2000}	Date	GRB	BATSE α_{2000}	δ_{2000}	Error	Date
1991X	195.2530	-14.5106	910505	107	193.4700	-8.3800	11.2614	910423
1991aa	191.2910	-6.3173	910507	107	193.4700	-8.3800	11.2614	910423
1991bg	186.2655	12.8711	911203	1120	183.9200	8.2400	8.5634	911125
1992Q	182.7567	-1.8808	920407	1506	181.6100	-1.9700	4.0706	920321
1992T	198.2500	-31.9000	920409	1485	202.0100	-29.9800	9.8353	920315
				1510	198.8400	-34.3500	7.5336	920321
1992aj	356.9500	-35.2000	920705	1649	353.2900	32.8400	7.3595	920615
1992al	311.4854	-51.3944	920727	1690	308.3400	-49.9100	3.8600	920708
1992am	21.2500	-4.6500	920726	1704	20.7500	-22.3200	9.0223	920717
				1708	22.4400	-3.9800	2.2158	920718
1992aw	286.7300	51.0486	920827	1742	286.4000	50.6900	2.6104	920804
				1791	280.0700	59.4800	9.8843	920811
1992bb	319.5619	-7.5843	920930	1962	322.7700	9.3800	21.5838	920928
1993K	96.1500	-22.8500	930328	2210	79.4400	-9.7700	18.6669	932028
1993R	356.0500	10.7667	930602	2352	351.1600	-6.0200	18.1099	930602
1993W	352.4117	-3.0405	930818	2483	351.9300	-1.3600	12.1791	930807
1994J	150.4032	54.5824	940305	2821	151.0800	53.1500	10.0610	940214
1994R	167.5000	-23.7167	940603	2998	163.6400	-25.3300	3.9386	940529
1994U	196.2339	-7.9476	940627	3046	192.5300	-155.3700	8.4660	940626
1995bc	147.9312	40.3141	951201	3901	148.8700	39.8500	4.4283	951107
1996bx	59.8186	-53.3740	961118	5648	59.3600	-52.6200	3.8252	961029
1997B	88.2624	-17.8732	970113	5718	97.7550	-21.7300	12.8314	961218
1997bs	170.0599	12.9721	970415	6145	169.7400	8.8000	5.0417	970329
				6166	167.9700	7.3900	9.1495	970408
1997bz	170.6061	1.1893	970427	6162	174.2800	2.6600	4.1949	970407
				6166	167.9700	7.3900	9.1495	970408
1997dp	38.5270	32.9371	971101	6439	51.7900	49.3200	17.9906	971016
1997dq	175.2329	11.4794	971102	6428	167.0300	2.6600	8.9832	971013
1998B	116.5000	18.7167	980101	6536	116.1700	16.6700	5.5269	971218
1998bw	293.7638	-52.8458	980428	6707	291.9100	-53.1100	2.5230	980425
1998ck	145.2750	-29.1049	980531	6796	148.3100	-27.2900	8.8953	980530
1999cs	181.1053	-20.1006	990414	7518	180.0000	-18.4400	2.1288	990413
1999ct	198.2690	46.2646	990613	7587	199.9300	49.3500	7.9504	990527
1999cu	198.3143	46.3336	990613	7587	199.9300	49.3500	7.9504	990527
1999dg	227.8684	13.4849	990723	7662	224.5100	10.8600	11.1333	990719
1999dn	354.0612	2.1524	990819	7706	358.0700	1.7300	4.7560	990810
1999dp	67.3194	69.5335	990902	7749	58.2300	62.4900	15.6259	990902
1999ds	358.4651	0.1564	990907	7706	358.0700	1.7300	4.7560	990810

can be retrieved from deep archival plates (some of the archival plate collections reach the limiting magnitudes of 20 and even 23 such as ROE Edinburg and TLS Tautenburg) but a systematic deep CCD patrol can provide more precise and much more complete database.

SUMMARY

The unrecognized OAs of GRBs may be among the SNe detected by various projects. A small fraction of them may be revealed by cross-correlation with GRB catalogues. Theories exist for SN emission preceding the GRB, depending on the beam position relative to the line of sight. There may be however also OAs without GRBs due to different beaming in gamma rays and in optical. The OAs may be revealed by their characteristic light curves and by their typical colors [9]. However, this information is not available for most of the detected Sne. Better optical data are urgently needed for the SNe recently detected which may be related to GRBs.

ACKNOWLEDGEMENTS

The work has been supported by the project of the Czech Ministry for Education and Youths, No. ES02, by grant of the Grant Agency of the Czech Republic No. 205/99/0145 and by the project CNR-AS CR.

REFERENCES

1. Woods, E. and Loeb, A., *ApJ* **508**, 760 (1998).
2. Li-Xin, L. and Paczynski, B., *ApJ* **507**, L59 (1998).
3. Patat, F. and Piemonte, A., *IAUC* 7017 (1998).
4. Sazonov, S. Y. et al., *A&ASS* **129**, 1 (1998).
5. Terekhov, O. V. et al., *Astron. Lett.* **20**, 265 (1994).
6. Kippen, R. M. et al., *ApJ* **506**, L27 (1998).
7. Wang, L. and Wheeler, J. C., *ApJ* **504**, L87 (1998).
8. Cen R., *ApJ* **507**, L131 (1998).
9. Simon, V., Pizzichini, G. and Hudec, R., these proceedings.

IV. THEORY

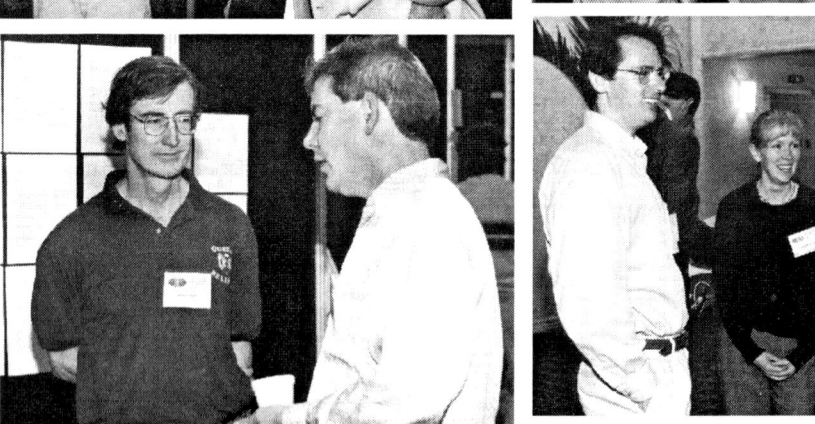

External Shock Model for Gamma-Ray Bursts during the Prompt Phase

Charles D. Dermer[1]

Naval Research Laboratory, Code 7653, Washington, DC 20375-5352

Abstract. The hard X-ray and γ-ray phenomenology of gamma-ray bursts (GRBs) can be explained by an external shock model where a *single* relativistic blast wave interacts with the surrounding medium. Besides reproducing the generic spectral behavior of GRB profiles, the external shock model provides quantitative fits to the peak flux distribution, the > 1 s t_{50} duration distribution, and the distribution of the peaks E_{pk} of the νF_ν spectra of GRBs measured with BATSE. The apparent paradox between a relativistic beaming scenario and the empirical finding that E_{pk} values are preferentially measured within the triggering range of a GRB detector is resolved by this model when blast wave physics and detector triggering criteria are taken into account. Some surprising implications follow, namely that the fireball event rate is ~ 1 per 10^4 years per Milky Way galaxy for unbeamed sources, and proportionally more if fireball outflows are collimated. This is ~ 3 orders of magnitude larger than normally quoted. Most of the clean and dirty fireball transients are undetected due to telescope sensitivity and design limitations.

Strongly variable GRB time histories with good radiative efficiencies are possible because of the strongly enhanced emissions when a blast wave interacts with density inhomogeneities located nearly along the line-of-sight to the observer. Arguments against short timescale variability in an external shock model are answered, and difficulties in an internal shock/colliding shell model are mentioned.

I INTRODUCTION

An important question in GRB studies is whether the GRB engine produces a single impulsive collapse and ejection event, or instead operates over a period of time much longer than the $\sim$ ms dynamical time scale of the central engine. In the external shock model [1,2], a single relativistic shell is ejected by the GRB engine and energized by interactions with the surrounding medium. Variability in the light-curves is attributed to interactions with an inhomogeneous surrounding medium. In the colliding shell (or internal shock) model [3,4], collisions between a succession of shells in a relativistic wind are thought to produce the variability observed in GRB light curves. If a conclusive resolution to this problem is obtained, then physical information can be extracted directly from GRB light curves. In the case

---

[1] Work supported by the Office of Naval Research.

of the external shock model, variations in GRB light curves reveal the distribution of circumstellar material near the sources of GRBs. In the case of the internal shock model, GRB light curves reflect the structure of, and accretion processes operating within, the putative disk of material that is accreted by the newly formed collapsed object to energize the relativistic wind.

Here we review work focusing on the external shock model in the prompt γ-ray luminous phase. We find that the extensive phenomenology of GRBs can be explained with this model, so that the addition of multiple relativistic shells and the numerous parameters associated with a hybrid internal/external shock model are unnecessary. The fewer number of free parameters in the external shock model places definite constraints on the number and type of fireballs needed to explain GRB statistics. The most important implication is that classes of clean and dirty fireballs with well-defined properties must exist, and that the fireball event rate is much larger than previously estimated on the basis of detected GRBs.

II NUMERICAL SIMULATION OF LIGHT CURVES

When a relativistic blast wave with Lorentz factor Γ encounters an external medium, charged particles will be captured by the blast-wave shell even if the shell has only a very weak entrained magnetic field. A captured particle in the shell frame gets Lorentz factor Γ. This internal energy derives from the directed energy of the relativistic shell, causing the shell to decelerate.

We have developed a numerical simulation model [5,6] for a GRB blast wave that interacts with an external medium. The model treats synchrotron, Compton, and adiabatic processes, and blast-wave deceleration is self-consistently calculated. The parameters that enter the numerical model are those of the standard blast wave model. The macroscopic variables are the implied isotropic energy release $E = 10^{54} E_{54}$ ergs/(4π sr) and the initial Lorentz factor $\Gamma_0 = 300\Gamma_{300}$ of the blast wave. The environmental variables are the external density $n(x) \propto n_0 x^{-\eta}$, where x is the distance from the center of the explosion. We let $n_0 = 10^2 n_2$ cm^{-3} and consider a uniform surrounding medium ($\eta = 0$). (Inhomogeneities in the external medium are considered in §V.) We also let the opening half-angle of the outflow $\psi = 10°$, corresponding to a beaming factor $f = 0.76\%$. As long as $\psi \gg \Gamma_0^{-1}$, the collimation has little effect on γ-ray emission during the prompt phase if the observer's line-of-sight falls within an angle $\theta \lesssim \Gamma_0^{-1}$ of the jet axis.

The microscopic variables are the fraction of energy ϵ_e that is transferred from the swept-up protons to the swept-up electrons, and the injection index p of the assumed power-law electron energy distribution. A parameter $\epsilon_{\max}$ is defined in terms of a maximum Lorentz factor $\gamma_{\max}$ obtained by balancing the minimum acceleration time scale and the synchrotron loss time scale, giving $\gamma_{\max} = 4 \times 10^7 \epsilon_{\max}/[B(G)]^{1/2}$. The magnetic field B is specified by a magnetic-field parameter ϵ_B through the relation $B^2/(8\pi) = 4\epsilon_B m_p c^2 n(x) \beta(\Gamma^2 - \Gamma)$, where $\beta c = (1 - \Gamma^{-2})^{1/2} c$ is the speed of the blast wave. Standard values used here are $\epsilon_e = 0.5$, $p = 2.5$, $\epsilon_{\max} = 1$,

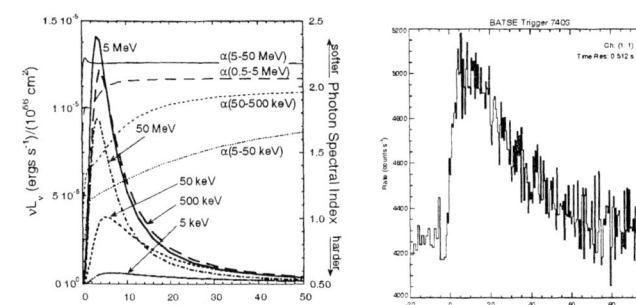 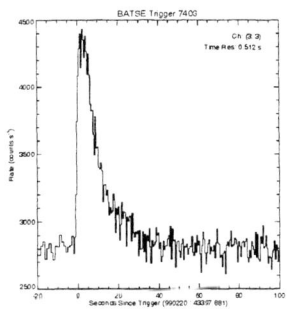

FIGURE 1. Calculated GRB light curves (thick curves) and photon spectral indices (thin curves) due to an external shock interacting with a uniform surrounding medium (left) are shown in the left figure. The 20–50 and 100–300 keV count rates of a typical GRB with a smooth profile (GRB 990220; BATSE trigger 7403) are shown in the middle and right figures, respectively.

and $\epsilon_B = 10^{-4}$. The low value of ϵ_B is required to avoid forming cooling spectra, which are not commonly observed in GRBs [7]. We also note that the microscopic variables are assumed to be constant in time.

The left panel in Fig. 1 shows calculations of light curves and spectral indices at different observing energies for a model GRB with standard parameters. For comparison, we also show a typical GRB with a smooth light curve. Several effects are apparent here. The first is that the generic Fast Rise, Exponential Decay (FRED) profile found in some 20–30% of all GRB light curves is reproduced (FRED is actually a misnomer, as the decay law is more closely approximated by a power law). The second is that the peaks are sharper at higher energies and broader at lower energies. Another is a hardness-intensity relation and a hard-to-soft evolution of the GRB light curves, so that the well-known correlations are reproduced. A prediction of the model is that the peaks are aligned at γ-ray energies, but lag at X-ray energies [8]. This prediction seems to be confirmed by observations with BeppoSAX [9] which has spectral coverage in the 2–700 keV range.

Fig. 2a shows model GRB spectra at different observing times, and Fig. 2b shows the calculated relationship between $E_{\rm pk}$, flux, and fluence. At X-ray energies, the photon spectral index approaches a value $\alpha \approx 2/3$, corresponding to the nonthermal synchrotron emissivity spectrum from an uncooled electron distribution with a low-energy cutoff. The spectrum turns over and approaches the value $1 + (p/2)$ associated with a cooling electron distribution at the highest energies. Fig. 2b shows that the qualitative behavior of the $E_{\rm pk}$-fluence relationship observed in GRBs [10] is reproduced. The spectral aging inferred from the decay of $E_{\rm pk}$ values in smooth GRB light curves is a natural consequence of the external shock model.

The external shock model therefore accounts for the best established phenomenological correlations of FRED-type GRBs [11,8].

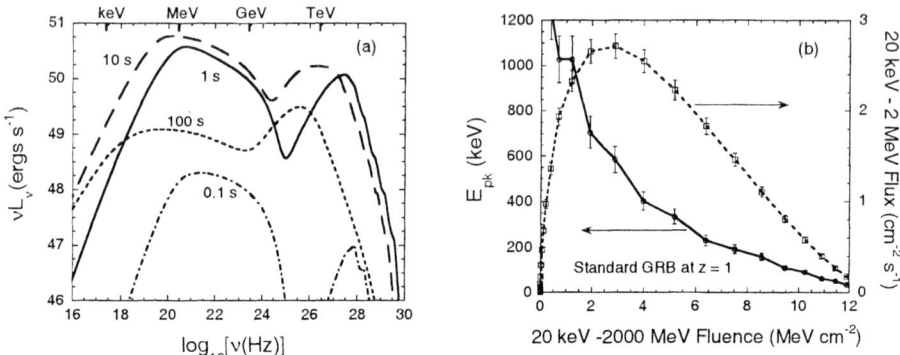

FIGURE 2. Generic behavior of a model GRB from an external shock energized by a uniform surrounding medium. The left panel shows the broadband X-ray and γ-ray spectra at different observing times. The right panel shows the dependence of $E_{\rm pk}$ and flux as a function of fluence.

III STATISTICAL PROPERTIES OF GRBS

Even if beaming is neglected, seven parameters enter into a blast-wave model calculation with an assumed uniform surrounding medium. We carried out a parameter study [12] showing that GRB observables are most sensitive to the value of the initial Lorentz factor (or baryon-loading parameter) Γ_0 of the explosion. The typical duration of a GRB in the prompt phase varies as $(E/\Gamma_0^8 n_0)^{1/3}$ at observing energies $\mathcal{E} \gtrsim \mathcal{E}_0$. The quantity $\mathcal{E}_0 = E_{\rm pk}(t=0)$ is the photon energy of the peak of the νF_ν spectrum at early times, and $\mathcal{E}_0 \propto q n_0^{1/2} \Gamma_0^4$, where q is a parameter related to the magnetic field and Lorentz factor of the lowest energy electrons. The power Π_0 at photon energy $\mathcal{E}_0$ varies as $(\Gamma_0^8 E_0^2/n_0)^{1/3}$. These relations show that the mean duration, peak photon energy, and peak power output of a GRB are most sensitive to the value of Γ_0.

A central criticism of a relativistic beaming scenario has been to explain the apparent paradox between a model involving relativistically beamed outflows, and observations showing that $E_{\rm pk}$ is narrowly confined to an energy range near a few hundred keV. Brainerd's Compton attenuation model [13], for example, was specifically designed to account for this fact, but the large column densities required by this model make it unable to explain rapid variability in GRB light curves [14]. The beaming paradox is resolved by the external shock model [8] when the spectral behavior implied by blast wave physics is convolved with detector response. A dirty fireball with $\Gamma_0 \ll 300$ will have a νF_ν peak at low energies, and will rarely be detected because the blast wave energy is radiated over a long period of time ($\propto \Gamma_0^{-8/3}$); thus its peak power is very weak ($\Pi_0 \propto \Gamma_0^{8/3}$). The flux in the BATSE range is even lower than implied by this relation because BATSE would be sensitive to only the soft, high-energy portion of the spectrum. A clean fireball, by contrast, would produce a brief, very luminous GRB, but BATSE would sample the very

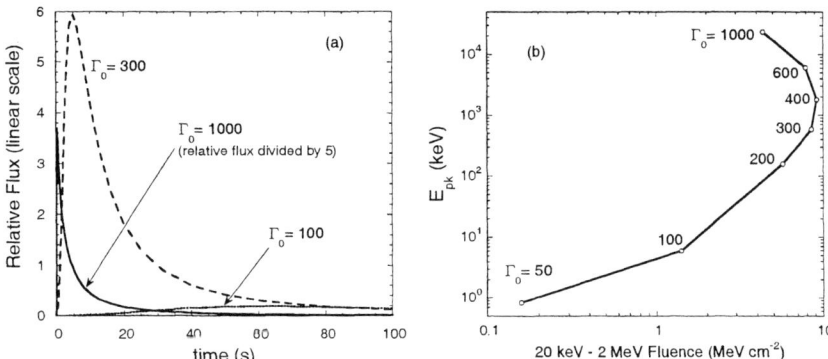

FIGURE 3. (a) Prompt 100 keV light curves for blast-wave Lorentz factors $\Gamma_0 = 1000$ (solid curve), 300 (dashed), and 100 (dotted). Other parameters are given in the text. (b) Dependence of $E_{\rm pk}$ measured at the deceleration time scale $t_{\rm dec}$ on the 20 keV – 2 MeV fluence, which is integrated from the start of the burst to $3t_{\rm dec}$.

hard portion of the spectrum below the νF_ν peak where the received flux is not so great. Fig. 3a illustrates this behavior. Dirty fireballs would rarely trigger BATSE because the flux is so weak in the BATSE triggering range, and clean fireballs with $\Gamma_0 \gg 300$ would be so brief that the total fluence measured within the BATSE window would not be sufficient to trigger it.

Fig. 3b shows the relationship between $E_{\rm pk}$ and fluence for a model calculation when only the parameter Γ_0 is varied. When $\Gamma_0 \lesssim 500$, the external shock model predicts a positive correlation between $E_{\rm pk}$ and fluence, as has been recently reported [16]. The dirty fireballs with $\Gamma_0 \lesssim 100$ would not normally be detected and, as just described, there would also be biases against detecting the clean fireballs with $\Gamma_0 \gg 300$. It is necessary, however, to convolve temporal and spectral model results through a simulation of the detector response before drawing conclusions about the viability of the model.

The BATSE instrument has provided the largest and most uniform data base on GRBs. It nominally triggers on 64, 256, and 1024 ms timescales when the flux in at least two detectors exceeds 5.5σ over background. The data points in Fig. 4 show the peak photon-flux size distribution, the t_{50} duration and the $E_{\rm pk}$ distributions measured with BATSE. The observable t_{50} is the time interval over which the integrated counts range from 25% to 75% of the total counts over background. For comparison with statistical data, we developed an analytic model for the temporally evolving GRB spectrum [12] based on the detailed numerical calculations. To make a valid comparison between the external shock model and the observed statistical properties of GRBs, we have modeled detector triggering criteria. Model results were integrated over time to determine if the peak 50–300 keV flux exceeded the BATSE threshold so that the simulated BATSE detector would be triggered [15]. Trigger efficiencies were explicitly taken into account, which is important for GRBs with fluxes near threshold. The underlying assumption of our statistical model is that the event rate of fireballs follows the star formation history of the universe [17].

We [15] found that it was not possible to fit simultaneously the size, t_{50} and $E_{\rm pk}$ distributions with a monoparametric model. Broad distributions of explosion energy E and initial Lorentz factor Γ_0 are needed to fit these distributions. The model fits shown in Fig. 4 are based upon power-law distributions of E and Γ_0, where $dN/dE \propto E^{-1.52}$ for $10^{48} \leq E({\rm ergs}) \leq 10^{54}$, and $dN/d\Gamma_0 \propto \Gamma_0^{-0.25}$ for $\Gamma_0 \leq 260$. The upper limit to Γ_0 corresponds to a density of $n_0 = 10^2$ cm^{-3}; the analytic model is degenerate in the quantity $n_0 \Gamma_0^8$.

As can be seen from Fig. 4, the model provides reasonable fits to the peak-flux, $E_{\rm pk}$ and t_{50} distribution of the long-duration ($\gtrsim 1$ s) GRBs. The short, hard GRBs must arise from a separate component. The implied redshift distribution of GRBs detected with BATSE is also shown in Fig. 4. We predict that most GRBs detected with BATSE lie in the redshift range $0.2 \lesssim z \lesssim 1.2$, with a tail of GRBs extending to high redshifts. The predicted number and distribution of high-z GRBs detected with BATSE is quite uncertain, because the star-formation rate at high redshifts is poorly known, and the fit depends on the unproven assumption that the comoving space density of fireball transients follows the star formation rate. Moreover, the distribution of explosion energies is assumed to be described by a

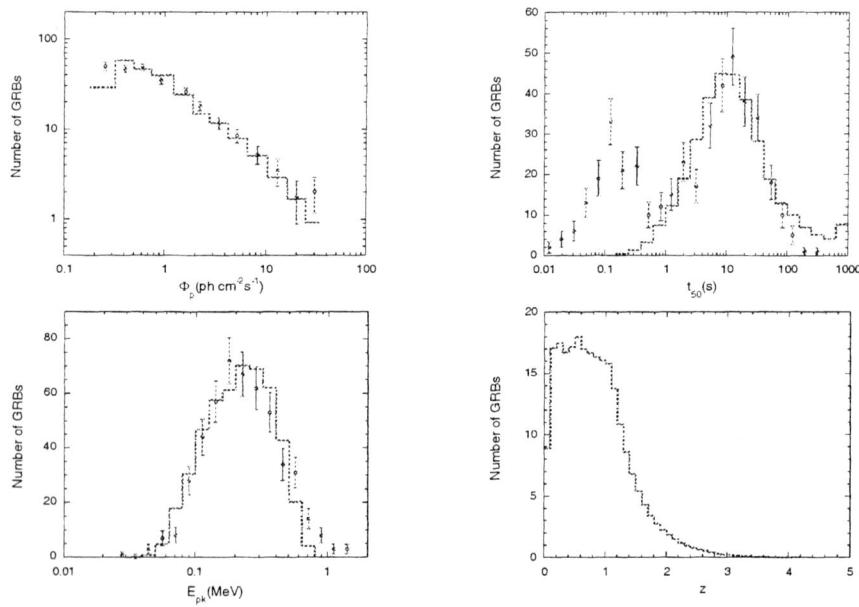

FIGURE 4. Data points give the 3B catalog peak 50–300 keV photon-flux (upper left), t_{50} duration (upper right) and $E_{\rm pk}$ (lower left) distributions of GRBs measured with BATSE [18,19]. Dotted histograms give fits from the external shock model [15] with a range of values of total explosion energy and baryon-loading. Lower-right panel shows the redshift distribution predicted by this set of parameters.

power-law function with a discrete cutoff. This distribution might instead have a tail extending to very high values.

IV DIRTY AND CLEAN FIREBALLS

An overall normalization factor for the fireball event rate per unit comoving volume is implied by the joint fits to the statistical properties of GRBs shown in Fig. 4. If no beaming is assumed, this normalization corresponds to a local event rate of $\cong 440$ yr^{-1} Gpc^{-3}, which is equivalent to a local GRB rate of $\cong 90$ Galactic events per Myr. This is a factor ~ 4000 greater than the result of Wijers et al. [20], who fit the combined BATSE/PVO peak-flux distributions only. This difference is due to an approach to GRB statistics where we abandon a standard candle assumption for the luminosity and rely on blast wave physics and detector response properties to determine whether a fireball transient would be detected with BATSE. Most crucially, we do not assume that there is a preference in nature to make fireballs with a specific energy release E and baryon-loading parameter Γ_0 (which would also entail a typical density of the surrounding medium) that would produce radiation that would trigger BATSE; any such assumption is highly artificial.

The consequence of this approach is that fireballs with a wide range of energies and baryon-loading parameters are formed in nature, the bulk of which are not detected and for which we have no evidence except for the limits implied by surveys [21,22]. Only a very few nearby fireball transients with low values of E would be detected, and fireballs with $\Gamma_0 \lesssim 10^2$ would be invisible to BATSE because most of the dirty fireball radiation is emitted at X-ray energies and below. The dirty fireball transients have longer durations and lower E_{pk} values than standard GRBs, and are difficult to detect because they are lost in the glow of the luminous diffuse X-ray background for wide field-of-view instruments. The X-ray transient events discovered with the Beppo-SAX WFC and reported at this meeting [24] might be fireball transients with a baryon load that is large enough that such events would not normally trigger a burst detector at hard X-ray energies. The number of clean fireball transients is not well constrained, but our results show that there must be a break or cutoff in the Γ_0-distribution at high values of Γ_0. Clean fireballs have shorter durations and E_{pk} values extending to MeV and GeV energies, and require sensitive, wide field-of-view gamma-ray telescopes to be detected [12,23].

If there are many more fireball events than implied by direct observations of GRBs, then a number of important implications follow:

- The hypothesis that ultra-high energy cosmic rays are produced by GRBs remains viable. This hypothesis has been questioned [25] in light of redshift measurements of GRB counterparts that suggest a much lower event rate within the GZK radius than formerly thought.

- The identification of X-ray hot spots in M101 with GRB remnants [26] appears more probable. These associations seemed unlikely given the event rate inferred directly from GRB observations.

- GRB explosions could leave many more observable Galactic remnants such as stellar arcs and HI holes, and produce greater biological effects than has been estimated [27].

V INHOMOGENEOUS EXTERNAL MEDIUM

Several arguments have been advanced to the effect that an external shock model cannot reproduce the short timescale variability observed in GRB light curves. We address these point by point.

1. *An external shock model will display short timescale variability only if the radiative efficiency is poor.* The analytic argument [28] assumes that density inhomogeneities (or "clouds") located at an angle $\theta \sim \Gamma^{-1}$ to the line-of-sight make the dominant contribution to variability. Clouds at $\theta \ll \Gamma^{-1}$ actually make much stronger contributions to large-amplitude flux variability because of the combined effects of Doppler beaming and the much shorter observer timescale over which on-axis clouds radiate their emission [2].

2. *A condition of local spherical symmetry in radiating blast wave produces pulses in light curves which spread with time, contrary to the observations* [29]. An external shock model breaks the condition of local spherical symmetry if clouds with radius $r \ll R/\Gamma_0$ are present, as must be assumed to make the short timescale variability. Here R is the distance of the cloud from the explosion center.

3. *A decelerating blast wave produces spreading pulses, contrary to the observations.* Only the portion of the blast wave that interacts with a cloud experiences strong deceleration, and its energy is dissipated by the interaction. The rest of the blast wave does not undergo significant deceleration until it intercepts another cloud, so no spreading from deceleration results. Thus, it is not surprising that there is no spreading of peaks in GRB 990123 [30], because different portions of the blast wave are producing the distinct pulses and peaks in the light curve.

4. *Gaps and precursors are not possible due to the interference between a large number of causally disconnected regions.* If there are shells of material from winds of GRB progenitor stars, as seems likely if GRB sources are associated with the collapse of massive stars, then gaps in the light curves can be formed.

5. *A low-density confining medium will produce a low level of emission unless the density contrast between the clouds and the confining medium is very large.* First, it is not necessary to have a confining medium if the massive star progenitor ejects material. Even if there is a low-density confining medium, the standard blast wave model implies that this residual emission will be radiated in a different energy band than the radiation emitted from the blast-wave/cloud interaction.

Finally, we note difficulties in a colliding shell scenario. The efficiency for dissipating internal energy in a relativistic shell is maximized for collisions between a shell and a stationary external medium, and is much poorer in collisions between relativistic shells. It is simple to get $\gtrsim 10\%$ radiative efficiency in the BATSE band for an external shock model, but efficiencies $\sim 1\%$ are more likely in an internal shock model [31], which calls into question the validity of the internal shock model for GRB 970508 [32]. A colliding shell scenario must contend with spreading profiles unless pairs of shells collide only once near the burst source, which would mean an additional loss of efficiency. GRBs with widely separated pulses generally have νF_ν peak photon energies within a factor of 2-3 of each other. This is natural for an external shock model, where a blast wave with a single Lorentz factor collides with different clouds within the Doppler cone, but requires fine-tuning of the speeds between pairs of shells in a colliding shell model.

VI SUMMARY

The original motivation for an external shock model was that it provided a simple explanation for the mean duration of GRBs [33]. This duration roughly corresponds to the time scale $t_{\rm dec}$ where the relativistic shell has swept up a sufficient amount of matter to cause the shell to decelerate. For a GRB source at redshift z,

$$t_{\rm dec} \approx 10(1+z)(\frac{E_{54}}{n_2 \Gamma_{300}^8})^{1/3} \text{ s} ,\qquad(1)$$

which is comparable to the mean duration of GRBs observed with BATSE (see Fig. 4). This equation does not explain, however, why $\Gamma_0 \approx 300$. We now know the answer to this problem — fireballs do not have to have $\Gamma_0 \approx 300$. But if the baryon-loading is significantly different from this value, then a detector like BATSE will not be triggered. Dirty fireballs with $\Gamma_0 \ll 300$ will make long duration X-ray transients that will, in general, be too weak to trigger BATSE, and clean fireballs with $\Gamma_0 \gg 300$ make brief high-energy γ-ray transients with insufficient fluence in the BATSE band to be detected. The implication is that there are many fireball transients that will be detected with more sensitive telescopes employing appropriate triggering properties and scanning strategies.

No explanation has been given within the context of the colliding shell/internal shock model as to why GRB durations should range from a fraction of a seconds to hundreds of seconds. There seems to be no reason why intermittent or delayed accretion of a massive ring of material around a collapsed star should not take place over long time scales, particularly given the unusual behavior that the accretion process must display if it is to produce the variability observed in GRB light curves. The observation of a single GRB that recurs after several hours, days, or months would falsify the external shock model. No convincing case of recurrence has been observed.

ACKNOWLEDGMENTS

I acknowledge discussions, collaborations, and joint work with M. Böttcher, J. Chiang, and K. Mitman.

REFERENCES

1. Mészáros, P., & Rees, M. J., *ApJ* **405**, 405 (1993).
2. Dermer, C. D., & Mitman, K. E., *ApJ* **513**, L5 (1999).
3. Rees, M. J., & Mészáros, P., *ApJ* **430**, L93 (1994).
4. Kobayashi, S., Piran, T., & Sari, R., *ApJ* **513**, 679 (1997).
5. Chiang, J., & Dermer, C. D., *ApJ* **512**, 699 (1999).
6. Dermer, C. D., Chiang, J., & Mitman, K. E., *ApJ*, submitted, astro-ph/9910240 (1999).
7. Preece, R. D., et al., *ApJ* **506**, L23 (1998).
8. Dermer, C. D., Böttcher, M., & Chiang, J., *ApJ* **515**, L49 (1999).
9. Frontera, F., et al., *A&A*, in press, astro-ph/9911228 (1999).
10. Crider, A., et al., *ApJ* **519**, 206 (1999).
11. Panaitescu, A., & Mészáros, P., *ApJ* **492**, 683 (1998).
12. Dermer, C. D., Chiang, J., & Böttcher, M. *ApJ* **513**, 656 (1999).
13. Brainerd, J. J. *ApJ* **428**, 21 (1994).
14. Böttcher, M., Dermer, C. D., Crider, A. W., & Liang, *A&A*, **343**, 111 (1999).
15. Böttcher, M., & Dermer, C. D., *ApJ* **529**, in press, astro-ph/9812059 (2000).
16. Lloyd, N., Petrosian, V., & Mallozzi, R. S., *ApJ*, submitted, astro-ph/9908191 (1999).
17. Madau, P., Pozzetti, L., & Dickinson, M., *ApJ* **498**, 106 (1998).
18. Meegan, C. A., et al., *ApJ Supp.* **106**, 65 (1996).
19. Mallozzi, R. S., et al., in *Gamma-Ray Bursts: 4th Huntsville Symposium*, eds. C. A. Meegan, R. D. Preece, & T. M. Koshut, AIP, New York, 1998, pp. 273.
20. Wijers, R. A. M. J., Bloom, J. S., Bagla, J. S., & Natarajan, P., *MNRAS* **249**, L13 (1992).
21. Grindlay, J. E. *ApJ* **510**, 710 (1999).
22. Greiner, J., Hartmann, D. H., Voges, W., Boller, T., Schwarz, R., & Zharikov, S. V., *A&A*, submitted, astro-ph/9910300 (1999).
23. Dermer, C. D., & Chiang, J., in *Proc. of the GeV-TeV Gamma-Ray Astrophysics Workshop: Toward a Major Atmospheric Cherenkov Telescope*, ed. B. L. Dingus, AIP, New York, in press, astro-ph/9912164 (2000).
24. Heise, J., these proceedings.
25. Stecker, F. W., *ApJ Lett.*, submitted, astro-ph/9911269 (1999).
26. Wang, Q. D., *ApJ* **517**, L27 (1999).
27. Scalo, J., & Wheeler, J. C. *ApJ*, submitted, astro-ph/9912564 (1999).
28. Sari, R., & Piran, T., *ApJ* **485**, 270 (1997).
29. Fenimore, E. E., Madras, C., & Nayakshin, S., *ApJ* **473**, 998 (1996).
30. Fenimore, E. E., Ramirez-Ruiz, E., & Bobing, W., *ApJ* **518**, L73 (1999).
31. Kumar, P., *ApJ* **523**, L113 (1999).
32. Paczyński, B., in *Supernovae & Gamma Ray Bursts*, eds M. Livio, K. Sahu, & N. Panagia, astro-ph/9909048 (1999).
33. Rees, M. J., & Mészáros, P., *MNRAS* **258**, 41P (1992).

Spectral Modeling of GRB Pulses

Hara Papathanassiou

International School for Advanced Studies (SISSA), Trieste, Italy

Abstract. The energy spectra of pulses of GRBs are modeled for synchrotron and multiple self-inverse Compton scattering from a population of thermal and non-thermal electrons. The contribution from pairs that result from annihilation is also taken into account. A high particle density (enhanced by the pairs) will increase absorption but, if the pairs are not accelerated, the absorption frequency cannot lie in the BATSE window. Pairs will contribute in upscattering and will most likely increase the population of cold particles that will downscatter hard photons and thus suppress hard (e.g., above 300 keV) emission.

INTRODUCTION

The properties of time-resolved BATSE spectra can probe the actual physical mechanisms responsible for the emission in this range. Time-integrated spectra, on the other hand, result from the integration of time-dependent quantities which tends to smooth out spectral features. In the case of GRBs, such an averaging is unavoidable due to the photon time delay and aberration inherent in a relativistically moving medium. Spectral evolution sequences or average spectra over the course of the whole burst will reflect the hydrodynamical evolution of the quantities involved in the radiation processes. Quantities like the bulk Lorenz factor Γ_b of the flow, or the available energy, will remain roughly constant during the generation of a pulse. Therefore, spectral modeling of pulses best addresses the radiative mechanisms. Spectral fits of a large number of bright time-resolved spectra [1] have furnished distributions of the three main spectral parameters: the low (α) and high (β) photon spectral indices and the frequency of the spectral break (E_b). These quantities are primarily linked with the emission processes. The surprising narrow E_b distribution probably holds the key answer to the physical mechanisms (be them the radiation process itself, or the mechanism for magnetic field amplification and particle acceleration). In the widely used synchrotron self-inverse Compton model for GRB spectra, E_b is a function of free parameters and can only be used to constrain them [2].

The low-energy spectral index α has a distribution [1] that can be approximated by a Gaussian centered at $\alpha = -1$, with a FWHM of ~ 0.9 plus a separate component of a small percentage of cases with $\alpha = 1$. The radiative mechanisms that would *intrinsically* result in these values are: Comptonization by moderate optical depth [3], or bremsstrahlung in $\alpha \approx -1$, optically thin synchrotron from a (cooled) electron power law in $\alpha = -2/3(-3/2)$, inverse Compton (IC) in $\alpha = 0$, optically thick emission by any mechanism in the Rayleigh-Jeans limit in $\alpha = 1$. More importantly, any steep intrinsic spectral slope portion of the spectrum will be observed flatter due to sampling over volume, energy, and time distributions (e.g., a self absorbed synchrotron spectrum with the self-absorption frequency, ν_{abs}, sweeping through the window of interest due to an increase in the thickness of the medium will result in a photon index of 1, rather than 1.5 [4]). Further inferred flattening of the value of α is introduced when fitting a spectrum through broadband model functions because, in practice, the BATSE window is too narrow [1,5]. The large number of cases with $\alpha > -2/3$, coined as the "death line of the synchrotron model" [6], suggests that any model employed will not be in a fully optically thin regime in the BATSE window.

The high-energy power law index β provides the best indication for the acceleration of particles to a power law (of index p). The distribution of the fitted values in the time resolved sample [1] is bimodal, its main component peaking at $\beta \approx -2.25$. If interpreted as synchrotron emission from particles that have cooled down, it gives $p = 2.5$. This is in agreement with the values derived from afterglows and is reproduced by particle acceleration calculations [7]. Attributing the full range of β values to synchrotron emission from a power law suggests that the radiation populations of e^- have $2 \leq p \leq 4$ — but invoking IC could narrow this range down to $2 \leq p \leq 3$. Roughly 10% of the spectra have no high-energy power law component. These spectra can be either those classified as no high-energy (NHE) pulses [8] or those with $\beta \leq -4$. Such values can be interpreted as representing the Wien part of spectrum (e.g., resulting from a thermal particle distribution) or a power law spectrum that has suffered some absorption at the high end.

Here, I report on spectral evolution sequences for GRB pulses. I discuss the spectra in the comoving frame only, since the purpose is to account for spectral slopes that are sufficiently steep, so that they can result in observed spectra consistent with the fits to BATSE data (e.g., [1]).

SYNCHROTRON SELF ABSORPTION

As long as a *high* α spectrum is produced locally, all lower α values can be reproduced by the integrations necessary to obtain the observed spectra and can be attributed to time dilation and aberration of photons, adiabatic dilution of the e^- distribution, and inhomogeneous conditions in the emitting region. Here, I examine whether a steep slope spectrum ($\alpha \geq 1$) can be generated in an internal shock flow (external shocks are produced further out in the flow and thus involve less dense material). I pick typical values for the global parameters: total available luminosity per unit solid angle of 10^{52} erg/s/sr ($L_{52}/\Omega = 1$), variability timescale

$t_{var} = 0.01$ s, and $\Gamma_b = 300$ (note that Γ_b cannot be arbitrarily large even for a very clean flow because of Compton drag [9]). Figure 1 shows the range in particle number density n_e and magnetic field B that would place ν_{abs} in the 30–300 keV range. This frequency is defined as the one where the optical depth to synchrotron is 1. The spectral slope below that will be 1 (or 1.5 for emission from a power law that is optically thick above the peak). The calculation is performed for a power law spectrum of electrons with index $p = 2.5$ (the results depend very weakly on p) peaking at $\gamma_{min0} = 10^3$, one peaking at $\gamma_{min0} = 10$, and a thermal distribution with $kT = 50$ keV. These constitute the simplest cases one would test. A more realistic case, where the e^- spectrum results from continuous ejection with constant density and cooling via synchrotron emission only, for the first 1/10th of the duration of the pulse, is calculated as well. To the left of the diagonal lines, the dominant cooling mechanism for the electrons at peak energy is bremsstrahlung (therefore, the usual assumption of the emission being due synchrotron is no longer valid). As shown, in order to get ν_{abs} in the BATSE range, for an e^- distribution peaking at $\gamma_{min0} = 10^3$ and for an equipartition magnetic field value, one would need the number density of particles to be roughly 10^3 to 10^5 of that of the protons in the flow. If one allows for the cooling of the e^- distribution, the requirement on n_e is more severe. While having a thermal particle distribution relaxes the constraint on n_e, it requires a flow that is strongly magnetically dominated. Having a shorter

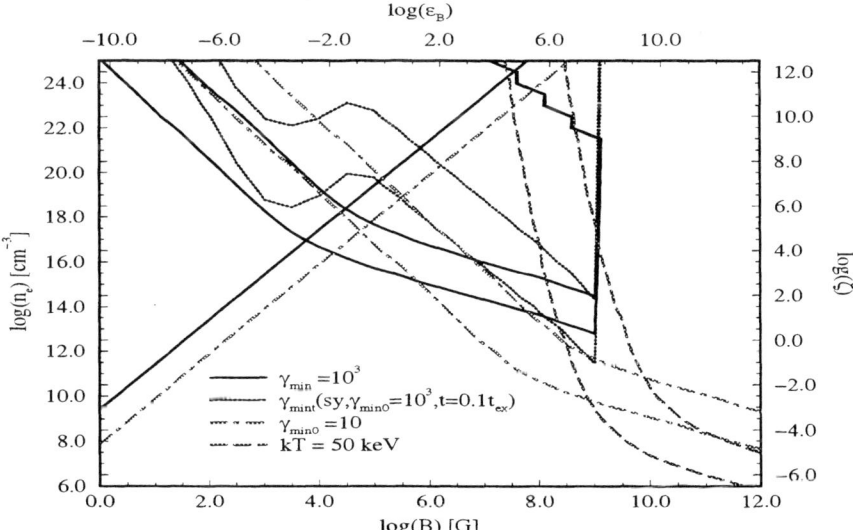

FIGURE 1. Values of the magnetic field and the particle number density for which the synchrotron spectrum turns optically thick in the 30–300 keV range. On the facing axes are given the corresponding equipartition values for a flow of $t_{var} = 0.01$ s, $L_{52} = 1$, and $\Gamma_b = 300$.

variability timescale (and/or lower Γ_b) implies a denser environment and this results in lower equipartition fractions (e.g., for $t_{var} = 1$ ms and $\Gamma_b = 200$, $\epsilon_B \approx 1$ would require $\zeta = 10^3$) but also makes bremsstrahlung the dominant cooling mechanism for the e^- peak. One could therefore conclude that it is not possible to have ν_{abs} in the BATSE window, if the emitting environment is that of internal shocks and the emission mechanism is synchrotron of the available electrons. There is a possibility though to substantially increase the number density of emitting particles by including the e^-e^+ pairs that are produced in the flow due to pair opacity of the interactions of hard IC photons [4,3]. But these pairs should be added with high densities and very hard spectra (i.e., produced abundantly and accelerated immediately).

SPECTRAL EVOLUTION SEQUENCES

I calculate spectral evolution sequences for continuous injection of particles with a prescribed power distribution (consisting of a relativistic Maxwellian at low energies, $\sim \gamma^2$, and a power law, $\sim \gamma^{-p}$ above the peak, γ_{min0}, and constant number density) and losses through adiabatic expansion, synchrotron and IC radiation in a region that is expanding at constant rate. The synchrotron and multiply (up to ~ 10) IC upscattered spectral components that result from the distribution at hand are calculated. The pair production is evaluated following the prescription of [10] (valid for scattering of photons with very different energies). This allows us to evaluate the attenuated hard spectrum, the pair distribution (which is a power law peaking at $\gamma \approx 1$ with index that of the hard photons). At each time step, the number of electrons with energies in the first bin are added to a population of cold electrons which are used in calculating the down-scattered spectrum. At the end of each cycle the pairs are added to the e^- distribution at an average constant rate (following two different prescriptions: (i) the aforementioned power law, (ii) a thermal distribution of 50 keV [3]). Details of the calculation will appear in a forthcoming paper.

Values of parameters close to equipartition and a low Γ_b turn the spectra optically thick to pair production. In the absence of re-acceleration, the pairs will contribute to absorption (ν_{abs} can reach up to a few keV), modest upscattering, and will provide abundant cold particles that down-scatter hard photons. Given a sufficient number density of cold electrons in the flow, the hard photons will lose energy in successive scatterings. If the optical depth of cold electrons is τ_c, and for scatterings in the Thompson regime, a cut-off will appear at around $\Gamma_b m_e c^2 / \tau_c^2$. Requiring such a cutoff at 300 keV introduces a lower limit on the number of cold electrons (and e^+) in the flow of $n_c \geq 6.5\, 10^{13}/\sqrt{\Gamma_b t_{var}}$. Even if all the electrons of the flow (those coming from the ionization of the explosion material) were cold they would not be able to account for the cut-off. This is possible in a situation where the flow is highly optically thick to pairs. Pair production interactions of photons eliminate all photons that in the comoving frame have energies above 0.5 MeV. This creates a large number of cold pairs that subsequently down-scatter the hard photons (all scatterings take place in the Thompson regime). A rough estimate shows that for a cut-off at 300 keV, the product of the e^- and IC radiative efficiency must be $\varepsilon_e \varepsilon_{ic} \approx 0.077 \eta_2 \Gamma_{b2}^{9/2} t_{var} \Omega / L_{52}$, where the specific entropy of the flow η and Γ_b are measured in units of 100.

The β distribution argues in favor of a particle acceleration mechanism resulting in a power law distribution in most of the pulses. A simple way (and an alternative one to the down-scattering by cold electrons) to explain the very steep high-energy spectra (or NHE pulses [8]) is by invoking a failed particle acceleration process. In this case, the e^- distribution is either thermal, or a narrow Gaussian centered at γ_{min0} and the power law extension is suppressed.

CONCLUSIONS

Observed synchrotron spectra can have *any* α value above $-2/3$. The $\sim 30\%$ of the spectra that require a higher value imply optically thick conditions in the BATSE window. For this, one infers high values of particle densities and a consistent spectral modeling calls for inclusion of bremsstrahlung emission. Pairs are present in the flow and they reach maximum density at around the pulse peak. They contribute to absorption, but since they are created with low energies they cannot push ν_{abs} into the BATSE window, *unless* they are accelerated. They may cause a changing spectral index through Comptonization (for very short pulses). They can provide a time varying population of cold particles that may be responsible for the lack of high-energy emission (or steep fall off) of about 10% of the spectra.

REFERENCES

1. Preece, R. D., et al., to appear in *ApJS*; astro-ph/9908119.
2. Papathanassiou, H. *Ph. D. Thesis*, UMI, 1998.
3. Ghisellini, G. & Celotti, A., *ApJ* **506**, L93 (1999).
4. Papathanassiou, H., *A&ASS* **138**, 525 (1999).
5. Lloyd, N. M. & Petrosian, V., these proceedings.
6. Preece, R. D., et al., *ApJ* **506**, L23 (1998).
7. Gallant, Y. A., et al., these proceedings.
8. Pendleton, G. N., et al., *ApJ* **489**, 175 (1997).
9. Mészáros, P. & Rees, M. J., *MNRAS* **269**, L41 (1994).
10. Bonometto, S. & Rees, M. J., *MNRAS* **152**, 21 (1971).

Time Profiles and Spectral Evolution of GRB Pulses

Edison Liang, Anthony Crider, Markus Böttcher, and Ian Smith

Rice University, Houston, TX 77005-1892

Abstract. We review the alternative scenarios of gamma-ray emission during the early GRB phase. A key uncertainty is the Thomson depth of the emitting shell. Whether it is Thomson thick or thin affects not only the emission mechanism, but also the overall energetics and beaming requirements. We argue that the transition time to the afterglow phase in the case of GRB990123 supports the proposal that GRBs are Thomson thick during the gamma-ray phase. We also propose a new approach to model the asymptotic pulse decay profiles.

INTRODUCTION

Currently popular scenarios of GRBs invoke relativistically expanding shells converting their bulk kinetic energy into internal energy, which is then radiated away. Such shells are energized either by "internal shocks" due to the collision of shells of slightly different Lorentz factors, or by "external shocks" due to the collision of a single shell with interstellar or circumburster matter. Here we adopt the popular hypothesis that the GRB gamma-ray emission is due to internal shocks [1,2], while the afterglow emission is due to a decelerating external shock [3,4]. A key question is the Thomson depth of the shell during the gamma-ray phase. The Thomson depth is important not only in determining the radiation mechanism and cooling rate, but also the overall energetics of the explosion [5,6]. Here we present evidence from several different directions that strongly suggests that the shell during the gamma-ray phase may not be Thomson thin as it is often assumed in the synchrotron shock model [7–9].

SPECTRAL EVOLUTION SIGNATURES

Two key signatures of GRB hard-to-soft spectral evolution are (a) the Liang-Kargatis [10] cooling law which states that during the decay phase of a GRB pulse the spectral break energy E_{pk} decays at a rate directly proportional to instantaneous luminosity [12]; (b) the low-energy (x-ray) photon spectral index α [11] often decreases with time, with initial $\alpha_{max} > -2/3$, the limit for optically thin synchrotron radiation [13-15]. As it has been suggested by Liang [5], both signatures are consistent with Compton cooling in an initially Thomson thick $\tau_T > 1$ plasma. This is in stark contrast with the popular scenario in which the shock cools via optically thin $\tau_T \ll 1$ synchrotron radiation [7-9] [1]. When $\tau_T > 1$, the overall energetics of the shell per unit solid angle is greater than that if $\tau_T \ll 1$, especially if the proton loading is significant [6]. Hence, to keep the total energy output of the explosion manageable, the beaming factor must be correspondingly increased in the Thomson thick case.

Another telltale signature of multiple Compton scattering is the existence of the terrace spectral shape at the x/gamma-ray energies, with a low-energy upturn in the soft x rays [6]. Such spectral behavior was clearly evident in the spectrum of GRB990123 [6], thanks to the broadband spectral coverage of this event from optical [16] to gamma rays [17]. While one single spectrum is not a proof of anything, this is tantalizing. We believe that future broadband observations by HETE-2 should help to confirm or dismiss such spectral behavior.

TRANSITION TO THE AFTERGLOW

If we identify the power-law decay of the afterglow of most GRBs with the decelerating self-similar phase of a relativistic blast wave [18], then the transition time (in the detector frame) T_{bw} from the GRB pulse to the power-law afterglow decay would correspond to the moment when the ejecta has swept up an amount of mass M_{csm} from the circumburster matter comparable to the ejecta mass M_{ej} divided by the bulk Lorentz factor Γ [19]. GRB990123 shows such a clear transition at $T_{bw} = 1000$ seconds [16]. Setting $M_{ej} = M_{csm}\Gamma$, we can relate the ejecta Thomson depth τ_{Tej} at earlier epochs T_{grb} to the transition time T_{bw}:

$$\tau_{Tej} = 1.3 \times 10^{-14} \mu_e n_{csm} \Gamma^3 T_{bw}^3 T_{grb}^{-2}$$

assuming conservation of the ejecta mass. In reality, the ejecta mass of a particular shell going through internal shock and radiating a gamma-ray pulse may be only a fraction of the total merged ejecta mass of the afterglow phase. But the above estimate should still be good to an order of magnitude, at least for the dominant GRB pulse. Here Γ is assumed to be constant prior to T_{bw}, n_{csm} is the circumburster density in units of baryons/cc and μ_e is the lepton (electron and pairs) number per baryon of the ejecta.

For GRB990123, if we set $T_{bw} = 1000$ s and $T_{grb} = 1$ s, we find that $\tau_{Tej} > 1$ if $\Gamma > 42 n_{csm}^{-1/3} \mu_e^{-1/3}$. In other words, during the early phase of the GRB pulse, for all reasonable values of the bulk Lorentz factor and circumburster density, the ejecta should be Thomson thick. This is consistent with the terrace spectral shape of GRB990123 during the first ROTSE [16] time bin [17,6].

GENERIC PULSE PROFILES

Motivated by the different cooling models, we have systematically studied the asymptotic decay profiles of many GRB pulses. We find that in most cases both the rise and decay more closely resemble power laws rather than exponentials in time, as proposed by Norris [20]. Here we propose a two-parameter analytic function for the generic pulse profile. The normalized profile is assumed to have the form:

$$F/F_{peak} = (1 + r/d)\,(t/t_{peak})^r \left[1 + (r/d)\,(t/t_{peak})^{r+d}\right]^{-1},$$

where t_{peak} is the time at peak flux. As $t \sim 0$, $F \sim t^r$, whereas as $t \gg t_{peak}$, $F \sim t^{-d}$. Thus, (r, d) are the (rise, decay) power-law indices. Figure 1 gives a sample pulse profile generated with the above function for $r = 20$ and $d = 10$, which strongly resembles typical narrow BATSE GRB pulses.

We have analyzed a number of long clean pulses with the above function using crude eyeball fits. Surprisingly, the decay indices d all lie in a narrow range (1.5–2.8), with a mean value of 2.1. This value is consistent with the 2.2 ± 0.2 ROTSE asymptotic early decay slope of GRB990123 [16] and the long tails of several pulses recently discovered by the BATSE team (see these proceedings). We plan to more systematically analyze a much larger sample of long, clean pulses in the near future.

SUMMARY

We find that both the spectral evolution signatures of GRB pulses and the transition time of the afterglow of GRB990123 are consistent with the ejecta shells being Thomson thick in the early phase. The terrace shape of the GRB990123

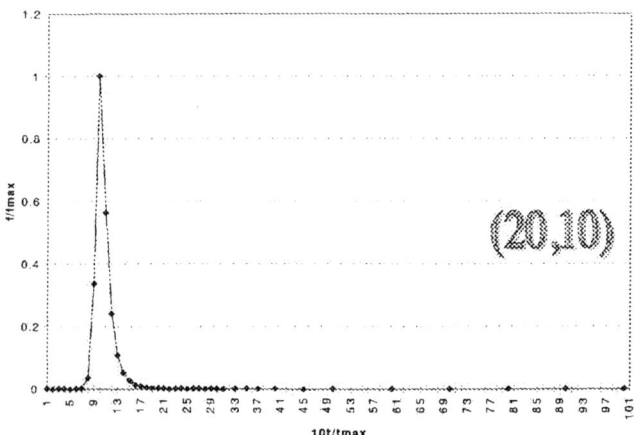

FIGURE 1. Pulse profile generated with the above equation for r=20 and d=10.

spectra also supports a Thomson thick interpretation. Finally, we propose a new analytic function to fit GRB pulse profiles. The average decay profile based on this functional form is consistent with a decay power-law index of around 2. The universality of this value and its physical interpretation remain to be studied.

ACKNOWLEDGMENTS

This work is partially supported by NASA grant NAG5-3824.

REFERENCES

1. Piran, T., and Sari, R., in *Gamma-Ray Bursts: 4th Huntsville Symp.*, eds. Meegan, C. A., Preece, R. D., & Koshut, T. A., AIP Conf. Proc. 428, New York, 1998, pp. 662.
2. Fenimore, E., Ramirez, E., and Sumner, M., in *Gamma-Ray Bursts: 4th Huntsville Symp.*, eds. Meegan, C. A., Preece, R. D., & Koshut, T. A., AIP Conf. Proc. 428, New York, 1998, pp. 657.
3. Meszaros, P., in *Gamma-Ray Bursts: 4th Huntsville Symp.*, eds. Meegan, C. A., Preece, R. D., & Koshut, T. A., AIP Conf. Proc. 428, New York, 1998, pp. 647.
4. Dermer, C., Chiang, J., and Böttcher, M., *ApJ* to appear (1999).
5. Liang, E., *ApJ* **491**, L15 (1997).
6. Liang, E., Crider, A., Böttcher, M., and Smith, I., *ApJ* **519**, L21 (1999).
7. Meszaros, P., and Rees, M., *ApJ* **405**, L67 (1993).
8. Katz, J., *ApJ* **432**, L107 (1994).
9. Tavani, M., *ApJ* **466**, 768 (1996).
10. Liang, E., and Kargatis, V., *Nature* **381**, 49 (1996).
11. Band, D., et al., *ApJ* **413**, 281 (1993).
12. Crider, A., et al., *ApJ* **519**, 206 (1999).
13. Crider, A., et al., *ApJ* **479**, L39 (1997).
14. Preece, R., et al., *ApJ* **496**, 849 (1998).
15. Preece, R., et al., these proceedings.
16. Akerloff, C., et al., *Nature* **398**, 400 (1999).
17. Briggs, M., et al., *ApJ* **524**, 82 (1999).
18. Blandford, R., and McKee, C., *Phys. Fluids* **19**, 1130 (1976).
19. Vietri, M., *ApJ* **478**, L9 (1997).
20. Norris, J., et al., *ApJ* **459**, 393 (1996). (1999)

Constraints on the Internal Shock Model from the Temporal Analysis of GRBs

Maddalena Spada*, Alin Panaitescu †, and Peter Mészáros†

*Dipartimento di Astronomia e Scienza dello Spazio, Università di Firenze,
L.E.Fermi 5, I-50125 Firenze, Italy [1]
†Department of Astronomy & Astrophysics, Pennsylvania State University,
University Park, PA 16802

Abstract. We analyze the power density spectrum (PDS) of Gamma Ray Bursts arising from multiple shocks in a relativistic wind. The factors to which the PDS is most sensitive are: the wind ejection features, which determinate the wind dynamics and its optical thickness, and the energy release parameters, which give the pulse 50–300 keV radiative efficiency. The model takes into account the photons down-scattering on cold electrons in the wind. For an almost optically thick wind we identify a combination of ejection features and wind parameters that yield bursts with an average power spectrum in agreement with the observations, and with an efficiency of converting the wind kinetic energy in 50–300 keV emission of order of 1%. The interval between peaks and pulse fluences have distributions consistent with the log-normal distribution observed in real bursts.

INTRODUCTION

The Gamma-Ray Burst (GRB) light-curves are complex and irregular, without any systematic temporal features [3], and an understanding of the origin of the temporal behavior of GRBs remains an open issue. Statistical studies are necessary in order to identify the physical properties of the emission mechanism existent in all or a group of GRBs. Recently [1], hereafter BSS98, have used the Fourier analysis of a sample of long GRB light-curves to study the statistical properties of their power density spectra (PDS). The PDS features together with other temporal properties of the observed GRBs, such as the distributions of the time interval between peaks and of the pulse fluence [5], can be used to constrain the physical characteristics of the GRB source. In the framework of the internal shock model, the rapid variability and complexity of the GRB light-curves is due to the emission from multiple shocks in a relativistic wind [10]. The shocks resulting from such instabilities heat the expanding ejecta, amplify pre-existing magnetic fields or generate a turbulent one,

---

[1] This research is supported by NASA NAG5-2857, NSF PAY94-07194 and the CNR.

and accelerate electrons, leading to synchrotron emission and inverse Compton scattering. Here we analyze the features of the GRB light-curves arising from internal shocks model, in order to identify the parameters that affect most strongly the GRB emission. By comparing the features of the simulated bursts with the observed burst PDS and the distributions of the interval time between peaks and of the pulse fluence, we constrain some of the physical properties of the ejecta.

OUTLINE OF THE MODEL

The Gamma-Ray Bursts' light curves are simulated by adding pulses radiated in a series of internal shocks that occur in a transient, unstable relativistic wind, as in the previous work presented by [2]. Moreover we have included in the modeling:

1. the collision hydrodynamics: the shock jump equations allow the calculation of the shocked fluid physical parameters, determine the shock velocity, the compression ratio, the thickness of the merged shells and the internal energy of the shocked fluid [7].

2. the reprocessing of the synchrotron photons by the interaction with the electrons inside the two colliding shells: the photons emitted can be up-scattered by the electrons which are relativistic in the co-moving frame of the source (hot-electrons) and down-scattered by the non-relativistic (cold-electrons) ones.

3. the optical thickness to Thomson scattering of the wind: while the photons propagate through the outer parts of the wind they can be down-scattered by the cold electrons in the shells.

4. the modifications in the shape of the emission spectrum, arising from the electrons cooling.

The wind is discretized as a sequence of shells, released by the central source during a time t_w and with an average interval t_v between consecutive ejections, so the number of shells is just $N = t_w/t_v$. The mass of each shell is drawn from a log-normal distribution determined by the average shell mass $\overline{M} = M_w/N$, M_w being the wind mass, and the dispersion $\sigma_M = \overline{M}$. The Lorentz factor (LF) η of each shell is randomly drawn from the interval $[\eta_m, \eta_M]$. Both η_m and η_M can be chosen constant during the entire wind ejection, so that η has a uniform distribution (we shall refer to this case as the "uniform wind"), or can vary on time-scale comparable to t_w and much larger than t_v, according to a certain law ("modulated wind"). After setting the dynamics of the wind ejection, we calculate the radii where the collisions take place, determine the relevant physical parameters in the shocked fluid and calculate the features of the emitted radiation: the energy of the synchrotron and inverse Compton spectral peaks, the Comptonization parameters, the optical depth to scattering on cold electrons, the 50–300 keV emission, and its duration. These quantities are necessary for the computation of the pulse features in the observed frame: fluence, duration, arrival time, taking into account relativistic and cosmological effects. Finally, assuming a given shape for the pulses [6], one can calculate the peak photon flux for each pulse and simulate the light curve.

RESULTS

An analysis of the PDS of real bursts was presented by BSS98. They calculated the Fourier transform of 214 long ($T_{90} > 20$ s) and bright burst, and have found that the average PDS is a power-law ($P_f \propto f^{-5/3}$, f is frequency) over almost two orders of magnitude in frequency, between 0.02 Hz and 2 Hz, where a break is observed, indicating a paucity of pulses with duration less than ≈ 0.5 s.

The distribution of intervals between peaks has been studied and by Li & Fenimore (1996), who showed that the distributions of the pulse fluence S_p and of the time interval δ_p between peaks are consistent with a log-normal distribution.

If the wind is optically thin and the ejection features are random, the pulse duration increases with the collision radius and the emission efficiency decreases during the wind expansion. The short inner collisions yield most of the 50–300 keV burst emission and the internal shock model predicts a flat PDS with equal power at low and high frequency. Thus, in order to explain the observed behavior, we need a configuration of the parameters which shifts power from the short to the long time-scales in the light-curves. Moreover, the δ_p distribution is not log-normal: in simulated GRBs arising from optically thin, uniform winds there are too many short intervals between peaks respect to a Gaussian $\log \delta_p$ distribution.

The analysis of the PDS [8] and δ_p distribution shows that there are three possible ways to explain the deficit of pulses with $\delta T < 1$ s:

1. a reduction in the electron injection fraction. This increases the photon energy, reducing the window efficiency of the short pulses (causing the high energy break) and increasing that of the longer ones. However the behavior of the PDS at lower frequency remains flat.

2. a modulation of the shell ejection Lorentz factor. This allows different configurations for the collisions series and a higher dynamical efficiencies for longer pulses.

3. an increase of the optical thickness of the wind. In this case the downscattering suffered by the photons as they propagate through the wind increases the pulse duration for the small radii collisions, which yield the shorter duration pulses.

In Figure 1 we show a simulated light-curve for a square-sine modulated wind. The burst 50–300 keV efficiency is 1%, and the 90% of the RS and 80% of the FS propagate in optically thick shells. The distributions are similar to a log-normal one, and the choice of N_{var} does not affect strongly their shape.

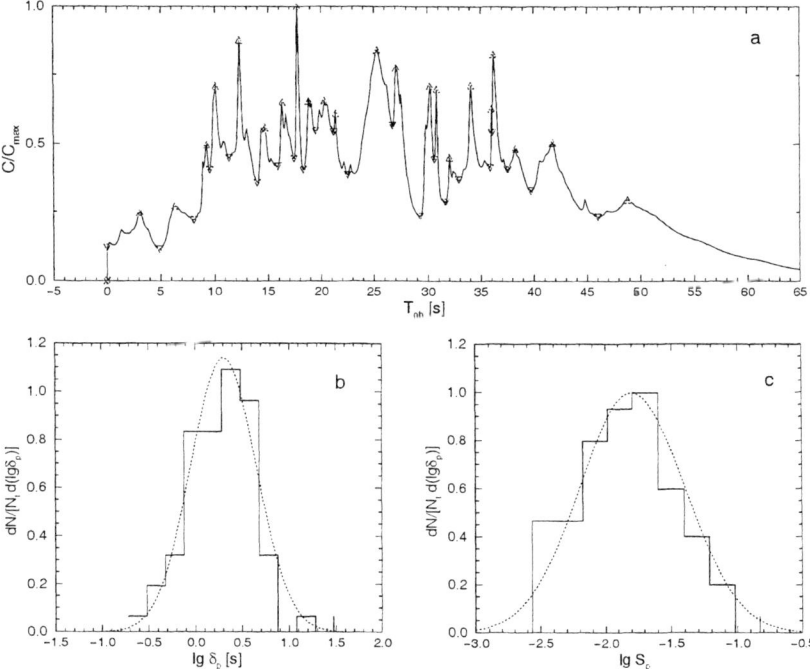

FIGURE 1. Top panel: light-curve for a modulated wind with period t_w and $\Gamma_m = 5$, $\Gamma_M = 150$, $L_w = 2 \times 10^{52}$ erg s^{-1}, $t_v = 25$ ms, $t_w = 30$ s, $z = 1$, $\zeta_e = 0.1$, $\epsilon_e = 0.25$, $\epsilon_B = 0.1$. The photon flux is normalized to its highest value. The triangle the peaks (38) and the valleys selected with the PFA ($N_{var} = 0.1$). Bottom: the distributions of the interval δ_p between peaks (b) and of the peak fluence S_p (c), the latter representing the fraction of the total fluence between two consecutive valleys. The distributions are calculated for four bursts with identical ejection features and wind parameters, total numbers of peaks is 80. The log-normal distributions that fit best the numerical results are shown with dotted curves.

In order to compare the PDS of the simulated bursts with the observed one, we consider an ensemble of cosmological GRBs. In this work, we use a GRB comoving rate density proportional to a power-law $n_c(z) \propto (1+z)^D$, mainly as a convenient parameterization. Given the rate density evolution, the GRB redshift is chosen from a probability distribution: $d\mathcal{P}/dz \propto n_c(z)/(1+z) \times dV/dz$, where dV/dz is the cosmological co-moving volume per unit redshift, with $q_0 = 0.5$ and $H_0 = 75$ km s^{-1}Mpc^{-1}. For the wind luminosity we use an un-evolving power-law distribution: $\Phi(L) \propto L^{-\beta}$ if $L_m \leq L \leq L_M$; and zero otherwise.

Figure 2b shows a burst-averaged PDS whose features are similar to that found by BSS98 in real bursts. The model parameters that led to the observed PDS shape yield bursts whose integral intensity distribution is consistent with the distribution found by Pendleton et al (1996): excluding the bursts dimmer than 1γ cm^{-2} s^{-1}, the model has $\chi^2 = 9.5$ for 9 degrees of freedom (Fig. 2a).

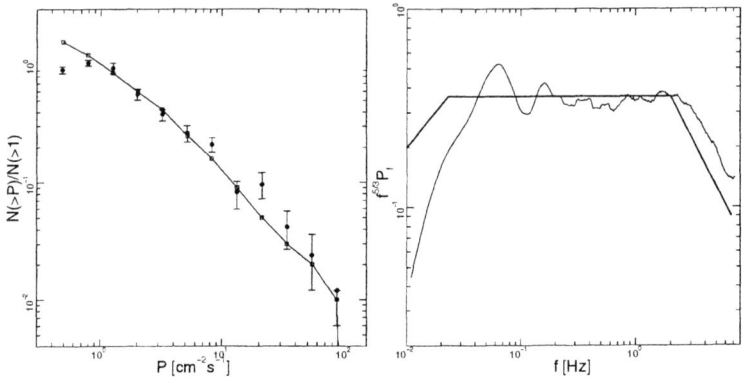

FIGURE 2. Left panel: intensity distribution of simulated bursts (open squares) compared to that of the observed ones, taken from Pendleton et al. (1996) (filled circles). Both distributions are normalized to the number of bursts brighter than 1γ cm^{-2} s^{-1}. Right panel: the averaged PDS of the simulated bursts, compared with the shape of the PDS (thick line) determined by BSS98 for real bursts. The average is done over 112 bursts with a square-sine modulated wind of random period between $t_w/4$ and t_w. Parameters: $L_m = 2 \times 10^{52}$ erg s^{-1}, $L_M = 100 L_m$, $\beta = 2$, $D = 3$, $t_w = 30$ s, $t_v = 25$ ms, $\epsilon_e = 0.25$, $\epsilon_B = 0.1$, $\zeta_e = 0.1$, $\Gamma_m = 5$, $\Gamma_M = 150$, $D = 3$.

In the internal shock model, the most efficient collisions, with a dynamical efficiency of 10–20% and a radiative efficiency of $10 - 30\%$, happen in the first part of the wind expansion where the wind optically thickness is higher and the angular spread time, the shell shock-crossing time and the electrons cooling time are shorter ($\ll 0.5$ s). In order to reproduce the observed PDS shape and interval between peaks distribution the features required to the simulated burst are: an optical thickness to scattering on cold electrons above unity, to increase the duration of pulses emitted by the first efficient collisions; and a modulated Lorentz factor of the ejected shells, in order to obtain wind efficiency of 1%, close to the maximum value admitted by the internal shock model [4,8].

REFERENCES

1. Beloborodov, A.M., Stern, B.E., & Svensson, R. (BSS98), *ApJ* **508**, L25 (1998).
2. Daigne, F. & Mochkovitch, R., *MNRAS* **296**, 75 (1998).
3. Fishman, G.J. & Meegan, C.A. *ARAA* **33**, 415 (1995).
4. Kumar, P., *ApJL* **523**, L113 (1999).
5. Li, H. & Fenimore, E.E., *ApJ* **469**, L115 (1996).
6. Norris, J.P., et al., *ApJ* **459**, 393 (1996).
7. Panaitescu, A. & Mészáros, P., astro-ph/9810258 (1999).
8. Panaitescu, A., Spada, M. & Mészáros, P., *ApJL* **522**, L105 (1999).
9. Pendleton, G.N., et al., *ApJ* **464**, 606 (1996).
10. Rees, M.J. & Mészáros, P., *ApJ* **430**, L93 (1994).

A Plasma Instability Theory of Prompt Gamma-Ray Burst Radiation

J. J. Brainerd

University of Alabama in Huntsville, Huntsville, AL 35899

Abstract. A new theory for converting the kinetic energy of a relativistic shell into radiation is discussed. In the theory, plasma instabilities arise from the relativistic streaming of two plasmas, producing a magnetic field and a relativistic distribution of electrons. The electrons produce synchrotron self-Compton radiation, which accounts for the prompt gamma-ray emission. For sufficiently high values of the bulk-motion Lorentz factor, the kinetic energy of the shell is efficiently converted into electromagnetic radiation. The processes present in this theory do not produce a shock.

THEORY

Much of the research to understand the prompt gamma-ray burst emission has focused on the formation of shocks within a relativistic flow. In this article, an alternative mechanism for creating the prompt emission from gamma-ray bursts is discussed: plasma streaming instabilities within a relativistic flow [1].

The source of the energy in this theory is a neutral relativistic shell traveling with a bulk Lorentz factor of $\Gamma > 100$. The gamma-ray emission occurs when the shell passes through the an external medium, such as a circumstellar or interstellar gas. It is assumed that both the shell and the external medium are ionized and composed solely of hydrogen. In the discussion that follows, the shell has a higher density than the external medium. Three aspects of the theory are discussed below: the plasma instabilities that convert the kinetic energy of the shell into electron thermal energy and magnetic field, the synchrotron self-Compton emission of prompt gamma-rays, and the efficiency of converting kinetic energy into prompt radiation.

For a relativistically streaming plasma, two plasma instabilities arise: the filamentation instability and the two-stream instability [3]. The filamentation instability is a magnetic pinch of the ion component with the lower proper density. It produces a toroidal magnetic field with a length scale set by the electron plasma frequency of the higher-density component. The two-stream instability heats the electrons so that their distribution spans the momentum space separating the ion streams.

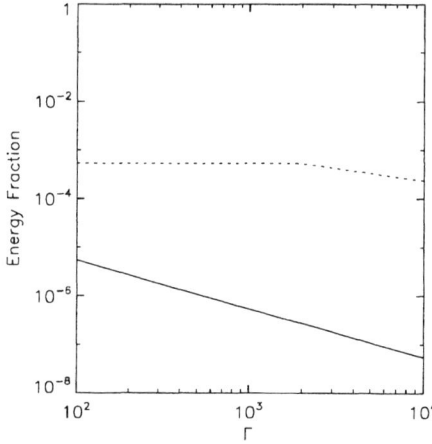

FIGURE 1. Energy release in the filamentation instability when the Rankine-Heugoniot equation for density is satisfied. The solid line gives, in the rest frame of the shell, the fraction of the external medium kinetic energy that goes into magnetic field, while the dashed line gives the fraction that goes into ion thermal energy.

Of these two instabilities, the filamentation instability has the higher growth rate for the conditions considered in the theory. Its growth rate is given by

$$\gamma'_{fi} = \frac{\beta \omega'_{p,i,em}}{\Gamma^{\frac{1}{2}}\sqrt{1 + c^{-2}k_\perp^{-2}\omega'^2_{p,e,shell}}} \approx \frac{1}{2}\sqrt{\frac{4\pi e^2}{m_p}} n_{em}^{\frac{1}{2}}, \qquad (1)$$

where the primes denote values measured in the shell rest frame, $\omega_{p,i,em}$ is the ion plasma frequency of the external medium, $\omega_{p,e,shell}$ is the electron plasma frequency of the shell, $\beta = v/c$, and $k_\perp$ is the instability wave number perpendicular to the velocity vector of the streaming plasma. The quantity n_{em} is the electron number density of the external medium. The length scale of the instability perpendicular to the velocity vector of the streaming ions is given by $x_f = k_\perp^{-1} = c/\omega'_{p,e,shell}$. In contrast, the two-stream instability has a maximum growth rate in the relativistic regime of

$$\gamma'_{2s} = \frac{1}{2}\omega'_{p,e,em}\Gamma^{-\frac{3}{2}} \approx \frac{1}{2}\sqrt{\frac{4\pi e^2}{m_e}} n_{em}^{\frac{1}{2}} \Gamma^{-1}. \qquad (2)$$

One sees that the first instability has the higher growth rate when $\Gamma > \sqrt{m_p/m_e} = 43$.

The filamentation instability saturates when the growth rate of the instability equals the bounce-frequency of the ions within the filament [4,2]. The bounce-frequency describes the sinusoidal oscillations of the ions across the toroidal magnetic field. The saturation condition sets the maximum strength of the magnetic

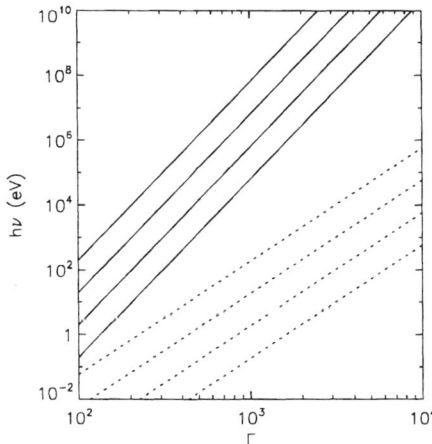

FIGURE 2. Maximum characteristic energy of radiation as a function of bulk Lorentz factor Γ in the external medium rest frame. The solid lines give the maximum energy of the Compton scattered radiation for a single scattering. The dotted lines give the maximum energy of the synchrotron radiation. For each set of curves, the values of n_{em} are, from bottom to top, $1\,\mathrm{cm}^{-3}$, $10^2\,\mathrm{cm}^{-3}$, $10^4\,\mathrm{cm}^{-3}$, and $10^6\,\mathrm{cm}^{-3}$. In calculating these curves, the shell density is set to $n'_{shell} = n_{em}\Gamma$, which gives the maximum photon energy for a given value of n_{em}.

field produced by the instability. The magnetic field strength at saturation is given in the rest frame of the shell by

$$B' = \left(8\pi m_e c^2 n_{em} \frac{n_{em}}{n'_{shell}}\right)^{\frac{1}{2}} \Gamma, \qquad (3)$$

where primes denote values measured in the shell rest frame, and n_{shell} is the electron density of the shell. For characteristic values of $n_{shell} = \Gamma n_{em}$, $n_{em} = 1\,\mathrm{cm}^{-3}$, and $\Gamma = 10^3$, one finds $B' = 0.143\,\mathrm{G}$.

Each instability converts only a small fraction of the kinetic energy of the plasma shell into thermal and magnetic field energy. Figure 1 shows as functions of Γ the energy fractions for the filamentation instability when the Rankine-Heugoniot equation for density is satisfied; less than 0.001 of the kinetic energy is released. For the two-stream instability, the fraction of energy that is thermalized is the ratio of the electron mass to the ion mass, with the electrons having a thermal Lorentz factor of order Γ. As a consequence, these two processes cannot mediate a shock, because the Rankine-Heugoniot equations require an energy conversion of order unity.

The radiative process that produces the gamma rays in this theory is synchrotron self-Compton emission. The synchrotron emission occurs as the electrons orbit within the instability-produced magnetic field. The low strength of the magnetic field makes the maximum frequency of the synchrotron photons fall below the x-ray band. Compton scattering of the synchrotron photons by the hot electrons boosts the photon energy into the gamma-ray range. Figure 2 shows the maximum energy of the photons produced by each of these radiative mechanisms for a variety of n_{em}

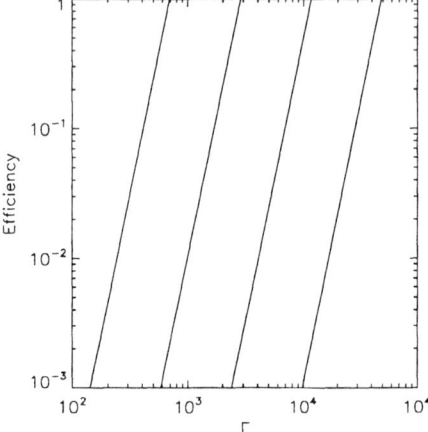

FIGURE 3. Radiative efficiency. The efficiency of converting relativistic-shell kinetic energy into electromagnetic radiation is given as a function of Γ (eq. [4]). In this figure, $\mathcal{M}_{27} = 1$, $\Gamma_3 = 1$, $f_{emis} = 1$, and $(R/R_0)^3 = m_p/m_e$. From left to right, the curves are for $n_{em} = 10^6 \, \text{cm}^{-3}$, $10^4 \, \text{cm}^{-3}$, $10^2 \, \text{cm}^{-3}$, and $1 \, \text{cm}^{-3}$.

values. One sees that Γ must be greater than 10^3 to produce gamma-ray emission. This subjects the theory to an important selection effect: bursts with low values of Γ are unobservable in this theory, since bursts are detected by their gamma-ray emission.

The efficiency with which the kinetic energy of the shell is radiated away in this theory depends strongly on the values of n_{em} and Γ. The radiative cooling of energy transferred to electrons through the two-stream instability can release close to 100% of the shell kinetic energy as electromagnetic radiation. An estimate of the efficiency is given by the ratio of the timescale for an electron to pass through the shell to the timescale for that electron to cool radiatively. This gives the efficiency as

$$g = 2.80 \times 10^{-7} \, \mathcal{M}_{27}^{\frac{2}{3}} \Gamma_3^{\frac{13}{3}} f_{emis}^2 \left(\frac{R}{R_0}\right)^{-1} n_{em}^{\frac{4}{3}}, \qquad (4)$$

where $\mathcal{M}_{27}$ is the mass of the relativistic shell in units of 10^{27}gm per ster radian, Γ_3 is the bulk Lorentz factor in units of 10^3, f_{emis} is the fraction of the electrons emitting synchrotron radiation, R is the actual distance traveled by the shell, and R_0 is the distance traveled by the shell if a shock arises in the interstellar medium. The external density n_{em} is expressed in units of cm^{-3}. Note that $\mathcal{M}_{27} = 1$ and $\Gamma_3 = 1$ corresponds to 1.1×10^{52} ergs. Generally, since the energy per particle lost to the external medium is of order $m_e c^2 \Gamma$, one has $(R/R_0)^3 \approx m_p/m_e$. Figure 3 shows the cooling efficiency as a function of Γ for several external-medium densities. One sees that to have efficient release of energy, the Lorentz factor must exceed a certain value that is strongly dependent on the density of the external medium.

This subjects the theory to a second selection effect.

SUMMARY

The plasma instability theory is an alternative to the shock-theory for the prompt emission of gamma rays in a gamma-ray burst. The plasma instabilities of this theory cannot mediate a shock, and the conversion of kinetic energy into electron thermal energy is more gradual than expected for a shock. It is found that the theory can efficiently convert the kinetic energy of a relativistic shell into prompt electromagnetic radiation for sufficiently large values of Γ. Conditions of emission at gamma-ray energies and efficient emission place lower limits on the value of Γ.

REFERENCES

1. Brainerd, J. J., *ApJ*, in press (1999).
2. Lee, R., & Lampe, M., *PRL* **31**, 1390 (1973).
3. Davidson, R. C., *Physics of Nonneutral Plasmas*, Addison-Wesley, Redwood City, 1997.
4. Davidson, R. C., Hammer, D. A., Haber, I., & Wagner, C. E., *Phys. Fluids* **15**, 317 (1972).

Emission of Cosmological Gamma-Ray Bursts in the GeV–TeV Energy Domain

E. V. Derishev[*], V. V. Kocharovsky[*], and Vl. V. Kocharovsky[*]

[*]*Institute of Applied Physics, Russian Academy of Science
46 Ulyanov Street, 603600 Nizhny Novgorod, Russia*

Abstract. The observed sub-MeV radiation from gamma-ray bursts (GRBs) arises most likely from synchrotron emission of a relativistic shock. We estimate the magnetic field strength and the Lorentz factor of accelerated electrons required by the synchrotron mechanism. We find that these values are furthermore sufficient to cause significant emission due to inverse Compton scattering. The Comptonized photons typically have energies well above 1 TeV, so that they are strongly absorbed by infrared background radiation and cannot be observed from a source at cosmological distance.

On the other hand, the ultra-high energy radiation may be absorbed in the close vicinity of the source due to interaction with ultraviolet and soft X-ray quanta originating from the same source and scattered in the ambient medium. In this case, almost all the energy initially contained in ultra-high energy radiation is reprocessed into a softer spectral range corresponding to the two-photon absorption threshold. Both the presence and the absence of absorption/reprocessing of the ultra-high energy photons places a rather strong limit on the density of interstellar medium near the GRB source and the Lorentz factor in fireball of a gamma-ray burst.

We discuss all types of GRB emission in GeV–TeV spectral range, including 100 GeV emission of GRB fireballs, which originates from the decay of neutral pions.

PROMPT 100 GEV EMISSION FROM GAMMA-RAY BURST FIREBALLS

In a typical cosmological gamma-ray burst fireball, a neutron component is initially present and has the energy of the order of the proton-electron component [1,2]. Velocity decoupling of the neutron component takes place if the temperature of the central engine, T_0, satisfies the following condition:

$$T_0 < 5 \text{ keV} \times \Gamma,$$

where Γ is the fireball Lorentz factor. Neutron flow decoupling is accompanied by pion production in inelastic proton-neutron collisions and generation of energetic gamma-quanta in subsequent decay of neutral pions. These quanta are blue-shifted so that their apparent energies are in the 100 GeV domain.

Emission of 100 GeV photons may precede the main pulse of gamma-radiation by a time interval approximately equal to the duration of the burst itself, contains $\sim 10^{-3}$ of the total GRB energy, and exactly traces activity of the central engine. The detection of prompt energetic gamma-rays will allow direct studying the physical conditions in fireballs.

THE ULTRA-HARD RADIATION FROM GRB SHOCKS

Assuming that the bulk of radiation from GRBs is due to synchrotron emission of electrons accelerated at the front of relativistic shock with the Lorentz factor Γ [3,4], it is possible to make definite conclusions about physical conditions in this shock. The following three requirements must be fulfilled. First, the observed spectrum has a maximum at $\varepsilon_m \sim 200$ keV. Second, the energy $E_{\rm GRB} \sim 10^{51} - 10^{52}$ erg is released during the burst, which lasts for $t_{\rm GRB} \sim 10$ s. (For definiteness, we take the external shock model.) Third, the optical depth for inverse Compton scattering is less than unity,

$$\frac{20}{3}\gamma_{\rm max}^2 \sigma_T \zeta n_e L \equiv \tau_{\rm ic} < 1. \qquad (1)$$

Here $\gamma_{\rm max}$ is the maximum value of the Lorentz factor of of the accelerated electrons, n_e is their number density, and ζ is the fraction of electrons with $\gamma_e \sim \gamma_{\rm max}$.

Let us suppose that the relativistic shock forms a spherical source of radiation with radius $\Gamma^2 c t_{\rm GRB}$ and thickness L. This approximation also works well for sources with a jet-like geometry unless the opening angle is less than $1/\Gamma$. As measured in the comoving frame, the bolometric luminosity of the source is $E_{\rm GRB}/\Gamma^2 t_{\rm GRB}$. Given the expression for the integral luminosity of a single particle, the second of the above requirements gives

$$\frac{E_{\rm GRB}}{\Gamma^2 t_{\rm GRB}} = \frac{4}{3}\gamma_{\rm max}^2 \sigma_T \zeta N_e c \frac{B^2}{8\pi}, \qquad (2)$$

where N_e is the total number of electrons in a GRB shell.

The emitting region formed by relativistic shock is inevitably geometrically thin, so that the value of $n_e L$ from eq. (1) is equal to the total number of electrons divided by the surface area of the GRB shell. Thus, the third requirement takes the following form:

$$\frac{20}{3}\gamma_{\rm max}^2 \sigma_T \zeta \frac{N_e}{4\pi(\Gamma^2 c t_{\rm GRB})^2} < 1. \qquad (3)$$

Here σ_T is the Thomson cross-section. Taken together, eqs. (2) and (3) limit the magnetic field strength,

$$B \gtrsim \left[\frac{10\,E_{\rm GRB}}{(\Gamma^2 c t_{\rm GRB})^3}\right]^{1/2} \sim \frac{5 \times 10^8}{\Gamma^3}\ {\rm G}, \qquad (4)$$

and the maximum Lorentz factor of accelerated electrons,

$$\frac{\gamma_{max}}{\Gamma} \lesssim \left(\frac{2m_e c\varepsilon_m}{e\hbar}\right)^{1/2} \left(\frac{c^3 t_{GRB}^3}{10\, E_{GRB}}\right)^{1/4} \sim 200. \tag{5}$$

Here m_e and e are the mass and the charge of the electron, respectively.

The total energy of the magnetic field in the emitting region (eq. 4) is equal to $E_B = B^2 (\Gamma^2 ct_{GRB})^2 L/2$, where the thickness of relativistically expanding emitting region grows during the burst up to $L \sim \Gamma ct_{GRB}$ (in the comoving frame). So, even the smallest required magnetic field has the energy $5E_{GRB}/\Gamma$, which is comparable with the thermal energy of plasma behind the shock. However, the magnetic field can only be generated at the expense of the latter energy. Hence, the expression in eq. 4 represents at the same time an upper limit of B, that is, provides the estimate of the actual magnetic field strength. Consequently, eq. 5 also gives a rough estimate of the actual value of γ_{max}, so that the optical depth for inverse Compton scattering is always not very far from unity.

It follows from the above arguments that at least several percent of the total GRB energy is converted via inverse Compton scattering into ultra-hard emission consisting of photons with typical energies $\Gamma\gamma_{max}m_e c^2 \sim 10^{-4}\Gamma^2$ TeV. Photons above 1 TeV are strongly absorbed by the infrared background radiation [6], so that direct observation of the ultra-hard emission is possible only for GRBs with $\Gamma \lesssim 200$. In this case the fluence of ultra-hard photons should be ~ 100 km^{-2}, even for the weakest bursts having total fluence $\sim 5 \times 10^{-7}$ erg/cm^2.

ABSORPTION AND REPROCESSING OF THE ULTRA-HIGH ENERGY PHOTONS

Part of the radiation from a GRB source is scattered by the surrounding plasma. From the point of view of an observer in the laboratory frame, the photons emitted from the front of relativistic shock move faster than the shock itself and the velocity difference is small, $\sim c/2\Gamma^2$. Therefore, from the time when scattering occurred in some point, till the arrival of the GRB shock in the same point, all scattered photons may be considered as frozen into the interstellar medium. The energy density w_r of such photons can be expressed via the GRB fluence, which is equal to $E_{GRB}/4\pi R^2$ at the distance R from the source, and the density of the surrounding plasma ρ:

$$w_r(R) = \frac{\sigma_T \rho}{m} \frac{E_{GRB}}{4\pi R^2}, \tag{6}$$

where m is the proton mass.

On its way to an observer, the ultra-hard radiation actually passes through a medium filled with soft X-ray photons. This may significantly alter the GRB spectrum in the ultra-high energy domain, provided the optical depth for the two-photon pair production exceeds unity at the distance $R = \Gamma^2 ct_{\text{GRB}}$. A projectile (photon) of energy ε has the largest cross-section of electron-positron pair production when it interacts with photons of energy $\sim 2m_e^2c^4/\varepsilon$, which make the main contribution to the opacity for the two-photon absorption. If absorbing photons are produced by the synchrotron mechanism, their spectrum is $I_\omega \propto \omega^{-1/2}$ and the effect of absorption is stronger for more energetic projectiles.

Let us introduce the new variable J, which shows what fraction of the total fluence is contained within a logarithmic interval of photon energies around ε, $J(\varepsilon) = \varepsilon I_\varepsilon / I$. This is an actual observable. The two-photon absorption optical depth for quanta with energy ε is equal to $\tau_a = J(2m_e^2c^4/\varepsilon)\sigma_{\text{pp}} w_r R\varepsilon/(2m_e^2c^4)$, where $\sigma_{\text{pp}} \simeq \sigma_T/8$ is the angle-averaged value of the pair-production cross-section. Assuming $J(\varepsilon) = \sqrt{\varepsilon/\varepsilon_m}$ one has

$$\tau_a \simeq \frac{\rho\sigma_T^2}{8m} \frac{\sqrt{\varepsilon/\varepsilon_m}}{\sqrt{2}m_ec^2} \frac{E_{\text{GRB}}}{4\pi\Gamma^2 ct_{\text{GRB}}} \sim \frac{0.3\, E_{51} n \sqrt{\varepsilon_{12}}}{\Gamma^2 t_{\text{GRB}}}. \tag{7}$$

Here E_{51} is the GRB energy in units 10^{51} erg, n the electron number density in surrounding plasma (in cm^{-3}), ε_{12} the photon energy (in TeV), and t_{GRB} the GRB duration in seconds. Photons with energy above

$$\varepsilon_t \sim \frac{10\,\Gamma^4 t_{\text{GRB}}^2}{E_{51}^2 n^2}\, \text{TeV} \tag{8}$$

are absorbed in the vicinity of the GRB source and cannot be observed. Because the relativistic shock should not decelerate significantly during the burst, the density of the interstellar medium has an upper limit. Therefore, for GRBs having $\Gamma \gtrsim 300$ a value of ε_t is larger than a characteristic energy of ultra-hard photons $\Gamma\gamma_{\text{max}} m_e c^2$, i.e., there is no absorption. At the same time, for short and powerful bursts located in relatively dense galactic environments, the absorption threshold ε_t may drop below 1 GeV. In these rare cases the effect of reprocessing may be observed by orbital telescopes operating in the appropriate spectral range.

Now, let us follow the fate of absorbed photons. Each of them produces an electron-positron pair where the daughter particles have roughly equal energies. Electrons and positrons have larger interaction cross-section than photons do, so that if the optical depth for interaction of initial ultra-hard quanta with scattered soft photons is larger than unity the same is true for daughter electrons (positrons). The first photon scattered off by one of these energetic electrons takes away about a half of its energy. Step by step, the electromagnetic radiation becomes softer until the absorption threshold ε_t is reached. It is this energy domain where the bulk of energy initially contained in the ultra-hard radiation is concentrated. If the absorption threshold is below 1 TeV, then the reprocessed radiation may reach the Earth. For a typical burst having the total fluence $\sim 10^{-6}$ erg/cm^2 the fluence of ultra-hard photons is of the order of $\sim 10^4\, \text{TeV}/\varepsilon_t$ km^{-2}. It is more than sufficient for detection by means of modern ground-based telescopes.

WHAT TO OBSERVE?

Prior to or during the GRB main pulse. At this time a flash of ~ 100 GeV photons from π^0 decay should be observed. This emission carries approximately 10^{-3} of the total energy and originates directly from the fireball as a result of pion production in inelastic proton-neutron collisions, which start when proton and neutron flows decouple [1]. The expected flux from a GRB at redshift ~ 1 corresponds to approximately 100 photons km^{-2} accumulated during the period of activity of the central engine. As the nature of central engine is still unknown, its period of activity may last from several milliseconds up to the full duration of the GRB main pulse. About one third of all GRBs should produce emission of this kind and they must be among the subset of short bursts.

During the GRB main pulse. For the inverse Compton radiation, the highest luminosity should be observed just below the IR background absorption cut-off (~ 1 TeV for $z \simeq 1$). Depending on the intrinsic properties of the GRB shell, which cannot be predicted beforehand, the ultra-hard radiation in the sub-TeV energy domain may carry about 5% of the total burst energy. For a source with redshift $z \simeq 1$ this corresponds to ~ 100 sub-TeV photons km^{-2}, as integrated over the GRB main pulse. *All bursts* should produce such emission.

At the afterglow stage. Depending on the density of surrounding interstellar gas and on the Lorentz factor of the GRB shell, the ultra-hard radiation may undergo absorption and reprocessing in the vicinity of the source. Almost all reprocessed photons have an energy close to the absorption threshold ε_t, which is a highly variable parameter. In few cases (perhaps, only in ~ 10 % of GRBs) ε_t falls into sub-TeV energy domain allowing ground-based observations of reprocessed emission. The sub-TeV emission from such bursts appears about an order of magnitude brighter than that discussed in the previous paragraph. Long (in usual sense) bursts are the prime candidates for observations of such kind.

Acknowledgments. This work has been supported by the Russian Foundation for Basic Research (the project 99-02-18244) and by the Russian Academy of Science through a grant for young scientists.

REFERENCES

1. Derishev, E.V., Kocharovsky, V.V., & Kocharovsky, Vl.V., *A&A* **345**, L51 (1999).
2. Derishev, E.V., Kocharovsky, V.V., & Kocharovsky, Vl.V., *ApJ* **521**, 640 (1999).
3. Mészáros, P. & Rees, M.J., *ApJ* **405**, 278 (1993).
4. Tavani, M., *ApJ* **466**, 768 (1996).
5. Ginzburg, V.L. & Syrovatskij, S.I., *The Origin of Cosmic Rays*, Moscow: USSR Academy of Science, 1963.
6. MacMinn, D. & Primack, J.R., *Sp. Sci. Rev.* **75**, 413 (1996).
7. Derishev, E.V., Kocharovsky, V.V., & Kocharovsky, Vl.V., *Spectra and radiation mechanisms in the sources of cosmological gamma-ray bursts.*, Preprint IAP RAS 492. Nizhny Novgorod, 1999.

Synchrotron Radiation as the Source of GRB Spectra, Part I: Theory

Nicole M. Lloyd and Vahé Petrosian

Stanford University, Stanford, California 94305

Abstract. We investigate synchrotron emission models as the source of gamma-ray burst spectra. We show that allowing for synchrotron self absorption and a "smooth cutoff" to the electron energy distribution produces a wide range of low-energy spectral behavior. We show that there exists a correlation between the value of the peak of the νF_ν spectrum, E_p, and the low-energy spectral index α as determined by spectral fits over a finite bandwidth. Finally, we discuss the implications of synchrotron emission from internal shocks for GRB spectral evolution.

INTRODUCTION

It has been suggested (e.g., [4]) that synchrotron emission is a likely source of radiation from GRBs, and later shown [11] that an optically thin synchrotron spectrum is a good fit to some bursts. However, some features seen in the low-energy portion of GRB spectra cannot be explained by the simple synchrotron model (SSM) — optically thin synchrotron emission from a power law distribution of relativistic electrons with a minimum energy cutoff. This model predicts that the asymptotic value of the low-energy photon index, α, should be a constant value of $-2/3$. However, the data show an α distribution with a mean of about -1.1 and a standard deviation of about 1. Furthermore, there are a significant fraction of bursts with $\alpha > -2/3$ — above the so-called "line of death" [10]. In addition, spectral evolution of α and the peak of νF_ν, E_p, are inconsistent with an instantaneous optically thin synchrotron spectrum in an external shock model [3]. Consequentially, other models — usually involving inverse Compton scattering [2,5] — were invoked to explain these "anomalous" spectral behaviors.

In this paper, we discuss how GRB spectra can be accommodated by synchrotron emission, including those spectra not explained by the SSM. We discuss the various spectral shapes from a general form for synchrotron emission, allowing for the possibility of self-absorption and a smooth cutoff to the electron energy distribution, and show that these models fit GRB spectra well. We show that there is a correlation between α and E_p as determined by fits using the Band [1] (and similar) spectral forms. Finally, we briefly discuss the variety of spectral evolution behaviors seen in GRBs in the context of synchrotron emission. In Part II (Lloyd et al., these proceedings, [7]), we compare our theoretical predictions with the data and show how synchrotron emission can explain the spectral behavior of GRBs.

SYNCHROTRON EMISSION

The general form for an instantaneous synchrotron spectrum for a power law distribution in the electron energy with a sharp cutoff, $N(E) = N_o E^{-p}$, $E > E_{min}$, is given by [9]

$$F_\nu = A\nu^{5/2}[\frac{I_1}{I_2}] \times [1.0 - \exp[-Q\nu^{-(p+4)/2} I_2]] \qquad (1)$$

$$I_1 = \int_0^{\frac{\nu}{\nu_{min}}} dx \, x^{(p-1)/2} \int_x^\infty K_{5/3}(z) dz, \; I_2 = \int_0^{\frac{\nu}{\nu_{min}}} dx \, x^{p/2} \int_x^\infty K_{5/3}(z) dz \qquad (2)$$

A is the normalization and contains factors involving the perpendicular component of the magnetic field, $B_\perp$, bulk Lorentz factor, Γ, and number of electrons, N_o. The frequency $\nu_{min} = (\Gamma E_{min}^2 B_\perp 3e)/(m^3 4\pi c^2)$. The parameter Q represents the optical depth of the medium (for example, if $\nu \gg \nu_{min}$, the photon spectrum will be absorbed at the frequency $\nu_{abs} \sim Q^{2/(p+4)}$). The high-energy asymptotic behavior is the usual $F_\nu \propto \nu^{-(p-1)/2}$. The low-energy asymptotic forms of the function depend on the relative values of ν_{min} and ν_{abs}: $F_\nu \sim \nu^{5/2}$ for $\nu_{min} < \nu \ll \nu_{abs}$, $F_\nu \sim \nu^2$ for $\nu \ll \min[\nu_{abs}, \nu_{min}]$, $F_\nu \sim \nu^{1/3}$, for $\nu_{abs} < \nu < \nu_{min}$.

Note that we do not address the case of cooling electrons, which will have the effect of increasing the electron power law distribution index p by 1, $p \to p+1$ at some characteristic cooling energy.

THE ELECTRON DISTRIBUTION

In most models of synchrotron emission, the electron distribution is modeled by a power law with a sharp cutoff at some minimum energy (as done in the previous section). This is not a realistic (and may even be an unstable) distribution. We characterize the electron distribution by the following equation:

$$N(E) = N_o \frac{(E/E_*)^q}{1 + (E/E_*)^{p+q}} \qquad (3)$$

where E_* is some critical energy that characterizes where the electron distribution changes. For $E \gg E_*$, $N(E) \propto E^{-p}$, while for $E \ll E_*$, $N(E) \propto E^q$. Hence, q characterizes the "smoothness" of the cutoff. This has a significant impact on the low-energy portion of the synchrotron spectrum. An optically thin synchrotron spectrum takes the form:

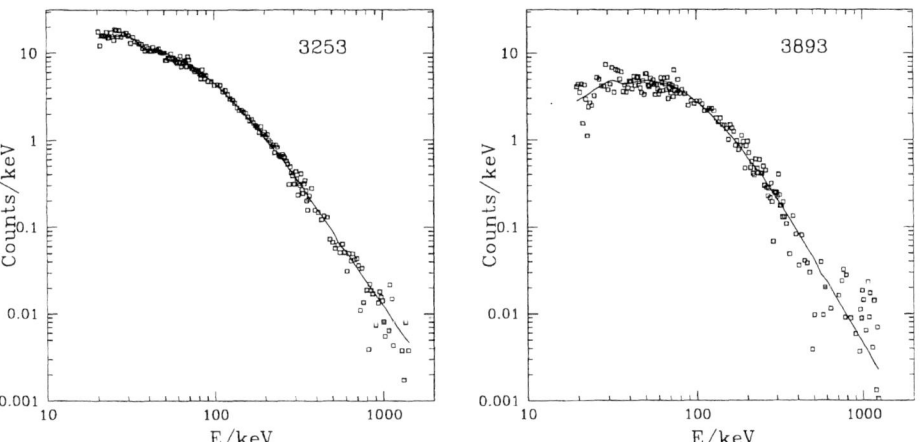

FIGURE 1. Spectra of two GRBs; the solid line shows the synchrotron model fit. An optically thin model is the best fit to 3253 (**left**), while and optically thick model best describes 3893 (**right**).

$$F_\nu = A(\nu/\nu_*)^{(q+1)/2} \int_0^\infty dx \frac{x^{-(q+p)/2}}{1 + ((\nu/\nu_*)^{(q+p)/2} x^{(p+q)/2})} \int_x^\infty K_{5/3}(z) dz \quad (4)$$

where $\nu_* = c_1 B_\perp E_*^2$. Depending on the smoothness of the cutoff, the spectrum of the emitted photons can change significantly. The peak of the spectrum is shifted to lower energies as the cutoff becomes smoother (q smaller), and the width of the spectrum increases, which implies that it takes longer for the spectrum to reach its low-energy asymptotic value.

SPECTRAL FITS

Synchrotron emission models fit the GRB data well. We fit 11 bursts with 256 channel energy resolution to the synchrotron spectral forms described above. Five of these bursts have a low-energy photon index (as determined by fitting a Band spectrum) above the "line of death" for optically thin synchrotron emission; that is, $\alpha > -2/3$. In all of these cases, we found that *including an absorption parameter will accommodate the hardness of the low-energy index*, and provided the best fit. Figure 1 shows the spectra for 2 GRBs in our sample (burst triggers 3893 and 3253). A self-absorbed spectrum in which the absorption frequency just enters the BATSE window best fits 3893, while an optically thin spectrum is the best fit to 3253. For a more complete discussion of the spectral fits, see [8].

A RANGE OF SPECTRA

Figure 2a shows the many types of low-energy spectral behavior one can obtain from the above synchrotron models, normalized to the peak of F_ν (at 500 keV). The vertical lines mark the approximate width of the BATSE spectral window.

Now, the α distribution will depend largely on how quickly the spectrum reaches its low-energy asymptote or how well spectral fits can determine the asymptote. As E_p moves to lower and lower energies, we get less and less of the low-energy portion of the spectrum; in this case, our spectral fits probably will not be able to determine the asymptote and will measure a lower (softer) value of α. [Preece et al. [10] pointed out this effect and attempt to minimize it by defining an effective α, which is the slope of the spectrum at 25keV (the edge of the BATSE window). However, a correlation between α_{eff} and E_p will still exist if the asymptote is not reached well before 25keV.] This difficulty becomes more severe the smoother the cutoff to the electron distribution, because the spectrum takes longer to reach its asymptote. To test this, we produce sets of data from optically thin synchrotron models with different parameters (ν_{min}, q, etc.), all of which have a low energy asymptote of $-2/3$. We fit a Band spectrum to this data (to be conservative, we extended the range of BATSE's sensitivity to 10 keV). Figure 2b shows the value of the asymptote as determined by the Band spectrum, as a function of E_p, for different degrees of the smoothness of the electron energy distribution cutoff. Not surprisingly, there is a strong correlation between the value of E_p and the value of the "asymptote", α, as determined by a Band fit to the data. We can use this relationship and knowledge of the E_p distribution to determine the resultant distribution for α. This is discussed in Part II [7].

SPECTRAL EVOLUTION

The behavior of the spectral characteristics with time throughout the GRB can give us information about the environment of the emission region and conceivably constrain the emission mechanism. Given the apparent correlation between α and E_p induced by the fitting procedure, we expect evolution of α (obtained from such fits) to mimic the behavior of E_p in time during a pulse or spike. Note, however, if each pulse in the time profile is a separate emission episode (as in an internal shock scenario), parameters such as q and the optical depth can vary from shock to shock; this can create a change in α from pulse to pulse, independent of E_p.

CONCLUSIONS

Synchrotron emission can produce a variety of GRB spectral shapes, particularly when one allows for a smooth cutoff to the electron distribution and includes effects of self-absorption. In addition, we expect a relationship between α and E_p, as a consequence of the fitting procedure; this will have implications for the observed

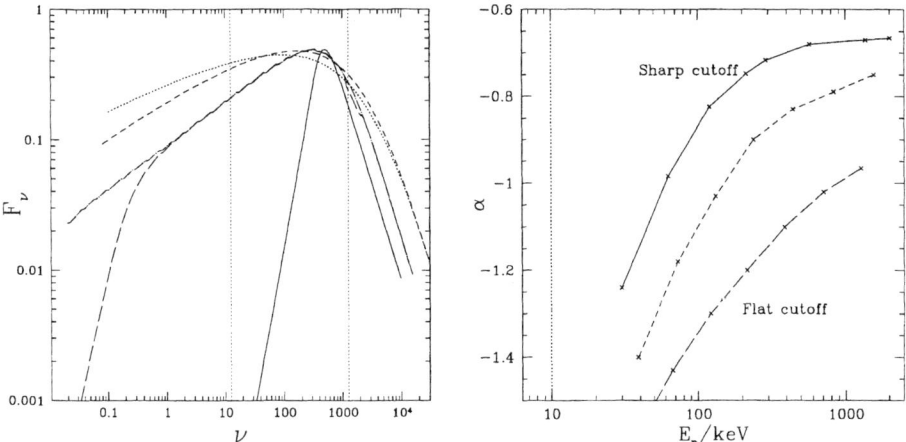

FIGURE 2. *Left Panel*: Synchrotron spectra for different values of the optical depth and smoothness of the electron cutoff. Optically thin spectra are shown by the the dot-dashed line, the dotted line, and the short-dashed line for a sharp ($q = \infty$, SSM), an intermediate ($q = 2$) and flat ($q = 0$) cutoff to the e^- distribution, respectively. The solid and long dashed lines show the self-absorption cutoff when $\nu_{abs} > \nu_{min}$ and $\nu_{abs} < \nu_{min}$, respectively. The vertical lines mark the BATSE window. *Right Panel*: The correlation between α and E_p for a spectrum with a sharp (solid line), intermediate (short dashed), and flat (long dashed) cutoff to the e^- distribution.

distributions and temporal evolution of spectral parameters. We test this model against the data in Part II [7].

REFERENCES

1. Band, D., et al., *ApJ* **413**, 281 (1993).
2. Brainerd, J.J., *ApJ* **428**, 21 (1994).
3. Crider, A., et al., *ApJ* **479**, L39 (1997).
4. Katz, J.I., *ApJ* **432**, L107 (1994).
5. Liang, E.P. & Kargatis, V.E., *Nature* **381**, 49 (1996).
6. Lloyd, N.M. & Petrosian, V., *ApJ* **511**, 550 (1999).
7. Lloyd, N.M. et al., (Part II) these proceedings.
8. Lloyd, N.M. et al., in preparation.
9. Pacholczyk, A.G., *Radio Astrophysics*, (Freeman, San Francisco) (1970).
10. Preece, R.D., et al., *ApJ* **506**, L23 (1998).
11. Tavani, M., *ApJ* **466**, 768 (1996).

Comparisons between Analytic and Numerical Calculations of GRB Spectra

Charles D. Dermer[*][1], Markus Böttcher[†], and James Chiang[‡]

[*]*Naval Research Laboratory, Code 7653, Washington, DC 20375-5352*
[†]*Rice University, Space Physics and Astronomy Department, MS 108
6100 S. Main Street, Houston, TX 77005-1892, USA*
[‡]*JILA, University of Colorado, Campus Box 440, Boulder, CO 80309-0444*

Abstract. Formulas of Sari, Piran, and Narayan for the spectral energy distributions of gamma-ray burst blast waves are extended to include synchrotron self-absorption and Compton scattering and, given a simple numerical integration, to apply to general radiative regimes. The expressions are compared with detailed numerical simulation results and are most accurate (within a factor of ~ 3) in the optical/X-ray regime during the afterglow phase. Comparison of spectral energy distributions are made for the parameters used by Wijers and Galama to fit GRB 970508.

INTRODUCTION

Analytic formulas [1,2] to model the spectra from gamma-ray burst blast waves energized by capturing material from the external medium are often used to model afterglow radiation at X-ray, optical, and radio frequencies. But just how good are these formulas? We extend the expressions of Sari, Piran, and Narayan [2] to apply to general radiative regimes and to include the effects of synchrotron self-absorption and synchrotron self-Compton (SSC) processes. The semi-analytic spectra are compared with detailed numerical simulation results [3].

We [4] find that the spectral and temporal breaks from the numerical model are much smoother than the analytic formulas imply, and that the discrepancies between the analytic and numerical results are greatest near the breaks and endpoints of the synchrotron spectra. The expressions are most accurate (within a factor of ~ 3) in the optical/X-ray regime during the afterglow phase, and are more accurate when ϵ_e, the fraction of the kinetic energy of protons swept-up by a blast wave with Lorentz factor Γ that is transferred to nonthermal electrons, is < 0.1. The analytic results provide at best an order-of-magnitude accuracy in the self-absorbed radio/infrared regime, and give poor fits to the SSC spectra due to complications from Klein-Nishina effects and photon-photon opacity.

---

[1] Work supported by the Office of Naval Research.

ANALYTIC FORMULAS

In the external shock model, nonthermal electrons are assumed to be energized and injected in the blast-wave frame with index p between Lorentz factor $\gamma_m \leq \gamma \leq \gamma_2 \cong \epsilon_{\max} \cdot 4 \times 10^7/[B(G)]^{1/2}$, where B is the magnetic field of the shocked fluid and $\epsilon_{\max}$ is a parameter which we set equal to unity in the calculations shown here. The value of γ_m is given numerically by a simple leapfrog calculation, but when $\gamma_2 \gg \gamma_m \gg 1$ and $p > 2$, $\gamma_m \cong \epsilon_e[(p-2)/(p-1)](\Gamma-1)m_p/m_e$. Electrons cool in the comoving frame by synchrotron and Compton losses to Lorentz factor

$$\gamma_c = \frac{3m_e}{16m_p n c \sigma_t \Gamma^3 t(1+f_{KN}u_{ph}/u_B)} \tag{1}$$

by observing time t [2]. The quantities $u_B = B^2/8\pi = 4nm_p c^2 \epsilon_B(\Gamma^2-\Gamma)$ and u_{ph} are the magnetic field and photon energy densities, respectively. When Klein-Nishina effects on Compton scattering are important, u_{ph} is corrected for the reduction in the scattering efficiency using the factor f_{KN} (≤ 1).

For an uncollimated blast wave expanding into a uniform medium with density n, $N_e = 4\pi r^3 n/3$ is the total number of nonthermal electrons. The electron spectrum $N(\gamma) \propto \gamma^{-s}$ for $\gamma_0 \leq \gamma \leq \gamma_1$, and $N(\gamma) \propto \gamma^{-(p+1)}$ for $\gamma_1 \leq \gamma \leq \gamma_2$. In the slow cooling limit, $\gamma_0 = \gamma_m$, $\gamma_1 = \gamma_c$ and $s = p$, whereas in the fast cooling limit $\gamma_0 = \gamma_c$, $\gamma_1 = \gamma_m$, and $s = 2$. The blast wave's radiative efficiency $\epsilon = \epsilon_e \epsilon_{rad}$, where $\epsilon_{rad} = (\gamma_m/\gamma_c)^{p-2}$ in the slow cooling limit, and $\epsilon_{rad} \cong 1$ in the fast cooling limit [6]. The evolution of Γ in radiative regimes bridging the adiabatic and radiative limits is easily obtained by numerically solving the equation of motion $d\Gamma/dm = -(\Gamma^2-1)/M$ (see [7] for analytic solutions). Here $dm = 4\pi r^2 m_p n dr$, r is the radial coordinate, and $M = \int dm \, [\epsilon + \Gamma(1-\epsilon)]$ is the total energy in the blast wave, including both rest mass and kinetic energy.

The synchrotron spectral luminosity $L_{\nu,\max}$ at frequency ν_0 is given by $L_{\nu,\max} = N_e m_e c^2 \sigma_T \Gamma B/(3e)$ [2]. Consequently, the synchrotron spectral power when the self-absorption frequency $\nu_a < \nu_0$ is given by

$$L_\nu = L_{\nu,\max} \times \begin{cases} (\frac{\nu_a}{\nu_0})^{1/3}(\frac{\nu}{\nu_a})^2 & \text{for } \nu < \nu_a \\ (\frac{\nu}{\nu_0})^{1/3} & \text{for } \nu_a < \nu < \nu_0 \\ (\frac{\nu}{\nu_0})^{-(s-1)/2} & \text{for } \nu_0 < \nu < \nu_1 \\ (\frac{\nu_1}{\nu_0})^{-(s-1)/2}(\frac{\nu}{\nu_1})^{-p/2} & \text{for } \nu_1 < \nu < \nu_2 \end{cases} \tag{2}$$

Here $\nu_i = \Gamma \gamma_i^2 eB/(2\pi m_e c)$, with i = 0, 1, 2, and ν_a is given by

$$\nu_a(\text{Hz}) = 9.87 \, \Gamma[\frac{(s+2)(s-1)}{(s+2/3)}]^{3/5} (rn)^{3/5} B^{2/5} \gamma_0^{-1}, \quad \text{when } \nu_a < \nu_0, \tag{3}$$

where all terms are in cgs units. See [4] for the case $\nu_0 < \nu_a < \nu_1$. We employ a very simple treatment for the SSC process by approximating the synchrotron emission by a δ-function distribution at the comoving frame frequency ν_1/Γ. This is the frequency of the peak of the power output if $2 < p < 3$. The relative amount of energy radiated in Compton and synchrotron photons is, neglecting Klein-Nishina effects, equal to $u_{\rm ph}/u_B \cong L_{\rm syn}/(u_B 4\pi r^2 c)$, where $L_{\rm syn}$ is the Compton power in the comoving frame. The synchrotron power is

$$L_{\rm syn} = \frac{4}{3} c\sigma_T u_B \int_{\gamma_0}^{\gamma_2} d\gamma N(\gamma)\gamma^2 =$$

$$\frac{4}{3}(s-1)c\sigma_T u_B N_e \gamma_o^{s-1} \left[\gamma_1^{3-s}\left(\frac{1}{3-s} - \frac{1}{2-p}\right) + \frac{\gamma_2^{2-p}\gamma_1^{p+1-s}}{2-p} - \frac{\gamma_0^{3-s}}{3-s}\right]. \quad (4)$$

Because of the strong reduction in scattering efficiency when photons are scattered in the Klein-Nishina limit, we shut off the integration when $\gamma h\nu_1/(\Gamma m_e c^2) > 1$, or when $\gamma = \bar{\gamma} = \Gamma m_e c^2/(h\nu_1)$. This implies that no photons with frequencies $\nu > \nu_{\rm KN} \equiv (\Gamma m_e c^2/h)^2 \nu_1^{-1}$ are observed. The energy density of photons that are efficiently scattered in the Thomson regime is obtained by repeating the exercise in equation (4), but integrating over the range $\gamma_0 < \gamma < \bar{\gamma}$ rather than over $\gamma_0 < \gamma < \gamma_2$. This yields a ratio f_{KN} of the Compton power which approximately takes into account Klein-Nishina effects to that without Klein-Nishina effects. This ratio is roughly given by $f_{\rm KN} = 1$ if $\bar{\gamma}/\gamma_1 > 1$, and $f_{\rm KN} = (\bar{\gamma}/\gamma_1)^{3-s}$ if $\bar{\gamma}/\gamma_1 < 1$. Because $\bar{\gamma}$ is the electron Lorentz factor for which Klein-Nishina effects become important, it is also necessary that $\bar{\gamma} > \gamma_0$ for significant SSC emission.

The spectral power of the Compton-scattered radiation at frequency $\nu_0^C = \gamma_0^2 \nu_1$ is

$$L_{\nu,\max}^C = \frac{4}{3} \frac{c\sigma_T u_{\rm ph} \gamma_0^2 \Gamma^2}{\nu_0^C} N_e = \frac{u_{\rm ph}}{u_B} \frac{\nu_0}{\nu_0^C} L_{\nu,\max} = \gamma_1^{-2} \frac{u_{\rm ph}}{u_B} L_{\nu,\max}. \quad (5)$$

The SSC component in this approximation is therefore given by

$$L_\nu^C = L_{\nu,\max}^C \times \begin{cases} (\frac{\nu}{\nu_0^C})^{1/3} & \text{for } \nu < \nu_0^C \\ (\frac{\nu}{\nu_0^C})^{-(s-1)/2} & \text{for } \nu_0^C < \nu < \nu_1^C = \gamma_1^2 \nu_1 \\ (\frac{\gamma_0}{\gamma_1})^{(s-1)}(\frac{\nu}{\nu_1^C})^{-p/2} & \text{for } \nu_1^C < \nu < \nu_2^C = \gamma_2^2 \nu_1 \end{cases}, \quad (6)$$

provided $\nu_{\rm KN} > \nu_2^C$. When this is not the case, the SSC emission is set equal to zero at $\nu > \nu_{\rm KN}$.

ANALYTIC AND NUMERICAL COMPARISONS

Figure 1 shows a comparison between the semi-analytic results and the numerical simulation model using a parameter set for the external blast wave model that yields a GRB with typical observed properties [3,5]. The agreement is reasonably good at optical and X-ray energies in the afterglow phase, but is quite poor whenever

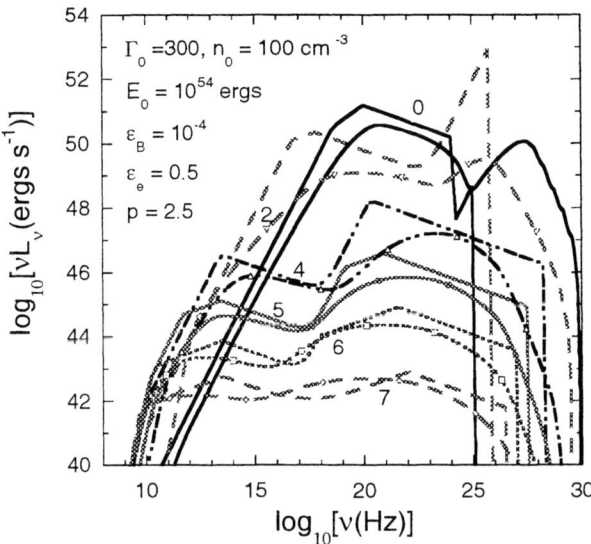

FIGURE 1. Calculations of νL_ν spectral energy distributions emitted by a relativistic blast wave with initial Lorentz factor $\Gamma_0 = 300$ that is energized by sweeping up material from a uniform surrounding medium with density n. Curves are labeled by the base 10 logarithm of the observing time in seconds. The analytic and numerical models are shown, with the latter curves identified by data points. Other parameters of the calculation are shown in the legend.

endpoint effects are important, as in the radio regime or in the hard X-ray/soft γ-ray regime during the prompt phase. This is due to the neglect of adiabatic losses on the electrons and the use of a δ-function approximation to model the synchrotron spectrum. The SSC component is also poorly modeled due to the simplicity of the underlying assumptions. Note, however, the importance of the SSC component at X-ray and optical energies in the afterglow phase if one assumes no temporal evolution of the blast-wave parameters (see [5] for details).

Figures 2a and 2b compare analytic and numerical spectral energy distributions and light curves for a model GRB with parameters derived by Wijers and Galama [8] to fit afterglow data from GRB 970508. Here the total energy injected into the fireball is $E = 3.5 \times 10^{52}$ ergs, and the density of the external medium is $n = 0.03$ cm^{-3}. The relative importance of the SSC component, which depends on the ratio ϵ_e/ϵ_B [9,6], is not as great as in Fig. 1. The analytic expressions provide a reasonable representation of the numerical model in the optical/X-ray regime, but the discrepancies at radio frequencies remain significant, as is clear from Fig. 2b. Although the times of the temporal breaks arising from synchrotron emission emitted by the lowest energy electrons are in good agreement with the numerical calculation, the analytic model can be discrepant in flux values by an order-of-magnitude or more. The cooling break, which is clearly defined in the V band

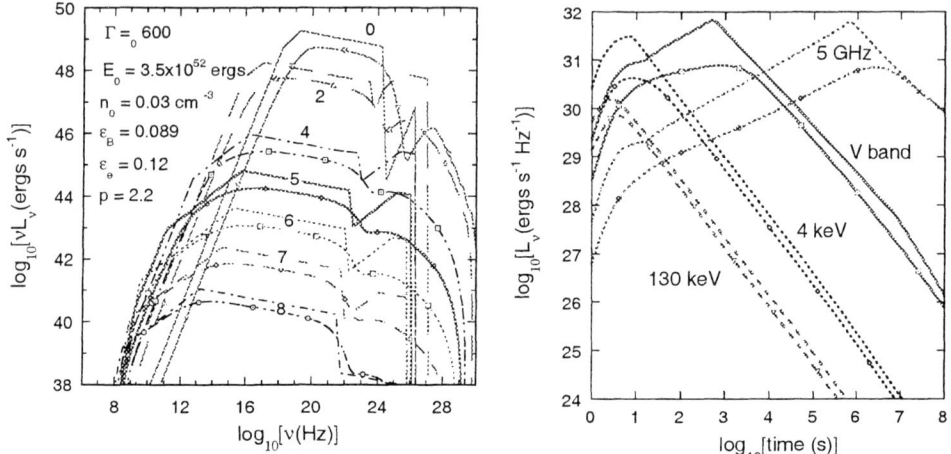

FIGURE 2. Left panel: Same as Figure 1, but with parameters taken from the fit to GRB 970508 by Wijers and Galama [8]. Right panel: light curves at different energies in an L_ν representation.

light curve at $t \approx 10^7$ s in the analytic model, is almost indistinct in the numerical calculation. This model also does not reproduce the flux values of GRB 970508 in its prompt phase, which are much brighter than calculated here.

To summarize, we [4] have extended some widely used formulas [2] and compared the resulting expressions with detailed, but numerically intensive, blast-wave model calculations [3] of GRB emissions. The semi-analytic model is quite accurate in the optical/X-ray afterglow regime, but is not very accurate when used to model the level of the radio emission or the hard X-ray and soft γ-ray emissions during the prompt phase.

REFERENCES

1. Dermer, C. D., Chiang, J., and Böttcher, M. *ApJ* **513**, 656 (1999).
2. Sari, R., Piran, T., and Narayan, R., *ApJ* **497**, L17 (1999)
3. Chiang, J., and Dermer, C. D., *ApJ* **512**, 699 (1999).
4. Dermer, C. D., Böttcher, M., and Chiang, J., *ApJ*, submitted, astro-ph/9910472 (1999).
5. Dermer, C. D., Chiang, J., and Mitman, K. E., *ApJ*, submitted, astro-ph/9910240 (1999).
6. Moderski, R., Sikora, M., and Bulik, T., *Astrophys. J.* in press, astro-ph/9904310 (1999).
7. Böttcher, M., and Dermer, C. D., *ApJ*, in press, astro-ph/9907270 (2000).
8. Wijers, R. A. M. J., and Galama, T. J., *ApJ* **523**, 177 (1999).
9. Sari, R., Narayan, R., and Piran, T. *ApJ* **473**, 204 (1996).

Evaluating Spectral Functions Used to Test the Synchrotron Shock Model

Anthony Crider

Naval Research Laboratory, Code 7653, Washington, DC 20375-5352

Abstract. Analyses of time-resolved BATSE spectra, deconvolved using the Band GRB spectral function, reveal inconsistencies with the synchrotron shock model (SSM) for gamma-ray bursts. However, we discover here that by instead deconvolving spectra using a smoothly-broken power law (SBPL) function, a vast majority of bursts are consistent with the SSM and do not violate the "line of death". To resolve these discrepant conclusions, we compare both empirical functions to true synchrotron spectra and conclude that the Band GRB function is better suited for testing the SSM. This suggests that without some form of absorption, the SSM is still in conflict with observations.

INTRODUCTION

Evidence for spectral evolution in gamma-ray bursts was found shortly after their discovery [1]. However, it has been generally assumed that the shape of the gamma-ray continuum remains constant throughout a burst's duration. Based on analyses of GRB spectra, integrated over their duration, theorists proposed the optically-thin synchrotron shock model (SSM) [3]. One prediction of this model, that the spectral photon slope α below the νF_ν peak would be $\leq -2/3$ seemed to be confirmed by the time-integrated spectra [3–5]. However, in the past few years, we have discovered that the continuum spectrum changes during a burst and that the time-resolved spectra of many bursts are inconsistent with the $\alpha \leq -2/3$ prediction of the SSM [6]. This finding has been dubbed the "line of death" problem [7] and its seriousness is well illustrated in Figure 1 of Preece et al. [7] with the distribution of $\Delta \alpha_\sigma = [\alpha - (-2/3)]/\sigma_\alpha$. In Figure 1a, we re-plot this using 4726 GRB spectra from the spectral catalog of Preece et al. [8]. Here we used the true asymptotic value of α rather than the spectral slope measured at 25 keV as was done in Preece et al. [7]. While extreme negative values of $\Delta \alpha_\sigma$ can be accommodated by changing the underlying lepton energy distribution [9], extreme positive values cannot be explained without invoking synchrotron self-absorption [9,10].

The spectra in the Preece et al. [8] catalog were fit using convenient empirical functions and *not* a true synchrotron spectrum. Most often used was the Band

GRB function [11]. Also used was the smoothly broken power law (SBPL) which has been suggested as a more flexible alternative to the Band GRB function [12]. It differs from the Band GRB function in that it introduces a fifth parameter which controls the sharpness of the spectral break. A small fraction of bursts were fit using a power law with an exponential break (PLEB). The PLEB function is equivalent to the Band GRB function when the Band parameter $\beta \equiv -\infty$. It is often used when E_{pk} (the energy where the peak in the νF_ν spectra occurs) is close to the upper limit of BATSE's sensitivity and β cannot be determined. For historical reasons, this function is called the COMP model within the BATSE WINGSPAN data analysis software.

NO "LINE OF DEATH" CRISIS WHEN USING SBPL

In Figure 1b-d, we plot the the distribution of $\Delta\alpha_\sigma$ as seen for each spectral function separately. Curiously, we find that while bursts from Preece's sample which were fit with the Band GRB function are inconsistent with $\alpha \leq -2/3$, this is not true for bursts which were fit instead with the SBPL function. If the bursts fit by Preece with the SBPL are not intrinsically different from those fit with the Band GRB function, then *the assumption of a spectral function must affect the resulting low-energy slope α.*

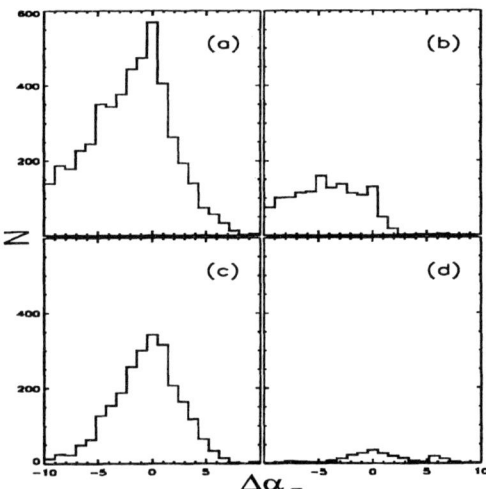

FIGURE 1. Histogram of the deviations of the low-energy asymptotic spectral slope α from the $\alpha \leq -\frac{2}{3}$ prediction of the synchrotron shock model for (a) all bursts, (b) bursts deconvolved using the smoothly broken power law, (c) bursts deconvolved using the Band GRB spectral function, and (d) bursts deconvolved using the Band GRB spectral function with $\beta \equiv -\infty$. While a substantial fraction of bursts fit with the Band GRB function violated the predictions of the SSM, this is not true for bursts fit with the SBPL.

TABLE 1. The ten bursts with spectra most inconsistent with the $\alpha \leq -2/3$ prediction of the synchrotron shock model. In all cases, fitting with the SBPL results in an α more consistent with the SSM prediction. We also note that discrepancies between columns 2 and 4 reveal the degree of inaccuracy stemming from the assumption of different initial parameters in the forward-folding fits.

BATSE Trigger	Preece et al. [8] max[$\Delta\alpha_\sigma$]	α	Band GRB max[$\Delta\alpha_\sigma$]	SBPL max[$\Delta\alpha_\sigma$]
2083	13.2	-0.05±0.05	13.3	10.6
5773	11.2	1.33±0.18	8.9	4.0
3734	10.9	-0.09±0.05	11.1	3.5
6630	8.0	0.12±0.10	11.6	5.8
1085	7.9	0.27±0.05	10.2	6.0
2329	7.7	-0.20±0.06	6.1	2.0
6350	7.6	0.31±0.13	5.4	4.6
3657	7.6	0.03±0.09	7.7	2.3
3138	7.4	0.12±0.11	5.9	3.5
6235	7.1	0.11±0.11	7.9	6.7

Before continuing, we must first convince ourselves that the distributions of $\Delta\alpha_\sigma$ in Figure 1b-d are different due to the choice of function and not because of some intrinsic differences in the bursts. To do this, we selected the 10 bursts from Preece et al. [8] most inconsistent with the SSM and determined α when we re-fit them with a Band GRB function and SBPL function. Our results appear in Table 1. Indeed, the SBPL provides a systematically lower $\Delta\alpha_\sigma$ such that only 5 of these 10 bursts are extremely inconsistent (max[$\Delta\alpha_\sigma$] > 4). In Figure 2, we show an example of the difference in fitting the Band GRB function and the SBPL. Fitting all 157 bursts with the SBPL would not likely provide many more inconsistent bursts since the remaining bursts all have lower values of max[$\Delta\alpha_\sigma$] when fit with the Band GRB function. Thus, if the SBPL were in fact more appropriate than the Band GRB function at representing true synchrotron spectra, then the "line of death" crisis would exist only for a handful of bursts rather than in one-sixth of bursts [7]!

APPROPRIATENESS OF SBPL FOR TESTING SSM

In the remainder of this paper we explore which of our two empirical functions is better suited for testing the synchrotron shock model. We produced our SSM spectra using the formulae of Crusius & Schlickeiser [13]. We assumed a spherical plasma with radius r and magnetic field B. The lepton distribution is a power law ($\frac{dN}{d\gamma} \propto \gamma^{-p}$) with low and high-energy cutoffs ($\gamma_{\min}$ and $\gamma_{\max}$).

Of our SSM parameters, only p, and to a lesser extent, B and $\gamma_{\min}$ significantly affect the shape of the spectrum near the νF_ν peak. We evaluated the synchrotron spectra at 36 (4 B × 3 p × 3 $\gamma_{\min}$) logarithmically-spaced points in the parameter space defined by Table 2. In 6 of these spectra, when B = 1 G and $\gamma_{\min} = 10^3$ or when B = 10^6 G and $\gamma_{\min} = 10^7$, E_{pk} was less than a factor of 100 from the boundaries of our code and for convenience we drop these. We fit our empirical spectral functions to the remaining synchrotron spectra evaluated at 33 logarithmically-spaced points within a factor of 100 of the νF_ν peak. We made an *ad hoc* assumption of 0.1 % confidence in our synchrotron spectral points. The resulting fits of the Band GRB and smoothly broken power law function all had a better than a 4% rms accuracy throughout the 4 decade range surrounding E_{pk}. The Band approximation was adequate until B became high enough that self-absorption occurred near E_{pk}. The fifth SBPL parameter for controlling the sharpness of the spectral break remained nearly constant (≈ 0.45) in our fits. Thus, for an infinitely short integration time, we conclude that this fifth parameter is superfluous.

One complication is that E_{pk} changes by $\sim 10\%$ within each bin [14]. By refitting our empirical functions to summed synchrotron spectra with a $\Delta\Gamma/\Gamma$ of 10%, we find this small degree of spectral evolution does not affect our conclusions. Thus, when approximating a synchrotron spectrum, the Band GRB function is better suited than the smoothly broken power law. Consequently, the "line of death" is still a hurdle for the synchrotron shock model.

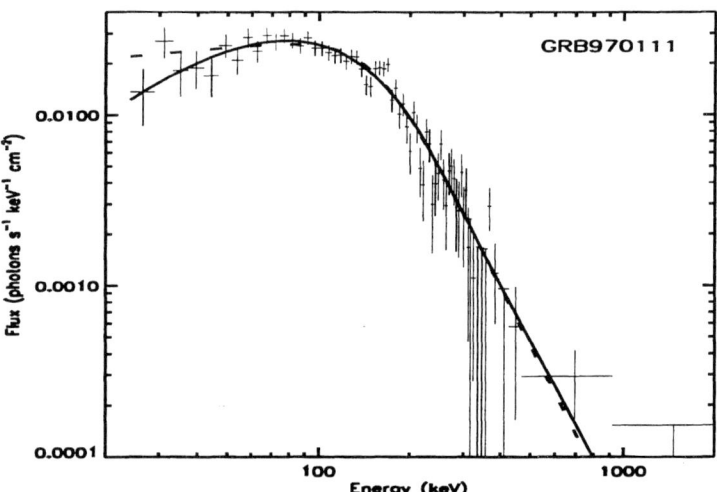

FIGURE 2. A Band GRB function (solid line) and a smoothly broken power law (dashed line) fit to BATSE LAD #0 data for GRB970111. For this time interval (3.904 to 4.672 s), $\alpha_{\text{Band}} = 1.65 \pm 0.26$, a value inconsistent with the SSM prediction of $\alpha \leq -\frac{2}{3}$. In contrast, $\alpha_{\text{SBPL}} = 0.162 \pm 0.25$. The curvature parameter for the SBPL fit was 0.18 ± 0.08, a much sharper break than the 0.45 curvature expected from the SSM. [Residuals in this figure are taken from the fit of the Band GRB function and are plotted for illustrative purposes.]

TABLE 2. Values of the parameters used in evaluating our synchrotron shock spectra for comparison with the Band GRB function and the smoothly broken power law. We fixed parameters which we found to not affect the shape of the spectra near E_{pk} unless self-absorption break was sufficiently high. In the parameter space described by the 3 varied parameters, we found the Band GRB function was good to within 4%.

Fixed Parameters				Varied Parameters			
Radius	r	$= 10^{16}$	cm	Magnetic Field	1 G	$< B$	$< 10^6$ G
Total Energy	E	$= 10^{50}$	ergs	Energy Index	4	$< p$	< 8
Bulk Lorentz Factor	Γ	$= 10^2$		Low Energy Cutoff	10^3	$< \gamma_{min}$	$< 10^7$
High Energy Cutoff	γ_{max}	$= 10^{10}$					

A second complication is that the low-energy edge of the lepton energy distribution could be smoother than the sharp cutoff we assume [9]. Presumably, this would lead to a smoother curvature at the νF_ν peak. However, fits with the SBPL instead suggest a *sharper* break. Thus, it seems improbable that modifications to the underlying energy distribution could circumvent this dilemma.

Acknowledgements. Many thanks to Charles Dermer for providing his SSM code and Rob Preece for his useful comments. This article was written while the author held a NRC/NRL Associateship.

REFERENCES

1. Golenetskii, S. V., Mazets, E. P., Aptekar, R. L., & Ilyinskii, V. N., *Nature*, **307**, 41
2. Katz, J., *ApJ* **432**, L107 (1994).
3. Tavani, M., *ApJ* **466**, 768-778 (1996).
4. Cohen, E. et al., *ApJ* **488**, 330-337 (1997).
5. Schaefer, B. E. et al., *ApJ* **492**, 696-702 (1998).
6. Crider, A. et al., *ApJ* **470**, L39-L42 (1997).
7. Preece, R. et al., *ApJ* **506**, L23-L26 (1998).
8. Preece, R., et al., astro-ph/9908119 (1999).
9. Lloyd, N. M. & Petrosian, V., these proceedings.
10. Crider, A. & Liang, E. P., *A & A Supp. Ser.* **138**, 405-406
11. Band, D. et al., *ApJ* **413**, 281-292 (1993).
12. Ryde, F., astro-ph/9811462 (1999).
13. Crusius, A. & Schlickeiser, R., *A & A* **164**, L16-L18 (1986).
14. Crider, A., et al., *ApJ* **519**, 206-213 (1999).

Redshifts and Compton Attenuation

J. J. Brainerd and R. D. Preece

University of Alabama in Huntsville, Huntsville, AL 35899, USA

Abstract. A number of gamma-ray bursts have spectra that are accurately fit by a Compton attenuated power-law spectrum. One free parameter in this spectral model is the cosmological redshift of the gamma-ray burst. From the fitting of this model to burst spectra observed with BATSE, values of z of order unity or greater were derived, a result that was later derived from optical observations. The model-derived z distribution contains many more bursts with $z > 2$ than does the optically derived distribution. Fits of the model to gamma-ray spectra of bursts with optically derived redshifts are consistent with the observations for the 2 bursts that have spectra with x-ray excesses. In cases without an x-ray excess, the model is unable to fully explain the observations. This suggests that other processes in addition to Compton attenuation are required to fully explain the prompt gamma-ray spectrum.

REDSHIFT ENSEMBLE

Prompt gamma-ray burst emission has two unusual properties: all burst spectra are characterized by approximately the same frequency, and many bursts have an x-ray excess. These characteristics are consistent with a model of a power-law spectrum attenuated through Compton scattering by an optically thick medium [1]. It has been shown that this model produces excellent fits to many of the observed spectra [2]. One of the free parameters in the model is the cosmological redshift of the gamma-ray burst, so if the theory is correct, one could derive a redshift for every gamma-ray burst with an x-ray excess, independently from optical observations.

A comparison of the observed optical redshifts to an ensemble of redshifts derived by fitting the Compton attenuation model to BATSE gamma-ray burst spectra is shown in Figure 1. As found with the optical redshifts, the model-derived redshifts have values of $z > 1$. The ensemble of values is systematically higher for the model-derived values than for the optically-derived values. The significance of the discrepancy is difficult to evaluate. The optical determination of redshift is difficult for $z > 2.5$, and there may be selection effects that make observation of afterglows from high-redshift bursts difficult to detect. The difficulty in deriving z for high-redshift bursts will produce a truncation in the optical redshift distribution relative to the model-determined distribution.

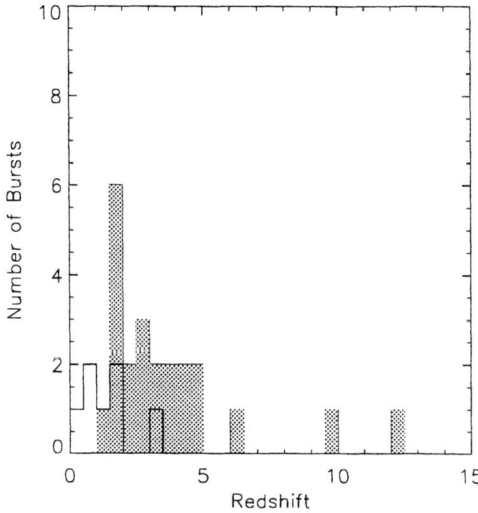

FIGURE 1. The shaded histogram shows the redshift distribution for 23 bursts as determined by fitting the Compton attenuation model to BATSE burst spectra [2]. The solid-line histogram shows the distribution of five GRB redshifts determined through optical observations [6-12].

INDIVIDUAL BURSTS

Three bursts with optically determined redshifts have been observed by BATSE. Only one of these bursts, GRB 990123, is strong enough to derive the redshift directly from the gamma-ray spectrum. When one fits the model to this burst with z as a free parameter, one finds $z = 1.20 \pm 0.05$, versus the optically-derived value of 1.61 [9]. Figure 2 shows this burst for a fit in which z is fixed to the value of 1.61. One sees in this figure the importance of the x-ray excess in the model for fitting the data.

Figures 3 and 4 show the model fit to GRB 980703 and GRB 970508 when z is set to their optically-derived values. The burst GRB 980703 is well-fit by the theory. This burst also has an x-ray excess that requires the upturn in the model spectrum. The burst GRB 970508, however, is poorly fit, with model parameters that make the spectrum nearly a power-law. The burst GRB 990510 has been analyzed using the combined data of BATSE and the BeppoSAX Wide Field Cameras [5], and it has been shown not to have an x-ray excess. This suggests that despite its successes in explaining many of the observations, the Compton attenuation theory does not provide the full answer to the question of what produces the observed spectral shape.

CONCLUSION

The Compton attenuation theory provides a means of deriving the burst redshift independent of an optical counterpart. The small number of gamma-ray bursts

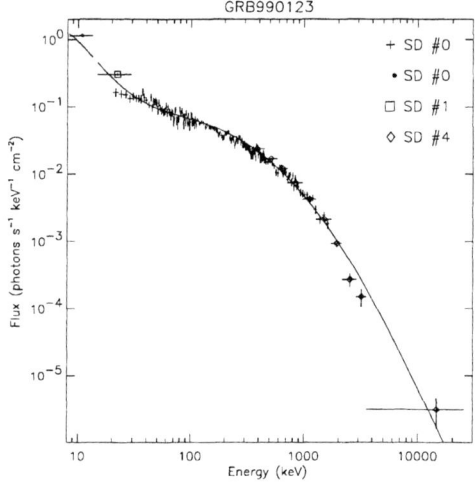

FIGURE 2. Photon spectra for GRB 990123. With the redshift set to the optically-determined value of $z = 1.61$ [9], a fit with $\chi^2 = 617$ for 411 degrees of freedom was found.

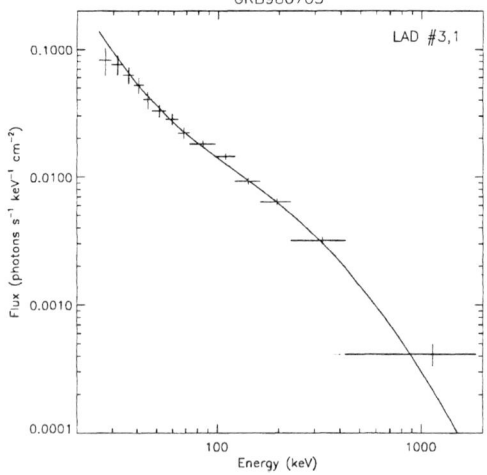

FIGURE 3. Photon spectra for GRB 980703. With the redshift set to the optically-determined value of $z = 0.966$ [7], a fit with $\chi^2 = 11.07$ for 11 degrees of freedom was found.

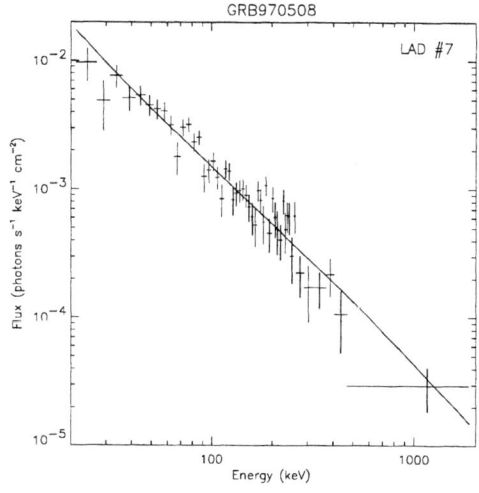

FIGURE 4. Photon spectra for GRB 970508. With the redshift set to the optically-determined value of $z = 0.835$ [11], a fit with $\chi^2 = 140$ for 112 degrees of freedom was found.

with optically derived redshifts show some consistency. To fully test the ability of the model to provide a redshift requires better observations of the burst spectrum between 1 keV and 100 keV than are currently available. Observations by both HETE-2 and SWIFT should enable one to fully test the theory.

The theory does not appear to fully explain the shape of the burst spectrum, as seen in GRB 990510. This suggests that the radiative processes are somewhat more complex than contemplated in the Compton attenuation theory. One possibility is that the Compton attenuation is occurring in the emission region, so that it is an optical depth effect. This would require a population of electrons that have a temperature of order 1 MeV and no bulk motion towards the observer. This would not be consistent with shock theories for burst emission, but may be consistent with a plasma instability theory [3,4].

REFERENCES

1. Brainerd, J. J., *ApJ* **428**, 21 (1994).
2. Brainerd, J. J., Preece, R. D., Briggs, M. S., Pendleton, G. N., & Paciesas, W. S., *ApJ* **501**, 325 (1998).
3. Brainerd, J. J., *ApJ*, submitted, astro-ph/9904040 (1999).
4. Brainerd, J. J., these proceedings.
5. Briggs, M. S., et al., these proceedings.
6. Djorgovski, S. G., Kulkarni, S. R., Bloom, J. S., Frail, D., Chaffee, F. & Goodrich, R., *GCN Circ.* 189 (1999).

7. Djorgovski, S. G., Kulkarni, S. R., Goodrich, R., Frail, D. A., & Bloom J. S., *GCN Circ.* 139 (1998).
8. Galama, T. J., Vreeswijk, P. M., Rol, E., Kaper, L., Masetti, N., Pian, E., Palazzi, E., Frontera, F., van Paradijs, J., & Kouveliotou, C., *GCN Circ.* 388 (1999).
9. Kelson, D. D., Illingworth, G. D., Franx, M., Magee, D., & van Dokkum, P. G., *IAU Circ.* 7096 (1999).
10. Kulkarni, S. R., et al., *Nature* **393**, 35 (1998).
11. Metzger, R. M., Djorgovski, S. G., Kulkarni, S. R., Steidel, C. C., Adelberger, K. L., Frail, D. A., Costa, E., & Frontera, F., *Nature* **387**, 878 (1997).
12. Vreeswijk, P. M., Galama, T. J., Rol, E., Stappers, B., Palazzi, E., Pian, E., Masetti, N., Frontera, F., van Paradijs, J., Kouveliotou, C., & Boehnhardt, H., *GCN Circ.* 324 (1999).

An Eigenfunction Method for Particle Acceleration at Ultra-relativistic Shocks

Axel W. Guthmann*, John G. Kirk*, Yves A. Gallant[†,‡],
and Abraham Achterberg[‡]

*Max-Planck-Institut für Kernphysik, Postfach 103980 , D-69029 Heidelberg, Germany [1]
[†]Astronomical Institute, Utrecht University, P.O. Box 80000, 3508 TA Utrecht, The Netherlands
[‡]Dublin Institute for Advanced Studies, 5 Merrion Square, Dublin 2, Ireland

Abstract. We adapt and modify the eigenfunction method of computing the power-law spectrum of particles accelerated at a relativistic shock front via the first-order Fermi process [6] to apply to shocks of arbitrarily high Lorentz factor. The power-law index of accelerated particles undergoing isotropic small-angle scattering at an ultra-relativistic, unmagnetized shock is found to be $s = 4.23 \pm 0.2$ (where $s = d\ln f/d\ln p$, with f the Lorentz-invariant phase-space density and p the momentum), in agreement with the results of Monte-Carlo simulations. We present results for shocks in plasmas with different equations of state and for Lorentz factors ranging from 5 to infinity.

THE METHOD

We study a stationary shock front in the $x - y$ plane. The accelerated particles are assumed to be test-particles without influence on the dynamics of the plasma or the jump conditions at the shock-front. The plasma flows along the z-axis, with constant velocities u_- in the upstream ($z < 0$) region and u_+ downstream ($z > 0$), and the velocities are related by the Rankine-Hugoniot jump conditions.

Test-particles are injected into the acceleration process and their interaction with the plasma flow is assumed to give rise to diffusion in the angle $\cos^{-1}\mu$ between a particle's velocity and the shock normal. In the frame of the shock front this leads to a stationary transport equation valid for the local plasma rest frame and given in mixed coordinates as [6]

$$\Gamma(u+\mu)\frac{\partial f}{\partial z} = \frac{\partial}{\partial \mu}D_{\mu\mu}(1-\mu^2)\frac{\partial f}{\partial \mu} , \qquad (1)$$

where the plasma speed u is measured in units of the speed of light, $\Gamma = (1-u^2)^{-1/2}$ is the Lorentz factor, $f(p,\mu,z)$ is the (Lorentz invariant) phase-space density as

---
[1]) homepage: http://www.mpi-hd.mpg.de/theory/

a function of the particle momentum p, direction μ and position. p and μ are measured in the local rest frame of the plasma, whereas z is measured in the rest frame of the shock front.

Equation (1) is solved using the separation *Ansatz* [6]

$$f(p, u, \mu, x) = \sum_{i=-\infty}^{+\infty} g_i(p) Q_i(\mu, u) \exp\left(\Lambda_i z / \Gamma\right), \qquad (2)$$

valid in each half-plane with Λ_i and Q_i the eigenvalues and eigenfunctions of the equation

$$\left\{ \frac{\partial}{\partial \mu} \left[D_{\mu\mu} \frac{\partial}{\partial \mu} \right] - \Lambda_i (u + \mu) \right\} Q_i(\mu, u) = 0 \ . \qquad (3)$$

The momentum distribution of particles with energy far above the injection energy range — those in which we are interested — takes the shape of a power-law $g_i(p) \propto p^{-s}$ with index s, since there is no preferred momentum scale in this range.

Matching the expansion (eq. 2) across the shock front according to Liouville's theorem and imposing physically realistic boundary conditions up and downstream leads to a nonlinear algebraic equation for the power-law index s.

In [6] and [4] only the eigenfunctions with $i < 0$ were used and the method was applied to mildly relativistic shock speeds ($\Gamma_- \leq 5$). Here, we use the eigenfunctions with $i > 0$ and calculate them directly with a numerical scheme. In the limit $u_- \to 1$, an analytic expression is available [7]. Four eigenfunctions ($i = 1, 3, 5, 7$) are shown in Fig. 1A as functions of the cosine $\mu_s = (\mu + u)/(1 + \mu u)$ of the angle between the particle direction and the shock normal, measured in the shock rest frame. For $i > 1$ they are oscillatory for $-1 < \mu_s < 0$ and for all $i > 0$ fall off monotonically in the range $0 < \mu_s < 1$.

RESULTS

The index s of the momentum spectra of the accelerated particles in different cases are shown in Fig. 1B. The jump conditions investigated are those for a relativistic gas both up and downstream: $u_- u_+ = 1/3$ and for a strong shock in a medium with adiabatic index $4/3$ [5].

Also we investigate two different scattering operators, $D_{\mu\mu 1} = 1 - \mu^2$ (isotropic small/angle scattering) and $D_{\mu\mu 2} = (1 - \mu^2) \times (\mu^2 + 0.01)^{1/3}$ corresponding to scattering in weak Kolmogorov turbulence, together with a rough prescription for avoiding the lack of scattering at $\mu = 0$ [4]. For high upstream Lorentz-factors the power-law index settles at a value around 4.23 for all equations of state, which is reproduced in the limiting case $u_- \to 1$. The scattering operator has only a minor effect.

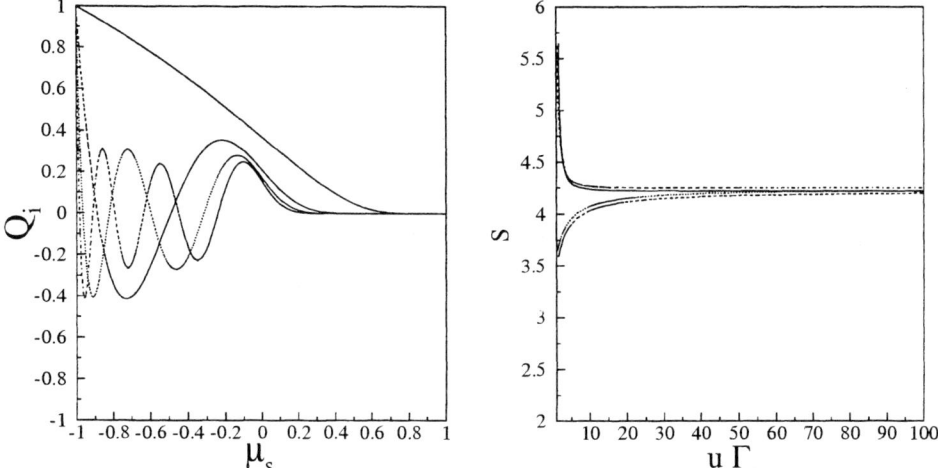

FIGURE 1. (**A**, *left*) The eigenfunctions Q_i for $i = 1, 3, 5, 7$ for $\Gamma_- = 223$, as a function of the (cosine of the) angle between the particle speed and the shock normal, measured in the shock frame, for a relativistic gas. (**B**, *right*) The power-law index s for relativistic gas with isotropic (solid, 2nd from top) and anisotropic (dashed-dotted, top) scattering operator and for a strong shock in a gas of adiabatic index 4/3, with isotropic (dashed, 4th from top) and anisotropic (dotted, 3rd from top) scattering.

SUMMARY

These results are in agreement with the asymptotic Monte Carlo results of Gallant et al. [3] and those of Bednarz & Ostrowski [1] for $\Gamma_- \approx 200$. Anisotropic scattering, which has not been treated by Monte Carlo simulations, leads to a slight steeping in the power-law spectrum, because fewer particles are able to cross the region $\mu \approx 0$ and return to the shock. From observations of GRB afterglows, Galama et al. [2] and Waxman [8] have found synchrotron spectral indices corresponding to $s \approx 4.25$, implying that the particles could indeed have been accelerated by the first order Fermi mechanism operating at an ultra-relativistic shock front.

ACKNOWLEDGMENTS

This work was supported by the European Commission under the TMR programme, contract number ERBFMRX-CT98-0168

REFERENCES

1. Bednarz, J., Ostrowski, M., *Phys. Rev. Lett.* **80**, 3911 (1998).

2. Galama, T. et al., *Astrophysical Journal* **500**, L101 (1998).
3. Gallant, Y.A., Achterberg, A., Kirk, J.G., *Astron. Astrophys. Suppl. Ser.* **138**, 549 (1999).
4. Heavens, A.F., Drury L.O'C., *MNRAS* **235**, 997 (1988).
5. Kirk, J.G., Duffy, P., *Journal of Physics G: Nucl. Part. Phys.* **25**, R163 (1999).
6. Kirk, J.G., Schneider, P., *Astrophysical Journal* **315**, 425 (1987).
7. Kirk, J.G., Schneider, P., *Astronomy & Astrophysics* **225**, 559 (1989).
8. Waxmann, E., *Astrophysical Journal* **485**, L5 (1997).

The Synchrotron Spectrum of Fast Cooling Electrons Revisited

Jonathan Granot*, Tsvi Piran*, and Re'em Sari[†]

*Racah Institute, Hebrew University, Jerusalem 91904, Israel
[†] Theoretical Astrophysics, California Institute of Technology, Pasadena, CA 91125

Abstract. We discuss the spectrum arising from synchrotron emission by fast cooling electrons when fresh electrons are continually accelerated by a strong blast wave into a power law distribution of energies. The fast cooling spectrum was so far described by four power law segments: $F_\nu \propto \nu^2, \nu^{1/3}, \nu^{-1/2}$ and $\nu^{-p/2}$, divided by three break frequencies: $\nu_{sa} < \nu_c < \nu_m$. This is valid for a homogeneous electron distribution. However, hot electrons are located right after the shock, while most electrons are farther down stream and have cooled. We show that this spatial distribution introduces a new break frequency, ν_{ac}, in the optically thick regime: $\nu_{ac} < \nu_{sa} < \nu_c < \nu_m$, and a new spectral slope: $F_\nu \propto \nu^{11/8}$ for $\nu_{ac} < \nu < \nu_{sa}$, while the familiar $F_\nu \propto \nu^2$ is obtained only for $\nu < \nu_{ac}$. While the above ordering of the break frequencies is relevant for afterglows with typical parameters in an ISM environment, other possibilities may also be relevant for internal shocks or afterglows in dense circumstellar winds. We discuss possible applications for gamma-ray bursts (GRBs) and their afterglows, in the context of the internal-external shock model. This may explain spectral slopes steeper than 1/3 seen in the 1 − 10 keV range in some bursts, if ν_{sa} reaches the X-ray.

INTRODUCTION

The spectrum of GRBs and their afterglows is well described by synchrotron and inverse Compton emission. It is better studied during the afterglow stage, where we have broad band observations. The observed spectrum is in a good agreement with the theory. Within the fireball model, both the GRB and its afterglow are caused by the slowing down of a relativistic flow through shock waves. According to the internal-external shock scenario, the GRB itself arises from internal shocks within the flow, while the afterglow is due to the external shock produced when the flow is decelerated as it sweeps the ambient medium. In this paper we consider fast cooling, where the electrons cool due to radiation losses on a time scale much shorter than the dynamical time of the system.

We consider the synchrotron spectrum of relativistic electrons which are continually accelerated by a strong blast wave into a power law energy distribution: $N(\gamma_e) \propto \gamma_e^{-p}$ for $\gamma_e \geq \gamma_m$. After being accelerated by the shock, the electrons cool down due to radiation losses. On the dynamical time of the system, all electrons cool to a Lorentz factor of γ_c [1]. The typical synchrotron frequency of an electron with a randomly oriented velocity is $\nu_{syn}(\gamma_e) = 3\gamma q_e B \gamma_e^2 / 16 m_e c$, where γ_e, m_e and q_e are its Lorentz factor, mass and electric charge, respectively, B is the magnetic field and γ is the bulk Lorentz factor of the flow. The two critical Lorentz factors, γ_m and γ_c, thereby define two critical frequencies: ν_m and ν_c, respectively. In the fast cooling regime, $\gamma_c \ll \gamma_m$ and therefore $\nu_c \ll \nu_m$. Another critical frequency is ν_{sa}, below which the system becomes optically thick to self absorption.

The spectrum of a homogeneous distribution of fast cooling electrons had been investigated for $\nu_{sa} < \nu_c < \nu_m$ [1,2]. It consists of four power law segments: $F_\nu \propto \nu^2, \nu^{1/3}, \nu^{-1/2}$ and $\nu^{-p/2}$, divided by these three critical frequencies. The spectral slope above ν_m is related to the electron injection distribution; the number of electrons with Lorentz factors of the order γ_e is $\propto \gamma_e^{1-p}$ and their energy $\propto \gamma_e^{2-p}$. As these electrons cool, they deposit most of their energy into a frequency range of the order of $\nu_{syn}(\gamma_e) \propto \gamma_e^2$ and therefore $F_\nu \propto \gamma_e^{-p} \propto \nu^{-p/2}$. In the frequency range $\nu_c < \nu < \nu_m$ we have contribution from all the electrons in the system as they all cool on the dynamical time. Since the electron's energy is $\propto \gamma_e$ and its typical frequency $\propto \gamma_e^2$, the flux per unit frequency is $\propto \gamma_e^{-1} \propto \nu^{-1/2}$. At $\nu_{sa} < \nu < \nu_c$ we see the synchrotron low frequency tail of the cooled electrons, which is given by $\nu^{1/3}$. At $\nu < \nu_{sa}$ the spectrum is self absorbed, and we see the Rayleigh-Jeans portion of the black body spectrum: ν^2. According to the Rayleigh-Jeans law, $F_\nu \propto \nu^2 \gamma_{e,typ}(\nu)$, where $\gamma_{e,typ}(\nu)$ is the typical thermal Lorentz factor of the electrons emitting at the observed frequency ν. The assumption leading to the ν^2 spectral slope below ν_{sa} is that the relevant temperature is that of the cooled electrons, $\gamma_{e,typ}(\nu) = \gamma_c \propto \nu^0$, as these are the electrons which radiate the low energy tail.

We show here that for fast cooling, the self absorbed flux deviates from the ν^2 power law due to spatial effects which have been ignored in previous works. We derive the fast cooling spectrum taking into account the inhomogeneous distribution of the electron effective temperatures behind the shock. We provide expressions for the break frequencies and maximal flux density, and discuss the relevance of the new spectral slope, $F_\nu \propto \nu^{11/8}$, both to the prompt emission and the afterglow.

FAST COOLING SPECTRUM

The shape of the fast cooling spectrum is determined by the relative ordering of the break frequencies: ν_{sa}, ν_c and ν_m. For fast cooling, $\nu_c \ll \nu_m$, leaving three possibilities for this ordering, each of which gives rise to a different spectrum.

We begin the "canonical" situation for which the homogeneous spectrum was described above: $\nu_{sa} < \nu_c < \nu_m$. We denote this by case (1). The optically thin ($\nu > \nu_{sa}$) part of the spectrum is the same for the homogeneous and inhomogeneous electron distributions behind the shock, since all the emitted radiation escapes the system, and the location of the emitting electrons becomes irrelevant. Immediately behind the shock there is a thin layer where most electrons have not yet cooled significantly, followed by a much wider layer where all electrons at a given distance, l', behind the shock (in the local frame) have cooled down to $\gamma_e \propto 1/l'$. In the optically thick regime, most of the photons which escape the system are emitted at an optical depth close to unity: $\tau_\nu \sim 1$. Therefore, $\gamma_{e,typ}(\nu)$ must be evaluated at the place where $\tau_\nu = 1$. An optical depth of unity is obtained at a distance l' behind the shock where the optically thin emission equals the optically thick emission:

$$n'l'\gamma^2(P_{\nu,max}/4\pi)[\nu/\nu_{syn}(\gamma_e)]^{1/3} = (2\nu^2/c^2)\gamma\gamma_e m_e c^2 , \qquad (1)$$

where n' is the proper number density and $P_{\nu,max} \propto \gamma_e^0$ is the peak spectral power of an electron. Since $\nu_{syn}(\gamma_e) \propto \gamma_e^2$, and within the wide layer of cooled electrons, $l' \propto 1/\gamma_e$, equation (1) implies that $\gamma_{e,typ}(\nu) = \gamma_e \propto \nu^{-5/8}$, and according to the Rayleigh-Jeans law, $F_\nu \propto \nu^{11/8}$.

The new spectral slope, $F_\nu \propto \nu^{11/8}$, begins at ν_{sa}, as the system becomes optically thick, and extends to lower frequencies. However, at sufficiently low frequencies ($\nu < \nu_{ac}$), $\tau_\nu = 1$ is reached within the thin layer just behind the shock where the electrons have not cooled significantly, implying $\gamma_{e,typ}(\nu) = \gamma_m \propto \nu^0$ and $F_\nu \propto \nu^2$. The spectrum in case (1) is shown in the upper frame of Figure 1, and the break frequencies and maximal flux density are given by

$$\nu_{ac} = (2^7 m_e 4 c^{13} \gamma^5 n'^3 / 3^7 \pi^3 B^4 q_e^4 \gamma_m^8)^{1/5} = 1.3 \times 10^8 \text{ Hz } n_2^{3/5} \gamma_{40} B_{10}^{-4/5} \gamma_{m,4}^{-8/5} ,$$

$$\nu_{sa}^{(1)} = (\sigma_T^8 B^{12} c^5 t_{dyn}^8 \gamma^{13} n'^3 / 3^{15} 2 \pi^{11} m_e^4 q_e^4)^{1/5} = 7.3 \times 10^{10} \text{ Hz } n_2^{3/5} \gamma_{40}^{13/5} B_{10}^{12/5} t_{dyn,3}^{8/5} ,$$

$$\nu_c = 27\pi^2 m_e c q_e / 4\sigma_T^2 \gamma B^3 t_{dyn}^2 = 4.9 \times 10^{13} \text{ Hz } \gamma_{40}^{-1} B_{10}^{-3} t_{dyn,3}^{-2} ,$$

$$\nu_m = 3 q_e B \gamma \gamma_m^2 / 16 m_e c = 1.3 \times 10^{17} \text{ Hz } \gamma_{40} B_{10} \gamma_{m,4}^2 ,$$

$$F_{\nu,max}^{(1)} = 2\sigma_T m_e c^2 N_e \gamma B / 9\pi^2 q_e D^2 = 100 \text{ mJy } N_{e,52} D_{28}^{-2} \gamma_{40} B_{10} , \qquad (2)$$

where $n_2 = n'/100 \text{ cm}^{-3}$, $\gamma_{40} = \gamma/40$, $B_{10} = B/10$ G, $\gamma_{m,4} = \gamma_m/10^4$, $t_{dyn,4} = t_{dyn}/10^4$ sec, $D = D_{28} 10^{28}$ cm is the distance to the source, and $N_e = N_{e,52} 10^{52}$ is the number of radiating electrons, assuming an isotropic flow. This parameterization reflects typical values during the early afterglow. The superscript $^{(i)}$ labels the specific case under consideration.

We now turn to case (2), where $\nu_c < \nu_{sa} < \nu_m$. Now, ν_c is unimportant as it is within the optically thick regime, and the spectrum is given in the middle frame of Figure 1. The maximal flux density $F_{\nu,max}^{(2)}$ is now reached at $\nu_{sa}^{(2)}$, and both quantities are given by

$$\nu_{sa}^{(2)} = (\nu_{sa}^{(1)})^{5/9} \nu_c^{4/9} , \qquad F_{\nu,max}^{(2)} = F_{\nu,max}^{(1)} (\nu_c/\nu_{sa}^{(1)})^{5/18} . \qquad (3)$$

The last case (3) occurs when $\nu_c < \nu_m < \nu_{sa}$. Now the system becomes optically thick due to absorption from a very thin layer behind the shock front, well within the thin layer where most electrons have not cooled significantly. Therefore, both ν_{ac} and ν_c are now irrelevant, since the inner parts where these frequencies are important are not visible. Furthermore, the initial electron distribution may be used to evaluate $\gamma_{e,typ}(\nu)$. Thus, $\gamma_{e,typ}(\nu) \propto \gamma_m \propto \nu^0$ at $\nu < \nu_m$, implying $F_\nu \propto \nu^2$, while at $\nu_m < \nu < \nu_{sa}$ the emission is dominated by electrons with $\nu_{syn}(\gamma_e) \sim \nu$, implying $\gamma_{e,typ}(\nu) \propto \nu^{1/2}$ and $F_\nu \propto \nu^{5/2}$. The resulting spectrum is shown in the lower frame of Figure 1, where ν_m remains unchanged while

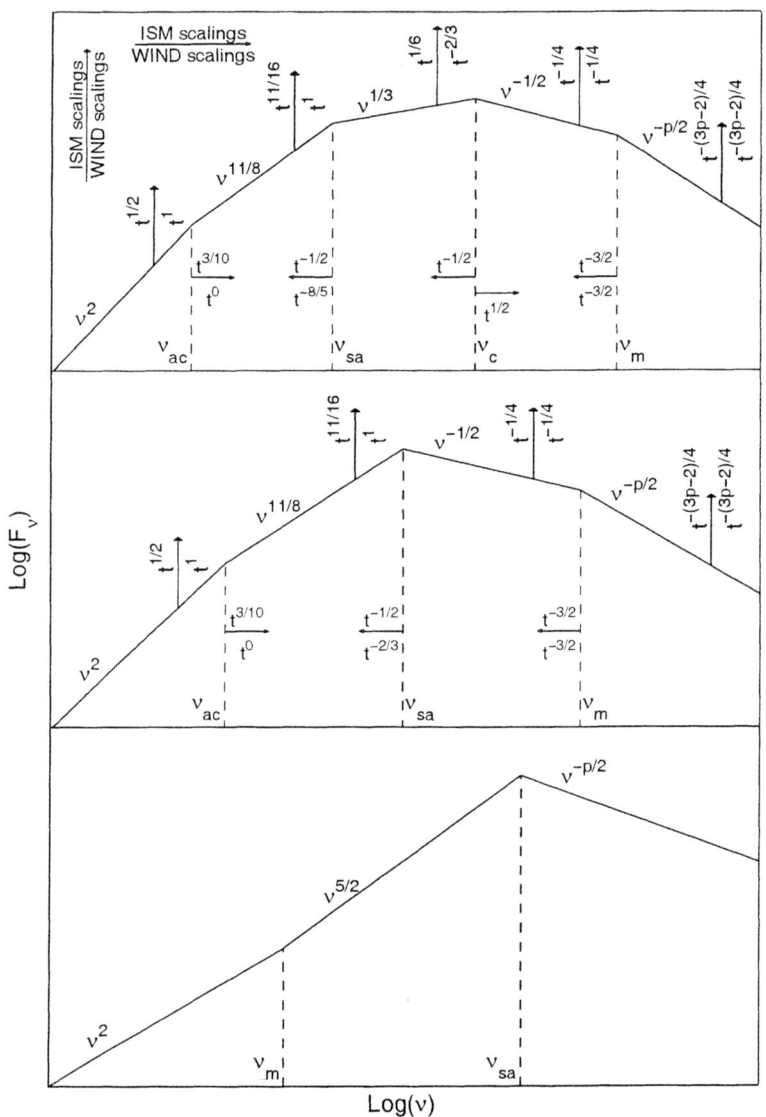

FIGURE 1. Fast cooling synchrotron spectra from a shock injected power law electron distribution. The shape of the spectrum is determined by the ordering of the self absorption frequency, ν_{sa}, with respect to $\nu_c < \nu_m$. There are three possible shapes for the spectrum, corresponding to $\nu_{sa} < \nu_c$, $\nu_c < \nu_{sa} < \nu_m$ and $\nu_{sa} > \nu_m$, from top to bottom. The temporal scalings for an external shock, both for an ISM and stellar wind (denoted WIND) environments, are given only in the first two frames, since the last (bottom) frame is unlikely.

$$\nu_{sa}^{(3)} = \left[(\nu_{sa}^{(1)})^{10/3} \nu_c^{8/3} \nu_m^{p-1}\right]^{\frac{1}{(p+5)}} \quad , \quad F_{\nu,max}^{(3)} = F_{\nu,max}^{(1)} \left[(\nu_{sa}^{(1)})^{-p/3} \nu_c^{\frac{(3-p)}{6}} \nu_m^{\frac{(p-1)}{2}}\right]^{\frac{5}{(p+5)}} . \quad (4)$$

APPLICATION TO GRBS

For afterglows in an ISM surrounding, fast cooling typically last for one hour, and the spectrum is of the first type (upper frame in Figure 1). If the surrounding is a stellar wind of a massive progenitor, the fast cooling stage typically lasts for about a day [3], during which the spectrum is of the second kind (middle frame in Figure 1) for the first $1-2$ hours, and then switches to the spectrum of the first kind. The time dependence of the break frequencies and flux density for an afterglow in an ISM or wind environment is given in Figure 1.

During the prompt gamma-ray emission, both the highly variable temporal structure of most bursts and the requirement of a reasonable radiative efficiency [4,5] suggest fast cooling. All three cases for the spectrum are possible. The spectra of the third type (lower frame of Figure 1) is likely to occur with slow cooling conditions ($\nu_m < \nu_c$), but this should not effect the spectrum as long as $\nu_c < \nu_{sa}$.

DISCUSSION

The synchrotron spectrum of fast cooling electrons is summarized in Figure 1. We find three possible spectra, depending on the ordering of ν_{sa} with respect to $\nu_c < \nu_m$. Two of these cases include a new self absorption regime where $F_\nu \propto \nu^{11/8}$.

During the initial fast cooling stage of the afterglow, the system is typically optically thick in the radio and optically thin in the optical and X-ray. We therefore expect the new feature, $F_\nu \propto \nu^{11/8}$, to be observable only in the radio band during the afterglow. This requires fast radio observations, $\lesssim 1$ hour (day) for an ISM (stellar wind) environment, which may become possible after the launch of HETE-2.

During internal shocks, fast cooling conditions are expected, and a spectrum of the first or second type (upper or middle frames of Figure 1) is most likely. If the self absorption frequency reaches the X-ray, this may explain the observed spectral slopes greater than $1/3$ (photon number slope larger than $-2/3$), in the $1-10$ keV range [6-8] seen in some bursts.

REFERENCES

1. Sari, R., Piran, T. & Narayan, R., *ApJ* **497**, L17 (1998).
2. Sari, R. & Piran, T., *ApJ* **520**, 641 (1999).
3. Chevalier, R.A., & Li, Z.Y., astro-ph9908272 (1999).
4. Sari, R. & Piran, T., *ApJ* **485**, 270 (1997).
5. Fenimore, E. et al., *ApJ* **473**, 998 (1996).
6. Preece, R.D. et al., *ApJ* **496**, 849 (1998).
7. Crider, A. et al., *ApJ* **479**, L39 (1998).
8. Strohmayer, T.E. et al., *ApJ* **500**, 873 (1998).

Flow-Field Dependent Variation Method for Complex Relativistic Fluids

G. A. Richardson[1], T. J. Chung[2], G. R. Karr[2], G. N. Pendleton[1,3]

[1] *Center for Space Plasma, Aeronomical, and Astrophysical Research, University of Alabama - Huntsville*
[2] *Department of Mechanical and Aerospace Engineering, University of Alabama - Huntsville*
[3] *Department of Physics, University of Alabama - Huntsville*

Abstract. Many current high-energy astrophysics problems, particularly those containing shock waves and high-speed flow, do not take advantage of new computational fluid dynamics (CFD) techniques available in such fields as aerospace engineering. We will present the flow-field dependent variation (FDV) method to accurately solve very high-speed flow problems, as well as capture relativistic shocks, all while allowing the user to apply their familiar finite difference method (FDM) or finite element method (FEM). This method is also versatile enough to apply the non-relativistic Naiver-Stokes equations to solve low speed flows. In the FDV method, numerical schemes are automatically adjusted from the current flow field information reflecting shock discontinuities and/or effects of viscosity in boundary layers. To demonstrate the validity of this theory, the shock tube using the relativistic hydrodynamic equations has been applied.

INTRODUCTION

The need to address situations where inviscid and viscous, compressible and incompressible, as well as laminar and turbulent flows may all occur within the same flow field is met by the flow-field dependent variation (FDV) method [1]. The physical complexities of such problems found in relativistic astrophysical flows are only outweighed by the computational complexities. Traditionally, solving applications containing a combination of these different types of flows required different techniques in order to avoid instabilities. Finite difference methods (FDMs) have been used predominantly in the past [e.g., 2-4]. The FDV method concentrates on each element (or spatial discretization) individually, automatically adjusting for the current flow field. This is important for flow fields with widely varying scales of length and time such as those found in turbulent flows and complex shock waves, and is what allows FDV to solve such a variety of problems. The FDV method is derived by introducing variational parameters into a form of the Taylor series expansion. The variational parameters are used to track the flow field information in the individual elements. The Taylor series is expanded in such a way that the method is second order accurate in both time and space. The Naiver-Stokes system of equations written in conservation form are applied to the series. Conservation variables are used since they result in continuous functions. Primitive variables are found from the conservation variables revealing the discontinuities such as shock waves. FDV is unique in that the geometrical discretization can be done using either a FDM or a finite element method (FEM). FDV contains all the physical information, so the user is not limited by the physics dictating the use of either a FDM or a FEM.

FLOW FIELD DEPENDENT VARIATION METHOD

Equations

The governing equations for three dimensional relativistic hydrodynamic flow can be written in conservation form as

$$\frac{\partial \mathbf{U}}{\partial t} = -\frac{\partial \mathbf{F}_i}{\partial x_i} - \frac{\partial \mathbf{G}_i}{\partial x_i}, \tag{1}$$

where U, F, and G are the conservation flow variables, convective flux, and viscous diffusion flux respectively:

$$\mathbf{U} = \begin{pmatrix} \rho\Gamma \\ \rho\mu\Gamma^2 V_j \\ \rho\mu\Gamma^2 - \frac{P}{c^2} \end{pmatrix}, \quad \mathbf{F}_i = \begin{pmatrix} \rho\Gamma V_i \\ \rho\mu\Gamma^2 V_i V_j + P\delta_{ij} \\ \rho\mu\Gamma^2 V_i \end{pmatrix}, \quad \mathbf{G}_i = \begin{pmatrix} 0 \\ -\tau_{ij} \\ -\tau_{ij}V_j + q_i \end{pmatrix} \tag{2}$$

$$\tau_{ij} = \mu(V_{i,j} + \frac{1}{c^2}V_{i,k}V_k V_j + V_{j,i} + \frac{1}{c^2}V_{j,k}V_k V_i) - \frac{2\mu}{3}(V_{k,k}\delta_{ij} + \frac{1}{c^2}V_i V_j V_{k,k})$$

$$q_i = -k\left(\delta_{ij} + \frac{V_i V_j}{c^2}\right)\left(T_{,j} + \frac{T}{c^2}\Gamma^2\frac{dV_j}{dt}\right)$$

$$\mu = 1 + \frac{\gamma}{\gamma+1}\frac{P}{\rho c^2}.$$

The primitive variables include ρ, V_i, P, μ, T, and Γ corresponding to the density, dimensional velocity, pressure, mass equivalent enthalpy, temperature, and Lorentz factor respectfully. Other parameters include the thermal conductivity k, and the polytropic index γ. The FDV equations are derived by introducing variational parameters into a form of the Taylor series expansion of the conservation flow variables:

$$\Delta U^{n+1} = \Delta t\left(\frac{\partial U^n}{\partial t} + s_a\frac{\partial \Delta U^{n+1}}{\partial t}\right) + \frac{\Delta t^2}{2}\left(\frac{\partial^2 U^n}{\partial t^2} + s_b\frac{\partial^2 \Delta U^{n+1}}{\partial t^2}\right) + \mathcal{O}(\Delta t^3). \tag{3}$$

Combining the above equations leads to

$$\left(I + \frac{\partial E_i}{\partial x_i} + \frac{\partial^2 E_{ij}}{\partial x_i \partial x_j}\right) \Delta U^{n+1} = Q^n, \tag{4}$$

where
$$E_i = \Delta t s_1 a_i + \Delta t s_3 b_i,$$
$$E_{ij} = \Delta t s_3 c_{ij} + \frac{\Delta t^2}{2} s_2 (a_i a_j + b_i a_j) + \frac{\Delta t^2}{2} s_4 (a_i b_j + b_i b_j),$$
$$Q = \Delta t \left(\frac{\partial F_i^n}{\partial x_i} + \frac{\partial G_i^n}{\partial x_i}\right) + \frac{\Delta t^2}{2} \left[\frac{\partial}{\partial x_i}(a_i + b_i)\left(\frac{\partial F_j^n}{\partial x_j} + \frac{\partial G_j^n}{\partial x_j}\right)\right],$$

and where the Jacobians are defined as
$$a_i = \frac{\partial F_i}{\partial U}, \quad b_i = \frac{\partial G_i}{\partial U}, \quad c_{ij} = \frac{\partial G_i}{\partial x_j}. \tag{5}$$

The variational parameters are given as
$$s_1 = \begin{cases} min(r,1) & r > \alpha \\ 0 & r < \alpha \\ 1 & \Gamma_{min} = 0 \end{cases} \quad r = \frac{\sqrt{\Gamma_{max}^2 - \Gamma_{min}^2}}{\Gamma_{min}} \tag{6}$$
$$s_2 = s_1^\eta \quad 0 < \eta < 1$$
$$s_3 = \begin{cases} min(s,1) & s > \beta \\ 0 & s < \beta \\ 1 & Re_{min} = 0 \end{cases} \quad s = \frac{\sqrt{Re_{max}^2 - Re_{min}^2}}{Re_{min}} \tag{7}$$
$$s_4 = s_3^\eta \quad 0 < \eta < 1$$

where Γ is the Lorentz factor (or Mach number for non-relativistic flow) and Re is the Reynolds number. α and η are user defined constants.

Variational Parameters

The variational parameters introduced in the above equations are used for a variety of purposes. All the variational parameters fall between 0 and 1, and are calculated locally at each element making them flow field dependent.

- s_1 and s_3 are the first order variation parameters.
- s_2 and s_4 are the second order variation parameters.
- s_1 and s_3 are calculated from local flow speeds (the Lorentz factor for relativistic speeds and the Mach number for non-relativistic speeds) and local Reynolds number. The contours of these parameters over time closely resemble that of the flow field. The parameters are large in regions of high gradients, and small in regions of small gradients. s_1 and s_3 provide computational accuracy to the system.

- s_2 and s_4 are exponentially proportional to the first order variation parameters. These parameters were introduced in the second order time derivative of the Taylor series approximation and act as an artificial viscosity term does to add stability to the solution.

- s_1 represents convection. If $s_1 \approx 0$ then the effect of convection is small and the computational scheme is automatically altered to take this into account by forcing the equations to be predominantly parabolic-elliptic.

- s_3 is associated with diffusion. If $s_3 \approx 0$ then the effect of viscosity or diffusion is small and the equations are automatically altered to predominantly hyperbolic equations.

- If s_1 and s_3 are both nonzero, then convection and diffusion are equally important, and a mixed form of hyperbolic, parabolic, and elliptic equations results. The result is incompressible, low speed flow. The ability of FDV to handle both low speed and high speed is what makes instrument design analysis possible.

- The transition to turbulence can be seen by the increase in the Reynolds number. This physical instability is detected by the increase of s_3 for incompressible flow and both s_1 and s_3 if the flow is compressible. This physical instability is also likely to cause numerical instability which will be countered by s_2 and/or s_4 ensuring numerical stability. The variational parameters are capable of capturing turbulent unsteady fluctuations.

- s_1 and s_3 can be used as error indicators for adaptive mesh generators. Large values indicate large gradients which need smaller mesh sizes. Since s_1 and s_3 are dependent on the primitive variables, they participate in resolving the adaptive mesh.

Element-by-Element GMRES Solver

Element by element (EBE) techniques are used to avoid assembling very large matrices [5]. It is common to have arrays with dimensions on the order of hundreds of thousands, even millions. In EBE, element parameters are stored individually and solved locally to that element. This works well with FDV since the physical parameters are calculated local to each element. Once the individual elements have been solved, the local element solutions are assembled into a large single column global matrix. EBE can be used with the generalized minimum residual method (GMRES). GMRES is an iterative technique that was developed specifically for solving large systems. It is a projection method based on minimizing the residual norm over a Krylov space. The GMRES solver is derived from adding a preconditioner to the Conjugate Gradient Method (CGM), the idea being that the proper preconditioner will reduce the convergence time and in come cases add stability.

RESULTS

The shock tube has traditionally been used to determine the accuracy and stability of a technique since an analytical solution is known. An element by element/generalized minimum residual method was used in solving the FDV equations for the relativistic shock tube. Results presented in [6] demonstrate the FDV method stability and accuracy in comparison to more traditional techniques used in relativistic astrophysical flow studies. Results indicate that the FDV accuracy, combined with the methods ability to handle difficult flows and the user's freedom to choose their own FDM or FEM technique, makes the FDV method not only a viable, but desirable option.

REFERENCES

1. Chung, T. J., *Inter. J. for Num. Methods in Fluids* **31**, 223-246 (1998).
2. Hawley, J. F., Smarr, L. L., and Wilson, J. R., *ApJS* **55**, 211-246 (1984).
3. Balsara, D. S., *J. Comput. Phys.* **114**, 284-297 (1994).
4. Panaitescu, A., Meszaros, P., *ApJ* **492**, 683-695 (1998).
5. Saad, Y., *Iterative Methods for Sparse Linear Systems*, PWS Publishing, Boston, pp.158-173, 377-379 (1996).
6. Richardson, G. A., Chung, T. J., "Flow-Field Dependent Variation Method for Relativistic Hydrodynamics" in *Finite Elements in Flow Problems 2000* (May 2000).

A Unified Picture for the Various Total Energies of GRBs

Tomonori Totani

National Astronomical Observatory, Theory Division
Mitaka, Tokyo 181-8588, Japan

Abstract. The observed total energies of GRBs show a wide dispersion by a factor of 10^{2-3}, and the most energetic GRBs emit $E_{\gamma,\rm iso} \sim 3 \times 10^{54}$ erg as the isotropic energy. Therefore, the true kinetic energy must exceed $E_{\rm iso} \sim 10^{55}$ erg, and may reach $E_{\rm iso} \sim 10^{56}$ erg if the efficiency of internal shocks is not so high. (Of course, the actual energy can be much smaller if GRBs are strongly beamed.) We propose a unified picture for the energetics of GRBs, in which all GRBs emit a true kinetic energy of $E_{\rm iso} \sim 10^{55-56}$ erg per 4π sr with a strong beaming implicitly assumed. Difference of the observed total energies in soft gamma rays is attributed to different efficiency of energy transfer from protons into electrons. This model predicts that, for GRBs weak in the sub-MeV emission, energy emission in very high energy (VHE) photons ($\gtrsim$ TeV) is even larger than the ordinary emission in the sub-MeV range. We try to interpret the suggestion for VHE emission from GRB970417a reported by the Milagro project in this conference in the context of this model.

INTRODUCTION

It has now been confirmed that there is a wide dispersion in the observed total energies of GRBs. Relatively weak GRBs such as GRB 970508 at $z = 0.835$ typically have a total energy of $E_{\gamma,\rm iso} \lesssim 10^{52}$ erg [1], where $E_{\gamma,\rm iso}$ is the isotropic total energy of observed gamma rays assuming isotropic radiation. On the other hand, there is a population of GRBs with quite a large amount of energy with $E_{\gamma,\rm iso} \sim 3 \times 10^{54}$ erg such as GRB 990123 at $z = 1.6$ [2]. Despite this large dispersion in total energy, there is no significant change in other overall properties of GRBs such as spectra or light curves. If all GRBs are triggered by a similar event, it is somewhat strange that GRB total energies have such a wide dispersion. The difference of beaming may explain this dispersion; however, in this case it is expected that the GRB luminosity and afterglow luminosity is correlated while observations suggest that there is almost no correlation between the two.

Currently the most popular explanation for the GRB phenomenon is dissipation of the kinetic energy of ultra-relativistic bulk motion with a Lorentz factor of $\Gamma \gtrsim 10^{2-3}$ in internal shocks generated by relative velocity differences of relativistic shells ejected from a central engine. All the total energy ejected as relativistic bulk motion cannot be dissipated in internal shocks, and hence the total energy truly emitted as kinetic motion ($E_{\rm iso}$) should be larger than the observed $E_{\gamma,\rm iso}$, at least by a factor of several. Therefore, some GRBs must emit quite a large amount of energy, $E_{\rm iso} \gtrsim 10^{55}$ erg. If the efficiency of the internal shock is not so high, we may have to consider an isotropic energy reaching $\sim 10^{56}$ erg. Therefore, if GRBs are produced by stellar death events, GRBs must be strongly beamed at least by a factor of 100. In this paper we refer to the isotropic energy $E_{\rm iso}$ for convenience, but a strong beaming is implicitly assumed to reduce the actual energy emission.

Here we propose a theoretical model to explain the wide dispersion of total GRB energies, in which all GRBs emit roughly the same amount of energy, $E_{\rm iso} \sim 10^{55-56}$ erg with the same beaming factor. The observed difference of total GRB energies is attributed to difference of efficiency of gamma-ray production. This model predicts strong emission above TeV energies from some fraction of GRBs, in which the total energy emitted as very high energy (VHE) gamma rays is comparable with or even larger than that in the ordinary sub-MeV range. In this paper we briefly describe this model, and then try to interpret the interesting report by the Milagro group in this conference, which suggests strong emission of TeV gamma rays from GRB 970417a with a TeV fluence at least 10 times greater than the sub-MeV fluence [3].

THE PROTON-SYNCHROTRON MODEL FOR THE ENERGETICS OF GRBS

Full description of this model has already been given in Ref. [6], and here I summarize the qualitative features of the model. Since the origin of the GRB energy is relativistic bulk motion, protons should carry a much larger amount of energy than electrons by a factor of $m_p/m_e \sim 2,000$, at least in the initial stage of the internal shock generation. It is uncertain what fraction of the proton energy is converted into electrons, but the simplest Coulomb interaction cannot transfer the proton energy into electrons within the time scale of GRBs. The soft γ-rays are generally considered to be generated by electrons, because of the short time variability of GRBs. Therefore it is not unreasonable that, in some GRBs, only $(m_e/m_p) \sim 10^{-3}$ of the total kinetic energy is carried by electrons and then emitted as soft γ-rays. On the other hand, it may also be possible that a physical process works as an energy conveyor from the hidden energy reservoir (i.e., protons) into electrons (or positrons). If the energy transfer is almost complete in a GRB, a significant fraction of $E_{\rm iso}$ can be radiated as soft γ-rays.

We suggest that this new energy transfer channel from protons into electrons is $e^{\pm}$-pair creation by very high energy photons of proton-synchrotron. When $E_{\rm iso} \gtrsim 10^{55}$ erg, synchrotron radiation of protons accelerated to 10^{20} eV can be a very efficient emission process because the cooling time of such protons is comparable with the typical GRB duration (~ 10 s). The energy of these synchrotron photons for an observer is about 1–10 TeV, and strong TeV emission from GRBs is possible. On the other hand, such proton-synchrotron photons may interact with low-energy electron-synchrotron photons and create $e^{\pm}$ pairs. The opacity of this reaction is of order unity, and strongly depends on the Lorentz factor of GRBs, as $\tau \propto \Gamma^{-5}$ in a simple model. The GRB luminosity is determined by the efficiency of energy transfer from protons into $e^{\pm}$ pairs, i.e., the opacity of pair-creation reaction. If the energy transfer is sufficient, we observe a strong GRB such as GRB 990123, while we observe a weak GRB such as GRB 970508 in case of almost no energy transfer. A modest dispersion in Γ from one GRB to another results in drastic change in the energetics of GRBs in sub-MeV range. It can also be shown that the photon energy band of the synchrotron radiation of the created pairs becomes $\sim$ MeV, giving an explanation for the sub-MeV GRB phenomenon. This mechanism may be similar to a see-saw between sub-MeV and TeV energies, in which the total kinetic energy of GRBs is roughly the same for all GRBs and difference of GRB energetics is whether the dominant emission is at TeV or MeV energies.

AN INTERPRETATION OF THE MILAGRITO RESULT

The Milagro group reported that they found an excess of gamma rays above several hundred GeV from GRB 970417a out of the 54 BATSE GRBs in the field of view of the Milagrito detector (proto-type of Milagro), and estimated the chance probability of this excess after examining 54 GRBs as 1.5×10^{-3} [3]. If this signal is truly from the GRB, the TeV fluence must be at least 10 times greater than the sub-MeV fluence of this GRB. The impact on the GRB energetics would be quite strong. Our model has predicted such an extreme phenomenon, if the energy transfer from protons into electrons is inefficient [4,5]. Here we discuss whether the Milagrito result can be explained by our model.

In order to observe such TeV emission much stronger than sub-MeV emission, there are two necessary conditions in our scenario: 1) GRB 970417a must be faint in sub-MeV range like the GRB 970508 rather than the most energetic GRBs such as GRB 990123, and 2) the redshift of GRB 970417a must be relatively low. If GRB 970417a is a strong GRB in the sub-MeV range like the GRB 990123, the even larger total energy emitted in the TeV range cannot be explained by stellar death models even with a strong beaming. If the redshift of GRB 970417a is too high, TeV gamma rays would be strongly absorbed by the interaction with the cosmic infrared background radiation. The fluence of GRB 970417a measured by BATSE (> 20 keV) is 3.9×10^{-7} erg cm^{-2} which is not especially bright, and hence it seems possible that this GRB has relatively low redshift and small total energy.

Quantitative analysis requires the absorption optical depth of VHE gamma rays due to the cosmic infrared background. We have calculated this optical depth as a function of the source redshift and observed photon energy, by using a model of luminosity density evolution of stellar lights in the universe [7]. The emission component of dust is calculated assuming that the dust emission spectrum is the same as that of the solar neighborhood and the fraction of stellar light absorbed by dust is determined to reproduce the observed far infrared background radiation measured by the COBE satellite [8]. This model of optical depth is quantitatively consistent with other publications within the model uncertainties (e.g., [9,10]).

Figure 1 shows the total energies emitted from GRB 970417a in MeV and TeV bands as a function of redshift assumed for this GRB (solid lines). The energy in the MeV band is simply calculated by the observed BATSE fluence. We have calculated the energy in the TeV band taking into account the absorption of TeV gamma rays in the intergalactic field, as mentioned above. We estimated the optical depth of absorption at the observed photon energy of 200 GeV, considering that the threshold energy of the Milagrito is a few hundred GeV [3]. The attenuation factor $e^{-\tau}$ by this absorption is also shown in Fig. 1, by the dashed line (see the right-hand axis of the figure). The estimate of the fluence of TeV gamma rays observed by the Milagrito depends on the higher cut-off energy and power-law index of the spectrum (see Fig. 5 of Ref. [3]). We assume the power-law index of -1.5 (standard synchrotron index corresponding to the particle power-law index of -2), and estimated the cut-off energy as a function of redshift by the energy at which the optical depth of the intergalactic absorption becomes the unity.

The fraction (1/54) of GRBs possibly detected in the TeV band in all the BATSE GRBs in the field-of-view of the Milagrito is also important information. We should consider that the GRB 970417a was the nearest one among the 54 GRBs, because the opacity of intergalactic absorption of TeV gamma rays rapidly increases with

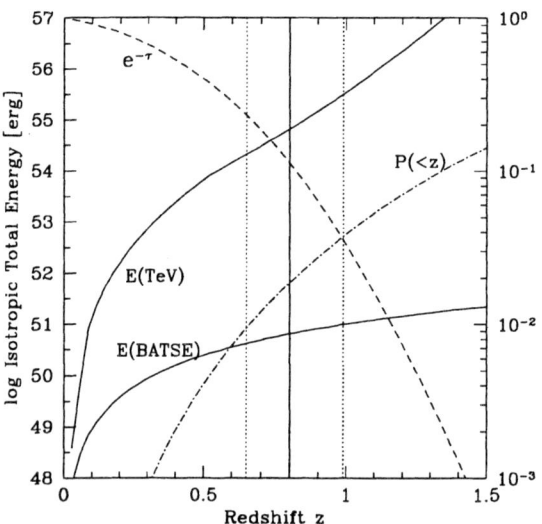

FIGURE 1. TeV/MeV energies of GRB 970417a and detection probability. See the right-hand axis for $e^{-\tau}$ (dashed) and $P(<z)$ (dot-dashed). (See text for detail.)

redshift (see dashed line of Fig. 1). Then we can estimate the likely redshift of this GRB from the event rate of $\sim 1/54$. In our model the total energy in the BATSE range is expected to widely distribute in $E_{\rm iso} \sim 10^{51}$–10^{54} erg, and hence we assume the total energy distribution in the $\log E_{\rm iso}$ space, $dN/d(\log E_{\rm iso})$, is constant in a range of $E_{\rm iso} = 10^{51}$–3×10^{54} erg. We also assume that the GRB rate is proportional to the star formation rate in the universe. Then we can calculate the probability that a GRB is nearer than a given redshift z, in the whole sample of the BATSE GRBs. (We have estimated the threshold fluence of the BATSE GRBs by the published BATSE catalog.) This is shown in Fig. 1 by the dot-dashed line (see the right-hand axis of the figure). We can infer from this result that the likely range of the redshift of GRB 970417a is $0.6 \lesssim z \lesssim 1$. (The vertical lines show the redshift corresponding to the detection rate $1/54$ with 1σ statistical uncertainties.)

It is interesting to note that the energy emitted in the TeV range is $E_{\rm iso} \sim 10^{54-55}$ erg for this likely redshift range. This energy scale is comparable with that of the most energetic GRBs such as GRB 990123, and hence this energy is not too large as the total energy budget of GRBs. In fact, our model assumes that all GRBs emit $E_{\rm iso} \gtrsim 10^{55}$ erg as the total kinetic energy, and considerable fraction of this energy is converted first into TeV gamma rays by proton-synchrotron radiation. If the opacity of the pair-creation reaction is not high, these TeV gamma rays are emitted without conversion into the MeV band. In this case the energy emitted in the MeV band by the synchrotron radiation of electrons originally loaded in the internal shock could be as low as $\sim 10^{-3}$ of the TeV band, because they carry only very small fraction of kinetic energy by the electron-proton mass ratio. We suggest that GRB 970417a with $E_{\gamma, \rm iso} \sim 10^{51}$ erg in the sub-MeV band was such a class of GRBs. Therefore, our model gives an explanation for the TeV and MeV fluence of GRB 970417a and the detection rate of BATSE GRBs by Milagrito.

REFERENCES

1. Metzger, M.R. et al., *Nature* **387**, 878 (1997).
2. Kulkarni, S.R., et al., *Nature* **393**, 35 (1999).
3. McEnery, J.E., et al., these proceedings (see also astro-ph/9910549).
4. Totani, T., *ApJ* **502**, L13 (1999).
5. Totani, T., *ApJ* **509**, L81 (1999).
6. Totani, T., *MNRAS* **307**, L41 (1999).
7. Totani, T., Yoshii, Y., & Sato, K., *ApJ* **483**, L75 (1999).
8. Hauser, M.G., et al., *ApJ* **508**, 25 (1998).
9. Salamon, M.H & Stecker, F.W., *ApJ* **493**, 547 (1998).
10. Primack, J.R., et al., *Astroparticle Phys.* **11**, 93 (1999).

Beaming and Jets in Gamma-Ray Bursts

Re'em Sari

Theoretical Astrophysics 130-33, California Institute of Technology, Pasadena CA 91125

Abstract. The origin of GRBs has been a mystery for almost 30 years. The afterglow observed in the last few years enabled redshift determination for a handful of bursts, and the cosmological origin is now firmly established. Though the distance scale is settled, there still remains orders of magnitude uncertainty in their rate and in the total energy that is released in the explosion due to the possibility that the emission is not spherical but jet-like. Contrary to the GRB itself, the afterglow can be measured up to months and even years after the burst, and it can provide crucial information on the geometry of the ejecta. We review the theory of afterglow from jets and discuss the evidence that at least some of the bursts are not spherical. We discuss the prospects of polarization measurements, and show that this is a powerful tool in constraining the geometry of the explosion.

I JETS? – A FUNDAMENTAL QUESTION

The study of γ-ray bursts was revolutionized when the Italian-Dutch satellite BeppoSAX delivered arcminutes positioning of some GRBs within a few hours after the event. This enabled other ground and space instruments to monitor the relatively narrow error boxes. Emission in X-ray, infrared, optical and radio, so called "afterglow", was observed by now for more than a dozen of bursts.

The current understanding of the GRBs phenomenon is that a compact source emits relativistic flow with Lorentz factor γ of at least a few hundreds. This flow emits, probably by internal shocks (see e.g., [2,3]), the GRB. After these internal shocks have produced the GRB, the ultra relativistic flow interacts with the surrounding medium and decelerates. Synchrotron radiation is emitted by the heated surrounding matter. As more and more of the surrounding mass is accumulated, the flow decelerates and the emission shifts to lower and lower frequencies. Excitingly, the afterglow theory is relatively simple. It deals with the emission on timescales much longer than those of the GRBs. The details of the complex initial conditions are therefore forgotten and the evolution depends only on a small number of parameters.

We begin by clarifying some of the confusing terminology. There are two distinct, but related, effects. The first, **"jets"**, describes scenarios in which the relativistic flow emitted from the source is not isotropic but collimated in a finite solid angle.

FIGURE 1. Schematic geometric description of jets in GRBs. The scheme shows the multiple shells before internal shocks have occurred. After that they all merge to one shell with typical width a factor of γ^2 thinner than their distance from the source.

The term jet refers to the geometrical shape of the relativistic flow emitted from the inner engine. The second effect is that of **"relativistic beaming"**. The radiation from any object that is radiating isotropically in its own rest frame, but moving with a large Lorentz factor γ in the observer frame, is collimated into a small angle $1/\gamma$ around its direction of motion. This is an effect of special relativity. It has nothing to do with the ejecta's geometry (spherical or jet) but only with the fact that the ejecta is moving relativistically. The effect of relativistic beaming allows an observer to see only a small angular extent, of size $1/\gamma$ centered around the line of sight. Unfortunately, the term beaming was also used for "jets" by many authors (including myself). We will keep a clear distinction between the two in this paper. Since we know the flow is ultra-relativistic (initially $\gamma > 100$), there is no question that the relativistic beaming is always relevant for GRBs. The question we are interested in is that of the existence of "jets".

The idealized description of a jet is a flow that occupies only a conical volume with half opening angle θ_0. In fact, the relativistic dynamics are such that the width of the matter in the direction of its propagation is much smaller than its distance from the source by a factor of $1/\gamma^2$. The flow, therefore, does not fill the whole cone. Instead it occupies only a thin disk at its base, looking more like a flying pancake [4] - see Figure 1. If the "inner engine" emits two such jets in opposite directions then the total solid angle towards which the flow is emitted is $\Omega = 2\pi\theta_0^2$. The question whether the relativistic flow is in the form of a jet or a sphere has three important implications.

The Total Emitted Energy. Optical observations of afterglows enabled redshift determination, and therefore a reasonably accurate estimate of the distance, D, to these events (the uncertainty is now in the cosmological parameters of the universe). The so called "isotropic energy" can then be inferred from the fluence F (the total observed energy per unit area at earth) as $E_{iso} = 4\pi D^2 F$ (taking cosmological corrections into account, $D = D_L/\sqrt{1+z}$ where D_L is the luminosity distance and z is the redshift). The numbers obtained in this way range from 10^{51}erg to 10^{54}erg with the record of 3×10^{54}erg held by the famous GRB 990123. These huge numbers approach the equivalent energy of a solar mass, all emitted in a few tens of seconds!

These calculations assumed that the source emitted the same amount of energy in all directions. If instead the emission is confined to some solid angle Ω, then the true energy is $E = \Omega D^2 F$. As we show later, Ω is very weakly constrained by the GRB itself and can be as low as 10^{-6}. If so, the true energy in each burst is $E \ll E_{iso}$. We will show later that interpretation of the multi-wavelength afterglow lightcurves indeed indicates that some bursts are jets with solid angles considerably less than 4π. The isotropic energy estimates may be fooling us by a few orders of magnitudes! Clearly this is of fundamental importance when considering models for the sources of GRBs.

The Event Rate. BATSE sees about one burst per day. With a few redshifts measured this translates to about 10^{-7} bursts per year per galaxy. However, if the emission is collimated to $\Omega \ll 4\pi$ then we do not see most of the events. The true event rate is then larger than that measured by BATSE by a factor of $4\pi/\Omega$. Again this is of fundamental importance. Clearly, the corrected GRB event rate must not exceed that of compact binary mergers or the birth rate of massive stars if these are to produce the majority of the observed GRBs.

The Physical Ejection Mechanism. Clearly, different physical models are needed to explain collimated and isotropic emission. For example, in the collapsar model (e.g., [1]), relativistic ejecta that is capable of producing a GRB is produced only around the rotation axis of the collapsing star with half opening angle of about $\theta_0 \cong 0.1$. Such models would have difficulties in explaining isotropic bursts as well as very narrow jets.

With these uncertainties we are therefore left with huge ignorance in how, how much and how many GRBs are produced. The question as to whether the emission of GRBs is spherical or collimated in jets is fundamental to almost all aspects of the GRB phenomenon.

II AFTERGLOW SPECTRUM – BASIC THEORY

When the ejecta interacts with the surrounding medium, a shock wave (the so called the forward shock) is going through the cold ambient medium and heating it up to relativistic temperatures. The basic afterglow model assumes that electrons are accelerated by the shock into a powerlaw distribution of their Lorentz factor γ_e: $N(\gamma_e) \sim \gamma_e^{-p}$ for $\gamma_e > \gamma_m$. The lower cutoff of this distribution is assumed to be a fixed fraction of equipartition. It is also assumed that a considerable magnetic field is being built behind the shock, again characterized by a certain fraction of equipartition. The relativistic electrons then emit synchrotron radiation and produce the observed afterglow. The broad band spectrum of such emission was given by Sari, Piran & Narayan [5] (see Figure 2).

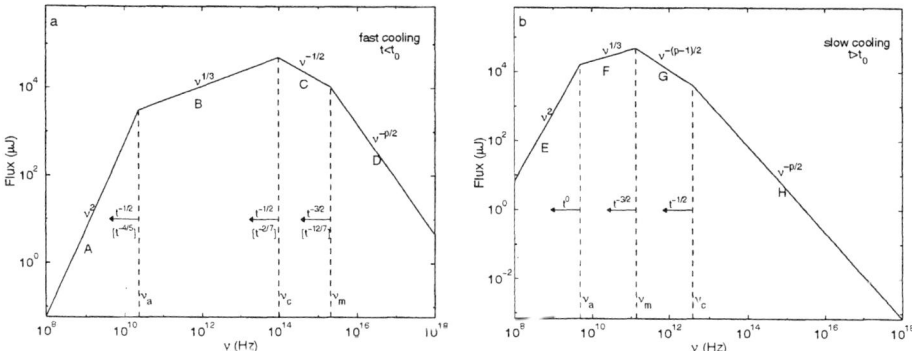

FIGURE 2. Theoretical spectra of synchrotron emission from fast cooling ($\nu_c < \nu_m$ left) and slow cooling ($\nu_m < \nu_c$ right) powerlaw distribution of electrons. This spectrum is robust and holds for jets as well as spherical ejecta. In general, the break frequencies change in time as well as the overall normalization. The arrows on the figure indicate the evolution of these break frequencies for a spherical emission in a constant density environment. $p = 2.2 - 2.4$ fits the observed spectra well.

At each instant, there are three characteristic frequencies. (i) ν_m is the synchrotron frequency of the minimal energy electron, having a Lorentz factor γ_m. (ii) The cooling time of an electron is inverse proportional to its Lorentz factor γ_e. Therefore, electrons with a Lorentz factor higher than some critical value $\gamma_e > \gamma_c$ can cool on the dynamical timescale of the system. This characteristic Lorentz factor corresponds to the "cooling frequency" ν_c. (iii) Below some critical frequency ν_a the flux is self absorbed and is given by the Rayleigh-Jeans portion of a black body spectrum. The broad band spectrum of the well studied GRB 970508 [6] is in very good agreement with the theoretical picture.

We stress that the spectrum given above is quite robust. The only assumption is synchrotron radiation from a powerlaw distribution of relativistic electrons. The same spectrum will hold whether the shock propagates into a constant density interstellar medium or a decreasing surrounding density produced earlier by the progenitor's wind. It will be valid whether the ejecta is spherical or jet-like, and whether the equipartition parameters are constant with time or not.

On the contrary, the temporal evolution of the spectrum is more subtle. The simplest evolution, which well describes the data of some bursts, is the spherical adiabatic model with a constant density ambient medium. In this scenario, $\gamma \sim R^{-3/2}$, or in terms of the observer time, $t = R/\gamma^2 c$, $\gamma \sim t^{-3/8}$. Given the evolution of $\gamma(t)$ one can derive the temporal evolution of the break frequencies, and the results are indicated in Figure 2. The peak flux in the adiabatic, spherical constant ambient density model is constant with time.

III HYDRODYNAMICS OF JETS

Interestingly, due to the effect of relativistic beaming (which is independent of jets) we are only able to see an angular extent of $1/\gamma < 0.01$ during the GRB itself where the Lorentz factor $\gamma > 100$. Moreover, it is only regions of size $1/\gamma$ that are causally connected. Therefore, each fluid element evolves as if it is part of a sphere as long as $1/\gamma < \theta_0$. Combining these two facts, we cannot distinguish a jet from spherical ejecta as long as $1/\gamma < \theta_0$.

However, as the afterglow evolves, γ decreases and it will eventually fall below the initial inverse opening angle of the jet. The observer will notice that some of the sphere is missing from the fact that less radiation is observed. This effect alone will produce a significant break, steepening the lightcurve decay by a factor of $\gamma^2 \sim t^{-3/4}$ even if the dynamics of each fluid element has not changed. The transition should occur at the time t_{jet} when $1/\gamma \cong \theta_0$. Observing this time can therefore provide an estimate of the jet's opening angle according to

$$t_{\rm jet} \approx 6.2(E_{52}/n_1)^{1/3}(\theta_0/0.1)^{8/3}\,{\rm hr}. \tag{1}$$

Additionally, Rhoads [7] has shown that at about the same time (see however [8–10]), the jet will begin to spread laterally so that its opening angle $\theta(t)' \sim 1/\gamma$. The ejecta now encounters more surrounding matter and decelerates faster than in the spherical case. The Lorentz factor now decays exponentially with the radius and as $\gamma \sim t^{-1/2}$ with observed time. Taking this into account, the observed break is even more significant. The slow cooling spectrum given in Figure 2 evolves now with decreasing peak flux $F_{\nu,m} \sim t^{-1}$, and the break frequencies evolve as $\nu_m \sim t^{-2}$, $\nu_c \sim t^0$ and $\nu_a \sim t^{-1/5}$. This translates to a temporal decay in a given frequency as listed in Table 1.

The jet break is a hydrodynamic one. It should therefore appear at the same time at all frequencies – an achromatic break. Though an achromatic break is considered to be a strong signature of a jet, one should keep in mind that any other

TABLE 1. The spectral index β and the temporal index α as function of p for a spherical and a jet-like evolution. Typical values are quoted using $p = 2.4$. The parameter free relation between α and β is given for each case (eliminating p). The difference in α between a jet and a sphere is always substantial at all frequencies.

	Spectral Index β, $F_\nu \propto \nu^{-\beta}$	Light Curve Index α, $F_\nu \propto t^{-\alpha}$ sphere	jet
$\nu < \nu_a$	$\beta = -2$	$\alpha = -1/2$	$\alpha = 0$
$\nu_a < \nu < \nu_m$	$\beta = -1/3$	$\alpha = -1/2$	$\alpha = 1/3$
$\nu_m < \nu < \nu_c$	$(p-1)/2 \cong 0.7$	$\alpha = 3(p-1)/4 \cong 1.05$ $\alpha = 3\beta/2$	$\alpha = p \cong 2.4$ $\alpha = 2\beta + 1$
$\nu > \nu_c$	$p/2 \cong 1.2$	$\alpha = (3p-2)/4 \cong 1.3$ $\alpha = 3\beta/2 - 1/2$	$\alpha = p \cong 2.4$ $\alpha = 2\beta$

hydrodynamic transition will also produce an achromatic break. To name a few, the transition from relativistic to non-relativistic dynamics, a jump in the ambient density, or the supply of new energy from slower shells that catch up with the decelerated flow. However, the breaks produced by the transition from a spherical like evolution (when $1/\gamma < \theta_0$) to a spreading jet has a well defined prediction for the change in the temporal decay indices. The amount of break depends on the spectral regime that is observed. It can be seen from Table 1 that the break is substantial $\Delta\alpha > 0.5$ in all regimes and should be easily identified.

Finally we note that if jet's opening angle is of order unity, the total energy may still be about an order of magnitude lower than the isotropic estimate. However, in this case the break will be "hidden" as it will overlap the transition to non-relativistic dynamics. It was suggested that this is the case for GRB 970508 [11]

IV OBSERVATIONAL EVIDENCE FOR JETS

Evidence of a break from a shallow to a steep powerlaw was first seen in GRB 990123 [12,13]. Unfortunately the break was observed only in one optical band while the infrared data were ambiguous. Yet, the strongest evidence for this burst being a jet does not come from this optical break but rather from radio observations, as explained below. A famous and exciting event this year was the first detection of a bright (9th magnitude) optical emission simultaneous with GRB 990123 [14]. Another new ingredient in GRB 990123 is a radio flare [15]. Contrary to previous afterglows, where the radio peaks after a few weeks and then decays slowly, this burst had a fast rising flare, peaking around a day and then decaying quickly. Sari and Piran [16] have shown that the bright optical flash and the radio flare are related. Within a day the emission from the adiabatically cooling ejecta that produced the 60s optical flash shifts into the radio frequencies. Given this interpretation, the regular forward shock emission should have come later, on the usual few weeks timescale. The fact that this "usual" forward shock radio emission did not show up is in agreement with the interpretation of this burst as a "jet" which causes the emission to considerably weaken by the time the typical frequency ν_m arrives to radio frequencies.

GRB 990510 had a very clear break simultaneously in all optical bands and in radio [17,18]. In GRB 990123 and GRB 990510 the transition times were about 2.1 days and 1.2 days reducing the isotropic energy estimate by a factor of ~ 200 and ~ 300, respectively. The total energy is now well below a solar rest mass!

Sari, Piran & Halpern [19] have noted that the observed decays in GRB afterglows that do not show a break are either of a shallow slope of $\sim t^{-1.2}$ or a very steep slope of $\sim t^{-2.1}$. They argued that the rapidly decaying bursts are those in which the ejecta was a narrow jet and the break in the light curve was before the earliest observation. Interestingly, evidence for jets are found when the inferred energy E_{iso} (which does not take jets into account) is the largest. This implies that jets may account for a considerable fraction of the wide luminosity distribution seen in

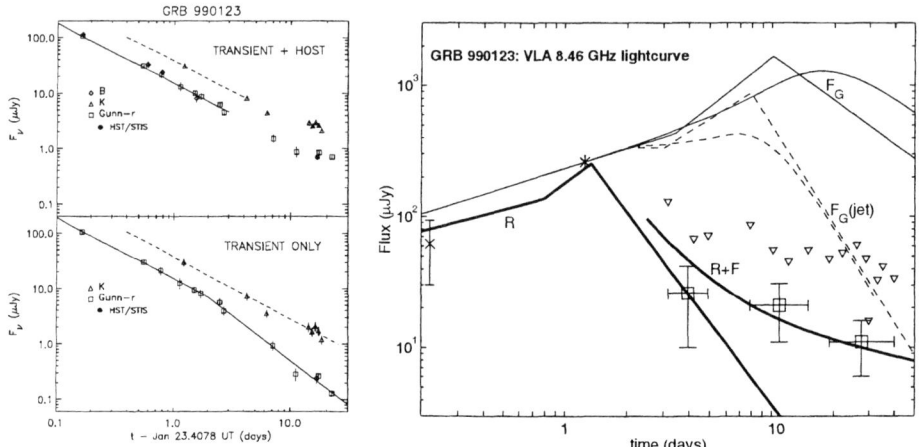

FIGURE 3. GRB990123: Optical data (left) shows some break in the light curve at Gunn-r band. K band seems to have no break but the contribution of the host galaxy is less certain. Radio "flare" (right) seen a day after the burst agrees with theoretical scaling of the optical flash (heavy solid line marked R). In the jet interpretation, only faint radio emission is expected at late times as given by the heavy solid line marked R+F, in agreement with observations. Thin and dashed lines indicate the theoretical expectations if the radio signal at day two is interpreted as the forward shock (independent of the optical flash) and if jets are not taken into account. These will largely over predict the late radio upper limits [15], marked by triangles (see however [20]).

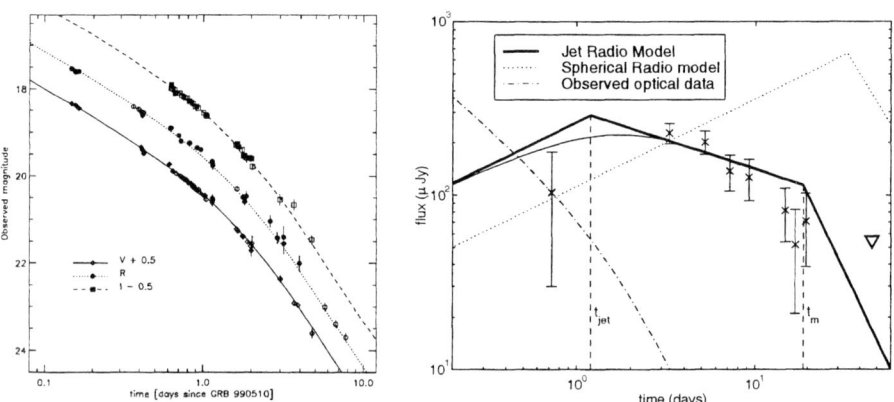

FIGURE 4. GRB 990510, the best evidence for a "jet": an achromatic break in optical and radio at $t_{jet} \cong 1.2$ days implying $\theta_0 \cong 0.08$. The temporal slope before and after the break agrees well with theory if $p \cong 2.2$. For this burst $E_{iso} = 2.9 \times 10^{53}$ erg but the true total energy is only $E = 10^{51}$ erg.

GRBs, and that the true energy distribution is less wide than it seems to be.

An alternative explanation for these afterglows with fast decline is propagation into a medium with decreasing density, i.e., a wind produced earlier by the progenitor [21]. We favor the jet interpretation for two reasons: (i) decreasing density only enhances the decay by $t^{-1/2}$ for $\nu_m < \nu < \nu_c$ and does not enhance the decay at all for $\nu > \nu_c$ (with typical parameters the optical and certainly the x-ray bands are above ν_c). The rest of the needed effect, in the wind interpretation, is associated with a higher value of the electron powerlaw distribution index p ($p \cong 3$ instead of $p \cong 2.2 - 2.4$). Why should the value of p be different for shocks propagating into winds? With the jet interpretation one can explain all afterglows with a single value of p, as in [19]. (ii) The jets interpretation makes the luminosity distribution of GRBs more narrow, since evidence for jets is found in bright events. Clearly, this is circumstantial evidence. A more clear cut between these two possible interpretations can be made with the use of early afterglow observations, preferably at radio frequencies (see [22]).

In summary, there are several kind of afterglows:
Shallow decline: $\sim t^{-1.2}$ for as long as the afterglow can be observed. These are probably spherical or at least have a large opening angle (e.g., GRB 970508).
Fast decline: $\sim t^{-2.1}$ (e.g., GRB 980519 and GRB 980326). These are either narrow jets, in which the break was very early, or they have high values of p and propagate into decreasing density medium.
Breaks: Initially slow decline that changes into a fast decline. These are the best candidates for jets (e.g., GRB 990510).

V POLARIZATION - A PROMISING TOOL

An exciting possibility to further constrain the models and obtain a more direct proof of the geometrical picture of "jets" is to measure linear polarization. High levels of linear polarization are usually the smoking gun of synchrotron radiation. The direction of the polarization is perpendicular to the magnetic field and can be as high as 70%. Gruzinov & Waxman [23] and Medvedev & Loeb [24] considered the emission from spherical ejecta by which symmetry should produce no polarization on the average, except for fluctuations of order of a few percent. Polarization is more natural if the ejecta is a "jet" and the line of sight from the observer is within the jet but does not coincide with its axis. In this case, the spherical symmetry is broken [25-27], and the natural polarization produced by synchrotron radiation should not vanish. For simplicity, let's assume that the magnetic field behind the shock is directed along the shock's plane (the results hold more generally, unless the magnetic field has no preferred direction). The synchrotron polarization from each part of the shock front, which is perpendicular to the magnetic field, is therefore directed radially.

As long as the relativistic beaming angle $1/\gamma$ is narrower than the physical size of the jet θ_0, one is able to see a full ring and therefore the radial polarization

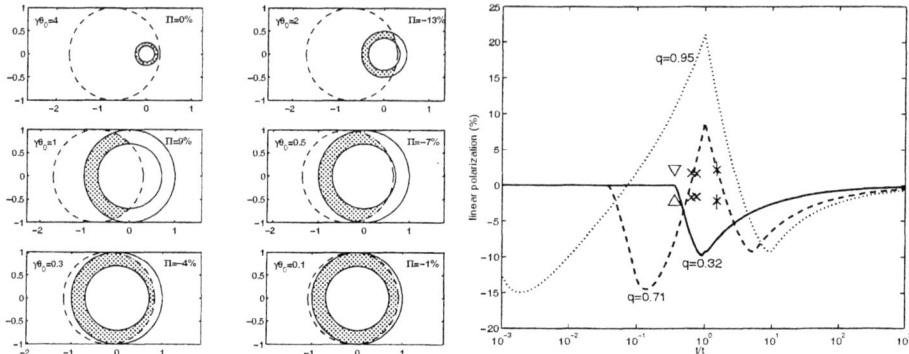

FIGURE 5. Left: Shape of the emitting region. Dash line marks the physical extent of the jet while solid lines give the viewable region $1/\gamma$. The observed radiation is coming from the gray region. On each frame, the percentage of polarization is given on the top right and the initial size of the jet relative to $1/\gamma$ is given on the left. The frames are scaled so that the size of the jet is unity. Right: observed and theoretical polarization lightcurve for three possible offsets of the observer relative to the jet axis. Observational data for GRB 990510 is marked by x, assuming $t_{jet} = 1.2\,\mathrm{d}$. The upper limit for GRB 990123 is given by a triangle, assuming $t_{jet} = 2.1\,\mathrm{d}$.

averages to zero (the first frame, with $\gamma\theta_0 = 4$ of the left plot in Figure 5). As the flow decelerates, the relativistic beaming $1/\gamma$ becomes comparable to θ_0 and only a part of the ring is visible; net polarization is then observed. Note that due to the radial direction of the polarization from each fluid element, the total polarization is maximal when a quarter ($\gamma\theta_0 = 2$ in Figure 5) or when three quarters ($\gamma\theta_0 = 1$ in Figure 5) of the ring is missing (or radiate less efficiently) and vanishes for a full and for a half ring. The polarization when more than half of the ring is missing is perpendicular to the polarization direction when less than half of it is missing.

At late stages the jet expands and since the offset of the observer from the physical center of the jet is constant, spherical symmetry is regained. The vanishing and re-occurrence of significant parts of the ring results in a unique prediction: there should be three peaks of polarization, with the polarization position angle during the central peak rotated by 90° with respect to the other two peaks. In case the observer is very close to the center, more than half of the ring is always observed, and therefore only a single direction of polarization is expected. A few possible polarization light curves are presented in Figure 5.

VI SUMMARY

Now when redshifts for GRBs are routinely measured, the largest uncertainty in their energy budget and event rate is the possibility that the emission is not spherical but jet-like. We discussed the theory of afterglow from jet-like event.

These should produce a substantial break at all frequencies. The time where this break occurs is an indication of the jet's opening angle. GRB 990510 seems to be a perfect example of this behavior. The inferred opening angle is about 0.1, consistent with upper limits from searches of orphan X-ray afterglows [28]. Several other candidates for jets are bursts with fast decline, where the break presumably took place before the earliest observation. This question will be settled when more frequent early observations are available. We have shown that afterglow from jets should show a unique signature of polarization at detectable levels. Observing such a signature will confirm the jet interpretation and the synchrotron model in general.

Acknowledgements. I thank Titus Galama for very useful comments, and the Sherman Fairchild foundation for support.

REFERENCES

1. MacFadyen, A. I., Woosley, S. E., *ApJ* **526**, 152 (1999).
2. Sari, R., & Piran T., *ApJ* **485**, 270 (1997).
3. Fenimore, E. E., Madras, C., & Nayakshine, S., *ApJ* **473**, 998 (1996).
4. Piran, T., in the proceedings of the Gräftåvallen workshop "Gamma Ray Bursts: The First Three Minutes", ed. Juri Poutanen (1999).
5. Sari, R., Piran, T. & Narayan, R., *ApJL*, **497**, L17 (1998).
6. Galama, T. J. et al., *ApJ* **500**, 101 (1998).
7. Rhoads, J. E., *ApJ* **525**, 737 (1999).
8. Panaitescu, A. & Mészáros, P., *ApJ* **503**, 314 (1999).
9. Mészáros, P., & Rees M. J., *MNRAS* **299**, L10 (1999).
10. Moderski, R., Sikora, M., Bulik, T., astro-ph/9904310 (1999).
11. Frail, D. A., Waxman, E. & Kulkarni, S. R., astro-ph/9910319 (1999).
12. Kulkarni, S. R. et al., *Nature* **398**, 389 (1999).
13. Fruchter, A. S. et al., *ApJ* **519**, L13 (1999).
14. Akerlof, C. et al., *Nature* **398**, 400 (1999).
15. Kulkarni, S. R. et al., *ApJL* **522**, L97 (1999).
16. Sari, R., & Piran T., *ApJL* **517**, L109 (1999).
17. Stanek, K. Z., Garnavich, P. M., Kaluzny, J., Pych, W. & Thompson, I., *ApJL* **522**, L39 (1999).
18. Harrison F. A. et al., *ApJL* **523**, L121 (1999).
19. Sari, R., Piran, T., & Halpern, J. 1999, ApJ, 519, L17
20. Galama, T. J. et al., *Nature* **398**, 394 (1999).
21. Chevalier, R. A. & Li, Z. Y., *ApJ* in press, and astro-ph/9908272.
22. Frail, D. A. et al., astro-ph/9910060 (1999).
23. Gruzinov A., & Waxman E., *ApJ* **511**, 852 (1999).
24. Medvedev, M. V., & Loeb A., astro-ph/9904363 (1999).
25. Gruzinov A., *ApJL* **525**, L29 (1999).
26. Ghisellini, G., & Lazzati, D., *MNRAS* **309**, L7 (1999).
27. Sari, R., *ApJL* **524**, L43 (1999).
28. Greiner, J. et al., these proceedings.

Photospheres, Comptonization and X-ray Lines in Gamma-Ray Bursts

P. Mészáros [1,2,3]

[1] Pennsylvania State University, 525 Davey, University Park, PA 16802
[2] California Institute of Technology, MS 105-24, Pasadena, CA 91125
[3] E-mail address: nnp@astro.psu.edu

Abstract. Steep X-ray spectral slopes, X-ray excesses and preferred spectral energy breaks in the 0.1–0.3 MeV range are discussed as the possible consequences of the photospheric component of the GRB relativistic outflow, and of pair breakdown in internal shocks leading to Comptonization on semi-relativistic electrons or MHD waves. We also discuss the X-ray and UV spectra of GRB afterglows occurring in a dense environment characteristic of massive stellar progenitors, including their ability to produce detectable Fe or other metallic line features.

PHOTOSPHERES, SHOCKS AND PAIRS

A significant fraction of bursts appear to have low-energy spectral slopes steeper than 1/3 in energy [1,2]. This has motivated consideration of a thermal or nonthermal [3,4] Comptonization mechanism, while leaving the astrophysical model largely unspecified. There is also evidence that the apparent clustering of the break energy of GRB spectra in the 50–500 keV range may not be due to observational selection [1,5,6]. Models using Compton attenuation [7] require reprocessing by an external medium whose column density adjusts itself to a few g cm^{-2}. More recently, a preferred break has been attributed to a blackbody peak at the comoving pair recombination temperature in the fireball photosphere [8]. In order for such photospheres to occur at the pair recombination temperature in the accelerating regime requires an extremely low baryon load. For very large baryon loads, a different explanation has been invoked [9] involving scattering of photospheric photons off MHD waves in the photosphere, which upscatters the adiabatically cooled photons up to the observed break energy.

Motivated by the above observations, these ideas have been synthesized [10] to produce a generic scenario in which the presence of a photospheric component as well as shocks subject to pair breakdown can produce steep low-energy spectra and preferred breaks (see Figure 1). In some of our previous work [11,12] considering photospheres and pair formation, their thermal character, the uncompensated photosphere redshift in the coasting phase, and the requirement of a power law extending to GeV energies were arguments in favor of a synchrotron and inverse Compton mechanism in shocks. The latter should, indeed, play a significant role in any model. However, a photosphere is always present, even if not always dominant. If the photosphere occurs in the accelerating regime where $\Gamma \propto r$, its energy is comparable to that of shocks which may occur further out, and the energy at which the black-body peak (T) is observed is in the "magic" range near 0.5 MeV, for $\eta \geq \eta_*$, where $\eta = L/\dot{M}c^2 \to \Gamma_f$ is the terminal bulk Lorentz factor and $\eta_* = (L\sigma_T/4\pi m_p c^3 r_o)^{1/4} \simeq 10^3 (L_{52} r_7^{-1})^{1/4}$. Both its peak energy and its total energy are lower if the photosphere occurs in the coasting phase ($\eta \leq \eta_*$). A steep low-energy spectral slope is provided by the Rayleigh-Jeans part of the photosphere, and a low-energy excess or terrace by its Wien part. A high-energy power law extending above this up to GeV requires, however, a separate explanation. One possibility is up-scattering of photospheric photons in the $\tau_T \gtrsim 1$ region by Alfvén waves, whose energy may be a fraction of the bulk kinetic energy [9]. This leads to a Comptonized broken power law spectrum (PHC) in xF_x ($x = h\nu/m_e c^2$) of slope 1 up to the "magic" break energy $x \lesssim 1$, and slope 0 up to $x \lesssim \eta$ above that (Fig. 1). The energy in this PHC wave-Comptonized component can be substantial relative to the photosphere, and equals the ratio of wave to bulk kinetic energy.

Above the photosphere, internal shocks are expected to occur [12], which would lead to a nonthermal Synchrotron/IC spectrum (S) in addition to the above. How-

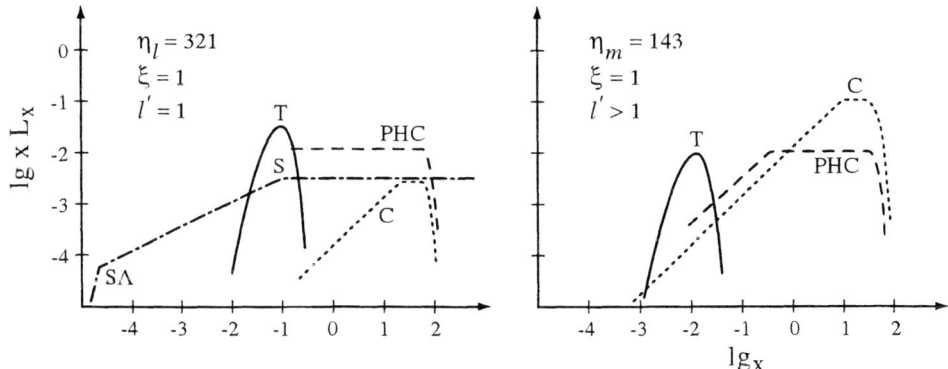

FIGURE 1. Luminosity per decade xL_x vs. $x = h\nu/m_e c^2$ for two values of $\eta = L/\dot{M}c^2$ and marginal (left) or large (right) pair compactness. T: thermal photosphere, PHC: photospheric Comptonized component; S: shock synchrotron; C: shock pair dominated Comptonized component [10]

ever, if the compactness parameter ℓ' (or comoving luminosity) is high, pair formation occurs which could produce a self-regulated low pair (comoving) temperature $\Theta'_p = kT'/m_e c^2 \sim 10^{-1}$ favoring Comptonization [13]. In this $\ell' \geq 1$ case, thermal Comptonization on the sub-relativistic electrons leads to another Comptonized component (C) of slope 1 up to an observer-frame energy $x \sim \Theta'_p \eta \sim 10^{-1} \eta$. Above this, if scattering off waves also occurs in the shocks, a second component of slope 0 would extend above it to $x \lesssim \eta$.

X-RAY AND UV LINE SPECTRA OF GRB

The environment in which a GRB occurs may also lead, in the afterglow phase, to specific spectral signatures from the external medium imprinted in the continuum, such as atomic edges and lines [14–16]. These may be used both to diagnose the chemical abundances and the ionization state (or local separation from the burst), as well as serving as potential alternative redshift indicators. (In addition, the outflowing ejecta itself may also contribute blue-shifted edge and line features, especially if metal-rich blobs or filaments are entrained in the flow from the disrupted progenitor debris [17], which could serve as diagnostic for the progenitor composition and outflow Lorentz factor). An interesting prediction [16] is that an Fe K-α X-ray *emission* line could be a diagnostic of a hypernova, since in this case one may expect a massive envelope at a radius comparable to a light-day where $\tau_T \lesssim 1$, capable of reprocessing the X-ray continuum by recombination and fluorescence (see also [18,19]). Detailed radiative transfer calculations have been performed to simulate the time-dependent X/UV line spectra of massive progenitor (hypernova) remnants [20], see Figure 2. Two types of hypernova environment geometries were considered, which are illuminated by a typical time-dependent broken power law afterglow continuum spectrum. One model consists of a dense shell, such as a supernova remnant, which could be the product of an inhomogeneous wind of variable velocity. This is essentially a transmission model, and produces initially an absorption X-ray line spectrum, turning later into an emission spectrum, in which for Fe abundances 10 or 100 times solar the Fe line luminosities are $\lesssim 10^{42} - 10^{43}$ erg s^{-1}. The other model assumes a funnel geometry and is essentially a reflection model, with an empty or low density region along an axis, such as would arise in a rotating stellar envelope or a wind. The fireball and the afterglow propagate inside this funnel, which acts as a channel that collimates and reflects the continuum. This results in an emission line spectrum (Fig. 2), where for 10 or 100 times solar abundances the Fe K-α line luminosity reaches $L_{Fe} \lesssim 10^{44}$ erg s^{-1}, with line and edge equivalent widths $EW \lesssim 1$ keV. This is comparable to the 3σ Fe features reported by two groups [21,22] in GRB 970508 and GRB 970828.

It is interesting that the Fe K-edge is significant in a funnel model such as shown in Figure 2. While the energy of an Fe line 6.7 keV feature in GRB 970508 agrees with its previously known redshift $z = 0.835$, the line feature of GRB 970828 would be in agreement with the 9.28 keV Fe K-edge energy at this object's newly reported

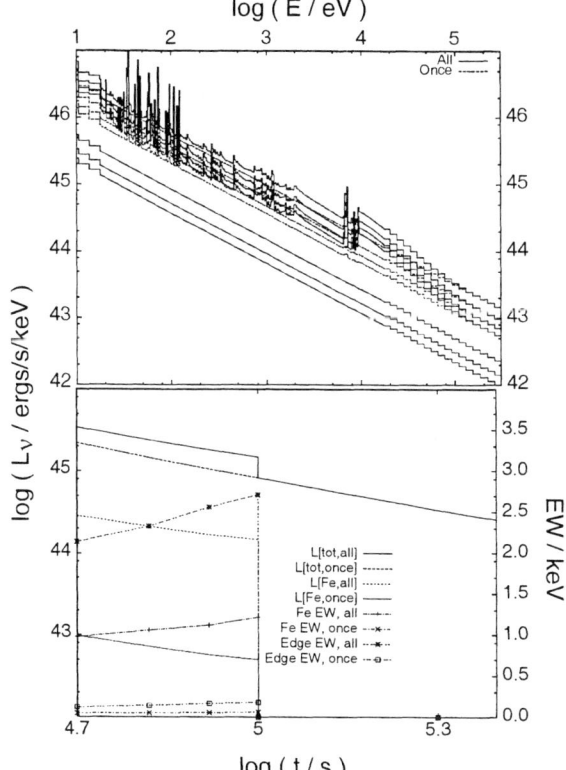

FIGURE 2. Model hypernova funnel spectrum (top) for observer times of 50, 66, 83, 100, 200, 300 ks top to bottom, and (bottom) the total and Fe light curves and equivalent widths [20], for $R = 1.5 \times 10^{16}$ cm, $n = 10^{10}$ cm^{-3}, and Fe abundance 10^2 times solar.

[23] redshift of $z = 0.958$. The line features in the 30–40 eV source-frame range seen in Figure 2 would be redshifted into the optical for $z \gtrsim 5$, but are likely to be blanketed by the Ly-α forest of intervening high-redshift galaxies. However, it may be possible to detect the soft X-ray metallic lines which become prominent soon after the Fe features, as the continuum softens and the gas cools, e.g., S and Si in the 2–3 keV source-frame range, or ~1–1.5 keV at $z \sim 1$.

ACKNOWLEDGMENTS

I am grateful to M.J. Rees, C. Weth and T. Kallman for stimulating collaborations, NASA NAG-5 2857, the Guggenheim Foundation and the Division of Physics, Math & Astronomy, the Astronomy Visitor and the Merle Kingsley funds at Caltech for support.

REFERENCES

1. Preece, R., et al., *ApJ* **496**, 849 (1998).

2. Crider, A., et al., *ApJ* **479**, L39 (1997).
3. Liang, E., et al., *ApJ* **491**, L15 (1997).
4. Liang, E., et al., *ApJ* **519**, L21 (1999).
5. Brainerd, J., et al., in *Abstr 19th Texas Symp*, Paris, astro-ph/9904039 (1999).
6. Dermer, C.D., et al., *ApJ* **515**, L49 (1999).
7. Brainerd, J., et al., *ApJ* **501**, 325 (1998).
8. Eichler, D. & Levinson, A., *ApJ*, submitted, astro-ph/9903103 (1999).
9. Thompson, C., *MNRAS* **270**, 480 (1994).
10. Mészáros , P. & Rees, M.J., *ApJ*, in press, astro-ph/9908126 (1999).
11. Mészáros , P., Laguna, P. & Rees, M.J., *ApJ* **415**, 181 (1993).
12. Rees, M.J. & Mészáros , P., *ApJ* **430**, L93 (1994).
13. Ghisellini, G. and Celotti, A., *ApJ* **511**, L93 (1999).
14. Bisnovatyi-Kogan, G. & Timokhin, A., *Astr. Rep.* **41**, 423 (1997).
15. Perna, R. & Loeb, A., *ApJ* **503**, L135 (1998).
16. Mészáros , P. & Rees, M.J., *MNRAS* **299**, L10 (1998).
17. Mészáros , P. & Rees, M.J., *ApJ* **502**, L105 (1998).
18. Ghisellini, G., et al., astro-ph/9808156 (1998).
19. Böttcher, M., et al., astro-ph/9809156 (1998).
20. Weth, C., Mészáros , P., Kallman, T. & Rees, M.J, *ApJ*, submitted, astro-ph/9908243 (1999).
21. Piro, L., et al., in *Proc. Rome GRB Conference, A&AS*, in press (1999).
22. Yoshida, A., et al., in *Proc. Rome GRB Conference, A&AS*, in press (1999).
23. Djorgovski, S.G., et al., *ApJ*, submitted (2000).

Modeling the Iron Line in GRB Afterglows

Markus Böttcher[1]

*Space Physics and Astronomy Department; Rice University, MS 108
6100 S. Main Street; Houston, TX 77005-1892; USA*

Abstract. The time and angle dependent yield of fluorescence line and continuum emission from a dense torus around a cosmological gamma-ray burst (GRB) source is simulated, taking into account photoionization, collisional ionization, recombination, and the self-consistent electron heating and cooling. A model calculation to reproduce the Fe Kα line emission observed in the X-ray afterglow of GRB 970508 indicates that $\sim 10^{-4}\,M_\odot$ of iron must be concentrated in a metal-enriched region of $R \lesssim 10^{-3}$ pc of extent. Similar scenarios from misaligned GRBs may result in observable X-ray flashes with strong emission line features.

INTRODUCTION

The recent marginal detection of a redshifted iron Kα emission line in the X-ray afterglow of GRB 970508 [1] has stimulated vital interest in the processes of photoionization and fluorescence line emission in gamma-ray burst (GRB) environments. X-ray absorption features and fluorescence line emission from the environments of cosmological GRBs have been investigated by several authors [2–4]. Motivated by suggestions that GRBs are caused by the death of a very massive star [5–7] (and are therefore likely embedded in the dense environment of a star-forming region), the influence of a dense, quasi-isotropic environment on the observable radiation, in terms of X-ray absorption features and fluorescence line emission, has been investigated [4,3].

However, the results of [8,3] clearly show that while a temporally varying Fe K absorption edge might be detectable, the luminosity and duration of the Fe Kα line observed in GRB 970508 [1], if real, is inconsistent with a quasi-isotropic environment.

A plausible way to solve this problem is the assumption of a dense torus surrounding the GRB source, which could be produced by anisotropic ejecta of the burst progenitor (e.g., [6,7]). An anisotropic geometry has recently been considered in [9]. In this geometry, the density of the torus may be so high that recombination and electron-impact ionization become important processes. In this paper, I

[1] Chandra X-Ray Observatory Fellow

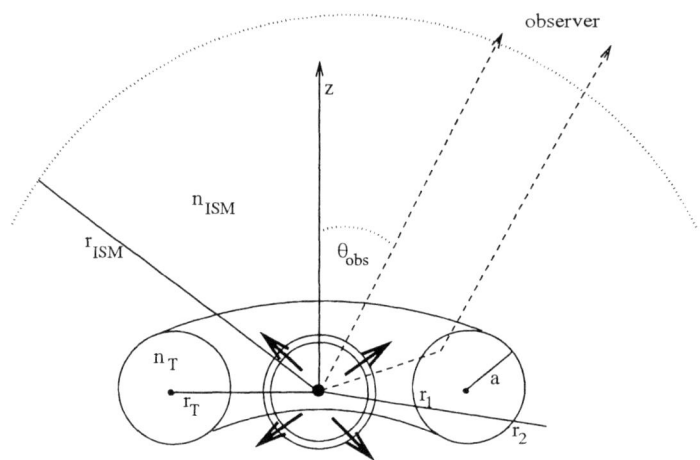

FIGURE 1. Sketch of the model geometry.

summarize the results of a detailed numerical study of the relevant processes in a dense torus illuminated by GRB radiation and hydrodynamically interacting with the relativistic blast wave triggered by the GRB. The physics included in the numerical study is first outlined, followed by a discussion of general considerations and a model calculation reproducing the (marginally) detected Fe Kα line in the X-ray afterglow of GRB 970508.

MODEL ASSUMPTIONS AND COMPUTATIONAL SCHEME

In [3], we have investigated the problem of time-dependent photoionization and radiation transport in the case of an isotropic, moderately dense GRB environment, where recombination and electron-collisional effects were negligible. In the situation investigated in this paper, such effects have to be considered carefully. Furthermore, the numerical problem is no longer isotropic. The code used in [3] has thus been modified in order to account for the anisotropy of the GRB environment.

The geometry assumed to treat this problem is illustrated in Fig. 1. The center of the coordinate system is the center of the GRB explosion. The GRB source is surrounded by a torus of dense material (particle density n_T), at a distance r_T from the center of the explosion. The radius of the cross-section of the torus is denoted a. The burst source and the torus are embedded in dilute ISM of density n_{ISM} which extends out to a radius r_{ISM} from the center of the explosion.

The ISM and the torus are illuminated by the time-dependent radiation field of a GRB, represented by the analytical expressions of [11]. The numerical scheme used to treat the time-dependent radiation transport problem is basically the same as described in [3], except that now the environment is anisotropic and more processes are included. To account for the anisotropy of the CBM, the expressions of [11] need to be modified because the blast wave will be decelerated much more efficiently in the dense torus than in the dilute ISM. When the blast wave begins to interact with the torus, it will produce an extremely short, extremely luminous flash of very high-energy radiation. The duration of this flash, as it would be measured by an observer located within or behind the torus, is $\Delta t_t \approx 1.1 \cdot 10^{-5}$ s $E_{54}/(n_{10}\,\Gamma_{300}^4\,r_{15}^2)$, where $E_{54} = E_0/(10^{54}\,\mathrm{erg})$, $n_{10} = n_T/(10^{10}\,\mathrm{cm}^{-3})$, $\Gamma_{300} = \Gamma/300$, and $r_{15} = r_1/(10^{15}\,\mathrm{cm})$.

As the parts of the blast wave interacting with the torus are decelerated to sub-relativistic velocities almost instantaneously, the material of the torus will be energized via shock heating as well as via illumination by the extremely luminous gamma-ray and X-ray flash. Heating by the sub-relativistic shock wave is taken into account using basic energy and momentum conservation arguments. Line and continuum emission resulting from atomic processes in the ISM or the torus is assumed to be emitted isotropically at each point. The time delay (due to the light travel time difference) of such radiation reaching the observer from directions misaligned with respect to the line of sight to the GRB source, is properly taken into account. The output spectra and light curves are sampled under different viewing angles θ_{obs} with respect to the symmetry axis of the torus.

In addition to the processes of photoelectric absorption, photoionization and fluorescence line emission following photoionization events, which had been included already in [3], now radiative and dielectronic recombination [12,13], electron-collisional ionization [13], electron heating and cooling due to bremsstrahlung, Compton scattering and Coulomb scattering, and continuum emission due to radiative recombination and bremsstrahlung emission are taken into account. 133 strong UV and X-ray lines due to radiative transitions following recombination into excited states of H, He, C, O, Ne, Mg, Si, S, Ca, Fe, and Ni have been included.

MODELING THE IRON LINE OF GRB 970508

GRB 970508 was a moderately bright burst with a peak flux of $\Phi_p \approx 3.4 \cdot 10^{-7}$ erg cm^{-2} s^{-1} and a duration of $t_\gamma \approx 15$ s in the energy band of the GRBM on board the BeppoSAX satellite (40 – 700 keV). The X-ray flux measured by the WFC exhibits a power-law decay ($F_\nu(t) \propto t^{-\chi}$) up to $t \sim 6 \cdot 10^4$ s after the burst with index $\chi = 1.17 \pm 0.1$, before a secondary outburst at X-ray energies occurs [10]. [1] have recently reported the marginal detection of a possible Fe Kα line with a line flux of $\Phi_L = (5\pm2) \cdot 10^{-5}$ cm^{-2} s^{-1} at the likely redshift of the burst, $2.0 \cdot 10^4$ s – $5.6 \cdot 10^4$ s after the burst trigger time.

When estimating the efficiency of reprocessing the illuminating GRB flux into Fe Kα line flux and thus estimating the amount of mass required to produce the observed fluorescence line, it is important to take the effects of anisotropy of the CBM into account, as described in the previous section. If the density anisotropy due to the torus is strong, the assumption of the dense torus being illuminated and

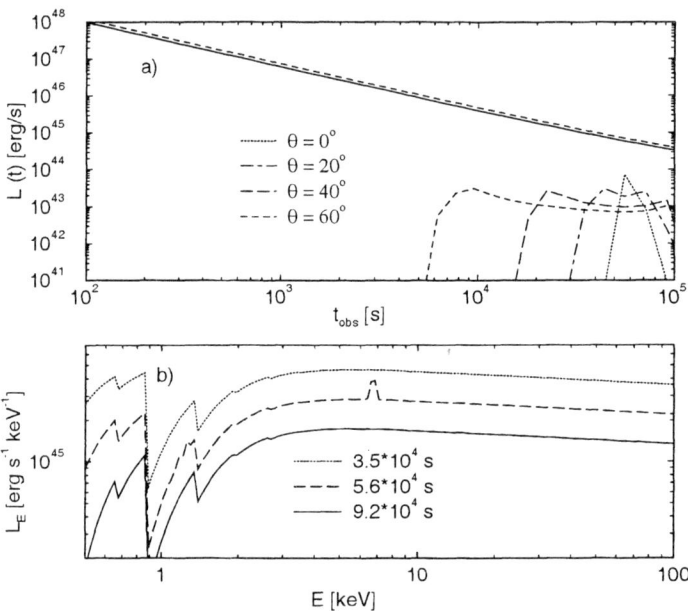

FIGURE 2. Results of a model calculation for GRB 970508. For parameters see text. a) Light curves in the iron line under different observing angles compared to the GRB continuum in two continuum energy channels. b) Simulated energy spectra as observed along the symmetry axis.

photoionized by radiation with the observed GRB characteristics will obviously yield a completely unrealistic picture.

The duration of the observed line emission is probably dominated by light-travel time effects rather than by the intrinsic time scale of the physical process involved, if the GRB and its afterglow are related to a relativistically expanding blast wave. A conservative estimate of the mass required to produce the observed iron line flux is based on the result of [3] that out to a radius of $R \sim 10^{20} \, (n_T[\text{cm}^{-3}])^{-1/3}$ cm iron in the CBM will be completely ionized. Taking into account the large number of Auger electrons ejected following photoionization of iron in low ionization states, a fiducial number of Kα line photons in the energy range 6.4 – 6.7 keV emitted in the course of complete ionization of an iron atom is ~ 5. Thus, for a line luminosity $L_{K\alpha} = 10^{44} L_{44}$ erg s^{-1} emitted over a time scale $\Delta t_L = 10^5 t_5$ s, a total of $N_{Fe} = 2 \cdot 10^{56} \, L_{44} \, t_5/f$ iron atoms is needed, where $f \geq 1$ is a correction factor accounting for the enhancement of the efficiency of line emission due to recombination and electron-collisional ionization. This yields a required mass of iron of $M_{Fe} = 0.16 \, (L_{44} \, t_5/f) \, M_\odot$. The line flux measured in the afterglow of GRB 970508 would thus lead to a mass estimate of $\sim 1 \, M_\odot$ of iron, which appears very unlikely to be realized in any well-understood astrophysical object.

Following [11], the spectral and temporal properties of GRB 970508 can be reasonably well reproduced assuming $\partial E/\partial \Omega = 3.6 \cdot 10^{52}$ erg/sr, $\Gamma_0 = 100$,

$n_{ISM} = 4.8 \cdot 10^5$ cm^{-3}, $q = 8.15 \cdot 10^{-5}$, $g = 1.6$, and photon spectral indices $\alpha_{le} = 2/3$ and $\alpha_{he} = 2.1$ below and above the peak, respectively. In a series of simulations with this model setup, the intensity of the iron line was found to be strongly positively correlated with the density of the torus, but only weakly dependent on the degree of anisotropy of the blast wave (before hitting the torus), as long as $\partial E/\partial\Omega(\text{torus}) \gtrsim 10^{-2}\, \partial E/\partial\Omega(\text{l.o.s.})$ and $\Gamma_{\text{torus}} \gtrsim 30$.

Fig. 2 shows the results of a model simulation assuming an intrinsically anisotropic blast wave, which is a factor of 100 less energetic in the direction of the torus than along the evacuated funnel. The torus has a total mass of 0.71 $M_\odot$, a density of 10^{12} cm^{-3}, and $r_T = 7 \cdot 10^{-4}$ pc. The material in the torus has a ten-fold iron overabundance with respect to solar-system abundances and contains a total of $2.2 \cdot 10^{-4}\, M_\odot$ of iron. The observed iron line intensity is very well reproduced with this model calculation.

In the scenario investigated here, the dominant cause of iron line emission from the torus results from its heating due to the X-ray and gamma-ray flash as the blast wave is decelerated in the torus, and the subsequent non-relativistic shock wave. Consequently, similar X-ray flashes with strong emission line features may be observable from such sources even when their symmetry axis is misaligned with respect to the line of sight, i.e., when they are not observed as GRBs. Such events may be observable with Chandra or with upcoming missions such as XMM or Astro-E.

ACKNOWLEDGMENTS

The work of M.B. is supported by NASA through Chandra Postdoctoral Fellowship grant PF 9-10007.

REFERENCES

1. Piro, L., et al., *ApJ* **514**, L73 (1999).
2. Mészáros, P., & Rees, M.J., *MNRAS* **299**, L10 (1998).
3. Böttcher, M., et al., *A&A* **343**, 111 (1999).
4. Ghisellini, G., et al., *ApJ* **517**, 168 (1999).
5. Woosley, S. E., *ApJ* **405**, 273 (1993).
6. Paczyński, B., *ApJ* **494**, L54 (1998).
7. Stella, L., & Vietri, M., *ApJ* **507**, L45 (1998).
8. Böttcher, M., et al., *A&AS* **138**, 543 (1999).
9. Lazzati,, D., et al., *MNRAS* **304**, L31 (1999).
10. Piro, L., et al., *A&A* **331**, L41 (1998).
11. Dermer, C. D., et al., *ApJ* **513**, 656 (1999).
12. Aldrovandi, S. M. V., & Péquignot, D., *A&A* **25**, 137 (1973).
13. Shull, J. M., & van Steenberg, M., *ApJS* **48**, 95 (1982).

Particle Acceleration at Ultra-Relativistic Shocks and the Spectra of Relativistic Fireballs

Yves A. Gallant[*,†], Abraham Achterberg[*],
John G. Kirk[‡], and Axel W. Guthmann[‡]

[*] *Astronomical Institute, Utrecht University, Postbus 80 000, 3508 TA Utrecht, Netherlands*
[†] *Dublin Institute for Advanced Studies, 5 Merrion Square, Dublin 2, Ireland*
[‡] *Max-Planck-Institut für Kernphysik, Postfach 10 39 80, 69029 Heidelberg, Germany*

Abstract. We examine Fermi-type acceleration at relativistic shocks, and distinguish between the initial boost of the first shock crossing cycle, where the energy gain per particle can be very large, and the Fermi process proper with repeated shock crossings, in which the typical energy gain is of order unity. We calculate by means of numerical simulations the spectrum and angular distribution of particles accelerated by this Fermi process, in particular in the case where particle dynamics can be approximated as small-angle scattering. We show that synchrotron emission from electrons or positrons accelerated by this process can account remarkably well for the observed power-law spectra of GRB afterglows and Crab-like supernova remnants. In the context of a decelerating relativistic fireball, we calculate the maximum particle energy attainable by acceleration at the external blast wave, and discuss the minimum energy for this acceleration process and its consequences for the observed spectrum.

INTRODUCTION

The spectrum of emission from GRB afterglows is well-accounted for by synchrotron emission from electrons accelerated at a decelerating relativistic blast wave. The mechanism responsible for the gamma-ray burst emission itself is less well-established, but it has been interpreted as synchrotron emission as well, from electrons accelerated either at internal, mildly relativistic shocks, or at the ultra-relativistic external shock as it runs into a clumpy medium.

In current models of afterglow emission, however (e.g., [1]), the particle acceleration physics is simply described by two parameters which are left to be adjusted to the observations: the shock is assumed to accelerate the electrons to a power-law spectrum of index p, with a lower cutoff $E_{\min}$ which is simply related to the efficiency of energy conversion into these accelerated electrons. In what follows, we first examine more closely the spectral index p that can be theoretically expected for

Fermi-type acceleration at relativistic shocks and compare it with observed values, and then consider the maximum and minimum energies over which this spectrum can extend, $E_{\max}$ and $E_{\min}$, and their consequences for observations.

FERMI ACCELERATION AT RELATIVISTIC SHOCKS

Shock-Crossing Kinematics and Energy Gain

In what follows, we restrict our attention to ultra-relativistic shocks, i.e., those with Lorentz factor $\Gamma_{\rm sh} \gg 1$ with respect to the upstream medium. For a weakly magnetized shock, the shock jump conditions then imply a relative Lorentz factor $\Gamma_{\rm rel} \approx \Gamma_{\rm sh}/\sqrt{2}$ between the downstream and upstream media, and a shock velocity of $c/3$ relative to the downstream medium.

Assuming a standard Fermi-type process at the shock, in which charged particles are deflected elastically by magnetic fluctuations in both the upstream and downstream media, the energy change of a particle in a single cycle of crossing and re-crossing the shock is given by

$$\frac{E_f}{E_i} = \Gamma_{\rm rel}^2 (1 - \beta_{\rm rel}\mu_{\to d})(1 + \beta_{\rm rel}\mu'_{\to u}). \tag{1}$$

Here E_i and E_f are the particle's initial and final energies, and $\mu_{\to d}$ and $\mu_{\to u}$ the cosine of its direction angle θ (between its velocity and the shock normal) upon crossing the shock into the downstream and upstream media, respectively. Throughout, primed and unprimed variables refer to quantities measured in the downstream and upstream rest frames, respectively.

Since kinematics require $1 \geq \mu'_{\to u} > 1/3$, the energy gain factor (1) will depend most sensitively on the distribution of $\mu_{\to d}$. For a pre-existing isotropic population of relativistic particles upstream, energy gains E_f/E_i of order $\Gamma_{\rm rel}^2$ can be achieved in the first shock crossing cycle. However, for particles having crossed the shock from downstream, realistic deflection processes upstream yield $\theta_{\to d} \lesssim 2/\Gamma_{\rm sh}$, so that for all subsequent shock crossing cycles, on average the particle energy is only roughly doubled by each cycle [2].

Numerical Simulations and the Spectral Index p

The power-law index of the accelerated particle spectrum depends on the average energy gain per shock crossing and on the return probability, the chance that a particle crossing downstream will eventually re-cross the shock upstream. Both these factors are strong functions of the angular distribution of particles crossing the shock, which, as suggested by the above considerations, is highly anisotropic. Thus, the quasi-isotropic approximations current in non-relativistic shock acceleration do not apply, and we turn to numerical simulations.

For simplicity, we focus our attention here on the case where both the upstream and downstream particle dynamics are dominated by scattering; in other words, we assume that magnetic fluctuations, possibly amplified by turbulence downstream, dominate the effect of the regular magnetic field for transport in the shock normal direction. Since the nature of the particle transport is by assumption independent of particle energy, we decoupled the dynamical problem from the energy gains, and first computed a numerical approximation to the function $f_d(\mu'_{\to u}; \mu'_{\to d})$, the distribution of downstream egress angles $\mu'_{\to u}$ for a given ingress angle, by Monte Carlo simulation of the downstream scattering process for a grid of $\mu'_{\to d}$ values. Along with a similarly obtained representation of the upstream dynamics, $f_u(\mu_{\to d}; \mu_{\to u})$, and the energy gain formula (1), these constitute the necessary ingredients for a Monte Carlo calculation of the accelerated particle distribution.

The results of such a calculation are shown in Figure 1(a). Particles are injected at E'_0 with $\mu'_{\to d} = -1$, but the influence of this highly anisotropic initial condition disappears after a little more than a decade in energy, where the self-consistent angular distribution is established with a smooth power-law dependence, $F(E', \mu') \propto F(\mu') E'^{-p}$. We obtained $p = 2.23$ for this case, in perfect agreement with the results of the semi-analytical eigenfunction method [3]. The angular distribution obtained with that method is compared with the simulation results in Figure 1(b), which shows both the asymptotic flux distribution $F(\mu')$ measured in the simulations and the corresponding density distribution, $n(\mu') \propto F(\mu')/(\mu' - 1/3)$. The agreement is excellent except near the loss cone, $\mu' \approx 1/3$, where the denominator amplifies Poisson noise in the measured Monte Carlo flux.

While the spectral index found above is valid only for pure scattering, simulations incorporating more complex transport dynamics, especially upstream where the effect of the regular magnetic field may well not be negligible, yield similar or only slightly steeper spectra in the same ultra-relativistic limit [4,5].

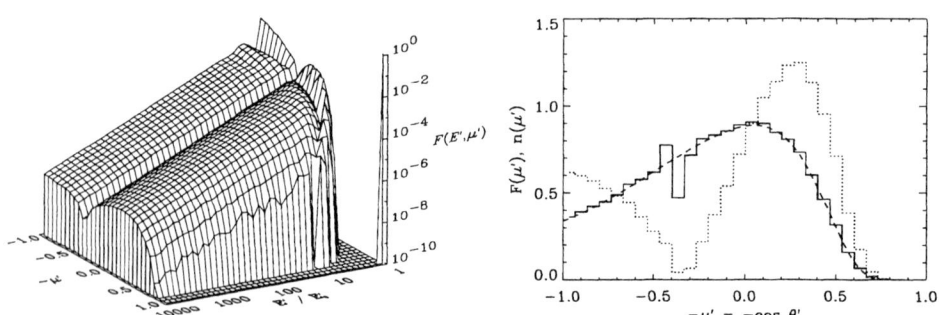

FIGURE 1. (a) Left panel: Steady-state flux distribution at the shock, $F(E', \mu')$, as a function of downstream particle energy E' and direction angle cosine μ'. (b) Right panel: Asymptotic angular distribution of the particles at shock crossing, expressed both in terms of flux $F(\mu')$ (dotted line) and density $n(\mu')$ (solid line), each normalized to unity. The dashed line shows the distribution $n(\mu')$ obtained by the eigenfunction method for the same case.

Comparison With Observations

Early observations of two GRB afterglows suggested $p = 2.3 \pm 0.1$ [6], and detailed analysis of the GRB 970805 afterglow spectrum yielded $p = 2.2$ [7]. While multi-wavelength spectral analyses have not been carried out in such detail for other afterglows, a value of $p \approx 2.2$ seems compatible with most [8]. Moreover, in another class of astrophysical objects where an ultra-relativistic shock is thought to accelerate particles, namely Crab-like supernova remnants, the inferred spectral indices are similar: the best-fit model for the Crab Nebula spectrum corresponds to p in the range 2.2–2.3 [9].

This spectral index might thus be considered a signature of ultra-relativistic shock acceleration, as the spectral index for acceleration at mildly relativistic shocks is expected to be different [3]. In this connection, it is intriguing to note that for GRB prompt emission, the average value of 2.12 for the high-energy spectral index [10], when interpreted as cooled synchrotron emission, corresponds to $p = 2.24$, in excellent agreement with the theoretical value found above. This could perhaps be viewed as spectral evidence that the prompt GRB emission also originates at an ultra-relativistic shock, such as the external shock, rather than a mildly relativistic one, such as internal shocks.

MAXIMUM AND MINIMUM ENERGIES

Age Limit and Ultra-high-energy Cosmic Rays

We first consider the maximum particle energy attainable by Fermi acceleration at a relativistic blast wave in the absence of energy loss processes; this is set by the constraint that the acceleration time $t_{\rm acc}$ be shorter than the age of the system. Since the fractional energy gain per shock crossing cycle is typically of order unity, the acceleration time is roughly the cycle time, which is the sum of the upstream and downstream residence times, t_u and t_d. When the downstream magnetic field is simply the shock-compressed upstream field, one can show that $t_d \sim t_u$; if the downstream field is amplified from this value, e.g., by turbulence, t_d is correspondingly shorter, so that the total cycle time is of order t_u. Comparing this with the age of the fireball yields a maximum energy which is attained at the beginning of the deceleration phase, and has value

$$E_{\rm max}^{\rm (age)} \simeq 5 \times 10^{15} B_{-6} \left(\frac{\mathcal{E}_{52} \Gamma_3}{n_0} \right)^{1/3} {\rm eV}, \qquad (2)$$

where B_{-6} is the upstream magnetic field, $\mathcal{E}_{52}$ the isotropic fireball energy, Γ_3 its initial Lorentz factor and n_0 the upstream density, respectively in units of μG, 10^{52} erg, 10^3, and cm^{-3}. This upper limit has important implications for the hypothesis that ultra-high-energy cosmic rays might be produced in GRBs [2].

Synchrotron Loss Limit and Spectral Upper Cutoff

While the above upper limit is appropriate for cosmic-ray protons, for the electrons responsible for the observed afterglow emission synchrotron losses must also be taken into account. This yields the additional criterion that the energy lost to synchrotron radiation in time t_u upstream and t_d downstream must be less than the energy gained per shock crossing cycle. If the downstream field is simply the compressed upstream one, synchrotron losses upstream and downstream will be comparable; otherwise, the downstream losses will dominate, so we need only consider the latter. Using this criterion yields another upper limit $E_{\max}^{(\mathrm{syn})}$, which is more stringent than (2) when the downstream magnetic field B' exceeds about $0.1\,\Gamma_3^{10/9}\xi^{2/3}$ G, with weak dependences on the other fireball parameters $\mathcal{E}_{52}$ and n_0, where ξ is the field amplification factor above simple compression, i.e., $B' \equiv \xi\sqrt{8}\,\Gamma_{\mathrm{sh}} B$.

The maximum synchrotron photon energy emitted by electrons of energy $E_{\max}^{(\mathrm{syn})}$ will be roughly 150 MeV in the proper (downstream) frame, independently of the value of the magnetic field. This B-independence is a generic result for acceleration times scaling like the gyro time, with longer t_{acc} yielding correspondingly lower maximum photon energies. Boosting to the observer's frame yields

$$h\nu_{c,\max}^{(\mathrm{syn})} \simeq 150\,\Gamma_3\ \mathrm{GeV}; \tag{3}$$

the same result also holds for synchrotron-limited acceleration at internal shocks. This suggests that establishing the presence of a cutoff to the synchrotron spectrum in the range of EGRET to TeV gamma-ray energies could place direct constraints on the fireball Lorentz factor.

Minimum Energy and Electron Pre-acceleration

The Fermi process requires that particles feel the shock as a discontinuity, which requires that their Larmor radius be larger than the shock thickness, which is in turn roughly the downstream thermal ion Larmor radius. This sets a minimum electron energy for Fermi acceleration $E'_{\min} \simeq \Gamma_{\mathrm{sh}} m_i c^2$, where m_i is the ion mass, corresponding to an observed synchrotron photon energy

$$h\nu_{c,\min} \simeq 160\,\xi\,B_{-6}\Gamma_3^4\ \mathrm{keV}. \tag{4}$$

It is interesting to note that for our fiducial parameters this falls in the BATSE break energy range, although it is unclear how the strong dependence on Γ_3, in particular, could keep $h\nu_{c,\min}$ constrained to a narrow range of values.

The energy $E'_{\min}$ exceeds by a factor m_i/m_e that resulting from randomization of the bulk upstream electron energy, and electrons must thus be pre-accelerated by some other process before Fermi acceleration can operate. One candidate for this pre-acceleration mechanism is the resonant ion cyclotron wave acceleration process

[11], which efficiently accelerates electrons over precisely the required energy range, and yields harder power-law spectra than those obtained above, providing a possible explanation for the BATSE low-energy spectral indices.

SUMMARY

We examined the particle energy gain per shock crossing cycle for Fermi-type acceleration at ultra-relativistic shocks, and found that while the initial shock crossing cycle can yield a very large energy gain, in all subsequent crossing cycles the particle energy on average roughly doubles. We used Monte Carlo simulations to obtain the spectrum of accelerated particles, and found a power-law spectral index $p = 2.23$ for the case of small-angle scattering, in agreement with the results of the semi-analytical eigenfunction method. This value is compatible with those inferred from observations of GRB afterglows and Crab-like supernova remnants, and might thus be considered a signature of ultra-relativistic shock acceleration.

For protons, the maximum energy attainable by Fermi acceleration at a relativistic blast wave is set by the age limit, and is of order 5×10^{15} eV for typical fireball parameters. For electrons, synchrotron losses can become the dominant limiting factor for moderately amplified downstream magnetic fields. In that case the maximum observed synchrotron photon energy is independent of the magnetic field, and is proportional only to the fireball Lorentz boost factor. Finally, electrons must be pre-accelerated to a minimum energy comparable with the thermal ions' before they can undergo Fermi acceleration; a good candidate for the pre-acceleration mechanism is the resonant ion cyclotron wave acceleration process.

Acknowledgments: This work was supported by the Netherlands Foundation for Research in Astronomy (ASTRON) under project 781-76-014, and by the European Commission under TMR programme contract ERBFMRX-CT98-0168.

REFERENCES

1. Sari, R., Piran, T., and Narayan, R., *ApJ* **497**, L17 (1998).
2. Gallant, Y.A., and Achterberg, A., *MNRAS* **305**, L6 (1999).
3. Guthmann, A.W., Kirk, J.G., Gallant, Y.A. and Achterberg, A., these proceedings.
4. Bednarz, J., and Ostrowski, M., *Phys. Rev. Lett.* **80**, 3911 (1998).
5. Gallant, Y.A., Achterberg, A., and Kirk, J.G., in *"Rayos Cósmicos 98"*, Proc. 16th European Cosmic Ray Symposium, ed. J. Medina, Alcalá de Henares, p. 371 (1998).
6. Waxman, E., *ApJ* **485**, L5 (1997).
7. Galama, T.J., Wijers, R.A.M.J., Bremer, M., et al., *ApJ* **500**, L101 (1998).
8. Frail, D.A., personal communication.
9. Kennel, C.F., and Coroniti, F.V., *ApJ* **283**, 710 (1984).
10. Preece, R.D., Pendleton, G.N., Briggs, M.S, et al., *ApJ* **496**, 849 (1998).
11. Hoshino, M., Arons, J., Gallant, Y.A., and Langdon, A.B., *ApJ* **390**, 454 (1992).

Afterglow Lightcurves From Beamed Outflows

A. Majczyna[1], T. Bulik[2], R. Moderski[2,3], and M. Sikora[2]

[1] *Astronomical Observatory, Warsaw University, Al. Ujazdowskie 4, 00478 Warsaw, Poland*
[2] *Copernicus Center, Bartycka 18, 00716 Warsaw, Poland*
[3] *JILA, University of Colorado, Boulder, CO 80309, USA*

Abstract. We study the dependence of relativistic outflow lightcurves on the initial parameters like the opening angle of the outflow, the initial bulk Lorentz factor, and varying the lateral flow velocity. To this end we use the numerical code by Moderski, Sikora, & Bulik (2000). We analyze the lightcurve of GRB 990510 and find that the data do not require a small opening angle of the outflow, $\theta_j > 0.3$ rad, while the initial bulk Lorentz factor must be above 30.

INTRODUCTION

Collimation of the relativistic outflow is frequently invoked in various models of GRBs in order to reconcile the observed luminosities with the available energy reservoirs. It has been noted by [4] that the lateral outflow of the ejecta can strongly influence the dynamics and the radiation of an afterglow. When the bulk Lorentz factor of the blast wave drops below the inverse of the initial opening angle of the outflow, the surface of the blast wave increases faster due to the lateral outflow than just due to the radial divergence. This causes faster deceleration of the blast wave and causes a faster decrease in the lightcurve.

This effect has been studied analytically and numerically [3,5,6]. Numerical calculations have shown that the break in the lightcurve could be very smooth and may even extend over almost two orders of magnitude in time. On the other hand, breaks in the lightcurves of GRB afterglows have been reported to occur at different times. Here we calculate afterglow lightcurves and find their dependence on the outflow parameters. We also confront the numerical model with observations of the lightcurve of GRB 990510.

OUTLINE OF THE MODEL

The deceleration of a blast wave is described by [1,6]

$$\frac{d\Gamma}{dm} = -\frac{\Gamma^2 - 1}{M}, \quad \frac{dM}{dr} = \frac{dm}{dr}[\Gamma - \xi_{rad}\xi_e(\Gamma - 1)], \quad \frac{dm}{dr} = \Omega_j r^2 \rho = 2\pi r^2(1 - \cos\theta_j)\rho,$$

where Γ is the bulk Lorentz factor of the blast wave, M is the total mass including internal energy, r is the distance from the central engine to the blast wave, dm is the rest mass swept up in the distance dr, ρ is the mass density of the external medium, ξ_e is the fraction of dissipated energy converted to relativistic electrons, ξ_{rad} is the fraction of electron energy which is irradiated, and θ_j is the angular size of the blast wave. This angular size is not constant but increases due to thermal expansion [4],

$$\theta_j \equiv \frac{a}{r} = \theta_{j0} + \frac{c_s}{c\Gamma},$$

where $c_s = c/\sqrt{3}$ is the sound speed in the relativistic plasma.

We assume that electrons are injected with a power law energy distribution $Q = K\gamma^{-p}$ and that the minimum energy of injected electrons is $\gamma_m = \frac{\xi_e \Gamma m_p}{m_e}$. The normalization of the injection function, K, is given by

$$\int_\gamma^{\gamma_{max}} Q \frac{dt'}{dr} \gamma m_e c^2 \, d\gamma = \frac{dE'_e}{dr} = \xi_e \frac{dm}{dr} c^2 (\Gamma - 1) = \xi_e \Omega_j r^2 \rho (\Gamma - 1) c^2,$$

where $dr = c\beta\Gamma dt'$.

Evolution of the electron energy distribution is given by continuity equation

$$\frac{\partial N_\gamma}{\partial r} = -\frac{\partial}{\partial \gamma}\left(N_\gamma \frac{d\gamma}{dr}\right) + Q,$$

where

$$\frac{d\gamma}{dr} = -f(r)\gamma^2 - \frac{\gamma}{r}, \tag{1}$$

are the electron energy losses. In both equations above, the derivatives over comoving time t' are replaced by derivatives over the distance r, according to the relation $\partial/\partial t' = c\beta\Gamma \partial/\partial r$,. The first term on the rhs of Eq. 1 represents synchrotron energy losses ($\to f(r) \propto B'(r)^2$), while the second one represents the adiabatic losses. B' is the magnetic field intensity as measured in the blast wave frame and is calculated from the formula $\frac{B'^2}{8\pi} = \xi_B^2 \kappa \rho c^2 \Gamma^2$, where κ is the compression ratio and ξ_B parameterizes the departure of the magnetic field intensity from its equipartition value.

We compute the evolution of the radiation spectrum in the blast wave comoving frame. This is given by

$$L'_{\nu'}(r) = \int N_\gamma(r) P(\nu', \gamma) d\gamma,$$

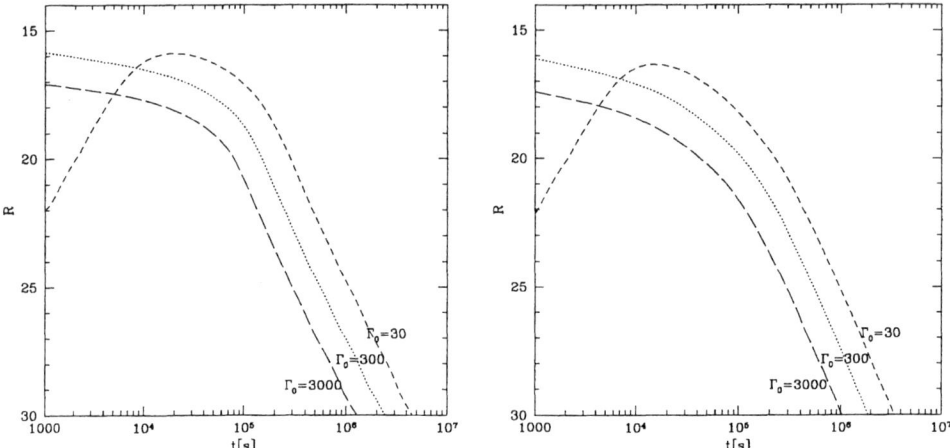

FIGURE 1. R-band lightcurves from outflows for three different values of the initial bulk Lorentz factor $\Gamma = 30, 300, 3000$ (short dashed, dotted, and long dashed lines, respectively), and with $\theta_J = 0.1$. The left panel corresponds to the case with no lateral outflow, and the right panel shows the case with lateral outflow velocity $c/\sqrt{3}$.

where $P(\nu', \gamma)$ is the power spectrum of synchrotron radiation of a single electron [2]. The apparent monochromatic luminosity as a function of time (a lightcurve) is calculated from

$$L_\nu(t, \theta_{obs}) = \iint_{\Omega_j} \frac{L'_{\nu'}[r(\theta)]\mathcal{D}^3}{\Omega_j} d\cos\theta d\phi,$$

where $\mathcal{D} = 1/\Gamma(1 - \beta\cos\theta)$ is the Doppler factor of the blast wave. The coordinates (θ, ϕ) are chosen so that the observer is located at $\theta = 0$ and the jet axis is at θ_{obs}. The integral is taken over the surfaces $t = \int \frac{(1-\beta\cos\theta)}{c\beta} dr = const$, enclosed within the blast wave boundaries.

RESULTS

We show the dependence of the afterglow optical lightcurve in the R band on Γ_0 in Figure 1. In the case when the lateral flow is neglected (left panel) the solutions tend to a power law when the outflow is decelerating. The maximum of the lightcurve, which corresponds to the transition from the coasting to the decelerating regime takes place the later for smaller initial bulk Lorentz factors. When the lateral outflow is taken into account (right panel of Figure 1) a break occurs in the lightcurves. As mentioned above this is due to faster deceleration of the blast wave because of the increase in the surface of the blast wave. It has been noted in [5] that the break is very smooth.

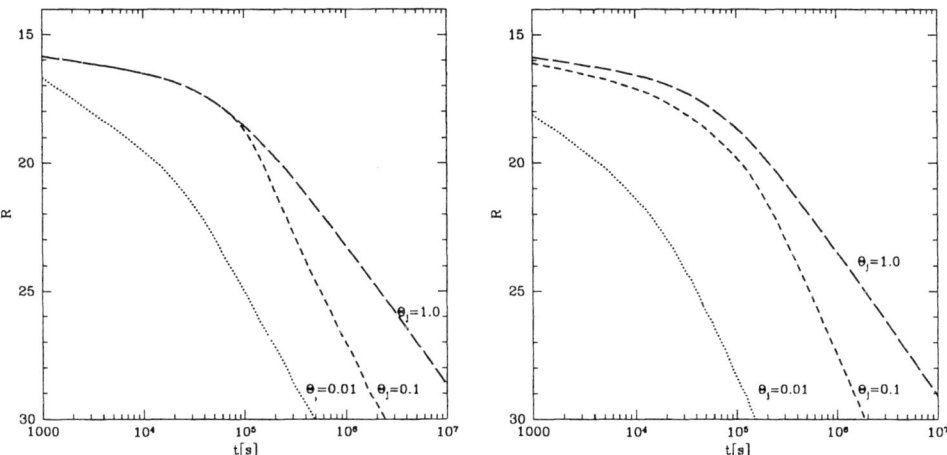

FIGURE 2. R-band lightcurves from outflows for three different values of the outflow initial opening angle: $\theta_j = 0.01, 0.1, 1.0$ rad. (long dashed, short-dashed and dotted lines, respectively), and for $\Gamma = 300$. The left panel corresponds to the case with no lateral outflow, and the right panel shows the case with lateral outflow velocity $c/\sqrt{3}$.

The dependence of the afterglow lightcurves on the jet opening angle θ_j is shown in Figure 2. When the lateral outflow is neglected (left panel) the shape of the lightcurve is determined by the time when the Doppler cone becomes larger than the opening angle of the jet. Until this time there is no observable difference between outflows with different collimation because we can not see outside the Doppler cone. Introduction of the lateral outflow smoothes out the curves and makes the cases with different opening angles different because of their different dynamics. The increasing area of the outflow in this case leads to the increase in the swept-up mass and to faster deceleration of the blast wave.

Observations of GRB 990510 provide a very good database to test dynamics and collimation of the outflow. We have collected data from different telescopes for this burst [7-9]. All observations have been rescaled to the R-band assuming a spectral index of -1. To analyze the data, we have calculated a two dimensional grid of models in the space spanned by the initial bulk Lorentz factor Γ_0, and the jet opening angle θ_j. We have considered a 9 by 9 grid with Γ_0 spaced in uniform logarithmic intervals between $\Gamma_0 = 10$ and $\Gamma_0 = 1000$; and θ_j spaced uniformly between $\theta_j = 0.01$ rad and $\theta_j = 1.0$ rad. The parameters of the blast wave used were: $E_0 = 10^{54}$ erg, $\xi_e = 0.03$, $\xi_B = 0.5$, and $\xi_{rad} = 1$. We follow the evolution of the blast wave from the initial radius of 10^{14} cm, and assume that the observer is located right on the axis of the outflow. The best fit was obtained for $\Gamma_0 = 31.6 \pm 10$, and $\theta_j = 0.3 \pm 0.1$ rad, where the error bars are given at the 1σ level. We present the data and our best fit model in Figure 3. The fit is quite good except for the

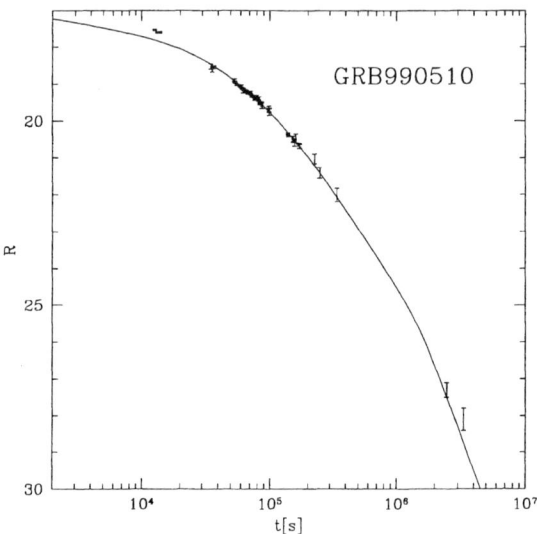

FIGURE 3. Our best fit model of the afterglow lightcurve of GRB 990510 in the R-band.

first observation at about $t = 1.5 \times 10^4$ s after the burst. After performing the fit we calculated the confidence contours in the Γ_0, θ_j plane and found that at the 2σ level the collimation angle of the outflow is not bounded from above. Also, the initial bulk Lorentz factor is not bounded from above: $\Gamma_0 > 30$ at the level of 2σ. Thus, we conclude that the data in GRB 990510 are consistent with a collimated jet. However, collimation is not required by the data. The data are also consistent with a model of a very wide, practically uncollimated jet.

Acknowledgements: This research has been supported by grants KBN 2P03D00415 and NASA 4690.

REFERENCES

1. Blandford, R.D., & McKee, C.F., *Phys. Fluids* **19**, 1130 (1976).
2. Chiaberge, M., & Ghisellini, G., *MNRAS* **306**, 551 (1998).
3. Mészáros, P., & Rees, M.J., *MNRAS* **306**, 39 (1999).
4. Moderski, R., Sikora, M., Bulik, T., *ApJ*, in press (2000).
5. Chiang, J., & Dermer, C.D., *ApJ* **512**, 699 (1999).
6. Rhoads, J.E. 1997, *ApJ* **487**, 1L (1997).
7. Israel, G. L., et al., *A&A* **348**, 5 (1999).
8. Stanek, K.Z, et al., *ApJ* **522**, 39L (1999).
9. Harrison, F.A., et al., *ApJ* **523**, 121 (1999).

The Patchy Shells Model

Tsvi Piran[†] and Pawan Kumar[*]

[†]*Racah Institute, The Hebrew University, Jerusalem 91904, Israel*
[*] *IAS, Princeton, NJ 08540, USA*

Abstract. We propose that the angular inhomogeneities within the relativistic flow cause the same GRB to look very different to different observers. Consequently the most energetic bursts do not correspond to an exceptional energy release but instead they correspond to accidental exceptionally bright spots along the line of sight on colliding shells. We describe the patchy shell model and calculate the distribution function of the observed fluence for bursts with random angular fluctuations of ejecta. We predict, according to this model, that the GRB luminosity function will be much wider than the afterglow X-ray luminosity function and only little correlation between the γ-ray fluence and the afterglow emission. These two predictions are confirmed by the GRB-afterglow data. We also predict that the early (minutes to hours) afterglow would depict large temporal fluctuations whose amplitude decreases with time. Finally we predict that there should be many weak bursts with average afterglow luminosity in this scenario.

The recent redshift determination of several GRBs revealed that the GRB luminosity function is very broad; the width of the fluence distribution is about two orders of magnitude in energy. Assuming isotropic emission, one finds that the energy of the most energetic bursts is larger than 10^{54} ergs. Beaming with an opening angle of a few degrees reduces the energy estimate by a factor of a hundred. However, when taking into consideration the relatively low efficiency of conversion of kinetic energy to γ rays [1–3], one finds that the total kinetic energy required is, after all, of order 10^{54} erg — even after the beaming correction. This energy is too large to be released by a compact solar mass object. According to the internal-external shock model, comparable amounts of energy should be released during the GRB phase and the afterglow. However, quite generally, only a fraction of the energy emitted as γ rays is seen in the afterglow (mostly as X rays). Moreover, there appears to be little correlation between the γ-ray fluence and the afterglow flux.

All the above mentioned phenomena could be unrelated. However, we suggest a possible connection within the relativistic fireball model (see [4] for a review). These properties could all be manifestations of large angular inhomogeneity of the relativistic ejecta. The size of a causally connected region of a shell of radius

R moving with Lorentz factor γ is $\sim R/\gamma$. Because of relativistic beaming this is also the size of the region visible to a distant observer. During the GRB the angular size of these regions (< 0.01 rad) is significantly smaller than the inferred angular width of the ejecta, $\delta\theta \sim$ a few degrees. There are therefore $(\gamma\delta\theta)^2$ causally disconnected regions within this cone. Thus, the observed γ-ray luminosity seen by different observers from the same burst could fluctuate strongly due to small-scale inhomogeneities in the emitting regions and one would over-estimate the energy release in γ rays in cases in which a hot spot has been observed if we simply multiply the observed fluence by the angular widths of the jet. The dispersion of the afterglow luminosity seen along different lines of sight should be smaller at a later time when the Lorentz factor of the ejecta has become smaller and the sizes of causally connected regions are larger. The emission at this stage should yield a better estimate of the overall total energy involved.

We have simulated, numerically, the radiation from a system of successive, randomly ejected blobs [6]. These blobs undergo multiple collisions. We compute the energy generated in the forward and the reverse shocks during each collision and the resulting synchrotron emission (including IC corrections). We keep the overall energy ejected into a given cone of opening angle $10°$ approximately the same — 10^{52} ergs. The total duration is 30 s. The Lorentz factor of each shell is assumed to be a random number and the energy distribution of blobs is taken to be either log-normal, with a mean of $10^{52}/N_b$ erg and width (FWHM) in $\log_{10}$(energy) of 1, or a delta function distribution; N_b is the number of blobs ejected in the explosion. The number of blobs ejected along a line of sight is also taken to be a random number uniformly distributed between N_{min} and N_{max}. We have considered two different values for the average number of shells ejected equal to 5 and 40. The computed emergent power spectrum is integrated between 10^1–10^3 keV, in the observer frame, to determine fluence along different lines of sight. From the 'observed' fluence we calculate the isotropic radiative energy in γ rays (it should be emphasized, however, that the radiation is highly anisotropic when shells are not uniform). The total energy in the explosion is obtained by adding the energy of all the blobs in a cone of $10°$ opening angle, which as stated previously is taken to be 10^{52} erg.

The resulting effective "isotropic" γ-ray ($10^1 - 10^3$ keV) fluence distribution is shown in Fig. 1. The full width at half maximum (FWHM) of fluence distribution function is about two orders of magnitude in the linear energy scale when the mean number of shells along the line of sight is 5. We have kept the total energy fixed. The distribution would have been wider if the total energy varies as well. However, such a variability would introduce a correlation between the GRB luminosity and the afterglow luminosity. The width of the fluence distribution function decreases with increasing number of blobs along the line of sight. The width of the distribution function is not sensitive to the burst duration; moreover, it is almost independent of the energy fraction in the magnetic field as long as ϵ_B is not so small that the shocks become almost adiabatic. It is also evident from Fig. 1 that there are a number of bright spots in the cone containing the ejecta for which the observed isotropic γ-ray fluence is 5×10^{53} erg.

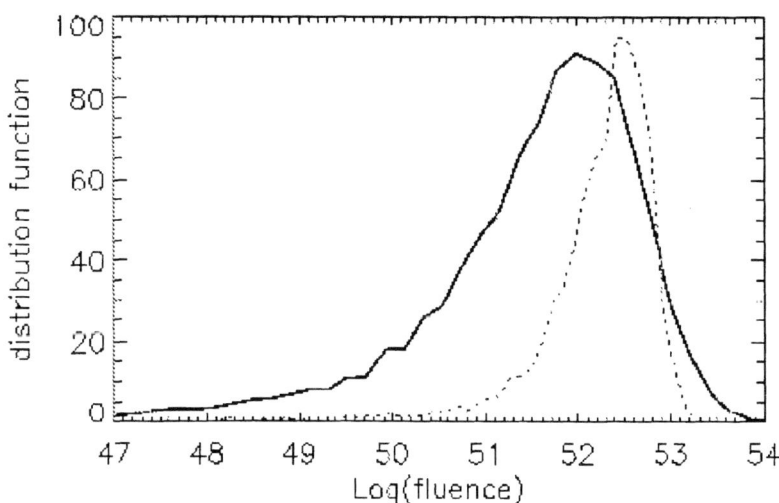

FIGURE 1. The distribution of GRB fluence in $10^1 - 10^3$ keV band. The mean number of shells ejected along a fixed direction is 5 for the solid curve and 40 for the dotted curve. The energy in the explosion, where material is ejected in a cone of opening angle $10°$, is taken to be 10^{52} erg. The burst duration is 30 s, and $\gamma_{min} = 5$ & $\gamma_{max} = 400$, in all cases shown in the figure.

This model suggests that γ-ray fluence is not a reliable measure of the total energy in the explosion. A more promising way to estimate the energetics of explosion is to consider emission at a later time when a number of causally disconnected regions have merged — but the shock is still radiative — thereby reducing the dispersion seen along different lines of sight. The ratio of the synchrotron cooling and the dynamical time is $t_s/t_d \approx 6\pi m_e c/(\sigma_T t_{obs} B^2 \gamma^2) \approx m_e/(\epsilon_B \sigma_T m_p c n_{ism} t_{obs} \gamma^4) \propto t_{obs}^{1/2}$, which becomes greater than 1 at $t_{obs} \sim 1$ hr. At this time $\gamma \sim 30$ and we expect about 20 disconnected regions to have merged, and the dispersion of the radiative flux along different lines of sights to have decreased by a factor of about 4. Note that the energy of the ejecta has also dropped by a factor of about 4 at this time.

To compare the model with GRB-afterglow observations we have estimated the GRB energy and the X-ray luminosity of bursts with observed afterglow and known redshift (see also [7]). Assuming a normal distribution of $log(E_\gamma)$ we find that the standard deviation of the logarithm of the (isotropic) energy emitted in γ rays, σ_γ, is 0.87. The average isotropic γ-ray fluence is 1.4×10^{53} ergs. The likelihood is larger than 0.05 of the maximal value within the range $0.45 < \sigma_\gamma < 2.4$. Similar analysis for the isotropic X-ray luminosity 5 hours after the burst for six bursts yields: $\sigma_x = 0.58$. The average (isotropic) X-ray luminosity is 1.3×10^{46} ergs/s. A variance range for likelihood larger than 0.05 of the maximal likelihood is: $0.35 < \sigma_x < 1.45$.

These results show that the FWHM of γ-ray energy distribution is wider by a factor of five than the X-ray afterglow luminosity distribution and is roughly consistent with the expected decrease in fluctuation amplitude by a factor of 7 based on the merger of causally disconnected regions (γ decreases by a factor of about 7 at 5 hr). If the γ-ray emitting surface were uniform (not highly patchy as considered here) and the large width of the isotropic γ-ray energy distribution were due to a wide distribution of the explosion energy (or the opening angle of the jet) then the distribution of afterglow luminosity in the X-ray and other wavelengths should have been wider than the γ-ray energy distribution since the afterglow flux $\log f_\nu = \frac{1}{4}(p+3)\log E + 0.5\log n - \frac{3}{4}(p-1)\log t_{obs} + constant$; where $p \approx 2.5$, n is the density of the circumstellar medium, E is the energy in the explosion per unit solid angle so long as the opening angle of the ejecta is larger than $\gamma^{-1}(t_{obs})$, and the *constant* term includes the dependence on ϵ_e and ϵ_B. Assuming that E and n are uncorrelated, we expect the width of the afterglow luminosity, $\log(f_\nu)$, to be larger than the width of the $\log E$ distribution by at least a factor of 1.4. Since the observed X-ray afterglow luminosity distribution is narrower by a factor of ~ 1.5 compared to γ-ray fluence (on log scale), this suggests that the distribution of E is very narrow[1] and the large width of the γ-ray fluence distribution arises as a result of angular fluctuation in the γ-ray emitting surface.

The $\sim 1\%$ radiative efficiency of internal shocks [3,5] in the $10^1 - -10^3$ keV energy band requires the total explosion energy to be larger than the observed energy in γ-ray photons by a factor of about 100. The finite opening angle for burst ejecta reduces the energy requirement by a factor of 10–100. The angular inhomogeneity of shells ejected in the explosion could further reduce the energy budget of the brightest bursts, such as GRB990123, by a factor of ~ 10 — thereby bringing down the total energy involved in the brightest observed bursts to a value of order the energy in weaker bursts (i.e., $\lesssim 10^{53}$ erg). It should noted that the energy requirement for an average GRB in the patchy shell and the uniform shell models are very similar; the largest difference, a factor of ~ 10 in the energy budget, arises only for the very brightest bursts.

An interesting result of this model is that in spite of the very wide observed luminosity function of GRB, the total energy in GRB explosions could be roughly comparable in all bursts. This could have interesting implications on the nature of the inner engine.

There are several predictions of our model. First, the width of the distribution function in the X-ray afterglow flux should be significantly smaller than the spread of fluence seen in γ rays. Moreover, the dispersion of optical luminosity should be smaller than the X-ray luminosity, and the late time radio afterglow should have the smallest dispersion which reflects the variation of energy in GRB explosions. The γ-ray, X-ray, and optical data for GRBs with known redshifts are consistent with these expectations. The observed decrease in the width of the isotropic luminosity distribution for the afterglow emissions, compared to the width of the isotropic γ-

---

[1] The ratio of E and the observed isotropic γ-ray fluence is a constant of order 100, almost independent of E, in the internal shock scenario when shells are uniform.

ray fluence distribution, is contrary to what is expected if the width of the fluence distribution were a consequence of a wide distribution of energy release in GRBs.

A second prediction is that the afterglow flux should show small-amplitude fluctuation with time if the energy distribution of blobs is not a delta function or the shells are not uniform. The early afterglow light curve should show larger fluctuations whose amplitude decreases in time. This prediction could be directly tested with the forthcoming quick-response GRB missions.

A third prediction is the existence of numerous weak bursts — which will arise from the low-energy tail of the GRB luminosity function. The afterglow flux from these week bursts should be comparable to the afterglow from stronger bursts — this may have implications to the rate of "orphan" afterglows.

A fourth prediction is that the fluence distribution of multi-peaked bursts (which arise due to numerous collisions) would be narrower than that of bursts with only a few peaks. Most of the bursts detected by BeppoSAX, for which afterglow emission and redshifts have been measured, show light curves consisting of a few peaks, which could arise as a result of collision of just a few shells. For such bursts we expect a very wide fluence distribution, as observed.

Finally, as the prompt optical and the prompt X-ray emissions arise in regions that are moving with very high Lorentz factor [8], we expect these emissions to also have a very wide luminosity function, whose width should be comparable to the GRB luminosity function (i.e., the prompt emission could be dominated by small hot spots and produce unusually large fluences in some cases). As mentioned earlier, we expect temporal fluctuations with a decreasing amplitude in time during this stage.

REFERENCES

1. Kobayashi, S., Piran, T., & Sari, R., *ApJ* **490**, 92 (1997).
2. Daigne, F. & Mochkovitch, R., *MNRAS* **296**, 275 (1998).
3. Kumar, P., *ApJ* **523**, L113 (1999).
4. Piran, T., *Physics Reports* **314**, 575 (1999).
5. Panaitescu, A., Spada, M., & Mészáros, P., *ApJ* **522**, L105 (1999).
6. Kumar, P. & Piran, T., *ApJ*, in press (2000).
7. Piran, T., Jimenez, R., & Band D., these proceedings.
8. Sari, R. & Piran, T., *ApJ* **520**, 641 (1999).

Hydrodynamics and Radiation from a Relativistic Expanding Jet with Applications to GRB Afterglow

Jonathan Granot*, Mark Miller[†], Tsvi Piran*, and Wai-Mo Suen[†]

*Racah Institute, Hebrew University, Jerusalem 91904, Israel
[†]Department of Physics, Washington University, St. Louis, MO, USA

Abstract. We describe fully relativistic three dimensional calculations of the slowing down and spreading of a relativistic jet by an external medium like the ISM. We calculate the synchrotron spectra and light curves using the conditions determined by the hydrodynamic calculations. Preliminary results with a moderate resolution are presented here. Higher resolution calculations are in progress.

INTRODUCTION

The level of beaming in GRBs is one of the most interesting open questions in this subject. The relativistic flow that drives a GRB may range from isotropic to strongly collimated into a narrow opening angle. The degree of collimation (beaming) of the outflow has many implications on most aspects of GRB research, such as the requirements from the "inner engine", the energy release, and the event rate. During the prompt emission the Lorentz factor of the flow is very high ($\gamma > 100$), and due to relativistic aberration of light, only a narrow angle of $\sim 1/\gamma$ around the line of sight (LOS) is visible. During the afterglow stage, the flow decelerates and an increasingly larger angle becomes visible. As long as $\gamma > 1/\theta$, an outflow collimated within an angle θ around the LOS produces the same observed radiation as if it were part of a spherical outflow. Once γ drops below $1/\theta$, the observer can notice that radiation arrives only from within an angle θ around the LOS, instead of $1/\gamma$ as in the spherical case. Sideways expansion is also expected to become important when $\gamma \sim 1/\theta$ [3,4]. These two effects combine to create a break in the light curve at $\gamma \sim 1/\theta$, but it is not quite clear whether this break is sharp enough to be detected [3-5]. To explore this question we have performed fully relativistic three dimensional simulations that follow the slowing down and the lateral expansion of a relativistic jet. We then calculate the synchrotron light curve and spectrum using the conditions determined by the hydrodynamical simulation.

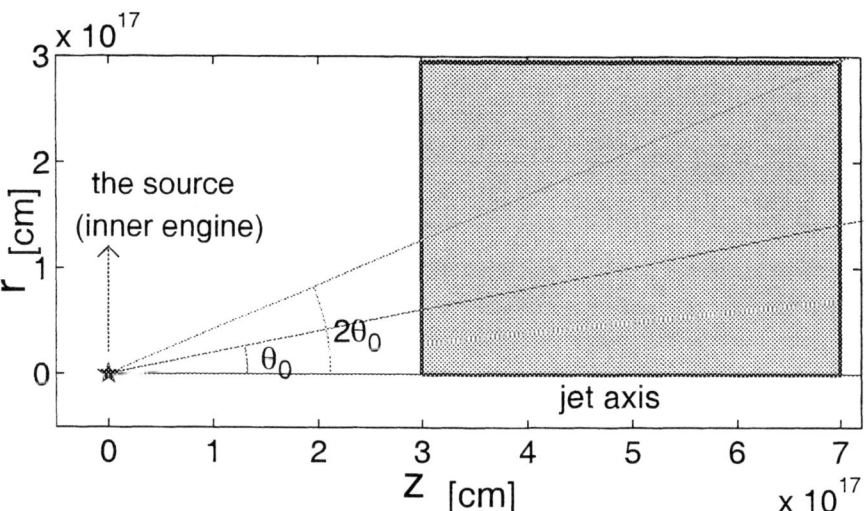

FIGURE 1. The region depicted in Figures (2–5) is within the shaded region. z and r are the cylindrical coordinates. The jet axis is in the z direction. Two lines which indicate opening angles of θ_0 and $2\theta_0$ are added to help follow the lateral expansion.

THE HYDRODYNAMICS

We use a fully relativistic three dimensional code for the hydrodynamical calculations. The initial conditions are a wedge with an opening angle $\theta_0 = 0.2$ taken from the Blandford McKee [1] (BM hereafter) self similar spherical solution, embedded in a cold and homogeneous ambient medium. For this initial opening angle, the jet is expected to show considerable lateral expansion when $\gamma \sim 1/\theta_0 = 5$, where γ is the Lorentz factor of the fluid. The BM solution used for the initial conditions was therefore at the time when $\Gamma = 10$, where $\Gamma = \sqrt{2}\gamma$ is the Lorentz factor of the shock. The total isotropic energy was $E = 10^{52}$ ergs, and the ambient number density was $n = 1$ cm^{-3}.

In Figures 2–5 we present snapshots of the number density, internal energy density, Lorentz factor and velocity field of the fluid, as the jet slows down and spreads sideways. An explanation of what is seen in these snapshots is given in Figure 1. The snapshots are taken at consecutive times, ranging $t = 137 - 282$ days in the rest frame of the ambient medium (corresponding roughly to observer times from one and a half to twenty days for an observer along the jet axis). The results shown in this work are still preliminary. The resolution is not sufficient to resolve the very thin initial shell. For this reason the maximal Lorentz factor of the matter is just over 3, instead of $10/\sqrt{2} \cong 7.07$ (see Figure 4). Higher resolution runs are in progress.

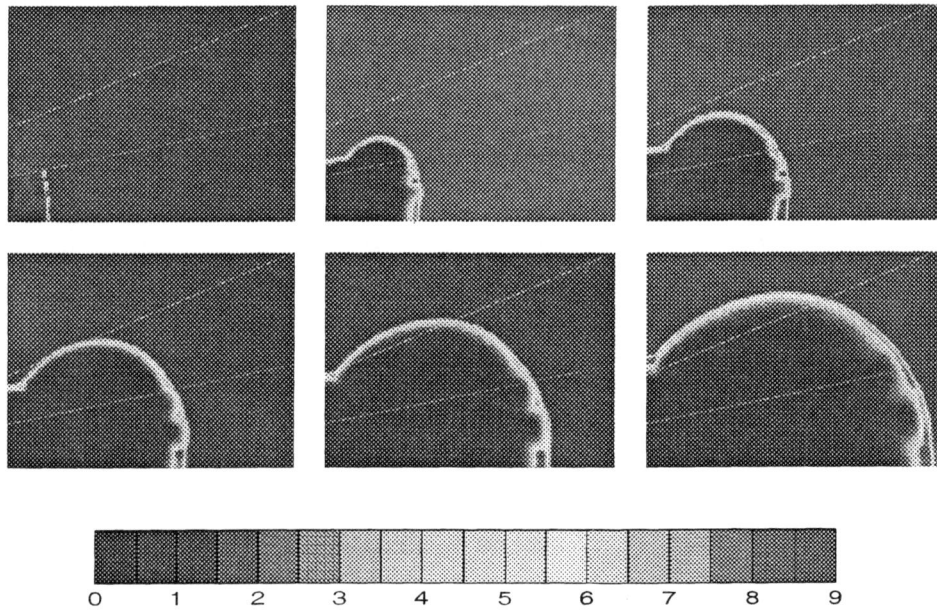

FIGURE 2. The proper number density, in cm^{-3}.

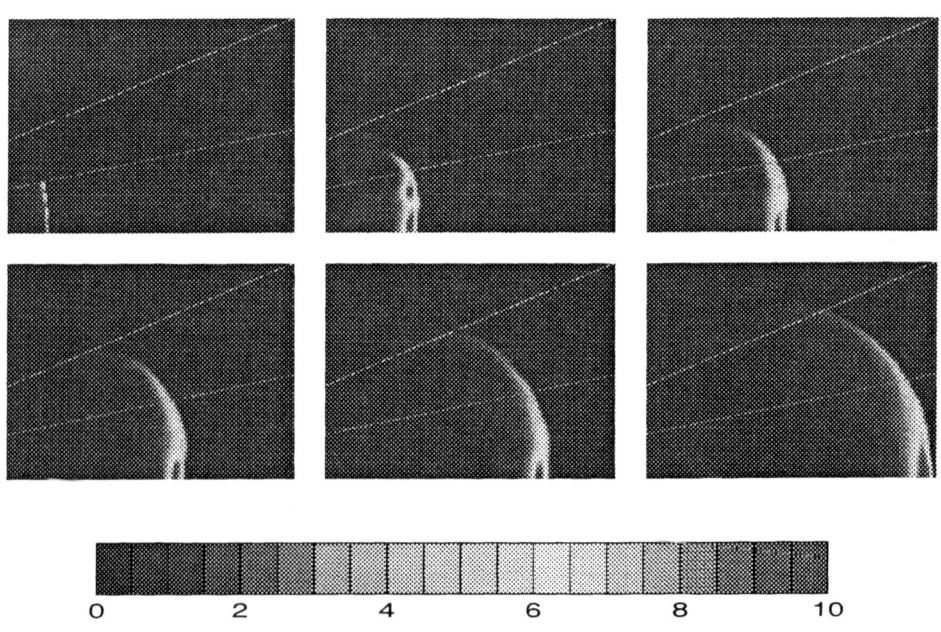

FIGURE 3. The internal energy density, in units of $m_p c^2/cm^3$.

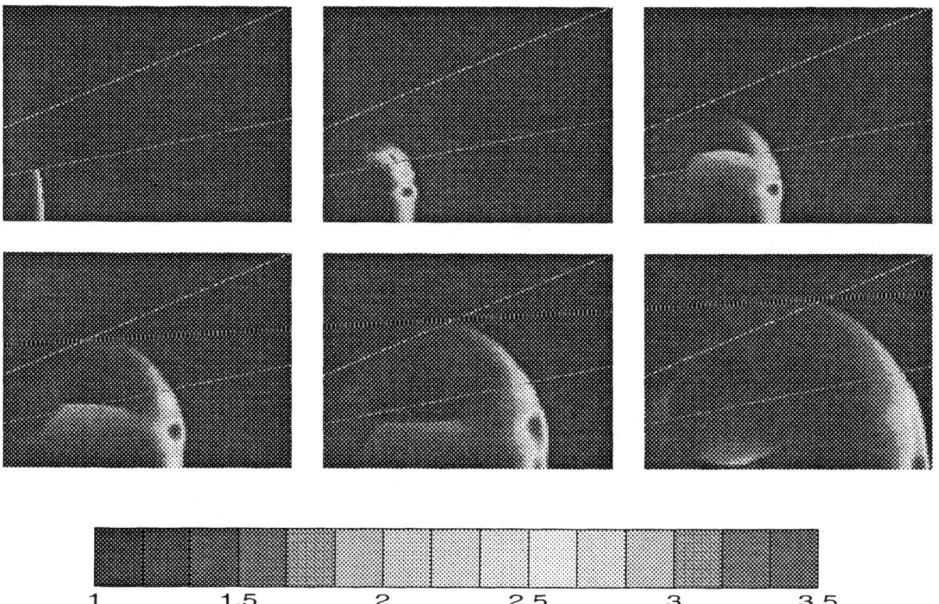

FIGURE 4. The bulk Lorentz factor of the fluid.

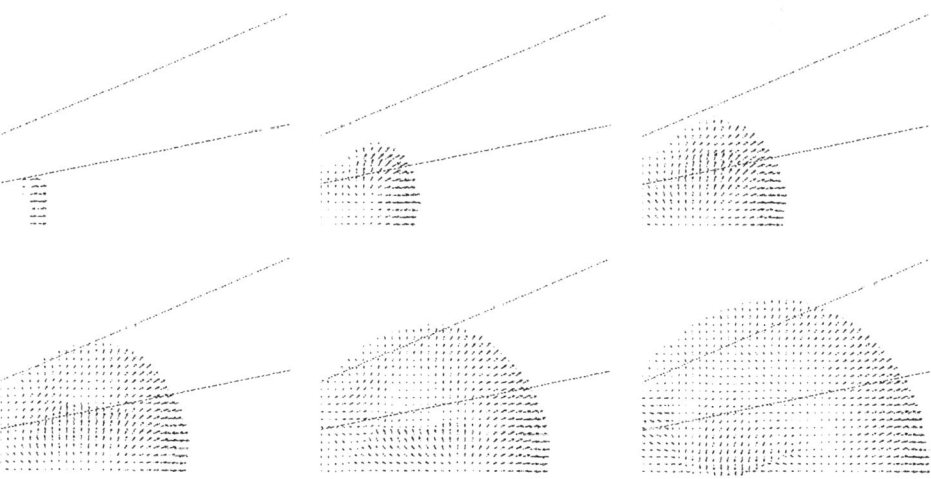

FIGURE 5. The velocity field of the fluid. Each arrow points at the direction of the velocity at the point where it begins, while its length is proportional to the size of the velocity.

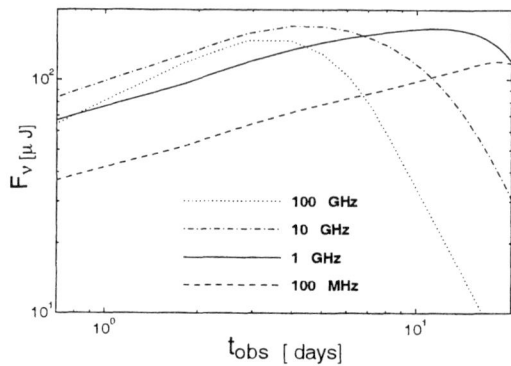

FIGURE 6. Light curves at different observed frequencies, as seen by an observer at a distance of 10^{28} cm along the jet axis.

CALCULATING LIGHT CURVES AND SPECTRA

The local emission coefficient at a given space-time point is calculated directly from the hydrodynamical quantities there. The magnetic field and electron energy densities are assumed to hold constant fractions, ϵ_B and ϵ_e, respectively, of the internal energy, while the electrons posses a power law energy distribution, $N(\gamma_e) \propto \gamma_e^{-p}$. The local emissivity is approximated by a broken power law: $F_\nu \propto \nu^{1/3}$ and $\nu^{(1-p)/2}$, below and above the typical synchrotron frequency, respectively [2].

The light curves and spectra are calculated for several viewing angles with respect to the jet axis. Once the emission coefficient is determined in the local frame, it is transformed to the frame of each observer. The time of arrival to each observer is then calculated, and the contributions are summed over space-time, producing the various light curves. This is done for several frequencies, simultaneously, so that the spectrum may be obtained, as well as the light curve at different frequencies.

A few light curves, calculated for $\epsilon_B = \epsilon_e = 0.1$ and $p = 2.5$, are shown in Figure (6). These light curves serve to demonstrate the potential of this approach. Future simulations are expected to achieve sufficient resolution to produce realistic light curves which could be compared with afterglow observations.

REFERENCES

1. Blandford, R.D. & McKee, C.F., *Phys. of Fluids* **19**, 1130 (1976).
2. Granot, J, Piran, T. & Sari, R., *ApJ* **513**, 679 (1999).
3. Rhoads, J.E., *ApJ*, submitted, astro-ph9903399 (1999).
4. Sari, R., Piran, T. & Halpern, J.P., *ApJ* **519**, L17 (1999).
5. Wei, D.M. & Lu, T., astro-ph9908273 (1999).

Early GRB Afterglows from Relativistic Blast Waves in General Radiative Regimes

Markus Böttcher[*][1] and Charles D. Dermer[†]

*Rice University, MS 108; 6100 S. Main Street
Houston, TX 77005 - 1892; USA
[†]Code 7653; NRL; 4555 Overlook Ave. SW
Washington, DC 20375-5352; USA*

Abstract. We present analytical expressions for the spectral evolution of the early afterglow radiation from gamma-ray bursts in radiative regimes intermediate between the adiabatic and the fully radiative solutions. Our expressions are valid as long as the relativistic electrons responsible for the observed synchrotron emission are in the fast cooling regime and the blast wave is relativistic. We show that even a slight deviation from a perfectly adiabatic evolution results in significant changes in the temporal characteristics of the afterglow emission.

INTRODUCTION

The relativistic blast wave model has met with considerable success in explaining the afterglows of cosmological GRBs. The dominant radiation mechanism responsible for the radio – X-ray emission from these relativistic blast waves is believed to be optically thin synchrotron emission from relativistic electrons accelerated behind the shock front [6,7,10]. The evolution of flux levels and characteristic frequencies of the synchrotron spectrum from electrons in a relativistic blast wave can be formulated in a very elegant way if the blast wave is either perfectly adiabatic or fully radiative [9]. However, while the comparison to afterglow data of several GRBs suggests that in the late afterglow phase GRB blast waves are well described by the adiabatic solution [4], they can obviously not be perfectly adiabatic in order to be observable, and they are believed to be strongly non-adiabatic in the early afterglow phase.

Self-consistent solutions of the blast wave dynamics under realistic assumptions on the energy transfer from protons to electrons and on the magnetic field evolution have until now only been possible numerically [3,5,8]. In this paper, we show that under certain conditions, the blast wave kinetic equation can be solved analytically,

---

[1] Chandra X-Ray Observatory Fellow

yielding a self-consistent analytical representation for the synchrotron emission from relativistic blast waves in general radiative regimes.

THE BLAST WAVE KINETIC EQUATION

In the relativistic blast wave model it is assumed that a total energy $E = 10^{52} E_{52}$ erg is deposited into a small volume, giving rise to a relativistic blast wave expanding with a bulk Lorentz factor Γ_0. The mass of the initial baryon-loaded ejecta is given by $M_0 = E_0/(\Gamma_0 c^2)$ in the comoving frame of the shocked material. As it expands into an external medium of density $n_{ext}(r) \propto r^{-\eta}$, it sweeps up matter at a rate $dm = \Omega r^2 n_{ext}(r) m_p dr$, where Ω is the solid angle element into which the blast wave is expanding. The kinetic equation governing the blast wave evolution is then given by [1,3]

$$\frac{d\Gamma}{dm} = -\frac{\Gamma^2 - 1}{M}, \qquad (1)$$

where M is the total, comoving mass (rest mass plus internal kinetic energy) of the ejecta + swept-up material. We assume that a fraction ϵ_e of the swept-up energy per unit time is transferred to relativistic electrons behind the shock front, and a fraction ϵ_{rad} of this energy is then radiated as synchrotron radiation. Thus, a fraction $\epsilon = \epsilon_e \epsilon_{rad}$ of the swept-up energy will be transformed into radiation. If the relativistic electrons are in the fast cooling regime, then $\epsilon_{rad} \approx 1$, and $\epsilon \approx \epsilon_e$ may be regarded as constant. In this regime, Eq. (1) can be solved analytically. As the blast wave propagates through the surrounding medium, its mass increases at a rate $dM = (\epsilon + \Gamma[1 - \epsilon]) dm$. Using this relation in Eq. (1), we find

$$\frac{m(r)}{M_0 A_0} = \int_{\Gamma}^{\Gamma_0} d\gamma \, \frac{(\gamma + 1)^\epsilon}{(\gamma^2 - 1)^{3/2}}, \qquad (2)$$

where $m(r) \equiv \Omega m_p \int_0^r d\tilde{r} \, \tilde{r}^2 n_{ext}(\tilde{r})$ and $A_0 \equiv \sqrt{\Gamma_0^2 - 1}/(\Gamma_0 + 1)^\epsilon$. The integral in Eq. (2) can be solved analytically in the relativistic limit. The asymptotic forms for the coasting and the deceleration phase, respectively, are given by

$$\Gamma_{rel}(r) \approx \begin{cases} \Gamma_0 & \text{if } (2 - \epsilon) m(r) \Gamma_0 \ll M_0 \\ a r^{\frac{3-\eta}{\epsilon - 2}} & \text{if } (2 - \epsilon) m(r) \Gamma_0 \gg M_0 \end{cases} \qquad (3)$$

where

$$a \equiv \left(\frac{[3 - \eta] E_0 \Gamma_0^{-\epsilon}}{[2 - \epsilon] \Omega n_{r_0} m_p c^2 r_0^\eta} \right)^{\frac{1}{2-\epsilon}}. \qquad (4)$$

Being interested in the spectral characteristics of the afterglow radiation, we will use the late-time asymptote of Eq. (3) in the remainder of this paper. Integrating

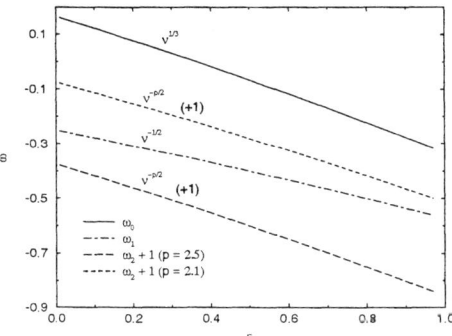

FIGURE 1. The temporal decay index ω_i in the three different spectral phases as a function of the radiative efficiency ϵ. The index ω_2 is shown for two different values of the electron injection spectral index p.

over time t in the observer's frame, using $dt = dr/(\Gamma^2 c)$, we find $R(t) = (a^2 c t/\beta)^\beta$ and $\Gamma(t) = a^\beta (ct/\beta)^\delta$, where $\beta = (2-\epsilon)/(8-2\eta-\epsilon)$ and $\delta = (\eta-3)/(8-2\eta-\epsilon) = (\beta-1)/2$. In the limits $\epsilon = 0$ and $\epsilon = 1$, these equations reproduce the well-known scaling laws for adiabatic and the radiative blast waves.

EVOLUTION OF SYNCHROTRON SPECTRA

We will derive the relevant scaling laws of the early afterglow emission here for a homogeneous external medium ($\eta = 0$) of density n_0 cm^{-3}. We are using the prescription for the evolution of the electron energy distribution and the magnetic field analogous to [9], which we are generalizing to arbitrary radiative regimes, using the scaling laws for $R(t)$ and $\Gamma(t)$ found in the previous section.

At any given time, the spectrum of electrons behind the shock will be given by a low-energy cutoff at γ_c, a power-law with index 2 for $\gamma_c < \gamma < \gamma_m$ and a second power-law with index $p+1$ for $\gamma > \gamma_m$. The corresponding synchrotron spectrum consists of a double-broken power-law given by

$$F_\nu = F_c \begin{cases} (\nu/\nu_c)^{1/3} & \text{for } \nu < \nu_c, \\ (\nu/\nu_c)^{-1/2} & \text{for } \nu_c < \nu < \nu_m, \\ (\nu_m/\nu_c)^{-1/2} (\nu/\nu_m)^{-p/2} & \text{for } \nu > \nu_m \end{cases} \quad (5)$$

where the characteristic synchrotron frequencies $\nu_i = (m_e c^2/h)(B/B_{cr})\gamma_i^2$ with $B_{cr} = 4.414 \cdot 10^{13}$ G. Now, normalizing the total number of swept-up, relativistic electrons in the shock to $N_e = (4/3)\pi n_{ext} R^3$, and the total radiative power to $L = (4/3)\Gamma c \sigma_T (B^2/[8\pi]) \int_{\gamma_c}^\infty N_e(\gamma) \gamma^2 d\gamma$, we find the relevant scaling laws for the flux normalization F_c and the break frequencies ν_c and ν_m:

$$F_c = f_{F_c}\, d_{28}^{-2}\, \epsilon_B^{1/2} \left(\frac{E_{52}}{\Omega}\right)^{\phi_E} \Gamma_0^{-\epsilon\phi_E}\, n_0^{\phi_n}\, t_d^{\phi_t}\ \mathrm{Jy}, \qquad (6)$$

$$\nu_m = \frac{f_{\nu_m}}{(1+z)}\, \epsilon_B^{1/2}\, \epsilon_e^2 \left(\frac{E_{52}}{\Omega}\right)^{\chi_E} \Gamma_0^{-\epsilon\chi_E}\, n_0^{\chi_n}\, t_d^{\chi_t}\ \mathrm{Hz}, \qquad (7)$$

$$\nu_c = \frac{f_{\nu_c}}{(1+z)}\, \epsilon_B^{-3/2} \left(\frac{E_{52}}{\Omega}\right)^{\psi_E} \Gamma_0^{-\epsilon\psi_E}\, n_0^{\psi_n}\, t_d^{\psi_t}\ \mathrm{Hz}. \qquad (8)$$

TABLE 1. Indices for time, energy, and density scaling of the synchrotron flux and break frequencies

	F_C	ν_m	ν_c
t	$\phi_t = -\frac{3\epsilon}{8-\epsilon}$	$\chi_t = -\frac{12}{8-\epsilon}$	$\psi_t = -2\frac{2-\epsilon}{8-\epsilon}$
E	$\phi_E = \frac{8}{8-\epsilon}$	$\chi_E = \frac{4}{8-\epsilon}$	$\psi_E = -\frac{4}{8-\epsilon}$
n	$\phi_n = \frac{3}{2} - \phi_E$	$\chi_n = \frac{1}{2} - \chi_E$	$\psi_n = -\frac{3}{2} - \psi_E$

The power-law indices of these relations are given in Table 1, $d_{28} = d_L/(10^{28}\,\mathrm{cm})$, and the normalization constants f_{F_c}, f_{ν_m}, and f_{ν_c} are given in [2].

At a given observing frequency ν the flux will decay according to power-laws in time. For $\nu < \nu_c$ we have $F_\nu \propto \nu^{1/3}\, t^{\omega_0}$ with $\omega_0 = (1/3)\,(4 - 11\,\epsilon)/(8-\epsilon)$. For $\nu_c < \nu < \nu_m$ we have $F_\nu \propto \nu^{-1/2}\, t^{\omega_1}$ with $\omega_1 = -2\,(1+\epsilon)/(8-\epsilon)$. For $\nu > \nu_m$, $F_\nu \propto \nu^{-p/2}\, t^{\omega_2}$ with $\omega_2 = -[2\,(1+\epsilon) + 6\,(p-1)]/(8-\epsilon)$. Eqs. (7) and (8) are easily inverted to find the sweep-through times t_c and t_m of the break frequencies ν_c and ν_m at a given observing frequency, which define the times of breaks in the light curves between the above power-law decay slopes.

Generally the break frequency ν_m decreases more rapidly than the cooling frequency ν_c. This means that a burst which starts out in the fast-cooling regime will become radiatively less efficient, with a time-dependent ϵ, after both break frequencies have become equal. After this transition, the radiative efficiency will steadily decrease as $\epsilon \approx \epsilon_e\,(\gamma_m/\gamma_c)^{p-2}$ [8] until the blast wave evolution approaches the adiabatic limit ($\epsilon \to 0$).

The dependence of the temporal rise/decay slopes ω_i on the radiative regime is illustrated in Fig. 1. Both the slopes ω_0 and ω_2 are rapidly decreasing with increasing ϵ. The typically observed power-law decays of X-ray and optical afterglows, $F_\nu \propto t^{-\alpha}$ with $1.1 \lesssim \alpha \lesssim 1.4$, if produced by a blast wave in the fast-cooling regime, imply constraints on a rather hard electron injection spectrum, if one allows for a finite radiative efficiency $\epsilon \gtrsim 0.1$.

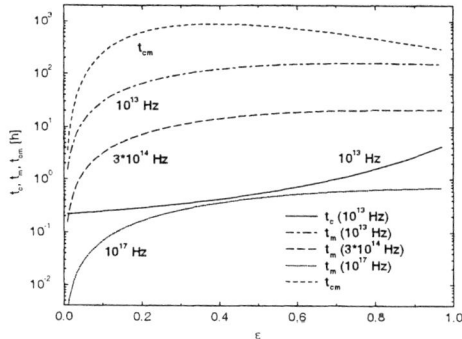

FIGURE 2. The sweep-through times of the cooling and the ν_m break as a function of the radiative efficiency ϵ for three different observing frequencies. Parameters: $E_{52}/\Omega = 100/(4\pi)$, $n_0 = 10$, $\Gamma_0 = 100$, $\epsilon_B = 0.2$, $\epsilon_e = \epsilon$, $p = 2.5$, $z = 1$. Also shown is the time t_{cm}, at which the transition from fast cooling to slow cooling occurs ($\nu_c = \nu_c$). The cooling-break sweep-through times t_c for $\nu_{obs} = 3 \cdot 10^{14}$ Hz and 10^{17} Hz are of the order $\lesssim 1$ s and thus irrelevant since this is shorter than the deceleration time scale.

In Fig. 2 we plot the sweep-through times of the break frequencies ν_c and ν_m for observations at optical, infrared, and soft X-ray frequencies for a standard set of burst parameters as a function of ϵ. In particular for low ϵ, the break time t_m depends very strongly on the radiative efficiency. Thus, any conclusions drawn from the observed sweep-through times based on the adiabatic solution of the blast wave dynamics, need to be taken with caution. Note that in calculating these break times we have set $\epsilon_e = \epsilon$, which is the reason for the sweep-through time t_m approaching 0 for $\epsilon \to 0$.

REFERENCES

1. Blandford, R. D., & McKee, C. F., *Phys. Fluids* **19**, 1130 (1976).
2. Böttcher, M., & Dermer, C. D., *ApJ* **532**, in press (2000).
3. Chiang, J., & Dermer, C. D., *ApJ* **512**, 699 (1999).
4. Galama, T. J., et al., *ApJ* **500**, L97 (1998).
5. Huang, Y.-F., et al., *CPL*, in press (1999).
6. Mészáros, P., & Rees, M. J., *ApJ* **418**, L59 (1993).
7. Katz, J. I., 1994, *ApJ* **432**, L107 (1994).
8. Moderski, R., et al., *ApJ*, in press (1999).
9. Sari, R., et al., *ApJ* **497**, L17 (1998).
10. Tavani, M., *PRL* **76**, 3478 (1996).

Optical Flashes and Radio Flares in GRB Afterglows: Numerical Study

Shiho Kobayashi[1] and Re'em Sari[2]

[1] *Department of Earth and Space Science, Osaka University, Toyonaka, Osaka 560, Japan*
[2] *Theoretical Astrophysics 130-33, California Institute of Technology, Pasadena, CA 91125, USA*

Abstract. We numerically study the evolution of an adiabatic relativistic fireball interacting with an ambient uniform medium, both in the initial energy transfer stage and in its late evolution. It is shown that the Blandford-McKee solution adequately describes the evolution of the shocked ejecta quite early on and for as long as the fireball material has relativistic temperatures. In the case where the reverse shock is only mildly relativistic, the shocked ejecta becomes cold almost immediately and the evolution deviates from the Blandford-McKee solution. We derive analytical expressions for the ejecta evolution in its cold regime. This solution gives a good approximation to the numerical results. We estimate the radiation from the fireball ejecta using the numerical hydrodynamic evolution in both cases: cold and hot ejecta. Surprisingly, we find that both evolutions give rather similar light curves, decaying approximately as t^{-2} in the optical and peaking after about one day in the radio, even though the hydrodynamics is different.

INTRODUCTION

A new clue to understand the nature of gamma-ray bursts was found in the event of GRB 990123 by ROTSE which detected a strong optical flash during the "gamma-ray" burst. The optical flash reached to a peak of 9th magnitude and then decayed with a slope of a power law index ~ 2 [1].

Such a strong, prompt optical flash was predicted [5,7]. The prompt optical flash observed by ROTSE is compatible with these predictions [6,8]. In the energy transfer stage of a fireball evolution, the forward shocked ISM and the reverse shocked fireball ejecta carry comparable amount of internal energy. The typical temperature in the shocked ejecta is considerably lower than that of the shocked ISM, the typical frequency of the emission from the shocked ejecta comes to the optical band with reasonable values of the parameters.

Besides the optical flash, GRB 990123 had another newly observed phenomenon, a radio flare [4]. Sari and Piran [8] have interpreted this flare as the emission from the ejecta particles earlier shocked by the reverse shock. The same particles producing the prompt optical flash have been cooled adiabatically and their emission shifts quickly towards lower frequencies while weakening. According to the analytical estimates, scaling the prompt optical emission to the epoch of the radio detection gives the right amount of radio emission.

Although the discussion of the prompt optical flash is quite robust, the estimate of the hydrodynamic evolution of the shocked ejecta is more fragile. In this paper we study numerically the evolution of the ejecta in order to estimate the decay rate of the optical flash as well as the light curve and timing of the radio flare.

THE SHOCKED EJECTA

Sari and Piran [7,8] used the Blandford-McKee (BM) solution to estimate hydrodynamics of the shocked ejecta. However, it is not clear whether the BM solution is applicable to the shocked ejecta for the following reasons: (i) The BM solution, as a self similar solution, describes the shocked ISM long after the energy transfer stage where the details of the initial conditions are no longer important. (ii) It assumes that the initial ejecta is irrelevant, however the evolution of that ejecta is what we are interested in. (iii) The initial ejecta contains many more particles than those collected by the forward shock. Its density therefore must be higher than that predicted by the BM solution. (iv) Though the BM solution assumes relativistic temperatures, a mildly relativistic reverse shock can not heat ejecta well.

It can be argued that since the shocked ejecta is located not too far behind the forward shock at the end of the energy transfer stage and it has a comparable amount of energy to that of the system, it roughly fits the BM solution. The fact that its rest mass density is much higher than that given by the BM solution should not play an important role as long as the temperature is relativistic. At relativistic temperature, the fluid inertia is due to its thermal energy rather than the rest mass.

The evolution of an adiabatic spherical fireball with an energy E, a dimensionless entropy η and a radius R_0 surrounded by a uniform medium with a low density ρ_1 can be classified into two categories by a dimensionless quantity $\xi \equiv (E/\rho_1 R_0^3)^{1/6}/\eta^{4/3}$ [2]. If $\xi > 1$ the reverse shock is initially Newtonian and becomes mildly relativistic when the reverse shock crosses the ejecta while the reverse shock becomes relativistic during the crossing if $\xi < 1$.

Using numerical calculation we verified that the BM solution adequately describes the evolution of the ejecta if the reverse shock is relativistic [3]. In this paper we discuss only the evolution of the cold shell (the case of $\xi > 1$). We can roughly estimate the evolution of cold ejecta as follows: When the ejecta arrives at radius R, the time in the local frame is $R/c\gamma$. The ejecta sound speed is sub-relativistic and can be estimated by $(p/\rho)^{1/2}$. The width of the ejecta will be $(p/\rho)^{1/2} R/c\gamma$. The ejecta density is therefore: $\rho \propto R^{-3}\gamma(\rho/p)^{1/2}$. Using the adiabatic expansion law and assuming a power law $\gamma \propto R^{-g}$, we get $p \propto R^{-8(3+g)/7}$ and $\rho \propto R^{-6(3+g)/7}$.

We compare numerical results with this estimate and the BM solution in Figure 1.

The numerical scaling of the ejecta (thick solid line) is very different from the BM one for a fluid element (dashed line). The power law estimates (thin solid line) are good approximation as long as the bulk motion is relativistic.

THE EJECTA EMISSION

We estimate in this section the light curve of the optical flash and the radio flare of GRB 990123. According to the internal shock model, the duration of the gamma-ray burst is determined by the thickness of the relativistic flow. We assume $R_0/c = 40$ s. The observations suggest that the typical synchrotron frequency of

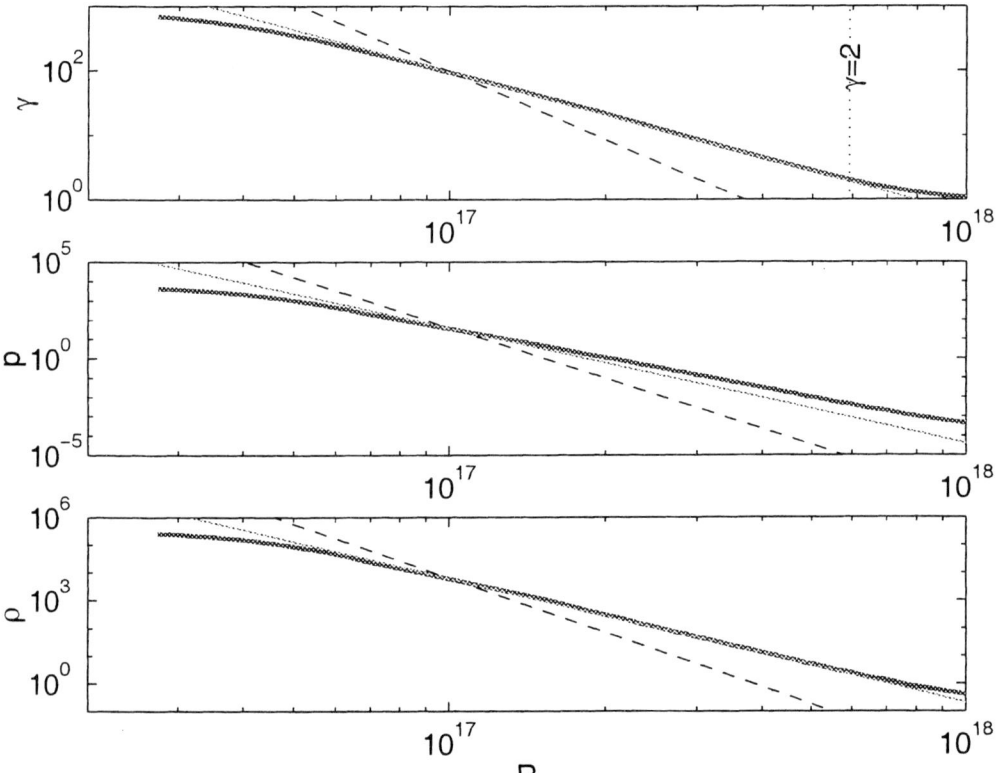

FIGURE 1. R vs Lorentz factor, pressure and density of the ejecta: Newtonian reverse shock case of $\xi = 44, E = 3 \times 10^{54}$ erg, $\rho_1 = 10$ proton/cm^3, $\eta = 10^3$ and $R_0 = 3 \times 10^7$ cm. The numerical results (thick solid lines) are compared with the Blandford-McKee scalings of a fluid element (dashed lines) and the power law estimates (thin solid lines). The vertical dotted line show the radius where the Lorentz factor of the ejecta is 2

the reverse shock is below the optical bands quite early on, then the initial Lorentz factor of the ejecta is a few hundred [8]. $\eta = 400$ is assumed here. The explosion energy and the ambient density are rather ambiguous. We assume $E = 1 \times 10^{54}$ ergs and $\rho_1 = 5$ protons cm^{-3}. The fireball of GRB 990123 is a marginal case of $\xi = 0.7$. However, such a marginal case behaves very much like a case of $\xi \gg 1$ as the ejecta becomes cold at the early stage.

In order to estimate the light curve from the thermodynamic quantities of the fireball ejecta, we assume that the energy of the magnetic field and the internal energy of electrons remain constant fractions of the internal energy of the fluid. If the shocked electrons in the shocked ejecta are accelerated into the power law distribution of the electron random Lorentz factor with index $\hat{p}$, the spectral flux at a given frequency above the typical frequency ν_m is $F_\nu \sim F_{\nu_m}(\nu/\nu_m)^{-(\hat{p}-1)/2}$.

Using the numerical evolution of γ, p and ρ, the optical light curve is plotted in Figure 2. The ROTSE observations are also plotted (stars), the triangles represent upper limits. The steepness of the light curve is about -2 at late time with a steeper slope at early times. The numerical light curve normalized at the radio flare reasonably fits the observation.

Emission from the reverse shock can also explain the radio flare, the radio detection one day after the burst. The shocked ejecta initially radiates in the optical band. As the ejecta expands, the temperature of the ejecta becomes lower. The emission frequency and the flux drop quickly; eventually the emission comes into the radio band. The flux at 8.5 GHz is plotted in Figure 2 in which we consider the upper limit by the self absorption (dashed line) [3]. The numerical light curves fit to both the optical flash and the radio flare.

DISCUSSION

We have seen that the Blandford-McKee solution is not applicable to the shocked ejecta if the reverse shock is Newtonian or mildly relativistic and the temperature of the shocked ejecta is not relativistic. The hydrodynamics of the cold shocked ejecta are very different from that of the hot shocked ejecta which is well described by the Blandford-McKee solution. The numerical scaling relations for a cold one were well approximated by the spreading assumption and the adiabatic expansion law, while the Blandford-McKee solution satisfies the relativistic version of it. We have estimated the radiation from the fireball ejecta. Surprisingly, we find that the evolution of both the cold and hot ejecta give rather similar light curves of optical and radio even though the hydrodynamics is very different.

REFERENCES

1. Akerlof, C.W., et al., *Nature* **398**, 400 (1999).
2. Kobayashi, S., Piran,T. and Sari, R., *ApJ* **513**, 669 (1999).
3. Kobayashi S. and Sari R., *ApJ* submitted (1999) astro-ph/9910241.

4. Kulkarni, S.R., et al., *Nature* **398**, 389(1999).
5. Mészaros, P. and Rees,M.J., *ApJ* **476**, 231 (1997).
6. Mészaros, P. and Rees,M.J., *MNRAS* submitted (1997) astro-ph/9902367.
7. Sari, R. and Piran,T., *ApJ* **520**, 641(1999).
8. Sari, R. and Piran,T., *ApJ* **517**, L109(1999).

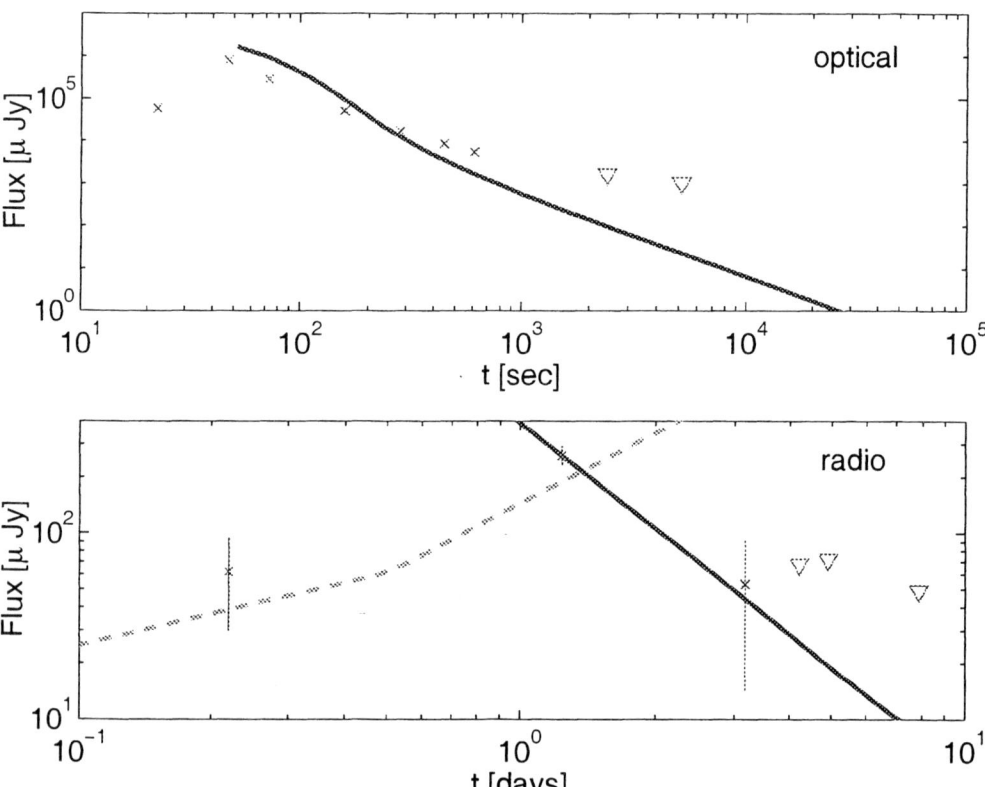

FIGURE 2. Light curves of optical flash and radio flare. The solid lines depicts the numerical light curves. The stars are the observations and the triangles are upper limits. The dashed line is self absorption upper limit.

Gamma-Ray Bursts: The Central Engine

S. E. Woosley

Department of Astronomy, University of California, Santa Cruz, CA 95064

Abstract. A variety of arguments suggest that the most common form of gamma-ray bursts (GRBs), those longer than a few seconds, involve the formation of black holes in supernova-like events. Two kinds of "collapsar" models are discussed, those in which the black hole forms promptly — a second or so after iron core collapse — and those in which formation occurs later, following "fallback" over a period of minutes to hours. In most cases, extraction of energy from a rapidly accreting disk (and a rapidly rotating black hole) is achieved by magnetohydrodynamical processes, although neutrino-powered models remain viable in cases where the accretion rate is $\gtrsim 0.05$ $M_\odot$ s^{-1}. GRBs are but one observable phenomenon accompanying black hole birth and other possibilities are discussed, some of which (long, faint GRBs and soft x-ray transients) may await discovery. Since they all involve black holes of similar mass accreting one to several $M_\odot$, collapsars have a nearly standard total energy, around 10^{52} erg, but both the fraction of that energy ejected as highly relativistic matter and the distribution of that energy with angle can be highly variable. An explanation is presented why inferred GRB luminosity might correlate inversely with time scales and arguments are given against the production of ordinary GRBs by supergiant stars.

I GENERAL GRB MODEL REQUIREMENTS

Given the locations afforded by x-ray and optical afterglows, redshifts have now been determined for approximately 10 GRBs so that we have at least a small sampling of GRB energies [5]. They are by no means standard candles. Even discounting the unusual case of GRB 980425 (8×10^{47} erg), energies range from about 5×10^{51} erg (GRB 980613) to 2×10^{54} erg (GRB 990123). In addition, GRB time profiles and spectra are very diverse and separate into at least two classes — the "short-hard" bursts (with average duration 0.3 s) and the "long-soft" bursts (average duration 20 s). The first challenge any model builder must confront is deciding just which GRBs, and which features, he or she is attempting to model, since it is increasingly doubtful that all GRBs are to be explained in the same way. Moreover, in all of today's models, the gamma-rays observed from a cosmological GRB are produced far from the site where the energy is initially liberated — presumably conveyed there by relativistic outflow or jets. How much of what we see in GRB reflects the central engine and how much the environment where the outflow dissipates its energy? So our first step is to define the problem we are attempting to address.

Even after the dramatic progress of the last two years, few definitive statements can be made about GRBs without provoking controversy. Still, in 1999, most people feel that the following are facets of a common GRB that the central engine must provide:

1. Highly relativistic outflow — $\Gamma \gtrsim 100$, possibly highly collimated.

2. An event rate that, at the BATSE threshold and in the BATSE energy range, is about 1/day. Beaming, of course, raises this number appreciably.

3. Total energy in relativistic ejecta $\sim 10^{53} - -10^{54} \epsilon_\gamma^{-1} f_\Omega$ erg where ϵ_γ is the efficiency for turning relativistic outflow into gamma-rays ($\sim 10\%$?), and f_Ω is the fraction of the sky into which that part of the flow having sufficiently high Γ ($\gtrsim 100$) is collimated ($\sim 1\%$?). For reasonable values of these parameters, the total energy required for a common GRB is 10^{52} erg. Fainter GRBs can result from the same 10^{52} erg event if the efficiency for producing relativistic matter is reduced (e.g., GRB 980425); brighter ones if the collimation is tighter.

4. A duration of relativistic flow in our direction no longer than the duration of the GRB. This constraint is highly restrictive for the short (0.3 s) bursts and may imply multiple models. For GRB models produced by internal shocks, the flow may additionally need to last *as long as* the GRB (*modulo* the relativistic time dilation). This makes a natural time scale ~ 10 s attractive.

5. In the case of long bursts, association with star forming regions in galaxies and, in perhaps three cases, with supernovae of Type I.

The near coincidence of 10^{52} erg with the energy released in the gravitational collapse of a stellar mass object to a neutron star (or, equivalently, the accretion disk of a black hole), has long suggested a link between GRBs and neutron star or black hole formation, a connection championed by Paczynski before cosmological models became fashionable. Viable models separate into three categories (Table 1), where $\epsilon_{\rm MHD}$ is the unknown efficiency for magnetohydrodynamical processes to convert either gravitational accretion energy at the last stable orbit ($\sim 0.1 \dot{M} c^2$) or neutron star rotational energy into relativistic outflow. Those using black hole accretion [6,7,15,9] typically employ $1 - 10\%$ for $\epsilon_{\rm MHD}$; pulsar advocates [17,18] need approximately 100%.

The collapsar model is incapable of producing relativistic jets of total duration less than a few seconds (hence, short-hard bursts are difficult — impossible unless the beam orientation wanders). Merging neutron stars and black holes, on the other hand, can produce short bursts if the disk viscosity is high (i.e., $\alpha \sim 0.1$), but cannot, with the same disk viscosity, produce long bursts. Merging neutron stars also lack the massive disks that help to focus the outflow in collapsars and it may be more difficult for them to emit highly collimated jets. Hence, their "equivalent

TABLE 1. Gravitational collapse models for GRBs

Model	Energy Source	Mass Reservoir	Possible Energy	Jet From	^{56}Ni /SN	Beaming
n*+n*,BH	BH accretion	0.01 – 0.5 M$_\odot$	10^{50} $10^{53}\epsilon_{MHD}$	$\nu\bar{\nu}$ MHD	No	~10%?
collapsar	BH accretion	1 – 5 M$_\odot$	10^{52} $10^{54}\epsilon_{MHD}$	$\nu\bar{\nu}$ MHD	Yes	0.1% – 10%
pulsar	n* rotation		$10^{52}\epsilon_{MHD}$	MHD	?	~1%?

isotropic energies" may be smaller (unless MHD collimation dominates). It seems more natural to associate the merging compact objects with short-hard bursts, but this conjecture presently lacks any observational basis. Hopefully future observations (e.g., with HETE-2) will clarify whether short bursts are associated with host galaxies in the same way as the long ones.

The pulsar based models have not been studied nearly as extensively as either the collapsar or merging compact objects, perhaps because the MHD phenomena they rely on are difficult to simulate numerically. The magnetic fields and rotation rates invoked for the pulsar models, though large (P $\sim$ ms; B $\sim 10^{15}$ gauss), are not much greater than employed for the disk in MHD collapsar models. However, it is not at all clear how such models would make the large mass of ^{56}Ni inferred for SN 1998bw or the highly collimated flow required to explain energetic events like GRB 990123. Also, the bare pulsar version of the model [17] ignores the effects of neutrino-powered winds and the supernova-based version [18] ignores the collapse of the massive star that would continue, at least at some angles, during the few seconds it takes the pulsar to acquire its large field. A complete calculation of the implosion of the iron core of a massive star, including the coupled effects of rotation, magnetic fields, and neutrinos has not been done, but could be in the next decade.

II COLLAPSARS – TYPE 1

We thus consider here a model that can, in principle, satisfy the five constraints above, at least for long bursts, and has the added virtue of being calculable, with a few assumptions, on current computers. A *collapsar* is a massive star whose iron core has collapsed to a black hole that is continuing to accrete at a very high rate. The matter that it accretes, that is the helium and heavy elements outside the iron core, is further assumed to have sufficient angular momentum ($j \sim 10^{16} - -10^{17}$ cm^2 s^{-1}) to form a centrifugally supported disk outside the last stable orbit. The black hole is either born with, or rapidly acquires a large Kerr parameter. It may also be possible to create a situation quite similar to a collapsar in the merger of the helium core of a massive star with a black hole or neutron star [4].

What follows has been discussed in the literature [7,8,1]. The black hole accretes matter along its rotational axis until the polar density declines appreciably. Accretion is impeded in the equatorial plane by rotation. The accretion rate through the disk is insensitive to the disk viscosity because a steady state is rapidly set up in which matter falls into the hole at a rate balancing what is provided by stellar collapse at the outer boundary. The mass of the accretion disk is inversely proportional to the disk viscosity and accretion rates 0.01 – 0.1 $M_\odot$/s are typical during the first 20 s as the black hole grows from about 3 $M_\odot$ to about 4 or 5 $M_\odot$. The accretion rate may be highly time variable down to intervals as short as 50 ms [7], and an appreciable fraction of the matter passing through the disk is ejected as a powerful "wind" that itself carries up to a few $\times 10^{51}$ erg and a solar mass [7,16]. Given the high temperature in the disk, this disk wind will, after some recombination, probably be mostly ^{56}Ni. This may be the origin of the light curve of SN 1998bw and other supernovae associated with GRBs.

Disk accretion also provides an energy source for jets. In the simplest, but perhaps least efficient version of the collapsar model, energy is transported from the very hot (~ 5 MeV) inner disk to the rotational axis by neutrinos. Neutrinos arise from the capture of electron-positron pairs on nucleons in the disk and deposit a small fraction of their energy, $\sim 1\%$, along the axis where the geometry is favorable for neutrino annihilation. The efficiency factor for neutrino energy transport is a sensitive function of the accretion rate, black hole mass and Kerr parameter, and the disk viscosity [14]. Only in cases where the accretion rate exceeds about 0.05 $M_\odot$ s^{-1} for black hole masses 3 – 5 $M_\odot$ and disk viscosities, $\alpha \sim 0.1$, will neutrino transport be significant. Using the actual accretion rate, Kerr parameter, hole mass as a function of time, and $\alpha \sim 0.1$, MacFadyen finds for a helium core of 14 $M_\odot$, a total energy available for jet formation up to $\sim 10^{52}$ erg. The typical time scale for the duration of the jet, and a lower bound for the duration of the GRB, is ~ 10 s, the dynamical time scale for the helium core.

In addition to any neutrino energy transport, one has the possibility of magnetohydrodynamical processes which could, in principle, efficiently convert a large fraction of the binding energy at the last stable orbit, up to 42% Mc^2, into jet energy. Adopting a more conservative value, $\epsilon_{\rm MHD} \sim 1\%$ [8], one still obtains 10^{52} – 10^{53} erg available for jet formation. Dumping this much energy into the natural funnel-shaped channel that develops when a rotating star collapses gives rise to a hydrodynamically collimated jet focused into $\sim 1\%$ of the sky [7,8,1]. Magnetic collimation though uncertain, could, in principle, increase the collimation factor still further.

Thus, jets of equivalent isotropic energy 10^{54}, and possibly 10^{55} erg (if, e.g., $\epsilon_{\rm MHD} \sim 0.1$) seem feasible in this model. The event rate of collapsars is also adequate [3].

The collapsar model also makes several "predictions" some of which have already been confirmed (these same predictions were inherent in the original 1993 model [19]. First, the GRB should originate from massive stars, in fact the *most* massive stars, and be associated with star forming regions. In fact, given the need for large helium core mass, collapsars may be favored not only by rapid star formation, but also by low metallicity. This reduces the loss of both mass and of angular momentum. Pre-explosive mass loss also provides a natural explanation for the surrounding medium needed to make the GRB afterglows and makes a prediction that the density decline as r^{-2}. The GRB duration, ~ 10 s, corresponds to the collapse time scale of the helium core. The explosion is expected to be highly collimated, though just how collimated was not realized until 1998 [7]. The jet blows up the star in which it is made so one expects some kind of supernova. Since the presence of a massive hydrogen envelope prohibits making a strong GRB, the supernova must be of Type I (a possible exception would be an extreme Type IIb supernova, one that had lost all but a trace of hydrogen on its surface). That the explosion might also produce a lot of ^{56}Ni from a disk powered wind was not appreciated until [7]. Without the ^{56}Ni, the supernova would have been very dim, which is why I originally referred to the collapsar model as a "failed supernova".

It also seems natural that both the variable accretion rate [7] and the hydrodynamical interaction of the jet with the star which it penetrates may introduce temporal structure into the burst. Implications for GRB diversity are discussed in §4.

III COLLAPSARS – TYPE 2

It is also possible to produce a collapsar in a delayed fashion by fallback in an otherwise successful supernova [8]. A spherically symmetric explosion is launched in the usual way by neutron star formation and neutrino energy transport, but the supernova shock has inadequate strength to explode the whole star. Over a period of minutes to hours a variable amount of mass, ~ 0.1 to 5 $M_\odot$, falls back into the collapsed remnant, often turning it into a black hole [20] and establishing an accretion disk. The accretion rate, ~ 0.001 to 0.01 $M_\odot$ s^{-1}, is inadequate to produce a jet mediated by neutrino annihilation [14], but MHD processes may still function with the same efficiency as in the Type 1 collapsar (or merging neutron stars, for that matter). Then the total energy depends not on the accretion rate, but the total mass that reimplodes. For 1 $M_\odot$ and $\epsilon_{MHD} = 1\%$, this is still 10^{52} erg.

A key difference is the time scale, now typically 10 – 100 times longer. Thus, the most likely outcome of a Type 2 collapsar in a star that has lost its hydrogen envelope is a less luminous, but longer lasting GRB. Indeed, there exist GRBs that have lasted hundreds of seconds and there may be a class of longer, fainter GRBs awaiting detection. Since black holes may be more frequently produced by fall back than by failure of the central engine [2], these sorts of events might even be more common than ordinary GRBs.

Both kinds of collapsars can also occur in stars that have *not* lost their envelopes. Stars with lower metallicity have less radiative mass loss so that solitary stars (or widely detached binaries) might also end their lives with both a rapidly rotating massive helium and a hydrogen envelope. Because the motion of the jet head through the star is sub-relativistic [1] and because fallback only maintains a high accretion rate for 100 – 1000 s, highly relativistic jets will not escape red supergiants with radii $\gtrsim 10^{13}$ cm. What happens in more compact blue supergiants is less certain. Generally speaking, the largest fallback masses will characterize the weakest supernova explosions and also have the shortest fallback time scales. With a jet head speed of 10^{10} cm s^{-1}, it would have taken 300 s, for example, to cross the blue progenitor of SN 1987A. The fallback mass in 87A is believed to have been $\lesssim 0.1$ M$_\odot$, probably inadequate to turn the neutron star into a black hole and certainly too little to make a powerful GRB, but perhaps enough to make a jet anyway — or at least cause some mixing. Larger mass helium cores (87A was 6 M$_\odot$) might have more fallback though, definitely making black holes and more energetic jets. Whether the jet can still have a large Lorentz factor remains to be calculated.

Even if they do not make GRBs, collapsar powered jets in blue and red supergiant stars may still lead to very energetic, asymmetric supernova explosions, possibly accompanied by large ^{56}Ni production and luminous soft x-ray transients due to shock breakout [8]. These transients may have luminosity up to $\sim 10^{49}$ erg s^{-1} times the fraction of the sky to which high-energy material is ejected (typically 0.01) and color temperatures of 2×10^6 K.

IV GRB DIVERSITY

As previously noted, the inferred total energy in gamma rays for those GRBs whose distances have been determined is quite diverse. One appealing aspect of the collapsar model is that its outcome is sufficiently variable to explain this diversity. The observed burst intensity is sensitive not only to the jet's total energy, but also to the fraction of that energy in the observer's direction that has Lorentz factor Γ above some critical value (~ 100). Most collapsars accrete about the same mass, 1 – 3 M$_\odot$, before accretion is truncated by the explosion of the star. For an efficiency factor of 1%, this implies a total jet (and disk wind) energy of $\sim$ few $\times 10^{52}$ erg. However, depending on the initial collimation of the jet, its internal energy (or equivalently the ratio of its pressure to its kinetic energy flux), and its duration, very different outcomes can result. A poorly collimated jet, or one that loses its energy source before breaking through the surface of the star, may only eject a little mildly relativistic matter and make, e.g., GRB 980425. A focused, low-entropy jet that lasts ~ 10 s after it has broken free of its stellar cocoon might make GRB 990123.

Duration can be affected by such things as the presupernova mass and angular momentum distribution. Internal energy depends on details of the jet acceleration. Neutrino powered jets, for example, have much higher internal energies than some

MHD jets and may be harder to focus. Hydrodynamical focusing of the jet also depends on the density distribution in the inner disk, which in turn depends on disk viscosity and accretion rate. And of course the efficiency factor need not always be 1%, e.g., for neutrino-powered models and MHD models. Calculations [8,1]

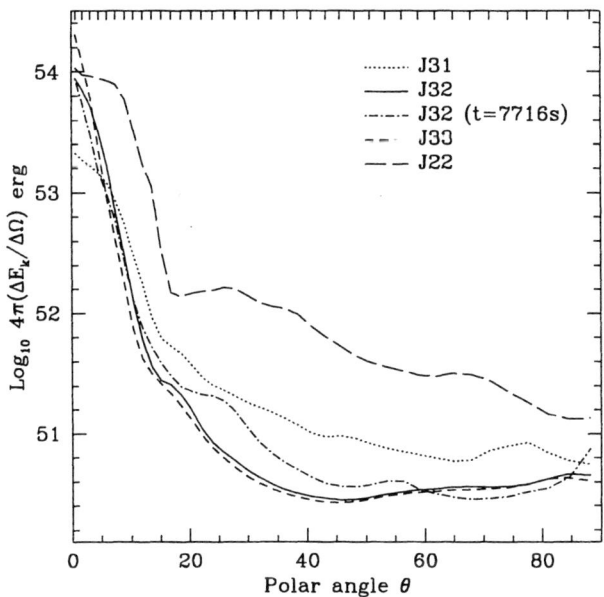

FIGURE 1. Equivalent isotropic energy as a function of angle for several models.

illustrate this. Fig. 1 shows the "equivalent isotropic kinetic energy" as a function of polar angle for three models having the same total jet energy, 3×10^{51} erg, at the base. All models except the dot-dash line for J22 are shown 400 s after the initiation of the jet, well after it has broken out of the helium core. The three models differ only in the ratio of internal energy to kinetic energy given to the jet at its base. Yet, even for a constant viewing angle, $\theta = 0$, the inferred isotropic energies vary by an order of magnitude. Larger variations are possible if one goes to other values of viewing angle — *not* because the GRB is being viewed "from the side", but because the material coming at the observer has both less energy and a lower Lorentz factor. Thus, it is also possible that GRB 980425 was a more typical GRB viewed off axis [10], but not in the sense of a single highly relativistic beam which emitted a few photons in our direction. Instead we saw emission from matter coming towards us with a lower Γ.

Fig. 2 is not the result of any current calculation, but just a sketch to illustrate what calculations may ultimately show. (See, for comparison, Fig. 4 of [1], a first pass at one collapsar model using a code with the necessary relativistic hydrody-

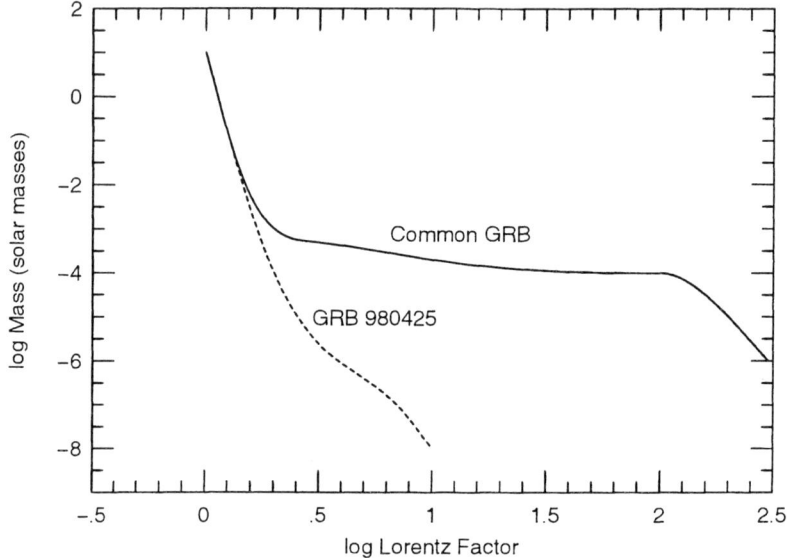

FIGURE 2. Mass ejected vs. Lorentz factor for two hypothetical models.

namics. Unfortunately this calculation has not yet been run long enough to show the final distribution of Lorentz factors). There is a large concentration of mass, ~ 10 M$_\odot$, moving at sub-relativistic speeds. This is the supernova produced by the jet passing through the star. Though the speed is "slow", most of the energy may be concentrated here if the jet did not last long enough or stay focused enough to become highly relativistic (dotted line). Then there is a relativistic "tail" to the ejecta. Even though it is a small fraction of the mass, this tail could, in some cases, namely the common GRBs, contain most of the energy in the explosion.

Table 2 indicates some of the diverse outcomes that might arise. Here R_{15} is an approximate radius in units of 10^{15} cm where the material might give up its energy. A typical Wolf-Rayet mass loss rate has been assumed for those cases where external shocks are clearly important (x-ray afterglows and GRB 980425). Supernovae also typically have a photospheric radius of 10^{15} cm. $\Omega/4\pi$ is the fraction of the sky into which the mass is beamed. The fractions sum to over 100% because the supernova is not beamed.

V TIME VARIABILITY, LAG TIME, AND LUMINOSITY

At this meeting we also heard of two fascinating results with important implications for the use of GRBs as calibrated "standard candles" for cosmology.

TABLE 2. Relativistic mass ejected in two artificial models

Γ	M/M$_\odot$	E(erg)	Ω/4π	R$_{15}$(10^{15} cm)	Comment
			Common GRB		
100	10^{-4}	10^{52}	< 1%	< 3	GRB
10	10^{-4}	5 × 10^{51}	10%	3	X-ray tail
1	10	5 × 10^{51}	100%	1	SN Ib/c
			GRB 980425		
7	10^{-7}	10^{48}	10%	0.01	GRB 980425
2	10^{-5}	10^{50}	20%	10	X-ray, radio afterglow
1	10	10^{52}	100%	1	SN 1998bw

Ramirez-Ruiz and Fenimore (Paper T-04) discussed a correlation between "variability" and luminosity. The more rapidly variable the light curve, the higher the absolute luminosity. Norris, Marani, & Bonnell [11,12] also showed data to support a high degree of (anti-)correlation between absolute luminosity and the "time lag", the delay time between the arrival of hard and soft-subpulses. The shorter the lag, the brighter the burst.

Both these effects may be understood as an outcome of Fig. 1. The bursts for which we infer the highest luminosities are those that are observed straight down the axis of the jet, $\theta = 0$. This is also the angle at which we see the largest Lorentz factors. Slightly away from $\theta = 0$, both the equivalent isotropic energy and Γ drop precipitously.

For larger Lorentz factors, the burst will be produced closer to the source. Ref. [13] gives a thinning radius where the GRB becomes optically thin to Thomson scattering that is proportional to $\Gamma^{-1/2}$. The distance where internal shocks form from two shells having Lorentz factors Γ_1 and Γ_2 is $\Gamma_1 \Gamma_2 c \Delta t$. For smaller radii and larger Γ, time scales will thus be contracted. That is larger Γ may imply more time structure on shorter scales and perhaps reduced lag times as well. Then variability and time lags would be related to the equivalent isotropic energy because both are functions of the viewing angle.

GRB 980425 is an exception since its GRB was produced by a external shock interaction between mildly relativistic matter and the presupernova mass loss.

ACKNOWLEDGMENTS

This research was supported by the National Science Foundation, (AST-97-31569) and by the NASA Theory Program (NAG5-8128).

REFERENCES

1. Aloy, M. A., Müller, E., Ibanez, J. M., Marti, J. M., & MacFadyen, A. I., *ApJ*, submitted, astro-ph/9911098 (1999).
2. Fryer, C. L., *ApJ*, in press, astro-ph/9902315 (1999).
3. Fryer, C. L., Woosley, S. E., & Hartmann, D. H., *ApJ*, in press, astro-ph/9904122 (1999).
4. Fryer, C. L., & Woosley, S. E., *ApJL* **502**, L9, astro-ph/9804167 (1998).
5. Galama, T. J., *Gamma-Ray Burst Afterglows*, PhD Thesis, Universititeit van Amsterdam, 1999.
6. Janka, H. -Th., Eberl, T., Ruffert, Mm., & Fryer, C., *ApJ*, submitted, astro-ph/9908290 (1999).
7. MacFadyen, A. I., & Woosley, S. E., *ApJ* **524**, 262, astro-ph/9810274 (1999).
8. MacFadyen, A., Woosley, S. E., & Heger, A., *ApJ*, submitted, astro-ph/9910034 (1999).
9. M'esz'aros, P., in *Proc. 19th Texas Symposium, Nucl. Phys B*, in press, astro-ph/9904038 (1999).
10. Nakamura, T., *ApJL* **522**, L101 (1999).
11. Norris, J. P., Marani, G. F. & Bonnell, J. T., these proceedings.
12. Norris, J. P., Marani, G. F. & Bonnell, J. T., *ApJ*, submitted, astro-ph/9903233 (1999).
13. Panaitescu, A., Wen, L., Laguna, P., & Mészáros, P., *ApJ* **482**, 942 (1997).
14. Popham, R., Woosley, S. E., & Fryer, C., *ApJ* **518**, 356, astro-ph/9807028 (1999).
15. Rosswog, S., Liebendoerfer, M., Thielemann, F.-K., Davies, M. B., Benz, W., Piran, T., *A&A*, in press, astro-ph/9811367 (1999).
16. Stone, J. M., Pringle, J. E., Begelman, M. C., *MNRAS*, in press, astro-ph/908185 (1999).
17. Usov, V., *MNRAS* **267**, 1035 (1994).
18. Wheeler, J. C., Yi, I., Höflich, P., & Wang, L., *ApJ*, in press, astro-ph/9909293 (1999).
19. Woosley, S. E., *ApJ* **405**, 273 (1993).
20. Woosley, S. E., & Weaver, T. A., *ApJS* **101**, 181 (1995).

2D Hydrodynamic Simulations of Relativistic Jets from Collapsars

Ewald Müller[†], Miguel Angel Aloy[*], José M$^{\underline{a}}$ Ibáñez[*],
José M$^{\underline{a}}$ Martí[*], and Andrew MacFadyen[‡]

[†] *Max-Planck-Institut für Astrophysik, Karl-Schwarzschild-Str. 1, 85748 Garching, Germany*
[*] *Dept. Astronomía y Astrofísica, Univ. Valencia, 46100 Burjassot (Valencia), Spain*
[‡] *Astronomy Department, University of California, Santa Cruz, CA 95064*

Abstract. Using a collapsar progenitor model of MacFadyen & Woosley we have simulated the propagation of an axisymmetric relativistic jet through a collapsing rotating massive star. The jet forms as a consequence of an assumed constant or variable energy deposition in the range $10^{50}\,\mathrm{erg\,s^{-1}}$ to $10^{51}\,\mathrm{erg\,s^{-1}}$ within a 30° cone around the rotation axis. The jet flow is strongly beamed ($\lesssim$ few degrees), spatially inhomogeneous, and time dependent. The jet reaches the surface of the stellar progenitor ($R_* = 2.98 \times 10^{10}$ cm) intact with a maximum Lorentz factor $\Gamma_{\max} = 33$. After breakout the jet accelerates into the circumstellar medium, whose density is assumed to decrease exponentially and then being constant. Outside the star the flow also expands laterally ($v \sim c$), but the beam remains very well collimated. At a distance of $2.54 \times R_*$, where we had to stop the simulation, the Lorentz factor has increased to 44.

MOTIVATION AND INITIAL MODEL

Various catastrophic collapse events have been proposed to explain the energies released in a gamma-ray burst (GRB) including mergers of compact binaries (Pacyński 1986; Goodman 1986), collapsars (Woosley 1993) and hypernovae (Pacyński 1998). According to the current view these models require a stellar mass black hole (BH) which accretes up to several solar masses of matter powering a pair fireball. Recently, it has been argued that the rapid temporal decay of several GRB afterglows is more consistent with the evolution of a relativistic jet after it slows down and spreads laterally than with a spherical blast wave (Sari, Piran & Halpern 1999; Halpern et al. 1999; Kulkarni et al. 1999a; Rhoads 1999). The lack of a radio afterglow in GRB 990123 provides independent evidence for jet-like geometry (Kulkarni et al. 1999b).

Using a 14 $M_\odot$ collapsar model from MacFadyen & Woosley (1999) we have performed axisymmetric relativistic simulations (see also Aloy et al. 1999b). When the central BH has acquired a mass of 3.762 $M_\odot$ we map the model to our computational grid. In a consistent collapsar model a jet will be launched by any process which gives rise to a local deposition of energy and/or momentum, as e.g., $\nu\bar{\nu}$–annihilation, or magneto-hydrodynamic processes. We mimic such a process by depositing energy at a prescribed rate homogeneously within a 30° cone around the rotation axis and for 200 km $\leq r \leq$ 600 km. We have investigated constant energy deposition rates of $\dot{E} = 10^{50}\,\mathrm{erg\,s^{-1}}$ and $\dot{E} = 10^{51}\,\mathrm{erg\,s^{-1}}$, and a varying deposition rate with a mean value of $10^{50}\,\mathrm{erg\,s^{-1}}$. The simulations were performed with the multidimensional relativistic hydrodynamic code GENESIS (Aloy et al. 1999a) using a 2D spherical grid with 200 logarithmically spaced radial zones ($2 \times 10^7\,\mathrm{cm} \leq r \leq R_* = 2.98 \times 10^{10}\,\mathrm{cm}$, and with 100 nonuniform angular zones providing a 0.5° angular resolution within a polar region of $0° \leq \theta \leq 30°$. The gravitational field is described assuming a Schwarzschild BH space–time. Effects due to the self-gravity of the star on the dynamics are neglected.

RESULTS

Constant small energy deposition rate (Model C50): For a constant $\dot{E} = 10^{50}\,\mathrm{erg\,s^{-1}}$ a relativistic jet forms within a fraction of a second and starts to propagate along the rotation axis with a mean speed of $7.8 \times 10^9\,\mathrm{cm\,s^{-1}}$. The jet exhibits all the morphological elements of the Blandford & Rees (1974) jet model: a terminal bow shock, a narrow cocoon, a contact discontinuity separating stellar and jet matter, and a hot spot. The propagation of the jet is unsteady, because of density inhomogeneities in the star. The Lorentz factor of the jet, Γ, increases non-monotonically with time, while the density drops to $\sim 10^{-6}\,\mathrm{g\,cm^{-3}}$. The density profile shows large variations (up to a factor of 100) due to internal shock waves. The mean density in the jet lies in the range $10^{-2} - 1\,\mathrm{g\,cm^{-3}}$. When the jet encounters a region along the axis where the density gradient is positive (at $\log r \approx 8.1$ and $\log r \approx 8.6$) the jet's head is decelerated, while a central channel in the beam is cleaned by outflow into the cocoon through the head, which accelerates the beam. The combination of both effects (deceleration of the head and beam acceleration) increases the strength of the internal shocks. We find that some results are strongly depending on angular resolution, the minimum acceptable one being 0.5° (at least near the axis). Within the uncertainties of the jet mass determination due to finite zoning and the lack of a precise numerical criterion to identify jet matter, the baryon load ($\eta \equiv Mc^2/E_{\mathrm{depos}}$ with $E_{\mathrm{depos}} = \int \dot{E}dt$) decreases with increasing resolution. We find an average $\eta \simeq 1.3$ at jet breakout.

Constant large energy deposition rate (Model C51): Enhancing $\dot{E}$ by a factor of ten (to $10^{51}\,\mathrm{erg\,s^{-1}}$), the jet flow reaches larger Lorentz factors. We observe transients during which the Lorentz factor becomes as large as 40. After 1.2 s the Lorentz factor steadily increases from 22 to 33. The jet propagates faster than in model C50. The time required to reach the surface of the star is 2.27 s instead of 3.35 s. The opening angle of the jet at breakout is $\sim 10°$, i.e., the jet is less collimated. Instead of the strong re-collimation shock present in model C50, several biconical shocks are observed within a very knotty beam and the Lorentz factor near the head of the jet is larger (~ 22 in the final model).

Varying energy deposition rate (Model V50): We have computed a model where the mean energy deposition rate ($10^{50}\,\mathrm{erg\,s^{-1}}$) randomly varies on time scales of a few milliseconds and the amplitude by a factor of ten. Compared to model C50 the jet structure is more knotty and also richer in shocks. At breakout $\Gamma_{\max} = 26.81$, which is *almost twice* as large as the one found in model C50. Thus, a variable $\dot{E}$ is more efficient in converting internal energy into kinetic energy, because the internal shocks are stronger and more numerous in model V50. The mean propagation speed is similar in both models, although the instantaneous velocity of the jet's head is clearly different. The opening angle behind the strongest re-collimation shock is quite small ($< 1°$) in both models.

Evolution after jet breakout: We endowed the star with a Gaussian atmosphere, which at $R_a = 1.8\,R_*$ passes over into an external uniform medium with a density $10^{-5}\,\mathrm{g\,cm^{-3}}$ and a pressure $10^{-8} p(R_*)$. The computational domain is extended to $R_t = 2.54 R_*$ with 70 additional zones. In models C50 and C51 the jet reaches R_t after $\sim 1.8\,\mathrm{s}$ (measured from breakout). Its mean propagation velocity is $\sim 0.85\,c$, which is almost three times faster than the velocity of the head inside the star ($0.30\,c$ in model C50; $0.44\,c$ in model C51). The evolution after jet breakout consists of three distinct epochs. The first one lasting $0.35\,\mathrm{s}$ is characterized by a head velocity of $0.48\,c$ and a small sideways expansion. During the second phase (of $0.3\,\mathrm{s}$) the jet head accelerates to $0.91\,c$, because of the steep external density gradient, and because the flux of axial momentum is still important compared to pressure. The sideways expansion is still sub-relativistic ($\approx 0.008\,c$), and the half opening angle of the beam increases to $\approx 10°$. During the final $1.2\,\mathrm{s}$ the bow-shock propagates within the uniform part of the ambient medium leading to a rapid lateral spreading. The Lorentz factor near the boundary of the cavity blown by the jet grows from ~ 1 (at jet breakout) to ~ 3 in both models decreasing with latitude. At the end of the simulation $\Gamma_{\max}$ is 29.35 (44.17) for model C50 (C51), which is still smaller than the ones required for the fireball model (Piran 1999). The shape of the expanding bubble is prolate during the post-breakout evolution until the jet reaches the uniform part of the circumstellar environment, when its shape changes appreciably. We expect a more isotropic expansion when most of the bubble is inside the uniform medium, and when it is pressure driven (in particular if the energy deposition is switched off).

DISCUSSION

Our simulations show that the jet has a small opening angle ($\sim 8°$) and possesses a highly collimated ($\sim 1°$), ultra-relativistic core in which the Lorentz factor reaches a value of $\Gamma_{\max} = 44$ (model C51) at the end of the simulation about $2\,\mathrm{s}$ after shock breakout. Due to the very good collimation of the outflow the equivalent isotropic kinetic energy (see MacFadyen, Woosley & Heger 1999) slightly exceeds $10^{54}\,\mathrm{erg}$

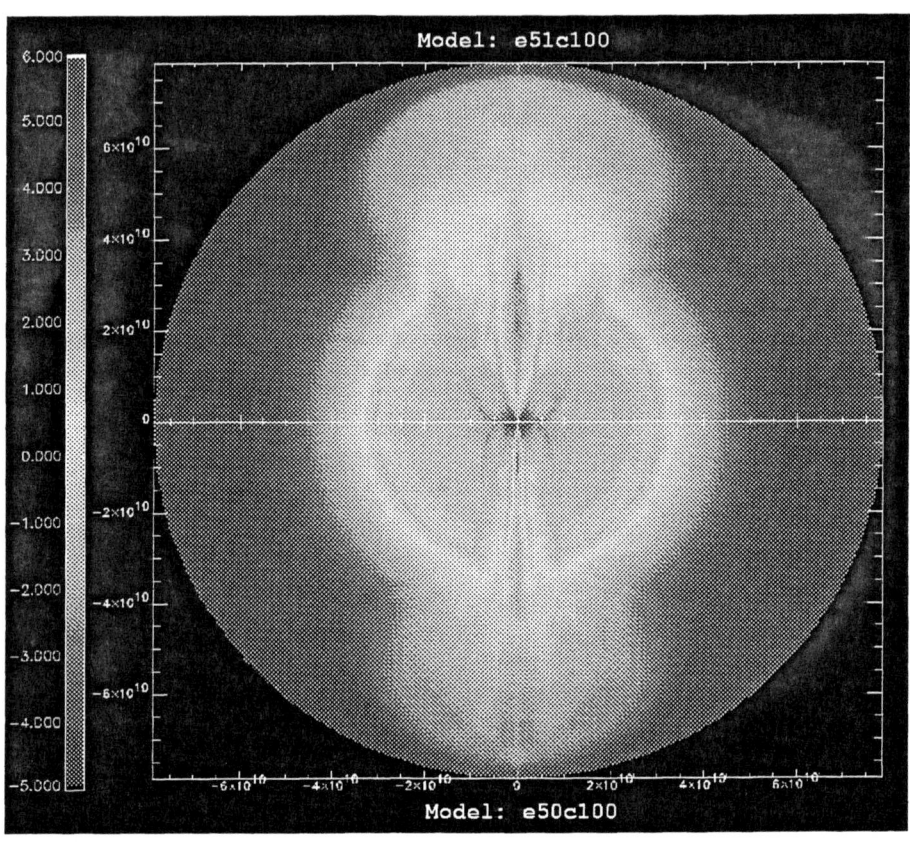

FIGURE 1. Contour maps of the logarithm of the rest–mass density ~ 1.8 s after shock breakout for model C51 (top) and C50 (bottom), respectively. The X and Y axes measure distance in centimeters.

for model C51 (10^{53} erg for model C50) within 2° (5°) of the rotation axis dropping by a factor of 10 within 17° (10°). The inner region contains $8 \times 10^{-4}\, M_\odot$ with $<\Gamma> \sim 4$. For a larger $\dot{E}$, the jet and in particular the cocoon are less collimated, because when the jet is driven harder it also expands stronger laterally (Fig. 1). The rest-mass density and the internal energy show strong spatial and temporal variations within the jet giving rise to a very inhomogeneous baryon load, i.e., the concept of η as a global parameter is useless. Instead it is more appropriate to discuss the efficiency of energy conversion in terms of incremental baryon loads considering only matter within a given range of Γ-values. Although we find an average baryon load of the jet of $\bar{\eta} \sim 1$, some parts of the flow have a baryon load as low as $\sim 10^{-5}$ or even less. After jet breakout $\bar{\eta}$ decreases by a factor 4 in less than 1.8 s. If this trend continues even $\bar{\eta} \sim 10^{-3}$ within 9 s.

At the end of the simulation $\sim 2\, M_\odot$ have a Lorentz factor of less than 3 in model C51, $3 \times 10^{-4} M_\odot$ move with $3 \leq \Gamma < 10$, and for $2 \times 10^{-6} M_\odot$ the Lorentz factor $\Gamma \geq 10$. The latter two masses reduce to $2 \times 10^{-5} M_\odot$ and $2 \times 10^{-7} M_\odot$ for model C50. The efficiency of energy conversion $\mathcal{E}$ (see Aloy et al. 1999b) is 44% (83%) for C50 (C51) at the end of the simulation. If only matter with $\Gamma > 10$ is considered, $\mathcal{E}$ drops to 0.6% (1%). Except for the early evolution ($t < 1$ s) the amount of matter moving at moderate ($3 \leq \Gamma < 10$) and highly ($\Gamma \geq 10$) relativistic velocities increases by a factor ~ 3 every second, i.e., if the central engine is active for another 5 s at the assumed $\dot{E}$, $\sim 10^{-4} M_\odot$ will move with $\Gamma \geq 10$. As the maximum Lorentz factor also rapidly increases, it is not unlikely that maximum Lorentz factors of several hundreds can be reached before the central engine is switched off. For models where the released total energy is equal (C50 and V50) the maximum Lorentz factor is by a factor of two higher for a time-dependent $\dot{E}$.

We plan to extend our simulations to later times and larger radii ($\sim 10^{13}$ cm) using adaptive mesh refinement techniques, in order to account for the observational properties of GRBs and their afterglows.

Acknowledgements: This work has been supported in part by the Spanish DGES (grant PB97-1432) and the CSIC. The calculations were performed on two SGI Origin 2000 at the CEPBA and at the SIUV.

REFERENCES

1. Aloy, M.A., Ibáñez, J.Mª, Martí, J.Mª, and Müller, E., *Astrophys. J. Suppl.* **122**, 151 (1999a).
2. Aloy, M.A., Müller, E, Ibáñez, J.Mª, Martí, J.Mª, and MacFadyen, A., *Astrophys. J. Lett.* submitted; and astro-ph/9911098 (1999b).
3. Blandford, R., and Rees, M.J., *Month. Not. Roy. Soc. Astron.* **169**, 395 (1974).
4. Goodman, J., *Astrophys. J.* **308**, L47 (1986).
5. Halpern, J.P, Kemp, J., Piran, T. and Bershady, M.A., *Astrophys. J. Lett.* **517**, L105 (1999).
6. Kulkarni, S.R., Djorgovski, S.G., Odewahn, S.C., Bloom, J.S., Gal, R.R., et al., *Nature* **398**, 389 (1999a).
7. Kulkarni, S.R., Frail, D.A., Sari, R., Moriarty-Schieven, G.H., Shepherd, D.S., Udomprasert, P., Readhead, A.C.S., Bloom, J.S., Feroci, M., and Costa, E., astro-ph/9903441 (1999b).
8. MacFadyen, A., and Woosley, S.E., *Astrophys. J.* **524**, 262 (1999).
9. MacFadyen, A., Woosley, S.E., and Heger, A., *ApJ* submitted; and astro-ph/9910034 (1999).
10. Pacyński, B., *Astrophys. J.* **308**, L43 (1986).
11. Pacyński, B., *Astrophys. J.* **494**, L45 (1998).
12. Piran, T., *Physics Reports* **314**, 575 (1999).
13. Rhoads, J.E., *Astrophys. J.* submitted; and astro-ph/9903399 (1999).
14. Sari, R., Piran, T. and Halpern, J.P., *Astrophys. J.* **519**, L17 (1999).
15. Woosley, S.E., *Astrophys. J.* **405**, 273 (1993).

Gamma-Ray Bursts via Pair Plasma Fireballs from Heated Neutron Stars

Jay D. Salmonson*, James R. Wilson*[†], and Grant J. Mathews[†]

Lawrence Livermore National Laboratory, Livermore, CA 94550
[†]*University of Notre Dame, Notre Dame, IN 46556*

Abstract. In this paper we model the emission from a relativistically expanding e^+e^- pair plasma fireball originating near the surface of a heated neutron star. This pair fireball is deposited via the annihilation of neutrino pairs emanating from the surface of the hot neutron star. The heating of neutron stars may occur in close neutron star binary systems near their last stable orbit. We model the relativistic expansion and subsequent emission of the plasma and find $\sim 10^{51} - 10^{52}$ ergs in γ-rays are produced with spectral and temporal properties consistent with observed gamma-ray bursts.

INTRODUCTION

It has been speculated for some time that inspiraling neutron stars could provide a power source for cosmological gamma-ray bursts [5,6]. However, previous Newtonian and post-Newtonian studies [1] of the final merger of two neutron stars have found that the neutrino emission time scales are so short that it would be difficult to drive a gamma-ray burst from this source. It is clear that a mechanism is required for extending the duration of energetic neutrino emission. A number of possibilities could be envisioned, for example, neutrino emission powered by accretion shocks, MHD or tidal interactions between the neutron stars, etc. The present study, however, has been primarily motivated by numerical studies of the strong field relativistic hydrodynamics of close neutron star binaries (NSBs) in three s-patial dimensions. These studies [11,12,4,2] suggest that neutron stars in a close binary can experience relativistic compression and heating over a period of seconds. During the compression phase released gravitational binding energy can be converted into internal energy. Subsequently, up to 10^{53} ergs in thermally produced neutrinos can be emitted before the stars collapse [4]. Here we briefly summarize the physical basis of this model and numerically explore its consequences for the development of an e^+e^- plasma and associated GRB.

In [4] properties of equal-mass neutron-star binaries were computed as a function of mass and EOS (Equation of State). From these studies it was deduced that compression, heating and collapse could occur a few seconds before binary merger.

Our calculation of the rates of released binding energy and neutron star cooling suggests that interior temperatures as hot as 70 MeV are achieved. This leads to a high neutrino luminosity which peaks at $L_\nu \sim 10^{53}$ ergs sec^{-1}. This much neutrino luminosity would partially convert to an e^+e^- pair plasma above the stars as is also observed above the nascent neutron star in supernova simulations [13].

NEUTRINO ANNIHILATION AND PAIR CREATION

Having outlined a mechanism by which neutrino luminosities of 10^{52} to 10^{53} ergs/sec may arise from binary neutron stars approaching their final orbits, we must calculate the efficiency of conversion of neutrino pairs into an electron pair plasma via $\nu\bar{\nu} \to e^+e^-$. Here we argue that the efficiency for converting these neutrinos into pair plasma is probably quite high. Neutrinos emerging from the stars will deposit energy outside the stars predominantly by $\nu\bar{\nu}$ annihilation to form electron pairs. A secondary mechanism for energy deposition is the scattering of neutrinos from the e^+e^- pairs. Strong gravitational fields near the stars will bend the neutrino trajectories. This greatly enhances the annihilation and scattering rates [8]. For our employed neutron-star equations of state the radius to mass ratio is typically between $R/M \sim 3$ and 4 just before stellar collapse (in units $G = c = 1$). In [8] it is shown that $\nu\bar{\nu}$ annihilation rates will be enhanced by a factor $\mathcal{F}(R/M) \sim 8$ to 28 due to relativistic effects. From Eq. 24 of [8] we obtain,

$$\frac{\dot{Q}}{L_\nu} \approx 0.03 \mathcal{F}(R/M) L_{53}^{5/4} \ . \tag{1}$$

Thus, the efficiency of annihilation ranges from ≈ 0.1 to $0.84 \times L_{53}^{5/4}$. For the upper range of luminosity the efficiency is quite large. Also, using the supernova code of Wilson and Mayle [13] we calculate the entropy per baryon of the plasma to be as high as 10^6, thus the resulting pair plasma will have low baryon loading.

PAIR PLASMA EXPANSION AND SHOCK WITH ISM

Having determined the initial conditions of the hot e^+e^- pair plasma near the surface of a neutron star, we wish to follow its evolution and characterize the observable gamma-ray emission. To study this we have developed a spherically symmetric, general relativistic hydrodynamic computer code to track the flow of baryons, e^+e^- pairs, and photons. For the present discussion we consider the plasma deposited at the surface of a $1.45 M_\odot$ neutron star with a radius of 10 km. Discussion of this code can be found in [9,7]. In those papers the emission from an expanding fireball was studied. In Figure 1 it is shown that the resulting emission spectrum and γ-ray emission efficiency E_γ/E_{tot} strongly depends upon the entropy per baryon of the plasma deposited near the surface of the neutron stars; entropies of $\lesssim 10^6$ resulted in weak emission with most of the original energy manifesting

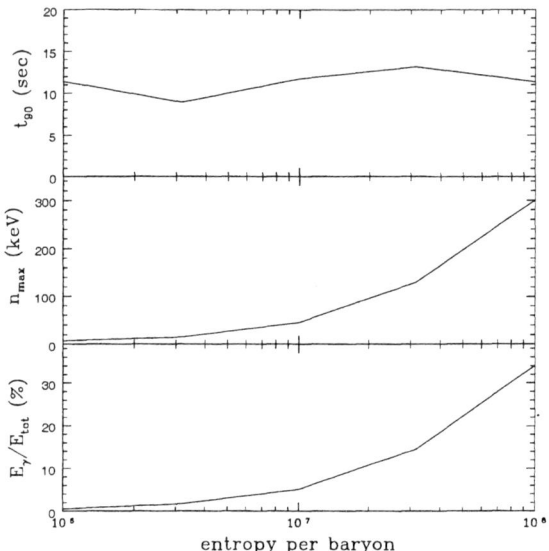

FIGURE 1. The duration, energy at the number spectrum peak, and gamma-ray efficiency are plotted for the emission from an expanding pair plasma with total deposited energy $E_{tot} = 10^{52}$ ergs over a range of entropies per baryon 10^6 to 10^8.

itself as kinetic energy of the baryons. Thus, for the low entropy per baryon fireballs ($s \sim 10^5 - 10^6$) produced by NSBs it is necessary to examine the emission due to the interaction of the relativistically expanding baryon wind with the interstellar medium (ISM). We find that these baryon winds typically have a Lorentz factor $\gamma \approx 300$ and have a total energy $\approx 10^{52}$ ergs.

After becoming optically thin and decoupling with the photons, the matter component of the fireball continues to expand and interact with the ISM via collisionless shocks. As the ISM is swept up, the matter decelerates. We model this process as an inelastic collision between the expanding fireball and the ISM as in, for example, [6]. We assume that the absorbed internal energy is immediately radiated away. From this we construct a simple picture of the emission due to the matter component of the fireball "snowplowing" into the ISM of baryon number density n.

We have constructed an analytic formula for the luminosity in time [7] of the fireball plowing into the ISM. We show a plot of this function in Figure 2 for a range of ISM densities. Defining t_{max} as the time of maximum luminosity

$$L(t) \propto \begin{cases} t^2 & \text{free expansion phase } (t < t_{max}) \\ t^{-10/7} & \text{deceleration phase } (t > t_{max}). \end{cases} \quad (2)$$

This luminosity curve has the so called "FRED" (Fast Rise, Exponential Decay) profile which is characteristic of real bursts.

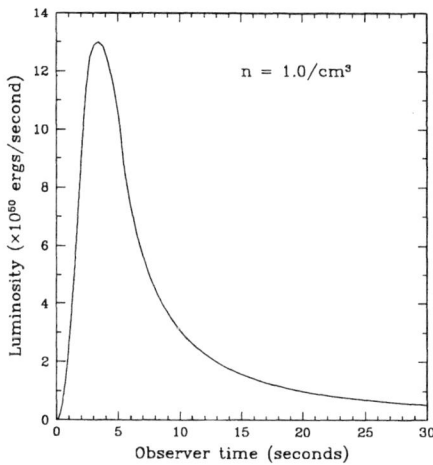

FIGURE 2. The light curve for a 10^{52} erg fireball expanding at $\gamma = 300$ into the interstellar medium with number density of one baryon per cm^3.

Synchrotron Shock Spectrum

Using the theory of synchrotron shocks (e.g., [10]) we can construct a spectrum as shown in Figure 3. To model the synchrotron spectrum there are three free parameters: ϵ_B, ϵ_e are the fractions of baryonic kinetic energy that is deposited into the magnetic field and the electrons respectively, and n is the number density of baryons in the ISM. In these calculations we assume $\epsilon_B = \epsilon_e = 1/4$. As shown in Figure 3, a reasonable ISM density of 1 baryon/cm^3 gives a peak in the νL_ν spectrum at ~ 100 keV in agreement with observations. Calculations of the efficiency show that 75 % of the energy is emitted at photon energies of 10 keV and above [7].

CONCLUSIONS

In these proceedings we have argued that heated neutron stars (perhaps by stellar compression of close neutron-star binaries) are viable candidates for the production of large, high entropy per baryon, e^+e^- pair plasma fireballs, and thus, for the creation of gamma-ray bursts. We find that fireballs of total energies $E \sim 10^{51}$ to 3×10^{52} ergs and entropies per baryon $s > 10^5$ are possible. Also, this model gives a power-law spectrum that peaks at hundreds of keV and has an overall efficiency of 10–20%.

Acknowledgments: This work was performed under the auspices of the U.S. Department of Energy by the Lawrence Livermore National Laboratory under contract

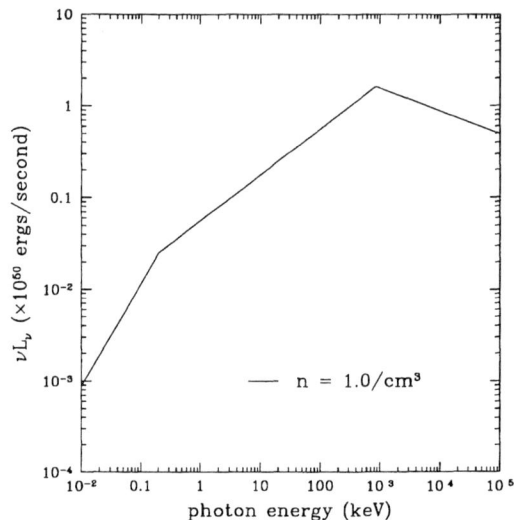

FIGURE 3. The peak synchrotron spectrum for a 10^{52} erg fireball expanding at $\gamma = 300$ into interstellar medium number density of one baryon per cm^3.

W-7405-ENG-48. J.R.W. was partly supported by NSF grant PHY-9401636. Work at the University of Notre Dame is supported in part by DOE grant DE-FG02-95ER40934, NSF grant PHY-97-22086, and by NASA CGRO grant NAG5-3818.

REFERENCES

1. Janka, H. T. and Ruffert, M., *A&A* **307**, L33 (1996).
2. Mathews, G. J., Marronetti, P. and Wilson, J. R., *Phys. Rev. D* **58**, 043003 (1998).
3. Mathews, G. J. and Wilson, J. R., submitted *Phys. Rev. D*, gr-qc/9911047 (2000).
4. Mathews, G. J. and Wilson, J. R., *ApJ* **482**, 929 (1997).
5. Mészáros, P. and Rees, M. J., *ApJ* **397**, 570 (1992).
6. Piran, T., astro-ph/9907392 (1999).
7. Salmonson, J. D. and Wilson, J. R. and Mathews, G. J., submitted *ApJ* (2000).
8. Salmonson, J. D. and Wilson, J. R., *ApJ* **517**, 859 (1999).
9. Salmonson, J. D. and Wilson, J. R. and Mathews, G. J., in *Gamma-Ray Bursts: 4th Huntsville Symposium*, AIP Conf. Proc. 428, eds. C. A. Meegan, R. D. Preece and T. M. Koshut, 1997, p. 788.
10. Sari R., Piran T., Narayan R., *ApJL* **497**, L17 (1998).
11. Wilson, J. R. and Mathews, G. J., *Phys. Rev. Lett.* **75**, 4161 (1995).
12. Wilson, J. R. and Mathews, G. J. and Marronetti, P., *Phys. Rev. D* **54**, 1317 (1996).
13. Wilson, J. R. and Mayle, R. W., *Phys. Rep.* **227**, 97 (1993).

Asymmetric Neutron Star Coalescences: Implications for GRBs

S. Rosswog[1], M. B. Davies[2], F.-K. Thielemann [3], and T. Piran[4]

[1] *Center for parallel computing (ZPR/ZAIK), Universität zu Köln, Germany,*
[2] *Dept. Physics and Astronomy, University of Leicester LEI 7RH, UK,*
[3] *Departement für Physik und Astronomie, Universität Basel, Switzerland,*
[4] *Racah Institute for Physics, Hebrew University, Jerusalem, Israel*

Abstract. The coalescence of two neutron stars in a binary is a prominent candidate for the central engine of GRBs. Previous calculations for realistic equations of state have focussed on symmetric systems, where both components possess the same spin and mass. Here we analyze systems that are slightly asymmetric with the asymmetry stemming from either different masses (1.4 and 1.6 $M_\odot$) or spins that are not aligned with the orbital angular momentum. Our calculations are performed using a 3D hydrodynamics code that is coupled to an elaborate equation of state for hot and dense nuclear matter. Implications for GRBs are discussed.

INTRODUCTION

A central black hole (BH) surrounded by a thick disk is at the heart of the most prominent models for the central engine to power Gamma Ray Bursts (GRBs). It is currently believed by most people working in the field that there should be (at least) two different scenarios to explain the enormous variety in the observed GRBs: one for the short time scale (~ 0.3 s) bursts and another one for the long bursts with a duration of ~ 30 s. The difference of both models lies in the formation mechanism of the BH and the corresponding masses of the disks: in one case it is believed to form in the coalescence of two neutron stars (or a neutron star with a low-mass BH) in the other it is suggested to form during the collapse of a massive rotating star [1,2].

Our focus here will be on the first scenario, especially on the consequences of slight asymmetries in the initial binary system coming from either a mass ratio different from unity or spins that are not aligned with the orbital angular momentum.

NUMERICAL METHODS AND INITIAL CONFIGURATIONS

The neutron stars are modeled as a fluid that is governed by the nuclear equation of state of Lattimer and Swesty [3] under the influence of self-gravity. The hydrodynamic equations are solved in 3D applying the well-known SPH method (see e.g., [4]). For the artificial viscosity (AV) a new hybrid scheme is used that benefits from two recently suggested improvements: the decrease of the viscosity parameters to low, minimum values in the absence of shocks [6] and the so-called Balsara-Switch [5] to suppress AV in pure shear flows. A detailed discussion may be found in [10].

Observed neutron star binary components have very similar masses [11], but it is known from calculations with polytropic equations of state [12] that the evolution of the final coalescence shows an enormous sensitivity to deviations of the mass ratio q from unity. Thus, we investigate the fate of unequal mass binary systems (1.4 and 1.6 $M_\odot$ of baryonic material) where the stars are either corotating, not rotating at all or rotating opposite to the orbital motion. In a second group of calculations we focus on equal mass systems (1.4 $M_\odot$) with one irrotational component and the second star's spin with or against the orbital motion or lying in the orbital plane. The spin period is for simplicity always chosen to be equal to the orbital period ($\sim$ 3 ms) at the initial separation ($a_0 = 45$ km). A detailed discussion of the results is given in [10].

RESULTS

We will focus here basically on two issues: the evolution of the distribution of mass in space and the baryonic loading of the polar regions of the central object. Results concerning the formation of heavy elements in the ejecta of neutron star coalescences may be found elsewhere [7–9].

Mass Distribution

The mass distribution of the corotating system with 1.4 and 1.6 $M_\odot$ is shown in Fig. 1. Prior to the merger the asymmetry in the distribution of mass is hardly visible. However, as soon as the dynamical instability sets in, the less massive component (secondary) gets stretched by the tidal forces of the heavier star (primary) and dynamic mass transfer sets in. The material approaches the primary at a subsonic speed ($v_{matter} v_{sound}^{-1} \sim 10^{-3}$) and quickly engulfs it in a way that it finally forms the central core of the merged configuration. The rest of the secondary forms a spiral arm like outflow. Around four percent of the tip of this spiral arm becomes ejected into space and probably forms r-process material [9]. By means of the mass gain from the secondary, the primary will probably become unstable to

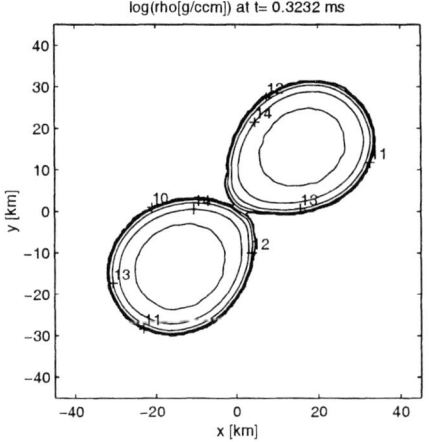 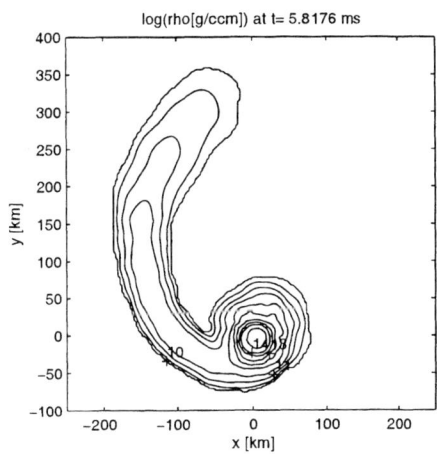

FIGURE 1. Density contours of a corotating system with 1.4 and 1.6 $M_\odot$ (10000 particles).

gravitational collapse. Though the morphology may be drastically different from equal-mass systems, the possible central engine to power the GRB is not: a fast spinning black hole of 2.5 to 3.0 $M_\odot$ surrounded by a thick, hot disk. The disk masses are very sensitive to the initial spins: the coalescence of the system where one star spins against the orbital angular momentum produces a disk of only ~ 0.05 $M_\odot$ while we find ~ 0.25 $M_\odot$ for corotation. The spins thus largely determine the available energy reservoir to produce a GRB.

Polar Region

For a long time the emergence of a GRB from a region "contaminated" with baryons has been thought to be impossible. If spherical symmetry is assumed, an amount as small as 10^{-5} $M_\odot$ of baryonic material injected into the fireball is enough to prevent a GRB from forming [14]. Most recently, more detailed calculations [1,13] have shown that under certain conditions still large relativistic gamma factors can be achieved despite considerable baryonic loading.

If the mechanism suggested for the emergence of the jet in the collapsar model also works for the coalescence scenario is at present still an open question. It is interesting to ask whether the region above the poles of the merged object remains baryon free if the neutron star spins are not aligned with the orbital angular momentum. However, even in the case where one star is spinning in the orbital plane still baryon free funnels emerge over the poles of the central object. The basic effect of this spin configuration is to tilt the disk by some small angle. Things might become more severe for the systems with a mass ratio $q \neq 1$. Especially in the corotating case more material is driven towards the rotational axis and might present a severe danger for the fate of a possibly emerging fireball.

SUMMARY

We have presented hydrodynamic calculations of the final stages of the inspiral and subsequent merger of a neutron star binary system. Our focus has been on the effects of slight asymmetries in the initial system coming either from a mass ratio deviating from unity or neutron star spins that are not aligned with the orbital angular momentum. We find that the dynamical evolution for $q \neq 1$ during the coalescence is drastically different from equal mass systems. However, the final BH-disk systems are similar to their equal mass counterparts: ~ 2.5 to ~ 3.0 $M_\odot$ for the BH and ~ 0.05 to ~ 0.25 $M_\odot$ for the disk. Non-aligned spins do not seem to present a danger for the emergence of a baryon free funnel above the central object. The problem might be more severe for the binaries with $q \neq 1$. Especially for the corotating case we find substantially more mass close to the rotational axis. However, this point needs further efforts: the present numerical resolution is too poor to draw conclusions on the precise amount of material that might contaminate the fireball. In addition, further investigations have to clarify if this would present a danger at all, or if mechanisms similar to the collapsar case become effective and lead to highly relativistic outflow in spite of the presence of baryonic matter.

REFERENCES

1. MacFadyen, A. & Woosley, S.E., astro-ph/9810274 (1999).
2. MacFadyen, A. & Woosley, S.E., Heger, A., astro-ph/9910034 (1999).
3. Lattimer, J. M., & Swesty, D., *Nucl. Phys.* **A535**, 331 (1991).
4. Benz, W., in *Numerical Modeling of Stellar Pulsations*, ed. J. Buchler, Kluwer Academic Publishers, Dordrecht, 1990, pp. 269.
5. Balsara, D., *J. Comp. Phys.* **121**, 357 (1995).
6. Morris, J.P., & Monaghan, J.J., *J. Comp. Phys.* **136**, 41 (1997).
7. Rosswog, S., Freiburghaus, C., & Thielemann, F.-K., in *Proc. Nuclei in the Cosmos V*, 1998.
8. Rosswog, S., Liebendörfer, M., Thielemann, F.-K., Davies, M.B., Benz, W., & Piran, T., *A&A* **341**, 499 (1999).
9. C. Freiburghaus, S. Rosswog, & F.-K. Thielemann, *ApJL* **525**, L121 (1999)
10. Rosswog, S., Davies, M.B., Thielemann, F.-K., & Piran, T., to be published (1999).
11. Thorsett, S.E. & Chakrabarti, *ApJ* **512**, 288 (1999).
12. Rasio, F.A., & Shapiro, S.L., *ApJ* **438**, 887 (1995).
13. Aloy, M.A., Müller, E., Ibanez, J.M., Marti, J.M., & MacFadyen, A., astro-ph/9911098 (1999).
14. Shemi, A., & Piran, T., *ApJL* **365**, L55 (1990).

Mass-Loss from a Magnetically Driven Wind Emitted by a Disk Orbiting a Stellar-Mass Black Hole

Frédéric Daigne[*,†] and Robert Mochkovitch[*]

*Institut d'Astrophysique de Paris, 98 bis boulevard Arago, 75014 Paris, France
†Max-Planck Institut für Astrophysik, Karl-Schwarzschild-Str. 1, 85748 Garching, Germany

Abstract. The source of cosmic gamma–ray bursts (hereafter GRBs) is usually believed to be a stellar mass black hole accreting material from a thick disk. The mechanism for the production of a relativistic wind by such a system is still unknown. We investigate here one proposal where the disk energy is extracted by a magnetic field amplified to very large values $B \sim 10^{15}$ G. Using some very simple assumptions we compute the mass loss rate along magnetic field lines and then estimate the Lorentz factor Γ at infinity. We find that Γ can reach high values only if severe constraints on the field geometry and the conditions of energy injection are satisfied. We discuss the results in the context of different scenarios for GRBs.

INTRODUCTION

Most of the sources which are now discussed to explain GRBs (the coalescence of two compact objects or the collapse of a massive star to a black hole (collapsar) [1–3]) lead to the same system : a stellar mass black hole surrounded by a thick debris torus. The release of energy by such a configuration can come from the accretion of disk material by the black hole or from the rotational energy of the black hole extracted by the Blandford-Znajek mechanism. The released energy is first injected into a relativistic wind and then converted into gamma–rays, via the formation of shocks probably within the wind itself [4,5]. The wind is finally decelerated by the external medium which leads to a shock responsible for the afterglow emission observed in the X-rays, optical and radio bands [6].

The production of the relativistic wind is a very complex question because of the very low baryonic load that has to be achieved in order to reach high values of the terminal Lorentz factor. Just a few ideas have been proposed and none appears to be fully conclusive. A first possibility to extract the energy from accretion is the annihilation of neutrino–antineutrino pairs emitted by the hot disk along the rotation axis of the system, which is a region strongly depleted in baryons due to centrifugal forces. The low efficiency of this process however requires high neutrino luminosities and therefore short accretion time scales [7]. Another possibility to extract the energy from accretion is to assume that the magnetic field in the disk is amplified by differential rotation to very large values ($B \sim 10^{15}$ G). A magnetically driven wind could then be emitted from the disk with a fraction of the Poynting flux being eventually transferred to matter. The energy can also be extracted from the rotational energy of the black hole by the Blandford-Znajek mechanism [8].

We present here an exploratory study of the case where a magnetically driven wind is emitted by the disk. Matter is heated at the basis of the wind (by $\nu\bar{\nu}$ annihilation, viscous dissipation, magnetic reconnection, etc.) and then escapes, guided along the magnetic field lines. We begin by describing a "toy model" to explore the behavior of such a wind. Despite its extreme simplicity, we expect that it can help to identify the key parameters controlling the baryonic load. Our results are then presented and discussed in the context of different scenarios for GRBs.

A "TOY MODEL"

We solve the wind equations with the following simplifications: (i) we assume a geometrically thin disk and a poloidal magnetic field with the most simple geometry (straight lines making an angle θ with the disk); (ii) we consider that a stationary regime has been reached by the wind; (iii) we use non–relativistic equations (to obtain the mass loss rate we just need to solve them up to the sonic point, where $v < 0.1c$) but we adopt the Paczyński-Wiita potential for the black hole

$$\Phi_{\rm BH} = -\frac{GM_{\rm BH}}{r - r_{\rm S}} \quad \text{with} \quad r_{\rm S} = \frac{2GM_{\rm BH}}{c^2} . \tag{1}$$

We write the flow equations (continuity, Euler and energy equations) in a frame co-rotating with the foot of the field line, anchored at a radius r_0 in the disk

$$\rho\, v\, s(x) = \dot{m} , \tag{2}$$

$$v\frac{dv}{dx} = g(x)r_0 - \frac{1}{\rho}\frac{dP}{dx} , \tag{3}$$

$$v\frac{d\epsilon}{dx} = \dot{q}(x)r_0 + v\frac{P}{\rho^2}\frac{d\rho}{dx} , \tag{4}$$

where $x = \ell/r_0$, ℓ being the distance along the magnetic field line, and ρ, P, ϵ and v are the density, pressure, specific internal energy and velocity in the flow.

The total acceleration $g(x)$ includes both gravitational and centrifugal terms. In this exploratory study, the power deposited per unit mass $\dot{q}(x)$ only takes into account the heating and cooling due to neutrinos. We assume that the inner part of the disk is optically thick (which is probably justified for compact object mergers but is more questionable for collapsars, except for low α–viscosity ($\alpha < 0.01$) [10]). We include the following processes: neutrino capture on free nucleons, neutrino scattering on relativistic electrons and positrons and neutrino–antineutrino annihilation (heating); neutrino emission by nucleons and annihilation of electron-positrons pairs (cooling). The temperature distribution in the disk corresponds to a geometrically thin, optically thick disk:

$$T_\nu(r) = T_* \left(\frac{r_*}{r}\right)^{3/4} \left(\frac{1 - \sqrt{\frac{r_{in}}{r}}}{1 - \sqrt{\frac{r_{in}}{r_*}}}\right)^{1/4} \quad (T_* \text{ is the temperature at } r_*) \; ; \quad (5)$$

The section of the wind $s(x)$ is easily related to the field geometry because the field and stream lines are coincident. We adopt the equation of state computed by [9] which includes nucleons, relativistic electrons and positrons and photons.

The acceleration $g(x)$ along a field line is negative up to $x = x_1$ for angles larger than $\theta_1 \simeq 60°$ (60° is the exact value for a Newtonian instead of a Paczyński-Wiita black hole potential). For $x > x_1$, $g(x)$ is dominated by the centrifugal force. The sonic point of the flow is located at a distance x_s just below x_1 (the relative difference never exceeds 1%). We solve the flow equations in a classical way by inward integration along the field line. We start at the sonic point by fixing trial values of the temperature T_s and the density ρ_s from which we get the velocity v_s and the position x_s (from the condition of regularity at $x = x_s$) and then the value of the mass loss rate $\dot m$. We observe that at some position x_{cr}, the velocity v begins to fall off rapidly, while T reaches a maximum $T_{max} \leq T_\nu(r_0)$. We adjust T_s and ρ_s so that x_{cr} is as close as possible to 0 and T_{max} to $T_\nu(r_0)$.

RESULTS

We have studied the dependence of the mass loss rate $\dot m$ on the different model parameters and found the following expression:

$$\dot m(r) \sim 3.8 \; 10^{13} \left(\frac{M_{BH}}{2.5 \; M_\odot}\right) \left(\frac{T_\nu(r)}{2 \; \text{MeV}}\right)^{10} f\left[\frac{r}{r_g}; \theta(r)\right] \text{ g/cm}^2/\text{s} \; . \quad (6)$$

The geometrical function f is normalized in such a way that it is equal to unity for $r = 4 \; r_g$ and $\theta(r) = 85°$. The very strong dependence of $\dot m$ with $T_\nu(r)$ (tenth power) is in agreement with what is found for neutrino driven winds in spherical geometry [11]. Figure 1 shows that $\dot m$ also strongly depends on the inclination angle. The other important parameters are the position in the disk and the mass of the black hole, while $\dot m$ depends only weakly on all other parameters like the size of the optically thick region (here $r_{in} = 3 \; r_g$ and $r_{out} = 10 \; r_g$). In the more general case where the source of heating is not restricted to neutrino processes but can also include viscous dissipation, magnetic reconnection, etc, we have obtained a very simple and general analytical approximation for $\dot m$ [12]:

$$\dot m \sim \frac{\dot e}{\Delta \Phi} \delta \; , \quad (7)$$

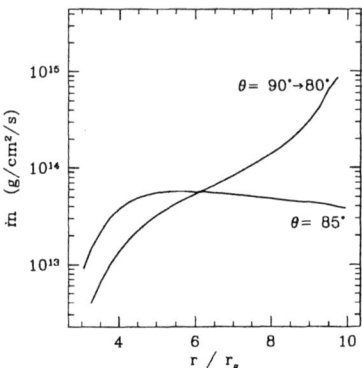

FIGURE 1. Mass loss rate $\dot{m}$ from the disk for a constant ($\theta = 85°$) and decreasing (from $90°$ to $80°$ between 3 and 10 r_g) inclination of the field lines. The disk temperature is $T_* = 2$ MeV at $r_* = 4\ r_g$. The mass of the black hole is $M_{\rm BH} = 2.5\ M_\odot$.

where $\dot{e}$ is the rate of energy deposition (in erg/cm^2/s) between the plane of the disk ($x = 0$) and the sonic point ($x = x_s \simeq x_1$), $\Delta\Phi$ is the difference of potential (gravitational+centrifugal) between $x = 0$ and $x = x_1$ and δ is a factor close to unity depending on the distribution of energy injection between $x = 0$ and $x = x_s$.

We can now estimate the average Lorentz factor $\bar{\Gamma} = \dot{E}/\dot{M}c^2$ at infinity. The total mass loss rate $\dot{M}$ and the power injected into the wind $\dot{E}$ are given by

$$\dot{M} = 2\int_{r_{in}}^{r_{out}} \dot{m} 2\pi r\, dr = 2.6\ 10^{26} \left(\frac{M_{\rm BH}}{2.5\ M_\odot}\right)^3 \left(\frac{T_*}{2\ {\rm MeV}}\right)^{10} F_{\rm geo}\ {\rm g/s} \quad (8)$$

$$\text{and } \dot{E} = 2\ 10^{51} \left(\frac{\Omega/4\pi}{0.1}\right) \left(\frac{f_\gamma}{0.05}\right)^{-1} \left(\frac{\dot{E}_\gamma}{10^{51}/4\pi\ {\rm erg/s/sr}}\right)\ {\rm erg/s}, \quad (9)$$

where $F_{\rm geo} = \int_{r_{in}/r_g}^{r_{out}/r_g} f[x;\theta(x)]\, x\, dx$ is a function of the field geometry only; $\dot{E}_\gamma$ is the burst power in gamma–rays, $\Omega/4\pi$ is the beaming factor and f_γ is the efficiency for the conversion of kinetic energy into gamma–rays. The wind is powered by accretion but at the same time the disk is heated by viscous dissipation and cools by emitting neutrinos. We assume that these losses represent a fraction α of the power $\dot{E}$ injected into the wind, so that we can estimate T_* at $r_* = 4\ r_g$:

$$\dot{E}_\nu = \alpha \dot{E} = 2\int_{r_{in}}^{r_{out}} \frac{7}{8}\sigma T_\nu^4(r)\, 2\pi r\, dr \quad (1$$

$$\text{and } T_* = 1.72\ \alpha^{\frac{1}{4}} \left(\frac{M_{\rm BH}}{2.5\ M_{\rm odot}}\right)^{-\frac{1}{2}} \left(\frac{\Omega/4\pi}{0.1}\right)^{\frac{1}{4}} \left(\frac{f_\gamma}{0.05}\right)^{-\frac{1}{4}} \left(\frac{\dot{E}_\gamma}{10^{51}/4\pi\ {\rm erg/s/sr}}\right)\ {\rm MeV}.\quad (1$$

From equations (8), (9) and (11), we can calculate the average Lorentz factor

$$\bar{\Gamma} = \frac{8500}{F_{\rm geo}}\alpha^{-\frac{5}{2}}\left(\frac{M_{\rm BH}}{2.5\ M_\odot}\right)^2 \left(\frac{\dot{E}_\gamma}{10^{51}/4\pi\ {\rm erg/s/sr}}\right)^{-\frac{3}{2}} \left(\frac{\Omega/4\pi}{0.1}\right)^{-\frac{3}{2}} \left(\frac{f_\gamma}{0.05}\right)^{\frac{3}{2}}. \quad (12)$$

The value of F_{geo} is 56 for a constant inclination $\theta = 85°$ and 250 if θ decreases from $90°$ to $80°$ between $r = 3$ and $10\ r_g$. We therefore conclude that large terminal Lorentz factors can be reached only if several severe constraints are satisfied: (i) low F_{geo} values, i.e., quasi-vertical field lines; (ii) low α values, i.e., good efficiency for energy injection into the wind with little dissipation; (iii) low value of $\Omega/4\pi$, i.e., necessity of beaming. With the more general equation (7) we can obtain another simple and useful constraint: if the power $\dot{e}$ deposited below the sonic point represents a fraction χ of the total power $\dot{e}_{tot}$ injected into the wind, we have

$$\Gamma \sim \frac{\dot{e}_{tot}}{\dot{m}c^2} \sim \frac{\Delta\Phi/c^2}{\delta\chi} \ . \qquad (13)$$

For $r = 4\ r_g$ and $\theta = 85°$, we obtain $x_1 = 2.182$ and $\Delta\Phi/c^2 = 0.18$ which implies that χ should not exceed 10^{-3} to have $\Gamma > 100$!

DISCUSSION

This study is clearly limited by its crude assumptions. However the severe constraints we get show how difficult it may be to produce a relativistic MHD wind from the disk. An optimistic view of our results would be to consider that this difficulty could just be a way to explain the apparent discrepancy between the observed rate of GRBs and the birthrate of sources in the collapsar scenario, most collapsars failing to give a GRB. A more pessimistic point of view would be to conclude that the baryonic load of such winds is never sufficiently low so that they remain non-relativistic. If one chooses to rely on the Blandford-Znajek mechanism to power the wind [8], it should be noted that this process must not be "contaminated" by frozen material carried along magnetic field lines coming from the disk and trapped by the black hole.

REFERENCES

1. Narayan, R., Paczyński, B. and Piran, T., *ApJ* **395**, L83 (1992).
2. Woosley, S.E., *ApJ* **405**, 273 (1993).
3. Paczyński, B., *ApJ* **494**, L219 (1998).
4. Rees, M.J. and Mészáros, P., *ApJ* **430**, L93 (1994).
5. Daigne, F. and Mochkovitch, R., *MNRAS* **296**, 275 (1998).
6. Wijers, R.A.M.J, Rees, M.J. and Mészáros, P., *MNRAS* **288**, L51 (1997).
7. Ruffert, M., Janka, H.T., Takahashi, K. and Schaeffer, G., *A&A* **319**, 122 (1997).
8. Lee, H.K., Wijers, R.A.M.J. and Brown, G.E., *ASTRO-PH 9906213* (1999).
9. Bethe, H.A., Applegate, J.H. and Brown, G.E., *ApJ* **241**, 343 (1980).
10. Popham, R., Woosley, S.E. and Fryer, C., *ApJ* **518**, 356 (1999).
11. Duncan, R.C., Shapiro, S.L. and Wasserman, I., *ApJ* **309**, 141 (1986).
12. Daigne, F. and Mochkovitch, R., in preparation.

Fireballs From Collapse of Neutron Stars Induced by Primordial Black Holes

E. V. Derishev*, V. V. Kocharovsky*, and Vl. V. Kocharovsky*

*Institute of Applied Physics, Russian Academy of Science
46 Ulyanov Street, 603600 Nizhny Novgorod, Russia*

Abstract. We outline a detailed analysis of a novel scenario in which gamma-ray bursts are of intergalactic origin and arise from the induced collapse of a lone neutron star triggered by a primordial black hole captured in its center. The energy released from the phase transition of accreted nucleon matter into the quark-gluon plasma is transferred by degenerate neutrinos to the star's surface, where neutrinos annihilate into an electron-positron plasma and produce an inverted temperature layer that preserves a fireball from undue baryonic pollution.

Possible observational tests include absence of apparent cosmological time dilation, primary location of γ-ray bursts outside of galaxies, specific shape of logN-logS dependence with a peak near redshift $z \sim 10$, emission of $10^{-4} - 10^{-3}$ of total energy in the form of 100 GeV photons, bimodal distribution of durations, very weak accompanying pulse of gravitational radiation, etc [1].

SCENARIO STEPS

0. Primordial black holes (PBHs) were formed in over-dense regions of the very early Universe [2-4]. Nowadays, their typical mass is $m_* \simeq 5 \cdot 10^{14}$ g, corresponding to a $0.8 \cdot 10^{-13}$ cm event horizon.

1. The massive protostellar cloud slowly collapses. Orbits of those PBHs that were gravitationally bound to it shrink, and the PBH density at the center rises.

2. Some of the captured PBHs have orbits completely inside the newly born star. A non-negligible fraction of them become bound to a neutron star (NS) when it forms in a supernova explosion.

3. The PBH on a close orbit loses energy by gravitational radiation and tidal friction. The fall-down time is less than age of the Universe if the periastron distance is $< 3R_{NS}$.

4. When a PBH enters beneath the neutron star crust, it quickly decelerates and comes to rest at the center of the NS. Accretion of nucleonic matter begins. A typical PBH swallows the entire NS at a time of several million years.

5. Compressed in an accretion flow, nucleonic matter undergoes a phase transition to quark-gluon plasma. Due to energy released in partial transmutation of d-quarks into s-quarks, this plasma is heated up to $T \sim 50$ MeV and emits hot neutrinos of all three flavors.

6. Neutrinos passing through the cold, outer part, of the NS become degenerate. All the heat is advected by accretion flow.

7. Dense nucleonic matter is essentially transparent for degenerate neutrinos, which annihilate at the NS surface, producing an inverted temperature layer.

8. Above the surface, a fireball with very low baryonic pollution is created, reaching a Lorentz factor of about 1000.

GRAVITATIONAL CAPTURE OF PRIMORDIAL BLACK HOLES

Almost all GRB progenitors should belong to the pre-galactic population of stars, and therefore bursts have predominantly intergalactic origin. However, the first Population III stars [5] were formed inside larger objects with masses of the order of Jean's mass at the time of radiation-matter decoupling, $M_0 \sim 10^6 M_\odot$. Consider a spherical protostellar cloud with homogeneous density ρ. As follows from conservation of adiabatic invariant for radial motion, $\int_{r_{min}}^{r_{max}} m v_r \, dr$, in the case of slow contraction a maximum orbital distance changes as $r_{max} \propto \rho^{-1/4}$, and the shape of an orbit (elliptical in this case) is preserved. These statements are also valid for any power-law potential [6]. A deviation from the adiabatic law will lead to more efficient trapping, up to freezing of ratio ρ_{PBH}/ρ_b in the free-fall case; index b refers to baryonic matter.

Given the protostellar ($\rho_* \sim 10^{-23}$g/cm^3) and the pre-supernova core ($\rho_c \sim 10^8$g/cm^3) densities, one may estimate the fraction of PBHs that cause GRBs:

$$\mathcal{F}_{GRB} \simeq 0.3 \left(\frac{R_i}{R_0}\right) \times \left(\frac{R_i}{R_*}\right)^3 \sim 10^{-13}, \qquad (1)$$

where $R_i = R_c(\rho_c/\rho_*)^{1/4}$ is a radius of the PBH orbit just after trapping, $R_c \simeq 3 \cdot 10^8$ cm a radius of the core, R_* a radius of part of protostellar cloud with $M \simeq 1.5 M_\odot$, and R_0 is a size of the large object with mass M_0. The probability that a NS becomes a GRB source is then

$$P_{GRB} = \mathcal{F}_{GRB} \frac{\Omega_{PBH}}{\Omega_b} \frac{M_{NS}}{m_*}, \qquad (2)$$

where Ω_{PBH} is the mass fraction of PBHs. This probability must be about few percent to explain the present day GRB rate, so we estimate $\Omega_{PBH} \sim 2 \div 5 \times 10^{-8} \Omega_b \sim 10^{-9}$ (an order of magnitude less than the current upper limit [4]). Small-scale clustering of dark matter prior to the formation of the first stars may significantly increase the probability of PBH capture.

PBHs that induce GRBs are spread over fall-down times in such a manner that there is a peak which traces the formation of Population III stars. PBHs on orbits passing through the NS were swallowed almost immediately at $z \sim 10$ and gave rise to a distinct population of GRBs.

The new scenario not only explains GRBs, but also provides a link between them and PBHs. We expect that it will become a sensitive indicator for PBH existence, 3–5 orders of magnitude more powerful than present day tests.

NEUTRINO PRODUCTION AND TRANSFER THROUGH DEGENERATE NEUTRONS

The energy density of compressed degenerate nucleonic matter grows more slowly than that of degenerate gas consisting of ultra-relativistic u- and d-quarks. When the two values become equal, a phase transition to quark-gluon plasma becomes inevitable. This happens at the matter density $\simeq 1.5 \cdot 10^{15}$ g/cm^3.

Quarks of all three flavors (u, d, and s) are relativistic and still degenerate with nearly the same Fermi energy $\varepsilon_F^{u,d,s} \simeq 390$ MeV. The thermal energy per quark is

$$E_{\text{thermal}} = \frac{\pi^2}{2}\frac{T^2}{\varepsilon_F} + 4\frac{11}{24}\frac{aT^4}{N_t}, \qquad (3)$$

where $N_t \simeq 8 \cdot 10^{38}$ cm^{-3} is the nucleon density at transition point and a is the radiation density constant. The above thermal energy E_{thermal} should be equal to 38 MeV after $d \to s$ transmutation is completed. This finally gives $T \simeq 50$ MeV.

The quark-nucleon interface is located at a distance of about 10 Schwarzschild radii from the central black hole.

We analyze quasi-stationary neutrino flow, in which the pressure gradient is equal to the rate of momentum density losses. In the case of spherical symmetry this gives

$$\frac{d\mu}{dR} = -\frac{3}{16\pi}\sigma_0 N_n \Sigma \lambda_c^2 h f, \qquad (4)$$

where μ is the neutrino Fermi energy, $\sigma_0 \simeq 1.8 \cdot 10^{-44}$ cm^2, N_n is the neutron density, $\lambda_c = h/(m_e c)$ is the Compton wavelength of the electron, and f is the neutrino flux density. Effective cross-section for scattering of degenerate neutrinos $\Sigma = (2m_e c^2/\mu)^2 \sigma_{eff}/\sigma_0$ has the thermal component [7]

$$\Sigma_T = \frac{1 + 2g_a^2}{4}\frac{\chi}{\mu}\frac{m_n c}{p_F^n}, \qquad g_a \simeq 1.254, \qquad (5)$$

(where $\chi = \pi^2 T^2/(4\varepsilon_F^n)$ is thermal energy per neutron, m_n the neutron mass, ε_F^n and p_F^n the Fermi energy and momentum of neutrons) and the non-thermal one

$$\Sigma_N = 0.24 \left(\frac{m_n c}{p_F^n}\right)^2 \frac{\mu}{p_F^n c} \left(\frac{\bar{\delta}}{\mu}\right)^2, \qquad (6)$$

which is calculated in the limit of heavy neutrons and zero temperature. Here $\bar{\delta}$ is a dipole moment of the deviation of the neutrino Fermi surface $\mu = \varepsilon_F^\nu + \bar{\delta}\cos\theta$ from the equilibrium level due to non-zero neutrino flux.

Equation (4) must be accompanied by two more continuity equations for functions f and χ. We assume $N_n = const$, constant ratio of radius of quark-gluon core to the radius of central black hole R_{in}/r_g, and zero neutrino losses. The value of $\partial \mu/\partial t$, as expressed via $dR_{in}/dt \neq 0$, provides an equation for flux f (the same for all neutrino flavors). Balancing the thermal neutron energy flux $v_n N_n \chi$ and all neutrino energy fluxes $6\mu f$, where v_n is a fall-down velocity of nucleonic matter, yields the last equation for χ. The boundary conditions are as follows: $\mu = \mu_{in}$ at $R = R_{in}$, $\chi = 0$ and $f = k_{tr}(4\pi/3)(\mu/hc)^3 c/4$ at the NS surface $R \simeq R_{NS}$. Here a factor $k_{tr} < 1$ includes all possible opaqueness. For definiteness, we take $\mu_{in} = 80\,\mathrm{MeV}$, though results are not very sensitive to this value.

Numerical analysis of the neutrino transfer equations shows that hot nucleonic matter is localized within hundreds of meters above the emitting quark surface, while the rest of the NS remains cold, allowing neutrinos to escape almost freely [1]. Inclusion of an outer opaque layer in our modeling *increases* the total energy output by several times if $k_{tr} \sim 0.1$. The energy of outgoing neutrinos rises from 15 MeV to 30 MeV during a $\sim 3\,\mathrm{ms}$ long pulse, which provides about 10^{51} ergs in $e^+ e^-$ outflow.

Neutrinos do not annihilate until they reach the surface layer of the NS where the Fermi level of electrons is low enough to permit the process $\nu + \bar{\nu} \to e^+ + e^-$. The latter begins in a degenerate neutrino gas and it takes some time (and distance) to heat the flow. Thus, an inverted temperature layer is formed that overlaps the base of the $e^- e^+$-wind. The Lorentz factor of an outflow may reach several thousands and is limited by the mass of nucleons laying above the sonic point.

DISCUSSION

We show that the induced collapse scenario may naturally resolve many problems typical for cosmological models of GRBs, including the absence of time dilation, lack of large host galaxies, and small baryonic pollution. The latter is possible thanks to the unusual combination of physical conditions in one and the same object (collapsing NS), where the hot neutrino-emitting quark core and the region of efficient neutrino annihilation are separated by a layer of cold (and hence transparent) neutron matter. Under these conditions, an inverted temperature layer in the $e^- e^+$-plasma outflow is formed which prevents baryonic pollution.

It should be noted that for two bursts, GRB971214 and GRB990123, the estimated value of energy release approaches 10^{54} ergs assuming isotropic emission. Since it is very divergent from the average value ($10^{51} - 10^{52}$ ergs) inferred from the analysis of logN-logS curve [8], these bursts may be of different origin.

For typical GRB sources, the surface temperature, $3 \div 10\,\mathrm{MeV}$, is higher than the nuclei dissociation threshold, $\simeq 1\,\mathrm{MeV}$. So, a hot wind initially carries away both protons and neutrons in roughly equal proportion [9]. This neutron stream crucially affects the widely accepted picture [10] of electron-proton shock deceleration in surrounding matter. Our considerations based on a two-flow model indicate that traditional scenarios face new difficulties explaining short events ($\tau < 5$ s). However, the neutron stream does not pose any danger to GRB models with high Lorentz factors, and may help to explain the bimodal distribution of burst durations [11].

TESTS FOR PBH-INDUCED COLLAPSE SCENARIO

1. The logN - logS dependence should have a large peak due to another, distinct, population of bursts at $z \sim 10$.
2. The fireball Lorentz factor ~ 1000, which can be directly measured due to neutron-proton flow decoupling [9].
3. Bimodal distribution of burst durations. Shorter GRBs correspond to decoupled flows, longer ones to sources where decoupling does not occur.
4. GRB sources are not tied to bright central part of host galaxy, as opposed to hypernova models. Instead, they are likely located either outside galaxies or in their extended halos.
5. Very weak (undetectable) accompanying pulse of gravitational waves in NS-PBH collapsing system, as opposed to the case of merging neutron stars [12].

If decoupling of a neutron flow takes place, then:
1. Fireball emits 100 GeV photons and 30 GeV neutrinos from decaying pions produced in inelastic proton-neutron collisions.
2. A power law Comptonized "tail" is present up to 1 MeV (in comoving frame) in the precursor spectrum. The tail contains about 10% of the energy stored in thermal blackbody radiation emitted from the fireball photosphere.
3. A cloud of positrons remains at the GRB location. These positrons annihilate during thousands of years and might be visible as an annihilation line.

REFERENCES

1. Derishev, E.V., Kocharovsky, V.V., and Kocharovsky, Vl.V., *Pis'ma v ZhETF* **70**, 642 (1999).
2. Zel'dovich, Ya.B., & Novikov, I.D., *Astron. Zhurn.* **43**, 758 (1966).
3. Hawking, S., *MNRAS* **152**, 75 (1971).
4. MacGibbon, J.H. & Carr, B.J., *ApJ* **371**, 447 (1991).
5. Ostriker, J.P. & Gnedin, N.Y., *ApJ* **472**, L63 (1996).
6. Derishev, E.V. & Belyanin, A.A., *A&A* **343**, 1 (1999).
7. Burrows, A. & Lattimer, J.M., *ApJ* **307**, 178 (1986).
8. Reichart, D.E. & Mészáros, P., *ApJ* **483**, 597 (1997).
9. Derishev, E.V., Kocharovsky, V.V., & Kocharovsky, Vl.V., *ApJ* **521**, 640 (1999).
10. Meszaros, P. & Rees, M.J., *ApJ* **405**, 278 (1993); *MNRAS* **269**, L41 (1994).
11. Kouveliotou, C., et al., *ApJ* **413**, L101 (1993).
12. Paczynski, B., *ApJ* **308**, L43 (1986).

Gamma-Ray Bursts from Rapidly Spinning Neutron Stars

H. C. Spruit

Max-Planck-Institut für Astrophysik
Box 1523, D-85740 Garching, Germany

Abstract. A neutron star in an X-ray binary (XRB) is spun up by mass transfer from its companion to rotation periods of the order of a millisecond. Evidence for this are the existence of ms pulsars, believed to descend from X-ray binaries. Direct evidence of such rotation rates in XRB is the of 2.5 ms rotation in SAX 1808. The rotation energy in a neutron star spinning at 1 ms is 10^{52} erg, in the range of that required for GRB. The environment also satisfies the baryon loading constraint, since the accretion disk contains only of the order $10^{-9} M_\odot$, and the companion star blocks only of the order 1% of the sky as seen from the neutron star.

INTRODUCTION

Extraction of the rotation energy from a spinning neutron star to power a GRB looks very implausible at first sight. The energy must be extracted in 0.3–30 s (e.g., Fishman & Meegan 1995). This requires an extremely efficient 'brake' which does not exist in an ordinary X-ray binary (XRB).

In the following I show, however, that such a brake can very elegantly be produced by the neutron star itself, through a sequence of straightforward events involving only known neutron star physics (Spruit 1999a):

1. The neutron star's spin slowly increases by accretion of angular momentum from the companion.

2. At a period of the order of a ms, the star becomes unstable to an r-mode oscillation, coupled to the emission of a gravitational wave.

3. The negative temperature dependence of the viscosity of neutron star matter makes this instability run away in a finite time of the order of 100 years. The last few e-foldings of the oscillation amplitude take place on a time scale of minutes.

4. The star loses angular momentum from its outer parts by the emitted gravitational wave, and starts rotating differentially by ~10–30%.

5. A weak initial magnetic field in the star (10^6–10^8G) gets wound up by the differential rotation into a field of $\sim 10^{17}$G on a time scale of days to months.

6. This field becomes buoyantly unstable at a threshold of $10^{16} - 10^{17}$G, rises through the surface on a time scale of ms to seconds, creating a non-axisymmetric field of the order 10^{13}G at the light cylinder. (From this point on, the model is like that of Kluźniak & Ruderman 1998).

7. The star spins down on the pulsar spindown time scale, which for these conditions is of the order of seconds. The angular momentum is carried by a relativistic MHD wind. This powers a GRB, and extracts essentially all the rotation energy of the star (as in Usov's 1992 model).

RUNAWAY GRAVITATIONAL WAVE INSTABILITY

Rotating neutron stars are naturally unstable to excitation of their oscillation modes by the emission of gravitational waves (e.g., Shapiro & Teukolsky 1983). A mode is excited like a squeaking brake. The deformation of the oscillating star provides surface roughness, and the gravitational wave provides the torque coupling the star to the vacuum outside, analogous to the frictional coupling of a brake lining.

Rotational modes (Rossby waves) turn out to be much more effectively excited than the pulsation modes of the star (Andersson 1997, Andersson et al. 1998). Modes with angular quantum numbers $l = m = 2$ grow fastest. The unstable displacements in these modes are mainly along horizontal surfaces. Their frequency in an inertial frame is $2/3\,\Omega$, and in the corotating frame $-4/3\,\Omega$ (retrograde). In the absence of viscous damping, the growth rate of the most unstable mode is (Andersson et al. 1998) $\sigma_{\rm GW} = 5 \times 10^{-2} P_{-3}^{-6}$ s^{-1}. Including viscous damping, the growth rate is $\sigma = \sigma_{\rm GW} - 1/\tau_{\rm v}$, where $\tau_{\rm v}$ is the viscous damping time $\tau_{\rm v} = (\Delta R)^2/\nu$, and $\Delta R \approx 0.5R$ is the width of the unstable eigenfunction. For the kinematic viscosity ν the standard value quoted (van Riper 1991) for a neutron superfluid is $\nu \approx 2 \times 10^7 T_7^{-2}$ cm^2 s^{-1} at an average density of 3×10^{14} g cm^{-3}, where $T = 10^7 T_7$ is the temperature. This expression holds at low temperatures; at temperatures above $\sim 3 \times 10^9$K the viscosity starts increasing strongly with T. During the spin-up by accretion, the temperature of the star is determined by the accretion rate, $T_7 \approx \dot{M}_{-9}^{1/4}$, where $\dot{M} = 10^{-9}\dot{M}_{-9}$ M$_\odot$/yr is the mass accretion rate [taking into account heating by nuclear reactions in the accreted atmosphere (Miralda-Escudé et al. 1990, Brown 1999), yields somewhat higher temperatures]. With the typical value $\dot{M}_{-9} = 1$, the critical rotation period for instability to set in is 3 ms.

A star slowly spun up by accretion (on a time scale of 10^9 yr) would become weakly unstable to the gravitational wave mechanism as its period decreases below the critical value. The unstable mode dissipates energy by viscous friction. This causes heating of the star, and a decrease of the viscosity. This upsets the delicate balance between the intrinsic growth and the viscous damping that exists at the time the marginal stability line is crossed, and the *growth rate* increases with the wave amplitude. This results in a runaway of the mode amplitude in a finite time.

A few hundred years after the rotation rate has crossed the instability threshold, the mode amplitude diverges. During the final runaway phase the temperature is so high (a few 10^9 K) that viscous damping is weak and the mode grows on a time scale of the order of hours to minutes (depending on the initial rotation rate).

DIFFERENTIAL ROTATION AND WINDING-UP OF THE MAGNETIC FIELD

The angular momentum loss by gravitational wave emission is largest in the outer parts of the star, where the unstable eigenfunction typically have their largest amplitudes, and where the coupling to the gravitational wave emitted is strongest. Depending on the saturation amplitude assumed for the r-mode oscillation, this creates differential rotation on a time scale of hours to days. The differential rotation, absent before the final runaway of the instability, winds the initial magnetic field into a toroidal (azimuthal) field. Because the rotation period is so short, the non-axisymmetric components of the initial field are quickly eliminated by magnetic diffusion (Rädler 1980, 1986; for a recent discussion see Spruit 1999b). The remaining axisymmetric and nearly azimuthal field is as sketched in Kluźniak and Ruderman (1998). To represent these processes in a simple model, I divide the star into two zones with boundary at $r \approx 0.5R$, such that the outer shell has approximately the same moment of inertia as the core, and rotating at rates Ω_s and Ω_c respectively. The angular momentum loss by gravitational waves is taken to be restricted to the outer shell for simplicity. The azimuthal field increases at a rate given by the difference in rotation between the two zones. The magnetic torque between the two zones is taken into account as it modifies the differential rotation during the winding process.

The growth of the unstable mode is limited at large amplitude by nonlinear mode couplings. Since these have not been computed yet, they are parametrized by adding a damping term which becomes effective at an assumed saturation level $\xi/R = 0.03$.

BUOYANT INSTABILITY OF THE TOROIDAL FIELD

The increasing azimuthal magnetic field inside the star does not manifest itself at the surface until it becomes unstable to buoyancy instability. For this to happen the field has to become strong enough (Kluźniak and Ruderman 1998), of the order $B_c = 10^{17}$G, to overcome the stable stratification associated with the increasing neutron/proton ratio with depth. The growth time τ_B of the instability is of the order of the Alfvén travel time: $\tau_B \sim R/(V_A - V_{Ac})$, where V_{Ac} is the Alfvén speed at the critical field strength B_c. Buoyancy instability is generally non-axisymmetric, forming loops of field lines which erupt through the surface (e.g., Matsumoto & Shibata 1992). The action of the radial velocities in the eruption process on the azimuthal field produces a strong small-scale radial field component, which couples to the differential rotation and links the different parts of the star together on an Alfvén crossing time scale. The whole process is likely to yield a complex time dependent field configuration, but important aspects can be illustrated with a simple model (Spruit 1999a). It takes into account the gradual crossing of the buoyancy stability limit, the time scale (of the order of a millisecond) of this instability, the magnetic torques coupling the star, and the external torque exerted by the spinning magnetic field.

SPINDOWN AND POWERING OF THE GRB

If the field emerging at the surface were a vacuum field, the energy output would be in the form of the Poynting flux of an electromagnetic wave, and would simply be given by the usual pulsar spindown formula. At a surface field strength of 10^{16}G and a spin period of 1 ms this yields a spin down time of seconds. In reality, the presence of some matter lifted into the atmosphere by the eruption of the magnetic field, and a pair plasma produced by the dissipation of some of this magnetic energy will enforce nearly ideal MHD conditions. The energy is then emitted in the form of a relativistic MHD wind instead of a pure EM wave. The energy output is still given approximately by the pulsar spindown formula, however (Usov 1992).

DISCUSSION

The model presented is related to that of Kluźniak and Ruderman (1998), but provides a plausible scenario for the development of the strong differential rotation assumed by these authors. It has a number of properties that make it attractive as a possible GRB engine. The baryon loading is intrinsically low. The natural time scale for variations in the energy output is the Alfvén travel time through the star, of the order of ms as observed. Burst durations as short as a second or as long as a minute can be accommodated by variation of the initial rotation rate. It invokes only properties of neutron stars and X-ray binaries believed to be known from current theory and observations.

The model also produces a reasonable energy output. For the largest observed energies, however, (such as the 23 January 1999 event), it is necessary to assume a significant beaming factor (as in other GRB engine models). Whether such a beaming is natural in the model depends on properties of the relativistic wind produced by a rotating neutron star which are not fully understood at present.

The model makes some predictions that differ from engine models based on collapsing neutron stars or black holes. In these models, the bulk of the gravitational wave energy is emitted at the time of the GRB itself. In the present model, the gravitational waves are emitted at the peak of the gravitational wave instability,

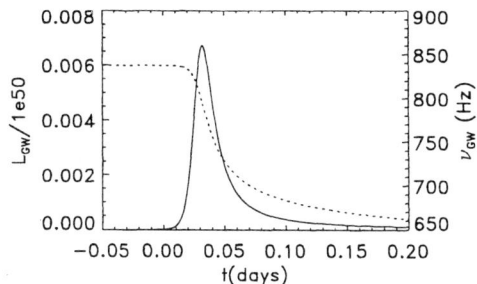

FIGURE 1. Typical variation of gravitational wave luminosity and frequency during the peak of the r-mode instability (which precedes the electromagnetic GRB emission). Assumed parameters are initial spin period of 1.5 ms, and saturation amplitude of the r-modes $\xi_{max}/R = 0.03$.

which precedes the GRB by days to months. This is because the differential rotation is generated only at this peak, and the winding up of the magnetic field that finally powers the GRB takes some time (depending on the initial field strength on the star). This difference should be obvious in gravitational wave observations whenever these become sensitive enough to detect GRB. The estimated amplitude and frequency of a gravitational wave produced near the peak of the r-mode instability are shown in Figure 1.

REFERENCES

1. Andersson, N., *ApJ* **502**, 708 (1997).
2. Andersson, N., Kokkotas, K., & Schutz, B., *ApJ* **510**, 846 (1999).
3. Brown, E., astro-ph/9910215 (1999).
4. Fishman, G.J., & Meegan, C.A., *ARAA* **33**, 415 (1995).
5. Kluźniak, W., & Ruderman, M., *ApJ* **505**, L113 (1998).
6. MacFadyen, A., & Woosley, S.E., *ApJ* **524**, 262 (1998).
7. Matsumoto, R., & Shibata, K., *PASJ* **44**, 167 (1992).
8. Miralda-Escudé, J., Paczyński, & B., Haensel, P., *ApJ* **362**, 572 (1990).
9. Paczyński, B., *ApJ* **494**, L45 (1998).
10. Rädler, K.-H., *Astron. Nachr.* **301**, 101 (1980).
11. Rädler, K.-H., in *Plasma Astrophysics* (Proceedings Joint Varenna-Abastumani Workshop), ESA SP-251, 1986, pp. 569.
12. Shapiro, S.L., & Teukolsky, S.A., *Black Holes, White Dwarfs and Neutron Stars*, Wiley, N.Y, 1983.
13. Spruit, H.C., *A&A* **341**, L1, astro-ph/9811007 (1999a).
14. Spruit, H.C., *A&A* **349**, 189, astro-ph/9907138 (1999b).
15. Usov, V.V., *Nature* **357**, 472 (1992).
16. van Riper, K.A., *ApJS* **75**, 449 (1991).

Red Hole Gamma-Ray Bursts: A New Gravitational Collapse Paradigm Explains the Peak Energy Distribution and Solves the GRB Energy Crisis

James S. Graber

407 Seward Square SE
Washington, DC 20003

Abstract. Gamma-ray bursts (GRBs) are still an enigma. In particular, the central engine, the total energy, and the very narrow distribution of peak energies challenge model builders. Motivated by recent theoretical developments (string theory, quantum gravity, critical collapse), which suggest that complete gravitational collapse can occur without singularities or event horizons, we explore how red-hole models (which lack singularities or event horizons) can solve these problems better than black-hole models.

KEY GRB MODEL BUILDING CHALLENGES

Gamma-ray bursts vary rapidly and therefore they must be compact. Because these compact gamma-ray bursts release enormous energy, they must form an intense fireball that is optically thick, pair-producing, and thermalized. But the spectrum is not thermal, and there is no sign of pair-production attenuation at the high end of the observed spectrum [1]. This seeming self-contradiction (the opacity problem) can be solved by having the fireball power a relativistic shell or jet that collides with something (perhaps itself) to produce the observed gamma rays [2]. This fireball/shock model is currently the leading candidate to explain GRBs [3]. It has already overcome several severe model-building challenges. But like almost all other published models, it fails to explain the observed spectroscopy of GRBs, particularly the narrowness of the observed peak energy distribution [4,5]. Furthermore, this model does not explain the high ratio of the energy of the GRB burst itself (caused by internal shocks) to the energy in the afterglow (caused by external shocks in the fireball/shock model) [6]. Nevertheless, the predictions of this model for the afterglows themselves are consistent with current observations [3].

Finally, there is the problem of the overall energetics of the GRB. The two leading candidates to produce the initial fireball or fireballs — the so-called central engine — are merging neutron stars and core-collapse supernovae [7,8]. Both these sources have over 10^{54} ergs of total energy available. This is more than enough energy for even the most energetic GRB, but it is not at all clear how to prevent most of it from falling into the newly created black hole which forms in the standard general relativity versions of these models.

There seems to be an inherent conflict between solving the opacity problem and solving the peak energy distribution problem. The only successful technique available to solve the transparency problem is to invoke highly relativistic bulk motion. In the relativistic frame, the gamma rays are below the pair-production threshold and so do not suffer pair-production attenuation. This definitively solves the opacity problem. But unless the Lorentz gamma factor of the bulk motion can be fine-tuned to a very narrow range for all GRBs, the resulting blueshift will not only relocate the peak of the photon energy distribution; it will also substantially widen it, inconsistent with the observed narrow E-peak distribution. Thus, one needs to find a way to fine-tune the Lorentz gamma factor or find some other way around this conflict. In the fireball/shock model, the gamma factor depends sensitively on the baryon loading, and hence will vary widely. Furthermore, the internal shocks model is dependent on shocks with varying Lorentz gamma factors colliding with each other. So fine-tuning is not a reasonable option for this model.

A generic solution to this problem is provided if the relativistic bulk motion results not from an initial explosion, but rather from the gravitational acceleration of matter falling into a deep potential well. An arbitrarily high Lorentz gamma factor can be attained, but the accompanying blueshift will be exactly cancelled when the matter and radiation are redshifted as they emerge from the potential well. (By that time, the matter and radiation will have separated, so the opacity problem has already been solved).

A black hole can provide the necessary deep potential well. But once matter or radiation is deep in the potential well of a black hole, it is almost impossible for it to escape. Therefore, we will consider an alternative gravitational collapse paradigm in which it is possible to escape from deep within the potential well of a gravitationally collapsed object.

WHY CONSIDER ALTERNATE GRAVITY MODELS?

The problems with constructing a GRB model might be sufficient motivation to consider alternate theories of gravity. However, a stronger motivation comes from the theory of gravitation. Recent theoretical developments in string theory, quantum gravity and critical collapse strongly suggest the possibilities of both gravitational collapse without singularities (and without loss of information) and also gravitational collapse without event horizons [9–14]. If these possibilities are correct, we are forced to consider the phenomenological consequences (such as different models for GRBs and core-collapse supernovae) of alternate paradigms for gravitational collapse in which black holes do not form [15].

RED HOLES — A NEW PARADIGM

Many authors have considered the alternative in which a hard core collapsed object similar to a smaller, harder, denser, neutron star forms in place of a black hole [16]. We here consider the alternative in which no such hard surface forms. Instead, the spacetime stretching that forms a black hole in the standard model occurs, but it does not continue to the extent necessary to form an event horizon or a singularity. Instead, spacetime stretches enormously, but not infinitely, and forms a deep, wide, potential well with a narrow throat. We call this a red hole.

This type of spacetime configuration was previously considered by Harrison, Thorne, Wakano and Wheeler (HTWW) in 1965, but only as a way station in the final collapse to a black hole (not yet then called by that name) [17]. In their version, part of the configuration is inside the event horizon, the collapse continues, and a singularity soon forms.

In the new alternate paradigm we call a red hole, no event horizon forms and no singularity forms. The gravitational collapse does not continue forever, but eventually stops. (Why? Perhaps due to quantum effects or string-theory dualities, but we cannot discuss this adequately here.) As the collapse proceeds, the collapsing matter becomes denser and denser until it reaches a critical point, after which the distortion of spacetime is so great that the density decreases. This happens because the spacetime is stretching faster than the collapsing material can fall inward. (This decreasing density effect was already noticed by HTWW in their analysis of gravitational collapse in the context of standard general relativity [17]. In general relativity, this expansion of spacetime is mostly hidden behind the event horizon and does not prevent the formation of a singularity in a finite time. This is not the case in several observationally viable alternate theories of gravity [18–20].) This is why we are confident that the center of a red hole resembles a low-density vacuum more than it resembles a high-density neutron star. The decrease in density due to this enormous stretching may also be a factor in halting the gravitational collapse of the red hole before the stretching becomes infinite.

As a result, even though the stretching of spacetime is enormous, it never becomes fast enough to exceed the speed of light and cause an event horizon to form. It stops before it reaches an infinite size or any other form of singularity. (Infinite density and infinite curvature also do not occur.) Nevertheless, it is very hard to escape from a red hole. First, there are trapped orbits inside the red hole for photons as well as massive particles, which allows permanent or nearly permanent trapping of mass and energy. Second, the Shapiro delay in crossing a red hole is very substantial, (in some cases, enormous) [21]. Hence, particles that are only crossing the red hole or passing through are in effect "temporarily" trapped.

In fact, most of the matter falling into a red hole will be trapped. However, radiation and highly relativistic matter that falls directly into the center of the red hole and does not re-scatter while inside the red hole, can travel straight through and emerge on the other side. This possibility is essential for our proposed new GRB models.

RED-HOLE MODELS FOR GRBS

In order to describe our new red-hole models for GRBs, which are based on modifying the existing fireball/shock model, we begin by resummarizing that model. In the fireball/shock model, some form of gravitational collapse deposits a large amount of energy in a very small region, (which is called the fireball, and also the central engine). The fireball has so much energy in such a small space that a relativistic expansion must occur. Part, or all, of this explosive expansion travels through a region with a very small critical number of baryons, which absorb essentially all of the energy and form a relativistic blast wave (either spherical or jetted). Multiple such relativistic shells (travelling in the same direction) are created by the central engine, perhaps by repeated explosions (possibly due to repeated accretion events). The faster relativistic shells overtake the slower relativistic shells and collide with them. The internal shocks convert the energy of the baryons to gamma rays (by synchroton emission, inverse compton scattering, or perhaps by both means). The shells eventually collide with external matter and generate the main afterglow. (Perhaps an early prompt afterglow is the result of a reverse shock [3].)

There are basically three important sites in this model. First, there is the central engine, or fireball site. (In the standard black-hole interpretation, this is probably near a newly forming black hole, perhaps at the pole of a Kerr black hole [22].) Second, there is the location of the internal shocks, where the main gamma-ray burst is generated. According to Piran, this is typically 10^{12}–10^{14} cm, or 30–3000 light seconds down stream from the central engine [3]. Third, there is the location of the external shock, where the relativistic matter collides with material that was not part of the original explosion, and the long-lasting (days to months), but weak (less total energy than the gamma rays) afterglow is generated. In the standard model, this is far from the central engine.

In our alternate red hole models we will relocate these three sites in, or near, a red hole instead of near the outside of a black hole.

In the first and most conservative red-hole model, we merely replace the black hole of the standard model with a red hole. The red hole can help the central engine by generating more energy than the corresponding black hole or by focusing the outgoing jet more narrowly, but the rest of the model is essentially the same as the standard one and there is no significant impact on the spectral issues. In other words, this first red-hole model can help solve the energy crisis, but does not help explain the broad spectrum, with its unusual slopes and narrowly distributed peak energy.

In the second — and more interesting — red-hole model, the central engine is located at the infalling bottleneck of the red hole, and the internal shocks that generate the primary gamma-ray burst are located at the outgoing bottleneck of the red hole (which is essentially the same place, but at a later time), and the external shocks and the afterglow still occur far away at the point where the ejecta encounter the interstellar material or some other external matter.

In this model, the great internal expansion of the red hole, along with the great acceleration of the gravitational infall, help to generate the relativistic jet that will later create the GRB. Then the focusing effects of the emerging bottleneck of the red hole help to create the internal shocks necessary for the final transformation of the energy into gamma rays, and to very substantially increase the efficiency of this process. Furthermore, since the blueshift of the infall should be exactly cancelled by the redshift of the outclimb, the gamma rays seen by the observer will have no net red or blue shift (on average). Therefore, the observed peak energy will be the same as the initial peak energy. Even if the internal transit involves enormous and substantially varying Lorentz gamma factors, they will not be observed as a net blueshift. So this model helps solve the narrow peak energy distribution problem, as well as the energy crisis. It can also help solve the spectral wideness and slope problems because of the tolerance for differing Lorentz gamma factors during the transit through the red hole.

REFERENCES

1. Band, D., et al., *ApJ* **413**, 281 (1993).
2. Rees, M. J., and Meszaros, P., *MNRAS* **258**, 41P (1992).
3. Piran, T., *Phys. Rep.* **314**, 575 (1999).
4. Preece, R. D. et al., *ApJS*, in press, astro-ph 9908119 (1999).
5. Brainerd, J. J., et al., *ApJ*, submitted.
6. Paczynski, B., and Rhoads, J.,*ApJL* **418**, L5 (1993).
7. Eichler, D., et al., *Nature* **340**, 126 (1989).
8. Woosley, S. E., *ApJ* **405**, 273 (1993).
9. Strominger, A., and Vafa, C., *Phys. Lett.* **B379**, 99 (1996).
10. Callan, C. G. and Maldacena, J. M., *Nucl. Phys.* **B472**, 591 (1996).
11. Maldacena, J. and Strominger, A., *Phys. Rev.* **D55**, 861 (1997).
12. Shapiro, S. and Teukolsky, S. A., *Phys. Rev. Lett.* **66**, 994 (1991).
13. Choptuik, M. W., *Phys. Rev. Lett.* **70**, 9 (1993).
14. Christodoulou, D., *Ann. Math.* **140**, 607 (1994).
15. Graber, J. S., to appear in *Largest Explosions Since the Big Bang*, Ed. M. Livio, astro-ph 9908113 (1999).
16. Robertson, S. L., *ApJL* **517**, 117 (1999).
17. Harrison, B. K., Thorne, K. S., Wakano, M., and Wheeler, J. A., *Gravitation Theory and Gravitational Collapse*, Chicago: University of Chicago Press, 1965, ch. 8, pp. 69-75.
18. Rosen, N., *Ann. Phys.* **84**, 455 (1974).
19. Yilmaz, H., *Ann. Phys.* **101**, 413 (1976).
20. Itin, Y., *Gen. Rel. Grav.* **31**, 187 (1999).
21. Shapiro, I. I., *Phys. Rev. Lett.* **13**, 789 (1964).
22. Kerr, R. P., *Phys. Rev. Lett.* **11**, 237 (1963).

Electric GRBs

László Körtvélyessy

Observatory Kleve, D-47533 Germany

Abstract. The gamma ray burst (GRB) is the electric explosion of a rest-star from a white dwarf in our Galactic arm. The GRB destroys all white dwarfs before they could be red. This is the first publication of a new model which is proven by known measured emissions of the GRBs compared with the emissions of the solar flares. The new model explains not only the gamma rays, their short pulses, the later peaks of continuous and redshifted afterglows, the emission of relativistic protons, the non-thermal radiation, the deficit of weak bursts (as usual goals of GRB models), but also a so far un-revealed deficit of strong GRBs and the missing cold (red) white dwarfs.

WHAT CAN GRBS NOT BE?

The missing gravity-lens-effect and the incredibly large amount of energy of a cosmological explosion — which is impossible in known physics and in pulses — suggest in my opinion a source in our proximity that is at least energetically clearly possible. Since the start with gamma rays and the later appearing visible photons and other weaker photons suggest a cold explosion, the GRBs are

- NOT ACCRETION DISCS,
- NOT HYPERNOVAE,
- NOT MERGING NEUTRON STARS,
- NOT FIREBALLS AT ALL.

All the 4 processes above would be not only hot, but also too big to send pulses shorter than one second. The source of a GRB must be smaller than 1 light-second.

The solar flares have exactly the same emission as the GRBs: gamma rays in pulses and then continual afterglow (X-ray, visible photons, etc.). Solar flares produce also a small redshift of the proximity in visible light. They also emit relativistic protons. The only difference is the synchrotron radiation. Stellar flares are cold, electric explosions of protons. If we want to understand the GRBs, we should start from all GRB-data. They show a flare without a star and in rotation. Many such bodies exist around us (Fig. 2).

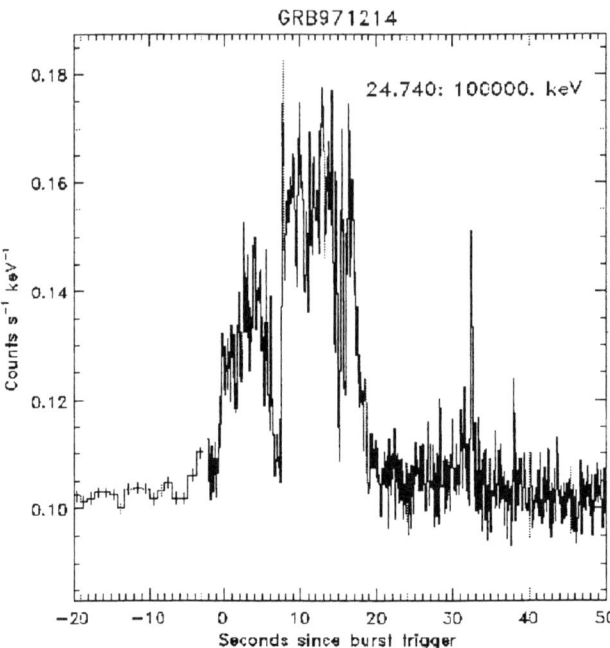

FIGURE 1. Sharp and short pulses show a nearby and small source. Solar flares produce such pulses of gamma rays, too (NASA/MSFC J. Horack and C.Kouveliotou 1998 May 6).

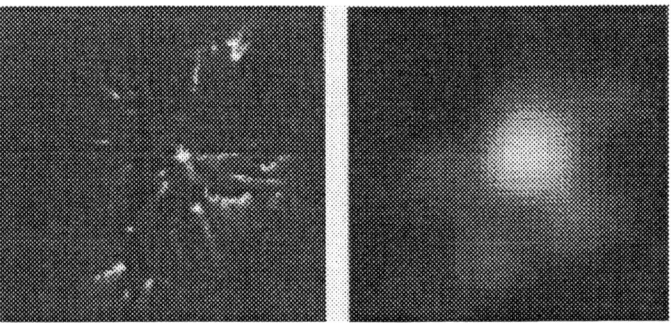

FIGURE 2. The last phase of a white dwarf (Abell 30) before its electric total-explosion: it pushes off its colder parts as positive helium-fingers (left: HST). X ray activity is shown by ROSAT on the right. Both sizes are 18″ × 18″ (S&T 1996 May).

A white dwarf contains a positive charge which — in rotation — produces its strong magnetic field. After the finger-emissions (Fig. 2), only a small (positive) rest-body of the white dwarf remains in fast rotation. Its positive charge imbalance cannot explode as long as the body is hot. The photons of the electrostatic repulsion run in a zig-zag course and are, therefore, ineffective. But, similar to solar flares, the rest-body of a white dwarf becomes transparent after the finger-emissions cool-down to, e.g., 7000 K, and explodes electrostatically at once. The ions accelerate each other and explode in diminishing rotation, emitting synchrotron radiation. The GRBs increase the huge mass of the X-ray active halo. These dispersed positive masses are the cemetery of supernovae and white dwarfs! They produce filaments of superclusters and the accelerated expansion of the Universe.

QUESTIONS

Why do GRBs emit gamma rays?

Positive ions accelerate positive ions. The GRBs and solar flares pulse similarly. (One of the three emissions of the electric explosion — i.e., fission of Uranium — was named gamma rays a century ago.)

Why is no object at the position of a GRB?

The rest-star of the white dwarf totally exploded by its own electric charge. This positive cloud is quickly cooled, but it emits X rays for some 104 years.

Will Chandra see something at the position of an older GRB?

Yes. It will see an expanding X ray cloud with radial filaments, similar to an SNR.

Why does a GRB send either no visible photons at all or only later?

This explosion is not hot. It produces heat, if at all, when the electrically exploded ions collide into a nebula around the white dwarf (Fig. 3).

Why do the X rays from a GRB show that these photons came through a cloud?

The X rays were released in the inside of the shell.

Why do we find GRBs from all directions?

White dwarfs are everywhere in our proximity and we cannot detect GRBs if they are far away.

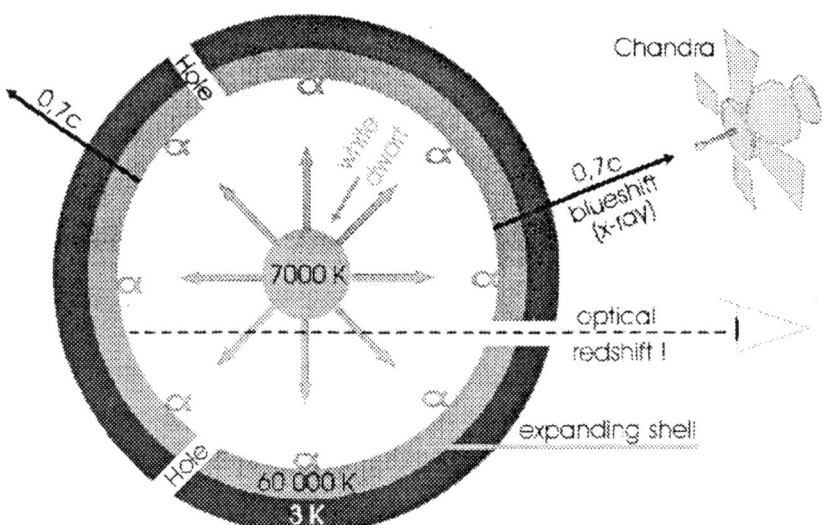

FIGURE 3. The whole body of the white dwarf is in collision with the nebula around it. Early measurements show a velocity of 0.9999c of charged particles and a brake-down to about 0.7c in some days. This brake is surely due to the swept away masses of the nebula. The inside of the shell glows, but the outside is still cold. We see only the rejected shell from inside (60 000K) with a redshift caused by 0.7c. The approaching shell-wall shows its cold side (3K) to us. The receding afterglow can only be seen through holes and gaps in the approaching wall of the shell. Only Chandra will see that the X ray from the inside of the shell shows a redshift and a blueshift, both caused by 0.7 c.

Why is there a deficit of weak and strong GRBs?

The mass of the white dwarfs have a lower limit, the Universe is still young. Big white dwarfs have the Chandrasekhar-limit.

Why is a white dwarf white?

It explodes electrically during its cool-down. It will never become red, see the white dwarf branch which is amputated at 7,000 K in the HRD.

ACKNOWLEDGEMENTS

Prof. S. N. Zhang (Univ. Al. Huntsville) not only confirmed that the temperature gradient separates the electric charges due to the different masses of the protons and electrons — in hydrogen burning stars — but also showed that the photons push the electrons outwards stronger than the heavier protons. This separation of the electric charges via temperature gradient is important for the resulting positive charge imbalance of the stellar core which is the later white dwarf.

A Ritzian Interpretation of Variable Stars

Robert S. Fritzius[1]

305 Hillside Drive, Starkville, Mississippi 39759

Abstract. A revived version of de Sitter's 1913 "binary stars" argument against Ritz's Galilean relativity general electrodynamic theory provides a means of explaining of the mechanism underlying the *apparent* variability of variable stars, pulsars, and gamma-ray bursters. Numerical code to compute $c + v$ induced intensity peaks and *blueshift chirps* for spectroscopic binaries provides graphical displays for comparison with observed light curves and spectra for different classes of variable stars.

INTRODUCTION

In 1908 Walter Ritz formulated an emission theory of general electrodynamics [1] in which the velocity of a light source is added to the velocity of the light emitted by it, i.e., the the velocity of light is $c+v$. In 1913 Willem de Sitter urged abandonment of Ritz's theory because visual binary stars failed to show its predicted $c+v$ effects [2]. His argument was along the following lines.

The addition of the source velocity of a visible component of a binary star to the velocity of its light emitted in the direction of an observer would allow slower light $c - v$ from one side of the orbit (when the component was traveling away from the observer) to be overtaken by the faster light $c + v$ from one half orbit later (when the component was traveling toward the observer). At a sufficient distance this effect could cause the visible component to periodically seem to be at two different locations simultaneously and generally would lead to observational departures from Keplerian motion.

These effects were not observed and Ritz's ideas on electrodynamic theory were largely laid to rest. (Ritz died in 1909 at age 31 and wasn't around to defend or modify his theory.) What follows represents recent re-examinations of the matter.

---

[1] The author works for the U.S. Air Force which does not endorse the views expressed herein.

DERIVATION OF DE SITTER'S OVERTAKING DISTANCE EQUATION

The expression to compute the de Sitter overtaking distance is derived as follows. (Circular binary orbits are assumed.)

Let the time interval $t1$, required for the slower light (visible component receding) to reach the observer at distance L, be

$$t1 = \frac{L}{c-v}. \tag{1}$$

and time interval $t2$ which is the sum of the time for one half orbit T plus the time for the faster light (component approaching) to travel the same distance L:

$$t2 = T + \frac{L}{c+v}. \tag{2}$$

For $t1 = t2$ we have:

$$\frac{L}{c-v} = T + \frac{L}{c+v}. \tag{3}$$

Solving for L we get:

$$L = T\frac{c^2 - v^2}{2v}. \tag{4}$$

For $v \ll c$ we can use

$$L = T\frac{c^2}{2v}. \tag{5}$$

RECENT RECONSIDERATIONS

Contrary to de Sitter's claim and to other arguments advanced more recently, John Fox [3] found that visible binary stars do *not* offer evidence against the Ritz theory. He takes this stand on the basis of Tolman's extinction theorem [4], i.e., the absorption and re-radiation of electromagnetic radiation by electrical charges in dispersive media. (The re-radiated light travels at c with respect to the medium.) The Oswald-Oseen extinction theorem is another version of this idea. One extinction length (an exponential process) in the interstellar medium is estimated to be on the order of one light year.

In 1987 Vladimir Sekerin [5] showed that when we consider the distances (binary to observer) required for de Sitter's "whimsical" images to manifest themselves that the atmosphere limited angular resolution (one arc second) of our best ground based telescopes were insufficient for us to resolve them. (This was before the advent of the Hubble Space Telescope and the growing family of Very Large Telescopes.)

Sekerin hypothesized that de Sitter's binary star scenario might provide an alternate explanation for the light and apparent radial velocity time histories of periodically varying stars. Instead of seeing a visible component at two locations simultaneously (because the images can't be resolved) we would get periodic variations in its light intensity and spectral signature.

MODELING THE PROCESS

Computer generated CRT photographs, furnished by Sekerin's colleague, M.S. Serbulenko, show light curves and radial velocity curves computed in accordance with Sekerin's hypothesis. The curves are for a binary system with one visible component. Sekerin was pursuing a mechanism to explain the characteristics of Cepheids but there is a problem. For Cepheids, the peak (maximum brightness) of the light curve is nearly coincident with the peak (approaching) *apparent* radial velocity curve. In Sekerin's modeling the radial velocity peak lags the light curve brightness peak by 90 degrees.

New numerical code was created to check Sekerin's hypothesis. The code models a binary system and remote observer. At selected positions in the orbit of one (or both) of the components, the distance from the component to the observer is "measured." The computed source-to-observer travel times, from these points, are added to the emission times to get arrival times, at the observer, which are then *scaled* in arrival time bins. The accumulating sums in these bins are displayed on a CRT time base. Doppler shifts for each "emission point" are used to create accompanying radial velocity curves. The new code also produced the 90 degree phase error.

Since the radial velocity curves in both computer models do not match observational data some other physical mechanism besides Doppler is needed in the modeling to bring the computed curves into some semblance of actual observations. The failure of this quest would constitute falsification of Sekerin's hypothesis.

Since Ritz's work mainly addressed electrodynamics, and light is electrodynamic in nature, we note that expressions for electrodynamic influences of charges on one another contain terms for positions, velocities, *and* accelerations. Heretofore accelerations had not been considered in either of our approaches.

If we consider a more or less circular binary star orbit with sinusoidally varying acceleration and velocity components (with respect to the remote observer), the acceleration effects are similar to Doppler shifts but lead them by 90 degrees. Thus, acceleration effects very nearly bring the Ritzian hypothesis into conformance with observational data.

The computer program was modified to use orbital accelerations, with respect to the observer, to produce the apparent radial velocity curves. (The light curve is the same regardless of whether Doppler or acceleration effects are used. The program will need to be modified to properly account for both effects.)

COMPARING NUMERICAL RESULTS TO OBSERVATIONS

For observer distances much less than the de Sitter overtaking distance the numerical code creates light and radial velocity curves which approximate those for simple Cepheid variables and similar objects. When both components are visible the program produces light and radial velocity curves for each component.

As the observer distance increases the Cepheid-like bumps evolve into sharp peaks which get higher and narrower as the de Sitter distance is approached.

For a binary star with both components visible we get a peak for each component. The double pulsed Crab pulsar may be considered as a candidate for a two component (fraternal twins) binary. A neutron star binary where the radii of the orbits are on the order of fourteen *Houstons* and their orbital speeds are on the order of $0.1c$ would suffice. There *are* intriguing spectral differences between the pulses.

If the Ritz hypothesis is valid, then we should expect to find lots of two-color, double pulsed, pulsars. AN Ursae Majoris, which is considered to be a possible slow pulsar [6] (period 1.914 hr), may be a candidate as a two-color pulsar. It is reported to have two different emitting sources.

At the de Sitter overtaking distance L, the sharp high peak splits into two peaks. (At L a telescope with sufficient resolving power would show two separate images pulsing in phase with one another with a smeared out faint bridge between them.) As the binary-to-observer distance increases beyond L the inter-peak time interval increases. The outer edges of the peaks become almost vertical and there is a saddle-like trough structure between the peaks. These peaks will flash alternately. At a binary-to-observer distance of $2L$ the computed light curve strongly resembles that of the Geminga Pulsar.

The $c + v$ effects produce arrival time modulation at the observer which further modifies the spectral content. The observed frequency modifications are in accordance with the relation $dt/d\tau$, where τ is the modulated arrival time.

APPLICATION OF THE HYPOTHESIS TO GAMMA-RAY BURSTERS

According to this Ritzian interpretation of variable stars, gamma-ray bursts could be caused by close encounters of stellar objects. The arrival time compressions that would accompany perihelion passages could produce blue shifted *observed* bursts of extremely high energies. (This means that the objects are not actually bursting, rather they just look that way from sufficient distances.) Short term elliptical orbits would produce repeating bursts such as the Geminga pulsar. Chance non-returning stellar encounters could produce non-repeating bursts.

In principle, any encounter of two stars could lead to perturbative accelerations which could produce a pulse or burst of light in the sense mentioned above.

Where a *capture* event takes place between two stars, we might expect to observe a series of pulses/bursts. The process may be accompanied by electromagnetic braking and perhaps inter-body electrical discharges. Both processes would produce changes in the settling down orbitals. The latter could produce radical energy changes in the orbits and abrupt changes in the observed light curves and spectra. When one of the participants happens to be a short period binary then an otherwise well behaved burst can take on a spiky or crenulated/serrated effect.

An unabridged version of this presentation, including graphics and the numerical code used, is available on the world wide web at:
http://www.shadetreephysics.com

REFERENCES

1. Ritz, W., *Ann. de Chim. et de Phys.* **13**, 145–275 (1908).
2. de Sitter, *Phys. Zeits.* **14**, 429 (1913).
3. Fox, J.G., *Am. J. Phys.* **33**, 1 (1965).
4. Tolman, R., *Phys. Rev.* **31**, 26 (1910); *ibid* **35**, 136 (1913).
5. Sekerin, V.I., *Contemporary Science and Regularity in its Development*, Tomsk University, **4**, 119–123, (1987).
6. Gilmozzi, R., Messi, R., & Natali, G., *Astrophys. J.* **245**, L119 (1981).

The Connection between Supernovae and Gamma-Ray Bursts: On the Distribution of the Circumstellar Matter

Roger A. Chevalier

Department of Astronomy, University of Virginia
P.O. Box 3818, Charlottesville, VA 22903

Abstract. Radio observations of supernovae can be interpreted as the interaction of the outer steep density profile of the supernovae with stellar winds from the progenitor stars. Modeling of the thermal supernova emission shows that the supernovae have massive star progenitors that are expected to have winds. Multiwavelength observations of the supernovae give additional constraints on the wind interaction which allow, in some cases, the determination of the parameters for the synchrotron emission. In the case of GRBs, thermal emission from a supernova was observed in one case and possibly in two others, but the nonthermal emission is generally the only diagnostic for determining the nature of the source. The distribution of the circumstellar matter is important for determining the progenitor object and there is some evidence for at least 2 types of progenitors among the observed afterglows. However, the density determination is complicated by a lack of knowledge of basic parameters for the synchrotron emission model. In addition, the possibilities of jet effects and a transition to sub-relativistic expansion further complicate the situation. Multiwavelength observations extended over time are needed to limit the possibilities.

I INTRODUCTION

The shock waves generated by both supernovae (SNe) and gamma-ray bursts (GRBs) give rise to synchrotron radiation from relativistic electrons in strong magnetic fields. The density of the surrounding medium and its distribution are important for this emission. The surroundings of an explosion provide clues about the progenitor of the explosion. A massive star is expected to have a substantial stellar wind, but compact objects may interact directly with the interstellar medium. The interpretation of observations of synchrotron radiation can thus be expected to provide critical information for determining the progenitor objects. Unfortunately, a lack of knowledge of the basic processes crucial for the synchrotron emission have made it impossible to fully exploit the implications of the synchrotron emission. The purpose of this paper is to review our understanding of the surrounding medium, and how it is influenced by uncertainties in the basic processes.

II SUPERNOVAE

The radio emission from supernovae can be interpreted as synchrotron emission from the interaction of the supernova with a progenitor wind. This emission has been observed from Type II, Type Ib, and Type Ic supernovae, which are thought to result from the explosion of core collapse supernovae. These supernovae are believed to have massive star progenitors, which is consistent with the presence of a stellar wind. The shock front moves at a nonrelativistic velocity and is in the dense part of the circumstellar wind at the time of observation. The energy density that is generated by the interaction is capable of giving the observed radio emission if the energy densities in relativistic electrons and magnetic fields are ~ 1 % of the total postshock energy density. For reference, the density of wind gas that is experienced by a 10,000 km s^{-1} shock front at an age of 100 days is 7×10^{-19} g cm^{-3} for a wind with mass loss rate $\dot{M} = 1 \times 10^{-5}$ $M_\odot$ yr^{-1} and velocity $v_w = 10$ km s^{-1} (red supergiant), and 7×10^{-21} g cm^{-3} for a wind with mass loss rate $\dot{M} = 1 \times 10^{-5}$ $M_\odot$ yr^{-1} and velocity $v_w = 1,000$ km s^{-1} (Wolf-Rayet star). These densities are much higher than a typical interstellar density, $\sim 1 \times 10^{-24}$ g cm^{-3}; a supernova in the interstellar medium is expected to be a faint radio source. In this context, the lack of detectable radio emission from Type Ia supernovae may indicate interstellar interaction, although interaction with a rarified stellar wind is also possible. The best limit on a Type Ia event is for SN 1986G in Centaurus A [1]. If more sensitive searches for radio emission from Type Ia supernovae are unsuccessful, this could be evidence for their progenitors being merging white dwarf binaries, which can interact directly with the interstellar medium.

In the model for radio supernovae initially proposed in [2], the shock front is driven by the outer steep power-law density distribution of the supernova and the energy densities in relativistic electrons and magnetic field are a constant fraction of the total postshock energy density. If the particle acceleration is rapid, the evolution of the optically thin radio flux is expected to be a power law in time. This has been observed in many cases, and the power-law index is compatible with interaction with a stellar wind. Additional evidence for wind interaction comes from the evolution of the free-free optical depth in SN 1979C and SN 1980K [2] and from observations of supernovae at optical and X-ray wavelengths [3]. The multiwavelength observations, together with the radio emission, allow estimates of the presupernova mass loss rate for Type II supernovae. For an assumed red supergiant progenitor with $v_w = 10$ km s^{-1}, $\dot{M}$ is typically in the range $10^{-5} - 10^{-4}$ $M_\odot$ yr^{-1}. These results are at the high end of the mass loss rates deduced for Galactic red supergiants. An interesting recent case is SN 1999em, which was detected as an X-ray source by *Chandra*, but at a relatively low luminosity [4]; radio emission from the source is also at low luminosity [5]. The implied $\dot{M}$ is $\sim 10^{-6}$ $M_\odot$ yr^{-1}. This supernova, which had a plateau type light curve, may be surrounded by a typical red supergiant wind.

The peculiar Type II supernova SN 1987A exploded as a blue supergiant and was a low-luminosity radio source, as expected for a low density wind. For $v_w = 500$ km s^{-1}, the estimated $\dot{M}$ is $\sim 10^{-8}$ $M_\odot$ yr^{-1} [6]. The radio emission rose sharply in 1990 as the supernova presumably began to interact with dense gas from a previous red supergiant phase. The radius at which this interaction began was $\sim 4 \times 10^{17}$ cm. SN 1987A is clearly interacting with a complex environment created by presupernova mass loss, but it probably represents a relatively rare event.

Type Ib and Ic supernovae are thought to be the explosions of massive stars that have lost their hydrogen envelopes, i.e., Wolf-Rayet stars. In these events, the radio emission provides the primary evidence for circumstellar interaction, and it does not contain sufficient information to determine the circumstellar density. The intensity of the emission is compatible with a typical Wolf-Rayet star wind only if the efficiencies of relativistic electron and magnetic field production are high [6].

In addition to the supernova and circumstellar density distributions, the basic parameters for the model of the synchrotron emission are the electron energy spectral index p, where $N(E)dE \propto E^{-p}$, the fraction of the total postshock energy density in relativistic electrons ϵ_e, and the fraction of the total postshock energy density in magnetic fields ϵ_B. An assumption of the standard model is that the value of p, which is determined from the observations, remains constant during the observed evolution. The existing observations are compatible with this assumption, but they clearly show that p can vary between supernovae. The values range from 2.1 (SN 1980K) to 3.2 (SN 1990B and SN 1987A) ([6,7] and references therein). The Type Ib and Ic supernovae typically have $p \approx 3$.

The most likely mechanism for accelerating the electrons is diffusive shock wave acceleration. In the test particle limit in a strong shock wave, a power-law spectrum is produced with $p = 2$ [8]. Steeper spectra are often implied by synchrotron sources and a plausible reason is that the shock wave is cosmic ray dominated. In this case, higher energy electrons experience a larger density jump than lower energy ones and the spectrum is concave [9]. If this is occurring, careful observations of an evolving source may show a flattening of the spectrum with time because higher energy particles are observed as the source expands and the magnetic field drops.

The values of ϵ_e and ϵ_B are also not well determined from basic theory. One suggestion for the magnetic field is that it is built up by Rayleigh-Taylor instabilities in the region where the supernova ejecta are decelerated by the surrounding medium [10]. The ejecta are being decelerated throughout the time of observation, so that the magnetic field is continually affected by the turbulent motions and the hypothesis of a constant ϵ_B is reasonable. Hydrodynamic simulations of the build up of the magnetic field show that the field is most amplified on the smallest scales; in one simulation, the magnetic field is locally amplified by up to a factor of 60, but the overall magnetic energy density is only ~ 0.3 % of the turbulent energy density [11]. However, the magnitude of amplification is sensitive to the numerical resolution. Although there have been suggestions that the magnetic field in radio supernovae is simply the shock compressed stellar wind field, the observed properties of synchrotron self-absorption in Type Ib and Ic supernovae show that this is very unlikely [6]. The magnetic field is stronger than can be explained in this way.

If there is sufficient information on a supernova, the values of ϵ_e and ϵ_B can be estimated from the observations. This is possible for SN 1993J because multiple physical processes play a role in the radio light curve, radio VLBI imaging is possible, and optical and X-ray observations lead to additional constraints [12]. Fransson and Björnsson [12] find that the evolution is consistent with constant efficiency factors, with $\epsilon_e \approx 5 \times 10^{-4}$ and $\epsilon_B \approx 0.14$; their model fits imply that $p = 2.1$ throughout the observed evolution. This model also requires that any stellar wind magnetic field be amplified in the postshock region.

III GAMMA-RAY BURSTS

The discovery of SN 1998bw in the error circle of GRB 980425 [13] provided the first strong observational evidence for a link between supernovae and GRBs. The radio emission from the supernova was unusually strong, implying relativistic motion [14,15]. In addition to the high velocity and consequent high energy in the synchrotron emitting electrons, the evolution of the radio source had properties which distinguish it from normal radio supernovae. With the assumption of constant particle and magnetic efficiency factors, the radio light curves implied constant blast wave energy over much of the observed evolution [15]. This behavior is distinct from normal radio supernovae in which the blast wave energy continually increases as a result of energy input from the steep supernova density profile. Another property of the SN 1998bw radio light curves was an increase in flux at an age of 20–40 days, which implied a sharp increase in the blast wave energy by a factor of ~ 2.5 [15]. The overall behavior of the source appeared to be more consistent with explosive release of energy from a central source, as in a GRB afterglow, as opposed to driving by the outer layers of a supernova envelope, as in a normal radio supernova. The model for the radio light curves also depends on the distribution of circumstellar matter and that was found to be consistent with the $\rho \propto r^{-2}$ distribution expected in a stellar wind. The supernova was of Type Ic [13], which are believed to have Wolf-Rayet star progenitors and strong stellar winds.

The electron energy spectral index in SN 1998bw could be determined from the radio observations, with the result $p = 2.5$ [14,15]. The model parameters for the synchrotron source could not be uniquely determined because the synchrotron cooling break, which is useful for determining ϵ_B, was not observed. In a model close to equipartition, $\epsilon_B = 0.15$ and $\epsilon_e = 0.1$ led to an explosion energy $E = 1.2 \times 10^{49}$ ergs and $A_\star = 0.04$, where $A_\star$ is the wind density ($\rho = Ar^{-2}$) in units of 5×10^{11} gm cm^{-1} [15]. This reference density is chosen to represent a wind with $\dot{M} = 1 \times 10^{-5}$ $M_\odot$ yr^{-1} and $v_w = 1,000$ km s^{-1}, which are typical parameters for a Wolf-Rayet star wind. In another model with $\epsilon_B \approx 10^{-6}$, the other parameters are $\epsilon_e \approx 1$, $E = 1.7 \times 10^{50}$ ergs, and $A_\star = 6$. In both cases, the mass loss rate is compatible with that expected for a Wolf-Rayet star. The energy is lower than that deduced for the supernova itself, which has $E_{SN} \approx (2-3) \times 10^{52}$ ergs [16,17], but it is unlikely that the outer layers of the supernova have sufficient energy to drive the radio source blast wave at a relativistic velocity. Most of the energy in this explosion is in relatively slow material.

If GRB 980425 was associated with SN 1998bw, as seems likely in view of the peculiar radio source, its distance of only 40 Mpc makes it a peculiar low-energy event among the GRBs. However, there is evidence that some of the cosmological GRBs are also associated with supernovae. Both GRB 980326 [18] and GRB 970228 [19,20] show supernova type emission in their optical light curves. If these GRBs are the explosions of massive stars, they must be expanding in the winds of the progenitor stars. However, the standard model for the afterglows of GRBs involves expansion into a uniform density interstellar medium [21–23]. This type of surrounding medium would be appropriate for a progenitor system involving a compact binary system. The determination of the distribution of circumstellar matter around a GRB thus gives crucial information on the nature of the progenitor. Unfortunately, this determination is affected by a number of uncertain parameters, including the values of p, ϵ_e, and ϵ_B, and the effects of jet expansion. The result is that there is considerable controversy surrounding the interpretation of GRB afterglows.

In the case of GRB 980326, the optical afterglow before the supernova light showed a rapid rate of decline, which can be interpreted in terms of adiabatic wind interaction with $p \approx 3$ [24]. However, an alternative interpretation is an expanding jet, above the cooling frequency, with $p \approx 2$ [25]. In this case, the interaction could be either with a wind or a uniform, interstellar medium.

A burst with a similar optical decline was GRB 980519, although a supernova was not observed in this case. There is again an ambiguity of wind or jet evolution. This source was observed at radio wavelengths over the first 63 days [26], and the emission was found to be compatible with a wind model prediction based on the first few days of radio data [24]. However, the uncertainties in the radio emission are sufficiently large that jet evolution is also possible. The X-ray afterglow from this source was moderately bright and there was no evidence for a cooling break between optical and X-ray wavelengths. In the wind model, this would require a low value of ϵ_B ($\lesssim 10^{-5}$) [24]. In the jet model, the optical emission is above the cooling break, so the observations are compatible with a higher value of ϵ_B.

The ambiguity in the emission from this type of source might be broken by more extensive observations of the afterglow, either before the jet effects have set in or more extensive radio observations. An alternative is to follow the late evolution, because the presence of a jet should eventually lead to spherical, nonrelativistic expansion [27]. This will appear as a flattening of the light curves.

In the case of GRB 970228, the subtraction of the supernova light leads to a moderately steep decline in the optical that is compatible with adiabatic wind interaction and $p \approx 2.6$ [28]. The expected drop of the peak flux with time in a wind model is also compatible with the upper limits on radio emission that have been placed on the source. However, it is possible that a jet model will also be capable of explaining the observed behavior.

The afterglow of GRB 970508 was the most extensively observed to date. Especially valuable are the multifrequency radio data, which extend to an age of 450 days [29]. In one interpretation, the emission is from a spherical explosion in a stellar wind; this model can reproduce most of the structure that is observed in the radio light curves [28]. In this case, optical observations do place constraints on the cooling frequency, so that all of the basic model parameters can be determined: they are $E = 3 \times 10^{51}$ ergs, $A_* = 0.3$, $\epsilon_e = 0.2$, $\epsilon_B = 0.1$, and $p = 2.2$. In this model the expansion is becoming sub-relativistic at the end of the observation period.

In another interpretation of GRB 970508, the source has interacted with a uniform density medium, but has gone through a phase of lateral jet expansion [29]. In this model, the first 25 days of evolution can be described by a spherically symmetric blast wave in a uniform medium because only a small part of the blast wave is observed due to relativistic effects. Parameters found for the blast wave in refs. [30,31] are the ambient density $n_o = 0.030, 5.3$ cm^{-3}, $E = (3.5, 0.53) \times 10^{52}$ ergs, $\epsilon_e = 0.12, 0.57$, and $\epsilon_B = 0.089, 0.0082$, respectively. The energy here assumes isotropic expansion. During ages 25–100 days, the GRB is inferred to have a jet that laterally expands leading to lower radio fluxes than otherwise expected, although this evolution is not evident at optical wavelengths. After 100 days, the lateral expansion is complete and the source expands as a nonrelativistic blast wave in a uniform medium. The parameters deduced for this phase are $n_o \approx 0.5$ cm^{-3}, $E \approx 5 \times 10^{50}$ ergs, $\epsilon_e \approx \epsilon_B \approx 0.5$ [29]. The lower energy deduced for this case can be attributed to the lateral expansion. However, the change in ϵ_B must be attributed to a change in conditions. The value of p remains constant (at 2.2) throughout the evolution in both this model and the previous one, which is interesting in view of the higher values of p inferred for rapidly expanding, but sub-relativistic, supernovae.

Although there is controversy surrounding the cases of possible wind interaction, there is general agreement that some afterglows are best described by interaction with a uniform medium. These include the afterglows of GRB 990123 and GRB 990510. In the case of GRB 990123, support for interstellar interaction comes from the relation between the optical spectral index and the rate of flux decline, the more rapid flux decline at X-ray wavelengths as compared to optical wavelengths, and the extended early emission that is apparently from a reverse shock wave [28,32]. In the case of GRB 990510, the relatively flat early optical decline and the relation between spectral index and decline rate indicate interstellar interaction [33]. This source also showed the best evidence for lateral jet expansion.

In addition to interpretations in terms of wind interaction and jets, the steep declines observed in some afterglow sources have been interpreted in terms of sub-relativistic expansion because of interaction with a high-density medium ($n_o \sim 10^5 - 10^6$ cm^{-3}) [34–36]. The model parameters for the afterglow of GRB 990123 are $n_o = 3 \times 10^6$ cm^{-3}, $\epsilon_e = 0.1$, and $\epsilon_B = 2 \times 10^{-8}$ when E is taken to be 1.6×10^{54} ergs [34]. The large luminosity that would be produced by the high ambient density is compensated by the low value of ϵ_B. The origin of the high-density medium is unclear. Mass loss from the progenitor can create a dense medium, but it would be unlikely to have the approximately uniform density that is inferred in this model.

An explosion in a dense interstellar clump is unlikely because of the small volume filling factor of such clumps; stars form out of dense matter, but a massive star is unlikely to still be embedded in the dense gas at the end of its life. Dai and Lu [34] propose that the gas is from a previous supernova explosion of the progenitor star. To have the proper density, a 10 $M_\odot$ supernova would have to have expanded to $R \sim 3 \times 10^{17}$ cm, which implies an age of about 10 years for a 10^{51} erg supernova explosion. The mechanism by which a GRB may follow a supernova at this time interval is unknown. Other evidence for a high density around a GRB includes the possible detection of the Fe K line in the X-ray afterglow of GRB 970508 [37], which would require a density of $n_o \sim 5 \times 10^9$ cm^{-3} at $R = 10^{16}$ cm for a solar abundance of Fe. However, the main part of the afterglow can be modeled by expansion in a low-density medium, as discussed above, and there is no unequivocal evidence for dense gas.

These considerations show that the determination of the density and distribution of circumstellar matter is intertwined with the energy spectral index p and other parameters of the synchrotron emission. The extensive observations of GRB 970508 imply $p = 2.2$. Numerical simulations of particle acceleration in ultrarelativistic shock waves in fact show that $p = 2.2$ is approached for large Lorentz factors for a range of turbulence amplitudes and magnetic field configurations [38]. However, the blast wave in GRB 970508 was observed to decelerate to the point that deviations from $p = 2.2$ are likely according to the numerical simulations. Although there is no evidence for a changing p in GRB 970508, some afterglows (e.g., GRB 980425/SN 1998bw, GRB 970228, GRB 990123) appear to have p in the range $2.5 - 2.6$. If sources like GRB 980326 and GRB 980519 are correctly described by non-cooling wind interaction models, $p \approx 3$ is implied. Thus, there is not clear evidence for a universal value of p in GRB afterglows, and the theory does not require one value. The value of ϵ_B is also poorly determined and can depend on the model for the circumstellar density. The instabilities due to deceleration of the ejecta invoked in the radio supernova case are not relevant to the GRB afterglow case because GRB afterglows are modeled as blast waves without any influence of the ejecta. The relativistic two-stream instability has been investigated as a possible source of magnetic field amplification [39], but a consensus has not yet been reached on this problem.

IV FUTURE PROSPECTS

The outstanding problem in this area and one on which progress can be expected in the next few years is the determination of the distribution of circumstellar gas around GRBs. If the gas has the $\rho \propto r^{-2}$ distribution expected in a stellar wind, the implication would be a massive star progenitor. If the gas has a low, uniform density, a massive star origin is unlikely and a progenitor involving compact objects is indicated. This issue can be explored in several ways [24,28]:

- The long term evolution of the radio afterglow of a GRB provides valuable constraints on a number of model parameters, including the self-absorption frequency, the "typical" frequency ν_m, and the peak flux. The observations of GRB 970508 [29] provide an excellent example. At high frequencies (optical, X-ray), the observations do not distinguish between the wind and interstellar cases if the observations are above the cooling frequency. However, observations at infrared through X-ray wavelengths over an extended time are likely to show the evolution of the cooling frequency, which does distinguish between wind and interstellar interaction.

- At an age of a few days, the density experienced by a shock front in a Wolf-Rayet star stellar wind is similar to a typical interstellar density. At earlier times ($t < 10^5$ sec), the density in the stellar wind case is substantially higher. One result of this is that a relativistic shell generated by the central engine is more rapidly decelerated in the stellar wind case. Thus, for a similar central engine, the interaction is more rapid and stronger in the stellar wind case. The reverse shock wave is expected to be in the strong cooling regime, which results in a cutoff of emission once the reverse shock has propagated through the initial shell, unlike the interstellar case, which is not in the strong cooling regime.

- There can be a problem in distinguishing the optical evolution of a jet source as compared to a wind interaction (e.g., the cases of GRB 980326 and GRB 980519). The late transition to spherical sub-relativistic expansion in the jet case provides a way of distinguishing between these [27].

- The observations of GRB 980326 and GRB 970228 provide tantalizing evidence for supernova type optical emission. The search for such emission in future sources is clearly of high priority. If the supernova light is indicating a massive star origin, the afterglow should be of the stellar wind type. Conversely, a source with a stellar wind type afterglow should be accompanied by a supernova. However, a range of supernova properties is likely, as is observed with normal Type Ib and Ic supernovae.

ACKNOWLEDGMENTS

I am grateful to Claes Fransson and Zhi-Yun Li for enjoyable collaborations on supernova and gamma-ray burst problems. This work was supported in part by NASA grant NAG5-8232.

REFERENCES

1. Eck, C. R., Cowan, J. J., Roberts, D. A., Boffi, F. R., and Branch, D., *ApJ* **451**, L53 (1995).
2. Chevalier, R. A., *ApJ* **259**, 302 (1982).

3. Chevalier, R. A., and Fransson, C., *ApJ* **420**, 268 (1994).
4. Fox, D. W., Lewin, W. H. G., et al., *IAUC* No. 7318 (1999).
5. Lacey, C. K., Van Dyk, S. D., Weiler, K. W., Sramek, R. A., and Panagia, N., *IAUC* No. 7336 (1999).
6. Chevalier, R. A., *ApJ* **499**, 810 (1998).
7. Weiler, K. W., Van Dyk, S. D., Montes, M. J., Panagia, N., and Sramek, R. A., *ApJ* **500**, 51 (1998).
8. Blandford, R. D., and Eichler, D., *Phys. Reports* **154**, 2 (1987).
9. Ellison, D. C., and Reynolds, S. P., *ApJ* **382**, 242 (1991).
10. Chevalier, R. A., Blondin, J. M., and Emmering, R. T., *ApJ* **392**, 118 (1992).
11. Jun, B.-I., and Norman, M. L., *ApJ* **465**, 800 (1996).
12. Fransson, C., and Björnsson, C.-I., *ApJ* **509**, 861 (1998).
13. Galama, T. J., et al., *Nature* **395**, 670 (1998).
14. Kulkarni, S. R., Frail, D. A., Wieringa, M. H., Ekers, R. D., Sadler, E. M., Wark, R. M., Higdon, J. L., Phinney, E. S., and Bloom, J. S., *Nature* **395**, 663 (1998).
15. Li, Z.-Y., and Chevalier, R. A., *ApJ* **526**, 716 (1999).
16. Iwamoto, K., et al., *Nature* **395**, 672 (1998).
17. Woosley, S. E., Eastman, R. G., and Schmidt, B. P., *ApJ* **516**, 788 (1999).
18. Bloom, J. S., et al., *Nature* **401**, 453 (1999).
19. Reichart, D. E., *ApJ* **521**, L111 (1999).
20. Galama, T. J., et al., *ApJ* submitted, astro-ph/9907264 (1999).
21. Mészáros, P., and Rees, M. J., *ApJ* **476**, 232 (1997).
22. Waxman, E., *ApJ* **489**, L33 (1997).
23. Sari, R., Piran, T., and Narayan, R., *ApJ* **497**, L17 (1998).
24. Chevalier, R. A., and Li, Z.-Y., *ApJ* **520**, L29 (1999).
25. Sari, R., Piran, T., and Halpern, J. P., *ApJ* **519**, L17 (1999).
26. Frail, D. A., Kulkarni, S. R., Sari, R., Taylor, G. B., Shepherd, D. S., Bloom, J. S., Young C. H., Nicastro, L., and Masetti, N., *ApJ* in press, astro-ph/9910060 (1999).
27. Livio, M., and Waxman, E., *ApJ* submitted, astro-ph/9911160 (1999).
28. Chevalier, R. A., and Li, Z.-Y., *ApJ* submitted, astro-ph/9908272 (1999).
29. Frail, D. A., Waxman, E., and Kulkarni, S. R., *ApJ* submitted, astro-ph/9910319 (1999).
30. Wijers, R. A. M. J., and Galama, T. J., *ApJ* **523**, 177 (1999).
31. Granot, J., Piran, T., and Sari, R., *ApJ* **527**, 236 (1999).
32. Kulkarni, S. R., et al., *Nature* **398**, 389 (1999).
33. Harrison, F. A., et al., *ApJ* **523**, L121 (1999).
34. Dai, Z. G., and Lu, T., *ApJ* **519**, L155 (1999).
35. Dai, Z. G., and Lu, T., *ApJ* in press, astro-ph/9906109 (1999).
36. Wang, X. Y., Dai, Z. G., and Lu, T., *MNRAS* submitted, astro-ph/9912492 (1999).
37. Piro, L., et al., *ApJ* **514**, L73 (1999).
38. Bednarz, J., and Ostrowski, M., *Phys. Rev. Lett.* **80**, 3911 (1998).
39. Medvedev, M. V., and Loeb, A., *ApJ* **526**, 697 (1999).

The Type Ib/c Supernova, Gamma-Ray Burst, Soft Gamma-ray Repeater, Magnetar Connection

J. Craig Wheeler*, Peter Höflich*, Lifan Wang*, and Insu Yi[†]

*Department of Astronomy, The University of Texas Austin, Texas 78751
[†]Korea Institute for Advanced Study, Seoul, Korea

Abstract. The polarization of core-collapse supernovae shows that many if not all of these explosions must be strongly bi-polar. The most obvious way to produce this axial symmetry is by the imposition of a jet as an intrinsic part of the explosion process. These jets could arise by MHD processes in the formation of pulsars and be especially strong in the case of magnetars. The jets will blow iron-peak material out along the axes and other elements from the progenitor along the equator, a very different composition structure than pictured in simple spherical "onion skin" models. In extreme cases, these processes could lead to the production of γ-ray bursts powered by strong Poynting flux.

POLARIZED SUPERNOVAE AND JETS

We have found that all core-collapse events, supernovae of Type II and Type Ib/c, are polarized at the 1% level and some much more so (Wang et al. 1996; Wang et al. 1999). Our data suggest a very important trend: the smaller the hydrogen envelope and the deeper within the ejecta we see, the larger the observed polarization. Polarization of the level we observe then forces us to abandon timid phrases like "asymmetric supernovae." For these events, it is appropriate to talk about "bi-polar supernovae."

The next issue is thus how to account for the observed high levels of polarization. An asymmetric impulse in an otherwise spherical configuration will tend to turn spherical before homologous expansion is reached by the propagation of lateral pressure gradients. What is needed is the directed flow of energy and momentum in a single direction for a time that is substantial compared to the dynamical timescale. We need a jet. This conclusion is independent of any connection to γ-ray bursts, but, of course, the potential for this connection is clear.

A preliminary study in which conditions were selected to represent the sort of MHD jet found by LeBlanc & Wilson (1970; see also Müller & Hillebrandt 1979; Symbalisty 1984) has been presented by Khokhlov et al. (1999; see also Höflich,

FIGURE 1. Composition structure of a jet-driven supernova. The axial jet (light lobes) contains jet material. The equatorial shell (darker region) shows the distribution of the oxygen layer from the initially spherical progenitor model (from Höflich et al. 2000).

Wheeler & Wang 1999). This study has been extended to explore a range of jet energies and stellar configurations, both bare helium cores and red supergiants.

The code employed, developed by Khokhlov (1998), is an Eulerian adaptive mesh code. The calculations are fully three dimensional. The adaptive mesh gives great resolution. The finest scale corresponds to a uniform grid of some 10^{10} cells. The adaptive mesh also allows substantial dynamic range. For the jet models this ranges from $2^{12} \sim 10^4$ to $2^{19} \sim 10^6$. The imposed jets are cylindrically symmetric and the initial stellar model is spherical. The resulting jets are thus highly cylindrically symmetric, but this is not imposed in the dynamics, only the initial conditions. The jet dynamics are sufficiently rapid for the models computed that Kelvin/Helmholz instabilities have little time to form.

Figure 1 shows the distribution of the jet matter (unspecified in the computation, but presumably rich in iron-peak elements), and of the oxygen layers of the star. The former reflects the bi-polar nature of the jet flow. The latter shows the effects of the lateral shocks that compress the oxygen into an equatorial shell. This will, in turn, affect the line profiles of the oxygen observed in the nebular phase. These profiles are presented after 4.84 seconds when the jet breaks through the surface of the helium core. They must be followed into homologous expansion before any direct connections to observations can be made. We have also studied models with red giant envelopes (Höflich, et al. 2000). The code allows us to follow the jet in a single calculation from the center of the star out through the extended envelope. We find that energetic jets can penetrate the hydrogen envelope, but that more modest jets cannot. The latter can still induce an asymmetric, bi-polar explosion.

The next issue is the origin of the jets. To account for normal supernovae we must have jets in routine circumstances, that is, associated with the formation of a neutron star and not restricted to the more rare circumstances of the possible formation of a black hole. This statement is independent of the likelihood that in rare cases or different circumstances such a jet might yield a γ-ray burst.

The obvious place to look for jets in frequent core collapse events is in the rotating, magnetic collapse of a neutron star with the equivalent dipole magnetic field ranging from "typical" values like the Crab pulsar to the extreme values associated with magnetars (Duncan & Thompson 1992) and soft γ-ray repeaters (Kouveliotou et al. 1998). This environment gives a framework in which to quantitatively address questions of physics that are germane to the nature of the core collapse process in general and to potential γ-ray production. The physics that could be at play in such a collapse has recently been considered by Wheeler et al. (1999).

Rotation and magnetic fields have a strong potential to create axial matter-dominated jets that will drive strongly asymmetric explosions for which there is already ample observational evidence in Type II and Type Ib/c supernovae, their remnants, and in the pulsar velocity distribution. The potential to also create strong flows of Poynting flux and large amplitude electromagnetic waves (LAEW) serves to reinforce the possibility to generate bi-polar explosions. These bi-polar explosions will, in turn, affect nucleosynthesis and issues such as fall-back that determine the final outcome to leave behind neutron stars or black holes. In addition, the presence of matter-dominated and radiation-dominated jets might lead to bursts of γ-rays of various strengths. The issue of the nature of the birth of a "magnetar" in a supernova explosion is of great interest independent of any connection to γ-ray bursts. Highly magnetized neutron stars might represent one out of ten pulsar births. Production of a strong γ-ray burst might be even more rare.

Wheeler et al. show that the contraction phase of a proto-neutron star could result in a substantial change in the physical properties of the environment. When the rotating magnetized neutron star first forms there is likely to be linear amplification of the magnetic field and the creation of a matter-dominated jet, perhaps catalyzed by MHD effects, up the rotation axis. The rotational energy of the proto-neutron star is $\sim 10^{51}$ ergs, sufficient to power a significant matter jet, but unlikely to generate a strong γ-ray burst. The matter jet could generate a smaller γ-ray burst as seems to be associated with SN 1998bw and GRB 980425 by the Colgate (1974) mechanism as it emerges and drives a shock down the stellar density gradient in the absence of a hydrogen envelope, e.g., in a Type Ib/c supernova.

As the neutron star cools, contracts, and speeds up, the rotational energy increases. The energy becomes significantly larger than required to produce a supernova and sufficient, in principle, to drive a cosmic γ ray burst if the collimation is tight enough and losses are small enough. For a neutron star with a period near 1 millisec the rotation energy becomes $\gtrsim 10^{52}$ ergs. If efficiently utilized and collimated, this energy reservoir could make a substantial γ-ray burst. The luminosity is estimated to be $\sim 10^{52}$ erg s^{-1} and to last for a few seconds.

The second important factor that accompanies the contraction and spin-up of the cooling neutron star is that the light cylinder contracts significantly, so that a stationary dipole field cannot form and the emission of strong LAEW occurs. Tight collimation of the original matter jet and of the subsequent flow of LAEW in a radiation-dominated jet is expected. The LAEW will propagate as intense low-frequency, long-wavelength radiation. The LAEW "bubble" could be strongly

Rayleigh-Tayor unstable, but still may propagate selectively with small opening angle up the rotation axis as an LAEW jet. If a LAEW jet forms, it can drive shocks which may selectively propagate down the axis of the initial matter jet or around the perimeter of the matter jet. The shocks associated with the LAEW jet could generate γ-rays by the Colgate mechanism as they propagate down the density gradient at the tip of the jet or there could be bulk acceleration of protons to above the pion production threshold. The protons could produce copious pions upon collision with the surrounding wind, thus triggering a cascade of high-energy γ-rays, pairs, and lower-energy γ-rays in an observable γ-ray burst. Yet another alternative is that the LAEW could eventually propagate into such a low-density environment that they directly induce pair cascade. The energy produced by the spin-down of the pulsar could emerge from the stellar surface along the axis of a low-density matter jet, or in an annulus surrounding a high-density jet. Either of these cases will give a Lorentz factor that depends strongly on the aspect angle of the observer.

CONCLUSIONS

Circumstantial evidence has accumulated that the γ-ray burst phenomenon is linked to Universal star formation and hence to massive stars. This alone does not say whether the product of the massive star is a black hole or a neutron star. Other evidence suggests that there may be a roughly canonical energy in γ-rays $\sim 10^{52}$ ergs that may appear as a larger "isotropic equivalent" energy in some cases because of collmation effects. If this remains a relevant number, then the possible association of some γ-ray bursts with neutron stars is still on the table. This energy is about what would expect in the rotation of a newly formed neutron star, and it could be delivered up in the form of Poynting flux in a few seconds if the neutron star is very highly magnetized; if it is a magnetar. Two key facts emerge that might support the connection of some supernovae that make neutron stars with some γ-ray bursts.
- Core collapse supernovae are strongly polarized.
- Magnetars exist!

This means that routine neutron star (not just black hole) formation requires the production of strong jets that may themselves explode the star. In addition, we must understand the birth event of strongly magnetized neutron stars.

Consideration of the systematics of the formation of magnetars suggests that the rotating collapse will first launch an MHD jet up the rotation axis. Later, after the neutron star cools and contracts, it will generate a Poynting flux that could be very intense for magnetar-type field strengths. That Poynting flux could emerge as a radiation-dominated jet following the path of the first matter-dominated jet. Because the radiation-dominated jet cannot form for several seconds as the neutron star contracts, spins up, and generates a large magnetic field, but then propagates faster than the matter- dominated jet, the matter jet could precede or follow the radiation dominated jet. In the former case an X-ray precursor could be generated. In the latter case the matter jet might not be conspicuous at all.

The process of neutron star spin-down has predictable properties if a simple dipole magnetic field is assumed. The bolometric luminosity declines like $L_{bol} \propto t^{-2}$ (Wheeler et al. 1999) and Blackman & Yi (1998) have estimated that for a synchrotron/Compton model the luminosity in the BATSE band might scale like $L_{BATSE} \propto t^{-1}$. The ratio of the energy emitted after several tens of seconds to the total energy is a few percent. These decline rates and the efficiency are very reminiscent of the "tails" of γ-ray bursts described at this meeting by Litvine, Connaughton, and Giblin. After the meeting, Giblin (private communication) reported that there is no sign of a periodic signal in the data from the especially bright tail source GRB 980923. It is not clear that the data were sampled in a way to reveal a rapid pulsar signal, so this issue might still be open.

The role of Poynting flux needs greater consideration, both in the context of the formation of magnetars and polarized supernovae and γ-ray bursts. For perspective, Usov (1999) has noted that a strong Poynting flux would not allow differential motion of particles and hence would tend to suppress internal shocks.

The interim conclusion is very clear. We need to explore the physics of rotating, magnetic core collapse in considerably more detail than has been done to date.

ACKNOWLEDGMENTS

This work is supported in part by NSF Grant 9818960, NASA grants LSTA-98-02 and HF-01085.01-96A and by a grant from the Texas Advanced Research Program.

REFERENCES

1. Blackman, E. G. & Yi, I., *ApJ* **498**, L31 (1998).
2. Colgate, S. A., *ApJ* **187**, 333 (1974).
3. Duncan, R. C. & Thompson, C., *ApJ* **392**, L9 (1992).
4. Müller, E. & Hillebrandt, W., *A&A* **80**, 147 (1979).
5. Höflich, P., Wheeler, J. C., & Wang, L., *ApJ* **521**, 179 (1999).
6. Höflich, P., et al., in preparation (2000).
7. Khokhlov, A.M., *J. Comput. Phys.* **143**, 519 (1998).
8. Khokhlov A.M., et al., *ApJ* **524**, L107 (1999).
9. Kouveliotou, C., et al., *ApJ* **510**, 115 (1998).
10. LeBlanc, J. M. & Wilson, J. R., *ApJ* **161**, 541 (1970).
11. Symbalisty, E. M. D., *ApJ* **285**, 729 (1984).
12. Usov, V. V., astro-ph/9909435 (1999).
13. Wang, L., Howell, D. A. Höflich, P., & Wheeler, J. C., *ApJ*, submitted (1999).
14. Wang, L., Wheeler, J. C., Li, Z. W., & Clocchiatti, A., *ApJ* **467**, 435 (1996).
15. Wheeler, J. C., Yi, I., Höflich, P., & Wang, L., *ApJ*, in press (1999).

Properties of Hypernovae: SNe 1997ef, 1998bw, and 1997cy

K. Nomoto[1,2], K. Maeda[1], T. Nakamura[1], K. Iwamoto[3],
T. Suzuki[1,2], P. A. Mazzali[2,4], M. Turatto[5],
I. J. Danziger[4], and F. Patat[6]

[1] *Department of Astronomy, University of Tokyo, Tokyo, Japan*
[2] *Research Center for the Early Universe, University of Tokyo, Tokyo, Japan*
[3] *Department of Physics, College of Science and Technology, Nihon University, Tokyo, Japan*
[4] *Osservatorio Astronomico di Trieste, via G. B. Tiepolo, Trieste, Italy*
[5] *Osservatorio Astronomico di Padova, vicolo dell'Osservatorio, Padova, Italy*
[6] *European Southern Observatory, Garching, Germany*

Abstract. We discuss the properties of the very energetic Type Ic supernovae (SNe Ic) 1998bw and 1997ef and Type IIn supernova (SN IIn) 1997cy. SNe Ic 1998bw and 1997ef are characterized by their large luminosity and the very broad spectral features. Their observed properties can be explained if they are very energetic SN explosions ($E_K \gtrsim 1 \times 10^{52}$ erg), originating probably from the core collapse of the bare C+O cores of massive stars ($\sim 30 - 40$ M$_\odot$). At late times, the light curve suggests that the explosion may have been asymmetric; this may help us understand the claimed connection with GRB's. SN IIn 1997cy is even more luminous than SN Ic 1998bw and the light curve declines more slowly than the ^{56}Co decay. We model such a light curve with circumstellar interaction, which requires an explosion energy of $\sim 5 \times 10^{52}$ erg. Because these kinetic energies of explosion are much larger than in normal core-collapse SNe, we call objects like these SNe "hypernovae".

INTRODUCTION

Recently, there have been an increasing number of candidates for the gamma-ray burst (GRB)/supernova (SN) connection, including GRB980425/SN Ic 1998bw (Galama et al. 1998; Iwamoto et al. 1998), GRB971115/SN Ic 1997ef (Wang & Wheeler 1998), GRB970514/SN IIn 1997cy (Germany et al. 1999; Turatto et al. 1999), GRB980910/SN IIn 1999E (Thorsett & Hogg 1999), and GRB991002/SN IIn 1999eb (Terlevich et al. 1999). Two other GRB's may also be associated with a SN: GRB980326 (Bloom et al. 1999) and GRB970228 (Reichart 1999; Galama et al. 1999). The optical transients of these GRBs showed that the decline of the light curve is slowed down at late phases, which can be fitted by the early power-law decay plus a later contribution of a SN.

Among the SNe with a possible GRB counterpart, we show here that SN Ic 1998bw (Iwamoto et al. 1998; Woosley et al. 1999), SN Ic 1997ef (Iwamoto et al. 2000), and SN IIn 1997cy (Turatto et al. 1999) are characterized by their extremely large kinetic explosion energy, $E_K \gtrsim 10^{52}$ erg. This is more than one order of magnitude larger than in typical SNe, so that these objects may be called "hypernovae". Also, SN IIn 1999E has a spectrum very similar to that of SN 1997cy (Cappellaro et al. 1999), and so it is probably a similar object.

SN 1997EF

SN 1997ef has been noticed for its unique light curve and spectra. At early times, the spectra were dominated by broad oxygen and iron absorption lines, but did not show any clear feature of hydrogen or helium (Garnavich et al. 1997; Filippenko et al. 1997), which led us to classify SN 1997ef as a Type Ic supernova (SN Ic). The most striking and peculiar feature of the spectra of SN 1997ef is the broadness of the line features, which suggests that SN 1997ef may have a large explosion energy.

In order to clarify whether SN 1997ef is indeed a hypernova, we constructed hydrodynamical models for an ordinary SN Ic and for a hypernova (Iwamoto et al. 2000). For the ordinary SN Ic (model CO60), a C+O star with a mass $M_{CO} = 6.0 M_\odot$ (the core of a 25 $M_\odot$ star) is exploded with a kinetic energy $E_K = 1.0 \times 10^{51}$ ergs and an ejecta mass $M_{ej} = 4.6 M_\odot$. For the hypernova model (CO100), a C+O star of $M_{CO} = 10.0 M_\odot$ (the core of a 30 – 35 $M_\odot$ star) is exploded with $E_K = 8.0 \times 10^{51}$ ergs and $M_{ej} = 8.0 M_\odot$. The mass of ^{56}Ni is set to be 0.15 $M_\odot$ for both models to explain the observed peak brightness.

The synthetic V light curves for models CO60 and CO100 show quite broad peaks and relatively slow tails, which reproduce the light curve of SN 1997ef reasonably well. Light curve modeling provides important constraints on M_{CO} and E_K. However, it is difficult to distinguish clearly between the ordinary SN Ic and the hypernova model from the light curve alone, since models with different values of M_{ej} and E_K can reproduce the observed light curve equally well if $M_{ej}^{3/4} E_K^{-1/4} = $ const. On the other hand, these models can be expected to produce different spectra, because of the different E_K. Therefore, spectrum synthesis can tell these models apart, taking advantage of their differences.

Spectra computed with model CO60 for the early epochs show narrow lines, much narrower than the observations. This clearly indicates a lack of material at high velocity in model CO60, and suggests that the kinetic energy of this model is too small. Synthetic spectra computed with model CO100 for the first three epochs are shown in Figure 1 (left). The spectra have much broader lines and are in good agreement with the observations. We thus conclude that SN 1997ef is a hypernova.

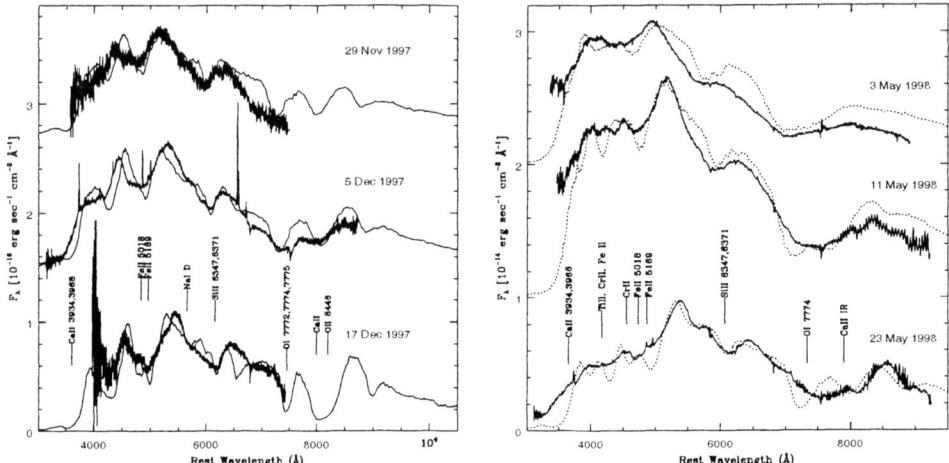

FIGURE 1. Left: Observed spectra of SN 1997ef (bold lines) and synthetic spectra computed using model CO100 (fully drawn lines). Right: Observed spectra of SN1998bw (full lines) and synthetic spectra computed using model CO138 (dashed lines).

SN 1998BW

The absorption lines are so broad in SN 1998bw that they blend together, giving rise to what could even be confused with an emission spectrum. Velocities in the Si II 6355Å line are as high as 30,000 km s^{-1}. Also, the SN was very bright for a SN Ic: the observed peak luminosity, $L \sim 1.4 \times 10^{43}$ erg s^{-1}, is almost ten times higher than that of previously known SNe Ib/Ic (Iwamoto et al. 1998; Woosley et al. 1999).

We calculate the light curves and spectra for various C+O star models with different values of E_K and M_{ej}. These parameters can be constrained by comparing the calculated light curves, the synthetic spectra, and the photospheric velocities with the data of SN 1998bw. Here we describe the best model, which is the explosion of a 13.8 $M_\odot$ C+O star with a large kinetic energy ($E_K = 6 \times 10^{52}$ erg). Such a massive C+O star progenitor is the product of a main-sequence star of $\sim 40\ M_\odot$ which lost its H/He envelope via a stellar wind or binary interaction. A ^{56}Ni mass of $\sim 0.63 M_\odot$ is necessary to reproduce the light curve maximum. Figure 1 (right) shows the observed spectra of SN1998bw (solid lines) and synthetic spectra computed using model CO138 (dashed lines). We can see that the broad lines are well reproduced because we have enough material at high velocities in this model.

Despite the success of our 'hypernova' model in reproducing the early light curve ($t \lesssim 60$ days), the model light curve tail declines more rapidly than the observations (Figure 2, left). After day ~ 200 the decline of the model light curve becomes slower, and it approaches the half-life of ^{56}Co decay around day 400. At $t \gtrsim 400$ days most γ rays escape from the ejecta, while positrons emitted from the ^{56}Co decay are mostly trapped and their energies are thermalized. Therefore, positron deposition determines the light curve at $t \gtrsim 400$ days. If the observed tail should follow the positron-powered light curve, the ^{56}Co mass could be determined directly.

The deviation of the model light curve from the observation at $t \gtrsim 60$ days suggests that there might be a high-density region in the ejecta of SN1998bw, where the γ rays are efficiently trapped. Another peculiarity appeared in the late phase spectra. Measuring the velocity of the various elements from the width of their emission lines (Patat et al. 1999), we note that iron expands faster than oxygen, which is contrary to expectations.

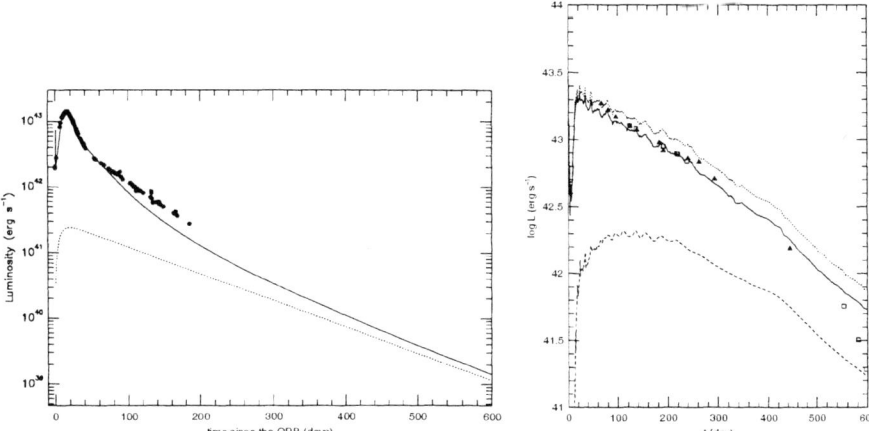

FIGURE 2. Left: The bolometric light curve of model CO138 compared with the observations of SN1998bw (a distance modulus is $\mu = 32.89$ mag). The dotted line indicates the energy deposited by the positrons. Right: The $uvoir$ bolometric light curve of SN 1997cy compared with the synthetic curve obtained with the CSM interaction model ($\mu = 37.40$ mag).

Both the need for a high-density region and the velocity inversion might indicate that the explosion is aspherical (Höflich et al. 1999; MacFadyen & Woosley 1999) as also indicated by polarization measurements (Patat et al. 2000). If the outburst in SN 1998bw took the form of a prolate spheroid, the explosive shock along the major axis was probably strong, ejecting material with large velocities and producing abundant ^{56}Ni. In directions away from the major axis, on the other hand, oxygen is not consumed and the density is high enough for γ rays to be trapped even at advanced phases, thus giving rise to the slowly declining tail. Our spherical models may overestimate E_K by a factor of a few, but still E_Ks are unusually large.

SN1997CY

SN 1997cy displayed narrow Hα emission, which lead to the classification of SN 1997cy as a Type IIn (Germany et al. 1999; Turatto et al. 1999). It is the brightest

SN II discovered so far. After about day 120 the SN light decline becomes slower than the lifetime of ^{56}Co, suggesting circumstellar interaction for the energy source.

In the interaction model, collision of the SN ejecta with the slowly moving circumstellar matter (CSM) converts the kinetic energy of the ejecta into light, thus producing the observed intense light display of the SN. Our exploratory model considers the explosion of a massive star of $M = 25 M_\odot$ with a parameterized kinetic energy E_K. The regions excited by the forward and reverse shocks emit mostly X rays. The density in the shocked ejecta is so high that the reverse shock is radiative and a dense cooling shell is formed (e.g., Suzuki & Nomoto 1995; Terlevich et al. 1992). The X rays are absorbed in the outer layers and in the core of the ejecta, and are re-emitted as UV-optical photons.

Figure 2 (right) shows the model light curve which best fits the observations. The model parameters are: $E_K = 5 \times 10^{52}$ erg and a mass-loss rate of $\dot{M} \sim 4 \times 10^{-4} M_\odot$ yr^{-1} for a wind velocity of 10 km s^{-1}). The large CSM mass and density are necessary to have large shocked masses and thus to reproduce the observed high luminosity, and so is the very large explosion energy. For models with low E_K and high CSM density, the reverse shock speed is too low to produce a sufficiently high luminosity. For high E_K or low CSM density, the expansion of the SN ejecta is too fast for the cooling shell to absorb enough X rays to sustain the luminosity. Thus, in this model E_K and $\dot{M}$ are constrained within a factor of ~ 3 of the reported values. Its large kinetic energy gains SN 1997cy enrollment in the family of 'hypernovae'.

NUCLEOSYNTHESIS IN ASYMMETRIC EXPLOSIONS

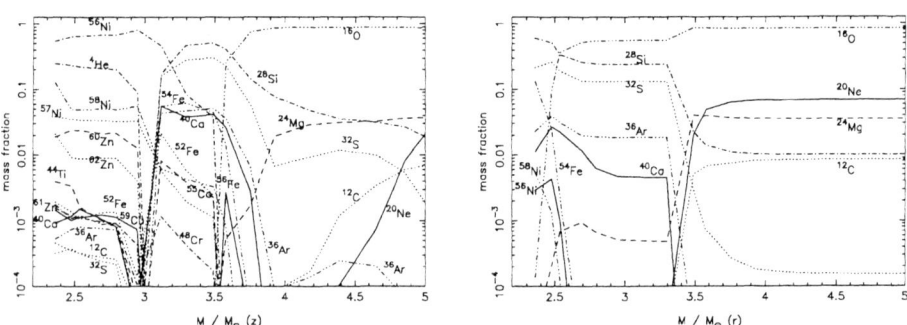

FIGURE 3. The isotopic compositions of the ejecta of the axisymmetric explosion in the direction of the jet (left) and the perpendicular to the jet (right).

We speculate above that the expansion velocity of Fe and O in SN 1998bw betrays the effect of the asymmetry in the explosion. To confirm this we calculate the explosive nucleosynthesis in an axisymmetric explosion. Figure 3 shows the isotopic compositions of the ejecta of the axisymmetric explosion along the jet direction (left) and perpendicular to it (right). The progenitor model is CO138. The explosion energy is $E_K = 1 \times 10^{51}$ erg. Starting the hydrodynamical simulation, we deposit the energy as 50% thermal energy and 50% kinetic energy in the direction of the jet (z) below the mass cut which divides the ejecta from the collapsing core.

The shock is stronger and the post-shock temperatures are higher along the jet direction (z), so that explosive nucleosynthesis takes place in a more extended, lower density region compared with the perpendicular direction (r). A larger amount of ^{56}Ni is produced in the jet direction. In addition, elements produced by α-rich freezeout are enhanced because nucleosynthesis proceeds at higher entropies than in the region away from the jet. Figure 3 clearly shows that in the jet direction a larger amount of ^{4}He is left after the shock decomposition. Hence, elements synthesized through α-particle capture, such as ^{44}Ti and ^{48}Cr (which decay into ^{44}Ca and ^{48}Ti, respectively) are more abundant (see also Nagataki et al. 1997). In contrast, little ^{56}Ni is produced in the r-direction. Also, the expansion velocities are lower than those in the z-direction. Therefore, the Fe velocities (mostly in the z-direction) can exceed the O velocities (in the r-direction), as observed in SN 1998bw. Oxygen in the z-direction has the highest velocities but the densities may become too low to be excited by γ rays.

REFERENCES

1. Bloom, J.S., et al., *Nature* **401**, 453 (1999).
2. Cappellaro, E., et al., *IAU Circ.* 7091 (1999).
3. Filippenko, A.V., et al., *IAU Circ.* 6783, 6809 (1997).
4. Galama, T.J., et al., *Nature* **395**, 670 (1998).
5. Galama, T.J., et al., *ApJ*, submitted, astroph/9907264 (1999).
6. Garnavich, P., et al., *IAU Circ.* 6778, 6786, 6798 (1997).
7. Germany, L., et al., *ApJ*, submitted, astroph/9906096 (1999).
8. Höflich, P., Wheeler, J.C., & Wang, L., *ApJ* **521**, 179 (1999).
9. Iwamoto, K., et al., *Nature* **395**, 672 (1998).
10. Iwamoto, K., et al., *ApJ* **534**, in press, astroph/9807060 (2000).
11. MacFadyen, A.I. & Woosley, S.E., *ApJ* **524**, 262 (1999).
12. Nagataki, et al., *ApJ* **486**, 1026 (1997).
13. Patat, F., et al., *Mem Sait*, in press (1999).
14. Patat, F., et al., *ApJ*, submitted (2000).
15. Reichart, D.E., *ApJ* **521**, L111 (1999).
16. Suzuki, T. & Nomoto, K., *ApJ* **455**, 658 (1995).
17. Terlevich, R., Tenorio-Tagle,G., Franco,J., & Melnick,J., *MNRAS* **255**, 713 (1992).
18. Terlevich, R., Fabian, A., & Turatto, M., *IAU Circ.* 7269 (1999).
19. Thorsett, S.E. & Hogg, D.W., *GCN Circ.* 197 (1999).
20. Turatto, M., et al., *ApJ*, submitted, astroph/9910324 (1999).
21. Wang, L.-F. & Wheeler, J.C., *ApJ* **504**, L87 (1998).
22. Woosley, S.E., Eastman, R.G., & Schmidt, B.P., *ApJ* **516**, 788 (1999).

Is Nova Sco 1994 (GRO 1655−40) a Relic of a GRB ?

G. E. Brown*, C.-H. Lee*, H. K. Lee[†], and H. A. Bethe[††]

*Department of Physics & Astronomy, SUNY, Stony Brook, NY 11794, USA
[†]Department of Physics, Hanyang University, Seoul 133-791, Korea
[††]Floyd R. Newman Laboratory of Nuclear Studies, Cornell University, Ithaca, NY 14853, USA

Abstract. We suggest Nova Sco 1994 (GRO 1655-40) as a possible relic of a Gamma Ray Burster (GRB) and Type Ib supernova (SN) explosion, showing that there is evidence both that the black hole was spun up by accretion and that there was a supernova explosion. We use the disc energy delivered from the rotational energy of the black hole to power the SN, and give arguments that roughly equal energy goes into the GRB and into the accretion disc to power the supernova.

INTRODUCTION

In this note we consider only the long term GRBs, of duration from several seconds up to several minutes. This is just the dynamical time of a He star, which we consider as progenitor.

The formation of black holes in single stars of Zero-Age Main Sequence (ZAMS) masses $20 - 35 M_\odot$ was proposed by Brown, Lee, & Bethe (1999). However, these do not lose their envelopes except in binaries. This latter case has been studied by Brown et al. (1999) who evolve the transient sources in this way. These have mostly low mass main sequence companions, although in two cases the companions are subgiants. In many more cases, binaries of $\sim 7 M_\odot$ black holes with companions up to nearly equal mass of the ZAMS mass of the black hole progenitors are predicted. These companions do not, however, fill their Roche Lobes, and consequently are not observed. None the less, the Wolf-Rayet progenitors of the $\sim 7 M_\odot$ black holes in these binaries offer a set of progenitors for GRB's. They are already somewhat in rotation because of the companion star. Note that in the Brown, Lee, & Bethe (1999) scenario, their envelopes are removed only following He core burning, in the supergiant stage, so there is only a short time left in their evolution for them to lose He by wind. Thus, a substantial amount of He should be left in the W.-R., although most of it will have been burned to carbon and oxygen (Woosley, Langer, & Weaver 1993), and the explosion we describe below would be Type Ib.

The conclusion of the Bethe & Brown (1999) paper was that in order for Wolf-Rayets followed by high-mass black holes like that in Cyg X-1 to be formed in single stars, a ZAMS mass of $\gtrsim 80 M_\odot$ was necessary. This was based on the calculation of Woosley, Langer, & Weaver (1993) who used a too large mass loss rate for the He winds. These stars have been reevolved by Wellstein & Langer (1999) with lower mass loss rates, but the evolution has not been carried beyond the carbon-oxygen core stage, so we do not yet know how much the lower winds will decrease the mass limit for evolving into a high-mass black hole. The carbon oxygen cores still have about 33% central carbon abundances, so they clearly will not skip the convective carbon burning stage and therefore may well end up as low mass compact objects. These high mass Wolf-Rayet stars are the progenitors of GRBs in Woosley's Collapsar model (MacFadyen & Woosley 1999).

In addition to these and the high-mass black holes in the transient sources, the coalescence of the low-mass black holes in the Bethe & Brown (1998) scenario of compact binary evolution with the companion He star (Fryer & Woosley 1996) offers another type of generator for the long term GRBs.

All three of the above possibilities involve a He star being accreted by a black hole. In this process the black hole will be spun up. The energy can be extracted by the Blandford-Znajek (BZ) 1977 process, as in Lee, Wijers, & Brown (1999). The BZ power supplied into the disc will halt the inflow, and later propel the matter outwards in a Type Ib supernova explosion (Brown, Lee, Lee, & Bethe 2000).

ENERGETICS OF GRBS

The maximum energy that can be extracted from the BZ mechanism (Lee, Wijers, & Brown 1999) is $E_{max} = 0.09 M_{BH} c^2$. For a $7 M_\odot$ black hole, such as is found in Nova Sco 1994, $E_{max} \simeq 1.1 \times 10^{54}$ ergs. The black hole is first formed with a mass of at least $1.5 M_\odot$ (Brown & Bethe 1994). The maximum energy, after these corrections, is still an order of magnitude greater than the 3×10^{52} ergs used by Iwamoto et al. in the supernova explosion. Presumably, the explosion will take place before the BZ mechanism can deliver full energy, leaving the black hole with substantial spin energy.

Without beaming, the estimate of the energy in the jet of the GRB (Anderson et al. 1999) is $E_{990123} = 4.5 \times 10^{54}$ ergs. The BZ scenario entails substantial beaming, so this energy should be multiplied by $d\Omega/4\pi$, which may be a small factor ~ 0.01. The BZ power can be delivered at a maximum rate of

$$P_{BZ} = 6.7 \times 10^{50} \left(\frac{B}{10^{15} G}\right)^2 \left(\frac{M_{BH}}{M_\odot}\right)^2 \text{ erg s}^{-1}. \tag{1}$$

IS NOVA SCO 1994 A RELIC OF A GRB ?

Several characteristics of Nova Sco 1994 (GRO 1655-40) can be understood if it is a relic of a GRB. First of all the high space velocity -150 ± 19 km/s can be understood if a supernova explosion is associated with black hole formation (Brandt et al. 1995, Nelemans et al. 1999).

In our scenario, first a GRB is initiated by the BZ-mechanism, following which a Type Ib supernova explosion is begun by the energy deposited in the thick accretion disc. Following Brandt et al. (1995) we note that a binary symmetric in the frame of the exploding star will be asymmetric in the center of mass of the binary. The amount of mass that can be ejected is constrained by the fact that if more than half of the total initial mass is ejected, the system will become unbound. Brandt et al. consider collapse to a $4M_\odot$ black hole; then at the time of collapse the collapsing He star must have a mass of $\sim 9M_\odot$ or greater.

In fact, the initial black hole needs to be no more than $\sim 1.5M_\odot$ according to Brown & Bethe (1994), but one would expect a substantial amount of the carbon-oxygen core to collapse with the Fe, and also substantial fallback of the original carbon-oxygen core, probably burned to Fe in the explosion. In order to obtain an $\sim 7M_\odot$ black hole, the initial He envelope would have to be $\sim 15M_\odot$ corresponding to a ZAMS mass of $35 - 40M_\odot$.

Israelian et al. (1999) find a large overabundance of oxygen, magnesium, silicon and sulphur in the F3-F8 IV/III companion star of $1.6 - 3.1M_\odot$ orbiting the companion. These are just the elements copiously produced in a Type Ib supernova explosion. Contrary to Iwamoto et al. (1998), who need $0.7M_\odot$ of ^{56}Fe to reproduce the brightness of SN 1998bw, Israelian et al. find no enhancement in the Fe. In our scenario we expect the jet of the GRB preceding the supernova explosion to go along the rotation axis of the black hole, and the supernova explosion to be initiated perpendicular to this in the accretion disc. The highly non-equilibrium processes in the jet would not initially affect the supernova, but might be expected to excite the He lines in later stages where the expanding supernova interacts with the jet.

There are indications that the black hole in Nova Sco 1994 is spinning rapidly (Sobczak et al. 1999, Gruzinov 1999). We would expect the BZ central engine to stop delivering energy following the supernova explosion which disrupts the magnetic fields.

In Appendix C of Brown, Lee, Wijers, & Bethe (1999) the observed GRB rate at the present time is estimated to be ~ 0.1 GEM (Galactic Event per Megayear). With a factor of 100 for beaming, this would require 10 GEM. Brown, Lee, & Bethe (1999) estimate the birth rate for visible transient sources in the Galaxy to be 8.8 GEM. However, including the high-mass black-hole binaries with companions which have not evolved to their Roche Lobes, and therefore would not be visible, they arrive at a number 25 times higher; namely 220 GEM. (Inclusion of the "silent" binaries effectively removes the q, the ratio of masses, in the calculation. The more massive companions are all inside of their Roche Lobes, except for the two subgiants in the systems V404 Cyg and XN Sco94.) Also, there are the other possible GRB progenitors, Collapsars and coalescing black hole and He star, mentioned above. Thus, there must be other severe criteria for GRBs; e.g., high magnetic fields in the rotating He envelopes, etc.

In the evolution of the transient sources, the hydrogen envelope of the massive star must be taken off only following He core burning, if collapse into a high-mass black hole is to be obtained (Brown, Lee, & Bethe 1999). In this last stage the companion star would try to spin the hydrogen envelope up towards co-rotation in the binary before expelling it. With the large viscosity from magnetic turbulence assumed in the Spruit & Phinney (1998) argument, the He core would be carried along, but probably in differential rotation because the common envelope time is very short, ~ 1 year.

Similar considerations follow for the Fryer & Woosley (1998) model of coalescence of black hole with companion He star. The common envelope evolution here need not happen as late as in the transient source scenario, but on the other hand the predicted merger rate is high, 380 GEM (Brown, Lee, Wijers, & Bethe 1999).

DISCUSSION

We suggested Nova Sco 1994 (GRO 1655-40) as a relic of a GRB and Type Ib SN explosion.

Our model begins from a black hole in a He star. It is assumed that in the hypercritical accretion, the disc has a high magnetic field. The black hole is spun up and the GRB, powered by the Blandford-Znajek mechanism, is driven along the nearly matter-free axis of rotation of the black hole. With high viscosity such as follows from magnetic turbulence, the swallowing of the He star by the black hole takes a dynamical time of the star. The BZ mechanism stops once the accretion disc disappears and the magnetic field disperses, leaving the black hole with some spin energy.

We point out that power roughly equal to that driving the GRB will be delivered into the disc made up out of hyperaccreting helium. The delivered energy first brings the accreting matter to rest, then drives it backwards through the accretion disc (and to the sides of it) in a Type Ib supernova explosion. From consideration of energetics, both the GRB and SN explosion can be powered by several times 10^{53} ergs, if we take the black hole to have mass $\sim 7 M_\odot$ typical of those in the transient sources, but the GRB may well take off and the explosion begin before maximum energy is delivered.

In a more detailed paper (Brown, Lee, Lee, & Bethe, 2000), we show that hypercritical accretion onto a black hole in the middle of a self-similar accretion disc of Narayan & Yi (1994) type will spin the black hole up to $\gtrsim 90\%$ of maximum, so that $\sim 10^{54}$ ergs is available. Running the plasma current through a current (Thorne 1986) around the black hole and through the accretion disc, we argue that roughly equal fraction of this energy are available to power the GRB and the Type

Ib supernova, the former by energy delivered in the loading area up the rotation axis of of the black hole and the latter in the accretion disc. In a more detailed paper (Lee, Brown, & Wijers 1999) we show how the plasma can pass through magnetosonic points, etc.

ACKNOWLEDGMENTS

We would like to thank Ralph Wijers and Stan Woosley for useful discussions. This work is supported partially by the U.S. Department of Energy Grant No. DE-FG02-88ER40388. HKL is supported also in part by KOSEF Grant No. 985-0200-001-2 and BSRI 98-2441.

REFERENCES

1. Andersen, M.I. et al. 1999, *Science* **283**, 2075
2. Baron, E. 1999, private communication
3. Bethe, H.A., and Brown, G.E. 1998, *ApJ* **506**, 780
4. Bethe, H.A., and Brown, G.E. 1999, *ApJ* **517**, 318
5. Bethe, H.A., Brown, G.E., Lee, C.-H. 1999, *ApJ*, submitted, astro-ph/9909132
6. Blandford, R.D. and Znajek, R.L. 1977, *MNRAS* **179**, 433
7. Brandt, W.N., Podsiadlowski, Ph., and Sigurdssen, S. 1995, *MNRAS* **277**, L35
8. Brown, G. E., & Bethe, H.A. 1994, *ApJ* **423**, 659
9. Brown, G. E., Lee, C.-H., and Bethe, H.A. 1999, *New Astronomy* **4**, 313
10. Brown, G. E., Lee, C.-H., Lee, H.K., and Bethe, H.A. 2000, in preparation
11. Brown, G. E., Lee, C.-H., Wijers, R.A.M.J., and Bethe, H.A. 1999, *Physics Reports*, to be published, astro-ph/9910088
12. Fryer, C.L., Woosley, S.E. 1998, *ApJL* **502**, 9
13. Gruzinov, A. 1999, *ApJ*, submitted, astro-ph/9910335
14. Israelian, G. et al., 1999, *Nature* **401**, 142
15. Iwamoto, K., et al. 1998, *Nature* **395**, 672
16. Lee, H.K., Wijers, R.A.M.J., and Brown, G.E. 1999, *Physics Reports*, to be published, astro-ph/9906213
17. Lee, H.K., Brown, G.E., and Wijers, R.A.M.J. 1999, *ApJ*, submitted
18. MacFadyen, A.I., and Woosley, S.E. 1999, *ApJ* **524**, 262
19. Mirabel, I.F., and Rodriquez, L.F. 1998, *Nature* **392**, 673
20. Nelemans, G., Tauris, T.M., and Van den Heuvel, E.P.J. 1999, *A&A*, accepted, astro-ph/9911054
21. Spruit, H., Phinney, E.S. 1998, *Nature* **393**, 139.
22. Sobczak, G.J., McClintock, J.E., Remillard, R.A., Bailyn, C.D., and Orosz, C.D. 1999, *ApJ* **520**, 776
23. Thorne, K.S., Price, R.H., and MacDonald, D.A. 1986, *Black Holes: The Membrane Paradigm* (Yale University Press, New Haven and London)
24. Woosley, S. E., Langer, N., & Weaver, T. A. 1993, *ApJ* **411**, 823
25. Wellstein, S., & Langer, N. 1999, *A&A* **350**, 148

Collapsars

Andrew MacFadyen

Astronomy Department, University of California, Santa Cruz, CA 95064

Abstract. A variety of stellar explosions powered by black hole accretion are discussed. All involve the failure of neutrino energy deposition to launch a strong supernova explosion. A key quantity which determines the type of high energy transient produced is the ratio of the engine operation time, t_{engine}, to the time for the explosion to break out of the stellar surface, t_{bo}. Stars with sufficient angular momentum produce collapsars – black holes accreting rapidly through a disk – in their centers. Collapsars can occur in stars with a wide range of radii depending on the amount of pre-collapse mass loss. The stellar radius and jet properties determine the degree of beaming of the explosion. In some cases the stellar envelope serves to focus the explosion to narrow beaming angles. The baryon loading of various models for classical GRBs formed in massive stars is examined and the consequences are explored. For $t_{engine} > t_{bo}$, highly relativistic outflow is possible and classical GRBs accompanied by supernovae can be produced. In other cases hyper-energetic, asymmetric supernovae are produced. Longer GRBs ($t \gtrsim 100$ s) can be produced by fallback following a weak neutrino-driven supernova explosion.

INTRODUCTION

Massive stars ($M_{ms} \gtrsim 25$ M$_\odot$) may not always launch successful neutrino-driven explosions [1]. We describe here the continued evolution of such stars after their cores collapse to black holes and accrete the surrounding stellar mantle. For the most massive stars (M$_{ms} \gtrsim 35$ M$_\odot$) with sufficient angular momentum, a collapsar – a rapidly accreting ($\dot{M} \approx 0.1$ M$_\odot$ s^{-1}) stellar mass black hole – forms promptly at the center of a collapsing star [2]. We refer to these as Type-I collapsars. Less rapidly accreting black holes can also form over longer time periods due to the fallback of stellar material which failed to escape during the initial supernova explosions (Type II collapsars). This probably happens for main sequence masses, $M_{ms} \gtrsim 20$ M$_\odot$. Stars with masses below this explode as normal supernovae and leave behind neutron star remnants. Collapsars power jetted explosions by tapping a fraction of the binding energy released by the accreting star through magnetohydrodynamical processes or neutrino annihilation, or possibly by extracting some of the black hole spin energy. The vast majority of stellar explosions do not make collapsars, only those which make black holes and have sufficient angular momentum.

TABLE 1. Observable phenomena resulting from collapsars.

	Prompt BH form. Type I Collapsar	Delayed BH form. Type II Collapsar
No H env. Wolf-Rayet *	$t_{engine} > t_{bo}$ GRB + TypeIb/c SN	$t_{engine} > t_{bo}$ long GRB + TypeIb/c SN
Small H env. Blue supergiant	$t_{engine} < t_{bo}$ asymmetric SN hard XRT	$t_{engine} \gtrsim t_{bo}$ long GRB? + Type II SN
Large H env. Red supergiant	$t_{engine} \ll t_{bo}$ Asym. hyper-energ. SNII Soft Xray transient	$t_{engine} \lesssim t_{bo}$ Asym. hyper-energ. SNII Soft Xray transient + tail

Further, not all collapsars make GRBs. Only those which happen in sufficiently small (in radius) stars and manage to accelerate a fraction of the explosion energy to sufficiently high Lorentz factor. Other collapsar explosions may be responsible for hyper-energetic and asymmetric supernovae like SN1998bw.

Table 1 shows a range of observable phenomena possible from collapsars in various kinds of stars. Prompt and delayed black hole formation can occur in massive stars with a range of radii depending on the evolutionary state of the star, its metallicity and its membership in a binary system. The key difference between the scenarios is the ratio of the time the engine operates, t_{engine}, to the time the explosion takes to break out of the surface of the star, t_{bo}. The breakout time is $t_{bo} = R_{star}/v_{jet}$ where v_{jet} is the propagation velocity through the star of the explosion shock or jet head. Typical velocities are 50,000 km s^{-1} [3]. If the engine operates for a sufficiently long time to continuously power the jet at its base after the explosion shock (the jet head) breaks out of the surface of the star, then highly relativistic outflow can be achieved for a fraction of the ejecta and a classical GRB can result. The column of stellar material pushed ahead of the jet, perhaps a few 0.01 M$_\odot$, escapes the star and expands sideways leaving a decreasing amount of material ahead of the jet.

The engine time, t_{engine}, is the time the star is able to feed the black hole at a sufficient rate and depends on the stellar mass and angular momentum distribution at collapse. Viscous entropy generation in the accretion disk and centrifugal bounce can eject significant fractions of the accreting mass flux in a wind [2]. This can effect the accretion rate onto the black hole and can shorten the accretion time by expelling the outer layers of the star and choking accretion onto the disk. The wind can also be important for ejecting radioactive nickel into the explosion. Until the star starts exploding, the engine time is given by the collapse time of the star onto the disk. It is not given by the initial disk mass divided by the accretion rate since the disk is simply an intermediate repository of mass that is coming from the collapsing stellar envelope. Helium cores can accrete for tens of seconds to minutes while the accretion time is longer for any size star if the star initially explodes then partially reimplodes (Type II collapsar) [4]. Not all of this time is available for producing an explosion since there is an initial collapse phase lasting several seconds when the disk forms, the deposited energy can be advected into the hole for several seconds until the density is sufficiently low for energy input to reverse the infall and drive an explosion, and the explosion takes several seconds to break out of the stellar surface. The star can also explode after a sound crossing time due to the lateral expansion of the jet shock especially for "hot" jets or due to the explosion of the star due to a disk wind.

A key characteristic of the model is the degree of spreading of the jet as it passes through the stellar envelope. GRBs may have an approximately common total energy, $E_\odot \approx 10^{52}$ erg, yet produce a variety of stellar explosions depending on the degree to which the explosion is focussed into a jet. Two characteristics of the explosion can determine the beaming of the jet: their entropy and their duration. In particular, explosions like SN1998bw can be explained by the jet being "hot" or "brief".

"HOT" JETS

"Hot" jets have large internal pressure compared to their ram pressure and the ambient stellar pressure. They expand laterally as they push through the star and share more of their energy with the stellar envelope. While it may be possible for a hot jet to escape the star and make a GRB, more of the star will participate in the explosion and the jet will take longer to penetrate the star. "Cold" jets on the other hand are capable of penetrating the star with relatively little sharing of energy with the stellar envelope. In some cases the star actually compresses the jet and helps to focus it [4,3]. In this case the supernova explosion is relatively weak and a large fraction of the energy goes into the narrow jet beam. These types of explosions can have long accretion episodes and leave large black hole remnants ($M_{bh} > 5 M_\odot$) since the lateral expansion of the jet is inefficient at exploding the star and choking the accretion feeding the hole.

"BRIEF" JETS

If the engine is only on for a short time ($t_{engine} < t_{bo}$), the power is lost before the jet head reaches the surface of the star even for a small star like the He and C/O cores thought to be responsible for SN1998bw [5,7] (Figure 1, right panel). In this case the jet expands quasi-isotropically into the star after its ceases to be energized at it's base (Figure 1, right panel). The resulting explosion is asymmetric since the explosion is initiated in the polar region. It could be distinguished from a conventional Type Ib/c supernova by high expansion velocities and large energy. SN1998bw may be an example of this with the weak GRB coming from a small amount of moderately relativistic material [5,7] interacting with the CSM.

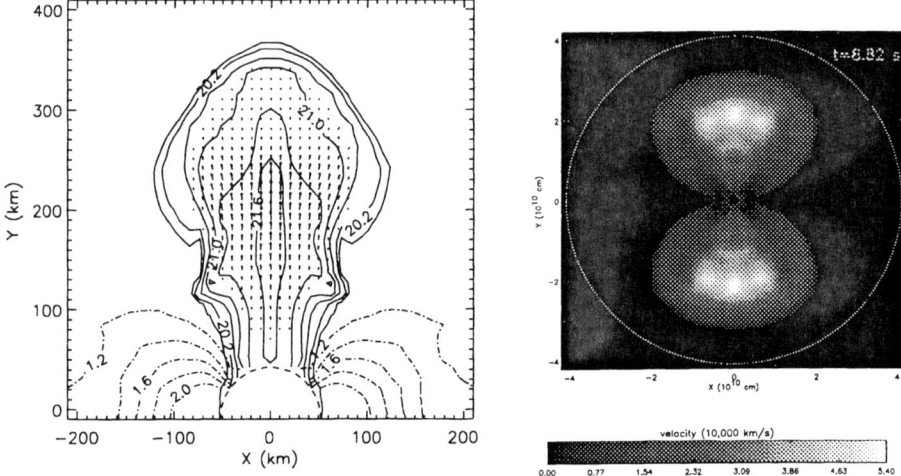

FIGURE 1. Left: The solid contour lines show the logarithm of total (kinetic plus internal) energy density (erg g^{-1}) 5.7 milliseconds after initiating energy input at 5×10^{50} erg s^{-1}. After 5.7 ms, 2.8×10^{48} have been deposited and a jet is starting to form, rising upward along the pole (steepest density gradient) as shown by the arrows representing the velocity field. The magnitude of the largest velocity arrows is $\sim c$ showing that relativistic expansion has begun. The dash-dot contours indicate the temperature in the accretion torus in MeV. Higher temperature exist interior to the inner boundary shown here as the dashed semi-circle with radius 50 km. Right: Asymmetric explosion for $t_{engine} = 5\text{ s} < t_{bo}$ in a Helium star. 10^{51} erg s^{-1} was deposited for 5 s as thermal energy near the inner boundary in the polar region. The explosion is shown 3.82 s later..

RELATIVISTIC OUTFLOW FROM MASSIVE STARS

Relativistic outflow can be achieved from an accreting black hole surrounded by a collapsing stellar mantle. The key is the prolonged deposition of energy into regions of the star which can expand due to their overpressure. A single impulsive release of energy $E_{dep}(t) = E_\circ \delta(t-t_\circ)$ will explode a star if the energy input exceeds the binding energy. But the explosion will be "baryon loaded", a "dirty fireball", i.e., a supernova. The maximum Lorentz factor will be $\Gamma \lesssim \frac{E_{dep}}{m_b c^2}$, where m_b is the mass of baryons in the region where the energy is deposited. The Lorentz factor will be less than the asymptotic limit because the expanding fireball will do work in accelerating the surrounding material. *For an impulsive energy deposition*, a clean environment with $m_b < \frac{E_{dep}}{\Gamma c^2}$ and $\Gamma \gtrsim 100$ is necessary to make a classical GRB or else the deposited energy will be shared with too many baryons (the "baryon pollution" problem).

The situation is different if the energy is deposited over a period long compared to the expansion time of the energy loading region. In this case, energy is injected into a region already expanding due to the previously deposited energy and the corresponding overpressure. Initially, $m_b \gg \frac{E}{\Gamma c^2}$ and the expansion is sub-relativistic, but as the gas expands the baryon density in the deposition region decreases and the energy per baryon increases (assuming constant energy deposition rate per unit volume). The expanding gas must do work in accelerating its surroundings so the deposited energy is shared with many baryons and extremely relativistic motion is initially impossible. Baryons can be mixed into the deposition region but centrifugal force and pressure gradients directed away from the pole can inhibit this (Figure 1, left panel). The amount of baryons which mix into the deposition region is important for determining the ultimate Lorentz factor that can be achieved. Current two-dimensional relativistic calculations for one particular model achieve $\Gamma \approx 40$ near the deposition region [3] and it is expected that higher Γ will result when the calculation is run longer or with greater and/or variable energy deposition. Detailed calculations are possible and will be performed soon. At late times after the baryons initially present in the deposition region have expanded away, the Lorentz factor depends on the energy flux and mass flux into the deposition region as $\Gamma \approx \frac{\dot{E}}{\dot{m}c^2}$. The mass flux depends on the rate that hydrodynamical instabilities mix baryons into the deposition region and the rate at which the engine injects baryons.

In the collapsar model for GRBs, or any other similar model involving a massive star, the key to obtaining relativistic motion is the escape of an energy loaded bubble from its surroundings (the stellar mantle). In the case of the toroidal density distribution as in the collapsar [2], a low density channel is left behind by regions of the star along the rotational axis which lacked centrifugal support and fell into the black hole. Recent simulations [2,3,6] have shown that energy deposition leads to expansion of gas along the pole, a jet. The key to achieving high Γ is for energy deposition to continue at the base of the jet even after the jet head has broken out of the surface of the star and begun free expansion into the low density circumstellar environment. Subsequent energy deposition at the base of the jet continues to load energy into an increasingly baryon-free region with the expanding gas continuously channelled along the rotation axis of the star.

REFERENCES

1. Fryer, C. L., *ApJ* (in press, and astro-ph/9902315).
2. MacFadyen, A. I., & Woosley, S. E., *ApJ* **524**, 262 (1999).
3. Aloy, M.A., Müller, E., Ibanez, J.M., Marti, J.M., MacFadyen, A.I., astro-ph/9911098 (1999).
4. MacFadyen, A.I., Woosley, S.E., & Heger, A., astro-ph/9910034 (1999).
5. Iwamoto, K. et al., *Nature* **395**, 672 (1998).
6. E. Müller, these proceedings.
7. Woosley, S. E., Eastman, R. G., Schmidt, B. P., *ApJ* **516**, 788 (1999).

Looking for GRB Progenitors

Krzysztof Belczyński[1], Tomasz Bulik[1], and Bronisław Rudak[2,3]

[1] *Nicolaus Copernicus Astronomical Center, Bartycka 18, 00-716 Warszawa, Poland*
[2] *Nicolaus Copernicus Astronomical Center, Rabiańska 8, 87-100 Toruń, Poland*
[3] *Toruń Center for Astronomy, Nicolaus Copernicus University, Gagarina 11, 87-100 Toruń, Poland*

Abstract. Using stellar binary population synthesis code we calculate the production rates and lifetimes of several types of possible GRB progenitors. We consider mergers of double neutron stars, black holes and neutron stars, black holes and white dwarfs, and helium stars. We calibrate the results with the measured star formation rate history. We discuss the viability of each GRB model, and alternatively, assuming that all bursts are connected with one model, we constrain the required collimation of GRBs. We also show the importance of widely used evolutionary parameters on the merger rates of calculated binary populations.

INTRODUCTION

It is surprising how little we know of GRB progenitors when we consider how much work is devoted to the subject. Different objects and models of progenitors were proposed, some already forgotten, while others still being intensively studied. Considering the diversity of gamma-ray bursts it is probable that they come from different types of astronomical objects, and this should promote the work on different types of proposed progenitors. Recently, much of the weight was placed on collapsars, and the connection between supernovae and GRBs, although other models, such as compact object mergers, are still on the stage.

Compact object binaries have recently drawn much attention in the astronomical community. Several models of compact object binaries were proposed as GRBs progenitors, namely: double neutron star mergers [11,18], black hole – neutron star mergers [9], black hole – white dwarf mergers [8] and helium star mergers [6]. Mergers of compact object binaries are also most often considered sources of gravitational waves. As gravitational wave detectors LIGO and VIRGO will soon be operational, the question of merger rates arises [1].

To predict compact object merger rates we use Monte Carlo simulations to produce populations of different compact object binaries. In this approach one generates massive binaries at Zero-Age Main Sequence (ZAMS) and evolves them through consecutive stages of single and binary star evolution, which may eventually lead to formation of a compact object binary. Final synthesis of large ensembles of compact object binaries allows then statistical studies and calculation of expected merger rates.

Several compact object merger rate estimates [12,13,21,17,2,3,7] have already been published. However, calculations of these rates are based on many assumptions and use parameters with very uncertain values. Whereas the evolution of single stars is reasonably well known [5], the distributions of initial binary conditions as well as some aspects of binary star evolution are uncertain. Finally, the population synthesis codes must deal with uncertainties in supernova explosion mechanisms, and in particular with the value of the kick a neutron star (or a black hole) receives at birth, and the mass formed in the explosion compact object [3]. Previous population synthesis calculations concerned double neutron stars and only few dealt with neutron star, black hole binaries. So far only in one case [7] were all kinds of the proposed GRBs compact object binary progenitors were considered. Our present contribution follows the same line of work, although we use a different population synthesis code and we would like to communicate our first results here. Preliminary comparison shows striking similarities of the results, which, taking into account different codes, points toward robustness of the population synthesis method in spite of the many uncertainties involved in the calculations.

THE MODEL

Our evolutionary code is primarily based on the prescriptions given by [20] and [5], but we use the number of many revised or newly developed specific evolutionary prescriptions from [4,16,15,14]. In our calculation of single and binary evolution we include stellar winds (normal and LBV wind), magnetic breaking, quasi-dynamic mass transfer, common envelope evolution, hyper-accretion onto compact objects, detailed supernova explosion treatment and gravitational wave energy loss in compact object binaries. We start the evolution of a given binary when both components are at ZAMS, then each star evolves through different stages of its life depending on its mass: main sequence, Hertzsprung gap, red giant branch, horizontal branch, asymptotic giant branch, and either supernova explosion which leads to formation of a neutron star (or a black hole) or a phase of enhanced mass loss and formation of a white dwarf. During each stage of the binary evolution the components may interact, which changes the consecutive component's evolution (either through rejuvenation, stripping the component off its outer layers, or even by swallowing one component by the other). Interaction of binary components may lead to one of the proposed GRB progenitors, the helium star merger, when the compact object is engulfed in the giant's envelope. If there is not enough orbital energy to eject the giant's envelope then the compact object spirals in through the giant's envelope, finally merging with the giant's core, which can be torn out in the process, and form an accretion disc around a compact object which is then already a black hole, due to the accretion during spiral in.

Other systems follow their evolutionary paths, and we collect information on the

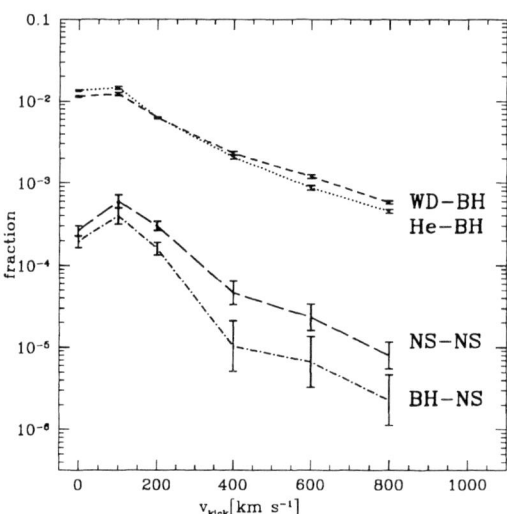

FIGURE 1. Relative production rates of different GRB progenitor types that merge within the Hubble time.

types of compact object binaries that are proposed for GRBs progenitors: double neutron stars, black hole – neutron stars and black hole – white dwarf systems. These systems interact only due to gravitational energy wave loss, finally merging and forming a black hole with massive thick accretion disc, which presumably may lead to a gamma-ray burst.

RESULTS

In Fig. 1 we show the relative numbers of 4 different GRB progenitor types that merge within the Hubble time (15 Gyr) as a function of the width of the distribution from which we draw kick velocity a compact object receives in a supernova explosion. Two things are clearly seen: First, the number of WD-BH binaries and Helium mergers (He-BH) is about the same and is more then an order-of-magnitude greater than the number of NS-NS and BH-NS binaries. Second, the number of a given progenitor type falls off approximately exponentially with the kick velocity. Relative production rates may be calibrated (e.g., see Eq. 14 in [2]). For example, assuming the width of kick velocity, say $v_{kick} = 200$ km s^{-1}, yields 1 merging event per Milky Way like galaxy per Myr for BH-NS systems, 3 events for NS-NS binaries, and 60 for WD-BH and Helium mergers.

In Fig. 2 we show the cumulative rate of different merging events. We have combined our relative numbers for different progenitors with the star formation rate function [10,19], and taking into account the evolutionary time delay of a given merging event we integrated our production rates to get the merger rates as

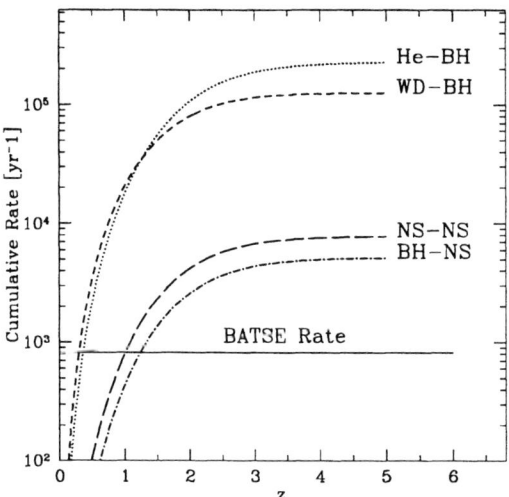

FIGURE 2. Cumulative merger rates for different GRB progenitor types as a function of redshift.

a function of redshift. In this example calculation we assumed $\Omega_M = 1.0$, $\Omega_\Lambda = 0.0$, and the Hubble constant $H_0 = 65$ km s^{-1}Mpc^{-1}. We used the kicks drawn from the distribution which is a weighted sum of two Gaussians: 80 percent with the width of 200 km s^{-1} and 20 percent with the width 800 km s^{-1}.

In Fig. 2 we also show the BATSE gamma-ray burst detection rate corrected for the uneven sky exposure. Comparison of the cumulative distributions for different progenitor types with the BATSE rate shows that if any of the progenitor types included here was to reproduce the BATSE rate, then we should not see GRBs from redshifts greater then unity! Of course this is not the case, as GRBs with higher redshifts have been observed. However, we haven't yet introduced the collimation factor into our results, which will certainly lower our predicted rates. To lower our calculated rates to the BATSE rate, for the average GRB redshift of ∼2, we would need collimation of about 4° for BH-WD and Helium mergers and about 12° for BH-NS and NS-NS mergers

As seen in Fig. 2 the curves flatten out for high redshifts ($z \geq 5$). In other words we do not expect binary mergers at high redshifts. This is a combined effect of the star formation rate function we have used, which decreases at high redshifts, and of the non-zero lifetimes of progenitors prior to the final merging event. Each of our binaries needs a specific time to evolve into a compact object binary (t_{evol}) and then needs time to merge due to gravitational wave energy losses (t_{merger}). These times are non-negligible and are specific for each group of proposed binary GRBs progenitors. For our sample of binaries we found that t_{life} ($t_{evol} + t_{merger}$) are, for NS-NS: ∼ 10^7–10^{12} yr, for BH-NS: ∼ 10^7–10^{10} yr, for BH-He: ∼ 10^6–10^9 yr, and for BH-WD: $\sim 10^7$–10^{12} yr.

CONCLUSIONS

- GRB binary progenitor production rates fall off exponentially with the width of natal kick velocity distribution.

- Evolutionary times can not be neglected in computations of GRB progenitor rates.

- Assuming that all GRBs are connected to NS-NS or BH-NS binaries, the collimation must be of order $\sim 3 \times 10^{-2}$ (12°). If we assume that all GRBs result from binary mergers, then the population is dominated by BH-WD and BH-He star mergers, and the collimation must be of order $\sim 3 \times 10^{-3}$ (4°).

- We do not expect binary mergers at high redshifts ($z \geq 5$).

Acknowledgments: We acknowledge the support of the following grants: KBN-2P03D01616, KBN-2P03D00415, and KBN2P03D02117.

REFERENCES

1. Allen, B., et al., *Phys. Rev. Lett.* **83**, 1498 (1999).
2. Belczynski, K. and Bulik, T., *A&A* **346**, 91 (1999).
3. Belczynski, K., Bulik, T., & Zbijewski, W., *A&A* accepted, astro-ph/9911435 (1999).
4. Bethe, H.A. & Brown G.E., *ApJ* **506**, 780 (1998).
5. Eggleton, P.P., Tout, C.A., & Fitchett, M.J., *ApJ* **347**, 998 (1989).
6. Fryer, C.L. & Woosley, S., *ApJ Lett.* **502**, L9 (1998).
7. Fryer, C.L., Woosley, S., & Hartmann, D.H., *ApJ* submitted, astro-ph/9904122 (1999).
8. Fryer, C.L., Woosley, S., Herant, M., & Davies, M.B., *ApJ* **520**, 650 (1999).
9. Lee, W.H. & Kluźniak, W., *Acta Astronomica* **45**, 705 (1995).
10. Madau, P., et al., *MNRAS* **283**, 1388 (1996).
11. Meszaros, P. & Rees, M.J., *ApJ* **476**, 232 (1997).
12. Narayan, R., Piran, T., & Shemi, A., *ApJ Lett.* **379**, L17 (1991).
13. Phinney, E.S., *ApJ Lett.* **380**, L17 (1991).
14. Podsiadlowski, P., Joss, P.C., & Hsu, J.J.L., *ApJ* **391**, 246 (1992).
15. Pols, O.R. & Marinus, M., *A&A* **288**, 475 (1994).
16. Portegies Zwart, S.F. & Verbunt, F., *A&A* **309**, 179 (1996).
17. Portegies Zwart, S.F. & Spreeuw, H.N., *A&A* **312**, 670 (1996).
18. Ruffert, M., Janka, H.T., Takahashi, K., & Schaefer, G., *A&A* **319**, 122 (1997).
19. Totani, T., *ApJ Lett.* **486**, L71 (1997).
20. Tout, C.A., Aarseth, S.J., Pols, O.R., & Eggleton, P.P., *MNRAS* **291**, 732 (1997).
21. Tutukov, A.V. & Yungelson, L.R., *MNRAS* **260**, 675 (1993).

Making Accretion Disks Around Black Holes: GRB Progenitors

Chris L. Fryer and Weiqun Zhang

*UCO/Lick Observatory, UC Santa Cruz
Santa Cruz, CA, 95064*

Abstract. We review the current progenitors of the rapidly accreting black holes which are thought to power gamma-ray bursts. Differences in the evolutionary scenarios of these progenitors lead to differences in the observed properties of the bursts (e.g., burst duration, burst location, and afterglow emission). We pay particular attention to the differences between helium merger and collapsar gamma-ray burst progenitors (the two leading progenitors for long-duration gamma-ray bursts).

INTRODUCTION

The current "favored" engine of classical gamma-ray bursts (GRBs) is thought to be powered by rapidly accreting black holes. This black hole accretion disk (BHAD) engine seems to be able to reproduce the large energies and high Lorentz factors necessary to drive the fireballs which produce GRBs (see [1,2] for reviews). But this is just the first step for GRB engine theorists. Current and future data are already placing more stringent constraints on the engine, and more specifically, on the formation of the BHAD systems. For example, the long-duration bursts (the only GRBs with accurate localizations, thus far) are clearly cosmological, occur within galaxies, and seem to trace star forming regions.

These constraints teach us about the formation processes of the BHAD GRB engines. So far, there are 5 distinct BHAD progenitors [1,2]: the merger of two neutron stars (DNS), of a neutron star and a black hole (BH/NS), of a white dwarf and a black hole (WD/BH), of a compact remnant and a helium core (He-merger), and the collapse of a rotating massive star (collapsar). The accretion disks and black holes in these systems are all slightly different, and these differences will produce a range of burst durations and energies.

Each BHAD system has its advantages and disadvantages. For instance, collapsars and He-mergers can use the remnants of their collapsing star to focus their energy into a narrow beam [3]. But DNS, BH/NS, and WD/BH mergers are the only engines in which a straightforward analogy to the Blandford-Znajek mechanism of active galactic nuclei is possible [4]. Thus, it is easier to imagine[1] magnetic fields playing a role in these mergers than with the He-merger or the collapsar. These differences are what theorists (e.g., [2]) use to separate short-duration (DNS,WD/BH) and long-duration engines (WD/BH, He-Merger, Collapsar). However, the burst afterglows provide insight into the environment in which GRBs explode, and this window can further separate burst engines.

Here we summarize the current understanding of the BHAD GRB formation scenarios and show that by knowing their environments, we can distinguish between the engines. Since nearly all of the BHAD GRB progenitors are formed in binaries, these results depend heavily upon binary population synthesis calculations.

POPULATION SYNTHESIS

The first four BHAD GRB progenitors listed above are all binaries, and hence it is no surprise that they are sensitive to the uncertainties in binary evolution. In addition, since the current collapsar model requires the removal of the massive star's hydrogen envelope [3], over 99% of collapsars are also formed in binaries. It is the binary companion which removes the collapsar's hydrogen envelope in what is called a "common envelope" phase. Unfortunately, stellar evolution is not well understood, and our understanding of binary evolution is even worse! Every study of the progenitors of GRBs must overcome these uncertainties and theorists studying binary population synthesis are just realizing the number of unknowns (a.k.a. free parameters) in the calculations. To make any reliable predictions about BHAD progenitors, one must study the effects caused by the uncertainties in all of these free parameters. These many dials, knobs, and twiddles make the building of any binary populations synthesis machine a daunting effort (Fig. 1).

Rates, Durations, and Burst Locations

Those brave (a.k.a. foolish) enough to pursue the study of BHAD progenitor formation (e.g., [2]) soon learn that any reliable formation rate for these systems is rather hopeless (Table 1). However, some distinguishing features can be determined via binary population synthesis calculations. For example, 17–39% of DNS and BH/NS will merge >1 Mpc from their host galaxy (assuming a galaxy mass equal to 1/4th that of the Milky Way — Table 1). Since all of the currently observed bursts with good localizations (long duration bursts only) are well within their host galaxy, our population synthesis calculations begin to rule out DNS and BH/NS mergers as viable GRB progenitors of long-duration bursts. To most theorists, the fact that DNS and BH/NS mergers do not produce long-duration bursts does not come as a surprise. The disks formed from these mergers are likely to accrete very quickly ($\lesssim 0.1$ s) and will only produce short bursts (of which we know very little).

---

[1] Since no rigorous solution for magnetic fields exists, however, anything is possible.

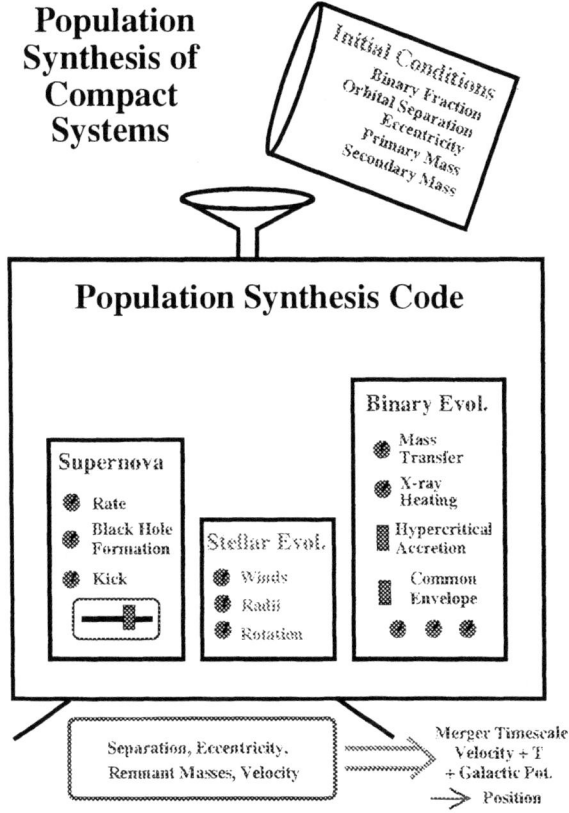

FIGURE 1. Binary population synthesis machine with some of its free parameters.

WD/BH binaries produce long-duration bursts which occur within their host galaxies, but the rate of WD/BH binaries is too low to explain all of the bursts. A typical rate for WD/BH binaries is 1 GRB day^{-1}. This rate seems to match the observed rate perfectly until one accounts for observational efficiencies and beaming (which must occur to get high enough energies) which make the merger rate of WD/BH binaries to be nearly a factor of 100 too low. However, the rates are uncertain, and if we include WD/NS mergers as well (which probably produce similar bursts), the rate is high enough. But then these binaries are more likely to occur outside their host galaxy [5].

He-mergers and Collapsars both produce long-duration bursts, occur in their host galaxies, and have rates that are sufficient to explain the well-localized gamma-ray bursts. In this meeting, collapsars certainly had the center stage, primarily because they are studied more than He-mergers. So it is worth briefly describing He-mergers here.

TABLE 1. BHAD GRB Models

Model	Form. Rate[a] (GRB/day)	Duration (s)	Fraction[b] >Mpc	$z_{\text{GRB}}^{\text{median}}$ ($z_{\text{SN}}^{\text{median}}$)
DNS	1–300	~0.1	3–8, 17–39%	0.5–0.8
BH/NS	1–1200	~0.1	3–8, 17–39%	0.5–0.8
WD/BH	0.1–50	15–150	0, 0%	$\lesssim 1$
He-Merg.	$<10 - 4\times 10^3$	15–500	0, 0%	1
Collapsar	$<100 - 5\times 10^4$	~10	0, 0%	$\gtrsim 1$

[a] These are the "most-likely" values of the rates using limited ranges for the uncertainties. If we include the full possible range of uncertainties, the rates are uncertain by another factor of 10 or 100 on both the upper and lower limits.

[b] We list values for galaxies with respective masses equal to and 1/4th that of the Milky Way, respectively.

He-mergers are binaries in which the primary ends its life as a neutron star or black hole. When the secondary evolves off the main sequence, it envelops the compact primary. The neutron star/black hole spirals into the secondary, ejecting the hydrogen envelope. If the secondary is large enough, the compact remnant will continue to spiral down to the core of the secondary, ejecting all of the hydrogen envelope and some of the helium core. The compact remnant will accrete as it inspirals, becoming a black hole if it is not one already. This spiral-in process also spins up the helium core, and eventually produces a torus around a black hole. These conditions are nearly identical to those produced in the collapse of a rotating massive star.

We have simulated this merging process and are just now starting to place numbers on the accretion rates, angular momenta of the tori, black hole spins and masses formed in this mergers (Fig. 2 [6,7]). The angular momenta are slightly higher than typical collapsar models, and hence, the accretion rates are lower, but He-mergers remain a viable GRB progenitor. They have one major advantage over collapsars in that they almost certainly occur. Collapsars require very specific conditions to form, and there is a portion of population synthesis parameter space which produces far too few Collapsars for Collapsars to be a major GRB progenitor.

Can we distinguish He-mergers and Collapsars? Fortunately, the observational constraints continue to grow, and the afterglows of GRBs can indeed provide further constraints on BHAD progenitors. Analyzing radio afterglows can tell us about the environments in which the fireball is moving [8]. Collapsars all will explode in an environment determined by the strong stellar wind that characterizes massive stars. He-mergers, on the other hand, may be in wind environments (the primary certainly had a strong wind). But, if the binary moves significantly from its birth site after the collapse of the primary, He-mergers may also occur in an environment which is more typical of the interstellar medium. The study of afterglows may be able to rule out one of these two models.

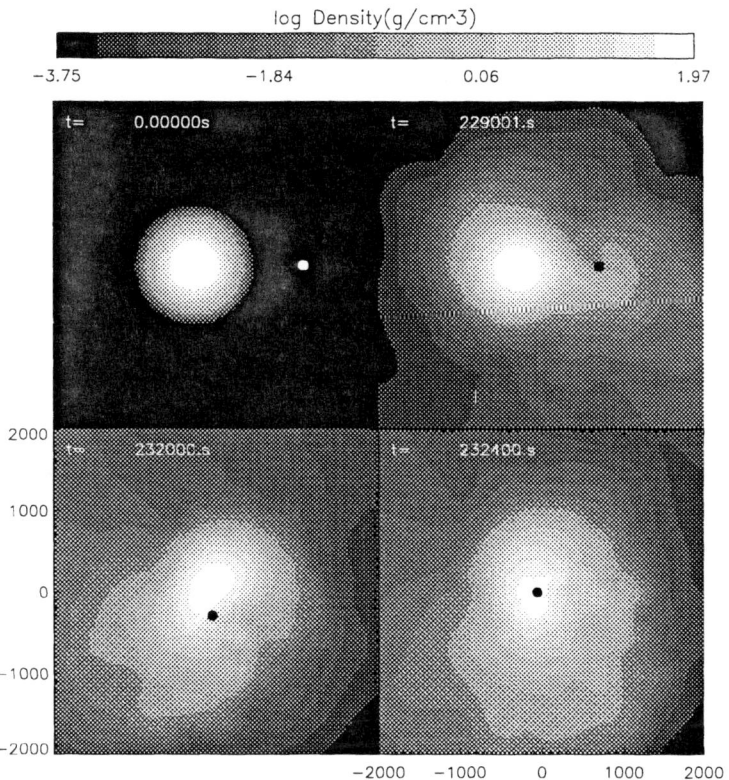

FIGURE 2. Merger of a $2 M_\odot$ compact remnant with a $8 M_\odot$ helium core. As the compact remnant merges, it spins up the helium core and accretes material, becoming a black hole if it is not one already. Note the asymmetric excretion disk.

Acknowledgments: This research was funded by NASA (NAG5-8128) and the US DOE ASCI program (W-7405-ENG-48).

REFERENCES

1. Popham, R., Woosley, S.E., & Fryer, C.L., *ApJ* **518**, 356 (1999).
2. Fryer, C.L., Woosley, S.E., & Hartmann, D.H., *ApJ* **526**, 152 (1999).
3. MacFadyen, A., & Woosley, S.E., *ApJ* **524**, 262 (1999).
4. Blandford, R.D., & Znajek, R.L., *MNRAS* **179**, 433 (1977).
5. Belczynski, K., & Bulik, T., in preparation.
6. Fryer, C.L., & Woosley, S.E., *ApJ* **502**, L9 (1999).
7. Zhang, W., Fryer, C.L., & Woosley, S.E., in preparation.
8. Chevalier, R., & Li, Z.-Y., *ApJ* **520**, L29 (1999).

Distribution of Binary Mergers Around Galaxies

Tomasz Bulik and Krzysztof Belczyński

Nicolaus Copernicus Astronomical Center, Bartycka 18, 00-716 Warszawa, Poland

Abstract. We use a stellar binary population synthesis code to find the lifetimes and velocities of several types of possible GRB progenitors: double neutron stars, black holes and neutron stars, black holes and white dwarfs, and helium stars. Assuming that they are born in different types of galaxies, we compute their spatial distribution and compare it with the observed locations of GRB afterglows within their hosts. We discuss constraints on the compact object merger model of GRBs imposed by this comparison and find that the observations of afterglows and their host galaxies appear inconsistent with the GRB compact object merger model.

INTRODUCTION

In the last few years the astronomical community has moved much closer to unveiling the nature of GRBs. The discovery of afterglows and identification of host galaxies for several bursts clearly links GRBs to some type of stellar events. Yet the nature of these events is unknown, and we still do not know what the GRB central engines are. Observations of GRB host galaxies and precise locations of GRBs within hosts provide a tool to test some of the possible central engine models. In this paper we discuss the consistency between the current observations and the results of binary population synthesis.

THE MODEL

The population synthesis code used here is described in [3]. One of the most important parameters determining the properties of the populations of binaries is the kick velocity a newly formed compact object receives at birth. However, several studies [4,7] indicate that the distribution of kick velocities consists of two components: a low-velocity term having a width of approximately 200 km s^{-1}, and a high-velocity term with a characteristic velocity around 800 km s^{-1}. About 80% of the kicks are drawn from the first component. It is also known that the production rate of compact object binaries falls off exponentially with increasing kick velocity, see Fig. 1 in [3]. Thus, the population of compact object binaries will be dominated by the objects formed in the systems that received the kicks drawn from the low-velocity component of the distribution. In the following we will consider the properties of the compact object binaries for the case when the kick velocity is drawn from a Gaussian distribution of width 200 km s^{-1}.

Little is known a priori about the masses and gravitational potentials of host galaxies where GRB progenitors reside. Therefore, to find the expected distribution of merger sites around galaxies we consider two extreme cases: propagation in a potential of a large galaxy like the Milky Way and propagation in empty space [2].

RESULTS

In Figure 1 we present the distribution of center-of-mass velocities gained by systems in supernova explosions in the galactic potential and binary lifetimes (the time a binary takes to evolve from Zero-Age Main Sequence [ZAMS] to the final merger of two components) for four types of compact object binaries.

For the case of propagation in a potential of a massive galaxy, only a fraction of NS-NS binaries will be able to escape from their host galaxies. The BH-NS binaries tend to stay in the galaxy. Here, we have assumed that the kick velocity does not depend on the nature of the compact object formed in the supernova. However, it has been argued that the kicks back holes receive should be smaller than those of the neutron stars. This is discussed in more detail in [1]. The helium star mergers stay in the host galaxies, while some of white dwarf black hole mergers have a chance of taking place outside the host, provided that the escape velocity from a given galaxy is not too large.

In the case of propagation in empty space quite a large number of binaries of any type will be able to escape from their birthplace.

In Figure 2 we present the cumulative distributions of the distances (projected on the sky) between merger sites and their host galaxies. In the case of propagation in empty space (left panel of Figure 2), most of the mergers take place far outside the host. In the case of propagation in the potential of a massive galaxy (right panel of Figure 2), black hole – neutron star mergers and helium star mergers take place inside the host, and only a small, but non-negligible, fraction of double neutron stars and black hole – white dwarf mergers happen outside the host.

DISCUSSION

We have learned at this conference [6] that the afterglow locations coincide with galaxies, and that typically there are intense star forming processes in these galaxies.

In the compact object merger model of GRBs one has to take into account the fact there are significant delays between the stellar formation and the time of merger. These delays consist of the stellar evolutionary time leading to supernovae explosions, the formation time of the compact object binary, and the evolution of

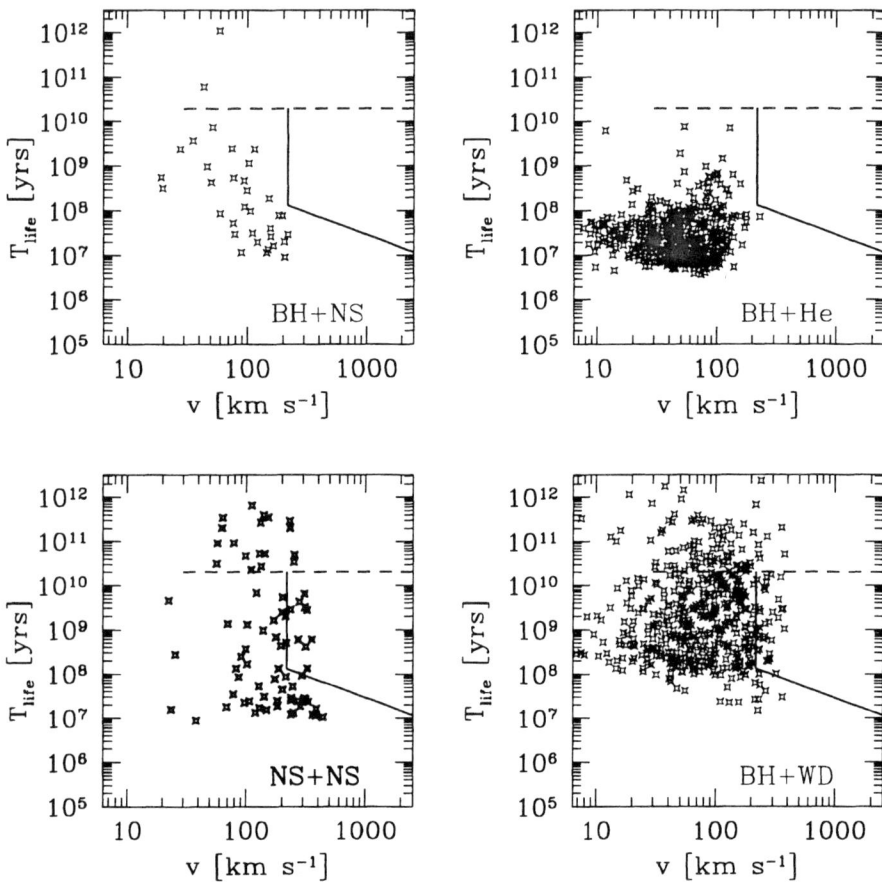

FIGURE 1. The distribution of different types of compact object binary mergers in the space spanned by the binary center-of-mass velocity and the binary lifetime (from ZAMS to final merger). The horizontal dashed line corresponds to the Hubble time (15 Myr). In the region for $t_{merge} < 15$ Myr we present two solid lines: the vertical corresponding to $v = 200\,\mathrm{km\,s^{-1}}$ (approximately the escape velocity from a galaxy) and the diagonal line corresponding to a constant value of $v \times t_{merge} = 30\,\mathrm{kpc}$. Together these lines define the region in the parameter space with systems that can escape from the large host galaxy.

the compact object binary due to gravitational wave energy loss. The distribution of the delay times is rather wide and varies for different types of binaries [3], see Figure 1. In the case of helium star mergers the delays can be as short as a few million years.

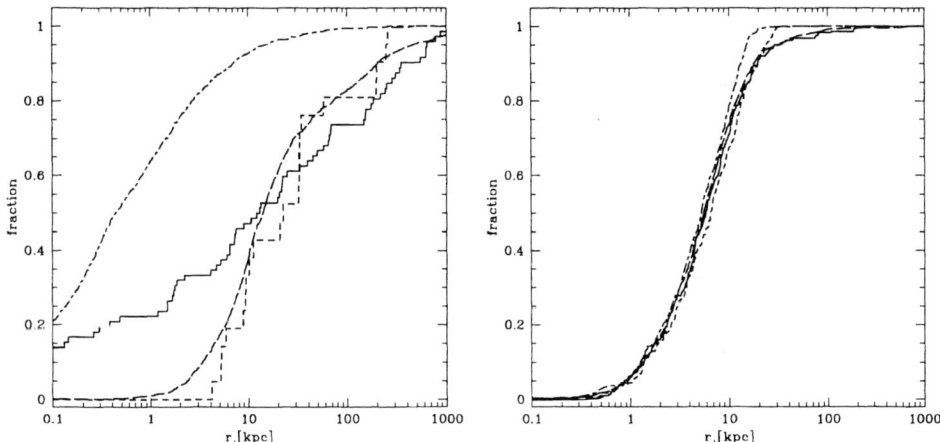

FIGURE 2. Distributions of different types of compact object binaries around their hosts. The left panel shows the case of propagation in empty space and the right panel shows the case of propagation in a potential of a massive galaxy. The solid, short-dashed, long-dashed and dashed-dotted lines correspond to NS-NS, BH-NS, WD-BH, and He-BH mergers, respectively.

Assuming that GRBs are related with NS-NS, BH-NS, or BH-WD mergers, we do not expect any correlation between the GRB sites and star formation because the star formation processes could have ceased by the time the merger happens. Thus, within this model, the GRB rate should be proportional to the luminous mass in the Universe. As most of the luminous mass is concentrated in massive galaxies we expect to find GRBs within such galaxies. However, this should be typical galaxies and we tend to find GRB hosts in small, star forming ones [6].

Let us now consider the case when the delay between the stellar formation and the merger event is shorter than the star forming episode itself. Naturally, a correlation between the GRB sites and star forming galaxies exists. However, the observed host galaxies in this case are typically small. Thus, as shown above, a significant fraction of the mergers should take place outside the host galaxies and we should be finding GRBs with no underlying host galaxies.

One can also argue that the GRBs that take place outside the host galaxies do not produce significant afterglows because of the low density of the ambient medium. This would make a strong selection effect against detecting GRBs happening outside of the hosts. At this conference we have heard, however, that the BeppoSAX data are consistent with all GRBs having X-ray afterglows [5]. This means that all GRBs have afterglows and that GRBs with no afterglows do not exist, at least within the sample observable by BeppoSAX.

The reasoning presented above strongly argues against the compact object merger model of GRBs. However, we must remember that afterglows have been detected only from the long bursts, and we do not know if short bursts also produce afterglows or what are the locations in relation to host galaxies of short bursts. Moreover, numerical models of compact object coalescences [8,9] agree with analytical estimates and show that the time scales of these events can not be stretched beyond a fraction of a second. Yet the long bursts have a median duration of approximately $\approx 20 - 30$ s.

Thus, we conclude that the compact object merger model appears to be inconsistent with the observations of afterglows and their locations within the host galaxies. Long bursts are therefore probably not connected with compact object mergers. However, it is quite likely that we will find that the short bursts are connected with mergers of compact objects.

Acknowledgments: We acknowledge the support of grants KBN-2P03D01616 and KBN-2P03D00415.

REFERENCES

1. Belczynski, K., Bulik, T., & Zbijewski, W., *A&A* accepted, astro-ph/9911435 (1999).
2. Belczynski, K., Bulik, T., & Zbijewski, W., *MNRAS* **309**, 629 (1999).
3. Belczynski, K., Bulik, T., & Rudak, B., these proceedings.
4. Dewey, R.J. & Cordes, J.M., *ApJ* **321**, 780 (1987).
5. Costa, E., these proceedings.
6. Fruchter, A., these proceedings.
7. Fryer, C.L., Burrows, A., & Benz, W., *ApJ* **496**, 333 (1998).
8. Lee, W.H. & Kluźniak, W., *Acta Astronomica* **45**, 705 (1995).
9. Ruffert, M., Janka, H.T., Takahashi, K., & Schaefer G., *A&A* **319**, 122 (1997).

The Most Distant Gamma-Ray Bursts

Dieter H. Hartmann[*], A. I. MacFadyen[†], and S. E. Woosley[†]

[*]*Department of Physics and Astronomy, Clemson University, Clemson, SC 29634*
[†]*Department of Astronomy and Astrophysics, UC Santa Cruz, Santa Cruz, CA 95064*

Abstract. GRBs have redshifts comparable to, perhaps even larger than those of quasars. Indeed, they are the most energetic explosions in the universe, with energies of order $M_\odot c^2$. Their host galaxies are faint, but are actively forming stars at rates typical of galaxies in the early universe. The current paradigm associates GRBs with the formation of black holes in massive, rotating stars, or with the mergers of compact binaries. GRBs thus trace the cosmic star formation rate and may be the most easily detectable emission of the earliest generation of stars.

INTRODUCTION

Cosmic gamma-ray bursts now appropriately hold the distinction of being the "largest explosions in the universe". Their afterglows are brighter than supernovae, and are thus sometimes referred to as "hypernovae" (Paczynski 1998). Photometric and spectroscopic observations of their afterglows provided a major breakthrough in our understanding of these explosions. These data place bursts at large distances and in association with faint host galaxies. But what is (or are) the underlying cause(s) of these violent events? The answer remains uncertain, but several theoretical arguments suggest the creation of hyperaccreting black holes with accretion rates in the range 10^{-3} to 10 $M_\odot$ s^{-1}. The accretion disks produce narrow jets of relativistically expanding plasma. As these jets propagate into the circum-burster material, internal and external shocks generate the GRB and also the afterglows.

With the launch of the Italian-Dutch Beppo-SAX X-ray satellite (Piro et al. 1998) the study of GRBs underwent a major transition. SAX established fading X-ray counterparts, GRB970228 being the first. The arc-minute accuracy locations allowed ground-based telescopes to successfully search for fading counterparts in the optical band (OTs) (van Paradijs et al. 1997). While extended emission was found for GRB970228, the nature of this host has never been established. Then followed GRB970508, for which there there was no evidence for an extended source. However, spectroscopy with Keck II found absorption lines suggesting a redshift of $z = 0.835$ (Metzger et al. 1997). After those breakthroughs, the study of GRBs became an international multiwavelength effort, involving a very large and growing group of astronomers who studied the emission at all wavelengths from radio to gamma-ray.

GRB971214 broke the redshift record with $z = 3.4$ (Kulkarni et al. 1998), which implies a total burst energy of over 10^{54} ergs (assuming isotropic emission) — clearly a challenge to theorists. GRB980425 surprised us by its association with a peculiar Type Ic supernova (SN1998bw), located in a nearby galaxy at $z = 0.008$ (Galama et al. 1998). Models (Woosley, Eastman, & Schmidt 1999; Iwamoto 1999; Iwamoto et al. 1998) required a kinetic energy in excess of 10^{52} ergs if the explosion was isotropic, but perhaps less if it was not (Höflich, Wheeler, & Wang 1998). High velocities were required to explain the SN spectrum, and about 0.5 $M_\odot$ of ^{56}Ni was needed to power the light curve. GRB980425 was likewise unusual. Assuming isotropic emission, a total energy of only 10^{48} ergs was released in about 20 s.

Several more bursts were detected at $z \sim 1$, and in essentially all cases, the candidate host galaxy was faint (R $\sim$ 22–26). Estimates of the star formation rate in these hosts are typically a few $M_\odot$ yr^{-1}. GRB990510 was found to be beyond $z = 1.6$, and observations with the VLT led to the discovery of polarization (1.7 %; Covino et al. 1999). GRB970508 was the first burst with an established radio afterglow (Frail et al. 1997), but there are now many such events, including the recent bright burst GRB991216 (Frail et al. 1999). Perhaps the most spectacular burst was GRB990123, at $z \sim 1.6$, associated with a group of (interacting?) galaxies (Fruchter et al. 1999). This burst established the existence of prompt optical emission. The ROTSE experiment observed a peak brightness of V $\sim$ 9 (Akerlof *et al.* 1999). Automated systems such as ROTSE and LOTIS (http://hubcap.clemson.edu/~ggwilli/LOTIS/) are ready to spring into action and catch prompt OTs, but so far this has only been successful for GRB990123.

MODELS AND THE LINK TO SUPERNOVAE

The measured redshifts suggest that GRBs compete in distance with quasars and the earliest galaxies. It is likely that GRBs trace the cosmic star formation history, which would provide a new and valuable tool of observational cosmology. So far, the detected OTs are close to their hosts and not associated with the galactic central regions. The high redshifts imply large energies, comparable in some cases to a solar mass rest. This is beyond the reach of most theoretical models, and geometric beaming into a small solid angle is often invoked as a way out of this energy crisis. But even if beaming is significant, the explosive release of more than 10^{52} ergs of energy (electromagnetic or in particle form) into a small volume inevitably leads to an opaque region (a "fireball") which will rapidly expand to relativistic velocities. The Lorentz factors achieved in this expansion depend on the amount of baryon loading. The escape of hard photons from the burst requires bulk Lorentz factors greater than $\Gamma \sim 100$. The resulting relativistic beaming of the photons reduces the optical depth in various ways, which is needed to let the high-energy photons escape.

The emerging picture is a relativistically expanding shell of electrons, positrons, and baryons: the fireball scenario (Meszaros & Rees 1997; Piran 1999). In this fireball scenario the GRB emission is believed to result from shocks in the expanding fireball, where shock amplified magnetic fields allow radiative cooling via synchrotron radiation. The shocks could be due to internal energy dissipation from colliding shells with different initial Lorentz factors, or due to external shocks occurring when the shell(s) run into an external medium. The observations and underlying physics of these afterglows are described by Meszaros (astro-ph-9904038) and Piran (1999).

Here we use the term "hypernova" to describe the observational manifestation of a supernova-like outburst that has a total of over 10^{52} ergs of isotropic equivalent energy or an optical brightness for a sustained period of over 10^{44} erg/s. The mechanism is not specified. If the hypernova is powered by a specific mechanism, one should refer to the model by this name. The collapsar model (Woosley 1993; MacFadyen & Woosley 1999) refers to the explosion of a massive star by the formation of a hyper-accreting black hole and a jet in its middle. The jet may be powered by neutrinos or by MHD processes. By hyper-accretion we mean accretion rates from 10^{-3} to 10 $M_\odot$ s^{-1}. Collapsars can power a GRB jet for hundreds of seconds as the collapsing stellar envelope continues to feed the accretion disk and power the jet. Short time scale variability can come both from variations in the disk accretion rate or from the interaction of the jet with the outer layers of the star. Merging compact objects, on the other hand, will have a time scale of only tens of milliseconds if neutrino losses are important (Janka & Ruffert 1996; Janka et al. 1999) and last tens of seconds only if the disk viscosity is very low.

The collapsar mechanism may also operate in stars which have extended envelopes, either because of lower mass loss or due to accretion from a binary companion. In this case the observed event is unlikely to be a classical GRB since the time for the jet to break out of the stellar surface (1,000 – 100,000 s) exceeds the time the engine is "on". The engine turns off because of starvation due to the decreasing accretion rate either because the envelope density is low or because shocks from the jet sweep around the star and blow up the star. A very powerful, asymmetric explosion will result in this case with possible soft X-ray transient due to shock acceleration along the steep density gradient at the stellar surface. SN1997cy might be an example of a GRB engine turning on in a star with an extended envelope.

There is growing evidence for a supernova component in afterglow light curves. In addition to direct evidence from GRB980425/SN1998bw, there is further evidence for "SN-contamination" from GRB970228 (Reichart 1999, Galama 1999), GRB980326 (Bloom et al. 1999), and GRB980519 (Chevalier & Li 1999; Halpern et al. 1999; Vrba et al. 2000). The total SN rate in the universe is about one event per second, while the GRB rate is ~ one event per day. If beaming of the gamma rays is into 1% of the sky, the rate is closer to one per 10^3 seconds, still far less than the supernova rate. This implies that only a very small fraction of supernovae leads to observable GRBs. In the collapsar model we are dealing with very massive progenitors and require a sufficiently large angular momentum. Apparently, these constraints select a tiny fraction of all supernovae for GRB formation.

GRBS AS COSMOLOGICAL PROBES

Whether GRBs are due to merging neutron star binaries or collapsars, the events trace the star formation rate. Collapsars trace it directly because the lifetimes of the progenitors are so short, while mergers involve a significant delay (e.g., Fryer, Woosley, & Hartmann 1999; Belczynski & Bulik 1999) of order 10^8 yrs. In the case of mergers we also expect a significant spatial separation between the GRB/OT and the host galaxy, because of the rather large systemic velocities of the progenitor systems (e.g., Fryer, Woosley, & Hartmann 1999). The observations indicate typical angular separations of less than 1 arcsec, which argues in favor of sites more closely associated with star forming regions in the host galaxies. The relative position distribution argues in support of the collapsar model, and disfavors the merger scenario.

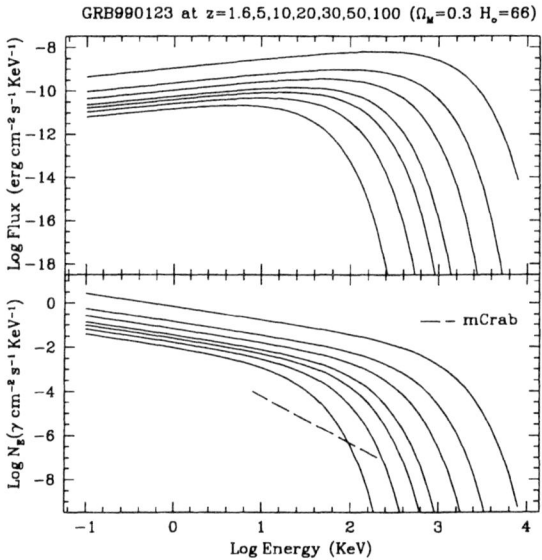

FIGURE 1. Spectrum of GRB990123 at its known redshift $z = 1.6$, and higher values.

If GRBs trace star formation, in particular that of the most massive stars, they provide a tool to trace the elusive Population III stars. The universe recombined when the CMB temperature fell below $\sim 3,000$ K, i.e., at a redshift $z_{\rm rec} \sim 1000$, leaving a residual freeze-out ionization of $n_e/n \sim 10^{-4}$, but was again almost fully ionized at a poorly known re-heating redshift $z_{\rm reh} \sim 5\text{-}10$, inferred from the Gunn-Peterson effect. The cause of this re-ionization is unknown, but early quasars and Population III stars are the most likely sources. GRBs might thus point us toward the first luminous objects formed in the early universe. Their intrinsic brightness is so large that it is easy to detect them at very large redshifts. For example, GRB990123 was one of the brightest GRBs ever recorded, yet its host galaxy is located at a redshift of $z = 1.6$. Figure 1 shows the flux GRB990123 would have had if located at $z = 5, 10, 20, 30, 50$, and 100, for a particular choice of cosmological parameters ($H_0 = 66$, $\Omega_M = 0.3$, $\Omega_\Lambda = 0.7$). For comparison we also show a line indicating a flux of 1 mCrab. GRBs as bright as GRB990123 are easily detectable in the X-ray band, even if located as far away as $z = 100$! Their afterglows are bright as well, but at such large redshifts the observations have to be carried out in the IR band. How GRBs and their afterglows probe the early universe is also discussed in greater detail by Lamb & Reichart (1999 and in these proceedings).

REFERENCES

1. Akerlof, C., et al., *Nature* **398**, 400 (1999).
2. Belczynski, K. & Bulik, T., *A&A* **346**, 91 (1999).
3. Bloom, J. S., et al., *Nature* **401**, 453 (1999).
4. Briggs, M. S., et al., *ApJ*, in press, astro-ph/9903247 (1999).
5. Chevalier, R. A., & Li, Z.-Y., astro-ph/9904417 (1999).
6. Covino, S. et al., *IAU Circ.* 7172 (1999).
7. Frail, D., et al., *IAU Circ.* 6662 (1997).
8. Frail, D., et al., *GCN Circ.* 451 (1999).
9. Fryer, C., Woosley, S. E., & Hartmann, D. H., *ApJ* **526**, 152 (1999).
10. Fruchter, A. S., et al., *ApJ*, in press, astro-ph/9902236 (1999).
11. Galama, T. J., et al., *Nature* **395**, 670 (1998).
12. Galama, T. J., *Gamma-Ray Burst Afterglows*, PhD Thesis, Universititeit van Amsterdam, 1999.
13. Halpern, J. P., et al., *ApJ* **517**, L105 (1999).
14. Höflich, P., Wheeler, J. C., & Wang, L., *ApJ* **521**, 179 (1999).
15. Iwamoto, K., *ApJ* **512**, L47 (1999).
16. Iwamoto, K., et al., *Nature* **395**, 672 (1998).
17. Janka, H.-T., & Ruffert, M., *A&A* **307**, L33 (1996).
18. Janka, H.-T., Eberl, T., Ruffert, M., & Fryer, C. L., astro-ph/9908290 (1999).
19. Kulkarni, S. R., et al., *Nature* **393**, 35 (1998).
20. Lamb, D. Q. & Reichart, D. E., *ApJ*, astro-ph/9909002 (1999).
21. MacFadyen, A. I., & Woosley, S. E., *ApJ*, in press, astro-ph/9810274 (1999).
22. MacFadyen, A. I., Woosley, S. E., & Heger, A., *ApJ*, astro-ph/9910034 (1999).
23. Meszaros, P., & Rees, M. J., *ApJ* **476**, 232 (1997).
24. Metzger, M. R., et al., *Nature* **387**, 878 (1997).
25. Paczyński, B., *ApJ*, **494**, L45 (1998).
26. Piran, T., *Physics Reports* **314**, 575 (1999).
27. Piro, L., et al., *A&A* **329**, 906 (1998).
28. Reichart, D. E., *ApJ Lett.*, in press, astro-ph/9906079 (1999).
29. van Paradijs, J., et al., *Nature* **386**, 686 (1997).
30. Vrba, F., et al., *ApJ* 528, in press (2000).
31. Woosley, S. E., *ApJ* **405**, 273 (1993).
32. Woosley, S. E., Eastman, R., & Schmidt, B., *ApJ* **516**, 788 (1999).

Gamma-Ray Bursts as a Probe of the Very High Redshift Universe

Donald Q. Lamb and Daniel E. Reichart

*Department of Astronomy & Astrophysics, University of Chicago,
5640 South Ellis Avenue, Chicago, IL 60637*

Abstract. We show that, if many GRBs are indeed produced by the collapse of massive stars, GRBs and their afterglows provide a powerful probe of the very high redshift ($z \gtrsim 5$) universe.

INTRODUCTION

There is increasingly strong evidence that gamma-ray bursts (GRBs) are associated with star-forming galaxies [1-4] and occur near or in the star-forming regions of these galaxies [2-6]. These associations provide indirect evidence that at least the long GRBs detected by BeppoSAX are a result of the collapse of massive stars. The discovery of what appear to be supernova components in the afterglows of GRBs 970228 [7-8] and 980326 [9] provides direct evidence that at least some GRBs are related to the deaths of massive stars, as predicted by the widely-discussed collapsar model of GRBs [10-15]. If GRBs are indeed related to the collapse of massive stars, one expects the GRB rate to be approximately proportional to the star-formation rate (SFR).

GRBS AS A PROBE OF STAR FORMATION

Observational estimates [16-19] indicate that the SFR in the universe was about 15 times larger at a redshift $z \approx 1$ than it is today. The data at higher redshifts from the Hubble Deep Field (HDF) in the north suggests a peak in the SFR at $z \approx 1 - 2$ [19], but the actual situation is highly uncertain. However, theoretical calculations show that the birth rate of Pop III stars produces a peak in the SFR in the universe at redshifts $16 \lesssim z \lesssim 20$, while the birth rate of Pop II stars produces a much larger and broader peak at redshifts $2 \lesssim z \lesssim 10$ [20-22]. Therefore, one expects GRBs to occur out to at least $z \approx 10$ and possibly $z \approx 15 - 20$ — redshifts that are far larger than those expected for the most distant quasars. Consequently

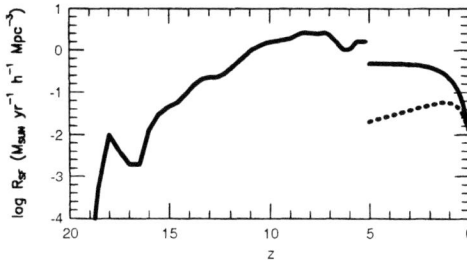

FIGURE 1. The cosmic SFR R_{SF} as a function of redshift z. The solid curve at $z < 5$ is the SFR derived by [23]; the solid curve at $z \geq 5$ is the SFR calculated by [21] (the dip in this curve at $z \approx 6$ is an artifact of their numerical simulation). The dotted curve is the SFR derived by [19].

GRBs may be a powerful probe of the star-formation history of the universe, and particularly of the SFR at very high redshifts (VHRs).

In Figure 1, we have plotted the SFR versus redshift from a phenomenological fit [23] to the SFR derived from submillimeter, infrared, and UV data at redshifts $z < 5$, and from a numerical simulation by [21] at redshifts $z \geq 5$. The simulations done by [21] indicate that the SFR increases with increasing redshift until $z \approx 10$, at which point it levels off. The smaller peak in the SFR at $z \approx 18$ corresponds to the formation of Population III stars, brought on by cooling by molecular hydrogen. Since GRBs are detectable at these VHRs and their redshifts may be measurable from the absorption-line systems and the Lyα break in the afterglows [24], if the GRB rate is proportional to the SFR, then GRBs could provide unique information about the star-formation history of the VHR universe.

More easily but less informatively, one can examine the GRB peak photon flux distribution $N_{GRB}(P)$. To illustrate this, we have calculated the expected GRB peak flux distribution assuming (1) that the GRB rate is proportional to the SFR[1], (2) that the SFR is that given in Figure 1, and (3) that the peak photon luminosity distribution $f(L_P)$ of the bursts is independent of z. There is a mis-match of about a factor of three between the $z < 5$ and $z \geq 5$ regimes. However, estimates of the star formation rate are uncertain by at least this amount in both regimes. We have therefore chosen to match the two regimes smoothly to one another, in order to avoid creating a discontinuity in the GRB peak flux distribution that would be entirely an artifact of this mis-match.

For a peak luminosity function $f(L_P)$ and for $dL_P/d\nu \propto \nu^{-\alpha}$, the observed GRB peak flux distribution $N_{GRB}(P)$ is given by the following convolution integration:

$$N_{GRB}(P) = \Delta T_{obs} \int_0^\infty R_{GRB}(P|L_P) f[L_P - 4\pi D^2(z)(1+z)^\alpha P] dL_P , \qquad (1)$$

where ΔT_{obs} is the length of time of observation, $D(z)$ is comoving distance,

$$R_{GRB}(P|L_P) \propto \frac{R_{SF}(z)}{1+z} \frac{dV(z)}{dz} \left| \frac{dz(P|L_P)}{dP} \right| , \qquad (2)$$

---

[1] This may underestimate the GRB rate at VHRs since it is generally thought that the initial mass function will be tilted toward a greater fraction of massive stars at VHRs because of less efficient cooling due to the lower metallicity of the universe at these early times.

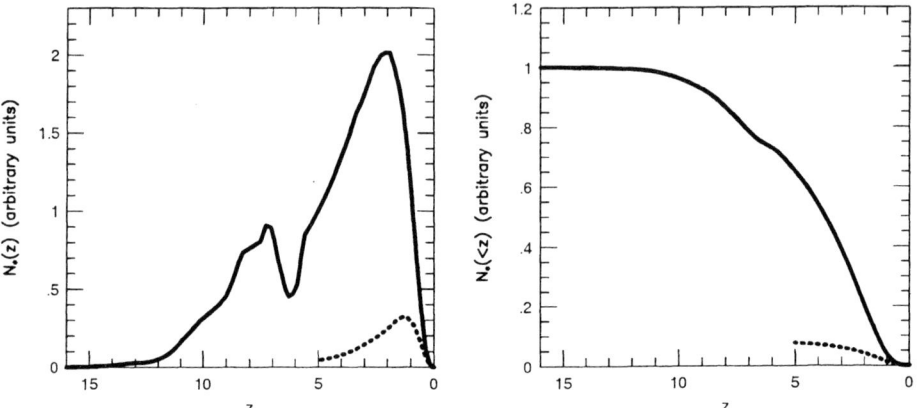

FIGURE 2. Left panel: The number N_* of stars expected as a function of redshift z (i.e., the SFR from Figure 1, weighted by the differential comoving volume, and time-dilated) assuming that $\Omega_M = 0.3$ and $\Omega_\Lambda = 0.7$. Right panel: The cumulative distribution of the number N_* of stars expected as a function of redshift z. Note that $\approx 40\%$ of all stars have redshifts $z > 5$. The solid and dashed curves in both panels have the same meanings as in Figure 1.

$R_{SF}(z)$ is the local co-moving SFR at z, and $dV(z)/dz$ is differential comoving volume [24].

The left panel of Figure 2 shows the number $N_*(z)$ of stars expected as a function of redshift z (i.e., the SFR, weighted by the co-moving volume, and time-dilated) for an assumed cosmology $\Omega_M = 0.3$ and $\Omega_\Lambda = 0.7$ (other cosmologies give similar results). The solid curve corresponds to the star-formation rate in Figure 1. The dashed curve corresponds to the star-formation rate derived by [19]. This figure shows that $N_*(z)$ peaks sharply at $z \approx 2$ and then drops off fairly rapidly at higher z, with a tail that extends out to $z \approx 12$. The rapid rise in $N_*(z)$ out to $z \approx 2$ is due to the rapidly increasing volume of space. The rapid decline beyond $z \approx 2$ is due almost completely to the "edge" in the spatial distribution produced by the cosmology. In essence, the sharp peak in $N_*(z)$ at $z \approx 2$ reflects the fact that the SFR we have taken is fairly broad in z, and consequently, the behavior of $N_*(z)$ is dominated by the behavior of the co-moving volume $dV(z)/dz$; i.e., the shape of $N_*(z)$ is due almost entirely to cosmology. The right panel in Figure 2 shows the cumulative distribution $N_*(> z)$ of the number of stars expected as a function of redshift z. The solid and dashed curves have the same meaning as in the left panel. This figure shows that $\approx 40\%$ of all stars have redshifts $z > 5$.

The upper panel of Figure 3 shows the predicted peak photon flux distribution $N_{GRB}(P)$. The solid curve assumes that all bursts have a peak (isotropic) photon luminosity $L_P = 10^{58}$ ph s^{-1}. However, there is now overwhelming evidence that GRBs are not "standard candles." Consequently, we also show in Figure 3, as an illustrative example, the convolution of this same SFR and a logarithmically flat photon luminosity function $f(L_P)$ centered on $L_P = 10^{58}$ ph s^{-1}, and having widths

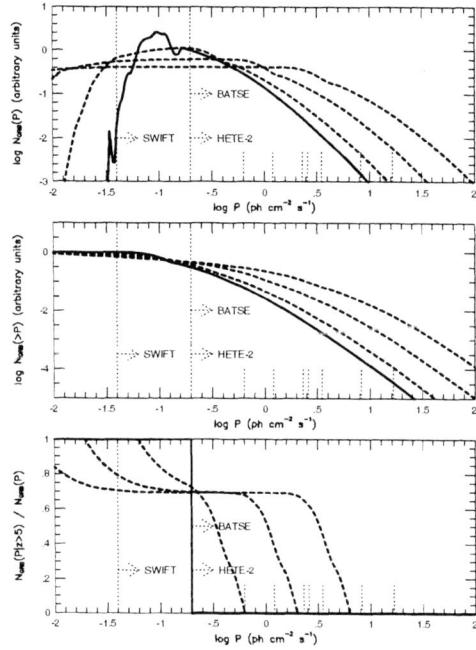

FIGURE 3. Top panel: The differential peak photon flux distribution of GRBs, assuming that (1) the GRB rate is proportional to the SFR, (2) the SFR is that shown in Figure 1; and (3) the bursts are standard candles with a peak photon luminosity $L_P = 10^{58}$ ph cm^{-2} s^{-1} (solid curve), or have a logarithmically flat peak photon luminosity function that spans a factor of 10, 100, or 1000 (dashed curves). Approximate detection thresholds are plotted for BATSE and HETE-2, and for *Swift* (dotted lines). Middle panel: The cumulative peak photon flux distribution of GRBs for the same luminosity functions. Lower panel: The fraction of GRBs with peak photon flux P that have redshifts of $z \gtrsim 5$ for the same luminosity functions. In all three panels, the dotted hashes mark the peak photon fluxes of the bursts with known peak photon luminosities and redshifts.

$\Delta L_P / L_P = 10$, 100 and 1000.[2] The actual luminosity function of GRBs could well be even wider [25].

The middle panel of Figure 3 shows the predicted cumulative peak photon flux distribution $N_{GRB}(> P)$ for the same luminosity function. For the SFR that we have assumed, we find that, if GRBs are assumed to be "standard candles," the predicted peak photon flux distribution falls steeply throughout the BATSE and HETE-2 regime, and therefore fails to match the observed distribution, in agreement with earlier work. In fact, we find that a photon luminosity function spanning at least a factor of 100 is required in order to obtain semi-quantitative agreement with the principle features of the observed distribution; i.e., a roll-over at a peak photon flux of $P \approx 6$ ph cm^{-2} s^{-1} and a slope above this of about $-3/2$. This implies that there are large numbers of GRBs with peak photon number fluxes below the detection threshold of BATSE and HETE-2, and even of *Swift*.

The lower panel of Figure 3 shows the predicted fraction of bursts with peak photon number flux P that have redshifts of $z > 5$, for the same luminosity functions. This panel shows that a significant fraction of the bursts near the *Swift* detection threshold will have redshifts of $z > 5$.

---

[2] The seven bursts with well-determined redshifts and published peak (isotropic) photon luminosities have a mean peak photon luminosity and sample variance $\log L_P = 58.1 \pm 0.7$.

CONCLUSIONS

We have shown that, if many GRBs are indeed produced by the collapse of massive stars, one expects GRBs to occur out to at least $z \approx 10$ and possibly $z \approx 15 - 20$, redshifts that are far larger than those expected for the most distant quasars. GRBs therefore give us information about the star-formation history of the universe, including the earliest generations of stars. The absorption-line systems and the Lyα forest visible in the spectra of GRB afterglows can be used to trace the evolution of metallicity in the universe, and to probe the large-scale structure of the universe at very high redshifts. Finally, measurement of the Lyα break in the spectra of GRB afterglows can be used to constrain, or possibly measure, the epoch at which re-ionization of the universe occurred, using the Gunn-Peterson test. Thus, GRBs and their afterglows may be a powerful probe of the very high redshift ($z \gtrsim 5$) universe.

REFERENCES

1. Castander, F. J., & Lamb, D. Q., *ApJ* **523**, 593 (1999).
2. Fruchter, A. S., et al., *ApJ* **516**, 683 (1999).
3. Kulkarni, S. R., et al., *Nature* **395**, 663 (1998).
4. Fruchter, A. S., *ApJ* **516**, 683 (1999).
5. Sahu, K. C., et al., *Nature* **387**, 476 (1997).
6. Kulkarni, S. R., et al., *Nature* **398**, 389 (1999).
7. Reichart, D. E., *ApJ* **521**, L111 (1999).
8. Galama, T. J., et al., *ApJ*, submitted, astro-ph/9907264 (2000).
9. Bloom, J. S., et al., *Nature* **401**, 453 (1999).
10. Woosley, S. E., *ApJ* **405**, 273 (1993).
11. Woosley, S. E., in *Gamma-Ray Bursts*, eds. C. A. Meegan, R. D. Preece, & T. M. Koshut, AIP, New York, 1996, pp. 520.
12. Paczyński, B., *ApJ* **494**, L45 (1998).
13. MacFadyen, A. I., & Woosley, S. E., *ApJ* **524**, 262 (1999).
14. Wheeler, J. C., et al., *ApJ*, submitted, astro-ph/9909293 (1999).
15. MacFadyen, A. I., Woosley, S. E., & Heger A., *ApJ*, submitted, astro-ph/9910034 (1999).
16. Gallego, J., *ApJ* **455**, L1 (1995).
17. Lilly, S. J., et al., *ApJ* **460**, L1 (1996).
18. Connolly, A. J., *ApJ* **486**, L11 (1997).
19. Madau, P., Pozzetti, L., & Dickinson, M., *ApJ* **498**, 106 (1998).
20. Ostriker, J. P., & Gnedin, N. Y., *ApJ* **472**, L63 (1996).
21. Gnedin, N. Y., & Ostriker, J. P., *ApJ* **486**, 581 (1997).
22. Valageas, P., & Silk, J., *A&A* **347**, 1 (1999).
23. Rowan-Robinson, M., *Astroph. & Space Sci.*, in press, astro-ph/9906308 (1999).
24. Lamb, D. Q., & Reichart, D. E., *ApJ*, in press, astro-ph/9912368 (2000).
25. Loredo, T. J., & Wasserman, I. M., *ApJ* **502**, 108 (1998).

Gamma-Ray Burst Lensing Limits on Cosmological Parameters

Robert J. Nemiroff[*], Gabriela F. Marani[†], Jay P. Norris[†],
Jerry T. Bonnell[†], Charles A. Meegan[§], and Kevin C. Hurley[‡]

[*]*Michigan Technoloyical University, Department of Physics,
1400 Townsend Drive, Houghton, MI 49931*
[†]*NASA Goddard Space Flight Center, Code 661, Greenbelt, MD 20771*
[§]*NASA Marshall Space Flight Center, SD 50, Huntsville, AL 35812*
[‡]*University of California, Space Science Lab., Berkeley, CA 94720*

Abstract. The lack of detection of a gravitational lens signature on gamma-ray bursts (GRBs) places limits on cosmological geometry and the mass fraction of compact objects in the universe. We report on two new searches we have done for macro- and millilensing of GRBs and their corresponding limits.

INTRODUCTION

Practically all classes of objects visible at cosmological distance are seen to undergo gravitational lensing except, so far, gamma-ray bursts (GRBs). GRBs carry unique qualities, though, that allow their lens statistics to yield information otherwise unobtainable by other forms of lensing and even other probes of cosmology. Sub-minute durations make GRBs capable of *time-resolving* not only galaxy lenses, but compact dark matter upwards of a million solar masses. Large redshifts make GRBs capable of probing lenses into the early universe. Because lensed GRB images are capable of being resolved in both angle and time, they hold promise of yielding relatively detailed information about the lenses and sources (Paczynski 1987).

As with all sources, the detection of GRB gravitational lensing is a "numbers game." One must inspect a high enough number of sources to find typically rare gravitational lensing signatures. Although thousands of GRBs have now been detected, no lens has yet been discovered. Indeed, there are densities and cosmological scenarios where GRB lensing would have been expected. Statistics published here and elsewhere show that in several popular cosmological scenarios GRB lensing is tantalizingly close to detection.

A NEW MACROLENSING SEARCH

We report here preliminary results and theoretical implications of our most recent search for galaxy lenses in GRB data. We believe this search to be the most comprehensive of its kind yet. First, selected GRBs were compared to those within their statistical error boxes in a preliminary visual search. Later, a complete and automated search involving 1892 of the brightest GRBs that triggered the Burst and Transient Source Experiment (BATSE) onboard the Compton Gamma Ray Observatory (CGRO) between April 1991 and July 1999 were considered for a match in light curve and spectral shape. No statistically significant matches were found. Details of this search will be reported published elsewhere.

Why aren't 1892 GRBs enough to win the lensing numbers game? For example, roughly 1 in 500 QSOs has been found to exhibit multiple image gravitational lensing. Are GRBs typically too close to make lensing likely? To test this assumption, we derived likely GRB redshift distributions from Monte-Carlo simulations that includes a fit to the BATSE detection rate, the BATSE Log N - Log P, $< V/V_{max} >$, and the amount of time dilation (Bonnell et al. 1997; Norris 1999 private communication), inclusive of statistical errors. Matching simulation results to BATSE bursts, we obtain GRB redshifts that are similar to those measured for many QSOs.

The answer therefore probably lies in areas unrelated to distance. In fact, GRB gravitational lens statistics are sensitive to blackout areas and times for BATSE, the number of GRBs bright enough for a statistically meaningful comparison, and the relative dynamic range between images to which BATSE is sensitive. Given a model for each of these and a preliminary estimate of the magnification bias, we compute the number of lenses expected in various cosmologies in Table 1.

TABLE 1. Preliminary

(Ω_M, Ω_L)	N_{lens} (No Evolution)	N_{lens} (Lens Evolution)
(0.3, 0)	0.13	0.088
(0.3, 0.7)	0.094	0.062
(1, 0)	0.24	0.16
(0, 1)	0.86	0.80

From Table 1, it is clear that the lack of GRB macrolensing in BATSE data was not very surprising. In the most "lensing-friendly" universe geometry, however, it was slightly more likely that a GRB lens should have been discovered.

To win the GRB lensing numbers game, higher GRB numbers are needed. Many models predict the existence of many sub-BATSE GRBs (see, e.g., Kommers et al. 1997, Nemiroff et al. 1998). As BATSE took up only a fraction of the space onboard CGRO, the barrier to building larger telescopes capable of detecting more GRBs is clearly more financial than technical. More GRBs are not just more GRBs, however. More GRBs likely allow the detection of GRB lensing. These, in turn, could allow astronomers to survey the universe at several interesting epochs and mass scales of lenses, and perhaps even allow for an independent determination of the geometry of the universe.

A NEW MILLILENSING SEARCH

Several prominent cosmologists have hypothesized that dark matter may be hidden in Jean's mass compact objects that condensed at recombination (see, e.g., Carr & Rees 1984; Gnedin & Ostriker 1992). Such compact objects (COs) could create two detectable lens-induced images separated in time by just seconds. The very short durations of some GRBs makes them good sources for the lens detections of such objects (Blaes & Webster 1992). Previously, Nemiroff et al. (1993) searched 44 GRBs for millilensing, and the null detection was used to show the universe was not filled to closure density of such lenses (Nemiroff et al. 1993). Using the same data, the (0.15, 0) cosmology of Gnedin & Ostriker (1992) was marginally ruled out Marani et al. (1999).

In our present search 774 GRBs were inspected for millilensing, a factor of over 15 increase in source number. Many GRBs were discarded from this purely automated search sample because of variable backgrounds or data gaps. Again, no millilensing candidates were found. This null detection when combined with the Monte-Carlo redshifts as described above yields likely limits on millilensing in cosmogenic scenarios.

To compute these limits, we used the detection volume formalism discussed in Nemiroff (1989) and Nemiroff et al. (1993). To be detectable, lenses must fall inside a volume where they would create a detectable gravitational lens magnification. Additionally, lenses must fall inside another volume where the two bright images would be separated in time by less than a maximum amount, and fall *outside* a third volume where the two bright images would be separated in time by less than a minimum amount. Details of this millilensing analysis will be reported in Nemiroff et al. (2000).

Results are shown for two of these cosmologies in Figures 1 and 2. In the figures, the number of expected lens detections is plotted as a function of lens mass and density. The contours represent 1, 5, and 10 expected lens detections, and hence roughly correspond to exclusions at the 1, 2, and 3 σ levels.

The expected number of lens detections in a (0.3, 0.7) universe is shown in Figure 1, on the left. Here one finds a wide range of masses is excluded for several interesting values of Ω_{CO}. In fact, Ω_{CO} dips below even 0.03 for a narrow range of masses near $M/M_\odot \sim 10^7$. Therefore, even a significant amount of baryonic dark matter is excluded from forming compact lenses in this mass range.

To test Gnedin & Ostriker's universe directly, Figure 2, on the right, shows similar results for a (0.15, 0) universe. It appears difficult for any significant fraction of the mass density to hide in super massive compact objects in the mass range $10^6 < M/M_\odot < 10^8$.

Cursory searches for picolensing (Nemiroff & Gould 1995; Marani et al. 1999) were made comparing GRBs detected by both Ulysees and BATSE. Since none were found, a universe filled near closure density with planet sized masses is excluded.

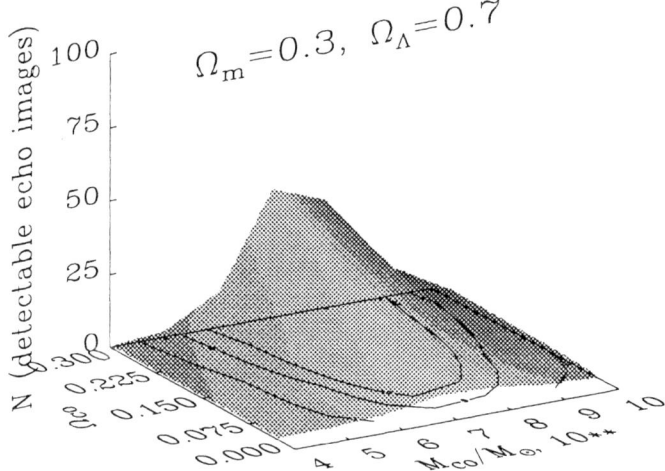

FIGURE 1. The (preliminary) number of expected millilenses as a function of millilens mass and universe geometry.

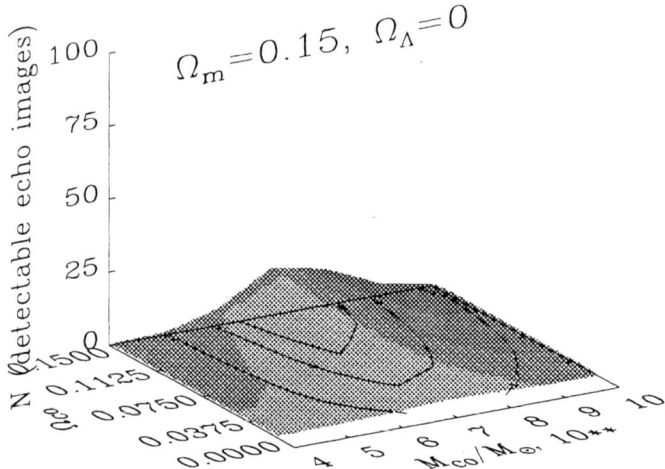

FIGURE 2. The (preliminary) number of expected millilenses as a function of millilens mass and universe geometry.

ACKNOWLEDGMENTS

This research was supported, in part, by grants by NASA. RJN acknowledges additional support from a grant by the NSF. GFM acknowledges additional support from the NRC.

REFERENCES

1. Blaes, O. M., & Webster, R. L., *Ap. J.* **391**, L63 (1992).
2. Bonnell, J. T., Norris, J. P., Nemiroff, R. J., & Scargle, J. D., *Ap. J.* **490**, 79 (1997).
3. Carr, B. J., & Rees, M. J., *Mon. Not. Roy. Astron. Soc.* **206**, 315 (1984).
4. Gnedin, N. Y., & Ostriker, J. P., *Ap. J.* **400**, 1 (1992).
5. Kommers, J., et al., *Ap. J.* **491**, 704 (1997).
6. Marani, G. F., Nemiroff, R. J., Norris, J. P., Hurley, K., & Bonnell, J. T., *Ap. J.* **512**, L13 (1999).
7. Nemiroff, R. J., *Ap. J.* **341**, 579 (1989).
8. Nemiroff, R. J., et al., *Ap. J.* **414**, 36 (1993).
9. Nemiroff, R. J. & Gould, A., *Ap. J.* **452**, L111 (1995).
10. Nemiroff, R. J., Norris, J. P., Bonnell, J. T., & Marani, G. F., *Ap. J.* **494**, 173 (1998).
11. Nemiroff, R. J., et al., *Phys. Rev. Lett.* (2000).
12. Paczynski, B., *Ap. J.*, **317**, L51 (1987).

V. INSTRUMENTATION

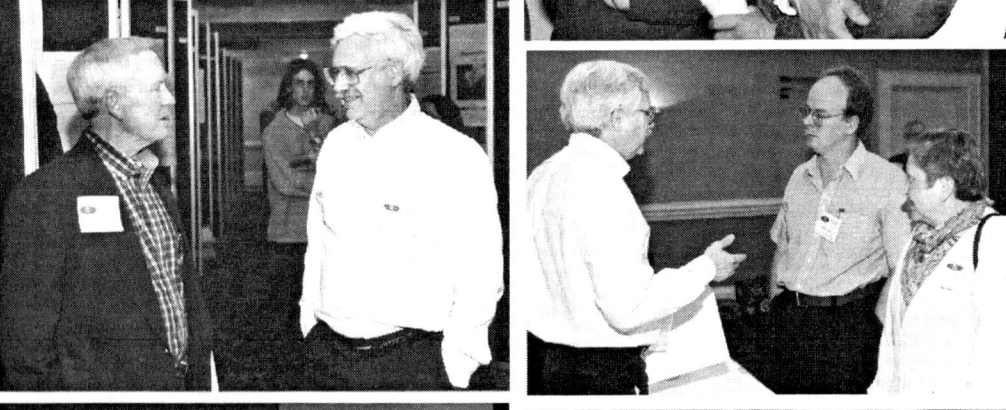

The Swift Gamma-Ray Burst MIDEX

Neil Gehrels

*NASA Goddard Space Flight Center
Greenbelt, Maryland 20771*

on behalf of the Swift Team

Abstract. Swift is a first of its kind multiwavelength transient observatory for gamma-ray burst astronomy. It has the optimum capabilities for the next breakthroughs in determining the origin of gamma-ray bursts and their afterglows as well as using bursts to probe the early Universe. Swift will also perform the first sensitive hard X-ray survey of the sky. The mission is being developed by an international collaboration and consists of three instruments, the Burst Alert Telescope (BAT), the X-ray Telescope (XRT), and the Ultraviolet and Optical Telescope (UVOT). The BAT, a wide-field gamma-ray detector, will detect $\sim$1 gamma-ray burst per day with a sensitivity 5 times that of BATSE. The sensitive narrow-field XRT and UVOT will be autonomously slewed to the burst location in 20 to 70 seconds to determine 0.3–2.5 arcsec positions and perform optical, UV, and X-ray spectrophotometry. On-board measurements of redshift will also be done for hundreds of bursts. Swift will incorporate superb, low-cost instruments using existing flight-spare hardware and designs. Strong education/public outreach and follow-up programs will help to engage the public and astronomical community. Swift has been selected by NASA for development and launch in 2003.

I INTRODUCTION

The discovery by BeppoSAX and ground observers of afterglow [1–3] from gamma-ray bursts (GRBs) has revolutionized our understanding of these enigmatic events. We now know that they are cosmological with redshift $z \gtrsim 0.5$ and involve the most powerful explosions known. These explosions are thought to create super-relativistic blast waves resulting in afterglow that fades from gamma rays to radio. A panchromatic approach to observations of GRBs is now essential for the next phase of discovery.

Much of the information on GRB afterglows is being lost due to the fact that it currently takes $\sim$8 hours or more to point X-ray and optical telescopes to look at burst sources. By this time the fluxes have dropped by a factor of thousands (Figure 1).

In addition, current instruments may not be sensitive to the full diversity of the GRB population. BATSE showed that the distribution of burst durations is bimodal [4,5], with the weaker bursts having, on average, longer durations than the stronger bursts [6]. Short bursts (<1 s) have not been observed by BeppoSAX, so we do not know the nature of this distinct class of GRB.

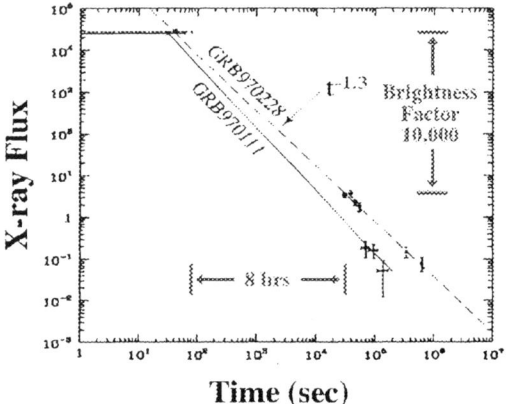

FIGURE 1. The Gap

Swift is a multiwavelength observatory that exploits the newly discovered afterglow characteristics to make a comprehensive study of ~1000 bursts. It will determine the origin of GRBs, tell us how the blast wave interacts with its surroundings, and identify classes of bursts and their associated physical processes. In addition, Swift will investigate how GRBs can be used to study the early Universe.

II SWIFT INSTRUMENTS

The Swift instrumentation was carefully chosen for GRB discovery. It incorporates a wide-field GRB detector, plus two sensitive narrow-field telescopes for identifying and observing the X-ray, UV, and optical afterglow (Figure 2).

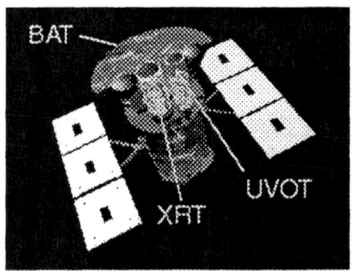

FIGURE 2. Swift Instruments

The *Burst Alert Telescope (BAT)* covers the 10–150 keV energy band and uses a coded aperture mask to provide positions of 1–4 arcmin accuracy depending on burst brightness. The telescope has a CdZnTe (CZT) detector array with an area of 5200 cm^2 and a coded aperture mask that covers 2 sr of the sky. The mask is positioned one meter away from the detectors. The BAT will detect $\sim$300 bursts per year. The large detector area and a sophisticated triggering system that takes advantage of the imaging capability of the instrument allows the BAT to detect bursts of all durations to a sensitivity 5 times better than BATSE (BAT threshold $\sim 10^{-8}$ erg cm^{-2} for a 1 s GRB). The BAT parameters are summarized in Table 1. The instrument is being developed at NASA's Goddard Space Flight Center.

TABLE 1. BAT Parameters

Energy Range	10–150 keV
Aperture	Coded mask
Detecting Area	5200 cm^2
Detector	CdZnTe
Detector Operation	Photon counting
Field of View	2 sr (half-coded)
Detection Elements	256 modules of 128 elements
Detector Size	$4 \times 4 \times 2$ mm^3
Telescope PSF	17 arcmin

The *X-Ray Telescope (XRT)* utilizes flight-spare X-ray optics from the JET-X instrument on the Spectrum X-Γ mission. This mirror has a 15 arcsec half-power diameter at 1.5 keV. The XRT will locate bursts to 2.5 arcsec accuracy. The detector is a 600 square pixel CCD from the XMM program, giving a field of view (FoV) of 24 arcmin square in an energy range of 0.2–10.0 keV. Compared to the BeppoSAX X-ray telescope, the XRT has twice the effective area ($\sim$110 cm^2 @ 1.5 keV) and four times better angular resolution. The XRT parameters are summarized in Table 2. The instrument is being developed at Leicester University, Osservatorio Astronomico di Brera, and Penn State University.

TABLE 2. XRT Parameters

Energy Range	0.2–10 keV
Telescope	JET-X Wolter 1
Detector	EPIC CCD
Effective Area	110 cm^2 @ 1.5 keV
Detector Operation	Photon counting, Integrated imaging, & Timing
Field of View	23.6×23.6 arcmin
Detection Element	600×600 pixels
Pixel Scale	2.36 arcsec
Telescope PSF	15 arcsec HPD @ 1.5 keV
Sensitivity	2×10^{-14} erg cm^{-2} s^{-1} in 10^4 s

The *UV/Optical Telescope (UVOT)* is a 30 cm diameter modified Ritchey-Chrétien equipped with an image intensified CCD covering 170 to 650 nm. It has a FoV 17 arcmin square and is based closely on the design of the XMM Optical Monitor (OM). The UVOT is capable of reaching $m_B = 24$ in 1000 s (open filter). A filter wheel provides 6 colors plus two grisms and a 4× magnifier. The grisms will obtain spectra with resolving power of $\lambda/\Delta\lambda = 200$–$400$ for sources brighter than $m_B = 17$. The optical point spread function of the telescope is 0.3 arcsec, allowing for excellent astrometry. By registering the field against foreground stars, the UVOT will provide <0.3 arcsec positions. The UVOT parameters are summarized in Table 3. The instrument is being developed at Mullard Space Science Laboratory and Penn State University.

TABLE 3. UVOT Parameters

Wavelength Range	170–650 nm
Telescope	Modified Ritchey-Chrétien
Aperture	30 cm diameter
F-number	12.7
Detector	Intensified CCD
Detector Operation	Photon counting
Field of View	17 × 17 arcmin
Detection Element	2048 × 2048 pixels
Telescope PSF	0.3 arcsec @ 350 nm
Colors	6
Sensitivity	B=24 in white light in 1000 s
Pixel Scale	0.5 arcsec

III SWIFT MISSION

The strategy of the Swift mission is to slew to each new GRB position as soon as possible and to follow the GRB afterglows as long as they are visible. To observe the earliest phase of the afterglow, new BAT positions will trigger an autonomous slew of the spacecraft followed by a programmed sequence of observations with the XRT and UVOT.

The initial GRB position is normally determined by the BAT, but positions can also be uploaded from other satellites through a real-time TDRSS uplink. Either case will trigger the spacecraft software to plan and execute an autonomous slew. All calculations of slew path and pointing constraints will be done on-board. Figure of Merit (FoM) software will determine when to slew to a new position. The FoM has a flexible design that can accommodate more focused studies of specific GRB questions as the mission progresses.

Each of the three Swift instruments rapidly produces alert messages after a GRB is detected. To ensure prompt delivery, these messages are sent through a real-time TDRSS downlink to the ground, and routed immediately to the GRB Coordinates Network (GCN) [7] for delivery to the community. Based on the similar process already in place for BATSE, we estimate the entire acquisition and delivery process will take eight seconds for BAT positions and 96 seconds for XRT positions.

The bulk of the data will be downloaded to the Malindi ground station and then sent to the Penn State Mission Operations Center (MOC). The Malindi service is contributed by the Italian Space Agency (ASI).

When Swift is not engaged in prompt observations of the most recent bursts, it will follow a one week schedule uploaded from the ground each working day and as needed. This schedule will provide for long term follow-up of GRB afterglows and other science. The MOC will be capable of generating a new schedule in <2 hours.

TABLE 4. Mission Characteristics

Autonomous slew decision capability
Fast Slew: 50° in <60 s
Low Earth Orbit; 22° Inclination
Launch Vehicle: Delta 7320 with 3 meter fairing
Mass: 1271 kg
Power: 1650 W
Launch Date: October 2003

IV SWIFT SCIENCE

Recent GRB discoveries have shown that X-ray, optical, and radio afterglows exist, continuing for days after the bursts, but fading quickly (t^{-1} to t^{-2} is typical). Better data on faster time scales for many more bursts is needed. (See ref. [8] for discussion of requirements for future GRB missions.) The Swift mission provides the needed capability to answer the following four key science questions: What are the progenitors of GRBs? How does the blast-wave evolve and interact with its surroundings? Are there different classes of bursts with unique physical processes at work? What can GRBs tell us about the early Universe?

A GRB Progenitors

To determine the origin of GRBs three parameters are needed, the total energy released, the nature of the host galaxy (if one exists), and the location within the host galaxy. The Swift mission is optimized to measure all three of these for many hundreds of bursts.

Obtaining the energetics requires a reliable redshift measurement. Ideally, this should be done independently for both the afterglow and any host galaxy to check that there has not been a chance coincidence [9]. The UV grisms and filters of Swift can make redshift determinations by searching for the Ly-α cutoff in the UV and eliminate the $1.3 < z < 2.5$ deadband of current observations [9] during the early phase of the afterglow. In addition, time varying optical, UV and X-ray lines and edges are expected within the first hour following a burst from the illumination of the immediate (100 pc) environment by the initial event [10,11]. The rapid response of Swift will enable a search for predicted X-ray lines (see next section) and again provide a direct redshift measure from the afterglow.

The UVOT will obtain <0.3 arcsec positions by using background stars to register the field. This will provide a unique host galaxy ID and also allow later comparison with HST fields to determine the position within the galaxy.

There will probably be events where no optical afterglow is detected because of dust extinction surrounding the site of the GRB. The position from the XRT will then be crucial. By obtaining 2.5 arcsec positions, the XRT will enable unique identification of candidate host galaxies down to $m_R \sim 26$. Follow-up observations with Chandra made within a couple of days for a selection of these events will give sub-arcsec positions within the Swift 2.5 arcsec error circle.

B Blast-Wave Interactions

Afterglow is thought to be produced by the interaction of an ultra-relativistic blast-wave with the interstellar or intergalactic medium. The blast-wave model [12] predicts a series of stages as the wave slows. A key prediction is a break in the spectrum that moves from the gamma-ray to optical band, and is responsible for the power law decay of the source flux. This break moves through the X-ray band in a few seconds, but takes up to 1000 s to reach the optical. Thus, observations within the first 1000 s in the optical and UV are crucial to see this early phase. While it now seems likely that all the GRBs have X-ray afterglow, not all have bright optical afterglow (at least after several hours). This may be due to optical extinction, but it is also possible that in some cases the optical (and X-ray) afterglow is present, but decays much more rapidly, [13] and is a function of the density of the local environment [14]. Prompt high-quality X-ray, UV and optical observations over the first minutes to hours of the afterglow (inaccessible without Swift) are crucial to resolve this question. Continuous monitoring is important since model-constraining flares can occur in the decaying emission.

Star forming regions are embedded in large columns of neutral gas and dust. The presence of extinction can be readily determined by multi-band photometry in the optical and IR. The simultaneous detection of high X-ray absorption, coupled with photometric $E(B-V)$ measurements with Swift, will determine whether dust and gas are present. Continuous monitoring over the first few hours to days will indicate whether dust is building up (due to condensation out of an expanding hot wind) or disappearing (due to ablation and evaporation).

C Classes of GRBs

Swift will determine whether sub-classes of GRBs exist and what fundamental differences in the source physics cause the classes. While some evidence of sub-classes has been obtained (e.g., bimodal duration distribution, possible correlation of hardness and $\log N$–$\log P$ shape, short-bursts having $V/V_{\max}$ consistent with a Euclidean distribution) it is not clear if these are real differences in physical phenomena or simply represent the distribution function of GRB properties such as beaming angle, density of the local medium, or initial energy injection. Swift data will determine locations, redshifts, and afterglow properties of the different classes and thus allow physical understanding of their nature. Central to the confusion regarding potential classes is that we do not have a reliable standard candle. Swift remedies that by directly measuring the distance through redshift. This will give an exact determination of the GRB luminosity function.

Since BeppoSAX does not presently detect bursts shorter than ~ 1 second, we have no idea of the nature or even existence of optical, X-ray, or radio afterglows for these objects. Swift will be sensitive to the shortest events, and will provide far better coverage of these events than has been possible.

In the interesting scenario that Swift discovers some GRBs with no X-ray or UV/optical afterglow, the BAT will still provide positions of 1–4 arcmin, which is sufficient to look for radio or IR counterparts. Only the rapid response of Swift will be able to identify such a new and elusive sub-class of GRB event.

If there are classes of GRBs that are the signal of conventional supernova explosions (e.g., refs. [17,18]), the UVOT will provide unique and unprecedented coverage of the optical and UV light curve during the early stage.

D Bursts as Astrophysical Tools

Since the lifetime of NS–NS or massive star progenitors is short compared to the Hubble time at $z < 5$, the cosmic GRB rate should be proportional to the star formation rate. The cosmic rate of massive star formation is at present controversial. Estimates that star formation peaks at $z \sim 1$–2 and declines sharply at high redshifts have been reported [19]. However, recent IR [20,21] and X-ray cluster [22] results show a considerably higher rate in dust enshrouded galaxies at higher redshifts. Swift, by obtaining a large sample of GRBs over a wide range of fluences and redshifts, will determine whether their evolution follows that of star formation in the Universe and, because the X-ray flux does not depend greatly on the line of sight column, these results will be independent of absorption.

Since GRBs are the most luminous objects in the Universe, they provide a unique opportunity to probe the intergalactic medium (IGM) and the ISM of the host galaxies via measurement of absorption along the line of sight [23]. Depending on evolution, GRBs might originate from redshifts up to ~ 15 and have a median redshift > 2, larger than that of any other observable population. By rapidly providing both accurate positions and optical brightness, Swift will enable the immediate follow-up of those GRBs bright enough for high resolution optical absorption line spectroscopy at redshifts large enough to study the reionization of the IGM [24]. This information on the high-z Ly-α forest will be unique because there are currently no known bright (m < 17) galaxies or quasars at $z > 4.8$ [23].

V GROUND SYSTEM AND DATA ANALYSIS

A layered data analysis approach will be used to achieve rapid dissemination of Swift results and data to the community. The most urgently needed results, namely GRB positions, are produced on the spacecraft. Quicklook results, including optical finding charts and multiwavelength light curves, are produced in the MOC in near real-time and distributed using the GCN. Definitive standard products, including spectra, multi-band light curves, and images, will be made into production FITS files. This data base of Swift results will be augmented with contributed results from other observers.

All the Swift data will be processed at the Swift data center at Goddard and will be made available to the general public through the HEASARC in the US, a data center in the UK, and a data center in Italy. The end result will be easy access for the entire community to a broad range of timely information on GRBs.

VI SCIENCE TEAM

The Swift science team is made up of world experts in GRB astronomy, space instrumentation, data analysis, theory, GRB follow-up observation, and outreach. The Co-Investigator list with roles and responsibilities is given in Table 5.

An essential Swift capability is providing the world-wide community of ground and space observers with rapid arcsec positions for hundreds of GRBs. Until now such work has been based on less than 10 GRBs/yr, with no information on the optical appearance of the field, and lagging at least 5–8 hours after the burst. We have assembled a Follow-up Team under the leadership of Kevin Hurley, to use their expertise on large facilities to guarantee systematic study of Swift GRBs. These, and other scientists, are Associate Scientists for the Swift mission (Table 6).

We will actively encourage follow-up observations of all kinds, regardless of membership in the Swift team, by providing precise positions and other data in real-time to the GCN. The GCN founder, Scott Barthelmy, is a member of the Swift team. Observers will be free to use the data with no restrictions. We will encourage all observers to make their data public by maintaining a web database.

ACKNOWLEDGMENTS

I thank John D. Myers for assistance in writing this paper. I gratefully acknowledge the Swift Science Team for preparing the Swift Proposal and Phase A study on which this paper is based, and the Swift Management and Engineering Team led by Tim Gehringer for the technical development of the mission.

TABLE 5. Swift Co-Investigators

Team Member	Role/Responsibility	Institution
Neil Gehrels	Principal Investigator	GSFC
John Nousek	Narrow-Field Instruments Lead	PSU
Nicholas White	Science Working Group Chair	GSFC
Scott Barthelmy	BAT Co-Lead, Development	GSFC
Louis Barbier	BAT Detector Electronics	GSFC
Ann Parsons	BAT Co-Lead, Fabrication-Calibration	GSFC
Ed Fenimore	BAT Flight Software	LANL
Francois Lebrun	BAT Detectors	CEN-Saclay
David Palmer	BAT Software	GSFC
David Burrows	XRT Lead	PSU
Alan Wells	XRT System	Leicester Univ
Richard Willingale	XRT Calibration	Leicester Univ
Oberto Citterio	XRT Mirrors	Brera (OAB)
Leisa Townsley	UVOT Lead	PSU
Keith Mason	UVOT Telescope	MSSL
Mark Cropper	UVOT Detectors	MSSL
Tim Sasseen	UVOT OM Liaison	UCSB
Scott Horner	UVOT DPU Heritage	Lockheed-Martin
Frank Marshall	GS Lead	GSFC
Lorella Angelini	GS Archive	GSFC
Patrizia Caraveo	GS Malindi	IFC/CNR
Margaret Chester	GS Penn State Ops	PSU
Paolo Giommi	GS Archive	BeppoSAX SDC
Keith Jahoda	GS Hard X-ray Survey	GSFC
Alan Smale	GS Science Center	GSFC
Gianpiero Tagliaferri	GS Italian Archive Center	Brera (OAB)
Jack Tueller	GS Hard X-ray Survey	GSFC
Peter Mészáros	STT Lead	PSU
Robin Corbet	STT	GSFC
Guido Chincarini	STT	Brera (OAB)
Tom Cline	STT	GSFC
France Cordova	STT	UCSB
Gordon Garmire	STT	PSU
Dale Frail	STT	NRAO/VLA
Rich Mushotzky	STT	GSFC
Jay Norris	STT	GSFC
Bohdan Paczyński	STT	Princeton Univ
Jacques Paul	STT	CEN-Saclay
Luigi Stella	STT	OSS di Roma
Martin Turner	STT	Leicester Univ
Mario Vietri	STT	Univ di Roma
Will Zhang	STT	GSFC
Kevin Hurley	Followup Team Lead	UC Berkeley
Laura Whitlock	Education and Outreach Manager	Sonoma State
Eric Feigelson	Education and Outreach Team	PSU
Lisa Brown	Education and Outreach Team	PSU
Lynn Cominsky	Public Relations	Sonoma State
GS = Ground System	STT = Science Theory Team	

TABLE 6. Swift Associate Scientists

Member	Association	Member	Association
Michel Boer	CESR	Hye-Sook Park	LLNL
Niel Brandt	PSU	Holger Pederson	Copenhagen Univ
Michael Busby	Tenn State	Martin Rees	IoA, Cambridge
Andrea Cimatti	OAA	Paolo Saracco	Brera (OAB)
Massimo Della Valle	Obs Astro, Padova	Bradley Schaefer	Yale Univ
Alex Filippenko	UC Berkeley	Don Schneider	PSU
Gabriele Ghisellini	Brera (OAB)	Ian Smith	Rice Univ
Piero Madau	STScI	Christopher Stubbs	Univ Washington
Bruce Margon	Univ Washington	Chris Thompson	Univ North Carolina
Mark Metzger	CalTech	Fred Vrba	USNO

REFERENCES

1. Costa, E., et al., *Nature* **387**, 783 (1997).
2. Van Paradijs, J., et al., *Nature* **386**, 686 (1997).
3. Frail, D.A., et al., *Nature* **389**, 361 (1997).
4. Kouveliotou, C., et al., *ApJ* **413**, L101 (1993).
5. Klebesadel, R.W., in *Proc. Los Alamos Workshop on GRBs*, eds. C. Ho, R. Epstein, & E. Fenimore, Cambridge Univ. Press, Cambridge, 1992, p. 161.
6. Norris, J.P., et al., *ApJ* **439**, 542 (1995).
7. Barthelmy, S., et al., in *Proc. Fourth Huntsville GRB Workshop*, eds. C. Meegan, R. Preece, & T. Koshut, AIP, New York, 1998, p. 99.
8. Gehrels, N., in *Cosmic Explosions*, eds. S. Holt & W. Zhang, AIP, New York, 2000, in press.
9. Hogg, D.W. & Fruchter, A.S., *ApJ* **520**, 54 (1998).
10. Perna, R. & Loeb, A., *ApJ* **501**, 467 (1998).
11. Mészarós, P. & Rees, M., *ApJL* **502**, L105 (1998).
12. Mészarós, P. & Rees, M., *ApJL* **418**, L59 (1993).
13. Groot, P.J., et al., *ApJL* **502**, L123 (1998).
14. Piran, T., astro-ph/9807253 (1998).
15. Fenimore, E., Madras, C., & Nayakshin, S., *ApJ* **473**, 998 (1996).
16. Fenimore, E., et al., astro-ph/9802200 (1998).
17. Bloom, J.S., et al., *Nature* **401**, 453 (1999).
18. Woosley, S.E., et al., *ApJ* **516**, 788 (1999).
19. Madau, P., et al., *MNRAS* **283**, 1388 (1996).
20. Blain, A.W., et al., astro-ph/9806062 (1998).
21. Rowain-Robinson, M., et al., *MNRAS* **289**, 490 (1998).
22. Mushotsky, R.F., & Lowenstein, M., *ApJL* **481**, L63 (1997).
23. Lamb, D.Q. & Reichart, D.E., astro-ph/9909002 (1999).
24. Miralda-Escide, J., *ApJ* **501**, 15 (1998).

Observing Gamma-Ray Bursts with INTEGRAL

Christoph Winkler[1]

Astrophysics Division, Space Science Department, ESA-ESTEC
2200 AG Noordwijk, The Netherlands

Abstract. The International Gamma-Ray Astrophysics Laboratory INTEGRAL is dedicated to the fine spectroscopy, accurate imaging, and arcminute location of celestial gamma-ray sources in the energy range 15 keV to 10 MeV. Simultaneous monitoring of the GRB site in X rays and in the optical V-band will be provided by two on-board monitors. Gamma-ray bursts will be discovered both during routine program observations as serendipitous sources in the large FOVs and as high-resolution time histories from the spectrometer anticoincidence system, forming part of the interplanetary network of GRB detectors. The observational capabilities of INTEGRAL further include the dissemination of INTEGRAL GRB alerts and associated data to the science community at large in order to facilitate rapid follow-up observations of GRB error boxes for counterpart searches, afterglow observations, and subsequent investigations.

INTRODUCTION

INTEGRAL [8] is ESA's next major 15 keV – 10 MeV gamma-ray mission, in collaboration with Russia (PROTON launcher) and NASA (ground stations). The observatory is scheduled for launch in 2001, with a lifetime of two years to up to five years. The scientific objectives focus around high energy resolution spectroscopy and fine imaging/location of gamma-ray sources and include: study of compact objects; explosive and hydrostatic nucleosynthesis; high-energy transients and GRBs; mapping of the Galactic structure (Galactic Center) and ISM; normal galaxies and clusters; AGN, Blazars and Seyferts; cosmic diffuse background and identification of high-energy sources. The payload consists of a Germanium spectrometer (SPI) optimized to perform high-resolution spectroscopy of gamma-ray lines and observations of the large scale diffuse emission, a CdTe/CsI imager (IBIS) optimized to provide fine point-source images with accurate locations and sensitive studies

---

[1] This paper is written on behalf of the INTEGRAL Science Working Team consisting of: T. Courvoisier, N. Gehrels, A. Gimenez, S. Grebenev, W. Hermsen, F. Lebrun, N. Lund, G. Palumbo, J. Paul, R. Sunyaev, P. Ubertini, V. Schönfelder, G. Vedrenne, and C. Winkler.

of continuum and broad lines, and two monitors in the X-ray (JEM-X) and optical (OMC) energy range that identify and monitor high-energy sources at the low end of the INTEGRAL energy band. All instruments on-board INTEGRAL are co-aligned with overlapping FOVs and they operate simultaneously. The three high-energy instruments utilize coded aperture masks. A particle radiation monitor complements the payload. The key features of the instruments are shown in Table 1. INTEGRAL[2] will be launched in 2001 by a Russian PROTON launcher into a High Earth Orbit (72 hr period, 10,000 km (initial) perigee, 153,000 km (initial) apogee, 51.6° inclination) allowing operation of the science instruments entirely outside the Earth's trapped proton belt, and largely outside the electron belts. Nominally, the orbit above 40,000 km altitude (90% of orbital time) will be used for real-time scientific observations (downlink science data-rate = 86 kbps).

GRB DETECTIONS AND OBSERVATIONS

From the scientific objectives and the instrument characteristics in Table 1 it follows that GRBs are one important scientific objective for INTEGRAL, but the payload is not specifically designed to support primarily a GRB mission. However, INTEGRAL is well-suited to provide important new observational data on the GRB phenomenon. In particular, INTEGRAL will provide GRB locations to arcminute accuracy or better; spectral coverage from 3 keV (JEM-X) up to 10 MeV (IBIS, SPI) with concurrent optical monitoring and high-resolution spectroscopy (2 keV FWHM @ 1.33 MeV). Recent observations of optical, X-ray and radio counterparts, as discussed during this Symposium, show, as expected, that major progress in understanding the GRB phenomenon comes from accurate locations, broad band coverage, and rapid response.

High-Energy Detections and Optical Follow-Up (On-Board)

SPI, IBIS and JEM-X would detect GRBs in their nominal operational modes (on-ground, photon-by-photon events). The expected number of GRBs per year detected within the fully coded field of view has been estimated to 13 (SPI), about 6 to 24 (IBIS) — where the lower and upper limits are given by the BATSE detection rate and extrapolation of the BATSE $\log N - \log P$ distribution [6] — and 2 to 6 (JEM-X). IBIS will be able to locate GRBs with its upper CdTe detector area (2621 cm^2) to $< 30''$ (Table 1). The IBIS GRB location will be determined on-ground (Figure 1) at the INTEGRAL Science Data Centre [1,4] from the near real-time data. Then the source position for the optical monitor OMC is calculated and the CCD co-ordinates are automatically uplinked to the OMC instrument. This automatic procedure allows centering a new CCD read-out window of the

---

[2] Further details on INTEGRAL can be found in [8] and
http://astro.estec.esa.nl/SA-general/Projects/Integral/integral.html

TABLE 1. Key Parameters of the INTEGRAL Scientific Payload

	SPI	IBIS	JEM-X	OMC
Energy range	20 keV – 8 MeV	15 keV – 10 MeV	3 keV – 35 keV	500 nm – 600 nm
Detector area (cm^2)	500	2600 (CdTe) 3100 (CsI)	1000 (2 units each 500)	CCD (2048×1024 pxl)
Spectral resolution (FWHM, keV)	2 @ 1.3 MeV	7 @ 100 keV 60 @ 1 MeV	1.5 @ 10 keV	–
Field of view (fully coded)	16°	9° × 9°	4.8°	5.0° × 5.0°
Angular resolution	2° FWHM	12′ FWHM	3′ FWHM	17.6″/pixel
10σ source location	20′	< 0.5′	< 20″	< 8″
Continuum sensitivity[a]	7×10^{-8} @ 1 MeV	4×10^{-7} @ 100 keV	1×10^{-5} @ 6 keV	19.7^{m_v} (3σ, 10^3 s)
Line sensitivity[b]	5×10^{-6} @ 1 MeV	1×10^{-5} @ 100 keV	2×10^{-5} @ 6 keV	–
Timing accuracy (3σ)	100 μs	67μs – 1000 s	128 μs	> 1 s

[a] Units are (ph cm^{-2} s^{-1} keV^{-1}) for 3σ detection in 10^6 s.
[b] Units are (ph cm^{-2} s^{-1}) for 3σ detection in 10^6 s.

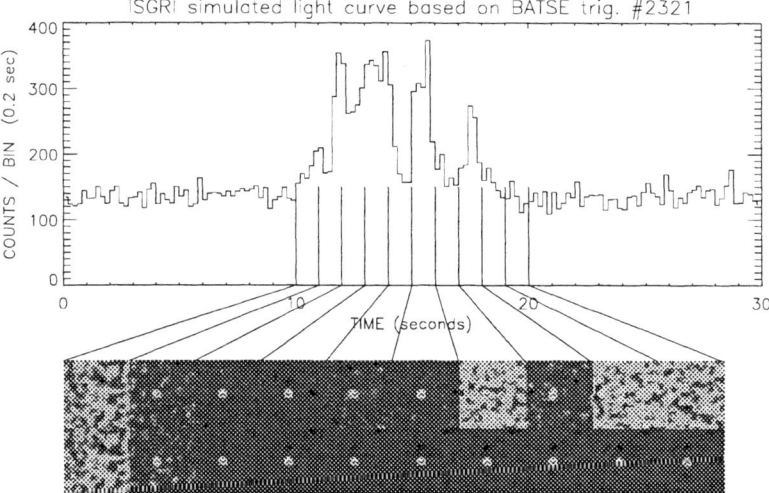

FIGURE 1. Simulated GRB in the IBIS (CdTe array ISGRI) fully coded FOV according to [4]. Photon arrival times are generated based on a BATSE light curve. The peak flux is 0.85 ph/(cm^2 s) (50–300 keV) and a double power-law spectrum is assumed ($\alpha = -0.7$ and E$_{break}$ = 100 keV). The upper row displays 15–300 keV images of 1 s integration each, while the images in the lower row are for increasingly longer time intervals (from 1 s to 10 s). With a monitoring in 50 ms time bins and a threshold at 7σ, this GRB triggered at T = 11.7 s. Less than 2 s later, the peak in the deconvolved image (third image from left in bottom row) has reached a S/N of ∼ 20.

OMC rapidly (few seconds up to ∼ minute) on the GRB position to allow for near-simultaneous optical observation of the GRB event and its possible afterglow.

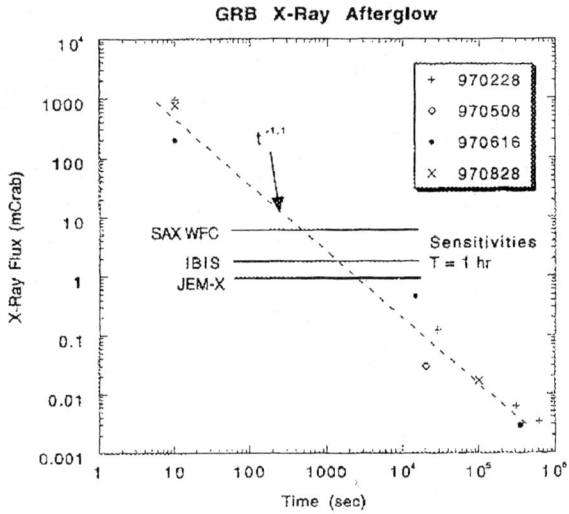

FIGURE 2. INTEGRAL (IBIS/JEM-X) sensitivity to GRB afterglow emission.

Studies of Afterglow Emission

Routine INTEGRAL observations (targeted pointings of up to typically 14 days) allow in principle the observation of pre- and post-GRB emission in the gamma-ray, X-ray and optical domains. Due to specific requirements by the SPI instrument to improve its coded mask imaging, the spacecraft will perform, during many pointings, a number (7–25) of small (2°) dithering steps around the nominal target position along a hexagonal or rectangular pattern. This increases the effective sky area viewed by the instruments considerably. Quiescent or afterglow counterparts can therefore be captured during nominal target pointings. Alternatively, the spacecraft can react to GRBs outside its FOV through the standard Target of Opportunity mechanism that allows a re-scheduling of the observing program and re-pointing of the spacecraft within ∼30 hours, on average. It is noted that INTEGRAL will perform, as part of its Core Program [9], weekly scans of a $b = \pm 20°$ wide strip along the entire Galactic plane to detect and monitor high-energy transients (e.g., GRS 1915+105, etc.), and GRBs would be covered due to their spatial and temporal serendipity. The prospects for observing the fading counterpart with INTEGRAL are promising: X-ray afterglow data for 4 GRB counterparts in 1997 indicate a generic $t^{-1.1}$ decay curve up to 10^6 s after the event. IBIS and JEM-X would have sufficient sensitivity to discover afterglow emission on ∼ mCrab level up to about one hour after the event (Figure 2, [7]).

Data for GRB Arrival Time Analysis

INTEGRAL will also use the SPI anticoincidence subsystem [2] to provide GRB event time histories (50 ms binning, $E > 100\,\text{keV}$) for ~ 350 GRBs/year (10σ, 4π FOV) contributing to the Fourth Interplanetary Network (IPN). A GRB event of moderate strength would produce a typical annulus width of $\sim 7''$, assuming Ulysses at a distance of 4.3 AU. These SPI-provided GRB time series would be made available to subscribers of the INTEGRAL GRB alert system described below.

INTEGRAL GRB ALERTS AND THE SCIENTIFIC COMMUNITY

Following the successful utilisation of the BACODINE/GCN network, it is planned to implement a similar GRB alert system for INTEGRAL as described in detail by [5]: Its purpose is to faciliate rapid follow-up observations. Based on INTEGRAL detections, a largely automatic system at the INTEGRAL Science Data Centre (ISDC, [1]) in Geneva monitors the real-time telemetry in order to detect, localize, and validate an event before an alert is distributed to the community within seconds after the GRB onset. The disseminations of various alerts (as a function of time) will contain increasingly more improved data on the location accuracy: the GRB location accuracy depends on GRB detections performed during (i) spacecraft slews (predicted pointing $\sim 10'$ using slew path simulator); (ii) at start of a stable pointing (predicted position based on previous slew, i.e., $\sim 30''$ for $\leq 2°$ slews and $\sim 5'$ for $\geq 2°$ slews); and (iii) from 10 minutes after start of stable pointing until the end of stable pointing (on-ground attitude reconstitution $\leq 30''$).

Dissemination of alerts are expected to occur in the few seconds to minutes time frame (see above), and we note that a significant number of GRBs, as deduced from the BATSE 3B distribution of durations [3], have durations longer than 14 s.

REFERENCES

1. Dubath, P., et al., in *Proc. 5th Compton Symposium*, AIP, in press (1999).
2. Lichti, G., et al., in *Proc. 5th Compton Symposium*, AIP, in press (1999).
3. Meegan, C., et al., *ApJSS* **106**, 65 (1996).
4. Mereghetti, S., et al., in *Proc. 3rd INTEGRAL Workshop*, in press (1999).
5. Mereghetti, S., et al., these proceedings.
6. Skinner, G., et al. in *Proc. 2nd INTEGRAL Workshop*, ESA-SP 382, 487 (1997).
7. Teegarden, B., *Minutes INTEGRAL Science Working Team Meeting*, #9 (1997).
8. Winkler, C., in *Proc. 3rd INTEGRAL Workshop*, in press (1999).
9. Winkler, C., et al., in *Proc. 3rd INTEGRAL Workshop*, in press (1999).

The INTEGRAL Burst Alert System

S. Mereghetti[1], S. Brandt[2], D. Jennings[3], J. Borkowski[3], and R. Walter[3]

[1] Istituto di Fisica Cosmica G.Occhialini - CNR, Milano, Italy
[2] Danish Space Research Institute, Copenhagen, Denmark
[3] Integral Science Data Center, Versoix, Switzerland

Abstract. ESA's INTEGRAL high-energy observatory is scheduled for launch in 2001. INTEGRAL carries two gamma-ray instruments, optimized respectively for spectroscopy and high-resolution imaging, complemented by an X-ray and an optical monitor. The high sensitivity of the INTEGRAL instruments will allow the detection and detailed studies of relatively faint gamma-ray bursts.

The INTEGRAL Burst Alert System (IBAS) is implemented as a ground-based system, working on the near-real time telemetry stream. It is expected that the system will detect more than one GRB per month in the field of view of the main instruments. Positions with an accuracy of a few arc-minutes will be distributed to the community for follow-up observations within a few tens of seconds of the event. Furthermore, the system will upload commands to optimize the detection of an associated transient with the INTEGRAL optical monitor.

INTRODUCTION

INTEGRAL is a major gamma-ray astronomy mission of the European Space Agency currently under development, with a launch scheduled for the end of 2001 [1]. INTEGRAL is devoted to high-resolution spectroscopy and high angular resolution imaging of the γ-ray sky, together with a broad coverage of the electromagnetic spectrum. These goals will be achieved by two main instruments (IBIS and SPI) optimized respectively for the angular and spectral resolution, complemented by two monitors in the X-ray (JEM-X) and optical (OMC) bands. The four instruments are co-aligned to simultaneously observe a large region of the sky. IBIS, SPI and JEM-X make use of the coded mask principle to obtain images at wavelengths where focusing techniques are impractical.

Thanks to its highly eccentric orbit with a 72-hour period, INTEGRAL will spend most of the time outside the radiation belts and will almost continuously downlink the collected data. To optimize the imaging capabilities of the SPI instrument and to reduce background systematic effects, the pointing strategy will usually consist of raster scan observations, covering predefined patterns of directions around the main target. INTEGRAL will also devote a large fraction of its Core Program to scans of the Galactic Plane [2]. As a consequence, a non-negligible fraction of the total observing time will consist of slews.

THE INTEGRAL INSTRUMENTS

IBIS [3] covers the energy range from 15 keV to 10 MeV thanks to two different position sensitive detection planes: a top layer composed of an array of 128×128 CdTe pixels (2 mm thick) operating in the lower energy range, followed by a second layer made of CsI scintillators (64×64 pixels, 3 cm thick) that extends the response to higher energies. A tungsten coded mask, placed $\sim$3.1 m from the detection plane, will provide an angular resolution of 12′ over a wide field of view (9°×9° full sensitivity, 19°×19° half sensitivity, 29°×29° zero response). The wide field of view and high angular resolution will make IBIS the most useful instrument for GRB studies. The data from its first layer (also known as ISGRI) will be in photon by photon mode, i.e., position, arrival time and energy information will be available for all the events. Due to telemetry limitations, the CsI layer (PICSIT) will normally work by integrating images (typically $\sim$1000 s long), and it is therefore less useful for the study of rapidly varying objects.

SPI [4] is optimized for detailed measurements of γ-ray lines and the mapping of diffuse sources with an angular resolution of $\sim$2°. The cooled Germanium detectors (19 pixels arranged in a hexagonal pattern) will allow precise measurements ($E/\Delta E \sim 500$ at 1 MeV) of the γ-ray energies over the 20 keV to 8 MeV range. Also, SPI will always operate in photon by photon mode. Its field of view is similar to that of IBIS, and even if its angular resolution is worse, it can be used to confirm GRB detected with IBIS, and possibly to discover events that for some reason did not trigger in IBIS.

Another important characteristic of SPI is related to the possibility of using its anti-coincidence shield (ACS) as a GRB detector. The SPI-ACS consists of about 512 kg of BGO scintillators, providing good sensitivity especially for GRB at large off-axis angles. The ACS will continuously produce a single light curve with the overall counting rate in time bins of 50 ms. Though no directional information will be available, the GRB detected in the ACS will allow the use of INTEGRAL as a near Earth spacecraft in the interplanetary network (IPN) for GRB location [5]. The expected rate of GRB detections in the ACS is $\sim$300 per year.

The JEM-X Monitor [6] will yield imaging with angular resolution of 3′ in the 3–120 keV band over a $\sim$5° fully coded field of view. It can be used to obtain a more precise localization of those GRB occurring in the central part of the IBIS/SPI field of view, that are bright enough in this energy range.

The OMC [7] will observe the optical emission from the prime targets of the other instruments. It consists of a 50 mm aperture telescope equipped with a passively cooled CCD. Its field of view is of 5°×5° and the pixel size is $\sim$17″. During normal operations only a subset of windows in the OMC field of view will be transmitted to the ground, corresponding to the position of preselected sources. Following the IBAS detection of a possible GRB in the central region of the IBIS/SPI field of view, a telecommand requesting a new window to cover the GRB location will be automatically sent to the OMC. The expected limiting magnitude is a function of the galactic latitude. For high latitudes it is of the order of V = 18–19 for a few hundred seconds exposure.

THE BURST ALERT SYSTEM

IBAS Overview

Since no onboard GRB detection system is foreseen on INTEGRAL, the search for GRBs will be performed on the ground by means of a near real time analysis done by the IBAS [8,9] at the INTEGRAL Science Data Center (ISDC) [10]. This gives the advantage of the greater computing power available on ground that allows the implementation of more complex, sensitive and flexible algorithms. Furthermore, the triggering criteria can be more easily tuned to minimize false alarms and to optimize the detection of GRBs with different characteristics.

Burst triggering shall result from the monitoring of ISGRI, SPI and JEM–X event data, e.g., by looking for significant excesses with respect to a running average, much like traditional on-board triggering. This will be done in parallel for different instruments, energy ranges and time bin durations. When a candidate event is detected, a process of image analysis shall start to verify the origin of the count rate variation and to assure that it was not caused by an instrumental malfunction or a background variation. Images shall be accumulated for different time intervals, deconvolved with very fast algorithms, and compared to the pre-burst reference images in order to detect the appearance of the GRB as a new source. Simulations have shown that in most cases it is possible to localize the GRB position in the ISGRI deconvolved images within less than one 1 s from the trigger time [8]. Of course this is dependent on the GRB light curve profile and on its intensity and location in the field of view. The image analysis will be based on the time intervals, derived from the GRB light curve, that optimize the signal to noise ratio.

After a candidate GRB has been detected, other automatic processes shall validate the event before an external alert is sent. If the event is genuine then the satellite attitude information is applied to derive a sky position that is then automatically transmitted to all the subscribed users (by e-mail and/or direct TCP/IP socket). In addition, if the GRB is located in the sky region covered by the OMC, a telecommand will be sent to the satellite to reprogram the CCD windows in such a way to cover the GRB error box. Because full event validation and localization might require a longer time, we foresee different levels of alert messages providing increasingly accurate and reliable information. For example, the errors on the GRB position can be reduced by analyzing images derived from the secondary offline processes.

Location Accuracy

The location accuracy of coded mask imaging systems depends on the signal to noise ratio of the source. For sources detected with a high statistical significance the accuracy can be a small fraction of the angular resolution. Theoretical evaluations, confirmed by several independent simulations, have shown that for ISGRI a source location accuracy smaller than 30" (90% confidence level) can be obtained for a signal to noise ratio of 30. In such cases, the final accuracy on the GRB location is also affected by the uncertainties on the satellite attitude (see below). For most of the detected GRB, however, we expect smaller signal to noise values, resulting in typical uncertainties, dominated by the photon statistics, of the order of $\sim$2-3'.

The expected INTEGRAL attitude accuracy depends on three cases: (1) attitude during slews, (2) attitude at the start of a stable pointing period, and (3) attitude from about ten minutes after the start to the end of a stable pointing period.

For case (1) a slew path simulator generates predictions of the pointing direction every 10 seconds based upon the expected slew start and end times. In most circumstances these predictions should be accurate to within 5% of the slew path. The average slew rate during the Galactic Plane scans and during the normal dithering observations will be of the order of only $\sim 1°$/min. Therefore, it will be possible to detect short GRBs even during slews, although their location accuracy will be worse.

For case (2) the attitude is based upon a prediction of the expected position with respect to the previously commanded slew. In most cases (i.e., slews shorter than $2°$) these predicted values will have an accuracy of the order of the star-tracker/instrument alignment uncertainty (30" or less). For larger slews the predicted attitude will have an error smaller than $5' \pm 1'$ (3σ).

Approximately 5 minutes after the start of a pointing (case 3) a "snapshot" attitude reconstruction will be completed based upon the first down-linked star-tracker map. The result shall be made available to the IBAS within a maximum delay of 10 minutes since the start of the stable pointing, yielding, as above, an attitude with accuracy of $\lesssim 30''$.

Time Performance

Under normal circumstances telemetry frames are expected at the ISDC within 30 s. In general, IBAS analysis might require between ~ 1 and ~ 20 s (the time for a GRB event triggering and validation is of course partially dependent upon the rise time of the event itself). Thus, in many cases, we foresee to be able to generate *first level* alerts while the GRB is still ongoing. Of course a trade-off has to be made between accuracy of the results (i.e., confidence of a real GRB vs. false trigger, dimensions of the error region, etc.) and the rapidity of the alert. For this reason IBAS will automatically generate, for the same GRB event, several alert messages with increasingly more accurate results. These messages will be configured in such a way as to allow an easy filtering by the users in order to react only to the situations that best fit their needs.

CONCLUSIONS

Although INTEGRAL was not specifically designed to study GRBs, its good imaging properties coupled with the almost continuous telemetry downlink, allow accurate localizations of GRBs in near real time [8]. Since no onboard GRB detection system will be available, the search for GRBs will be performed on the ground by means of an automatic analysis performed at the INTEGRAL Science Data Center. We expect to detect and localize accurately $\gtrsim 1$ GRB per month [11], and to quickly disseminate its coordinates for rapid follow-up observations at other wavelengths. Furthermore, the greater flexibility affordable with a ground based system will allow the detection of GRBs with different properties (e.g., slow rise time) that up to now failed to trigger GRB detectors.

REFERENCES

1. Winkler C., et al., *Proc. 3rd INTEGRAL Workshop "The Extreme Universe"* A.Bazzano ed., *Astroph. Letters & Communications*, **38**, 309, (1999).
2. Gehrels N. et al., *Proc. 2nd INTEGRAL Workshop "The Transparent Universe"*, C.Winkler, T.Courvoisier & Ph. Durouchoux eds., **ESA-SP 382**, 587 (1997).
3. Ubertini P. et al., *Proc. 2nd INTEGRAL Workshop "The Transparent Universe"*, C.Winkler, T.Courvoisier & Ph. Durouchoux eds., **ESA-SP 382**, 599 (1997).
4. Mandrou P. et al., *Proc. 2nd INTEGRAL Workshop "The Transparent Universe"*, C.Winkler, T.Courvoisier & Ph. Durouchoux eds., **ESA-SP 382**, 591 (1997).
5. Hurley K., *Proc. 2nd INTEGRAL Workshop "The Transparent Universe"*, C.Winkler, T.Courvoisier & Ph. Durouchoux eds., **ESA-SP 382**, 491 (1997).
6. Westergaard N.J., et al., *Proc. 2nd INTEGRAL Workshop "The Transparent Universe"*, C.Winkler, T.Courvoisier & Ph. Durouchoux eds., **ESA-SP 382**, 605 (1997).
7. Gimenez, A. et al., *Proc. 2nd INTEGRAL Workshop "The Transparent Universe"*, C.Winkler, T.Courvoisier & Ph. Durouchoux eds., **ESA-SP 382**, 613 (1997).
8. Mereghetti S., et al., *Proc. 3rd INTEGRAL Workshop "The Extreme Universe"*, A.Bazzano ed., *Astroph. Letters & Communications*, **38**, 301, (1999).
9. Kretschmar B. et al., *Astron. & Astrophisics Suppl. Series* **138**, 571 (1999).
10. Courvoisier T., *Proc. 2nd INTEGRAL Workshop "The Transparent Universe"*, C.Winkler, T.Courvoisier & Ph. Durouchoux eds., **ESA-SP 382**, 581 (1997).
11. Pedersen H., Jennings D., Mereghetti S. and Teegarden B., *Proc. 2nd INTEGRAL Workshop "The Transparent Universe"*, C.Winkler, T.Courvoisier & Ph. Durouchoux eds., **ESA-SP 382**, 433 (1997).

Observations of Gamma-Ray Bursts with MAXI on the International Space Station

N. Kawai[1,2], A. Yoshida[1,2], T. Mihara[1,2], H. Negoro[1],
Y. Shirasaki[1,2], I. Sakurai[1], M. Matsuoka[2], K. Torii[2],
S. Ueno[2], M. Sugizaki[2], H. Tomida[2], W. Yuan[2],
H. Tsunemi[3,2], E. Miyata[3,2], and M. Yamauchi[4]

[1] *RIKEN (The Institute of Physical and Chemical Research), Wako, Saitama, 351-0198 Japan*
[2] *National Space Development Agency of Japan, Tsukuba, Ibaraki, 305-8505, Japan*
[3] *Osaka University, Earth and Space Science, Toyonaka, Osaka, 560-0043, Japan*
[4] *Miyazaki University, Faculty of Engineering, Miyazaki, 880-2155, Japan*

Abstract. MAXI is an X-ray all-sky monitor mission which will be deployed on the Japanese Experiment Module of the International Space Station in 2004. It is designed to monitor X-ray sources by scanning the entire sky every 90 minutes, and can detect X-ray sources of 3 mCrab in one day. Although its instantaneous field of view is limited to narrow fan beams, one can expect ∼5 GRBs to occur in the field of view and ∼25 afterglows to be detected per year.

INTRODUCTION

Monitor of All-Sky X-Ray Image (MAXI) is the first astrophysical payload for the Japanese Experiment Module (JEM) on the International Space Station (ISS). The main scientific objective of MAXI is to study medium or long term variations of large number of X-ray sources in the entire sky. The prime target is the activity of active galactic nuclei, most of which are too faint to study with the current all-sky monitor missions. It will also monitor and detect bright Galactic transients [1–3].

In addition to these, its sensitivity is sufficiently high to detect GRB afterglows in the early phase of the decay. This paper describes the MAXI experiment and discusses its expected capability for gamma-ray burst observations.

INSTRUMENTS

MAXI is box-shaped with a size of 1.85 m (L) × 0.8 m (W) × 1.0 m (H) and the mass of ~500 kg (Figure 1). It is equipped with two scientific instruments: Gas Slit Cameras (GSC) and Solid-state Slit Cameras (SSC) [4].

The ISS will rotate synchronously with its orbit so that one side will always face towards the center of the Earth and the opposite side will see the sky all the times. Figure 2 shows how it scans across the sky in the course of an orbit. Since observations are not possible at particular regions of the orbit with high charged particle flux, two fan beams are used to simultaneously cover two perpendicular fields of view, one centered on the zenith ("zenith view"), and the other centered at the forward direction of motion of the space station ("horizon view"). This will enable MAXI to achieve a more uniform coverage of the sky.

GSC consists of twelve 1-D position-sensitive proportional counters. The design of the proportional counter is based on the WXM/HETE-2 counters. They are filled with 1.4 atm gas composed of 97 % Xe and 3 % CO_2, and have 100 μm Be window. They are sensitive to X-rays in the 2–30 keV range. Each proportional counter is equipped with a slat collimator and a slit aperture on top, defining a narrow and long field of view (1.5° FWHM × 80°). Three identical GSC detectors are combined to cover a fan-beam field of view with 160° × 1.5°, which is mapped one-dimensionally in the long direction. There are four such detector sets, of which two cover the zenith view, and the other two cover the horizon view. The combined geometrical area of GSC is ~5340 cm^2 with the angular resolution of better than 1.5° for 3–8 keV photons. SSC uses X-ray CCD to achieve a soft X-ray response

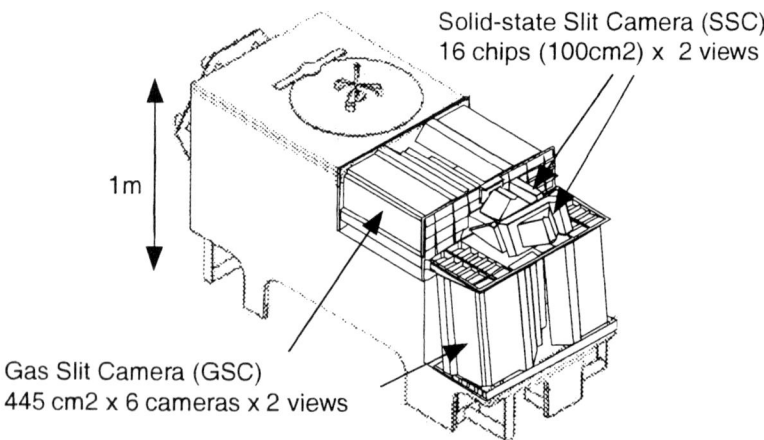

FIGURE 1. A layout of the two X-ray instruments, GSC and SSC, in MAXI. Each has two sets of cameras looking at the horizon and the zenith directions. Radiators and thermal shields are removed in this figure to show the cameras.

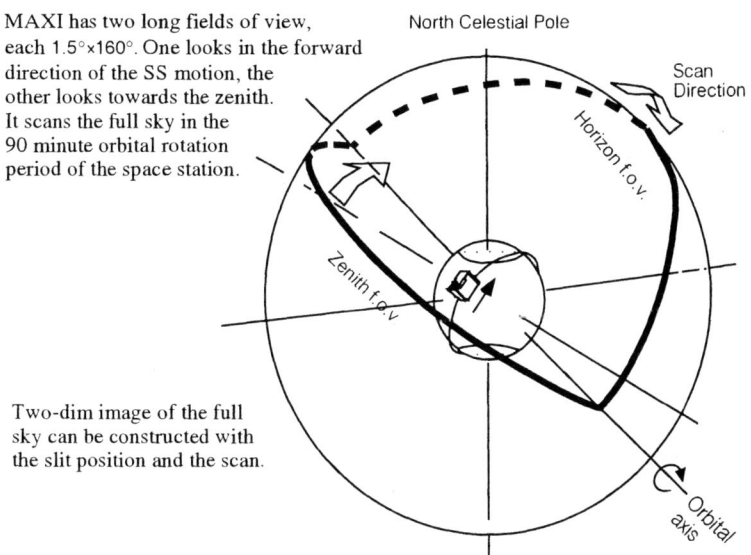

FIGURE 2. The fan-beamed fields of view of MAXI, which scan the entire sky in one orbit.

and good energy resolution. The sensitive energy range is 0.5–10 keV. The CCD chip is 1-inch-square full-frame transfer type, has 1024×1024 pixels with a pixel size 24 μm square, and is made by Hamamatsu Photonics [5]. We have two SSC cameras, horizon and zenith. Each contains 16 chips. The total number is 32 chips with an effective area of 200 cm^2. The CCD is used as a one-dimensional position sensitive X-ray detector with fast readouts. The energy resolution of the test chip was 152 eV FWHM at 5.9 keV.

GRB OBSERVATION CAPABILITY OF MAXI

MAXI scans the entire sky with the narrow fan-beamed fields of view. Due to its instantaneous small field of view, the fraction of GRBs caught at their onset is small, but it is not negligible.

Moreover, MAXI can detect an afterglow of a GRB which started outside the fields of view, if they are brighter than the GSC's detection limit when one of its fields of view passes over the source. The SSC is less sensitive than the GSC; its detection limit is $\sim$1/3 of GSC. Therefore we concentrate on GSC in the following discussions. We estimate the detection capability with very simple assumptions. As the property of GRBs is very diverse, the estimation here is inevitably crude.

Detection rate of GRBs at their onset

The rate at which GRBs occur in the field of view of MAXI are crudely estimated with the following assumptions.

- **Threshold:** Set to that of BATSE, i.e., 0.2 photons s^{-1} cm^{-2} in the 50–300 keV energy range. It is equivalent to 60 mCrab (2–10 keV) for MAXI assuming a photon index of ~ -1.

- **Number of GRBs:** 666 for the full sky per year, based on the BATSE 4B catalog [6].

- **Observation efficiency:** 0.6, considering the South Atlantic Anomaly region in which GSC cannot be operated due to high charged particle background.

- **Sky coverage:** $160° \times 1.5° \times 2 = 1/80$ of the sky. It is not negligible. It is approximately half of that of BeppoSAX WFC.

Multiplying the full sky number, sky coverage, and observing efficiency, we obtain 5.0 GRBs/year as the detection rate. As the burst profile is convolved with the slit passage pattern, the localization is affected in the scanning direction. We estimate the localization accuracy to be $\sim 0.2°$–$0.5°$.

Detection rate of GRB afterglows

MAXI detects only a small fraction of GRBs at their onset, as shown above. However, MAXI can detect substantially larger number of GRB sources if its field of view arrives on GRB sources while their afterglows are still sufficiently bright.

The expected rate of detection is estimated using the following simplified assumptions.

- **Threshold:** 7 mCrab ($\approx 1.4 \times 10^{-10}$ erg cm^{-2} s^{-1} in 2–10 keV) when the field of view passes over the GRB source. It is the 5σ detection limit of GSC for a single transit (dwell time ~ 30 seconds) [7].

- **Number of GRBs:** Brighter half of the bursts in the 4B catalog are assumed to have X-ray afterglows, i.e., 333 per year. The threshold in the 50–300 keV fluence roughly corresponds to 1×10^{-6} erg cm^{-2}.

- **Observing efficiency:** 0.6.

- **Afterglow flux:** Independent of GRB flux, starting at $t=10$ s at 500 mCrab, and follows a power-law decay:

$$F_x = 500 \text{ mCrab} \times (t/10 \text{ s})^{-1.2},$$

where t is the time after the burst onset.

- **Sky coverage:** With the above assumptions, the afterglow flux decays to the threshold (7 mCrab) in 350 s. The two fan-beamed fields of view sweeps ~1/7.5 of the sky within this period.

The initial afterglow flux of 500 mCrab is consistent for a case in which

- a GRB has a gamma-ray fluence (50–300 keV) $F_{50-300} = 1 \times 10^{-6}\,\mathrm{erg\,cm^{-2}}$,
- the profile is a flat-top box car with a 10 second duration,
- the X-ray flux (2–10 keV) of the burst is 10% of the gamma-ray flux,
- the X-ray afterglow is the smooth continuation of the X-ray flux in the burst, starting at 10 seconds from the onset.

Combining these factors, we obtain ~25 per year as the expected rate of GRB afterglow detection.

With a more optimistic assumption for the afterglow flux, i.e., the X-ray afterglow flux being proportional to the gamma-ray fluence in the burst, the average duration for the afterglow to be above the MAXI detection limit becomes 600 s, significantly longer than the previous estimation for a constant afterglow normalization. We obtain 44 per year for the detection of afterglows in this case.

Both of the estimation above is obviously based on over-simplified assumptions, and the expected rate of detection is highly uncertain. However, it is certain that MAXI has the capability not only to observe daily variation of weak sources such as AGN, but also to detect GRB afterglows in the early phase within several minutes after the burst onset.

REFERENCES

1. Matsuoka, M., *Proc. All-sky X-ray Observations in the Next Decade*, Matsuoka M., Kawai N. (eds.), RIKEN Cosmic Radiation Lab. Special Report IPCR CR-100, pp. 275 (1997).
2. Matsuoka, M. et al., SPIE Proc. 3114, p.414 (1997).
3. Kawai, N. et al., *Proc. All-sky X-ray Observations in the Next Decade*, Matsuoka M., Kawai N. (eds.), RIKEN Cosmic Radiation Lab. Special Report IPCR CR-100, pp. 279 (1997).
4. Mihara, T., *Broad Band X-ray Spectra of Cosmic Sources, 32nd COSPAR Scientific Assembly Symposium E1.1*, Makishima K., Piro L. (eds.), Advances Sp. Res., in press
5. Miyata, E. et al., *NIM A* **436**, 91 (1999).
6. Paciesas, W.S. et al., *ApJS* **122**, 465 (1999).
7. Rubin, B.C. et al., *Proc. All-sky X-ray Observations in the Next Decade*, Matsuoka M., Kawai N. (eds.), RIKEN Cosmic Radiation Lab. Special Report IPCR CR-100, pp. 285 (1997).

MARGIE: A Gamma-Ray Burst Ultra-Long Duration Balloon Mission

D. Band,[1] J. Matteson,[2] M. Cherry,[3] J. G. Stacy,[3,4] P. Altice,[3]
T. G. Guzik,[3] S. C. Kappadath,[3] J. Buckley,[5] P. Hink,[5] J. Macri,[6]
M. McConnell,[6] J. Ryan,[6] T. O'Neill,[7] and A. Zych[7]

[1] X-2, LANL, Los Alamos, NM 87545
[2] CASS, UC San Diego, La Jolla, CA 92093
[3] Louisiana State University, Baton Rouge, LA 70803
[4] Southern University, Baton Rouge, LA 70813
[5] Washington University, St. Louis, MO 63130
[6] University of New Hampshire, Durham, NH 03824
[7] UC Riverside, Riverside, CA 92521

Abstract. We are designing MARGIE as a 100 day ULDB mission to: a) detect and localize gamma-ray bursts; and b) survey the hard X-ray sky. MARGIE will consist of one small field-of-view (FOV) and four large FOV coded mask modules mounted on a balloon gondola. The burst position will be calculated onboard and disseminated in near-real time, while information about every count will be telemetered to the ground for further analysis. In a 100-day mission we will localize ~ 40 bursts with peak photon fluxes from 0.14 to ~ 5 ph cm^{-2} s^{-1} using 1 s integrations; the typical localization resolution will be better than ~ 2 arcminutes.

INTRODUCTION

An ultra-long duration balloon (ULDB) flight is ideal for a gamma-ray burst mission since the ~ 100 day duration is long enough for the detection and localization of a significant burst sample using new technologies which may still be considered too risky for a spacecraft. Consequently we are designing the Minute of Arc Resolution Gamma-ray Imaging Experiment (MARGIE) as a possible ULDB payload. This project is a synthesis of advanced mask construction, detector material and gondola control technologies. Funded by NASA, we are currently in the midst of a 2 year multi-institution design study of the MARGIE concept, and we will propose to build MARGIE when ULDB missions become feasible.

The MARGIE mission will consist of a central imager with a small field-of-view (FOV) and 4 large FOV detectors whose orientations will be offset from that of the central detector. The detectors will be coded mask systems with a detector plane of either Cadmium-Zinc-Telluride (CZT) or pixellated CsI scintillators. The detectors will be sensitive over ~50–200 keV, with the upper cutoff determined by the detector material and the lower cutoff by the atmosphere. The central detector will have a 43.5 cm × 43.5 cm detector plane (total area $A_{cen} = 1892.25$ cm^2) with 0.8 mm × 0.8 mm pixels. Separated from the detector plane by 150 cm, the mask will consist of two cycles of a 521×523 URA (the mask cells and detector pixels are currently planned to be the same size). The fully-coded FOV will be 16.5° across and the geometric angular resolution (a mask cell projected on the sky) will be 1.8'. The detector plane of each side-looking detector will be 44 cm × 44 cm (total area $A_{side} = 1936$ cm^2) with 0.5 mm × 0.5 mm pixels. The mask will consist of two cycles of a 881×883 URA (once again, the mask cells and detector pixels will be the same size) separated from the detector plane by 45 cm. The fully-coded FOV will now be 52.1° across and the geometric angular resolution 3.8'. Surrounding these detectors will be ~ 10^5 cm^2 of non-imaging NaI shields which may also be used as burst detectors. An active pointing system is currently not planned, but sub-arcminute aspect information will be provided by fiber optic gyroscopes, CCD star cameras and a GPS receiver. MARGIE will have a total mass of 990 kg and will consume 760 W of power.

Previous reports have outlined the details of the proposed instrument and discussed the technologies which will be implemented [1,2]. Here we discuss MARGIE's burst detection and localization capabilities, as well as the possibility of surveying the sky while waiting for a burst to occur.

BURST DETECTION THRESHOLD

Bursts will be detected by a statistically significant increase (e.g., $\kappa_b \sim 5\sigma$) in the count rate accumulated over a time bin of Δt (e.g., 1 s). We consider only bursts in the fully-coded FOV.

A number of factors are included in the calculation of the burst count rate. First, the mask blockage is assumed here to be 0.5. Second, we include the atmospheric attenuation by both photoelectric absorption and Compton scattering; not only the source photons but also the X-ray background photons are attenuated. Finally, we model the detector efficiency.

The background consists of a number of components. First, UCSD's experience with CZT leads us to approximate the internal background as 10^{-4} ct s^{-1} cm^{-2} keV^{-1} across the energy band of interest. This number will be refined as the project progresses. Second, we use the Gruber parameterization of the diffuse X-ray background [3], attenuated by both the mask and the atmosphere. Third, we include the hard X-rays originating in the atmosphere which consist of both scattered X-ray background photons and cosmic-ray induced emission. We use the flux measured by a UCSD balloon at 3 g cm^{-2} overburden [4]; this is a direct measurement of the atmospheric flux which should not be attenuated for atmospheric absorption.

The calculation of the diffuse background components requires the average solid angle Ω_{mask} viewed by the detector plane. Using relevant formulae [5] we find Ω_{cen}=0.296 sr for the central detector and Ω_{side}=1.594 sr for each side detector. Note that the atmospheric attenuation, detector efficiency and mask opacity are all energy dependent.

Translated into the BATSE 50–300 keV trigger band using a typical burst spectrum, the threshold peak flux for the central detector should be $\psi_{cen} = 0.14\Delta t^{-1/2}(\kappa_b/5\sigma)$ ph s^{-1} cm^{-2} and for the side detectors $\psi_{side} = 0.30\Delta t^{-1/2}(\kappa_b/5\sigma)$ ph s^{-1} cm^{-2}. Using 1) the burst rate, 2) the FOV, 3) a cumulative burst distribution $N(>\psi) \sim \psi^{-0.8}$ at threshold and 4) a mission duration of $M \sim 100$ days, we find that MARGIE should detect $N_{cen}(>\psi_{cen}) = 1.5\Delta t^{0.4}(\kappa_b/5\sigma)^{-0.8}(M/100$ d) bursts in the fully coded FOV of the central detector and $N_{side}(>\psi_{side}) = 8.6\Delta t^{0.4}(\kappa_b/5\sigma)^{-0.8}(M/100$ d) bursts in each of the side detectors. Accounting for overlapping FOVs, ~ 25 bursts should occur in some detector's fully coded FOV. Because the detectors will also detect bursts in the partially coded FOVs, we should detect at least 40 bursts. We assume that each detector triggers independently.

For the fully-coded FOVs, the brightest burst expected in 100 days (i.e., ψ_b such that $N(>\psi_b) \sim 1$) will have a peak flux of $\psi_{b,cen} = 0.25$ ph cm^{-2} s^{-1} and $\psi_{b,side} = 4.36$ ph cm^{-2} s^{-1} in the central and each of the side detectors, respectively. Of course, with the much larger area covered by all 4 side detectors, including the partially coded regions, the "burst of the mission" will be somewhat larger.

As an aside, note that the all sky burst rate is less than the often quoted 800 bursts yr^{-1}. BATSE's detection threshold is a peak flux of $\psi_B \sim 0.3$ ph s^{-1} cm^{-2} in the 50–300 keV energy band accumulated on the 1.024 s timescale. Above ψ_B 772 bursts were detected in the 3B catalog for which the latest BATSE team live-time is 5.16×10^7 s (C. Meegan 1998, personal communication), resulting in $n(>\psi_B) = 470$ yr^{-1} all sky. Two effects diminish the burst rate: the greater live-time than initially calculated and the threshold of ψ_B=0.3 ph cm^{-2} s^{-1} on the 1.024 s timescale.

BURST LOCALIZATION

The ability to localize bursts and the techniques to perform this localization depend on the burst and background count numbers. For these calculations we use bursts with durations of $T_d = 10$ s and average fluxes $\xi = 1/3$ of the peak flux. At the detection threshold there should be 190 and 400 burst counts in the central and side detectors, respectively, while for the brightest expected burst 250 and 4525 burst counts are expected in the central and side detectors, respectively. During the burst 1275 central detector and 5745 side detector background counts should also be accumulated. With this small number of counts distributed over a much larger number of pixels ($\sim 8 \times 10^5$ for the side detectors), back-projection will be the most efficient image reconstruction technique.

There are two localization issues: a) will the burst counts back-project to a statistically significant cluster on the sky; and b) how well can the burst be localized.

To determine the significance of the cluster of counts back-projected onto the burst's position, consider a grid of possible source positions superimposed on the sky. The pixels of this grid will be the mask cells projected onto the sky. For a mask which is half-open, half-closed, a given count on the detector plane can be back-projected through the mask onto half the grid's pixels. Thus the average count number projected into a non-burst sky pixel will be $(n_{GRB} + n_{back})/2$, and the standard deviation of this count number will be $(n_{GRB} + n_{back})^{1/2}/2$ (the total number of counts is known and therefore the counts per pixel follows the binomial distribution, and *not* the Poisson distribution; this changes the standard deviation). The expected number of counts in the burst pixel will be $n_{GRB} + n_{back}/2$. Thus, the expected difference between the number of counts in the burst and non-burst pixels should be $n_{GRB}/2$. Consequently, for a burst at threshold the expected sensitivity of a source detection will be 5.0σ for the central detector and 5.1σ for a side detector, while for the brightest expected burst the sensitivities are 6.4σ and 44.7σ for the central and side detectors, respectively. This sensitivity gives the probability that a fluctuation in a non-source pixel will equal or exceed the counts in the source pixel. Since only positive fluctuations would be mistaken for a source, we use the positive tail of the Gaussian distribution. The probability per pixel is less than 5×10^{-7} at threshold for both detectors, but less than 10^{-13} for the brightest expected burst for both detector types. However, the number of pixels on the sky is $\sim 3 \times 10^5$ for the central detector and $\sim 8 \times 10^5$ for each side detector, and therefore there is a probability of ~ 0.13 of a source-like fluctuation somewhere in the FOV at threshold, but an infinitesimal probability for the brightest bursts.

Persistent astrophysical sources are probably not an issue. Bursts at threshold will be 0.40 $\Delta t^{-1/2}(\kappa_b/5\sigma)(\xi/\frac{1}{3})$ Crab for the central detector and 0.83 $\Delta t^{-1/2}(\kappa_b/5\sigma)(\xi/\frac{1}{3})$ Crab for the side detectors. Nonetheless, we will identify steady sources in our FOV to avoid misidentifying bursts.

The rule-of-thumb is that a source can be localized to within the point-spread-function divided by the square-root of the source significance. Our point-spread-function is the size of a mask cell projected on the sky: $1.8'$ for the central detector and $3.8'$ for the side detectors. Thus the side detectors should localize threshold bursts to better than $\sim 1.7'$.

SURVEY CAPABILITIES

Between bursts MARGIE will survey the sky. However, the balloon gondola will be an unstable platform. The gondola will swing with a period of ~ 17.5 s and a likely amplitude of $\sim 0.5°$. In addition, a source will move relative to the earth at $15° \cos|\lambda|$ hr^{-1} (λ is the source's declination), or $2.5' \cos|\lambda|$ over 10 s, a typical burst duration. However, our aspect determination system should provide sub-arcminute resolution positions.

A given field will be visible for only a fraction of a day but observations from a number of days may be stacked. Techniques exist for compensating for shifts in the FOV, although systematic effects may be introduced.

Using the methodology developed for bursts, we find MARGIE's sensitivity to sources is $4.46(\kappa_s/5\sigma)(T_s/1 \text{ day})^{-1/2}$ mCrab for the central detector and $9.18(\kappa_s/5\sigma)(T_s/1 \text{ day})^{-1/2}$ mCrab for a side detector. At the detection threshold the number of source counts will typically be $5 \times 10^{-2}(\kappa_s/5\sigma)(T_s/1 \text{ day})^{1/2}$ counts/detector pixel on top of $50(T_s/1 \text{ day})$ background counts/detector pixel. The image reconstruction will be computationally expensive given the large number of counts accumulated while the FOV shifts by at least the source resolution on short timescales.

The expected number of survey sources is uncertain because the hard X-ray sky has not yet been well surveyed, and extrapolations from less sensitive, lower energy surveys depend on the assumed source spectrum. Galactic sources usually have spectra steeper than $\psi \propto E^{-2}$, but type II AGN may be heavily absorbed at lower energies, and thus will be more numerous at higher energies. In addition, we can point MARGIE at regions which are likely to have a higher hard X-ray density, such as the Galactic Center or the Magellanic Clouds. Nonetheless, it is worth attempting an extrapolation.

We assume a cumulative source distribution of $n(>\psi) \propto \psi^{-\gamma}$, where $\gamma = 1$ for Galactic and 3/2 for extragalactic sources. We normalize these distributions using the $HEAO\text{-}1$/A4 catalog [6]—65 Galactic and 8 extragalactic sources in an all-sky survey to a flux limit of 13mCrab above 13 keV—which is consistent with the HELLAS ($Beppo\text{-}SAX$) 5–10 keV survey [7] for extragalactic sources. We extrapolate these fluxes to higher energies with a Crab spectrum.

The mission duration of M days is broken into N_{FOV} pointings of T_s days each. Clearly, with enough pointings, the whole sky will be covered, which limits $T_s > 0.66$ d for the central detector and $T_s > 6.6$ d for each side detector. Thus we expect in 100 days $39.5(T_s/10 \text{ d})^{-1/2}$ Galactic and $14.8(T_s/10 \text{ d})^{-1/4}$ extragalactic sources for the central detector and $192(T_s/10 \text{ d})^{-1/2}$ Galactic and $50(T_s/10 \text{ d})^{-1/4}$ extragalactic sources for each side detector.

Acknowledgment: This study is supported by NASA grant NAG5-5208.

REFERENCES

1. Stacy, J. G., et al., in *Proc. 19th Texas Symp. on Relativistic Astrophysics*, Paris, in press (1999).
2. Cherry, M. L., et al., in *Proc. SPIE*, submitted (1999).
3. Gruber, D. E., Matteson, J. L., Peterson, L.E., and Jung, G. V., *Ap. J.* **520**, 124 (1999).
4. Matteson, J. L., et al., *Sp. Sci. Inst.* **3**, 491 (1977).
5. Sullivan, J. D., *Nucl. Inst. Meth.* **95**, 5 (1971).
6. Levine, A., et al., *Ap. J. Suppl.* **54**, 581 (1984).
7. Fiore, F., et al., in *First XMM Workshop*, in press (1999).

A New X-ray Telescope for Monitoring the X-ray Afterglows of GRBs

Rene Hudec[1], Adolf Inneman[2], Ladislav Pina[3], and Paul Gorenstein[4]

[1]*Astronomical Institute Ondrejov, Czech Republic, e-mail: rhudec @ asu.cas.cz*
[2]*Faculty of Mechanical Engineering, Czech Technical University Prague, Czech Republic*
[3]*Faculty of Nuclear Engineering, Czech Technical University Prague, Czech Republic*
[4]*SAO, Cambridge MA, USA*

Abstract. The first prototypes of innovative very wide field X-ray telescopes of Lobster-Eye type confirm the feasibility to develop such instruments for flight in the near future. These devices are expected to allow very wide field (more than 1000 square degrees) monitoring of the sky in X-rays (up to 10 keV and perhaps even more) with high sensitivity. We discuss the recent status of the development of very wide field X-ray telescopes as well as related scientific questions including expected major contributions such as monitoring and study of X-ray afterglows of Gamma Ray Bursts.

INTRODUCTION

The lobster-eye (LE) geometry X-ray optics offer an excellent opportunity to achieve very wide fields of view, while the widely used classical Wolter grazing incidence mirrors are limited to roughly 1 deg FOV. Wide field X-ray telescopes with imaging optics are expected to represent an important tool in future space astronomy projects, especielly those for deep monitoring and surveys in X-rays over a wide energy range including hard X-rays up to 10 keV (and possibly even more).

Wide field X-ray optics have been suggested in the 70s by Schmidt (orthogonal stacks of reflectors [1]) and by Angel (array of square cells [2]) but have not been constructed yet. Up to 180 deg FOV may be achieved. Possible solution is offered by the replication technology as confirmed by first prototypes presented here for both Schmidt as well as Angel arrangements.

SCHMIDT OBJECTIVES

One dimensional lobster-eye geometry was originally suggested by Schmidt [1]. The device consists of a set of flat reflecting surfaces. The plane reflectors are arranged in an uniform radial pattern around the perimeter of a cylinder of radius R. X rays from a given direction are focussed to a line on the surface of a cylinder of radius R/2. Focussing is not perfect and the image size is finite. Angular resolution of the order of one tenth of a degree or better may be achieved.

The one dimensional focusing device offers a wide field of view, up to maximum of 2π with the coded aperture. Two such systems in sequence, with orthogonal stacks of reflectors, form a double-focusing device. There is potential for extending the wide field imaging system to higher energy (perhaps up to 100 keV) by the use of additional coatings.

ANGEL OBJECTIVES

The idea of two dimensional lobster-eye type wide-field X-ray optics was first mentioned by Angel [2]. The full lobster-eye optical grazing incidence X-ray objective consists of numerous tiny square cells located on the sphere and is similar to the reflective eyes of macruran crustaceans such as lobsters. The field of view can be made as large as desired. It is possible to achieve good efficiency for photon energies up to 10 keV and/or even more if additional coatings are applied. Spatial resolution of a few seconds of arc over the full field is possible, in principle, if very small reflecting cells can be fabricated. This idea was, however, never further developed because of difficulties with production of numerous polished square cells of very small size (with aperture/length ratios of 30 or more i.e. about 1 mm × 1 mm or smaller at lengths of order of tens of mm).

LOBSTER TELESCOPE PROTOTYPES

The first Lobster-eye X-ray telescope prototypes have been finished. The prototype of the Schmidt geometry represents one module and consists of two perpendicular arrays of double-sided X-ray reflecting flats (36 and 42 double-sided flats 100 mm × 80 mm each). The flats are 0.3 mm thick and gold-coated. The microroughness is below 1 nm. The focal distance is 400 mm from the midplane. The FOV of one module is about 6.5°. More such modules may create an array with substantially larger FOV. The optical and X-ray tests indicate performance close to that calculated and expected (e.g., by ray tracing).

For the Angel geometry, numerous square cells of very small size (about 1 mm × 1 mm or less at lengths of order of tens of mm, i.e., with the size/length ratio of 30 and more) are to be produced. This demand can be also solved by modified innovative replication technology. First test modules with LE Angel cells have been succesfully produced. First test module has 47 cells 2.5 mm × 2.5 mm, 120 mm long (i.e., size/length ratio of almost 50), surface microroughness 0.8 nm, and f = 1.3 m. A second test module with 5 × 5 cells was finished recently. A third test module with 96 × 96, i.e., 9216 cells is in development. The surface microroughness of the replicated reflecting surfaces is better than 1 nm.

The first prototypes of lobster eye X-ray lenses of both the Schmidt as well as Angel geometries demonstrate the feasibility that the wide-field telescopes may be constructed in the future based on this type of reflective X-ray optics.

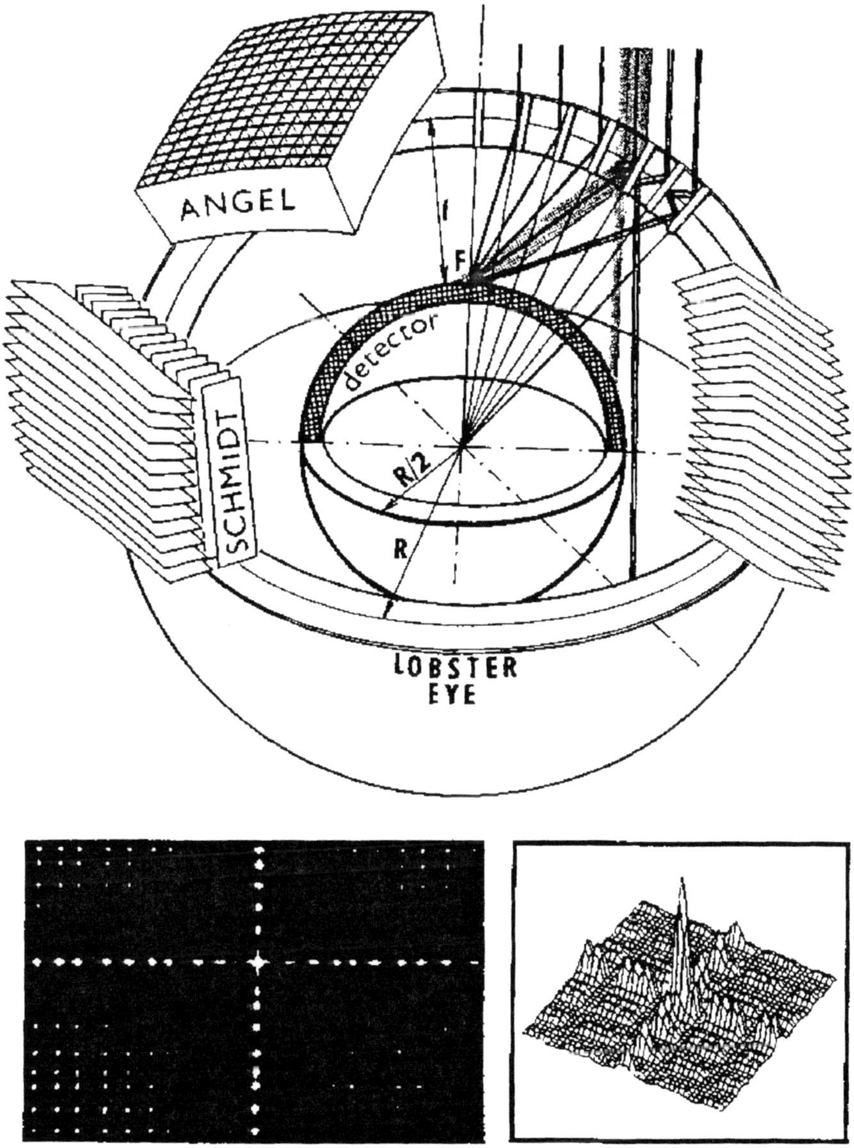

FIGURE 1. Schematic arrangement of the Lobster-Eye X-ray optics (top) with real image of the Schmidt objective prototype in optical light (bottom left) and distribution of intensity in the focal sphere for a point-like source (computer ray-tracing, bottom right)

FIGURE 2. Lobster-Eye prototypes: Schmidt (left) and Angel (right).

LOBSTER TELESCOPES AND GRB X-RAY AFTERGLOWS

Lobster-Eye X-ray telescopes are extremely important since the discovery of X-ray afterglows of Gamma Ray Burst (GRBs) sources in 1997. The expected rate of GRBs is ~1 per day. However, the theoretical prediction assumes a larger beaming angle in X-rays compared to gamma rays. Hence, the actual rate of X-ray afteglows is expected to be substantially larger (nearly 10×, or even more) than the rate of GRBs; about 10 X-ray afterglows are expected daily. The sensitivity of the LE telescopes is sufficient to detect the recently discovered X-ray GRB afterglows. The localization accuracy of the LE telescopes is of order of 1 arcmin, substantially exceeding the recent localization accuracy of most gamma-ray instruments (2° and more). The LE telescopes are expected to provide a substantial contribution to the science and statistics of GRBs. The sensitivity of $\sim 10^{-14}$ erg cm^{-2}s^{-1}, or better, may be achieved (0.5–3 keV energy range). This is sufficient to position the fading X-ray GRB counterpart to 1 arcmin or better, as well as to obtain the light curve.

MORE SCIENCE WITH LE TELESCOPES

The additional science of LE X-ray telescopes includes supernova explosions, high energy binary sources, AGNs, blazars, X-ray novae, X-ray flares on stars, X-ray transients etc. The use of LE telescopes will allow these objects to be detected and studied by sky patrol monitoring.

DISCUSSION

The use of very wide field X-ray imaging systems could be very valuable for many areas of X-ray and gamma-ray astrophysics. Results of analyses and simulations of lobster-eye X-ray telescopes indicate that they will be able to monitor the X-ray sky at an unprecedented level of sensitivity, an order of magnitude better than any previous

X-ray all-sky monitor. Limits as faint as 10^{-12} erg cm^{-2} s^{-1} for daily observation in the soft X-ray energy range are expected to be achieved, allowing monitoring of all classes of X-ray sources, not only X-ray binaries, but also fainter classes such as AGNs, coronal sources, cataclysmic variables, as well as fast X-ray transients including gamma-ray bursts and nearby type II supernovae. For pointed observations, limits better than 10^{-14} erg cm^{-2} s^{-1} (0.5–3 keV) could be obtained, sufficient to detect X-ray afterglows of GRBs. The production of corresponding optical elements can be reasonably achieved by methods of electroforming and composite replication as an alternative to other methods.

The first prototypes of Schmidt as well as Angel arrangements have been produced succesfully for the first time, demonstrating the possibility to construct these lenses by innovative but feasible technologies. This makes the proposals for space projects with very wide field lobster eye optics possible.

FUTURE STEPS TOWARD A REAL LOBSTER EYE TELESCOPE

For both Schmidt and Angel arrangements, the further steps should involve: (1) application of additional layers to extend the energy range to higher energies, perhaps up to 100 keV, (2) improving the surface quality (microroughness, slope errors) to further improve the reflectivity and the angular resolution, (3) construction of larger or multiple modules to achieve a larger FOV of order of at least 30 deg or more, (4) for Angel objective, to further reduce the cell apertures and to enhance the length/aperture ratios to achive a better angular resolution.

ACKNOWLEDGEMENTS

The development of double-sided reflecting X-ray foils was supported by a grant within the US-Czech Science and Technology program, No. 930 37. The development of the Angel objective is supported by the grant provided by the Grant Agency of the Czech Republic No. 106/97/1223. The design and development of innovative X-ray telescopes for future space projects is supported by the grant provided by the Grant Agency of the Czech Republic, No. 105/99/1546. We also acknowledge the support provided by the UK X-ray astronomy group at the University of Leicester in X-ray tests of the LE telescope Schmidt prototype.

REFERENCES

1. Schmidt, W. H. K., *Nucl. Instr. and Methods* **127**, 285 (1975).
2. Angel, J. R. P., *Astrophs. J.* **233**, 364 (1979).

Estimation of GRB Detection by FiberGLAST

S. Phengchamnan[1], K. Aisaka[2], M. Atac[2], W.R. Binns[3],
J.H. Buckley[3], M.L. Cherry[4], D. Cline[2], P. Dowkontt[3],
J.W. Epstein[3], M.H. Finger[5,9], G.J. Fishman[5], T.G. Guzik[4],
P.L. Hink[3], M.H. Israel[3], S.C. Kappadath[3], G.R. Karr[1],
R.M. Kippen[1], J. Macri[6], R.S. Mallozzi[1], M.L. McConnell[6],
Y. Pischalnikov[2], W.S. Paciesas[1], T.A. Parnell[1], G.N. Pendleton[1],
R.D. Preece[1], G.A. Richardson[1], K. Rielage[3], J.M. Ryan[6],
J.G. Stacy[2,3], T.O. Tümer[9], D.B. Wallace[1], and R.B. Wilson[5]

[1] *University of Alabama in Huntsville, Huntsville, AL 35899, USA*
[2] *Department of Physics and Astronomy, University of California,
Los Angeles, CA 90095, USA*
[3] *Department of Physics & McDonnell Ctr. for Space Sciences,
Washington Univ., St. Louis, MO 63130, USA*
[4] *Department of Physics, Louisiana State Univ., Baton Rouge, LA 70803, USA*
[5] *NASA/Marshall Space Flight Center, Huntsville, AL 35812, USA*
[6] *Space Science Center, Univ. of New Hampshire, Durham, NH 03824, USA*
[7] *Southern Univ., Baton Rouge, LA 70813, USA*
[8] *Institute of Geophysics & Planetary Physics, Univ. of California,
Riverside, CA 92521, USA*
[9] *Universities Space Research Association*

Abstract. FiberGLAST is one of several instrument concepts being developed for possible inclusion as the primary Gamma-ray Large Area Space Telescope (GLAST) instrument. The predicted FiberGLAST effective area is more than 12,000 cm^2 for energies between 30 MeV and 300 GeV, with a field of view that is essentially flat from $0°-80°$. The detector will achieve a sensitivity more than 10 times that of EGRET. We present results of simulations that illustrate the sensitivity of FiberGLAST for the detection of gamma-ray bursts.

INTRODUCTION

Understanding the nature of gamma-ray bursts (GRBs) is among the most important science objectives of NASA's GLAST (Gamma-ray Large Area Space Telescope) mission. Observations by CGRO/EGRET have shown that the high-energy gamma-ray emission can be a significant fraction of the burst fluence, and that burst spectra commonly extend to GeV energies with no evidence for a high-energy cut-off [1]. Even more exciting is the observation by EGRET of high-energy burst emission after lower energy gamma rays were no longer detectable, with time delays of as much as 90 minutes [2,3]. Clearly there is a need for high-energy observations of more bursts and with greater sensitivity, and this has been considered in defining the GLAST scientific requirements.

FiberGLAST is one of several concepts being developed for possible inclusion as the primary GLAST instrument. The predicted response of FiberGLAST has been estimated using Monte Carlo simulations. We use extrapolations of BATSE burst spectra along with a simulated response database to estimate the number of burst photons that will be detected by FiberGLAST. The predicted FiberGLAST effective area is more than 12,000 cm^2 for energies between 30 MeV and 100 GeV, with a field of view that is essentially flat from $0° - 80°$. The FiberGLAST instrument concept is described in more detail by Rielage et al. [4].

FIRST METHOD

This method is a rough estimation of the number of GRBs that will be detected by FiberGLAST. The energy range that FiberGLAST will observe is from 10 MeV to 300 GeV, and the effective area is ~12,000 cm^2 up to an 80° viewing angle. We also assume Band's GRB spectral model with $\alpha = -1$, $\beta = -2$ and $E_0 = 200$ keV. Since the effective area of FiberGLAST depends on energy, we divide our peak flux calculation into 10 different energy bins. Using 5 counts $\cdot$ s^{-1} as the minimum detection criteria and by comparing our calculation value with the $LogN - LogP$ of GRBs that BATSE has detected [5], we can estimate how many GRBs we would expect to detect. $LogN - LogP$ on the 1024 ms time scale is used to extrapolate the number of GRBs. Our calculation yields the result of 549 bursts per years that FiberGLAST might detect.

We checked the validity of this simulation by repeating the same procedure using the EGRET response [8]. During its first 3.4 years, EGRET detected five bursts, each with at least seven spark chamber events above 30 MeV (during the gamma-ray active phase as defined by BATSE). Assuming 30% observing efficiency for EGRET, this methodology predicts twenty-four detected bursts in 3.4 years, but EGRET observed only 5 GRBs. We can see that this simple method overestimates the number of GRBs that might be detected by FiberGLAST. We will show another method in the next section that yields a more reliable estimate.

SECOND METHOD

In this method, in order to simulate the response of FiberGLAST to GRBs, we use the measured distribution of BATSE burst spectra, and extrapolate them to higher energies. The high-energy gamma-ray flux from a GRB of a given BATSE peak flux or fluence was estimated using the catalog of Preece et al. [6,7], which includes time-resolved spectral fits for a large sample of bright BATSE bursts. We chose for our spectral templates 102 events from this catalog that also had BATSE peak flux and fluence measurements. For most burst sub-intervals in the catalog, the Band GRB function, or other broken broken power-law models, provided an acceptable fit. For each of these, we extrapolated the best-fit high-energy power-law spectral model to the FiberGLAST energy range and computed a fluence for each burst by summing over all such sub-intervals. For a few sub-intervals no high-energy power-law could be determined, so the FiberGLAST fluence was set to zero. Fluences were computed in five energy bins between 10 MeV and 300 GeV. We consider this method to be more robust than extrapolating a single spectrum averaged over each burst. However, due to the finite BATSE energy bandwidth, there is some concern that the high-energy power law spectral index determined from BATSE spectra is systematically too hard. Some evidence for this comes from the comparison of the distribution of BATSE high-energy spectral indices with COMPTEL observations of GRB spectra in the 0.75–30 MeV range. The mean COMPTEL spectral index is -2.53 [9], whereas the BATSE high-energy spectral index distribution peaks around -2.25, though with a pronounced skewness toward softer spectra. For better consistency, we arbitrarily softened each BATSE spectrum by adding -0.2 to the spectral index before extrapolation.

We generated two sets of simulated bursts by first picking a peak flux P_B (1.024 s timescale) or fluence S_B (25–2000 keV) randomly from the observed BATSE $LogN - LogP_B$ and $logN - logS_B$ distributions and then computing the FiberGLAST fluence S_F that each template burst would have if it were scaled to the chosen flux or fluence:

$$S_F = \frac{P_B}{P_B^t} S_F^t \qquad \text{and} \qquad S_F = \frac{S_B}{S_B^t} S_F^t, \qquad (1)$$

respectively, where P_B^t and S_B^t are the BATSE peak flux and fluence of a template burst, and S_F^t is its extrapolated FiberGLAST fluence. A random direction of incidence (within 90° of the FiberGLAST primary axis) was chosen for each such event, and the extrapolated fluence S_F was convolved with the simulated FiberGLAST response to determine the total detected counts in each of five energy bins: 10–30 MeV, 30–100 MeV, 0.1–1 GeV, 1–10 GeV, and 10–300 GeV. For each simulated set, 1,292 BATSE values were used, so the total number of simulated bursts in each set is $1.292 \times 102 = 131{,}784$.

We checked the validity of these simulations by repeating the same procedure using the EGRET response (as in the first method). This second methodology predicts nine EGRET detections if we use peak flux scaling, and six if we use fluence scaling, whereas five bursts were actually detected. Given the small numbers, both techniques produce reasonably consistent results — giving us confidence in the FiberGLAST predictions using this method.

TABLE 1. Burst detectability using peak flux scaling

Energy (GeV)	Fraction of unocculted bursts (percent)					
	<3 photons	3-10 photons	$10\text{-}10^2$ photons	$10^2\text{-}10^3$ photons	$10^3\text{-}10^4$ photons	$>10^4$ photons
0.01-0.03	37	12	32	16	2.5	0.1
0.03-0.1	39	12	31	15	2.6	0.1
0.1-1	48	15	25	10	1.6	0.1
1-10	77	10	10	2.5	0.2	0
10-300	93	4.0	2.4	0.3	0	0
0.01-300	34	7.5	28	23	6.4	0.6

TABLE 2. Burst occurrence rate using peak flux scaling

Energy (GeV)	Rate (bursts per year)				
	3-10 photons	$10\text{-}10^2$ photons	$10^2\text{-}10^3$ photons	$10^3\text{-}10^4$ photons	$>10^4$ photons
0.01-0.03	52	143	73	11	0.4
0.03-0.1	54	136	69	12	0.6
0.1-1	68	110	46	7.2	0.3
1-10	46	45	11	1.1	0
10-300	18	11	1.4	0	0
0.01-300	33	126	103	29	2.6

TABLE 3. Burst detectability using fluence scaling

Energy (GeV)	Fraction of unocculted bursts (percent)					
	<3 photons	3-10 photons	$10\text{-}10^2$ photons	$10^2\text{-}10^3$ photons	$10^3\text{-}10^4$ photons	$>10^4$ photons
0.01-0.03	46	14	26	12	1.6	0
0.03-0.1	48	14	25	11	1.6	0
0.1-1	58	14	20	7.1	0.9	0
1-10	83	8.5	7.4	1.5	0.1	0
10-300	96	2.7	1.5	0.1	0	0
0.01-300	41	11	26	17	4.3	0.3

TABLE 4. Burst occurrence rate using fluence scaling

Energy (GeV)	Rate (bursts per year)				
	3-10 photons	$10\text{-}10^2$ photons	$10^2\text{-}10^3$ photons	$10^3\text{-}10^4$ photons	$>10^4$ photons
0.01-0.03	64	115	52	7.1	0
0.03-0.1	64	109	49	7.2	0
0.1-1	63	87	32	4.1	0
1-10	38	33	6.7	0.4	0
10-300	12	6.7	0.6	0	0
0.01-300	50	117	77	19	1.3

DISCUSSION

Our first method is a rough estimation that most likely overestimates the number of GRBs that FiberGLAST should detect. The second method gives us better estimation. From Tables 1–4, it can be seen that FiberGLAST will detect at least 10 photons each from roughly half of the unocculted bursts above BATSE's threshold, and roughly one in five bursts should produce at least 100 photons in FiberGLAST. Bursts that produce at least 10 photons above 10 GeV should be detected once every month or two, and there is a good chance of detecting at least 100 photons above 10 GeV from a burst during the GLAST mission. The photons above 10 GeV are important not only for their direct physical implications, but also because their directions can be more accurately determined.

These simulations assume no spectral cut-offs and do not consider additional photons from the delayed/extended emission observed by EGRET. The large predicted numbers of bursts with more than 10 photons per energy bin should allow measurement of spectral breaks or cut offs up to at least 10 GeV if they exist, even for the "wide-field" bursts for which FiberGLAST has worse energy resolution.

REFERENCES

1. Dingus, B. L., Catelli, J. R., & Schneid, E. J., in *Gamma-Ray Bursts: 4th Huntsville Symposium*, AIP CP428, Ed. C. A. Meegan et al., New York: AIP, 1998, pp. 349.
2. Dingus, B. L., in *Gamma-Ray Bursts: 2nd Huntsville Symposium*, AIP CP307. Ed. G. J. Fishman et al., New York: AIP, 1994, pp. 22.
3. Hurley, K., *Nature* **372**, 652 (1994).
4. Rielage, K., in *Proc. 26th Intl. Cosmic ray Conf.*, Salt Lake City, paper OG4.2.21, in press (1999).
5. Pendleton, G. N., *ApJ* **464**, 606 (1996).
6. Preece, R. D., 1998, *ApJ* **496**, 849 (1998).
7. Preece, R. D., *ApJS* **496**, submitted (1999).
8. Thompson, D. J., *ApJS* **86**, 629 (1998).
9. Kippen, R. M., *Adv. Sp. Res.* **22**, 1097 (1998).

A Robust Filter for the BeppoSAX Gamma-Ray Burst Monitor Triggers

M. Feroci[1], C.L. Bianco[1], F. Lazzarotto[1], A. Mattei[1],
G. Ventura[1], E. Costa[1], and F. Frontera[2,3]

[1] Istituto di Astrofisica Spaziale - CNR, Via del Fosso del Cavaliere, Roma, Italy
[2] Istituto Tecnologie E Studio Radiazioni Estraterrestri - CNR, Via Gobetti 101, Bologna, Italy
[3] Dipartimento di Fisica, Università di Ferrara, Via del Paradiso 12, Ferrara, Italy

Abstract. The BeppoSAX Gamma Ray Burst Monitor (GRBM) is triggered any time a statistically significant counting excess is simultaneously revealed by at least two of its four independent detectors. Several spurious effects, including highly ionizing particles crossing two detectors, are recorded as onboard triggers. In fact, a large number of false triggers is detected, on the order of 10/day. A software code, based on an heuristic algorithm, was written to discriminate between real and false triggers. We present the results of the analysis on an homogeneous sample of GRBM triggers, thus providing an estimate of the efficiency of the GRB detection system consisting of the GRBM and the software.

THE GAMMA RAY BURST MONITOR

The GRBM [1,3,2,4] is a gamma-ray detection system onboard the BeppoSAX satellite. It is the secondary function of the anticoincidence shields of the PDS experiment. It is composed of four ~1100 cm^2 slabs of CsI(Na) scintillators operating in the 40–700 keV range. Each detector provides time series of the detected counts in the above energy range, with 1 s resolution as a continuous housekeeping and with better than 8 ms resolution upon trigger. An onboard trigger is active whenever a statistically significant counting excess is simultaneously detected in at least two of the four detectors.

AIMS, METHOD AND LIMITS

Besides being gamma-ray detection devices, the scintillating crystals composing the GRBM are sensitive to highly ionizing particles that, leaving a large amount of energy (~GeV) in just one shot, result in a phosphorescence phenomenon, with a consequent detection by the electronics of a large number of counts in few tens of ms, i.e., a *spike* in the counting rate. When the same particle crosses two detectors, it causes a trigger in the onboard logic that is electronically indistinguishable from a cosmic gamma-ray transient trigger (no particle anticoincidence is available to the GRBM). The number of spikes that trigger the onboard electronics is of the order of 9–10/day. Therefore, the aim of this work is to develop a software filter that provides a "safe" first-order discrimination of the "instrumental" triggers from those of "cosmic-origin". This will reduce of the huge number of onboard triggers recorded so far ($\sim$10,000 in the first three years of BeppoSAX operations), and will apply a more refined program to the generated sample of triggers. A first drawback of the "roughness" of our filter is that it does not include criteria to distinguish between gamma-ray bursts, solar flares, and soft gamma repeaters. This task will be carried out by the "second order" filter.

Our software filter is based on the automatic on-ground analysis of the high-resolution time series, according to criteria established on the basis of known detector/electronics behavior and an extended study of GRBM time series. Usually, a manual inspection of a GRBM light curve is sufficient to discriminate cosmic gamma-ray events from spurious events. However, when an archival search for real GRBs is carried out this becomes impractical, and an automatic filter is needed. We therefore developed IDL-based software that applies a number of discrimination criteria to the GRBM light curves. This is a first approach to the problem and we did not make use of all the information that the GRBM provides for each onboard trigger. In fact, 1-s resolution data are available in two energy ranges (40–700 keV and >700 keV) and time-averaged energy spectra, but they are not used in the analysis presented here.

The criteria implemented in the program are based on the knowledge of the instrument operational principles and on experience with the observed light curves. It is likely that particular cases exist that were not taken into account in our code. At this time, the following parameters are computed for each of the four GRBM detection units: duration, rise time, simultaneity, shape, and full width at half maximum. The basic criterium of the code is to assign a "score" to each of these parameters, based on their comparison between pairs of detectors, with the goal of having the smallest score to the most-likely-cosmic events. The individual scores for the different parameters then combine together to give a total score, accounting for all the measured event characteristics. For the final score, the higher the value, the smaller the probability of being a cosmic event.

CALIBRATION OF THE FILTER

In order to test and calibrate the efficiency of our software filter, we created a sample of selected events whose origin was known by different methods (typically BATSE or IPN events). To these, a number of manually-screened false triggers were added. In Fig. 1 we present the results of the analysis carried out with our filter on such a sample of events. The x-axis reports the score assigned to each event by the filter. The texture classifies the events, based on the comparison with data

TABLE 1. Score calibration for the software filter.

Score	Burst	Spike	Doubt
<2	$(83.3 \pm 5.8)\%$	$(9.6 \pm 4.5)\%$	$(7.1 \pm 4.0)\%$
2–3	<4.5% (68% Conf.)	$(16.0 \pm 7.3)\%$	$(84.0 \pm 7.3)\%$
≥4	$(1.7 \pm 1.0)\%$	$(94.0 \pm 1.8)\%$	$(4.2 \pm 1.6)\%$

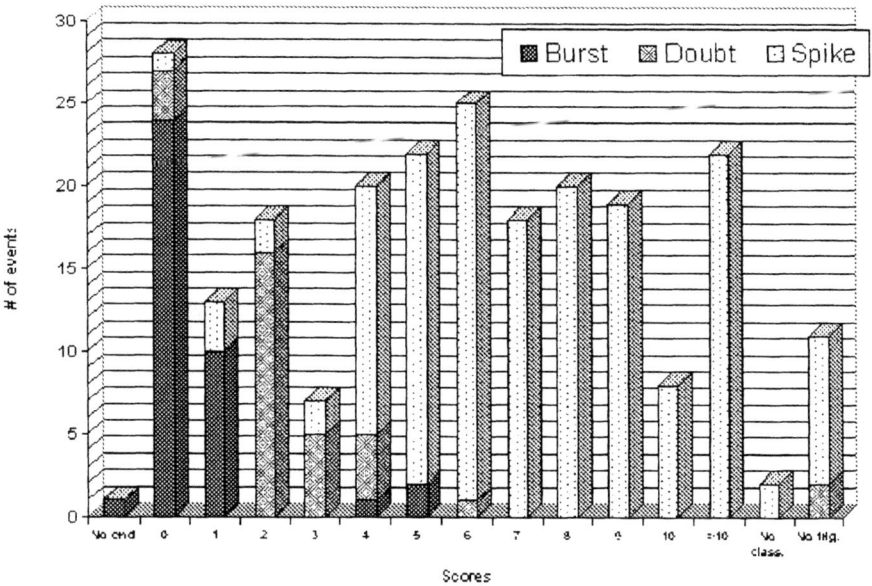

FIGURE 1. Efficiency diagram of the software filter: probability of true cosmic events is inversely proportional to the score value on x-axis.

from other satellites (BATSE, Ulysses, etc.). The sample includes several short GRBs in order to test the ability of the software to distinguish between them and spikes, but many long GRBs are also included. Thus, we can define 3 score classes (Burst, Spike, and Doubt), with the confidences given in Table 1 (the sum of every row is 100%), computed assuming a multinomial distribution. The software works on single peaks even during the evolution within a multi-peaked event. Therefore, the results reported in the Table should be applicable to individual peaks in each light curve. Examples of events belonging to these three categories are presented in Fig. 2.

There are 3 additional special classes in the plot: No-End, No-Class, No-Trig. No-End: the signal does not return to noise level before the end of the light curve (GRBM high time resolution data have a maximum coverage of 106 s). This can be due to long bursts, or to a variable background (see an example in [2]). No-Class: the signal has not been analyzed. This can happen for weak events if the software is not able to estimate the duration of the signal. No-Trig: the software is unable to find anything other than noise in the light curve, possibly due to very weak signals.

APPLICATION TO THE GRBM DATABASE

The software filter has been applied to a fraction of the BeppoSAX GRBM data archive, covering more than 3 years of elapsed time, but consisting of about 603 days of satellite observing time. The goal was to test our software on an homogeneous sample of onboard triggers. A noticeable by-product of this operation is an estimate of the GRB detection efficiency. As stated above, the filter operates on individual peaks. The result of the analysis gives: 440 peaks from cosmic events, 510 doubt cases, 4648 spikes, 58 No End, 86 No Class. If we apply the filter efficiency that was defined in the previous section, then the number of cosmic peaks become (367 ± 26). They belong to about 340 individual events, of which only $\sim$180 have been post-facto verified to be real cosmic events.

SUMMARY AND CONCLUSIONS

The results of the analysis on an unselected portion ($\sim$40%) of the BeppoSAX/GRBM data archive presented in this paper allow us to draw the following conclusions:

- A number of $\sim$180 cosmic events detected by the BeppoSAX/GRBM were identified (including a small number of Solar Flares and events from Soft Gamma Repeaters), over a net exposure time of $\sim$545 days, leading to an estimation of the *GRBM efficiency for triggering cosmic events of* $\sim$*0.33/day* (additional events are detected, but without an onboard trigger).

- The automatic filter has a $\sim$50% efficiency for selecting real events out of false triggers (i.e., any event selected by the program has a $\sim$50% probability of being of cosmic origin). This reduced efficiency is likely due to the incompleteness of the sample on which the code was calibrated.

- The automatic filter has a >90% efficiency for discarding false triggers (i.e., any event not selected by the program has a less than 10% probability of being a real event, based on the software calibration). Thus, the filter allows us to reduce the number of light curves to be analyzed to about 10% of the total, with an expected efficiency of more than 90%.

REFERENCES

1. Frontera, F., et. al., *A&ASS* **122**, 357 (1997).
2. Feroci, M., et. al., *Proc. SPIE* **3114**, 186 (1997).
3. Amati, L., et. al., *Proc. SPIE* **3114**, 176 (1997).
4. Costa, E., et. al., *Adv. Space Res.* **22**, 1129 (1998).

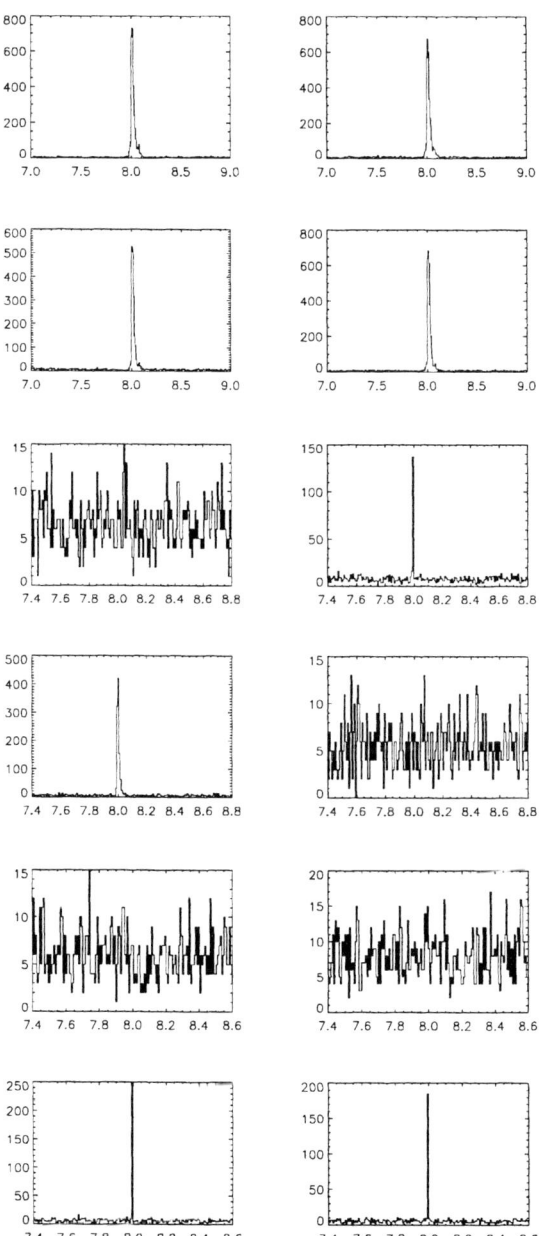

FIGURE 2. 8-ms resolution light curves in the four GRBM detectors, for a real short gamma ray burst (Top 4 panels), a spike (Middle 4 panels) and a dubious event (Bottom 4 panels). X-axis is time in seconds, Y-axis gives the counts/bin.

GRB Localization with the BeppoSAX Gamma-Ray Burst Monitor

B. Preger[1], E. Costa[1], M. Feroci[1], L. Amati[2], and F. Frontera[2,3]

[1] *Istituto di Astrofisica Spaziale - CNR, Via del Fosso del Cavaliere, Roma, Italy*
[2] *Istituto Tecnologie E Studio delle Radiazioni Extraterrestri - CNR, Via Gobetti 101, Bologna, Italy*
[3] *Dipartimento di Fisica, Università di Ferrara, Via Paradiso 12, Ferrara, Italy*

Abstract. A method to evaluate the capability to localize Gamma-Ray Burst (GRB) sources with the Gamma-Ray Burst Monitor (GRBM) onboard BeppoSAX is presented. The instrument is composed of four differently oriented detectors, so that their relative counting rates can be used to reconstruct the incoming direction of the GRB photons. The method involves the minimization of an energy- and direction-dependent χ^2-like variable related to the difference between the intensities observed by the four GRBM detectors and their expected values, as derived from the instrument's response functions. The method is applied to a sample of GRBs simultaneously detected and localized by other experiments, in order to compare the achieved results with locations determined independently.

INTRODUCTION

Source localization can be performed with non-imaging gamma-ray experiments by using the relative fluxes registered by individual detectors oriented in different directions. The error boxes obtained with this method strongly depend (both in shape and in dimension) on the intensity of the GRB, on its incoming direction with respect to the instrument's rest frame and on the accuracy the response function of the instrument is known with. The BATSE experiment onboard CGRO (see e.g., [1]), as an example, uses this method to localize approximately 1 GRB per day with a typical uncertainty of a few degrees (statistical plus systematic [2]). In the case of the GRBM (see e.g., [3-5]) onboard BeppoSAX [6] larger error boxes and fewer events are expected, due to the smaller total area and to the partial obscuration of the detectors' fields of view. However, even a rough determination of the angular position can be used to choose between the two intersections of annuli derived by the IPN for bursts that BATSE does not detect. In order to test the capability of the GRBM to localize GRB sources (a preliminary study of this problem can be found in [7]), we have chosen a sample of events whose angular positions are known independently and compared our results with these locations.

THE INSTRUMENT

The GRBM is a secondary function of four 1 cm thick CsI(Na) slabs (LS1, LS2, LS3 and LS4) primarily operating as an active anticoincidence of the Phoswich Detection System experiment. It operates in the range 40–700 keV and has a total geometric area of about 4500 cm^2, but its location in the inner part of the spacecraft causes a strong dependence of the effective area on the energy and the direction of the incoming photons. Housekeeping data include 1 s rate meters for the four detectors separately in two energy bands (40–700 keV, GRBM band, and >100 keV, AC band) and a trigger function is enabled for the detection of GRBs.

On-ground calibrations have been performed [8] after the full integration of the spacecraft, except for the solar panels, in order to provide the response of each detector as a function of the incoming direction and spectrum of the GRBs. Each LS unit suffers from the presence of different parts of the satellite in front of the detector, and the response functions strongly differ from one another. In particular, units LS1 and LS3 are located behind the carbon-fiber tubes of the LECS and MECS X-ray telescopes, showing a good agreement with the geometrical dependence of the effective area on the incoming direction. The field of view of unit LS2 is partially obscured, especially in the lower part, by the High Pressure Gas container, while the two Wide Field Cameras strongly disturb unit LS4.

LOCALIZATION METHOD AND RESULTS

We have developed a method for localizing GRB sources that is based on the comparison between the intensities registered by the four LS units in the two available energy ranges and their expected values, as derived from the instrument's response functions. In the following we will use a GRBM rest frame where ϕ is the azimuthal angle and θ the elevation of the source with respect to the spacecraft xy plane and we assume a power law energy spectrum for the GRBs ($\propto E^{-\alpha}$).

Based on the GRBM detectors' response functions, we computed the expected number of counts in the four detectors and the two energy bands for each incoming direction (ϕ, θ) and for three values of the spectral index $\alpha = 1, 2, 3$. For each GRB we assume that these values, N_i ($i = 1...8$), are related to the observed ones, O_i ($i = 1...8$) with experimental uncertainties σ_i ($i = 1...8$), through the expression $N_i = K \cdot O_i$.

For each value of ϕ, θ and α we find through a least-square minimization technique the value of K that best describes the observed phenomenology, $\tilde{K}(\phi, \theta, \alpha)$, and define an energy- and direction-dependent variable

$$\Phi(\theta, \phi, \alpha) = \sum_{i=1}^{8} \frac{(N_i(\theta, \phi, \alpha) - \tilde{K}(\theta, \phi, \alpha) \cdot O_i)^2}{\sigma_i^2} \quad (1)$$

whose numerical values are related to the probability that a given set (θ, ϕ, α) corresponds to the source parameters: lower values of $\Phi(\theta, \phi, \alpha)$ have higher prob-

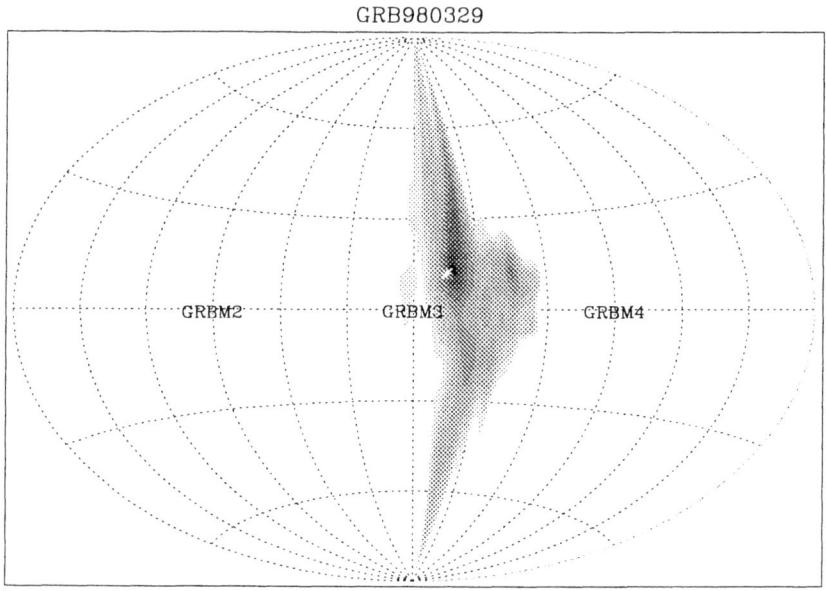
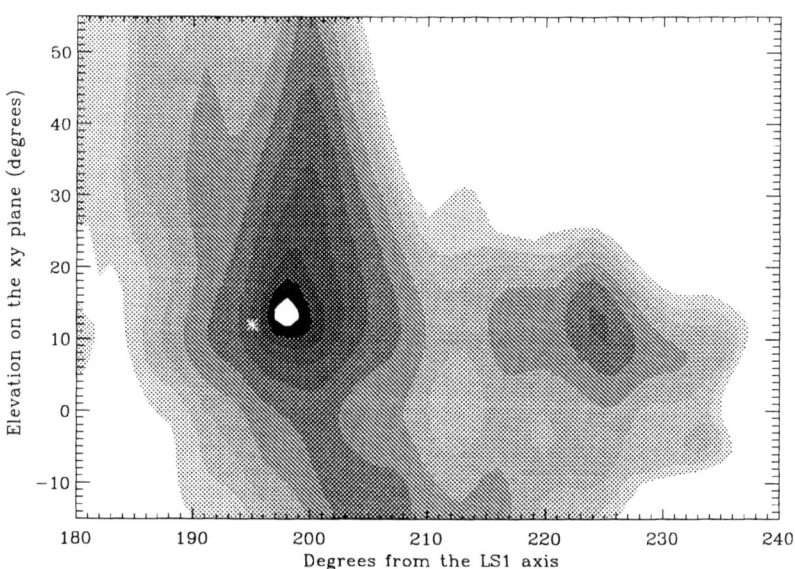

FIGURE 1. Contour plot of the $\Phi_{min}(\theta,\phi)$ variable for GRB980329. The upper panel shows the entire sky in the Aitoff projection, centered in the LS3 detector axis. The lower panel shows a zoom of the minimum zone: the x axis is the azimuth and the y axis is the elevation in a spacecraft rest frame centered in the LS1 axis. The asterisk shows the WFC position of the source and the contours are shown in the range $200 < \Phi_{min} < 400$.

TABLE 1. Results obtained for 19 events with high-resolution positions.

EVENT	FLUX[a]	FLUENCE[b]	ϕ_1[c]	θ[d]	INST.[e]	OFFSET[f]
GRB961026	2098 ± 55	6329 ± 182	173°	-22°	BATSE+IPN	≈ 37°
GRB961211	1027 ± 48	1863 ± 83	65°	42°	BATSE+IPN	≈ 3°
GRB970111	5547 ± 80	78050 ± 348	193°	16°	SAX/WFC	≈ 42°
GRB970402	868 ± 40	12243 ± 341	2°	4°	SAX/WFC	≈ 4°
GRB970508	420 ± 49	2141 ± 115	174°	1°	SAX/WFC	≈ 16°
GRB970517	3547 ± 76	11789 ± 137	133°	13°	BATSE+IPN	≈ 7°
GRB970816	5643 ± 86	18349 ± 168	35°	3°	BATSE+IPN	≈ 12°
GRB971214	803 ± 51	8622 ± 180	7°	-14°	SAX/WFC	≈ 22°
GRB971227	651 ± 49	1521 ± 77	177°	-9°	SAX/WFC	≈ 9°
GRB980203	8101 ± 99	102645 ± 162	135°	15°	BATSE+IPN	≈ 8°
GRB980306	2322 ± 64	33423 ± 510	135°	29°	BATSE+IPN	≈ 27°
GRB980310	1356 ± 55	1530 ± 64	188°	29°	BATSE+IPN	≈ 7°
GRB980329	5902 ± 87	64018 ± 332	195°	12°	SAX/WFC	≈ 3.5°
GRB980425	384 ± 46	4854 ± 159	203°	1°	SAX/WFC	≈ 14°
GRB980519	1524 ± 49	11765 ± 194	174°	0°	SAX/WFC	≈ 4°
GRB980703	198 ± 43	4734 ± 283	240°	-57°	XTE/ASM	≈ 7°
GRB981220	1538 ± 60	6776 ± 164	222°	-11°	XTE/ASM	≈ 5°
SGR1900+14	636 ± 49	627 ± 46	143°	35°	GRBM+IPN	≈ 37°
Solar Flare	4412 ± 82	200736 ± 450	97°	-28°	–	≈ 5°

[a] Peak flux in counts · s^{-1}
[b] In counts
[c] Approximate azimuth with respect to LS1 axis
[d] Approximate source altitude on the satellite xy plane
[e] Instrument providing the accurate position
[f] Distance between the accurate position and the absolute minimum of the Φ_{min} variable

abilities and vice versa. Then we build a new variable, which is only a function of the incoming direction, by choosing for each (ϕ, θ) the value of α that corresponds to the minimum $\Phi(\theta, \phi, \alpha)$

$$\Phi_{min}(\theta, \phi) = \min_{\alpha=1,2,3}(\Phi(\theta, \phi, \alpha)) \qquad (2)$$

The $\Phi_{min}(\theta, \phi)$ is formally equal to a χ^2, but since the variables involved are not completely independent (the two energy bands are partially overlapped) and systematic effects dominate the statistical ones (in particular, atmospheric backscattering is not considered at this stage), we can't use the χ^2 statistics to assign a numerical probability to each set of parameters on the basis of the value assumed by the $\Phi_{min}(\theta, \phi)$. The distribution of the $\Phi_{min}(\theta, \phi)$ variable can be studied applying the method to a sample of GRBs whose locations have been determined by other experiments and comparing our results to these reference locations. As a first attempt, we have chosen a sample of GRBs whose locations are known with high angular resolution, i.e., with uncertainties less than ≈1°. These can be BeppoSAX/WFC, RossiXTE/ASM or IPN events. A typical contour plot of the $\Phi_{min}(\theta, \phi)$ variable is shown in Fig. 1, while the results obtained for 17 GRBs, one solar flare and one event from a Soft Gamma Repeater (SGR1900+14) are

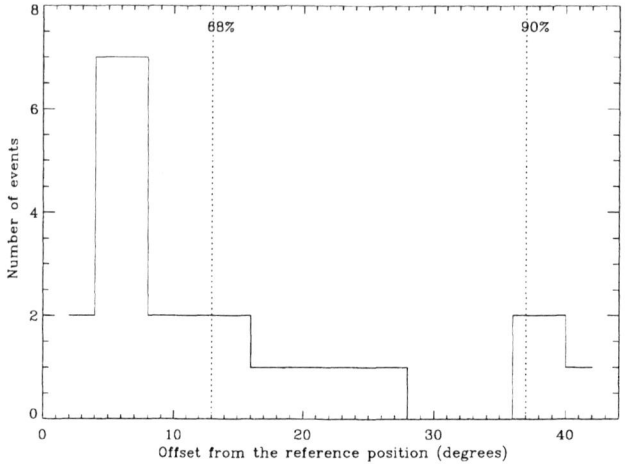

FIGURE 2. Distribution of the offsets between the reference positions and the Φ_{min} minima

shown in Table . The distribution of the offsets between the reference position and the $\Phi_{min}(\theta,\phi)$ minima, i.e., the positions that have the maximum probability to correspond to the incoming direction of the GRBs, is shown in Fig. 2. In this preliminary phase 68% of the events lie within 13° of the reference position and 90% of the events lie within 36°, while there is no evident dependence of the offset on the incoming direction (ϕ, θ).

CONCLUSIONS AND FUTURE DEVELOPMENTS

The application of a χ^2-like minimization technique as a GRB localization method with the BeppoSAX/GRBM has shown good agreement with angular positions determined independently. Further improvements of the method include the estimation of the main systematic effects (above all atmospheric backscattering) and the standardization for a quick look use.

REFERENCES

1. Fishman, G.J., et al., *ApJ Suppl. Ser* **92**, 229 (1994).
2. Briggs, M.S., et al., *ApJ Suppl. Ser.* **122**, 503 (1999).
3. Frontera, F., et al., *A&A Suppl. Ser.* **122**, 357 (1997).
4. Costa, E., et al., *Adv. Space Res.* **22**, 1129 (1998).
5. Feroci, M., et al., *Proc. SPIE Conf.* **3114**, 187 (1997).
6. Boella, G., et. al., *A&A Suppl. Ser.* **122**, 299 (1997).
7. Preger, B., et al., *A&A Suppl. Ser.* **138**, 559 (1999).
8. Amati, L., et. al., *Proc. SPIE Conf.* **3114**, 176 (1997).

Response Function of the Gamma-Ray Burst Monitor (GRBM) Onboard the BeppoSAX Satellite

F. Calura[*], M. Rapisarda[†], F. Frontera[*‡], E. Montanari[*], C. Guidorzi[*], L. Amati[‡], M. Feroci[‖], E. Costa[‖], and P. Collina[*]

[*] *Physics Department, University of Ferrara, Italy*
[†] *ENEA, Frascati, Italy*
[‡] *ITESRE, CNR, Bologna, Italy*
[‖] *IAS, CNR, Roma, Italy*

Abstract. Preliminary results of a study aimed to the determination of the response function, in direction and energy, of the Gamma-Ray Burst Monitor (GRBM) onboard the BeppoSAX satellite are presented. The study has been carried out by means of Monte Carlo techniques, taking into account on-ground calibration before launch, and the Crab spectrum measured during the flight. The derived response matrix will be used to evaluate properties (e.g., direction, spectrum, fluence) of all Gamma-Ray Bursts detected with the GRBM.

GRBM EXPERIMENT ONBOARD BEPPOSAX

The BeppoSAX GRBM [1–4] is an experiment based on four CsI(Na) scintillators that are also used as active lateral shields of the PDS experiment [3]. They form a square box around the main PDS detectors. The GRBM has specific electronics for GRB recognition. The GRBM data continuously transmitted to ground include 1 s rate meters in two different energy ranges: the so called GRBM band (nominal 40–700 keV) and the AntiCoincidence (AC) band (>100 keV). In the case of an on-board trigger, high time resolution rate meters (0.488 ms for 10 s from the trigger time and 7.8125 ms for 8 s before and for 88 s after the trigger time) are stored. Spectra are continuously accumulated in the GRBM band with a time integration of 128 s, independent of the trigger time.

MONTE CARLO MODEL

The goal of the present investigation is to reconstruct direction, fluence and spectral properties of all the GRBs detected by the GRBM. To achieve this, the

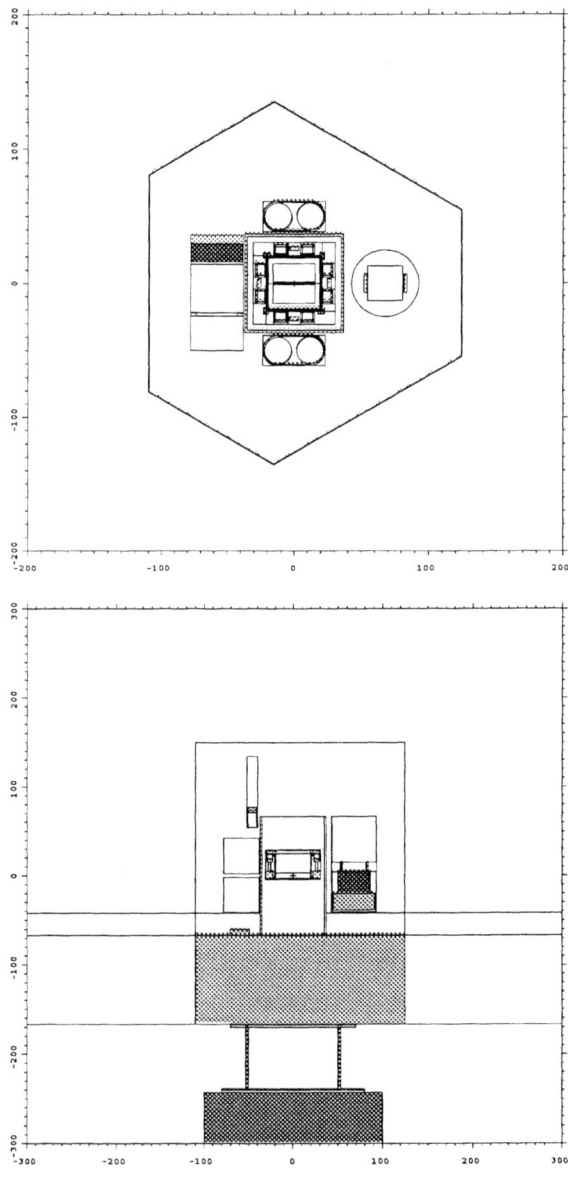

FIGURE 1. Different views of the BeppoSAX MCNP model; top: x-y plane at z=10.8 cm, bottom: y-z plane at x=0 cm.

response matrix of each GRBM detector is needed. Given the presence of other instruments around the monitor, the response function has been determined by means of Monte Carlo techniques. For that we have developed a geometric model of the BeppoSAX payload that describes the geometry and material of all the instruments and electronics (see Figure 1). The code used is MCNP4b [5], which allows the transport of photons, electrons and neutrons through matter. Photons with energies from 1 keV to 1 GeV, can be transported. All interactions of photons in this energy range with matter (photoelectric absorption, Compton and Thomson scattering and pair production) are treated in detail. The Monte Carlo model has been successfully tested by using the on-ground GRBM calibration results [1] and the reconstruction of the Crab spectrum measured with GRBM. Preliminary work has been done by using only the on-ground calibrations [6].

RESPONSE MATRIX

The GRBM response matrix has been evaluated as a function of photon direction and energy. A spherical reference frame has been used with the origin in the satellite, z-axis co-aligned with the Narrow Field Instruments [7], and x-axis directed as the GRBM unit-2 axis. A grid of 72 directions (12 azimuthal angles ϕ in the range $[0°, 360°]$, and 6 polar angles θ in the range $-50°, +75°$) and 10 logarithmically-spaced energies in the range 40–1000 keV, have been chosen. For each direction and energy a run of the MCNP code has been performed assuming a plane wave source of 5×10^6 photons. The number of counts $N_i(E_j, \theta_m, \phi_l)$ detected by the GRBM unit-i ($i = 1, \ldots, 4$), in the GRBM 40–700 keV energy range, for a photon plane wave source with energy E_j, and incident direction (θ_m, ϕ_l) ($m = 1, 12; l = 1, 6$), is evaluated. The grid of energies has been refined by interpolating the derived $N_i(E_j, \theta_m, \phi_l)$ by means of cubic splines. In this way the detected events have been evaluated with energy steps of 1 keV. Assuming a power law spectrum of the incident radiation that is suitable to describe GRB spectra (order zero approximation), the expected counts in the GRBM unit-i is given by:

$$C_i^k(\theta_m, \phi_l) = \sum_{j=1}^{960} \Delta E_j \, E_j^{-\alpha_k} \, N_i(E_j, \theta_m, \phi_l) \qquad (1)$$

where α_k is the photon index of the power law. The following values for the spectral index have been used: $\alpha_1 = 1$, $\alpha_2 = 1.5$, and $\alpha_3 = 2$. The best direction (θ, ϕ) of the GRB is determined by minimization of the following χ^2 statistics:

$$\chi_k^2(\theta_m, \phi_l) = \sum_{i=1}^{4} \frac{\left(n_i - \frac{n \, C_i^k(\theta_m, \phi_l)}{C^k(\theta_m, \phi_l)}\right)^2}{\sigma_i^{k\,2}} \qquad k = 1, 2, 3 \qquad (2)$$

where n_i represents the total number of GRB counts measured in detector unit-i; σ_i^k is the variance of n_i plus $C_i^k(\theta_m, \phi_l)$; $n = \sum_i n_i$; and $C^k(\theta_m, \phi_l) = \sum_i C_i^k(\theta_m, \phi_l)$.

TABLE 1. Comparison between GRBM and other (BATSE, BeppoSAX/Wide Field Cameras (WFC)) position determinations. Δ is the angular discrepancy.

	BATSE		WFC		GRBM		
GRB	α_{2000}	δ_{2000}	α_{2000}	δ_{2000}	α_{2000}	δ_{2000}	Δ
960720			262.6	49.1	249.9	49.5	8.3
961211	281.7	54.8			273.7	43.0	12.9
970420	213.0	-15.9			210.1	-10.9	5.7
970816	91.6	44.9			94.4	40.7	4.7
971110	241.6	50.3			227.9	48.9	9.0
971220	33.2	-38.6			29.3	-36.2	4.0
980306	7.0	-45.5			6.9	-46.7	1.2
980329			105.7	38.8	108.6	32.3	6.9
980810	350.0	24.6			350.8	13.6	11.1

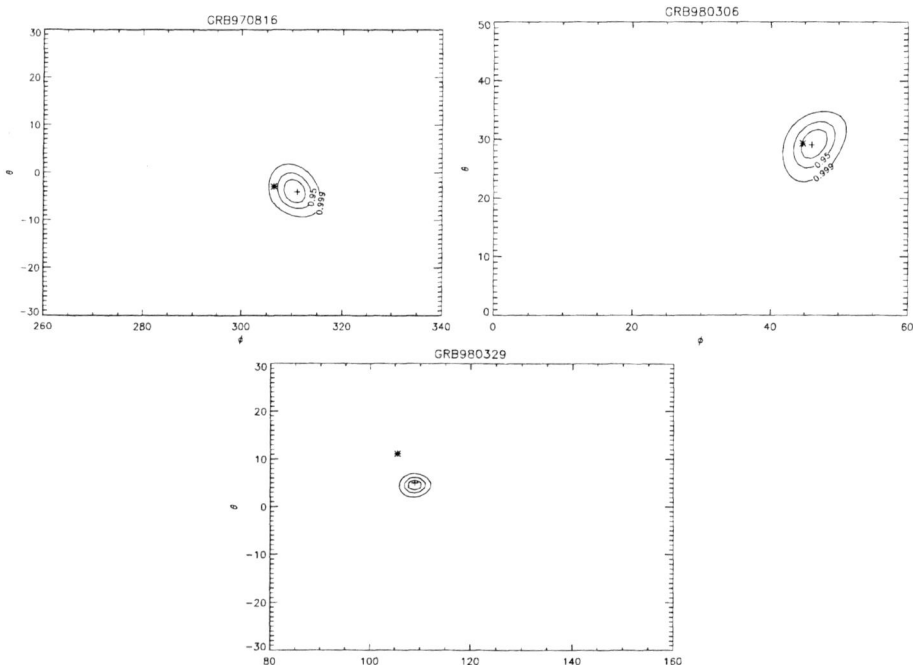

FIGURE 2. Contour plots of the chi-square of some of the GRBs of Table 1 are shown. The curves denote error regions at different confidence levels (0.70, 0.95, 0.999). Asterisk: estimated position by either BATSE or BeppoSAX/Wide Field Cameras; Plus sign: GRBM estimated position. Local coordinates θ and ϕ are used.

For a given value of α_k, the the best (θ_m, ϕ_l) has been evaluated. The estimate is refined by extracting a small region around the direction of the minimum. In this region an interpolation of the expected counts $C_i^k(\theta_m, \phi_l)$ with bicubic splines is carried out; finally the chi-square is calculated once again and another minimum is found. The direction of the GRB is given by the (θ, ϕ) associated with this minimum.

RESULTS

In Table 1 results regarding the determination of the incoming directions of some GRBs are summarized.

In Figure 2 we show the contour plots of the recalculated χ^2 for some of the above GRBs. The position of the source as determined either by BATSE or BeppoSAX/Wide Field Cameras (see Table 1) is denoted by an asterisk, while the GRBM centroid position is represented by a plus sign. The curves around the centroid denote the error regions at different confidence levels (0.70, 0.95, 0.999). As can be seen, for some GRBs the agreement between our position and that estimated by other instruments is very satisfactory. However, in other cases (e.g., GRB980329) we found a statistically significant discrepancy that could be due to other effects (e.g., albedo from the terrestrial atmosphere) still not included in our code. Further improvements are in progress.

REFERENCES

1. Amati, L., *PhD Thesis* (1999).
2. Costa, E., Frontera, F., Dal Fiume, D., et al., *Adv. Sp. Res.* **22**, 1129–32 (1998).
3. Frontera, F., Costa, E., Dal Fiume, D., et al., *A&A Supp.* **122**, 357 (1997).
4. Pamini, M., Natalucci, L., Dal Fiume, D., et al., *Il Nuovo Cimento* **13C**, 337 (1990).
5. Briesmeister, J. F., *MCNP Manual*,
   http://www-xdiv.lanl.gov/XCI/PROJECTS/MCNP/manual.html
6. Preger, B., et. al., *A&A Suppl.* **138**, 559 (1999).
7. Boella, G., et al., *A&A Suppl.* **122**, 327 (1997).

Progress Incorporating the NEAR Mission into the Interplanetary GRB Network

T. L. Cline[a], S. Barthelmy[a], P. Butterworth[b], T. McClanahan[a], D. Palmer[c], J. Trombka[a], K. Hurley[d], R. Gold[e], R. M. Kippen[f], C. Kouveliotou[f], D. Frederiks[g], S. Golenetskii[g], and E. Mazets[g]

[a] *NASA's Goddard Space Flight Center, Greenbelt, MD, 20771*
[b] *Raytheon Systems Company at NASA's GSFC*
[c] *Universities Space Research Association at NASA's GSFC*
[d] *Space Sciences Laboratory, UC Berkeley, Berkeley, CA 94720-7450*
[e] *JHU Applied Physics Laboratory, Laurel, MD 20723*
[f] *NASA's Marshall Space Flight Center, SD-50, Huntsville AL 35812*
[g] *Ioffe Physical-Technical Institute, St. Petersburg, 194021, Russia*

Abstract. The present gamma-ray burst (GRB) network consists of the Ulysses and the Near Earth Asteroid Rendezvous missions in deep space, with the BATSE experiment on Compton-GRO, the Konus experiment on GGS-Wind, and the BeppoSAX mission all near to the Earth. The NEAR spacecraft, built and launched without any GRB capability, was modified in flight to provide 1-second GR count rates, creating the first 3-cornered long-baseline IPN since the early 1990s. The arc-minute precision of this IPN was confirmed with the known locations of SGRs and GRB afterglows. After the Eros orbital insertion maneuver was postponed until February 2000, the NEAR spacecraft was placed in a low bit-rate, dormant mode, but was dedicated to GR data by command. Most events are now saved, although data recovery is necessarily delayed up to several days. Restoration to the active mode is expected in January 2000. Note: since the Huntsville Symposium, the first all-IPN alert to result in radio and/or optical afterglow observations followed GRB991208.

INTRODUCTION

The remarkable developments of the last few years have revolutionized the gamma-ray burst field. BeppoSAX's discovery of a persistent soft x-ray component enabled the discoveries of longer-lasting but hitherto elusive counterparts in the optical and radio domains, with alerts given only hours after the events. These observations in turn yielded direct evidence for cosmological GRB sources. The first detection of a GRB-associated transient during the gamma-ray profile itself was recently made [1] with a wide-angle telescope. This dramatically justified prior investments in rapid-alert systems for astronomy, in particular the GCN [2], a system created at Goddard to provide observers with all available GRB information, beginning in real time as the burst is taking place. The HETE-2 mission, built to give rapid GRB alerts with great directional precision, has been modified to include a soft x-ray trigger; HETE-2 launch is expected in early 2000. Most recently, the excitement in the GRB discipline and the emphasis on rapid alert concepts have culminated with the NASA selection of the next-generation Swift mission.

THE GRB INTERPLANETARY NETWORK

The interplanetary GRB network has been in place in its various manifestations from 1976 to the present. The earliest network was successful in its goal of defining GRB source locations with near arc-minute accuracy; it found that the transient source population(s) must be distinct from known x-ray emitters or other identifiable astronomical populations, that one repeater-associated and anomalous event had N49 in the LMC as its source direction, and that repeaters (SGRs) are a separate population of transients from 'classical' GRBs.

However, rapid GRB source determinations were not then very likely, since data from deep space were not collected continuously, but delayed by the processing efforts that followed the daily or less frequent contacts. Nevertheless, the network arrangements that episodically survived during these two decades (often with only one distant vertex, i.e., Pioneer-Venus Orbiter in the 1980s or Ulysses in the 1990s), continued to advance the field. The IPN localization of SGR sources made possible other discoveries showing their association with distant supernova remnants, confirming the N49 identity, and additional SGR sources and new kinds of events have also been found. An IPN with only one distant vertex, providing only GRB source annuli, can serve to refine source fields found independently, reducing the delayed counterpart search labors.

Finally, that the existing IPN can be rapid and precise enough to prompt GRB afterglow studies (even with NEAR in the 'dormant' mode) was demonstrated recently with GRB991208. We can expect an increased rate of IPN-prompted GRB and GRB afterglow studies to follow the NEAR full-scale reactivation.

NEAR

After the decades that had seen the loss of four Mars missions (Phobos-1 and -2, Mars Observer and Mars-96), the IPN now includes NEAR as a second distant vertex, finally completing the fully long-baseline triangle that is required for precise GRB error box definition. NEAR had been designed and flown without a GRB capacity and had been in interplanetary space for some time before a heroic, inflight, software modification was successfully attempted by command. ('Heroic' in the sense that the possibility of courting disaster was great.) The only useful data that could be extracted consisted of a 1-second GR rate, a housekeeping item from the anticoincidence shield of the XGRS spectrometer. Many resulting GRB profiles have since been posted on the Web during the last year. An unexpected discovery that the renewed IPN enabled was the detection [3] of the giant SGR flare of 1998 August 27 — the first event of its kind since the 1979 March 5 event. It was localized to a known SGR source direction [4] and studied in detail [5].

NEAR was initially intended to be maneuvered into Eros orbit in early 1999 and to remain there for some time, but the maneuver attempt failed. Later, after a rescheduling of the next entry maneuver attempt for mid-February 2000, NEAR was placed in a passive cruise mode using low bit-rate telemetry. However, a system was designed to store GR data continuously and provide a rapid and out-of-sequence download, which was ground tested and sent by command to the spacecraft data system. This change has successfully enabled nearly all the GRB and SGR events in the last several months to be saved and processed.

The NEAR sensitivity to several-hundred-keV gamma rays is somewhat greater than that of Ulysses, and is roughly omnidirectional except for the shielding provided by other spacecraft mass. Thus, most of the events seen with Ulysses and with BATSE or Konus or RXTE also are detected with NEAR. The temporal precision of the NEAR 1-second GR rate is its limiting item. For very brief events, a 1-second binning results in a limiting precision of about 4 arc minutes, for a spacecraft distance of 2 AU. However, given that BATSE and Konus profiles are usually available with 64-millisecond binning, it is possible to obtain greater time accuracy for complex and extended GRBs.

Figure 1 shows the result of comparing the Konus profile of the 1999 May 10 event (rebinned at 1 second, phased at 64-ms intervals) with the NEAR data. The least-squares study illustrates that a 0.1-second accuracy can be obtained. In this case, the result is verified (to within 0.05 s) with the agreement of the curve minimum with the value of 40.27 seconds calculated from the precisely known source location of the afterglow. Thus, the 0.1-second precision at 2 AU can translate into a source localization as precise as ± 0.5 arc-minute.

SUMMARY

This manuscript was intended to publicize our expectation that adding the NEAR mission to the IPN should make possible IPN alerts adequate to generate GRB afterglow studies, despite their necessary delays. The event of GRB991208 has been the first such event to fulfill that expectation. A GCN notice [6] sent a day or more after that event prompted a radio detection that, in turn, prompted optical transient observations that mutually agree in precise location [7]. Figure 2 indicates the initially distributed 'error box' and the location of the radio/optical afterglows (the same on this scale), in which the proximity of the afterglow to the center of the approximately $4' \times 4'$ source field testifies to our initial overestimate of the IPN's locational uncertainty. The event of GRB991208, in fact, successfully prompted a renewal of afterglow-study activity. We hope that this result will increase the enthusiasm of astronomers to follow up more GRB alerts from the IPN, the number of which opportunities should increase considerably after NEAR returns to its full-scale data mode, expected before the next asteroid encounter in mid-February 2000. A period of many months with a full IPN with optimum characteristics should then follow throughout 2000.

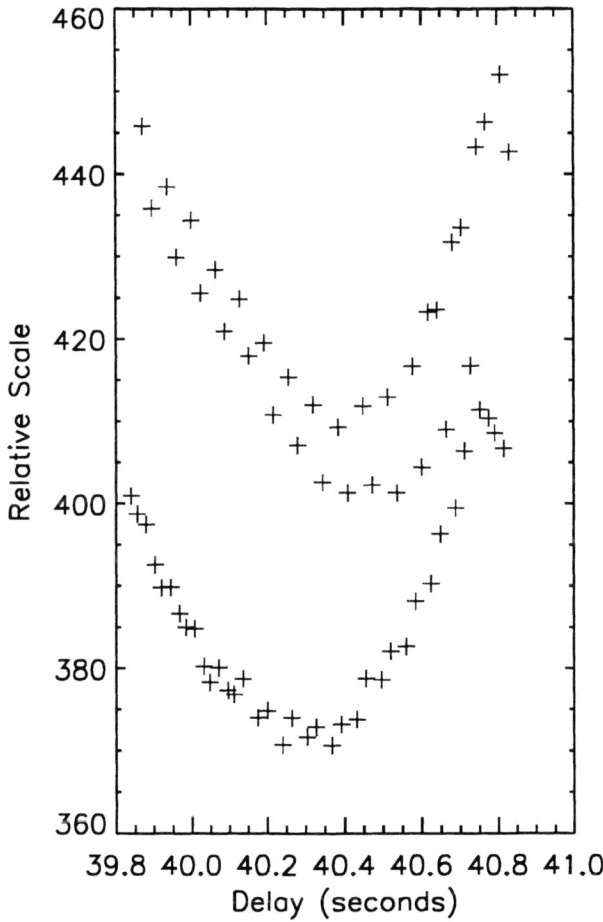

FIGURE 1. A calibration of the clock accuracy, indicating the consequent precision of the Ulysses/NEAR/Konus network. Despite its necessarily coarse, 1-second binning, the NEAR event profile can be fitted to phased Konus 64-ms profiles to better than 0.1-second accuracy, and the resulting fit agrees with the delay value calculated from the observed GRB990510 optical transient location, i.e., 40.273 seconds.

REFERENCES

1. Akerlof, C., et al., *Nature* **398**, 400–402 (1999).
2. Barthelmy, S. D., et al., *SPIE Conf. Proc.* **3768**, 444–449 (1999).
3. Cline, T. L., et al, *IAU Circular* 7002 (1998).
4. Hurley, K., et al., *IAU Circular* 7004 (1998).
5. Hurley, K., et al., *Nature* **397**, 41–43 (1999)
6. Hurley, K., and Cline, T., *GCN Circular* 450 (1999).
7. *GCN Circulars* 451–462, and more (1999).

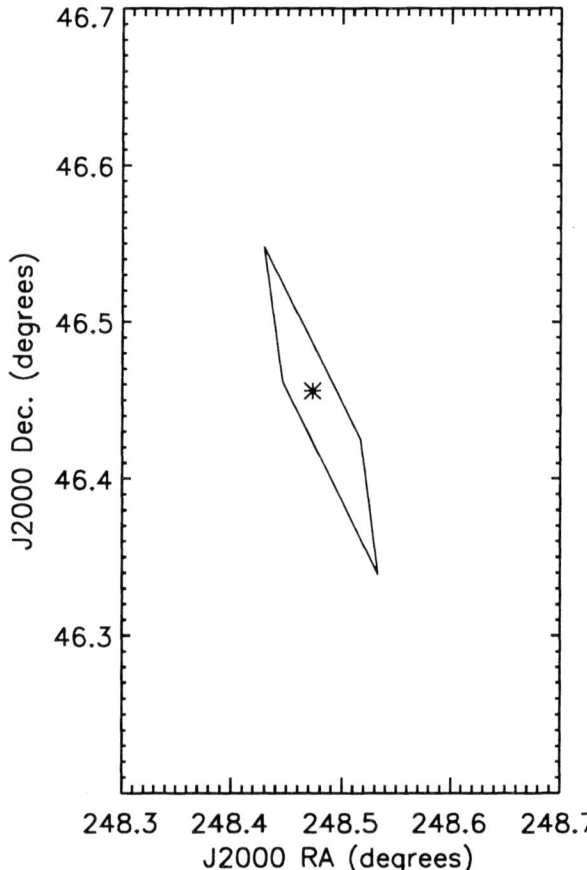

FIGURE 2. A plot of the GRB991208 source field corners alerted by GCN Circular 450, compared with the location of the associated transient found in both the radio and optical domains. This was the first purely-IPN GRB to promote such counterpart observations.

GRB Coordinates Network (GCN): A Status Report

S. D. Barthelmy*, T. L. Cline*, P. Butterworth*, R. M. Kippen[†], M. S. Briggs[†], V. Connaughton[†], and G. N. Pendleton[†]

*NASA-GSFC, Greenbelt,MD 20771
[†]NASA-MSFC, Huntsville,AL 35812

Abstract. The GRB Coordinates Network (GCN) was designed to deliver locations of GRBs to instruments and observers in real-time (a few seconds) — while the burst is still bursting — so that they can make multi-band simultaneous follow-up observations. This goal has been realized with the optical detection of the burst counterpart for GRB990123 by the ROTSE instrument [1]. A brief review of the function and capabilities of the GCN system is given; the types of GRB location information available plus the distribution methods are described. Complementing the real-time location Notices, the GCN Circulars allow follow-up observers to rapidly share the results of their observations with the community. A status report on recent improvements to the GCN system and a list of future improvements is given. One of the key improvements will be the conversion of the semi-manual RBR LOCBURST locations into the fully automated LOCFAST Notices.

INTRODUCTION

The original BATSE Coordinates Distribution Network (BACODINE) started operations in 1993 by distributing GRB locations calculated from the real-time telemetry of the CGRO-BASTE instrument [2]. It has since grown into the GRB Coordinates Network (GCN) by distributing location and light curve information for GRBs detected by all spacecraft capable of detecting GRBs. There are four basic components of the GCN system (the details of GCN are described in Barthelmy et al. [3]). The GCN is a system of computers and programs which: 1) Monitor the CGRO-BATSE real-time telemetry for bursts, calculate their locations, and distribute the locations to interested parties within 5 s after the BATSE trigger. 2) Distribute raw CGRO data to BATSE, COMPTEL, and auto-IPN. 3) Collect GRB location information from other sources (other spacecraft) and distribute them to interested parties. 4) Collect reports from burst follow-up observers and distribute them to the GRB community (the Circulars).

The GCN system is all encompassing and all automatic. GCN collects all the information on GRB locations from all the various sources into a single point and transmits that information to all the various sites. Each site need only develop and maintain one connection for all their GRB needs. There are no humans involved (within the GCN system proper), so there is minimal delay (3–5 s for BATSE information and only a 1–60 s delay after receipt of the information from the other sources (e.g., BeppoSAX, RXTE, COMPTEL, etc.).

The original goal of the GCN (BACODINE) system of observing a GRB in a non-gamma-ray bandpass while the GRB is still busting has been realized. At 9:46:59 UT ($T_0 + 3$ s) on 23 Jan 99, the GCN system detected a GRB in the CGRO-BATSE telemetry stream, calculated a rough location, and distributed that location to the ROTSE instrument located at LANL [1]. The ROTSE instrument is a fully computer-controlled, fast slewing, wide FOV CCD camera instrument connected to the GCN system. ROTSE received the GCN burst location and was on-target and integrating the first of a series of exposures at $T_0 + 22$ s. During the 90-s duration of the burst, ROTSE recorded open-filter magnitudes of 11.8, 8.9, and 10.1. The idea of making observations of GRBs in bandpasses other than the gamma-ray while the burst is still bursting has been proven.

NOTICE TYPES AND DISTRIBUTION METHODS

Table 1 shows all the Notice types available through the GCN system. The time delay after the burst, size of the error box, occurrence rates, and a comment about which types of instruments are suitable to make use of the type are given.

TABLE 1. Sources of GRB locations within GCN.

SOURCE	TIME DELAY	ERROR BOX SIZE	RATE	COMMENTS
Original	5 s	6°–20° dia	1/day	Dedicated, automated, wide FOV
Final	37 s	6°–18° dia	1/day	Dedicated, automated, wide FOV
LOCFAST	2 min	3°–5° dia	50/year	Medium response, medium FOV
Light Curve	5 min	n/a	1/day	Useful for burst assessment
MAXBC	10 min	5°–20° dia	1/day	Fast response, wide FOV
LOCBURST	15–30 min	4°–8° dia	10/month	Medium response, medium FOV
COMPTEL	15–30 min	3°–5° dia	5/year	Medium response, medium FOV
RXTE-ASM	1–2 hr	4' × 15'–150'	8/year	Small FOV telescopes
SAX-WFC	2–3 hr	6'–20' dia	8/year	Small FOV telescopes
RXTE-PCA	2–5 hr	6'–40' dia	6/year	Small FOV telescopes
SAX-NFI	12–48 hr	100" dia	4/year	Small FOV telescopes
IPN	0.5–3 day	4' × 4°–8°	2/month	Small FOV with tiling
ALEXIS	12 hr	0.6° dia	20/year	Small FOV telescopes

All of the Notice types listed in Table 1 are available via the 5 distribution methods: Internet socket, phone/modem, e-mail, pagers/cellphones, and the GCN web site; plus the light curves are also on the web site. The Internet socket connection directly couples the site's instrument control program to the GCN program with a TCP/IP socket connection. It is fast — 0.1–2 s. All the software required at the "site"-end of the connection has been developed and is available [9]. The dedicated phone line is slightly faster (0.3 sec) than the socket method, but seldom worth the expense. The e-mail method uses the standard e-mail protocols with the body of the e-mail having a "TOKEN:value" format suitable for human or computer parsing. The delivery times are typically less than 30 s, which is fine for human destinations. There are several forms of pager and cell-phone methods — you can get beeped by the Universe. There are normal and terse formats, subject-line-only formats, and decimal degrees vs hh:mm:ss formats. The alpha-numeric pagers are convenient because you can get the RA, Dec, Time, and Intensity of the burst no matter your location. They have the same time delays as the e-mail method.

The web site [9] is an archive facility. There are pages within the site that contain all the Notices for each Source instrument (BATSE, COMPTEL, RXTE-PCA/-ASM, BeppoSAX-WFC/-NFI, IPN, and ALEXIS). There are also pages that contain the lightcurves from the CGRO-BATSE, Wind-KONUS, and NEAR-XGRS instruments in alpha-numeric text, JPEG, GIF, and Postscript formats. All of the various web archive pages are updated within 60 s of the Notice being distributed to the community. The web site also archives the published GCN Circulars. It also has the detailed technical description of the system.

GCN FILTERING METHODS

There is a wide range of filters available to sites wishing to customize the quantity and quality of the notices they receive. There are 4 main filtering categories and each filter category acts independent of the other 3 categories. 1) Source type — sites can elect to receive each of the following Source types: BATSE, COMPTEL, RXTE, BeppoSAX, IPN, ALEXIS. Within each Source type, there are one or more sub-types for the instrument and/or the Original, Updated, Final, Will-observe, Won't-observe, Saw-something, Didn't-see-something sub-types. 2) There is a filter category based on the location of the burst on the sky: ALL, VISIBLE, or NIGHT. The VISIBLE filter requires that the RA,Dec location of the bust have an elevation angle at the site greater than $10°$. The NIGHT filter further requires nautical twilight at the site. There are also customized versions which can include declination-based, Sun angle, UT window, and proximity to previous GRBs constraints. 3) There is also a filter based on the size of the uncertainty in the locations, and 4) a filter based on the amount of time from burst to Notice availability — the details for both are discussed in the Improvements section.

SOME GCN STATISTICS

Currently, the GCN system distributes Notices to 112 "sites" involving ∼290 researchers. There are 45 locations with 65 instruments (24 optical, 12 radio, 16 gamma-ray, 7 x-ray, 3 gravity wave and 3 neutrino). There are 9 fully automated instruments; the rest are manual. There are 64 heads-up and/or cross-instrument correlation operations. Approximately 500 burst follow-up observations have been made by the GCN sites on about 380 GRBs. As of Dec. 99 the afterglows from 42 GRBs have been detected using GCN Notices and Circulars, plus the truly simultaneous detection of GRB990123 by ROTSE [1]. As of 30 Dec. 99, 512 Circulars were distributed to a list of 416 recipients.

IMPROVEMENTS TO THE GCN SYSTEM

The GCN system is an evolving and improving system. It is continuously adding new capabilities and new sources of GRB information to provide a better system to the clients. The GCN system is funded by NASA through the CGRO Guest Investigator and SR&T programs. It is envisioned that it will have a long and productive future, and therefore can be counted on for long-range plans of the follow-up community. The following are improvements made within the last 2 years — the dates in parentheses are when the feature was added to the system.

In the past, the BeppoSAX-WFC/NFI burst location notices had been distributed semi-automatically, such that the last step in the sequence involved the GCN Operator formulating a message to be imported into the GCN system for distribution. Now this last step has been automated too (Dec. 99), resulting a decrease in the time delay of 0.5 to a few hours.

The GCN system captures the periodic telemetry down-loads from the Wind and NEAR spacecraft and extracts the data from the KONUS (Oct. 98) and XGRS (Mar. 99) [4] instruments, respectively. It then scans these data looking for sudden increases in the background counting rates. A 2-min section of this count-rate lightcurve is extracted and sent to the IPN system operated by K. Hurley [5]. These lightcurves are cross-correlated with the Ulysses burst detections to produce automated IPN annuli and IPN boxes. The IPN solutions are then sent back to the GCN system for distribution to those sites requesting this type of Notice. The time delay on these IPN solutions is dependent on the rate of telemetry dumps from the spacecraft; typically 1-25 hours for Wind and 1 hour to 3 days for NEAR. (Note that NEAR will enter into orbit around the EROS asteroid in Jan. 2000 and the telemetry dump rate will increase to once per day.) There have been a total of 4 IPN solutions using KONUS data and 3 solutions using the XGRS data, including the IPN solution for GRB991208 which resulted in a radio and optical counterpart detections based on the small error-box of the IPN solution [6]. These IPN solutions based on the Wind-KONUS and NEAR-XGRS contributions are significant because they provide the third node in the IPN triangulation technique to yield true error boxes, instead of the annuli segments using only the Ulysses and BATSE data.

The ASM instrument on RXTE scans about 80 percent of the sky every 90 minutes. D. Smith et al. [8] have created a real-time data analysis system to scan the count-rate data from the ASM to look for hard x-ray transients. When a burst is detected, the location and error box are sent to the GCN for distribution to those sites requesting this Notice type (Jan. 99). It is expected to generate about 8 bursts/year with $4' \times 15'$–$150'$ locations with 5–60 min delays.

Since the number of sources producing small error-box locations is increasing and these error boxes are comparable to the FOV of traditional telescopes, a filter based on the size of the location uncertainty was added. At first, the filters were set at fixed values of $1°$ and $2'$ (Feb. 99 and Oct. 99, respectively), and they were exclusive of the standard ALL, VISIBLE, and NIGHT filters. However, this error-box-based filtering was later changed (Dec. 99) to be continuously adjustable ($1''$ to $360°$) and to be independent of all other filtering criteria.

A filter was added (Dec. 99) that allows sites to filter the Notices they receive based on the time delay between the start of the burst and the time the location Notice is available for distribution. This is useful for the non-automated sites that have a sensitivity limit such that they would not be able to detect an afterglow N hours after the burst. It is continuously adjustable from 1 s to 1000 days.

The GCN has had a system in place since May 97 that sends the BATSE data to the BATSE team at MSFC where they quickly produce a location within 15–30 minutes, which is significantly better than the "ideal physics" locations calculated by the GCN system alone. This LOCBURST procedure [7] (which includes the full response function for the BATSE LADs) will be fully automated and incorporated into the GCN system, thus reducing the time delay to 1–2 minutes. The LOCFAST errors are smaller by a factor of 3 compared to the Original, Final, and MAXBC locations. The manual LOCBURST procedure has been automated into computer code by MSFC and it will soon (Feb. 00) be incorporated into the GCN program.

HETE-2 will be launched in late May of 2000, work is already in progress to incorporate the GRB locations detected by the Wide-field X-ray Monitor and Soft X-ray Camera instruments into the GCN system. There will be about 40 bursts/yr and with error box sizes of $0.2'$–$30'$. And when INTEGRAL launches (Apr. 2001), the GRB locations imaged by the SPI and ISGRI instruments (20/yr at few arcmin with delays of less than 60 s) will be distributed by GCN. And when Swift launches in late 2003, the locations from the BAT, XTR, and UVOT instruments will be also distributed (150–300/yr at $4'$ at $T_0 + 10$ s and $1''$–$2''$ at $T_0 + 60$ s).

REFERENCES

1. Akerlof, C., et al., *Nature* **398**, 400 (1999).
2. Barthelmy, S.D., et al., in *Proc. 3rd Huntsville GRB Symp.*, AIP 384, 1996, pp. 580.
3. Barthelmy, S.D., et al., in *Proc. 4th Huntsville GRB Symp.*, AIP 428, 1998, pp. 99.
4. Barthelmy, S.D., et al., *Proc. SPIE* **3768**, 444 (1999).
5. Hurley, K., et al., these proceedings.
6. Hurley, K., et al., *GCN Circ.* 450 (1999).
7. Kippen, R.M., et al., in *Proc. 4th Huntsville GRB Symp.*, AIP 428, 1998, pp. 119.
8. Smith, D.A., et al., *ApJ* **526**, 683 (1999).
9. The GCN Web Pages URL: http://gcn.gsfc.nasa.gov/gcn.

Super-LOTIS Early Time Optical Counterpart Measurements

H. S. Park[1], R. A. Porrata[1], G. G. Williams[2], E. Ables[1],
D. L. Band[5], S. D. Barthelmy[3], R. M. Bionta[1], T. L. Cline[3],
D. H. Ferguson[6], G. J. Fishman[4], N. Gehrels[3], D. Hartmann[2],
K. Hurley[7], C. Kouveliotou[4], C. A. Meegan[4], R. Nemiroff[8],
and W. Pereira[8]

[1] *Lawrence Livermore National Laboratory, Livermore, CA 94550*
[2] *Dept. of Physics and Astronomy, Clemson University, Clemson, SC 29634-1911*
[3] *NASA/Goddard Space Flight Center, Greenbelt, MD 20771*
[4] *NASA/Marshall Space Flight Center, Huntsville, AL 35812*
[5] *Los Alamos National Laboratory, Los Alamos, NM 87545*
[6] *Dept. of Physics, California State University at Hayward, Hayward, CA 94542*
[7] *Space Sciences Laboratory, University of California, Berkeley, CA 94720-7450*
[8] *Dept. of Physics, Michigan Technological University, Houghton, MI 49931*

Abstract. We present an update on our ongoing effort to establish a dedicated observation program with an automated 0.6 meter telescope system that can detect GRB optical signals from 30 s to many hours after the start of the burst. The Super-LOTIS telescope has a $0.8° \times 0.8°$ field-of-view, is sensitive to V $17 \sim 19$ objects, depending on the integration times, and will be placed at the Kitt Peak National Observatory. This paper presents technical aspects of this telescope and first results from initial operations at LLNL. Utilizing real-time coordinates from BATSE, BeppoSAX, XTE, IPN, HETE-2 and INTEGRAL, our LOTIS and SLOTIS systems will measure prompt GRB optical light curves that will enhance our understanding of GRBs.

INTRODUCTION

Nearly thirty years after the discovery of GRBs, x-ray, optical and radio afterglows now have been observed for a dozen bursts during the last two years [1–4]. These observations finally determined that GRBs are at cosmological distances and have established some of the GRB parameters such as energy, ambient environment and dynamics. However, there is still very little understanding of the nature of the GRB progenitors. Recent detection of a prompt optical signal [5] is inconsistent with the brightness and spectrum of the later time afterglows [6]. This amplifies our needs to measure more simultaneous optical counterparts associated with GRBs. Prompt optical activity measurements will provide clues to understanding of the GRB production mechanism [7].

To search for simultaneous optical counterparts of GRBs, we are operating an automated wide field-of-view telescope at Lawrence Livermore National Laboratory (LLNL) to rapidly image GRB coordinate error boxes distributed by the Gamma-ray burst Coordinate Distribution Network (GCN) [8]. The LOTIS results are given in many papers including the one presented at this conference [9–12]. In this paper, we describe our next generation prompt optical measurement experiment, Super-LOTIS, that will search for quasi-simultaneous GRB optical signals starting 30 s to many hours after the burst with a sensitivity of V 17 $\sim$ 19.

SUPER-LOTIS

The telescope is a Boller and Chivens 0.6-meter reflective telescope of $f/3.5$. Figure 1 shows the telescope. In order to make it a dedicated and automated GRB follow-up telescope, we added computer controllable drives. These drives can point to any part of the sky within 30 s upon receipt of a GCN trigger. We also designed and fabricated a custom 4-element coma corrector to match the point spread function to the pixel scale at the corners of the imaging CCD. The sensor is a LOTIS CCD camera utilizing a Loral 442A 2048×2048 CCD (15×15 μm pixels) with LLNL-built readout electronics. The CCD has thermo-electric cooling (to

FIGURE 1. The Super-LOTIS Boller and Chivens 0.6-meter reflective telescope. We converted it into a dedicated prompt GRB counterpart search telescope by adding computer controlled motors, a 2K × 2K CCD camera and an automated data acquisition system.

−30°C) to minimize dark current and readout noise. Super-LOTIS has a 0.84° × 0.84° field-of-view (1.5″/pixel), which is sufficient for most GRB satellite triggers distributed by the GCN.

Our data acquisition system includes custom readout electronics, a custom hardware power control unit, a weather station and a housing control unit. Extensive on-line scheduling software has been written to handle various triggers. Priority is given to the most recent trigger that has the smallest error box. For example, in response to a GCN/BATSE "Original" trigger, which has only a 5 s delay but a large 15° error box, the telescope begins to systematically acquire a mosaic of images covering the error box. When refined positions are received, i.e., BATSE-LOCBURST, XTE, BeppoSAX, or HETE-2 triggers, the telescope moves to that region and stays at that location the rest of the night. Our scanning strategy and automation allows us to record GRB optical activity as early as 30 s after receipt of the trigger.

FIRST LIGHT AND FIRST EVENT OF SUPER-LOTIS

We have completed installing the motor drivers, coma corrector, CCD camera and the data acquisition system. We imaged the night sky successfully the first

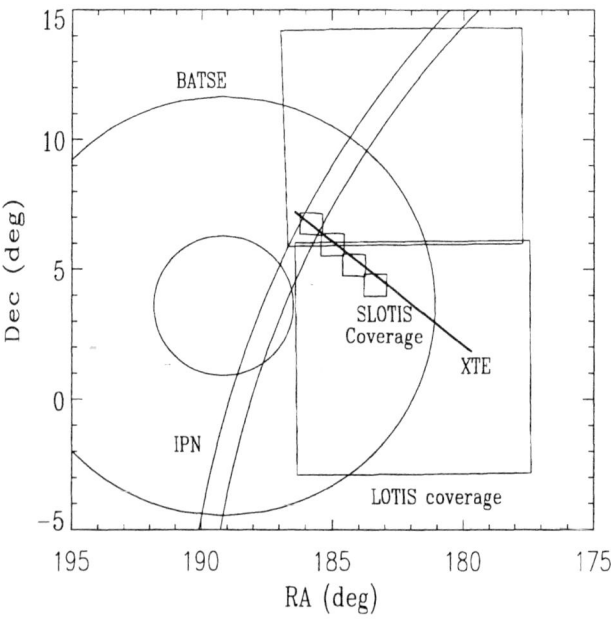

FIGURE 2. Super-LOTIS coverage of GRB990308.

time on Feb. 25, 1999.

While we were testing the on-line software Super-LOTIS obtained early-time observations of the error box of GRB 990308 (BATSE Trig. 7457). This event was detected and localized by the XTE/All-Sky Monitor. Super-LOTIS began a raster scan around the BATSE-LOCBURST GCN coordinates 1700 s after the start of the burst. Four Super-LOTIS images (30 s integration; $t = 1694$ s, 1809 s, 2620 s, and 3923 s) covered most of the XTE/ASM error box within the BATSE 2-σ error circle.

The coverage of this event is shown in Figure 2. Here the circles represent the BATSE 1- and 3-σ errors; the arc represents the IPN error, and the thin line represents the XTE localization. The large boxes represent LOTIS coverage of this event at 132 s after the burst; and the small boxes are the Super-LOTIS coverage during the raster scan. We searched for an optical transient in the area where the error boxes overlap. Even though an optical afterglow for this event has been reported at 3.28 hr and 3.47 hr after the burst [13], Super-LOTIS detected no fading or flaring objects brighter than $V > 15.3$ at 28.2 minutes. The weather conditions and the usage of an uncooled prototype CCD camera at the time prevented us from reaching deeper limits.

We have shown that the Super-LOTIS is already operating and capable of responding to GRB triggers. The Super-LOTIS telescope is currently at LLNL for integration before final installation. We will place it at Kitt Peak National Observatory in early 2000 and will be ready for the HETE-2 Coordinate Distribution. With the LOTIS and Super-LOTIS systems, we will be able to cover GRB optical activity from 10 s to many hours to a magnitude level of V 14 $\sim$ 19. With HETE-2 and other prompt GRB coordinate distributing satellites, we will be able to measure early-time optical activity. Once we detect an optical transient, we will also be able to promptly alert other telescopes.

ACKNOWLEDGMENTS

This work was supported by the U.S. Department of Energy, under contract W-7405-ENG-48 to the Lawrence Livermore National Laboratory .

REFERENCES

1. Costa, E., et al., *Nature* **387**, 783 (1997).
2. Heise, J., et al., *IAU Circ.* 6654 (1997).
3. van Paradijs, J., et al., *Nature* **386**, 686 (1997).
4. Frail, D., et al., *Nature* **389**, 261 (1997).
5. Akerlof, C., et al., *Nature* **398**, 400 (1999).
6. Briggs, M., et al., *ApJ* **524**, 82 (1999).
7. Sari, R., et al., *ApJ* **517**,L109 (1999).
8. Barthelmy, S., et al., *AIP Conf. Proc.* **428**, 99 (1998).

9. Williams, G., et al., these proceedings.
10. Park, H., et al., *ApJ* **490**, L21 (1997).
11. Park, H., et al., *ApJ* **490**, 99 (1997).
12. Williams, G., et al., *ApJ* **519**, L25 (1997).
13. Schaefer, B., et al., *ApJ* **524**,L103 (1999).

Rapid, Deep GRB Observations with The U.S. Naval Observatory 1.3-m Wide-Field Telescope

F. J. Vrba[1], H. C. Harris[1], B. Canzian[1], A. A. Henden[1,5],
S. E. Levine[1], C. B. Luginbuhl[1], D. H. Hartmann[2],
M. C. Jennings[3], and R. M. Kippen[4]

[1] *U.S. Naval Observatory, P.O. Box 1149, Flagstaff, AZ 86002-1149*
[2] *Dept. of Physics and Astronomy, Clemson University, Clemson, SC 29634-0978*
[3] *IGPP, University of California Riverside, Riverside, CA 92521*
[4] *CSPAAR, University of Alabama in Huntsville, Huntsville, AL 35899*
[5] *Universities Space Research Association*

Abstract. An automated, wide-field, 1.3-m aperture optical telescope was recently commissioned at the U.S. Naval Observatory, Flagstaff Station. Employing a multi-filter camera with six SITe 2048×4096 CCDs, it has a field of view of approximately $1.4° \times 1.0°$ with $0.6''$ pixelization. It is capable of imaging at V = 21 mag with S/N = 10 in less than 20 s of integration. While it will be employed primarily for astrometric observations, it is planned to be used also for GRB counterpart searches. The combination of automated operation, moderate aperture, large field of view, and small pixelization should allow for systematic and deeper searches than have previously been attempted, for example, of LOCBURST positions. Examination of GRB localizations that are not x-ray selected may reveal new counterpart properties.

INTRODUCTION

Enormous advances have been made recently in the detection of optical radiation from classical gamma-ray bursts (GRBs). Working mainly from the small x-ray localizations of the *BeppoSAX* satellite [1], standard optical telescopes of moderate to large aperture have been used to identify afterglow radiation from at least 12 GRBs beginning with GRB 970228 to the most recent success, GRB 991208. The 12 successful detections, however, represent less than one percent of the more than 1500 BATSE-detected bursts which occurred during that interval. While approximately 20 other x-ray derived localizations were examined without revealing an afterglow, it is clear that current techniques allow only a small fraction of GRBs to be intensively examined for lower energy radiation. The HETE-2 satellite, if launched successfully, will provide additional small localizations, but likely at a rate no greater than that of *BeppoSAX*. Small Earth-based telescopes with large fields of view show that bright counterparts might be found for at least some GRBs, such as GRB 990123 ($V \leq 9$) [2]. However, the current observational statistics from these telescopes indicate that the brightness of the prompt optical radiation from GRB 990123 may have been a relatively rare event [3,4].

Taken together, these results indicate the usefulness of a moderate aperture telescope with a large enough field of view to be able to utilize the numerous BATSE observations that are reduced and distributed in the form of LOCBURST localizations. In this paper we describe a new 1.3-m telescope at the USNO, Flagstaff Station and its potential for observing GRB optical counterparts.

THE USNO 1.3-M WIDE-FIELD TELESCOPE

The U.S. Naval Observatory, Flagstaff Station's 1.3-m telescope provides a moderately wide field observing capability for a variety of astrometric and photometric survey programs. Among these are: astrometry tied to the Hipparcos/Tycho catalog, photometry and astrometry supplements to other surveys such as the Sloan Digital Sky Survey, variable star surveys, astrometry of artificial satellites, and various other wide-field projects such as the search for GRB optical counterparts.

The 1.3-m telescope (Figure 1) was built by DFM Engineering and was commissioned during mid-November 1999. The telescope employs a modified Ritchey-Chrétien design and a Gascoigne-type corrector lens giving good quality images

FIGURE 1. The 1.3-m telescope at the U.S.N.O. Flagstaff Station.

and a flat focal plane over a 1.7 degree diameter field of view. The telescope is enclosed in an 18.5 foot Ash dome with louvered ventilation to minimize dome seeing (Figure 2).

The telescope will employ a camera, currently nearing completion, which uses six SITe 2K×4K, thinned, three-side buttable CCDs with 15 micron pixels. With a telescope focal plane scale of $40''$/mm this provides a scale of $0.6''$/pixel and a total camera field of $1.0° \times 1.4°$. The camera filter changer has positions for seven 6-inch square filters. Initially, the filters used will be U,B,V,R,I, SDSS r$'$, and a narrow-band, low-transmission red filter for bright stars. The 1.3-m camera uses an ICCD camera for guiding and a high-speed pneumatic shutter for accurate exposure timing. The system can be used in either stare- or scan-mode.

Table 1 provides estimates of the exposure times needed to reach various S/N ratios for a point source detection in the V-band. These estimates are based on typical $1.2''$ FWHM seeing and a sky background of $V = 21.4$ mag/arcsec^2 (measured during dark time at the Flagstaff Station).

Due to the nature of its work, telescope operation is automated. Fields observed will normally be selected during the night by software from several program lists depending on observing conditions and program requirements. The observing queue can be autonomously interrupted at any time to respond to LOCBURST, *BeppoSAX*, HETE-2, or other GRB notifications. Due to the enormous quantity of data that will be generated both in normal- and interrupt-mode observing, data will be processed immediately after being taken.

FIGURE 2. The 1.3-m telescope dome at the U.S.N.O. Flagstaff Station.

TABLE 1. Estimated exposure times (in minutes) needed to reach given S/N ratios.

V (mag)	S/N = 10	S/N = 30	S/N = 100
20	0.1	0.8	8
21	0.3	3	35
22	1.5	15	
23	10	100	
24	55		

PROSPECTS FOR GRB COUNTERPART DETECTION

With a slew rate of approximately 3 deg per second, a full CCD array readout time of 90 seconds, and the ability to interrupt the observing queue, the 1.3-m telescope will be capable of acquiring multiple images of a GRB position, reaching well below 20th magnitude, within minutes of its reporting. Combining the accessible area of the sky and the weather statistics of the Flagstaff site with the fraction of the sky not obscured by the Sun, we estimate that approximately 20% of all GRBs can be observed by the 1.3-m telescope within 24 hours of the burst.

It is estimated that the HETE-2 satellite will localize as many as 30 GRBs per year with accuracies between 10 arcsec – 10 arcmin and distribute them in near real time. Although not taking advantage of its wide field capabilities, the 1.3-m telescope will be capable of rapid follow up observations of about six HETE-2 GRBs per year, using the above statistics.

The wide-field capabilities of the 1.3-m can be employed to search for optical radiation from BATSE-detected GRBs when LOCBURST positions are available. The current LOCBURST system provides coordinates within 30 minutes of the burst. The automated LOCFAST system, expected to be in operation soon, will provide coordinates within one minute, thus allowing a 1.3-m response similar to that for HETE-2 data. LOCBURST positional uncertainties are very close to those of the final locations published in the BATSE catalogs for which an empirical model has been determined [5]. Based on the combined BATSE systematic and statistical errors, we have computed probabilities for observing the true GRB position as a function of field of view and BATSE statistical errors. In Figure 3 we show the number of true GRB positions that would actually be observed with the 1.3-m telescope per year as a function of statistical error and for several sensible telescope pointing rasters. At the bottom of the graph in parentheses are the number of such events per year expected to be accessible from Flagstaff. For instance, if we chose to limit ourselves to GRB events with a statistical error radius of $\leq 2.0°$ (8.4 events per year), we would expect to include the GRB position within our imaged raster for about 4.5 of these events using a 3×4 imaging pattern. These statistics include the 20% visibility and weather factor from above. Thus, with modest observing efforts, approximately four true GRB positions should be observed each year, based on LOCBURST localizations alone.

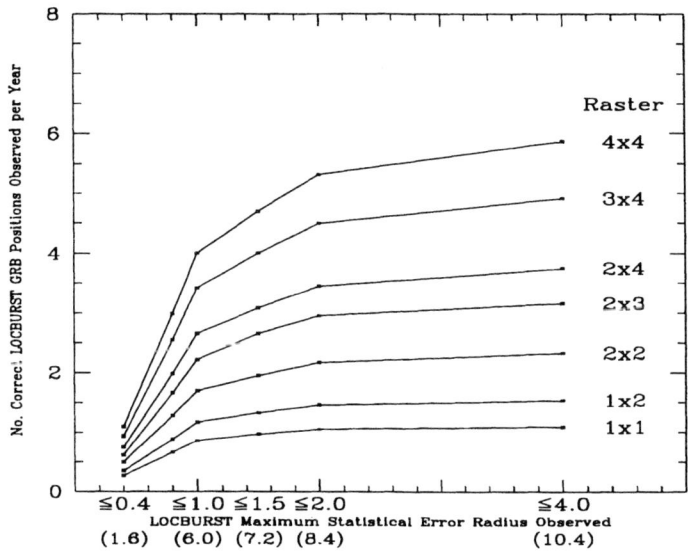

FIGURE 3. Estimated number of correct GRB positions observed per year with the 1.3-m telescope based on LOCBURST positions.

CONCLUSIONS

The new USNO wide field 1.3-m telescope in Flagstaff is capable of providing optical observations for about four LOCBURST/LOCFAST and six HETE-2 localized GRBs per year, in addition to localizations that may still be provided by BeppoSAX. Coverage of four additional GRBs from LOCBURST/LOCFAST is a small, but important, addition to those events likely to be observed in the post-HETE-2 era, since these events will not be x-ray selected as have most of the other localizations observed successfully thus far.

REFERENCES

1. Boella, G., et al., *A&A Supp. Ser.* **122**, 299 (1997).
2. Akerlof, C., et al., *Nature* **398**, 400 (1999).
3. Kehoe, R., et al., *Astro-Ph/9909219*, in press (1999).
4. Park, H.S., et al., *A&A Supp. Ser.* **138**, 577 (1999).
5. Briggs, M.S., et al., *ApJ Supp.* **122**, 503 (1999).

TAROT-2: A Versatile Large Observatory for Optical Transients

Michel Boër

Centre d'Etude Spatiale des Rayonnements (CESR/CNRS),
BP 4346, 31028 Toulouse Cedex 4 France

Abstract. TAROT-2 (Télescope à Action Rapide pour les Objets Transitoires — Rapid Action Telescope for Transient Objects) represents a major step towards a relatively large observatory dedicated to the rapid observation of optical transients or variable sources. This completely autonomous observatory will feature a 1.5 meter telescope, with an imagery mode. A spectroscopic mode will be implemented, possibly on another telescope, linked to the imaging telescope as well as with TAROT-1. TAROT-2 will be fully autonomous, i.e., no human intervention will be required from the telescope schedule to the processing of the observations. The instrument will be networked with other instruments. New materials will be heavily used for all the optical and mechanical parts. The imaging telescope will have a 2° field of view, and a maximum slew speed of 60 deg/s. A 8000×8000 pixel CCD camera will give a pixel size of 0.8″. If a new, interesting, source is detected by the imaging telescope, its position will be automatically sent to the spectroscopic telescope for immediate observation, i.e., for a GRB, while the source still has a large optical flux level.

INTRODUCTION

There are three main categories of scientific objectives which may benefit from the use of autonomous, automatic telescopes with large fields of view:

- Survey of a large number of objects

- Serendipitous discoveries based on wide field surveys

- Target of opportunity observations

Conventional observatories are not well designed for these objectives. Survey tasks require a significant amount of time, both for observing and for reducing the data, resulting in a considerable investment in people for doing these tasks. A large number of TOO observations will result in a competition for observing time with "normal" programs, and they will be scheduled in general with large delay, and usually not in the best instrumental configurations.

Automated telescopes may be an answer to these problems. They may respond quickly to TOOs, while spending a major part of their observing time on long, time consuming, survey programs. Their instrumentation is always in the same configuration, simplifying the problem of the data reduction software and pipeline. Many of them have been built to respond to a particular scientific project: this is the case of the Télescope à Action Rapide pour les Objets Transitoires (TAROT, Rapid Action Telescope for Transient Objects [1,2]), already in use at the Calern observatory (France).

Prompted by the need to reach better sensitivity, we decided to study a new and larger instrument, called TAROT-2. We present here the specific aspects of TAROT-2, and the new materials and techniques we plan to use.

SCIENTIFIC OBJECTIVES

Cosmic Gamma-Ray Bursts

Cosmic Gamma-Ray Bursts (hereafter GRBs) are among the most prominent scientific objectives of TAROT-2. TAROT-2 will be able to reach at least magnitude 22 in several seconds, and to slew to any target at a speed of 60 deg/s. In addition, we plan to have a specific instrument, linked to TAROT-2, able to perform broad-band spectroscopy of selected sources. With only one detection [3] over more than 20 observed sources at optical wavelengths, it is unclear whether optical transients associated with GRBs are rare or faint (or both). This point applies also to the afterglow at optical and radio wavelengths. The specific GRB objectives for TAROT-2 will be the following:

1. TAROT-2 will be able to detect and localize the optical burst associated with GRBs up to sources of magnitude 22 for a 10 second observation. These observations have to be performed with any of the dedicated satellites in orbit at that time: HETE-II, CGRO/BATSE, SWIFT, INTEGRAL, BALLERINA, GLAST, etc.

2. Follow-up observations of the GRB afterglows for several days, with the same instrument. Specifically, the transition between the burst and afterglow regime will be studied, and eventually correlated with data taken at other energies.

3. Sample the optical transient light curve with good temporal resolution, and correlate it with the gamma-ray data.

4. Connected (or not) with the topic of jets in GRBs, TAROT-2 should be able to detect "orphan" optical transients, i.e., OTs presumably associated with undetected GRBs. Probably, both for observational and physical reasons, TAROT-2 will only be able to detect the afterglow coming from these objects.

5. Provide good, sub-arcsecond, localizations of detected events to other instruments within about one minute

6. Derive the distance of detected events as quickly as possible

We note that objective 1 implies a large telescope with fast slewing capabilities, and a good connection to the INTERNET. Also, the field of view has to match the worst accuracy provided by the above mentioned satellites. Objective 2 and 3 requires a good sensitivity, and a reliable photometry. In addition, objective 3 requires a large field of view in order to span a large area of the sky at an acceptable rate, and a high duty cycle, as well as low dead-time. Objective 4 is achievable if the optical PSF and the pixel size is small enough, and if the telescope site has a sub-arcsecond seeing. Finally, objective 5 requires an associated spectroscopic capability, though with low resolution. We note also that many of the above mentioned items are achievable only if the data processing and the tools to recognize the sources are rapid enough, i.e., if within a time on the order of one minute, the system is able to process the images, to recognize the sources of "interest", and to disseminate their positions.

Other Objectives

In addition to the above goals, and because of the low rate of GRB observations (about 1 every 15 days with BATSE, a comparable rate with Swift [4]), TAROT-2 will be able to address many scientific objectives. They are in general connected with celestial variability. TAROT-2 will try to catch several exo-planets, using the transit method. We plan to have also simultaneous observation of several objects in conjunction with high-energy detectors, both on Earth and in space. TAROT-2 will also be able to study Kuiper-type objects. Finally, TAROT-2 will probably catch quite a lot of Earth-crossing asteroids and space debris.

TECHNICAL ASPECTS

The above mentioned objectives lead to a telescope about 1.5 m in size. One of the problems is to be able to move such an instrument in a time compatible with the objectives. In order to be able to slew to any target in a time of 1 or 2 seconds, the telescope has to reach a speed of 60 deg/s. We decided to select a ceramic, Silicon Carbide (SiC), to achieve these objectives. The advantages of using SiC for TAROT-2 are the following:

- SiC has a large Young modulus resulting in a high stiffness. The telescope elements will be almost unaffected by gravity.

- SiC has an excellent thermal conduction coefficient, resulting in reduced polishing times, avoiding thermal gradients and the need of any thermal control, since all the telescope will be at the same (outside) temperature.

TABLE 1. TAROT-2 main technical characteristics.

Optics	
Aperture	1.50 m, f/2.7
Field of view	2° × 2°
Optical Resolution	20 μm
Mount	
Mount type	Alt-Azimuthal
Slew speed	Max. 60 deg/s
Traking speed	Adjustable 0–40 deg/s
Weight of moving parts	≤ 150 kg (with instrumentation)
Detectors	
Detector type	CCD mosaic
Size	2000 × 2000 pixel, 12 × 12 cm
Pixel size	15 μm, 0.8″
Readout speed	2 s (full mosaic)
Readout noise	≤ 8 e$^-$
Spectrometer	
Slew speed	10 deg/s
Detector size	2000 × 2000 pixels
Readout	1sec, ≤ 8 e$^-$
Data flow and links	
Required link data rate	128 kbps (no remote archiving)
Image size	128 MB
Image rate	500 frame/night
Data amount	15–20 TB/year

- The structure and all the optical parts will be made of the same material. Because of the above mentioned points, it will be largely simplified, and we will not have any barrel, actuators, etc. Thus, the weight and system complexity will be further reduced.

Our goal is to have a weight of the telescope moving parts of about 100 kg, or 150 kg including the instrumentation, keeping the moment of inertia quite low.

In addition to the main imaging telescope, we plan to have a second instrument dedicated to the spectroscopy of "interesting" sources, though with low resolution. This instrument, as autonomous as TAROT-2, will be linked to TAROT-2, and will get the source positions from it.

TAROT-2 will have also a large field of view, 2 degrees, together with a good sampling: at this time we plan to have a pixel size of 0.8″. This results in a 8000 × 8000 pixel camera, probably a mosaic of CCD chips, read out in a time ∼1 s. The resulting data rate will be 128 MB/s, and for a mean exposure time of 1 min in routine mode, and 10 s in alert mode, about 15–20 TB will have to be archived and scientifically exploited per year. The challenge here is not only to develop the tools needed to cope with the tremendous amount of data, but also to operate the telescope scientifically, i.e., to automatically recognize the interesting phenomena (e.g., the burst source, or the extrasolar planet), to send the position around the globe within minutes, and to acquire spectra of the interesting objects. This means a far higher degree of autonomy than that reached already with TAROT-1.

This huge amount of data and the high degree of system complexity, as well as the networking aspects and the dissemination to remote users, implies that TAROT-2 will have to be accessible from the world wide web in a very flexible way, and that high level tools will have to be available to navigate the data.

Table 1 gives a summary of the foreseen characteristics of TAROT-2.

CONCLUSIONS

TAROT-2 will feature a large number of developments and will be a prototype of a new generation of autonomous instrument. From its conception it will be the first networked observatory, linking a photometric/imaging telescope, a spectroscopic instrument, and the already existing TAROT-1 experiment. Other automatic telescopes may be added, e.g., to perform continuous, round the Earth monitoring of a GRB afterglow. Another possibility will be the efficient use of the instrument network, the scheduling algorithm being responsible for sending the right requests to the right instrument. As an example, if a new source is discovered, the scheduler may send the follow-up observations to TAROT-1 while the source is bright enough, then to TAROT-2. At the same time, spectra of the source will be taken. When the source starts to be too far from the transit, another instrument may handle the observations. The TAROT-2 project is currently under study and review by several institutions in France, and we hope to begin the development by the year 2000.

ACKNOWLEDGMENTS

TAROT-2 is at present supported by the Ministère de l'Education, de la Recherche et de la Technologie and the Université Paul Sabatier, Toulouse.

REFERENCES

1. Boër, M., et al., *A&ASS* **138**, 579 (1999).
2. Boër, M., et al., these proceedings.
3. Akerlof, C., et al., *Nature* **398**, 400 (1999).
4. Gehrels, N., these proceedings.

Milagro: A TeV Observatory for Gamma-Ray Bursts

B.L. Dingus[1], R. Atkins[1], W. Benbow[2], D. Berley[3,10],
M.L. Chen[3,11], D.G. Coyne[2], D.E. Dorfan[2], R.W. Ellsworth[5],
D. Evans[3], A. Falcone[6], L. Fleysher[7], R. Fleysher[7], G. Gisler[8],
J.A. Goodman[3], T.J. Haines[8], C.M. Hoffman[8], S. Hugenberger[4],
L.A. Kelley[2], I. Leonor[4], M. McConnell[6], J.F. McCullough[2],
J.E. McEnery[1], R.S. Miller[8,6], A.I. Mincer[7], M.F. Morales[2],
P. Nemethy[7], J.M. Ryan[6], B. Shen[9], A. Shoup[4], C. Sinnis[8],
A.J. Smith[9,3], G.W. Sullivan[3], T. Tumer[9], K. Wang[9], M.O. Wascko[9],
S. Westerhoff[2], D.A. Williams[2], T. Yang[2], and G.B. Yodh[4]

(1) University of Utah, Salt Lake City, UT 84112, USA
(2) University of California, Santa Cruz, CA 95064, USA
(3) University of Maryland, College Park, MD 20742, USA
(4) University of California, Irvine, CA 92697, USA
(5) George Mason University, Fairfax, VA 22030, USA
(6) University of New Hampshire, Durham, NH 03824, USA
(7) New York University, New York, NY 10003, USA
(8) Los Alamos National Laboratory, Los Alamos, NM 87545, USA
(9) University of California, Riverside, CA 92521, USA
(10) Permanent Address: National Science Foundation, Arlington, VA 22230, USA
(11) Now at Brookhaven National Laboratory, Upton, NY 11973, USA

Abstract. Observation of prompt TeV γ-rays from GRBs requires a new type of detector to overcome the low duty factor and small field of view of current TeV observatories. Milagro is such a new type of very high energy (> a few 100 GeV) gamma-ray observatory, which has a large field of view of > 1 steradian and 24 hours/day operation. Milagrito, a prototype for Milagro, was operated from February 1997 to May 1998. During the summer of 1998, Milagrito was dismantled and Milagro was built. Both detectors use a 80m×60m×8m pond of water in which a 3m×3m grid of photomultiplier tubes detects the Cherenkov light produced in the water by the relativistic particles in extensive air showers. Milagrito was smaller and had only one layer of photomultipliers, but allowed the technique to be tested. Milagrito observations of the Moon's shadow and Mrk 501 are consistent with the Monte Carlo prediction of the telescope's parameters, such as effective area and angular resolution. Milagro will have improved flux sensitivity over Milagrito due to larger effective area, better angular resolution and cosmic-ray background rejection.

TeV gamma-rays are a logical consequence of fireball models of gamma-ray bursts [1]. In these models the gamma rays can be created by synchrotron radiation of protons [2] or inverse Compton scattering of electrons accelerated in internal [3] or external [4] shocks of the ultra-relativistic energy flow. In most of these scenarios, determining the upper energy end of the spectra provides strong constraints on the physical conditions of the emitting region. EGRET observations of GRBs show no indication of a high-energy cut off, and the average spectrum of the four brightest bursts detected by EGRET has a differential spectral index of 1.95±0.25 extending up to 10 GeV [5].

However, gamma-ray emission above a few 100 GeV may not be observable for sources at redshifts much greater than 0.5 because of pair production with infrared extragalactic background photons [6,7]. Alternatively, if the distance to a gamma-ray burst source can be identified by optical redshift measurements, then the absorption of high-energy gamma-rays due to pair production can be used to measure the infrared intergalactic photon density. The spectrum produced by interactions with intergalactic photons will have a sharp exponential cutoff, and can be distinguished from the attenuation due to photons in the source, which result in a slower bend in the spectrum [8]. TeV gamma rays from GRBs can also be used to constrain theories of quantum gravity by measuring the constancy of the velocity of light for different energy photons [9].

At energies greater than 30 GeV, gamma-ray fluxes become too small for current satellite-based experiments to detect because of their small sensitive areas. Only ground-based experiments have large enough areas [10]. These instruments detect the extensive air showers produced by the high energy photons in the atmosphere, thus giving them a much larger effective area at high energies. These showers can be observed by detecting the Cherenkov light emitted by the cascading relativistic particles as they traverse the atmosphere, or by detecting the particles which reach ground level.

TeV gamma-ray emission from several astrophysical sources has been detected using atmospheric Cherenkov telescopes. These instruments have extremely large collection areas ($\sim 10^5$ m^2) and good hadronic rejection. Unfortunately, they have relatively narrow fields of view (a few degrees) and can operate only on dark clear nights, resulting in a low duty cycle. They are therefore ill suited to search for transient sources such as GRBs. Searches for GRBs at energies above 300 GeV have been made by slewing these telescopes within a few minutes of the notification of the GRB location [11]. No detections have been reported. However, because of the narrow field of view, coupled with the delay in slewing to the correct position, there have not been any prompt TeV gamma-ray observations at the GRB location.

A NEW TYPE OF TEV γ-RAY OBSERVATORY

Milagro, a new type of TeV gamma-ray observatory, is ideally suited to observe TeV emission from gamma-ray bursts due to its large field of view of > 1 sr and > 90% duty factor. Milagro is located in the Jemez Mountains near Los Alamos,

NM (106.7° W, 35.9° N, 2650 m above sea level). The observatory just became operational in fall 1999, but a smaller the prototype, called Milagrito, was operated at the same site from February 1997 to May 1998 [12].

Both detectors used the large pond of water 80m×60m×8m which can be seen in the photograph of Figure 1. The pond has a light-tight cover. Milagro contains 723 photomultipler tubes (PMTs) which are placed on a 3m×3m grid in 2 layers at 1.5m and 6m below the surface. The prototype Milagrito had only one layer of 228 PMTs on a 3m×3m grid spread over the smaller area of 30m×50m.

FIGURE 1. Aerial photograph of the 60m×80m×8m pond and cover used by Milagro and Milagrito. The pond as instrumented for Milagro contains 5 million gallons of water.

An extensive air shower is detected when the relativistic particles radiate Cherenkov light in the water causing several tens of PMTs to observe the light within a few 100 nsec of each other. From the relative timing of the photomultiplier tube signals, the direction of the particle or gamma-ray initiating the shower can be determined to ∼1 degree depending on the number of PMTs hit. The field of view is such that 50% of the showers detected are within 20 degrees of zenith and 90% are within 50 degrees. Almost all of these triggers are due to the background of showers that are initiated by charged cosmic rays. Monte Carlo simulations correctly predict the observed rate and zenith angle distribution of cosmic-ray initiated showers. Simulations of γ-ray initiated showers show sensitivity to γ-rays as low as ∼100 GeV with the effective area increasing as $\sim E^2$, where E is the γ-ray energy, and flattening near 10 TeV.

MILAGRITO OBSERVATIONS

The expected performance of Milagrito has been confirmed by observations. The Moon blocks cosmic rays and a deficit of showers has been detected, which is a 10σ deviation from the background (Figure 2) [13]. The shape and size of the deficit is consistent with the Moon's angular size and Milagrito's angular resolution, and the deflection in right ascension (R.A.) is consistent with the Earth's magnetic field and the energy of the cosmic rays detected by Milagrito.

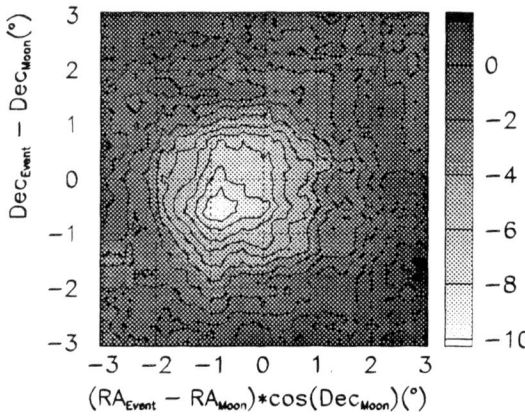

FIGURE 2. The shadow of the Moon due to the blockage of cosmic rays. The scale is in standard deviations of the Gaussian distributed background. The deficit is not located at the direction to the Moon, but is deflected in R.A. because the trajectories of the charged cosmic rays are bent by the Earth's magnetic field.

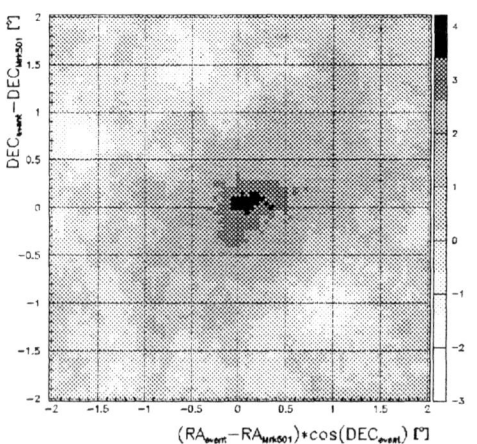

FIGURE 3. The number of showers near Mrk501 plotted in standard deviations of the background.

The simulations of gamma-ray initiated showers were verified by Milagrito observations of Mrk501, an x-ray selected BL Lac, which was a bright TeV source during 1997. The Milagrito detection of Mrk501 shown in Figure 3 was a $3.7sigma$ deviation from the cosmic-ray background [14]. The flux and spectrum were well measured by several atmospheric Cherenkov telescopes and the significance of the Milagrito detection agrees with this spectrum folded with the Milagrito effective area. Simulations also indicate that the sensitivity of Milagrito was too low to detect the Crab nebula, the standard candle of TeV astronomy.

New observations have been performed by Milagrito that atmospheric Cherenkov telescopes have not been able to do. Specifically, an all sky survey of the Northern Hemisphere was performed and no steady sources brighter than 5 times the Crab nebula flux were detected [15]. An all-sky search for 10 second duration transients was also done and none were found [16], which places limits on the local density of evaporating primordial black holes. However, the most interesting result was the evidence of TeV emission from GRB970417a. During T90, 18 events were observed when 3.46±0.11 were expected for a direction 3.8° from the position determined by BATSE. The probability is 2.8×10^{-5} that such an excess is due to a fluctuation of the background anywhere within the BATSE 90% confidence interval (statistical

+ systematic) of 9.4° radius. The probability for any of the 54 bursts that were within Milagrito's field of view to have such an improbable excess is $54 \times 2.8 \times 10^{-5} = 1.5 \times 10^{-3}$. More details were reported separately at this conference [17].

MILAGRO EXPECTATIONS

Milagro has a larger effective area and better angular resolution than Milagrito. Milagro is also be able to reject some of the cosmic ray background due to the addition of a lower layer of photomultiplier tubes, which Milagrito did not have. Milagro began taking data in December 1999 and will soon be operating at at a rate of $\sim$2000 showers per second (Milagrito triggered on $\sim$300 showers per second) resulting in more than 100 GBytes of data per day.

If the Milagrito result from GRB970417a is not a random fluctuation of the background, then the improvements of Milagro over Milagrito will produce exciting results about gamma-ray bursts. The increased effective area will result in observations of weaker but still low-redshift bursts, and more TeV γ-rays in brighter bursts will allow correlations between TeV and sub-MeV light curves. The improved angular resolution will yield localizations for rapid TeV, x-ray, optical, and radio afterglow observations. And the improved energy resolution will measure the highest energy γ-rays from a gamma-ray burst.

Milagro is supported in part by National Science Foundation, Department of Energy, Los Alamos National Lab, University of California, and CalSpace.

REFERENCES

1. Meszaros, P., Rees, M. J. & Papathanassiou, H., *ApJ* **432**, 181 (1994).
2. Totani, T., *ApJL* **509**, L81 (1998).
3. Pilla, R. P. & Loeb, A., *ApJL*, **494**, L167 (1998)
4. Dermer, C.D., Chang, J. & Mitman, K.E., astro-ph/9910240 (1999) and these proceedings
5. Dingus, B.L., Catelli, J.R., & E.J. Schneid, "4th Huntsville GRB Symposium", *AIP* **428**, 349 (1998).
6. Primack, J.R. et al., *Astroparticle Physics* **11**, 93 (1999).
7. Salamon, M.H. & Stecker, F.W., *ApJ* **493**, 547 (1998).
8. Baring, M.G., "Towards a Major Atmospheric Cherenkov Detector V", (South Africa), 113 (1997).
9. Amelino-Camelia et al., *Nature* **393**, 763 (1998).
10. Hoffman, C. M., Sinnis, C., Fleury, P., Punch, M. *Rev. of Mod. Phys.* **71(4)**, 897 (1999).
11. Connaughton, V. et al., *ApJ* **479**, 859 (1997).
12. Atkins, R. et al., *Nucl. Inst. and Methods* (2000) (in press) and astro-ph/9912456.
13. Wascko, M. et al., *26th ICRC Proceedings* (Salt Lake City,UT) (1999).
14. Atkins, R. et al., *ApJL* **525**, L25 (1999).
15. Smith, A.J. et al., *AIP Proc. GeV-TeV γ-Ray Astrophysics* (Snowbird, UT) (2000).
16. Sinnis, C. et al., *26th ICRC Proceedings* (Salt Lake City,UT) (1999).
17. McEnery, J. et al., astro-ph/9910549 (1999) and these proceedings.

Searching for Optical and High-Energy Transients from Gamma-Ray Bursts Simultaneously With Čerenkov Telescopes

A. Piccioni[1], C. Bartolini[1], G. Beskin[2], S. Biryukov[3], D. Eichler[3], D. Faiman[3], and A. Guarnieri[1]

[1] *Dipartimento di Astronomia, Università di Bologna, Italy*
[2] *Special Astronomical Observatory, Nizniji Arhkiz, Russia*
[3] *Ben-Gurion University, Beer-Sheva, Israel*

Abstract. The observation of OTs connected to GRBs seems to have confirmed the cosmological hypothesis. However, a detailed knowledge of the source behavior in different spectral bands, with high time resolution and maximum possible bandwidth, before and during the GRB emission is necessary to understand the nature of these sources. In order to obtain this goal, the use of very large, wide field *bad* mirrors is proposed and justified. The dependence of detection limits from size, beam aperture, transparency, background level, time exposition, and detector quantum efficiency are considered for huge Čerenkov telescopes; the possibility of a contemporaneous observation in the optical band and in the VHE band with the same mirror, by using two different mirror/detector configurations, is suggested.

INTRODUCTION

To understand the physics that underlies the GRB events, the observational efforts must be extended in time and up in frequencies to obtain a full description of the development of the GRB light curves

1. before and during the GRB emission phase,

2. in all possible spectral bands,

3. with high time resolution.

The first condition was never fulfilled; only in the case of GRB 990123 a very close optical observation was performed by ROTSE [1]; however, this time delay does not allow one to firmly state a correlation between the optical transient and the gamma activity phase. Starting the observations *before* the event, at a temporal resolution similar to or better than that of gamma instrumentation would allow us to compare the optical with the gamma-ray behavior.

The second condition that has to be considered a priority owing to the non thermal characteristic of the energy flux associated to the GRB events. Contemporaneous observations in all possible bands should be performed, simultaneously with the gamma-ray measurements. In this context, no special effort should be done for the identification of the optical counterpart, leaving this target to standard observations of the optical afterglow. The fast evolution of the OT of GRB 990123 could be enough to justify the third request. However, additional interest in very high temporal resolution comes in principle because the minimal observable time structure in the frame of the fireball model is related to the dimension of the fireball. If $\Delta\tau$ is the smallest observed time structure and ΔT is the smallest time structure in the fireball rest frame, then $\Delta\tau \sim \Delta T \gamma^{-2}$; so by measuring $\Delta\tau$ it is possible to have information about the γ factor, and from the maximum energy of photons $E_{max} \sim \gamma$ it is possible to infer information about the energy release, connected to the maximum energy of photons, independently on the distance of the parent body. We already suggested the possibilities to solve the mentioned problems by big telescopes with *bad* mirrors and wide fields [2]. We develop these ideas here.

UNCONVENTIONAL INSTRUMENTATION

From the point of view stressed in this paper, we propose a new approach to the observational problem. This requires

1. large mirrors, to have high photon fluxes,

2. photoelectric, position sensitive cameras, to have high temporal resolution,

3. a combination of different measure techniques, to have contemporaneous observations of bursts in different energy bands,

4. use of existing instrumentation, to reduce costs and times,

5. continuous tracking of the field observed by a satellite with a few wide field instruments, or continuous scanning of the same field at a speed high enough to finish the cycle before the OT complete decay.

Suitable existing instruments for these kind of observations are paraboloidal mirror arrays used for Čerenkov light pulse measurements [4].

DETERMINATION OF DETECTION THRESHOLD

An estimate of the sensitivity of these kind of instruments in the optical band can be easily performed [2] by analyzing the detection limit corresponding to a set of mean experimental conditions. We will take into account a mirror array diameter of D meters, a pixel size of $d \times d$ arcsec^2, a detector quantum efficiency η, a transparency of the atmosphere τ, a typical local normalized background m_{bg} (referred to a 1 m^2 mirror surface and to a 1 arcsec^2 beam aperture) and a measurement time T. Assuming as R-band photon flux calibration $F_R = 2.5 \times 10^{-26}$ erg·cm^{-2}s^{-1}Hz^{-1}, for R = 15 [3] the corresponding photon flux is $N_R = F_R \Delta\nu D^2/2h\nu = 4900 D^2$ phot·s^{-1}. From this value, the count number for a background magnitude m_{bg} is

$$N_{bg} = 30.9 D^2 d^2 \eta \tau T 10^{-0.4(m_{bg}-20.5)} \text{cnts} \cdot \text{px}^{-1}.$$

Assuming Poisson statistics of background photocounts, we find for the minimum detectable signal (optical transient) $\alpha\sigma_{bg} = \alpha\sqrt{N_{bg}}$, where α is the detection threshold level. We find that the faintest detectable magnitude of a flare is

$$m_p = 2.5 \log (30.9 \eta \tau T)^{1/2} D(\alpha d)^{-1} + 0.5 m_{bg} + 10.25. \quad (1)$$

In the case of non-Poisson noise, the background variance σ_b depends on the long-time variability of background intensity, i.e., $\sigma_b = \zeta N_{bg}$, where ζ is much smaller than 1, and is the relative amplitude of this variability. As a magnitude limit in this case we find

$$m_n = -2.5 \log(\alpha \zeta d^2) + m_{bg}. \quad (2)$$

We use the non-Poisson limit under the condition $\sigma_b = \sigma_{bg}$, or

$$\zeta d D (30.9 \eta \tau T)^{1/2} = 1 \quad \text{for} \quad m_{bg} = 20.5. \quad (3)$$

To do a comparison between the restriction coming from the Poisson detection threshold and that resulting from background variability, formula (1) and formula (3), computed for $\eta = 0.6$, $\tau = 0.7$, $\alpha = 3$, $m_{bg} = 20.5$, different T (1 s, 1 ms) and ζ (10^{-3}, 10^{-4}) are reported in Figure 1. The two line families show how the two conditions concur to the observability of the sources, depending on the mirror diameter and beam aperture. Any object can be observed only if the point corresponding to the used telescope is placed under the background line (negative slope) and the detectivity line (positive slope) computed for the chosen integration time. In the area of non-Poisson noise (above the background line), the level of detection does not depend on the mirror size (formula 2). However, since the background lines move up for short expositions (formula 3) it is necessary to use huge mirrors to observe with high time resolution. The points corresponding to the mirror size and to the electronic camera pixel size (beam aperture) of MAGIC and PETAL (the two largest Čerenkov single telescopes) are reported in the graph.

POSSIBLE CAMERA/MIRROR CONFIGURATIONS

It is possible to get contemporaneous Čerenkov and optical measurements with two different arrangements:

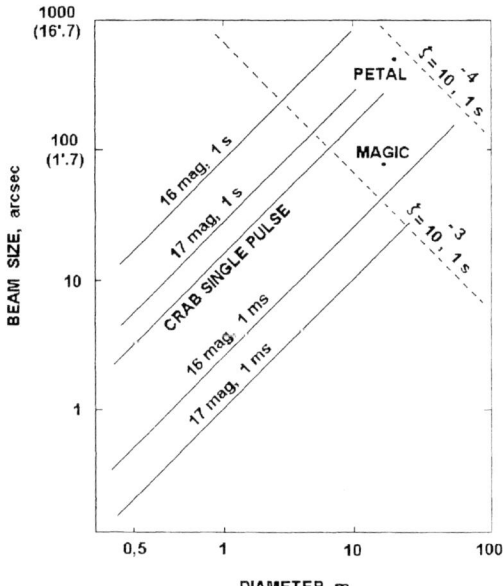

FIGURE 1. Flare detection level. Two line families, computed in the R band assuming $m_{bg} = 20.5$, $\eta = 0.6$, $\tau = 0.7$, for $T = 1$ and $T = 10^{-3}$ (positive slope) and $T = 1$, $\zeta = 10^{-3}$ and $\zeta = 10^{-4}$ (negative slope) represent the two limiting conditions; any detectable signal must be under the lines corresponding to the two limitations.

1. by splitting the mirror surface in two sections, with different focal planes and with two different imaging systems,

2. by using all the mirror surface and only one Čerenkov electronic camera, splitting the signals by a twin trigger system.

The first configuration requires two different sets of submirrors. The external part of the mirror array should be composed by the elements with the larger focal length; the inner part, having the focal plane nearer to the mirror surface, should be composed by good elements with fine mounting supports to minimize the aberrations and to allow the use of panoramic photoelectric detectors. The second configuration is based on the use of the existing Čerenkov electronic camera. It requires only minimal changes to obtain contemporaneous output as an optical and Čerenkov imager. With the huge mirrors of the largest Čerenkov telescopes, a very high background counting rate is expected and a comparable or larger photon flux is expected from field stars; therefore very fast low gain PMT and high speed electronics are required.

CONCLUSION

Multifrequency observations are the next step to arrive at a full understanding of the GRB puzzle. High time resolution observations are important as probes to check the fireball model. Čerenkov telescopes offer very large collecting areas and could be used with good sensitivity to perform contemporaneous observations in the VHE and optical bands. In conditions of large background noise, a reduction of the mirror surface can increase the sensitivity, but to increase the measure speed larger telescopes are required. The feasibility of a special electronic camera for contemporaneous optical/Čerenkov observations seems to be good; this opens new exciting perspectives in the field of the OT observations.

ACKNOWLEDGEMENTS

This investigation was supported by the Italian Ministry of Foreign Affairs, by the University of Bologna (Funds for Selected Research Topics) and by the Russian Fund of Fundamental Research.

REFERENCES

1. Akerlof C.W. and McKey T.A., I.A.U. Circ. 7100 (1999).
2. Beskin G.M. et al., *A&AS* 138, 589 (in press).
3. Fukugita M., *PASP* 107, 945 (1995).
4. Weeks T.C., *Physics Reports*, 160 (1988).

VI. SOFT GAMMA REPEATERS

The 4.5 ± 0.5 Soft Gamma Repeaters in Review

K. Hurley

UC Berkeley
Space Sciences Laboratory
Berkeley, CA 94720-7450

Abstract. Four Soft Gamma Repeaters (SGRs) have now been identified with certainty, and a fifth has possibly been detected. I will review their X-ray and gamma-ray properties in both outburst and quiescence. The magnetar model accounts fairly well for the observations of SGR1806−20 and SGR1900+14, but data are still lacking for SGR1627−41 and SGR0525−66. The locations of the SGRs with respect to their supernova remnants suggest that they are high-velocity objects.

I INTRODUCTION

The Soft Gamma Repeaters are sources of short, soft-spectrum ($\leq$ 100 keV) bursts with super-Eddington luminosities. They undergo sporadic, unpredictable periods of activity, sometimes quite intense, which last for days to months, often followed by long periods (up to years or decades) during which no bursts are emitted. Very rarely, perhaps every 20 years, they emit long duration *giant flares* which are thousands of times more energetic than the bursts, with hard spectra ($\sim$ MeV). The SGRs are quiescent, and in some cases periodic, 1–10 keV soft X-ray sources as well. They all appear to be associated with supernova remnants, and a good working hypothesis is that they are all *magnetars*, i.e., highly magnetized neutron stars for which the magnetic field energy dominates all other sources, including rotation [6,29]. Figure 1 shows the time histories of bursts from SGR1900+14, and Figure 2 shows a typical energy spectrum.

In this paper, I will mainly review the radio, X-ray, and gamma-ray properties of the SGRs in outburst and in quiescence, and indicate how the magnetar model accounts for these properties.

II SGR1806−20

Kulkarni and Frail [22] suggested that this SGR was associated with the Galactic supernova remnant (SNR) G10.0−0.3, based on its localization to a $\sim 400\,\text{arcmin}^2$

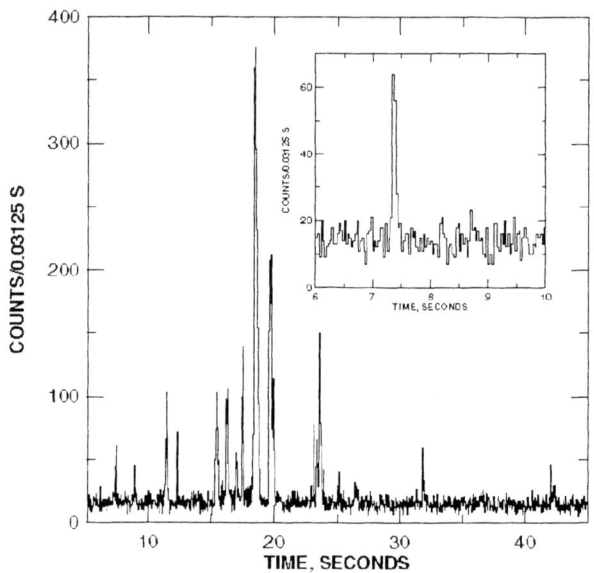

FIGURE 1. Inset: a typical burst from SGR1900+14 as observed in the 25–150 keV range by *Ulysses*. Main figure: bursts during a period of intense activity. (Reprinted from [12])

error box by the old interplanetary network (IPN) [1]. This was confirmed when ASCA observed and imaged the source *in outburst*, localizing it to a 1' error circle [25]. A quiescent soft X-ray source was also detected by Cooke [5] using the ROSAT HRI. Based on more recent observations, Kouveliotou et al. [20] have found that the quiescent source is periodic ($P = 7.48$ s) and is spinning down rapidly ($\dot{P} = 2.8 \times 10^{-11}$ s/s). If this spindown is interpreted as being due entirely to magnetic dipole radiation, the implied field strength is $B = 8 \times 10^{14}$ G. The 2–10 keV X-ray luminosity of the source is 2×10^{35} erg/s, and the low-energy X-ray spectrum may be fit by a power law with index 2.2.

The SNR G10.0−0.3 has a non-thermal core, and Frail et al. [8] have detected changes in the radio contours of the core on $\sim$ year time scales. Van Kerkwijk et al. [30] have found an unusual star at the center of this core, which they identify as a luminous blue variable (LBV). The presence of this object has been a mystery up to now, because it was thought that the SGRs were single neutron stars. Recent work from the 3rd IPN has shed some light on this issue [13]. Figure 3 shows the location of the SGR superimposed on the radio contours of the SNR. It can be seen that the SGR is in fact offset from the LBV. The LBV may be powering the non-thermal core of the SNR, and causing the changes in the radio contours. It is also possible that the SGR progenitor was once bound to the LBV, but that it became unbound when it exploded as a supernova. A transverse velocity of $\sim$100 km/s would then be required to explain the displacement between the two.

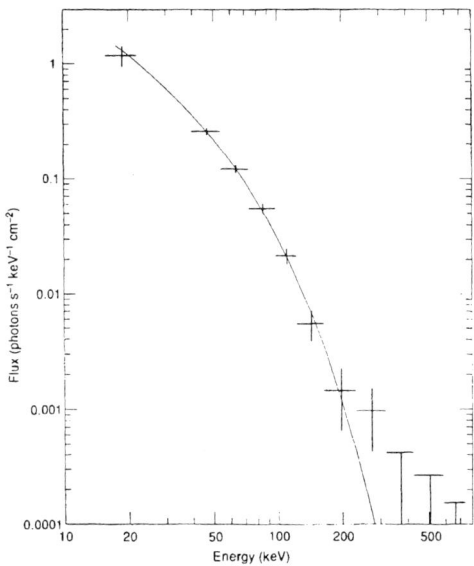

FIGURE 2. Typical spectrum of a burst from SGR1900+14 as observed by BATSE. The spectrum is fit here with an optically thin thermal bremsstrahlung function, with $kT = 39$ keV. (Reprinted by permission from Nature [19], copyright 1993 Macmillan Magazines Ltd.)

Alternatively, it is possible that the apparent SGR-SNR association is due to a chance alignment of these two objects along the line of sight.

III SGR1900+14

SGR1900+14 was discovered by Mazets et al. [23] when it bursted 3 times in two days. A precise localization by the IPN [12] showed that this source lay just outside the Galactic SNR G42.8+0.6, with an implied proper motion >1000 km/s. The SGR is associated with a quiescent soft X-ray source [32,14,21]. The quiescent source has a period of 5.16 s, and a period derivative of 6.1×10^{-11} s/s; again, assuming purely dipole radiation, $B \sim 8 \times 10^{14}$ G. The 2–10 keV luminosity is 3×10^{34} erg/s, and the spectrum may be fit with a power law of index 2.2.

On 1998 August 27, the SGR emitted a giant flare which was probably the most intense burst ever detected at Earth [15]. Its luminosity was 2×10^{43} erg/s in >25 keV X-rays, or $10^5 L_E$ (the Eddington luminosity). The time history of this burst clearly displayed the 5.16 s periodicity of the quiescent source (Figure 4). The magnetic field strength required to contain the electrons responsible for the X-ray emission is $> 10^{14}$ G; this constitutes an independent argument for the presence of strong fields in SGRs. From measurements of the ionospheric disturbance that this

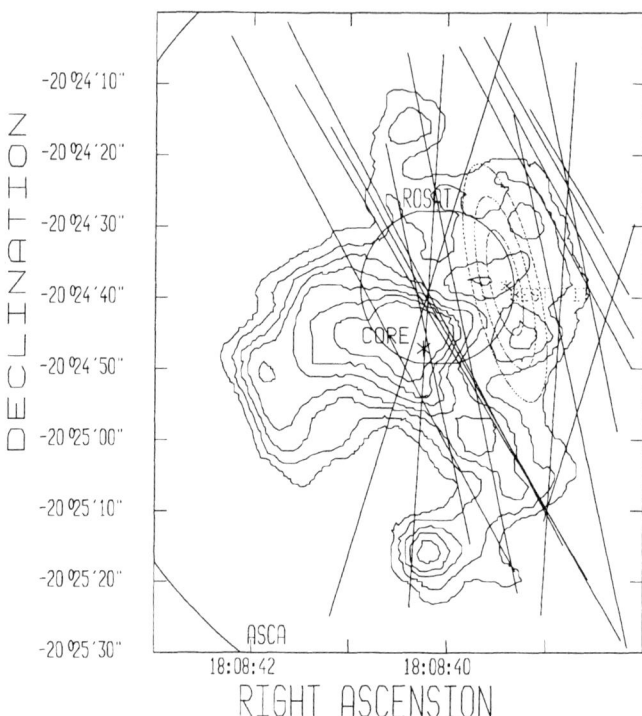

FIGURE 3. Eight IPN annuli (lines), and the 1, 2, and 3σ equivalent confidence contours (ellipses) for SGR1806−20. The best fit position and the position of the non-thermal core are indicated. The ASCA error circle is just visible in the lower left and upper left hand corners [25]. The ROSAT PSPC error circle is at the center; its radius is 11″ [5]. The 3.6 cm radio contours of G10.0−0.3 are also shown, from [33]. (Reprinted from [13])

burst caused, Inan et al. [18] have estimated that there must have been one order of magnitude more energy in 3–10 keV X-rays than in >25 keV X-rays, bringing the total energy to $\sim 4 \times 10^{44}$ erg. Frail et al. [9] detected a transient radio source with the VLA at the SGR position following the giant flare. This is the only case where a radio point source is present at an SGR position.

IV SGR0525−66

This SGR was discovered when it emitted the giant flare of 1979 March 5 [3,10]. It was localized by the IPN to a 0.1 arcmin² error box within the N49 supernova remnant [7]. For an LMC distance of 55 kpc, this burst had a luminosity of 5×10^{44} erg/s in X-rays >50 keV, or $2 \times 10^6 L_E$; the total energy emitted was $\sim 7 \times 10^{44}$ erg in >50 keV X-rays. The time history displayed a clear 8 s periodicity [2]. Duncan and Thompson [6] and Paczynski [26] suggested a strongly magnetized

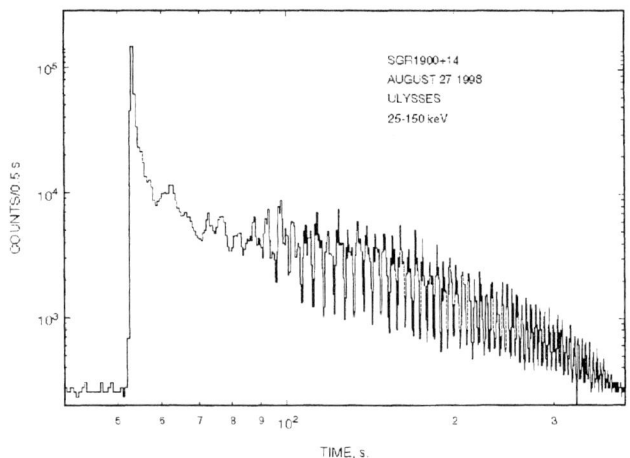

FIGURE 4. The *Ulysses* 25–150 keV time history of the 1998 August 27 giant flare from SGR1900+14. Note the 5.16 s periodicity. (Reprinted by permission from Nature [15], copyright 1999 Macmillan Magazines Ltd.)

neutron star as the origin of this burst. Although the source remained active through 1983 [11], it has not been observed to burst since then.

Rothschild et al. [27] found a quiescent soft X-ray point source in the SGR error box with a ROSAT HRI observation. As no energy spectra are obtained from the HRI, the soft X-ray luminosity can only be estimated by assuming various spectral shapes. The 0.1–2.4 keV luminosity is in the range $10^{36} - 10^{37}$ erg/s, depending on the assumed spectrum. No periodicity was detected in this observation, but the upper limit to the pulsed fraction is only 66%. If the age of the N49 SNR is taken to be 5 kyr [31], the implied transverse velocity of the SGR is several thousand km/s. *Chandra* observations of the SNR are scheduled, and are bound to reveal more about this interesting object.

V   SGR1627−41

SGR1627−41 bursted about 100 times in June–July 1998, and has not been observed to burst since then. During that period, observations by BATSE [34], *Ulysses* [16], KONUS-*Wind* [24], and RXTE [28] led to a precise source localization. The SGR lies near the SNR G337.0−0.1, at a distance of ~ 11 kpc. The implied transverse velocity of the SGR is in the range 200–2000 km/s. Although no giant flare has been observed from this source, there is a KONUS-*Wind* observation of an extremely energetic event [24]. The luminosity and total energy of the burst in the >15 keV range were $\sim 8 \times 10^{43}$ erg/s and $\sim 3 \times 10^{42}$ erg/s, respectively.

Like the other SGRs, this one also appears to be a quiescent soft X-ray source.

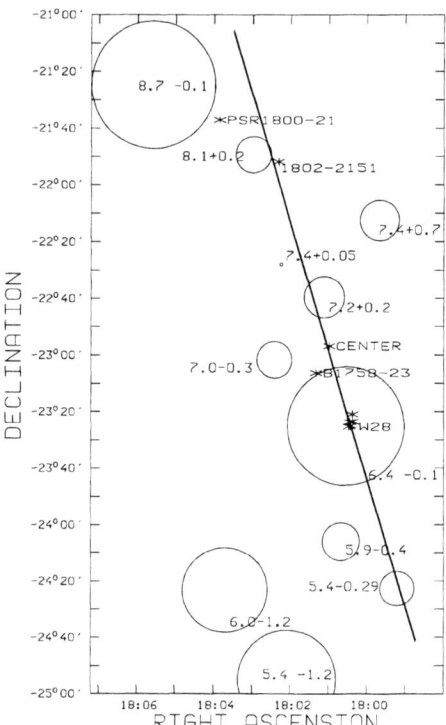

FIGURE 5. IPN error box for SGR1801−23 (the lines are too closely spaced to distinguish). The center is indicated with an asterisk. Circles give the approximate locations of confirmed and suspected SNRs; the radii have been taken as half the size given in the catalogs. Asterisks give the positions of ROSAT X-ray sources, and two pulsars, PSR1800−21 and B1758−23, probably associated with SNRs 8.7−0.1 and 6.4−0.1. Coordinates are J2000. (Reprinted from [4])

BeppoSAX observations revealed a variable source with spectral index 2.1 and luminosity $\sim 10^{35}$ erg/s [35]. Although the *BeppoSAX* observations gave weak evidence for a possible 6.4 s periodicity, this was not confirmed in later ASCA observations of the source with better statistics [17].

VI SGR1801−23

The latest SGR to be discovered is 1801−23 [4]. It was observed to burst just twice, on June 29, 1997, by *Ulysses*, BATSE, and KONUS-*Wind*. The burst spectra were soft, and could be fit by an optically thin thermal bremsstrahlung function with $kT \sim 25$ keV. The time histories were short. In both respects, then, the source properties resemble those of the other SGRs. However, because only two bursts were observed, and they occurred on the same day, the IPN localization is not very precise. The error box is 3.8° long, and has an area of ∼ 80 arcmin². The source lies in the general direction of the Galactic center, and the error box crosses numerous possible counterparts (Figure 5). The source would have a super-Eddington luminosity for any distance >250 pc; at the approximate distance of the Galactic center, its luminosity would be 1200 L_E. At present, the best hypothesis is that this source is indeed an SGR; recall that SGR1900+14 was similarly detected when it bursted 3 times in two days, and it remained quiescent for many years. Like SGR1900+14, the identification of SGR1801−23 may have to await a new period of bursting activity.

Table 1 summarizes the essential properties of the SGRs.

TABLE 1. Essential properties of the SGRs

SGR	Super-Eddington Bursts?	Giant Flare?	Periodicity Observed in Burst?	Quiescent Soft X-ray Source? (erg/s)	Periodicity in Quiescent Source?	$\dot{P}$ 10^{-11} s/s
1806−20	1000×	No	No	2×10^{35}	7.47 s	2.8
1900+14	1000×	270898	5.16 s	3×10^{34}	5.16 s	6.1
0525−66	20000×	050379	8 s	10^{36-37}	No	—
1627−41	400000×	No	No	10^{35}	6.4 s?	—
1801−23	?	No	No	?	—	—

VII THE MAGNETAR MODEL

Briefly, the magnetar model [6,29] explains the short, soft bursts by localized cracking on the neutron star surface, with excitation of Alfven waves which accelerate electrons. Every 20–100 yr, a massive, global crustquake takes place. Regions of the neutron star with magnetic fields of opposite polarity suddenly encounter one another, resulting in magnetic field annihilation and energization of the magnetosphere, giving rise to a giant flare. Magnetars are thought to be born in ∼ 1 out of 10 supernova explosions, and remain active for perhaps 10,000 yr. Thus, there should be about 10 active magnetars in the Galaxy at any given time. So far, we have found 4.5 ± 0.5. Stay tuned for more!

ACKNOWLEDGMENTS

We are grateful for JPL support of *Ulysses* operations under Contract 958056 and to NASA for support of the IPN under NAG5-7810.

REFERENCES

1. Atteia, J.-L., et al., *Ap. J.* **320**, L105 (1987).

2. Barat, C., et al., *Astron. Astrophys.* **79**, L24 (1979).
3. Cline, T., et al., *Ap. J.* **237**, L1 (1980).
4. Cline, T., et al., *Ap. J.*, accepted, astro-ph/9909054, (1999).
5. Cooke, B., *Nature* **366**, 413 (1993).
6. Duncan, R., and Thompson, C., *Ap. J.* **392**, L9 (1992).
7. Evans, W. D., et al., *Ap. J.* **237**, L7 (1980).
8. Frail, D., et al., *Ap. J.* **480**, L129 (1997).
9. Frail, D., et al., *Nature* **398**, 127 (1999).
10. Golenetskii, S., et al., *Sov. Astron. Lett.* **5**, 340 (1979).
11. Golenetskii, S., et al., *Sov. Astron. Lett.* **13(3)**, 166 (1987).
12. Hurley, K., et al., *Ap. J.* **510**, L107 (1999).
13. Hurley, K., et al., *Ap. J.* **523**, L37 (1999).
14. Hurley, K., et al., *Ap. J.* **510**, L111 (1999).
15. Hurley, K., et al., *Nature* **397**, 41 (1999).
16. Hurley, K., et al., *Ap. J.* . **519**, L143 (1999).
17. Hurley, K., et al., *Ap. J.*, in press, astro-ph/9909355 (1999).
18. Inan, U., et al., *GRL*, in press (1999).
19. Kouveliotou, C., et al., *Nature* **362**, 728 (1993).
20. Kouveliotou, C., et al., *Nature* **393**, 235 (1998).
21. Kouveliotou, C., et al., *Ap. J.* **510**, L115 (1999).
22. Kulkarni, S., and Frail, D., *Nature* **365**, 33 (1993).
23. Mazets, E., et al., *Sov. Astron. Lett.* **5(6)**, 343 (1979).
24. Mazets, E., et al., *Ap. J.* **519**, L151 (1999).
25. Murakami, T., et al., *Nature* **368**, 127 (1994).
26. Paczynski, B., *Acta Astronomica* **42**, 145 (1992).
27. Rothschild, R., et al., *Nature* **368**, 432 (1994).
28. Smith, D., et al., *Ap. J.* **519**, L147 (1999).
29. Thompson, C., and Duncan, R., *Mon. Not. R. Astron. Soc.* **275**, 255 (1995).
30. van Kerkwijk, M., et al., *Ap. J.* **444**, L33 (1995).
31. Vancura, O., et al., *Ap. J.* **394**, 158 (1992).
32. Vasisht, G., et al., *Ap. J.* **431**, L35 (1994).
33. Vasisht, G., et al., *Ap. J.* **440**, L65 (1995).
34. Woods, P., et al., *Ap. J.* **519**, L139 (1999).
35. Woods, P., et al., *Ap. J.* **519**, 139 (1999b).

BeppoSAX and Ulysses Data on the Giant Flare From SGR 1900+14

M. Feroci[1], K. Hurley[2], R. Duncan[3], C. Thompson[4],
E. Costa[1], and F. Frontera[5,6]

[1] *Istituto di Astrofisica Spaziale - CNR, Via del Fosso del Cavaliere, 00133 Roma, Italy*
[2] *University of California, Space Sciences Laboratory, Berkeley, CA 94720-7450*
[3] *University of Texas, Department of Astronomy, Austin, TX 78712, USA*
[4] *University of North Carolina, Department of Physics and Astronomy,
Philips Hall, Chapel Hill, NC 27599-3255*
[5] *TESRE/CNR, Via Gobetti 101, Bologna, Italy*
[6] *Dipartimento di Fisica, Università di Ferrara, Via del Paradiso 12, Ferrara, Italy*

Abstract. The extraordinary giant flare of 1998 August 27 from SGR 1900+14 was the most intense event ever detected from this or any other cosmic source (even more intense than the famous 1979 March 5th event). It was longer than any previous burst from SGR1900+14 by more than one order of magnitude, and it displayed the same 5.16-s periodicity in hard X-rays that was detected in the low-energy X-ray flux of its quiescent counterpart. The event was detected by several gamma-ray experiments in space, among them the *Ulysses* gamma-ray burst detector and the *BeppoSAX* Gamma Ray Burst Monitor. These instruments operate in different energy ranges, and a comparison of their data shows that the event emitted a strongly energy-dependent flux, and displayed strong spectral evolution during the outburst, itself. Here we present a joint analysis of the *BeppoSAX* and *Ulysses* data, in order to identify the energy-dependent features of this event and understand some of the physical conditions in the environment of the neutron star which generated this flare.

INTRODUCTION

After several years of quiescence, SGR 1900+14 became extremely active in May 1998. ASCA and RXTE observations after the 1998 May activity episode revealed a periodicity of 5.16 s in the quiescent X-ray emission (2–10 and 2–20 keV, respectively) with period derivative $\sim 1 \times 10^{-10}$ s·s^{-1} [1,2]. This spindown rate is consistent with a magnetar-strength magnetic field [3,4].

On 1998 August 27 a giant flare from SGR1900+14, lasting more than five minutes, was detected by Konus-*Wind*, *Ulysses*, *BeppoSAX*, and NEAR [5–7]. Gamma rays during the first second were extraordinarily intense, overwhelming detectors on several other spacecraft as well. The Compton Gamma Ray Observatory was

Earth-occulted for this flare. The 5.16 s neutron star rotation period was strongly detected during the giant flare. Indeed, the periodic signal was intense enough to produce a marked 5.16-second modulation in the height of the Earth's ionosphere, which affected long-wavelength radio transmissions [8]. After $\sim$ 40 seconds, the flare evinced a $\sim$1.03 s repetitive pattern that is unlike any emission previously detected from any source [6,7]. A radio afterglow was found with the Very Large Array [9]. This source was apparent in the error box of SGR1900+14 on 1998 September 3, but it faded away in less than one week, providing evidence for an abrupt outflow of relativistic particles during the flare.

In this paper we show preliminary results of a comparative analysis of *BeppoSAX* and *Ulysses* observations of the August 27th event. Additional analysis, results and their interpretation may be found in [10].

THE INSTRUMENTS

The *BeppoSAX* Gamma Ray Burst Monitor (GRBM [11,12]) consists of the four anticoincidence CsI(Na) detectors of the Phoswich Detection System (PDS [11]), forming a square box, surrounding the main PDS detectors and located in the core of the *BeppoSAX* payload. Each shield is 1 cm thick and has dimensions 27.5×41.3 cm. The GRBM electronics record data from each shield with both low (1 s) and high (<8 ms) time resolution. The 1 s data consist of count rates in the 40–700 keV and >100 keV energy ranges. Additional details on the GRBM may be found in [12].

The *Ulysses* GRB detector [13] consists of two 3 mm thick hemispherical CsI(Tl) scintillators with a projected area of about 20 cm^2 in any direction. The detector is mounted on a magnetometer boom far from the body of the spacecraft. The energy range is $\sim$ 25–150 keV. The lower energy threshold is set by a discriminator, and is in practice an approximate one; photons with energies > 10 keV can penetrate the housing and be counted either because of the rather poor energy resolution at low energies, or, in the case of very intense events, due to pulse pile-up. The instrument takes time history data with time resolutions of 31.25 ms for 64 s, and 0.5 s for the full duration of the event.

DATA ANALYSIS

Low Time Resolution Data

In Figure 1 (top panel) we show the event light curve as detected by the two above experiments. The 5.16-s pulsation is evident in both energy ranges for the entire duration of the event ($\geq$300 s). Even if the low-resolution data are independently synchronized with the onboard clocks of the two spacecraft, and the event was detected at different UT times due to spatial separation, the relative timing

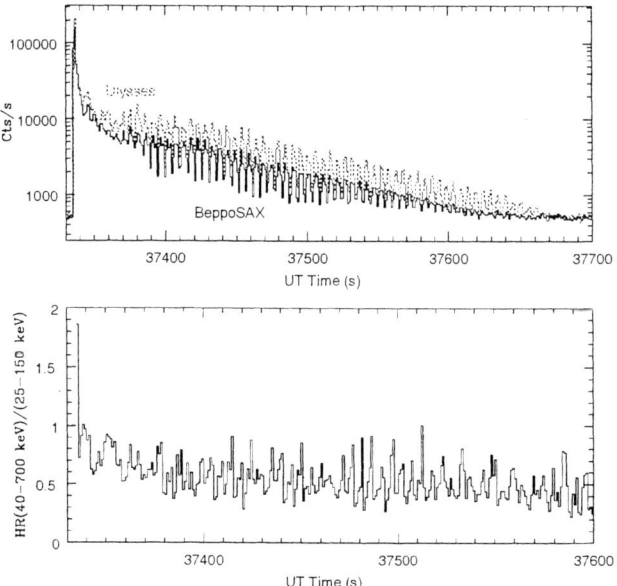

FIGURE 1. Top panel: *BeppoSAX*/GRBM (40–700 keV) and *Ulysses*' GRB detector (25–150 keV) 1-s light curve of the giant outburst. A constant value of 500 cts/s has been added to the background-subtracted GRBM data for display purposes. Time reference is seconds of day August 27 1998. Bottom panel: Hardness Ratio between the light curves in the two energy ranges (typical 1-σ error $\sim$0.01–0.02). The *Ulysses* time has been shifted to account for its position with respect to the Earth.

between the two light curves turned out to be, purely by chance, synchronized within approximately 100 ms. This allows one to use the light curves in the two energy ranges (25–150 and 40–700 keV) to produce a hardness ratio as a function of time, with a time resolution of 1 s. The result is shown in Figure 1 (bottom panel).

It is clear from the hardness ratio plot that the energy spectrum of the emitted radiation significantly changes during each 5.16-s pulse. In fact, performing a Fast Fourier Transform of the hardness ratio curve we obtain the power density spectrum presented in Figure 2, where the fundamental harmonics at $\sim$0.2 Hz is clearly detected. The minor peaks are easily identified as well. The two peaks around 0.4 Hz derive from aliasing of the higher order harmonics, due to the limited frequency span. The peak at $\sim$0.08 Hz is actually spurious, deriving from the spinning motion of the *Ulysses* spacecraft at about 5 rpm. The modulation in the *Ulysses* count rate is likely due to a partial occultation by the carbon fiber magnetometer boom.

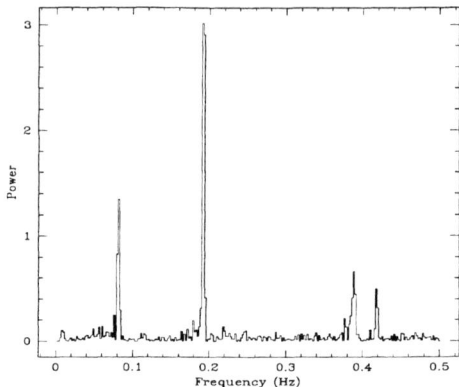

FIGURE 2. Power density spectrum of the hardness ratio curve shown in the bottom panel of Figure 1

High Time Resolution Data

High time resolution data are available from both the instruments for a limited lapse of time. In particular, these *Ulysses* data stop about 60 s after the event onset, whereas the *BeppoSAX*/GRBM recorded the high-resolution light curve for approximately 106 s after the trigger. In Figure 3 they are shown with a time resolution of 31.25 ms, over the time interval in which they are both available.

CONCLUSIONS

The data from experiments operating in different energy ranges (the *Ulysses* count rate is dominated by ~10–40 keV photons, whereas only photons above ~40 keV contribute to the *BeppoSAX* count rate) allow for a time-resolved study of the giant outburst from SGR 1900+14.

From the low time resolution data it appears that the energy spectrum of the emitted X-ray flux is strongly modulated by the 5.16-s rotation of the neutron star. From the high time resolution data in Figure 3, we see that the relative contribution of soft (i.e., *Ulysses*) and hard (i.e., GRBM) X-rays changes over the spin phase, in a different way in subsequent pulses, at least during the initial ~60 s of the event. It is also interesting to note that the soft counts usually exceed numerically the hard ones, except near the 5.16-s pulse minima, where they are basically equal. This indicates that harder emission persists during the occultation of a softer radiation beam.

Hence, both data sets indicate significant spectral evolution over the spin phase and from one pulse to another. These data likely reflect the complex behavior of bubbles of relativistic plasma trapped by the star's magnetosphere following a transient outflow that gave rise to the hard initial spike in the event light curve. The

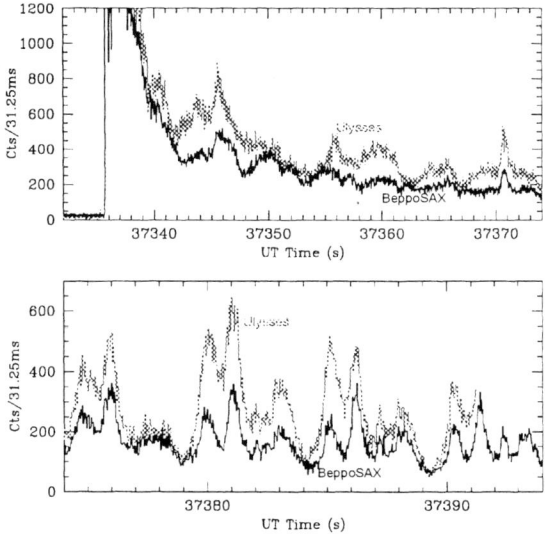

FIGURE 3. High time resolution (31.25 ms) light curves of an initial ~60 s portion of the giant outburst from the *Ulysses* (25-150 keV) and *BeppoSAX* (40-700 keV) GRB experiments. The *Ulysses* time has been shifted to account for its position with respect to the Earth.

late-time stability of the four-peaked pulse shape (see also [6,7]) strongly suggests a multi-polar structure for the magnetic field. These facts, in addition to the harder emission at the minima, suggest the existence of two spatially distinct emission components dominating the soft and hard energy domains, undergoing separate histories both in time and space.

REFERENCES

1. Hurley, K., et al., *ApJ* **510**, L111 (1999).
2. Kouveliotou, C., et al., *ApJ* **510**, L115 (1999).
3. Duncan, R.C. & Thompson, C., *ApJ* **392**, L9 (1992).
4. Thompson, C. & Duncan, R.C., *MNRAS* **275**, 255 (1995).
5. Hurley, K., et al., *Nature* **397**, 41 (1999).
6. Feroci, M., et al., *ApJ* **515**, L9 (1999).
7. Mazets, E.P., et al., preprint, astro-ph/9905196 (1999).
8. Inan, U., et al., *Geophys. Res. Lett.* **26**(22), 3357 (1999).
9. Frail, D., Kulkarni, S., & Bloom, J., *Nature* **398**, 127 (1999).
10. Feroci, M., et al., in preparation (1999).
11. Frontera, F., et al., *Astron. Astrophys. Suppl. Ser.* **122**, 357 (1997).
12. Feroci, M., et al., *Proc. SPIE* **3114**, 186 (1997).
13. Hurley, K., et al., *Astron. Astrophys. Suppl. Ser.* **92**(2), 401 (1992).

The Hard Side of SGR 1900+14

P. M. Woods[†,*], C. Kouveliotou[#,*], J. van Paradijs[†,§],
M. S. Briggs[†,*], K. Hurley[¶], E. Göğüş[†,*], R. D. Preece[†,*],
T. W. Giblin[†,*], C. Thompson[‡], and R. C. Duncan[∥]

[†]*Department of Physics, University of Alabama in Huntsville, Huntsville, AL 35899;*
peter.woods@msfc.nasa.gov
[#]*Universities Space Research Association*
[*]*NASA Marshall Space Flight Center, SD50, Huntsville, AL 35812*
[§]*Astronomical Institute "Anton Pannekoek", University of Amsterdam, 403 Kruislaan, 1098 SJ Amsterdam, NL*
[¶]*University of California at Berkeley, Space Sciences Laboratory, Berkeley, CA 94720-7450*
[‡]*Department of Physics and Astronomy, University of North Carolina, Philips Hall, Chapel Hill, NC 27599-3255*
[∥]*Department of Astronomy, University of Texas, RLM 15.308, Austin, TX 78712-1083*

Abstract. We present evidence for burst emission from SGR 1900+14 with a power-law high-energy spectrum extending beyond 500 keV. Not only is the emission hard, but the spectra are better fit by Band's GRB function rather than by the traditional optically-thin thermal bremsstrahlung model. We find that the spectral evolution within these hard events obeys a hardness/intensity *anti*-correlation. Temporally, these events are distinct from typical SGR burst emissions in that they are longer ($\sim$ 1 s) and have relatively smooth profiles. Despite a difference in peak luminosity of $\gtrsim 10^{11}$ between these bursts from SGR 1900+14 and cosmological GRBs, there are striking temporal and spectral similarities between the two kinds of bursts, aside from spectral evolution. We outline an interpretation of these events in the context of the magnetar model.

INTRODUCTION

Soft gamma repeaters (SGRs) constitute a group of high-energy transients named for the observed characteristics that set them apart from classical Gamma-Ray Bursts (GRBs). SGRs emit brief ($\sim$ 0.1 s), intense (up to $10^3 - 10^4$ L_{Edd}) bursts of low-energy γ rays [1]. The vast majority of SGR burst spectra ($\gtrsim$ 20 keV) can be fit by an OTTB model with temperatures between 20 and 35 keV [2,3].

Hard burst emission has been detected during the brightest bursts recorded from three of the four SGRs (0526−66, 1627−41, 1900+14 [see, e.g., [1]]). Further evidence for hard emission from SGRs comes from RXTE observations of SGR 1806−20. For a small fraction of the more common short events, high OTTB spectral temperatures in the range 50 – 170 keV were measured [4,5]. SGR burst spectra show little or no variation over a wide range of timescales [2], both within individual bursts, and between source active periods that cover years. There have been some exceptions, however, where modest hard-to-soft evolution within bursts from SGR 1806–20 was detected [4].

Here, we present strong evidence for spectrally hard burst emission from SGR 1900+14 during its recent active episode which started in May 1998. Two events recorded with BATSE have temporal and spectral signatures quite distinct from typical SGR burst emissions. We show that although the time-integrated spectrum of each event resembles a classical GRB spectrum, spectral evolution is found that shows a hardness/intensity anti-correlation, never before seen in GRBs. For a more complete description of this work, see [6].

BATSE OBSERVATIONS

On 22 October 1998, during a period of intense activity of SGR 1900+14, BATSE triggered at 15:40:47.4 UT on a burst with a smooth, FRED-like temporal profile (Figure 1a). Such light curves are commonly seen in GRBs, but are rare for SGR events. This burst was located near SGR 1900+14, but was longer (~ 1 s) than typical bursts from this source and its spectrum was much harder. Using Ulysses and BATSE, an IPN annulus was constructed and the joint BATSE/IPN error box contained the source location of SGR 1900+14 [7]. We estimate the probability of a GRB error box of this size overlapping a known SGR at 2×10^{-3}.

Ten weeks later on 10 January 1999 at 08:39:01.4 UT, BATSE detected a strikingly similar burst (Figure 1b), which again located near SGR 1900+14. This burst also triggered Ulysses, so an IPN annulus was constructed, which again contained the position of SGR 1900+14. We find the probability that this burst and any known SGR are related by chance coincidence of 3×10^{-3}. The combined probability that these two events are GRBs with BATSE/IPN error boxes that are consistent with a known SGR by chance coincidence is 6×10^{-6}. We, therefore, conclude that these two bursts originated from SGR 1900+14 during its recent active episode of 1998 – 1999.

Using HERB data (128 energy channels covering 20 – 2000 keV), we fit the time-integrated background-subtracted burst spectra of each event to three models: an OTTB (dN/dE $\propto E^{-1}$ exp[-E/kT]), a simple power law, and Band's GRB function. We find that for each burst, the spectrum is not well characterized by the OTTB model based upon the large value of χ_ν^2 (Table 1). Using a $\Delta\chi^2$ test between the Band and OTTB models [8], we find the Band function is strongly favored over the OTTB, with chance probabilities of 4.6×10^{-7} and 7.2×10^{-11}, respectively. The peak fluxes and fluences of these two events are not exceptional when compared to other burst emissions from this source (Table 1). However, the spectral form is clearly different from typical SGR burst emissions (typical OTTB $kT \sim 25$ keV).

Using data with coarser spectral resolution, but finer time resolution, we fit

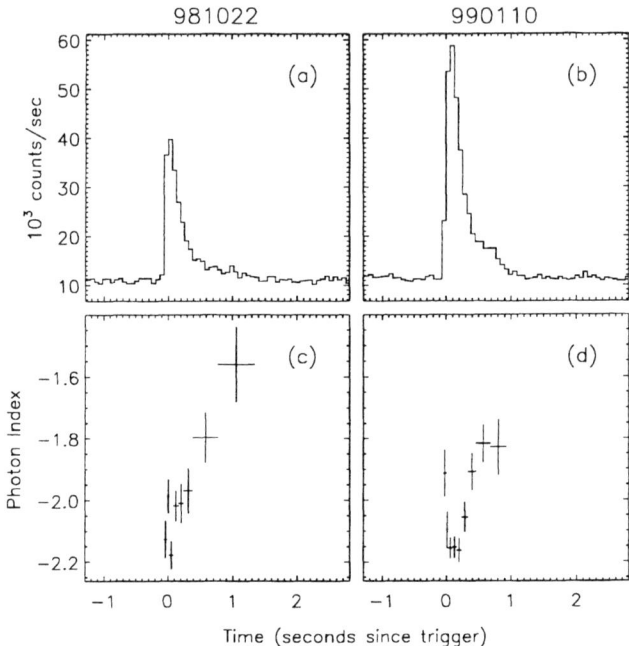

FIGURE 1. Light curves (25 – 2000 keV) of two bursts from SGR 1900+14 BATSE trigger 7171 (981022) is shown in panel (a) and trigger 7315 in panel (b). Panels (c) and (d) give the photon index as a function of time to illustrate the spectral evolution observed during these events.

multiple segments of each burst in order to search for spectral evolution. Guided by the results of our fits to time-integrated spectra and our limited number of energy channels, we chose to fit the power-law model to these spectra. We find significant spectral evolution through each burst, as the power-law photon index

TABLE 1. Spectral Fit Summary

SGR Burst	Model	χ^2/dof	kT or E_{peak} (keV)	Photon index[a]	Peak flux[b] (10^{-6} ergs cm^{-2} s^{-1})
981022	OTTB	119.4/96	102 ± 5	–	
	Power law	92.7/96	–	1.91 ± 0.06	
	Band's GRB	90.2/94	54 ± 50	1.96 ± 0.08	2.94 ± 15
990110	OTTB	139.8/95	94 ± 4	–	
	Power law	106.4/95	–	2.06 ± 0.03	
	Band's GRB	93.1/93	59 ± 11	2.19 ± 0.06	4.96 ± 18

[a] For Band's GRB function, the high-energy index β is given here.
[b] Given over the energy range 25 – 2000 keV on the 0.064 s timescale.

varies between −1.5 and −2.4 (Figures 1c and 1d). We find that these events obey an intensity/hardness anti-correlation (Spearman coefficient $\rho = -0.86$; chance probability $= 8.3 \times 10^{-6}$).

A COMPARISON WITH GRBS

For each event, we have calculated two quantities that are traditionally used to delineate between the two classes [9] of GRBs in the BATSE catalog, specifically t_{90} and spectral hardness (Ch 3/Ch 2). We find t_{90} durations of 1.2 ± 0.2 and 0.9 ± 0.2 s, and fluence hardness ratios of 1.9 ± 0.1 and 2.1 ± 0.1 for the bursts of 981022 and 990110, respectively. When plotted together with the reported values of the 4B (revised) catalog [10], we find that these two bursts fall outside the main concentrations of each distribution (Figure 2), but nearer the centroid of the short, hard class. These observations demonstrate that a Galactic source—probably a strongly magnetized neutron star with a large velocity—is capable of producing a burst of γ rays whose time-integrated spectrum resembles that of a classical GRB. This similarity is remarkable in light of the extreme difference in peak luminosities of $\gtrsim 10^{11}$!

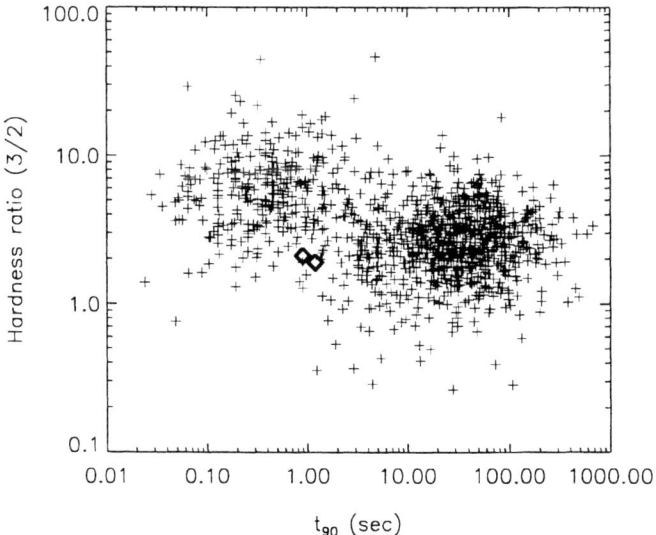

FIGURE 2. Duration (t_{90}) versus hardness (100 – 300 keV/50 – 100 keV) for 1191 gamma-ray bursts (crosses) and the two SGR 1900+14 events presented here (asterisks). Typical durations and hardness ratios for the more common SGR bursts are 0.1 s and 0.08, respectively.

Given the spectral similarities between these two events and a fair fraction of GRBs, the time-integrated spectrum is not sufficient to distinguish these bursts from GRBs. Many GRBs show a strong hardness/intensity correlation within individual bursts [11], but to the best of our knowledge, a consistent *anti*-correlation, as shown for these two events, has never been seen throughout a GRB. If this behavior is inherent to all *hard* SGR events, it would be a useful diagnostic (but secondary to location) with which to select them.

DISCUSSION

The proposed identification of SGRs with very strongly magnetized neutron stars has received strong support from the discovery that both SGR 1806−20 and S-GR 1900+14 are X-ray pulsars spinning down at rapid rates [12,13]. In this magnetar model, diffusion of the magnetic field from the core can cause a sudden fracture of the rigid neutron star crust that injects a pulse of Alfvén radiation directly into the magnetosphere. The Alfvén waves cascades to high wavenumber and create a trapped fireball [14]. The soft spectrum arises from a combination of photon splitting and Compton scattering in the cool, matter-loaded envelope of the fireball. The relative hardness of the two reported events, combined with luminosities in excess of 10^{40} erg s^{-1}, points directly to an emission region of shallower scattering depth, $\tau_{es} < 1$, situated outside $\sim 10^3 (L_X/10^{40}$ erg s$^{-1})$ neutron star radii. In certain circumstances, one expects that Alfvén radiation will disperse rapidly throughout the magnetosphere: if the initial impulse occurs on extended dipole field lines; or if it involves a buried fracture of the crust. This rapid dispersal is made possible by the strong coupling between external Alfvén modes and internal seismic waves [15].

REFERENCES

1. Hurley, K., these proceedings.
2. Fenimore, E., Laros, J., & Ulmer, A., *ApJ* **432**, 742 (1994).
3. Göğüş, E., et al., *ApJ* **526**, L93 (1999).
4. Strohmayer, T. & Ibrahim, A., in *Proc. 4th Huntsville GRB Symp.*, AIP **428**, 1997, p. 947.
5. Marsden, D., PhD thesis, UCSD (1999).
6. Woods, P.M., et al., *ApJ* **527**, L47 (1999).
7. Frail, D., Kulkarni, S., & Bloom, J., *Nature* **398**, 127 (1999).
8. Band, D., et al., *ApJ* **485**, 747 (1997).
9. Kouveliotou, C., et al., *ApJ* **413**, L101 (1993).
10. Paciesas, W., et al., *ApJS* **122**, 465 (1999).
11. Ford, L.A., et al., *ApJ* **439**, 307 (1995).
12. Kouveliotou, C., et al., *Nature* **393**, 235 (1998).
13. Kouveliotou, C., et al., *ApJ* **510**, L115 (1999).
14. Thompson, C., & Duncan, R., *MNRAS* **275**, 255 (1995).
15. Blaes, O., Blandford, R., Goldreich, P., & Madau, P., *ApJ* **343**, 839 (1989).

Spin Period and Burst Rate History of SGR 1900+14

Peter M. Woods[†,*], Chryssa Kouveliotou[#,*], Jan van Paradijs[†,§],
Mark H. Finger[#,*], Christopher Thompson[‡], Robert C. Duncan[||],
Kevin Hurley[¶], Tod Strohmayer[#,@], Jean Swank[@],
and Toshio Murakami[&]

[†] *Department of Physics, University of Alabama in Huntsville, Huntsville, AL 35899*
[#] *Universities Space Research Association*
[*] *NASA Marshall Space Flight Center, SD50, Huntsville, AL 35812*
[§] *Astronomical Institute "Anton Pannekoek", University of Amsterdam, 403 Kruislaan, 1098 SJ Amsterdam, NL*
[¶] *University of California at Berkeley, Space Sciences Laboratory, Berkeley, CA 94720-7450*
[‡] *Dept. of Physics and Astronomy, Univ. of North Carolina, Philips Hall, Chapel Hill, NC 27599-3255*
[||] *Department of Astronomy, University of Texas, RLM 15.308, Austin, TX 78712-1083*
[@] *NASA Goddard Space Flight Center, Greenbelt, MD 20771*
[&] *ISAS, 3-1-1, Yoshinodai, Sagamihara-shi, Kanagawa 229, Japan*

Abstract. We have analyzed *RXTE* PCA observations of the pulsed emission from SGR 1900 + 14 during September 1996, June – October 1998, and early 1999. Using these measurements and results reported elsewhere, we construct a period history of this source for 2.5 years. We find significant deviations from a steady spin-down trend during quiescence and the burst active interval. BATSE observations of the burst emission are presented and correlations between the burst activity and spin-down rate of SGR 1900 + 14 are discussed. We find an 80 day interval during the summer of 1998 when the average spin-down rate is larger than the rate elsewhere by a factor ~ 2.3. This enhanced spin-down may be the result of a discontinuous spin-down event or "braking glitch" at the time of the giant flare on 27 August 1998.

INTRODUCTION

Soft gamma repeaters (SGRs) form a rare class of persistent X-ray sources that are associated with young ($\sim 10^4$ yr) supernova remnants (SNRs; see [1] for a review). SGR 1806−20 and SGR 1900+14 were recently found to spin down on long time scales at a rate $\sim 10^{-11}$ to 10^{-10} s s^{-1} [2,3]. This spin-down has been interpreted as evidence that they are neutron stars with very intense magnetic fields in the $10^{14} - 10^{15}$ G range, i.e., magnetars [4,5]. Except for their emitting brief

($\sim$ 0.1 s), intense ($\sim 10^{39} - 10^{42}$ ergs s^{-1}) bursts of low-energy γ-rays and having harder persistent emission spectra, the characteristics of SGRs are similar to those of the anomalous X-ray pulsars (AXPs; [6,7]).

The spin-down histories of at least some AXPs have both a steady spin-down component and a variable perturbing component. Two AXPs with well-sampled period histories, 1E 1048.1−5937 and 1E 2259+586, have shown evidence for such perturbations ([8], and references therein). Following the discovery of pulsations from SGR 1900+14 during an observation in April 1998 with *ASCA* [13], subsequent analysis of both archival and more recent observations with *RXTE*, *BeppoSAX*, and *ASCA* have shown that the spin-down rate of this magnetar is not constant, varying from 5×10^{-11} to 14×10^{-11} s s^{-1} [3,9,10,11]. Furthermore, it was noted that the observed changes in period derivative appeared to be associated with the burst reactivation of the source in May 1998.

Here, we present an analysis of *RXTE* PCA observations of SGR 1900+14 from 1996, 1998 and 1999, and construct a period history for this source over 2.5 yr. We also report on the burst rate history of SGR 1900+14 as seen with BATSE and discuss possible correlations between the burst activity and the changes observed in the spin-down. For further details on this work, see [12].

PCA AND BATSE OBSERVATIONS

Using event mode data, we energy-selected all observations for 2 − 10 keV photons, binned the data at 0.125 s time resolution and barycenter-corrected the bin times. Adopting a constant period phase model, we calculated the phase at multiple points during each observation. The phases for the observations from 1996 and 1998 could not be well fit with a linear phase model due to their long baseline, so we included a second-order term (i.e., frequency derivative). These fits yielded good χ_ν^2 for the observations of September 1996 and June 1998, however, the χ_ν^2 was larger (2.4) for the data during the fall of 1998. The measured period and period derivatives are given in Table 1.

The period derivative measurement for September 1996 has been extrapolated for 250 days in Figure 1 (dotted lines represent $\pm 2\sigma$) to clearly indicate the discrepancy between the local slope and the long-term trend. The precision of the period derivative measurement within this observation and the inclusion of the *BeppoSAX* period measured from May 1997 allows us to conclude that there are significant deviations from a constant spin-down rate during quiescence. The phase-folded profile remains constant between September 1996 and June 1998, in spite of the enhanced burst activity [3,13,12]. The phase-folded profile in the fall of 1998 is consistent with the profile found in January 1999 [12,3,9]. This shows that the pulse shape changes observed during the tail of the August 27th flare [17] persist for months.

Between May 1998 and January 1999, BATSE triggered on 63 bursts from SGR 1900+14. In order to obtain a more complete database, we performed an

TABLE 1. X-ray Period and Period Derivative Measurements for SGR 1900+14

Time of Observation mm/dd/yy	Expos. ksec	Epoch MJD TDB	Period s	Period Derivative 10^{-11} s s^{-1}	Instrument	Refs
09/04/96–09/19/96	100.9	50337.0	5.15581568(19)	8.27(14)	RXTE	12,11
05/12/97–05/13/97	45.7	50580.5	5.157190(7)	–	BeppoSAX	10
04/30/98–05/02/98	84.6	50935.0	5.1589715(8)	–	ASCA	13
05/31/98–06/09/98	43.5	50970.0	5.15917011(55)	8.2(6)	RXTE	12,3
08/28/98–10/08/98	146.9	51070.0	5.16026572(12)	5.93(3)	RXTE	12
09/15/98–09/16/98	33.2	51071.5	5.160262(11)	–	BeppoSAX	10
09/16/98–09/17/98	39.0	51073.3	5.160295(3)	–	ASCA	9
01/03/99–01/04/99	31.9	51181.5	5.160934(56)	–	RXTE	12
03/21/99–03/21/99	9.1	51259.0	5.16145(18)	–	RXTE	12
03/30/99–03/30/99	8.4	51268.0	5.16156(11)	–	RXTE	12

off-line search for un-triggered events from SGR 1900+14 which is described in [12]. Between 24 May 1998 and 3 February 1999, we detected most of the SGR 1900+14 events that triggered BATSE in addition to 137 untriggered events. Based upon our experience with classifying BATSE triggers, we expect the number of false triggers within our sample to be less than 5%. Figure 1 displays the burst rate (per 3-day interval) over the time period searched. No emission from SGR 1900+14 triggered the BATSE instrument between August 1992 [14] and 25 May 1998. We note that the most recent trigger from this SGR was recorded on 29 April 1999, so it appears the burst activity has ceased for the time being.

DISCUSSION

The period history of SGR 1900+14 appears to be divided into two sections during which the spin-down rate is nearly, but not exactly, constant. Before 9 June 1998 and after 27 August 1998, the average rate is 6.1×10^{-11} s s^{-1}. These two sections are separated by 80 days during which the period increased by 1 ms, which implies an average rate of 14.0×10^{-11} s s^{-1}. It appears that the period history during this interval allows for two obvious descriptions: (i) a gradual increase of the nominal spin-down rate, perhaps due to particle outflow from the stellar surface caused by the burst activity [11,15,16] and (ii) a discontinuous spin-down event associated with the 27 August 1998 flare.

The number of bursts recorded with BATSE between the onset of activity in May and 26 August 1998 was 40. Following the 27 August 1998 flare, 123 bursts were recorded up to 3 February 1999. If one assumes that the burst rate or the burst energy released is correlated with the increase in spin-down, then one would expect to see an even steeper spin-down rate between 28 August 1998 and 3 January 1999, which is not the case. The increased braking torque on the star must then be

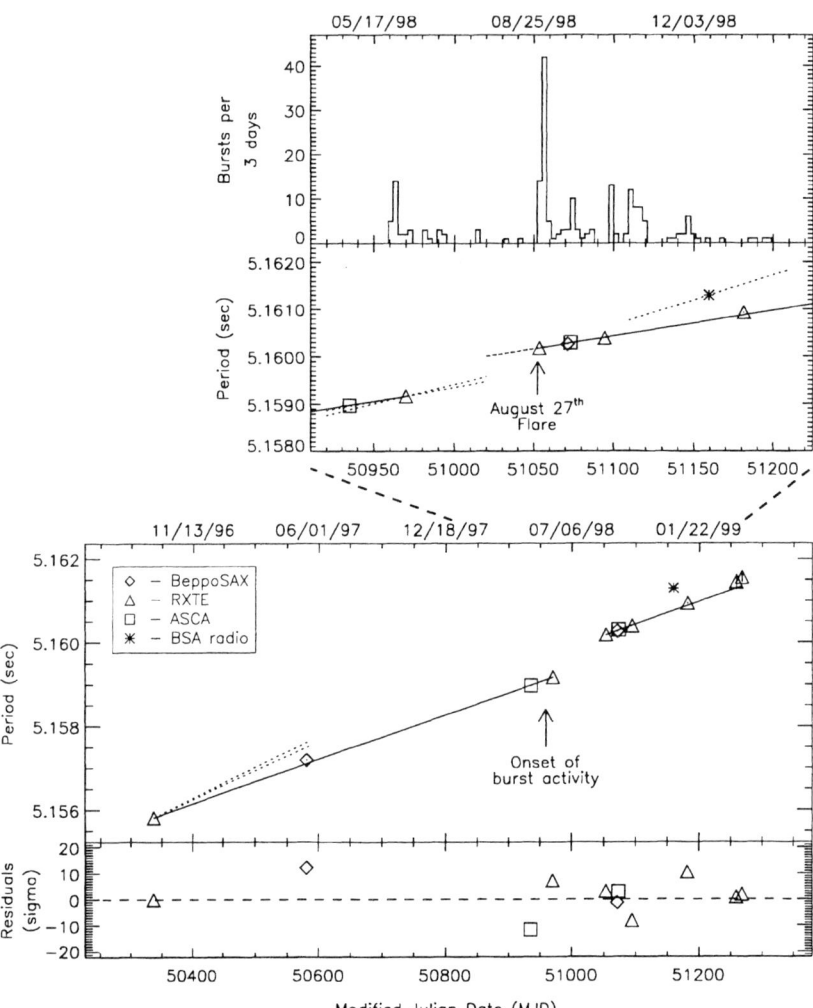

FIGURE 1. *Bottom* – Period history of SGR 1900+14 from September 1996 through March 1999. Lower axis label is modified Julian date and upper axis is mm/dd/yy. The solid lines indicate least square fits to the period measurements found in two separate intervals (September 1996 – June 1998 and August 1998 – March 1999; $\sim 6.1 \times 10^{-11}$ s s^{-1}). Due to the long series of observations with *RXTE* from 28 August to 8 October 1998, two period measurements from the beginning and end are shown. Residuals of fit are shown in lower panel. Dotted lines represent extrapolation of local period derivative measurement ($\pm 2\sigma$; see Table 1) found in September 1996 *RXTE* observation. *Upper Right* – Inset of lower figure showing burst rate history (upper panel) and period history (lower panel) of SGR 1900+14 from 7 April 1998 through 16 February 1999. Dotted lines represent extrapolation of local period derivative measurement ($\pm 2\sigma$).

attributed to something other than the burst activity of the smaller, more common bursts.

An alternative scenario to account for the rapid spin-down during the period June – September 1998 is that the star underwent a more or less steady spin-down at a long-term average rate of $\sim 6 \times 10^{-11}$ s s^{-1} from June through most of August. The star then suffered a discontinuous upward jump in period or a 'braking glitch', which is attractive to link with the occurrence of the very energetic flare of 27 August 1998. Extrapolating the long-term trends found before and after 27 August, we find that this braking glitch would have a magnitude ΔP = 5.72(14) $\times$ 10^{-4} s. This corresponds to a rotational energy loss for the star $\Delta E_{\rm rot} \approx 2 \times 10^{41}(M_*/1.4M_\odot)(R_*/10$ km$)^2$ ergs if the whole star participates, or 0.5% of the energy released in high-energy photons during the 27 August flare. Physical mechanisms that may lead to a sudden braking of the neutron star rotation at the time of the giant flare are discussed in [15].

REFERENCES

1. Hurley, K., these proceedings.
2. Kouveliotou, C., et al., *Nature* **393**, 235 (1998).
3. Kouveliotou, C., et al., *ApJ* **510**, L115 (1999).
4. Thompson, C., & Duncan, R., *MNRAS* **275**, 255 (1995).
5. Thompson, C., & Duncan, R., *ApJ* **473**, 322 (1996).
6. Mereghetti, S. & Stella, L., *ApJ* **442**, L17 (1995).
7. van Paradijs, J., Taam, R.E. & van den Heuvel, E.P.J., *A&A* **299**, L41 (1995).
8. Heyl, J.S. & Hernquist, L., *MNRAS* **304**, L37 (1998).
9. Murakami, T., et al., *ApJ* **510**, L119 (1999).
10. Woods, P.M., Kouveliotou, C., van Paradijs, J., Finger, M.H., & Thompson, C., *ApJ* **518**, L103 (1999).
11. Marsden, D., Rothschild, R.E., & Lingenfelter, R.E., *ApJ* **523**, L97 (1999).
12. Woods, P.M., Kouveliotou, C., van Paradijs, J., Finger, M.H., Thompson, C., Duncan, R.C., Hurley, K., Strohmayer, T., Swank, J., Murakami, T., *ApJ* **524**, L55 (1999).
13. Hurley, K., et al., *ApJ* **510**, L111 (1999).
14. Kouveliotou, C., et al., *Nature* **362**, 728 (1993).
15. Thompson, C., Duncan, R., Woods, P.M., Kouveliotou, C., Finger, M.H., & van Paradijs, J., *ApJ*, in press (2000).
16. Harding, A.K., Contopoulos, I., & Kazanas, D., *ApJ* **525**, L125 (1999).
17. Mazets, E.P., et al., *Astron. Lett.* **25**, 635 (1999).

Deep Searches for Pulsations in SGR 1627−41 and 0526−66

T. Murakami[a,1], M. Ando[a], K. Hurley[b], P. Li[b], C. Kouveliotou[c],
P. Woods[d], D. Hartmann[e], I. Smith[f], M. Tsujimoto[g],
T. Strohmayer[h], and A. Yoshida[i]

[a] Institute of Space and Astronautical Science,
3-1-1, Yoshinodai, Sagamihara, Kanagawa 229-8510, Japan
[b] Space Science Laboratory, University of California, Berkeley CA, USA
[c] University Space Research Association, MSFC, ES-84, Huntsville AL, USA
[d] Marshall Space Flight Center, ES-84, Huntsville AL, USA
[e] Dept. of Physics, Clemson University, Clemson SC, USA
[f] Dept. of Space Physics and Astronomy, Rice University, Houston TX, USA
[g] Dept. of Physics, University of Kyoto, Kyoto, Japan
[h] Goddard Space Flight Center, Greenbelt MD, USA
[i] Physical and Chemical Research, Wako, Saitama, Japan

Abstract. Deep searches for X-ray pulsations from SGR 1627−41 and SGR 0526−66 were performed with ASCA. SGR 1806−20 and 1900+14 were previously established as X-ray pulsars. SGR 0526−66 is suspected to be a strong candidate for a magnetar, but X-ray pulsation has not been confirmed yet. Detection of X-ray pulsations from SGR 1627−41 and SGR 0526−66, is the key issue to the unified magnetar model of the SGRs. Although ASCA has done the deepest observations of $\sim$ 80 ks for both of them, we did not detect pulsations. The long-term period history of SGR 1900+14 is compared with a very similar change of the AXP pulsar 1E1048.1−5937.

INTRODUCTION

SGR 1900+14 and SGR 1806−20 are well established as X-ray pulsars. The rapid spin-downs of the order of 10^{-10} s/s found with RXTE and ASCA suggest a magnetar origin of the energy, with a magnetic field of $\sim 10^{15}$ G [1,3,5]. Although SGR 0526−66 showed pulsations of 8.0 ± 0.05 s during the decay of the 1979 March 5th event [6], stable pulsations were not detected from the point-like X-ray source, which was discovered by ROSAT [7], in the supernova remnant (SNR) N49. SGR 1627−41 was discovered independently by the three experiments, BATSE,

---

[1] E-mail: murakami@astro.isas.ac.jp

RXTE and Konus and was localized to a small error region [1,9]. The IPN3 errorbox coincided with the SNR G337.0−0.1 [11] and the BeppoSAX observation of this region discovered an X-ray source located inside the SNR [9,10]. The BeppoSAX observation suggested a probable pulsation period of 6.41 s [9]. As two of the SGRs show stable pulsations and rapid spin-downs, if the same physics is at work in all the SGRs, we also expect to detect pulsations from the others.

OBSERVATIONAL RESULTS

SGR 1627−41

The observation of SGR 1627−41 was done on 1999 Feb. 26–28 with a total of 73 ks using ASCA. The ASCA observation found at least three X-ray sources in the field of view (FOV) as shown in Figure 1 and the brightest one coincided with the reported BeppoSAX position which is also within the IPN3-RXTE error box. We searched for pulsations using the FFT and pulse folding methods at around 6.41 s, but no pulsations were detected. The upper limit to the signal power at this period is 1.8% (rms) in 1–10 keV [4]. However, the spectrum is best represented by a power-law of photon-index −2.2, which is similar to the plerionic X-ray pulsar SGR 1806−20. So still there is a possibility for detection of pulsation with a deeper observation.

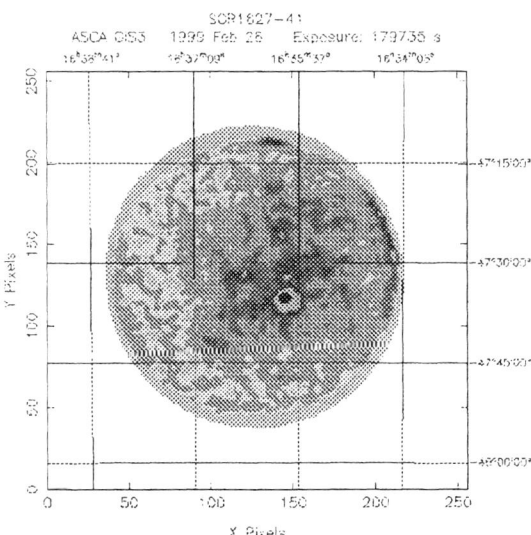

FIGURE 1. A raw ASCA image of the SGR 1627−41 region. At least three X-ray sources are detected. The brightest one is consistent with the IPN3-RXTE error.

SGR 0526−66

SGR 0526−66 was observed once in 1993 with ASCA. We made a new long observation in 1999 with 80 ks. The GIS-mode was set to a pulsar search mode in this observation. The PSF of the ASCA X-ray mirrors is poor, so we could not separate the SNR emission from the X-ray point source. Thus, we focused our pulsar search to the higher energy band which is less contaminated with the SNR emission. Since an 8.0 ± 0.05 s pulsation was reported before during the decay of the 1979 March 5th burst, we searched extensively for pulsations at around this period. However, we did not find any pulsation. Moreover, the SIS X-ray spectrum of N49 can be explained without any hard component and therefore the presence of a plerionic pulsar like SGR 1806−20 is doubtful [12].

SGR 1900+14 and 1E1048.1−5937

Deep ASCA observations to search for pulsations from both SGRs gave negative results. A hard and power-law type spectrum of SGR 1627−41 still suggests a pulsar origin. All reported pulse periods of SGR 1900+14 are shown in Figure 2. Three epochs can be seen in this figure. A change in the spin-down rate was first detected with the RXTE observation in June 1998 [2], before the giant burst on August 27. Since the first change at June 1998 coincided with the start of SGR activity, it is clear that the change was not a sudden jump. The change might

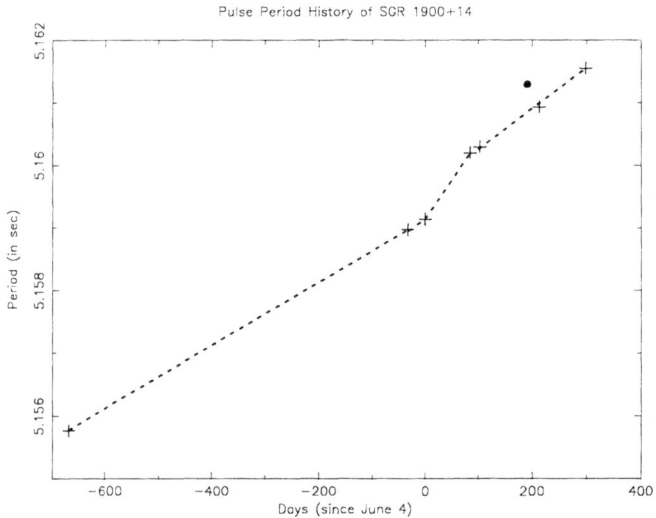

FIGURE 2. Pulse period history of SGR 1900+14. The filled circle is the radio observation by Shitov (IAUC 7110). Three different epochs can be seen. One coincided with the start of SGR activity.

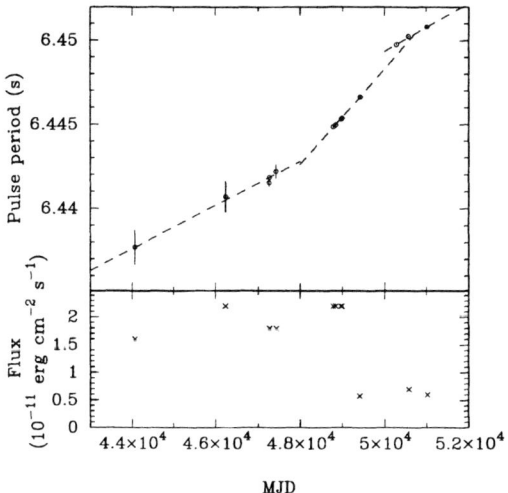

FIGURE 3. Pulse period and flux history of 1E1048.1−5937. The lines show three different epochs. (Adapted from Paul et al. [13].)

be a change in the spin-down rate. However, this fluctuation in the pulse period raised a doubt about the magnetar origins, which fully rely on the magnetic-dipole radiation of a single neutron star [14]. They alternatively suggest the possibility of a pulsar-wind with normal magnetic fields for the spin-down. The fast spin-down can be explained by a strong pulsar-wind, but we think that it is very difficult to explain the large amount of internal energy required for the SGR activity, of an order of 10^{40-44} ergs, by the pulsar-wind mechanism. Moreover, the recent report by the Parkes multibeam radio observations, which discovered three radio pulsars with $B \sim 10^{14}$ G [15], supports the existence of super-strong magnetic fields in neutron stars. The period history of 1E1048.1−5937 with ASCA is interesting (Figure 3). Paul et al. [13] claim that the rapid spin-down of 1E1048.1−5937 lack of correlation between the X-ray flux and the spin-down rate suggest the absence of accretion and the changing in spin-down rate does not support the magnetar. We think that the lack of correlation is also very hard to support in pulsar-wind models.

Acknowledgment: TM thanks B. Paul for his critical reading and discussion.

REFERENCES

1. Kouveliotou, C., et al., *Nature* **393**, 235 (1998).
2. Kouveliotou, C., et al., *ApJL* **510**, 115 (1999).
3. Hurley, K., et al., *ApJL* **510**, 111 (1999).
4. Hurley, K., et al., *ApJL* submitted, Astroph/9909355(1999).

5. Murakami, T., et al., *ApJL* **510**, 119 (1999).
6. Mazets, P., et al., *Nature* **282**, 587 (1979).
7. Rothschild, R.E., et al., *Nature* **368**, 432 (1994).
8. Kouveliotou, C., et al., *GCN* **107**, (1998).
9. Woods, P., et al., *ApJL* **519**, 139 (1999).
10. Smith, D., et al., *ApJL* **519**, 147 (1999).
11. Hurley, K., et al., *ApJL* **519**, 143 (1999).
12. Murakami, T., et al., *Nature* **368**, 127 (1994).
13. Paul, B., et al., *ApJ* submitted , (1999).
14. Marsden, D., et al., *ApJL* submitted, Astroph/9904244 (1999).
15. Manchestr, R.N., et al., *Pulsar Astronomy 2000 and Beyond* ASP conference series, astro-ph/9911319 (1999).

Soft Gamma Repeaters as Relaxation Systems

David M. Palmer

*Universities Space Research Association,
Goddard Space Flight Center/NASA Code 661, Greenbelt, MD 20771*

Abstract. SGR bursts come from systems that accumulate a continuous energy input and release it as discrete bursts. This behavior is analogous to the tectonic systems that produce earthquakes. The fact that these systems can be detected by examining the relationship between burst times and fluences has additional implications, e.g., that burst emission is isotropic.

INTRODUCTION

A *relaxation system* [1] is a system that continuously accumulates some quantity (e.g., energy) in a reservoir and releases it in sudden bursts. The quantity of energy in the reservoir is the upper limit to the burst size at any given time, and a finite-sized reservoir sets the maximum possible burst size. If a burst completely empties the reservoir then another big burst cannot occur until the continuous input has time to replenish it. Multiple moderate-sized bursts can occur in rapid succession if the first bursts only partially deplete the reservoir's contents. Thus, the bursts from a relaxation system need not show a simple relationship between burst size and the time since the previous burst, but may reveal a pattern under more detailed analysis.

A familiar example of a relaxation system is a tectonic zone that continuously accumulates energy as strain in the rocks on either side of a fault, and releases it as earthquakes. Tsuboi [2] found that, during 1885–1963 in the region surrounding Japan, the cumulative energy released by earthquakes is close to, but never higher than, a linear function of time. The slope of this linear function, $r = 2.24 \times 10^{23}$ erg year^{-1}, would represent the rate of energy input into the system (adjusted by an efficiency factor). The deficit of the cumulative earthquake energy below this line represents the stored strain energy available for earthquakes at any given time. The maximum deficit is comparable to the energy of the largest observed earthquakes.

This behavior is also found in bursts from SGR 1806−20, in the data recorded by the International Cometary Explorer (*ICE*). The most compelling interval consists of the 33 bursts during 1983 Nov. 10–15, which I will hereafter refer to as 'Interval B', its designation in [1]. The burst sequence during Interval B is inconsistent with chance at the $P = 2.1 \times 10^{-5}$ level, based on a bootstrap that shuffles the burst intensities while preserving their absolute times.

The statistic used is in this analysis is the 'Sum Of Residuals' (SOR). This is the total of the energies remaining in the reservoir at the ends of the bursts, assuming a best-fit model. The model assumes a constant rate of energy input and is constrained so that the energy in the reservoir is never negative. The SOR statistic measures how closely the cumulative burst energy can follow a linear function of time (representing the cumulative energy input) without exceeding it.

Figure 1 shows the burst intensities, cumulative burst energy, and estimated reservoir energy for this time period. The best-fit model implies that energy is flowing into a reservoir at a rate of 2×10^{36} erg s$^{-1}/\epsilon$, where ϵ is the efficiency factor for turning this energy into γ-ray photons, using the distance and spectral analysis of Fenimore et al. [3] to convert detector count rates to source luminosity. The maximum of the reservoir contents during this interval is also the size of the largest burst, $\sim 1.7 \times 10^{41}$ erg, and may approach the limiting capacity of this reservoir.

Further statistical analysis of the *ICE* data shows that multiple relaxation systems are active simultaneously. For example, the SGR produced a strong sequence

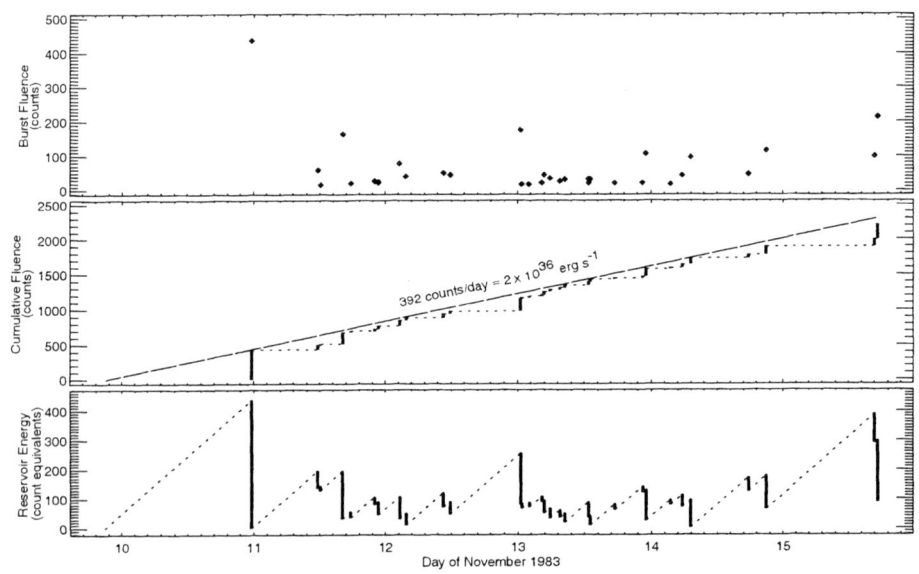

FIGURE 1. SGR 1806−20 burst history and relaxation system model for 1983 Nov. 10–15 (Interval B). (*top*) Burst times and fluences. (*middle*) Cumulative burst fluence and best-fit input rate. (*bottom*) Best-fit energy remaining in reservoir.

of 19 bursts on 1983 Nov. 16, the day following the end of Interval B. After that, from late Nov. 16 through Nov. 19 ('Interval C' of [1]), a dozen bursts resumed the pattern and energy accumulation rate seen in Interval B. The most likely explanation is that the bursts from Intervals B & C came from a single system, with the intervening bursts coming from an independent system or systems.

BURST ISOTROPY

A relaxation system can be present, yet undetectable, in the available data. There are many ways in which a relaxation system can be hidden, such as inadequate observational coverage, non-proportional conversion between stored energy and detector signal, variations in the relaxation system's parameters, and others listed in [1]. To make a relaxation system evident, the data must accurately and completely reflect all energy releases from the system over a long, uninterrupted time period.

If, for example, some of the bursts released during Interval B had not been observed by the detector, the reservoir energy estimate would not reflect those releases, the SOR statistic would be considerably worse, and the statistical significance of the detection would be decreased or even eliminated. The detection of a relaxation system therefore implies that all bursts from this system were detected.

This is *a priori* surprising because this SGR is known to be a neutron star with quiescent emission modulated at its $\sim$7.4 s rotation period [5]. The tail emission from bright bursts and super-bursts can be heavily and intricately spin-modulated, such as the 1999 Aug. 27 giant burst of SGR 1900+14, which showed 4 strong main peaks and finer structure with each rotation over many minutes of decay [6].

The detection of a relaxation system therefore leads us to an unexpected and independently testable prediction: the main emission from a burst is isotropic[1]. The test is to look for variations in the number or strength of bursts as a function of spin phase—this will show modulation if bursts are not viewable or appear weaker when they occur on the far side of the neutron star.

Extrapolating the period history of SGR 1806-20 in the 1990's [5] back to 1983, with generous allowance for variation in $\dot{P}$, gives a period between 7.40 and 7.48 s. Examining the burst times in Interval B shows no variation with spin phase at any period in this range [1]. This indicates that the relaxation system energized a photosphere that was large enough to be viewed equally from all angles, even though the energy reservoir involved is probably local to a small part of the neutron star. It rules out burst emission from a hot spot on or near the surface of the neutron star.

This test can be performed more directly using bursts during periods when the rotation phase is directly known from simultaneous measurements of the quiescent

---

[1] More rigorously, the burst emission is independent of viewing longitude at our latitude relative to the spinning neutron star. However, since any emission anisotropy would be likely to be related to the magnetic rather than the spin-axis coordinates, this rigor is probably superfluous.

flux modulation. Observations of SGR 1900+14 on 1998 Aug. 29 provide this. The sensitive *RXTE* PHAs measured the phase and period of the spin-modulated quiescent emission while detecting many bursts. Most of these bursts were small, but one was rather large and showed modulation in its tail, although it was well below the 'super-burst' category [7]. Figure 2 shows that burst time is independent of spin phase. Although this measurement could be explained if different bursting sites are causing the bursts seen at different phases, the agreement with the *ICE* measurement makes this explanation unnecessary.

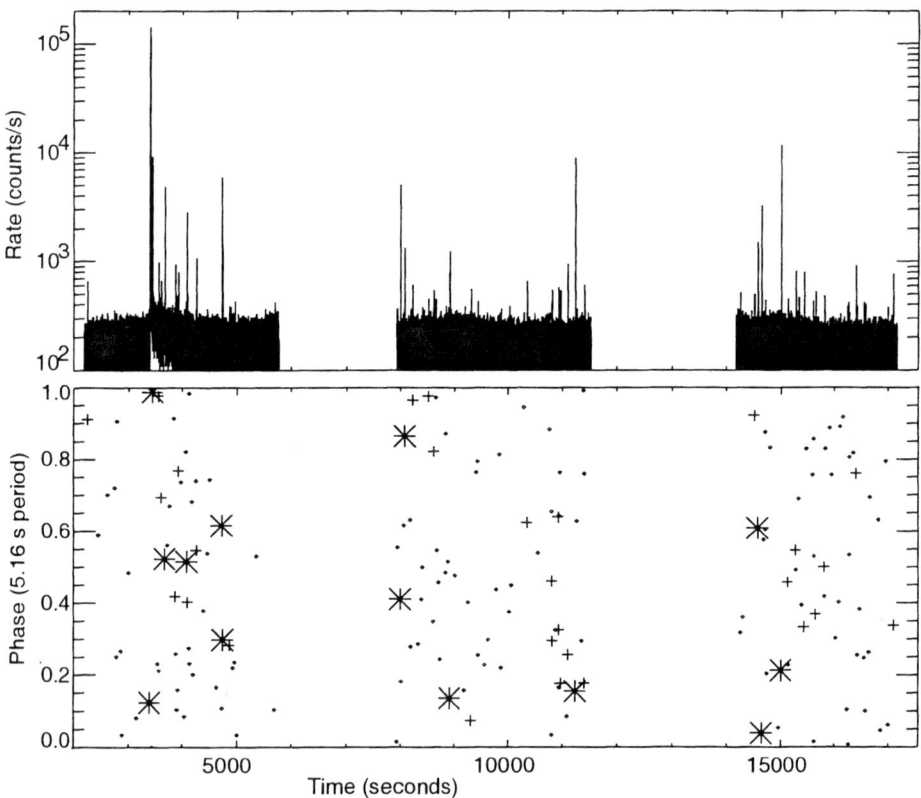

FIGURE 2. *RXTE* PHA observations of SGR 1900+14 on 1998 Aug. 29. (*top*) Light curve showing bursting activity. (*bottom*) Plot of rotation phase during burst, with different symbols representing strong ($* > 10\sigma$), medium ($+ 4.5 - 10\sigma$) and weak ($\cdot\ 3 - 4.5\sigma$) bursts. There is no apparent phase-dependence on bursts overall, nor on any subclass of burst strength.

CONCLUSIONS

The presence of relaxation systems in SGR 1806−20, and by extrapolation on all SGRs, is statistically compelling and consistent with the leading 'starquake' models.

The observability of these relaxation systems imposes further restrictions on their behavior, resulting in the prediction that burst emission is isotropic.

The predicted isotropy of the burst emission is confirmed by spin-sensitive and spin-resolved observations of burst times and strengths. This constrains emission models by requiring that the radiating region that emits the burst photons either surrounds or is much larger than the neutron star. This is in contrast to the regions that emit quiescent and burst-tail flux, which must be localized near the surface or otherwise beamed to produce spin modulation.

REFERENCES

1. Palmer, D.M., *ApJ* **512**, L113 (1999).
2. Tsuboi, C., *Proc. Jap. Acad.* **41**, 392 (1965).
3. Fenimore, E.E., Laros, J. G., & Ulmer, A., *ApJ* **478**, 624 (1994).
4. Laros, J.G, et al., *ApJ* **320**, L111 (1987).
5. Kouveliotou, C., et al., *Nature* **393**, 235 (1998).
6. Mazets, E.P., et al., *ALett* **25**, 635 (1999).
7. Ibrahim, A., et al., presented at AAS Meeting #194, Chicago (1999).

Statistical Properties of SGR 1900+14 Bursts

Ersin Göğüş*, Peter M. Woods*, Chryssa Kouveliotou[†],
Jan van Paradijs*, Michael S. Briggs*, Robert C. Duncan[‡],
and Christopher Thompson[||]

*Department of Physics, University of Alabama in Huntsville, Huntsville, AL 35899
[†] Universities Space Research Association
[‡] Department of Astronomy, University of Texas, RLM 15.308, Austin, TX 78712-1083
[||] Department of Physics and Astronomy, University of North Carolina, Philips Hall, Chapel Hill, NC, 27599-3255

Abstract. BATSE detected 200 bursts from SGR 1900+14 between May 1998 and January 1999. The reactivation of the bursts initiated a series of RXTE observations within which more than 800 bursts were observed. Here we study statistical characteristics of SGR 1900+14 bursts using BATSE and RXTE observations.

INTRODUCTION

Soft Gamma Repeaters (SGRs) are characterized by the recurrent brief (~ 0.1 s), emission of gamma-ray bursts with relatively soft spectra (well described by optically-thin thermal bremsstrahlung at $kT \sim 20-40$ keV) (Kouveliotou 1995).

Cheng et al. (1996) observed that particular statistical properties of a sample of 111 SGR events from SGR 1806−20 are quite similar to those of earthquakes. These properties include the distribution of event energies, which follow a power law $dN \propto E^{-\gamma} dE$ with an exponent, $\gamma = 1.6$. A similar distribution was obtained empirically by Gutenberg and Richter (1956a) for the distribution of EQ energies, with power-law index $\gamma_{EQ}=1.6 \pm 0.2$ The distribution of time intervals between successive SGR 1806−20 events is well described by a log-normal distribution analogous to the waiting times distribution of microglitches seen in the Vela pulsar (see Hurley et al. 1994). Cheng et al. (1996) also showed that cumulative waiting time distributions of SGR 1806−20 and earthquake events are similar. These results support the idea that SGR bursts are caused by starquakes, as expected to occur in the crusts of magnetically-powered neutron stars, or "magnetars" (Duncan & Thompson 1992; Thompson & Duncan 1995).

BATSE AND RXTE OBSERVATIONS

In the period from May 1998 until January 1999 a total of 200 events were detected (Woods et al. 1999) with BATSE. Out of these 200 events, 63 led to an on-board trigger. We determined fluences of 187 BATSE events (triggered and untriggered) which had DISCLA data. We find that the fluences of SGR 1900+14 bursts observed with BATSE range between 2×10^{-8} and 2.5×10^{-5} ergs cm^{-2}. For an estimated distance to SGR 1900+14 of 7 kpc (Vasisht et al. 1994), and assuming isotropic emission, the corresponding energy range is $1.1 \times 10^{38} - 1.5 \times 10^{41}$ ergs.

The sudden change in source activity initiated a series of RXTE observations between May 31 and December 21, 1998. During these observations, 837 bursts from SGR 1900+14 were detected with the Proportional Counter Array (PCA). We determined the count fluence of each burst, then using a conversion factor we derived between PCA counts and BATSE fluences for SGR 1900+14, we estimated PCA burst fluences which extends from 1.2×10^{-10} to 3.3×10^{-7} ergs cm^{-2} and the burst energies range from 7×10^{35} to 2×10^{39} ergs.

STATISTICAL ANALYSIS

(i) Energy distributions: The fluences of BATSE and RXTE bursts were binned in equally spaced logarithmic fluence steps $(dN/d\log E)$ (Fig. 1). We employed a maximum likelihood analysis to fit a power-law model to the unbinned fluence values between 5.0×10^{-8} and 2.5×10^{-6} ergs cm^{-2}. This method yields $\gamma = 1.66 \pm 0.13$ for the energy exponent. The unbinned RXTE fluences between 1.6×10^{-10} and 3.3×10^{-7} ergs cm^{-2} were then fit to a power-law model using the using the same method obtaining 1.66 ± 0.05 for the power-law exponent.

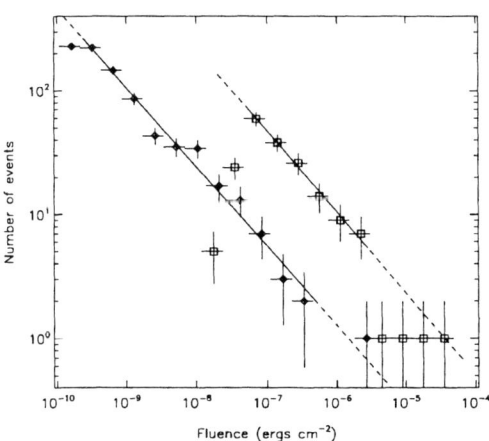

FIGURE 1. Differential distribution of the fluences of bursts from SGR 1900+14 as measured with RXTE (diamonds) and BATSE (squares).

Combined RXTE and BATSE fluences range from 1.2×10^{-10} to 2.5×10^{-5} ergs cm^{-2} (Fig. 1) which demonstrates that power-law distribution of energies with an exponent $\gamma \approx 1.66$ is valid for SGR 1900+14 over 4 orders of magnitude.

(ii) Waiting times statistics: We have measured the waiting times (ΔT) between successive bursts for 779 events. Fig. 2 shows the distribution of waiting times which range from 0.25 to 1421 s. We fit the (ΔT)-distribution to a log-normal function and found a peak at $\sim$ 49 s.

To investigate any relations between waiting times till the next burst (ΔT^+) and the intensity of the bursts, we divided the 779 events sample into 8 intensity inter-

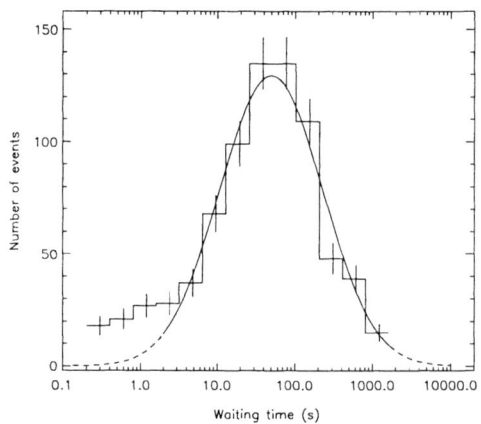

FIGURE 2. Distribution of the waiting times between successive RXTE PCA bursts from SGR 1900+14.

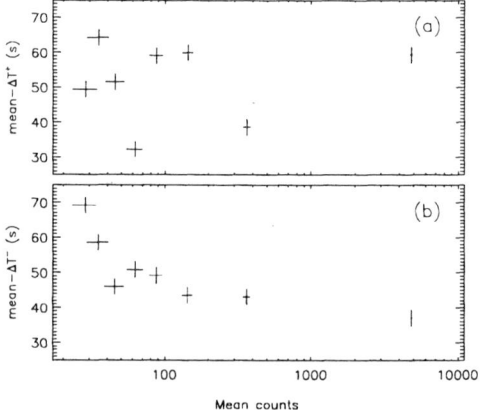

FIGURE 3. (a) Plot of mean waiting times till the next burst (ΔT^+) vs mean counts, (b) plot of mean waiting times since the previous burst (ΔT^-) vs mean counts.

vals each of which contains approximately 100 events. We fit the ΔT^+-distribution to also a log-normal function and determined the mean-ΔT^+ (i.e., where the fitted log-normal distribution peaks), and the mean counts for each of the 8 groups. We show in Fig. 3a that there is no correlation between ΔT^+ and energy of the bursts (the Spearman rank-order correlation coefficient, $\rho = 0.05$ and the probability that this correlation occurs by a random data set, P = 0.91). Similarly, we searched for the relation between the elapsed times since the previous burst (ΔT^-) and the intensity of the bursts. Fig. 3b shows that there appears to be an anti-correlation between mean-ΔT^- and burst energy ($\rho = -0.93$, P $= 8 \times 10^{-4}$).

(iii) Burst durations: Gutenberg and Richter (1956a; 1956b) demonstrated that there is a power-law relation between the magnitude, or energy of the EQ events and the durations of the strong motion at short distances from an EQ region. In order to investigate if a similar correlation exists for SGR events, we obtained t_{90} durations for 281 of the bursts using event mode PCA data with 1 ms time resolution. Fig. 4 shows that burst energies and durations are correlated ($\rho=0.54$, P$\sim 10^{-24}$), although there is a significant spread of fluences at a given duration.

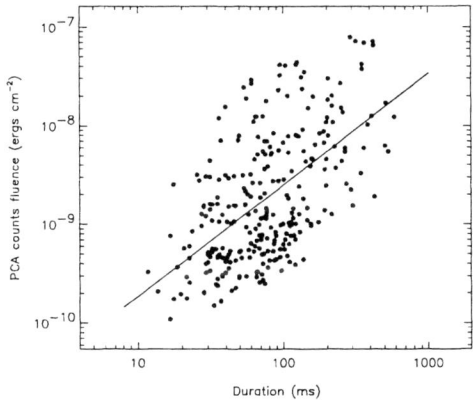

FIGURE 4. Scatter plot of the PCA fluence vs. duration for 281 SGR 1900+14 bursts which shows a correlation between them ($\rho = 0.54$). The solid line is a power law with an exponent 1.13 obtained using least squares fitting.

DISCUSSIONS

We find that the size distribution of SGR 1900+14 bursts follows a power law of index 1.66 over more than four orders of magnitude in burst fluence. Power-law energy distributions have also been found for earthquakes with $\gamma = 1.4$ to 1.8 (Gutenberg & Richter 1956a; Chen et al. 1991), and solar flares, $\gamma = 1.53$ to 1.73 (Crosby et al. 1993).

There is no correlation between the intensity of the burst and the waiting time until the following burst. This result agrees well with the results of Laros et al. (1987) for SGR 1806−20 and distinguishes the physical mechanism of SGR 1900+14 bursts from that of type II X-ray bursts from the Rapid Burster (Lewin et al. 1976) in which the burst energy is proportional to the waiting time until the next burst.

Power-law energy distribution, along with a log-normal waiting time distribution and energy-correlated burst durations, are characteristics of self-organized critical systems in general, and earthquakes and solar flares in particular. The concept of self-organized criticality states that many composite systems will self-organize to a critical state in which a small perturbation can trigger a chain reaction that affects any number of elements within the system (see Bak et al. 1988). In the magnetar model, the triggering mechanism for SGR bursts is a hybrid of crustquakes and magnetically-powered flares (Thompson & Duncan 1995).

ACKNOWLEDGMENTS

We acknowledge support from the cooperative agreement NCC 8-65 (EG); NASA grants NAG5-3674 and NAG5-7060 (JvP); Texas Advanced Research Project grant ARP-028 and NASA grant NAG5-8381 (RCD).

REFERENCES

1. Bak, P., Tang, C. & Wiesenfeld, K., *Phys. Rev. A* **38**, 364 (1988).
2. Chen, K., Bak, P. & Obukhov, S.P., *Phys. Rev. A* **43**, 625 (1991).
3. Cheng, B., et al., *Nature* **382**, 518 (1996).
4. Crosby, N.B., et al., *Solar Phys.* **143**, 275 (1993).
5. Duncan, R.C. & Thompson C., *ApJL* **392**, L9 (1992).
6. Gutenberg, B. & Richter, C.F., *Bull. Seis. Soc. Am.* **46**, 105 (1956a).
7. Gutenberg, B. & Richter, C.F., *Ann. Geophys.* **9**, 1 (1956b).
8. Hurley, K.J., et al., *A & A* **288**, L49 (1994).
9. Kouveliotou, C., *Astrophys. & Space Sci.* **231**, 49 (1995).
10. Laros, J.G., et al., *ApJ* **320**, L111 (1987).
11. Lewin, W.H.G, et al., *ApJ* **207**, L95 (1976).
12. Thompson, C. & Duncan, R.C., *MNRAS* **275**, 255 (1995).
13. Vasisht, G., et al., *ApJ* **431**, L35 (1994).
14. Woods, P., et al., *ApJ* **519**, L139 (1999).

Optical Imaging of the SGR 1627−41 Error Box During the SGR Activity in June 1998

A. J. Castro-Tirado[1,2], N. Lund[3], D. Pinfield[4], and S. Covino[5]

[1] *Laboratorio de Astrofísica Espacial y Física Fundamental (LAEFF-INTA), P.O. Box 50727, E-28080, Madrid, Spain*
[2] *Instituto de Astrofísica de Andalucía (IAA-CSIC), P.O. Box 03004, E-18080 Granada, Spain*
[3] *Danish Space Research Institute, Copenhagen, Denmark*
[4] *Queens University, Belfast, Ireland*
[5] *Osservatorio Astronomica di Merate, Italy*

Abstract. I-band imaging of the G337.0-0.1 region [1] limited by the IPN annulus reported by Hurley [2] were obtained on 25 June 1998, just 7.2-hr prior to a burst detected by Ulysses and Konus. A comparison image was obtained in late August 1998. No variable source was detected above a limiting magnitude of I $\sim$ 20.

INTRODUCTION

SGR 1627-41 was discovered by BASTE/CGRO on 15 June 1998. Eight bursts occurred between 15-18 June [3,4]. The detection by the GBD/Ulysses led to determine a diamond-shape error box of $\sim$ 1700 arcmin^2 [5]. The SGR activity was also recorded by the GRBM/BSAX instrument [6] and Konus-Wind [7].

The supernova remnant G337.0-0.1 (= CTB 33) is possibly associated to SGR 1627-41 [8]. VLA observations of the SNR revealed no point sources detected with fluxes densities F > 345 μJy [9]. Further RXTE pointed observations centered on the SN remnant resulted in the detection of one burst, thus confirming the proposed association [10–12]. The SNR-SGR association was definitively confirmed by seven events detected by both Konus and Ulysses between 17-22 June 1998 [13].

An X-ray counterpart, SAX J1635.8-4736, was detected within the IPN arc by BSAX [3,4].

OBSERVATIONS

We carried out optical observations in the I-band at the 1.54-m Danish telescope at ESO´s La Silla Observatory on 25 June 1998, just 7.2-hr prior to the burst

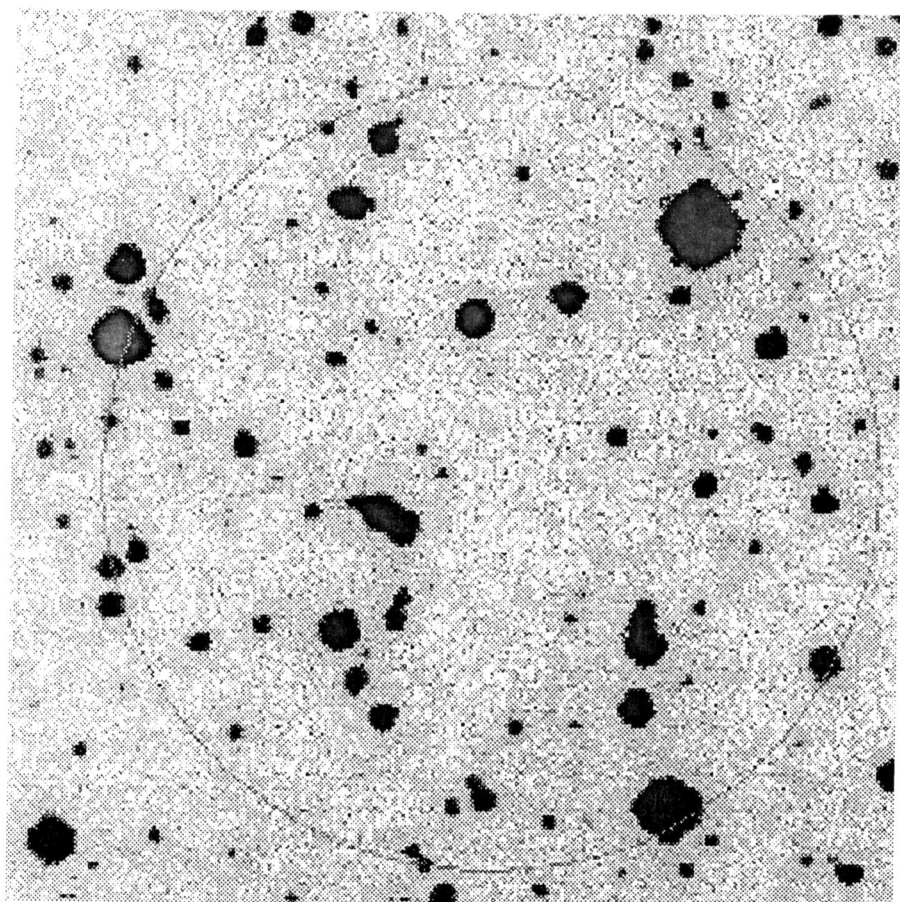

FIGURE 1. A fraction of the I-band image obtained at the 1.5-m Danish telescope at La Silla on 25 June 1999, ~ 7.2-hr prior to a burst detected by Ulysses. No variable source is seen within the error box for SAX J1635.8-4726, presumably the X-ray counterpart to SGR 1627-41, when comparing to an image taken one month later. The field is 2.5′ × 2.5′. North is up and east to the left.

detected by Ulysses and Konus at 39739 s U.T. Our first frame was centered at AR(2000) = 16h 35m 50s, Dec(2000) = −47° 37′ 00″ and the other two frames were shifted by 3′ in declination. Another 4 frames were obtained for comparison purposes using the same set-up on 25 July 1998 (Table 1).

TABLE 1. Log of optical observations for SGR 1627-41.

Date of 1998	Telescope	Band	Exp. time (s)	frames	lim. mag.
June 25	1.54-m Danish	I	300	3	I ~ 20.5
June 28	1.54-m Danish	I	300	1	I ~ 20.5

RESULTS AND DISCUSSION

After a careful analysis when comparing both sets of images, no variable source is detected within the intersection of the IPN arc and the SNR region (including the BSAX error box shown on Figure 1), above a limiting magnitude of I ~ 20.5.

This is not surprising, as no optical/IR variable has been ever found in the X-ray error boxes of the other SGRs, and this result is consistent with the high column density derived from the BSAX observation ($N_H \sim 8 \times 10^{22}$ cm^{-2} [4]), thus placing SGRs beyond the reach of optical telescopes.

Acknowledgements. This work has been partially supported by a Spanish CICYT grant ESP95-0389-C02-02.

REFERENCES

1. Woods, P. M. et al., GCN 113 (1998).
2. Hurley, K., GCN 110 (1998).
3. Kouveliotou, C. et al., IAU Circ. 6944 (1998).
4. Woods, P. M. et al., *ApJ* **519**, L139 (1999).
5. Hurley, K. et al., *ApJ* **519**, L143 (1999).
6. Feroci, M. et al., GCN Circ. 111 (1998).
7. Mazets, E. et al., *ApJ* **519**, L151 (1999).
8. Woods, P. M. et al. , IAU Circ. 6948 (1998).
9. Frail, D. & Kulkarni, S., GCN Circ. 121 (1998).
10. Smith, D. & Levine, A. M., IAU Circ. 6950 (1998).
11. Dieters, S. et al., IAU Circ. 6962 (1998).
12. Smith, D., Bradt, H. V. & Levine, A. M., *ApJ* **519**, L147 (1999).
13. Hurley, K., Mazets, E. & Golenetskii, S., IAU Circ. 6966 (1998).

Rapid Optical Follow-up Observations of SGR Events with ROTSE-I

R. Balsano[1], C. Akerlof[2], S. Barthelmy[3], J. Bloch[1], P. Butterworth[3],
D. Casperson[1], T. Cline[3], S. Fletcher[1], G. Gisler[1], J. Hills[1],
R. Kehoe[2], B. Lee[4], S. Marshall[5], T. McKay[2], A. Pawl[2],
W. Priedhorsky[1], N. Seldomridge[1], J. Szymanski[1], and J. Wren[1]

[1] *Los Alamos National Laboratory, Los Alamos, New Mexico 87545*
[2] *University of Michigan, Ann Arbor, Michigan 48109*
[3] *NASA/Goddard Space Flight Center, Greenbelt, MD 20771*
[4] *Fermi National Accelerator Laboratory, Batavia, Illinois 60510*
[5] *Lawrence Livermore National Laboratory, Livermore, California 94550*

Abstract. The primary mission of the Robotic Optical Transient Search Experiment (ROTSE) is to search for contemporaneous optical emission from GRBs. Among the triggers ROTSE receives via the GRB Coordinates Network (GCN), there are a number from Soft-Gamma Repeater (SGR) events. Since beginning operations in March 1998, ROTSE-I has triggered on 16 observable SGR events. Ten of these events had useful data, eight events from SGR 1900+14 and two events from SGR 1806−20.

The error regions for these SGRs are a small fraction of the ROTSE 16° × 16° field of view and have been searched for new or variable objects. Limits on optical transient counterparts are in the range $m_{ROTSE} \approx 12.5 - 15.5$ during the period 10 seconds to 1 hour after the observed SGR events.

INTRODUCTION

There is currently nothing known about the emission of Soft Gamma-ray Repeater (SGR) bursts at energies below a few keV. In this paper we present the first known attempts to detect optical emission from SGRs in the period just following SGR bursts. The Robotic Optical Transient Search Experiment (ROTSE) [1] is configured to respond to transient events from the Gamma-Ray Burst Coordinates Network (GCN) and is capable of rapidly slewing to the coordinates of a transient event such as an SGR burst. Since beginning operations in March 1998, the first generation system, ROTSE-I, has triggered on 16 SGR events, ten of which had useful data.

TABLE 1. Summary of ROTSE-I SGR Observations

SGR	Date	BATSE trigger	Delay before first usable exposure (s)	Number of usable images	Comments[a]
1900+14[b]	980530	6798	168	4	tiles
	980607	6809	19	13	direct
			1023	4	tiles
	980719	6932	226	1	tiles
			669	4	tiles
	980720	6934	316	23	direct
	980921	7107	682	22	direct
	980927	7124	153	4	tiles
			617	16[c]	direct
	990429	7536	933	2	tile
	990429	7537	831	4	tile
1806−20	980908	7073	425	6	direct, first images cloudy
	980922	7109	174	4	tiles

[a] "Direct" means the SGR was in direct exposures, "tiles" that it was only in tiles.
[b] For three triggers, a second GCN trigger was received at a new location.
[c] This does not include 6 frames in which only half of the SGR error region was covered.

OBSERVATIONS

A typical ROTSE-I response to a GCN trigger consists of a series of *direct* exposures centered at the trigger coordinates followed by a series of *tiled* exposures with the mount shifted by ±8° (half the total FOV) in both right ascension and declination to extend the ROTSE-I coverage of the GCN error box. ROTSE-I SGR responses are summarized in Table 1. The durations of the SGR bursts were all < 1 s with the exception of BATSE trigger 6798 which was a series of bursts lasting 350 s. ROTSE generally begins observations ∼ 10 s after receiving a GCN trigger. However, the first useful image may be taken up to several minutes later if the SGR position is only in tiled images. Six of the 16 SGR responses have no useful data, either because they occured during twilight or in cloudy conditions (3 of 6), because the GCN trigger positions differed from the SGR location by more than 16 degrees (2 of 6) or because a software failure occurred (1 of 6).

Data reduction

All images have been dark-corrected and flat-fielded. Examples of corrected images are shown in Figure 1. Parameters for all objects in an image are measured using SExtractor [2]. The object lists are photometrically and astrometrically calibrated against the Tycho Reference Catalogue [3]. Since ROTSE-I uses an unfiltered CCD, the photometry is color-corrected using Tycho $B - V$ to produce a ROTSE-I equivalent V-band magnitude, $m_{\rm ROTSE}$.

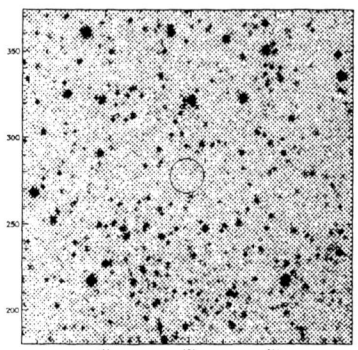 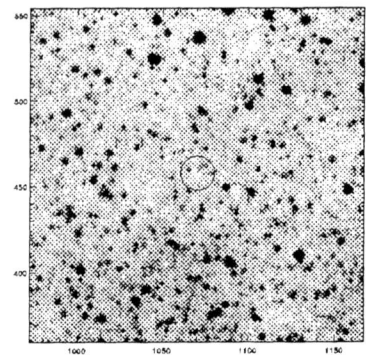

FIGURE 1. Sample images for SGR 1900+14 (left) and SGR 1806−20 (right). Each image is ≈ 50′ wide. The adopted search region, which is 5′ in diameter, is circled.

DISCUSSION

No new or variable objects were detected in any images for either SGR. Limits, shown in Figure 2, were obtained by determining the magnitude at which the efficiency for detecting artificial objects falls to 50%. For SGR1806−20, we have used objects found with SExtractor. For SGR1900+14, we performed this estimate visually, although a cross-check performed with SExtractor on a subset of the data indicates the limits agree to within 0.25 mag for the two methods.

Extinction

Since both SGR 1900+14 and SGR 1806−20 are at low galactic latitudes, extinction is highly uncertain. Unfortunately, no direct measurements of the extinction to either SGR exist. However, infrared observations covering the IPN localization for SGR 1900+14 provide estimates of $A_V = 15.4 \pm 1.2$ mag at $2.2 - 6.6$ kpc and $A_V = 19.1 \pm 1.2$ mag at $12 - 15$ kpc [4]. To get an alternate estimate for the extinction to these SGRs, we turn to the X-ray data.

Both SGR 1900+14 and SGR 1806−20 have been detected as X-ray pulsars so the hydrogen column density measured from the X-ray spectra may be related to optical extinctions. SGR 1900+14 was detected as a pulsar by ASCA and spectral fits gave a hydrogen column density of $n_h = (2.16 \pm 0.07) \times 10^{22}$ cm^{-2} and an estimated distance to the SGR of ~ 5 kpc [5] which agrees with the distance estimate to the supernova remnant G42.8+0.6 in which the SGR appears to be embedded [6]. The value of n_h can be converted to extinction via: $A_V = R_V \times n_h/E(B-V)$ where $R_V \equiv A_V/E(B-V) = 3.1$ (see e.g. [8]) and $n_h/E(B-V) = 5 \times 10^{21}$ cm^{-2} mag^{-1} [7]. Since ROTSE-I uses an unfiltered CCD which is sensitive out to ~ 1 μm, A_V can be used as only a rough guide to the expected extinction. Using these values gives an estimate of the total visual band extinction

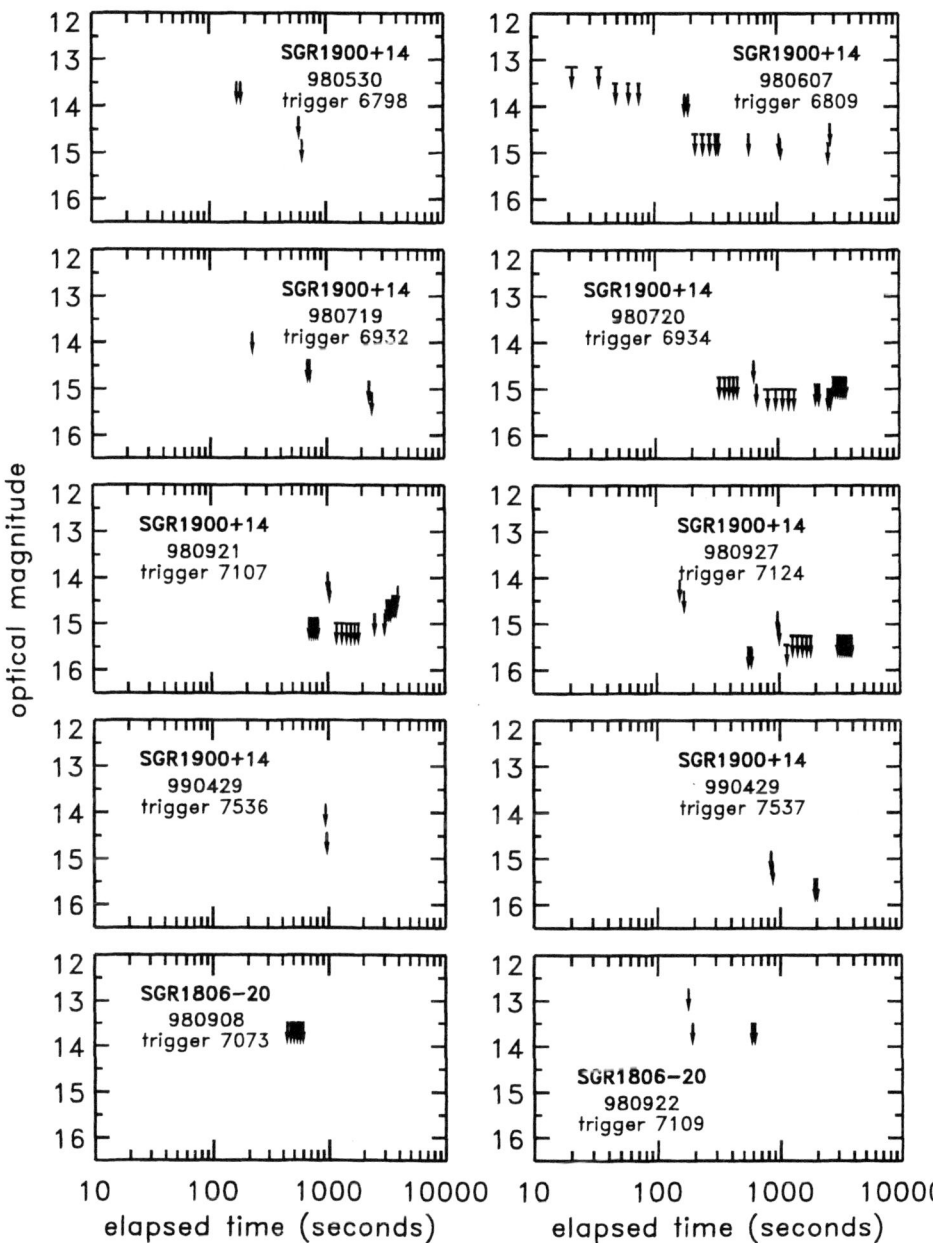

FIGURE 2. Limits for the 10 triggers with useful data. Each plot gives the limits, $m_{\rm ROTSE}$, as a function of delay since the SGR trigger.

to SGR 1900+14 of $A_V \approx 13$ mag. This agrees well with the extinction value found above for a distance of $2.2 - 6.6$ kpc, so we adopt $A_V \approx 13$ mag. For SGR 1806−20, $n_h \approx 6 \times 10^{22}$ cm^{-2} [9,10] gives $A_V \approx 30$. This large value, if indicative of the true extinction to SGR 1806−20, would make it impossible to see this source in the optical.

With the calculated extinction to SGR 1900+14, we can now address how bright an optical transient would have to be for ROTSE-I to detect it. Accepting the distance of 5 kpc for SGR1900+14, the ROTSE-I limits of $m_{\rm ROTSE} \approx 14$ mag give absolute magnitude limits of $M_{\rm ROTSE} \approx -1$. The extinction of $A_V \approx 13$ mag reduces this to $M_V \approx -14$, roughly between a nova and supernova in brightness.

With the large extinction in the direction of the known SGRs, a campaign specifically designed to detect SGRs would utilize a rapid-response detector sensitive in the $1 - 10$ μm region of the spectrum. However, ROTSE-I will continue to observe SGR triggers since doing so is a simple extension of ROTSE's main GRB response program. To improve the chances that ROTSE-I observes an SGR event in the future, we may point to the coordinates of a particular SGR based on classification information in the GCN trigger and the position of the event as given by GCN. Furthermore, the ROTSE collaboration is developing several 45 cm aperture telescopes which should reach several magnitudes deeper than ROTSE-I. It is possible that this increased sensitivity will be enough to overcome the very large extinction in the direction of SGR 1900+14.

CONCLUSION

We have presented limits on optical emission in the period immediately following SGR bursts for a total of ten events. Limits on optical transient counterparts are in the range $m_{\rm ROTSE} \approx 12.5 - 15.5$ during the period 10 seconds to 1 hour after the bursts.

REFERENCES

1. Kehoe, R., et al., in *Proc. of the 1999 STSci May Symposium*, astro-ph 9909219 (1999).
2. Bertin, E. & Arnouts, S., *Astr. Astrophys. Suppl. Ser.* **117**, 393–404 (1996).
3. Høg, E., et al., *Astr. Astrophys.* **335**, L65–L68 (1998).
4. Vrba, F. J., et al., *Astrophys. J.* **468**, 225 (1996).
5. Hurley, K., et al., *Astrophys. J.* **510**, L111–L114 (1999).
6. Vasisht, G., et al., *Astrophys. J.* **431**, L35 (1994).
7. Bohlin, R. C., Savage, B. D., & Drake, J. F., *Astrophys. J.* **224**, 132–142 (1978).
8. Savage, B. D. & Mathis, J. S., *Ann. Rev. Astr. Ap.* **17**, 73–111 (1979).
9. Murakami, T., et al., *Nature* **368**, 127–129 (1994).
10. Sonobe, T., et al., *Astrophys. J.* **436**, L23–25 (1994).

Search for Photometric Variability in the Vicinity of SGR 1900+14 and Discovery of a High-Mass Cluster

F. J. Vrba[1], C. B. Luginbuhl[1], A. A. Henden[1,3], H. H. Guetter[1], and D. H. Hartmann[2]

[1] *U.S. Naval Observatory, P.O. Box 1149, Flagstaff, AZ 86002-1149*
[2] *Dept. of Physics and Astronomy, Clemson University, Clemson, SC 29634-0978*
[3] *Universities Space Research Asociation*

Abstract. A pair of spectroscopically nearly identical M supergiant stars were proposed [1] as possible counterparts to SGR 1900+14 based on positional coincidences with gamma-ray and x-ray localizations. Whether these stars or the nearby fading radio source [9] are associated with SGR 1900+14 is not yet clear. We obtained I- and J-band photometric observations of the M stars with 10 to 120 second time resolution obtained on 16 nights before and during the active phase of SGR 1900+14 beginning in June 1998. Despite more than 54,000 seconds of integration time, no observations were obtained within an hour of a gamma-ray burst event. No short term variability was found for any object within the ROSAT error circle to I ≤ 24.5, although an unrelated eclipsing binary was discovered just outside the ROSAT circle. PSF subtraction of the bright M stars reveals a previously hidden compact star cluster.

INTRODUCTION

An imaging survey [1] of the original Network Synthesis Localization (NSL) of SGR 1900+14 [2] found a pair of nearly identical M5 supergiant stars, separated by 3.4 arcsec, and at an estimated distance of 12–15 kpc. While just outside of the original NSL, they lie within the ROSAT HRI localization of an X-ray source thought to be associated with SGR 1900+14 [3]. On the basis of the small probability that even one supergiant would lie within the ROSAT error circle and that a supergiant had been associated with SGR 1806−20 [4,5], it was proposed that the M star pair may be associated with the SGR 1900+14 source [1].

The position of the M star pair has continued to be consistent with recent X-ray and gamma-ray observations which have narrowed considerably the actual location of SGR 1900+14 from the original NSL area of 5 arcmin^2: ASCA X-ray observations, which discovered the 5.16 s period pulsar [6,7]; a new IPN localization using the extraordinary activity of SGR 1900+14 during 1998 [8]; and the ROSAT HRI localization of the quiescent X-ray source RX J190717+0919.3 mentioned above [3]. In addition, a variable radio source [9] has been found only a few arcseconds from the M stars (see section below). These positional coincidences, the lack of a plerionic radio source, and, despite arguments for SNR G42.8+0.6 in the literature, the lack of a coincident supernova remnant, indicate that this proximate, high-mass system of stars should not yet be dismissed as a possible evolutionary companion to the X-ray pulsar associated with the SGR. How such a companion fits into the SGR magnetar model [10] is not clear, though we note that the case for a SGR 1900+14 magnetar may still be in question [11,12].

Finding direct evidence that the M star pair is associated with SGR 1900+14 has thus far proven elusive [13]. In this paper we report on an attempt to detect photometric variability before and during the bursting activity of 1998 and the results of deep I-band imaging of the SGR 1900+14 region.

VARIABILITY OBSERVATIONS AND RESULTS

The 89 second period X-ray pulsar XTE J1906+09 discovered by RXTE [14] was found to be consistent with the position and estimated distance of the M stars and, hence, a potential counterpart to SGR 1900+14. In early May 1998 we began an observational program to search for the 89 second period in J-band observations of the M stars. Shortly thereafter, the reactivation and extraordinary outburst

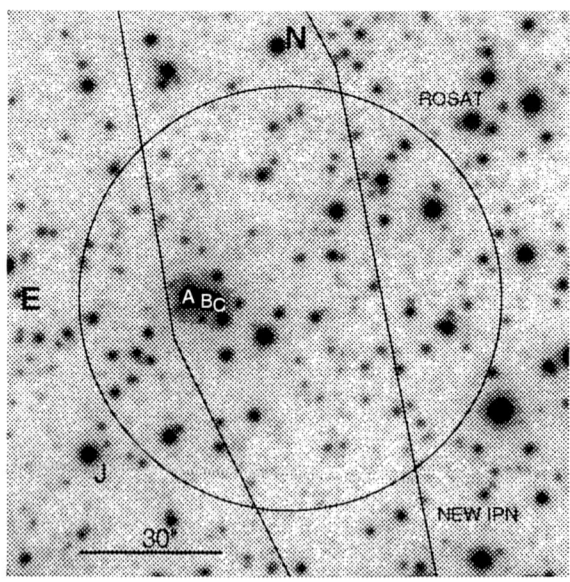

FIGURE 1. An I-band image of the SGR1900+14 region to I ≈ 26.5 showing the ROSAT error circle, IPN localization, the M stars ABC, and star J, the probable eclipsing binary.

activity of SGR 1900+14 prompted us to continue and expand the observations through mid-July 1998, hoping to find optical activity from the M stars or other sources within the ROSAT error circle. Observations were made in the I- and J-bands at the 1.55-m and 1.0-m telescopes of the Flagstaff Station of the U.S.N.O., comprising a total of 2025 frames of Tek 1K, Tek 2K, and NICMOS III data with 54,460 seconds of open shutter time during 16 nights. The exposure times and observing frequencies were intended to sample variability timescales from a few seconds to weeks. Figure 1 is a median-filter composite of about 6.5 hours of I-band integration at the 1.55-m telescope with limiting detection magnitude of about I ≈ 26.5 covering the approximate region searched for variable objects. The MI stars (A,B) and a foreground MIII star (C) are shown along with the ROSAT HRI error circle and the new IPN localization, for reference.

Only one object was found with significant evidence of variability amongst a total of 149 stars within and near the ROSAT error circle to a limiting magnitude of about I = 24.5. This object is marked "J" in Figure 1 and is a probable eclipsing binary, whose lightcurve is shown in Figure 2, but which is likely unrelated to SGR 1900+14.

Most of the observations were short integrations aimed at finding short time scale light variations in the M stars, particularly around the 89-s period. Periodogram analysis was carried out for all the fast data for periods between 10 and 300 seconds, but no significant power peaks were found. With 5.0 second integration times, our

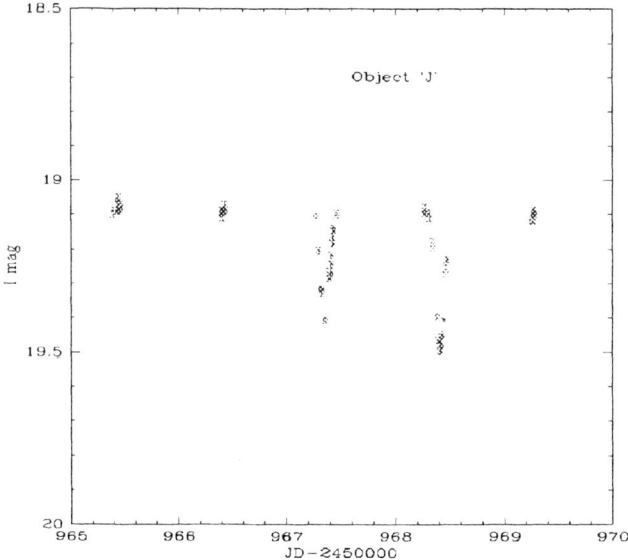

FIGURE 2. I-band measurements for the star labelled "J" in Figure 1, showing possible eclipses.

fast data were not well suited for finding the 5.16 sec period of the X-ray pulsar [6]. Nonetheless, we phased the data at this period to look for variations, again without success.

While no obvious counterpart candidates were revealed via variability, it is possible that excess optical/infrared radiation is only correlated to gamma-ray fluence. Despite a fairly intense effort, none of our observations were nearer in time than one hour before or 3 hours after an SGR 1900+14 burst. These null results indicate that significant excess low-energy radiation might not be a general characteristic of SGRs in their active state.

SPATIAL RELATION OF M STARS TO FADING RADIO SOURCE

A few days after the strong SGR 1900+14 outburst of UT 1998 August 27 a variable, probably fading, radio source was discovered at $\alpha = 19^h\ 07^m\ 14.33^s$, $\delta = +09^d\ 19'\ 21.1''$ (J2000), with positional accuracy of $\pm 0.15''$ in each coordinate [9].

We have performed a formal astrometric solution for the positions of the M stars ($\pm$ 0.1 arcsec) based on 21 USNO-A2.0 stars with the result:

Star A: $\alpha = 19^h\ 07^m\ 15.35^s$, $\delta = +09^d\ 19'\ 21.4''$ (J2000)
Star B: $\alpha = 19^h\ 07^m\ 15.13^s$, $\delta = +09^d\ 19'\ 20.7''$ (J2000)
Star C: $\alpha = 19^h\ 07^m\ 14.97^s$, $\delta = +09^d\ 19'\ 20.0''$ (J2000)

Thus, the M stars have formal offsets from the radio position of 15.2'', 11.8'', and 9.5'' ($\pm 0.2''$) for A, B, and C, respectively.

In Figure 3 we show a closeup of the Figure 1 I-band image, but with M stars A, B, C removed via PSF subtraction, and the radio source position indicated. There is no object seen at the radio position to I < 26.5, consistent with near infrared non-detections at J,H,K of 18.1, 16.6, and 15.3, respectively [15].

DISCOVERY OF AN EMBEDDED CLUSTER

Figure 3 also shows a cluster of stars, and possibly nebulosity, centered on the position of the M stars. This probable cluster was discovered due to the fortuitous situation that the 6.5 hours of total integration time was comprised of many short exposures thus allowing the non-saturated M star PSFs to be subtracted successfully.

The SGR 1900+14 cluster appears to be similar to the cluster discovered near SGR 1806−20 [16]. In the SGR 1806−20 case, the cluster is located about 20'' from a "luminous blue variable" (LBV) supergiant [4,5] and is interpreted as a young cluster of massive star formation. The new SGR 1900+14 cluster is discussed in [17]. Now that similar clusters have been found at or near the positions of two SGRs, the possiblity that young SGR neutron stars have their origins in massive young clusters should be considered seriously.

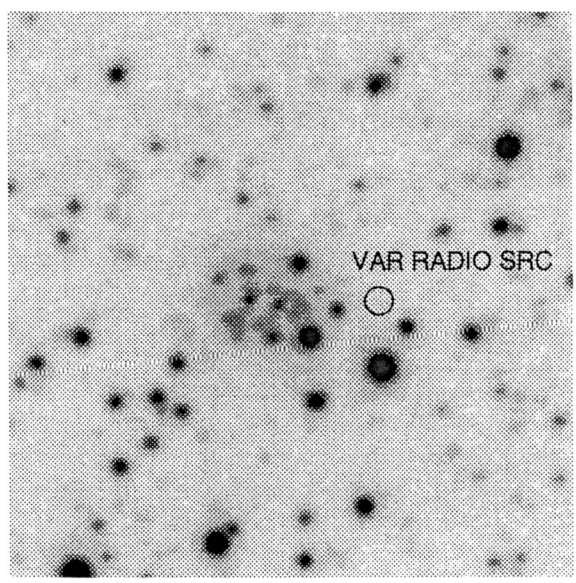

FIGURE 3. A portion of the same image as Figure 1, but with stars A, B, and C removed by PSF subtraction, showing the star cluster otherwise hidden by the bright stars.

REFERENCES

1. Vrba, F.J., et al., *ApJ* **468**, 225 (1996).
2. Hurley, K.H., et al., *ApJ* **431**, L31 (1994).
3. Hurley, K.H., et al., *ApJ* **463**, L13 (1996).
4. van Kerkwijk, M.H., et al., *ApJ* **444**, L33 (1995).
5. Kulkarni, S., et al., *ApJ* **440**, L61 (1995).
6. Hurley, K.H., et al., *ApJ* **510**, L111 (1999).
7. Murakami, T., et al., *ApJ* **510**, L119 (1999).
8. Hurley, K.H., et al., *ApJ* **510**, L107 (1999).
9. Frail, D.A., Kulkarni, S.R., & Bloom, J.S., *Nature* **398**, 127 (1999).
10. Koveliotou, C., et al., *ApJ* **510**, L115 (1999).
11. Marsden, D., Rothschild, R.E., & Lingenfelter, R.E., *ApJ* **520**, L107 (1999).
12. Rothschild, R.E., Marsden, D., & Lingenfelter, R.E., these proceedings.
13. Guenther, E.W., Klose, S., & Vrba, F., these proceedings.
14. Marsden, D., et al., *ApJ* **502**, L129 (1998).
15. Eikenberry, S.S. & Dror, D.H., *ApJ*, in press (1999).
16. Fuchs, Y., et al., *A&A* **358**, 891 (1999).
17. Vrba, F.J., et al., *ApJ*, submitted (2000).

Soft Gamma-Ray Repeaters in Clusters of Massive Stars

I. Félix Mirabel*[†], Yaël Fuchs*, and Sylvain Chaty[‡]

*Service d'Astrophysique, CEA Saclay,
Bat. 709, Orme des Merisiers, 91191 Gif sur Yvette cedex, France
[†]Instituto de Astronomía y Física del Espacio, cc67, suc 28. 1428 Buenos Aires, Argentina
[‡]Department of Physics and Astronomy, The Open University
Walton Hall, Milton Keynes, MK7 6AA, United Kingdom

Abstract. Infrared observations of the environment of the two Soft Gamma-ray Repeaters (SGRs) with the best known locations on the sky show that they are associated with clusters of massive stars. Observations with ISO revealed that SGR 1806−20 is in a cluster of giant massive stars, still enshrouded in a dense cloud of gas and dust [1]. SGR 1900+14 is at the edge of a similar cluster that was recently found hidden in the glare of a pair of M5 supergiant stars [2]. Since none of the stars of these clusters has shown in the last years significant flux variations in the infrared, these two SGRs do not form bound binary systems with massive stars. SGR 1806−20 is at only ~ 0.4 pc, and SGR 1900+14 at ~ 0.8 pc from the centers of their parental star clusters. If these SGRs were born with typical neutron star runaway velocities of ~ 300 km s^{-1}, they are not older than a few 10^3 years. We propose that SGR 1806−20 and SGR 1900+14 are ideal laboratories to study the evolution of supernovae explosions inside interstellar bubbles produced by the strong winds that prevail in clusters of massive stars.

INTRODUCTION

Neutron stars and stellar mass black holes are the last phase of the rapid evolution of the most massive stars, which are known to be formed in groups. In this context, it is expected that the most recently formed collapsed objects should be found near clusters of massive stars, still enshrouded in their placental clouds of gas and dust. This should be the case for SGRs, if they indeed are very young neutron stars [3]. Among the four SGRs that have been identified with certainty, SGR 1806−20 and SGR 1900+14 are the two with the best localizations with precisions of a few arcsec ([4,5] and references therein). Both are on the galactic plane at distances of ~ 14 kpc, beyond large columns of interstellar material; $A_v \sim 30$ mag in front of SGR 1806−20 [6] and $A_v \sim 19$ mag in front of SGR 1900+14 [7]. Because of these large optical obscurations along the lines of sight, and in the immediate environment of the sources, infrared observations are needed to understand their origin and nature.

INFRARED OBSERVATIONS

Mid-infrared (5–18 μm) observations of the environment of SGR 1806−20 were carried out with the ISOCAM instrument aboard the Infrared Space Observatory (ISO) satellite [1]. By chance, the ISO observations were made in two epochs, 11 days before, and 1–4 hr after a soft gamma-ray burst detected with the Interplanetary Network on 1997 April 14 [4].

We also observed[1] SGR 1806−20 in the J (1.25 ± 0.30 μm), H (1.65 ± 0.30 μm) and K′ (2.15 ± 0.32 μm) bands on 1997 July 19, and SGR 1900+14 in the J (1.25 ± 0.30 μm), H (1.65 ± 0.30 μm) and Ks (2.162 ± 0.275 μm) bands on 1999 July 25, at the European Southern Observatory (ESO), using the IRAC2b camera on the ESO/MPI 2.2-m telescope for SGR 1806−20, and using the NTT/SOFI for SGR 1900+14. In the near infrared, SGR 1806−20 was monitored by us during the last four years, and SGR 1900+14 by Vrba et al. [2].

SGRS IN CLUSTERS OF MASSIVE STARS

The results of the infrared observations of SGR 1806−20 and SGR 1900+14 are summarized in Figures 1 and 2, respectively. Figure 1 shows a cluster of massive stars deeply embedded in a dense cloud of molecular gas and dust. Using the ISO fluxes as a calorimeter, Fuchs et al. [1] show that each of the four stars at the centre of the cluster could be equaly, or even more luminous than the LBV identified in the field by Kulkarni et al. [8].

van Paradijs et al. [9] reported the possible association of SGRs to strong IRAS sources. The IRAS fluxes listed in Table 1 suggest that the infrared emission at longer wavelengths detected by IRAS does arise in clouds of gas and dust that enshroud these two clusters of massive stars.

TABLE 1. Association between SGRs and IRAS sources. The IRAS position and 12, 25 and 60 μm fluxes for 18056−2025 and 19048+0914 come from van Paradijs et al. [9].

SGR	SGR Position (J2000)	IRAS source	IRAS position (J2000)	IRAS flux densities (Jy)		
				12	25	60 μm
SGR 1806−20	α 18 08 39.5 δ -20 24 40	18056−2025	α 18 08 40.4 δ −20 24 41.6	0.98	35	29
SGR 1900+14	α 19 07 14.33 δ +09 19 20.1	19048+0914	α 19 07 15.3 δ +09 19 20.0	2.5	6.3	12.3

---

[1] Based on observations collected at the European Southern Observatory, La Silla, Chile under proposal numbers 59.D-0719 and 63.H-0511.

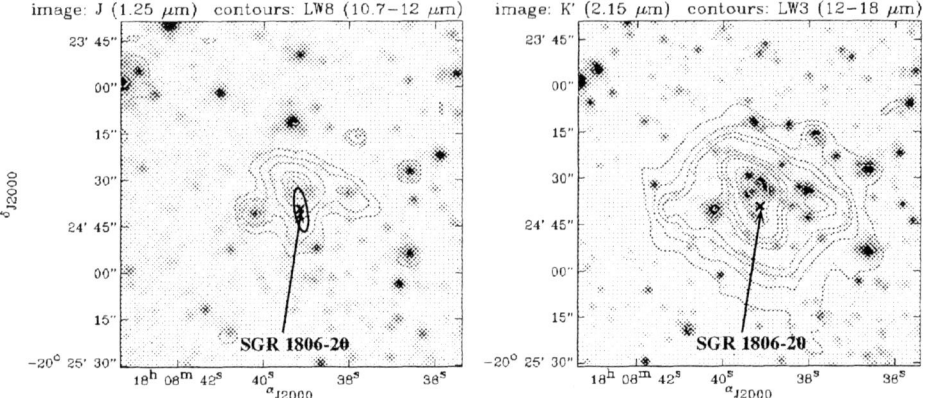

FIGURE 1. J ($1.25 \pm 0.32\,\mu$m) and K' ($2.15 \pm 0.32\,\mu$m) band images of SGR 1806−20 with 0.507″/pixel, together with the ISO fluxes ($11.35 \pm 0.65\,\mu$m) and ($15 \pm 3\,\mu$m) in contours with 3″/pixel. The best fit position of SGR 1806−20 [4] is marked as a small cross and ellipse. These images show that SGR 1806−20 is $\leq 5''$ (≤ 0.4 pc at a distance of 14 kpc) from the centre of a cluster of hot, giant, massive stars, which are still partly embedded in their "placental" cloud of gas and dust.

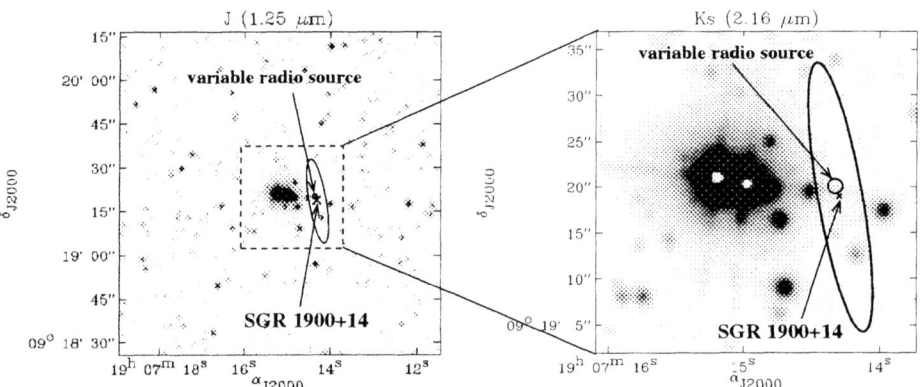

FIGURE 2. J ($1.25 \pm 0.32\,\mu$m) and Ks ($2.162 \pm 0.275\,\mu$m) band images of SGR 1900+14 with 0.292″/pixel. The two white points on the Ks image are due to detector saturation. The position of a fading radio counterpart [10] is indicated by a small circle, and the best fit position of SGR 1900+14 [4] is marked as a small cross and ellipse. SGR 1900+14 is ~ 0.8 pc from a pair of M5 supergiant stars [7] whose glare hides a cluster of stars [2]. SGR 1900+14 is located near the edge of this cluster of massive stars.

CONCLUSIONS

1. SGR 1806−20 and SGR 1900+14 are associated with clusters of massive stars. From ISO observations we find evidence that the cluster associated with SGR 1806−20 is enshrouded and heats a dust cloud that appears very bright at 12–18 μm. Although we did not make ISO observations of SGR 1900+14, it is, as SGR 1806−20, a strong IRAS source [9], and very likely it is also enshrouded in a dust cloud.

2. These SGRs cannot be older than a few 10^3 years. At the runaway speeds of neutron stars this is the time required to have moved away from the centers of their parental clusters of stars.

3. J, H and K′ band observations of the massive stars close to the SGRs positions show no significant flux variations [1,2]. Therefore, these SGRs do not form bound binary systems with any of these massive stars.

4. There is strong excess emission at 12–18 μm associated with SGR 1806−20. However, there is no evidence of heating by the high-energy SGR activity, although observations were made only 2 hours after a soft gamma-ray burst reported by Hurley et al. [4].

ACKNOWLEDGMENTS

The authors are grateful to F.J. Vrba for communicating his results on SGR 1900+14 prior to publication.

REFERENCES

1. Fuchs, Y., Mirabel, I. F., Chaty, S., et al., *A&A* **358**, 891–899 (1999)
2. Vrba, F. J., Henden, A. A., Luginbuhl, C. B., and Guetter, H. H., *ApJ*, in press (2000)
3. Kouveliotou, C., Strohmayer, T., Hurley, K., et al., *ApJ* **510**, L115–L118 (1999)
4. Hurley, K., Kouveliotou, C., Cline, T., et al., *ApJ* **523**, L37–L40 (1999a)
5. Hurley, K., astro-ph/9912061 (1999b)
6. Corbel, S., Wallyn, P., Dame, T. M., et al., *ApJ* **478**, 624–630 (1997)
7. Vrba, F. J., et al., *ApJ* **468**, 225–230 (1996)
8. Kulkarni, S. R., Mathews, K., Neugebauer, G., et al., *ApJ* **440**, L61–L64 (1995)
9. van Paradijs, J., et al., *A&A* **314**, 146–152 (1996)
10. Frail, D. A., Kulkarni, S. R., and Bloom, J. S., *Nature* **398**, 127–129 (1999)

ISO Observation of a Fraction of the SGR 1801−23 Error Box

A. J. Castro-Tirado[1,2], L. Metcalfe[3], and R. Laureijs[3]

[1] *Laboratorio de Astrofísica Espacial y Física Fundamental (LAEFF-INTA), P.O. Box 50727, E-28080, Madrid, Spain*
[2] *Instituto de Astrofísica de Andalucía (IAA-CSIC), P.O. Box 03004, E-18080 Granada, Spain*
[3] *ESA Astrophysics Division, VILSPA Satellite Tracking Station, P.O. Box 50727, E-28080, Madrid, Spain*

Abstract. We present the result of a target of opportunity observation performed by the ESA's Infrared Space Observatory (*ISO*) following the detection of the possible new SGR 1801-23 in June and September 1997. Four 60 μm sources are detected in a $10' \times 10'$ region, but the association of any to SGR 1801-23 remains open due to the small fraction of the revised IPN error observed by *ISO*.

INTRODUCTION

A possibly new soft gamma-ray repeater, SGR 1801-23, was discovered by *Ulysses* on 29 June 1997. The first two short, soft events were detected by KONUS-*WIND*. The third one was also recorded by BATSE on 12 Sep 1997 [1], and a region about 0.5° long was initially provided by the IPN [2]. This third burst was also detected by *RXTE*, leading to a preliminary combined IPN-ASM error box centered at AR(2000) = 18h 14m 49s, Dec(2000) = −13° 40′ 00″ of 5′ radius [3].

A strong near-IR source was detected about 2′ from the center of the IPN which was coincident with the source IRAS 18119-1342 [4], but a near-IR spectrum revealed an M-type star showing no apparent emission lines. It is likely to be an obscured supergiant unrelated to the SGR [5].

A refined *RXTE*/ASM position (a circle with a 5′ radius), centered at AR(2000) = 18h 14m 25s, Dec(2000) = −14° 06′ 42″ was released, intersecting the IPN arcs at the southern tip of the diamond [6].

SGR 1900+14 and SGR 1806-20 appear to be associated with sources with spectra peaking in the IR [7], hence we requested a target-of-opportunity observation of the new SGR field by the European Space Agency's Infrared Space Observatory (*ISO*).

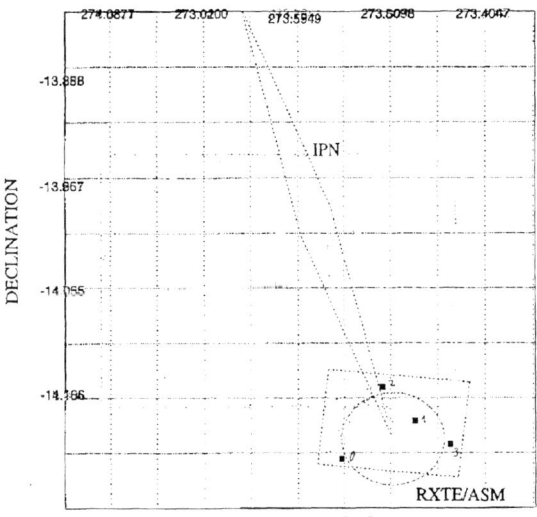

FIGURE 1. A sketch displaying the two error boxes for SGR 1801-23: the preliminary IPN region (diamond, from [2]) and the RXTE/ASM error box (circle, from [6]). The squared region is the field observed with ISO PHT, showing the location of the four sources found at 60 μm.

OBSERVATIONS

An observation at 60μm was performed by means of *ISO* on 13 Oct 1997, during a 2-hr observing window. The PHT instrument and the P22 filter (60μm) were used, as the latter provides the highest resolution (69″) and the best positional determination in case of a detection.

A raster of 11 steps was followed, with 69″ step size in both directions and 20 s integration time per raster point. An area of $10' \times 10'$ centered at the refined *RXTE*/ASM position was imaged.

RESULTS

Four sources were detected at 60μm within the *RXTE*/ASM error box, with one of them being close to the IPN arc (see Figure 1). The brightest one has a density flux of 3.6 mJy, and $\sim$ 1 mJy for the other three (details are given in Table 1). No extended emission is seen.

Follow-up optical/infrared observations were obtained at the 1.54 m Danish telescope at La Silla (Gunn-z filter) in Oct 1997 and at the 3.5 m telescope at Calar Alto (JHK-filters) in Sep 1999. Data reduction is in progress at the time of this writing, and the final results will be reported elsewhere.

TABLE 1. ISO 60μm sources detected in the SGR 1814-14 field.

source ID	AR(2000)	DEC(2000)	60μm flux (mJy)
0	18 14 45	-14 03 54	3.6
1	18 14 16	-13 59 46	1
2	18 14 29	-13 56 04	1
3	18 14 02	-14 02 19	1

In any case, the association of any of the four *ISO* sources to SGR 1801-23 is an open question, due to a somewhat larger 3.8° IPN error box later reported [8,9] that was not totally imaged by the *ISO* observation.

Acknowledgements. We are grateful to *ISO* staff in VILSPA for having rapidly scheduled the SGR 1814-14 observations just before the end of the observing window. This work has been partially supported by a Spanish CICYT grant ESP95-0389-C02-02.

REFERENCES

1. Kouveliotou, C. et al., IAU Circ. 6743 (1997).
2. Hurley, K., Cline, T. and Mazets, E., IAU Circ. 6743 (1997).
3. Smith, D. et al., IAU Circ. 6743 (1997).
4. Henden, A. et al., IAU Circ. 6744 (1997).
5. Djorgovski, G. et al., private communication 1997).
6. Smith, D. et al., private communication (1997).
7. van Paradijs, J. et al., *A&A* **314**, 146 (1996).
8. Hurley, K. et al., *ApJ*, in press (2000).
9. Hurley, K. et al., these proceedings (2000).

Optical/Near-IR Observations of SGR 1900+14 During the May-June and Aug-Sep 1998 Active Periods

A. J. Castro-Tirado[1,2], S. Beckwith[3], D. Kelson[4], T. Kerr[5], C. Lázaro[6], and S. Madruga[6]

[1] *Laboratorio de Astrofísica Espacial y Física Fundamental (LAEFF-INTA), P.O. Box 50727, E-28080, Madrid, Spain*
[2] *Instituto de Astrofísica de Andalucía (IAA-CSIC), P.O. Box 03004, E-18080 Granada, Spain*
[3] *Space Telescope Science Institute, Baltimore, MA, USA*
[4] *Carnegie Institute, Washington, USA*
[5] *Joint Astronomy Centre, Hilo, Hawaii 96720, USA*
[6] *Instituto de Astrofísica de Canarias, La Laguna, Tenerife, Spain*

Abstract. We present the results of optical/IR observations obtained in the period June-October 1998. Images in the Gunn-z band were serendipitously taken just 3 days after the giant burst that took place on 27 Aug 1998.

INTRODUCTION

The Soft Gamma-ray Repeater 1900+14 (hereafter SGR 1900+14) was first detected by the Konus experiment on *Venera 11* and *12* on 24-27 March 1979, when it produced three outbursts [1,2] coming from the same direction of the sky.

Thirteen years later (1992), a series of events detected by BATSE on *CGRO* was tentatively attributed to SGR 1900+14. The first event arising from the vicinity was observed on 19 June 1992 [3]. Two additional bursts were detected on 8 July and 19 August, thus confirming a SGR nature, and suggesting that SGRs can become active again after many years. This has been supported by the new activity period observed by BATSE in SGR 1806-20 [4].

At that time, an improved position was obtained when combining the data from the *Ulysses* gamma-ray burst detector and BATSE. Very close to this tiny error box, an X-ray source was found with the *ROSAT* HRI detector [5]. This was tentatively suggested as the long sought X-ray counterpart to SGR 1900+14.

Within the 10" radius *ROSAT* error box a double IR source was suggested as the potential counterpart for this galactic soft gamma-ray repeater [6]. The stars were found to be variable but otherwise normal supergiant stars at a distance of

TABLE 1. Log of optical/near-IR observations for SGR 1900+14.

Date of 1998	Telescope	Band	Exp. time (s)	frames	lim. mag.
June 3.06	1.0-m JKT	I	300	1	I ~ 20
June 3.08	1.0-m JKT	V	300	1	V ~ 20
June 5.13	0.8-m IAC80	I	240	1	I ~ 20
June 8.35	3.5-m CAHA	J	60	5	J ~ 20
June 9.11	0.8-m IAC80	I	120	75	I ~ 19
June 9.92	3.5-m CAHA	J	60	1	J ~ 20
June 9.94	3.5-m CAHA	2.166	60	9	K ~ 16
June 9.96	3.5-m CAHA	2-248	60	9	K ~ 16
June 9.99	0.8-m IAC80	I	120	3	I ~ 17
June 10.00	0.8-m IAC80	I	120	3	I ~ 17
June 11.07-11.19	1.0-m JKT	I	40	233	I ~ 20
Aug 30.05	1.5-m Danish	Gunn-z	180	3	z ~ 19
Sep 7.73	3.5-m CAHA	K	45	5	K ~ 18
Sep 7.75	3.5-m CAHA	2.166	60	5	K ~ 17
Oct 5.75	3.5-m CAHA	2.166	60	5	K ~ 17
Oct 22.74	0.8-m IAC80	I	300	1	I ~ 18
Oct 22.74	0.8-m IAC80	I	300	1	I ~ 18
Nov 7.02	1.5-m Danish	Gunn-z	180	3	z ~ 19

TABLE 2. Log of photoelectric photometric observations.

Date of 1998	Telescope	Band	Int. time (s)	measurements	lim. mag.
June 9.04-9.19	1.5-m TCS	K	5	864	K ~ 10

12-15 kpc (i.e., distinguished by $A_v = 19$ mag). At that time, SGR 1900+14 was in the so-called off state that began in 1992, with no bursts being detected until then. JHK spectroscopy only revealed strong CO absorption at 2.3 μm, and did not show any emission line that would have revealed the presence of an unseen compact object.

The very recent BATSE data indicate that SGR 1900+14 entered in 1998 a new prolific period, the third one since 1979, as reported in [7]. A giant outburst was detected on 27 August, similar to the famous 5 March event for SGR 0526-66, with oscillations detectable with a period of 5.16 s [8,9] and a transient radio counterpart [10].

Observations of the field by *Asca* on 30 April–1 May discovered a pulsar with a period of 5.15897 s from the known *ROSAT* source. The detection of several bursts on 16-17 September provided an accurate position for the SGR [11].

OBSERVATIONS

Optical/near-IR imagery of the SGR 1900+14 field was obtained in June–October 1998 (Table 1), but the photoelectric photometry and JHK-band spectroscopy of the two bright reddened stars were only acquired in June 1998 (Tables 2 and 3).

RESULTS

With the exception of the double IR source (the bright M stars), no other object is detected with a variation > 0.2 mag within the *ROSAT* error box. No source is firmly detected at the position of the radio transient [10] and not even on the Gunn-z image obtained 2.7 d after the giant outburst of 27 August 1998 (see Figure 1).

Regarding the aperture photoelectric photometry, no IR outbursts (peaking at $K > 10$) are detected within the *ROSAT* error box during a our 1-hr monitoring on 9 June (with time resolution of 10 s).

Our JHK-band spectra are consistent with those of a highly-reddened M5 I-III stars, in agreement with the results given in [6].

Acknowledgements. One of us (AJCT) is grateful to C. Kouveliotou for useful discussions, and to N. Walton (INT Group, La Palma) and to S. Kemp and M. Kidger (IAC) for the images taken at the JKT on 3 June and and at the IAC-80 telescope on 5 June 1998. This work has been partially supported by a Spanish CICYT grant ESP95-0389-C02-02.

REFERENCES

1. Mazets, E. P., Golenetskii, S. V., Guryan, Yu. A., *SvA Lett* **5(6)**, 343 (1979).
2. Mazets, E. P. et al., *ApSS* **80**, 1 (1981).
3. Kouveliotou, C. et al., *Nature* **362**, 728 (1993).
4. Kouveliotou, C. et al., *Nature* **368**, 125 (1994).
5. Hurley, K. et al., AIP Conf. Proc. **366** (1995).
6. Vrba, F. J. et al., *ApJ* **468**, 225 (1996).
7. Hurley, K. et al., *ApJL* **510**, L107 (1999).
8. Hurley, K. et al., *Nature* **397**, 41 (1999).
9. Feroci, M. et al., *ApJL* **515**, L9 (1999).
10. Frail, D. A., Kulkarni, S. R. & Bloom, J. S., *Nature* **398**, 127 (1999).
11. Murakami, T. et al., *ApJL* **510**, L119 (1999).

TABLE 3. Log of spectroscopic observations for the M-type stars.

UT (10 June 1998)	Telescope	Central wavelength (μm)
11:10-11:26	3.8-m UKIRT	1.15
11:47-12:00	3.8-m UKIRT	1.30
12:32-12:45	3.8-m UKIRT	1.70
13:15-13:32	3.8-m UKIRT	2.18

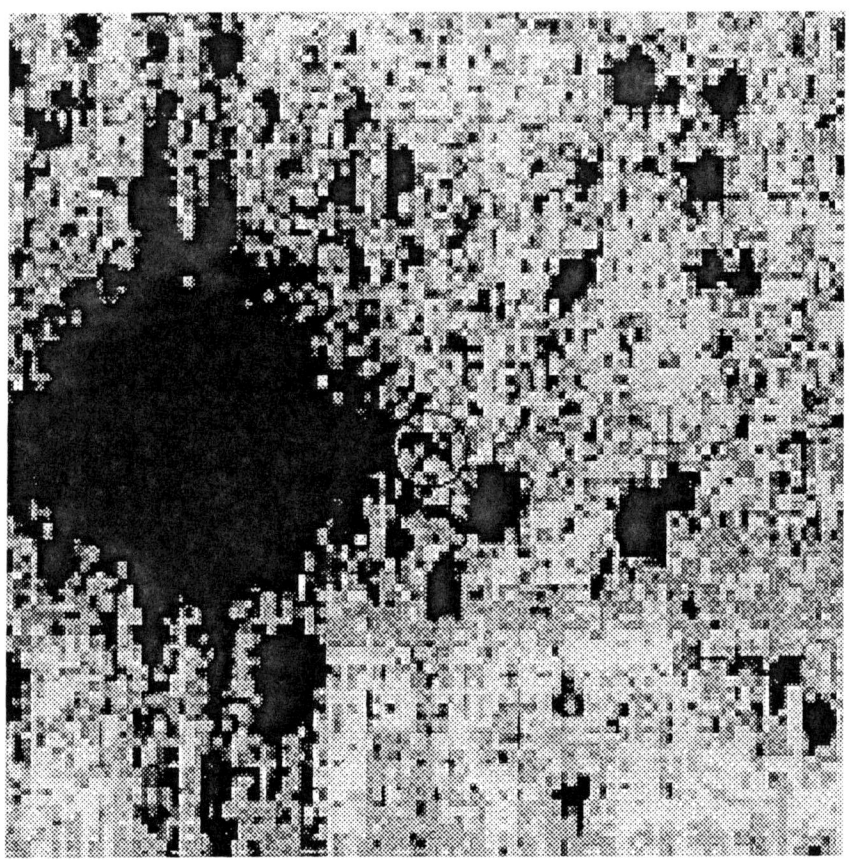

FIGURE 1. A deep frame obtained on 8 June 1998 at the 3.5-m telescope at the German-Spanish Calar Alto Observatory. A narrow band filter centered at 2.25 μm (continuum K) was used. The image covers a fraction of the *ROSAT* error box for the X-ray source RX J190714.2+0919, the X-ray counterpart to SGR 1900+14 and is centered at the position of the VLA radio transient [11] (circle). The brightest object is the pair of M5 I-III stars. North is up and east to the left.

NIR Spectroscopic Observations of the SGR 1900+14 M Stars

E. W. Guenther[1], S. Klose[1], and F. J. Vrba[2]

[1] Thüringer Landessternwarte Tautenburg, D-07778 Tautenburg, Germany
[2] U.S. Naval Observatory, Flagstaff Station, P.O. Box 1149, Flagstaff, AZ 86002

Abstract. We report on medium-resolution near-infrared spectroscopic observations of the M5 super-giant binary that may be related to the Soft Gamma Repeater S-GR 1900+14. The observations were performed with CGS4 at UKIRT and cover the wavelength range from 1.9 to 2.5μm.

INTRODUCTION

The most unusual object that is located in the error box of SGR 1900+14 is a pair of super-giant M stars (Vrba et al. 1996). Since the super-giants themselves are unlikely to be the source of the gamma-rays, it has been suggested that the gamma-rays originate from an unseen neutron star which is a companion of one of the super-giants. However, there is little evidence for a link between the source of the gamma-rays and the super-giants, as photometric monitoring of the supergiants do not show any evidence for a relation between the optical emission from the stars and the gamma-rays (Oppenheimer et al. 1998; Vrba et al. 2000).

Here we undertake another approach to search for a hypothetical neutron star companion of the super-giants. The basic idea is that the presence of high-excitation optical/infrared emission lines in one of the super-giants implies the presence of a nearby compact object. In the case of the presence of a neutron star companion close to one of the stars, such lines may arise either from a hot accretion disk, from a funneled accretion flow, or from reprocessing of high-energy photons in the atmosphere of one of the super-giants (cf. Castro-Tirado, Geballe, & Lund 1996; Dhillon & Marsh 1995; Eikenberry et al. 1998; Smith 1995). For this project, we have selected Brδ (1.945 μm), He I (2.058 μm), and Brγ (2.166 μm), in the K-band, because of the very high extinction of the object, and because these lines are prominent in hot accretion discs and in the case that high-energy photons are reprocessed in a stellar atmosphere. Although near-infrared spectroscopy of SGR 1900+14 has been carried out before (Vrba et al. 1996), the spectra presented here have a significantely higher signal-to-noise-ratio and resolution.

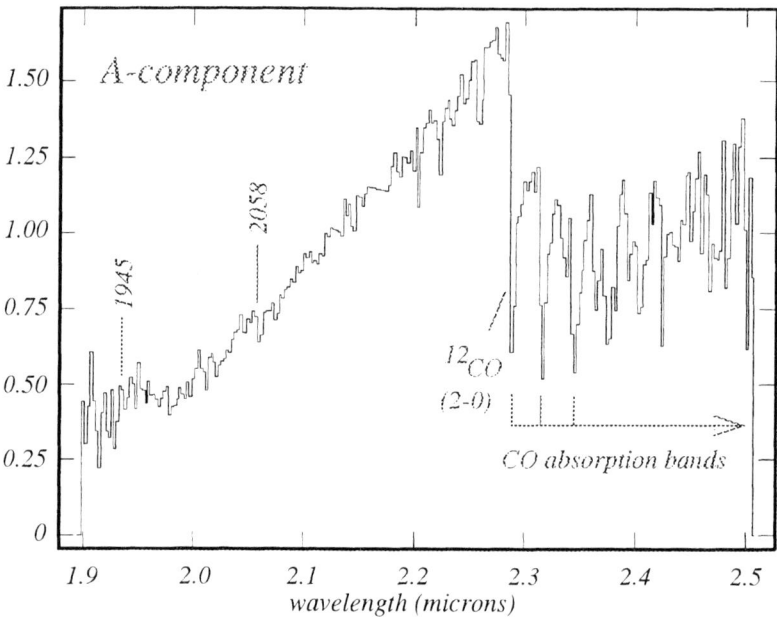

FIGURE 1. K-band spectrum of the A-component after correction for telluric features. There is no evidence for potential coronal emission lines. Note the depth of the (2-0) CO overtone band. The ordinate is given in arbitrary units.

OBSERVATIONS

The spectra were taken in service time, on 1999 May 19, with the cooled-grating spectrometer CGS4 (Davies 1991) of the United Kingdom Infrared Telescope, nearly 1 year after the recent activity of SGR 1900+14 (cf. Hurley et al. 1999). The 40 line mm^{-1} grating was used, providing a resolving power of ~ 800 (see "The CGS4 Online Handbook" at the UKIRT WWW-homepage). The standard star observed was HD 181383, which is of spectral type A2V.

The angular separation of the M stars is only about 3″.5. Therefore, the slit of the spectrograph was placed so that both stars were observed simultaneously.

The spectra were reduced in a standard manner using CGS4dr. Subsequent data reduction was performed using standard IRAF routines. To correct for telluric absorption, the spectra of both stars were divided by the standard star. Unfortunately, bad pixels affected the spectral region around the Brγ (2.166 μm) line, so this line could not be examined in our spectra.

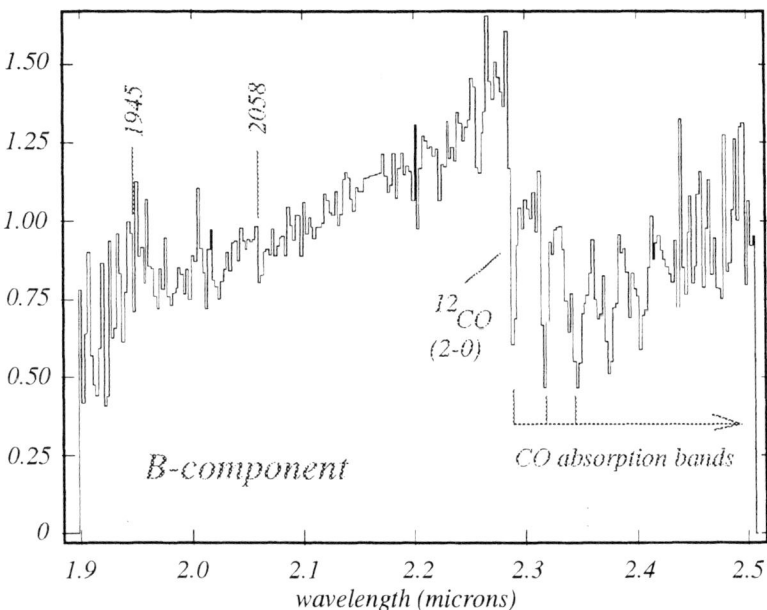

FIGURE 2. The same as Fig. 1, but for the B-component.

RESULTS

Figures 1 and 2 show the spectra of the binary stars A and B (we follow the designation in Figure 1 of Vrba et al. 1996). The spectra are not flux calibrated, but this is not crucial for our purposes. We expect that either star A or star B should show emission lines in its spectrum if it has an X-ray bright neutron star in a close orbit around it. Because of the similar spectral type of the stars and their simultaneous spectroscopy, a calculation of the A/B spectral ratio is the best strategy to search for emission lines. Within the observational errors the A/B spectral ratio shows no evidence for high-excitation emission lines in one of the stellar spectra (Fig. 3). Thus, we see no evidence for a neutron star in our spectra, although based on these findings we cannot exclude that a neutron star is in orbit around one of these stars.

In spite of this negative result, various other conclusions can be deduced from our observations.

1. No significant difference in the radial velocities between the two super-giants were found ($\Delta v = 12 \pm 21$ km/s). It is thus likely that the two super-giants are in fact a physical pair.

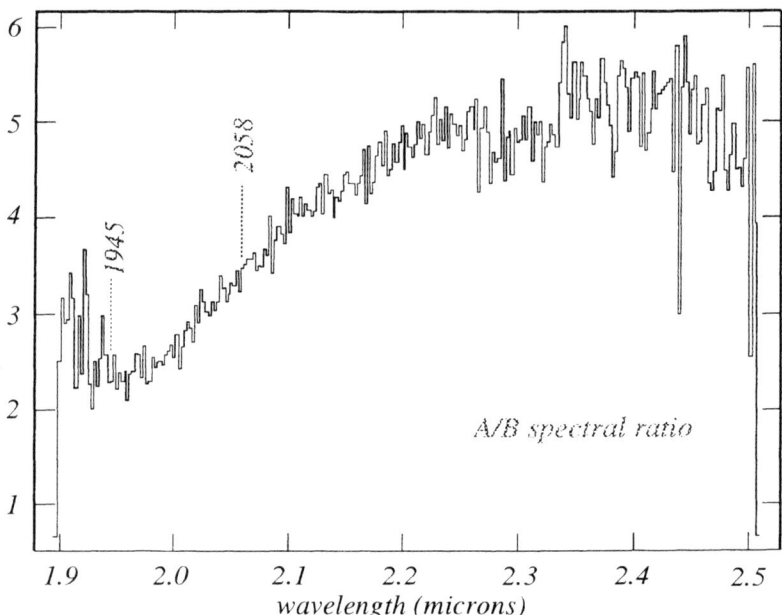

FIGURE 3. The A/B spectral ratio shows no evidence for emission lines in one or both stellar spectra.

2. The stars do not have the same spectral type. We find that star A is cooler than star B (Fig. 4). This is in agreement with the conclusion drawn by Vrba et al. (1996) based upon optical spectra.

3. We can constrain the luminosity class of the stars based on the drop in intensity at the short-wavelength side of the ^{12}CO (2-0) band (Kleinmann & Hall 1986; Wallace & Hinkle 1997). The drop in intensity we measure for star A is more than a factor of 2.5, and for star B more than a factor of 2.3 which, in both cases, excludes luminosity class III. This again confirms the conclusion drawn by Vrba et al. (1996) based on proper motion arguments that the stars must be more luminous than class III objects.

ACKNOWLEDGMENTS

The United Kingdom Infrared Telescope is operated by the Particle Physics and Astronomy Research Council (PPARC) of the U.K. We would like to thank the UKIRT service staff, especially Dr. John Davies, for performing the observations and providing reduced data files.

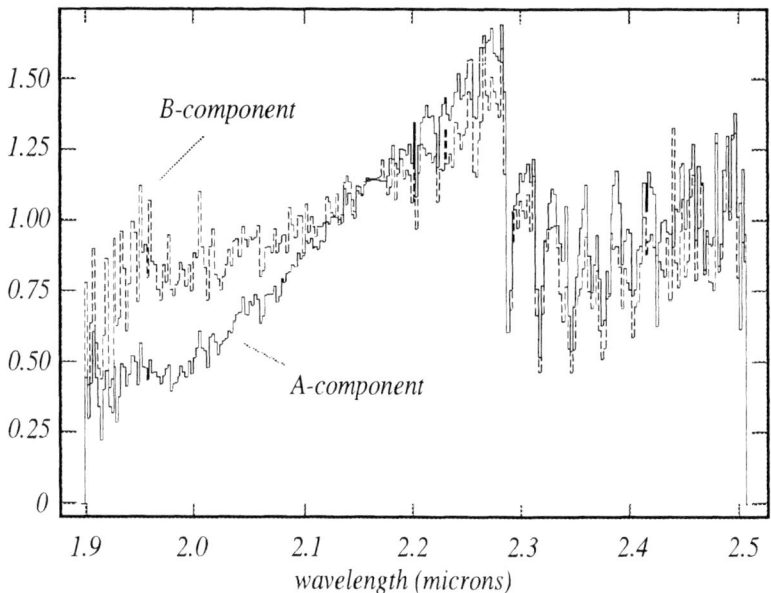

FIGURE 4. The spectra of stars A and B, arbitrarily normalized to 1 at about 2.15 μm. Star B shows more emission at shorter wavelengths, indicating an earlier spectral type compared to star A.

REFERENCES

1. Castro-Tirado, A., Geballe, T. R., & Lund, N., *ApJ* **461,** L99 (1996).
2. Davies, J. K., UKIRT Observer's Manual (1991).
3. Dhillon, V. S., & Marsh, T. R., *MNRAS* **275,** 89 (1995).
4. Eikenberry, S. S., et al., *ApJ* **506,** L31 (1998).
5. Hurley, K., et al., *Nature* **397,** 41 (1999).
6. Kleinmann, S. G., & Hall, D. N. B., *ApJ Suppl. Ser.* **62,** 501 (1986).
7. Oppenheimer, B. R., et al., *Astron. Telegram* 23 (1998).
8. Smith, R. C., *ASP Conf. Proc.* **85,** 417 (1995).
9. Vrba, F. J., et al., *ApJ* **468,** 225 (1996).
10. Vrba, F. J., et al., these proceedings.
11. Wallace, L., & Hinkle, K., *ApJ Suppl. Ser.* **111,** 445 (1997).

Physics in Ultra-strong Magnetic Fields

Robert C. Duncan

Dept. of Astronomy and McDonald Observatory
University of Texas at Austin

Abstract. In magnetic fields stronger than $B_Q \equiv m_e^2 c^3/\hbar e = 4.4 \times 10^{13}$ Gauss, an electron's Landau excitation energy exceeds its rest energy. I review the physics of this strange regime and some of its implications for the crusts and magnetospheres of neutron stars. In particular, I describe how ultra-strong fields

- render the vacuum *birefringent* and capable of distorting and magnifying images ("magnetic lensing");
- change the self-energy of electrons: as B increases they are first slightly lighter than m_e, then slightly heavier;
- cause photons to rapidly split and merge with each other;
- distort atoms into long, thin cylinders and molecules into strong, polymer-like chains;
- enhance the pair density in thermal pair-photon gases;
- strongly suppress photon-electron scattering, and
- drive the vacuum itself unstable, at extremely large B.

In a concluding section, I discuss the spindown of ultra-magnetized neutron stars and recent soft gamma repeater observations.

I ELECTRONS AT $B > B_Q$

The significance of the quantum electrodynamic field strength, B_Q, can be understood via a simple, semi-classical argument. A classical electron gyrating in a magnetic field satisfies $\dot{p} = ev \times B/c$, where $p = \gamma m_e v$ is the momentum. Substituting $\dot{p} = \omega p$ and $v = \omega r$ in this equation and cancelling factors of ω (along with orbital phase factors), one finds a radius of gyration $r = cp/eB$, where p is the transverse momentum ($\perp \mathbf{B}$). Quantum mechanics implies $r \cdot p \sim \hbar$ in the ground state, thus the semi-classical gyration radius is $r_{\rm gyr} \sim \lambda_e (B/B_Q)^{-1/2}$, where $\lambda_e \equiv \hbar/m_e c$ is the electron Compton wavelength. The associated momentum is $p \sim (\hbar/r_{\rm gyr}) \sim m_e c\, (B/B_Q)^{1/2}$.

This shows that electrons gyrate *relativistically* in fields $B > B_Q$. One thus expects excitation energies in excess of $m_e c^2$. This is borne out by the solution to the Dirac equation for an electron in a homogeneous magnetic field. The Dirac spinors are proportional to Hermite polynomials, and the energy levels or "Landau levels" are

$$E_n = [m_e^2 + p_z^2 + m_e^2 n\,(2B/B_Q)]^{1/2}, \qquad (1)$$

in units with $\hbar = c = 1$ (adopted also in many equations that follow). The first term in the square brackets is the rest energy. The p_z-term gives the energy of motion parallel to the field, which can take a continuum of values. The discrete energy levels are given by $n = 0, 1, 2\ldots$ These states are also eigenstates of spin, with the $n = 0$ ground state always having $s = -\frac{1}{2}$. For $p_z = 0$, the ground state energy is $E_o = m_e$, independent of B. In a semi-classical picture, one could say that the negative spin-alignment energy in the ground state cancels with the zero-point gyration energy. Excited Landau levels are two-fold degenerate in s. The first Landau-level excitation energy is $\omega_B(1) = E_1 - E_o \approx (2B/B_Q)^{1/2}\, m_e$ for $B \gg B_Q$. Because this energy is so large, electrons almost always remain in the ground state for processes thought to occur near the surfaces of ultra-magnetized neutron stars.

Electron *self-interactions* resolve the degeneracies of the Landau levels, and shift the ground state energy. This was first demonstrated by Schwinger, who estimated the "anomalous" magnetic moment of the electron [1]. The relevant Feynman diagram is shown in Figure 1: a free electron (traveling upward on the page) emits a virtual photon, interacts with the magnetic field, then reabsorbs the photon. The electron's effective spin magnetic moment is enhanced by $(1 + \alpha/2\pi)$ to first order in $\alpha = e^2/\hbar c = 1/137$, the fine-structure constant. This results in a ground-state energy shift

$$E_o = m_e\,[\,1 - (\alpha/2\pi)\,(B/B_Q)\,]^{1/2}\,. \qquad (2)$$

If extrapolated to $B > B_Q$, this formula would imply that the ground-state energy of an electron goes to *zero* at $B = (2\pi/\alpha)B_Q \approx 4 \times 10^{16}$ Gauss. For stronger B, the vacuum would become unstable to pair production, with dramatic astrophysical consequences [2]. But eq. (2) is actually only valid in the sub-B_Q regime. More generally, the electron's self-energy is determined by the sum of Feynman diagrams shown in Figure 2 (ref. [3]). The triple line on the left-hand side represents the physical electron propagator (i.e., the probability amplitude for an electron to move from point A to point B). The double lines on the right are bare propagators for an electron in the presence of a magnetic field, corresponding to basis states with

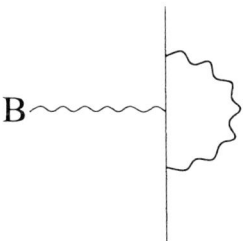

FIGURE 1. Anomalous magnetic moment diagram.

energies given by eq. (1).[1]

When the calculation is done, it is found that the electron ground-state energy diminishes according to eq. (2) as B increases within the Schwinger domain $B \ll B_Q$, but it reaches a *minimum value* of $(1 - 4.6 \times 10^{-5})m_e$ at $B = 0.25\,B_Q$ and then rises [4,5]. At $B > 0.65\,B_Q$ the electron grows heavier than m_e, but only slowly. The asymptotic fractional enhancement, valid at very large B, is [6]

$$(E_o - m_e)/m_e = (\alpha/4\pi)\left(\left[\ln(2B/B_Q) - \xi - \tfrac{3}{2}\right]^2 + \beta\right) \qquad B \gg B_Q \qquad (3)$$

to first order in α, where $\xi = 0.577$ is Euler's constant, and $\beta \approx 3.9$ is a numerical constant (estimated here from the numerical integrations of ref. [4]).

Thus, an electron's ground-state energy is doubled, $E_o \sim 2m_e$, only at $B \sim 10^{32}$ Gauss. (Higher-order corrections might change this result somewhat.) Of course, the maximum fields attained in neutron stars fall far short of this. The dynamical saturation field for convective motions in nascent neutron stars is $\sim 10^{16}$ G; and $B \sim 3 \times 10^{17}$ G is possible if the free energy of differential rotation in a rapidly-rotating, newborn neutron star is efficiently converted by a post-collapse dynamo [7,8]. But if $B > (8\pi P Y_e)^{1/2} \sim 10^{17}$ G, where P is the pressure and Y_e the electron fraction in the liquid interior of a neutron star, then buoyancy overcomes stable stratification and an inhomogeneous field is dynamically lost [9,8].

For $B \sim 10^{17}$ G, eq. (3) implies $E_o - m_e \approx 0.03\,m_e$. Thus, magnetic self-energy corrections for electrons and positrons are probably not important over the range of magnetic fields and at the level of accuracy typically attained in neutron star astrophysics.

---

[1] Self-interactions also occur when $B = 0$. The double-line propagators of Fig. 2 are then replaced with single-line, free-electron propagators (plane-wave states), and the resultant energy shifts—formally divergent—are absorbed into the electron's known rest mass by the renormalization of quantum electrodynamics. A strong magnetic field changes the self-energy when the same renormalization prescriptions are used. Note that the Schwinger diagram of Fig. 1 is included in the second diagram on the right of Fig. 2: when $B < B_Q$, the double-line propagators can be approximated as single-line, free electron propagators undergoing discrete, perturbative interactions with the magnetic field. Positron intermediate states are included; they correspond to a subset of the vertex time-orderings which are summed over. The lowest-order tadpole diagram gives no contribution in a homogeneous magnetic field.

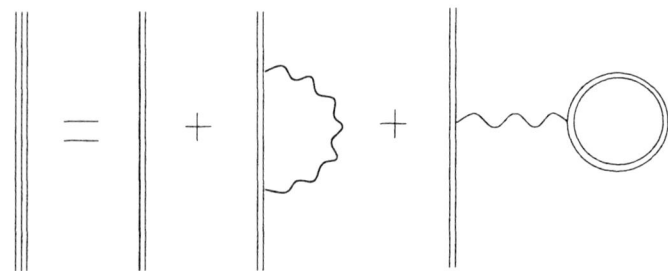

FIGURE 2. Electron self-energy in a magnetic field: lowest-order diagrams.

II ATOMS AND MOLECULES AT $B > B_Q$

At sufficiently low temperatures, a magnetar's surface will be covered with atoms and molecules. This surface structure can have consequences for the star's quiescent X-ray emissions, because it determines the work function for removing charged particles from the surface, as necessary for maintaining currents in the magnetosphere [10]. Such currents may result from magnetically-driven crustal deformations such as *twists* of circular patches of the crust.[2] If a bundle of field lines, describing an arch in the magnetosphere, has one footpoint twisted (with the motion driven from below by the evolving field), then a current must flow along the arch to maintain the twisted exterior field, since $\oint \mathbf{B} \cdot d\boldsymbol{\ell} = 4\pi I/c$. Surface impacts of the flowing charges create hot spots at the arch's footpoints and ultimately dissipate the exterior magnetic energy of the twist, with implications for SGR and AXP X-ray light curves and their time-variations [10]. Here we focus on the atomic and molecular physics that comes into play, following a paper by Ruderman [11] and extending the arguments to $B > B_Q$.

The Bohr radius of a hydrogen atom is $r_o = \lambda_e/\alpha$. The quantum gyration radius, $r_{\text{gyr}} = \lambda_e (B/B_Q)^{-1/2}$, is smaller than r_o for $B > \alpha^2 B_Q = 2.4 \times 10^9$ G. This is the characteristic field strength at which magnetism radically alters the atomic structure of matter.[3] At $B > \alpha^2 B_Q$, an atomic electron is constrained to gyrate along a cylinder which lies entirely within the spherical volume that the unmagnetized atom would occupy. Electrostatic attraction binds the electron strongly to the central nucleus. At $B \gg \alpha^2 B_Q$ the cylinder becomes very long and narrow, and atomic binding energies are adequately given by eigenvalues of the one-dimensional Schrödinger equation. A simple, intuitive estimate—which gives a good estimate of the ground state energy despite its lack of rigor—involves idealizing the atom as a line-charge of length 2ℓ. For linear charge density $e/2\ell$, the electrostatic energy is $\varepsilon = -(e^2/\ell) \ln[\ell/r_{\text{gyr}}]$. A lower cutoff r_{gyr} is necessary because the charge distribution does not resemble a line when you get within $\sim r_{\text{gyr}}$ of the nucleus. It is more like a sphere, contributing an energy $\sim -qe/r_{\text{gyr}}$ where $q = er_{\text{gyr}}/\ell$; but this contribution can be neglected in the limit $\ell \gg r_{\text{gyr}}$ or $B \gg \alpha^2 B_Q$. Thus, the ground state energy, including the energy of non-relativistic motion parallel to **B**, is $\mathcal{E}_o(\ell) = (\hbar^2/2m_e\ell^2) - (e^2/\ell) \ln[\ell/r_{\text{gyr}}]$. Minimizing this according to $d\mathcal{E}_o/d\ell = 0$, we find $\ell \simeq r_o [\ln(r_o/r_{\text{gyr}})]^{-1}$. This shows that the *length* of the thin cylindrical atom is less than the Bohr diameter, but only by a modest, logarithmic factor. The ground state hydrogen binding energy is then

---

[2] Crustal twists, with spiral patterns of shear strain, may be a common type of magnetically-driven deformation. The pressure in the crust is due mostly to degenerate particles (relativistic electrons, and free neutrons at densities above neutron drip), but the shear modulus is due only to relatively weak Coulomb forces of the lattice. Hence the crust is relatively incompressible, and pure shear deformations allow the largest range of motion, with the greatest energy transfer between the crust and the magnetic field.

[3] The largest field you are ever likely to encounter personally is $\sim 10^4$ G if you have an medical MRI scan. Fields $\gtrsim 10^9$ G would be instantly lethal.

$$\mathcal{E}_o \simeq -(\epsilon_o/4) \, [\ln(B/\alpha^2 \, B_Q)]^2 \quad \text{for} \quad B \gg \alpha^2 B_Q, \tag{4}$$

where $\epsilon_o = \alpha^2 m_e/2 = 13.6$ eV is one Rydberg. Note that $E \propto [\ln B]^2$ energy scalings are ubiquitous in ultra-magnetized systems (cf. eqs. 3,4,5).

As B increases beyond $B \sim B_Q$, the radius of the atomic cylinder shrinks to less than the Compton wavelength but eq. (4) remains a reasonably good approximation. This is because the electron's inertia for longitudinal motion ($\parallel \mathbf{B}$) stays close to m_e in the ground-state Landau level even at $B > B_Q$. Equation (4) would become invalid if the longitudinal motion became relativistic. But this would require $\ell < \lambda_e$, which occurs only at $B > \alpha^2 \exp(2/\alpha) \, B_Q \approx 10^{115}$ G. Magnetic fields can never get this strong. We will show that the vacuum breaks down at smaller B.

Equation (4) implies that the binding energy of hydrogen near the surface of a magnetar with $B \simeq 10 \, B_Q$, is $\mathcal{E}_o \simeq 0.5$ keV. This is comparable to the surface temperatures of some young magnetar candidates [12,13].

There are two classes of *hydrogenic excitations*. Longitudinal excited states are well-approximated as multi-nodal eigenfunctions of the 1-D Schrödinger equation; e.g., the first excited state has a node at the position of the nucleus. Transverse excited states involve transverse displacements of the center of electron gyration away from the nucleus. Semi-classically, the electron then experiences $\mathbf{E} \times \mathbf{B}$ drift, and its center of gyration moves in a circular orbit around the nucleus. (See ref. [11] for details.) Of course, Landau-level excitations are also possible, but the excitation energy is enormous for $B > B_Q$. Atoms generally become unbound when such free energies are present.

Longitudinal excitations tend to require more energy than transverse, so in ultra-magnetized *multi-electron atoms*, orbitals corresponding to transverse hydrogenic states fill up before longitudinal. In fact, for $B > Z^3 \alpha^2 B_Q \approx (Z/26)^3 \, B_Q$, where Z is the electron number, no orbitals with longitudinal nodes are occupied, and the atomic structure is very simple [11,14]. Note that Fe^{56}, which is likely to be the dominant nuclear species on a clean neutron star surface, has $Z = 26$. Thus, this condition is satisfied on magnetars, but not on radio pulsars with fields $\sim 10^{12}$ G. The atomic binding energy is then

$$\mathcal{E}_o(Z) \simeq -(7/24) \, Z^3 \, \epsilon_o \, [\ln(B/Z^3 \, \alpha^2 \, B_Q)]^2. \tag{5}$$

When $B \gg Z^3 \alpha^2 B_Q \approx (Z/26)^3 \, B_Q$, atoms on a neutron star's surface form long polymer-like molecular chains parallel to $\mathbf{B}$, bound by the electrostatic attraction of shared electrons. The molecular binding energy per nucleus is [11,15,16]

$$\Delta \mathcal{E} \simeq -(3/2) \, Z^3 \, \epsilon_o \, (B/Z^3 \, \alpha^2 \, B_Q)^{0.37}. \tag{6}$$

Together, these results determine the approximate work function for ionic emission from a magnetar's surface [10].

III VACUUM POLARIZATION AND RADIATIVE PROCESSES

Photon modes in the magnetized vacuum include the extraordinary mode or E-mode, with oscillating electric vector $\mathbf{E_E} \perp \mathbf{B}$, and the ordinary mode or O-mode, with $\mathbf{E_O} \perp \mathbf{E_E}$. Both electric vectors are also orthogonal to $\mathbf{k}$, the direction of propagation.[4] Due to the process shown in Fig. 3, where the double-lines are propagators for a magnetized, virtual $e^+ e^-$ pair, the indices of refraction of the two modes are very different at $B > B_Q$:[5]

$$n_O = 1 + (\alpha/6\pi) \sin^2\theta_{kB} (B/B_Q)$$
$$n_E = 1 + (\alpha/6\pi) \sin^2\theta_{kB}.$$

If $n_O - n_E \gtrsim (k\ell_B)^{-1}$, where k is the wavenumber and ℓ_B is the scale-length of variation of the magnetic field, then the modes adiabatically track: E stays E and O stays O as photons move through the changing field geometry. This condition is generally satisfied for X-rays in the magnetospheres of magnetars. Shaviv, Heyl & Lithwick [19] used geometrical optics to model *magnetic lensing* in the vicinity of a magnetar with a pure dipole field and a uniformly bright surface. They found O-mode image distortion and amplification, varying with viewing angle. This remarkable effect may be hard to observe in practice because fields $B \gtrsim (6\pi/\alpha) B_Q \sim 10^{17}$ G are required to produce strong lensing effects. Observations may also be complicated by non-uniform surface brightness, gravitational lensing [20], higher-order magnetic multipoles, photon splitting (see below) and X-ray emission from a magnetar's diffuse, Alfvén-heated halo.

When the excitation energy of the first Landau-level is much greater than the photon energy, $\omega_B(1) \equiv m_e[(1 + 2B/B_Q)^{1/2} - 1] \gg \omega$, then *photon scattering off electrons* is strongly suppressed in the E-mode. Semi-classically, this is easy to understand: the radiation electric field ($\mathbf{E_E} \perp \mathbf{B}$) is unable to significantly drive electron recoil. Paczyński first noted [21] that this greatly accelerates

---

[4] Photon eigenmodes are linearly polarized, as described here, except in narrow zones of $\mathbf{k}$-space where the angle between $\mathbf{k}$ and $\pm\mathbf{B}$ satisfies $\theta_{kB} \lesssim (\omega/m_e)^{1/2} (B/B_Q)^{-1/2}$ and $\hbar\omega$ is the photon energy. For propagation along $\pm\mathbf{B}$ within these zones, the E and O modes are elliptically polarized; and circularly polarized for $\mathbf{k} \parallel \pm\mathbf{B}$.

[5] The affect of a magnetized *plasma* on the eigenmodes and indices of refraction is small in comparison to the magnetic vacuum polarization so long as $\omega \gg \omega_{c2} \equiv (3\pi/\alpha)^{1/2} (B/B_Q)^{-1/2} \omega_p$ for $B \gg B_Q$, where $\omega_p = (4\pi N_e e^2/m_e)^{1/2}$ is the plasma frequency and N_e is the electron density (see [18] and references cited therein). This is satisfied in many or most applications to observable phenomena in magnetar magnetospheres since $\omega_{c2} = 0.13 (N_e/10^{23} \text{cm}^{-3})^{1/2} (B/10\, B_Q)^{-1/2}$ keV.

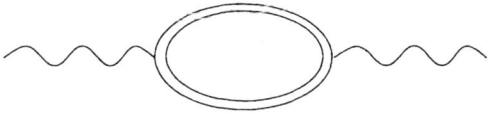

FIGURE 3. Vacuum polarization diagram.

X-ray diffusion in the vicinity of magnetars, facilitating hyper-Eddington burst and flare emissions. The E-mode scattering cross section, relative to Thomson, is $\sigma(E)/\sigma_T \sim (\omega/m_e)^2 (B/B_Q)^{-2}$ in the regime of possible relevance for soft gamma repeater (SGR) bursts; see §3.1 of ref. [17] for more details.

Photon splitting and merging, another important radiative effect, is depicted in Figure 4, with time advancing from left to right for splitting, and right to left for merging. These processes are kinematically forbidden in free space, but they operate at $B > B_Q$ because the field acts as an efficient sink of momentum. (Note the double-line, *magnetized* e^- and e^+ propagators in Fig. 4.) The dominant splitting channel is $E \to O\ O$. The rate for $B > B_Q$ and $\omega < m_e$ is [22,23]

$$\Gamma_{\rm sp} = (\alpha^3/2160\pi^2)\ \sin^6\theta_{kB}\ (\omega/m_e)^5\ m_e. \tag{7}$$

(Splitting $E \to O\ E$ also occurs, but at a lower rate.) Note that $\Gamma_{\rm sp}$ increases steeply with increasing photon energy; but it is independent of B for $B > B_Q$. At $B < B_Q$ the process shuts down abruptly: $\Gamma_{\rm sp} \propto (B/B_Q)^6$.

How does this process affect SGR burst spectra? Simple splitting cascade models [24] are illustrative but not realistic since O-mode photons do not split. Realistically, one must consider the subtle interplay of splitting/merging and Compton scattering [17]. In particular, E-mode splitting outside the E-mode scattering photosphere produces O-mode photons which are isotropized by rapid Compton scattering. Subsequent mergers $O\ O \to E$ yield a quasi-isotropic E-mode source function. Only at $B < B_Q$ and outside the O-mode photosphere do the modes truly decouple and all photons stream outward; see §6 of ref. [17] for many more details.

IV THE ULTRA-MAGNETIZED PAIR GAS

Ultra-strong magnetic fields also have profound *thermodynamic* effects. The magnetized photon-pair gas gives an example. Such a gas may be created during an SGR burst or flare, when a crust fracture or other magnetically-driven instability suddenly injects a large quantity of energy into the magnetosphere [17]. The result is an optically-thick *trapped fireball*, confined by closed field lines, anchored to the star's surface. The gas inside this fireball has remarkable properties, as illustrated in Fig. 5. (This figure is included here courtesy of A. Kudari [25].) The figure shows the ratio of pair energy density to the photon energy density, $\Lambda \equiv U_{e^+e^-}/U_\gamma$,

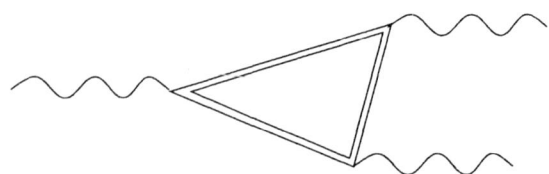

FIGURE 4. Photon splitting and merging in a strong magnetic field.

as a function of T and B. For $T \gg m_e$ and $T \gg \omega_B(1)$, the magnetic field has little effect on the ultra-relativistic pairs: $U_{e^+e^-} = 2 \cdot (7/8)aT^4$, so $\Lambda = (7/4)$. This should hold true across the whole right-hand side of Fig. 5, but only 1000 Landau levels were used in making this graph, so the ratio falls artificially below $(7/4)$ at high T and low B.

The striking peak in Fig. 5 is real, however. It occurs for pairs with non-relativistic longitudinal motion, $T \ll m_e$, and $B > B_Q$. In this regime, only the first Landau level is occupied: $T \ll \omega_B(1)$. The peak occurs because electrons and positrons are strongly localized in directions transverse to the field: $r_{\text{gyr}} = \lambda_e (B/B_Q)^{-1/2}$. This allows more of them to be packed into a given volume of ultra-magnetized gas. Formulae for thermodynamic parameters of a pair-photon gas in various limits are given in ref. [25] and in §3.3 of ref. [17].

Note that the trapped fireball of a common SGR burst, with energy $\Delta E \sim 10^{41}$ ergs confined within a volume of order $(\Delta R)^3 \sim (10\,\text{km})^3$ at $B \sim 10\,B_Q$ has a temperature $T \sim 160$ keV [17,25]. This puts it right on the peak in Fig. 5!

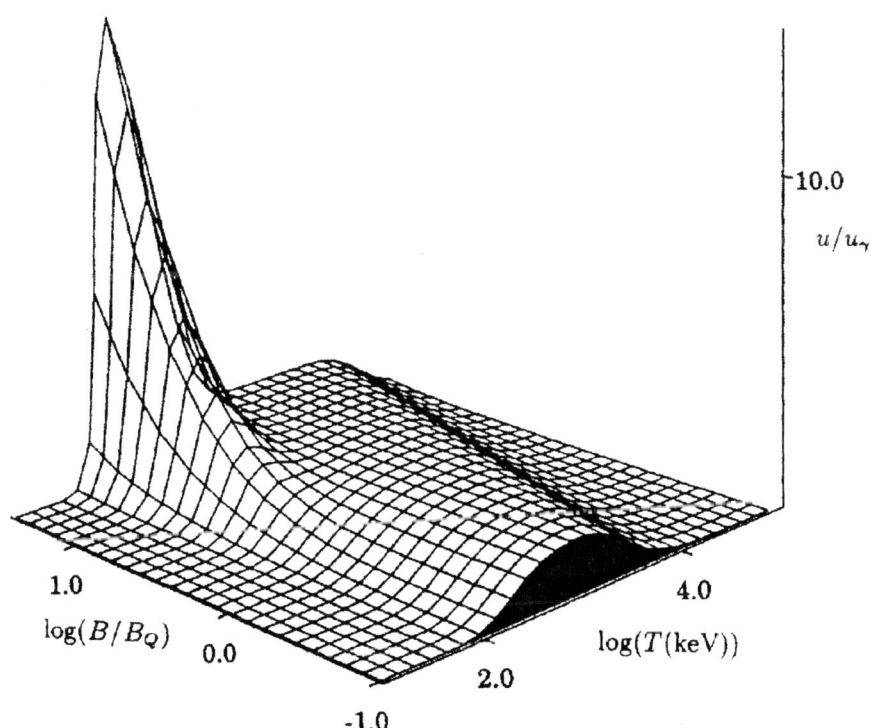

FIGURE 5. The ratio of pair energy density to photon energy density in a pair-photon gas, as a function of temperature and magnetic field strength.

V MAGNETIC VACUUM BREAKDOWN

We argued above that a uniformly magnetized vacuum is stable against spontaneous electron-positron pair production. Nevertheless, at sufficiently high B the vacuum must break down. Magnetic monopoles with mass m_η and magnetic charge η are spontaneously created when the energy they acquire in falling across a monopole Compton wavelength, $\varepsilon \sim \eta B \cdot (\hbar/m_\eta c)$, exceeds their rest energy $m_\eta c^2$. Dirac showed that a monopole charge is an integral multiple of $\eta = (\hbar c/2e)$ from the condition that an electron wavefunction must be single-valued in the field of a monopole [26]. Thus, magnetic fields can never get stronger than

$$B_{\max} \sim \alpha \, (m_\eta/m_e)^2 \, B_Q \ . \tag{8}$$

A firm upper bound is $B_{max} \simeq 10^{55}$ G for Planck-mass monopoles, $m_\eta = 10^{19}$ GeV. GUT theories predict $m_\eta = 10^{16}$ GeV or $B_{\max} \simeq 10^{49}$ G. Superstring/M-theory predicts intermediate values: $m_\eta = \alpha_s^{-1/2} = 10^{17}$–$10^{18}$ GeV, where α_s is the string tension, thus

$$B_{\max} \simeq 10^{51} - 10^{53} \text{ G}. \tag{9}$$

New work shows that the energy scale for quantum gravity could be as low as $M_o \sim 1$ TeV if there exist "large" extra dimensions to space [27]. The extra dimensions are wrapped in closed geometries (e.g., circles) of size $L \sim 1(M_o/1\,\text{TeV})^{-2}$ millimeter for two extra dimensions, or $L \sim \ell_p(M_o/M_p)^{-(n+2)/n}\,\ell_p$ for n extra dimensions; where M_p the Planck mass and ℓ_p is the Planck length. This would imply a small limiting field strength: $B_{\max} \simeq 10^{23}\,(m_\eta/1\,\text{TeV})^2$ G. However, there is no experimental evidence for large extra dimensions at the present time. The most plausible upper limit is given by eq. (9).

Thus, *a vast range of tremendous field strengths are possible in Nature*. We don't yet know any objects that generate such fields, but some possibilities have been suggested. For example, superconducting cosmic strings—if they exist—could generate fields $\gtrsim 10^{30}$ G in their vicinities [28]. Perhaps future astrophysicists will regard neutron star magnetic fields as mild!

VI MAGNETAR SPINDOWN

In this final section, I consider a topic of great current interest, namely recent observations of soft gamma repeater spindown histories [29–34], and their interpretation in the context of the magnetar model. At present, the most promising scenario involves *episodic, wind-aided spindown* (§4 in ref. [10]). This is based upon several background developments. In 1995 Thompson and I proposed that frequent, small-scale fractures in the crust of a young magnetar produce quasi-steady seismic and magnetic vibrations, energizing the magnetosphere and driving a diffuse, relativistic outflow of particles and Alfvén waves (§7.1.2 in ref. [17]). A year later we made a first estimate of this outflow's power [12]. Thompson & Blaes subsequently noted that a magnetar's rate of spindown is greatly accelerated by such a wind

(§VII B in ref. [35]). All of this work pre-dated the discovery of X-ray pulsations from SGRs [29].

How strong is the wind? The wind luminosity, L_W scales roughly with the magnetic energy density in the deep crust,[6] $\propto B_{\text{crust}}^2$ [12]. But if $B_{\text{crust}} > (4\pi\mu)^{1/2} \sim 6 \times 10^{15}$ G, where μ is the shear modulus in the deep crust, then evolving magnetic stresses overwhelm lattice stresses and the crust deforms plastically instead of fracturing, choking off the Alfvén-powered wind. This suggests an upper limit $L_W \lesssim 5 \times 10^{36}$ erg s^{-1} for a $\sim 10^4$-year-old magnetar [12]. In 1996, we proposed that a wind operating near this upper limit could account for radio synchrotron nebula that seemed to surround SGR 1806−20 [36]. However, we now know that the SGR is not coincident with this nebula [37]. There is no direct observational evidence for a quasi-steady wind from any SGR. Magnetar winds must be mild enough to produce no detected radio emission, with L_W probably much less than the theoretical upper limit of ref. [12], because this limiting value assumed optimal conditions, including the dubious application of a formula at the edge of the regime where it breaks down (i.e., $B_{\text{crust}} \sim B_\mu$). It is likely that L_W is comparable to the steady X-ray luminosity emitted by the hot stellar surface and Alfvén-energized halo: $L_W \sim 10^{35}$–10^{36} erg s^{-1}.

Rothschild, Marsden and Lingenfelter have plotted two graphs, included in this volume, which nicely elucidate constraints on SGR spindown for constant L_W and B_{dipole}, based upon formulae derived independently in refs. [38,10].[7] These plots show that wind luminosities $L_W \lesssim 10^{36}$ erg s^{-1} imply $B_{\text{dipole}} \gtrsim 10^{14}$ G in order to match the observed values of P and $\dot{P}$; but the implied stellar ages are then moderately shorter than the estimated ages of the putative associated supernova remnants (SNRs). Of course, the SNR associations or ages may be unreliable, since the SGRs lie far from the SNR centers, and the ages are only rough order-of-magnitude estimates. However, we favor a different interpretation: the wind is probably *episodic*, so the effective spindown age of the star is less than the SNR age. In particular, we proposed (in §4.1–4.2 of ref. [10]) that strong winds and rapid spindown prevail only during limited episodes of a young magnetar's life, when it is magnetically active and observable as an SGR. This fits in nicely with observations of anomalous X-ray pulsars (AXPs). These objects have spindown ages $P/2\dot{P}$ that are comparable or *longer* than the ages of their associated SNRs [39,40], suggestive of young magnetars observed during their non-windy, inactive episodes. A fully consistent scenario is possible [10].

Note, incidentally, that $B_{\text{dipole}} \lesssim B_Q$ is possible in a magnetar if the lowest-order magnetic moment decays quickly, e.g., via the Flowers-Ruderman instability [41] (see §14.2 and 15.2 in ref. [8]; §7.1.2 in ref. [17]). This is because a magnetar is a *magnetically-powered* neutron star: its emissions depend upon the *total free*

---

[6] Most of the free energy in magnetars is stored in the *internal* magnetic field, probably in toroidal and high-multipole components, so B_{crust} is usually much greater than B_{dipole}.

[7] These references correct an inaccuracy in the original wind-aided spindown rate given by ref. [35].

energy and configuration of its magnetic field, not simply upon its exterior dipole moment. The light curve of the 1998 August 27 giant flare gives evidence for strong higher-order multipole moments in SGR 1900+14 [42]. SGR bursts give evidence for magnetically-powered activity (e.g., refs. [17,42–44]).

Marsden et al. [33] also suggested that the spindown rate of SGR1900+14 was enhanced by a factor ~ 2 during the summer of 1998. In the context of the magnetar model, this could mean that L_W increased by a factor ~ 4. Possible evidence for this comes from $\dot{P}$ measurements during RXTE runs immediately preceding and following the interval in question [31]; but it should be noted that RXTE was observing the SGR at those times as a "target of opportunity" because the star was emitting hundreds of bursts [31,34]. Transient accelerated spindown during episodes of vigorous bursting can occur in the magnetar model, because the relativistic outflow may be enhanced. But only a handful of bursts were detected by BATSE during mid-summer of 1998 [34], and the RXTE/ASCA determination of $\dot{P}$ between 1998 Aug. 28 and Sept. 17 was 6.2×10^{-11} s/s [34] (a number that was rounded up to $1. \times 10^{-10}$ in ref. [32]). Furthermore, spindown rates measured over short time intervals, such as during single RXTE runs, can be affected by other transient or periodic effects (e.g., free precession: see §4.3 in ref. [10]; also ref. [45]).

Although an increase in $\dot{P}$ by ~ 2 during the summer of 1998 cannot be ruled out, we suspect that the average spindown rate was similar to that which prevailed at other times during the past few years, and the observed shift in the spindown history was due to an abrupt *spindown glitch* during the extraordinary giant flare of 1998 August 27th. Such a glitch could be caused by the unpinning of crustal superfluid vortices in a magnetar with a crust that has been deformed by evolving magnetic stresses [10].

In conclusion, we have come a long way from the days of ref. [7] when simple magnetic dipole radiation seemed to be an adequate idealization for SGR spindown! More observations are needed to determine whether glitches really occur in SGRs and AXPs [12,46,10] and what sign they may have; whether these stars exhibit free precession (which could give us the first direct measure of a magnetar's *internal* field [45,10]); and to further test and constrain models of these fascinating stars.

Acknowledgments: This work was supported by NASA grant NAG5-8381 and Texas Advanced Research Project grant ARP-028.

REFERENCES

1. Schwinger, J., *Phys. Rev.* **73**, 416L (1948).
2. O'Connel, R.F., *Phys. Rev. Lett.* **21**, 397 (1968); Chiu, H.Y. & Canuto, V., *Ap.J.* **153**, L157 (1968).
3. Demeur, M., *Acad. Roy. Belg., Classe Sci., Mem.* **28** (1953).
4. Constantinescu, D.H., *Nuclear Phys* **B44**, 288 (1972).
5. Geprägs, R., Riffert, H., Herold, H., Ruder, H. & Wunner, G., *Phys. Rev. D* **49**, 5582 (1994).

6. Jancovici, B., *Phys. Rev.* **187**, 2275 (1969).
7. Duncan R.C., & Thompson C., *Ap.J.* **392**, L9 (1992).
8. Thompson C., & Duncan R.C., *Ap.J.* **408**, 194 (1993).
9. Goldreich P., & Reisenegger A., *Ap.J.* **395**, 250 (1992).
10. Thompson, C., Duncan, R.C., Woods, P., Kouveliotou, C., Finger, M.H. & van Paradijs, J., *ApJ*, submitted, astro-ph/9908086, (2000).
11. Ruderman, M., in *The Physics of Dense Matter*, I.A.U. Symp. No. 53, ed. C.J. Hansen, Reidel, Dordrecht, 1974, pp. 117.
12. Thompson C. & Duncan R.C., *Ap.J.* **473**, 322 (1996).
13. Heyl, J.S. & Hernquist, L., *Ap.J.* **489**, L67 (1997).
14. Lieb, E.H., Solovej, J.P., & Yngvason, J., *Phys. Rev. Lett.* **69**, 749 (1992).
15. Neuhauser, D., Koonin, S.E. & Langanke, K., *Phys. Rev. A* **36**, 4163 (1987).
16. Lai, D., Salpeter, E.E., & Shapiro, S.L., *Phys. Rev. A* **45**, 4832 (1992).
17. Thompson, C., & Duncan, R.C., *M.N.R.A.S.* **275**, 255 (1995).
18. Bulik, T. & Miller, M.C., *M.N.R.A.S.* **288**, 596 (1997).
19. Shaviv, N.J, Heyl, J.S. & Lithwick Y., *M.N.R.A.S.* **306**, 333 (1999).
20. Page, D., *Ap.J.* **442**, 273 (1995).
21. Paczyński, B., *Acta Astron.* **42**, 145 (1992).
22. Adler, S.L., *Ann. Phys.* **67**, 599 (1971).
23. Thompson, C. & Duncan, R.C., in *Compton Gamma-Ray Observatory*, ed. M. Friedlander et al., AIP, New York, 1993, pp. 1085.
24. Baring, M.G., *Ap.J.* **440**, L69 (1995).
25. Kudari, A., Master's Thesis, University of Texas (1996).
26. Dirac, P.A.M., *Proc. Roy. Soc.* **A133**, 60 (1931).
27. Arkani-Hamed, N., Dimopoulos, S., Dvali, G., *Phys. Lett. B* **429**, 263 (1998).
28. Ostriker, J.P., Thompson, C. & Witten, E., *Phys. Lett. B* **180**, 231 (1986).
29. Kouveliotou, C., et al., *Nature* **393**, 235 (1998).
30. Hurley, K., et al., *Ap.J.* **510**, L111 (1999).
31. Kouveliotou, C., et al., *Ap.J.* **510**, L115 (1999).
32. Murakami, T., et al., *Ap.J.* **510**, L119 (1999).
33. Marsden, D., Rothschild, R.E. & Lingenfelter, R.E., *Ap.J.* **520**, L107 (1999).
34. Woods, P., et al., *Ap.J.* **524**, L55 (1999).
35. Thompson, C. & Blaes, O., *Phys. Rev. D* **57**, 3219 (1998).
36. Kulkarni, S., et al., *Nature* **368**, 129 (1994).
37. Hurley, K., et al., *Ap.J.* **523**, L37 (1999).
38. Harding, A.K., Contopoulos, I. & Kazanas, D., *Ap.J.* **525**, L125 (1999).
39. Gotthelf, E.V., Vasisht, G. & Dotani, T., 1999, *Ap.J.* **522**, L49 (1999).
40. Kaspi, V.M., Chakrabarty, D. & Steinberger, J., *Ap.J.* **525**, L33 (1999).
41. Flowers, E. & Ruderman, M., *Ap.J.* **315**, 302 (1977).
42. Feroci, M., et al., *Ap.J.* (2000). [see preliminary paper in these proceedings]
43. Palmer, D.M., *Ap.J.* **512**, L113 (1999).
44. Göğüş, E., et al., *Ap.J.* **526**, L93 (1999).
45. Melatos, A., *Ap.J.* **519**, L77 (1999).
46. Heyl, J.S. & Hernquist, L., *M.N.R.A.S.* **304**, L37 (1999).

Magnetic Field Limits on SGRs

R. E. Rothschild [1], D. Marsden[1,2], and R. E. Lingenfelter[1]

[1] *Center for Astrophysics and Space Sciences, University of California, San Diego, La Jolla, CA USA,*
[2] *presently NAS/NRC Research Associate at Goddard Space Flight Center*

Abstract. We measure the period and spin-down rate for SGR 1900+14 during the quiescent period two years before the recent interval of renewed burst activity. We find that the spin-down rate doubled during the burst activity, which is inconsistent with both magnetic dipole driven spin-down and a magnetic field energy source for the bursts. We also show that SGRs 1900+14 and 1806−20 have braking indices of ~1, which indicates that the spin-down is due to wind torques and not magnetic dipole radiation. We further show that a combination of dipole radiation and wind luminosity, coupled with estimated ages and present spin parameters, imply that the magnetic fields of SGRs 1900+14 and 1806−20 are less than the critical field of 4×10^{13} G and that the efficiency for conversion of wind luminosity to x-ray luminosity is <2%.

SPIN-DOWN HISTORY OF SGR 1900+14

The spin-down of SGR 1900+14 from 1966 September to 1999 April (Fig. 1) is characterized by three intervals of time for which the spin-down rate was essentially constant within the interval. This characterization of the SGR 1900+14 spin-down is based upon both direct measurements of $\dot{P}$ as part of the period determination, and upon differences in measured spin periods between two different observations. The first interval begins with the RXTE observation in September of 1996 and ends with the ASCA observation at the beginning of 1998 May. The mean spin-down is $\dot{P} \sim 6\times10^{-11}$ s/s [1-3]. The second interval begins with the onset of bursting on 1998 May 26 and continues until mid-September 1998. The mean spin-down during this time was $\dot{P} \sim 13\times10^{-11}$ s/s [4,1,5]. The third interval begins in mid-September 1998 and continues at least until 1999 March 30. The mean spin-down rate at that time was again $\sim 6\times10^{-11}$ s/s [8].

Woods et al. [8] have suggested that the data may be consistent with a discontinuous spin-down event during the second interval as a result of the giant burst of 1998 August 27, as opposed to a doubling of $\dot{P}$ during the entire second interval [1]. This suggestion, however, appears to be at odds with the measurement of $\dot{P}=(11.0\pm1.7)\times10^{-11}$ s/s in early 1998 June [4], approximately 3 months before the Superburst and with the RXTE/ASCA determination of $\dot{P} \sim 10\times10^{-11}$ s/s just

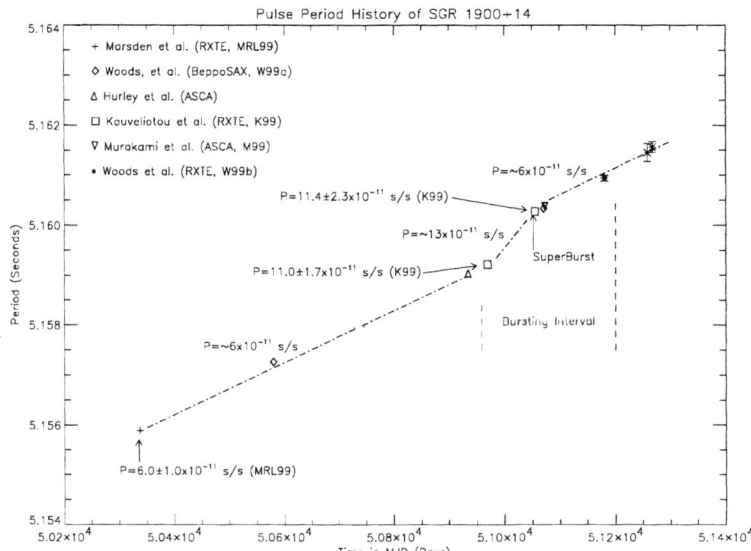

FIGURE 1. The pulse period history for SGR 1900+14 versus time. All of the published values of the pulse period are given along with the three measurements of $\dot{P}$ made as part of the period determination analysis. The mean spin-down in the three time intervals are also given.

after the event [5]. A $\dot{P}$ variation of more than a factor of 3 was also indicated in the timing measurements of SGR 1806−20 [9].

Very similar variations in the spin down rate $\dot{P}$ have been observed in AXPs 1E1048.1−5937 and 1E2259+58 (for a review, see [6]). Such behavior argues strongly against SGR spin-down by magnetic dipole radiation, since that would require large increases (e.g., ∼100%) in the total magnetic field energy of the star. The SGR 1900+14 observations also strongly argue against magnetic fields as the source of energy for SGR bursts, since that should lead to a decrease in the field energy, and hence a decrease in the spin-down rate during bursting periods. The observations show the spin-down rate increases.

SGR 1900+14 & 1806−20 SPIN-DOWN DUE TO RELATIVISTIC WINDS

The association of SGRs 1900+14 and 1806−20 with radio supernova remnants provides additional evidence as to the origin of their spin-down torques. Assuming that the spin-down torque is given by $\dot{\Omega} \propto \Omega^n$, the age of a pulsar with period P and spin-down torque $\dot{P}$ is given by

$$t_{age} = P/[(n-1)\dot{P}]$$

where the spin-down braking index, $n = 3$ for pure magnetic dipole radiation and

$n \sim 1$ for wind torques. Taking a nominal age of 10 kyr for a detectable [7] supernova remnant, inverting the age equation yields
$$n = 1 + (P/\dot{P})(t_{age})^{-1}.$$
Using $P = 5.16$ s and the long-term $\dot{P} = 6\times10^{-11}$ s/s measured for SGR 1900+14, we find that
$$n = 1 + 0.27/(t_{age}/10^4 \text{yr}),$$
and $P = 7.47$ s and the long-term $\dot{P} = 8.3\times10^{-11}$ s/s measured [9] for SGR 1806−20
$$n = 1 + 0.29/(t_{age}/10^4 \text{yr}).$$
This indicates that the braking index for SGRs 1900+14 and 1806−20 must be ~1, and that the spin-down of SGR 1900+14 is dominated by torques due to the relativistic wind and not magnetic dipole radiation.

SPIN-DOWN TORQUES OF SGRS

Addressing these problems, Thompson et al. [10] considered a magnetar driven Alfvén wind. The torque provided by the emission of such a relativistic wind is
$$I_* \dot{\Omega}_w = -\Lambda (L_w/c^2) R_A^2 \Omega$$
where I_* is the neutron star moment of inertia, L_w is the mechanical luminosity of the wind, $\Omega \equiv 2\pi/P$ is the spin frequency, $\dot{\Omega}_w$ is the spin-down rate due to the wind, and R_A is the Alfvén radius. The constant Λ is equal to 2/3 for a magnetic dipole field aligned with the rotation axis. The Alfvén radius is given by:
$$\frac{L_w}{4\pi R_A^2 c} = \frac{B_*^2(R_A)}{8\pi}$$
where B_* is the magnetic field of the neutron star. When the Alfvén radius is inside the light cylinder radius ($R_A < R_{lc}$, where $R_{lc} = c/\Omega$),
$$I_* \dot{\Omega}_w = -\Lambda B_* R_*^3 \left(\frac{L_w}{2c^3}\right)^{1/2} \Omega$$
where R_* is the radius of the neutron star and dipole geometry is assumed. When the Alfvén is outside the light cylinder radius, the torque is limited to
$$I_* \dot{\Omega}_w = -\Lambda L_w \Omega^{-1}$$
The transition frequency between these two wind spin-down regimes is
$$\Omega_{tr} = 8.572 \left(\frac{L_w}{10^{36} \text{ergs/s}}\right)^{1/4} \left(\frac{B_*}{10^{14} \text{G}}\right)^{-1/2} \text{ radians/s}.$$
The torque due to a rotating magnetic dipole is:
$$I_* \dot{\Omega}_{mdr} = -k \frac{B_*^2 R_*^6}{6c^3} \Omega^3$$
where k = 1 [11].

Once the total spin-down torque is specified as a function of Ω, the age of the SGR can found by the integral of $d\Omega$ over the total torque divided by I_*, where the integration is performed from an initial frequency to the present-day angular frequency.

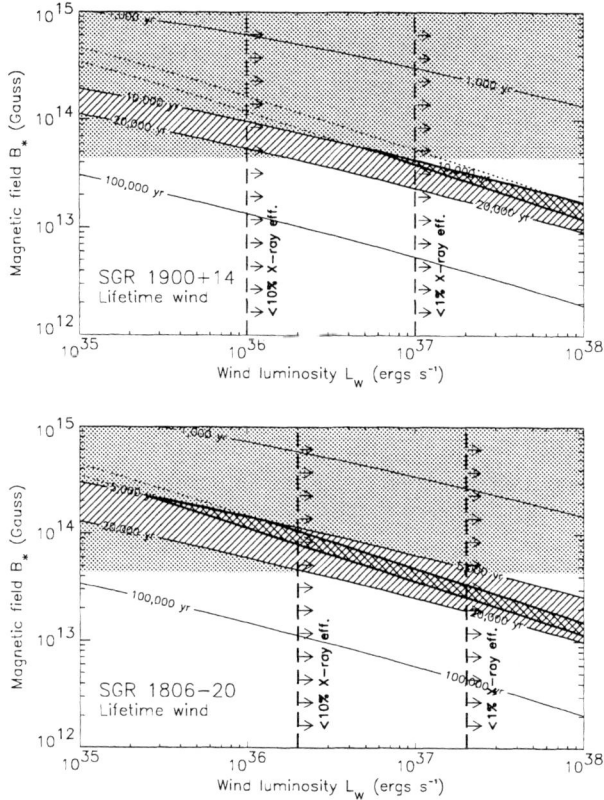

FIGURE 2. Age contours for SGRs 1900+14 and 1806−20 for constant magnetic fields and Alfvén wind luminosities. The cross-hatched areas denote the allowed regions of parameter space given the constraints provided by the age of the associated supernova remnant (solid lines) and long term present-day spin-down rate (dotted lines). The vertical dashed lines denote the 10% and 1% efficiencies of the wind in producing the observed x-ray flux of ~10^{35} ergs/s for SGR 1900+14 and ~2×10^{35} ergs/s for SGR 1806−20.

MAGNETIC FIELD AND WIND LUMINOSITY LIMITS

Using the above model we explore a wide range of magnetic fields B_* and wind luminosity L_w, shown in Fig. 2. We see in the upper panel that the presently observed period of $P = 5.157$ s, the spindown rate of $\dot{P} = 6 \pm 1 \times 10^{-11}$ s/s (dotted lines) of SGR 1900+14, and the 10 to 20 Kyr range of ages (solid lines) of its associated supernova remnant G42.8+0.6 [12], tightly constrain the allowable magnetic field to $B_* < 6 \times 10^{13}$ G and wind luminosities $L_w > 5 \times 10^{36}$ erg/s. Compared to the quiescent 2–10 keV x-ray luminosity of ~10^{35} erg/s [5], this wind luminosity implies a < 2% conversion efficiency of wind energy to x-rays in that band, which is quite consistent with theoretical calculations [11,13,14]. We also see in the lower panel of Fig. 2, that a similar set of constraints can be obtained for SGR 1806−20 and its supernova remnant G10.0−0.3, using the present spin period [9] P=7.47 s and $\dot{P} = 8.3 \times 10^{-11}$s/s. Again, a wind luminosity with <2% conversion efficiency to x-rays yields a sub-critical (4×10^{13} G) magnetic field. The limit for SGR 1806−20 is very similar that found from a comparable analyses by Harding et al. [11]. Thus, we see that with such winds the magnetic field limits are quite consistent with the limiting values inferred for normal radio pulsars, but not with those expected for magnetars.

SUMMARY

We show that the large variations in the spin down rate $\dot{P}$ measured in SGRs and AXPs argue strongly against spin-down by magnetic dipole radiation, since that would require large increases (e.g., ∼100%) in the total magnetic field energy of the star. The SGR 1900+14 observations also strongly argue against magnetic fields as the source of energy for SGR bursts, since that should lead to a decrease in the field energy, and hence a decrease in the spin-down rate during bursting periods rather than the increase observed. We further show from the braking indices that the spin-down torque of the two SGRs is in fact due to winds, not magnetic dipole radiation, and that with such, the magnetic fields are $< 6 \times 10^{13}$ G, which is quite consistent with normal pulsars, but not with magnetars.

REFERENCES

1. Marsden, D., Rothschild, R.E., & Lingenfelter, R.E., *Ap. J.* **520**, L107 (1999a).
2. Hurley, K., et al., *Ap. J.* **510**, L111 (1999).
3. Woods, P.M., et al., *Ap. J.* **518**, L103 (1999a).
4. Kouveliotou, C., et al., *Ap. J.* **510**, L115 (1999).
5. Murakami, T., et al., *Ap. J.* **510**, L119 (1999).
6. Stella, L., Israel, G.L., & Mereghetti, S., *Adv. Space Res.* **22**, 1025 (1998).
7. Braun, R., Goss, W.N., & Lyne, A.G., *Ap. J.* **340**, 355 (1989).
8. Woods, P.M., et al., astro-ph/9907173 (1999b).
9. Kouveliotou, C., et al., *Nature* **393**, 235 (1998).
10. Thompson, C., et al., *Ap. J.*, submitted (1999).
11. Harding, A.K., Contopoulos, I., & Kazanas, D., *Ap. J.*, in press (1999).
12. Vasisht, G., et al., *Ap. J.* **431**, L35 (1994).
13. Tavani, M., *Ap. J.* **431**, L83 (1994).
14. Harding, A.K., in *High Velocity Neutron Stars and Gamma-ray Bursts*, eds. R.E. Rothschild & R.E. Lingenfelter, AIP Press, New York, 1995, pp. 118.

Environmental Influences in SGRs and AXPs

David Marsden[*,1], Richard Lingenfelter[†], Richard Rothschild[†], and James Higdon[‡]

[*]*NASA/Goddard Space Flight Center, Greenbelt, MD 20771*
[†]*Center for Astrophysics and Space Sciences, University of California at San Diego, La Jolla, CA 92093*
[‡]*W. M. Keck Science Center, Claremont Colleges, Claremont, CA 91711*

Abstract. Soft gamma-ray repeaters (SGRs) and anomalous x-ray pulsars (AXPs) are young (<100 kyr), radio-quiet, x-ray pulsars which have been rapidly spun-down to slow spin periods clustered at 5–12 s. Nearly all of these unusual pulsars also appear to be associated with supernova shell remnants (SNRs) with typical ages < 20 kyr. If the unusual properties of SGRs and AXPs are due to an innate feature, such as a super-strong magnetic field, then the pre-supernova environments of SGRs and AXPs should be typical of neutron star progenitors. This is *not* the case, however, as we demonstrate that the interstellar media which surrounded the SGR and AXP progenitors and their SNRs were unusually dense compared to the environments around most young radio pulsars and SNRs. Thus, if these SNR associations are real, the SGRs and AXPs can not be "magnetars", and we suggest instead that the environments surrounding SGRs and AXPs play a controlling role in their development.

INTRODUCTION

Soft gamma-ray repeaters (SGRs) are neutron stars whose multiple bursts of gamma rays distinguish them from other gamma-ray burst sources [1]. SGRs are also unusual x-ray pulsars in that they have spin periods clustered in the interval 5–8 s, and they all appear to be associated with supernova remnants (SNRs), which limits their average age to approximately 20 kyr [2]. The angular offsets of the SGRs from the apparent centers of their associated supernova remnant shells indicates that SGRs are endowed with space velocities > 500 km s^{-1}, which are greater than the space velocities of most radio pulsars [3]. Anomalous x-ray pulsars (AXPs) are similar to SGRs in that they are radio quiet x-ray pulsars with spin periods clustered in the range 6–12 s, and have similar [4] persistent x-ray luminosities as the SGRs ($\sim 10^{35}$ ergs s^{-1}). Most of the AXPs appear to be associated with

---
[1)] NAS/NRC research associate

supernova remnants, and therefore they are also thought to be young neutron stars like the SGRs. Here we present a new look at environmental evidence which shows that the SGRs and AXPs can not be due to a purely innate property, such as superstrong magnetic fields [5].

THE ENVIRONMENTS OF SGRS AND AXPS

If the unusual properties of SGRs and AXPs were due solely to an intrinsic property of the neutron star, that developed independently of the external environment, then the characteristics of the interstellar medium which surrounded the AXP and SGR progenitors should be typical of that around the massive O and B stars which are progenitors of all neutron stars. Observations clearly show that the majority of neutron stars are formed in "superbubbles" — evacuated regions of the ISM which surround the OB associations in which the massive progenitors of most neutron stars live. The supernovae from the massive O and B stars which form SGRs and AXPs are heavily clustered in space and time and form vast (> 100 pc) HII regions/superbubbles [6] filled with a hot ($> 10^6$ K) and tenuous ($n \sim 10^{-3}$ cm^{-3}) gas. The occurrence of most supernovae in the hot phase of the ISM is confirmed from observations of nearby galaxies [7] and from studies of Galactic SNRs [8]. It is estimated that $90 \pm 10\%$ of all core-collapse supernova should occur in this hot and tenuous environment [9].

The environments of SGRs and AXPs are probed by the blastwaves of their associated supernova remnants, and from the size of the remnant shell as a function of the age we can constrain the external density. In Table 1 we have listed the 12 known SGRs and AXPs and their associated supernova remnant shells [10]. The identification of the associated remnants are based on both positional coincidences of the remnant and the SGR/AXP, and on similar distances of the SGR/AXP and its associated remnant. We include the new tentative [11] SGR candidate 1801−23, which appears to be associated with the SNR W28. The thin SGR error box passes roughly through the center of the SNR and through the compact, nonthermal x-ray source [12] within the remnant. No associated remnants can be found for AXPs 0720−3125 and 0142+615, which is not surprising given their close distance (~ 0.1) of 0720−3125 [13], and the molecular clouds associated with 0142+615 [14]. A more detailed discussion and reference list for the sources in Table 1 will be published elsewhere [10].

Most of the SGR/AXP positions are significantly displaced from the apparent centers of their associated SNRs, as can be seen in Table 1 from the ratio of the neutron star angular displacement θ_* divided by the angular radius θ_{SNR} of the remnant shell. These displacements clearly indicate that the SGR/AXPs have large transverse velocities. There is considerable uncertainty in the actual velocities, however, because the estimated remnant ages are probably uncertain by a factor of two in most cases, which introduces a corresponding uncertainty in the transverse velocities. In addition, the actual space velocities of the SGR/AXPs are larger by an

TABLE 1. The Supernova Remnants of SGRs and AXPs

Object	Period (s)	SNR	Age[a] (kyr)	Dist. (kpc)	Rad.[b] (pc)	$\frac{\theta_*}{\theta_{SNR}}$	Vel.[c] (km s^{-1})
SGR 1627−41	6.4?	G337.0−0.1	5	11	2	2.3	1000
AXP 1841−045	11.8	Kes 73	2	6.5	4	0.1	200
AXP 1845−0258	6.97	G29.6+0.1	10	12	8	0.1	200
SGR 0526−66	8	N49	5.5	55	8	0.8	1200
AXP 1709−40	11.0	G346.6−0.2 [d]	20	10	12	1.7	1000
SGR 1900+14	5.16	G42.8+0.6	10	5	15	1.4	2000
AXP 2259+586	6.98	CTB 109	10	4	16	0.2	300
SGR 1806−20	7.47	G10.0−0.3	10	14.5	16	0.5	800
SGR 1801−23	—	W28 [d]	60	3	19	0.1	30
AXP 1048−5937	6.45	G287.8−0.5 [d]	10	3	11	2.2	1100
AXP 0720−3125	8.39	— [e]	50 [f]	0.1	—	—	—
AXP 0142+615	8.69	— [g]	60 [f]	1	—	—	—

[a] SNR age
[b] Radius of radio shell
[c] Transverse velocity of SGR/AXP
[d] "Tentative" remnant identification (see text)
[e] Too close to identify remnant
[f] MDR timing age used since there is no SNR
[g] In/behind molecular cloud (no remnant)

unknown factor dependent on the viewing angle. Nonetheless, the data suggest that the typical SGR/AXP velocities are of the order of 1000 km s^{-1}. Such velocities, while much larger than the typical neutron star velocities, are not unprecedented, as $\sim$ 10% of radio pulsars may have space velocities of 1000 km s^{-1} or greater [3]. We conclude, therefore, that the SGRs and AXPs are a *high velocity* subset of young neutron stars.

In Figure 1 we have plotted the SNR shell radii as a function of the estimated age of each remnant. Overplotted in solid lines are simple approximations of the evolutionary tracks [15] of supernova remnant expansion in the wide range of the external ISM densities, and we see that these SNRs are all in the denser (> 0.1 cm^{-3}) phases of the ISM which slow their expanding shells to < 2000 km s^{-1} in < 10 kyr. Also overplotted are the tracks of neutron stars born at the origin of the supernova explosion with varying velocities, showing the times required for fast (e.g., > 500 km s^{-1}) neutron stars to catch up with the slowing supernova ejecta and swept-up matter.

DISCUSSION

From the discussion above, we saw that neutron stars should preferentially reside in the diffuse ($n < 0.01$ cm^{-3}) gas which constitutes the hot phase of the interstellar medium. As seen from Figure 1, however, the SGRs and AXPs tend form in denser regions of the ISM. Given the entire sample of AXPs and SGRs, the probability

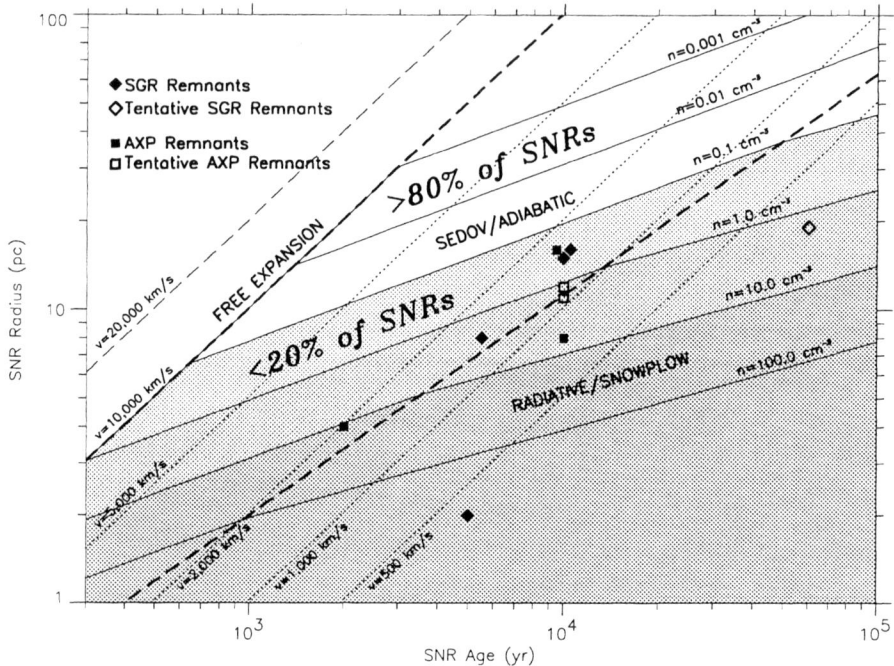

FIGURE 1. The radius of the SGR and AXP supernova remnant shells as a function of their age. The solid lines denote SNR expansion trajectories and the dotted lines denote the tracks of neutron stars born at the origin of the supernova explosion with varying space velocities. We see that essentially all of these sources were formed in the denser phase of the interstellar medium (ISM), which clearly indicates that the environment, and not a purely intrinsic property such as a superstrong magnetic field, is the controlling factor in the development of the SGRs and AXPs

that this is merely due to chance depends on the the ability to detect supernova remnants in the different phases of the interstellar medium. For the SGRs, the detection sensitivity is independent of the interstellar medium, because they are detected via their bright gamma-ray/x-ray bursts. Therefore, using only the SGRs yields a chance probability of less than $(0.2)^5 \sim 10^{-4}$, if one accepts the tentative W28/SGR 1801−23 association, and $\sim 10^{-3}$ if one excludes SGR 1801−23 from the SGR sample. The AXPs are also preferentially in the dense phase, which further lowers the chance probability for the class as a whole. The evidence then suggests that the environments surrounding SGRs and AXPs are significantly different than otherwise normal neutron stars in a way which is *inconsistent* with the hypothesis that the properties of these sources are the result of an innate characteristic such as a superstrong magnetic field.

These observational facts imply instead that the environment is crucial in the development of SGRs and AXPs. One plausible scenario is that the rapid spin-down of the SGR/AXPs may result from their interaction with co-moving ejecta and swept-up ISM material [16,17]. Calculations [10] indicate that such an *interaction* scenario, involving the formation of accretion disks by fast (> 500 km s^{-1}) neutron stars from co-moving ejecta of supernova remnants slowed to < 2000 km s^{-1} by the denser (> 0.1 cm^{-3}) phases of the ISM, could spin-down SGRs and AXPs to their present-day spin periods in ~ 10 kyr — consistent with the estimated ages of these sources — without requiring the existence of a population of neutron stars with ultrastrong magnetic fields. In addition, such a scenario can explain the clustering of spin periods, present-day spin-down rates, and the number of SGRs and AXPs in our galaxy [10].

REFERENCES

1. Hurley, K., these proceedings.
2. Braun, R., Goss, W. M., & Lyne, A. G., *ApJ* **340**, 355 (1989).
3. Cordes, J. M., & Chernoff, D. F., *ApJ* **505**, 315 (1998).
4. Mereghetti, S., in *The Neutron Star – Black Hole Connection*, 1998, in press (astro-ph/9911252).
5. Thompson, C., & Duncan, R. C., *MNRAS* **275**, 25 (1995).
6. Mac Low, M. M., & McCray, R., *ApJ* **324**, L776 (1988).
7. van Dyk, S. D., Hamuy, M., & Filippenko, A. V., *AJ* **111**, 2017 (1996).
8. Higdon, J. C., & Lingenfelter, R. E., *ApJ* **239**, L867 (1980).
9. Higdon, J. C., Lingenfelter, R. E., & Ramaty, R., *ApJ* **509**, L33 (1998).
10. Marsden, D., et al., *ApJ*, submitted, astro-ph/9912207 (2000).
11. Cline, T., et al., *ApJ*, in press, astro-ph/9909054 (1999).
12. Andrews, M. D., et al., *ApJ* **266**, 684 (1983).
13. Haberl, F., et al., *A&A* **326**, 662 (1997).
14. Israel, G. L., Mereghetti, S., & Stella, L., *ApJ* **433**, L25 (1994).
15. Shull, J. M., Fesen, R. A., & Saken, J. N., *ApJ* **346**, 860 (1989).
16. Corbet, R. H. D., et al., *ApJ* **443**, 786 (1995).
17. van Paradijs, J., Taam, R. E., & van den Heuvel, E. P. J., *A&A* **299**, L41 (1995).

Relativistic Compton Scattering in Ultra-Strong Magnetic Fields

Peter L. Gonthier[*], Rachel M. Costello[†], Cassandra L. Mercer[‡],
Alice K. Harding[§], and Matthew G. Baring[§]

[*]*Hope College, Department of Physics and Engineering, 27 Graves Place Holland, MI 49422-9000*
[†]*The College of Wooster, Department of Physics, 1189 Beall Ave, Wooster, OH 44691*
[‡]*The Colorado College, Department of Physics, 14 E. Cache La Poudre Colorado Springs, CO 80903*
[§]*NASA - Goddard Space Flight Center, Laboratory for High Energy Astrophysics Greenbelt, MD 20771*

Abstract. Recent observations of soft gamma-ray repeaters and anomalous X-ray pulsars are furnishing greater evidence for the existence of a class of neutron stars with surface magnetic fields exceeding the critical field of 4.4×10^{13} Gauss. The main effort of this study is to understand the role of relativistic Compton scattering as it operates with other QED processes, to develop approximate expressions of the exact rate and to incorporate them in a full acceleration-cascade model. Previous modeling of Compton scattering in magnetic fields has assumed that the scattering cross section can be adequately described by the nonrelativistic Compton scattering cross section (Thomson limit) below resonance and at the resonance, and the Klein-Nishina cross section above the resonance. Consequently, these studies have not included the effects of the strong fields of pulsars and magnetars. The study of the strong field effects on the inverse Compton scattering process will provide insight into the particle-photon interactions associated with a variety of pulsar phenomena, burst spectra of soft gamma-ray repeaters and the transport of thermal radiation through neutron star atmospheres. This paper will summarize the role of inverse Compton scattering and the effects of strong magnetic fields upon the integrated cross section for Compton scattering. An analytical approximation to the exact QED cross section will be discussed.

INTRODUCTION

As we have heard in this workshop, recent observations are providing evidence for the existence of isolated neutron stars having ultra-strong magnetic fields. Growing support for a new class of isolated neutron stars with ultra-strong magnetic fields ($B > 10^{14}$ G) has come from the observations of Soft Gamma-Ray Repeaters (SGRs) and Anomalous X-ray pulsars (AXPs). Assuming dipole radiation torques, the measured period derivatives of SGRs and AXPs imply surface magnetic fields between $10^{14} - 10^{15}$ G, well above the quantum critical field, $B_{crit} = 4.4 \times 10^{13}$ G. The AXPs are bright X-ray sources with luminosities, $L_X \sim 10^{35}$ erg/s, far exceeding their spin-down luminosity. This energetics issue has motivated Thompson & Duncan [1] and Kulkarni & Thompson [2] to suggest that, unlike rotation-powered pulsars, the X-ray and particle emission in AXPs is powered by a decay of the magnetic field in the stellar interior.

Various studies indicate that inverse Compton scattering (ICS) plays a significant role in the magnetospheric physics of strongly magnetized neutron stars. Relativistic electrons accelerated above the polar cap can Compton scatter off thermal radiation from the neutron star surface, producing high-energy gamma rays that can power pair cascades. Daugherty & Harding [3] found that in the presence of a strong magnetic field, resonant scattering greatly increases electron energy losses over those of non-resonant scattering, making Compton scattering efficient even at lower temperatures. Harding & Muslimov [4] considered ICS by the trapped, back flowing positrons and found that the pairs from the ICS photons may cause surface acceleration gaps to be unstable, forcing them to higher altitudes.

Compton scattering is also very important in SGR radiation models. The highly super-Eddington luminosities of the bursts ensures high densities of both photons and particles, so that scattering will be a critical factor. Paczyński [5] proposed that the lower scattering cross section below the resonance in the strong magnetic field (for photons in the $\perp$-polarization mode) could allow super-Eddington luminosities. However, Miller et al. [6] argued that scattering into the $\parallel$-mode keeps the radiation pressure high, and thus the effective Eddington luminosity lower, in hydrostatic atmospheres. In Thompson & Duncan's model [7] for the radiation from SGR bursts, Compton scattering plays a critical role in establishing equilibrium between pairs and photons and in the spectral formation. Compton scattering may also be important in the photon splitting cascade model for SGR burst emission [8-10].

SCATTERING OF RELATIVISTIC ELECTRONS

While full details have been recently submitted [11], here in this talk, we present some findings regarding the particular problem of scattering of photons from relativistic electrons, common to a variety of astrophysical phenomena. We have developed expressions for scattering of ultra-relativistic electrons with $\gamma \gg 1$ moving parallel to the magnetic field lines. The photon may have any angle of incidence, ψ_i, in the laboratory frame with respect to the magnetic field. Due to the large γ's, the laboratory angle, ψ_i, gets Lorentz contracted to $\theta = \psi_i/2\gamma \sim 0$ deg in the electron rest frame.

In Figure 1, we display the QED, exact angle-integrated cross section (solid curves) for the indicated magnetic fields, in units of B_{crit}, as a function of the incident photon energy, ω/ω_B, in cyclotron energy units. We have averaged over the final spin of the electron and over the polarization of the scattered photon. For this particular case in the scattering of relativistic electrons, there is only one resonance occurring at the fundamental cyclotron resonance of ω_B. We scale the photon energy by the cyclotron energy so that the resonance occurs at the same place independent of the magnetic field, B. For comparison, we also plot in the figures the nonrelativistic Thomson limit (dot-dashed curves) and the Klein-Nishina (dotted curves) predictions.

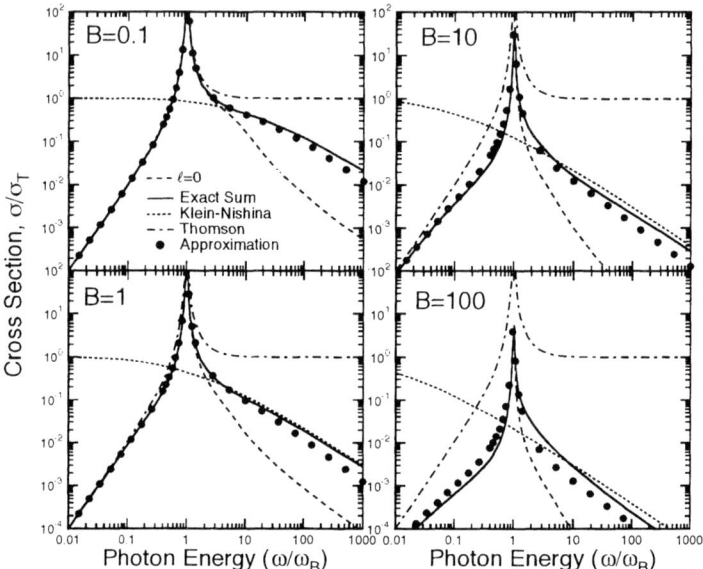

FIGURE 1. Total Compton scattering cross section (in Thomson units) as a function of the incident photon energy (in units of the cyclotron energy) for the indicated magnetic field strengths (in units of the critical field, B_{crit}). The exact QED scattering cross section, summed over all contributing final electron Landau states is indicated as a solid curve. The non-relativistic magnetic Thomson cross section is plotted as a dot-dashed curve (labeled Thomson), while the Klein-Nishina cross section is plotted as a dotted curve. The cross section for only the final Landau state $\ell = 0$ is plotted as a dashed curve.

Above the resonance, the exact cross section approaches the Klein-Nishina cross section. As expected for smaller fields in the case of $B = 0.1$, the convergence occurs at lower photon energies. At this field strength, typical of radio pulsars, there are no significant deviations from the Thomson limit below the resonance and from the Klein-Nishina limit above the resonance. The main discrepancy occurs right above the resonance where the two limiting cases do not match the exact cross section. As the field strength increases, the exact cross section below the resonance drops significantly beneath the Thomson limit by over a factor of ten in the case of $B = 100$. For scattering above the resonance at these high fields, there are deviations between the exact cross section and the Klein-Nishina cross section. However, as the energy increases, the exact cross section and the Klein-Nishina cross section appear to converge as seen in the cases of $B = 0.1$ and $B = 1$.

In Figure 1, the solid circles represent the polarization-averaged cross section obtained from an analytical approximation [11]. In the region of validity, $\omega < 1$ (corresponding to $\omega/\omega_B < 1/B$), the approximation agrees very well with the exact $\ell = 0$ cross section. Above the region of validity, the approximation over estimates the exact $\ell = 0$ cross section. However, the approximation does surprisingly well, when compared to the exact cross section, extrapolating above the region of validity. While the analytical approximation is a result of integrating the approximation to the exact $\ell = 0$ differential cross section, it remains close to the total cross section at high ω above the resonance ($\omega > \omega_B$) even for high field strengths.

In Figure 2, we present the polarization-dependent, QED Compton scattering cross section as a function of energy for magnetic fields of 0.1 and 10 times B_{crit}. For this particular case of scattering of relativistic electrons, there are two polarization scattering channels, in which the scattering leads to photons with ||-polarization (dashed curve) and with ⊥-polarization (dotted curve). In the $B = 10$

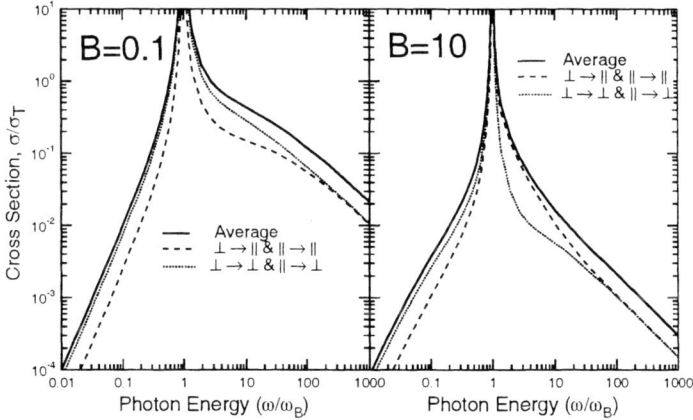

FIGURE 2. Total Compton scattering cross section (in Thomson units) as a function of the incident photon energy (in units of the cyclotron energy) for the indicated magnetic field strengths in units of the critical field, B_{crit}. The exact QED scattering cross section, summed over all contributing final electron Landau states and averaged over photon polarization states is indicated as a solid curve. The QED cross section leading to parallel and perpendicular polarizations are plotted as dashed and dotted curves, respectively.

case, above the resonance, the scattering process preferably produces photons with ||-polarization, whereas below the resonance, the channel producing ⊥-photons dominates. This behavior, where ⊥-polarized scattered photons dominate below the resonance and ||-polarized scattered photons dominate above the resonance, is characteristic of the magnetic-relativistic cross section. In the nonrelativistic case, the ⊥-polarization channel will be three times larger than the ||-polarization channel, but has the same shape at all photon energies. As can be seen in Figure 2 for the $B = 0.1$ case, $\|$-polarization in the QED cross section dominates at low fields, thus the switching to $\perp$-polarization dominance at high fields is a relativistic effect.

The polarization dependence of the scattering cross section in high fields will have significant implications for other polarization-dependent mechanisms such as pair production and, especially, photon splitting. As Baring & Harding [12] noted, photon splitting could dominate pair production at supra-critical magnetic fields, thereby suppressing pair creation and possibly accounting for the radio quiescence of SGRs and AXPs. However, kinematic selection rules [13,14] allow only one splitting mode to operate in the limit of weak dispersion, that in which photons with $\perp$-polarization split into two photons, each with $\|$-polarization. Under such restrictions, photon splitting would occur only once, and then pair production would take over as the dominant attenuation mechanism. It is possible that the dispersion characteristics of the ultra-strong field environment, or perhaps plasma properties present during outburst mode of SGRs, may permit the two other splitting modes allowed by CP (charge-parity) invariance to operate, providing $\|$-mode photons the opportunity to split. Nevertheless, even if other modes do not become operational in high fields, Compton scattering below the resonance is able to convert the photons with $\|$-polarization into $\perp$-polarization, refueling photon-splitting cascades. Optical depths for such scattering could be quite significant in SGRs during their high-luminosity gamma-ray outbursts, provided that the photon field does not dominate the SGR energetics.

REFERENCES

1. Thompson, C. & Duncan, R. C., *Astrophys. J.* **473**, 332 (1996).
2. Kulkarni, S. R., & Thompson, C., *Nature* **393**, 215 (1998).
3. Daugherty, J. K., & Harding, A. K., *Astrophys. J.* **336**, 861 (1989).
4. Harding, A. K. & Muslimov, A. G., *Astrophys. J.* **500**, 862 (1998); Harding, A. K. & Muslimov, A. G., *Astrophys. J.* **508**, 328 (1998).
5. Paczyński, B., *Acta Astro.* **42**, 145 (1992).
6. Miller, M. C., *Astrophys. J.* **448**, L29 (1995).
7. Thompson, C. & Duncan, R. C., *Mon. Not. R. Astr. Soc.* **275**, 255 (1995).
8. Baring, M. G., *Astrophys. J.* **440**, L69 (1995).
9. Harding, A. K., Baring, M. G. & Gonthier, P. L., *Astron. & Astr. Supp.* **120(4)**, 111 (1996).
10. Harding, A. K., Baring, M. G. & Gonthier, P. L., *Astrophys. J.* **476**, 246 (1997).
11. Gonthier, P. L. et al., *Astrophys. J.*, submitted (1999).
12. Baring, M. G. & Harding, A. K., *Astrophys. J.* **507**, L55 (1998).
13. Adler, S. L., *Ann. Phys.* **67**, 599 (1971).
14. Shabad, A. E., *Ann. Phys.* **90**, 166 (1975).

Nuclear Equation of State and Internal Structure of Magnetars

In-Saeng Suh and G. J. Mathews

Department of Physics, University of Notre Dame, Notre Dame, IN 46556, USA

Abstract. Recently, neutron stars with very strong surface magnetic fields have been suggested as the site for the origin of observed soft gamma repeaters (SGRs). We investigate the influence of a strong magnetic field on the properties and internal structure of such strongly magnetized neutron stars (magnetars). The presence of a sufficiently strong magnetic field changes the ratio of protons to neutrons as well as the neutron appearance density. We also study the pion production and pion condensation in a strong magnetic field. We discuss the pion condensation in the interior of magnetars as a possible source of SGRs.

INTRODUCTION

Recently, observations of the soft gamma repeaters, SGR 0526−66, SGR 1806−20, SGR 1900+14, SGR 1627−41 and SGR 1801−23 (see [1]) with BATSE, RXTE, ASCA, and BeppoSAX have confirmed the fact that these SGRs are a new class of γ-ray transients corresponding to strongly magnetized neutron stars (magnetars). Magnetars [2,3] are newly born neutron stars with a surface magnetic field of $B \sim 10^{14} - 10^{15}$ G, probably created by a supernova explosion.

As relics of stellar interiors, the study of the magnetic fields in and around degenerate stars should give important information on the role such fields play in star formation and evolution. However, the origin and evolution of stellar magnetic fields remains obscure. The strength of the internal magnetic field in a neutron star in principle could be constrained by any observable consequences of a strong magnetic field. For example, rapid motion of neutron stars may be due to anisotropic neutrino emission induced by a strong magnetic field [4]. One could also consider the effect of magnetic fields on the thermal evolution [5] and mass [6] of neutron stars. Recently, Chakrabarty et al. [7] have investigated the gross properties of cold nuclear matter in a strong magnetic field in the context of a relativistic Hartree model and have applied their equation of state to obtain masses and radii of magnetic neutron stars.

Since strong interior magnetic fields modify the nuclear equation of state for degenerate stars, their mass-radius relation also will be changed relative to that of non-magnetic stars. Recently, we have obtained a revised mass-radius relation for magnetic white dwarfs [8]. For strong internal magnetic fields of $B \sim 4.4 \times (10^{11} - 10^{13})$ G, we have found that both the mass and radius increase significantly and the mass-radius relation of some observed magnetic white dwarfs may be better fit if strong internal fields are assumed.

If ultrastrong magnetic fields exist in the interior of neutron stars as well, such fields will primarily affect the behavior of the residual charged particles. Standard internal properties such as the nuclear equation of state, neutron appearance, and the threshold density of muons and pions, would be modified by the magnetic field. Under charge neutrality and chemical equilibrium conditions, we calculate the ratio of protons to neutrons as well as the pion condensate equation of state in the presence of a sufficiently strong magnetic field. Here we briefly describe and summarize the results. The details of this work will be published elsewhere [9].

INVERSE β-DECAY AND NEUTRON APPEARANCE IN A STRONG MAGNETIC FIELD

Let us consider a homogeneous gas of free neutrons, protons, and electrons (npe) in β-equilibrium [10] in a uniform magnetic field. At high densities above 8×10^6 g cm^{-3}, protons in nuclei are converted into neutrons via inverse β-decay: $e^- + p \longrightarrow n + \nu$. Since the neutrinos escape a star, energy is transported away from the system. Thus, the composition and structure of the star will be modified mainly by inverse β-decay. This reaction can proceed whenever the electron acquires enough energy to balance the mass difference between protons and neutrons, $Q = m_n - m_p = 1.293$ MeV. β-decay is blocked if the density is high enough that all energetically available electron energy levels in the Fermi sea are occupied.

In order to determine the equilibrium composition and equation of state, the coupled equations for chemical equilibrium and charge neutrality should now be solved simultaneously. Figure 1 shows the proton fraction $Y_p = n_p/n_B$, where n_B is the baryon density, as a function of the neutron density ρ_n for given field strength, $\gamma_e = B/B_c^e$, where $B_c^e \simeq 4.4 \times 10^{13}$ G. If the charged particles are in the lowest Landau level, inverse β decay is not suppressed in magnetic fields. This means that rapid neutron star cooling can occur in a strong magnetic field through the direct URCA process [11]. However, electrons and protons, actually, are not in the lowest Landau level for higher densities above a critical density from which higher Landau levels begin to contribute to the chemical potential of electrons and protons, and hence, particle number densities. Therefore, discrete Landau levels become continuous and thus the proton concentration Y_p goes back to the non-magnetic case as the neutron density increases. As a result, inverse β decay is still suppressed at high densities in strong magnetic fields.

Finally, the equation of state, the mass-energy density $\rho = (\mathcal{E}_e + \mathcal{E}_p + \mathcal{E}_n)/c^2$, and the pressure $P = P_e + P_p + P_n$ (see [12] for a field strength less than $\log \gamma_e = 2$), are straightforwardly determined. Figure 2 shows the equation of state for a npe gas

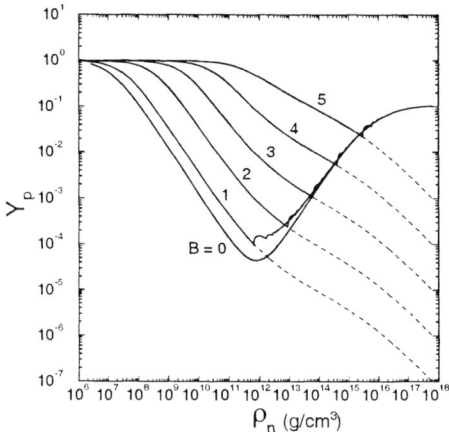

FIGURE 1. The proton fraction $Y_p = n_p/n_B$ as a function of the neutron density ρ_n for given $\gamma_e = B/B_c^e$'s. The $B = 0$ line is the non-magnetic case. The dashed lines occur if charged particles are restricted in the lowest Landau level. Numbers 1,2,...,5 correspond to the values of $\log \gamma_e$ (From [9].)

in various magnetic fields. In this figure we can see that the neutron appearance density for an ideal npe gas increases linearly with magnetic field strength.

PION PRODUCTION AND CONDENSATION IN A STRONG MAGNETIC FIELD

At very high density ($\rho \gtrsim \rho^*$), neutron-rich nuclear matter is no longer the true ground state of neutron star matter. It will quickly decay by weak interactions into chemically equilibrated neutron star matter. Fundamental constituents, besides neutrons, may include a fraction of protons, hyperons, and possibly more massive baryons. In particular, if pion condensation exists in a magnetic field [13], charged pion production and condensation through $n \to p + \pi^-$ is possible.

Figure 3 shows the equation of state for an ideal magnetic $npe\pi$ gas with pion condensation. We can see that magnetic fields reduce the pion condensation. However, we still have a distinguishable pion condensate equation of state in strongly magnetized neutron stars.

DISCUSSION

In this work, we have studied the nuclear equation of state for an ideal npe gas in a strong magnetic field. Here, we show that the higher Landau levels are significant at

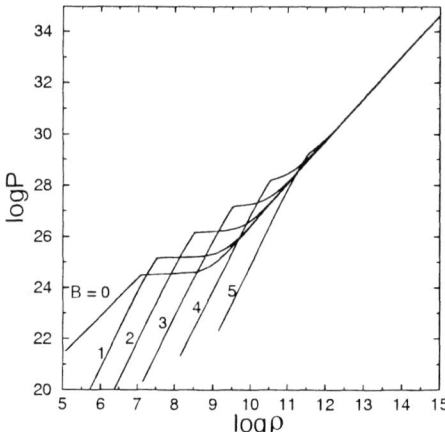

FIGURE 2. The equation of state for *npe* gas in various magnetic fields. The $B = 0$ line correspond to the non-magnetic case. Numbers 1,2,...,5 mean the values of log γ_e. (From [9].)

high density in spite of the existence of very strong magnetic fields. In particular, at high density, the proton concentration approaches the same non-magnetic limit. As a result, the inverse β decay is still suppressed in intense magnetic fields. Therefore, neutron star rapid cooling is not affected by the direct URCA process which is enhanced in strong magnetic fields. Finally, we see that the magnetic field reduces the amount of pion condensation. However, we have distinguishable effects of a pion condensate equation of state in strongly magnetized neutron stars.

It is generally accepted that neutrons and protons in a *npe* gas are superfluid [14,15]. The charged pion condensate is also superfluid and superconductive [14]. This pion formation and condensation in dense nuclear matter would have the significant consequence [9] that the equation of state would be softened. First of all, softening the equation of state reduces the maximum mass of the stars. This softening effect with pion condensation also leads to detectable predictions [14]. These are: (i) the rate of neutron star cooling via neutrinos would be enhanced, (ii) a possible phase transition of the neutron star to a superdense state, and (iii) sudden glitches in the pulse period. In particular, if pion condensation occurs in a strong magnetic field, it may significantly affect starquakes.

According to the magnetar model by Duncan and Thompson [2,3], SGRs are caused by starquakes in the outer solid crust of magnetars. In addition, Cheng and Dai [16] recently suggested that SGRs may be rapidly rotating magnetized strange stars with superconducting cores. Although such models can explain some crucial features, there are still several unsettled issues [17]. Therefore, superconducting cores with a charged boson (pion, kaon) condensate in magnetars might be an alternative model to explain the energy source of soft gamma rays from magnetars.

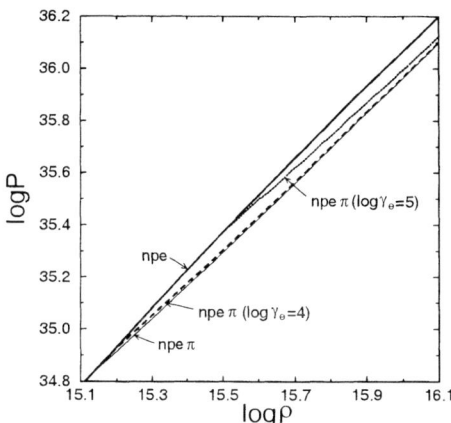

FIGURE 3. The equation of state for an ideal magnetic $npe\pi$ gas with pion condensation. The thick and thin solid lines are non-magnetic cases ($B = 0$). (From [9].)

ACKNOWLEDGMENTS

This work is supported in part by NSF Grant-97-22086 and DOE Nuclear Theory Grant DE-FG02-95ER40934.

REFERENCES

1. Hurley, K., these proceedings.
2. Duncan, R. C. & Thompson, C., *Astrophys. J.* **392**, L9 (1992).
3. Thompson, C. & Duncan, R. C., *Mon. Not. R. Astron. Soc.* **275**, 255 (1995).
4. Kusenko, A. & Segre, G., *Phys. Rev. Lett.* **77**, 4872 (1996).
5. Heyl, J. S. & Hernquist, L., *Astrophys. J.* **489**, L67 (1997).
6. Vshivtsev, A. S. & Serebryakoba, D. V., *Sov. Phys. JETP* **79**, 17 (1994).
7. Chakrabarty, S, Bandyopadhyay, D., & Pal, S., *Phys. Rev. Lett.* **78**, 2898 (1997).
8. Suh, I.-S., & Mathews, G. J., *Astrophys. J.*, in press, astro-ph/9906239 (2000).
9. Suh, I.-S. & Mathews, G. J., *Astrophys. J.*, submitted, astro-ph/9912301 (1999)
10. Shapiro, S. L. & Teukolsky, A. A., *Black Holes, White Dwarfs, and Neutron Stars*, Wiley-Interscience, New York, 1983.
11. Leinson, L. B., & Perez, A., *JHEP* **09**, 020 (1998).
12. Lai, D. & Shapiro, S. L., *Astrophys. J.* **383**, 745 (1991).
13. Rojas, H. Perez, *Phys. Lett.* **B379**, 148 (1996).
14. Migdal, A. B., et al., *Phys. Rep.* **192**, 179 (1990).
15. Yakovlev, et al., *Physics - Uspekhi*, accepted, astro-ph/9906456 (1999).
16. Cheng, K. S. & Dai, Z. G., *Phys. Rev. Lett.* **80**, 18 (1998).
17. Liang, E. P., *Astrophys. Space Sci.* **231**, 69 (1995).

New QED Calculations for Processes in Strong Magnetic Fields

D. Leahy* and L. Semionova[†]

Department of Physics and Astronomy, University of Calgary, Alberta, Canada
[†]*Department of Physics, Universidad National, Heredia, Costa Rica*

Abstract. The results of some new QED calculations are presented. We consider two-photon emission by electrons, and determine magnetic field, spin and polarization dependence of the transition rates. We utilize the electron wave functions of Sokolov and Ternov (1968, Synchrotron Radiation, Berlin: Akademie), rather than those of Johnson and Lippmann (1949, Physical Review D, 76, 828). As pointed out by Graziani (1993, Astrophysical Journal, 412, 351), use of the former wave functions is necessary to obtain results valid outside the weak field limit. The results are of particular interest for processes in magnetars, for which the magnetic field is near, or can exceed, the critical value ($B_{cr} = 4.414 \times 10^{13}$ G).

INTRODUCTION

The emission of two photons by an electron in a strong magnetic field is a second-order QED process. We present new calculations which are valid for any value of magnetic field. Previous calculations assumed $B << B_{cr}$. With the observational discovery of magnetars [1,2], which are pulsars with magnetic fields in the range $\sim 10^{14} - 10^{15}$ G, calculations which are valid for such high fields are of great interest.

Previous work related to photon emission by electrons in strong magnetic fields includes the following. The first-order emission process is discussed by [3–8]. [9] calculates first and second order Compton scattering, in which there is one photon in both the initial and final states. They use an S-matrix formalism which is applicable to arbitrarily strong fields, but the photon normal modes they use to include the effects of vacuum polarization and plasma apply only in the weak field limit ($B << B_{cr}$). They also use the Johnson-Lippmann electron wavefunctions [10] which makes their calculation valid only in weak fields, as shown by [11]. [11] also has given the correct treatment of the resonant line width for strong field cyclotron scattering using the appropriate electron wavefunctions, which are those of [12]. [13] use the results of [9] to calculate opacities and source functions, thus are also valid only in the weak field limit.

ELECTRON ENERGY STATES AND CALCULATION OF THE TRANSITION RATE

We assume a constant external magnetic field. The electron energy is given by:

$$E_N = [p_z^2 + m^2 + 2NeB]^{1/2}, \qquad (1)$$

and is characterized by the principal quantum number $N = l + \frac{1}{2}(s+1) = 0, 1, 2, \ldots$, where ($l = 0, 1, 2, \ldots$). For each N (called a Landau state), the electron may have spin-up ($s = +1$) or spin-down ($s = -1$) along the field direction, except in the ground state ($N = 0$), where only the spin-down state is allowed [3]. The critical magnetic field is $B_{cr} = \frac{m^2 c^3}{e\hbar} = 4.414 \cdot 10^{13}$ G, which is the field strength when the cyclotron energy is equal to the electron's rest energy.

The electron energy levels, E_N, consist of a discrete and continuous spectrum, since the perpendicular momentum $p_\perp = m\sqrt{2N\gamma}, N = 0, 1, 2 \ldots$, is discrete, whereas p_z is continuous (with $\gamma = \frac{B}{B_{cr}}$). Figure 1 presents the dependence of E on p_z/mc for values of $N = 0, 4, 8, 12$ and for two values of magnetic field, $B = 0.05 B_{cr} (\gamma = 0.05)$ and $B = 0.05 B_{cr} (\gamma = 1)$. The weak field limit ($\gamma \ll 1$) reduces to the $N = 0$ line. It is seen that increasing principal quantum number, N, and increasing magnetic field, γ, have the same effect in increasing the electron energy levels.

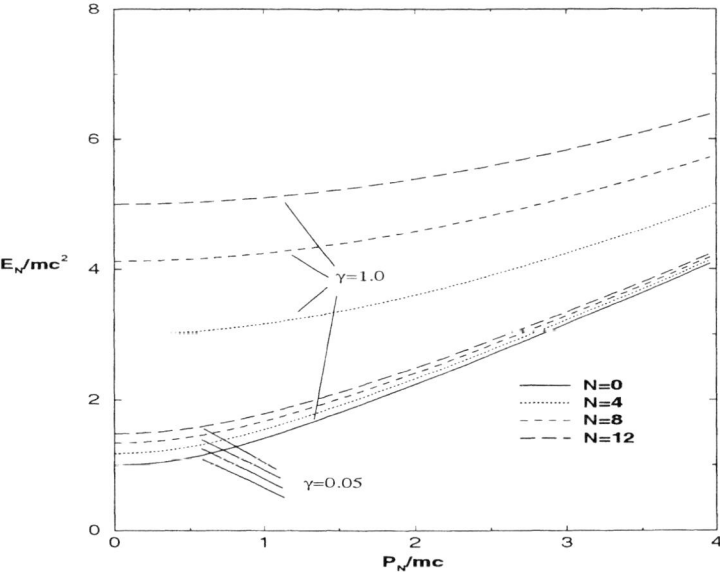

FIGURE 1. Electron energy divided by the electron mass energy, E/mc^2, vs. p_z/mc for $N = 0, 4, 8, 12$ and for two values of magnetic field, $\gamma = 0.05$ and $\gamma = 1$.

The non-relativistic approximation of E_N is: (defining $E^{NR} = E_N - m$):

$$E^{NR} \approx p_z^2/2m + NeB/m \qquad (2)$$

It is seen that the energy splitting between two successive Landau levels, with same p_z, is $E^{NR}{}_{N+1} - E^{NR}{}_N = eB/m = \omega_B$. The critical magnetic field is seen as the value of B for which this splitting is equal to the electron rest mass energy, m.

Here we study the transition between the states (N', s', p') and (N, s, p) where $N' > N$ which results in the emission of two photons of energies ω and ω', at angles θ and θ' with respect to the magnetic field. The Feynman diagrams for the process are given in Figure 2. The double lines represent electron propagation in the magnetic field and the single lines represent photons.

We use the S-matrix formalism (e.g., see [15]) and the electron wavefunctions of Sokolov, which are also used by [14]. These wave functions correctly describe the spin of the electron in the presence of a magnetic field. We also explicitly keep track of photon polarization, which can be described by the polarization unit vectors for linear polarization or for circular polarization. The details of the calculation are are to be published in Physical Review D. The four possibilities for the spins of the incoming and outgoing electrons are: (up, up), (up, down), (down, up) and (down, down). The transition probability into the states $(N' \to N)$ is evaluated from the S-matrix, with a summation over the quantities not observed in the final state, yielding the expression for the transition rate from the initial state.

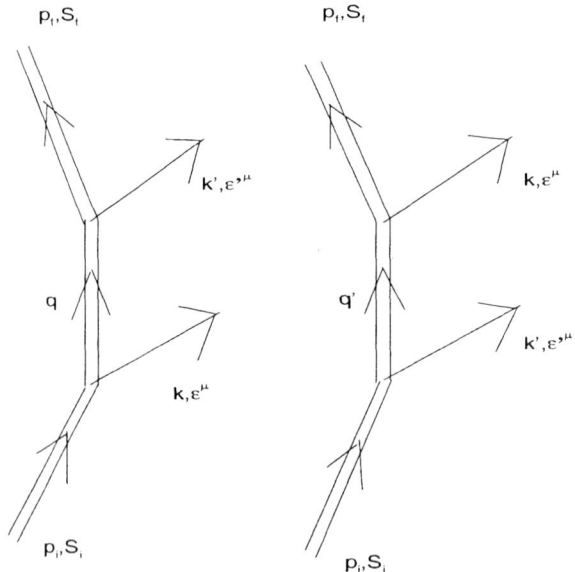

FIGURE 2. Feynman diagrams for the second order emission process.

TABLE 1. Cyclotron Resonance Energies $\omega_{res}^{(N',n)}(\theta)$ in Units of ω_B

N'	$\theta = 20°$		$\theta = 40°$		$\theta = 60°$		$\theta = 80°$	
	(1) [a]	(2)	(1)	(2)	(1)	(2)	(1)	(2)
2	0.9811	0.4526	0.9825	0.4674	0.9841	0.4870	0.9852	0.5018
3	1.9447	0.7690	1.9502	0.8067	1.9565	0.8610	1.9607	0.9066
4	2.8914	1.0203	2.9035	1.0804	2.9175	1.1716	2.9267	1.2543
5	3.8220	1.2328	3.8429	1.3135	3.8673	1.4407	3.8836	1.5634

[a] ω_{res} is given for two values of B: (1) $\frac{B}{B_{cr}} = 0.01$, (2) $\frac{B}{B_{cr}} = 1$, assuming that the intermediate state $n = 1$.

RESULTS

Resonances occur where the denominator of the expression for the transition rate vanishes. Generally, the resonance energies are smaller than ω_B. Table 1 gives the first resonance energies for two values of the magnetic field and for four angles. The expected resonance energies (from simple considerations) for the case of final state $N = 0$, intermediate state $n = 1$ and $N' = 2, 3, 4$ and 5, are 1, 2, 3 and 4 times ω_B. For $B = 0.1 B_{cr}$, the resonance energies are just slightly less than the expected energies, but for $B = B_{cr}$ they are far below the expected energies: by factors of $\simeq 2$ to $\simeq 4$.

Figure 3 presents the dependence of the resonance frequency ω_{res}/m on the angle of emission of the photon, θ, for $B = B_{cr}$. The direction of the magnetic field is the horizontal axis in the figure; ω_{res} is independent of azimuthal angle. The curves have monotonically increasing resonance frequency in the $\theta = 0$ direction as p_z' increases, and monotonically decreasing resonance frequency in the $\theta = 180°$ direction. The resonance energy is seen to be strongly dependent on both the photon emission angle and on p_z'.

In practice the most common final state is the ground state ($N = 0$), and the most common transitions have the electron in initial state $N' = 2$, spin-down, and

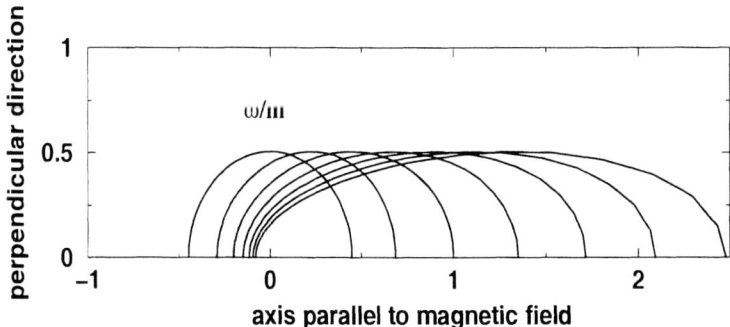

FIGURE 3. The dependence of the resonance frequency on the angle of emission of the photon, θ, for $p_z'/m = 0, 1, 2, 3, 4, 5, 6$ and for $B = B_{cr}$.

have intermediate state $n = 1$. We take this case in what follows. The differential transition rate is a strong function of the magnetic field strength (details to appear in the Phys. Rev. D paper) due to the presence of the resonances. For example, for the case of one photon emitted with energy 58 keV the transition rate is peaked near $0.12 B_{cr}$. This is expected, since that is the field strength for which the resonant energy is 58 keV. We have also calculated the angular dependence of the differential transition rate. For one photon emitted at a specific energy and angle to the magnetic field, and for given parallel momentum of the incoming electron, the rate is a function of the polar angle of the other photon. (details to be in the Phys. Rev. D paper). The rate is maximum when the second photon is emitted in the opposite direction to the first photon.

The differential transition rates also depend on the two photons' polarizations. The rates for the two linear polarizations are nearly equal. The rate for right-circular is much lower than for left-circular, e.g., by a factor of $\sim 10^{11}$ for the case of one photon emitted with energy 58 keV at angle 5° to the field, the incoming electron with parallel momentum of 0, and the second photon emitted at 5° to the field. For this example, the conservation laws give the second photon emitted with energy 53.7 keV, and the final electron parallel momentum as $p_z = -111.2$ keV, i.e., opposite to the magnetic field. The angular dependences of the differential transition rates for different photon polarizations have also been calculated for a number of cases. The resulting polar diagrams are very different for the different photon polarizations (details to be in the Phys. Rev. D paper).

REFERENCES

1. Vasisht, G. & Gotthelf, E., *ApJ* **486**, L129 (1997).
2. Kouveliotou, C., et al., *ApJL* **510**, L115 (1999).
3. Harding, A. & Preece, R., *ApJ* **319**, 939 (1987).
4. Latal, H., *ApJ* **309**, 372 (1986).
5. Semionova, L., M.Sc. Thesis, University of Costa Rica (1983).
6. Herold, H., Ruder, H., & Wunner, G., *A&A* **115**, 90 (1982).
7. White, D., *Phys. Rev. D* **18**, 2166 (1978).
8. White, D., *Phys. Rev. D* **9**, 868 (1974).
9. Bussard, R., Alexander, S., & Meszaros, P., *Phys. Rev. D* **34**, 440 (1986).
10. Johnson, M. & Lippmann, B., *Phys. Rev. D* **76**, 828 (1949).
11. Graziani, C., *ApJ* **412**, 351 (1993).
12. Sokolov, A. & Ternov, I., *Synchrotron Radiation*, Akademie, Berlin, 1968.
13. Alexander, S. & Meszaros, P., *ApJ* **372**, 554 (1991).
14. Herold, H., *Phys. Rev. D* **19**, 2868 (1979).
15. Bjorken, J. & Drell, S., *Relativistic Quantum Mechanics*, McGraw-Hill, New York, 1964.

Electric Magnetars

László Körtvélyessy

Observatory Kleve, D-47533 Germany

Abstract. All magnetic fields are produced electrically, also the field of magnetars. Their new electric model was partly published by the author in the book The Electric Universe in 1998. The new model is based on the positive charge which is collected in the pre-supernovas core via the different masses and sizes of proton and electron. The supernova implosion produces neutrons, but only from proton-electron pairs. The protons in overbalance will be fixed on the surface of the neutron body and create a magnetic field and synchrotron-radiation via rotation. The enclosed proton-bubbles swim upwards. The biggest proton bubbles surface first, and explode into space. This "proton-volcano" kicks the neutron star from the center of the SNR. The later proton-explosions produce only proton-jets.

ELECTRIC NEUTRON STARS

Why does a neutron star have a magnetic field? One of the old answers was: the neutron star inherited its field from the field of the pre-supernova. *Chandra*, however, showed the Crab pulsars two jets in its rotational axis, clearly without precession. The magnetic and rotational axes are identical, but the pulsar pulses. This old model cannot explain this identity and also the non-axial jet (Fig. 1) that was recently discovered by *Chandra*!

PROTON SKIN, WHAT IS IT?

The proton has a 1836 times larger mass than the electron. Because of this asymmetry, the protons have a 43 times lower velocity than the electrons in the same temperature due to the Boltzmann-equation. The core will be positively charged, since more electrons than protons leave the core. (See additionally the Zhang-effect below.) During the forming of the neutron body, already 1 cm^3 of protons in overbalance form a "proton-skin". This mono-proton-layer is fixed to the neutron body by the strong nuclear force (Fig. 2).

A second proton layer is inhibited by the first one. This mono-proton layer also inhibits many black holes, e.g., the total-collapse of the implosion of 20 solar masses. It repulses the inner positive star-masses of the implosion after its birth.

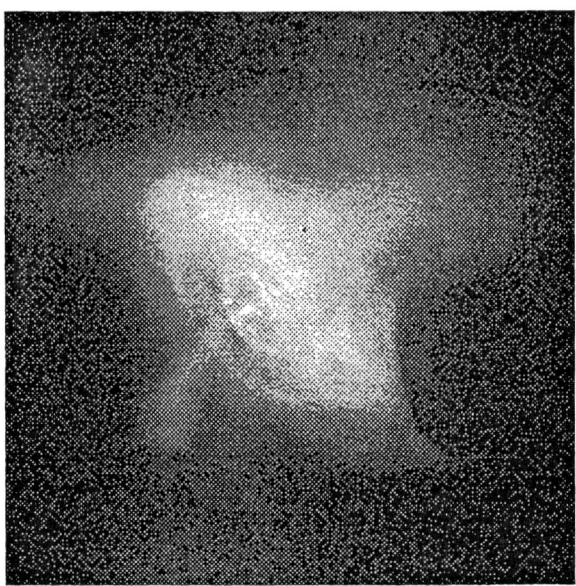

FIGURE 1. *Chandra* shows two jets in the rotational axis and a third, non-axial, jet with circular but unequal cross sections. Only the new electric model explains this X-ray picture. A magnetar with dynamo or with frozen-in fields should produce inducted currents of order of 10^{20} A, which would stop its rotation via Lenz' law and resistance-heat immediately. This more recent idea is also not clear. However, the magnetar's magnetic field is simply produced by its rotation and proton "skin".

The electric charge of the proton skin is $+4 \cdot 10^{20}$ Coulomb, easy to calculate. In rotation, all skin-protons produce a total magnetic field 30 GT, i.e., $3 \cdot 10^{14}$ Gauss for a magnetar. The calculation is simple if we substitute the sphere with a fitting cylinder which is 10 km high:

$$B = \frac{\pi \bullet 4 \cdot 10^{-7} \bullet 3 \cdot 10^{20} A}{10^4 m} = 3 \cdot 10^{10} \text{T} = 30 \,\text{GT}$$

The electric current of a neutron star is on order of 10^{20} A, which does not flow in the body but it rotates with the body in this model! Therefore, this star does not need power for the sustaining of its magnetic field. Only the rotation of the protons emits synchrotron radiation which calculably consumes the rotational energy very slowly and very evenly. This spin-down due to synchrotron radiation is well known for decades without an exact model. Now it became calculable! This positive star produces the cosmic-ray particles by its high voltage of $+10^{25}$ V. (Easy to calculate via sphere-condensator and charge of 1 cm^3 protons!) This highest voltage of the Universe strips the electrons from falling atoms, attracts and fixes them. It repulses the positive atomic nuclei electrostatically along lightyears to relativistic velocities. These are the cosmic-ray particles.

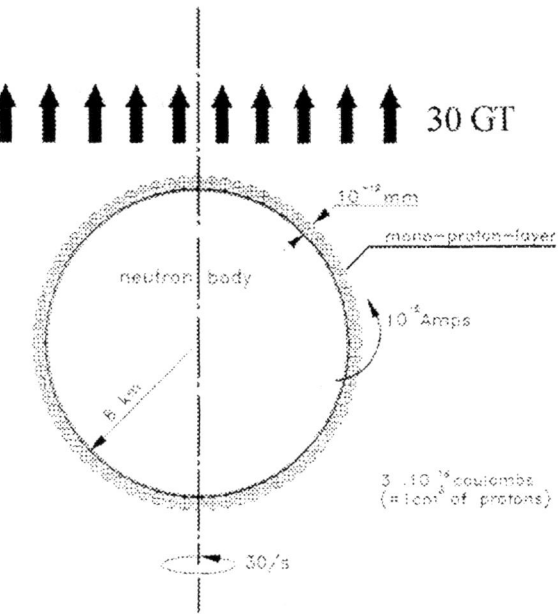

FIGURE 2. The electric model of the neutron stars: the neutron body is covered by a mono-proton-layer which has a simply calculable charge if the protons have the maximal density as shown in this sketch. The star-surface must be divided by the area of one proton. No electric current flows in this star. The rotation of the charge of this proton-skin gives the magnetars field and its synchrotron radiation.

WHAT ARE PROTON VOLCANOES?

The neutron body contains not only 1 cm^3 of free protons on the surface, but often many m^3 of them in the depth. These are partly fixed to the neutrons and to each other, but the strong electrostatic repulsion somewhat expands their volume, therefore they swim slowly upwards. I name them proton "bubbles". They arrive at the surface one after another in thousands of years. Naturally, such proton "bubbles" elevate quicker if they are larger and during a star-quake. A surfaced proton bubble explodes into the empty space. The erupted protons form a filament and fly parallel in it (see the non-axial, cold filament of the Crab above). There are three such filaments in N 49 and 4 in Vela. The proton filament will be repulsed by the proton skin along a distance of lightyears. The result of the strong acceleration is the emission of hard X rays. The rotation of the magnetar shows this proton volcano repeatedly. (Solar proton bubbles explode as X-ray bright points into coronal filaments.)

ACKNOWLEDGEMENT

Prof. S. N. Zhang (Univ. Al. Huntsville, USA) not only confirmed that the temperature gradient separates the electric charges due to the different masses of the protons and electrons — in hydrogen burning stars — but he showed that also the photons push outwards the larger electrons stronger than the smaller protons. This separation of the electric charges is important for all of astronomy! The positive charge overbalance of the stellar core automatically explains all properties of the magnetars.

VII. TECHNIQUES AND MISCELLANY

Attributes of GRB Pulses: An Improved Bayesian Blocks Algorithm for Binned Data

Jeffrey D. Scargle,[1] Jay Norris,[2]
Gabriela Marani,[2,3] and Jerry Bonnell[4]

[1] NASA/Ames Research Center, Moffett Field, CA 94035
[2] NASA/Goddard Space Flight Center, Greenbelt, MD 20771
[3] National Research Council
[4] Universities Space Research Association

Abstract. A procedure to estimate temporal locations, amplitudes, widths, rise and decay times of pulses occurring within a large sample of GRB light curves, using BATSE 64-ms concatenated data, is based on an improved version of Bayesian Blocks. It determines the maximum likelihood value of the number of blocks by marginalizing over all the other parameters (block locations and sizes). The blocks are then used to obtain objective, automatic estimates of pulse parameters, which can either be used to study pulse-attribute correlations or as the starting solution for iterative, nonlinear fits of parametric models.

THE PULSE PARADIGM

Our goal is to characterize burst light curves [6] in order to illuminate the underlying physical mechanisms. One common characteristic of GRB light curves that rises above their bewildering heterogeneity is the frequent presence of one or more sub-structures. Such *pulses* are distinct features, localized in time (possessing well-defined onset, a monotonic rising portion, peak, and monotonic declining portion), and with no apparent fine-time-scale structure. It seems natural that these structures signal the presence of underlying, physically independent events.

A previous Huntsville paper [10] dealt with the BATSE TTE data. Here we discuss four-channel concatenated data, consisting of photon counts in fixed-length time intervals -- bins of length 1.024 s before trigger and 64 ms after. These data can be found at the Compton Gamma Ray Observatory Science Support Center (http://cossc.gsfc.nasa.gov/cossc/batse/). We ignore dead time and other possible departures from the ideal, essentially perfect Poisson counting process model.

BAYESIAN BLOCKS

To address the fundamental difficulties of extracting overlapping pulses from noisy data, we have developed an analysis scheme called Bayesian Blocks [11]. The procedure determines the most likely partitioning of the data into segments, or *blocks*, in which the signal rate is sensibly constant[1]. The transition points between the pieces are conventionally called *changepoints* in the large literature on this topic [1,4,5,7-9,14-17].

The sizes and locations of the blocks are determined by the data in a rigorous statistical way that (1) imposes no limits on time resolution, (2) is semi-parametric (i.e., no parametric model of pulse shape), (3) effectively de-noises the data, and (4) yields immediately useful quantities, such as pulse locations, widths, and heights.

The fundamental step is determining whether a given interval of data should be subdivided into two subintervals or kept as one. The details are given in [11], where the problem of dealing with a whole data series was solved by applying this fundamental step first to the whole data interval, and then recursively to each subinterval until the Bayes criterion favors keeping all the subintervals whole. While this somewhat *ad hoc* procedure gives useful results, it is more correct to treat the multiple changepoint problem directly. Below is an outline of this approach; details will be described elsewhere [12].

THE ALGORITHM

Here or in the iterative "divide and conquer" method of [11], we use a Poisson Model with a constant photon rate for various intervals of time, or blocks. Denote such a model $M(\theta)$, with an array of parameters θ representing the Poisson rate parameters and the change point locations. The fundamental measure of the probability of a model, given the data is [11]

$$p(D|M) = \int p(D|\theta, M) p(\theta|M) \, d\theta . \quad (1)$$

The integration in this formula implements the underlying idea of the straightforward Bayesian prescription for estimating the number of model parameters (e.g., [13], especially §4.2). This *marginalizing of nuisance parameters* provides an Occam razor that remarkably and automatically takes model complexity into account — allowing direct comparison of models of very different form, and with different numbers of parameters.

This *Bayes factor* is useful when comparing models of different form. That is, the relative probability of models M_1 and M_2, using Bayes theorem, is

$$\frac{p(M_1|D)}{p(M_2|D)} = \frac{p(D|M_1)}{p(D|M_2)} \frac{p(M_1)}{p(M_2)} , \quad (2)$$

---

[1] This means, not that the signal is actually believed constant over the block, but that the data at hand are statistically consistent with a signal rate that is constant.

where $p(M_1)$ and $p(M_2)$ are the prior probabilities of the two models. The idea is to marginalize all of the model parameters except N_p, the number of parameters, and then find the maximum likelihood as a function of N_p.

Two major simplifications result in a powerful, general purpose algorithm. First, the likelihood for a series of blocks is just the product of those for the blocks. Therefore *the only likelihood computation needed is that for a single block.* Second, the Poisson rate parameters can be marginalized exactly — not just for time intervals, but for an arbitrary volume B in a space of any dimension.

Assuming a constant-rate Poisson process throughout B (thus distributing points independently and uniformly), the counts in B — denoted X — follow the probability distribution

$$P(X) = \frac{(\lambda V)^X e^{-\lambda V}}{X!} \qquad (3)$$

where λ is the event rate per unit volume, V is the volume of the set B. To specify any prior information about the rate parameter λ, we adopt the *conjugate prior* [2]

$$p(\lambda) = \frac{\beta^\alpha}{\Gamma(\alpha)} \lambda^{\alpha-1} e^{-\beta\lambda} \qquad (4)$$

The parameters α and β can be adjusted to represent the constraints placed on λ by the instrument. This prior yields the Bayes factor

$$p(D|M) = \frac{\beta^\alpha V^X}{\Gamma(\alpha) X!} \int_0^\infty \lambda^{X+\alpha-1} e^{-\lambda(V+\beta)} d\lambda , \qquad (5)$$

or

$$p(D|M) = \frac{\beta^\alpha V^X}{\Gamma(\alpha)} \frac{\Gamma(X+\alpha)}{(V+\beta)^{X+\alpha}} , \qquad (6)$$

where as usual the model-independent factor $X!$ is discarded.

The marginalization of the location parameters (i.e., the discrete bin indices) requires a sum over a high-dimensional space. A simple implementation of the standard tool in such cases, Markov Chain Monte Carlo [3], works surprisingly well — probably because these data, integer counts with low dynamic range and noise exactly describable by the Poisson distribution, are relatively well behaved.

The resulting posterior probability, as a function of N_p (here equal to the number of changepoints), peaks sharply at a good estimate of the number of changepoints — so sharply that any reasonable prior distribution for N_p suffices. Determining the values of the other parameters, namely block locations and heights, requires little more than keeping track of the maximum likelihood computed during the Monte Carlo exploration of parameter space. *Integration and optimization are the same problem, computationally.* More sophisticated procedures, such as averaging the model parameters with the likelihood acting as a weighting function, involve only more manipulations of the same computed quantities.

We will report elsewhere [12] on an extensive program of GRB modeling.

REFERENCES

1. Chib, Siddhartha, "Estimation and Comparison of Multiple Change Point Models," *Journal of Econometrics* **85**, in press (1998).
2. Gelman, A., Carlin, J., Stern, H., and Rubin, D., *Bayesian Data Analysis*, Chapman & Hall, New York, 1995.
3. Gilks, W. R., Richardson, S., and Spiegelhalter, D. J., editors, *Markov Chain Monte Carlo in Practice*, Chapman and Hall, London, 1996.
4. Gustafsson, F., "Segmentation of signals using piecewise constant linear regression," *IEEE Transactions on Signal Processing* **46**, in press (1998).
5. Gustafsson, Fredrik, "A Change Detection and Segmentation Toolbox for Matlab," Linköping University Technical Report LiTH-ISY-R-1669, http://www.control.isy.liu.se/cgi-bin/reports?author~Gustafsson (1998).
6. Norris, J., Nemiroff, R., Bonnell, J., Scargle, J., Kouveliotou, C., Paciesas, W., Meegan, C., and Fishman, G., *ApJ* **459**, 393 (1996).
7. Ogden, R. Todd, "Wavelets in Bayesian Change-Point Analysis," preprint (1997).
8. Ogden, R. Todd and Parzen, Emanuel, "Data Dependent Wavelet Thresholding in Nonparametric Regression with Change-point Applications," preprint (1997).
9. Ogden, R. Todd and Parzen, Emanuel, "Change-point Approach to Data Analytic Wavelet Thresholding," preprint (1997).
10. Scargle, J. Norris, J, and Bonnell, J., in *Gamma-Ray Bursts: 4th Huntsville Symposium*, ed C. A. Meegan *et al.*, AIP, New York, 1998, pp. 181.
11. Scargle, J., "Studies in Astronomical Time Series Analysis V. Bayesian Blocks, A New Method to Analyze Structure in Photon Counting Data," *ApJ* **504**, http://xxx.lanl.gov/abs/astro-ph/9711233 (1998).
12. Scargle, J., Norris, J, Bonnell, J., and Marani, G., "Attributes of GRB Pulses: Bayesian Blocks and Pulses for a Sample of 1661 Long Gamma Ray Bursts," in preparation (2000).
13. Sivia, D. S., *Data Analysis: A Bayesian Tutorial*, Clarendon Press, Oxford, 1996.
14. Smith, A. F. M., "A Bayesian approach to inference about a change-point in a sequence of random variables," *Biometrika* **62**, 407 (1975).
15. Stark, J. Alex, Fitzgerald, William J., and Hladky, Stephen B., "Multiple-order Markov Chain Monte Carlo Sampling Methods with Application to a Change-point Model," Cambridge University Technical Report CUED/F-INFENG/TR, 302 (1997).
16. Sugiura, N. and Ogden, R. T., "Testing Change-points with Linear Trend," preprint (1997).
17. West, R. Webster, and Ogden, R. Todd, "Continuous-time Estimation of a Change-point in a Poisson Process," preprint (1997).

A GRB Tool Shed

David J. Haglin[†] Richard J. Roiger[†], Jon Hakkila[†],
Geoffrey Pendleton[‡], and Robert Mallozzi[‡]

[†]*Minnesota State University, Mankato, MN 56001*
[‡]*University of Alabama, Huntsville, AL 35899*

Abstract. We describe the design of a suite of software tools to allow users to query Gamma Ray Burst (GRB) data and perform data mining expeditions. We call this suite of tools a shed (**SH**ell for **E**xpeditions using **D**atamining). Our schedule is to have a completed prototype (funded via the NASA AISRP) by February, 2002. Meanwhile, interested users will find a partially functioning tool shed at http://grb.mankato.msus.edu.

INTRODUCTION

We are implementing a suite of software tools to aid Gamma-Ray Burst (GRB) researchers in working with the GRB data. The major features of the tool shed — a **SH**ell for **E**xpeditions using **D**atamining — are a web-based data query facility, web-based data visualization capability, and a web-based interface to data mining software tools. The tool shed maintains a database of users allowing each user to store their own work at the tool shed site. Each user's data will not be visible to other users of the system.

Our GRB tool shed is populated with a standard set of preprocessed GRB data such as the basic table data, flux/fluence data, and duration [1]. These data are stored in tabular form as rows (burst instances) and columns (attributes). Each burst has the same attributes as all other bursts, with a provision for indicating "missing" attributes. Users may augment this database by uploading their own table or performing SQL database queries for data selection and calculations.

The data mining tools can be given any of the queried data and produce either rules for classification of bursts or an identification of classes of bursts by identifying which bursts belong to which class. Note that these "identified" classes are not necessarily classes based on physical properties of bursts; they may be due to instrumental bias, or even statistical fluctuations from the small numbers of instances (c.f. [2]). This step can lead to defining new data in new tables for further exploration. An example of applying these tools is given in [3].

Data Storage

Once a registered user has logged in, the data can be manipulated in a variety of ways. The user may simply query the existing data, selecting a subset of the bursts and/or attributes for later processing. They may augment the data by uploading their own attributes for existing bursts. Or they may augment the data by uploading information for new bursts.

As an example database query, consider selecting all bursts from the 4B catalog with high relative measurement errors on the channel 1 fluence measurement. To determine what might be considered "high", it may be necessary to see all of the bursts sorted by relative measurement error, easily done with this SQL query:

```
SELECT burstnum, channel1FluenceError FROM 4BFluxTable
WHERE channel1Fluence <> 0
ORDER BY (channel1FluenceError/channel1Fluence);
```

Now, after viewing the results of this query, one may decide that a "quality" threshold of 1.0 standard deviations would be required on the fluence data. The quality data can be extracted using this query:

```
SELECT burstnum, channel1Fluence, channel1FluenceError
FROM 4BFluxTable WHERE channel1Fluence <> 0
AND (channel1FluenceError/channel1Fluence) > 1.0;
```

For those users unfamiliar with SQL, the web application will provide point-and-click, fill-in-the-box forms for generating a database query. These web pages should provide the user with most of the features they would be interested in. There will also be a blank form for users who may wish to enter the SQL query directly. Either way, the names of the tables and attributes will be shown on the web browser so the user need not memorize them.

Data Visualization

At any time the user may decide to invoke visualization tools to help "see" the data in their scratch area or in the system tables. The web application will send these requests to ION (IDL On the Net) to produce graphical views of the data. The display of these graphs will be done on the user's web browser window.

Data Mining Tools

Several data mining tools will be available as part of the web application. Initially, there will be the classic classification tool, C4.5 [4], a tool developed by a member of our group, ESX [5], and at least one Neural Network package. These software packages run on the web server and the web application will guide the user, requesting information needed by the specific data mining tool being invoked.

The output from the data mining tools differ from one tool to the next. Our web application will be able to capture the output and transform it to an internal "rule" format. Once captured, the user may view the rules, or go even one step further by applying the rules to a database in their scratch area. The application of the rules to the data is a significant feedback component of the tool shed.

Online Help

There will be extensive online help to guide the user through the data mining process. The help system will be written in a hyper-linked book format complete with a table of contents, an index, and a search engine. There will also be context-sensitive help in the sense of hyper-links from web forms to relevant pages of the help system. The help system will address how to use the tool. The GRB data will be minimally documented, with pointers (hyper-links) to existing help in understanding the meaning of the data attributes.

A tutorial will be provided that brings the user through a data mining session. This tutorial will provide (scientific) rationale for selecting options along with way. Although the tool shed need will not be specific to GRB data, the tutorial will be.

GRB TOOL SHED SYSTEM PLATFORM

The GRB tool shed will initially run on a pentium-class computer system running RedHat Linux. To maximize portability, the Java language will be used. Since this tool is web-based, all of the Java code will run as Java Servlets. The Apache web server will be used along with its companion Java Servlet Apache-JServ. The GRB data will be stored in a PostgreSQL database with access to the Java code provided via JDBC. All of these software packages are available on many platforms, including Unix, Windows, and Mac, and they are all freely available. The flow of information through these packages is shown in Figure 1.

This whole process is initiated by aiming a web browser at the appropriate URL (http://grb.mankato.msus.edu/) where the Apache web server is configured to start a Java Servlet via Apach JServ. The Java Servlet will then make requests to the PostgreSQL database via the JDBC package that comes with PostgreSQL. The major development effort for this application is in the creation of the Java Servlet code, which we call "the web application."

GRB TOOL SHED DATA FLOW

The Java Servlet application is a complex set of Java code with many data structures and interfaces. There are two major data formats that will be used: Standard User Interface Format (SUIF) and Standard Internal Classifier Format (SICF). The SUIF will be used when presenting data to the user on the client

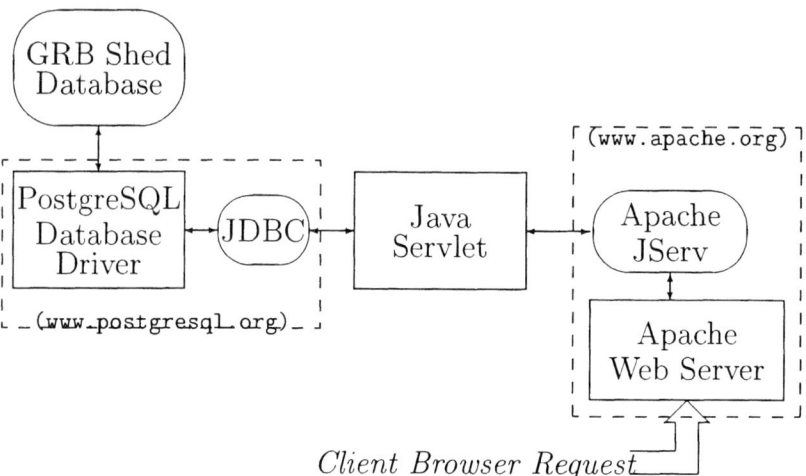

FIGURE 1. Information Flow

machine (web browser) and SICF will be used when presenting data to any of the supported classifiers (data mining tools). The SICF format is proposed as a tool-independent representation format for holding all information necessary to conduct a classification/data mining run.

We expect to provide a user interface on the web browser that looks very much like a spreadsheet program. This familiar view will allow the user to inspect and possibly update data easily.

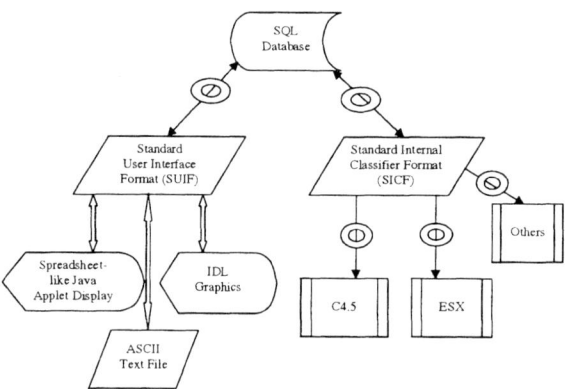

FIGURE 2. Data Flow

Note that a "donut" object indicates a data format conversion is needed at that point. Some of these data format conversions may require additional information, which would require the use of web pages to interact with the user. And the wide arrows indicate data flowing across the internet from server to client (browser).

The control of when and where the data flows is completely independent of this diagram. Imagine a controller sitting above this page directing the data to flow along the various paths of the diagram. This controller is a web-based menu system under the direction of the user.

WEB PAGES (CONTROL FLOW)

The design of this perspective of the GRB Tool Shed is least developed at this time. It is clear there needs to be a login menu screen. Once that information has been verified, the user will be placed in a main menu/dispatcher web page. This page will allow the user to click on various functions on a menu bar either along the side, the top, or the bottom. The menu selections/actions might be:

- graph/visualize data
- show data in spreadsheet form
- initiate upload/download of data
- work with the data (query database)
- initiate classification
- view documentation

Each of these items constitutes a large implementation effort. For example, the "work with the data" selection initiates a sequence of web pages/forms where the user is required to make selection criteria for the various rows and columns of the data. The "initiate classification" selection will cause the user to be prompted for which classifier tool to invoke, then guide the user through selecting parameter values that are specific to that tool.

CONCLUSIONS

With the creation of this software comes a powerful research tool capable of automating many aspects of manipulating GRB data. Our goal is to go beyond the tool creation and build up the collection of attributes about the GRBs, emerging as a significant repository of GRB information with a built-in efficient methodology.

Our implementation strategy is to incrementally develop components. That way, as the development progresses, the web application will contain some usable software and data.

REFERENCES

1. Paciesas, W.S., et al., *ApJ* **519**, 206 (1999).
2. Hakkila, J., et al., *APJ* submitted.
3. Hakkila, J., et al., these proceedings.
4. Quinlan, J.R., *C4.5: Programs for Machine Learning*, San Francisco: Morgan Kaurmann, 1993.
5. Roiger, R.J., et al., "ESX – A Tool for Knowledge Discovery", in *Proc. of the Federal Data Mining Symp. & Exposition '99*, edited by W.T. Price, AFCEA International, publisher, Fairfax VA., 1999, pp. 109–120.

Properties of Karhunen-Loeve Expansion of Astronomical Images in Comparison with Other Integral Transforms

Petr Páta*, Martin Bernas*, A. J. Castro-Tirado[†], and René Hudec[‡],

*Czech Technical University Prague, Department of Radioelectronics,
16627 Prague, Czech Republic,
[†]Laboratorio de Astrofísica Espacial y Física Fundamental (LAEFF-INTA),
P.O. Box 50727, E-28080, Madrid, Spain,
Instituto de Astrofísica de Andalucía (IAA-CSIC), P.O. Box 03004, E-18080 Granada, Spain,
[‡]Astronomical Institute, Academy of Sciences of Czech Republic,
25165 Ondrejov, Czech Republic

Abstract. This work deals with the use of Karhunen-Loeve Expansion (KLE) for astronomical image compression. A mathematical algorithm of KLE has been derived using characteristics of $(N \times N)$-dimensional Hilbert space. The KLE frequency domains and their reductions leading to image compression have been studied in comparison with other integral transforms. The following important results have been obtained during accuracy and suitability testing of KLE in astronomical photometry measurements. The achieved results show that KLE is suitable and probably the best method of astronomical image compression.

INTRODUCTION

New technical parameters of CCD image sensors bring not only high accuracy of optical measurements, but also a huge amount of data. Because of the limited capacity of transmission channels and archive storage media, it is important to find compression algorithms appropriate for supposed image data processing. Standard methods, like the well-known JPEG algorithm, reach an excellent compression ratio, but for our purposes loose too much important information. These standards were developed with respect to physiological characteristics of the human eye and are not appropriate for coding of scientific image data. Such a lossy reduction of an image is not noticeable to the eye, but introduces significant errors in mathematical processing. The solution of these problems usually involves using a lossless compression method, but these methods have much lower compression ratios (usually 1:2) [1]. Another method makes use of an integral transformation followed by image processing, which is more appropriate for scientific image data.

Examples of such a possibility involve using wavelet, fractal, or Karhunen-Loeve (KLE) transforms. In this work, the characteristics of KLE are compared with two other integral transforms based on Hadamard and cosines orthogonal bases.

IMAGE MATRIX SPLITTING

The image data from BOOTES experiment were used for this work [2]. The experiment uses a ST8 CCD camera equipped with two types of lenses — D = 50 mm, f/1.2 (Nikkor Japan) and D = 0.3 m, f/3.3 (Meade LX 200).

To reduce computational costs it is suitable to split the input image into sets of square submatricies. In the following, we describe an input image $\overleftrightarrow{X}$ with dimensions $N_1 \times N_2$. The distribution operator R transforms $\overleftrightarrow{X}$ into the set of M submatricies with dimensions $N \times N$

$$\overleftrightarrow{X} \xrightarrow{R} \{|x_j^i]^\alpha\}_{i,j=1,\alpha=1}^{N,M} , \tag{1}$$

where $N_1 \cdot N_2 = N \cdot N \cdot M$. Index i is the column number, j the line number and Greek letters are used for the submatrix index (i.e., the realization).

Submatricies $\{|x_j^i]^\alpha\}_{i,j=1,\alpha=1}^{N,M}$ are elements of abstract Hilbert space V_N^N dimension equal N^2 over the body of complex numbers $\Im$.

INTEGRAL DISCRETE TRANSFORM

Let us define scalar multiplication in the space V_N^N of image submatrix

$$\sum_{i,j=1}^{N} [y_j^i\|x_j^i] = \lambda , \tag{2}$$

where $[y_j^i|$ are elements of dual vector space (formalism of description is known from quantum theory or from linear algebra). The known orthonormal relations are valid in the space V_N^N and in our case it is possible to write for dual vectors

$$[x_j^i| = |x_j^i]^H . \tag{3}$$

In the space V_N^N can also be found by the standard way the basis — $N \times N$ basis vectors — which confirm the orthonormal relation

$$\sum_{i,j=1}^{N} [\{\phi_j^i\}_\alpha^\beta \| \{\phi_j^i\}_\gamma^\delta] = \delta_\gamma^\beta \delta_\alpha^\delta . \tag{4}$$

Using these relations it is possibile to define a two dimensional integral transformation of the vector $|x_j^i]$ in space V_N^N by the following set of equations:

$$|y_k^l| = \sum_{i,j=1}^{N} [\{\phi_l^k\}_j^i \| x_j^i] \quad l, k = 1, \ldots, N \tag{5}$$

$$|x_j^i| = \sum_{l,k=1}^{N} |\{\phi_l^k\}_j^i\| y_k^l] \quad i, j = 1, \ldots, N. \tag{6}$$

BASIS OF SPACE V_N^N

The Karhunen-Loeve basis [3] is the only orthogonal basis that allows full decorrelation of an image. This transformation is sometimes referred to as Hotteling transform or PCA (Principal Component Analysis).

Transformation is based on statistical properties of image data. By the distribution R, we split the image matrix into a set of M submatricies — elements of space V_N^N. Let every such submatrix be one realization of an $N \times N$ dimensional random process. The statistical mean value of each realization is

$$|\bar{x}_j^i| = E\{|x_j^i|^\rho\}_{\rho=1}^M = \sum_{\rho=1}^{M} |x_j^i|^\rho p(\rho) \quad i, j = 1, \ldots, N, \tag{7}$$

where $p(\rho)$ is the probability of realization. We can write for the correlation between realizations

$$\Xi_{in}^{jm} = E\{(^\rho[x_j^i] - [\bar{x}_j^i|)(|x_n^m|^\rho - |\bar{x}_n^m|)\}_{\rho=1}^M \quad i, j, m, n = 1, \ldots, N. \tag{8}$$

In terms of the mean square error (MSE), the best orthonormal basis of the space V_N^N can be built from the eigenvectors

$$\sum_{i,j=1}^{N} \Xi_{nj}^{im} |\{(\Phi_i^j)_s^r\}| = \beta_s^r |\{(\Phi_n^m)_s^r\}| \quad m, n, r, s = 1, \ldots, N. \tag{9}$$

In this work, the comparison of KLE properties to those of the discrete cosine transform (DCT) and the Hadamard transformation has been done [3]. Both transformations are separable, and have a symmetric transform kernel. The basis functions of DCT are built from $\cos(ax)$, whereas the Hadamard transform is generated from rectangular pulses of different width, and KLE has the basis composed of eigenvectors of the covariance matrix (eqn. 9).

COMPARISON OF PROPERTIES OF INTEGRAL TRANSFORMATIONS

During the comparison we reduced a number of quantization steps in the spectral plane of the integral transforms [3], starting with 10^5 possible quantization levels (16-bit quantization depth) and during the reductions we measured following parameters:

1. *Mean square error between original and reconstructed image δ (%).*
   This parameter expresses the absolute error of reconstruction.

2. *Peak signal-to-noise ratio PSNR (dB).*
   PSNR describes well the quantity of noise in measured signal (Figure 1, left).

3. *Photometric measurement of magnitude m (mag).*
   It shows, how the reduction changed the photometric measurement of magnitude (Figure 1, right).

For this analysis we choose the image M42.d03 taken 19/3/1999 21:01:21 UT by the BOOTES system. The CCD camera was placed in the focal plane of the Meade telescope [2] for a total exposure time of 100 s in 1:1 binning mode. The image matrix was split into blocks with dimension $N = 32$, so the number of realizations of random effect is $M = 1536$. For photometric measurement we choose the dimension of the mask to be 13 pixels.

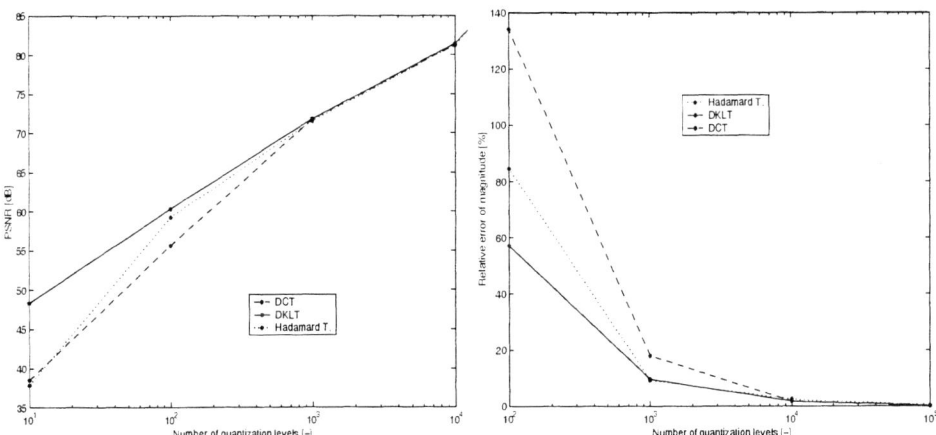

FIGURE 1. Left: PSNR of reconstructed image as a function of quantization levels number. **Right:** Relative error of magnitude of reconstructed image as a function of quantization levels number.

CONCLUSIONS

The comparison was made for Hadamard, DCT and KLE integral transforms. The MSE parameter does not show any significant difference between these transformations. Above 10^4 quantization levels, the measurement results are practically identical. The peak signal-to-noise ratio clearly shows the advantages of KLT, where the difference for low numbers of quantization levels is over 10 dB. The relative error of photometric measurement also shows the advantages of KLT.

The results of this work show two possible ways of solving the problem of scientific image data compression. Limitation of spectral coefficients to the $\gtrsim 10^4$ quantization levels does not introduce any significant error. For higher compression rates it is possible to use KLT with lower quantization depth and for coding to use also differential image matrix. This solution will bring a higher compression ratio and lower processing errors.

ACKNOWLEDGMENTS

A part of this work has been conducted at the Department of Radio Engineering of the Faculty of Electrical Engineering of The Czech Technical University in Prague in the frame of the research project "Digital Signal Processing Application in Radio Engineering" and has been supported by grant No. 102/98/1464 of the Grant Agency of Czech Republic. This research work also has been partially supported by the research program No. J04/98:212300014 "Research in the Area of Information Technologies and Communications".

REFERENCES

1. Bernas M., Páta P., Hudec R., Rezek T., 3rd INTEGRAL Workshop, Taormina, in press, (1998).
2. Castro-Tirado, A. J., Soldán J., Bernas M., Páta P., Rezek T., Hudec R., Sanguino T., M., De La Morena B., Berná B., Rodriguez J., Pena A., Gorosabel J., Ms-Hesse J., Giménez A., *A&AS* **138**, 583C (1999).
3. Pratt W. K., *Digital Image Processing*, 2nd editon, John Wiley & Sons (1991).

Fast and Simple Data Compression

Martin Nekola and Martin Bernas

*Faculty of Electrical Engineering,
Czech Technical University, Prague, Czech Republic
e-mail: M.Nekola@atlas.cz, bernas@feld.cvut.cz*

Abstract. Digital data transmission or data storage are branches often using data compression. In this paper we describe a simple and fast lossless compression algorithm for CCD images of star fields which reduces the amount of data by roughly 50%.

INTRODUCTION

Data compression is a process often used for digital transmission (e.g, from IN-TEGRAL to ground base) and storing information. Its purpose is to reduce the amount of data without loss of either any or important informations.

Our purpose was to find a fast and low-memory algorithm for lossless data compression of astronomical CCD images of star fields. The algorithm was developed for use with INTEGRAL OMC, but is also useful for other applications.

We tested several combinations of different methods such as RLE, packed bits, and distinguishing bits. The highest level of data reduction (roughly 50%) was reached by using the algorithm described below.

THE PRINCIPLE

This simple and fast compression algorithm is based on the DPCM of 12-bit input words sequence. After the DPCM, the small numbers ranging from -2^{N-1} to $+2^{N-1}-1$ (N-bit numbers) are stored using only N bits, other data are introduced by the label and then split into $\lceil 16/N \rceil$ N-bit words[1]. The label must be one number from the range -2^{N-1} to $+2^{N-1}-1$, mostly the minimum from this range is used (for $N = 5$, the label is 16), then also this number must be stored adding the label. See Figure 1.

For astronomical images, the best choice for N is usually $N = 5$. The reduction approaches 50% for images of star fields. See Table 1. The values in the table were

---

[1] $\lceil x \rceil$ - rounding up of x

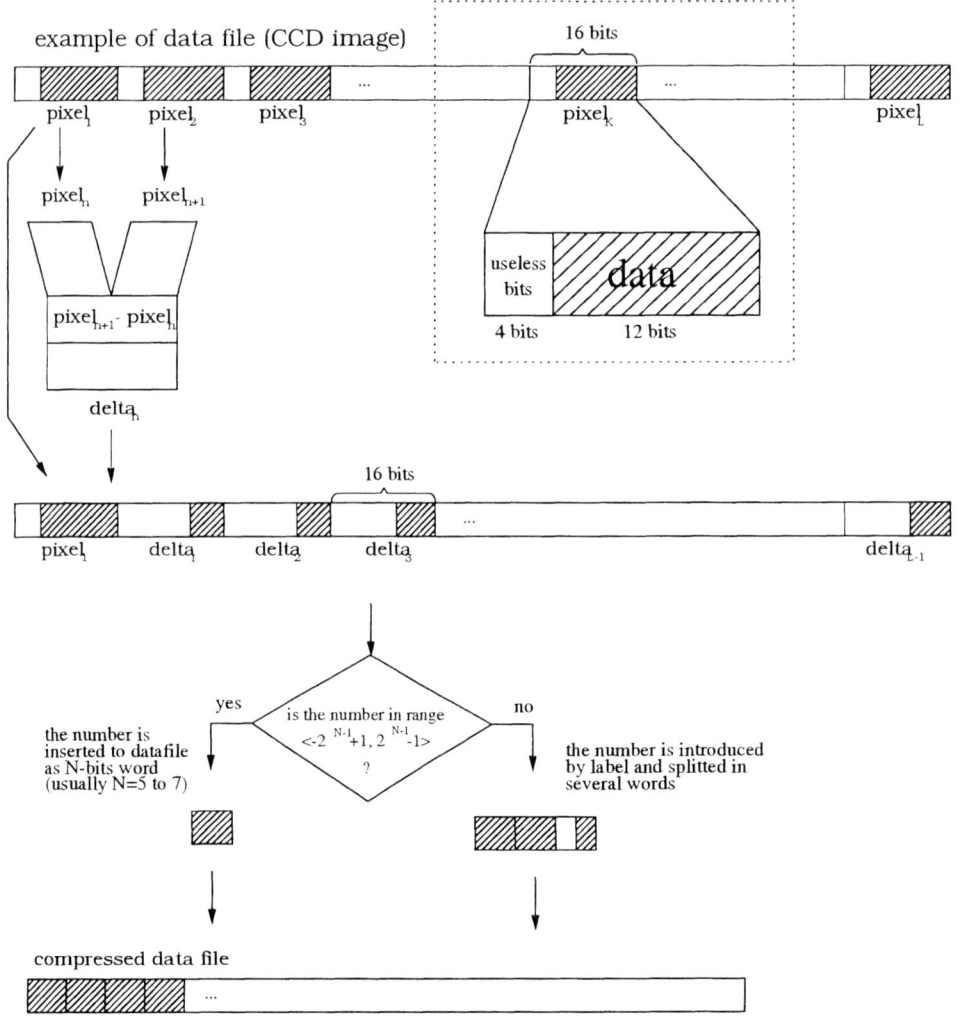

FIGURE 1. Processing of a data file by the compression algorithm.

TABLE 1. Percent reduction of data files depending on N.

filename	3b	4b	5b	6b	7b	8b	9b
Califo_12.fit	116.6	72.2	52.8	53.7	60.3	67.4	75.4
M87_an_12.fit	92.4	49.4	45.0	51.5	59.2	67.0	75.1
North_12.fit	129.1	89.3	67.7	58.7	62.7	68.0	75.5
com_h_12.fit	115.2	71.6	54.5	54.3	60.4	67.4	75.4
omcc1_12.fit	101.9	57.9	48.3	52.6	59.7	67.1	75.2
t100s_12.fit	83.6	46.0	46.0	51.7	59.3	67.0	75.2
t200s_12.fit	107.0	63.2	51.2	53.7	60.3	67.3	75.3
t20s_12.fit	37.3	34.8	42.4	50.3	58.5	66.7	75.0
t50s_12.fit	58.5	37.8	43.8	50.9	58.8	66.8	75.1
t5s_12.fit	26.7	33.6	41.8	50.1	58.4	66.7	75.0
test_12.fit	107.0	63.2	51.2	53.7	60.3	67.3	75.3
average:	88.7	56.3	49.5	52.8	59.8	67.2	75.2

obtained using the following equation:

$$C = \frac{12S_c}{16S_o}100[\%] \ ,$$

where S_c is the size of the compressed file and S_o is the size of the input file. The pixels in tested input files had a depth of 12 bits per pixel, but each was stored in two bytes.

All programs were written in C and tested under Linux.

Accuracy of Press Reports on Gamma-Ray Astronomy

Bradley E. Schaefer[a], Robert J. Nemiroff[b], and Kevin Hurley[c]

[a] *Yale University, Physics Department, 260 Whitney, New Haven CT 06511*
[b] *Michigan Tech. Univ., Department of Physics, 1400 Townsend Dr., Houghton MI 49931*
[c] *Space Sciences Laboratory, University of California, Berkeley CA 94720-7450*

Abstract. Most Americans learn about modern science from press reports, while such articles have a bad reputation among scientists. We have performed a study of 148 news articles on gamma-ray astronomy to quantitatively answer the questions "How accurate are press reports of gamma-ray astronomy?" and "What fraction of the basic claims in the press are correct?" We have taken all articles on the topic from five news sources (UPI, New York Times, Sky & Telescope, Science News, and five middle-sized city newspapers) for one decade (1987–1996) We found an average rate of roughly one trivial error every two articles, while none of our 148 articles significantly mislead the reader or misrepresented the science. This quantitative result is in stark contrast to the nearly universal opinion among scientists that the press frequently butchers science stories. So a major result from our study is that reporters should be rehabilitated into the good graces of astrophysicists, since they actually are doing a good job. For our second question, we rated each story with the probability that its basic new science claim is correct. We found that the average probability over all stories is 70%. Since the reporters and the scientists are both doing good jobs, then why is 30% of the science you read in the press wrong? The reason is that the nature of news reporting is to present front-line science and the nature of front-line science is that reliable conclusions have not yet been reached. The combination of these two natures forces fast breaking science news to have frequent incorrect ideas that are subsequently identified and corrected. So a second major result from our study is to make the distinction between textbook science (with reliabilities near 100%) and front-line science which you read about in the press (with reliabilities near 70%).

INTRODUCTION

"How accurate are press reports?" This is an important question since most Americans get most of their science from the press. Yet most scientists have a low opinion regarding the accuracy of popular science reporting. For example, Tankard & Ryan [1] found that 70–80% (depending on the exact question being asked) claim that science reporting is generally bad. Previous attempts [2–4] to quantitatively answer this question are old. A typical (and valid) procedure was to send press clippings to the quoted scientist for evaluation and error spotting [5].

"What fraction of the basic new science claims in the press are correct?" This second question is independent of the first, and is more to the point for newspaper readers. The reader should know the average confidence level for their new knowledge. We have quantified this confidence level as a probability that the basic science claim being presented by the article is correct, based on our evaluation of all evidence. Science is a good field for evaluating the overall correctness of the press because (a) science has objective standards for evaluating claims, (b) scientists aggressively test claims, and (c) political or social biases are minimized.

STUDY DESIGN

Each of the three authors independently analyzed each of 148 papers on the topic of "Gamma-Ray Astronomy" from one decade (1987–1996) from each of five popular news sources (Sky & Telescope, Science News, the New York Times, UPI, and newspapers in five middle-sized cities). The analyzer did not analyze any article for which for which they were involved or for which they felt they had inadequate knowledge. Articles primarily discussing 'instrumentation' were not used.

We evaluated each article on many issues, such as importance, how 'new' the result was, whether logic/evidence was presented, was the report placed in context, and whether the primary uncertainty was noted. Two primary sets of evaluations were identifying any error made by the reporter (important or 'trivial') and estimating a confidence level that the basic science claim being made by the article is correct (both as evaluated at the time the article was written and now).

The study reported here is just one of three identical studies, each with different topics. In addition, the topics of "supernovae" (with reviewers David Branch, Saul Perlmutter, and Brad Schaefer) and "Mars" (with reviewers Hap McSween, Guy Consolmagno, Robert Strom, and Martha Schaefer) have analyzed 402 articles. All results from our study of gamma-ray astronomy articles are confirmed by these two extensions.

One way to measure the consistency of evaluations (and indeed on the existence of a coherent 'community opinion') is to find the correlation coefficient between evaluations from independent experts. That is, if the correlation is poor then a community opinion does not exist, while if the correlation is good then our evaluations are likely to be close to some objective weighting of the available evidence. The overall correlation coefficient between all pairs of evaluations is $r=0.73$. The differences in evaluated confidence levels has a standard deviation of 25% (implying an underlying uncertainty for each person's evaluations of 18%). These values should be $r=0$ and a standard deviation of 41% if our evaluations were random. We take this as evidence that a consistent community opinion does exist and our evaluations are a reasonable measure of it. For our combined estimates from three reviewers, the result should have a typical uncertainty of 10%.

ACCURACY OF THE PRESS

What fraction of the articles have significant errors that are the fault of the reporter? Did the article not represent the basic claim of the scientist correctly? Was the report confused or give such a poor basic explanation that the reader cannot grasp the basic claim? Did the reporter make some grievous error of fact that changes the result?

Out of our 148 articles, none significantly misrepresented the science and none presented any grievous factual errors and none were muddled. This is to say that the reporters basically did a good job in 148-out-of-148 articles. This systematic and quantitative result is in contrast to the expectations of scientists that $\sim 30\%$ of press science articles are significantly bad. We believe that much of this discrepancy arises because (a) the few truly bad articles are well-remembered while the many good articles are not noted, and (b) scientists often react to the article negatively when what they really don't like is the science that the reporter is faithfully reporting.

A 'trivial' error is one that does not change the science or the result in any significant manner. Examples include calling the constellation Chamaeleon a galaxy, saying that the detectors were cooled to near absolute zero, and attributing data to the wrong GRO instrument. While these errors are insignificant for the purposes of a lay reader, a knowledgeable scientist will harp on them and take them as indicators of overall accuracy. In all, we identified 46 'trivial' errors in 31 articles. Both Sky & Telescope and Science News had virtually a zero rate of error, while the newspapers all had rates close to 0.5 errors per article. This is astoundingly good given the pressure of deadlines.

Many of the complaints about science reporting are actually about errors of omission. Characteristic complaints are that co-authors are not mentioned, detectors are not explained, and that the original discoverer is not identified. But these omissions are fully understandable from the point of view of the lay reader for whom these details are irrelevant. But there are other more serious errors of omission. These are where the reporter does not present the logical/evidential basis for the claim, does not place the claim in context, does not identify the primary uncertainties, and does not indicate the size of the uncertainties. To quantify each of these omissions, we have graded each article on a 0–1 scale. So an article that gives adequate details of the basis for a new science result would get a '1' while an article that discusses relevant historical background but not the meaning of a result might get a '0.5'. Roughly, our average values will be the fraction of articles doing good.

From the results tabulated in Table 1, we see a steady progression in errors of omission from sources that have significant space for detailed reporting to sources where perhaps 300 words must describe a complex science issue. We also see a steady progression of errors of omission from sources with monthly deadlines to sources with daily deadlines.

TABLE 1. Fraction of Articles with Errors of Omission.

	Logic/evidence given?	Placed in context?	Uncertainty discussed?	Error bars quoted?
OVERALL	0.65	0.66	0.26	0.04
Sky & Telescope	0.82	0.69	0.34	0.09
Science News	0.64	0.69	0.28	0.00
New York Times	0.55	0.60	0.17	0.05
UPI	0.56	0.64	0.21	0.06
5 newspapers	0.37	0.60	0.11	0.00

WHAT FRACTION OF THE BASIC CLAIMS ARE CORRECT?

Each article has had its basic science claim evaluated as to the confidence level (a probability of the basic result being correct) both at the time the article was written and currently. Our results are tabulated in Table 2.

Our conclusions from these results are: (1) About 30% of what you read in the press on gamma-ray astronomy is wrong. (2) All magazines and newspapers are equally reliable, likely because they all choose stories primarily from the same set of stories which are made prominent by various groups [chiefly AAS, Nature, and NASA]. (3) Contrary to our strong expectations, we found that the highest average confidence levels were given for articles based on press conferences and press releases, while the lowest average confidence levels were given for articles based on refereed journal papers (with ApJ at the bottom).

DISCUSSION

Our key results can be summarized as "Science reporters are doing a remarkably good job" and "About 30% of the basic science claims that you read about in the press are wrong." But this gives a paradox, since how can published results be 30% wrong if both the reporters and the scientists are doing good jobs?

The key to this paradox is your expectations. Commonly, the public (and scientists) expect that the newspapers should be 100% reliable. This expectation is perhaps based on an assumed authority, the examples of textbooks, and hope. However, the press is not reporting textbook science (this is not 'news'), instead the press reports late-breaking front-line science. Front-line science, by its nature is highly uncertain, with an overall reliability of perhaps 70%.

The public should be aware that what they read regarding science is $\sim 70\%$ confident (as appropriate for front-line science) and is not $\sim 100\%$ confident (as generally expected). This lack of awareness often leads to confusion and disillusionment among the public when they read claims and counterclaims.

TABLE 2. Confidence Levels that Basic Science Claim in Article is Correct.

Category	When Written	Currently	Comments
OVERALL:	70%	68%	
SOURCE:			
Sky & Telescope	66%	65%	No dependence
Science News	77%	71%	on source
New York Times	68%	69%	
UPI	74%	67%	
5 Newspapers	68%	70%	
TYPE OF NEW RESULT:			
Observational	75%	74%	Strong effect
Model proposal	32%	19%	on type of claim
Theoretical	50%	45%	
Review	90%	91%	
OUTSIDE COMMENTATOR:			
Pundit quoted	70%	71%	Quoting pundit does
No pundit quoted	70%	67%	not give reliability
IMPORTANCE OF CLAIM (if true):			
Most important half	66%	61%	Important claims
Least important half	74%	75%	tend to be daring
HOW NEW OR ORIGINAL IS CLAIM:			
'New'	67%	65%	No dependence on
'Old'	70%	67%	'newness'
CITED SOURCE FOR CLAIM:			
Journal article	61%	55%	Very surprising,
Conference paper	74%	78%	journals are
Press conference	81%	85%	reputed to be better
No cited source	74%	69%	than press releases

REFERENCES

1. Tankard, J., & Ryan, M., *Journalism Quarterly* **51**, 219–225 (1974).
2. Tichenor, P., et al., *Journalism Quarterly* **47**, 673–683 (1970).
3. Broberg, K., *Journalism Quarterly* **50**, 763–767 (1973).
4. Moore, B., & Singletary, M., *Journalism Quarterly* **62**, 816–823 (1985).
5. Pulford, D., *Journalism Quarterly* **53**, 119–121 (1976).

Symposium Participants

Anfimov, Dmitrij, Institute for Space Research – Russia, dima@cgrsmx.iki.rssi.ru

Balsano, Rick, Los Alamos National Laboratory, balsano@lanl.gov

Band, David, Los Alamos National Laboratory, dband@lanl.gov

Barthelmy, Scott, NASA/GSFC, scott@milkyway.gsfc.nasa.gov

Beathley, Louis Jr., Southern University, ljbeath@grant.phys.subr.edu

Belczyński, Krzysztof, Copernicus Center – Poland, kabel@camk.edu.pl

Belli, Bianca, IAS/CNR Rome – Italy, bianca@saturn.ias.rm.cnr.it

Beloborodov, Andrei M., Stockholm Observatory – Sweden, andrei@astro.su.se

Bhat, P. N., Tata Institute of Fundamental Research – India, phbhat@tifr.res.in

Bloom, Joshua, California Institute of Technology, jsb@astro.caltech.edu

Boër, Michel, CESR/CNRS – France, michel.boer@cesr.fr

Böttcher, Markus, Rice University, mboett@spacsun.rice.edu

Bonnell, Jerry, NASA/GSFC, bonnell@grossc.gsfc.nasa.gov

Borozdin, Konstantin, Los Alamos National Laboratory, kbor@lanl.gov

Brainerd, J. James, University of Alabama in Huntsville, jim.brainerd@msfc.nasa.gov

Brandt, Soren, Space Research Institute – Denmark, sb@dsri.dk

Briggs, Michael, University of Alabama in Huntsville, michael.briggs@msfc.nasa.gov

Bulik, Tomek, Centrum Astronomiczne – Poland, bulik@camk.edu.pl

Castro-Tirado, Alberto, LAEFF-INTA – Spain, ajct@iaa.es

Chevalier, Roger, University of Virginia, rac5x@virginia.edu

Cline, Thomas L., NASA/GSFC, cline@apache.gsfc.nasa.gov

Clingempeel, Richard, The Amateur Sky Survey, oerlicon@intelos.net

Connaughton, Valerie, University of Alabama in Huntsville, vc@msfc.nasa.gov

Costa, Enrico, IAS-CNR Rome – Italy, costa@saturn.ias.rm.cnr.it

Crider, Anthony, Naval Research Laboratory, acrider@gamma.nrl.navy.mil

Curry, Charles, The Curry Foundation, ragtime100@aol.com

Dai, Zigao, Nanjing University – China, zdai@bohr.physics.hku.hk

Daigne, Frederic, Institut d'Astrophysique de Paris – France, daigne@iap.fr

Deng, Ming, Yale University, ming.deng@yale.edu

Dermer, Charles, Naval Research Laboratory, dermer@gamma.nrl.navy.mil

Dingus, Brenda, University of Utah, dingus@physics.utah.edu

Duncan, Robert, University of Texas, duncan@astro.as.utexas.edu

Elsner, Ronald, NASA/MSFC, elsner@avalon.msfc.nasa.gov

Espley, Jared, University of Virginia, jared@virginia.edu
Fenimore, Ed, Los Alamos National Laboratory, efenimore@lanl.gov
Feroci, Marco, IAS/CNR Rome – Italy, feroci@ias.rm.cnr.it
Finger, Mark, USRA/MSFC, mark.finger@msfc.nasa.gov
Fishman, Gerald, NASA/MSFC, fishman@msfc.nasa.gov
Frail, Dale, National Radio Astronomy Observatory, dfrail@nrao.edu
Fritzius, Robert, Shade Tree Physics.com, rsf1@ebicom.net
Fruchter, Andrew, Space Telescope Science Institute, fruchter@stsci.edu
Fryer, Chris, University of California - Santa Cruz, cfryer@uoclick.org
Galama, Titus J., California Institute of Technology, titus@astro.uva.nl
Gallant, Yves, Astronomical Institute Utrecht – The Netherlands, gallant@cp.dias.ie
Gehrels, Neil, NASA/GSFC, gehrels@gsfc.nasa.gov
Giblin, Timothy, University of Alabama in Huntsville, tim.giblin@msfc.nasa.gov
Göğüş, Ersin, UAH/MSFC, ersin.gogus@msfc.nasa.gov
Gomez, Enrique, University of Alabama, gomez002@bama.ua.edu
Gonthier, Peter, Hope College, gonthier@physics.hope.edu
Gotthelf, Eric, Columbia University, evg@astro.columbia.edu
Graber, James, unaffiliated, jgra@loc.gov
Granot, Jonathan, Hebrew University Jerusalem – Israel,
    jgranot@merger.fiz.huji.ac.il
Greiner, Jochen, Astrophysical Institute Potsdam – Germany, jgreiner@aip.de
Guarnieri, Adriano, Bologna University – Italy, adraino@astbo3.bo.astro.it
Gupta, Varsha, Delhi University – India, varsha@ducos.ernet.in
Guthmann, Axel, Max-Planck-Institut für Kernphysik – Germany,
    axel.guthmann@mpi-hd.mpg.de
Haglin, David, Mankato State University, david.haglin@mankato.msus.edu
Hakkila, Jon, Mankato State University, jon.hakkila@mankato.msus.edu
Hardy, Stephen, Max-Planck-Institut für Astrophysik – Germany,
    stephen@mpa-garching.mpg.de
Harmon, Alan, NASA/MSFC, alan.harmon@msfc.nasa.gov
Hartmann, Dieter, Clemson University, hartmann@grb.phys.clemson.edu
Heise, John, Space Research Organization – The Netherlands, j.heise@sron.nl
Henden, Arne, U.S. Naval Observatory, aah@nofs.navy.mil
Henze, William, Teledyne-Brown Engineering, william.henze@msfc.nasa.gov
Holland, Stephen, University of Aarhus – Denmark, holland@obs.aau.dk

Horack, John, NASA/MSFC, john.horack@msfc.nasa.gov

Horan, Deirdre, Smithsonian Astrophysical Observatory, deirdre@ferdia.ucd.ie

Horváth, István, BJKMF – Hungary, hoi@bjkmf.hu

Hudec, René, Astronomical Institute Ondrejov – Czech Republic, rhudec@asu.cas.sz

Hurley, Kevin, University of California - Berkeley, khurley@sunspot.ssl.berkeley.edu

Johnson, Audress, University of Alabama - Tuscaloosa, audress@hera.astr.ua.edu

Kassin, Susan, Ohio State University, kassin@astronomy.ohio-state.edu

Kawai, Nobuyuki, RIKEN – Japan, nkawai@postman.riken.go.jp

Kippen, R. Marc, University of Alabama in Huntsville, marc.kippen@msfc.nasa.gov

Klose, Sylvio, TLS Tautenburg – Germany, klose@tls-tautenburg.de

Kobayashi, Shiho, Osaka University – Japan, shiho@vega.ess.sci.osaka-u.ac.jp

Kocharovsky, Vladimir, Academy of Science – Russia, kochar@appl.sci-nnov.ru

Körtvélyessy, Laszlo, Observatory Kleve – Germany, drlaky@aol.com

Koshut, Thomas, USRA/MSFC, thomas.m.koshut@msfc.nasa.gov

Krimm, Hans, USRA/GSFC, krimm@milkyway.gsfc.nasa.gov

Kulkarni, Shri, California Institute of Technology, srk@astro.caltech.edu

Kumar, Pawan, Institute for Advanced Study, pk@sns.ias.edu

Lamb, Donald, University of Chicago, lamb@oddjob.uchicago.edu

Lazzati, Davide, Osservatorio Astronomico Brera – Italy, lazzati@merate.mi.astro.it

Leahy, Denis, University of Calgary – Canada, leahy@iras.ucalgary.ca

Lee, Chang-Hwan, State University of New York - Stony Brook, chlee@silver.physics.sunysb.edu

Lestrade, J. Patrick, Mississippi State University, lestrade@ra.msstate.edu

Liang, Edison, Rice University, liang@spacsun.rice.edu

Lindsay, David, Hitachi, david.lindsay@hds.com

Litvak, Maxim, Institute for Space Research – Russia, max@cgrsmx.iki.rssi.ru

Litvine, Dmitri, Institute for Space Research – Russia, litvin@space.ru

Lloyd, Nicole, Stanford University, nicole@urania.stanford.edu

Luginbuhl, Christian, U.S. Naval Observatory, cbl@nofs.navy.mil

MacFadyen, Andrew, University of California - Santa Cruz, andrew@ucolick.org

Mallozzi, Robert, University of Alabama in Huntsville, robert.mallozzi@msfc.nasa.gov

Marsden, David, NASA/GSFC, dmarsden@milkyway.gsfc.nasa.gov

Marshall, Stuart, Lawrence Livermore National Laboratory, stuart@igpp.llnl.gov

Matthey, Christina, CERN – Switzerland, christina.matthey@cern.ch

Mattox, John, Boston University, mattox@bu.edu

McCollough, Michael, USRA/MSFC, Michael.McCollough@msfc.nasa.gov
McEnery, Julie, University of Utah, mcenery@mail.physics.utah.edu
McKay, Timothy, University of Michigan, tamckay@umich.edu
Meegan, Charles, NASA/MSFC, charles.meegan@msfc.nasa.gov
Mereghetti, Sandro, IFC-CNR Milan – Italy, sandro@ifctr.mi.cnr.it
Mészáros, Peter, Pennsylvania State University, pmeszaros@astro.psu.edu
Mészáros, Attila, Charles University Prague – Czech Republic, meszaros@mbox.cesnet.sz
Miller, Richard, University of New Hampshire, richard.miller@unh.edu
Mirabel, Felix, CE Saclay – France, mirabel@discovery.saclay.cea.fr
Mitrofanov, Igor, Institute for Space Research – Russia, imitrofa@space.ru
Müller, Ewald, Max-Planck-Institut für Astrophysik – Germany, emueller@mpa-garching.mpg.de
Montanari, Enrico, University Ferrara – Italy, montana@fe.infn.it
Murakami, Toshio, ISAS – Japan, murakami@astro.isas.ac.jp
Nekola, Martin, Czech Technical University Prague – Czech Republic, m.nekola@atlas.cz
Nemiroff, Robert, Michigan Technological University, nemiroff@mtu.edu
Nomoto, Ken'ichi, University of Tokyo – Japan, nomoto@astron.s.u-tokyo.ac.jp
Norris, Jay, NASA/GSFC, norris@groax0.gsfc.nasa.gov
Paciesas, Bill, University of Alabama in Huntsville, bill.paciesas@msfc.nasa.gov
Palmer, David, USRA/GSFC, palmer@lheamail.gsfc.nasa.gov
Papathanssiou, Hara, SISSA – Italy, hara@sissa.it
Park, Hye-Sook, Lawrence Livermore National Laboratory, hpark@llnl.gov
Parsons, Ann, NASA/GSFC, parsons@milkyway.gsfc.nasa.gov
Páta, Petr, Czech Technical University Prague – Czech Republic, oata@feld.cvut.cz
Patel, Sandy, University of Alabama in Huntsville, sandeep.patel@msfc.nasa.gov
Pelaez, François, Mississippi State University, fpelaez@ra.msstate.edu
Peterson, Burl, USRA/MSFC, burl.peterson@msfc.nasa.gov
Petrosian, Vahe, Stanford University, vahe@astronomy.stanford.edu
Phengchamnan, Surasak, University of Alabama in Huntsville, surasak.phengchamnan@msfc.nasa.gov
Piccioni, Adalberto, Universita' Degli Studi Di Bologna – Italy, piccioni@ermione.bo.astro.it
Piran, Tsvi, Hebrew University – Israel, tsvi@nikki.fiz.huji.ac.il
Piro, Luigi, IAS/CNR Rome – Italy, piro@ias.rm.cnr.it

Pizzichini, Graziella, TESRE/CNR Bologna – Italy, pizzichini@tesre.bo.cnr.it
Porrata, Rodin, Lawrence Livermore National Laboratory, porrata@llnl.gov
Pozanenko, Alexei, Space Research Institute – Russia, apozanen@iki.rssi.ru
Preece, Robert, University of Alabama in Huntsville, rob.preece@msfc.nasa.gov
Quilligan, Fergus, University College Dublin – Ireland, fquillig@bermuda.ucd.ie
Ramirez-Ruiz, Enrico, Los Alamos National Laboratory, enrico@nis.lanl.gov
Rattenbury, Nicholas J., University of Auckland – New Zealand, nrat001@phy.auckland.ac.nz
Reichart, Daniel, University of Chicago, reichart@oddjob.uchicago.edu
Rhie, Sun Hong, University of Notre Dame, srhie@condor.phys.nd.edu
Richardson, Georgia, University of Alabama in Huntsville, georgia.richardson@msfc.nasa.gov
Ricker, George, Massachusetts Institute of Technology, grr@space.mit.edu
Roiger, Richard J., Mankato State University, roiger@krypton.mankato.msus.edu
Rol, Evert, University of Amsterdam – The Netherlands, evert@astro.uva.nl
Rosswog, Stephan, University of Cologne – Germany, rosswog@quasar.physik.unibas.ch
Rothschild, Richard, University of California San Diego, rrothschild@ucsd.edu
Ryde, Felix, Stockholm Observatory – Sweden, felix@astro.su.se
Sahi, Maityree, USRA/MSFC, maitrayee.sahi@msfc.nasa.gov
Salmonson, Jay, Lawrence Livermore National Laboratory, salmonson@llnl.gov
Sanin, Anton, Institute for Space Research – Russia, asanin@cgrsmx.iki.rssi.ru
Sari, Re'em, California Institute of Technology, sari@tapir.caltech.edu
Scargle, Jeff, NASA/AMES, jeffrey@sunshine.arc.nasa.gov
Schaefer, Bradley, Yale University, schaefer@grb2.physics.yale.edu
Schilling, Govert, unaffiliated, goverts@casema.edu
Schmidt, Maarten, California Institute of Technology, mxs@astro.caltech.edu
Scott, Matt, USRA/MSFC, matt.scott@msfc.nasa.gov
Simpson, Morgan, Mississippi State University, mss8@ra.msstate.edu
Smith, Ian, Rice University, ian@spacsun.rice.edu
Soderberg, Alicia M., Los Alamos National Laboratory, alicia@nis.lanl.gov
Spada, Maddalena, University of Florence – Italy, spada@arcetri.astro.it
Spruit, Helmut C., Max-Planck-Institut für Astrophysik – Germany, henk@mpa-garching.mpg.de
Stacy, J. Gregory, Louisiana State University, gstacy@phys.lsu.edu
Suh, In-Saeng, University of Notre Dame, isuh@nd.edu

Sun, Xuejun, USRA/MSFC, xuejun.sun@msfc.nasa.gov
Svensson, Roland, Stockholm Observatory – Sweden, svensson@astro.su.se
Swartz, Douglas, USRA/MSFC, Doug.A.Swartz@msfc.nasa.gov
Tajima, Toshi, Lawrence Livermore National Laboratory, tajima1@llnl.gov
Takahashi, Yoshiyuki, University of Alabama in Huntsville, takahyx@sslab.msfc.nasa.gov
Terrell, James, Los Alamos National Laboratory, jterrell@lanl.gov
Thompson, Christopher, University of North Carolina, act@physics.unc.edu
Tikhomirova, Yana Yu, Lebedev Physical Institute – Russia, jana@dpc.asc.rssi.ru
Totani, Tomonori, National Astronomical Observatory – Japan, totani@th.nao.ac.jp
Tovmasian, Gagik, OAN UNAM – Mexico, gaghik@bufadora.astrosen.unam.mx
Ventura, Joseph, University of Crete – Greece, ventura@physics.uch.gr
Vrba, Frederick, U.S. Naval Observatory, fjv@nofs.navy.mil
Vreeswijk, Paul, University of Amsterdam – The Netherlands, pmv@astro.uva.nl
Wheeler, J. Craig, University of Texas, wheel@astro.as.utexas.edu
Wijers, Ralph, State University of New York - Stony Brook, rwijers@astro.sunysb.edu
Williams, G. Grant, Clemson University, ggwilli@hubcap.clemson.edu
Wilson, James R., Lawrence Livermore National Laboratory, jimwilson@llnl.gov
Wilson, Robert B., NASA/MSFC, robert.b.wilson@msfc.nasa.gov
Wilson, David, Soundprint, wilsondb@euclid.colorado.edu
Woods, Peter, University of Alabama in Huntsville, peter.woods@msfc.nasa.gov
Woosley, Stan, University of California - Santa Cruz, woosley@ucolick.org
Yan, Yuan, Mississippi State University, yy1@ra.msstate.edu
Ye, Lu, University of Hong Kong, ly66@bohr.physics.hku.hk
Yoshida, Atsumasa, RIKEN – Japan, ayoshida@postman.riken.go.jp
Young, Chad, Mississippi State University, chy1@ra.msstate.edu
Zhang, Shuang Nan, University of Alabama in Huntsville, shuang.zhang@msfc.nasa.gov

AUTHOR INDEX

Note, **bold** page number indicates author is first author.

A

Ables, E., 250, 736
Achterberg, A., 485, 524
Aisaka, K., 706
Akerlof, C., 804
Altice, P., 696
Amati, L., 18, 716, 721
Ando, M., 786
Anfimov, D. S., 68, **73**, 92, **140**, 230, 235
Angel Aloy, M. A., 565
Antonelli, L. A., 375
Arefiev, V. A., **245**
Atac, M., 706
Atkins, R., 240, 751
Atteia, J.-L., 255

B

Bagoly, Z., 102, 200
Balázs, L. G., 102
Balsano, R., **804**
Band, D. L., 87, 185, 250, **696**, 736
Baring, M. G., 852
Barthelmy, S. D., 250, 726, **731**, 736, 804
Bartolini, C., 355, 756
Beathley, Jr., L. J., **270**
Beckwith, S., 821
Belczyński, K., **638**, 648
Belli, B. M., **160**
Beloborodov, A. M., **205**
Benbow, W., 240, 751
Benítez, N., 334
Bennett, K., 28, 170
Berezhnoy, A., 13
Berger, E., 277
Berley, D., 240, 751
Berná, J. A., 260
Bernas, M., 260, 265, 882, 887
Beskin, G. M., **355**, 756
Bethe, H. A., 628
Bhargavi, S. G., 107
Bhat, P. N., 215

Bianco, C. L., 711
Binns, W. R., 706
Bionta, R. M., 250, 736
Biryukov, S., 756
Bloch, J., 804
Bloom, J. S., 277
Boër, M., **255, 746**
Boller, T., 380
Bond, I. A., 313, 342, 347
Bonnell, J. T., 78, **210**, 663, 873
Bontekoe, T. R., 270
Borgonovo, L., **130**, 180
Borkowski, J., 686
Borozdin, K. N., 245
Böttcher, M., 446, 470, **519, 545**
Brainerd, J. J., **150, 455, 480**
Brandt, S., 686
Briggs, M. S., 68, 73, 92, **125**, 140, 150, **165**, 175, 195, 230, 235, 394, 731, 776, 796
Bringer, M., 255
Brown, G. E., **628**
Broz, O., 265
Buckley, J. H., 696, 706
Bulik, T., 530, 638, **648**
Butterworth, P., 726, 731, 804

C

Calura, F., **721**
Campos, A., 334
Canzian, B., 741
Casperson, D., 804
Castander, F. J., 414
Castro-Tirado, A. J., **260**, 265, **313**, 334, 342, 347, 409, **801, 818, 821**, 882
Celidonio, G., 23
Chaffee, F., 277
Chaty, S., 255, 814
Chen, M. L., 240, 751
Cherry, M. L., 696, 706
Chevalier, R. A., **608**
Chiang, J., 470
Chung, T. J., 494
Cline, D. B., **97**, 706

Cline, T. L., 250, **726**, 731, 736, 804
Coletta, A., 23
Collina, P., 721
Collmar, W., 28, 170
Connaughton, V., **385**, 394, 731
Connors, A., **28**, 170
Cosentino, G., 355
Costa, E., 18, 23, 313, **365**, 375, 711, 716, 721, 771
Costello, R. M., 852
Covino, S., 318, 801
Coyne, D. G., 240, 751
Crider, A., 446, **475**
Cunningham, S. J., 298

D

Daigne, F., **579**
Danziger, I. J., 622
Das Gupta, P., 215
Davies, M. B., 575
de la Morena, B., 260
Deng, M., **63**, 419
Derishev, E. V., **460, 584**
Dermer, C. D., **431, 470**, 545
de Ugarte, A., 260
Di Ciolo, L., 23
Diercks, A., 277
Dingus, B. L., 240, **751**
Djorgovski, S. G., 277
Dorfan, D. E., 240, 751
Dowkontt, P., 706
Duggan, P., 190
Duncan, R. C., 771, 776, 781, 796, **830**

E

Eichler, D., 756
Ellsworth, R. W., 240, 751
Epstein, J. W., 706
Espley, J. R., 175
Evans, D., 240, 751

F

Faiman, D., 756
Falcone, A., 240, 751

Fenimore, E. E., 200
Ferguson, D. H., 250, 736
Feroci, M., 18, 23, 313, 347, 375, **711**, 716, 721, **771**
Finger, M. H., 706, 781
Fischer, O., 323
Fishman, G. J., 230, 235, 250, 394, 706, 736
Fletcher, S., 804
Fleysher, L., 240, 751
Fleysher, R., 240, 751
Florian, J., 265
Frail, D. A., 277, **298**
Frederiks, D., 726
Fritzius, R. S., **603**
Frontera, F., 18, 23, 313, 375, 711, 716, 721, 771
Fryer, C. L., **643**
Fuchs, Y., 814

G

Galama, T. J., 277, **303**, 326
Gallant, Y. A., 485, **524**
Gandolfi, G., **23**, 375
Gehrels, N., 250, **671**, 736
Ghirlanda, G., 185
Ghisellini, G., 318
Giblin, T. W., **394**, 399, 776
Giménez, A., 260, 265
Gisler, G., 240, 751, 804
Göğüş, E., 776, **796**
Gold, R., 726
Golenetskii, S., 726
Gonthier, P. L., **852**
Goodman, J. A., 240, 751
Goodrich, R. W., 277
Gorenstein, P., 701
Gorosabel, J., 260, 313, **334, 342**, 347, **409**
Graber, J. S., **594**
Granot, J., **489, 540**
Greiner, J., 107, 313, **380**
Groot, P. J., 326
Guarnieri, A., 313, 355, 756
Guenther, E. W., **825**
Guetter, H. H., 809
Guidorzi, C., **18**, 721
Gupta, V., **215**

Guthmann, A. W., **485**, 524
Guzik, T. G., 696, 706
Guziy, S., 313, **339**

H

Haglin, D. J., 33, 38, 48, **877**
Haines, T. J., 240, 751
Hakkila, J., 33, 38, 43, **48**, 53, **83**, **399**, 877
Hanlon, L., 190
Harding, A. K., 852
Harris, H. C., 741
Harrison, F. A., 277
Hartmann, D. H., 250, 380, **653**, 736, 741, 786, 809
Hearnshaw, J., 342, 347
Heise, J., 125, 375
Henden, A. A., 313, 741, 809
Hermsen, W., 28, 170
Higdon, J., 847
Hills, J., 804
Hink, P. L., 696, 706
Hoffman, C. M., 240, 751
Höflich, P., 617
Horváth, I., 102, **200**
Hroch, F., 265
Hudcová, V., 265, 424
Hudec, R., 260, **265**, 313, 329, 339, **352**, **360**, **424**, **701**, 882
Hugenberger, S., 240, 751
Hurley, K., **3**, 250, 726, 736, **763**, 771, 776, 781, 786, 890
Hurley, K. C., 663
Hurley, K. J., 190

I

Inneman, A., 701
in't Zand, J., 125
Israel, M. H., 706
Iwamoto, K., 622

J

Jackson, P. D., 270
Jennings, D., 686

Jennings, M. C., 741
Jimenez, R., 87
Johnson, A., 43, **53**

K

Kappadath, S. C., 696, 706
Karr, G. R., 494, 706
Kawai, N., **691**
Kehoe, R., 804
Kelley, L. A., 240, 751
Kelson, D., 821
Kerr, T., 821
Kilmartin, P. M., 342, 347
Kippen, R. M., 28, 170, 250, 706, 726, 731, 741
Kirk, J. G., 485, 524
Klose, S., 313, **323**, 825
Klotz, A., 255
Kobayashi, S., **550**
Kocharovsky, V. V., 460, 584
Kocharovsky, Vl. V., 460, 584
Kolaczyk, E. D., 170
Kompaneets, D., 13
Körtvélyessy, L., **599, 867**
Koshut, T. M., 230, 235
Kosyrev, A. S., 390
Kouveliotou, C., 125, 250, 326, 394, 726, 736, 776, 781, 786, 796
Kuiper, L., 28, 170
Kulkarni, S. R., **277**, 298
Kumar, P., 535
Kuulkers, E., 125

L

Lamb, D. Q., 414, **658**
Laureijs, R., 818
Lázaro, C., 821
Lazzarotto, F., 711
Lazzati, D., **318**
Leahy, D., **862**
Lee, B., 804
Lee, C.-H., 628
Lee, H. K., 628
Leonor, I., 240, 751
Lestrade, J. P., 230, 235
Levine, S. E., 741

Li, P., 786
Liang, E., **446**
Lingenfelter, R. E., 842, 847
Litvak, M. L., 68, 73, 92, 140, **195, 230**, 235
Litvine, D. A., **390**
Lloyd, N. M., **145, 155, 465**
Lodi, S., 355
Loznikov, V. M., 220
Luginbuhl, C. B., 313, 741, 809
Lund, N., 801

M

MacFadyen, A. I., 565, **633**, 653
Macri, J., 696, 706
Madruga, S., 821
Maeda, K., 622
Ma Ibáñez, J., 565
Majczyna, A., **530**
Mallozzi, R. S., 33, 38, 43, 48, 145, 150, 706, 877
Ma Martí, J., 565
Marani, G. F., 78, 210, 663, 873
Marsden, D., 842, **847**
Marshall, S., 804
Masetti, N., 265, 313, 424
Más-Hesse, J. M., 260, 265
Mathews, G. J., 570, 857
Matsuoka, M., 691
Mattei, A., 711
Matteson, J., 696
Matthey, C., 97
Mazets, E., 726
Mazzali, P. A., 622
McBreen, B., 190
McClanahan, T., 726
McConnell, D., 298
McConnell, M. L., 28, 170, 240, 696, 706, 751
McCullough, J. F., 240, 751
McEnery, J. E., **240**, 751
McKay, T., 804
Meegan, C. A., 33, **43**, 48, 53, 68, 73, 92, 140, 195, 230, 235, 663, 736
Mercer, C. L., 852
Mereghetti, S., **686**
Mészáros, A., **102**
Mészáros, P., 200, 450, **514**

Metcalfe, L., 818
Metcalfe, N., 334
Mihara, T., 691
Miller, M., 540
Miller, R. S., 28, 170, 240, 751
Mincer, A. I., 240, 751
Mirabel, I. F., **814**
Mitrofanov, I. G., **68**, 73, 92, 140, 195, 230, **235**, 390
Miyata, E., 691
Mochkovitch, R., 579
Moderski, R., 530
Mohan, V., 313
Montanari, E., 721
Morales, M. F., 240, 751
Morand, F., 255
Moriarty-Schieven, G. H., 298
Müller, E., **565**
Muller, J. M., 23
Murakami, T., 781, **786**
Muraki, Y., 342, 347

N

Nakamura, T., 342, 347, 622
Negoro, H., 691
Nekola, M., 265, **887**
Nemethy, P., 240, 751
Nemiroff, R. J., 210, 250, **663**, 736
Noda, S., 342
Nomoto, K., **622**
Norris, J. P., **78**, 200, 210, 663, 873

O

Ohnishi, K., 342, 347
O'Neill, T., 696
Otwinowski, S., 97

P

Paciesas, W. S., 68, 73, 92, 140, 195, 230, 235, 706
Palazzi, E., 265, 313, 424
Palmer, D. M., 726, **791**
Panaitescu, A., 450
Paolino, A., 23

Papathanassiou, H., **441**
Park, H. S., 250, **736**
Parnell, T. A., 706
Páta, P., 260, 265, **882**
Patat, F., 622
Pawl, A., 804
Pedersen, H., 255
Pendleton, G. N., 33, 38, 43, 48, 68, 73, 83, 92, 140, 150, 195, 230, 235, 494, 706, 731, 877
Pereira, W., 736
Petrosian, V., 145, 155, 465
Phengchamnan, S., **706**
Piccioni, A., 355, **756**
Pina, L., 701
Pinfield, D., 801
Piran, T., **87**, 489, **535**, 540, 575
Piro, L., 23, 313, 375
Pischalnikov, Y., 706
Pizzichini, G., 265, 329, 424
Pollas, C., 255
Porrata, R. A., 250, 736
Pozanenko, A. S., **220**
Preece, R. D., 68, 73, 83, 92, **115**, 125, 140, 150, 155, 165, **175**, 195, 230, 235, 394, 399, 480, 706, 776
Preger, B., **716**
Priedhorsky, W. C., 245, 804

Q

Quilligan, F., **190**

R

Rapisarda, M., 721
Rattenbury, N. J., 313, 342, **347**
Reichart, D. E., **414**, 658
Reid, M., 342, 347
Richardson, G. A., **494**, 706
Rielage, K., 706
Roiger, R. J., **33**, 38, 48, 877
Rol, E., 326
Rosswog, S., **575**
Rothschild, R. E., **842**, 847
Rudak, B., 638
Ryan, J. M., 28, 170, 240, 696, 706, 751
Ryde, F., 130, **135, 180**

S

Sagar, R., 313
Saito, T., 342, 347
Saizar, P., 342
Sakurai, I., 691
Salmonson, J. D., **570**
Sánchez, E., 334
Sanguino, T. M., 260
Sanin, A. B., 68, **92**, 230, 235
Sanin, A. L., 73
Sari, R., 277, 489, **504**, 550
Sasseen, T., 250
Scargle, J. D., **873**
Schaefer, B. E., 63, **404, 419, 890**
Schmidt, M., **8, 58**
Schönfelder, V., 28, 170
Schwarz, R., 380
Seldomridge, N., 804
Semionova, L., 862
Sethi, S. K., **107**
Shen, B., 240, 751
Shepherd, D. S., 298
Shirasaki, Y., 691
Shlyapnikov, A., 313, 339
Shoup, A., 240, 751
Sikora, M., 530
Šimon, V., **329**
Sinnis, C., 240, 751
Smith, A. J., 240, 751
Smith, I. A., **326**, 446, 786
Smith, M. J. S., 23
Soffitta, P., 375
Soldán, J., 260, 265
Spada, M., **450**
Spruit, H. C., **589**
Stacy, J. G., 270, 696, 706
Stecklum, B., 323
Stepanov, M., 13
Stern, B. E., **13**, 205
Stratta, G., **375**
Strohmayer, T., 781, 786
Subrahmanyan, R., 298
Suen, W.-M., 540
Sugizaki, M., 691
Suh, I.-S., **857**
Sullivan, G. W., 240, 751
Suzuki, T., 622
Svensson, R., 13, 135, 180, 205
Swank, J., 781

Szymanski, J., 804

T

Tanvir, N., 326
Tarei, G., 23
Tassone, G., 23
Tavani, M., **185**
Taylor, G. B., 298
Thielemann, F.-K., 575
Thompson, C., 771, 776, 781, 796
Tikhomirova, Y., 13
Tilanus, R. P. J., 326
Tomida, H., 691
Torii, K., 691
Totani, T., **499**
Toublanc, D., 255
Trapero, J., 334
Trombka, J., 726
Tsujimoto, M., 786
Tsunemi, H., 691
Tümer, T. O., 240, 706, 751
Turatto, M., 622

U

Ueno, S., 691

V

van Paradijs, J., 125, 326, 394, 776, 781, 796
Vavrek, R., 102
Ventura, G., 711
Voges, W., 380
Vrba, F. J., 313, **741, 809,** 825
Vreeswijk, P., 326

W

Wallace, D. B., 706
Walter, R., 686
Wang, K., 240, 751
Wang, L., 617
Wark, R. M., 298
Wascko, M. O., 240, 751
Watson, D., 190
Westerhoff, S., 240, 751
Wheeler, J. C., **617**
Wieringa, M. H., 298
Wijers, R. A. M. J., 326, 394
Williams, D. A., 240, 751
Williams, G. G., **250**, 736
Williams, O. R., 28, 170
Wilson, J. R., 570
Wilson, R. B., 706
Winkler, C., 28, 170, 270, **681**
Woods, P. M., **776, 781,** 786, 796
Woosley, S. E., **555**, 653
Wren, J., 804

Y

Yamauchi, M., 691
Yan, Y., **225**
Yang, T., 240, 751
Yi, I., 617
Yock, P., 313, 342, 347
Yodh, B. G., 751
Yodh, G. B., 240
Yoshida, A., 691, 786
Yost, S. A., 277
Young, C. A., 28, **170**
Yuan, W., 691

Z

Zapatero-Osorio, M. R., 313, 334
Zhang, W., 643
Zharykov, S. V., 380
Zhu, J., 313
Zych, A., 696